Natural resources research   XIV

Other titles in the series
Natural resources research

Tropical forest ecosystems

A state-of-knowledge report
prepared by
Unesco/UNEP/FAO

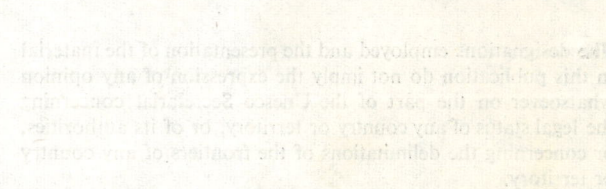

Published in 1978 by the United Nations
Educational, Scientific and Cultural Organization,
7 Place de Fontenoy, 75700 Paris
Printed by the Imprimerie des
Presses Universitaires de France, Vendôme

ISBN 92–3–101507–9 ✓
French edition: 92–3–201507–2
Spanish edition: 92–3–301507–6
UNEP Project 0202–73–01

Preface

The organization and promotion of research on the natural resources of the tropical world, including the periodical assessment and evaluation of the state of knowledge in this field, has always figured prominently in Unesco's co-operative international scientific programmes. In 1956, at a time when Unesco's Arid Zone Research Programme was in full swing, a more modest programme of research on the humid tropics was launched. Apart from the intrinsic usefulness of establishing a synthesis and disseminating the results of research carried out in different parts of the world and covering a very wide range of topics, these efforts demonstrated the value of integrated resource research as opening the way to an interdisciplinary approach to the planning of development projects founded on a sound ecological basis; they also brought fresh recognition of the vital interplay between fundamental research and land-use requirements, in which human aspects are of the utmost importance.

Works such as *A Review of the Natural Resources of the African Continent* (1963) and *Natural Resources of Humid Tropical Asia* (1974) bear witness to Unesco's desire to present as full a picture as possible of world knowledge of natural resources in different regions and continents. These studies, however, were still sectoral in their approach and followed the lines of the major scientific disciplines and, in particular, maintained the classic separation between the social and natural sciences. Between 1961 and 1972, with the International Biological Programme, a serious effort was made to promote research on the productivity of terrestrial and aquatic ecosystems without regard to disciplinary boundaries, at least in the realm of biology; thus ecology and ecological methods came to the forefront of research at the regional and the international level; the most important results were obtained in the temperate zones, the tropical world having been the subject only of limited and specific research undertakings. There remained, however, the task of linking the knowledge acquired on the functioning of natural ecosystems to the various ways of development and management.

The time had therefore come to consider the whole question of making rational use of and of conserving the resources of the biosphere on the basis of scientific knowledge already possessed or likely to be acquired in the future. The intergovernmental *Biosphere Conference*, convened by Unesco in 1968, marked a turning point in this respect since it confirmed the view that social and economic development was intimately linked with the rational use of

renewable resources and that the conservation of these resources should be seen as an element of their rational use on a sustained yield basis, rather than as an obstruction to it. Thus the Conference gave a clear call for a new conception of man's relationship with his environment based on an approach to development that did not compromise resource renewal. This was the underlying theme that led to the launching, in 1970, of the intergovernmental Man and the Biosphere (MAB) Programme. The Biosphere Conference bore witness to the spread of ecological awareness throughout the world and to the need, first and foremost, to establish resource inventories and to broaden and to synthesize our basic knowledge. In this domain, Unesco, in co-operation with other United Nations Agencies, was actively preparing and publishing a series of thematic maps relating to the environment and its resources. This action was accompanied by the development of comparable methodologies and agreed nomenclatures, practical examples of which are to be found in regional vegetation maps, the FAO/Unesco *Soil Map of the World* and the WMO/Unesco climatic atlases. Direct assistance was given to various research institutes, such as those created under the Arid Zone Research Programme, and to scientists in developing countries not only to enable them to acquire the techniques of integrated natural resource survey, but also to ensure that the development of natural resources was more closely linked to socio-economic conditions whilst taking into account biological productivity and other ecological constraints.

Unesco's concerns and activities with regard to research and to the synthesis of knowledge relating to natural resources and their rational management were put before the United Nations Conference on the Human Environment (Stockholm, June 1972), which culminated in the creation of the United Nations Environment Programme (UNEP). In pursuit of its role of co-ordination and stimulation, UNEP gave priority to the study of the functioning of terrestrial ecosystems with a view to their rational management. In the UNEP plan of action, arid lands, rangelands and pastures, forests and tropical wooded areas were picked out for special study and urgent action. Thus there was complete accord with the themes of the MAB Programme the objective of which is 'to develop the basis within the natural and social sciences for the rational use and conservation of the resources of the biosphere and for the improvement of the global relationship between man and the environment; to predict the consequences of today's actions on tomorrow's world and thereby to increase man's ability to manage efficiently the natural resources of the biosphere'. Two of these themes are particularly concerned with the tropical world.

1. Ecological effects of increasing human activities on tropical and subtropical forest ecosystems.
2. Impact of human activities and land use practices on grazing lands: savanna and grassland (from temperate to arid areas).

Thus the joint action of Unesco and UNEP in preparing this state-of-knowledge Report on tropical forest ecosystems can be seen as a natural outcome of the desire to provide better guidance to research that is vital for the satisfactory management of these ecosystems and to contribute to the training of specialists in the various scientific fields involved. Not only does this initiative follow logically from the priority accorded by Unesco and UNEP in their respective programmes to tropical forest ecosystems, it also provides another opportunity for Unesco to underline its preoccupation with the synthesis and diffusion of knowledge. For this reason the Report attempts to summarize the results of research undertaken over the past two decades, to point out gaps in knowledge and to sketch out the lines of future research. It is therefore intended as a tool for action while at the same time offering teachers and students a valuable source of information and documentation.

Another similar report on tropical grazing land ecosystems is in preparation.

The elaboration of this world Report was a complex operation in which many experts from the tropical regions as well as the non-tropical world took part. It also involved the invaluable co-operation of the United Nations Specialized Agencies, in particular FAO with whom Unesco also collaborates closely in a number of activities under the MAB Programme. Unesco, UNEP and FAO wish to express their gratitude to all those who helped in the preparation of this work. Final editing and the many revisions required to ensure a homogeneous presentation were carried out by A. Sasson (Division of Ecological Sciences, Unesco) with the assistance of B. Hopkins (New England College, Tortington Park, Arundel, Sussex, United Kingdom).

The designations employed and the presentation of the material in this work do not imply the expression of any opinion whatsoever on the part of the Secretariats of Unesco and UNEP concerning the legal status of any country, territory, city or area or of its authorities, or concerning the delimitation of its frontiers or boundaries.

The views expressed in this book reflect the opinions of their authors and not necessarily those of Unesco, UNEP and FAO.

Errata

Page 33, Table 1:   line Sea: 13 358 *instead of* 1 358

line Percentage land: *the numbers of the whole line are changed as follows:*

35	29	23	24	22	
24	23	20	24	25	26

page 37, Table 3:   line Total, 6th column, *delete* 58.

page 57, 2nd column, line 37: Buch, M. Von *instead of* Buch, M. V.

page 58, 1st column, line 27: 1974 *instead of* 1973.

page 210, 1st column, line 44: vol. 11 *instead of* vol. 2.

page 266, 2nd column, line 58: 1961 *instead of* 1965.

page 269, 2nd column, line 9: *Fournier, F. *instead of* *——.

page 281, 1st column, line 20: is from *instead of* in from.

page 303, 2nd column, line 38: These include *instead of* These include of.

page 376, 1st column, line 8: 390 *instead of* 392.

page 495, 1st column, line 57: vol. 154 *instead of* vol. 54.

Tropical forest ecosystems. A state-of-knowledge report prepared by Unesco/UNEP/FAO. Paris, Unesco. Natural Resources Research XIV, 1978.

Contents

Acknowledgements

Unesco and UNEP are deeply grateful to the contributions made by the various consultants who wrote the preliminary chapters of this Report, to those who contributed at regional levels and to the participants in the two synthesis workshops, as well as to the specialists of the sister United Nations organizations for their close co-operation. The following names need to be mentioned (the number preceding the name refers to the number of the chapter to which the contribution was directed).

1. Mr R. G. FONTAINE, (past director of the) Forest Resources Division of FAO, Via Lidia 73, Rome, Italy, and
Mr J. P. LANLY, Forestry Department, FAO, Via delle Terme di Caracalla, Rome 00100, Italy.

2. Prof. A. BAUMGARTNER, Meteorologisches Institut, Amalien-strasse 52, Forstliche Forschungs anstalt, 8 München 40, Federal Republic of Germany, and
Prof. E. F. BRÜNIG, Institut für Weltforstwirtschaft, Bundes-forschungsanstalt für Forst und Holzwirtschaft, Leuschner-strasse 91, 2050 Hamburg 80, Federal Republic of Germany.

3. Prof. D. A. LIVINGSTONE, Department of Zoology, Duke University, Durham, North Carolina 27706, U.S.A., and
Prof. T. VAN DER HAMMEN, Hugo de Vries Laboratorium, Universiteit van Amsterdam, Afdeling Palynologie, Sar-phatistraat 221, Amsterdam 4, The Netherlands.

4. Dr R. LETOUZEY, Laboratoire de Phanérogamie, Muséum National d'Histoire Naturelle, 16, rue Buffon, 75005 Paris, France.

5. Dr B. ROLLET, 7, Villa Cœur de Vey, 75014 Paris, France.

6. Prof. G. H. ORIANS, University of Washington, Department of Zoology, Seattle, Washington 98195, U.S.A.
Dr C. R. CARROLL and Dr B. CARROLL BENTLEY, Department of Ecology and Evolution, Division of Biological Sciences, State University of New York at Stony Brook, Stony Brook, N.Y. 11794, U.S.A.
Dr T. W. SCHOENER, Harvard University, The Biological Laboratories, 16 Divinity Avenue, Cambridge, Massachu-setts 02138, U.S.A., and
Dr T. ZARET, Department of Zoology, University of Washington, Seattle, Washington 98105, U.S.A.

7. Prof. O. J. SEXTON, Washington University, Department of Biology, St Louis, Missouri 63130, U.S.A.

8. Dr P. S. ASHTON, University of Aberdeen, Department of Botany, St Machar Drive, Aberdeen AB9 2UD, United Kingdom.

Drs M. J. HOPKINS, L. J. WEBB and W. T. WILLIAMS CSIRO, Division of Plant Industry, Rain Forest Ecology Section, Long Pocket Laboratories, Private bag no. 3, Indooroopilly, Queensland 4068, Australia, and
Mr J. PALMER, Commonwealth Forestry Institute, South Parks Road, Oxford, United Kingdom.

9. Mr R. G. FONTAINE and
Prof. A. GÓMEZ-POMPA and Mrs B. LUDLOW, Instituto de Biología, Departamento de Botánica-Universidad Nacional Autonoma, Apartado Postal 70-268, Mexico, 20 D.F., Mexico.

10 and 11. Dr F. B. GOLLEY, Institute of Ecology, The Rockhouse, University of Georgia, Athens, Georgia 30602, U.S.A.

12. Dr F. FOURNIER, ORSTOM, 24, rue Bayard, 75008 Paris, France (and Unesco's Division of Ecological Sciences).

13. Dr F. GOLLEY
Prof. B. HOPKINS, New England College, Tortington Park, Arundel, Sussex BN18 ODA, United Kingdom, and
Mme F. BERNHARD-REVERSAT, Laboratoire de Botanique, ORSTOM, B.P. 1386, Dakar, Senegal.

14. Dr B. GRAY, P.O. Box 61, Beerwah, Queensland 4519, Australia.

15. Dr. P. KUNSTADTER et al., East West Population Institute, 1777 East West Road, Honolulu, Hawaii 96822, U.S.A.

16. Dr J. V. G. A. DURNIN, Institute of Physiology, University of Glasgow, Glasgow, United Kingdom.

17. Prof. R. S. DESOWITZ, Department of Tropical Medicine and Medical Microbiology, School of Medicine, University of Hawaii, 3675 Kilauea Avenue, Honolulu - Hawaii, U.S.A.

18. Dr O. G. EDHOLM, School of Environmental Studies, University College of London, Gower Street, London WC1E 6BT, United Kingdom.

19. Prof. G. SAUTTER, Département de Géographie, Université de Paris I, 12, place du Panthéon, 75231 Paris Cedex 05, France.
Prof. J. BARRAU, Laboratoire d'Ethnobotanique, Muséum National d'Histoire Naturelle, 57, rue Cuvier, 75231 Paris Cedex 05, France.
Prof. P. GOUROU, 13, place Constantin Meunier, 1180 Bruxel-les, Belgique.

20. Mr R. G. FONTAINE
Prof. D. J. GREENLAND, Department of Soil Science, The University of Reading, London Road, Reading RG1 5AQ, United Kingdom.

Dr R. Herrera, Instituto Venezolano de Investigaciones Científicas, Centro de Ecología, Apartado 1827, Caracas 101, Venezuela.

Dr G. W. Ivens, Agronomy Department, Massey University, Palmerston North, New Zealand, and

Mr J. Palmer

21. Mr R. G. Fontaine
Dr J. P. Milton, Threshold, Suite 113, 1785 Mass. Ave., N.W., Washington, D.C. 20036, U.S.A.
Mr J. Palmer and Mr S. L. Pringle, Forestry Department, FAO, Via delle Terme di Caracalla, Rome 00100, Italy.

Regional contributions

Africa

Office de la recherche scientifique et technique outre-mer (ORSTOM), 24, rue Bayard, 75008 Paris, France.

Centre technique forestier tropical, 45 bis, avenue de la Belle Gabrielle, 94130 Nogent-sur-Marne, France.

Land Resources Division, Ministry of Overseas Development, Tolworth Tower, Surbiton, Surrey KT6 7DY, United Kingdom.

Asia and Oceania

Dr S. K. Seth, Ministry of Agriculture and Irrigation, Department of Agriculture, Krishi Bhavan, New Delhi, India, and
Dr O. N. Kaul and Dr P. B. L. Srivastava, Forest Research Institute and Colleges, P.O. New Forest, Dehra Dun, Uttar Pradesh, India.

Dr S. Sabhasri and Dr S. Aksornkoae, National Research Council, Paholyothin Road, Bangkok 9, Thailand.

Prof. P. Ho, Université de Saïgon, Faculté des Sciences, Département de Botanique, Boîte postale A-2, Ho Chi Minh (Saïgon), Viet Nam.

Dr T. C. Whitmore (for the Malay and Melanesian archipelagos), Commonwealth Forestry Institute, South Parks Road, Oxford OX1 3 RB, United Kingdom.

Dr B. L. Lim (Malaysia), Institute of Medical Research, Department of Medical Ecology, Jalan Pahang, Kuala Lumpur 02-14, Malaysia.

Dr J. A. Bullock (for Malaysia), Department of Zoology, School of Biological Sciences, University of Leicester, Adrian Building, University Road, Leicester LE1 7RH, United Kingdom.

Dr F. L. Dunn (for Malaysia), Department of International Health, University of California, San Francisco, California 94143, U.S.A.

Dr D. Sastrapradja and Dr K. Kartawinata, Indonesian Institute of Sciences, Botanical Garden, Djl. Teuku Tjhik Ditiro No. 43, Jakarta, Indonesia.

Prof. T. Shidei, Department of Forestry, Faculty of Agriculture, Kyoto University, Kyoto 606, Japan.

Prof. J. D. Ovington, Atmospheric, Marine and Living Resources Division, Department of the Environment and Conservation, P.O. Box 1937, Canberra City, A.C.T. 2601, Australia.

Prof. R. W. Hornabrook (for South-East Asia and Melanesia), Institute of Medical Research, P.O. Box 60, Goroka E.H.D., Papua New Guinea.

Latin America

Dr S. Capote, Laboratorio de Fisiología Vegetal, Instituto de Botánica de la Academia de Ciencias de Cuba, Calzada del Cerro nº 1257, La Habana, Cuba.

Prof. E. F. Brünig, Institut für Weltforstwirtschaft, Bundesforschungsanstalt für Forst und Holzwirtschaft, Leuschnerstrasse 91, 2050 Hamburg 80, Federal Republic of Germany, and

Dr H. Klinge (for South America), Abteilung Tropenökologie, Max-Planck-Institut für Limnologie, Postfach 165, Plön 232, Federal Republic of Germany.

Dr J. Murça-Pires, Instituto de Pesquisas Agropecuárias do Norte, Caixa Postal 48, Belém, Pará 66000, Brasil.

Dr W. A. Rodrigues, Instituto Nacional de Pesquisas de Amazonia, Rua Guilhermo Moreira no. 112/116, Manaus, Brasil.

Prof. R. Crist, Department of Geography, University of Florida, Gainesville, Florida, U.S.A.

Participants to the synthesis workshops

Besides some of the authors already cited, the following consultants attended the two synthesis workshops and their comments and contributions were very helpful.

Prof. J. N. Anderson, Department of Anthropology, University of California, Berkeley, California 94720, U.S.A.

Prof. G. Aubert, Office de la recherche scientifique et technique outre-mer (ORSTOM), 70/74, route d'Aulnay, 93140 Bondy, France.

Dr J. Ave, Rijksmuseum Voor Volkenkunde, Steenstraat 1, Postbus 212, Leiden, The Netherlands.

Prof. P. T. Baker, Department of Anthropology, Social Sciences Building 511, University Park, Pa. 16802 U.S.A.

Prof. F. Bourlière, INSERM, Centre de Gérontologie, Unité de recherche 118, 29, rue Wilhem, 75016 Paris, France.

Prof. N. A. Burges, New University of Ulster, Coleraine, County of Londonderry, United Kingdom.

M. R. Catinot, Centre technique forestier tropical (CTFT), 45 bis, avenue de la Belle Gabrielle, 94130 Nogent-sur-Marne, France.

Dr. M. GODELIER, Collège de France, 11, place Marcelin Berthelot, 75005 Paris, France.

Dr R. M. LAWTON, Land Resources Division, Ministry of Overseas Development, Tolworth Tower, Surbiton, Surrey KT6 7DY, United Kingdom.

Prof. G. LEMÉE, Université de Paris XI, Laboratoire d'Ecologie, Faculté des Sciences, 15, rue Georges Clemenceau, 91405 Orsay, France.

Dr J. MOUCHET, Office de la recherche scientifique et technique outre-mer (ORSTOM), Services scientifiques centraux, 70-74, route d'Aulnay, 93140 Bondy, France.

Prof. P. W. RICHARDS, University College of North Wales, School of Plant Biology, Memorial Buildings, Bangor LL57 2UW, Now: 14 Wootton Way, Cambridge CB39LX, United Kingdom.

Dr A. P. VAYDA, Department of Human Ecology, Cook College, P.O. Box 231, Rutgers University, New Brunswick, N.J. 08903, U.S.A.

Foreword

Forests and woodlands form the natural type of vegetation cover in most of the humid and subhumid parts of the tropical world. They are important for many reasons, not least because of their role in meeting man's needs for timber, food and other economic, environmental and socio-cultural values. For many developing countries, the wise use of the resources provided by tropical forests, and the balanced rural development of zones covered by them, lie at the heart of their national development strategies and programmes. One aspect of these strategies and pro-grammes is the search for improved sylvicultural and affor-estation techniques, as well as ways and means for more effective use of tropical hardwoods, including the marketing of tropical timbers and pulping. Another aspect concerns the conversion of tropical forests on a sustained yield basis to tree crop plantations and to mixed agricultural systems, particularly at the village level.

Some progress has been made in certain of these prob-lem areas in particular parts of the world. In most areas, however, sound scientific information on which optimal development can be based is lacking or is not readily available in useable form. In many areas of the humid and subhumid tropics, there is a clear need for much greater information on the quality and availability of ter-restrial biological resources, and on the ecological and social constraints to successful change in land use and enhanced productivity on a sustained yield basis. Assess-ments are particularly needed on the ecological and social implications of different land-use options and management practices in specific situations.

At the same time, better use should be made of the information that is already available on tropical forest eco-systems. Mechanisms are needed for the effective and rapid communication of relevant scientific information, and its timely insertion in the key policy-making processes. These mechanisms must be developed primarily at the local and national levels, since it is at these scales that the problems of resource use and ecosystem management have essentially to be tackled.

In addition, the synthesis and exchange of information between and among countries which share similar ecological conditions and comparable socio-economic and land-use problems, must be encouraged if more efficient use is to be made of existing knowledge. It is here that international organizations have a traditional role to play and a particular function to fulfil. Recognition of this role and function led Unesco, UNEP and FAO to agree to undertake a collation and synthesis of existing information on tropical forest ecosystems, and to make this information available in a single volume. The present publication represents the result of this co-operative venture, which was undertaken in consultation with other international intergovernmental and non-governmental organizations and with a large num-ber of national institutions and individual scientists.

This Report aims to provide a clear summary of know-ledge of the structure, functioning and evolution of tropical forest ecosystems and of the human populations that live within and around these ecosystems. It also describes some of the main patterns of use by man of these ecosystems. In presenting this information, the Report attempts: to identify gaps in knowledge; to present recommendations for future research; to indicate appropriate methodologies for problem-oriented studies on tropical forest ecosystems; to describe examples of land and resources management, and to examine some of the reasons for success or failure in specific concrete situations; and thus to highlight needs for future orientation in land-use development and man-agement strategies.

Even a report as large as this one cannot be an exhaus-tive and encyclopaedic work. In the areas of management, marketing, economic development, etc., the main facts have been summarized. It was neither possible nor desirable to enter into great detail; this is especially so as FAO is preparing a publication on these aspects, which will comp-lement this state-of-knowledge report; it is anticipated that the two studies will together adequately cover the subject of tropical forest ecology, exploitation, management and development.

The Report does not give equal emphasis to the various types of tropical forest ecosystems. It is mainly concerned with the lowland evergreen rain forests of the humid tropics, but whenever possible other types of dense tropical hu-mid forest ecosystems are included. The so-called semi-deciduous and deciduous or seasonal moist forests are therefore mentioned in several parts of the Report. Apart from a case study on the African miombo, the dry tropical forests—whether open or closed in structure—are generally excluded. They will to a certain extent be treated in another state-of-knowledge report on tropical grazing land ecosys-tems, to be published by Unesco in 1978.

There are some departures from the planned shape. Initially, the idea was to base the Report on information from integrated studies on man's relationships with tropical forests in specific geographic areas. At an early stage, it became apparent, however, that there were very few published studies on tropical forest ecosystems which had economic, social, cultural, biological, and physical dimensions. There was, and is, a general lack of case histories of assessments of land-use options and management strategies which incorporate short- and long-term ecological and social constraints and which operate at ecologically and socially acceptable scales in time and space. This lack of published information from interdisciplinary research meant that a somewhat traditional and sectoral structure was adopted for this Report—a structure which reflects the main groupings of disciplines involved.

The Report comprises three principal parts. Part I chiefly contains the research results of biologists and other natural scientists. Fourteen chapters deal with the description, functioning and evolution of tropical forest ecosystems, covering both disturbed and undisturbed situations. It includes information on composition, structure, biomass, primary and secondary productivity, water budgets, nutrient cycling, energy flow, stability, species interactions, succession, growth, regeneration, pests and diseases. In certain chapters, some comparative data are presented for food and tree crops which have replaced tropical forests in many lowland areas.

Part II will be of particular interest to demographers, social anthropologists, ethnologists and other scientists, as well as resource managers and land planners. It is mainly concerned with the biological behaviour and socio-cultural aspects of the human populations living in and around tropical forest ecosystems and with their patterns of use and management. In several chapters, attention is drawn to the role that the understanding of traditional resource use strategies of local populations can play in the development of new management techniques. Emphasis is also given to the key relationship that exists between traditional land-use systems and their associated social systems on the one hand, and the ecological constraints of the biological and physical environments in which they exist on the other. Part II ends with an exposé of the policies of forest protection, conservation, management and development, as they are related to scientific and technical knowledge and to socio-economic considerations.

Part III of the Report contains eight regional case studies which describe specific tropical forest ecosystems from various viewpoints (either basic research or utilization and management) to illustrate the kind and orientation of research.

The preparation of the Report was started in 1974. Following discussions between the international organizations concerned, contacts were made with a wide range of national and regional institutions and individual scientists concerned with various aspects of research and management of tropical forest ecosystems, and their inputs to the Report were solicited. These contributions were passed on to consultants who, working at world level, drafted the various chapters of the Report. Subsequently two workshops were held in June 1975, at Unesco's headquarters in Paris, to assess and review the various contributions and to try to achieve an integrated and balanced presentation of the information which would meet the requirements of the Report. The final editing of the work started in 1976.

In editing contributions to the Report, an attempt was made to achieve harmony in presentation and consistency in the use of terms and abbreviations, without suppressing individual styles. Each chapter bears the individual stamp of its author. In some cases, authors have inclined towards particular viewpoints which may prove provocative. The primary editorial aim has been to ensure accuracy of facts while retaining the privilege of authors to draw inferences. As in any multi-author book, there is inevitably a certain unevenness in the comprehensiveness and detail of the various chapters. In order to help the specialist, an effort has been made to give each chapter some autonomy, while linking it to the general requirements of the Report. Each chapter ends with a concise conclusion, an identification of gaps in knowledge, recommendations for future research and a selective bibliography. For several chapters, the major references are indicated in the bibliography with an asterisk; the absence of asterisks in other bibliographies indicates that these are already selective. Furthermore, for documentation purposes, several references of general interest, not quoted in the text, are added in the bibliographies.

It is hoped that this Report will provide, for a number of years, a useful source of facts and ideas for all those interested in tropical forest ecosystems. In particular, it is hoped that the Report will prove of practical value to those concerned with the design and conduct of research and management programmes in the countries and regions concerned. The Report should also be helpful to a variety of national and international activities in the fields of training, education and information exchange.

Part I

Description, functioning and evolution of tropical forest ecosystems

Part 1

1 Inventory and survey: international activities

Introduction

The determination of the extent and the growing stock of the world's forests has been the subject of continuous interest since the beginning of the century not only because of their significance in the estimation of potential raw material but also because of their stabilizing role in the biosphere. This is why this question received so much attention during the United Nations Conference on the Human Environment at Stockholm in 1972 and why it was made the subject of a resolution. However, in taking into account the requirements for a continuous monitoring of the world's forests, the Secretariat of UNEP has limited its proposal to that of tropical forest which is particularly threatened at the moment by increasing clearing for cultivation, by overgrazing, by intensive forestry and by the development of road networks. One other justification for this choice could equally well be the interest shown at the present time in tropical climates and the possibility of establishing correlations between regression of forested areas and climatic change.

Actual trends, both in reduction of the wooded area of the tropics and in its quality and quantity, are strikingly portrayed by the statistics. In one province of Venezuela, Veillon (in Hamilton, 1976) has shown that the ratio of forest cover to total land area decreased from 56 per cent in 1956 to 37 per cent in 1970. In Indonesia more than two thirds of the forestry exploitation concessions have been granted during the last ten years and out of 40 million exploitable hectares of forest, 12.5 million are already under exploitation, 4 million are under concession and 10 million are the subject of inventory for future concessions.

It is not a question of conserving forest at any price and for sentimental reasons, or of opposing their transformation into sylvicultural, agricultural or pastoral land for the economic and social development of countries in which the ratio of forest area to total land surface is still relatively high. The aim of forest conservation is their rational use as well as the satisfaction of the needs of mankind. It is simply a question of drawing attention to the necessity of maintaining under woodland not only the areas necessary to supply industries, but also for the stability of the environment in areas where ecological and economic studies have shown that transformation into agricultural or pastoral land is not profitable in the long term.

It must also be stressed that the inventory techniques to which reference is made are founded on the basis of widespread present uses of wood, such as timber, industrial wood and fuel-wood. New ideas, e.g. on the complete use of the tree from roots to leaves, or on 'material engineering', could change the terms of inventory.

Inventory and mapping syntheses raise the problem of defining the forests in question. This Report concerns mainly the humid tropical forests without a marked dry season, especially rain forest and semi-deciduous dense humid forest; open woodland and wooded savanna are excluded. This chapter extends somewhat beyond this definition by treating tropical forests as a whole, so as to take into account international projects in progress, which are not limited to humid forest but deal with all forests between the tropics, with some adjustments for countries located on the tropics. The chapter deals specifically with efforts made since the second world war, especially by FAO and UNEP, to gain a better understanding of the forest cover through the continuous monitoring of tropical forests. This is followed by examination of the mapping syntheses now in progress at world level, especially those of Unesco, but including those at regional or sub-regional level which could usefully complete these inventories by locating the different types of forest and making valid comparisons possible. Classification is extremely important for a mapping synthesis. But synthesis of different environmental factors is hampered by methodological problems as well as the absence of available data. Physiognomic and structural criteria could be used for definition of major divisions within which a more exhaustive ecological classification can be undertaken.

The utility of remote sensing techniques for continuous monitoring of tropical forest is also discussed, as is air photography, including new remote sensing techniques such as air-borne radar imagery (e.g. SLAR radar) and satellite photography in the visible and near infra-red spectra.

Finally, mention will be made of efforts to carry out model programmes for data processing for forest inventory, and to facilitate education and training in this field.

Area covered by tropical forests

At the end of the last century and before the first world war various attempts were made to assess world forest resources, including tropical forests, particularly on the occasion of forestry or universal congresses. One of these assessments was presented to the Exposition universelle in Paris in 1900 where an attempt was made to assess forest potential vis-a-vis present and future needs. However, the first comprehensive attempt to estimate the forest resources of the world was made by Sparhawk (in FAO, 1963) of the United States, immediately after the first world war. The results of this enquiry were published in 1922 and served as a standard source of reference throughout the inter-war period. Illevesallo and Jalava (in FAO, 1963) of Finland made further investigations in 1928 and 1931. Streyffert (in FAO, 1963) of Sweden, in 1930, published an important study of the coniferous resources of the world and in 1956 a *World Geography of Forest Resources* was published by the American Geographical Society. In the meantime the International Institute of Agriculture in Rome published annual estimates

supplied by national governments between 1933 and 1938. In 1946 available information on forest cover was summarized by the FAO Secretariat in a report submitted to the Second Session of the Conference of this Organization under the title *Forestry and Forest Products — World Situation, 1937–1946*. In the same year Sir Hugh Watson of United Kingdom made an independent study which appeared in the *(British) Empire Forestry Handbook*. The sources used in these studies were national reports, published material, correspondence and travel questionnaires.

FAO world forest inventory

At the First Session of the FAO Conference in 1945, a recommendation was made that a world forest inventory be undertaken as soon as possible. The first FAO world forest inventory was carried out in 1947–1948 and the results were published in the 1948 FAO report entitled *Forest Resources of the World*. At the Sixth Session of the FAO Conference in 1951, it was recommended that available information on the forest resources of all countries of the world be collected and published, at five year intervals, and FAO world forest inventories were published subsequently, covering 1953, 1958 and 1963. All these inventories were based on the replies from national services. When insufficient or no information was provided, the best possible estimates were made. During the same period the FAO Secretariat, in liaison with UN Regional Economic Commissions, published regional studies on timber trends and prospects in which the resource situation and the projected demand was analysed. All these regional studies have been updated and summarized in a special issue of *Unasylva* (FAO, 1966).

However, while the FAO World Forest inventories have been largely used for analysing world forest resources trends and capabilities and for defining broad guidelines for regional forest policies, they have proved to be inadequate for forest policy formulation and forest development planning at regional, national and local levels. Furthermore, the validity of the questionnaire approach, based on definitions (forestland, management plans, growing stock, increment, etc.) which were interpreted differently by the various governments, has been questioned. The fact that these world inventories were based on data supplied by individual countries means that their accuracy and reliability varies according to the quality of the information available to the compiler in the country. Consequently, a new regional approach was considered by FAO. The inventory efforts were to be carried out for a series of regions and on a continuing basis. For this reason, the World Forest Inventory, which was started in 1968, was planned to cover one region at a time. Over a five year period the whole world should have been covered. However, the whole idea of regional questionnaires was subsequently abandoned. In the meantime, a mapping approach, combining vegetation maps and local detailed inventories, was discussed, while a pilot-study on the forest resources of Africa was carried out by Persson (1975); this could help to elaborate a new method to be applied for forest resources appraisals in tropical countries. This method may be summarized as follows: identification of available information; elaboration of statistical tables showing some of the main features, such as forest areas, ownership, management plans, etc.; and preparation of maps, for each of the important forestry countries. These maps should if possible show the location of forests, forest types, the location of inventories already undertaken, the location of forest resources, the exploited areas and areas under concession, the transport systems, the location of plantations, etc.

The FAO/UNEP project of monitoring tropical forest cover

In view of the above mentioned difficulties in carrying out a world forest inventory, FAO presented to the UN Conference on the Human Environment held in Stockholm in 1972 an action proposal for a World Forest Appraisal Programme, with the main purpose of a continuous monitoring of the world forest cover. This was endorsed by the Conference. As a result of this recommendation UNEP requested FAO to formulate a global project for the monitoring of tropical forest cover. The project was limited to this ecological zone since threats to the forest cover are much more severe here. This project has to be viewed within the framework of the Global Environmental Monitoring System (GEMS) and will have the following objectives:

— obtaining data on the present forest cover of each forest type and forest condition class for each tropical country as a whole and for parts of it;
— determining the quantitative and qualitative changes which have taken place in the tropical forest cover during the last 20 years and which are considered to have a significant impact on the stability of the biosphere (e.g. on the heat balance at the surface of the earth);
— making proposals for a periodic assessment of the forested area and consequently of the changes in forest cover for tropical areas.

This project is also to be considered within the whole framework of FAO activities in the monitoring of natural resources (water, soil, vegetation, fish, etc.). FAO intends to maintain close contact with Unesco, particularly regarding its related activities, such as Project 1 (*Ecological effects of increasing human activities on tropical and subtropical forest ecosystems*) of the Man and Biosphere Programme, and the work on small-scale vegetation mapping. The following concepts were included:

— changes in forest cover are the most important results that this programme can provide, it being understood that the error to be expected in their estimation could be larger than the estimate of forest cover at a given time; it is therefore all the more essential to use objective measurements and not rely upon questionnaires;
— on the statistical side, continuous forest inventory (CFI) methods should be used;
— stratification would include some kind of land-use classification; reference was made in this connection to the recent Unesco vegetation classification;
— remote sensing techniques are to be used, including extensive study of very small-scale imagery (especially

satellite imagery) over the whole vegetation cover of the tropical belt (first phase), supplemented by a detailed assessment and analysis of changes in the vegetation cover in several endangered areas (second phase).

Total area, forest area and population in the main tropical countries as compiled by the FAO Secretariat are reproduced in table 1. An attempt has also been made to collect for some countries in Africa, Asia and tropical America data concerning the moist tropical forest areas only, and to break down these figures into evergreen and semi-deciduous forests. The results are given in table 2.

A tentative FAO estimate (1976) indicates that the total area of tropical moist forest is close to 935 million ha, while the entire tropical moist forest climax area is *ca.* 1 234 million ha.

Pilot-project for monitoring tropical forest cover

A pilot-project on tropical forest monitoring has been included in UNEP's activities and falls within the general functional task of environmental assessment (Earth Watch) and more particularly within the scope of the Global Environment Monitoring System (GEMS).

The main objective of the project is to implement in four adjoining countries of tropical Africa (Togo, People's Republic of Benin, Nigeria, Cameroon) the monitoring of the forest cover with a view to:

— obtain for these four countries data on the present forest cover and on the quantitative and qualitative changes of the forest cover which have occurred in critical areas;
— refine, test and possibly correct the general methodology, as outlined in an earlier project, by applying it to these four countries;
— prepare the extension of the tropical forest cover monitoring project to the whole African tropical and subtropical region and to the Latin America and Asia regions.

These four African countries were selected for the UNEP/ FAO project mainly because of the existence of related UN activities and similar ecological conditions.

National forest inventories

The FAO has carried out a large number of national inventories within the framework of the UNDP and has prepared publications on the methodology used. A new manual on forest inventory has been published (FAO, 1973). This manual is especially intended for foresters who have the task of evaluating and managing mixed tropical forests. Matters such as the accessibility of forests and the treatment of data are also considered. The importance of accessibility in the resource inventory must be stressed as the costs of felling and transportation play an ever-increasing role in determining the cost of the finished product.

Very large countries are, of course, particularly concerned with the accessibility of their forest resources. In the forest inventories of the USA, the wooded areas are assigned a coefficient of accessibility; in the USSR, the forest literature of the 1950s is particularly rich in articles dealing with the accessibility of Siberian forests with reference to supplying factories in European Russia with wood from this region. In tropical countries, especially in Africa, one speaks of forests of the first and second zones; the role of the ex-

tension of the 'trans-gabonais' railway in the development of the second zone is of interest.

At the international level, the United Nations Conference on Science and Technology (1963) proposed limits beyond which the cropping of certain products was no

TABLE 1. Total area, forest area and population in tropical countries[1].

I. AFRICA

Country	Total land area in 1 000 km² (whole country or part within tropics)	Approximate area of closed forest, open woodland and scrub in 1 000 km²	Population in 1 000	Approximate population density per km²	Approximate percentage of population which is urban
Angola	1 250	430	5 670	5	
Botswana	400	20	500	1	4
Burundi	30	—	3 400	113	3
Cameroon	480	300	5 950	12	20
Central African Empire	620	400	1 670	3	27
Chad	1 280	50	3 790	3	8
Congo (People's Republic of)	340	200	980	3	26
People's Republic of Benin	110	50	2 870	26	15
Equatorial Guinea	30	20	290	15	?
Ethiopia	1 220	740	25 930	21	10
Gabon	270	200	500	2	32
Gambia	10	—	380	38	15
Ghana	240	190	9 090	38	33
Guinea	250	200	4 110	16	12
Guinea Bissau	40	—	630	16	?
Ivory Coast	320	170	4 530	14	?
Kenya	580	90	12 070	21	10
Liberia	110	80	1 620	15	26
Madagascar	590	360	6 900	12	15
Malawi	120	20	4 670	39	10
Mali	1 240	150	5 260	4	15
Mauritania	1 170	—	800	1	?
Mauritius	20	20	850	43	?
Mozambique	780	40	8 230	11	?
Niger	1 270	—	4 210	3	10
Nigeria	920	310	58 020	63	10
Rhodesia	390	220	5 690	15	20
Rwanda	30	—	3 900	130	5
Senegal	200	120	4 120	21	26
Sierra Leone	70	70	2 630	38	?
Somalia	640	360	2 940	5	?
Sudan	2 510	130	16 490	7	12
Tanzania	950	700	14 000	15	10
Togo	60	50	2 070	35	15
Uganda	240	30	10 460	44	8
Upper Volta	270	50	5 610	21	?
Zaire	2 350	1 200	22 860	10	18
Zambia	750	370	4 420	6	30
Total	22 150	7 340	267 110		

1. Extracted from the FAO/UNEP report on the formulation of a tropical forest cover monitoring project (FAO, 1975).

II. SOUTH AND CENTRAL AMERICA

Country	Total land area in 1 000 km² (whole country or part within tropics)	Approximate area of closed forest, open woodland and scrub in 1 000 km²	Population in 1 000	Approximate population density per km²	Approximate percentage of population which is urban
Bolivia	1 100	450	5 190	5	35
Brazil	7 900	4 660	90 000	11	55
Belize	20	20	120	6	?
Colombia	1 140	650	22 490	20	53
Costa Rica	50	40	1 840	37	35
Cuba	110	20	8 750	80	60
Dominican Republic	50	20	4 300	86	40
Ecuador	280	220	6 510	23	60
El Salvador	20	—	3 760	188	40
French Guyana	90	70	50	1	80
Guatemala	110	70	5 600	51	34
Guyana	210	190	750	4	30
Haiti	30	—	5 670	169	20
Honduras	110	70	2 690	24	32
Jamaica	10	—	1 930	193	37
Mexico	850	160	25 000	29	60
Nicaragua	150	70	1 990	13	45
Panamá	80	50	1 520	19	48
Paraguay	200	150	1 200	6	36
Peru	1 290	700	14 460	11	53
Puerto Rico	10	—	2 810	281	?
Surinam	160	120	410	3	80
Trinidad and Tobago	10	—	1 040	104	20
Venezuela	910	480	10 970	12	72
Total	14 890	8 210	218 450		

III. ASIA, OCEANIA (p.p.)

Country	Total land area in 1 000 km² (whole country or part within tropics)	Approximate area of closed forest, open woodland and scrub in 1 000 km²	Population in 1 000	Approximate population density per km²	Approximate percentage of population which is urban
Australia	2 500	150	2 000	1	83
Brunei	10	10	100	10	16
Burma	470	270	9 000	19	20
Democratic Kampuchea (Cambodia)	180	130	6 700	37	15
India	1 720	470	300 000	174	20
Indonesia	1 900	1 220	121 300	64	18
Laos	240	150	3 110	13	15
Malaysia	340	240	10 800	32	29 (West) 16 (East)
Papua-New Guinea	480	330	1 700	35	?
Philippines	300	180	39 040	130	32
Sri Lanka	70	40	13 030	186	20
Thailand	510	310	36 290	71	20
Viet-Nam	330	140	41 390	125	20–30
Total	9 050	3 600	584 460		

IV. SUMMARY

Region	Total land area in 1 000 km²	Percentage	Closed forest, open woodland and scrub in 1 000 km²	Percentage	Closed forests in 1 000 km²	Percentage of forest which is closed	Population	Percentage
Tropical Africa	22 150	48.1	7 340	38.3	2 100	29	267	25.0
South and Central America	14 890	32.3	8 210	42.9	5 900	72	218	20.4
South-East Asia and Australia (without Oceania)	9 050	19.6	3 600	18.8	3 000	83	584	54.6
Total	46 090	100.0	19 150	100.0	11 000		1 069	100.0

Sources :
Various maps and tables of the *World forest atlas* (Federal Research Institute of forestry and forest products, Reinbek, Federal Republic of Germany).

Fischer's Weltalmanach (1974).
World forest inventory (FAO, 1963).

onger profitable. For timber the situation has changed markedly with new technology, such as containerized transport and new uses for wood chips and particles. In national and local planning, it is often necessary to identify those forests which should be inventoried first and, among these, the units which will permit the highest level of profit. It is this need which induced FAO to try to develop a method

TABLE 2. Area of moist tropical forest types in Africa, America and Asia.

	Evergreen rain forests	Semideciduous forests	Total area of the tropical humid forests (in million ha)
AFRICA[1]			
Cameroon	6.5	6.5	13
Ivory Coast	4.5	4.5	9
People's Republic of the Congo	3	7	10
Gabon	17	5	22
Central African Empire	0.75	3	3.75
Zaire	50	25	75
Madagascar	1.5	4.5	6
TROPICAL AMERICA[2]			
Bolivia	25	10	35
Brazil	362	101	463
Colombia	49	22	71
Ecuador	14	4	18
Guyana	14	—	14
French Guyana	7	—	7
Peru	62	6	68
Surinam	7	—	7
Venezuela	17	25	42
ASIA[3]			
India	4.5	1.8	6.3[4]
Sri Lanka	0.2	0.1	0.3
Burma	—	—	6[5]
Thailand	—	—	10[6]

	Evergreen rain forests	Semideciduous forests	Total area of the tropical humid forests (in million ha)
Indonesia	89.2	1.4	90.6
Malaysia	21[7]	—	21
Philippines	10[7]	—	10
Democratic Kampuchea (Cambodia)	—	—	6[8]
Laos	—	—	5[8]
Viet-Nam	—	—	6[8]

1. According to Catinot (personal communication).
2. Calculated from the vegetation map by Hueck (1966).
3. Mangroves and various kinds of swamp forests represent 100 000 ha in Kampuchea, 14 000 000 ha in Indonesia, 2 000 000 ha in Malaysia, 500 000 ha in the Philippines and 1 500 000 ha in Burma.
4. Add 23 000 000 ha of moist deciduous forests.
5. Add 14 000 000 ha of moist deciduous forests.
6. Add 5 500 000 ha of moist deciduous forests (monsoon forests).
7. Including the semi-deciduous forests.
8. Mainly moist tropical broad-leaved and mixed deciduous forests, also called monsoon forests.

for grading forests according to their cost of exploitation; however, it is sometimes difficult to define the exploitable volume per hectare, taking into account the influence of both the international and local market on profitability.

Accessibility for a particular stage of inventory, at a given degree of precision, can perhaps be defined from answers to the following questions: what quantity of wood, described in terms of dimensions and quality, can be delivered to a determined market within certain cost limitations per unit volume? The method adopted must also answer the following:

— what are the parameters to collect?
— how can these parameters be qualified or graded?
— which quantitative equations must be used to infer from the parameters of accessibility, the cost of administration and those of hauling and transport?

An accessibility index which represents the cost of exploitation per cubic meter could be calculated for each unit under consideration. Three types of data are needed to determine this index:

— data concerning the condition of the forest, i.e. on trees, stands and terrain;
— socio-economic data on the available labour force and its cost, equipment (cost, maintenance, fuel, etc.) and the institutional framework (transport legislation, railway tariffs, etc.);
— general data concerning exploitation, and drawn from studies on the forest work and the control of the production and its costs.

A number of methodological problems concerning the collection of data remain unsolved or require certain modifications:

— Methodology. It is possible to subdivide the accessibility index into: mean cost of exploitation per unit volume, delivered on road, and mean cost of transport to the mill, thus permitting consideration of various locations for the mill. This underlines the importance of a good description of the terrain conditions in the inventory.
— Exploitation systems. The costs of exploitation have been established on the basis of exploitation models derived from studies of forest work and the outcome of applying different technologies which together allow the establishment and quantification of various formulae. One must take into account not only the technical aspects but also the forest conditions under which these operations take place. As long as sufficient data on these operations are available, it is possible to define the limiting factors for each felling system and to develop formulae for the costing of partial operations of each system.
— Terrain classification. The aim of a terrain classification is to determine the topographic and environmental aspects of the terrain in quantitative terms in a manner that relates to the costs of exploitation. In this study, slope angle has been considered as the principal topographic factor and great efforts have been made to find a slope factor which has a real influence on felling and transportation. There are in fact two limiting slope values which are accompanied by a change in the technology used: the first (generally slopes of 50–60 per cent) necessitates the replacement of hauling on the soil by pulling by cable; the second (around 70 per cent) makes road construction difficult and increases the risk of erosion and land-slides. It is assumed that a correlation exists between slope and cost but further studies are needed.
— Production and costs. Problems arise from the availability of data concerning production and costing and their relationship to other variables covered by the inventory. Data are generally inadequate and, even when available, must be recalculated for adjustment to this exercise. This is especially the case for data on costs of felling, hauling, road construction and improvement of transport and handling.

Ecological questions are at last taking on an increasing importance and it will be interesting to link all inventories to an ecological classification system which enables comparisons to be made. This necessity was acknowledged at a meeting of experts on forest inventory (FAO, 1967) and the Holdridge system (1967) was selected as it can be applied to the whole range of global environments. It differs from other purely descriptive systems in that it is above all a classification of the existing relationship between climate and vegetation type. It was recommended, however, that this system must be applied only at the higher levels of classification of vegetation types.

Mapping syntheses

Although forest maps do not exist for all tropical countries, maps of vegetation, land-use or ecological maps exist for practically all these countries. The position is shown, at least for Africa and Asia, in two Unesco publications (1963, 1974). Similarly, a list of land-use maps and vegetation maps, catalogued by continent and by country, is given in the catalogue of maps published by the FAO/Unesco *Soil Map of the World* (FAO, 1973).

Criteria and classification adopted

The FAO/UNEP group on the monitoring of the tropical forests (FAO, 1975) defined tropical forest cover as that which falls between the tropics, with some adjustments in the case of countries crossed by the tropics, particularly on the Indian subcontinent (Bangladesh, Burma, India and Pakistan). The group examined the following useful classifications:

— the Yangambi classification (CCTA, 1956, see chapter 4) based principally on forest physiognomy but which also includes ecological distinctions;
— the ITC classification of humid tropical forests based chiefly on soil drainage conditions;
— Unesco's international classification (Unesco, 1973) based on the physiognomy and structure of the vegetation, linked to "habitats or ecologically important environments" and intended for vegetation mapping at a scale of 1/1 000 000 or smaller;

these three classifications, although based partly on physiognomic criteria, are not wholly suited to the work of air photograph interpretation;

— the key to vegetation of uncultivated areas (Blair Rains, in FAO, 1975) for interpretation of conventional air photographs.

The group concluded that:

— the value of height for vegetation classification intended for use primarily in interpretation of aerial or space imagery appears to be questionable since no agreement exists as to the effect of this height on the albedo;
— stand density appears to be the most important criterion to retain;
— a wide-ranging classification of density vegetation classes, for example 0–25 per cent, 25–75 per cent, 75–100 per cent, can be used for visual interpretation of satellite imagery;

— the first two levels of the Unesco classification could probably be used but some modification would be necessary beyond the third level.

For tropical America, and in particular for Central America, the ecological maps which give information on forest cover, most frequently use the Holdridge (1967) system based on the relationship between climate and vegetation type, taking into account both the physiognomic and structural characteristics as well as life-form and the macro-climate. In tropical humid Asia the classifications adopted at regional or national level for the establishment of a hierarchy of vegetation types use the relationship between temperature and rainfall. However, some countries have used other criteria, for example those proposed by Burtt Davy (1938) or Beard (1955), the latter correlating a physiognomic classification with the diagrammatic profile method derived from the work of Davis and Richards (1933, 1934, see chapter 5). In tropical Africa, maps have generally used the Yangambi classification (CCTA, 1956) which distinguishes between:

— forest formations controlled by climate (seasonal, or altitudinal);
— forest formations controlled by edaphic conditions (mangroves, swamp forests, periodically flooded forests and riverine forests);
— forest-savanna formations and grassland formations (open woodland, wooded savanna). See chapter 4.

The scales of vegetation or forest maps may be 1/5 000 000 for large countries such as Angola, Australia, Indonesia or Zaire or 1/800 000. e.g. for Gambia or Salvador. A scale of 1/1 000 000 is widely used. Maps of provinces, regions or districts are often at scales of 1/20 000 or 1/50 000.

A tendency is evident, especially in Africa, of using the phyto-sociological methods developed in Europe by the Zurich-Montpellier School, but there are other approaches for the definition of ecological groups which provide information on the structure and evolution of the forest. See chapter 4.

Tentative mapping syntheses

In Africa the most recent maps at 1/10 000 000 are those published by AETFAT (Association pour l'étude taxonomique de la flore de l'Afrique tropicale, 1958) and by FAO (1960). The AETFAT map uses the Yangambi classification; it has been revised by Unesco (1977) and will appear shortly at a scale of 1/5 000 000 using the Unesco classification (1973). FAO's map indicates grasslands and gives information on wooded savannas. Maps of the vegetation of western Africa at a scale of 1/5 000 000 and of eastern Africa at a scale of 1/4 000 000 are also available.

For Latin America, a vegetation map is included in the world agricultural atlas (Anon., 1969). This constitutes a valid attempt at a synthesis of forest vegetation, but the extensive forest plantations are not included. Unesco is currently preparing a vegetation map of Latin America at a scale of 1/5 000 000, using its own classification. A map at the scale of 1/20 000 000 was also published in the FAO/Unesco series on soils of the world and added to the soil

map published at a scale of 1/5 000 000 (1968). In the former map, the vegetation cover of South America has been divided into 10 principal ecological units; their distribution is linked to a limited number of vegetation types occurring throughout the Latin American subcontinent; these vegetation types are based on habitat (climatic or edaphic), physiognomy and vegetation structure. A phyto-ecological map of the Amazon Basin (Project RADAM) is also in progress and several sheets have already been published at 1/5 000 000; see Snobohn (1973).

In Asia the situation is more complex. Vegetation maps at small scales do not exist for the entire region but a map of forest types is available at a scale of 1/30 000 000. Nevertheless, maps exist for the Indian subcontinent (a map of the total vegetation at a scale of 1/19 000 000 revised by Champion and Seth (1968) representing 16 vegetation types at a scale of 1/15 000 000), for Malaysia and Indonesia at a scale of 1/1 500 000 and for other major areas, such as the Mekong valley and other parts of South-East Asia, 1/1 500 000 sheets are available. Maps at a scale of 1/1 000 000, showing vegetation and ecological conditions, were produced by the Institut français at Pondichéry in collaboration with the forest services and the research organizations of the countries concerned. These so far cover Sri Lanka and India up to 24° N. Maps of Democratic Kampuchea and Viet-Nam are in preparation.

On a world scale map no. 1 of the *World forest atlas* distinguishes between dense tropical forests and degraded tropical forests but the definition of dense forest is not clear and is not widely acceptable. A map of humid tropical forests was published under the aegis of Unesco in 1961 in *The Geographical Review* (see Fosberg *et al.*, 1961) but this is a synthesis of existing maps. Dasmann (1973) has published a report on a system for defining and classifying natural regions for conservation purposes. This report does not only contain definitions of the biotic provinces of the world but also gives maps, although at a very small scale, showing the limits of these provinces and, more especially, the different types of dense humid forests.

It must be stressed that mapping at a small scale (1/1 000 000, 1/5 000 000, etc.) does not give sufficient information either for the scientist or the decision maker and must be revised to take into account the inventories completed during the last 10 years. These maps have, however, considerable value in stimulating national activity and in providing a framework and a methodology.

The detailed distribution of tropical forest, and in particular that of humid tropical forest, is not yet well known and current attempts at regional synthesis should be encouraged in order to achieve a synthesis at a global scale. It is this task which has been assigned to the Unesco group in vegetation mapping at a small scale.

The use of remote sensing techniques for the continuous monitoring of tropical forests

The use of electromagnetic radiation for the purposes of civil and military observation and detection is already well established and foresters have been to some extent the forerunners in the use of panchromatic air photography in the evaluation and development of natural resources.

Continuous monitoring of tropical forests by conventional air photography

Two principal approaches are commonly used in the evaluation of forest resources: the estimation of the extent of forest stands and the evaluation of the average features of the stands (number of stems, volumes, etc.) per unit of surface. Visual interpretation of conventional air photographs (eventually completed by forest mapping and corrected by ground and air reconnaissance) has been the principal technique for estimation of the areas of stands in almost all tropical forest inventories. This applies equally to inventories at national and subnational level as well as to preinvestment surveys for the granting of large concessions, the setting up of processing plants, and even to local inventories for working out harvesting regulations in small forest areas. The interpretation of air photos is also used for the classification of forest stands according to criteria linked to their development, such as the different types of forest (on firm ground, on marsh; mixed or homogeneous), accessibility classes, density and the height of the dominant canopy trees, etc. Some of these criteria are equally useful in stratification which permit a more precise estimation of the average unit area characteristics for both total and partial forest inventories. This interpretation can be supplemented by additional information on boundaries separating administrative units and on types of tenure.

Almost all tropical forest inventories undertaken to date indicate the condition of forest stands at a given time. Very few tropical forests have been inventoried more than once, and current information on many large areas where rapid changes are taking place, such as the Amazon and Congo basins, is difficult to obtain. One notable exception is Thailand where a forest inventory at national level was started in 1971; shortly before this the whole region had been covered by inventories undertaken between 1957 and 1970. Nevertheless, in the face of the extent to which clearing of forests has taken place due to shifting cultivation and other activities, forest services and sometimes even individual foresters have taken the initiative in quantifying and sometimes mapping the *regression of the forest cover*, but generally without trying to assess the changes in stands by a continuous inventory in the field.

The most suitable method consists of comparing two air photo coverages, taken at two different times, with specifications (scale, emulsion) as similar as possible. If these two coverages are complete, it is then possible to map the forest limits on the two dates and assess, as accurately as possible by use of planimetry or gridding, the differences between them. This mapping exercise can also be applied to different forest strata and it is thus possible to describe the evolution of the forest cover according to the criteria of such a stratification (type of forest, accessibility, height, density, etc.). Work of this kind was undertaken by the Centre technique forestier tropical (CTFT) in Panamá (FAO, 1972). The forest cover of the Azuero peninsula was compared between 1956 and 1964: in an area of 215 000 ha of forest in 1954, a total of 55 000 ha of the edges (this being nearly 26 per cent) was cleared for agriculture. A similar study was completed in Sumatra by Schwaar (1971) in two small sample sites: the site which had never officially come within the zone of colonization, by 1969 had nothing left of the 2 000 ha of forest measured in 1954.

Fairly rough statistical estimates of changes occurring in forest cover may be sufficient for the initiation of appropriate remedial or other steps, or at least to draw attention to the magnitude of changes that have taken place since a certain date. It is sufficient to compare an objective sample of photographed land units taken at two or more intervals to give a quantitative assessment of these changes. If complete air photo coverage exists for the earlier period, estimates of changes in the forest area should, all things being equal, be more accurate. A frequently cited example is the study of the entire forest zone of Ivory Coast made between 1956 and 1966: the change in the forest cover was estimated by a regression sampling method applied to complete air photo coverage in 1956 and to a sample of photographic strips taken in 1966. 2.8 million ha of forest had disappeared between these two dates, this being *ca.* 30 per cent of the area under forest in 1956 (Lanly, 1969).

In some cases an existing air photographic record showing the condition of the forest at an earlier period is compared with observations made by air or ground reconnaissance. These methods are not always devoid of subjectivity and errors of observation, but can nevertheless be very useful when it is not possible to obtain new air photo coverage. This method was used for estimation of a 2.15 million ha area of forest cleared between 1954 and 1965 in the valleys of Magdalena and Sinu, Colombia (FAO, 1970). This clearing represented a reduction of *ca.* 40 per cent of the forest present in 1954. The particularly rapid depletion rate is directly attributable to colonization resulting from the construction of the railway connecting Bogotá to Santa Marta on the Atlantic coast. A similar evaluation of the decline of the forest in the Azuero peninsula, Panamá, between 1964 and 1972, showed that 30 000 ha could be added to the 55 000 ha which disappeared between 1954 and 1964; the forest of 1972 represented not more than 60 per cent of that which was present in 1954[1] (FAO, 1972).

1. The relative decline in rate of deforestation (5 500 ha per year between 1954 and 1964 as compared with 37 000 ha between 1964 and 1972) is probably not attributable to any measures taken to protect the forest but probably arises from the fact that accessible forests had been cleared and remaining forests were on steeper slopes or inaccessible.

Change does not always take place in the direction of forest regression. In some cases forests regenerate themselves, although this occurs less frequently because of the shorter fallow periods in shifting agriculture. Letouzey (1967) observed spontaneous growth of forest in the grasslands and tree savannas on the margins of the semideciduous dense forest of the Cameroon.

All studies reported to date have assessed forest cover two or three times, each assessment being separated by periods of from 7 to 15 years. These intervals are fairly large in view of the rapidity of change in many places. The rapidity of these changes is due to various factors. In the Ivory Coast, for example, deforestation has been associated with the development of a road network for forest exploitation: it has been estimated that deforestation was equal to or less than 100 000 ha per year in 1956 and had risen to or surpassed 500 000 ha in 1966. It is evident that monitoring will only be considered as continuous if the state of the forest cover is observed at intervals of time which are shorter when the rate of change increases. When applied to tropical forest, this time interval should be less than 5 years, at least in the critical areas.

Conventional air photography, as long as coverage exists, lends itself very well to such periodic evaluations. As such conventional coverage will continue to be used in topographic mapping, civil engineering, land registry surveys and cartography, as well as for forestry, and since it is always possible to take a large-scale photographic sample coverage with an ordinary camera from a light aircraft at relatively low cost, the information supplied by air photographs can generally be used for continuous monitoring. These methods are, however, rapidly being supplanted by the more recent techniques of remote sensing.

Continuous monitoring of tropical forests by new techniques of remote sensing

New remote sensing techniques

Panchromatic convential air photographs reproduce on a layer of *emulsion* of silver salts the relative intensities of *natural electromagnetic radiation within the visible spectrum* (the range being also limited by a filter), the *camera* being mounted on an *aircraft*. Many technical innovations have been made in one or other of the features italicized in the above definition.

The range and flight altitudes of aircraft used have kept pace with the development of remote sensing techniques. Thus, air photos of a very small scale (less than 1/100 000) can be obtained from aircraft flying at altitudes of around 20 000 m. At the other end of the scale, helicopters are being used to take views at scales of 1/1 000 or even larger. Nevertheless, the main innovation is the use of satellites, and to a lesser extent the use of rockets, these having sufficient height to take pictures over several hundreds of kilometres (see American National Research Council, 1970; Billings *et al.*, 1976).

The addition of the *near infra-red* band (0.78 to 1 μm approximately) to the other wave-lengths of the visible spectrum is a photographic technique which has been perfected for a long time, whether applied to black and white infra-red photography or to infra-red colour photography, the latter also being known as 'false colour'. Radiation in the *thermal infra-red* band (3-14 μm) has been used more recently, particularly in the detection of heat-emitting surfaces (hence its role in preventive detection of plant disease and forest fires). A further innovation, adapted recently to the study of natural resources, is the use of *radar* radiation (centimetric wave-lengths). In the short term at least, radar photography is perhaps the most useful remote sensing technique for continuous monitoring of tropical forests (see IUFRO, 1971, 1973; Latham *et al.*, 1972; Lanly, 1973, 1974; Howard and Lanly, 1975).

Emulsion covered film is now very often replaced by television systems or by electronic scanners. These consist of an optical apparatus which sweeps over the area to be observed and transmits the emitted or reflected radiation to a receiver which converts the variations in radiation intensity into electrical impulses. These scanners can be grouped so as to continuously measure a number of radiation bands (the recording apparatus used on the Earth Resources Technology Satellite, ERTS-1, was a multichannel scanner gathering data simultaneously from four neighbouring wave-lengths of the visible spectrum and the near infrared, the range of wave-lengths being from 0.5 to 1.1 μm).

One difference between conventional air photography, which records only the solar radiation reflected by the objects observed, and some remote sensing systems, is that the latter can also record 'artificial' radiation emitted by themselves and reflected by the objects observed. This is the case in the 'active' radar systems which are now widely used.

The principle of band selection is not new, as filters have been used for a long time in air photography to eliminate certain wave-lengths in the violet and blue bands. The principle of wave-length selection has also been used in the development of colour or 'false-colour' photography, since each layer of the emulsion receives a specific band of radiation.

It has been noted that the multichannel scanning apparatus used on board the ERTS-1 satellite made it possible to take separate images in the green and red bands and in the two near infra-red bands. Alternatively, several cameras can be grouped, or cameras with multiple lenses can record the reflection of defined spectral bands, this being done by each camera in the first case and each lens in the second case.

Conventional black and white or colour pictures can be obtained from non-camera detection systems which operate in the visible or infra-red bands of the spectrum or with radar radiation, and can be reproduced on a *television screen*. The coded measurements of radiation intensities received in any given band can be stored on magnetic tape for subsequent automatic data processing, or can be transmitted for analysis directly. Other tools for remote sensing data analysis include *simultaneous projection* which consists of obtaining a composite image derived from many images of the same scene taken in different spectral bands. The composite image can be adapted to the needs of interpretation by modulating the lighting intensity and choosing the appropriate filter for each image. Another method is the *densiometric analysis* of images. Here colour variations detected by spot-scanning an image are reproduced graphically, the detailed study of these curves permitting the identification of different elements (trees of a given species for example) after automatic comparison with sample curves.

The most important aspect of the new automatic information analysis techniques is their objectivity and uniformity, two qualities which cannot be completely guaranteed from visual analysis of air photographs.

Application of new remote sensing techniques to the continuous monitoring of tropical forests

Two new types of remote sensing which can be applied to the continuous monitoring of tropical forests are:
— air radar photography, in particular SLAR (Side Looking Air-borne Radar);
— photographs taken in the visible and near infra-red spectrum from satellites or rockets.

Although the second method has had more publicity, it is not yet certain that it is, in the immediate future, the most operational approach, especially for use in the equatorial zone.

Wave-lengths used in radar are between 0.5 m and 1 m long. Wave-lengths of the order of a centimetre are the most useful for the study of vegetation, as longer wave-lengths pass through the vegetative cover and are therefore more useful in the study of soils. Radar waves are not intercepted by cloud formations and are thus useful in equatorial zones where the cloud cover, if not permanent, is at the least both frequent and extensive.

In the SLAR system (ESRO, 1973), radar waves are emitted by an apparatus with a scanning antenna and are reflected by the soil or vegetative cover; no natural radiation is used. The radiation emitted is monochromatic, that is, only one frequency band is involved. The SLAR system works by a continuous transverse scanning of a strip of ground parallel to the line of flight and on one side of it; the inclination of the observation axis with respect to the vertical makes for better determination of the distances to the ground by use of chronometric measurement. The electric signals corresponding to radiation reflected along the scanning lines are continuously reproduced on a cathode ray tube and filmed. The resulting black and white radar imagery has the appearance of a black and white infra-red photograph but its interpretation is entirely different. As the strip is photographed continuously, there is no end lap. It is possible to produce side lap and to obtain a stereoscopic view of at least part of the area covered. Simultaneous recording of radiation reflected with horizontal and with vertical polarization, produces a slight pseudo-stereoscopic effect which helps in the perception of relief (Newton, 1973). All the characteristic features of the terrain such as crests, thalwegs and water-ways, can be easily identified and are enhanced by the shadows cast by the high points on the surrounding areas, in the opposite direction to that of the

plane (although the shadows hamper the detailed interpretation of the photograph). The original scale varies from 1/150 000 to 1/400 000. The resolution of the images (in other words, the smallest permissible distance between two objects giving signals of equal intensity) is a complex matter resulting from the interaction of many factors such as the wave-length of the radiation, the length of the antenna and the type of aperture. In practice these systems are capable of a resolution down to 10 m which may be considered as sufficient for the study of the vegetative cover.

The first example of tropical vegetation mapping over an extensive area by use of radar imagery was completed in 1965 in the province of Darien, in south-east Panamá (RAMP project; Crandall, 1969; Viksne *et al.*, 1970). The area covered was *ca.* 17 000 km² and the entire flight lasted only four hours. An acceptable planimetric map was prepared, first at a scale of 1/250 000 by use of 13 ground reflectors located at sites with known co-ordinates. Many thematic maps were produced, including a vegetation map at 1/250 000, which identified the following features:
— evergreen and semi-deciduous forests,
— palm forests,
— mangroves,
— swamps with low trees,
— swamps with trees,
— crops,
— plantations,
— logging areas,
— built-up areas.
Similar mapping exercises have been completed in other tropical countries, in particular in Nicaragua, New Guinea, Indonesia and Ecuador (FAO, 1974). In Brazil, radar mapping was carried out over 4.2 million km² within the framework of the RADAM project. The original scale of the pictures was 1/400 000 and the photographic strips, 37 km wide, present a lateral overlap of 25 to 50 per cent. This immense radar flying programme covering half the country was completed within the relatively short time of 6 months. It is self evident that the results of such a programme are of interest both for planimetric mapping at a small scale, and for vegetation mapping (Snobohn, 1973).

The real advantages of SLAR are of special interest: its use is possible in all climates, costs are relatively low (from 3 to 5 US $ per km² over large areas), very rapid completion; its planimetric value is sufficient for small-scale work and this feature makes it suitable for continuous monitoring of forest cover or land use. Some improvements could be made to permit mapping at a larger scale and to stratify forest-land more precisely and more objectively. Such improvements could result from greater resolution by the use of 'synthetic apertures', combined use of polarization and greater wave-length spacing (so as to reduce mottling of the images), as well as overlap of flight paths to permit stereoscopic study, and the development of programmes for automatic data processing.

No comprehensive programme of remote sensing which combines the advantages of radar with the repetitive coverage of orbital satellites has been devised to date. The completion of air-borne radar cover at regular intervals of several years should be considered very seriously by many tropical countries for the monitoring of forests and other natural resources. This would appear to be the best solution to the problem of continuous monitoring of forests in areas with considerable cloud cover.

The LANDSAT programme (follow-up of ERTS), and to a lesser extent the SKYLAB programme, have very rapidly popularized the observation of natural resources from satellites. Other experiments using rockets (Petrel or Skylark) have also been used for the assessment of natural resources from platforms located at heights of over 100 km. See Wilson (1970), Howard (1973), NASA (1973).

The first ERTS satellite was launched in July 1972 and the second in January 1975, both having a synchronous orbit with the sun at an altitude of 910 km, giving a total coverage of the earth in 18 days. Observations were made in zones ('scenes') in the form of parallelograms, with 185 km long sides, generally lying in an east-west direction and having a slight slope with a north-south orientation. The area of each scene is about 34 000 km² but, because of a small overlap in two directions, of around 10 per cent at the equator, the true surface area appearing in each negative is about 26 000 km². Each series is rephotographed at the same time every 18 days. The sensing system used on ERTS-1 was a multiple scanner (MSS, multispectral scanner) recording the following wave-lengths separately:

band 4: 0.5-0.6 μm green (visible)
band 5: 0.6-0.7 red (visible)
band 6: 0.7-0.8 } near infra-red.
band 7: 0.8-1.1 }

Information sent to the ground is reproduced in various forms, of which three are particularly useful for analysis:

— positive transparencies 23 cm × 23 cm black and white (photographed area: 18 cm × 18 cm; scale: *ca.* 1/1 000 000) for each of the 4 bands;
— positive transparencies 70 mm × 70 mm black and white (at a scale a little less than 1/2 500 000) for each of the 4 bands;
— magnetic tapes for use in a computer and containing the information on the intensity of each *area element* in the 4 bands.

The size of an interpretable area element is determined by the ground resolution governed by the scanning system. This is in the order of 70 m; the interpretable area element is therefore a square of *ca.* 0.45 ha.

The cost of satellite photography to a user is equal to the very low cost of reproduction: a transparency 23 cm × 23 cm in a given band costs $3 and a 70 mm × 70 mm, $2.50. In principle, any individual or legal entity can obtain transparencies of their choice (upon request to Eros Data Centre, 10th Street and Dakota Avenue, Sioux Falls, South Dakota 57198, U.S.A.).

ERTS-1 output, whether in digital or visual form, produces a synoptic view of very large regions under uniform light conditions, with particularly good resolution (considering the size of the area photographed), and with negligible image distortion. It is thus eminently suitable as a base for planimetry and for use in continuous monitoring. The major disadvantage for tropical zones is that the radiation used is unable to penetrate the frequent cloud cover of these regions. Two types of ERTS-1 data analysis can be carried out:

— *visual* interpretation of the images in each band, or of composites in false colour, or colour images enlarged up to a scale of 1/200 000 (beyond which interpretation is hampered by the scanning line pattern). Such an interpretation is comparable to that of conventional air photographs without stereoscopy; the false colour negatives can be reconstituted by *simultaneous projection* devices which make it possible to obtain a colour relationship which is best adjusted to the requirements of interpretation;
— *automatic* computer interpretation, using data on magnetic tape. Multivariable statistical analysis of the intensity of each of the 4 bands for each area element (0.5 ha), followed by automatic classification, makes it possible to assign each elementary area to a defined class; by use of ground information it would be possible to relate this classification to one already in use in a particular field (soils, vegetation, etc.) for the management of the corresponding resources; an automatic mapping by 0.5 ha elements, of the area covered by the image, will thus be obtained.

Automatic interpretation of information obtained by the ERTS-1 satellite, as presently available is impressive. Nevertheless, one should stress the *amount of preliminary work in collection and interpretation of ground data* which would be required if one wished to prepare a map which was of use in forestry. This kind of work is still scarce in the forested tropical countries and the application of automatic data interpretation in these countries is, therefore, at present limited.

In 1974 FAO sponsored two pilot-studies on the use of ERTS-1 data in the forest zones of Colombia and Nigeria, where ground inventories had been carried out. Another study undertaken by FAO is at a more advanced stage in the Sudan (Mitchell, 1973). Although the pilot-study area does not form part of the tropical forest zone, the following results are relevant to the study of vegetation: visual analysis of the transparencies of each sector permits identification of vegetation cover in so far as some association exists between these and particular land-forms; visual interpretation of false colour enlargement on a scale of about 1/200 000 confirms the preceding interpretation, namely the possibility of distinguishing major vegetation types associated with major physiographic units. The first stage of automatic interpretation, undertaken without ground data, shows only certain major features of the area surveyed, such as the hydrographic system and bare areas. Nevertheless the interest of this interpretation lies not only in the fact that visual interpretation of satellite photographs appears to be worth while in this zone, where cloud cover is rare, but also because, until recently, it has only been possible to obtain very approximate thematic maps of this zone at a very small scale.

What can be learnt from these preliminary observations in the field of continuous tropical forest monitoring? There is no doubt that in developed countries, at temperate latitudes (and especially in the United States) a powerful and extremely complex tool has been developed to evaluate the state of natural resources at regular intervals. This capability will spread rapidly enough to the less developed countries, especially to those within the tropical forest zone. Already committees of users and groups of experts are working actively in this field in some countries. However, the difficulties which could hold back the judicious and effective use of this tool in its application to forestry, e.g. absence of extensive knowledge of the environment and lack of scientific ground data, must not be overlooked.

A forest inventory data processing system for use in tropical countries

Introduction

During the last twenty years forest inventory projects have been carried out in many parts of the developing world, often with bilateral or multilateral aid. To date, however, inventories have existed for only a very small part of the potential tropical forest cover; thus the forest resources of huge areas remain to be assessed in the near future. Moreover, very few permanent inventory programmes have been set up in these countries, hence inventory activities are likely to show a sharp increase in the future. These forest surveys will be executed increasingly by local experts using Electronic Data Processing (EDP), since computers are available in many developing countries. FAO therefore intends to create a general EDP system, for two main purposes:
— to provide developing countries with pre-programmed computer routines for use in forest inventory;
— to transmit some knowledge of electronic data processing to local forest services as an activity in education and training.
By the use of this EDP system, it should then be possible to both coordinate the inventory work and facilitate future comparison between countries, in order to generate regional and inter-regional assessments.

Short review of the present state of EDP systems for forest inventories

Many EDP systems have been developed for both national forest and smaller-scale inventories.

In the tropics, only very few smaller-scale inventories have been carried out with EDP, often through FAO/UNDP projects or bilateral assistance. Those developed in East and West Malaysia, Surinam and Nigeria are of particular interest (FAO/UNDP, 1973, 1974). These systems have special components for the quality assessment of standing volume for which more sophisticated data treatment is required. The first attempt to develop a generalized EDP system for tropical forest inventory was made by Haller (1968) for the Liberia Forest Inventory. A sophisticated and complex EDP system was also developed several

years ago by North-eastern Forest Experiment Station of the US Forest Service. This 'FINSYS system' can be truly classified as a *General EDP System*, as it covers several sampling designs and has a high flexibility in INPUT and OUTPUT procedures (Frayer *et al.*, 1968). There is no provision, however, for 1-stage ratio or regression sampling, or for 2-stage sampling. The system has, however, been used successfully for the data processing of forest inventory in Paraguay, where a 2-phase or double-sampling method was employed. Careful appraisal of the system by FAO suggests that the FINSYS system cannot cope satisfactorily with the needs of an EDP system for tropical forest inventory. See Marsh (1974).

Special requirements of an EDP system for use in tropical forest inventory

Three main sources of information are commonly used in forest inventories:
— data from *photointerpretation*, for stratification of the area into forest types or land-use classes;
— data from *field sampling*, containing information on plots and trees (including *quality* for assessment of extractable volumes);
— data from the *re-measurement* of *permanent plots* to assess the growing potential of the forests.
In addition to the above, tropical forest inventory should also aim at giving information on *extractable volumes* and *accessibility*. A particular problem arises in the inventory of mixed tropical hardwoods, due to the large number of species. This necessitates a prior investigation of the grouping of the species before volume tables can be generated and the final stand and growing stock tables derived. Since the requirements of the information differ from one country to another, it may well be necessary to produce highly flexible yield tables, in addition to the basic tables required from each inventory.

The choice of an appropriate sampling design for a given inventory can only be made according to the objectives of the inventory and the local situation. The following considerations have to be taken into account when selecting a field sampling design in a tropical forest inventory:
— systematic designs are generally more adequate for forest conditions in tropical countries;
— for logistic reasons 2-stage sampling may be desirable in tropical forests;
— fixed area plot sampling is—at least under difficult conditions in tropical high forests—generally more efficient and more reliable than point sampling (PSS sampling).
It should be pointed out that only a few Forest Services in developing countries have access to medium-sized or large computers. A questionnaire on computer facilities sent out by FAO at the end of 1972 to 25 tropical forest countries led to the following conclusions:
— about 70 per cent of the computers available to the Forest Services are IBM machines, the remaining 30 per cent are installations of nine different makes;

— about half the available computers have a memory of less than 12 K bits. Only 15 per cent have a memory of more than 40 K.

In view of the above situation, the proposed generalized EDP system is written in FORTRAN IV, which is the most universally used language, and is suitable for small-sized computers.

Design of FAO's system of EDP

The EDP system for forest inventory compromises between the need for an effective, generalized and flexible programme system, and the limitations imposed due to special conditions in tropical countries. Due to the relatively small computers available in developing countries, the FAO Forest Inventory Data Processing System (FIDAPS) has been broken down into several phases to carry out the following functions:

Phases	Functions
INPUT	To convert field data into a standardized form
EDITING	To edit the new data and to create clean data-files for further processing
PARAMETER VALUES	To convert the basic data at the level of the sampling unit
MEANS/ERRORS	To calculate stand tables and stock tables with precision estimates
OUTPUT	Printing of the tables.

The main features of the system will be:
— maximum flexibility for the *input* and *editing* part;
— restrictions in the phase of generation and error calculations;
— certain flexibility in the production of output.

Conclusions: research needs and priorities

Support for the continuous monitoring of tropical forests, and humid tropical forests in particular, appears to be indispensable and of the first priority. This project will facilitate the evaluation of the impact of man's activities on the functioning and structure of tropical forest ecosystems, along with the study of various stages of regression and biological recovery; it is conceived within the framework of the Global Environmental Monitoring System (GEMS). Links to the ETGAO project (the tropical component of the Global Atmospheric Research Programme, GARP, in the Atlantic) might also prove useful.

Continuous forest monitoring should also indicate trends in regional and world wood production and could give an overview of the roles of these forests as raw material producers and as protectors of the environment.

National forest inventories, on the other hand, must not be limited to the inventory of available growing stock and growth rate, but must link these to the ecology and floristic composition of the forest stands; to be effective, inventories must not neglect ground operations, which are often complex. For instance, the importance of the accessibility indices has already been stressed. Problems which require further study, include sampling methodology, extraction systems, terrain classification, and production and costs.

Small-scale vegetation maps can play an important role in defining the major forest types on an ecological basis and in locating sample stands. However, preparation of these maps requires improved definition and quantification of classification criteria which are ecologically meaningful and readily identifiable on both air photographs and on the ground. Studies in this field should be pursued.

The FAO Committee on forest development in the tropics at its third session in May 1974 has recommended the increased use of satellite remote sensing methods, but has warned against the overestimation of their utility at their present stage of development (FAO, 1974). The need for more in depth research, and for collection of ground data, has been recognized. Repetitive programmes of standard air photography, possibly under the auspices of several national institutions within a country are still valuable for continuous monitoring of forest resources. The need for coordinating international assistance to tropical countries in the rapidly changing field of remote sensing, has been recognized. This co-ordination should preferably be undertaken at a regional level.

Bibliography

Area covered by tropical forests

ANON. *World Forest Inventory*. Rome, FAO, 1963, 113 p.
——. *Le bois. Évolution et perspectives mondiales*. Rome, FAO, 1967, 133 p.
——. *Report of the Headquarters meeting of forest inventory experts on UNDP/SF Projects*. Rome, FAO, sept. 1967, 260 p.
——. *Environmental aspects of natural resources management. Forestry* (a basic paper prepared for the UN Conference on Human Environment by FAO as UN Interagency focal point with contribution from Unesco and WHO). Rome, FAO, 1971, 33 p.
ANON. *Evaluation of accessibility of forest resources. A pilot study on logging costs from inventory data*. Rome, FAO, 1972, 26 p.
——. *Manual of forest inventory with special reference to mixed tropical forest*. Rome, FAO, 1973, 200 p.
——. *Formulation of a tropical forest cover monitoring project, FAO/UNEP*. Rome, FAO, 1975.
FAO. *Attempt at a global appraisal of the tropical moist forest*. Rome, FAO, 1976, 15 p. multigr.

HAMILTON, L. S. *Tropical rain forest use and preservation: a study of problems and practices in Venezuela.* San Francisco, Sierra Club Special Publication, International Series no. 4, 1976, 72 p. + appendices.

HEINSDIJK, D. *Forest assessment.* Wageningen, Centre for agricultural publishing and documentation, 1976, 359 p., 1800 references.

PERSSON, R. *World forest resources. Review of World's forest resources in the early 1970's.* Stockholm, Department of Forest Survey, Royal College of Forestry, no. 17, 1974, 261 p., 3 maps.

——. *Forest resources of Africa.* Stockholm, Royal College of Forestry, no. 18, 1975.

VON SEGELADEN, G. Studies of the accessibility of forest and forest land in Sweden. *Studia Forestalia Technica,* no. 76, 1969, 64 p.

Mapping syntheses

ANON. *World atlas of agriculture.* 3 vol. Novara (Italia), Istituto geografico De Agostini, 1969, 527, 671, 497 p.

ASHTON, P. and M. *The classification and mapping of Southeast Asian ecosystems. Transactions of the fourth Aberdeen-Hull Symposium on Malesian ecology.* Univ. Hull, Dept. of Geography, miscellaneous series no. 17, 1976, 103 p.

BEARD, J. S. The classification of tropical American vegetation types. *Ecology,* 36, 1955, p. 89–100.

BURTT Davy, J. *The classification of tropical woody vegetation types.* Oxford, Imperial Forestry Institute, no. 13, 1938, 85 p.

CHAMPION, H. G. ; SETH, S. K. *Forest types of India.* Dehra Dun Forest Research Institute, 1965, multigr. Revised ed. 1968.

DASMANN, R. F. *Classification and use of protected natural and cultural areas.* Morges, IUCN occasional paper no. 4, 1973.

FAO. *Catalogue des Cartes. Carte mondiale des sols.* Projet FAO/Unesco. Rome, 1973.

FOSBERG, F. R.; GARNIER, B. J.; KUECHLER, A. W. Delimitation of the humid tropics. I-III. *The Geographical Review,* vol. 51, no. 3, 1961, p. 353–357.

HOLDRIDGE, L. R. *Life zone ecology* (revised edition). San José (Costa Rica), Tropical Science Center, 1967, 206 p.

HUECK, K. *Die Wälder Sudamerikas.* Stuttgart, Gustav Fischer Verlag, 1966, 422 p.

Unesco. *A review of the natural resources of the African continent.* Paris, Unesco, Natural resources research no. 1, 1963, 437 p.

——. *International classification and mapping of vegetation.* Ecology and Conservation no. 6. *Classification internationale et cartographie de la végétation.* Écologie et Conservation nº 6. Paris, 1973, 93 p.

——. *Ressources naturelles de l'Asie tropicale humide.* Paris, Unesco, Recherches sur les ressources naturelles nº 12, 1974, 490 p. *Natural resources of humid tropical Asia,* Natural resources research no. 12, 1974, 456 p.

——. *Carte de la végétation de l'Afrique.* Paris, Unesco, 1977, in press.

Welt Forst Atlas. Hamburg, Berlin, Paul Parey, 1945, 59 maps.

The use of remote sensing techniques for the continuous monitoring of tropical forests

Application of the new remote sensing techniques to forestry

American National Research Council. *Remote sensing with special reference to agriculture and forestry.* Washington, D.C., National Academy of Sciences, 1970, 424 p.

BILLINGS, W. D.; GOLLEY, F. B.; LANGE, O. L.; OLSON, J. S. (eds.). *Remote sensing for environmental sciences.* Berlin and New York, Springer Verlag, 1976, 367 p.

IUFRO. *Application of remote sensors in forestry.* Joint report of the working group 'Application of remote sensors in forestry'. Section 25, 1971, 189 p.

——. *Proceedings of a symposium* held in Freiburg (17–21 September 1973). Subject Group 56.05, 1974.

LATHAM, R. P.; MCCARTY, T. M. Recent developments in remote sensing for forestry. *Journal of Forestry,* vol. 70, no. 7, 1972, p. 398–402.

WILSON, R. C. *Potentially efficient forest and range applications of remote sensing using earth orbital spacecraft—circa 1980.* Special report in series 'Remote sensing application in forestry' of University of California, Berkeley, 1970, 199 p.

Continuous monitoring of tropical forests by conventional air photography

FAO. *Inventario forestal por N. Henning* (FO: SF/COL 14). *Estudio de preinversión para el desarrollo forestal en los valles del Magdalena y del Sinu.* Rome, FAO, Informe técnico 4, 1970.

——. *Reconocimiento general de los bosques e inventario detallado de Azuero, por Centre technique forestier tropical,* FO: SF/PAN 6 (Inventariación y demonstraciones forestales). Rome, FAO, Informe técnico 12, 1972, vol. 4, documentos cartográficos.

LANLY, J. P. Régression de la forêt dense en Côte d'Ivoire. *Bois et Forêts des Tropiques* (Nogent-sur-Marne), 127, 1969, p. 45–59.

——. L'utilisation des photographies aériennes et des cartes pour l'évaluation des ressources forestières. Application au cas de l'Afrique francophone. In: *3e Conférence régionale de cartographie pour l'Afrique* (Addis Abéba, oct.-nov. 1972).

LETOUZEY, R. Photointerprétation en forêt dense camerounaise. In: *Colloque sur le rôle des recherches techniques dans le développement de l'emploi des bois tropicaux en Europe* (Centre technique forestier tropical, Nogent-sur-Marne), 1967, I/E, p. 1–12.

PARKER, R. C.; JOHNSON, W. E. Small camera aerial photography — The K-20 system. *Journal of Forestry,* vol. 68, no. 3, 1970, p. 152–155.

SCHWAAR, D. C. *Land-use dynamics and transmigration in southern Sumatra.* Bogor, Soil Research Institute, 1971, 16 p.

Continuous monitoring of tropical forests by new techniques of remote sensing

Radar

CRANDALL, C. J. Radar mapping in Panamá. *Photogrammetric Engineering,* vol. 35, no. 7, 1969, 6 p.

FAO. *Committee on forest development in the tropics. Report of the third session* (May 1974). Rome, 1974, 65 p. plus appendices.

HOWARD, J. A.; LANLY, J. P. Remote sensing for tropical forest surveys. *Unasylva* (Rome, FAO), 108, 1975 II, p. 32–37.

LANLY, J. P. Monitoring tropical forest resources through remote sensing. *Span,* vol. 17, no. 3, 1974, p. 114–115.

——. Techniques récentes en matière d'évaluation des ressources forestières. *Bois et Forêts des Tropiques* (Nogent-sur-Marne), 147, 1973, p. 35–45.

NEWTON, A. R. Pseudostereoscopy with radar imagery. *Photogrammetric Engineering,* vol. 39, no. 11, 1973, 4 p.

European Space Research Organization (ESRO). *Side-looking radar systems and their potential application to earth resources surveys.* Summary volume. ESRO CR-141, 1973, 135 p.

SNOBOHN, A. J. Empleo de sensores remotos — radar en la inventariación de la selva amazónica y levantamiento de los recursos naturales. In: *7th World Forestry Congress* (Buenos Aires), 1973, 4 p.

VIKSNE, A.; LISTON, T. G.; SAPP, C. D. SLAR reconnaissance of Panamá. *Photogrammetric Engineering,* vol. 36, no. 3, 1970, p. 253–259.

Satellites and rockets

HOWARD, J. A. *Satellite sensing of land resources.* Rome, Document FAO-AGS: Misc/73/25, 1973, 6 p.

MITCHELL, C. W. *The application of ERTS-1 imagery to the FAO Sudan Savanna Project.* Rome, FAO, 1973, 23 p.

NASA. *Symposium on significant results obtained from ERTS-1* (Goddard Space Flight Center, Greenbelt, Maryland, 5–9 March 1973).

Data processing

ANON. *Data Processing Report, Pre-investment survey of forest resources, East Godavari.* New Delhi, Ministry of Agriculture, 1972, 43 p.

FAO/UNDP. *A description of the data processing system used in the inventory of selected areas of mixed dipterocarp forest in Sarawak* (by K. D. Singh). Rome, FAO, FO: DP/MAL/72/009, 1973, 70 p.

——. *Electronic data processing and computer programming. Forest development in Surinam* (by M. Blakstad; R. De Milde). Rome, FAO, 1974, 82 p. and 7 appendices.

FRAYER, E.; WILSON, R.; PETERS, C.; BICKFORD, C. FINSYS: an efficient data processing system for large forest inventories. *Journal of Forestry,* 12, 1968, p. 902–905.

HALLER, K. E. Inventory of natural tropical forests; a computer programme for the processing of data. Inventaire des forêts tropicales naturelles; programme de traitement par ordinateur des résultats. *Unasylva* (Rome, FAO), 89, 1968, p. 22-28.

MARSH, H. E. *Some views on a forest inventory data processing system for uses in tropical countries.* Rome, FAO, 1974, 8 p.

Tropical forests and the biosphere

Introduction

Tropics

The tropical zone between 23° 27′ North and 23° 27′ South contains about 40 per cent of the earth's surface. The distribution of land and sea surfaces by continents and 5° latitudinal zones is given in table 1.

TABLE 1. Areas of tropical regions in thousand km² (Baumgartner and Reichel, 1975).

	N 25°	20°	15°	10°	5°	0
Africa	2 976	3 214	3 562	3 646	2 260	
Asia	3 527	2 143	1 066	400	742	
Australia		.	.	.	.	
N. America	636	805	274	94	.	
S. America	.	.	179	1 200	1 761	
Land	7 139	6 162	5 081	5 340	4 763	
Sea	1 358	14 992	16 553	16 614	17 360	
Percentage land	53	41	31	32	27	

	5°	10°	15°	20°	25° S	Total
Africa	1 854	1 566	1 712	1 796	1 436	24 022
Asia	711	332	.	.	.	8 921
Australia	437	508	374	1 258	2 046	4 623
N. America	.	.	.	.	.	1 809
S. America	2 340	2 651	2 331	1 946	1 565	13 973
Land	5 342	5 057	4 417	5 000	5 047	53 348
Sea	16 785	16 899	17 218	16 150	15 435	149 364
Percentage land	32	30	26	31	33	36

Climatically, the tropics are a belt of varying width on either side of the climatic equator which deviates from the geographic equator as a result of the distribution of land and ocean surfaces and orographic influences. The climatic equator is the line of maximum uniformity of humidity and temperature; it is characterized by unpredictable variation of weather and climate. Seasonality increases with distance from the climatic equator. Tropical cyclone systems carry tropical air masses north (India, Mexico, Florida) or south (East Africa, Madagascar) and extend tropical climatic conditions beyond the geographic tropics.

TABLE 2. The climatic gradient from the climatic equator towards higher latitudes and the corresponding gradients of plant habitus and of the main plant formations. The habitus is expressed as the type of climatic adaptation (above the line) and the corresponding morphological adaptation type (below the line). The formation refers to formation group and formation levels of ecological-structural classification. The climatic data refer to tropical lowlands below 300 m altitude (from Brünig, 1972).

Amount of annual insolation on ground (kcal/cm²)	South-East Asia Congo basin Amazon basin	140-160 120-130 100-120	*ca.* 160
Mean annual temperature T_m (°C) annual variation (°C) diurnal variation (°C)	28 3 9		25 15-20 20
Wind	Predominance of tropical low pressure trough, low velocities except in convectional squalls and local tornadoes		High velocities during summer (typhoon, hurricane, cyclone), low during dry season, strong effect of tropical convergence zone. Local frontal storms toward end of dry season (habub in Africa)
Relative humidity (mean %)	95/100 night, 60/70 daytime, little seasonal variation		90/100 wet season, 60/80 dry season
Ratio of potential evaporation to actual precipitation, $\dfrac{E_p}{P_o}$	< 1		< 1
Precipitation (mm/a) mean, min. or max.	2 000; min. = 50 $(T_m + 12)$		1 300-3 000; min. = 25 $(T_m + 12)$
Distribution of precipitation	Even, < 2 dry months		2 or 4 seasons, 3-5 dry months
Location	Equatorial belt and areas with constant moist air-masses outside this belt		Subequatorial to outer tropics with influence of trades, monsoons and monsoonlike alternating winds
Climate type	Tropical perhumid (wet), isotherm, non-seasonal, diurnal variation > annual variation		Tropical humid (moist), isotherm, seasonal with predominantly summer rainfall
Growing period (months)	11-12		7-10
Characteristic habitus of climatic climax formation	Megatherm-hydrophilous		Megatherm-tropophytic
	Hygromorph-mesomorph ⟶		tropomorph
Main climatic climax formations	Superhumid to humid, ombrophilous, evergreen tropical forest and semi-evergreen wet forest		Humid to subhumid, semi-deciduous and deciduous tropical forest
Edaphic climax formations	Littoral forest Mangrove forest and woodland Freshwater swamp forest and grassland Peat swamp forest Riparian forest Single-dominant forests on certain soils Sclerophyll forest		Littoral forest Mangrove forest and woodland (less luxuriant) Freshwater swamp forest and grassland Riparian forest (often relic gallery) Evergreen forest (on moister well-drained soils) Sclerophyll forest (sandy terraces and skeletal soils)
Physiographic climax formations	Submontane forest (locally rich in oaks and laurels, simpler structure than the climatic climax) Montane forest (moist) Alto-montane forest (moist) Alto-montane moss forest (wet, misty) Alto-montane woodland and scrub (moist)		Similar to the perhumid zone, except for species composition, conifers increase in southern and northern hemisphere, bamboo species become more frequent in the northern hemisphere
Degraded formations	Subseral secondary forest Disclimax secondary forest Disclimax pine forest to pine savanna (mostly at higher altitude) Disclimax grassland (*Imperata cylindrica* usually common) Disclimax karst-woodland Disclimax sclerophytic savanna		Subseral secondary forest Disclimax secondary forest Disclimax savanna Disclimax pine forest, pine woodland or pine savanna (higher altitudes) Disclimax karst-woodland Disclimax sclerophytic or xerophytic savanna

ca. 180	ca. 200	ca. 220	> 220
1-32 0 0	20-33 35 30	Extreme variation	Extreme variation
easonally dry tropic (harmattan) or moist tropic air (trades, monsoon), velocities moderate except during passage of tropical cyclones	Predominance of tropic high pressure cell, average wind speeds low to moderate, occasionally high velocities in advective storms	As before, dry and hot storms more frequent	As before, dust storms common
0/90 wet, 40/60 dry season	60/80 wet, 20/50 dry season	Usually 50 %	Average very low but locally very high for short periods
> 1	> 2	> 4	> 8
00-1 500; min. $= 20 (T_m + 12)$	350-1 000; min. $= 10 (T_m + 12)$	100-.00; min. $= T_m + 70$	< 100; max. $= T + 70$ mm
or 4 seasons, 6-8 dry months	2 seasons, 9-10 dry months	11 dry months	12 dry months
ubequatorial to subtropical ummer-rain belt	Outer tropical belt summer-rain belt	Outer tropical to subtropical belt of descending air-masses	Outer tropical to subtropical belt of descending air-masses
ropical dry seasonal ith summer rainfall	Tropical very dry seasonal with summer rainfall	Tropical arid	Tropical arid
-6	2-3	1-2	
Megatherm-tropophytic	Megatherm-tropophytic-sclerophytic-xerophytic		bare
→ xeromorph			bare
ubhumid to semi-arid, eciduous tropical forest	Semi-arid deciduous tropical thorn woodland	Arid thorn scrub and semi-desert	Perarid desert
ittoral woodland Mangrove forest to scrub iparian fringing forest nd grassland clerophytic thorn woodland eciduous moist forest	Littoral scrub Mangrove woodland to scrub Riparian fringing woodland and grassland Xerophytic semi-desert scrub	Littoral scrub Mangrove scrub Riparian fringing scrub and grassland	As in the arid zone, but rarer and poorer
imilar to the humid zone, ut relative effect of exposition nd barrier-effect more ronounced	Similar to dry zone, very strong and noticeable effect of elevation, frost occurs regularly even at lower altitudes, particularly in hollows, and creates local dwarf vegetation. Generally more open and scrub-like. Barrier-effect very pronounced	As before	As before
isclimax xerophytic savanna isclimax thorn scrub isclimax semi-desert scrub	Disclimax xerophytic thorn scrub Disclimax semi-desert	Disclimax desert	bare

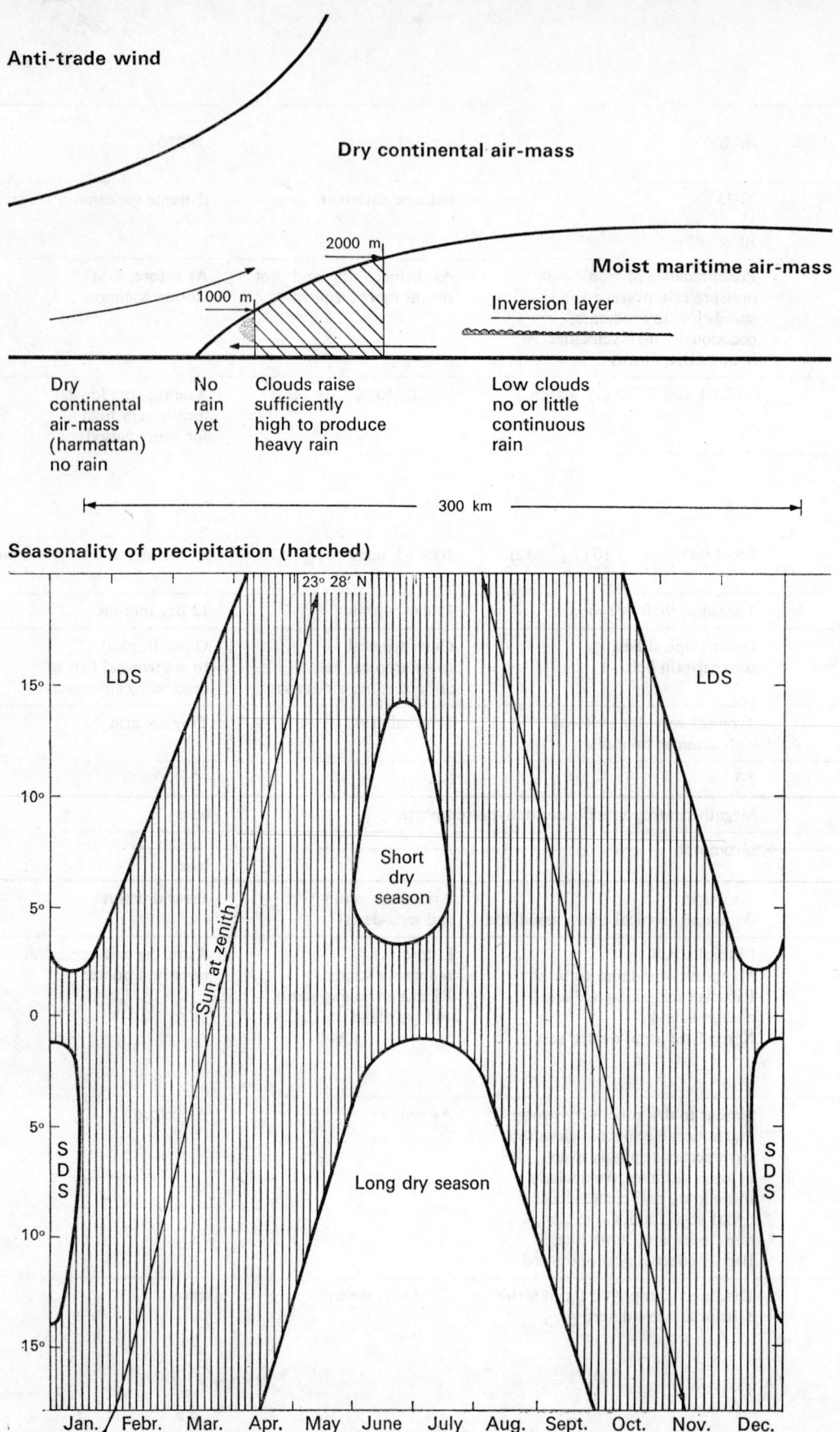

Fig. 1. Above: sectional view of the tropical convergence zone, the front approaching to the left or receding to the right following the zenithal position of the sun; below: the resulting assymetrical pattern of dry and wet seasons with a continuously wet equatorial rain belt.

Climates of tropical forest regions

Köppen (1930) empirically defines tropical lowland climates by monthly averages of air temperatures above 18 °C. Typical averages are between 24 and 18 °C. Mean monthly temperature in excess of 32 °C are rare and local. Seasonal temperatures variations are very small and less than the diurnal variation. Both variations increase with distance from the climatic equator but at different rates so that eventually seasonal variation exceeds diurnal variation. The corresponding climatic gradients are demonstrated in table 2. Seasonality in the tropics is governed by the annual march of the tropical convergence zone. This front follows the zenithal position of the sun with a time lag. The corresponding annual march of wet and dry seasons is illustrated in fig. 1. The coupling of seasons and sun position is close in continental Africa. In other areas it is strongly masked by the effects of unstable equatorial westerlies or trade-winds. Examples are the monsoonal wind systems in continental South and South-East Asia and West Africa.

Annual and seasonal humidity and temperature characteristics have been used by many authors to subdivide the tropical climate into ecologically meaningful classificatory units (e.g. Köppen, 1923, 1931; Fosberg, Garnier and Kuechler, 1961). The more widely adopted classifications are:
— Köppen's (1931) classification using temperature first and humidity second;
— Thornthwaite's (1931) classification emphasizing evapotranspiration for assessing the humidity situation;
— Gaussen's (1954) classification adding to the aerial humidity sources the water reservoirs in the soil and probability of drought conditions;
— Emberger's (1955) classification, essentially based upon the rhythm of seasonal climatic variations.
Of these the best known and most widely used is the Köppen classification which divides the humid tropical climate into:
A_f permanently wet rain forest, all months have sufficient precipitation;

A_m seasonally humid or subhumid, evergreen rain forest with months with arid characteristics;
A_w dry period in the winter of the corresponding hemisphere, subhumid or xeromorphic forests, woodlands or shrublands, savannas.
The distribution of the Köppen climatic types is shown in fig. 2. The pattern does not conform to the latitudinal zones as a result of the effects of land and ocean distribution and global atmospheric circulation. Shifts in the regional and global atmospheric circulation systems also seem to affect the extent of the various climatic types. However, very little is known about this and less on the effects of these shifts on the humid tropical forests. The effects are more obvious and spectacular at the subhumid and semi-arid margins of the tropical forest region but even there they are far from being fully understood.

A serious limitation of Köppen's approach is that the moisture regime and the availability of water to the vegetation is not adequately defined by the water input from rain. In addition evaporation losses must be considered. Holdridge (1967) attempts this in his life zone scheme. Finally, transpiration by the vegetation should be considered, as well as fog drip and dew, which locally may account for a substantial proportion of the total water turnover. Whitmore (1975) reviews the problem and summarizes experiences with various classification systems for the tropical forests of the Far East. Also, very rare climatic events, such as prolonged rainless periods with bright sunshine, freak storms or cold fronts, can be more critical for plant life than the averages calculated from long-term meteorological observations.

Tropical forests

Roughly 50 per cent of the total forested land area of the world is tropical forest. Fig. 3 shows the distribution of the world forest area, and table 3 the area of tropical forests and its human population.

The forest ecosystems are sensitive to the total annual precipitation, its reliability and its seasonal distribution.

TABLE 3. Total area, forest area and human population in the continents containing tropical forests.

Region	Land area × 1 000 km²	Forests					Population	
		Total* × 1 000 km²	%	Closed × 1 000 km²	%		× 10⁶	%
Tropical Africa	22 150	7 340	38	2 100	29		267	25
South and Central America	14 890	8 210	43	5 900	72		218	20
South-East Asia and Australia**	9 050	3 600	19	3 000	83		584	55
Total	46 090	19 150	100	11 000	58		1 069	100

* Closed forests, open woodlands and shrublands;
** Without Oceania.

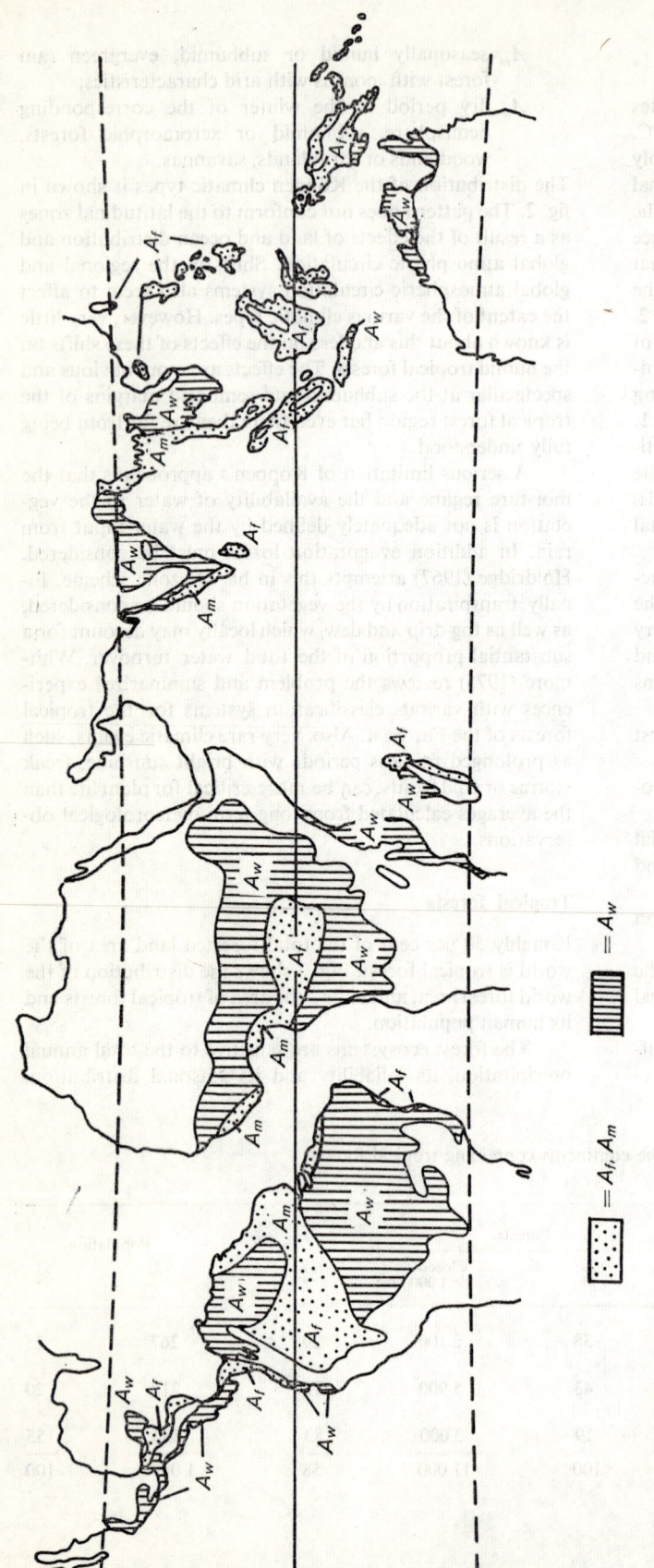

= A_w

= A_f, A_m

FIG. 2. Scheme of the world distribution of tropical climates (Köppen, 1930).

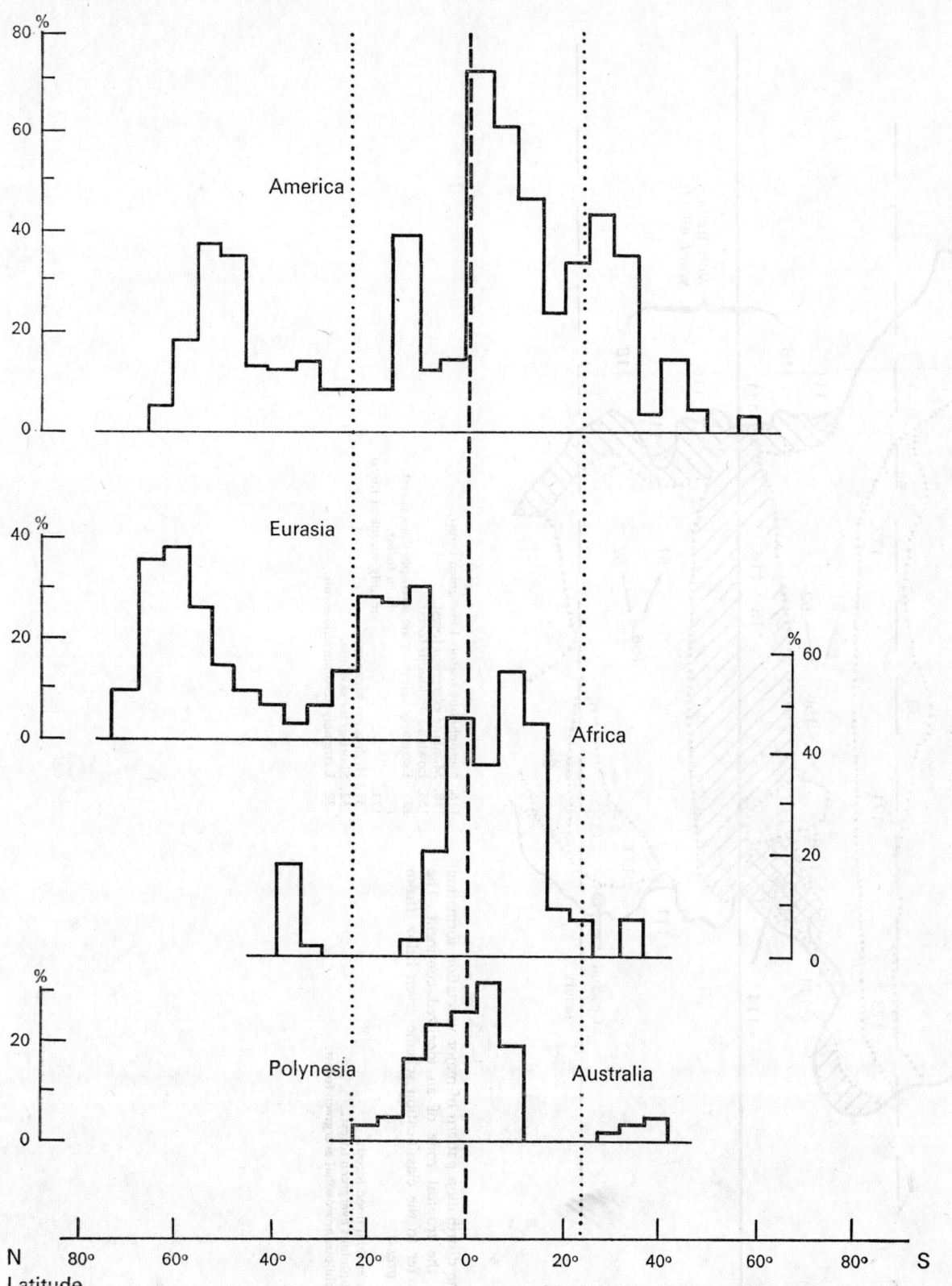

Fɪɢ. 3. Distribution of forested area expressed as a percentage of total land area in 5° latitudinal zones of the continents. Note the large proportion of forested land within the tropics (Baumgartner, in *Forstw. Centralbl.*, 76, 1976, in press).

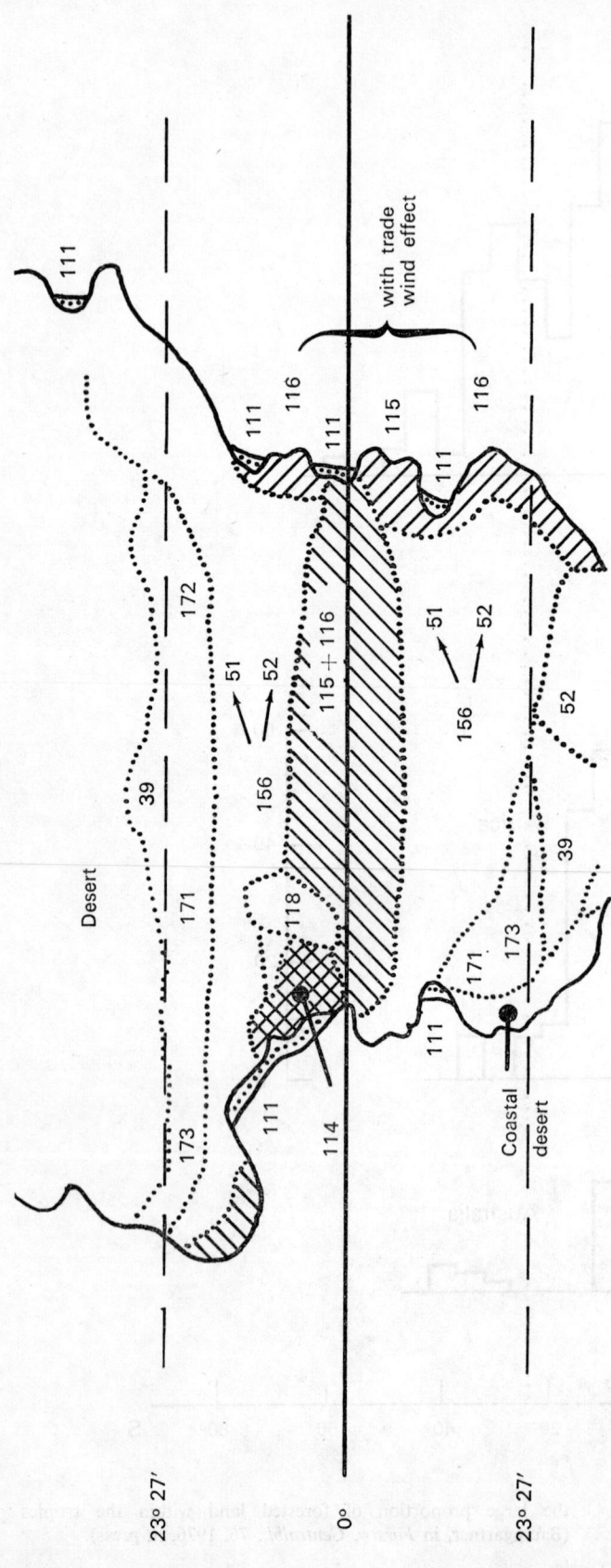

FIG. 4. The distribution pattern of major vegetation formation groups in the tropical zone of an idealized continent. The numbers refer to the classification scheme given below (from Brünig, in press).

111. Saltwater tidal flood forest.
114. Ombrophilous evergreen forest.
115. Mesophilous unseasonal evergreen forest.

116. Mesophilous seasonal evergreen forest.
118. Montane evergreen forest.
156. Lowland deciduous forest.
171. Extremely xeromorphic sclerophyllous forest.
172. — — thorn forest.
173. — — mainly succulent forest.
 51. High-grass savanna.
 52. Short-grass savanna.
 39. Extremely xeromorphic scrub.

Their floristics, architecture and structure, and their phenology vary with rainfall pattern. This variation is further modified by edaphic, orographic, biotic and historical influences. While considerable knowledge is available on the conditions under which certain structural forest types at a high level of classificatory hierarchy occur, little is known on the causal factors which determine the characteristics of forest structure at lower levels. Table 2 shows in a generalized way the association of major tropical vegetation types at formation group level of the Unesco classification (Ellenberg and Mueller-Dombois, 1966; Ellenberg, 1973) with the types of climate. Fig. 4 shows the corresponding distribution pattern at formation group level on an idealized continent.

The non-seasonal perhumid predominantly evergreen forests straddle the climatic equator. With increasing seasonality away from this line the forests gradually change through seasonal evergreen, semi-evergreen, semi-deciduous into deciduous forests, and finally into xeromorphic sclerophyllous forest, thorn forest or mainly succulent forest or woodlands. In the deciduous forest zone, the natural forest is to a large extent replaced by anthropogenic pyroclimax high-grass (moist) or short-grass (dry) savannas; in the extremely xeromorphic woodlands, by scrub or semi-desert (fig. 4).

In addition to this obvious large-scale variation of structure and function at formation group and formation level, there is more subtle variation at two lower levels. At subformation level, variation is between stands of mature natural forest. This is largely determined by differences in habitats. At small-scale level, variation is at the size of groups of trees and controlled by reproductive pressures and soil differences within a topographically uniform site. For recent reviews of the subject, see Ashton and Brünig (1975), Brünig (1975) and Whitmore (1975). Whitmore states that the small-scale variation is the most subtle mode of vegetation variation. There is such paucity of precise data that very opposite views are still held by different ecologists and foresters on its nature and significance.

Structural and functional variation are also related to the development phase of stands (medium level) or groups of trees (small-scale level). This variation is largely controlled by destructive forces causing openings in the forest. Gaps caused by lightning and wind in lowland rain forest usually occupy between 0.5 and 3 per cent of the area. They may be small (the size of one or a few emergent trees) or large (larger than the diameter of emergent tree height). The size of the gap and the regeneration determine the nature and sequence of floristic and architectural structure which ensues. The sequence ideally includes gap phase, building phase, mature phase and overmature phase. The gap phase may occupy between a fraction of a percent and 1–2 per cent and the building phase between 5 and 20 per cent of the forest area, but they account for substantially more if the destructive forces are severe (see also chapter 8). Knowledge of the kind and complexity of variation at medium and small-scale levels in relation to the available flora, disturbance, habitat, soil and reproductive pressure is essential for the assessment of the corresponding variation of functions and potentials of the tropical forest ecosystems.

This variation pattern is in addition subject to changes in the environment, such as variation of rainfall and of animal populations, particularly at medium- and small-scale patterns of variation. Hardly any reliable data are available.

The problems of enumerating and classifying, which arise out of the very complex variation patterns in the tropical forests, have been exhaustively studied, particularly by British and French ecologists and foresters. Adequate methods are now available for sampling, analysing and classifying tropical forests at large and medium scales of variation (see chapters 1 and 4), but not yet at the small-scale level.

Related to the variation problem is that of stability and fragility of the tropical forest ecosystems. These problems are of the utmost importance to land use planning and practice, but factual knowledge is sparse and views conflict. The conventional dogma of a simple and positive relationship between diversity and stability is contradicted by observation of tropical ecosystems and by recent theoretical work which suggests 'that communities with a rich array of species and a complex web of interactions (the tropical rain forest being the paradigm) are likely to be more fragile than relatively simple and robust temperate ecosystem' (May, 1975). The reproductive mechanisms of tropical rain forest plant and animal species are more adapted to biological competition than to large-scale environmental disturbance. The tropical ecosystems tend to store nutrients in the vegetation rather than in the soil. Both make the systems less resilient to change than temperate forests. Tropical forests are stable only within a comparatively small domain of parameter space. The humid tropical environment permits the systems to persist in spite of their fragility because the perturbations are relatively small and restricted to small areas. Natural catastrophic events tend to recurr with little variation. Thus tropical forests are in a constant turmoil of phasic developments. Each area exhibits a complex pattern of intricately mixed elements: less diverse but rapidly changing immature phases, more diverse persistant mature phases, and less diverse degrading overmature phases (see chapter 8). This confers to the whole system a certain but limited resilience to change within a narrow amplitude of fluctuation within which the dynamic stability can be maintained. This dynamic stability of the system is dependent on two important provisions. First, that the forces acting on the system do not exceed certain threshold values. If they do, then distinct and lasting changes may occur. Secondly, that the environmental conditions are themselves relatively constant if not necessarily uniform. The massive (in magnitude, scale of duration) interference by man in recent times has exceeded the capacity of the self-maintaining regulating processes and as a result the dynamic stability has been destroyed over large parts of the humid tropics (see chapter 9).

Biosphere

The biosphere is the portion of the earth and its environment in which life exists and sustains itself. It includes parts of the lithosphere, hydrosphere and atmosphere. The biosphere contains ecosystems of different natural and hierarchical levels. There is interaction and interchange between the different ecosystems. Fig. 5 illustrates this for the functional hierarchy of ecosystems from the natural base to the cultural top. Similar interrelationships exist between ecosystems of increasing size: from the smallest unit such as a part of a tree, a tree, a group of trees, a stand (subformation) and finally to the formation class (macro-ecosystem of Ellenberg, 1973).

The multidirectional exchanges within the biosphere involve energy in radiative, thermal, chemical and kinetic form and matter in gaseous, liquid or solid form. The various circulation systems at micro-, meso-, and macro-scales are interdependent and regulate inputs and outputs between ecosystems. Internal and external circulation processes of a system are coupled. The whole biosphere is a complex of interdependent structures and processes. Changes in one part of an ecosystem or of the biosphere will eventually cause corresponding adjustments in others.

Tropical forest ecosystems as part of the biosphere

Major forest functions

The forest ecosystems have protective, regulative and productive functions at the ecosystem level (primary ecosystem in fig. 5). These functions can acquire utility value for man and become functions of the cultural ecosystem (level of the secondary ecosystem in fig. 5). The effects of these functions on the environment are called forest influences. The more important of these are:

Protective functions
— soil protection by absorption and deflection of radiation, precipitation and wind;
— conservation of humidity and carbon dioxide by decreasing wind velocity;
— sheltering and providing required conditions for plant and animal species;

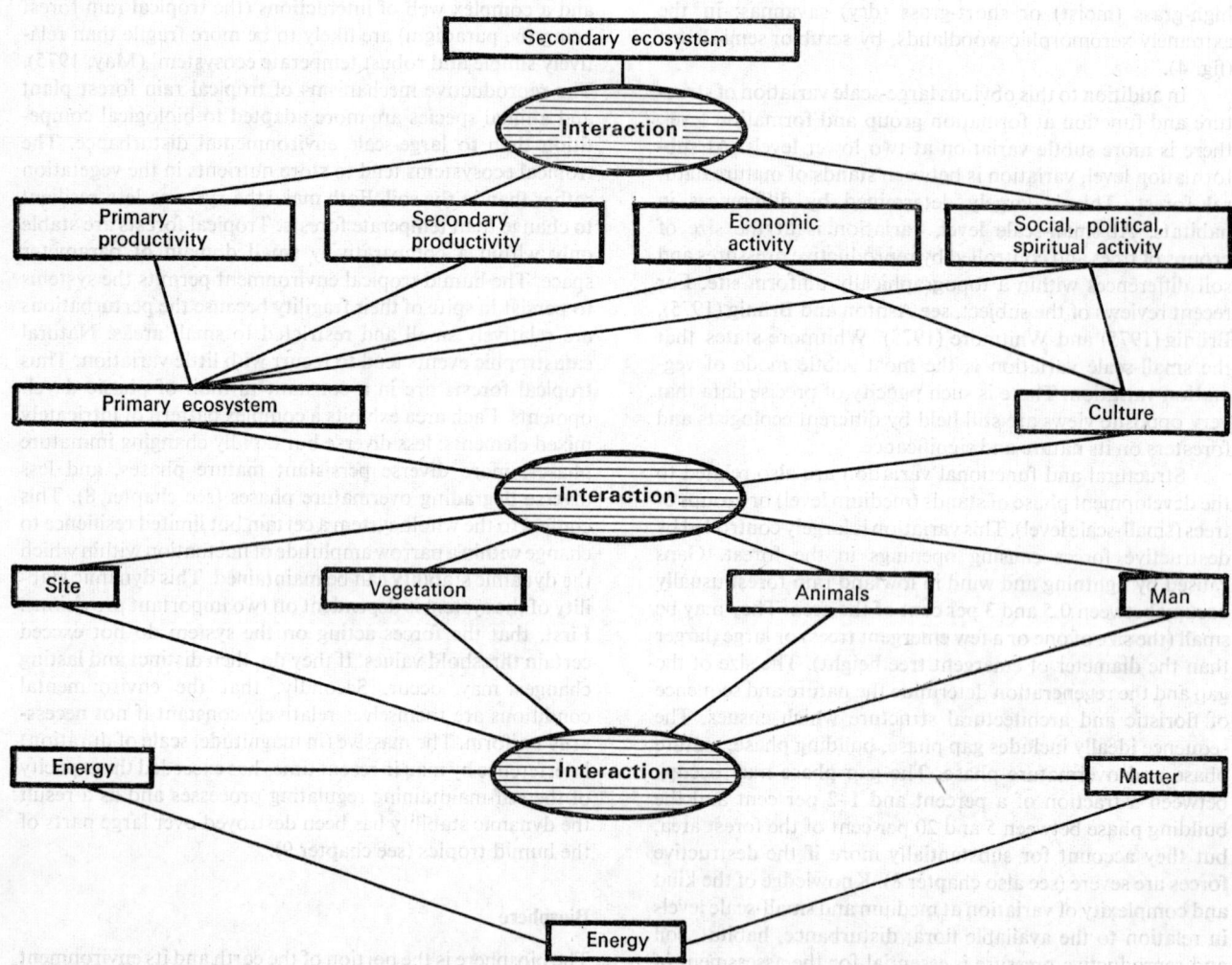

FIG. 5. Interactions and interrelationships within a functional hierarchy of ecosystems (from Brünig, 1971b).

Regulative functions
— absorption, storage and release of CO_2, O_2 and mineral elements;
— absorption of aerosols and sound;
— absorption, storage and release of water;
— absorption and transformation of radiant and thermal energy;

Productive functions
— efficient storage of energy in utilizable form in phyto- and zoomass;
— self-regulating and regenerative processes of wood, bark, fruit and leaf production;
— production of a wide array of chemical compounds, such as resins, alcaloids, essential oils, latex, pharmaceuticals, etc.

These functions can be utilized by man for:

Protection
— sheltering agricultural crops against drought, wind, cold, radiation;
— conserving soil and water;
— shielding man against nuisances (noise, sights, smells, fumes).

Regulation
— improvement of atmospheric conditions in residential and recreational areas;
— improvement of temperature regimes in residential areas (road-side trees, parks);
— improvement of the biotope value and amenity of landscapes.

Production
— supply of a wide array of raw materials to meet man's growing demands;
— supply of employment;
— creation of wealth.

Forest structure and climate

Interrelationships

Forest structure, including its phenological changes, is dependent on macro-climate, modified by physiographic and edaphic conditions, and in turn determines the micro-climate within the stand and immediately above and around it. Structural variation along the macro-climate gradient away from the equator is shown in table 2.

The changes of architectural structure and seasonality affect partioning of inputs and outputs, and the dynamics of the forests generally. The kinds of inputs and outputs and the directions of flows are shown in fig. 6. Practically all are influenced by stand structure, but there are hardly any data to quantify the interrelationships. The few available apply to specific stands and there is yet no adequate base for their generalization and extrapolation to different stands. For example, from data meas... region near Dar-es-Salam, Jac... local variation in rainfall is very high... erential local storm tracks rather than to relie... adjacent forest stands of the same type in a topogra... homogeneous area may possess very different micro climates and possibly differences in floristic and architectural structure which will again modify the micro-climate. But there is no way of ascertaining this by simple methods of field sampling.

The predominantly evergreen perhumid forest

Rainfall variation in space and time in the tropics seems to be considerably larger than in temperate climates. Associated is a corresponding variation in hours of sunshine and net radiation which accentuates the effects of rainfall extremes. For example, the rainfall statistics for the perhumid equatorial climate of Kuching, Sarawak, for the period 1896–1957 show a mean annual rainfall of 4 020 mm; the lowest annual amount was 2 740 mm and the highest 5 760 mm. The monthly extremes were 17 mm and 1 580 mm. 30 days running sums of rainfall frequently drop below the threshold level for a dry-month even if the calendar-monthly rainfall is well above 100 mm. The ecological significance of these masked dry spells are discussed by Brünig (1971c). His preliminary hypotheses are:
— the tropical humid evergreen forests frequently experience unpredictable periods of water deficiency and the probability of occurrence increases with decreasing water storage capacity of the rooting zone in the soil; consequently forest structure is accordingly adapted to avoid overheating during stress;
— on sites with ample water supply the structure is adapted to maximize evapotranspiration, but avoid damage to leaves during midday peaks of radiative influx;
— the various phases of stand development (from gap to over-mature) exhibit different structures to maximize the utilization of site potentials at a minimum risk of damage through excessive stress.

Sajise (1974) observes that the 'beautiful interplay of the absorbance and emittance characteristics—of leaves (good absorption in the visible fraction, poor in near infra-red, good absorption and emission in the far infra-red)—is not an adaptive or evolutionary consequence but is purely fortuitous; the spectral reflectance of aquatic plants is the same as that of land plants'. We know near to nothing about the significance of what looks to the observer to be something like an 'adaptative strategy' of tree and stand geometry to meet the challenges of the environment. It is also unclear to what degree each of the major determinants listed by Collier *et al.* (1973) (the dispersal mechanisms, the physical environment, and biotic influences such as shading, shelter and competition) contribute to the vegetative structure.

There are few data available on energy and matter exchange processes and those which exist are of localized value. Behn and Duffee (1965), reviewing micrometeorological studies with large towers in tropical forests in Colombia (BENDIX study), near Dakar (Dakar University

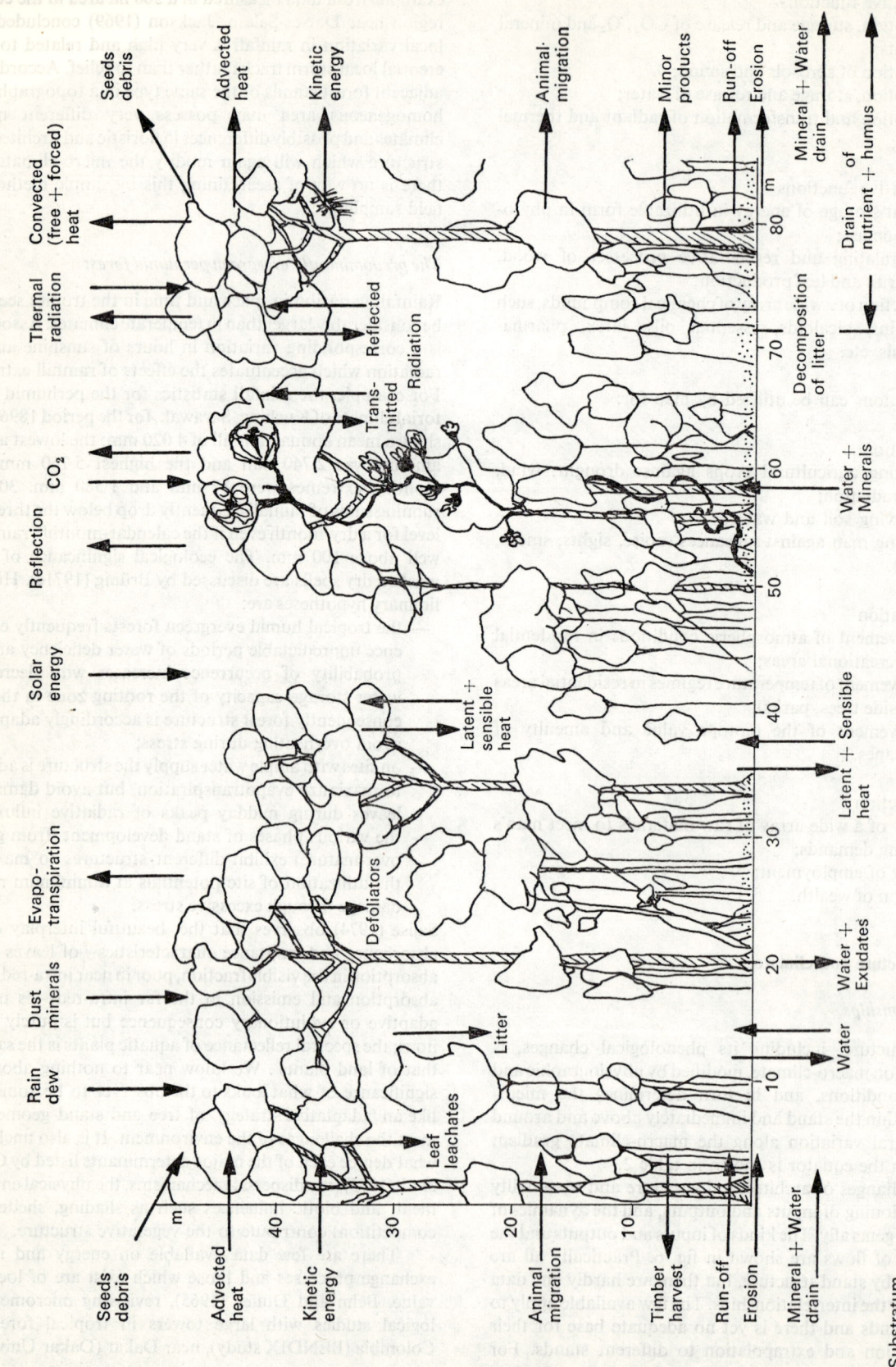

FIG. 6. Kinds and directions of inputs and outputs in a tropical forest ecosystem

study) and in Panamá (U.S. Army), confirm the general nature of radiation, temperature and humidity gradients which were already summarized by Richards (1952). Evaporation increases from the ground upwards; only moderately at first but the rate increases nearer to the upper canopy. This accords with the trends of the radiation and wind speed gradients. Short rainless periods during the major wet season drastically increase evaporation at all levels (by three times at ground level compared with twice at crown level).

Measured values from the IBP plot in Pasoh Forest Reserve, Malaysia, confirm that the water vapour pressure is generally very high within tropical rain forest and slightly decreases from the bottom to the top. At night, the water vapour content of the air within the canopy is slightly higher than in the atmosphere above the canopy. This indicates that transpiration occurs from the forest canopy surface even at night (Aoki, Yabuki and Koyama, 1974). The high water vapour pressure within the tropical rain forest damps temperature fluctuations. Maximum variation occurs in the top canopy stratum. Here maximum temperatures shortly after midday coincide with minima of humidity. Consequently large water vapour pressure deficits occur at this level (Evans, 1939; Aoki *et al.*, 1974). Under conditions of high saturation pressure deficit and high Bowen ratio an increase of turbulence is particularly effective on heat dissipation and transpiration rates. For example, 1 m/s increase in mean wind speed increases evapotranspiration about 3 per cent (Mock, 1974). Ventilation of the canopies confers ecological advantages; cooling is improved either by increased transpiration, if water is still available, or by rapid mixing of the atmospheric boundary layer and removal of sensible heat from the leaf surfaces (see fig. 7, right hand section) (Brünig, 1976).

Average figures of evapotranspiration from humid tropical forests are between 1 200 and 1 500 mm (Penman, 1970). They are deduced from very few meteorological measurements. It can be assumed that there is considerable variation between sites, forests of different structure, and species (Brünig, 1971c). See also chapter 12.

Kenworthy (1971) measured 450–500 mm water loss from interception of 2 500 mm annual rainfall in a natural mixed dipterocarp forest in Malaysia. Stem flow was negligible except in very heavy rainfall, especially if rain drift occurred in strong winds. Transpiration was calculated as 1 350 mm, soil evaporation as about 25 mm. Run-off was 650–700 mm, equal to 26–28 per cent of the rainfall. Loe Kwaisin (1974) measured 42 per cent interception in the IBP plot at Pasoh and Teoh (1974), 12–32 per cent in rubber plantations. Widely varying rates of stem flow are reported; this is not surprising because of the variation in plant habitus, rainfall pattern and wind speed.

Throughfall and ground-level radiation decline as secondary successions progress. For a continuous rainfall of 40 mm/d the data of Snedaker (1970) from lowland Guatemala indicate more than 90 per cent throughfall in a 14-year-old secondary succession, about 80 per cent at 32 years and roughly the same in undisturbed lowland forest. The incident radiation at ground level was 10 per cent in the 14-year-old stand and 3.9 per cent in the 32-year-old

stand. In addition to age, site quality also influences the trend of interception, the sites of high quality producing a denser and more complex cover much faster.

Light extinction in the canopy and light intensity in the undergrowth differ between seasons in forests with some degree of seasonality of leaf-fall. The improved illumination during periods of leaflessness is important for the regeneration of light-demanding species (Ashton and Brünig, 1975). Few measurements such as those by Brinkmann (1971) are available. In relatively open, sclerophyllous and xeromorphic stands radiation extinction in the top canopy is much weaker. Consequently more radiation penetrates to the ground layer. The generally poor ventilation in the ground layer, combined with the relatively high radiation of the ground surface, tends to produce high surface temperatures on leaves and the soil. The temperatures may equal those in the open on calm days and may even be higher on windy days. Examples and discussions of the ecological significance are given for pine forests in Cuba by Fojt (1969) and for some types of *kerangas* forests in Sarawak by Brünig (1974). Potential evaporation rates at the ground may be considerable under such conditions.

Brinkmann (1970) measured light intensity at short time intervals during 3 days in a dense secondary forest and in comparable cleared areas. In the forest relative light intensity was low (0.7–1.9 per cent), ultra-violet very low, but red and infra-red proportion somewhat higher than in the open. He concludes in accord with Smith (1936) and Walton (1936) that, for the undergrowth, root and stem competition seem to be more important than light. Stoutjesdijk (1972) relates the higher near-red/far-red ratio in tropical rain forest in comparison to temperate broad-leaved forests to the more irregular canopy structure of the former.

The illumination from the soil surface to a height of about 5 m under complex mesophyll ombrophilous forest is often less than 1 per cent of the illumination in the open and most of the energy is contributed by sunfleck light. Light saturation conditions extend through the upper, middle and top part of the lower tree strata and about 90 per cent of the light extinction is affected by these. The curve of relative illumination on the IBP plot at Pasoh shows a gradual decline from 100 to 25 per cent between 58 and 47 m height, then no further decline above a dense intermediate canopy at about 30 m. Below this there is a gradual logarithmic decline to 5 per cent at 20 m, 2 per cent at 10 m and 1 per cent at 5 m in the undergrowth (Yoda, 1974). Odum (1970) prepared an estimate of the budget of insolation and heat flows in a tropical rain forest (fig. 7) which illustrates the importance of evapotranspiration and eddy diffusion for maintaining a balanced energy budget in a tropical humid forest. Heat storage in the phytomass of a mature rain forest is possibly substantial. Yabuki and Aoki (1974) estimate the heat storage in the biomass (800 t/ha=8 g/cm², diurnal temperature range 7 °C) of the mixed dipterocarp forest in Pasoh Forest Reserve, Malaysia, as 56 cal/cm² and equivalent to 40 per cent of the incoming net radiation during the half day to noon.

The temperature change down the stand profile is small and amount to few degrees Celsius only, except around noon on bright days when a marked temperature peak develops in a narrow zone at the canopy surface (Yoda *et al.*, 1975). The significance of this peak in creating high saturation deficits of the water vapour in the atmospheric boundary layer of the emergent canopy trees has been discussed by Richards (1952) and Brünig (1971c, 1974). On average temperature gradients are weak, but the proportion of far-red light increases significantly in the lower strata. The relative humidity increases downward from midday values of 50–60 per cent at the top to above 90 per cent at ground level. Obviously leaves and plant geometry must be adapted to very different climatic conditions in the various forest strata, but reliable data on micrometeorological

conditions are very few and experimental data on the eco-physiological properties of the plants in these environments almost non-existent.

The chemical composition of the air within the stand also differs from the outside as shown by Freise (1936), by possessing:

— high carbon dioxide content often exceeding 1 000 ppm;
— $N_x O_x$: 44 mg/m³; NH_4: 2 mg/m³; $H_2 O_2$: 1 mg/m³; CH_4: 15 mg/m³;
— low fatty acids, 65 mg/m³, especially formic acid, and odorous organic coumpounds;
— tiny plant parts such as fibres, hairs, scales, but relatively few pollen grains.

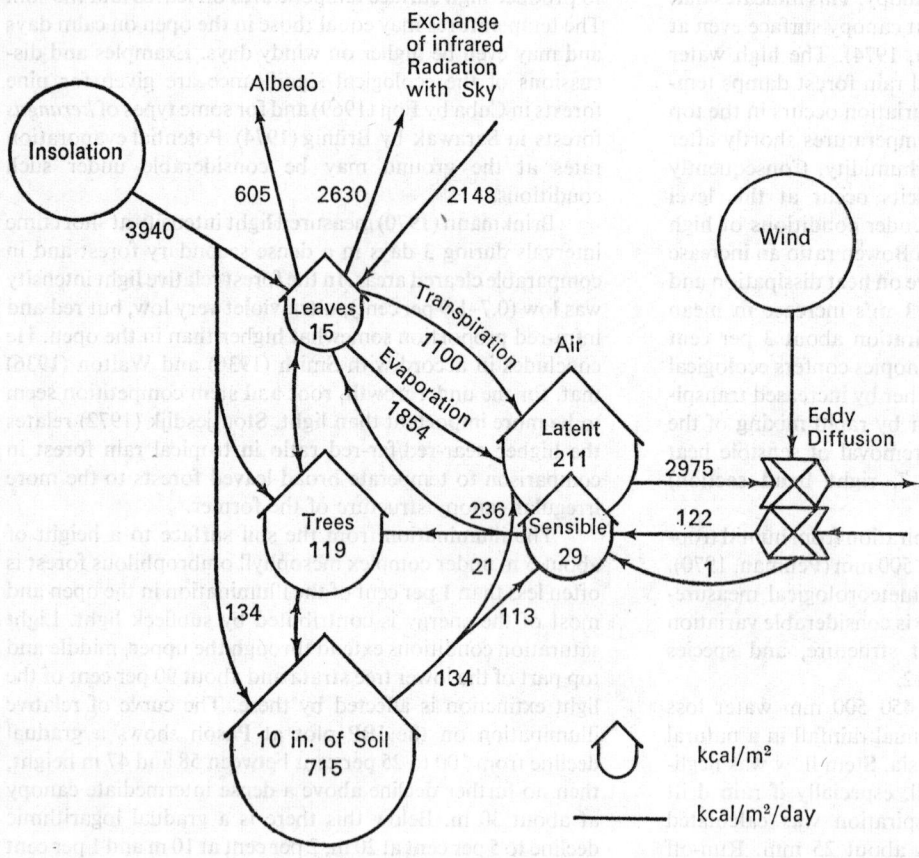

FIG. 7. Estimated budget of insolation and heat flows in the forest at El Verde, Puerto Rico, omitting photosynthetic deposition and rain cooling (from Odum, 1970).

The build-up of a large phytomass and the rapid turnover of organic matter, energy and water in the ecosystem causes a characteristic allocation of mineral nutrients to the various compartments of the system which is basic to its dynamic equilibrium. Klinge (1976) summarized and compared available data on biomass and nutrient content in tropical and temperate forest ecosystems. He concluded that there is a trend for tropical forest ecosystems to accumulate a larger proportion of the nutrients in the living biomass, but that the ranges of variation in the tropical and temperate forests are wide and overlapping. Any trend of the tropical ecosystems to store a large proportion of the system's

nutrients in the biomass would contribute to the sensitivity of the system to man-induced changes. Any change which involves the removal or destruction of the phytomass would cause a relatively large export of nutrients from the system into the surroundings in the form of solutes, gases and particles. See also chapters 10 and 13.

The deciduous forest

Away from the climatic equator rainfall declines whilst evapotranspiration and seasonality of the climate increase. Along this gradient the forests become increasingly deciduous, beginning with the top canopy and on sites of medium water-holding capacity, and progressively extending to include the undergrowth on deep soils with more ample water store. As a result of seasonality of climate and of leaf-fall pattern, micro-climatic conditions are distinctly different from those in the evergreen forest, but comparable measurements are rare.

During the wet season the micro-climate is similar to that in predominantly evergreen forest if the stature and structure are similar, but on average the deciduous forest possesses lower tree height, smaller leaf area and smaller biomass. Consequently the extinction rates of radiation and precipitation differ. The different means and variations of the macro-climatic parameters fully exert their influences and strongly affect vegetation structure and phenology. Malaisse *et al.* (1972) report a 35 per cent rainfall interception in miombo woodland during the peak of the growing season when the monthly rainfall is 250 mm. Almost all throughfall and stem flow is stored in the soil. Stem flow is estimated as 50 per cent of the intercepted water. The annual discharge into the drainage system is a mere 50 mm or about 4 per cent of the total rainfall. By changes of soil water content and rainfall interception, direct evapotranspiration was estimated to be 200 mm evaporation and 850–900 mm transpiration per year (see chapter 12 and annex).

Regional influences of tropical forests

Effect of partitioning on the regional climates

The micro-climate of forests is characterized by the peculiar partitioning of incoming matter and energy. Forest stands of different architectural structure and optical properties partition in different ways. Output flows will vary and this will affect regional climates. The effects will be modified by corresponding effects emanating from urban sites, grasslands, croplands, plantations or water surfaces. Each of these land-cover types has typical energy and mass exchanges and balances, and its effects extend for some distance in horizontal or vertical directions. The radiation exchange influences stability or lability of air stratification, water vapour content transfer influences cloudiness or cloud types or the internal water cycle, and the surface roughness changes the direction of air flow, aerosol fall-out and the distribution of pollutants. None of these effects can yet be quantified.

Range of forest influences

The range of the environmental effects of forests increases with the size of the source area and depends on the kind of the transported factor. Atmospheric contents such as water vapour, CO_2, pollen grains or spores may affect the biosphere at continental scale. Water exchanges influence the functioning of watersheds and the properties of the river sources through overland run-off, quick-seepage flow, erosion and leaching. The mechanisms of vertical radiation exchanges are well known for temperate conditions, but less information is available for tropical conditions, especially on temperatures and humidities, and albedo. The turbulent transfer of heat, vapour, gases or solids under certain conditions of lability of air stratification or of inversions is insufficiently understood. Large-scale atmospheric circulation systems may transfer environmental effects of tropical forests far beyond the tropical regions.

Examples of regional studies of the range to which forest influences make themselves felt, such as precipitation changes after reforestation or forest depletion, are available from Argentina, the Federal Republic of Germany, India, Spain and the USSR. The few studies conducted in tropical countries have been much less successful because it is very difficult to separate the effects of climatic fluctuations from the effects of changes of land use. Research in this field is difficult and requires sophisticated and often expensive experimental and evaluation methods.

Energy considerations

Energy exchange is a valuable means of quantifying the local and regional effects of specific vegetation cover and land use. The local conditions of the earth surface, such as the roughness and optical properties of vegetation, and of the sun radiation regime produce specific features of radiation balance, hydrological balance, and of individual thermal or hydrological features. Forests are very efficient absorbents of visible and infra-red radiation. The reflectivity of forests is 5–10 per cent lower than that of other soil covers. The low effective radiative temperature of forests damps infra-red emission. The high energy intake of forests is used primarily in the evapotranspiration process. Consequently water vapour flow to the air above the stands is intensive, and evaporation cooling avoids overheating of forested land and reduces the tendency to convective air exchanges. Moistening of air without destruction of the atmospheric stratification seems to be characteristic for moist forest areas (table 4).

TABLE 4. Energy balance terms in tropical regions (Baumgartner and Reichel, 1975).

Latitude (°)	North			South		
	25	15	5	5	15	25
kcal/cm²/a	Land surfaces only					
Q, net radiation	69	71	72	72	73	70
H, heat flux in the air	48	35	9	4	21	42
L, latent heat flux	21	36	63	68	52	28
$\dfrac{100\,L}{Q}$	31	51	88	94	71	40
$\dfrac{100\,H}{L}$	224	96	14	6	40	150

Net radiation (Q) increases toward the equatorial zone. The heat flux in the air (H) and the ratio H/Q is relatively small near the equator and increases toward the border of the tropics. The latent heat flux (L), which is proportional to evaporation, decreases from the equator polewards. The terms of the energy balance are sensitive to changes in the plant cover of the earth surface. Therefore the conversion of tall forests into perennial plantations, annual field crops or grasslands will modify the energy balance. These changes should be pre-assessed by comparative energy balance investigations before large-scale conversion of natural forest is implemented. The effects of the subsequent changes in aerodynamic surface roughness on the partitioning of kinetic energy (wind direction, speed and turbulence) must also be considered.

Hydrological considerations

Components of the hydrological balance are precipitation (P), evaporation (E), run-off or discharge (D), storage or reserves (R) and consumption (C). A rough approximation of the water balance of tropical regions is given in table 5.

TABLE 5. Water balances of tropical regions. Latitudinal averages for land surfaces (Baumgartner and Reichel, 1975).

Latitude	N	25°	20°	15°	10°	5°	0	5°	10°	15°	20°	25°	S
P		616	670	1 022	1 485	1 962	2 093	1 818	1 431	937	603		mm
E		389	455	792	1 042	1 118	1 214	1 118	1 002	778	517		mm
D		227	215	230	443	844	879	700	429	159	86		mm
E/P		63	68	77	70	57	58	61	70	83	86		%
D/P		37	32	23	30	43	42	39	30	17	14		%

Precipitation (P) in the equatorial rain belt is more than three times as large as the global average of 746 mm. Global average evaporation (E) is 460 mm for the land surfaces, but in the moist tropics it is at least 1 200 mm. Global run-off (D) from land surfaces averages 266 mm, but in the equatorial region it is equal to 879 mm. This causes the huge tropical rain forest rivers: the Amazon carries 18 per cent, the Congo 3 per cent and the Orinoco 2 per cent of all surface water discharge of the earth. The run-off pattern and the water quality depend on the effect of forested land on water yield, floods, erosion and sedimentation (Lull, 1971). In the perhumid regions, the influence of forests on the water yield is generally less important than on the run-off pattern. Forests prevent floods and high sediment loads.

Available knowledge is insufficient to quantify the climatic effects of different vegetation types on similar sites. This is particularly true with respect to energy and water balance where quantitative studies are almost completely lacking. There is an urgent need for intensifying investigations on the bio-ecological and human-ecological effect of forestry and alternative land use in the tropical regions. Also more precise and more extensive hydrographic information on river basins are needed. Comparative investigations are needed of catchments under different land uses, but equal climatic conditions.

Global influences of tropical forests

Evaluation problems

The share of tropical forests in the global resources is the scale by which their influences on the global biosphere may be assessed. Their areal (A_t) shares as a percentage of the surface area of the globe (A_g) are as follows:

tropical zone (land and ocean surfaces)	40 per cent
tropical land surface	12 per cent
tropical forest area	4 per cent

If C_t is the concentration (g/cm²) of a parameter in the tropical biosphere, then the source power is given by the product $A_t C_t$. Under this assumption the efficiency (E_t) of the influences of the tropical areas may be defined as:

$$E_t = \frac{A_t}{(A_g - A_t)} \cdot \frac{C_t}{C_g}.$$

The term C_g stands for the mean concentration of the same parameter in the biosphere of the globe. Using the above percentages of areal representation, then the efficiency may be quantified:

whole tropical zone	$= 0.67(C_t/C_g)$
tropical land surfaces	$= 0.14(C_t/C_g)$
tropical forest areas	$= 0.04(C_t/C_g)$.

Influences on the radiation and energy budgets, on water balance, on constitution of the atmosphere of the globe may be calculated on this basis. The concentration ratios C_t/C_g are > 1 for radiation emission, for water vapour content, for carbon dioxide content and for surface roughness, but < 1 for albedo and for some pollutants. The reason is that the infra-red emission, saturation vapour pressure and plant productivity increase with temperature, but albedo of tropical forests is relatively low and the emission of pollutants from the tropical forests generally smaller than from other land use types or areas outside the tropics.

For the evaluation of the dispersion of qualities outside the tropics, the persistence of the qualities must be recognized. Atmospheric circulation is effectively dispersing only some properties of the atmosphere such as gaseous components while radiation is limited to the source areas. Water vapour is eventually condensed and precipitated, pollutants eventually will fall or be washed out; however, gases such as CO_2 have a long turnover time so that their dispersion may have an effect on the biosphere of the whole globe. In the following section the main questions are considered in the light of the present situation; later, the problem will be examined in the light of the man-made changes of land use.

Global energy balance and thermal equilibrium

As shown in table 4 the net radiation input at the earth surface of tropical zone is high. This is due to the sun's elevation being at or near zenith at noon all year round, to the low albedo of forests and to the low effective temperature of forest canopies. The discrepancy between high net radiation and low effective temperature is not a contradiction as the absorbed radiation is used for the vaporization of water. The vaporization energy is transported to the atmosphere outside the tropics as latent heat and there becomes available in the thermal balance after condensation of the water vapour. The tropical forests are intensive converters of energy and of water and, therefore, an important link in the thermodynamic system of the atmosphere.

There are regional differences of the energy exchanges and heat flow components, governed by the atmospheric circulation systems (trade-winds, monsoons, cloudiness, precipitation) or by the properties of soil covers (effective emission temperature, albedo). It is unknown whether the tropical forests exert a substantial effect on the intensity of the equatorial westerlies or the direction of the trade winds. The effect of net radiation (Q) on climate and vegetation strongly depends on the amount of precipitation. The relationship can be expressed by the so-called Budyko number Q/P', where P' is the amount of energy used in evaporation. The Budyko number is 0.3 for tundra; 0.3–1.1 for forests; 1.1–2.3 for savanna, steppe or prairie; 2.3–3.4 for semi-deserts and 3.4 for deserts. In general, $Q < P'$ indicates humid, $Q/P' \simeq 1$–3 semi-arid and $Q/P' > 3$ arid conditions. It is open to argument whether the vegetation cover type depends on the local Q/P', or the Q/P' is caused by the local vegetation cover. However, changes in the vegetative cover will be reflected by changes in the Budyko number, in other words the partitioning of energy will be affected by a change of vegetation type.

Global water cycle and hydrological balance

The intertropical convergence zone receives very high precipitation which sustains the enormous tropical river systems. Most rivers discharge into the oceans within the humid tropics; but in some cases, such as the Nile, rivers carry surface water supply into arid zones. The major influence of the humid tropics on the global hydrological balance, however, is not by the rivers but through the water vapour transported within the atmosphere. The volumes of water vapour are given in table 6.

TABLE 6. Water vapour volumes originating from land and sea surfaces (Baumgartner and Reichel, 1975).

	Tropical areas	Global areas	Tropical as a percentage of global
	$\times 10^3$ km³		
Land surface (L)	44	71	62
Sea surfaces (S)	243	425	57
Total (G)	287	496	58
L/S (%)	18	17	—
L/G (%)	15	14	—
S/G (%)	85	86	—

The tropical regions which represent 40 per cent of total earth surface contribute 58 per cent of water vapour into the global water cycle. The share of the tropical ocean surface to the evaporation volume of the globe is 49 per cent and that of the tropical land surfaces 9 per cent. Water cycling in the atmosphere is rapid. The average residence time is 9 to 10 days (Penman, 1970) but possibly less in the humid tropics. Tropical forests occupy about one third of tropical land surfaces, so tropical forests influence the global water cycle by the order of 3 per cent. The assessment of the effects of changes in tropical land use on the global water balance must be in relation to this figure.

The assessment of the forest influences on the water balance must consider continental gradients of precipitation, evaporation and run-off. For example piedmonts in tropical America have precipitation rates of 6 000 mm and above with evaporation of 1 200 and run-off of 5 000 mm, while adjacent plains such as the Amazon basin have only 3 500, 1 500 and 2 000 mm respectively. Considerably smaller amounts occur in certain parts such as the notoriously dry North-East of Brazil. In Africa also, a steep gradient of diminishing rainfall away from the equator and some coasts leads to almost zero amounts in the deserts.

Carbon cycle and carbon balances

The carbon cycle and balance in the biosphere are governed by two main processes: the photosynthetic assimilation of CO_2 by plants, and the metabolism and release of CO_2 by living organisms, and by the release of CO_2 by burning fossil or recent fuels. The influence of the tropical forests on the atmospheric carbon balance is indicated by the amounts stored in the biomass and the amounts annually cycled.

The forests and woodlands of the world are estimated to contain between 400–700×10^9 t of carbon. The amount of dry organic matter would be about twice this figure. This compares with 700×10^9 t of carbon present in the form of carbon dioxide in the atmosphere. About one third of the organically bound carbon occurs in forests, one fourth in oceans, one fourth in atmosphere, the rest occurs in grasslands, tundra or other surface covers. Just over half of the world phytomass is stored in tropical forests and woodlands. These forests produce roughly 69 per cent of the total net primary production of the world's forests on about half of the world's forest area (Bolin, 1970; Whittaker, 1970; Brünig, 1971b). See also chapter 10.

Estimates of the earth's primary production range between about 120 and 160×10^9 tons of dry organic matter per year (Woodwell, 1970; Lieth and Box, 1972). Agricultural crops fix about 5 per cent of this, forests and woodlands about 45 per cent and other vegetation 13 per cent, the remaining 37 per cent, which may be an underestimate, being fixed by phytoplankton. The tropical forests covering about 20×10^6 km² or 4 per cent of the surface of the earth account for at least 25 per cent of the global terrestrial carbon fixation. They release through photosynthesis about 55.5×10^6 t O_2 per year which however is balanced by their simultaneous oxygen consumption in respiration (Brünig, 1971b). It is assumed that fossil fuel burning increases the CO_2 concentration of the atmosphere by 0.64 ppm/a. This is partly buffered in the oceans, but partly also fixed by photosynthesis in forests. Accordingly the highly productive tropical forest will serve as effective regulators of the CO_2 balance of the atmosphere.

The importance of tropical forests for the maintenance of climate cannot yet be quantified with any degree of reliance, because neither the changes of climate following changes in the atmospheric composition can be predicted, nor can the intensity of the forest influences be related to structural or site properties or to the productivity of the various natural or modified forest types.

Influences on atmospheric pollution

Aerosol particles from the tropical regions as polluting factors in the atmosphere originate above all from shifting cultivation and savanna burning. Small amounts are released by the exploitation of forest (forest fires, forest clearing), permanent agriculture (burning of crop wastes), and by the release of hydro-carbons (terpens), pollen grains or spores. The world total of aerosol production has been estimated 1 670 × 10⁶ t/a. Agriculture and forestry contribute 13 per cent of this. Forest fires, excluding shag burning, contribute less than 0.1 per cent. These calculations are tentative and lack a scientific data base. Similarly uncertain is the global assessment of photochemical oxidants, ozone, peroxyacetyl nitrates, ethylene, sulphur dioxide, fluorides, chlorines, ammonia, ions and radioactive materials.

The methodology of monitoring and evaluating regional and global environmental influences of forests is reasonably well known, but this methodology has hardly been used to quantify the environmental effects of tropical forests. The application of available methods is lagging well behind demands for information. Urgently required are: expansion and intensification of studies of the structure and functions of tropical forests, investigations into the concentrations and dynamics of the various constituents of the atmosphere, studies and continued monitoring of the horizontal and vertical fluxes of energy and masses (vapour, CO_2, O_2, aerosols, etc.) within forests and between the forests and the surroundings, and of the interactions between forest structure, functions and environmental effects in the humid tropical forest zone.

The consequences of modifying the vegetative cover

The problem

The tropical forest occupies a great variety of edaphically and climatically heterogeneous sites. The forest itself exhibits a great diversity of architectural structure and of species and chemical composition (see chapters 4, 5 and 8). The high degree of diversity is possibly a major condition for the sustained functioning of the tropical forest ecosystem (Klinge and Fittkau, 1972). There is very good reason to assume that forest structure is closely related to the energy, water and nutrient regime of the site (Brünig, 1971c, 1974).

This structure, in terms of phytomass distribution, leaf and crown geometry, and aerodynamic surface roughness, is probably essential for safeguarding sustained productivity and preventing the development of excessive stress. At present large areas of tropical moist forest are being destroyed by logging, shifting cultivation or agricultural development (see chapters 9 and 20). Invariably structural diversity and complexity will be reduced and modifications will occur in the processes of energy and matter exchange, and circulation in the boundary layers between atmosphere, vegetation and soil. In many cases, this will unbalance the functioning of the ecosystem. As the area of humid tropical forest is rapidly reduced, the effects will be felt at the regional and finally at the global scale.

Through recent centuries man has reduced the area of natural humid tropical forest by at least about one third. The current reduction rate is accelerating rapidly in South-East Asia, West Africa and Amazonia. The growing demand for food and raw materials will tend to force all arable land into use. The world resource of potential arable land is 3.2×10^9 ha or roughly 25 per cent of the land surface of the earth. 1.4×10^9 ha are already agriculturally and sylviculturally utilized (Nimlos, 1971). The balance of 1.8×10^9 ha potential arable land mostly lies within the tropics (1.1×10^9 ha = 61 per cent of the arable land reserve and 55 per cent of the remaining tropical forest area). At the present rate of forest destruction in the tropics this reserve area will have been converted before the end of this century, leaving natural forest only on inaccessible or extremely poor sites.

In 1960–62 the commercially harvested industrial timber was 6.1 per cent of the estimated tropical growing stock in Central America, 0.3 per cent in South America, 5.6 per cent in Africa and 2.7 per cent in Asia. Except for South America, these could not be sustained from the present forests unless very intensive management were introduced. The annual area of forest which is destroyed by shifting cultivation and settled agriculture is estimated to be in the order of 30–50 million ha (excluding the area affected by the Brazilian *Transamazonica* project). Timber exploitation and shifting cultivation alone therefore will remove all primary tropical forest on easily accessible sites within the next 20–30 years. This will change the meteorological and edaphic processes on about $1.2–1.5 \times 10^9$ ha of tropical lands, or on 6–8 per cent of the tropical land surface and 2–3 per cent of the global land surface. If this proportion seems small, it must be rembered that this is the region where climatic processes are most destructive and from where climatic processes in other regions are being initiated, and that the balance of the climate of the earth is exceedingly delicate. However, investigations into the effects of changes of plant cover on the soil, on the atmospheric boundary layer and on the ecosystem and its environment as a whole are very rare in the tropics. Also most available data are not strictly comparable due to deficiencies in scale and design of the observations. The following treatment and evaluation of the presented evidence must therefore be judged with these severe limitations in mind. A further great deficiency is that few humid tropical forest areas have yet been covered by detailed and efficient site, soil and vegetation surveys. For example, little more than 10 per cent of the Amazonian basin have been soil-surveyed even in a cursory manner. Therefore, any estimate of the natural scale of the effects of modifying the plant cover on the local and regional climate, and on soil and site conditions and productivity, must remain speculative.

Impact at the ground surface

Changes in the features of the earth's surface will cause changes of meteorological and climatic processes at various scales of space and time. Changes after forest clearing for agriculture with annual field crops or for engineering works have been severe. Less destructive on soil and climate seem to have been natural or plantation sylviculture, perennial field crops or rough grazing. Supplementary activities, such as irrigation, drainage, fertilization and pest control affect the environment directly and indirectly and usually accentuate the adverse effects of forest destruction, but adequate data are lacking (see chapter 20).

One of the most direct and immediate effects of forest conversion is the change of albedo. Initially this is the result of the different optical and geometrical properties of the smoky atmosphere during the season of burning. Later, the albedo will change as a result of the different optical properties of the new surface which may increase the amount of radiation reaching the ground by more than 10 per cent.

According to Tanner (1968) a 0.15–0.25 variation of the albedo of vegetation surfaces corresponds to a variation of the net radiation in the order of 5 per cent. Forest removal normally reduces the net radiation for the land surface because almost all substitute surface covers possess greater reflectivity. The net radiation on the exposed soil surface however increases as a result of the lack of protective cover and the greater impact of radiation. On moist soils, evap-

oration from the soil is increased and the soil surface may become cooler. On dry soils absorbed radiation will make the surface hotter. Lethal temperatures for life may occur, and a number of atmospheric processes are affected.

Hill (1966) measured soil temperatures at 3 to 50 cm depths in bare soil, grass, and disturbed mixed dipterocarp forest in Singapore. He found that relative temperatures at all depths were consistently higher in the bare soil, slightly lower under grass, but considerably lower and less variable under forest. At 50 cm depth all three soil locations showed little diurnal change, absolute temperatures were about equal in the bare soil and under grass (*ca.* 28 °C) but consistently lower under forest (*ca.* 24 °C).

Higher soil temperatures increase the rate of mineralization of the organic matter. This impairs the stability of soil crumb structure. As a result the shear resistance of the soil surface is reduced and the soil becomes more easily erodible. Hardening of the surface of some soil types on exposure also reduces infiltration which increases run-off and erosion.

Table 7 shows the scale of the differences of net radiation (Q), the fluxes of convective heat (H) and latent heat (L), Bowen ratio (H/L), corresponding evaporation (E) and short wave albedo (Rk) which are related to land-cover type. The figures give only a rough approximation but indicate that the climatic parameters vary considerably between land-cover types. Deforestation initiates changes towards aridity. Some climatologists are of the opinion (SMIC Report) that changes in land use may also affect the dynamics of atmospheric circulation systems. An example is the generation or the dissipation of tropical easterly waves or of local development above forest and absence above clearings of convective clouds. These effects are due to the change of aerodynamic surface roughness, albedo, water vapour flow and stability of thermal air stratification, as a result of the change of surface cover type.

TABLE 7. Energy balance terms for earth surfaces with different land uses (Baumgartner, 1965, 1970).

	Q kcal/cm²/a	H	L	E mm	Rk %	H/L
Coniferous forest	60	20	40	1 000	10	0.5
Deciduous forest	50	15	35	900	15	0.4
Open land, moist	60	15	45	1 000	20	0.3
Savanna	50	20	30	800	25	0.6
Grassland	45	15	30	750	20	0.5
Cropland	50	20	30	800	25	0.7
Bare sand	35	20	15	600	30	1.3
Urban area	35	20	10	600	30	2
Desert	35	30	15	600	30	6

Another important direct effect of forest conversion is loss of soil protection against wind and rain. This also has very serious consequences, which are relatively well known for temperate areas but cannot yet be reliably quantified for the tropics; results from temperate areas are not directly transferable and applicable to the tropics. The effectiveness of interception, radiation, precipitation and wind is directly related to stand height, basal area, biomass and number of

individuals (see chapter 12). This means that sylvicultural or agricultural modification will change, and usually reduce the rates of interception. Consequently, the impact of rain and radiation on the soil surface will increase if structural complexity of the vegetation is reduced. It is possible to calculate the energy of free falling raindrops of various sizes and densities. But it is not yet possible to calculate the effects of modifying the forest cover on the kinetic energy of the raindrops which fall through the canopy on to the ground. Odum (1970) tentatively assessed the rainfall intensities at the canopy surface and at the ground of the rain forest at El Verde. The results indicate that considerable differences exist. The erosive power of the impacting raindrops is further enhanced if burning has created hydrophobic substances which reduce wettability and infiltration rates. The net results of the changes in the soil cover and in the soil itself are accelerated overland flow, rapid seepage run-off and increased erosion by surface wash and internal leaching and mass wasting. Consequently the water content of the soil and the sediment and solute load of the discharged water will fluctuate more widely.

High rates of lateral water flow in humid tropical forest soils have been reported by many authors, but few measurements have been made. Leigh (1974) in a preliminary report on results of investigations on the IBP plot at Pasoh Forest Reserve, Malaysia, reviews available information on concentrated and diffuse throughflow of laterally percolating water in the soil and confirms previous opinions that 'until a substantial body of quantitative data is accumulated, the importance of lateral eluviation to slope development, and its significance in relation to other degradational processes, in particular surface wash and soil creep, remains conjectural'.

Available data for whole watersheds show that erosion under dense natural perhumid and seasonal humid forest is usually less than 1 t/ha/a. The rate increases up to 20–30 t/ha/a if the land is cultivated. The erosion problem will be discussed further in chapter 12, but the following information delineates the scale of erosion with respect to the functioning of the affected ecosystem and to the biosphere environment. Grazing on a 1:4 slope at Nurpur in Kangra (Himachal Pradesh) (Sinha cited by Ghosh, 1975) caused 22 t/ha soil loss compared to 0.3 t/ha from ungrazed grass and shrub during a rainstorm of 249 mm in 19 h; 95 per cent of the rainfall ran off from the grazed area, but only 20 per cent from the ungrazed area. At Dehra Dun, agriculturally used watersheds had significantly more soil loss than watersheds under mixed use by agriculture and forestry. Erosion rates on a 7 per cent slope in the Ivory Coast were between 138 and 90 t/ha/a on bare soils and under annual field crops respectively, but negligible (0.03 t/ha/a) under secondary evergreen forest (Roose, 1976: see chapter 12). Available experience indicates that less than 10 t/ha/a are normally eroded under natural tropical forest on undulating land and 10–40 t/ha/a on steep slopes, as compared to 20–160 t/ha/a under different annual or perennial agricultural or sylvicultural crops and up to 1 200 t/ha/a on slopes during the first year after burning for shifting cultivation. Kellman (1969) studied surface run-off characteristics during 227 days in upland Mindanao. Rates in a 10-year-old abaca plantation were twice as high as in natural mixed dipterocarp forest, 4 times as high in a young rice field and about 50 times in a 12-year-old rice field. Run-off in a well-developed mixed broad-leaved tree fallow was about the same as in natural forest. He concluded that erosion will increase if traditional shifting cultivation is continued. It has been observed elsewhere, e.g. in teak and pine plantations in Trinidad and Malaysia, that erosion rates in secondary tree stands are closely related

to the complexity of the stand structure, especially to the presence or absence of an understorey. This seems to be a general law; particularly erosion-susceptible are dense, uniform tall crops with a single-layered lofty canopy and without a well-developed understorey to absorb the kinetic energy of raindrops.

Williams (1969) established a significant correlation between rainfall momentum and rainsplash erosion in seasonal tropical northern Australia. Very similar conditions of high-intensity seasonal rainfall, poorly structured sandy top soils, gentle slopes, sparse savanna vegetation, and seasonal grass-fires are widespread in tropical Asia, Africa and South America. Monthly evaluation of soil loss and rainfall momentum for specific soil types and vegetation cover, coupled with a knowledge of the probable rainfall intensity distribution in each month, would allow splash losses from different sites to be predicted. Given a knowledge of run-off erosion, the effects of varying forms of land use on different soils could then be predicted, and land use planning accordingly adjusted to reduce soil degradation.

In conclusion we can generalize from the little and inadequate information available that under dense forest, surface flow is negligible or absent and seepage quick, and therefore storm-flow in streams will be moderate, except in extreme cloud-bursts (Wicht, 1972). Temporary or permanent exposure of the soil surface leads to dramatic degradational processes, which are difficult, costly, slow and uncertain to reverse. The effect of substituting complex forest with more simply structured forest stands is intermediate.

In a succession the standing phytomass increases very rapidly; leaf mass and surface reaches maximum levels within 10 or 20 years, but the woody phytomass increases much more gradually (see chapter 10). Ground protection against rain and radiation therefore increases very rapidly, as does the build-up of the litter cover on the soil surface. But the protective capacity of the stand continues to increase beyond the maximum level for leaf phytomass as a result of the increasing complexity of the stand architecture and of the increase in twig, branch and stem surface as the stand develops toward maturity.

Impact on the micro-ecosystem

Burning

Fire is used extensively in the tropics for the destruction of forests. The objectives of burning are site clearing for shifting cultivation, bush and weed control, hunting, grazing, and often simply for fun. Burning is a simple and effective method of land clearing but it has serious consequences for the ecosystem which is being burnt and for its surroundings.

Burning emits large amounts of matter into the atmosphere. Large aerosol particles from burning vegetation spend a short time in the atmosphere but may effectively increase the infra-red re-radiation from the lower atmosphere. Small particles remain for periods which may be as long as several weeks in the upper troposphere even years in long-distance transport. The extensive savanna fires during the dry season lead to heavy dust concentrations in the atmosphere which may even spread into the region of the perhumid equatorial forest. The net effect of smoke pollution at ground surface may be cooling or warming depending on the direction of changes of the surface albedo and on the absorption coefficient of the particles in the atmosphere (Flohn, 1973b). Burning also releases nutrients, especially nitrogen, into the atmosphere and into the soil water. Eventually part of the latter will enter the drainage system and be lost. The fate of the former is more complicated and little understood.

Simplification of architectural complexity

An important consequence of burning and replacing natural forests by man-made vegetation or secondary successions is the reduction of aerodynamic roughness of the vegetation surface. This reduction affects the wind field within and above the canopy, the net radiation and the thermal budget, and the water balance.

According to Tanner (1968) roughness variation does not cause drastic changes in potential evaporation, if the different surfaces have the same humidity and temperature. The differences between areas of different roughness is more pronounced if the upwind area is very dry and the air travelling across the evaporating surface must make a large adjustment. But changes in aerodynamic roughness will affect the exchanges of energy and matter within the forest ecosystem. The penetration of radiation, the absorption of kinetic energy and the turbulent transfer processes will be more intensive in canopies which possess an aerodynamically rougher canopy geometry. Leaf geometry and orientation and the optical properties of the leaves may amplify or compensate the effects of canopy surface roughness. This must be carefully considered in agricultural and sylvicultural planning, but there are very few data. Fränzle estimates from the few available data that converting Amazonian natural forest to cultivation would increase the Bowen ratio from 0.33 to 0.50 (table 8). Consequently, run-off and erosion rates and surface temperatures (sensible temperature) will increase and the humidity of the atmosphere will decrease.

TABLE 8. Change of micro-climatic parameters as a result of rain forest conversion in Amazonia (Fränzle, 1974).

	Natural rain forest	Cultivated land	Unit
Net radiation	80	75	kcal/cm²/a
Evapotranspiration	60	50	—
Sensible heat flux	20	25	—
Bowen ratio	0.33	0.50	ratio ET/SH
Precipitation	2 100	2 100	mm/a
Evapotranspiration	1 000-1 100	850- 950	—
Run-off	1 000-1 100	1 150-1 250	—

The effects of occasional drought conditions on vegetation with high exchange intensity, tall and complex structure, deep roots, and a relatively large water store in the biomass are damped by the available water reserve. Conversely, water excess is partly absorbed by intensive interception and

evapotranspiration (Brünig, 1971c). Shallow-rooted, simple-structured field crops will be subject to extremes. Fluctuations of the floristic, architectural, phenological and structural features, and of the functions of the ecosystem will consequently be more intensive. The simple field-crop system as a whole will under these conditions acquire a state of increased instability. This will create critical conditions if fluctuations of regional climate occur which expose the vegetation to long-lasting stresses which may exceed the power of resilience and recovery of the ecosystem. The result may be further reduced stability. Spectacular examples are the effects of droughts in the seasonal tropics.

Circumstantial evidence of lasting and profound changes being initiated by forest clearing is provided by long-term observations of rainfall in private rubber plantations in Malaysia. Large-scale forest clearing was followed by substantial changes of rainfall intensity. While the total annual rainfall appeared to remain unaffected, the number of rainfall incidents decreased and the intensity per rainfall increased. The ecological disadvantages of high rainfall intensity have already been described.

Bare soil and land covered with withered plants practically cease to evaporate after about 3 rainless days, while green-leafed forests continue to transpire for some time depending on the amount of water reserve available. In most soils this reserve equals about 15-20 per cent of the depth of the rooting zone (Mock, 1974). The plant-available water varies with soil type and rooting depth from between 60 to over 500 mm plus about 5-15 mm available in the phytomass. Consequently complex stand architecture and species diversity can be sylviculturally maintained on well water-supplied soils even in seasonally dry climates, but on more xeric soils such complex and diverse ecosystems would be extremely fragile (Brünig, 1971b, c, 1974). Holdridge (1967) calculates different transpiration rates for forests of different height and number of storeys; this is theoretical because comparable measurements are lacking. The replacement of tall, complex natural forest by secondary or plantation forest of smaller stature, simple architecture and sometimes more xeromorphic habit, will alter evaporation and transpiration rates. Babalola and Samie (1972) using neutron techniques found that a natural forest stored more, but also evapotranspired more water than a 10-year-old eucalypt plantation.

Smith (1936) and Walton (1936) measured light intensities with a photo-cell in sylviculturally treated mixed dipterocarp forest in Malaysia and found that for an effective improvement of light conditions for regeneration, drastic opening of the top canopy layer was essential. Understorey treatment alone is relatively ineffective for reducing competition for light. Possibly also root competition, which may be more important, is relatively unaffected by understorey treatment. Drastic top-canopy opening however, will expose the soil and cause the adverse effects described earlier. These may be ephemeral and therefore tolerable if the regeneration develops rapidly into a complex and dense stand. Tolerance limits may be exceeded if these sylvicultural measures are preceded by timber logging with heavy soil disturbance and damage to the regenerating stock.

Water consumption will be substantially reduced when

complex, multi-aged aerodynamically rough crops are replaced by uniform, simple-structured even-aged crops unless the new crop consists of plant species which use water profusely. In the case of the choice between deciduous or evergreen forest, the more simple and seasonally resting deciduous forest will produce greater total stream-discharge, owing mainly to reduced interception and transpiration during leafless periods (Wicht, 1972).

Intact natural forest ecosystems by their floristic and architectural diversity also support more diverse decomposer and consumer systems at the various trophic levels than substituted man-made vegetation. Most cultivated or degraded ecosystems have truncated decomposer and food chain pyramids and possess low diversities within each of the remaining trophic levels. The result is greater risk of pests and diseases (see chapter 14). These lead to further degradation, loss of structural complexity and slow-down of recovery, unless man intervenes with artificial correctives, which are costly and may contribute to a long-term imbalance. Genetical diversity is reduced, breeding systems are interrupted and potentially useful genes are lost if tropical rain forests are cleared, even if relics are kept as nature or biosphere reserves for the purposes of amenity, protection and preservation. Chemical diversity is similarly reduced.

The following consequences of man-made modification (usually in the direction of simplification) of the floristic and architectural complexity of the natural rain forest vegetation can be expected. The albedo, the intensity of infra-red radiation, the emission and transfer of sensible heat are substantially altered and the vertical air stratification is consequently shifted towards instability. Cloudiness and the internal water cycle of the ecosystem may in perhumid climates be reduced and may result in increased surface flow. Conversely, forest establishment may act in the opposite direction in the more xeric and seasonal climates. However, little information exists and there remain many uncertainties. The processes following forest destruction in the equatorial region may in the short term improve the living conditions for man, but long-term land use and health effects are usually unfavourable. There are good and fact-supported reasons to believe that generally forest destruction will increase aridity and put the natural ecosystem out-of-balance with a corresponding loss of dynamic equilibrium and reduction of power of recovery by intrinsic control mechanisms. Eventually effects will spread to regional and larger scales. As the area of vegetational change expands, turbulent exchange and Bowen ration increase, the air humidity over larger areas declines and the dust load of the atmosphere rises. The rates of these processes differ between the wet and dry seasons and between humid and arid climates.

Other particularly critical points in tropical land use development involving replacement of natural forest are:
— the soils with low absorption capacity in wet climates require continued large amounts of fertilizer to obtain and sustain adequate yields;
— maintenance of a stable soil structure, avoidance of crust formation and of erosion require effective natural soil cover or very expensive technological inputs;
— maintenance of adequate soil structure and nutrient cycling requires diverse perennial vegetation of suitable chemical properties, or very expensive and sophisticated land use technologies;
— insect and plant pests are very severe and persistent and require costly control;
— adequate infrastructure for large-scale development is almost invariably lacking and consequently adequate levels of sustained management are not ensured.

Urbanization

The transformation of rural lands into urban areas is connected with increase in albedo, heat storage capacity and aerodynamic roughness and reduced evaporating surface and water for evaporation. While the scale of such transformation is relatively small in tropical areas, the effects are profound and noticeable within the urbanized area itself. They are on a whole detrimental to man's well-being. The direction of change is the same as in cities in temperate climates, including increases of air and water pollution, cloudiness, intensity of precipitation, day-time temperature, while generally relative humidity, global radiation and wind speed decrease. The resulting harsh environment may be mollified by trees and green areas. Trees have numerous effects: they provide amenity and shade, they filter and absorb pollutants along traffic ways (gases, dust or noise), they are relatively temperate spots in hot environments, they are psychologically soothing and balancing components in an otherwise provocative and irritating urban environment.

Impacts on the global biosphere

Forests produce impacts of value to man in five fields: energy exchanges; water vapour exchanges; fixing and release of carbon dioxide; release, dilution and absorption of aerosols, gases and solutes; and the production of goods and cultural services.

The energy exchange complex can be divided into radiation and energy balance. It is interdependent with the other factors. The increase or reduction of water vaporization will influence the thermal radiation of the atmosphere and the thermal balance of the planet. The same is true for carbon exchange and aerosol production. Even if, on a area basis, the effects appear marginal, the delicate nature of the balance and interplay of global climatic processes makes effects at a larger scale appear likely.

Barinov (1972) speculates on the possibility of the interrelationship between phytomass fluctuations and geophysical phenomena by geologic periods, but seems to exaggerate the scale to which modern man is able to upset the global energy balance by thermal pollution. Man-made changes of the vegetation alter the albedo and roughness of the earth surface, the soil humidity, Bowen ratio, and direction and intensity of the water cycle. Flohn (1973b) estimates that at least 30 per cent of the continental land surfaces of the world are subject to such man-made changes. Mathematical models are being developed to estimate the size and direction of the variable effects on the climate, but fully satisfactory models are not yet available, because of the complex conditions and the lack of data.

Potter *et al.* (1975) used a two-dimensional (zonal) atmospheric model to simulate some of the effects of converting all humid tropical forest between 5° N and 5° S to a vegetation type with an increased albedo (0.25 instead of 0.07), increased run-off and decreased evaporation. They found an average global cooling of the surface by 0.2° K and a reduction in precipitation of less than 1 per cent. The probable results can be summarized as follows: deforestation → increased surface albedo → reduced surface absorption of solar energy → surface cooling → reduced evaporation and sensible heat flux from the surface → reduced convective activity and rainfall → reduced release of latent heat, weakened Hadley circulation and cooling in the middle and upper tropical troposphere → increased tropical lapse rates → increased precipitation in the latitude bands 5 to 25° N and 5 to 25° S and a decrease in the equator-pole temperature gradient → reduced meridional transport of heat and moisture out of equatorial regions → global cooling and a decrease in precipitation between 45 and 85° N and at 40 and 60° S.

Burning is a common method of forest clearing and of waste and debris disposal. Burning releases carbon dioxide and smoke into the atmosphere and subsequent cultivation causes soil destruction, erosion, water pollution, peak floods, and the silting and eutrophication of watercourses on a massive scale. The world CO_2 production by vegetation burning amounts to 1×10^9 t/a of carbon and is increasing. The tropical zone contributes by far the largest share. As the areas cleared for cultivation increase, this share continues to rise. The scale of the problem is illustrated by the following figures. The forests of the world contain *ca.* 1.6×10^{12} t dry biomass or 0.8×10^{12} t carbon. The moist tropical closed forests at the beginning of this decade covered *ca.* $1\,000 \times 10^6$ ha (Persson, 1974) and contained *ca.* 450×10^9 t dry above-ground biomass. At the present rate of forest destruction, all accessible tropical forests wil lhave disappeared by the end of this century and *ca.* 300×10^9 t dry matter will have been burnt or decomposed, using 400×10^9 t oxygen (0.03 per cent of the oxygen store in the atmosphere) and releasing 550×10^9 t carbon dioxide (equal to one third of the atmospheric store) into the atmosphere. The released carbon dioxide will eventually be fixed again by the substitute vegetation (10–20 per cent) or buffered in the oceans (*ca.* 40 per cent), but *ca.* 40 per cent will remain in the atmosphere and increase its carbon dioxide by 10–15 per cent. The consequences are yet unpredictable, but climatic changes are certain. Plants may react to the increased CO_2 content of the atmosphere by a greater rate of photosynthesis (Fairhall, 1973), but the climatic changes associated with the increase of CO_2 and vegetation burning may counteract or reduce plant production.

The dust concentration in the lower atmosphere has trebled during the past 60 years. This is due to burning vegetation and wind erosion especially in the tropics rather than traffic, industry and domestic heating. Burning and removing tropical forest is a major factor affecting atmospheric conditions in the world. Another and equally poorly under-stood factor is the spread of disease vectors, especially flying insects, along new roads. The potential threat to general human welfare from this source should not be ignored and underestimated.

The importance of the water cleansing and preserving capacity of tropical forests is emphasized by the fact that according to WHO estimates less than 30 per cent of the population in the tropics obtain and consume unpolluted water. The Amazon river with its more than 1 000 major tributaries contains 20 000 km of navigable water-ways, two-thirds of all river water of the world, and the world's richest and most diverse freshwater fish fauna with more than 2 000 species. Increased pollution, eutrophication and silting in this aquatic ecosystem may cause unassessable changes and degradational processes in the rivers.

Forests influence the local and global biosphere in an indirect manner by the low energy input required by the use of wood as an industrial and domestic raw material. If total energy input into one ton of sawn timber is unity, paper made from wood is 12, plastic 3–4, cement 4–5, glass 13, steel 23 and aluminium 126 (Fischer, 1973). Preserving existing growing stock and growing timber in new forests is therefore of major environmental importance to mankind as a whole.

Information gaps, research needs and priorities

Information gaps

Very little research has been carried out in natural and modified tropical forests on the interrelations between the floristic and architectural structure, the fractioning of matter and energy and their relevance to ecosystem stability and functioning. Available information refers to one or few factors, but the interrelationships and interactions are usually omitted.

Exceedingly little precise and reliable information is available on the ecology of the agricultural and sylvicultural ecosystems and their susceptibility to degradational processes. While these systems usually produce a large amount of useful biomass, their capacity for self-regulation is small. In many cases little understood complex processes have caused losses of productivity and a general decline of utility. Especially on poor soils, this can lead to the rapid formation of arrested and impoverished disclimaxes (e.g. savannas on bleached sands) with little ecological or economic value. Such degradation is accompanied by radical shifts in the matter and energy exchanges which are known only in a very general way. The few precise, detailed long-term observations usually include only a few of the effects of ecosystem modification on the surrounding biosphere. The conversion of natural tropical forest carries unknown and unacceptable ecological, economical and social risks.

Until very recently climatic research in the tropical zone has been wholly inadequate and even today meteorological observations are based on an exceedingly sparse network of stations. This and the short periods of observation make it impossible to distinguish natural climatic fluctuations and man-induced changes. The present unsatisfactory situation is indicated by the land area in km² per meteorological station:

South-East Asia	60 000
Africa	40 000
South America	220 000
European Community	10 000

The density of stations in the area of humid tropical forest is considerably lower than the figures suggest. It is therefore unlikely that changes of local climatic conditions in response to forest conversion will be detected in time for effective adjustments in land use techniques. The situation is aggravated by the absence of standardization, the lack of recording continuity, and the failure of most countries to supply the data to international storage and processing facilities. The state of knowledge in the field of local and regional hydrology is similar.

Research needs

There is an urgent need for coordinated and integrated research into the exchange processes of matter and energy in relation to the ecosystem as a whole. Such research should start of the lowest level of the individual plant and its parts and proceed to include successively larger units. Research is particularly urgent in the following fields.

— The intensity and direction of exchanges and turnovers of water, water vapour, carbon, carbon dioxide, oxygen, elements, radiant and thermal energy, air and pollutants; including micro-meteorological studies on mass transfer, radiation and heat balance, production studies by direct methods and by CO_2 exchange and balance estimations, hydrometeorological measurements of interception, stemflow, soil water content and transport, amount and quality of surface water run-off and water yield.

— The effect of modifying the vegetative cover on these exchanges and turnover rates, and consequently on the resilience and dynamic stability of the system.

— The specific effect of different technologies which are used to modify and manage the vegetation cover.

— The possibility of predicting these effects by simulating essential structural and functional features of ecosystems including the relationship and interaction between diversity, stability and adaptive features at species and ecosystem levels. These models are intermediate between the very general and descriptive models which have been characterizing community ecology in the past and the very specialized models which are commonly used in vegetation management to describe in detail the properties and growth reactions of one particular species. This intermediate level of analysis and synthesis

may prove to be very powerful in solving problems which are posed by the severe and extensive perturbations resulting from the unprecedented human activities in the ecosystems.

— The determination of the horizontal range of effects of local climatic and hydrological changes in response to land use by means of theoretical models and experimental data.

— The chemical composition of the atmosphere above natural and managed forest, and above deforested intensively managed and unmanaged lands, especially with respect to photochemical oxidants, ozone, hydrocarbons, nitrates, ethylene, sulphur dioxide, fluorides, chlorides, ammonia, ions, pollen grains, spores and radio-active materials.

— The criteria and techniques for efficient bioclimatic and ecological land classifications, and for mapping sites with respect to the vulnerability of soils, vegetation and the elements of the biosphere.

Research priorities

In order to achieve the general objective of filling the information gaps fast and efficiently, integrated and coordinated research projects are required. This precludes the creation of a multitude of individual research projects. Instead a limited number of regional and long-term research projects should be established at carefully selected sites to study the various aspects listed above in an ecosystem and biosphere context. The Unesco Man and the Biosphere Programme Project 1 is recommended as the most suitable vehicle for planning and conducting these researches. The research projects should include the following aspects which involve the exchanges between the vegetation ecosystem and the biosphere: data on floristic and architectural structure; productivity; storage, cycling and exchange of energy, water, organic matter and minerals; dynamics of plant and animal populations; quantification of the effects of man-made modifications on the structural and functional features of the ecosystem and its exchanges with the biospheric environment. Unless these fundamental aspects are understood, assessment and prediction of the influence of different land-cover types on the biosphere at local, regional and global level will remain speculative (see also the following chapters for specific research needs and priorities).

In addition, the network of standard meteorological stations should be increased to monitor long-term climatic changes in key areas of critical developments such as the Amazon basin. Reference in this respect is made to the list of projects in forest meteorology submitted by Reifsnyder (1973) to the inter-agency group (FAO-UNESCO-WMO) on Agricultural Biometeorology.

Bibliography

ALLEE, W. C. Measurements of environmental factors in the tropical rain forest of Panama. *Ecology*, 7, 1926, p. 273–302.

ALLISSON, I. K.; HERRINGTON, L. P.; MORTON, I. D. Turbulent diffusion under a jungle canopy. *Bull. Am. Meteor. Soc.* (Boston), 49, 1968, 302 p.

AOKI, M.; YABUKI, K.; KOYAMA, H. Micrometeorology of Pasoh Forest. In: *IBP Synthesis Meeting* (Kuala Lumpur, 12–18 Aug. 1974), 14 p. multigr.

ASHTON, P. S. Light intensity measurements in rain forest near Santarem (Brazil). *J. Ecol.*, 46, 1958, p. 65–70.

——; BRÜNIG, E. F. The variation of tropical moist forest in relation to environmental factors and its relevance to land use planning. *Mitt. Bundesforsch. anst. Forst -u. Holzwirtsch.* (Hamburg), 109, 1975, p. 59–86.

BABALOLA, O.; SAMIE, A. G. The use of a neutron technique in studying soil moisture profiles under forest vegetation in the northern guinea zone of Nigeria. *Tropical Science*, 14, 1972, p. 159–168.

BARINOV, G. V. Biosphere rhythmus and the problems of retaining oxygen balance. *Z. Obsc. Biol.*, 33, 1972, p. 771–778.

BARTLETT, H. H. *Fire in relation to primitive agriculture and grazing in the tropics. Annotated bibliography*, vol. II and III. Ann Arbor, University of Michigan, vol. II, 1957, 873 p.; vol. III, 1961, 216 p.

BAUMGARTNER, A. Energetical bases for differential vaporization from forest and agricultural land. In: *Proc. Int. Symp. on Forest Hydrology*, p. 381–389. London, Pergamon Press, 1965, 813 p.

——. Sauerstoffumsätze von Bäumen und Wäldern. *Allgem. Forstzeitschr.* (München), 25, 1970, p. 482-483.

——. Water and energy balances of different vegetation covers. In: *Proc. IASH-UNESCO Symp. on World water balance*, p. 56–65. Reading, 1970.

——. The influence of energetic factors on climate, production and water circulation in forested water-shed areas. In: *Proceedings 15th IUFRO Congress* (Gainesville), 1971, p. 75–90.

——; REICHEL, E. *The world water balance*. München, Oldenbourg; and Amsterdam, Elsevier, 1975, 179 p.

BAYTON, H. W. Wind structure in and above a tropical forest. *J. Appl. Meteor.* (Boston), 4, 1965, p. 670–675.

——; HAMILTON, H. L.; WORTH, I. I. Temperature structure in and above a tropical forest. *Quart. J. Roy. Meteor. Soc.* (London), 91, 1965, 225 p.

BEHN, R. C.; DUFFEE. *The structure of the atmosphere in and above tropical forest*. Batelle Mem. Inst. (Columbus, Ohio), Rep. no. BAT–171–8, 1965.

BIROT, P. *The cycle of erosion in different climates*. Berkeley, University of California Press, 1968, 144 p.

BOLIN, B. The carbon cycle. *Scientific American* (New York), vol. 223, no. 3, 1970, p. 124–132.

BRINKMANN, W. L. F. Relative light intensity measurements in a secondary forest (Capoeira) near Manaus (Amazonia, Brazil). *Bol. INPA, Pesq. Florestais* (Manaus), 17, 1970, 6 p.

——. Light environment in a tropical rain forest of central Amazonia. *Acta Amazonica*, vol. 1, no. 2, 1971, p. 37–49.

——; WEINMAN, J. A.; RIBEIRO, M. N. G. Air temperatures in central Amazonia. I. The daily record of air temperatures in a secondary forest near Manaus under cold front conditions (July 4th to 13th, 1969). *Acta Amazonica*, vol. 1, no. 2, 1971, p. 51–56.

——; RIBEIRO, M. N. G. Air temperatures in central Amazonia. II. The effect of near-surface temperatures on land use in the Tertiary region of Central Amazonia. *Acta Amazonica*, vol. 1, no. 3, 1971, p. 27–32.

BROWNE, L. R. Human food production as a process in the biosphere. *Scientific American* (New York), vol. 223, no. 3, 1970, p. 161–170.

BRÜNIG, E. F. Die Sauerstofflieferung aus den Wäldern der Erde und ihre Bedeutung für die Reinerhaltung der Luft. *Forstarchiv* (Hamburg), 42, 1971a, p. 21–23.

——. *Forstliche Produktionslehre*. Europ. Univ. Schriftenr. XXV, Forst- und Holzwirtsch., Bd. I, 1971b, 318 p. Herb. Lang Verl. Bern, Peter Lang Verl. Frankfurt/M.

——. On the ecological significance of drought in the equatorial wet evergreen (rain) forest of Sarawak (Borneo). In: *Trans. First Aberdeen-Hull Symp. on Malesian Ecology*, p. 66–79. Univ. Hull, Dept. Geography, Misc. ser. no. 11, 1971c, 97 p.

——. A physiognomic-ecological classification of tropical forests, woodlands and scrubland. In: *Silvics and silvicultural management in humid tropical forests*, in press (1978); 1972, multigr.

——. Species richness and stand diversity in relation to site and succession of forests in Sarawak and Brunei. 3rd Symp. Biogeogr. and Landscape Res. in South America (Max Planck Inst. Plön, 3.5.1972). *Amazoniana*, vol. 4, no. 3, 1973, p. 293-320.

——. *Ecological studies in the kerangas forests of Sarawak and Brunei*. Kuching, Borneo Lit. Bureau, 1974, 237 p.

——. Tropical ecosystems: state and targets of research into the ecology of humid tropical ecosystems. Plant research and development. *Inst. Science Co-operation* (Tübingen), vol. 1, 1975, p. 22–38.

——. Classifying for mapping of kerangas and peatswamp forest as examples of primary forest types in Sarawak (Borneo). In: *Trans. 4th Aberdeen-Hull Symp. on Malesian Ecology*, p. 57-75. Univ. Hull, Dept. Geography, Misc. ser. no. 17, 1976, 103 p.

——; BUCH, M. V.; HEUVELDOP, J.; PANZER, K. F. Stratification of the tropical moist forest for land-use planning. *Mitt. Bundesforsch. anst. Forst- u. Holzwirtsch.* (Hamburg), 109, 1975, p. 1–57.

CACHAN, P. Signification écologique des variations microclimatiques verticales dans la forêt sempervirente de basse Côte d'Ivoire. *Ann. Fac. Sci. Univ. Dakar*, 8, 1963, p. 87–155.

——; DUVAL, I. Variations microclimatiques verticales et saisonnières dans la forêt sempervirente. *Ann. Fac. Sci. Univ. Dakar*, 8, 1963, p. 1–87.

CLEGG, A. G. Rainfall interception in a tropical forest. *Caribbean Forester*, 24, 1963, p. 74–79.

COLLIER, B. D.; COX, G. W.; JOHNSON, A. W.; MILLER, P. C. *Dynamic ecology*. Englewood Cliffs, N.J., Prentice Hall, 1973, 563 p.

DORDIK, I. L.; THURONYI, G. Selective annotated bibliography on climate of the forest. *Meteor. Abstracts and Bibl.* (Boston), 8, 1957, p. 515–539.

DUVIGNEAUD, P. (ed.). *Productivity of forest ecosystems*. Paris, Unesco, 1971, 707 p.

ELLENBERG, H. Versuch einer Klassifikation der Ökosysteme nach funktionalen Gesichtspunkten. In: Ellenberg, H. (ed.). *Ökosystemforschung*, p. 235–265. Berlin, Heidelberg, New York, Springer Verlag, 1973, 280 p.

——; MUELLER-DOMBOIS, D. Tentative physiognomicecological

classification of plant formations of the earth. *Forsch. Inst. Ruebel*, 37, 1966, p. 21–55.

EMBERGER, L. Une classification biogéographique des climats. *Recl. Travaux Lab. Bot. Geol. Zool. Univ. Montpellier*, série Botanique, n° 7, 1955, p. 3-43.

EVANS, G. C. Ecological studies on the rain forest of southern Nigeria. II. The atmospheric environment condition. *J. Ecol.*, 27, 1939, p. 436–482.

——. Temperature gradients in a tropical rain forest. *J. Ecol.*, 54, 1966, p. 20–21.

——. *The quantitative analysis of plant growth*. Oxford, Blackwell, 1972, 734 p.

FAIRHALL, A. W. Accumulation of fossil CO_2 in the atmosphere and the sea. *Nature*, 245, 1973, p. 20–23.

FAO. *Shifting cultivation in Latin America* (by Watters, R. F.). Rome, Forestry development paper no. 17, 1971, 305 p.

——. *Environmental aspects of natural resources management. Forestry*. Report for the UN Conference on Human Environment (Stockholm, 1972), 1971, 33 p.

——. *Heat stress in forest work* (by Axelson, O.). Rome, TF-INT 74 (SWE), 1974, 31 p.

——. *The environmental aspects of forest land use*. FAO/SIDA Seminar on forest social relations for English-speaking countries in Africa and the Caribbean. Rome, 1975, 184 p.

FARNWORTH, E. G.; GOLLEY, F. B. (eds.). *Fragile ecosystems. Evaluation of research and applications in the Neotropics*. Berlin, Heidelberg, New York, Springer Verlag, 1973, 258 p.

FISCHER, F. Gedanken über die Zukunft von Wald und Holz. *Schweiz. Z. Forstwirtsch.*, 124, 1973, p. 174–179.

FITTKAU, E. J.; KLINGE, H. On biomass and trophic structure of the central Amazonian forest ecosystem. *Biotropica*, 5, 1973, p. 2–14.

FLOHN, H. Fluctuations of the global water balance due to man-made modification. *Die Naturwissenschaften*, 60, 1973a, p. 304–348.

——. Klimaschwankungen und Klimamodifikationen, Fakten und Probleme. *Deutsche Forsch. Gemeinsch.*, Mitt. 2, 1973b, p. 31-40.

FOJT, V. Estudio microclimatológico de los pinares de la Sierra de Nipe, Cuba. *Silvicultura trópica et subtrópica*, 1, 1969, p. 19-37.

FOSBERG, F. R.; GARNIER, B. J.; KUECHLER, A. W. Delimitation of the humid tropics. I-III. *Geogr. Rev.*, vol. 51, no. 3, 1961, p. 353–357.

FRÄNZLE, O. Der Wasserhaushalt des tropischen Regenwaldes und seine Beeinflussung durch den Menschen. Paper to: *5. Symp. Biogeogr. und landschafts-ökol. Probl. Südamerikas* (Plön, 2–3 May, 1974), unpublished.

FREISE, F. Das Binnenklima von Urwäldern im subtropischen Brasilien. *Petermanns Geogr. Mitt.*, 82, 1936, p. 301-304; p. 346-348.

GAUSSEN, H. Théories et classification des climats et des microclimats. In: *8ᵉ Cong. int. Bot.* (Paris), Section 7, 1954, p. 125-130.

GEIGER, R. *Das Klima der bodennahen Luftschicht*. 4 ed. Braunschweig, Vieweg Verlag, 1961, 646 p.

GÉRARD, P. Une année d'observations microclimatiques en forêt secondaire. *Florenz*, 1959, p. 206-209.

GHOSH, R. C. The protective role of forestry to the land. *Indian Forester*, vol. 101, no. 1, 1975, p. 28–38.

GUSINDE, M.; LAUSCHER, F. Meteorologische Beobachtungen im Kongo-Urwald. *Sitz. ber. Akad. d. Wiss.* (Wien), 150, 1941, p. 281-347.

HALES, W. B. Micrometeorology in the tropics. *Bull. Am. Meteor. Soc.* (Boston), 30, 1949, p. 81–89.

HILL, R. D. Microclimatic observations at Bukit Timah Forest Reserve, Singapore. *Malayan For.*, vol. 24, no. 2, 1966, p. 78–86.

HOLDRIDGE, L. R. *Life zone ecology*. San José, Costa Rica, Tropical Science Center, 1967, 206 p.

——; GRENKE, W. C.; HATHAWAY, W. H.; LIANG, T.; TOSI, J. A. Jr. *Forest environments in tropical life zones. A pilot study*. London, Pergamon Press, 1971, 731 p.

JACKSON, I. J. Tropical rainfall variations over small areas. *J. Hydrol.*, 8, 1969, p. 99–110.

JORDAN, C. F. A world pattern in plant energetics. *Amer. Sci.*, 59, 1971, p. 425–433.

——; DREWRY, G. E. *The rain forest project*. Ann. Rep. U.S. Atomic Energy Commission (USAEC), Rep. PRNC (Puerto Rico Nuclear Center), 129, 1969.

KELLMAN, M. C. Some environmental components of shifting cultivation in upland Mindanao. *J. Trop. Geogr.*, 28, 1969, p. 40–56.

KENWORTHY, J. B. Water and nutrient cycling in a tropical rain forest. In: *Trans. 1st Aberdeen-Hull Symp. Malesian Ecology*, p. 49–65. Univ. Hull, Dept. Geogr., Misc. ser. no. 11, 1971, 97 p.

KLINGE, H. Climatic conditions in lowland tropical podzol areas. *Trop. Ecol.*, 10, 1969, p. 222–239.

——. Struktur und Artenreichtum des zentral-amazonischen Regenwaldes. *Amazoniana*, vol. 4, no. 3, 1973, p. 283–292.

——. Bilanzierung von Hauptnaehrstoffen im Oekosystem tropischer Regenwald (Manaus)-vorlaeufige Daten. *Biogeographica*, 7, 1976, p. 59-77.

——; FITTKAU, E. J. Filterfunktionen im Ökosystem des zentral-amazonischen Regenwaldes. *Mitt. deutsche bodenkundl. Gesellsch.*, 16, 1972, p. 130-135.

KÖPPEN, W. *Die Klimate der Erde*. Berlin, 1923.

——. Das geographische System der Klimate. In: Köppen, W.; Geiger, R. (eds.). *Handbuch der Klimatologie*, vol. 1, part C. Berlin, Borntraeger, 1930.

——. *Grundriss der Klimakunde*. Berlin, Borntraeger, 1931.

KULLENBERG, B. Quelques observations microclimatiques en Côte d'Ivoire et Guinée française. *Bull. IFAN* (Dakar), 17, 1955, p. 755-768.

LAMPRECHT, H. *Importance of the tropical forest from the viewpoint of forest ecological-environmental relationships*. Paper multigr. for the 7th World Forestry Congress (Buenos Aires), 1972, 20 p.

LARCHER, W. *Okologie der Pflanzen*. Stuttgart, Ulmer Verlag, 1973, 320 p.

LEMON, E. R.; ALLEN, L. H.; MÜLLER, L. Carbon dioxide exchange of a tropical rain forest. Part II. *Bio-Science*, vol. 20, no. 19, 1970, p. 1054–1059.

LEVINS, R. *Evolution in changing environments. Some theoretical explanations*. Princeton, N.J., Princeton University Press, 1968, 120 p.

LIETH, H.; MALAISSE, F. Cartographie de la productivité primaire mondiale par ordinateurs. *Rev. Univ. Nat. du Zaïre* (Lubumbashi), ser. B. Sci., 26, 1971, p. 78-87.

——; Box, E. Evapotranspiration and primary productivity: C. W. Thornthwaite memorial model. In: Mather J. R. (ed.). *Papers on selected topics in climatology*, vol. 2, p. 37–46. N.Y., Elmer, Thornthwaite memorial volume 2, 1972.

LONGMAN, K. A.; JENÍK, J. *Tropical forest and its environment*. London, Longman, 1974, 196 p.

LUGO, A. E. Energy flow in some tropical ecosystems. *Soil Crop. Sci. Soc.* (Florida), 29, 1970, p. 254–264.

LULL, H. W. Run-off from forest lands. In: *Man's impact on terrestrial and oceanic ecosystems*, p. 240–251. Cambridge, Mass., MIT Press, 1971, 540 p.

MALAISSE, F.; ALEXANDRE, J.; FRESON, R.; GOFFINET, G.; MALAISSE-MOUSSET, M. The miombo ecosystem: a preliminary study. In: Golley, P. M.; Golley, F. B. (eds.). *Tropical ecology, with an emphasis on organic production*, p. 363–405. Athens, Univ. of Georgia, 1972, 418 p.

MARGALEF, R. On certain unifying principles in ecology. *Am. Naturalist*, 97, 1963, p. 357–374.

MATTEWS, W. H.; SMITH, F. E.; GOLDBERG, E. D. (eds.). *Man's impact on terrestrial and oceanic ecosystems*. Cambridge, Mass., MIT Press, 1971, 540 p.

MAY, R. M. *Diversity, stability and maturity in natural ecosystems, with particular reference to the tropical moist forests*. Rome, FAO, 1975, 9 p. multigr.

MEGGERS, B. J.; AYENSU, E. S.; DUCKWORTH, W. D. (eds.). *Tropical forest ecosystems in Africa and South America: a comparative review*. Washington, D.C., Smithsonian Institution, 1973, 350 p.

MOCK, F. S. Hydrological balance and climatic elements in the tropics (Indonesia). *Wasser und Boden*, 4, 1974, p. 88-92.

MONTEITH, J. L. (ed.). *Vegetation and the atmosphere*. London and New York, Academic Press, 1976, vol. 1 *(Principles)*, 278 p.; vol. 2 *(Case studies)*, 440 p.

MÜLLER, D. Kreislauf des Kohlenstoffs. In: *Handbuch der Pflanzenphysiologie*, vol. 12, 2, p. 934-948. Berlin, Springer Verlag, 1960, 1421 p.

NIMLOS, T. J. Soils and mankind. In: Behan, R. W.: Weddle, R. M. (eds.). *Ecology, economics, environment. Montana forest and conservation*. Univ. of Montana, Exp. Sta., School of Forestry, 1971.

ODUM, H. T. An emerging view of the ecological system at El Verde. In: Odum, H. T.; Pigeon, R. F. (eds.). *A tropical rain forest. A study of irradiation and ecology at El Verde, Puerto Rico*, p. I-191 - I-289. Division of Technical Information, US Atomic Energy Commission (USAEC, Washington), 1970, 1678 p.

OLSON, J. S. Geographic index of world ecosystems. In: Reichle, D. E. (ed.). *Analysis of temperate forest ecosystems*, p. 297–304. Berlin and New York, Springer Verlag, Ecological Studies no. 1, 1972, 304 p.

PAIJMANS, K. An analysis of four tropical rain forest sites in New Guinea. *J. Ecol.*, vol. 58, no. 1, 1970, pp. 77–101.

PARKHURST, D. F.; LOUCKS, O. L. Optimal leaf size in relation to environment. *J. Ecol.*, vol. 60, no. 2, 1972, p. 505–537.

PENMAN, H. L. The water cycle. *Scientific American* (New York), vol. 223, no. 3, 1970, p. 99–108.

PEREIRA, H. C. *Land use and water resources in temperate and tropical climates*. London, Cambridge University Press, 1973, 246 p.

PERSSON, R. *World forest resources*. Stockholm, Royal College of Forestry, 1974, 261 p., 3 maps.

POTTER, G. L.; ELLSAESSER, H. W.; MAC CRACKEN, M. C.; LUTHER, F. M. Possible climatic impact of tropical deforestation. *Nature* (London), 258, 1975, p. 697–698.

READ, G. Evaporation power in the tropical forest of the Panama Canal. *J. Appl. Meteo.*, 7, 1968, p. 417–424.

REIFSNYDER, W. E. *Research needs in forest meteorology. A report submitted to the Interagency Group on Agricultural Biometeorology*. FAO, Unesco, WMO, nov. 1973, 45 p.

RICHARDS, P. W. *The tropical rain forest: an ecological study*. Cambridge, Cambridge University Press, 1952, 450 p. 4th reprint with corrections, 1972.

ROSS, R. Ecological studies on the rain forest of southern Nigeria. III. *J. Ecol.*, 42, 1954, p. 259–282.

SAJISE, P. E. Energy transfer and production in a tropical rain forest. In: *IBP Synthesis Meeting* (Kuala Lumpur, 12–18 Aug. 1974), 10 p. multigr.

SARLIN, P. Evapotranspiration et végétation forestière tropicale. *Bois et Forêts des Tropiques* (Nogent-sur-Marne), 133, 1970, p. 17-26.

SCHULZ, I. P. *Ecological studies of the rain forest in northern Surinam*. Amsterdam, Noord. Hollandsche, Uitg. Mij., 1960, 267 p.

SIOLI, H. Zur Okologie des Amazonas-Gebietes. In: *Biogeography and ecology in South America*, p. 137–170. The Hague, Junk, Monographiae Biologicae no. 18, vol. 1, 1968, 446 p.

SLANAR, H. *Das Klima des östlichen Kongo-Urwaldes*. Mitt. Geogr. Ges. Wien, 1945.

SLESINGER, E. *Forest influences*. Rome, FAO, Forestry and Forest Product Studies, no. 15, 1962, 307 p.

SMITH, J. S. Light and the forest canopy. *Malayan Forester*, 5, 1936, p. 7–12.

SNEDAKER, S. C. *Ecological studies on tropical moist forest succession in eastern lowland Guatemala*. Univ. of Florida (Gainesville), Ph. D. thesis, 1970, 131 p. University Microfilms, High Wycomb (England), 1975.

STEPHENS, G. R.; WAGGONER, P. E. Carbon dioxide exchange of a tropical rain forest. Part I. *Bio-Science*, vol. 20, no. 19, 1970, p. 1050–1053.

STERNBERG, H. O'Reilly. Man and environment change in South America. In: Fittkau, E. J.; Illies, J.; Klinge, H.; Schwabe, G. H.; Sioli, H. (eds.). *Biogeography and ecology in South America*, p. 413–445. The Hague, Junk, Monographiae Biologicae no. 18, vol. 1, 1968, 446 p.

STOCKER, Q. Ein Beitrag zur Transpirationsgrösse im javanischen Regenwald. *Jahrb. Wiss. Bot.*, 81, 1953, p. 464-496.

STOUTJESDIJK, P. A note on the spectral transmission of light by tropical rain forest. *Acta Bot. Neerl.*, vol. 21, no. 4, 1972, p. 346–350.

SZESZTAY, K. *The hydrosphere and the climatic changes*. Nr. 2. New York, United Nations, Natural Resources Forum.

TANNER, C. B. Evaporation of water from plants and soil. In: *Water deficits and plant growth*, vol. 1, p. 73–106. New York and London, Academic Press, 1968.

THOMAS, P. K.; CHANDRASEKHAR, K.; HALDORAI, B. An estimate of transpiration by *Eucalyptus globulus* from Nilgiris watersheds. *Indian For.*, 98, 1972, p. 168–172.

THORNTHWAITE, C. W. The climates of North America according to a new classification. *Geogr. Rev.*, Oct. 1931, p. 634–655.

——. The water balance in tropical climates. *Bull. Am. Meteor. Soc.* (Boston), 32, 1951, p. 166–173.

TROLL, C. Die tropischen Gebirge. *Bonner Geograph. Abhandlungen*, Heft 25, 1959, 93 p.

Unesco. *Symposium on the impact of man on humid tropics vegetation* (Goroka, New Guinea). Unesco Science Cooperation Office for South-East Asia (Djakarta), 1960, 402 p.

——. *Ecological effects of increasing human activities on tropical and subtropical forest ecosystems* (Rio de Janeiro, 1974). MAB Report series, no. 16. Paris, Unesco, 1974, 96 p.

VARESCHI, V. Der Wasserhaushalt von Baümen, welche zur Aufforstung entwaldeter Gebiete Venezuelas verwendet werden. *Angew. Pflanzensoziologie* (Klagenfurt), II, 1954, p. 721–729.

WAGNER, H. Flächenausdehnung der Klimagebiete. *Petermanns Geogr. Mitt.*, 1921, 216 p.

WALTER, H. *Vegetationszonen und Klima*. Stuttgart, Ulmer Verlag, 1970, 244 p.

——. *Ecology of tropical and sub-tropical vegetation*. Van Nostrand Reinhold, 1971, 539 p.

WALTER, H. *Klimadiagramm-Karten/der einselnen Kontinente und die ökologische Klimagliederung der Erde*. Stuttgart, Gustav Fischer, 1975 36 p., 9 maps.

——; LIETH, H. *Klimadiagramm-Weltatlas*. Jena, Gustav Fischer, 1960–1967, 245 p.

WALTON, A. B. More investigations on light intensity. *Malayan Forester*, 5, 1936, p. 111–115.

WHITMORE, T. C. *Tropical rain forests of the Far East*. Oxford, Clarendon Press, 1975, 278 p., 550 references.

WHITTAKER, R. H. *Communities and ecosystems*. London, Macmillan, 1970, 162 p.

WICHT, C. L. Timber and water. *South Afric. For. J.*, 85, 1972, p. 3–11.

WILLIAMS, M. A. J. Prediction of rainsplash erosion in the seasonally wet tropics. *Nature*, 222, 1969, p. 763–764.

WILM, H. G. The influence of forest vegetation on water and soil. *Unasylva* (Rome, FAO), vol. 11, no. 4, 1957, p. 160–164.

WILSON, C. L.; MATTEWS, W. H. (eds.). *Man's impact on the global environment. Assessment and recommendations for*

action. *Report of the study of critical environmental problems (SCEP)*. Cambridge, Mass., MIT Press, 1970, 319 p.

WILSON, C. L.; MATTEWS, W. H. (eds.). *Inadvertent climate modification. Report of the study of man's impact on climate (SMIC)*. Cambridge, Mass., MIT Press, 1971, 308 p.

WILSON, J. W. Stand structure and light penetration. III. Sunlit foliage area. *J. Appl. Ecol.*, 4, 1967, p. 59–165.

WOODWELL, G. M. The energy cycle of the biosphere. *Scientific American* (New York), vol. 223, no. 3, 1970, p. 64–94.

YABUKI, K.; AOKI, M. Micrometeorological assessment of primary production rate of Pasoh Forest. In: *IBP Synthesis Meeting* (Kuala Lumpur, 12–18 Aug. 1974), 14 p. multigr.

YODA, K. Three-dimensional distribution of light intensity in a tropical rain forest of West Malaysia (light distribution under canopy in Pasoh). IBP Synthesis Meeting (Kuala Lumpur, 12–18 Aug. 1974). *Japanese J. Ecol.*, vol. 24, no. 4, 1974, p. 247–254.

——; OGAWA, H.; KIRA, T. *Structure and productivity of a tropical rain forest in West Malaysia*. Paper presented to the 12th Int. Cong. Botany (Leningrad, 1975), 23 p. multigr.

3 Palaeogeography and palaeoclimatology

Introduction

Geologic history has had profound influence on the evolution and distribution of plants and animals. The former connections and subsequent separation of the continents, the upheaval of mountain systems and the formation or disappearance of connections between land masses by tectonic and volcanic forces are important for a general understanding of the processes and rate of evolution. However, they act too slowly to be important for the last 10 000 years or for human history or for forest ecosystems during the next few thousand years.

The separation of Africa and South America commenced somewhere during the Cretaceous, more than 100 million years ago. The pollen flora of the Middle and part of the Upper Cretaceous of both continents is amazingly similar (Herngreen, 1974). After separation, the floras diverged evolutionarily and became more dissimilar in the course of the Tertiary. During the Tertiary, which began some 70 million years ago, modern taxa began to make their appearance. Certain Bombacaceae appear in the Palaeocene, one of the principal mangrove components, *Rhizophora*, appeared in the Oligocene, the Compositae and genera like *Symphonia* and *Cuphea* in the Miocene. These are only a few examples. A fuller account of tropical Tertiary pollen analysis is given by Germeraad, Hopping and Muller (1968). Relatively marked changes of climate seem to have occurred during the Tertiary (Van der Hammen, 1961, 1964) with a general cooling during the Miocene and Pliocene. It was, however, not until the beginning of the Quaternary, some 2.5 million years ago, that a continuous series of very conspicuous changes of climate, the Pleistocene ice ages, began. There is no reason to suppose that these changes of climate have come to an end.

The Quaternary glacials and interglacials have changed the face of the earth. They caused extinction, speciation and profound changes in the geographical distribution of plants and animals. This has long been known to have happened in temperate latitudes; it is only recently that undeniable evidence has been found for the tropics and subtropics. These climatic changes may be partly cyclic. They have had and will have a profound influence on ecosystems. Studies of palaeogeography and palaeoclimatology as related to tropical and subtropical forest ecosystems are therefore of considerable importance.

The largest store of information about the palaeogeography and palaeoclimatology of the tropics is to be found in the lake sediments. Several of these have been in existence for some millions of years and a few, like Lake Tanganyika, are among the oldest lakes in the world. A fine rain of sedimentary particles, some originating in the lake itself, some carried in by rivers, some by wind, falls slowly and continuously through the water to rest on the bottom. The accumulating mud is dark, anoxic and frequently acid. Under these conditions organic matter does not decay readily, and even such normally unstable compounds as sugars and plant pigments may persist for tens of thousands of years. Plant cell walls and animal exoskeletons are splendidly preserved, and a large part of the sediment is usually composed of pieces of plants and animals that have lived and died in the lake. Smaller components consist of fragments of organisms, such as pollen grains, from the land around the lake. A single milliliter of lake mud commonly contains 1–10 million fossils. Lake sediments, therefore, provide a record of the organisms that lived in and around the lake since its inception. The grain size, chemical composition and mineralogy of the sediment provide additional information concerning such features as the past depth, chemical composition and concentration of the lake water. All this information constitutes an extraordinary archive of palaeogeography.

Under a wide range of conditions about 0.5–1 mm/a of sediment accumulates in a lake but extremes of 0.1–10 mm/a may occur. After deposition the upper several centimeters of sediments are disturbed by water currents and to a larger extent by the burrowing bottom fauna of molluscs, worms and insect larvae. This blurs the record so that it is seldom easy to separate two events which occurred within ten years of each other. This places a limit on the temporal resolving power of the sedimentary record.

Lakes that are perennially stratified, however, with a light surface layer riding over denser anoxic water, usually lack a deep bottom fauna and the strata of separate years, or even separate seasons, may be preserved intact. Tropical Africa is very well provided with such stratified lakes, and it seems likely that a deep old lake, such as Tanganyika (P in fig. 1), may contain in its sediment a record of each separate month during the past 10–20 million years. No one has attempted such fine-grained resolution of a record so long, but in other parts of the world annually-laminated sediments have been put to valuable use in working out lake history over a few decades or centuries. There is no reason to doubt that such fine temporal resolution is attainable in Africa and other parts of the tropics.

Although the number of lakes is relatively large, and they are well distributed over at least the eastern half of the African and the western half of the South American continents, the number of investigators of their sediment is small. The completed work has yielded some very significant conclusions about the vegetational and climatic history but it is no more than a beginning.

A well-developed understanding of tropical and subtropical palaeoecology would provide three kinds of information for laying sound, far-sighted plans for the management of natural resources.

1. It would demonstrate the amount of variation from present mean conditions of climate and the many aspects of natural geography influenced by climate. In few parts of the world is the instrumental record long enough to provide a satisfactory data base for resource planning. For example, during the 1960s, the port facilities of the African Great Lakes were severely impaired by a historically unprecedented rise in lake levels. Palaeoecological study could have demonstrated the probability of such level change, and the harbour facilities might have been designed to cope with it.

2. It would permit the plotting of long-term trends of climatic change. If, as some believe, the recent Sahelian drought is part of such a trend, it cannot be handled by international pooling of food resources. Long-term resettlement and alteration of the economic base of the affected people will be required. Without good palaeoecological data we cannot say whether such a disaster should be met with temporary relief or permanent adjustment.

3. It would provide analogues to current events such as large-scale continent-wide felling of evergreen forest or to such possible future events as deliberate weather modification. Such shifts in the distribution and abundance of forest as are currently being produced by man have occurred previously only in response to major climate change. By studying the earlier natural events we may be able to predict the effects of modern land use on such disparate aspects of natural geography as genetic diversity, ecological community structure and the chemistry of soil formation.

General aspects of tropical pollen analysis

The reliable application of pollen analysis to a new region involves three stages. First, it must be possible to identify the fossil pollen grains. Secondly, one must develop a regionally consistent pollen stratigraphy. Thirdly, one must be able to interpret the stratigraphic changes at least in terms of the vegetation that produced them, and preferably in terms of the climate and other environmental factors that controlled the vegetation.

Pollen morphology

Much effort has gone into tropical pollen morphology. Of particular note are the monographs on African pollen being prepared at Montpellier (Van Campo, 1957, 1958, 1960; Van Campo and Hallé, 1959; Van Campo et al., 1964; Guinet, 1969; Lobreau et al., 1969; Guers et al., 1971), the pollen flora of Madagascar being prepared at Kiel (Straka, 1964a, 1964b, 1965, 1966; Straka and Simon 1967; Straka et al., 1967; Keraudren-Aymonin et al., 1969) and the pollen flora of southern Africa being prepared at Bloemfontein (Van Zinderen Bakker, 1953; Van Zinderen Bakker and Coetzee, 1959; Welman, 1970). In addition there are studies on a smaller scale of the pollen flora of local areas (Maley, 1970; Bonnefille, 1971a, b). For South

America the most important larger preparations are at Sao Paulo (*Pollen grains of plants of the cerrado* by Salgado-Labouriau, Melhem and collaborators) and at Rio de Janeiro (*Catalogo sistematico dos polens das plantas arboreas do Brasil Meridional*) by Barth and collaborators. Salgado-Labouriau (1973) also produced a more comprehensive review of the pollen morphology of the plants of the cerrado, with a bibliography containing most of the publications mentioned above. For Asia there is the series *Studies of Indian pollen grains* (Vishnu-Mittre and Sharma, 1962, 1963; Vishnu-Mittre and Gupta, 1964), and *Pollen grains of Indian plants* (Nair, 1961, 1962a, b; Nair and Rehman, 1962, 1963; Nair and Khan, 1965; Nair and Rastogi, 1966). For the Hawaiian islands, there is the classic study of Selling (1946, 1947). A series of publications on the pollen morphology of Colombian plants is being prepared in Amsterdam and Bogotá (Murillo and Bless, 1974). Although mainly outside the tropics proper, the pollen flora of Chile (Heusser, 1971) is of importance for the Southern Andean area in general.

Some taxonomists have provided monographic accounts of the pollen of particular plant groups. Major works such as those by Punt (1962), Sowunmi (1968), Walker (1971) and Lobreau-Callen (1972) are of value to analysts in the tropics in general. It is customary also for fossil analysts to convey in their publications some idea of the basis for their identifications (see, for example, Livingstone, 1967, and Kendall, 1969). Photographs and short descriptions of the main pollen types that are frequently found as fossils in certain areas are sometimes presented (see, for example, Van der Hammen and González, 1960a and 1963; Van der Hammen, 1963; Wijmstra and Van der Hammen, 1966). Hulshof and Manten (1971) provide a bibliography on recent pollen covering the entire world and examples of the pollen morphology of most families of Angiosperms, Gymnosperms, Pteridophyta and Bryophyta can be found in Erdtman (1957, 1965, 1966) and Erdtman and Sorsa (1971).

In order to make use of the morphological insights embodied in any of these papers, however, it is necessary to have a good working collection of identified pollen grains made up from herbarium material. For Africa, major collections exist at Montpellier, Bloemfontein, Duke University, and Kiel University. Smaller collections exist in the Centre national de la recherche scientifique (CNRS) at Bellevue, France, at Makerere University in Uganda and at the Polish National Academy. For South America relatively large collections exist in Rio de Janeiro and Sao Paulo and at the University of Amsterdam. Duplicates of a part of the Amsterdam collection are at National University and Ingeominas in Bogotá. Smaller collections exist in several places (Lima, Buenos Aires, Mexico D.F., Trinidad, Caracas, etc.). For Australia and New Guinea there is the collection of the National University in Canberra; duplicates of a part of this collection may be found, because of exchange of slides, in several other laboratories of the world (e.g. Amsterdam). No complete data can be given at this moment on Asia, but collections exist at Birbal Sahni Institute and the National Botanical Gardens in Lucknow, the French Institute in Pondichery, at the University of Hull

and the Rijksherbarium at Leiden. A Hawaiian collection is at the Swedish Museum of Natural History in Stockholm.

The more abundant pollen types in fossil assemblages from many parts of the tropics can be identified to genus or to family. Identifications may be somewhat uncertain and there almost always remains a residue, usually about five per cent, of many very rare and unidentifiable pollen types; few of these are represented by more than a few grains in an entire pollen profile. Tropical grass pollen so far cannot be distinguished with certainty. Grass pollen is very abundant, not only in assemblages from grassland, but also from a variety of savanna and woodland vegetation types. Identifications to genus are needed to carry out useful pollen analyses over wide areas of the tropics, but especially of tropical Africa. So far, some progress has been made at indicating the presence of certain tribes by means of size statistics (Bonnefille, 1972; Hamilton, 1972), and a few morphological types can be distinguished by scanning electron microscopy (Tsukada and Rowley, 1964).

Pollen stratigraphy

The second requirement for successful pollen analysis is a consistent regional stratigraphy. In Africa, so far this exists only for Uganda (N, M and south of it in fig. 1) and upland Kenya (G and H in fig. 1) (Van Zinderen Bakker, 1962, 1964; Coetzee, 1964, 1967; Livingstone, 1967; Kendall, 1969; Morrison, 1961, 1968; Morrison and Hamilton, 1974) where an initially grass-dominated pollen assemblage, with some pollen of shrubs and small trees, was supplanted *ca.* 10 000–12 000 B.P. by an assemblage of forest tree pollen. Within the past few thousand years grass pollen has regained something of its early prominence. In this still rather circumscribed area of tropical Africa we are dealing with a stratigraphic sequence of regional generality, a phenomenon like the shift from 'late-glacial' to 'post-glacial' vegetation in the temperate zone.

In South America, a consistent regional stratigraphy has been established in the mountainous regions of Colombia, above *ca.* 2 000 m (Van der Hammen and González, 1960a, b, 1965a, b; González, Van der Hammen and Flint, 1965; Van der Hammen, 1962, 1974; Van Geel and Van der Hammen, 1973). Paramo grasslands dominated much of this area before *ca.* 13 000. After *ca.* 9 500 most of the area below the level of the present tree line was forested whilst between *ca.* 13 000–9 500 there are several fluctuations corresponding in time reasonably well with fluctuations in the northern hemisphere. The regional stratigraphy in this area, however, is much longer and includes most of the Quaternary (Van der Hammen and González, 1964; Van der Hammen, Werner and Van Dommelen, 1973; unpublished diagrams from deep bore-holes). For the tropical lowlands of Colombia a long stratigraphy is being established and a reasonable stratigraphy has been recognized in the coastal lowlands of Guyana and Surinam (Van der Hammen, 1963; Wijmstra, 1969, 1971).

A regional stratigraphy also seems evident in Mexico, New Guinea, Australia and some parts of India and may be expected soon from Sumatra.

Interpretation

The third requirement for successful pollen analysis is the ability to interpret stratigraphic zones at least in terms of the vegetation that produced them, and consequently, in terms of the environmental factors such as climate that moulded the vegetation.

Most of the classical work in Europe and America relied on intuitive assessment of pollen diagrams by botanists with considerable ecological field experience. It rested on implied assumptions about the processes of pollen production, dissemination and preservation that went untested and unjustified for a generation. It is encouraging that now, when the broad correspondence of pollen assemblages and vegetational zones is being tested (Davis and Goodlett, 1960; Davis, 1963; Davis and Deevey, 1964; Livingstone, 1968; Ogden, 1969; Webb and Bryson, 1972; Webb, 1973, 1974) the assumptions are turning out to be at least broadly justified for temperate and subarctic North America. Modern objective statistical methods, and work with absolute pollen frequency rates, seem to yield conclusions that are not very different from those of the classical investigators. Investigators who have used modern methods in the tropics (Kendall, 1969; Hamilton, 1972) obtained results similar to those of the intuitive approach. Extensive but still unpublished studies on the quantitative relations of vegetation and pollen rain in the altitudinal zones of the Colombian Andes show that the interpretation of pollen diagrams from that area has been basically right, but that more precise interpretations will be possible.

The conventional, mainly intuitive method of interpretation of pollen diagrams, seems to give the most important facts about the history of vegetation and climate. Modern methods of interpretation may give much more exact data on the quantitative aspects of both vegetation and climatic change.

To summarize:

— The present specific composition and geography of tropical and subtropical forest ecosystems are the result of a long geologic history, including the drifting apart of continents by plate tectonics and the upheaval of mountain ranges.
— The Quaternary sequence of many glacial and interglacial periods, covering the last *ca.* 2.5 million years, has had a very pronounced effect on the geography of these ecosystems.
— Climatic changes of different lengths and amplitudes are still acting on these ecosystems and it is most necessary to know these changes and their effects for the purposes of conservation and management.
— Pollen analytical and palaeolimnological research (and to a certain extent also several other methods and disciplines) can give information on the palaeogeography and palaeoclimatology of tropical and subtropical ecosystems; the richest store of information lies in the sediments of lakes, which may contain large quantities of pollen grains and other micro-fossils per milliliter.
— For pollen analytical research it is first necessary to identify fossil pollen grains; identification should be based on morphological studies of recent pollen. More regional collections should be established and existing ones extended. Morphological studies on pollen of larger taxonomic groups should be stimulated in specialized centres.

Secondly, regional pollen stratigraphies should be established, based on pollen diagrams especially from lake sediments.

Thirdly, it is necessary to interpret these data in terms of quantitative change of vegetation and climatic environment.

Palaeogeography and palaeoclimatology of tropical Africa

Pollen analysis

The pre-Quaternary

Middle Cretaceous sediments from tropical Africa contain a very characteristic pollen flora, with *Galeacornea* and the elatere-bearing species *Elaterocolpites*, *Sofrepites*, *Senegalosporites*, etc. This association appears to be confined to a phytogeographical zone extending from northern South America via Central and North Africa to the Middle East, probably corresponding to the tropical zone of that time. The African continent was still connected with or very close to South America. In this floral belt, the first real Angiosperms, like *Hexaporotricolpites* (? Euphorbiaceae), *Triorites* and *Cretacaeiporites*, appear. It is, however, difficult to know with certainty the close botanical affinity of these pollen genera (Jardiné and Magloire, 1965; Jardiné, 1967; Herngreen, 1974). The physiognomy of the vegetation of that time is largely unknown.

Studies of Upper Cretaceous sediments (Van Hoeken-Klinkenberg, 1964, 1966; Jardiné and Magloire, 1965; Boltenhagen, 1967) show a much more diverse flora of Angiosperms with elements belonging with more or less certainty to Proteaceae, *Nypa*, *Ctenolophon* and other taxa. In the Palaeocene, new taxa such as Olacaceae and Palmae of the *Proxapertites*-type may be recognized. Pollen of *Pelleciera*, *Amanoa* and *Crenea* appear in the Eocene. Another pollen type of unknown affinity is *(Reti)brevitricolpites*, it probably belongs to plants of a mangrove-type vegetation (Van Hoeken-Klinkenberg, 1966).

Later grass pollen, *Alchornea*-type, Malpighiaceae and *Symphonia*-type appear (Germeraad *et al.*, 1968). In the early Miocene, *Rhizophora* occurs and the mangrove forests of that time on must have been very similar to the present ones. Somewhat later appear Compositae and Acanthaceae. The late Tertiary tropical forest and savanna communities may have been very similar to the present ones. Little, however, is known about possible changes in the distribution and extension of forest and grassland ecosystems during the Upper Tertiary. In Nigeria, clear fluctuations of the grass pollen curves were found in the Neogene, probably reflecting the shifting boundary between forest and savanna in the lowlands (Germeraad *et al.*, 1968).

The Quaternary

The longest pollen record that has been published so far from tropical Africa is that of Coetzee (1964, 1967) from Sacred Lake at an altitude of 2 400 m on Mt. Kenya (H in fig. 1). The lower part of the core contains many facies changes, and it is not clear from the published account that the record of deposition has been continuous since the time of deposition of the oldest radiocarbon date (33 350 B.P.). It is clear that a very long record is present, one in which grass pollen was very abundant, together with some pollen of high-altitude herbs and small trees, until *ca.* 10 500 B.P. At that time forest pollen became much more important in the record, and has continued so until the present day.

A somewhat shorter record was obtained at Kaisungor at an altitude of 2 900 m in the Cherangani Hills (G in fig. 1) of Kenya by Van Zinderen Bakker (1962, 1964). At Cherangani the record of changes is similar, except that the shift to forest pollen dominance seems to have occurred later. At Mahoma Lake at 2 900 m on the Uganda side of the Ruwenzori Mountains (M in fig. 1) Livingstone (1962, 1967) found a similar though shorter record, with the shift occurring at 12 500 B.P. Morrison (1961, 1968) and Morrison and Hamilton (1974) produced a diagram from 2 256 m in the Kigezi Highlands of western Uganda (just south of M in fig. 1) that seems similar to that from Cherangani, although the lower part of the core is not well radio-carbon dated. In Kigezi, probably because of vegetational disruption by prehistoric agriculturists, the vegetational zonation is less clear, but it is in accordance with the other localities in suggesting a relatively open, grassy vegetation prior to 11 000 B.P., and a more closed vegetation with more forest taxa since that time.

There are differences of opinion about the climatic interpretation of these stratigraphic changes, although all investigators seem in fair agreement about their vegetational significance. Bakker and Coetzee (1959) tend to interpret their vegetational sequences in terms of changing temperature, and proposed long-range correlations with the better established cool and warm periods of Europe, while Morrison (see especially his 1966 paper) and Livingstone have been more cautious about attributing vegetation changes to shifts in only one of the constellation of factors that influence African montane vegetation. Such differences of opinion are unavoidable whilst there are so few data on the modern pollen rain in regions of known climate and vegetation, and Hamilton has tried to fill the gap by careful analysis of surface spectra from 76 localities in highland Uganda. His re-interpretation (Hamilton, 1972) of the data is to some extent a compromise between the two earlier schools of thought. His most important contribution has been the attempt to separate fossil pollen taxa according to how far they have been dispersed from the sites where their parents grew.

It is known from geomorphological studies of glacier retreat stages (Osmaston, 1965) that the climate of the East African mountains was at least 4 °C colder than it is now during the height of the last glaciation, so the lack of trees during that time on Mt. Kenya, the Cherangani and Kigezi Highlands must be due at least partly to cold conditions. Livingstone (1967) has stressed that on Ruwenzori during the stage of glacial retreat the strictly botanical evidence for temperature change is mixed, whereas there are clear botanical indications, such as the presence of *Artemisia* pollen, that the climate was dry. The main botanical indication that conditions were not very cool on Ruwenzori prior to 12 500 B.P. is abundant *Myrica* pollen. This indicator seems to have survived Hamilton's (1972) attempt at an alternative explanation, but there are no other positive indications of such warmth. The vegetation on Mt. Kenya at that time was very similar to the vegetation of a somewhat earlier time for which the glacial record gives clear evidence of a cold climate. Taking into account the still-puzzling abundance of *Myrica* on Ruwenzori, and the wide spread of ages of appearance of tree pollen dominance on the various mountains, it is still not possible to say when the change from cold ice age climate to the warm climate of the present day occurred at high altitudes. It is not even clear from the pollen evidence that the temperature change was synchronous on these mountains, although it is meteorologically reasonable to presume that it was.

The firmest vegetational and climatic conclusions from Africa have been drawn by Kendall (1969), using a core from Pilkington Bay (N in fig. 1) on the north shore of Lake Victoria at 1 100 m. All other analyses from Africa had to plot each taxon as a percentage of the total assemblage of pollen grains occurring at the same depth, but Kendall had such accurate radio-carbon determinations of the sedimentation rate that he was able to express his results in terms of numbers of pollen grains of each kind sedimented per unit area per year. This removed one of the fundamental uncertainties of most pollen analyses. Kendall showed that the vegetation around Pilkington Bay from at least 15 000–12 500 B.P. was one that produced mostly grass pollen. The presence of traces of mimosaceous pollen strongly suggest that it was a savanna. After 12 500 B.P. forest trees became much more abundant, declining temporarily about 10 000 B.P. This initial forest appears to have been largely evergreen, and to have been displaced about 6 000 B.P. by a somewhat different forest in response to conditions that were drier, more seasonal, or both. During the past 3 000 years sedimentation of all forest pollen types at Pilkington Bay has declined whilst the percentage of grass grains increased, but due to excellent radio-carbon control and careful quantitative analyses Kendall was able to show that there was little absolute increase in the sedimentation rate for grass pollen. The main feature of the past few thousand years has been a decline in forest pollen without a corresponding increase in pollen of any other kind. Presumably the change represents the conversion of forest land to agriculture, because the indigenous crops of Uganda produce too little pollen to register clearly in the fossil record.

A similar suggestion of prehistoric human influence on the vegetation, though not so abundantly supported by radio-carbon dates, has been made by Bolick (1974) from Momela Lake at 1 500 m near Mt. Ujamaa (Mt. Meru) in Tanzania (K in fig. 1).

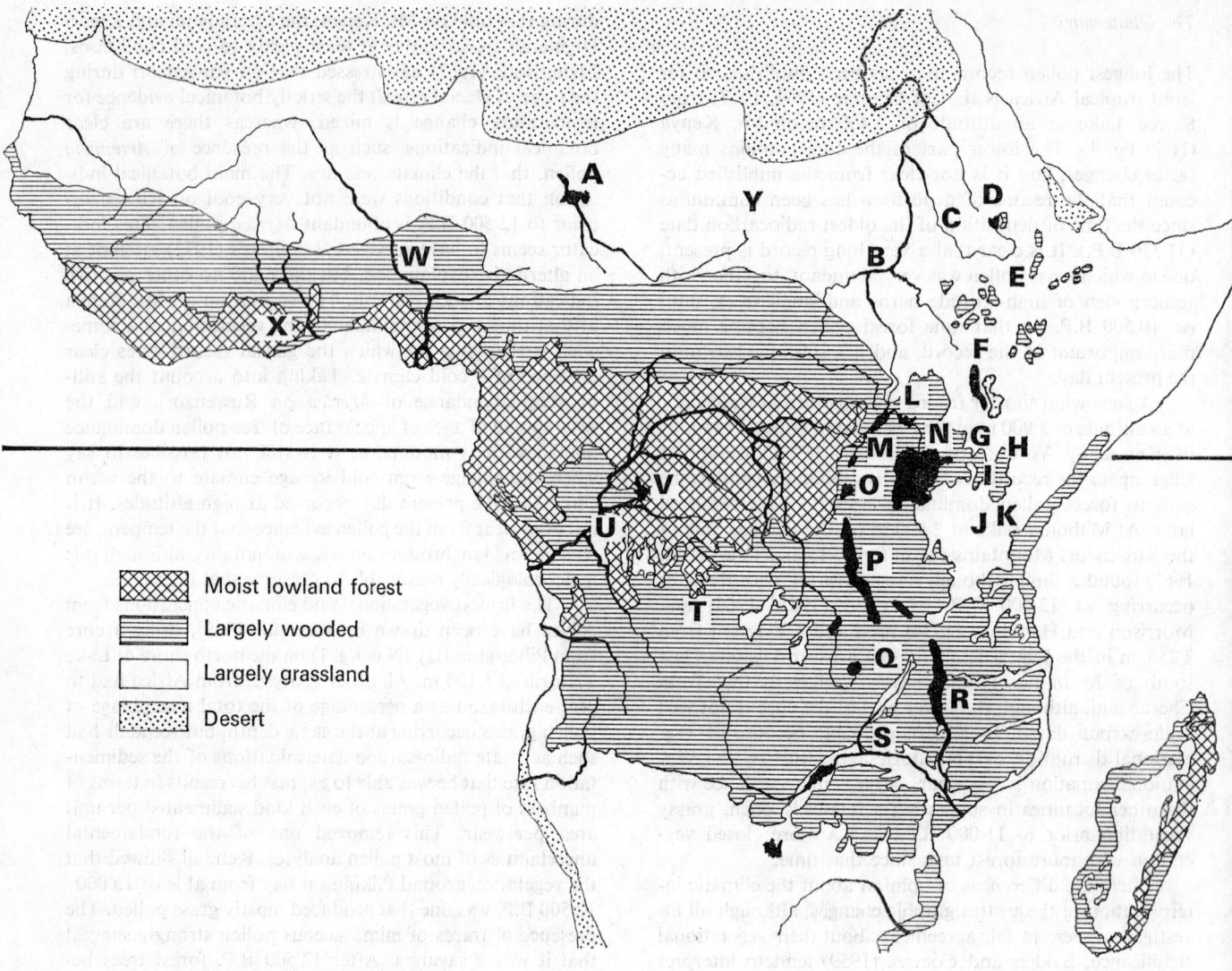

Moist lowland forest

Largely wooded

Largely grassland

Desert

FIG. 1. Synopsis of the vegetation of Africa.

A. Lake Chad
B. White Nile
C. Blue Nile
D. Lake Tana
E. Upper Awash River
F. Omo Delta, Lake Turkana (Rudolf)
G. Cherangani hills
H. Mount Kenya
I. Lakes Nakuru, Elmenteita and Naivasha
J. Lake Manyara
K. Momela Lakes, Mount Ujamaa (Mt. Meru)
L. Lake Mobutu Sese Seko (Albert)
M. Ruwenzori Mountains

N. Pilkington Bay, Lake Victoria
O. Lake Kivu
P. Lake Tanganyika
Q. Ishiba Ngandu
R. Lake Malawi (Nyasa)
S. Zambezi River
T. Luembe Valley
U. Stanley Pool on the Zaire River
V. Lake Leopold II
W. River Niger
X. Lake Bosumtwi
Y. Jebel Mara

Coetzee (1967) reports on short cores from several other localities. Bonnefille (1969a, 1969b, 1970, 1972) has succeeded in providing a number of isolated pollen spectra from beds of archaeological and palaeontological significance by pollen extraction from very poor material in Ethiopia. She has shown that changes of the same nature and extent as those of the past 15 000 or 20 000 years also occurred between two and three million years ago. Van Zinderen Bakker has obtained indications of vegetational, and by inference climatic, change from an altitude of 1 200 m at Kalambo Falls on the border between Tanzania and Zambia (Clark and Van Zinderen Bakker, 1964) and between 700 and 800 m in Angola (Van Zinderen Bakker and Clark, 1962), but the spectra are from river terraces and there are no comparable surface spectra to use in assessing the conclusions. Smit (1962) has demonstrated significant changes in the pollen assemblages that have been deposited in Lake Bosumtwi in Ghana (X in fig. 1) and Lawton (1963) has attempted a pollen analysis of the peats of the Bangweulu Swamp, which proved singularly poor in useful micro-fossils. Quézel and his co-workers have demonstrated vegetation less xeric than the modern over wide areas in the southern Sahara (Quézel, 1960, 1962; Quézel and Martinez, 1961, 1962). Dubois (Dubois and Dubois, 1939; Dubois and Jaeger, 1948), Hedberg (1954) and Osmaston (1958) have also carried out pioneering investigations. Livingstone (1971) provided a pollen record covering 22 000 years from Ishiba Ngandu in Zambia (Q in fig. 1), but the record is almost devoid of significant change, apparently because the vegetation types for hundreds of kilometers around the coring site are all wooded grasslands which produce indistinguishable grass-dominated pollen assemblages.

The stratigraphic consistency of the Uganda-Kenya region may now be summarized. Forest genera were very much scarcer in this area from at least 35 000 B.P. until a time that seems to vary with locality from 12 500 to *ca.* 9 000 B.P. Grasses were more important in the early fossil pollen assemblages, and presumably in the vegetation as well, than they have been since. The first forest elements to appear in abundance include taxa such as *Olea* that are often pioneers in seral succession, but they were quickly joined by an assemblage of trees suggesting a relatively moist evergreen forest, which prevailed until about 6 000 B.P., when conditions became at least seasonally drier. During the past few thousand years most forest taxa have declined to some extent, and grasses have become more important as a percentage of the total pollen assemblage. At the one site where there is adequate radio-carbon dating, the rise of grass pollen is mostly relative, and the decrease of forest pollen seems due more to displacement of trees by cultivated plants of low pollen productivity than to their displacement by grassland.

Throughout the record, and especially during the past ten or twelve thousand years, the vegetation of eastern Africa has been extremely dynamic. There has been no time when the vegetation has been stable for a thousand years and the various taxa are continually rising and falling in abundance. The amplitude of these changes has been great,

comparable with vegetational changes of the temperate zone during the same time span. The major change at 9 000–12 500 B.P. is very reminiscent of a similar change from open to closed forest vegetation in the well-watered parts of the temperate zone at the same time, and there is no reason to doubt that the change in both regions was due to world-wide changes in climate associated with the last ice age which was too cold for optimal forest growth in the temperate zone and on high tropical African mountains and too dry for optimal growth on both the mountains and the plateau of tropical eastern Africa.

Resource planners managing tropical African vegetation have to deal with a dynamic resource which has been continuously changing throughout its known history and seems likely to go on doing so. Static plans will not suffice for its exploitation and conservation. There is an immediate need for pollen analysis of surface samples of modern lake mud collected from the widest possible variety of modern vegetational environments. Such an array of analyses should be compared with existing vegetation maps, forest inventory records and satellite imagery to establish as closely as possible the correspondence of modern pollen assemblages and vegetation. This would provide a more secure basis for the interpretation of pollen diagrams. At the same time, the pollen results should be compared with climatic parameters by multivariate statistical analysis (Webb and Bryson, 1972) to establish direct correlations. This approach would give quantitative suggestions about past changes in climate if the sampling array covered as wide a range of conditions as are represented in the fossil record. It is particularly important to include in this investigation a large number of stations surrounded by evergreen forest, because such data are sparse and stratigraphic work has uncovered many fossil assemblages dominated by genera of the evergreen forest. Deforestation makes such work more difficult and less certain of success.

The area of stratigraphic investigation also needs to be extended into the lowland evergreen forest of Zaire basin and western Africa. The East African evidence shows that evergreen forest was much reduced during the last ice age, and we need to discover the locations and sizes of the areas in which it survived. Refuge size has a bearing on how one views the tropical forest: is it a recent reexpansion of a very much reduced remnant, or has it always been perhaps half as large as it is now? This is a fundamental question about the palaeogeography, not just of tropical Africa, but of the world, and must be answered before we can deal realistically with the general problem of latitudinal clines in ecological species diversity. Tumba and Leopold II (V in fig. 1), the large existing lakes of the Zaire basin, seem to contain a sediment of clean sand unsuitable for pollen analysis, or perhaps smaller lakes may be found that are suitable. In western Africa both Bosumtwi (X in fig. 1) in Ghana and the crater lakes of Cameroon hold great promise for extending the area of known vegetational history into the forested zone.

The area of stratigraphic investigation needs to be extended north and south into the tropical arid parts of the continent. The current investigations of Maley (1972) will, when completed, provide a useful record backed up by understanding of the relation between vegetation and pollen assemblages for Lake Chad (A in fig. 1), but there is no immediate prospect of data from Jebel Mara in Sudan (Y in fig. 1), the lakes of which appear eminently suitable for pollen analysis. Some information is available for arid northern Africa (see, for example, Van Campo and Coque, 1960; Van Campo *et al.*, 1966; Beucher, 1967; Saad, 1967; Beucher, 1971; Schulz, 1974) but in the southern part of the continent we have long cores only from coastal sediments in temperate areas (Martin, 1953, 1968; Schalke, 1973) and records that are shorter or more difficult to interpret from extra-tropical dry localities (Van Zinderen Bakker, 1955; Coetzee, 1967). There is a complete lack of information from the dry sub-tropical areas of southern Africa. Until these gaps are filled it will be difficult to know to what extent each ice age either brought generally drier conditions to Africa, or involved only a southward shift ot the well-watered area. This is a matter of the greatest importance in laying long-term plans for land use, especially if we are, as there is some reason to believe, moving again in the direction of ice age conditions.

There is a complete lack of detailed studies of laminated sediments which is not due to the lack of lakes that are likely to contain suitable deposits. This work might best begin in Ethiopia, which is well-supplied with suitable small stratified lakes and has had a long and important agricultural history.

There is reason to expect a record covering at least a million years in lakes Bosumtwi, Malawi, Tanganyika, Kivu, Mobutu Sese Seko, Turkana (Rudolf) and Natron. In some of these lakes the record is likely to cover ten million years. Some of the Cameroon lakes may also be very old, and a long record is expected in most of the major lakes associated with the Rift Valley. Some of these, such as Turkana and Natron, are unpromising for pollen work because they lie in regions where grass dominates the pollen rain, and the problems of grass identification are great and perhaps insuperable. Malawi and Tanganyika are apparently both very old (Brooks, 1950; Degens, von Herzen and How-Kin Wong, 1971) and at least some palynologically promising forest grows around them, but they are deep and rough. Perhaps the best compromise is Bosumtwi, which is of manageable size and situated in an excellent vegetational context, although its age is likely to be only a little over a million years (Bampo, 1963).

No pollen analytical prospect in the world presents a greater challenge or a greater potential for understanding palaeogeography than the ancient lakes of tropical Africa. If a core could be raised, analysed and interpreted from one of the very old lakes in the Rift Valley it would provide an unparalleled source of tropical palaeoclimatic information, a very long perspective on the changing natural geography of Africa, and an ecological background for understanding the evolution of human culture, the human species, and possibly the rich African bovid fauna as well.

The following previously reviewed gaps in knowledge exist:

— relations between climate, vegetation and modern pollen rain;

— cores from the ancient lakes, long enough to embrace at least several ice ages, need to be raised and analysed;

— the extent of contraction of the evergreen tropical forest needs to be determined from cores raised within the Zaire basin if suitable lakes can be found, and from the Guinea forest;

— the nature, timing, and extent of climatic changes in arid southern Africa are very poorly known;

— the timing and extent of immigration of temperate mediterranean trees into the northern Sahara should be more firmly established;

— data are needed from the Atlas mountains;

— palaeoclimatic information is required from especially the northern part of the Ethiopian plateau;

— high-resolution study of annually-laminated sediments in an agricultural area are needed to provide a temporal perspective on human land use;

— in no part of Africa except northern Lake Victoria is there an adequate density of radio-carbon dates for the palaeoclimatic sequence.

Palaeolimnology

Introduction

Richardson (1968), Kilham (1971), Hecky and Kilham (1973) and Kilham and Kopczynska (1974) have provided information about the mainly chemical factors which influence the distribution and abundance of diatoms in African lakes. Hecky (1971) and Richardson and Richardson (1972) have shown how to use such information to reconstruct changes in productivity, chemistry and climate. It is not to be expected, however, that all lakes affected by the same climatic change will react in exactly the same way. As Hutchinson (1957) showed, the response of a lake to a particular change in climate is a function of its shape and the relative importance of run-off and direct precipitation and evaporation in its water budget. The greatest changes in water chemistry result from closure of a lake, so an increase in aridity will have the greatest effect on the chemical composition of lakes in such delicate hydrologic poise that they are easily changed from open lakes to closed ones in which evaporative concentration of salts can occur. For this reason one does not find such great regional consistency in diatom stratigraphy as in pollen stratigraphy.

The mineralogy of lake sediment sometimes gives clear indications of climatic change, where, for example, the water has been sufficiently concentrated to precipitate crystalline salts (Kendall, 1969) or to form autochtonous zeolites (Hay, 1963, 1966; Stoffers and Holdship, 1975). If the water level of a lake falls, even without marked chemical concentration of the water, the episode is likely to register in the sedimentary record as an erosional unconformity, or as a stratum of coarse material, perhaps sand with shells, reflecting the high wave-energy of a shallow water environment.

Saharan lakes

The widely scattered literature on former lakes of the Sahara has been ably summarized by Faure (1969), who showed that large lakes existed in that desert region repeatedly during past geological time and especially during the Quaternary. The last of these lacustrine episodes began about 12 000 B.P. and reached its maximum about 8 000 B.P. Most of the lakes shrank quite quickly after that, and were dry by 7 000 B.P. At least locally there was another lacustrine episode between 5 500 and 3 000 B.P. There is a rough correspondence with sea-level change and temperature change in the temperate zone, with warm periods tending to be wet and cool ones dry, but the correspondence is far from exact, even during the past 12 000 years. For earlier episodes the correspondence seems even poorer: there is a strong suggestion of conditions moister than the present between 20 000 and 40 000 years ago. Although that was not the period of intensest glaciation, few palaeoclimatologists believe it to have been warmer than now.

Scattered lacustrine conditions existed during the early Holocene pluvial period, from 14 to 22° N, and from the Atlantic Ocean to the Nile. The drying up of a lake may not involve a great decrease in local rainfall. The evaporation-precipitation ratio must have changed, but the larger Saharan lakes were all nourished by large rivers such as the Chari and the Niger that drew their water from moister regions that are not desert even today. The lakes of arid and semi-arid regions are extremely evanescent; little additional rain is needed to make them expand, little additional evaporation to make them disappear. The palaeolimnological evidence tells us much about conditions of life for the fishes and diatoms whose remains are found in the lacustrine deposits, but we need more pollen evidence, such as that being provided by Maley, to know what the conditions of life were like for people and other terrestrial organisms.

Diatoms and stratigraphy of the Chad region (A in fig. 1) have been investigated by Servant and Servant (1970), whose paper, along with Faure's review, provides an entry to the large French literature on desert palaeolimnology.

Nilotic lakes

Kendall (1969) used gross stratigraphy, exchangeable cations, minerals, diatoms and other algae to show that Lake Victoria (N in fig. 1) was without an outlet from before 14 500 to *ca.* 12 000 B.P. During the driest part of this period the level of the lake fell by at least 26 m, the water was more alkaline than now, and its phytoplankton community differed from the modern one; *Stephanodiscus astraea* was more abundant than now, indicating a lower concentration of dissolved silica in the water (Kilham, 1971).

Livingstone (in press) sampled in deeper water, and showed that the level of the lake during the dry episode fell at least 75 m below the modern datum, but no one has cored in the deepest part of the lake to determine if it dried up completely.

After the outlet was established there was apparently a short period *ca.* 10 000 B.P. when the lake was closed again. Sinde then it has flowed continuously out through the Nile, and has been higher than it is now, for there are well marked horizontal strand-lines at 3,12 and 18 m

above the 1960 lake level. Only the lowest, which has been dated at 3 700 years B.P. is likely to have climatic significance. The older and higher two probably represent stages in down cutting of the Nile outlet at Jinja. Three wet years in the early 1960s caused the lake to rise almost 3 m suggesting that the lowest strand-line might represent a climatically-controlled level and not a down cutting stage.

Kendall's pollen and palaeolimnological study was important in establishing clearly and unequivocally for equatorial Africa the recent age of the last pluvial period and the glacial age of the interpluvial that preceded it, but this study did not establish the age of the basin or the length of time during which it has held water continuously. Victoria is the home of a rich fish fauna, including the endemic cichlids now known to number about 175 species (P. H. Greenwood, personal communication). It would be of both practical and theoretical interest to establish the age of such an ichthyological phenomenon.

Hecky and Degens (1973) and Thomas Harvey (personal communication) have taken cores from Lake Mobutu Sese Seko (Lake Albert) and shown that it, like Victoria, was dry *ca.* 13 000 B.P. The lake again fell briefly to at least —51 m, but may not have dried completely. Harvey's core, the longest, represents 28 000 years. This is a small part of the total history of the basin, which contains some 2 700 m of sediment and was probably first formed in Miocene time (Cahen,1954; Bishop, 1967).

These results show that the White Nile ceased to flow for a short time at least once prior to 12 000 years ago. Although its modern annual discharge is much less than that of the Blue Nile, it is White Nile water that sustains the river during the Ethiopian dry season. Loss of the White Nile must have had disastrous consequences for the lives of aquatic species and of terrestrial species, including man, that depended on the river for water.

Lake Turkana (Lake Rudolf) has a strongly nilotic fauna, and was connected to the Nile through the Pibor and Sobat during humid episodes. Studies of lake muds are being carried out by R. Yuretich, and there has been much study of the strand-lines around the lake and of the sedimentary deposits of rivers flowing into it, especially in the delta of the Omo. This work is summarized by Butzer *et al.* (1972). The lake was low from 35 000 to 9 500 B.P. when it rose to fluctuate between 60 and 80 m above the modern level until 7 500 B.P. It was high again from 6 200 to 4 400 B.P., when it fell temporarily, only to rise to +70 m a little before 3 000 B.P. Since then it has been relatively low, fluctuating rapidly through *ca.* 40 m; between 1897 and 1955 it fell 20 m. There have been frequent reoccupations of strand levels at 70 and 80 m above the modern level. These represent overflow levels into the Lotigipi mud flats west of the lake and into tributaries of the Nile respectively.

Turkana, being an arid closed lake receiving variable run-off from the much moister headwaters is extremely unstable in level. It is also gradually growing more concentrated, and is now so brackish as to be of borderline potability. It supports an indigenous fishery that has recently been supplemented by one experimental trawler. Before heavy capital and social investments are made it might be advisable to consult the palaeolimnological record of salinity change and the salinity tolerance of the commercial fish species to estimate the time that the lake will remain habitable by fish.

Other lakes in and near the Eastern Rift

Richardson (1966) and Richardson and Richardson (1972) investigated the palaeolimnology of Lake Naivasha in Kenya and found an elaborate history of diatom, chemical and climatic changes. The lake was high and fresh when their record began 9 000 B.P.; it possessed an outlet until *ca.* 5 600 B.P. After that the level fell and the lake dried up completely for a brief time *ca.* 3 000 B.P. Since then it has oscillated irregularly in both water-level and chemical concentration. This behaviour extends into the short historic period, to the annoyance of people who would exploit the resources of either the Rift lake fish populations or the dry land around them. Richardson and his co-workers are extending the known history of the Rift Valley in Kenya by analysing a series of cores from lakes Nakuru, Elmenteita and Naivasha that represent about 30 000 years of time.

Hecky (1971) studied the palaeolimnological record of the Momela lakes (K in fig. 1) formed 6 000 years ago by a mud-slide off the slopes of Mt. Ujamaa (Mt. Meru) in Tanzania. He used techniques of numerical taxonomy and multivariate analysis to compare fossil diatom assemblages with the modern diatom communities in lakes of known chemical concentration and productivity (Hecky and Kilham, 1973). The climatic results were in accordance with those of Richardson for Kenya, and Hecky showed that the composition and productivity of the diatom community were both extremely sensitive to climate and drainage changes. He found no tendency for the lakes to become more productive with time, and found productivity to be inversely correlated with species diversity.

The longest continuous lacustrine record that has been studied in Africa comes from Lake Manyara (J in fig. 1) in Tanzania. Stoffers and Holdship (1975) measured changes in the abundance of diatoms and autochtonous minerals, principally zeolites, in a 55 m long core that appears to represent about 60 000 years. The core presents the change from dry to wet conditions that is becoming the standard finding of African palaeolimnology. The last 5 000 years were similar to modern times, with episodes that were even drier. The greatest interest of the core is for the period prior to 15 000 B.P., which is poorly represented at most coring localities. This period was generally dry, rather than humid like the Early Holocene, but its dryness was interrupted at least three times by transitory periods of moderate wetness. There does not seem to be any detailed correspondence between the times of these wetter episodes and the wet periods between twenty and forty thousand years ago that have been reported from Chad (Servant and Servant, 1970).

Deep lakes of the Western Rift

Degens and his co-workers (Degens *et al.*, 1971; Degens *et al.*, 1973; Hecky and Degens, 1973) have taken cores from lakes Kivu, Tanganyika and Malawi (O, P, R, respectively in fig. 1) and provided seismic profiles from the first two of these.

The seismic profiles are of special interest for the detailed information they give about sedimentary structure over extensive areas of the lakes. Alternating beds of different sonic reflectivity are visible throughout the accessible part of the section in Tanganyika and to the crystalline basement in Kivu, and are interrupted by faulting and uncomformities. The stratigraphy seems comparable in complexity with that of the sea, and will demand extensive coring and correlation.

Although the seismic profiler penetrated over a kilometer of sediment it did not reach the crystalline basement under Lake Tanganyika. Such a sedimentary thickness is in accordance with the great age of the lake as deduced, in the absence of geologic data, from the rich fauna of endemic species and genera.

It was, however, surprising to find that there was *ca.* 1 km of sediment in Kivu. This lake has a very poor fauna of fish, and limnologists had concluded from this and from the apparent youth of some of the volcanoes that have contributed to the impounding of Kivu water, that the lake was young, perhaps no more than 10 000 years in age. Degens *et al.* (1973) are certainly more nearly correct in claiming a Pliocene age for the basin, and explaining the poverty of its fish fauna by the highly anoxic bottom water probably mixing with the surface frequently enough to keep the fish fauna modest.

Hecky and Degens (1973) found evidence suggesting very great changes in lake level, as much as 400 m in Kivu and 600 m in Tanganyika, with the low levels corresponding to cold times in the temperate zone. They also correlated pluvial conditions with hydrothermal activity.

Even though its water-level has fluctuated widely, the volume of the Lake Tanganyika is so enormous, and the water income so small, that evaporative concentration during an interpluvial period lasting for thousands of years will not make the lake so concentrated as to be inhospitable to fishes. The basin in which the lake lies is old, but more important from a biological point of view, the lake is old as well.

Palaeolimnological outlook

All the potentialities, the accomplishments and the difficulties that have already been discussed with respect to pollen analysis have palaeolimnological analogues. Palaeolimnologists are perhaps somewhat advanced with respect to understanding the modern ecology of their most useful group of organisms, the diatoms, but they face the same technical problems of coring and social organization before they can exploit the enormous palaeoecological potential of such lakes as Turkana or Tanganyika.

Summary and conclusions

Both pollen and palaeolimnology, together with work such as that of de Ploey (1963, 1965) on soils and geomorphology lead to the same conclusions: prior to 12 000 B.P. the climate of tropical Africa was dry. This dry period commenced before 60 000 B.P. and has been repeatedly interrupted by transitory periods of less severe drought, but was seldom, perhaps never, as dry as it was near its end *ca.* 13 000 B.P.

The past 12 000 years have been comparatively moist, but broken, in at least some tropical African localities, by transitory dry phases *ca.* 10 000 and 6 000 B.P. Maximal moistness seems to have prevailed between 8 000 and 7 000 B.P. and the past 3 000 years have been at least intermittently the driest time during the past twelve millenia.

The evidence for these conclusions is not equally good from all places. Especially with respect to pollen, the record of the past few millenia is complicated by changes in human land use that might produce vegetational changes similar to those of a dry climate. Agricultural deforestation since the beginning of the iron age might be mistaken for a drought-induced retreat of forest in the pollen record, but it should not produce a falling water-level. Agriculturalists would rather enhance run-off and the lake level would rise. Despite such local difficulties the general picture seems to be rather well established although it is conceivable that further work may demonstrate an even more detailed array of continent-wide changes in climate.

The dynamic nature of African natural geography is well established. For all kinds of information, in almost all places where they are available, the record is one of restless change over time scales of tens, hundreds, thousands, and tens of thousands of years. We do not yet know what happened near the middle of large blocks of rain forest when their peripheries were expanding and contracting in response to climatic change, but the findings of de Ploey near Stanley Pool (U in fig. 1) and the presence of windblown sand under forest and woodland in the lower Zaire basin (Cooke, 1964) suggest that not even the lowland evergreen forest escaped the vicissitudes of changing climate. Certainly the montane evergreen forest did not.

It has not been possible to treat the palaeogeography of evergreen forest, woodland and grassland separately in this chapter. Most of the major lakes, such as Tanganyika and Victoria, have some of all three vegetation types growing around their shores, and it is not possible, until better methods of identifying fossil grasses are developed, to make a distinction between woodland and grassland. The conclusion of climatic variability, however, seems to be generally valid, and applicable to all three vegetation types.

This chapter has not dealt with modern biogeographic patterns, partly for lack of space, partly because it is seldom possible to date the events that have led to any particular pattern of plant or animal range. It is clear, however, from the work of such biogeographers as Moreau (1966) and Lawton (1963, 1972) that the present discontinuities in distribution of plants and animals bespeak great environmental changes in the not-too-distant past. This evidence of modern distributions is in accord with the more direct palaeogeographic data discussed here.

Palaeogeography and palaeoclimatology of tropical America

Pollen analysis

Most pollen analysis data available for this area are from the northern Andes. The upheaval of the Cordillera de los Andes has been the latest major tectonic event. It occurred mainly during the Pliocene (probably more than 3 million years ago).

The pre-Quaternary

During the Middle Cretaceous, most tropical South America apparently belonged to the same phytogeographical zone as Central and North Africa, characterized by a typical association of pollen with *Galeacornea* and the elatere-bearing species *Elaterocolpites*, *Sofrepites* and *Senegalosporites*. South America may have been still connected to Africa or at least the two continents were not yet separated by a great distance (Herngreen, 1974). In this floral belt pollen of the first real Angiosperms appears (*Hexaporotricolpites*, *Triorites*, *Cretacaeiporites*, *Psilatricolpites*, *Retitricolpites*, *Striatricolpites*, *Psilatricolporites* which are mostly of uncertain or unknown botanical affinity). *Classopollis* which, since earlier Mesozoic time was an important element of the coastal flora, is still present.

The Upper Cretaceous pollen flora still shows a close affinity with that of Africa, an affinity that is still noticeable in the Lower Tertiary (see the earlier section on the pre-Quaternary of Africa).

In the Upper Senonian of Brazil (Herngreen, 1972) a group of very characteristic pollen grains occurs (*Crassitricolporites* and *Crassitriapertites*) which have not been found in the Maestrichtian of Colombia. This might be an indication for the existence of a western and an eastern floral province at that time. An important element in the Maestrichtian vegetation is the *Psilamonocolpites medius* group of palms. A considerable number of new taxa, like the Bombacaceae, appear near the boundary of the Cretaceous and Tertiary. Several of these (besides the *Psilamonocolpites medius* group, which was already present), became important elements of the vegetation of the lowlands such as *Proxapertites* (probably a palm and a close relative of *Mauritia*) (Van der Hammen, 1957; Van der Hammen and Wijmstra, 1964). During the Eocene, pollen of the *Brevitricolpites*-type appears in appreciable quantities and there is clear differentiation of the coastal vegetation into belts apparently parallel to the coast. The *Brevitricolpites*-type probably represented the mangrove belt, next was a *Mauritiidites* palm zone and inland from this a *Psilamonocolpites* palm zone (González, 1967). These zones shifted frequently according to changes of the relative sea-level.

During the Middle Eocene grass pollen appears in some quantity, associated with *Jussiaea* and Malvaceae. This possibly indicates the presence of some open, savanna-like vegetation. In the Oligocene, *Rhizophora* has definitely replaced *Brevitricolpites* in the mangrove zone. In the Miocene and Pliocene grass pollen became a very important part of the pollen associations and the diagrams show conspicuous maxima. This was established in the Caribbean area (Germeraad *et al.*, 1968) and seems to indicate the importance and variability of extensive open grasslands (probably savannas) at that time.

The upheaval of the Andes fundamentally changed the drainage system of the continent towards the Atlantic, leading finally to the present situation of the Amazon. The influence of this phenomenon on the tropical vegetation of the continent is still largely unknown (for the Northern Andes, see below).

The Quaternary

The Andes and the montane forest

In and partly around the high plains in the Colombian Eastern Cordillera, is a thick series of Pliocene lacustrine and fluviatile sediments. Pollen has been analysed from several parts of this sequence, from *ca.* 2 600 m altitude (Van der Hammen, Werner and Van Dommelen, 1973). The older part was deposited in a mainly forested tropical lowland environment with e.g. Bombacaceae, *Humiria*, *Alchornea*, *Hieronymus*, *Ilex* and palms like *Mauritia* and *Iriartea*. The genera *Podocarpus* and *Weinmannia* indicate that somewhat higher mountains may have been present in the area but other traces of high mountain species are lacking. The pollen content of the middle and upper part of this series of sediments shows that the area of deposition was successively in lower and higher montane forest belts: the tropical elements disappeared and were replaced by montane ones. Finally, some 3 million years ago, the Cordillera seems to have reached approximately its present elevation and high mountain elements, like *Polylepis* or *Acaena*, *Aragoa*, *Plantago* and *Valeriana* appear.

At the same time that the upheaval of the Andes took place, the Isthmus of Panamá came into existence, opening an easier way for migration between North and South America. During and after the upheaval of the Andes, species migrated into the 'tropical Andes' from the Antarctic and Holarctic regions. The present day montane and high mountain flora evolved from the local neotropical flora and the immigrants. The early higher montane vegetation types were still very poor in species and probably because of slow adaptation of woody species, the altitudinal tree-line was at a relatively lower level than today.

The high plain of Bogotá (Sabana de Bogotá, 2 600 m altitude; *a* in fig. 2) was a lake during most of the Pleistocene and several hundreds of meters of sediments were deposited in it. The results of the pollen analysis of these sediments, representing some 3 million years of history, have only partly been published (Van der Hammen and González, 1960a, 1964; Van der Hammen, Wijmstra and Zagwijn,

1971; Van der Hammen, Werner and Van Dommelen, 1973; Van der Hammen, 1974). The pollen diagrams show a long sequence of alternating forest and paramo vegetation, representing glacial and interglacial periods. The altitudinal forest limit, now *ca.* 3 500 m, was depressed during glacial times by *ca.* 1 500 m to altitudes that were, at least locally, below 2 000 m. The sequence can be dated by radiometric methods, because thin layers of volcanic ash are intercalated in the lake sediments. The sequence also shows the successive first appearance of new elements, so that the dates of arrival of *Alnus* and *Quercus* can be established with relative precision, at *ca.* 2 000 000 and 900 000 B.P. respectively. Further and more detailed study of this material may eventually reveal data on the time involved in the evolution of a number of taxa.

The Upper Pleistocene of the Sabana de Bogotá (*ca.* 2 580 m altitude) was studied in greater detail (Van der Hammen and González, 1960a; and unpublished). The terminal stage of an earlier glacial period apparently was very cold and dry with an open grass-paramo containing such high-paramo species as *Malvastrum acaule*. Virtually no traces of forest elements remain. During the last interglacial, which probably started *ca.* 13 000 B.P., the area around the high plain was forested once more; the frequency of some of the elements from the uppermost Subandean forest seems to indicate that the climate became slightly warmer than it is now. The first part of the last Glacial, with its much wetter climate, shows several interstadials and stadials comparable to those of the north temperate latitudes. During the forest phases of this early Glacial, *Quercus* becomes a dominant element. The following Pleniglacial was much colder and paramo vegetation dominated. Elements of the uppermost forest and shrub and subparamo became more important during some of the minor interstadial of the middle Pleniglacial. The coldest phase of the last Glacial started *ca.* 26 000 B.P. About that time the Pleistocene lake of the Sabana de Bogotá had dried up. Sedimentation continued, however, in the Laguna de Fuquene (*b* in fig. 2) on the next high plain (also at *ca.* 2 580 m) to the north.

The pollen diagram from Laguna de Fuquene (Van Geel and Van der Hammen, 1973) provides a complete record of the vegetational history of the last 32 000 years. From this diagram one may deduce that fluctuations in temperature and humidity occurred. The first kind of fluctuation may be estimated by the vertical displacement of vegetation belts, and the second can be deduced from the extension and retraction of the marshy zone of the hydrosere as reflected in the pollen diagram. Approximately 30 000 B.P. *Polylepis* scrub and *Acaena* were very common, their occurrence indicating conditions slightly above the limit of the proper Andean forest. Shortly afterwards the climate became progressively colder so that open paramo started to dominate. About 21 000 B.P. the climate became extremely cold and dry (the water in the lake dropped to a very low level and extreme paramo conditions prevailed). This lasted till the beginning of the late Glacial, 13 000 B.P. The lowering of vegetation belts during this period was of the order of 1 200–1 500 m. Taking the influence of aridity into account,

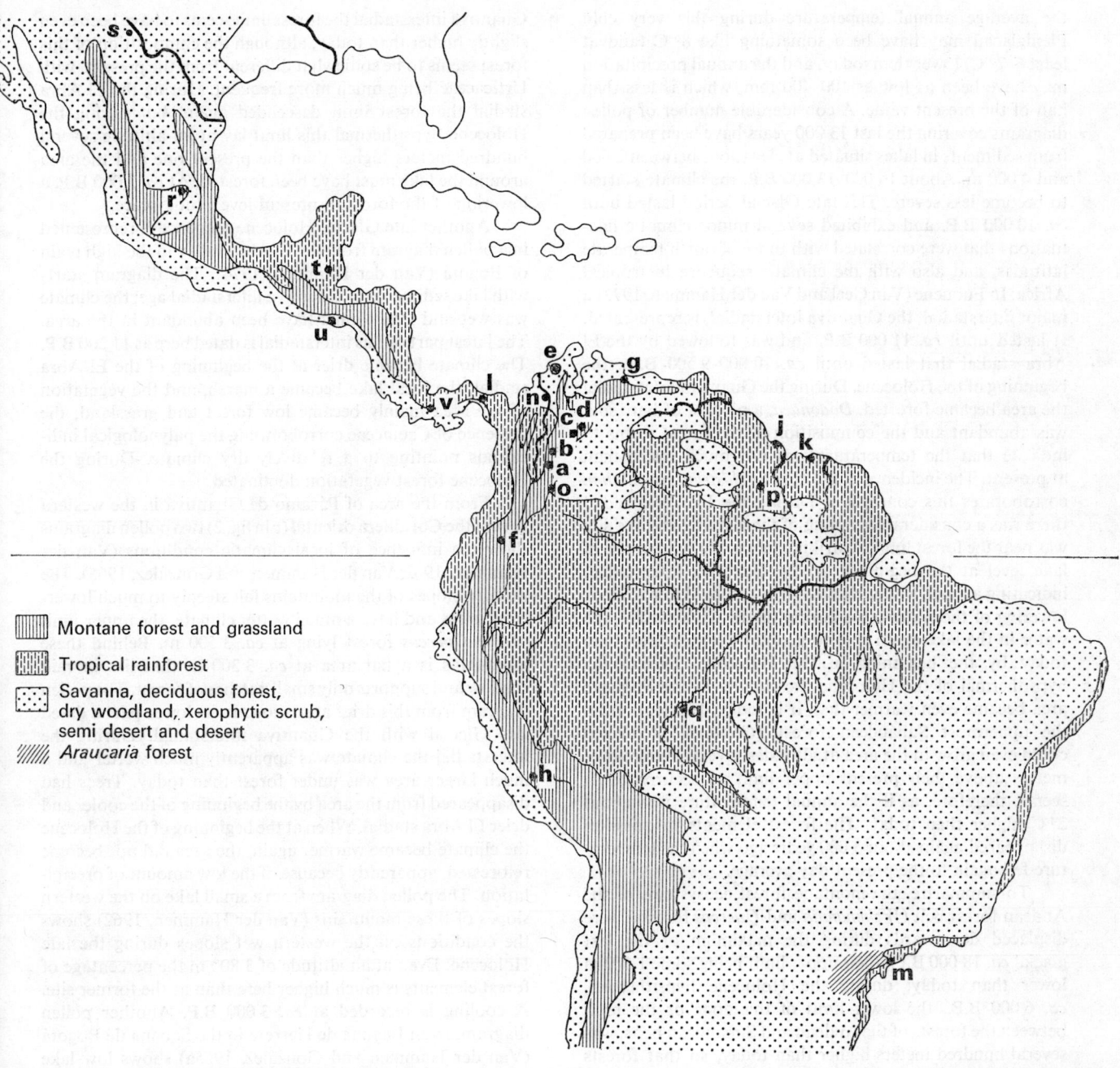

FIG. 2. Synopsis of the vegetation of tropical America, generalized after different sources. Letters mark the following features:

Legend:
- Montane forest and grassland
- Tropical rainforest
- Savanna, deciduous forest, dry woodland, xerophytic scrub, semi desert and desert
- *Araucaria* forest

a. Sabana de Bogotá
b. Laguna de Fuquene
c. Páramo de Guantiva
d. Sierra Nevada del Cocuy
e. Sierra Nevada de Santa Marta
f. Laguna de la Cocha
g. Lago Valencia
h. Ayacucho
j. Georgetown
k. Alliance
l. Gatun basin

m. Santos area
n. Ciénaga de Morrocoyal
o. Laguna Agua Sucia
p. Lake Moreiru
q. Rondonia
r. Valley of Mexico
s. Chihuahuan desert
t. Laguna Petenxil
u. Laguna Cuscachapa
v. Vicente Lachner bog.

the average annual temperature during this very cold Pleniglacial may have been something like 8° C (and at least 6–7° C) lower than today, and the annual precipitation may have been as low as 100–400 mm, which is less than half of the present value. A considerable number of pollen diagrams covering the last 13 000 years have been prepared from sediments in lakes situated at elevations between 2 000 and 4 000 m. About 14 000–13 000 B.P. the climate started to become less severe. This late Glacial period lasted until *ca.* 10 000 B.P. and exhibited several minor climatic fluctuations that were correlated with those of north temperate latitudes, and also with the climatic sequence in tropical Africa. In Fuquene (Van Geel and Van der Hammen, 1973) a major interstadial, the Guantiva interstadial, is represented. It lasted until *ca.* 11 000 B.P. and was followed by the El Abra stadial that lasted until *ca.* 10 000–9 500 B.P., the beginning of the Holocene. During the Guantiva interstadial the area became forested. *Dodonaea*, a pioneer of bare soil, was abundant and the composition of the forest seems to indicate that the temperature was not much lower than at present. The incidence of the alga *Coelastrum reticulatum* corroborates this conclusion. During the El Abra stadial there was a considerable cooling of the climate, and the site was near the forest limit. A striking fact is the sudden rise of lake level at the beginning of the Guantiva interstadial, indicating a much wetter climate than before. After a minor lowering of the water-table during the El Abra stadial, the water in the lake again rose to a slightly higher level than today by the beginning of the Holocene. During the hypsithermal of the Holocene the climate in the Fuquene area became even warmer than it is today. Elements from the uppermost Subandean forest (*Cecropia*, *Acalypha*) could even grow in the area at altitudes several hundreds of meters above their present upper limit of occurrence. It seems, therefore, as if the annual temperature was about 2° C higher than today. The forest elements in question disappeared again about 3 000 years ago when the temperature fell again to the present day average.

To summarize, the montane forest ecosystems of the Andean forest belt (*Weinmannia* and *Quercus* forests) were displaced downwards during the maximum of the last glacial *ca.* 18 000 B.P., their upper limit being some 1 500 m lower than today; during the Holocene hypsithermal, *ca.* 6 000 B.P., the lower limit of this zone or the limit between the forests of the Andean and Subandean belts, was several hundred meters higher than today, so that forests with *Cecropia* and *Acalypha* grew on the high plain. During the last 12 000 years, forest ecosystems around the Laguna de Fuquene changed in quantitative and specific composition because of changes in annual temperature and rainfall, and because of gradual development of the soil.

An example of the succession at higher elevation is provided by the diagram from Páramo de Palacio (*ca.* 3 500 m; near *a* in fig. 2), north-east of Bogotá (Van der Hammen and González, 1960b). The site lies at 200–300 m above the present day limit of the *Weinmannia* forest. The late Glacial clayey sediments, deposited after the retraction of the glaciers from the lake area, show fluctuations in pollen similar to those noted in the late Glacial of Fuquene. During the

Guantiva interstadial the forest limit seems to have been even slightly higher than today, although the composition of this forest seems to be somewhat different from the present, with Urticaceae being much more frequent. During the El Abra stadial the forest limit descended again, but during the Holocene hypsithermal this limit lay at an altitude several hundred meters higher than the present one and the area around the lake must have been forested. At *ca.* 3 000 B.P. a lowering of the forest to present levels took place.

Another late Glacial-Holocene sequence is represented in a pollen diagram from the El Abra valley in the high plain of Bogotá (Van der Hammen, 1974). The diagram starts with lake sediments of Guantiva interstadial age; the climate was wet and *Alnus* must have been abundant in the area. The latest part of this interstadial is dated here as 11 200 B.P. The climate became drier at the beginning of the El Abra stadial, the local lake became a marsh, and the vegetation of the area mainly became low forest and grassland; the presence of Cactaceae corroborating the palynological indications pointing to a relatively dry climate. During the Holocene forest vegetation dominated.

From the area of Páramo de Guantiva in the western part of the Cordillera oriental (*c* in fig. 2) two pollen diagrams show the influence of local climatic conditions (Van der Hammen, 1962; Van der Hammen and González, 1965). The western slopes of the mountains fall steeply to much lower-lying areas and have a much wetter climate, the upper limit of the *Quercus* forest lying at *ca.* 3 500 m. Behind these mountains is a flat area at *ca.* 3 300 m that lies in rain shadow and supports only small patches of forest. The pollen diagram from this drier area shows in its lower part a dated late Glacial with the Guantiva interstadial. During the interstadial the climate was apparently much wetter and a much larger area was under forest than today. Trees had disappeared from the area by the beginning of the cooler and drier El Abra stadial. When at the beginning of the Holocene the climate became warmer again, the area did not become reforested, apparently because of the low amount of precipitation. The pollen diagram from a small lake on the western slopes of these mountains (Van der Hammen, 1962) shows the conditions on the western wet slopes during the late Holocene. Even at an altitude of 3 800 m the percentage of forest elements is much higher here than at the former site. A cooling is recorded at *ca.* 3 000 B.P. Another pollen diagram, from Laguna de Herrera in the Sabana de Bogotá (Van der Hammen and González, 1965a) shows low lake levels (i.e. dry phases) *ca.* 5 000, 2 000, 760 and 500 B.P.

From Laguna de Pedro Palo (west of the Sabana de Bogotá) at *ca.* 2 000 m in the upper part of the Andean forest belt, a most elucidating diagram of the late Glacial was obtained (Van der Hammen, 1974). Sedimentation started in the early late Glacial or late Pleniglacial when the area was covered with open grassland vegetation and the forest must have been below 2 000 m. The presence of grains of *Isoetes* and of *Myriophyllum* (and of non-aquatic herbs), which are today mainly found above 3 000 m, corroborates the conclusion that this open vegetation very closely resembled paramo vegetation, so that the tree-line must have been at least 1 300 (and probably 1 500 m) lower than today.

Some 12 000 years ago the area was invaded by Andean forest, and towards the end of the Guantiva interstadial by Subandean forest.

A number of lakes in the Sierra Nevada del Cocuy (*d* in fig. 2) provided pollen diagrams for the last *ca*. 13 000 years (González, Van der Hammen and Flint, 1965); unpublished diagrams represent the last *ca*. 26 000 years. These and unpublished diagrams for the last *ca*. 10 000 years from the Sierra Nevada de Santa Marta (*e* in fig. 2) are correlated with endmoraines. From these data a considerable higher tree-line for the Holocene hypsithermal can be deduced.

Most of our knowledge of vegetation history of the Northern Andes is based on pollen analysis of lake sediments in the Colombian Eastern Cordillera. It may be summarized as follows:

The tropical-montane, northern Andean climatic belts originated during the late Pliocene upheaval of the Cordillera. These newly created belts gradually became populated by processes of evolutionary adaptation of elements from the local Neotropical flora and by the arrival of elements which immigrated from the Holarctic and Antarctic flora areas. This process continued during the entire Pleistocene. In the Andean forest belt local elements are frequent, but *Weinmannia* originally came from the south and *Myrica*, *Alnus* and *Quercus* from the north.

In the transitional and subparamo forest and scrub, local genera are still abundant, but elements from more remote areas gradually become more frequent to reach their highest frequencies of occurrence in open paramo vegetation (*Gentiana*, *Bartschia*, *Valeriana*, *Draba*, *Hypericum*, *Berberis*, etc., from the Holarctic; *Muehlenbeckia*, *Acaena*, *Azorella*, etc., from the Antarctic). The most characteristic paramo genus, *Espeletia*, is of local Andean origin as are such other endemics as *Aciachne*, *Distichia*, *Puya* and *Rhizocephalum*.

The Pleistocene shows a sequence of several glacials and interglacials comparable with those of the northern hemisphere and the contemporaneity of the changes of temperature could be substantiated for the last 50 000 years, i.e. for the period within the reach of reliable ^{14}C dating. The tree-line is now between *ca*. 3 200 and 3 500 m. Its depression was of the order of 1 200–1 500 m although this limit must have been as irregular as it is today owing to local climatic and microclimatic conditions. It seems as if a probable average value of 2 000 m for its altitudinal position during periods of maximum glaciation can be established. Under extreme arid conditions the lowermost limit of dry paramo vegetation in the Eastern Cordillera today is at 3 000 m. If such conditions prevailed during the upper Pleniglacial, a lowering of the temperature by 6–7° C is necessary to explain the total depression of the forest.

During the coldest period of the last Glacial (upper Pleniglacial), the climate on the high plains was much drier than today and for the areas of Fuquene the annual precipitation may have been less than half that at present. As we shall see later the temperature in the tropical lowlands during glacial time may only have been about 3° C lower than today; thus the temperature gradient in the Northern Andes was much steeper than it is at present.

During glacial times the surface area occupied by paramo vegetation was several times its present extent; many current 'islands' were linked together. The Páramos of Cocuy and Sumapaz formed part of a large and continuous area of paramo that covered the entire central portion of the Cordillera oriental. The superparamos of these two areas were never in direct contact with one another. This seems to be reflected in the relatively high degree of endemism, especially in the superparamo of the Sierra Nevada del Cocuy.

The distribution patterns in the Northern Andes may be partly explained by long-distance dispersal and partly by the erstwhile continuity of areas of paramo during glacial times. Conditions for immigration by long-distance dispersal (or from one 'island' to another) were certainly much more favourable for elements of the high Andean vegetation groups during glacial times. Speciation of both the local and the alien elements must have been stimulated by the successive periods of separation and union of populations.

Together with the study of actual vegetation this knowledge of the history seems sufficiently wide to form the base for a much wider study on tropical Andean ecosystems and their history. This is in preparation by members of botanical and other institutes of the universities of Amsterdam, Utrecht and of the National University in Bogotá. For this it will be necessary to collect lake sediments in the central and western Cordilleras. One of the most promising lakes is the Laguna de la Cocha (*f* in fig. 2). It will be especially important to find good lakes at elevations below 2 000 m, for they are scarce.

In the Venezuelan Andes comprehensive pollen analytical study of lake sediments by Salgado Labouriau in collaboration with Schubert has just begun, at the Instituto Venezolano de Investigaciones Científicas in Caracas. One of the most promising sites is the Laguna de Valencia, at low altitude (*g* in fig. 2). Peeters (1968, 1971) studied the origin and evolution of this basin and his conclusions (although only supported by two provisional pollen spectra) are that the present lake was preceded by an earlier one, whose level was much lower although it fluctuated to a higher level than the present one. There has recently been a lowering of the lake level. The fluctuations are interpreted as the result of precipitation changes. As a sample of a regressive phase contained more Chenopodiaceae-type pollen than a sample from a transgressive phase, it would be worth while to obtain a deep core from a carefully selected spot in this lake.

No pollen diagrams have been published from the Andes of Ecuador, Peru, Bolivia and northwestern Argentina and northernmost Chile but work is in progress. McNeish (e.g. 1971) and collaborators started archaeological research in the Ayacucho area, Peru (*h* in fig. 2). Graf has collected samples from peat bogs in Bolivia and d'Antoni has started work on pollen analysis of cave sediments in northwestern Argentina. From further to the south in temperate Chile, Heusser's (e.g. 1966, 1972) pollen diagrams and data for the last *ca*. 14 000 years show a climatic sequence very similar to the one in the Colombian Eastern Cordillera.

Data from the Peruvian Andes (especially from the dry western Andes and the Altiplano) are needed to understand the changing climate of the world and the history of vegetation of this area. There seem to be good possible sites for lakes are relatively abundant. For the longer sections, however, lakes have to be found that were not glaciated. The study of sediments of the *salares* of the Altiplano and surroundings may be very interesting as they might represent phases of a considerably wetter climate.

To summarize, the Andean tropical montane forest ecosystems originated after the upheaval of these mountains in the Pliocene. Elements evolved from the local neotropical flora and from elements that migrated into the area from the Antarctic and Holarctic floral areas. Their specific composition was slowly enriched during the entire Quaternary; they have been displaced many times, and often suddenly, during this period. Much is known from the Colombian Eastern Cordillera, but information from the remainder of the area is extremely scarce or lacking. Pollen analytical studies in the Colombian central and western Cordillera, in Ecuador and in the Venezuelan Andes are necessary (Lago Valencia being a very promising site for the lower belts) and are urgently needed for the Peruvian Andes.

Middle America and Mexico

The tropical flora of this area is a neotropical one derived from the south. In the temperate montane vegetation, however, there are a number of holarctic elements related to the flora of the eastern United States. A recent account mainly of the Tertiary history is found in Graham (1973). A progressive southward migration could be established by pollen analysis. Fourteen genera (like *Abies, Alnus, Betula, Carya, Fagus, Liquidambar, Myrica*, etc.) were present in the southeastern United States in the Eocene. Ten of these reached southern Mexico by the Middle Miocene. During the Miocene only three genera (*Alnus, Juglans, Myrica*) are recorded for Central America (Panamá). In northern South America these same three genera are first encountered in the Plio-Pleistocene.

Pollen analytical studies on the Quaternary in the valley of Mexico (*r* in fig. 2) started more than two decades ago (Clisby and Sears, 1955; Sears and Clisby, 1955; Foreman, 1955). Periods with increased percentages of *Alnus* and *Quercus* or *Abies, Alnus* and *Quercus* pollen, alternate with periods of dominating *Pinus* pollen or pollen of herbaceous plants. These fluctuations are interpreted as the result of changes in humidity, and partly as the result of volcanic activity. González Quintero (personal communication) found considerable changes in the vegetation in and around the valley and fluctuations of lake levels. The climate was considerably drier and colder between *ca.* 12 000 and 20 000 B.P. and woody vegetation mainly consisted of *Pinus*. There were wetter and warmer periods before and after these dates. Further to the north the history of the deserts and other dry areas (Sonoran desert, Chihuahuan desert, *s* in fig. 2) must have been similar to their continuation in the dry southwestern United States. Martin and Mehringer (1965) review the available evidence and give a map of the vegetation of the southwestern United States 17 000–23 000 B.P. The climate was then much wetter and deserts

were replaced by sage-brush, chaparral and woodland (*Pinus-Juniperus*). A change to drier conditions took place about 12 000 B.P. Watts (1969, 1971 and personal communication) found evidence of a long dry phase before 10 000–12 000 B.P. in Florida.

Further south, in the Maya area of Guatemala and El Salvador, a number of lakes were studied by Tsukada and Rowley (1966, 1967) including Laguna de Petenxil (*t* in fig. 2) and Irabal y Cuscachapa (*u* in fig. 2). They are in the lowlands or below *ca.* 1 200 m. The pollen record for the last 4 000 years for two lakes in the rain forest area shows that the vegetation changes were mainly induced by man; agriculture heavily affected the forest, especially during classic Maya time (*ca.* 200–900 A.D.).

A longer pollen diagram is available for a sequence of lake muds and peats from Parque Vicente Lachner in Costa Rica (*v* in fig. 2) at an altitude of 2 400 m (Martin, 1964). There are two ^{14}C dates from the upper part of the section. The present continuous upper limit of *Quercus* forest is at 3 100 m or slightly higher. During the last 8 000–10 000 years montane *Quercus* forest has predominated. Before that it was open grass paramo which commenced before 36 000 B.P. A minimum lowering of the forest limit by 650 m may be calculated. The sedimentation rate seems slow (0.1 mm per year) and it is difficult to know if the thin intercalation of clay at 300 cm represents a hiatus; so changes of lake level and annual precipitation are unknown.

Still further south data from tropical lowland deep core sediments in the Gatun basin, Panamá (*l* in fig. 2) are available on the history of vegetation, climate and sea-level during the last 12 000 years (Bartlett and Barghoorn, 1973). During the earliest part of this period plants now growing at higher altitudes were apparently found near sea-level (e.g. *Iriartea*, Ericaceae and *Symplocos*). They might have been growing some 500–1 000 m below their present habitat and temperatures may have been at least 2.5° C lower than now. By *ca.* 7 300 B.P. temperatures had risen to equal those of the present time. For the period between 7 300 and 4 200 B.P. a drier, more seasonal climate is suggested. Pollen of probably cultivated plants (*Zea, Manihot*) is found in samples from the last few thousands of years. While mangrove vegetation dominates the basin in the early Holocene, it later was replaced by freshwater swamps.

To summarize, in Middle America and Mexico the specific composition of tropical forest ecosystems is neotropical. During the Tertiary there was a progressive southward immigration via the montane belt of Holarctic forest species. While the northern Mexican deserts were wetter 18 000 years ago and locally supported chaparral or *Pinus-Juniperus* communities, in the valley of Mexico climate was apparently drier and colder. *Quercus* forests that are now found in the surrounding areas were then probably partly replaced by *Pinus* forest. In Costa Rica there is clear evidence that the montane *Quercus* forest at 2 400 m was replaced by open paramo vegetation, the upper forest limit being depressed by at least 650 m. In Panamá the temperature decreased by at least 2.5° C; some species from the hills may have been growing at sea-level.

After the main Holocene sea-level rise, mangrove swamp forests were replaced by freshwater communities in Panamá. Later the forest ecosystems in Guatemala and Yucatan were heavily affected by agricultural activities in classic Maya time.

Further pollen analytical studies of deep sections in the valleys of Mexico are necessary and there is a need for data (for at least the last 20 000 years) from the entire area of Middle America and Mexico.

The coastal lowlands, deltas and lowland marshes

For the interpretation of pollen diagrams from coastal sediments, the study of recent pollen sedimentation in the Orinoco delta and the adjacent shelf area is of considerable importance (Muller, 1959). Most available pollen analytical data from coastal lowlands are from the Caribbean coast in Guyana, Surinam and French Guyana (Van der Hammen, 1963; Van der Hammen and Wijmstra, 1964; Wijmstra, 1969, 1971).

A 30 m section from the coastal plain near Georgetown, Guyana (*j* in fig. 2) provided a pollen diagram that shows something of the history of this area during the last interglacial, the last Glacial and the Holocene. The lower part of the diagram, with a ^{14}C date of 45 000 B.P., indicates that at the time of deposition the site was within the mangrove belt. There follows an extension of freshwater swamp forest elements and the sea seems to have retired; the site then apparently lay behind the coast-line. Mangrove elements disappeared completely from the diagram in the next phase and open grass-savanna elements became completely dominant and the sediment showed clear signs of soil formation. The site is now well above sea-level. This situation lasted until the beginning of the Holocene, when the sea gradually invaded the area; first (*ca.* 8 600 B.P.) the *Avicennia* community and later *Rhizophora*. The presence of microforaminifera in this part of the section indicates that at that time the coast-line proper lay farther inland than today. The later Holocene part of the diagram shows that the coastline moved northwards again and *Avicennia* forest, swamp forest and open swamps were frequent in the Georgetown area. The last pollen spectra show the present situation and the influence of man on the vegetation. It is evident that the diagram not only shows the glacial/interglacial eustatic fluctuations of sea-level, but also shows that grass-savanna dominated in the area during at least a part of the last Glacial. Deep bore-holes from Surinam and Guyana show that in Quaternary sediments from the coastal plain the type of sequence recorded from near Georgetown is repeated many times and probably reflects the sequence of Quaternary glacial-interglacial eustatic movements of sea-level. An example is a 120 m bore-hole section from Alliance, Surinam (*k* in fig. 2). The age of the series can be deduced from the presence of a number of taxa such as *Alnus* (from a source area in the Andes). Although *Byrsonima* occurs locally in the late Holocene part of the sections, the overall composition of the spectra indicates that extensive grass-savannas were not developed at that time in this area. On the other hand, many of the recorded older, low sea-level intervals are similar to those from the last Glacial in Georgetown. They show a dominance of grass-savanna with *Byrsonima* and *Curatella*. Therefore, it can be concluded that savanna vegetation dominated the present coastal area of Guyana and Surinam during glacial times, or at least during a part of each glacial time. Although the influence of edaphic factors cannot be ruled out, it seems as if the dominance of grass-savannas over so large an area cannot be explained by edaphic factors alone and requires a lower annual precipitation and/or more pronounced dry and wet seasons than at present. The deduced vegetational succession during an eustatic climatic cycle is: *Rhizophora-Avicennia*-palm swamp forest, grass-savanna with *Byrsonima* and *Curatella* scrub, and vice versa.

Changes of the coastal vegetation in Surinam during the last *ca.* 2 000 years are shown by Laeyendecker-Roosenburg (1966). Minor transgressions and regressions of the sea are apparently the cause of sudden and profound changes of the vegetation. Mangrove forest was suddenly replaced by open herbaceous swamps and there is a sudden increase of *Avicennia* pollen during a minor transgression *ca.* 700 A.D.

Bartlett and Barghoorn (1973) reconstructed eustatic sea-level changes in Panamá on the basis of pollen diagrams from the Gatun basin (*l* in fig. 2). Mangrove vegetation established itself in the area during the postglacial sea-level rise, but was later replaced by freshwater marshes.

Pollen analytical study of coastal sediments near Santos, Brazil (*m* in fig. 2) has been started by Absy.

González Guzman is analysing 6 m cores from the marshes of the lower Magdalena and Cauca valleys. The changes of vegetation show a very marked alternation of dry and more humid phases during the last Glacial and Holocene. From the same area of the lower Magdalena (*n* in fig. 2) 7 m of sediment representing less than 2 000 years were deposited in a large lake (Ciénaga) in connection with river water (Wijmstra, 1967). The lake level fluctuated considerably and was so low *ca.* 1 100 A.D. and 1 500 A.D. that peat was formed. In the dry phases there was an increase of herb pollen and a decrease of tree pollen. Amongst the woody elements there is a relative increase of *Byrsonima* and *Curatella* during these dry phases, and of *Cecropia*, *Ficus* and Ulmaceae types. This seems to indicate an extension of savanna and savanna woodland vegetation into the formerly open marshes and the levee forest along the rivers and *caños*. The pollen diagrams from the lower Magdalena-Cauca seem to show considerable changes of the vegetation, caused apparently by cyclic changes of precipitation in the capture area of the rivers. In addition to major ones there are also cycles, with a period of *ca.* 250 years. This type of data is certainly of considerable importance for the management and control of this huge area.

To summarize, in the Guyanan coastal area the Pleistocene changes of sea-level and probably of climate caused the repeated replacement of the mangrove swamps and other lowland forest ecosystems by savanna and savanna woodland. In the marshes of the lower Magdalena valley

(Colombia) minor changes in rainfall during the Holocene caused considerable changes in the vegetation: during the drier phases savanna and savanna woodland vegetation extended in the formerly open marshes and levee forest extended further along the rivers and *caños*.

Data from the coastal lowlands and deltas are still very scarce and studies along the Pacific coast of Colombia and Ecuador, the Caribbean coast of Colombia and Venezuela and the Atlantic coast of Brazil are required. The Lago de Maracaibo and the Amazonas and Orinoco deltas may be of special importance.

Savanna woodlands and savannas

A number of pollen diagrams are available from lakes in the savannas of the Llanos orientales of Colombia and the Rupununi savannas in Guyana (Wijmstra and Van der Hammen, 1966). Most of these are from the Holocene but one reaches back to the last Glacial. Two of the more informative diagrams show the variation in time of the percentage of pollen grains of three groups: forest elements; savanna woodland and scrub (*Byrsonima* and *Curatella*); and open savanna (Gramineae, Cyperaceae and other savanna herbs). The first diagram is from Laguna de Agua Sucia, south of San Martin and not far from the Ariari river in the Colombian Llanos orientales (*o* in fig. 2). Today the area is dominated by grass-savanna with some swamp forest or gallery forests in low-lying places and in the Ariari valley. Two layers of peat, intercalated in the lake sediments, represent low lake-levels and were dated at *ca.* 4 000 and *ca.* 2 200 B.P. The lower part of the diagram, from possibly *ca.* 6 000–5 000 to *ca.* 4 000 B.P., shows the complete dominance of grass-savanna; it seems that the lake became seasonally dessicated. After *ca.* 3 800 B.P. the lake-level rose and open water remained throughout the year. At the same time *Byrsonima* woodland invaded the open savanna and became dominant. There is also an increase of the other forest elements. All this suggests a wetter climate. Subsequently open savanna increased gradually until *ca.* 2 200 B.P. when the formation of peat again indicates a very low lake-level; at this time *Mauritia* swamp forests invaded part of the lake. A little later the lake-level rose again and a final sharp increase of grass-savanna elements can only be interpreted as the effect of human influence (especially burning). From the data provided by this diagram we may conclude that, during the Holocene, changes in the annual or seasonal precipitation occurred and resulted in appreciable changes in the proportion of savanna compared to savanna woodland. A major period of open savanna lasted from about 6 000–5 000 B.P. to *ca.* 3 800 B.P.

The second diagram is from Lake Moreiru in the Rupununi savannas of Guyana (*p* in fig. 2). The lower part of the diagram shows approximately equal proportions of savanna woodland and open savanna, followed by a major extension of open savanna coinciding with a very low lake-level. When the lake-level rose again, the area around the lake was completely invaded by *Byrsonima* woodland, so that virtually no open grass-savanna was left. The age of the lower limit of this woodland period was not established

directly but may be calculated on a basis of sedimentation rates as *ca.* 13 000 B.P. or *ca.* 10 000 B.P. However, the sediments below that limit (at *ca.* 340 cm) are from the later part of the last Glacial period. Towards the end of this savanna woodland period, open savanna was dominant again *ca.* 7 300 B.P. Then a minor increase of *Byrsonima*, followed by an increase of open savanna, the lower part of which is dated *ca.* 6 000 B.P. Later, there was a slight final increase of trees. If the calculated date of *ca.* 13 000 B.P. is correct, the extreme grass-savanna period immediately before it, associated with very low lake-levels, would be of upper Pleniglacial age and the effective precipitation was low, probably lower than today, especially when we take into consideration that the temperature was probably lower. The previous period must have been wetter and the subsequent period, corresponding with the late Glacial and possibly the early Holocene, much wetter than today. This would mean that the curve for the effective precipitation, corresponding with the lower part of the diagram, is apparently in phase with that from Lake Fuquene in the Andes. There is no doubt that we urgently need more sections and more ^{14}C dates. However, the available data prove the existence of drier and wetter phases in the Holocene and in the late Pleistocene and are highly suggestive of a very dry period in the upper Pleniglacial and a wet period in the late Glacial.

All these changes of effective precipitation caused major or minor changes in the areas of especially savanna and savanna woodland ecosystems, savanna extending during the drier and woodland during the more humid phases.

Pollen analysis of the best and longest records from the lakes of the Llanos orientales, Orinoco savannas and the inland savannas of Guyana, Surinam and the Rio Branco area is necessary and exploration for suitable sediments in the area of the campos cerrados andrelat ed open vegetation types in Brazil is highly recommendable for nothing is known of the vegetation history of that area.

The Amazonian forest

Almost nothing is known of the Quaternary history of the Amazonian forests, although the occurrence of dry periods has been deduced from geomorphological and biogeographical data (see the following section). Recent palynological data strongly support the former existence of savannas in areas at present covered with tropical forest (Van der Hammen, 1972, 1974). The evidence comes from Rondonia (Brazil) in the southern part of the Amazon basin (*q* in fig. 2). The general diagram is composed of two sections from the same area. The uppermost part corresponds with 2 m of recent river sediments, and the lower part (representing *ca.* 13 m of sediments) is from the bottom of a small valley. The cumulative diagram shows the variation in time of the percentages of three groups of pollen grains: humid tropical forest trees; *Mauritia* (mainly occurring in swamp forest); and open grass-savanna (Gramineae, *Cuphea* and a few other herbs). The lower part of the lower section is a darker type of humid clay where the pollen of humid tropical and swamp forest dominates. In its upper part the clays are of a lighter colour and show intercalations of reddish and sandy material, apparently correlating with slope deposits washed down from the sides of the valley during a period of instability. This part of the section shows a complete dominance of open savanna elements (mainly grasses, but also herbs like *Cuphea*).

Elements of the humid tropical forest completely dominate the upper section. Here again savanna elements are much more abundant in the lower portion. The Holocene age of the uppermost recent sediments seems certain; their pollen content is entirely in agreement with the present vegetation. The age of the lower part of these river-valley sediments might be Holocene or late Glacial. Although we do not know the exact age of the other section, it is clear that it represents Quaternary or at least Cenozoic sediments from an earlier phase. We cannot yet be certain that they represent the last Glacial, but the data presented here show that there were periods during the Pleistocene when savannas locally replaced part of the forest.

An extensive exploration of the entire Amazon basin for suitable sediments for pollen analytical studies is urgently needed. If such important changes as described above took place so recently, we need to know about them for the rational management of these ecosystems. Moreover, if we want to understand the continuously changing climate of the world, we have to know what happened in this most important area.

Palaeolimnology and other data on palaeoclimates

Limnological studies are infrequent in the area and palaeolimnological studies still scarcer. Deevey (1957) studied recent lakes in Middle America and Loeffler (1968) in the Andes. Several studies on the Amazon basin appeared in *Amazoniana*. The studies of recent waters in the Amazon basin are most important for both management and historical aspects of the Amazonian forest (Sioli, Schwabe and Klinge, 1969). Palaeolimnological studies in Middle America were carried out by Cowgill *et al.* (1966) and Cowgill and Hutchinson (1966), covering the last few thousand years. Most changes seem to be related to agricultural practices, others with minor climatic changes or earthquakes. Palaeolimnology of lakes in South America is still very undeveloped.

Biogeography, geomorphology and even the study of deep sea sediments have contributed to the study of changes of vegetation and climate. An excellent review on this topic was published by Simpson (1971), while three of the more interesting original publications that refer to past climates are those of Haffer (1969) on Amazonian forest birds, Damuth and Fairbridge (1970) on deep-sea sediments, and Bigarella and Andrade (1965) on Quaternary geology. The results are of special importance for those areas from which no pollen analytical data are available, e.g. Peru and the Amazon basin. From both speciation patterns and geomorphological phenomena, it is concluded that considerable changes of climate and vegetation took place in these areas.

The most striking is the reduction of the Amazonian forest to a number of refuges during dry climates in the recent past. This view seems to be confirmed by pollen analysis from at least one place (Rondonia).

Hastenrath (1968) studied from records of precipitation and temperature the very recent and short-term climatic fluctuations in the Central American area. Several trends are evident all over the area which, when compared with the recent fluctuations of lake-levels and tree-ring sequences, show the effects on the vegetation. Further study of these short-term changes seems to be of considerable importance for the understanding of the influence of changing climatic factors on the vegetation.

Summary and conclusions

Since the separation of Africa and South America in the upper Cretaceous, the flora of both continents evolved separately. The upheaval of the Cordillera de los Andes, more than 3 million years ago, had a major impact on the climate, geography and vegetation of the continent. The montane forests came into existence by evolutionary adaptation of elements of the local neotropical flora and by elements immigrating from the Antarctic and Holarctic floral areas. The Isthmus of Panamá came into being around the same time, allowing the passage of neotropical elements to the north through Middle America into Mexico and of Holarctic elements to the south into South America.

In South America faunal evolution during the Cenozoic had been taking place in considerable isolation. The entrance of predators like puma and jaguar in the late Cenozoic is an important fact that may have contributed to the extinction of other animals. The entrance of *Mastodon* may have had a considerable effect on certain vegetation types.

The highest zone of tropical montane South America (the Northern Andes) during the Quaternary exhibited changes in the average annual temperature apparently contemporaneous with those of the north temperate latitudes. For the last Glacial the maximal decrease of temperature was at least 6–7° C. The upper limit of the Andean forest proper was lowered to an altitude of *ca.* 2 000 m (today it is at 3 200–3 500 m), and large areas of montane forest ecosystems were replaced by open paramo. In the last Glacial an extremely dry period coincided with the period of maximum glaciation (*ca.* 21 000–13 000 B.P.). The glacial climate before that time was wetter, the upper limit of the Andean forest somewhat higher and the paramo vegetation richer in *Polylepis-Acaena*. The late Glacial Guantiva interstadial (in time corresponding with the European Bølling and Allerød) was wet, the El Abra stadial (corresponding mainly with the late Dryas time) was drier. The forest ecosystems that at that time had invaded the areas formerly occupied by paramo, were different from the present ones. Their soil was still mainly undeveloped and some pioneer species played an important role. It seems that a major part of the last Glacial (between *ca.* 90 000 and 21 000 B.P.) had a climate with, at least at intervals, a higher effective precipitation than the Holocene. During the period 21 000–13 000 B.P. the effective precipitation was much lower than during the Holocene.

The tropical lowlands experienced a decrease of temperature of 2.5° C 12 000 B.P.; this seems to be in agreement with the calculated value of superficial sea-water in the Caribbean of 2–3° C (corrected $0^{16}/0^{18}$ data and data on foraminiferal assemblages) for the last Glacial maximum.

As at present the temperature of superficial sea-water in the Caribbean seems to correspond approximately with the average annual temperature, it seems reasonable to accept a lowering of *ca.* 3° C for the tropical lowlands of northern South America during the Glacial period. This figure was at least 6–7° C in the higher Andes so that the altitudinal temperature gradient must have been steeper during glacial time than it is now.

The changes of vegetation registered in diagrams from the tropical lowlands are changes from forest or woodland to savanna and vice versa. They can be explained by changes in effective precipitation. During at least a part of the last Glacial, savannas existed in the coastal lowlands of Guyana and the climate was favourable for the development of dry savanna in the Rupununi area. A period of savanna vegetation is also recorded in an area in the southern part of the Amazon basin. The fact that there are clear indications in northern South America of a dry period between *ca.* 21 000 and 13 000 B.P. (and such a period is also known from many places in Africa, see above summary) renders it probable that this was the case in many parts of the tropics and at least part of such dry periods may correspond with periods of the maximum extension of glaciation in northern latitudes. Many parts of the world (the southwestern United States of America forming one of the exceptions) seem to have had a drier climate during that period.

The repeated lowering of the tree-line in the Andes led to direct connections between groups of paramo islands and the possibility of species exchange. In interglacial times populations were isolated again and conditions for speciation through isolation were favourable. Immigration by species from abroad must have been much greater during glacial times when paramo islands joined up and were much larger. The repeated extension in the tropical lowlands of savanna vegetation (and locally perhaps of more xerophytic types of vegetation) in areas that now support forest vegetation (or a savanna vegetation, respectively), may have led to an exchange of savanna (and possibly of xerophytic) species meridionally through the Amazon basin and latidudinally along the Caribbean coast. It may also have led to the formation of forest refuges, temporarily separating populations of forest animals and plants and leading to speciation in isolation. Much more pollen-analytical work in the tropical lowlands is needed before the extent of these phenomena can be ascertained.

From pollen diagrams of the younger Holocene of different environments it is clear that climatic change is continuously and cyclically acting on the tropical and sub-tropical ecosystems. Dry periods are known both from the tropical lowlands and the montane belts, e.g. dated *ca.* 5 000, 2 000 and 700 and 500 B.P., causing the replacement of open swamp vegetation by savanna, savanna woodland by

savanna, and some changes in the quantitative composition of montane forests. Some climatic cycles were calculated to have a length of *ca.* 250 years and no doubt shorter ones exist. These changes often had a considerable effect on different ecosystems and for management planning it is most important to take into account the amplitude of these changes, especially in climatically critical areas.

There are still many gaps in the knowledge of tropical America and it may be said that the only area reasonably well studied is the Colombian Eastern Cordillera for the only really long continuous sections, covering several millions of years, are from this area. Other long sections might be found e.g. Lago Valencia (Venezuela) and one of the larger Peruvian lakes. Main gaps in our knowledge are in the Peruvian, Ecuadorian and Venezuelan Andes and the entire area of tropical lowlands. Extensive research is needed, especially in the Amazonian rain forest, north-western Brazil and the area of the cerrados. Studies in the Colombian-Venezuelan-North Brazilian savanna areas will equally have to be continued and extended.

While pollen analysis can provide us with much information on the flora and vegetation, it is much more difficult to obtain direct information on the evolution and palaeogeography of the fauna. While palaeontological knowledge is slowly advancing, recent biogeographical studies may provide useful information for the palaeogeography of the tropical forest ecosystems.

Palaeogeography and palaeoclimatology of tropical Asia, Australia and Oceania

Pollen analytical data from this area are still scarce, but a few important publications may be expected soon. This is especially so for the work of Walkers' group in New Guinea and northeastern Australia. Factual data on the pre-Quaternary history of vegetation are, with some exceptions, still scarce. Little attention is given to the theories that explain present distribution patterns (e.g. by plate-tectonics) as there are as yet little or no factual palynological-palaeobotanical data to support them.

Pollen analysis and other data

India and surrounding areas

Vishnu-Mittre (1963, 1969) has commented on some general evolutionary aspects of the Indian flora. He mentions that the noticeable change in the floristic history of India is due to the gradual recession of tropical evergreen forests, the relics of which are found today in the Western Ghats and Assam, their change to semi-evergreen forest and their replacement by deciduous forest and savanna. This process took place after the Eocene period. More general data on the present day distribution of plants and animals (Mani, 1973), are in agreement with what is known on this history.

The more recent history of vegetation in Rajasthan (supported by a study of pollen rain in northwestern India) was investigated in a number of playa deposits (Singh, 1963, 1967, 1971, 1973; Singh *et al.*, 1972, 1973). Pollen analysis, stratigraphy and radio-carbon dating showed considerable changes of the environment which is now semi-arid to arid. During phase one, before 10 000 B.P. there were severe arid environments; the now stabilized sand dunes were active. About 10 000 B.P. the deposition of lacustrine sediments starts; the vegetation was an open steppe rich in grasses, *Artemisia* and sedges, and poor in halophytes. Species (like *Typha angustata*, *Mimosa rubicaulis*) which now grow in areas of higher rainfall appear to have flourished in the semi-arid belt. *Artemisia* and *Typha* even grew in what is now the arid belt. This suggests that a general westward shift of the rainfall belts took place. The influence of man on the vegetation starts as early as 9 500 B.P. Between 5 000 and 3 000 B.P. an increase in rainfall is suggested by the increase of swamp vegetation, the intensification of the vegetation cover inland and the pollen maxima of all mesophytic elements. This more humid period was interrupted by a short relatively drier interval between about 3 800 and 3 500 B.P. This short dry phase correlated with the decline of the Indus culture in northwestern India. 3 000 B.P. is the approximate date for the onset of aridity which seems to have been widespread. About 300 A.D. the climate ameliorated, this phase lasting to the present. In that time the Rangmahal culture flourished in Rajasthan. In comparing the history of Rajasthan with known data from Iran (Van Zeist, 1967), one may suppose that the early dry phase (before 10 000) corresponded to the Pleniglacial and late Glacial. The monsoon summer rains later apparently brought sufficient water to the area for the development of one of the oldest cultures of the world, but then they suddenly failed to do so. Such fluctuations would have had an equal influence on the forested areas, but no further data from the lowlands are available.

Further to the north, data on the history of montane forests are known from the Kashmir valley (Vishnu-Mittre *et al.*, 1962; Singh, 1963; Vishnu-Mittre, 1963, 1966; Vishnu-Mittre and Sharma, 1966; Sharma and Vishnu-Mittre, 1968). The lower Karewa deposits were apparently deposited near the Plio-Pleistocene boundary. After a period of *Quercus-Cedrus-Alnus* forests, a period with dominating *Pinus* forest or open vegetation might represent the first (or an early) glacial period. The postglacial history (last *ca.* 10 000 years) is better known from deposits in mires. Amongst the forest elements *Pinus* dominates in the early Holocene, but in a later warmer phase *Quercus* and *Ulmus* dominated the forest. In the last part of the Holocene, *Quercus* and *Ulmus* disappeared completely or almost completely, being replaced by *Pinus* and *Abies*.

It will be clear that data from the Quaternary of tropical and subtropical India are still very scarce. For a better understanding of the changing monsoon climate and its influence on vegetation and agriculture, a further search and study of sediments in the entire area seems most important. While some data are known from Iran, from the area east of India (from Bangladesh to Vietnam) virtually no pollen diagrams are available. Knowledge of the vegetation history of the last 15 000 years from at least a few places is urgently needed.

Taiwan

Tsukada and Rowley (1966) published one pollen diagram from Jih Tan in the central part of Taiwan, representing more than 50 000 years of vegetation history at an altitude of *ca.* 750 m. The site is in the lower part of the warm temperate forest (but only some 250 m above the altitudinal limit of the subtropical rain forest). The present warm temperate forest reaches 1 800 m, above which is the cool temperate forest and above 2 400 m boreal forest. Cool temperate forest (*Cyclobalanopsis*, *Quercus*, *Ulmus*, *Zelkova*, *Juglans*, *Carpinus*) dominated completely from *ca.* 35 000 to 12 000 B.P. Then there was an increase of warmer elements and *ca.* 10 000 B.P. these elements increased considerably. Somewhat later the subtropical elements tended to dominate and reached a maximum *ca.* 4 000 B.P. After that the subtropical elements (mainly *Liquidambar*) decrease somewhat but there seems to be human influence on the forest from considerably before that date.

During the Pleniglacial (between 35 000 and 12 000 B.P.) the average annual temperatures may have been between 2–6° C lower than today. Holocene hypsithermal temperatures may, however, have been 2–3° C higher than today. No finite dates are available from before 35 000 B.P. and the sediment in that lower part is not lake mud but mainly peat, indicating a lower lake-level. The lowest part is silty clay and shows dominance of *Pinus* and boreal elements when the annual temperature was no less than 8–11° C lower and winter temperature probably dropped to freezing point in lowland Taiwan. Tsukada thinks that this interval corresponds to the Tali glacial, which he supposes to be the counterpart of the early Wisconsin in North America (lower Pleniglacial in northwestern Europe). If this is true, then the lower Pleniglacial was much colder in Taiwan than the upper Pleniglacial. However, some caution is still warranted, as the strong changes in lithology might represent stratigraphic gaps. Further study in Taiwan is needed to solve this problem.

Indonesia and Malaysia

The biogeography of this area is strongly influenced by plate-tectonics, sea level changes and the formation and disappearance of mountains. Principal plates are the Asian (including the Sunda Shelf) and the Australian-New Guinean one (including the Sulu Shelf). The islands lying on either of these shelves were connected and with the mainland of Asia or Australia, during the Pleistocene glacial low sea-levels, so that direct migrations of plants and animals were possible.

The history of the first appearance of several recent families or genera during the Tertiary was studied by pollen analysis in Borneo (Germeraad *et al.*, 1968; Muller, 1972). In another pollen analytical study Muller (1964) contributes to the Tertiary history of the mangrove vegetation in Borneo. A gradual change of composition must have taken place. In the Eocene only the *Nypa* and *Brownlowia* types are found, suggesting that mangrove vegetation as we know it today was hardly developed. In the Oligocene *Rhizophora* appears, but only in the Miocene is evidence of *Sonneratia* and *Avicennia* found. Only then did the mangrove forest reach a degree of complexity comparable to the present one. Pollen analysis of a Miocene coal-bed showed the presence of a mixed swamp forest very similar to today's which was replaced by a mangrove forest.

Muller (1972) summarized what is known from palynological data on the pre-Quaternary history of the major vegetation units. Of the beach forest species, *Barringtonia* and *Casuarina* are found throughout the Tertiary, while *Thespesia* and *Hibiscus* appear in the Upper Miocene and Pliocene.

For the dipterocarp forest, *Dipterocarpus* and *Dryobalanops* were recognized in Tertiary sediments and *Lithocarpus/Castanopsis* pollen type is frequent throughout the Tertiary. Myrtaceae pollen which is rare in pre-Miocene deposits, shows a distinct rise in the Lower Miocene. Most of these elements, however, are not restricted to the dipterocarp forest. Ashton (1972) assumes that all major existing genera immigrated in Western Malaysia, initially in Late Tertiary time, followed by rapid dispersal through the main hill systems, while continental conditions prevailed. A first phase of diversification leading to the evolution of a large group of species of the mixed dipterocarp forest is postulated for the Early Pleistocene with a moister climate. This type is still preserved in the inland of Western Malaysia. At the end of this period a second episode of diversification took place into the lowland hills. A third episode might correspond to the late Middle Pleistocene high sea-levels (Ashton, 1972).

In the kerangas or heath forests, the pollen of *Dacrydium* and *Casuarina* and the spores of *Lycopodium cernuum* could represent this forest type. There is a marked increase of this group in the lower part of the Miocene (Muller, 1972).

Muller (1972) reports also short maxima of pollen of the palms *Arenga*, *Nenga* and *Eugeissona* in the Miocene, which probably indicate ecological disturbances in the lowland forests. Similar phenomena are shown by Compositae and grass pollen (Miocene and Pliocene), indicating probably local disturbances of the vegetation cover, possibly due to volcanic activities.

Muller (1966, 1972), also reports on montane pollen from the Tertiary of northwestern Borneo. He concludes that *Podocarpus* and *Phyllocladus* arrived in the montane area of Borneo from the east and not until the late Pliocene. In Oligocene and Miocene sediments, however, an asiatic montane element is present (*Pinus*, *Picea*, *Tsuga*, *Alnus*, *Ephedra*) that must have arrived in the nearby highlands from mainland Asia via the Formosa-Luzon track. The later disappearance of these elements from the area seems to be related to erosion lowering these mountains to below 1 500 m.

Anderson and Muller (1975) studied the Holocene sequence in a raised bog in the coastal area of northwestern Borneo. They found that the lower part of the sequence was formed in a mangrove forest with *Rhizophora*, *Sonneratia* and *Nypa*. In the course of the Holocene this vegetation developed into a mesotrophic mixed swamp forest and finally into an oligotrophic stunted forest.

The relation of recent pollen rain and vegetation was studied by Flenley (1973) on Mt. Kinabalu in Borneo. He is now working on the analysis of sediments from swamps and lakes in the Central Sumatran highlands (see also Morley *et al.*, 1973), Java and Malaysia. Radio-carbon dates are already available, the oldest being from Sumatra (*ca.* 11 000 B.P.). Interesting changes were established, some due to hydroseral changes, others possibly to forest clearing. Two of the sections that date back to the late Glacial and early Holocene, show an early dominance by *Myrica* pollen, but the significance of this is uncertain.

Samples from the Sunda Shelf, now below sea-level, are being studied by Wijmstra and Van der Hammen; forests prevailed there in a period that might correspond to the early Holocene.

Van Zeist (personal communication) has started the study of some middle to lower Pleistocene samples from Java.

Thus the study of vegetation history is making a good start in the area of Indonesia and Malaysia. Although very few data are available as yet from the Quaternary, soon we will know much more about the last 12 000 years. Longer sequences, sections from other islands, and study of earlier Quaternary sequences will be necessary in the near future.

Papua-New Guinea and Irian Jaya

Studies on the vegetation, pollen production and dissemination from this area were started by Walker and collaborators (Flenley, Powell, Hope). Very little has been published as yet (Walker, 1970; Flenley, 1972) and the following results are based on personal information (Walker).

In the area between Mt. Hagen town and Laiagan in the highlands of Papua-New Guinea, at altitudes from 1 500 to 4 400 m, the belts of montane forests are *Quercus* (*ca.* 1 500–2 400 m), beech (*ca.* 2 400–2 700 m), mixed montane (*Podocarpus*, *Pittosporum-Quintinia*, etc., *ca.* 2 700–3 300 m), subalpine (*ca.* 3 000–3 900 m). Subalpine grasslands may commence at *ca.* 3 200 m and alpine grassland at *ca.* 3 800–4 100 m.

One pollen diagram already extends to 30 000 B.P. Others are correlated with archaeological material and document the impact of agriculturists on the vegetation during the past 5 000 years. Another series of diagrams demonstrate the recovery of the vegetation on Mt. Wilhelm following the melting of its glaciers which began *ca.* 14 000 B.P.

The altitudinal forest limit seems to have been as low as 1 500 m *ca.* 30 000 B.P. The most active period of vegetation change since then was probably between *ca.* 15 000 and 6 000 B.P., corresponding with the main period of deglaciation. The forest limit probably rose to *ca.* 4 000 m by 8 000 B.P. and subsequently fell to its present level *ca.* 100 m lower. The cause of this fall is uncertain, particularly because human activity has now replaced more of the subalpine forest.

At *ca.* 1 500 m correlations between pollen diagrams and

archaeological remains of agriculture are established after 2 300 B.P. Agriculture certainly predates this, however, and the pollen diagram show evidence of disturbance of the forest back to *ca.* 5 000 B.P. At 2 500 m, near the present altitudinal limit for farming, human interference is much later.

If the forests of New Guinea were compressed below 1 500 m during the last glaciation, it is probable that the modern lowland forests are ecologically less stable than is usually assumed. The synthesis of the upland forests can be traced; on the present data they do not seem to have migrated as coherent units.

Hope has begun work on the mountains of Irian Jaya (specifically Mt. Cartensz) where lowland sites have already been selected by Walker. Garrett-Jones has begun work on the history of the forest-grassland margin in the lower Markham valley. It seems most important that this work be continued, especially on lowland sites, in order to test the theory of instability mentioned above.

Northeastern Australia

Australia has been isolated since its separation from the other southern continents, nevertheless the tropical and subtropical forests have affinities with New Guinea and Malaysia. An important boundary, especially to the spread of animals, is Wallace's line which is related to the zone of deep water between the Asian Shelf and the New Guinea-Australia Shelf.

Pollen analysis in this area has been carried out on sites in northeastern Queensland (Atherton-Tableland near Cairns); these were in rain forest before European settlement, but close to the rain forest/sclerophyll woodland boundary, on a steep rainfall gradient. Published diagrams comprise the last 10 000 years (Kershaw, 1970, 1971), but current work covers at least 100 000 years (Walker, personal communication).

The pollen diagrams from Lake Euramoo and Quincan show that dry sclerophyll forest dominated from *ca.* 10 000 B.P. to *ca.* 7 500 B.P. when it was replaced for a period of *ca.* 1 000 years by a warm temperate rain forest, probably as a result of an increase in temperature and possibly also of rainfall. Rain forest then remained at its maximum areal extent for 3 000–4 000 years when it changed to a subtropical kind with increase in temperature and possibly also precipitation. At *ca.* 2 000 B.P. a decrease in effective precipitation caused the subtropical rain forest to become drier and allowed a partial return of the sclerophyll forest (Kershaw, 1971).

A sequence from Lynch's Crater, covering 100 000 years or more, is being studied by Kershaw. It shows that the boundary between rain forest and sclerophyll woodland was to the east (i.e. coastward) of its present position, or perhaps that rain forest was limited to small refuges. Throughout Pleniglacial times sclerophyll vegetation was more widespread in the region than before or since. Before that the rain forest of the Atherton Tableland was floristically distinct from what is there at present, having something in common with more southern rain forests of today (Walker, personal communication).

This work is of considerable importance, especially as it demonstrates the instability of what has formerly been supposed to be an ancient and changeless pattern.

Hawaii

The Late Quaternary history of the Hawaiian vegetation was studied by Selling (1948), who also made an extensive survey of the pollen morphology (Selling, 1946, 1947). Peat from montane mires at altitudes between *ca.* 1 100 and *ca.* 1 800 m might represent the last *ca.* 10 000–12 000 years and show three main periods. During the first, which may correspond to the late Glacial or early Postglacial, rain forest was restricted and drier vegetation types more widespread, at least in the higher levels. The summit parts of West Maui and the corresponding levels in the other islands, were covered by subalpine forests; these changed to more humid types of vegetation towards the very end of the period. During the second period, which may correspond to the postglacial warm period (the Holocene hypsithermal), the rain forest expanded considerably and the dry subalpine forests retreated. A succession (*Cheirodendron→Myrsine →Metrosideros*) occurs, possibly corresponding to an increasingly pronounced anticyclonic precipitation (trade-wind rains). During the third period, which may correspond to the climatic deterioration of the late Holocene of other parts of the world (subatlantic of Europe, last *ca.* 3 000 years) there is a marked increase of *Chenopodium oahuense*. It seems that the zone for maximum rainfall was lower and on the higher levels drier vegetation types advanced. Towards the end of this period the inversion limit rose again, but probably to below its level during the second period. It is believed that the later climatic changes affected Polynesian life and migrations and Selling believes that there is a connection between the nadir of the climatic fluctuations in the third period and the end of the last great Polynesian migration.

Summary and conclusions

Data are still relatively rare for the areas treated in this section, but may be expected to increase quickly. A considerable depression of vegetation zones and a lower temperature before 10 000 B.P. may be deduced from the study in montane areas (Papua-New Guinea, Taiwan).

In Kashmir and Hawaii the data do not go back that far, but the early Holocene was cooler (and eventually drier) than during the later hypsithermal. A late Holocene lowering of the upper forest limit has sometimes been interpreted as a lowering of temperature and eventually rainfall; human influence on the vegetation was considerable in New Guinea from 5 000 B.P.

Drier conditions before *ca.* 7 000 B.P. caused a considerable displacement of forests in northeastern Australia; vegetations were more sclerophyllous during most of the Pleniglacial. In Rajasthan the climate was very dry before the Holocene, but a shift of climate (increase in rainfall) and vegetation brought circumstances favourable for the development of early agriculture.

The longest continuous sequences known today are from Taiwan and northeastern Australia.

Further pollen analytical study in the entire area is urgently needed, especially in those countries from where no data have been obtained. It seems important to complete our knowledge from areas where much work has been done in the montane zones (like Papua-New Guinea), with data from the lowlands, so as to obtain a better understanding of the phenomena of climatic and vegetational change.

Conclusions: research needs and priorities

There is a clear general conclusion from what actually is known about the palaeogeography and palaeoclimatology of the tropical and subtropical forest ecosystems and grasslands: they are not stable for thousands of years. This is mainly due to the continuously changing climate. These changes may or may not be periodic. Large areas of the tropics were subjected to a much drier (and somewhat cooler) climate between 20 000 and 12 500 B.P. Some parts of the subtropics seem, however, to have had a more humid climate than today during that interval. The last 12 500 years were often more humid, but with rather marked drier phases locally. In parts of the dry subtropics part of this interval may have been considerably more humid than today. These changes of climate caused drastic changes of the vegetation, replacing evergreen forest and drier woodlands by savannas and deserts and vice versa. These continuous changes make the separate treatment of the palaeogeography of evergreen forest, drier woodlands and grasslands unmeaningful.

It is the belief of most ecologists that the rain forest is a stable community, at least over periods of 50–500 years and with respect to non-human perturbations. The evidence for this belief was not examined in detail, but the existence of descriptive vegetational evidence for such stability could be questioned, at least over intervals of hundreds of years. The pollen diagrams show no indications of rain forest communities that have been stable for periods as long as 500 years. Long records are not available from the central part of any large area of rain forest, but all the evidence of palaeogeography indicates that the tropical rain forest, at least around its periphery, is as sensitive to the continual changes of Quaternary climate as any other forest community.

The present knowledge of the history of tropical forest ecosystems is still small, but it has demonstrated their instability under changing environmental conditions. Data are still very scarce or absent over large areas and somewhat more abundant data are mainly concentrated in a few areas: tropical Andean Colombia, East African uplands and highlands, and Papua-New Guinean highlands.

More general information is needed, especially from pollen analytical and palaeolimnological studies, of all the remaining tropical areas, to show the extension of vegetation and the amplitude of climatic change in the different areas. This will help the understanding of climatic change in the tropics and indicate approximate limits of the resulting vegetational change. It will also help the understanding of the vegetational change in tropical forest ecosystems resulting from soil development or degradation.

To accomplish this, regional collections of recent pollen are needed. Larger herbaria are needed to provide the raw material and because of the time involved in the building-up of such collections, regional collaboration is highly desirable. Pollen slides might be produced near one or more of the larger herbaria and distributed to collaborating national centres, or they might be produced in every centre and an international exchange of slides organized. In order to do the work as economically as possible, the material for pollen and other analysis in each area or country should be carefully selected after a period of extensive search for suitable series of sediments.

While most of this work will have to cover the last 15 000–20 000 years, effort should be made to collect a few promising very old sections, in order to determine the period of the long-term climatic changes. Very deep ones were recovered near Bogotá and others are known from the valley of Mexico and northeastern Australia. In America it seems desirable to obtain such cores from a Peruvian Andean lake, from Lago Valencia (Venezuela) and a new one from Mexico. It is not certain if the Amazon basin could provide such a sequence, but it should be looked for. In Africa, there are abundant possibilities for a very long core: lakes Tanganyika, Nyassa, Mobutu, Meru and Bosumtwi among others. A promising site is badly needed in tropical Asia.

Collection of deep cores is more expensive than the usual hand borings and needs special equipment.

There seems to be a special need for knowledge of vegetation history in areas with less disturbed ecosystems, like the Amazon or the Congo basins, that may suffer increasing disturbance or exploitation in the near future. Other areas of special need for information are climatically critical areas (northwestern India, Sahel countries, northeastern Brazil, areas of tropical deciduous forests, etc.) and hilly mountainous areas (the Andes, Indonesian mountains and volcanoes, East African uplands and mountains, etc.), especially when these combine with climatically critical circumstances, where erosion may become important.

For purposes of conservation and the proper management of tropical forest ecosystems, integrated studies on vegetation, animal communities, ecology, hydrology, climate and history will be of great importance. It seems highly advisable to start these integrated studies in a few pilot areas, where the basic knowledge already available seems to justify this more detailed work and where a wide variety of vegetation types can be found in a relatively small area. Three of these pilot areas might be selected, one in each of the main areas. In America this could be the Colombian Northern Andes and the adjacing lowland (with sufficient general information on vegetation history, an increasing knowledge of vegetation and soils, and an incredible variety of environments and vegetation types). Drafts for such a project have already been made (as a collaboration of the Universidad Nacional in Bogotá with the Universities of Amsterdam and Utrecht). In Africa this could be an area from the East African Mountains westward over the Rift Valley to the Congo basin, and northward to the Sahel area. In the southeastern Asian-Australian area it could be New Guinea or Sumatra.

In these pilot areas palynologists and palaeoecologists should work with specialists in many other fields on the same ecosystems. These integrated studies in pilot areas will give much fundamental knowledge and insight that should be applicable to wide areas of the same continent.

The first effort should be in areas that are climatically critical, like rather steep rainfall gradients. Large virgin areas about to be exploited should also have priority. Also of high priority should be areas where the relief and climate are critical for eroding soils. It should be kept in mind, however, that palynological (and palaeolimnological) research depends on the presence of suitable sediments and that, if these are lacking in the critical areas, one should go to the nearest suitable sites.

Bibliography

In addition to the papers listed below, the reader is referred to detailed research summaries and bibliographies presented in: *Palaeoecology of Africa* (E. M. van Zinderen Bakker, ed.). Cape Town, Balkema (volumes 5 and 8 deal with Antarctica and adjacent regions rather than Africa). Another series of publications is: *The Quaternary of Colombia* (T. van der Hammen, ed.), Amsterdam.

*ANDERSON, J. A. R.; MULLER, J. Palynological study of a Holocene peat and a Miocene coal deposit in N. W. Borneo. *Rev. Palaeob. Palyn.*, vol. 19, 1975.

ASHTON, P. S. The Quaternary geomorphological history of western Malesia and lowland forest phytogeography. In: Ashton, P. S. and M. (eds.). *The Quaternary era in Malesia*, p. 35–49. Transact. 2nd Aberdeen-Hull Symposium on Malesian Ecology, Aberdeen 1971. University of Hull, Dept. of Geography, Misc. series no. 13, 1972, 122 p.

AUBRÉVILLE, A. *Contribution à la paléohistoire des forêts de l'Afrique tropicale.* Paris, Soc. Éd. géographiques, maritimes et coloniales, 1949, 98 p.

——. Savanisation tropicale et glaciations quaternaires. *Adansonia*, 2, 1962, p. 16–84.

——. Les origines des Angiospermes. *Adansonia*, 14, 1974, p. 5–27 et p. 145–198.

BAMPO, S. O. Kumasi conference on the Lake Bosumtwi crater. *Nature*, 198, 1963, p. 1150–1151.

*BARTLETT, A. S.; BARGHOORN, E. S. Phytogeographic history of the Isthmus of Panama during the past 12 000 years. In: Graham, A. (ed.). *Vegetation and vegetation history of northern Latin America*, p. 203–299. Amsterdam, London, New York, Elsevier, 1973.

BARTSTRA, G. J.; CASPARIE, W. A. *Modern quaternary research in South-East Asia.* Rotterdam, Balkema, 1975, 86 p.

BEUCHER, F. Une flore d'âge Ougartien (seconde partie du Quaternaire Moyen) dans les Monts d'Ougarta (Sahara nord-occidental). *Rev. Palaeob. Palyn.*, 2, 1967, p. 291–300.

——. *Étude palynologique de formations néogènes et quaternaires au Sahara nord-occidental.* Paris, Faculté des sciences, thèse, 1971, 796 p. + 23 pl.

BIGARELLA, J. J.; ANDRADE, G. G. Contribution to the study of the Brazilian Quaternary. *Geol. Soc. America*, special paper 84, 1965, p. 433–451.

BISHOP, W. W. The later Tertiary in East Africa volcanics, sediments and faunal inventory. In: Bishop, W. W.; Clark, J. D. (eds.). *Background to evolution in Africa*, p. 31–56. Chicago, Univ. of Chicago Press, 1967.

BOLICK, M. R. *A vegetational history of the Mt. Meru Lahar, Tanzania.* Durham, N. C., USA, Duke University, Zoology Dept., M.A. thesis, 1974, VII+96 p.

BOLTENHAGEN, E. Spores et pollen du Crétacé supérieur du Gabon. *Pollen et Spores*, vol. 9, n° 2, 1967, p. 335–355.

BONNEFILLE, R. Analyse pollinique d'un sédiment récent : vases actuelles de la Rivière Aouache (Éthiopie). *Pollen et Spores*, 11, 1969a, p. 7–16.

——. Indication sur la paléoflore d'un niveau du Quaternaire moyen du site de Melka Kontouré (Éthiopie). *C.R. Soc. Géol. Fr.*, 7, 1969b, p. 238–239.

——. Premiers résultats concernant l'analyse pollinique d'échantillons du Pléistocène inférieur de l'Omo (Éthiopie). *C.R. Acad. Sci. Paris*, 270, 1970, p. 2430–2433.

——. Atlas des pollens d'Éthiopie. *Adansonia*, 2, 1971a, p. 463–518.

——. Atlas des pollens d'Éthiopie, principales espèces des forêts de montagne. *Pollen et Spores*, 13, 1971b, p. 15–72.

——. *Associations polliniques actuelles et quaternaires en Éthiopie (vallées de l'Awash et de l'Omo).* Paris, Faculté des sciences, thèse, 1972, 513 p.

BROOKS, J. L. Speciation in ancient lakes. *Quart. Rev. Biol.*, 25, 1950, p. 30–60, p. 131–176.

*BUTZER, K. W.; ISAAC, G. L.; RICHARDSON, J. L.; WASHBOURN-KAMAU, C. Radiocarbon dating of East African lake levels. *Science*, 175, 1972, p. 1069–1076.

CAHEN, L. *Géologie du Congo Belge.* Liège, Vaillant-Carmanne, 1954, 577 p.

CLARK, J. D.; VAN ZINDEREN BAKKER, E. M. Prehistoric culture and Pleistocene vegetation at the Kalambo Falls, Northern Rhodesia. *Nature*, 201, 1964, p. 971–975.

CLISBY, K. H.; SEARS, P. B. Palynology in southern North America. Part III. Microfossil profiles under Mexico City correlated with the sedimentary profiles. *Bull. Geol. Soc. Amer.*, 66, 1955, p. 511–520.

COETZEE, J. A. Evidence for a considerable depression of the vegetation belts during the Upper Pleistocene on the East African mountains. *Nature*, 204, 1964, p. 564–566.

*——. Pollen-analytical studies in East and Southern Africa. In: Van Zinderen Bakker, E. M. (ed.). *Palaeoecology of Africa*. Cape Town, Balkema, 1967, 146 p.

COLE, S. *The prehistory of East Africa.* London, Weidenfeld and Nicolson, 1963, 383 p.

COOKE, H. B. S. Pleistocene mammal faunas of Africa, with particular reference to Southern Africa. In: Howell and Bourlière (eds.). *African ecology and human evolution*, p. 65–115. Viking Publ. in Anthropology, 1964.

COWGILL, U. M.; GOULDEN, C. E.; HUTCHINSON, G. E.; PATRICK, R.; RAČEK, A.; TSUKADA, M. *The history of Laguna de Petenxil.* Conn. Acad. Arts Science, Mem. 17, 1966, 126 p.

——; HUTCHINSON, G. E. La Aguada de Santa Ana Vieja (The history of a pond in Guatemala). *Arch. Hydrobiol.*, vol. 62, no. 3, 1966, p. 355–372.

* Major reference.

CROIZAT, L. L'âge des Angiospermes en général et de quelques Angiospermes en particulier (considérations sur l'âge des Angiospermes). *Adansonia*, 6, 1966, p. 239.

DAMUTH, J. E.; FAIRBRIDGE, R. W. Equatorial Atlantic deep-sea arkosic sands and ice-age aridity in tropical South America. *Geol. Soc. Am. Bull.*, 81, 1970, p. 189–206.

DAVIS, M. B. On the theory of pollen analysis. *Amer. J. Sci.*, 261, 1963, p. 897–912.

——; GOODLETT, J. C. Comparison of the present vegetation with pollen spectra in surface samples from Brownington Pond, Vermont. *Ecology*, 41, 1960, p. 346–357.

——; DEEVEY, E. S. Jr. Pollen accumulation rates: estimates from late-glacial sediment of Rogers Lake. *Science*, 145, 1964, p. 1293–1295.

DAVIS, T. A. W. On the island origin of the endemic trees of the British Guiana peneplain. *J. Ecol.*, 29, 1941, p. 1–13.

DEGENS, E. T.; VON HERZEN, R. P.; HOW-KIN WONG. Lake Tanganyika: water chemistry, sediments geological structure. *Naturwissenschaften*, 58, 1971, p. 229–240.

——; VON HERZEN, R. P.; HOW-KIN WONG; DEUSER, W. G.; JANNASCH, H. W. Lake Kivu: structure, chemistry and biology of an East African Rift Lake. *Geologischen Rundschau*, 62, 1973, p. 245–277.

DUBOIS, G.; DUBOIS, C. Caractères micropaléobotaniques d'une tourbe du Togo. *C.R. Acad. Sci. Paris*, 208, 1939, p. 1421–1422.

——; ——; JAEGER, P. Sol tourbeux d'*Eriospora* dans les monts Loma en Afrique occidentale. *C.R. Acad. Sci. Paris*, 227, 1948, p. 217–218.

EDMONDSON, W. T. Cultural eutrophication with special reference to Lake Washington. *Mitt. Internat. Verein. Limnol.*, 17, 1969, p. 19–32.

ERDTMAN, G. *Pollen morphology and plant taxonomy. I. Angiosperms*. New York, Haffner, 1966, 553 p.

——. *Pollen and spore morphology/plant taxonomy. II. Gymnospermae, Pteridophyta, Bryophyta* (illustrations). New York, Ronald, and Stockholm, Almqvist and Wiksell, 1957, 151 p.

——. *Pollen and spore morphology/plant taxonomy. III. Gymnospermae, Bryophyta* (text). New York, Ronald, and Stockholm, Almqvist and Wiksell, 1965, 191 p.

——; SORSA, P. *Pollen and spore morphology/plant taxonomy. IV. Pteridophyta* (text and additional illustrations). Stockholm, Almqvist and Wiksell, 1971, 302 p.

*FAURE, H. Les lacs quaternaires du Sahara. *Mitt. Internat. Verein. Limnol.*, 17, 1969, p. 131–146.

FITTKAU, E. J.; ILLIES, J.; KLINGE, H.; SCHWABE, G. H.; SIOLI, H. *Biogeography and ecology in South America*. Vol. 1 and 2. The Hague, Junk, 1968, 1969, 946 p.

FLENLEY, J. R. Evidence of Quaternary vegetational change in New Guinea. In: Ashton, P. S. and M. (eds.). *The Quaternary era in Malesia*, p. 99–120. University of Hull, Dept. of Geography, Misc. series no. 13, 1972, 122 p.

——. The use of modern pollen rain samples in the study of the vegetational history of tropical regions. In: Birks, H. J. B.; West, R. G. (eds.). *Quaternary plant ecology* (14th Symp. Brit. Ecol. Soc.), p. 131–141. London, Blackwell, 1973.

*——; WALKER, D.; WILLIAMS, J. Vegetation history of the Wabeg region, New Guinea Highlands. *New Phytol.*, 72, 1973.

FLORIN, R. The distribution of Conifer and Taxad genera in time and space. *Acta Horti Bergiani*, 20, 1963, p. 121–312.

FOREMAN, F. Palynology in southern North America. Part II. Study of two cores from lake sediments of the Mexico City Basin. *Bull. Geol. Soc. Amer.*, 66, 1955, p. 475–510.

FREY, D. G. Remains of animals in Quaternary lake and bog sediments and their interpretation. *Ergebnisse der Limnologie*, vol. 2, no. 1–2, 1964, p. 1–114.

*GERMERAAD, J. H.; HOPPING, C. A.; MULLER, J. Palynology of Tertiary sediments from tropical areas. *Rev. Palaeob. Palyn.*, 6, 1968, p. 189–348.

GONZÁLEZ, A. E. *A palynological study on the Upper Los Cuervos and Mirador Formations (Lower and Middle Eocene; Tibú area, Colombia)*. Thesis, Amsterdam. Leiden, J. Brill, 1967, 68 p.

——; VAN DER HAMMEN, T.; FLINT, R. F. Late Quaternary glacial and vegetational sequence in Valle de Lagunillas, Sierra Nevada del Cocuy, Colombia. *Leidse Geol. Meded.*, 32, 1965, p. 157–182.

*GRAHAM, A. (ed.). *Vegetation and vegetational history of northern Latin America*. Amsterdam, London, New York, Elsevier, 1973, 393 p.

GUERS, J.; LOBREAU-CALLEN, D.; DIMON, M. T.; MALEY, J.; CAMBON-BOU, G. Palynologie africaine, IX. *Bull. IFAN*, vol. 33, sér. A, 1971, pl. 215–238.

GUINET, P. *Les Mimosacées, étude de palynologie fondamentale, corrélations, évolution*. Travaux Section Scientifique et Technique, IX, 1969, 293 p.

*HAFFER, J. Speciation in Amazonian forest birds. *Science*, 165, 1969, p. 131–137.

*HAMILTON, A. C. The interpretation of pollen diagrams from highland Uganda. In: Van Zinderen Bakker, E. M. (ed.). *Palaeoecology of Africa, the surrounding islands and Antarctica*, vol. 7, 1972, p. 45–149.

HASTENRATH, S. Recent climatic fluctuations in the Central American area and some geo-ecological effects. In: Troll, C. (ed.). *Geo-ecology of the mountainous regions of the tropical Americas* (Colloquium Geographicum 9), p. 131–138. Bonn, 1968.

HAY, R. L. Stratigraphy of Beds I through IV, Olduvai Gorge, Tanganyika. *Science*, 139, 1963, p. 829–833.

——. Zeolites and zeolitic reactions in sedimentary rocks. *Geol. Soc. Amer. Spec. Pap.*, 85, 1966, 130 p.

HECKY, R. E. *The paleolimnology of the alkaline, saline Lakes on the Mt. Meru Lahar*. Ph. D. thesis, Duke University, 1971, 209 p.

——; DEGENS, E. T. Late Pleistocene-Holocene chemical stratigraphy and paleolimnology of the Rift Valley lakes of Central Africa. Technical Report WHOI-73-28, Woods Hole Oceanogr. Inst., 1973, 93 p. (unpublished manuscript).

——; KILHAM, P. Diatoms in alkaline saline lakes: ecology and geochemical implications. *Limnol. and Oceanogr.*, 18, 1973, p. 53–71.

——; MOPPER, K.; KILHAM, P.; DEGENS, E. T. The amino acid and sugar composition of diatom cell-walls. *Marine Biol.*, 19, 1973, p. 323–331.

HEDBERG, O. A pollen analytical reconnaissance in tropical East Africa. *Oikos*, 5, 1954, p. 137–166.

HEIM, M. R. Quelques pollens pléistocènes nouveaux pour le Hoggar. *C.R. Acad. Sci. Paris*, 258, 1964, p. 1297–1299.

HERNGREEN, G. W. F. Some new pollen grains from the Upper Senonian of Brazil. *Pollen et Spores*, vol. 14, n° 1, 1972, p. 97–112.

——. Middle Cretaceous palynomorphs from north-eastern Brazil. *Sci. Geol. Bull.* (Strasbourg), 27, no. 1–2, 1974, p. 101–116.

HEUSSER, C. J. Late-Pleistocene pollen diagrams from the province of Llanquihue, Southern Chile. *Proc. Amer. Philos. Soc.*, vol. 110, no. 4, 1966, p. 269–305.

——. *Pollen and spores of Chile*. Tucson, Univ. Arizona Press, 1971, 167 p.

HEUSSER, C. J. On the occurrence of *Lycopodium fuegianum* during Late-Pleistocene interstades in the province of Osorno, Chile. *Bull. Torrey Bot. Club*, vol. 99, no. 4, 1972, p. 178–184.

HOEKEN-KLINKENBERG, P. J. M. Van. A palynological investigation of some Upper Cretaceous sediments in Nigeria. *Pollen et Spores*, vol. 6, n° 1, 1964, p. 209–231.

——. Maestrichtian, Paleocene and Eocene pollen and spores from Nigeria. *Leidse Geol. Meded.*, 38, 1966, p. 37–48.

HULSHOF, O. K.; MANTEN, A. A. Bibliography of actuopalynology, 1671–1966. *Rev. Palaeob. Palyn.*, 12, 1971, p. 1–243.

HUTCHINSON, G. E. *A treatise on limnology*, vol. 1. New York, Wiley, 1957, 1 015 p.

JARDINÉ, S. Spores à expansions en forme d'élatères du Crétacé moyen d'Afrique occidentale. *Rev. Palaeobot. Palyn.*, vol. 1, n° 1–4, 1967, p. 235–258.

——; MAGLOIRE, L. Palynologie et stratigraphie du Crétacé des bassins du Sénégal et de Côte d'Ivoire. *Mém. BRGM*, 32, 1965, p. 187–245.

KELLOGG, W. W.; SCHNEIDER, S. H. Climate stabilization: for better or for worse? *Science*, vol. 186, no. 4170, 1974, p. 1163–1173.

*KENDALL, R. L. An ecological history of the Lake Victoria basin. *Ecol. Monogr.*, 39, 1969, p. 121–176.

KERAUDREN-AYMONIN, M.; STRAKA, H.; SIMON, A. Palynologia Madagassica et Mascarenica, Fam. 184–188. *Pollen et Spores*, 11, 1969, p. 299–332.

KERSHAW, A. P. A pollen diagram from Lake Euramoo, north-east Queensland, Australia. *New Phytol.*, 69, 1970, p. 785–805.

*——. A pollen diagram from Quincan Crater, north-east Queensland, Australia. *New Phytol.*, 70, 1971, p. 669–681.

KILHAM, P. A hypothesis concerning silica and the freshwater planktonic diatoms. *Limnol. and Oceanogr.*, 16, 1971, p. 10–18.

——; KOPCZYNSKA, E. E. African lake types and their specific diatom associations (abstract). In: *37th Meeting of the Amer. Soc. Limnol. Oceanogr.*, Seattle, Wash., 1974.

LAEYENDECKER-ROOSENBURG, D. M. A palynological investigation of some archaeologically interesting sections in north-western Surinam. *Leidse Geol. Meded.*, 38, 1966, p. 31–36.

LAWTON, R. M. Palaeoecological and ecological studies in the northern Province of Northern Rhodesia. *Kirkia*, 3, 1963, p. 46–47.

——. A vegetation survey of Northern Zambia. *Palaeoecology of Africa*, 6, 1972, p. 253–256.

LIVINGSTONE, D. A. Age of deglaciation in the Ruwenzori Range, Uganda. *Nature*, 194, 1962, p. 859–860.

——. Postglacial vegetation of the Ruwenzori Mountains in equatorial Africa. *Ecol. Monogr.*, 37, 1967, p. 25–52.

——. Some interstadial and post-glacial pollen diagrams from eastern Canada. *Ecol. Monogr.*, 38, 1968, p. 87–125.

——. A 22 000-year pollen record from the plateau of Zambia. *Limnol. and Oceanogr.*, 16, 1971, p. 349–356.

——. The Nile. Paleolimnology of headwaters. In: Rzoska Julian (ed.). *Biology of the Nile*, in press.

LOBREAU, D.; GUERS, J.; ASSEMIEN, P.; BOU, G.; GUINET, P.; POTIER, L. Palynologie africaine, IX. *Bull. IFAN*, 31, 1969, pl. 167–190.

LOBREAU-CALLEN, D. Pollen des Icacinaceae. I. Atlas (1). *Pollen et Spores*, 14, 1972, p. 345–388.

LOEFFLER, H. Tropical high mountain lakes. In: Troll, C. (ed.). *Geo-ecology of the mountainous regions of the tropical Americas* (Coll. Geogr. 9), 1968, p. 57–75.

MALEY, J. Contributions à l'étude du Bassin tchadien. Atlas de pollens du Tchad. *Bull. Jard. Bot. Belg.*, 40, 1970, p. 29–48.

*MALEY, J. La sédimentation pollinique actuelle dans la zone du lac Tchad (Afrique Centrale). *Pollen et Spores*, 14, 1972, p. 263–307.

MANI, M. S. *Ecology and biogeography in India*. Monographiae Biologicae, 23. The Hague, Junk, 1973, 774 p.

MARTIN, A. R. H. Some possible materials for pollen analysis in South Africa. *S. Afr. J. Sci.*, 50, 1953, p. 83–88.

*——. Pollen analysis of Groenvlei lake sediments Knysna (South Africa). *Rev. Palaeob. Palyn.*, 7, 1968, p. 107–144.

*MARTIN, P. S. Paleoclimatology and a tropical pollen profile, 1964. *Rep. VIth Int. Congr. Quat.* (Warsaw), 1961, 2, p. 319–323.

*——; MEHRINGER, P. J. Pleistocene pollen analysis and biogeography of the Southwest. In: Wright, H. E ; Frey, D. G. (eds.). *The Quaternary of the United States*, p. 433–451. Princeton Univ. Press, 1965.

McNEISH, R. S. Lack of correlation of glacial sequences in South America with those in Northern Hemisphere. *Abstr. of papers at Soc. Amer. Archaeol. meetings*, 31, 1971.

MOREAU, R. E. Pleistocene climatic changes and the distribution of life in East Africa. *J. Ecol.*, 21, 1933, p. 415–435.

*——. *The bird faunas of Africa and its islands*. New York, London, Academic Press, 1966, 424 p.

MORLEY, R. J.; FLENLEY, J. R.; KARDIN, M. K. Preliminary notes on the stratigraphy and vegetation of the swamps and small lakes of the Central Sumatran Highlands. *Berata Kajian Sumatera* (Sum. Res. Bull.), vol. 2, no. 2, 1973, p. 50–60.

MORRISON, M. E. S. Pollen analysis in Uganda. *Nature*, 190, 1961, p. 483–486.

*——. Low-latitude vegetation history with special reference to Africa. *Roy. Meteorol. Soc. Proc. Int. Symp. on World Climate from 8 000 to 0 B.C.*, 1966, p. 142–148.

*——. Vegetation and climate in the uplands of south-western Uganda during the later Pleistocene period. I. Muchova Swamp, Kigezi District. *J. Ecol.*, 56, 1968, p. 363–384.

——; HAMILTON, A. C. Vegetation and climate in the uplands of south-western Uganda during the later Pleistocene period. II. Forest clearance and other vegetational changes in the Rukiga highlands during the last 8 000 years. *J. Ecol.*, 62, 1974, p. 1–32.

MULLER, J. Palynology of recent Orinoco delta and shelf sediments. *Micropaleontology*, vol. 5, no. 11, 1959, p. 1–32.

——. A palynological contribution to the history of the mangrove vegetation in Borneo. In: *Ancient Pacific Floras*, p. 33–42. Univ. of Hawaii Press, 1964.

——. Montane pollen from the Tertiary of N.W. Borneo. *Blumea*, 14, no. 1, 1966, p. 231–235.

——. Palynological evidence for change in geomorphology, climate and vegetation in the Mio-Pliocene of Malesia. In: Ashton, P. S. and M. (eds.). *The Quaternary era in Malesia*, p. 6–16. Transact. 2nd Aberdeen-Hull Symposium on Malesian Ecology, Aberdeen 1971. University of Hull, Dept. of Geography, Misc. series no. 13, 1972, 122 p.

MURILLO, M. T.; BLESS, M. J. M. Spores of recent Colombian Pteridophyta. I. Trilete spores. *Rev. Palaeob. Palyn.*, 18, 1974, p. 223–269.

NAIR, P. K. K. Pollen grains of Indian plants, 1. *Bull. Natl. Bot. Gardens (India)*, 53, 1961, p. 1–35.

——. Pollen grains of Indian plants, 2. Cochlospermaceae, Moringaceae, Phytolaccaceae, Portulacaceae, Tamaricaceae. *Bull. Natl. Bot. Gardens (India)*, 63, 1962a, p. 9–30.

——. Pollen grains of Indian plants, 3. Malvaceae, Bombacaceae. *Bull. Natl. Bot. Gardens (India)*, 63, 1962b.

NAIR, P. K. K.; REHMAN, K. Pollen grains of Indian plants, 5. Verbenaceae. *Bull. Natl. Bot. Gardens (India)*, 76, 1962, p. 1–23.

——; ——. Pollen grains of Indian plants, 6. Convolvulaceae. *Bull. Natl. Bot. Gardens (India)*, 83, 1963, p. 1–16.

——; KHAN, H. A. Pollen grains of Indian plants, 7. Nyctaginaceae. *Bull. Natl. Bot. Gardens (India)*, 111, 1965, p. 1–13.

——; RASTOGI, K. Pollen grains of Indian plants, 8. Amaranthaceae. *Bull. Natl. Bot. Gardens (India)*, 118, 1966, p. 1–18.

NEWMAN, J. E.; PICKETT, R. C. World climates and food supply variations. *Science*, vol. 186, no. 4167, 1974, p. 877–881.

OGDEN, J. G. Correlation of contemporary and late Pleistocene pollen records in the reconstruction of Postglacial environments in northeastern North America. In: Frey, D. G. (ed.). *Symposium on Paleolimnology* (Mitteilungen Internationale Vereinigung für Theoretische und Angewandte Limnologie), 17, 1969, p. 64–71.

OSMASTON, H. A. *Pollen analysis in the study of the past vegetation and climate of Ruwenzori and its neighbourhood.* Oxford, B. Sc. thesis, 1958, 44 p.

*——. *The past and present climate and vegetation of Ruwenzori and its neighbourhood.* Oxford, Ph. D. thesis, 1965, unpublished.

*PLOEY, J. de. Quelques indices sur l'évolution morphologique et paléoclimatique des environs du Stanley Pool (Congo). *Studia Univ. Lovanium, Fac. des Sci.*, 1963, p. 1–16.

——. Position géomorphologique, genèse et chronologie de certains dépôts superficiels au Congo occicental. *Quaternaria.* VII, 1965, p. 131–154.

PUNT, W. Pollen analysis of the Euphorbiaceae with special reference to taxonomy. *Wentia*, 7, 1962, p. 1–116.

*QUÉZEL, P. Flore et palynologie sahariennes. Quelques aspects de leur signification biogéographique et paléoclimatique. *Bull. IFAN*, 22 A, 1960, p. 353–359.

——. A propos de l'olivier de Laperrine de l'Adrar. *Missions Berliet Ténéré-Tchad* (Paris), 1962, p. 329–332.

*——; MARTINEZ, M. C. Le dernier interpluvial au Sahara central. Essai de chronologie palynologique et paléoclimatique. *Libyca*, 6-7, 1961, p. 211–227.

——; ——. Premiers résultats de l'analyse palynologique de sédiments recueillis au Sahara méridional à l'occasion de la mission Berliet-Tchad. *Missions Berliet Ténéré-Tchad* (Paris), 1962, p. 313-327.

RICHARDSON, J. L. Plankton and fossil plankton studies in certain East African lakes. *Verh. Int. Ver. Limnol.*, 15, 1964, p. 993–999.

——. Changes in level of Lake Naivasha during postglacial time. *Nature*, 209, 1966, p. 290–291.

——. Diatoms and lake typology in East and Central Africa. *Int. Rev. Gesamten Hydrobiol. Hydrogr.*, 53, 1968, p. 299–338.

*——; RICHARDSON, A. E. History of an African Rift Lake and its climatic implications. *Ecol. Monogr.*, 42, 1972, p. 499–534.

SAAD, S. I. Studies of pollen and spores content of Nile delta deposits (Berendal region). *Pollen et Spores*, 9, 1967, p. 467–503.

SALGADO LABOURIAU, M. L. *Contribuçao à palinologia dos Cerrados.* Rio de Janeiro, Acad. Brasil. Cienc., 1973, 291 p.

*SCHALKE, H. J. W. G. The upper Quaternary of the Cape Flats area (Cape Province, South Africa). *Scripta Geologica*, 15, 1973, p. 1–57.

SCHNELL, R. *Le problème des homologies phytogéographiques entre l'Afrique et l'Amérique tropicales.* Mém. Mus. nat. Hist. nat., N. sér. B, Botanique, vol. XI, n° 2, 1961, 104 p.

SCHULZ, E. Pollenanalytische Untersuchungen quartärer Sedimente des Nordwest-Tibesti. In: *Forschungsstation Barda-Fu-Geologen in der Zentral Sahara (Pressedienst Wisseni schaft FU Berlin)*, 5, 1974, p. 59–69.

SEARS, P. B.; CLISBY, K. H. Palynology in southern North America. Part IV. Pleistocene climate in Mexico. *Bull. Geol. Soc. Amer.*, 66, 1955, p. 521–530.

SELLING, O. H. Studies in Hawaiian pollen statistics. Part I. The spores of the Hawaiian Pteridophytes. *Bishop Museum, special publ.*, 37, 1946, 87 p.

——. Studies in Hawaiian pollen statistics. Part II. The pollens of the Hawaiian Phanerogams. *Bishop Museum, special publ.*, 38, 1947, 430 p.

*——. Studies in Hawaiian pollen statistics. Part III. On the late Quaternary history of the Hawaiian vegetation. *Bishop Museum, special publ.*, 39, 1948, 154 p.

*SERVANT, M.; SERVANT, S. Les formations lacustres et les diatomées du Quaternaire récent du fond de la cuvette tchadienne. *Rev. Geogr. phys. et Géol. dynam.*, vol. 12, n° 2, 1970, p. 63–76.

SHARMA, B. D.; VISHNU-MITTRE. Studies of post-glacial vegetational history from the Kashmir valley. 2. Baba Rishi and Yus Maidan. *Palaeobot.*, vol. 17, no. 3, 1968, p. 231–243.

*SIMPSON, B. Pleistocene changes in the fauna and flora of South America. *Science*, 173, 1971, p. 771–780.

SINGH, G. A preliminary survey of the post-glacial vegetational history of the Kashmir valley. *Palaeobot.*, vol. 12, no. 1, 1963, p. 73–108.

——. A palynological approach towards the resolution of some important desert problems in Rajasthan. *Indian Geohydrology*, 3, 1967, p. 111–128.

——. The Indus valley culture seen in the context of post-glacial climatic and ecological studies in northwest India. *Archaeology and Physical Anthropology in Oceania*, 6, 1971, p. 177–189.

——. Late Quaternary changes in vegetation and climate in the arid tropics of India. In: *Abstracts Ninth Cong. Int. Union Quat. Res.*, 1973, 9, 332 p.

*——; JOSHI, R. D.; SINGH, A. B. Stratigraphic and radiocarbon evidence for the age and development of three salt lake deposits in Rajasthan, India. *Quaternary Research*, vol. 2, no. 4, 1972, p. 496–505.

——; CHOPRA, S. K.; SING, A. B. Pollen rain from the vegetation of northwest India. *New Phytol.*, 72, 1973, p. 191–206.

SIOLI, H.; SCHWABE, G. H.; KLINGE, H. Limnological outlooks on landscape ecology in Latin America. *Trop. Ecol.*, vol. 10, no. 1, 1969, p. 72–82.

SMIT, A. F. J. The origin of Lake Bosumtwi and some other problematic structures. *Ghana J. Sci.*, 2, 1962, p. 176–196.

SOWUNMI, M. A. Pollen morphology in the Palmae, with special reference to trends in apperture development. *Rev. Palaeob. Palyn.*, 7, 1968, p. 45–54.

STEENIS, C. G. G. J. Van. The land bridge theory in Botany with particular reference to tropical plants. *Blumea*, 11, 1962, p. 235–272.

*STOFFERS, P.; HOLDSHIP, S. Diagenesis of sediments in an alkaline lake: Lake Manyara, Tanzania. In: *IXth International Congress of Sedimentology* (Nice), 1975 (in press).

STRAKA, H. Palynologia Madagassica et Mascarenica, Fam. 126. Sarcolaenaceae (Chlaenaceae). *Pollen et Spores*, 6, 1964a, p. 289–301; p. 641–643.

——. Palynologia Madagassica et Mascarenica. *Pollen et Spores*, vol. 6, n° 1, 1964b, p. 239–288.

——. Palynologia Madagassica et Mascarenica, Fam. 212. Didiereaceae. *Pollen et Spores*, 7, 1965, p. 27–33.

STRAKA, H. Palynologia Madagassica et Mascarenica, Fam. 50-59. *Pollen et Spores*, 8, 1966, p. 241–264.

——; SIMON, A. Palynologia Madagassica et Mascarenica, Fam. 122-125. *Pollen et Spores*, vol. 9, n° 1, 1967, p. 59–70.

——; ——; CERCEAU-LARRIVAL, M. T. Palynologia Madagassica et Mascarenica, Fam. 155-166. *Pollen et Spores*, vol. 9, n° 3, 1967, p. 427–466.

*TEMPLE, P. Evidence of lake level changes from the northern shoreline of Lake Victoria, Uganda. In: Steel, R. W.; Prothero, R. M. (eds.). *Geographers and the tropics: Liverpool essays*, p. 31–56. London, Longmans, 1964.

TSUKADA, M.; ROWLEY, J. R. Identification of modern and fossil maize pollen. *Grana Palynol.*, 5, 1964, p. 406–412.

*——. Late Pleistocene vegetation and climate in Taiwan (Formosa). *Proc. Nat. Acad. Sciences USA*, 55, 1966, p. 543–548.

——. The pollen sequence. In: Cowgill *et al.* *The history of Laguna de Petenxil, a small lake in northern Guatemala*, p. 63–66. Conn. Acad. Arts Sci., Mem. 17, 1966, 126 p.

*——. Pollen analysis from four lakes in the southern Maya area of Guatemala and El Salvador. Quaternary palaeoecology. In: *Proc. VII Congr. Int. Ass. Quat. Res.*, 7, p. 303–331. Yale Univ. Press, 1967.

VAN CAMPO, M. Palynologie africaine. I. *Bull. IFAN*, 19 A, 1957, p. 659–726.

——. Palynologie africaine. II. *Bull. IFAN*, 20 A, 1958, p. 753–808.

——. Palynologie africaine. IV. *Bull. IFAN*, 22 A, 1960, p. 1165–1199.

——; HALLÉ, N. Les pollens des Hippocratéacées d'Afrique de l'Ouest. *Pollen et Spores*, 1, 1959, p. 191–272.

*——; COQUE, R. Palynologie et géomorphologie dans le Sud tunisien. *Pollen et Spores*, 2, 1960, p. 275–284.

——; BERTRAND, L.; BRONCKERS, F.; DE KEYSER, B.; GUINET, P.; ROLAND-HEYDACKER, F. Palynologie africaine. V. *Bull. IFAN*, 26 A, 1964, p. 105–120.

——; BRONCKERS, F.; GUINET, P. Palynologie africaine. VI. *Bull. IFAN*, 27 A, 1965, p. 795–842.

——; GUINET, P.; COHEN, J. Fossil pollen from late Tertiary and Middle Pleistocene deposits of the Kurkur oasis. In: Butzer; Hansen. *Desert and river in Nubia*, p. 515–521. Madison, University of Wisconsin Press, 1966.

VAN DER HAMMEN, T. Climatic periodicity and evolution of South American Maestrichtian and Tertiary floras. *Boletín Geológico* (Bogotá), vol. 5, no. 2, 1957, p. 49–91.

——. Upper Cretaceous and Tertiary climatic periodicities and their causes. *Ann. New York Acad. Sc.*, 95, 1961, p. 440–448.

——. Palinologia de la región de 'Laguna de los Bobos'. *Revta. Acad. Colomb. Cienc. Exact. Fis. Nat.*, vol. 11, n° 44, 1962, p. 359–361.

*——. A palynological study of the Quaternary of British Guiana. *Leidse Geol. Meded.*, 29, 1963, p. 125–180.

——. Paläoklima, Stratigraphie und Evolution. *Geol. Rundschau*, 54, 1964, p. 428–441.

*——. Changes in vegetation and climate in the Amazon basin and surrounding areas during the Pleistocene. *Geol. Mijnb.*, 51, 1972, p. 641–643.

*——. The Pleistocene changes of vegetation and climate in tropical South America. *J. Biogeogr.*, 1, 1974, p. 3–26.

*VAN DER HAMMEN, T.; GONZÁLEZ, A. E. Upper Pleistocene and Holocene climate and vegetation of the 'Sabana de Bogotá' (Colombia, South America). *Leidse Geol. Meded.*, 25, 1960a, p. 126–315.

——; ——. Holocene and late glacial climate and vegetation of Páramo de Palacio (Eastern Cordillera, Colombia). *Geol. Mijnb.*, 39, 1960b, p. 737–746.

VAN DER HAMMEN, T.; GONZÁLEZ, A. E. Historia de climá y vegetación del Pleistóceno superior y del Holoceno de la Sabana de Bogotá. *Bol. Geol.*, vol. 11, n° 1-3, 1963, p. 189–266.

——; ——. A pollen diagram from the Quaternary of the Sabana de Bogotá (Colombia) and its significance for the geology of the Northern Andes. *Geol. Mijnb.*, 43, 1964, p. 113–117.

——; ——. A pollen diagram from Laguna de la Herrera (Sabana de Bogotá). *Leidse Geol. Meded.*, 32, 1965a, p. 183–191.

——; ——. A late glacial and Holocene pollen diagram from 'Ciénaga del Visitador' (dept. Boyacá; Colombia). *Leidse Geol. Meded.*, 32, 1965b, p. 193–201.

——; WIJMSTRA, T. A. A palynological study on the Tertiary and upper Cretaceous of British Guiana. *Leidse Geol. Meded.*, 30, 1964, p. 183–241.

——; WIJMSTRA, T. A.; ZAGWIJN, W. H. The floral record of the late Cenozoic of Europe. In: Turekian, K. K. (ed.). *The late Cenozoic Glacial Ages*, p. 391–424. 1971.

*——; WERNER, J. H.; VAN DOMMELEN, H. Palynological record of the upheaval of the Northern Andes: a study of the Pliocene and lower Quaternary of the Colombian Eastern Cordillera and the early evolution of its high-Andean biota. *Palaeogeogr. Palaeoclim. Palaeoecol.*, 16, 1973, p. 1–24.

*VAN GEEL, B.; VAN DER HAMMEN, T. Upper Quaternary vegetational and climate sequence of the Fuquene area (Eastern Cordillera, Colombia). *Palaeogeogr. Palaeoclim. Palaeoecol.*, 14, 1973, p. 9–92.

VAN ZEIST, W. Late-Quaternary vegetation history of Western Iran. *Rev. Palaeob. Palyn.*, 2, 1967, p. 301–311.

VAN ZINDEREN BAKKER, E. M. *South African pollen grains and spores*. Part I. Cape Town, Balkema, 1953, 72 p.

——. A pollen analytical investigation of the Florisbad deposits (South Africa). In: *Proc. Third Pan-African Congress on Prehistory*, 1955, p. 56–67.

——. A late-glacial and post-glacial climatic correlation between East Africa and Europe. *Nature*, 194, 1962, p. 201–203.

*——. A pollen diagram from Equatorial Africa, Cherangani, Kenya. *Geol. Mijnb.*, 43, 1964, p. 123–128.

——. Palynology and stratigraphy in sub-Saharan Africa. In: Bishop, W. W.; Clark, J. D. (eds.). *Background to evolution in Africa*, p. 371–374. Univ. of Chicago Press, 1967.

——; COETZEE, J. A. *South African pollen grains and spores*. Part III. Cape Town, Balkema, 1959, 200 p.

——; CLARK, J. D. Pleistocene climates and cultures in north-eastern Angola. *Nature*, 196, 1962, p. 639–642.

VISHNU-MITTRE. The ice ages and the evolutionary history of the Indian Gymnosperms. *Journ. Indian Bot. Soc.*, vol. 42, no. 2, 1963, p. 301–307.

——. Oaks in the Kashmir valley with remarks on their history. *Grana Palynologica*, vol. 4, no. 2, 1963, p. 306–312.

——. Some aspects concerning pollen analytical investigations in the Kashmir valley. *Palaeobot.*, vol. 15, no. 1-2, 1966, p. 157–175.

——. Some evolutionary aspects of Indian flora. In: *J. Sen Memorial volume*, p. 385–394. Calcutta, 1969.

——; SINGH, G.; SAKSENA, K. M. S. Pollen analytical investigations of the Lower Karewa. *Palaeobot.*, vol. 11, no. 1-2, 1962, p. 92–95.

——; SHARMA, B. D. Studies of Indian pollen grains, 1. Leguminosae. *Pollen et Spores*, 4, 1962, p. 5–45.

——; ——. Studies of Indian pollen grains, 2. Ranunculaceae. *Pollen et Spores*, 5, 1963, p. 285–296.

——; GUPTA, H. P. Studies of Indian pollen grains, 3. Caryophyllaceae. *Pollen et Spores*, 6, 1964, p. 99–111.

VISHNU-MITTRE; SHARMA, B. D. Studies on post-glacial vegetational history from the Kashmir valley, 1. Haijam Lake.
Palaeobot., 15, 1966, p. 185–212.

*WALKER, D. The changing vegetation of the montane tropics.
Search, vol. 1, no. 5, 1970, p. 217–221.

WALKER, J. W. Pollen morphology, phytogeography and phylo-
geny of the Annonaceae. *Contrib. Gray Herbarium*, 202,
1971, p. 3–130.

WATTS, W. A. A pollen diagram from Mud Lake, Marion
County, northcentral Florida. *Geol. Soc. Amer. Bull.*, 80,
1969, p. 631–642.

——. Post-glacial and interglacial vegetation history of southern
Georgia and central Florida. *Ecology*, vol. 52, no. 4, 1971,
p. 676–690.

WEBB, T. A comparison of modern and presettlement pollen
from southern Michigan (USA). *Rev. Palaeob. Palyn.*,
16, 1973, p. 137–156.

——. Corresponding patterns of pollen and vegetation in lower
Michigan: a comparison of quantitative data. *Ecology*, 55,
1974, p. 17–28.

WEBB, T.; BRYSON, R. A. Late- and post-glacial climatic change
in the northern Midwest, USA: quantitative estimates derived from fossil pollen spectra by multivariate statistical
analysis. *Quaternary Res.*, 2, 1972, p. 70–115.

WELMAN, W. G. *South African pollen grains and spores.* Part VI.
Cape Town, Balkema, 1970, 110 p.

*WIJMSTRA, T. A. A pollen diagram from the Upper Holocene
of the lower Magdalena valley. *Leidse Geol. Meded.*, 39,
1967, p. 261–267.

——. Palynology of the Alliance Well. *Geol. Mijnb.*, 48, 1969,
p. 125–133.

*——. *The palynology of the Guyana coastal basin.* University
of Amsterdam, Dissertation, 1971.

*——; VAN DER HAMMEN, T. Palynological data on the history
of tropical savannas in northern South America. *Leidse
Geol. Meded.*, 38, 1966, p. 71–90.

Floristic composition and typology

Floristic composition

Floras

In the tropical forest ecosystem, the relations which exist between living organisms, and between these and their environment, are sometimes manifest, sometimes difficult to perceive, and sometimes unknown. It follows that, because of its role in the past, at present, or in the future, no living organism should be unconsidered and that the analysis of an ecosystem has to be as complete as possible. The inventory should be exhaustive and precise; the use of scientific names is essential. Most such analyses tend to omit a great many living elements; in practice this is unavoidable for reasons of feasibility, cost and effectiveness.

Floristic inventories are concerned with the knowledge of the flora in general and emanate more or less directly from herbaria, which are mostly on a national level but tend to work on an international scale. All necessary information will be found in Lanjouw's (1964) *Index Herbariorum*. These herbaria can provide information on the published floras and those in preparation. Some countries or regions are well supplied with fairly complete modern floras; others either lack them completely or they are in preparation but are rarely completed. Regional syntheses are sometimes obsolete and far from complete; in other cases they are modern, but unfinished. Complete, modern continental floras are virtually non-existent.

Richness and heterogeneity

It is thus hardly surprising that the estimates of the total number of higher plants in the world varies from 200 000–250 000 to 300 000–400 000 or even 700 000. The smaller numbers are usually found in earlier works; the larger figures are more realistic but still approximations.

The position is generally worst in the tropics and especially in tropical forests. This is due to the richness of their floras and the relative paucity of studies. The practical difficulties in identification of forest trees present serious obstacles to a rapid progress of knowledge.

An approximate idea of the numbers of taxa of all higher plants in tropical forests may be gathered from the following examples (some of which include non-forest species):

New Caledonia: 3 000 species (80 per cent are endemic) in over 680 genera (Schmid, 1975, see annex); Malesia: 4 000 tree species (600 genera) and 25 000 species (Van

Steenis, 1958b); Borneo: at least 10 000 species (Merrill, 1915); Malayan Peninsula: 3 000 tree species (DBH$\geqslant$30 cm) and 7 500 species (Whitmore, 1972); Java: 6 000 species in over 240 families and 2 000 genera (Bakhuizen van der Brink, 1963); Madagascar: 2 000 tree species (DBH$\geqslant$10 cm) in over 450 genera and 100 families (Capuron, 1971); West Africa: 7 000 species (Hepper, 1972); Ivory Coast: 600 tree species (DBH$\geqslant$10 cm) in dense forest in over 275 genera and 60 families (Aubréville, 1938); Nigeria: 4 500 species (Hall and Lowe, 1975); Cameroon: 8 000 species in over 1 800 genera and 220 families (Letouzey, 1976); Zaire: 11 000 species (Leonard, 1963); Costa Rica: 6 000 species (Aubréville, 1966); Panama Canal Zone: 2 000 species (Standley, 1928); Trinidad: 400 tree species and 1 300 species (Beard, 1946); Amazonia: 2 500 tall tree species (suggested by Huber, 1902).

The following are examples from much smaller areas:

On *ca.* 60 ares in the Malayan Peninsula, 183 species (among a total of 379 trees of DBH$\geqslant$10 cm) (Cousens, 1951); on *ca.* 45 ha in Brunei, 760 species among 30 000 trees of 10 cm (Ashton, 1964); on 1 ha in Sumatra, 60 species (among 320 trees of DBH$\geqslant$10 cm—where one or two species were represented by 20–50 stems while 30 species were each represented by a single individual (Anwary Dilmy, in Anon., 1958a); on 18 ha in Nigeria, 170 species of DBH$\geqslant$10 cm, of which 18 species were represented by >100 trees and 90 by <10 (Jones, 1955); on 1 are in Cameroon, 230 species among a total of 1 300 individual plants, including one species represented by 150 individual plants and 125 species represented by a single plant (Letouzey, 1968); on 8 ares in Gabon, 122 species of shrubs and trees $\geqslant$3 m high, among 418 individual plants, and 26 species of lianas among 96 individual plants (Hallé, Le Thomas and Gazel, 1967); on 500×500 m quadrats in Amazonia (Manaus region), *ca.* 50 species of trees and shrubs among 180 individual plants, 12 species of palm among 82 trees, 20 species of lianas among 40 individual plants, among plants above 1 m in height (Aubréville, 1961); on 5 ha of *terra firme* forest in Amazonia (Belém region) 224 species representing 136 genera and 52 families among 2 607 trees with GBH$\geqslant$30 cm (Murça Pires, see annex); for comparison Murça Pires cites corresponding figures of 180, 118, 47, 2 792 for adjoining areas of a swamp forest and 196, 124, 45, 2 912 for a periodically inundated forest respectively; only 83 species were common to all three forest communities.

These examples which mostly relate to trees and shrubs give an idea of the richness and heterogeneity which is almost universal in tropical forests. Richards (in Lowe-McConnell, 1969) stresses rightly that the richness in plant species per unit area is largely attributable to the simultaneous presence of many synusiae, something that is not found in other forests of the world; however, other vegetation types, such as for instance the maquis of New Caledonia, are even richer in species.

Forests with an extremely large number of trees of one predominating species are rare. Such forests, in which there is little heterogeneity, occur under special topographic or edaphic conditions, as for instance mangroves, African *Raphia* communities, *Mora excelsa* forests of Guyana, etc.

Nevertheless, the dry-land stands of *Gilbertiodendron dewevrei* in Zaire contain 347 species (Gérard, 1960), and the *Brachystegia laurentii* forests 535 species (Germain and Evrard, 1956). Other African forests, where certain Caesalpiniaceae appear to be absolutely predominating and, Malesian forests, where on small areas certain Lauraceae (*Eusideroxylon zwageri*) and Dipterocarpaceae (*Dryobalanops aromatica*) predominate, also contain many species. Richards (1952) gives examples of this tendency in some virgin forests.

Plant community diversity is of two kinds: intrinsic or α diversity—the number of species composing a community, and extrinsic or β diversity—the variation with habitat. Extrinsic diversity can occur on different scales, from shade to sun bryophyte floras on bark or boulders to the major climatic zones, regions, or continents. Communities of a few or single species such as stands of *Rhizophora apiculata* mangrove exist in close proximity and share a common climate with the most species-rich forests in the world in lowland Malaysia; these intrinsically diverse mixed dipterocarp forests only vary at the subtlest level in relation to changes in soil and topography, yet 90 per cent of the flora may change over a short distance with a change in soils in lowland Sarawak. The causes of this variation are complex and for the most part little studied and often defying critical scientific investigation. They are likely to be principally the following.

1. History (see also chapter 3). Angiosperms are thought to have originated at least 120 million years ago, in the late Jurassic or early Cretaceous periods. These periods were marked by major tectonic activity and coincided with the early stages of the fragmentation of the great northern Laurasian and southern Gondwana continents. These seem to have become linked for a period through North Africa-Lusitania in the Cretaceous but otherwise remained isolated until, during the Tertiary period and within the last 30 million years, one southern fragment migrated northwards to unite with Laurasia forming peninsular India and, later, North and South America became joined and the southern continents of Africa and Australasia took up their present positions. These events isolated major floras and provided periodic opportunities for migration and must clearly be partially the cause of the global patterns of contemporary plant geography. Raven and Axelrod (1974) have provided a recent review. The migration and eventual collision of these crustal plates led to the massive mountain building which characterized the Tertiary period and gave rise to the Himalayas, Andes, mountains of New Guinea, Sumatra and Kinabalu in Borneo, enhancing the diversity of climates and allowing the evolution within the tropics of floras that may later have migrated and further evolved into the temperate zone as it expanded at the close of the Tertiary. The Pleistocene period of the last 1.5–3 million years, with its drastic climatic changes throughout the world, has clearly led to extinction of whole floras which were either sedentary or had no migration route, and to rapid expansion and short-term evolution and diversification. Fossil evidence for the evolution of tropical floras, especially since the Cretaceous, is rapidly growing through the science of palynology (Van der Hammen, 1957; see chapter 3).

If natural selection operates in the evolution and maintenance of floristic diversity in species-rich vegetation such as tropical rain forests (see chapter 8 for further discussion) then intrinsic diversity should increase with the age. This would explain, for instance, the greater richness of the lowland rain forests of Malaysia, an ancient land surface that appears to have remained in a tropical and probably moist climate at least since the early Cretaceous, in comparison with those of South America and especially Africa which have experienced major eras of mountain building and erosion and were even more affected by changes in rainfall seasonality during the Pleistocene.

2.   The age and stability of land surfaces, by affecting geomorphological processes, will also affect habitat diversity though rates of geomorphological change vary also with climate. Young landscapes are more rapidly eroded and are characterized by more abrupt and extreme topography; their surfaces are covered with shallow and more diverse soils. Tectonically unstable areas may have experienced several cycles of uplift and erosion and the resultant sedimentary rocks may be chemically impoverished, leading to limiting soil conditions for plant growth. These factors could explain why the mixed dipterocarp forests of northwestern Borneo, which grow on shallow infertile soils and the immature physiography of a neogeosyncline, are so much more extrinsically diverse than those of Malaysia where residual soils on the rolling terrain can exceed 15 m depth.

Many tropical angiospermous tree families show a preference for certain soils and climates: these include Myrtaceae on leached and Meliaceae on fertile soils, Leguminosae in seasonal and Myristicaceae in aseasonal climates. These preferences are generally global and must therefore have an ancient evolutionary origin.

3.   The area occupied by a forest will influence the opportunities for evolutionary diversification and the chance extinction of small populations and hence floristic impoverishment. The characteristically high endemism and low intrinsic diversity of many island ecosystems can thus be partially explained; this has received elaborate theoretical treatment by MacArthur and Wilson (1967; see chapter 7). The same considerations apply to floras consisting of populations of species with narrow ecological ranges and/or occurring in fragmented habitats in a physically diverse landscape.

4.   Increased predictability of the physical habitat appears to be associated with an increase in biotic interactions and decline of physical limiting factors in the process of selective evolution. Biotic interactions, by providing a theoretically limitless range of evolutionary opportunities, can partially explain the high intrinsic diversity of forest ecosystems in the aseasonal humid tropics, and its decline as seasonality increases, in particular as the rains become increasingly unpredictable, and where other unpredictable factors such as volcanism and typhoons are frequent on an evolutionary time scale.

5.   These four physical factors interact with the plant species, leading through natural selection to the evolution of breeding systems which are best adapted to and therefore characteristic of a particular habitat, and circumscribe the potential of its species for further evolution and diversification. Though little studied, present evidence from both the monsoon and everwet tropics suggests that tree populations are genetically variable and that they are more or less free or obligate outcrossers (see chapter 8). This provides the potential for rapid evolution when populations are fragmented by environmental barriers but inhibits speciation where such barriers are absent. The breeding systems, through co-evolution of long or short range pollen and fruit dispersal systems (mainly by animals), affect the ease by which populations are fragmented. Both intrinsic biotic and extrinsic environmental factors frequently vary in the same direction when tropical rain forest and monsoon forest or savanna are compared and could help explain the often great difference in intrinsic diversity.

Thus, though it is unlikely that any forest ecosystem has reached an evolutionary zenith, it is possible that some have evolved further than others and, though floristic parallels exist within the three main regions of the tropics, equally remarkable alternative evolutionary pathways have been followed. This helps to explain why the cultivation of a crop species outside its region of origin frequently leads to higher yields. This is largely due to the initial absence of specific pathogens and predators, but these will inevitably increase and may eventually reduce productivity to the level in the country of origin. This emphasizes the vital importance of conserving wild stocks and indigenous cultivars as genetic material which will always be needed to build up disease resistance.

Thus, in general, the tropical forest is characterized by floristic heterogeneity, and the afore examples show that in addition to certain species represented by fairly large numbers of individuals, there exist great numbers of species represented by a very small number, not infrequently by only one specimen.

The development of knowledge of tropical floras may be illustrated by the three following examples (which underline the uncertainties inherent in the use of old sources): the numbers of known species of *Ocotea* in Madagascar rose from 17 to 35 between 1917 and 1957, and that of known species of *Cryptocarya* from 18 to 33; the numbers of families, genera and species known in Zaire in 1896 were respectively, 103, 479, and 854; in 1940 the corresponding figures had risen to 170, 1 631, and 9 705; the inventory of the flora for the whole of Africa south of the Sahara was increased by 194 genera and 3 550 species in the period 1954 to 1962 (to which have to be added 980 infraspecific taxa and 2 803 changes in nomenclature).

The tropical forest floras of Oceania are not very well known; those of Australia, Asia and Africa are reasonably well known; and those of the Americas not at all well known. There are many species of which only the genus is known (and sometimes only the family) and many other plants only known by their vernacular names—many of which refer to more than one species. Even in better inventories many uncertainties tend to remain. Time-consuming studies, sometimes on rather small areas, were often needed to achieve higher standards. A few examples might be cited:

— for Panamá, sample inventories made under the auspices of UNDP and FAO by the Centre technique forestier tropical at a sampling density of 0.1 °/₀₀ and covering trees of DBH⩾10 cm, contained only 147 identified species and 69 taxa identified to the genus and 338 vernacular names;

— for Colombia, Espinal and Montenegro (1963) describing a forest type by reference to its trees and shrubs, supply a list of 130 names, of which 75 are either doubtful or identified as to genus only;

— for Surinam, Schulz (1960) published lists of 35 and 33 species, of which 24 and 25 respectively were fully determined;

— for an area of 3.5 ha in Amazonia, Murça Pires, Dobzhansky and Black (in Cain and Oliveira Castro, 1959) list 179 tree species with a DBH⩾10 cm, of which 117 are identified, 46 identified to the genus and 16 only to the family;

— in another Amazonian study near Manaus, Aubréville (1961) from 1 652 woody and herbaceous plants listed *ca*. 100 plants that were entirely unknown (20 plants without even vernacular names); 45 species were identified to the family, 31 to the genus, and only 30 were completely identified;

— Takeuchi (1961) recorded 125 trees of DBH⩾10 cm from 1 850 m² of *terra firme* forest near Manaus, of which 20 were completely unknown, 10 were only identified to family and 19 only to genus; on a depressed terrain nearby, he found 112 trees (DBH⩾10 cm) of which 14 were completely unknown, 10 were identified as to family and 15 as to genus only;

— Heinsdijk (1960) carried out five large-scale inventories on the right bank of the Amazon river along a stretch of 1 300 km parallel to the equator. These inventories, commissioned by FAO, covered 15 000 000 ha containing 126 517 trees with a DBH⩾25 cm. Though identification was more precise, only 240 species of 374 known by vernacular names were identified, another 93 were identified to the genus (sometimes grouping several species together) and 41 to the family only;

— Heinsdijk and De Miranda Bastos (1965) summarized the inventories made in the Amazon region (covering 20 000 000 ha by representative sampling); for 348 vernacular names (some of which are conventional) they provided a list of 260 identified species, whilst for 88 names they gave the genus only.

Our ignorance of the taxonomy of lianas, at least as regards their rapid identification, is even greater; Rollet (1974) working on a list of 191 vernacular names from Venezuelan Guyana, could only provide immediate identifications for 42 species; 64 were identified to the genus, and 112 only to the family. It is important to stress that vernacular names are insufficient for the compilation of botanical inventories, since one single name often groups together several (sometimes entirely unrelated) species and one species may have several names.

In Africa and Asia the taxonomic position is perhaps somewhat more satisfactory, but even there the existing uncertainty is still quite considerable:

— on *ca*. 0.5 ha in Cameroon, Letouzey (1968) encountered 60 species of trees with a DBH⩾20 cm, 10 of which had not been satisfactorily identified to the level of species, genus or family;

— on *ca*. 0.5 ha in northern Borneo, Nicholson (in Anon., 1965) found 198 species of trees (DBH⩾10 cm) of which more than a third are either unknown or doubtful.

Inventories

There is all too often a lack of time available for compilation and interpretation, as well as an absence of literature bringing the scientific findings to the actual users. Not infrequently the floristic inventories only include the larger woody plants. Many such inventories have been carried out, often over large areas, but all adopting a minimum diameter: 10 cm, 15 cm (Centre technique forestier tropical, sample surveys over 100 000 ha in Cameroon); 20 cm (30 000 ha in the Central African Empire); 25 cm (Heinsdijk in Brazil; Surinam Forest Service, sample surveys over 200 000 ha); 100 cm/π (The Brazilian RADAM project); 40 or 50 cm (in Amazonia, FAO); or even 60 cm (Centre technique forestier tropical in Gabon). It is a universal observation that the numbers of plant individuals and species rapidly increase below the 25 cm-diameter class, and hence the identification of all species becomes correspondingly more arduous.

These forest inventories were initially carried out for economic reasons, and in many cases were limited to merchantable or potentially merchantable species; they represent therefore no more than a first step towards a scientific analysis of the forest ecosystems.

A better knowledge of tropical forest ecosystems requires a greater knowledge of the entire flora and, indirectly, more knowledge of the whole tropical flora. As Heinsdijk (1960) pointed out, the means available for the science of floristics are deplorably small and few botanists devote themselves to these studies, especially in the field. Murça Pires (see annex) similarly insisted on the necessity for close co-operation between foresters and taxonomists. The regional and international bodies operating in these fields will have to play an important part in the promotion of these botanical studies and the compilation of floras on local, national, regional and, if possible, continental levels, as well as the preparation of monographs on important taxa. The theoretical and applied botanical training of forest survey staff should also be greatly extended and improved.

Phytosociology

A comprehensive floristic inventory ought to enumerate all the phanerogamic and cryptogamic species occurring in the various synusiae: trees, shrubs, herbs, climbers, epiphytes, saprophytes, etc. Such complete enumerations have so far only been carried out on very small areas. Most of few carried out were on non-forest sites. Moreover, the idea behind them has been (e.g. in Brazil, the Ivory Coast and Zaire) to prove the existence of plant communities, 'plant associations' *sensu lato*, and to justify the application to the flora of the tropics of the principles and methods of the Zurich-Montpellier school of plant sociology or any other system; this was the approach of Mangenot (1955) for the

Ivory Coast, and of Lebrun and Gilbert (1954) for Zaire, notwithstanding the incontestable difficulties involved in the application of concepts such as abundance and dominance, overlap, characteristic species, etc., to a tropical forest. Most authors now admit that basic taxonomic studies are with few exceptions of greater importance in the tropical environment than purely phytosociological studies *sensu stricto*.

However, an important step in this direction appears to have been taken with the formation of the concept of leading species or dominant (preponderant) species which has been well defined by Schulz (1960). His work in Surinam is based on enumerations of 560 quadrats of 10×10 m. Leading species mean species that are well represented and show a marked vitality, thereby characterizing the forest in which they are found. Foresters or plant geographers tend to rely on some such concept, more or less clearly conceived and expressed, when they distinguish, say, *Pouteria, Pouteria-Eschweilera, Carapa-Eschweilera, Licania-Byrsonima*, or *Dacryodes-Sloanea* forests in America; *Brachystegia* and *Isoberlinia, Aucoumea klaineana* and *Sacoglottis gabonensis, Brachystegia laurentii*, and *Antiaris-Chlorophora* forests in Africa; *Dryobalanops aromatica, Koompassia*-Burseraceae, or *Shorea glauca* forests in Asia; or *Agathis macrophylla* forest in Oceania, etc. However the foundations on which analysis and evaluation are based vary enormously, and in most cases such designations must be taken to be no more than practical and provisional expedients; they frequently express only very vaguely the floristic composition of the forests in question.

Moreover these concepts and designations are mainly based on the fact that in all tropical forests there occur within a given, often rather narrow perimeter, one or two species of reasonably large trees that are really abundant, a few species that are frequent and a great number that are represented by only one or two individuals. Aubréville (1964), when carrying out 1 per cent sample inventories of 26 550 ha in the Central African Empire, found that in a population of 44 000 trees (DBH≥20 cm) belonging to 170 species, 54 per cent of all large trees belonged to 12 out of a total of 115 species and that 50 per cent of all small trees belonged to 13 out of a total of 150 species present; from the proportion of small trees to large trees within the four main species common to both types he was able to infer the important role of these four species in the dynamism of this forest (though admittedly they also provide evidence of earlier anthropogenic disturbances). Murça Pires (see annex) showed that in the Amazon region, near Belém, on three adjoining 5 ha plots, one situated on *terra firme*, one on swamp ground (*igapó*) and one on periodically flooded ground (*várzea*), there were respectively 224, 180 and 196 species, of which 21, 11 and 8 respectively accounted for 50 per cent of individuals of GBH≥30 cm; the 5 most common species of each plot contained *ca.* 30–40 per cent of the total number of trees present. On an area of nearly 100 ha in Guyana, Rees (1963) found that 20 per cent of all trees with a DBH≥10 cm belonging to *Eschweilera sagotiana*, and that 10 species comprised 60 per cent of all stems; in another forest, *Eperua falcata*, accounted for 25 per cent and 5 species for 70 per cent of all stems.

The flora

Despite the considerable dearth of information on the floristic composition of tropical forests, it is possible to give a rough picture of the distribution of certain families and genera of higher plants on a continental or intercontinental scale, and of certain species on a continental or regional scale.

The strictly intertropical families are of minor importance as regards the number of genera and species. Many have only one genus and their systematic position is not clear. Often they occur only in very restricted areas (the Strasburgeriaceae in New Caledonia, the Gonystylaceae in Malesia, the Diegodendraceae in Madagascar, the Medusandraceae in western and central Africa, the Lissocarpaceae in Brazil, etc.), sometimes they straddle two continents (the Ctenolophonaceae and Irvingiaceae in Asia and Africa, the Rapateaceae in Africa and America, the Schizandraceae in Asia and America, etc.). Many families which very often occur in all tropical regions and consist of many genera and species (such as the Annonaceae, Bombacaceae, Caesalpiniaceae, Ebenaceae, Lauraceae, Moraceae, Myrtaceae, Palmae, Sapotaceae, Zingiberaceae, etc.) also have representatives in nontropical areas. The origin of these dispersed elements should be traced to palaeogeographic phenomena.

Another aspect of the problem is concerned with the contact and occasional mixture of tropical and temperate floristic elements; with the exception of northern Africa, where continuity clearly exists, such contacts can be observed on various points of the globe and they have been the subject of special studies.

Problems of distribution

The distribution of certain sub-families and certain genera over different continents is a problem closely linked with palaeogeographic phenomena. It has so far only been tackled and quantified for a few families. For others, possibly equally important in terms of number of genera and species and distribution, but sometimes less well known taxonomically, our information is fragmentary and uncertain; this is true of the Annonaceae, Ebenaceae, Euphorbiaceae, Sapindaceae, Rubiaceae, Violaceae, etc. Often one largely deals with shrubs of the understorey where whole large genera need detailed studies (e.g. *Psychotria*, a pantropical genus with 800 species; *Ixora*, pantropical with 400 species; *Rinorea*, pantropical with 350 species; *Ouratea*, pantropical with 300 species; *Drypetes*, pantropical with 200 species; *Symplocos*, pantropical but not African with 400 species; *Pavetta*, palaeotropical with 400 species; *Memecylon*, palaeotropical with 300 species; *Elaeocarpus*, Oceanic and Asiatic with 250 species; *Saurauia*, Asiatic and American with 300 species; *Miconia*, neotropical with 900 species; *Cuphea*, neotropical with 250 species; and many others); in addition there are climbing, herbaceous terrestrial and epiphytic plants, the genera of which sometimes contain many species (e.g. *Alpinia, Anthurium, Begonia, Peperomia, Piper, Tillandsia*); all these frequently constitute important components of forest ecosystems.

At the present state of our knowledge it is difficult to

make an inventory, even if restricted to the level of families and genera, of the components of intertropical forest. The modern works available for such an enterprise, such as Willis' *Dictionary of the flowering plants and ferns* (Airy Shaw, 1973), Engler's *Syllabus der Pflanzenfamilien* (Melchior, 1964) as well as the *Traité de Botanique systématique* by Chadefaud and Emberger (Emberger, 1960), naturally do not distinguish between intertropical or subtropical forest species and non-tropical or non-forest species. The information regarding the number of species per genus or of genera per family sometimes vary quite widely, occasionally being over 20 per cent; this results partly from new descriptions, and from views on the delimitations of genera. The higher numbers of species have as a rule been included here as they are preferably closer to reality; in the case of marked disparities the minimum and maximum numbers are indicated. As far as possible, only those families and genera that can be found in intertropical forests have been listed; the number of species per genus which is frequently mentioned in only meant to serve as an indicator of the 'weight' attributable to these genera within the world of tropical forests; the absence of such a number, on the other hand, indicates that the genus is much better represented in temperate environments. Sometimes, the absence of species numbers suggests that the genus in question, though less important from the point of view of species, is yet interesting from the point of view of plant geography or sociology; such genera may be of special significance, as they frequently contain leading species or species of commercial importance.

The information on the distribution of genera and certain species is given in terms of rather large geographical units. The expression Oceania includes in general the eastern part of Australia; New Guinea means the island of that name; Malesia is used in the sense defined by Van Steenis; Asia frequently includes South-East Asia and India; the islands near Madagascar are attached chiefly to the eastern region of this large island; the term Africa covers the African rain forest regions and sometimes the southern African dry forests as well as the north of the continent; Central America, the Caribbean and South America are often included in the general denomination of America.

In view of the difficulties discussed above, this attempt at presenting a detailed picture of the distribution of families and genera has to be considered with certain reservations as to its real value. The presentation of the material, covering some hundred families (many small monogeneric or monospecific families of secondary importance had to be left out) cannot be ordered in a way that is both logical and practical; hence this presentation remains open to criticism.

Distribution of families and genera

Gymnospermae

Gymnosperms in the tropics occur at high altitudes and have restricted or endemic areas of distribution, or constitute transitory populations establishing themselves under special conditions (volcanic sites, dry soils, or clearing and burning) where they form only open woodlands with well developed graminaceous ground cover which will not be considered here.

Conifers are absent from the lowland forests of the Amazon region and Africa, but are found almost everywhere else, sometimes even at low altitudes. This is particularly the case for certain *Pinus* spp.: *P. merkusii* in South-East Asia; *P. strobus* var. *chiapensis* in Mexico (some 40 other species occur in Mexico at higher altitudes where they form rather open stands; they tend to disappear south of Mexico: only two are found in Nicaragua and one in Costa Rica); *P. caribaea* in the northern part of Central America; and various species in the Caribbean. Above an altitude of 600–800 m the conifers tend to increase in importance; they frequently reach their maximum extension between altitudes 1 000 and 2 000 m, just below or at the level of the cloud forests; with rising altitude, they quickly disappear, though sometimes they reach an altitude of 4 000 m, for instance in New Guinea.

Apart from Pinus and the other Pinaceae, the Araucariaceae play an important role with their two genera *Araucaria* (18 species) and *Agathis* (20); *Araucaria* is represented in New Guinea, Australia, and New Caledonia as well as in South America (*Araucaria angustifolia* in southeastern Brazil, especially between 800 and 1 800 m), and the area of *Agathis* extends from South-East Asia to Malesia, Australia and New Zealand. These *Araucaria* and *Agathis* species often occur as large isolated trees dominating a canopy of evergreen broad-leaved trees, but rarely in groups or stands; in Australia they occur at low altitudes, mixed with the broadleaved species of dense humid tropical forests.

The Podocarpaceae's main genera are *Dacrydium* (25) in Malesia and Oceania (there is also one species in Chile), *Phyllocladus* with some species in Malesia, and *Podocarpus*. This genus comprises 100 species and occurs throughout the southern hemisphere. In Madagascar and Africa, it is represented by some ten species only which inhabit the mountains of southern, central and eastern Africa, and reach as far north as Ethiopia. In Central America, it reaches the northern hemisphere by way of the central mountain chains especially in Colombia and Costa Rica; surprisingly it has also been reported from the Amazon region near Belém (Heinsdijk and De Miranda Bastos, 1965); another occurrence in the northern hemisphere is in South-East Asia.

The Cupressaceae include the genus *Callitris* (20), occurring in Australia, Tasmania, New Caledonia and New Zealand; its role in the rain forest environment is only a minor one; the genus *Widdringtonia* (5) is restricted to the mountains of southern Africa; *Juniperus procera* is the only tropical representative of its genus and occurs only between 2 000 and 3 000 m, from Ethiopia to Malawi. Finally there is the genus *Libocedrus* which is often split up into several genera; its distribution is large but discontinuous, it occurs in California, southern China, New Guinea, New Caledonia, and New Zealand.

Subtropical and extra-tropical families

These families and sub-families (presented in alphabetical order) are mainly represented within the tropics (often in the mountains) by trees and shrubs.

— Aceraceae ($3^{1}/200^{2}$, northern hemisphere, chiefly in the mountains): *Acer*[3], Malesia and South-East Asia above 1 000 m altitude.

— Asteraceae (Compositae, 900/13 000–20 000, cosmopolitan): *Brachylaena*, some trees in Madagascar and Africa; *Vernonia* (600–1 000), pantropical and non-forest, but some forest trees included.

— Cunoniaceae (26/350, southern hemisphere, especially Australia, Malesia and Central America, some species in Madagascar): *Weinmannia* (190), Oceania, Malesia, America (chiefly the Andes), some species in Madagascar.

— Epacridaceae (30/400, mainly New Zealand and Australia, some species in New Caledonia, Malesia, as well as in Hawaii, and in non-tropical South America).

— Ericaceae (50–80/1 300–2 500, cosmopolitan, on mountains in the tropics but not well represented in India and Sri Lanka): *Erica* (500–800, mainly South Africa), East African mountains; *Rhododendron* (and *Azalea*) (600–1 300, northern hemisphere), Malesia; *Vaccinium* (200–400, northern hemisphere), Malesia, Madagascar, South Africa and the Andes.

— Fagaceae (8/900, cosmopolitan except South Africa): *Castanopsis*, New Guinea and South-East Asia; *Lithocarpus* (and *Pasania*), Malesia and South-East Asia; *Nothofagus*, New Zealand, New Caledonia, southeastern Australia, New Guinea and southern Andes; *Quercus*, South-East Asia, Malesia and Central America (where certain species occur in low altitudes) to northern Colombia.

— Hamamelidaceae (25/120, warm-temperate and sub-tropical regions), not found in India and centred on eastern Asia; they occur in a non-contiguous manner on all continents.

— Juglandaceae (8/60, northern hemisphere except India, Sri Lanka and Africa): *Engelhardtia*, Malesia, South-East Asia and Central America; *Juglans*, Central America and northern Andes.

— Magnoliaceae (12/230, cosmopolitan, especially warm-temperate montane): *Michelia* (60), Malesia and South-East Asia; *Talauma* (50), Malesia, South-East Asia and America;

— Monimiaceae (and Siparunaceae) (25–34/300–450, mainly southern hemisphere but with extensions as far as the Tropic of Cancer in the Far East and in America; many gaps): *Siparuna* (160), America; *Tambourissa*, Madagascar.

— Myrsinaceae (35/1 000, hot and warm-temperate regions): *Ardisia* (and *Afrardisia*) (400), pantropical; *Embelia* (130), palaeotropical; *Maesa* (200), palaeotropical; *Rapanea* (200), pantropical.

— Myrtaceae (100/3 000, hot and warm-temperate regions especially in tropical America). Sub-family Myrtoideae: *Eugenia* (600–1 000), pantropical; *Psidium* (140), America; *Syzygium* and *Jambosa* (500), palaeotropical. Sub-family Leptospermoideae (Australia): *Bakhousia*, Australia and New Guinea; *Eucalyptus* (600, Australia, chiefly open *Eucalyptus* forests not considered here), some species in humid regions of Australia and Malesia: *Eucalyptus alba*, *E. deglupta*, *E. gigantea*, *E. obliqua*, *E. platyphylla*, *E. regnans*; *Melaleuca* (100), Oceania and Malesia; *Tristania*, Oceania and Malesia.

— Oleaceae (30/600, temperate and tropical regions, chiefly in eastern Asia): *Olea* (20–60), palaeotropical.

— Pittosporaceae (9/240, hot and warm-temperate palaeotropical regions, chiefly in Australia): *Pittosporum* (200), palaeotropical but with only a few species in Africa.

— Proteaceae (60/1 000–1 400, with 600 species in Australia and 300 in South Africa, widespread over the southern hemisphere with extensions towards the north in Africa and Asia; these are however frequently non-forest species): *Banksia*, *Grevillea* and *Personia* in Oceania and *Protea* in the dry African forests.

— Theaceae (Ternstroemiaceae) (35/600, intertropical and subtropical regions, mostly mountainous): *Ternstroemia* (130), Malesia, Asia and America and some species elsewhere; *Thea* (and *Camellia*) (80), Asia (including the mountainous regions of Malesia).

— Winteraceae (7/120, warm-temperate regions, discontinuous distribution).

Cosmopolitan families

These families are important in the intertropical zones where they are mainly represented by trees and shrubs.

— Boraginaceae (and Ehretiaceae) (100/2 000): *Cordia* (*Sebestena*) (250), pantropical.

— Celastraceae (60/850), not very well represented in forests in the intertropical regions: *Lophopetalum*, South-East Asia and Malesia; *Goupia*, South America.

— Euphorbiaceae (300/5 000–?10 000): *Antidesma* (170), Malesia, some species in Madagascar and Africa; *Baccaurea* (80), Oceania and Malesia; *Croton* (750), pantropical both moist and dry regions; *Drypetes* (200), pantropical; *Hevea*, America; *Hura*, America; *Macaranga* (280), palaeotropical; *Mallotus* (150), palaeotropical (some species only in Madagascar and Africa); *Sapium* (120), pantropical; *Uapaca* (50), Madagascar and Africa.

— Gramineae (Poaceae) (700/8 000–10 000): represented in the forests of the tropics by only a few herbaceous genera and species and by the sub-family Bambusoideae (40–100/500) which, particularly in mountainous or dry regions, can form stands in all intertropical zones, but are of minor importance in Africa; they play an important role in the understoreys of dense humid forest communities especially in Asia, and particularly in semi-deciduous or degraded forests.

— Leguminosae.

The Leguminosae (or Fabaceae) are often divided into three world-wide families: Papilionaceae, Mimosaceae, Caesalpiniaceae.

• Papilionaceae (or Faboideae) (400–600/9 000–12 000) is essentially represented by two tribes (the Dalbergieae and Sophoreae) and to a minor extent by trees, shrubs and lianas belonging to other tribes (*Milletia* (180) pantropical but not

1. World number of genera.
2. World number of species.
3. Species number not indicated either because there are only a few, or because the genus is better represented outside the tropics.

well represented in America); the number of these Papilionaceae species does not exceed 1 200. Among the Dalbergieae: *Dalbergia* (300), pantropical, with many lianas; *Derris* (100), mainly palaeotropical, some species in Brazil; *Lonchocarpus* (150), Africa and America, some species in Australia; *Machaerium* (150), America, with many lianas; *Pterocarpus* (100), pantropical. Among the Sophoreae: *Baphia* (60), Madagascar and Africa (with one species in Borneo). In addition there are smaller genera. *Pericopsis* (*Afrormosia*) palaeotropical, and 8 other genera, all American: *Andira* (and *Vouacapoua*) (35 with one species reaching Africa), *Brya, Centrolobium, Diplotropis, Dipteryx* (*Coumarouna* and *Taralea*), *Hymenolobium, Myroxylon* and *Platymiscium* (30).

• Mimosaceae (or Mimosoideae) (40–50/2 000) occupies only a minor place in the humid zones, for the large genera *Acacia* (800) and *Mimosa* (500) are hardly ever found in them at all. The tribe Ingeae is represented in America by the genera *Inga* (250), *Pithecellobium* (and *Samanea*) (200, including some African species), *Calliandra* (150) (this occurs also in Asia and Madagascar); the genus *Albizia* (150) is chiefly palaeotropical. There are also trees in the tribes Piptadenieae (better represented in Africa than in America) and Parkieae. There are also some smaller genera of various distributions: palaeotropical (*Xylia*, chiefly Madagascar and Africa), some African (*Calpocalyx, Cylicodiscus* with *C. gabunensis*), some American (*Dinizia* with *D. excelsa*, *Enterolobium*), some in both these continents (*Pentaclethra*).

• Caesalpiniaceae (or Caesalpinioideae) (100–150/ 2 200) is particularly well represented in forests of the humid or dry tropics, chiefly by trees and shrubs. There are not many large genera: *Bauhinia* (300, with many lianas especially in America but none in Africa), *Caesalpinia* (120); *Cassia* (600, particularly in America), *Cynometra* (70), all pantropical; many genera contain only a few species and monospecific genera are frequent.

This family and the Leguminosae in general are well represented in Africa and America and much less so in Oceania and Asia. Aubréville (1961) has compared the flora of the Guinea-Congo region (95 genera, of which 12 are Papilionaceae, 13 Mimosaceae, and 64 Caesalpiniaceae) with that of the Guyana-Amazon region (106 genera, of which 47 are Papilionaceae, 19 Mimosaceae and 40 Caesalpiniaceae) which may be taken to be the most representative floras of the two continents for large trees. He showed that among the Caesalpiniaceae the Cynometreae-Amherstieae predominate in Africa and the Sclerolobieae in America, but that some genera are shared (*Cassia, Copaifera, Crudia, Cynometra, Guibourtia*); he also emphasized that certain genera preponderate in Africa (*Dialium*, 25 species in Africa versus 1 species in America) and others in America (*Copaifera*, 30 American versus 5 African; *Swartzia*, 100 versus 2).

The Amazon region is an area of concentration for the American Caesalpiniaceae: *Cynometra* (actually pantropical with 70 species), *Dimorphandra* (25), *Hymenaea* (30), *Macrolobium* (80), *Peltogyne* (25), *Sclerolobium* (30), *Swartzia* (100), *Tachigalia* (25). Humid central Africa plays a similar role; there are 62 genera and 175 species in the floristic region of the Congo, and 42 genera and 81 species in that of West Africa (Aubréville, 1968). But it should be noted that in humid central Africa, the large genera among the Caesalpiniaceae contain far fewer species than the large American genera; for each of the genera *Anthonotha, Berlinia, Brachystegia, Dialium, Gilbertiodendron, Hymenostegia, Monopetalanthus* only 10 to 30 species can be cited, notwithstanding the fact that for some of them a combination with closely related small genera seems possible.

On the other hand, Africa possesses characteristic dry forests of which there are more in the southern than in the northern hemisphere; the limited numbers of genera found there are also rather poor in species; the most important of these are *Brachystegia, Cryptosepalum, Daniellia, Julbernardia*, all of which are also found in humid forests, and in addition, *Baikiaea, Isoberlinia* and some small genera.

The Caesalpiniaceae contains a number of smaller genera: *Peltophorum*, pantropical; *Afzelia* (and *Intsia*), *Erythrophloeum*, both palaeotropical; *Sindora*, essentially a genus of Malesia and Asia, with one African species; *Koompassia*, Malesian; *Burkea* (*B. africana*) and *Colophospermum* (*C. mopane*) from the dry forests of Africa; *Distemonanthus* (*D. benthamianus*), *Gossweilerodendron, Oxystigma, Scorodophloeus* and *Tessmania* from the humid forests of Africa; *Dicorynia, Eperua, Haematoxylon, Mora, Prioria* (*P. copaïfera*), *Schizolobium* and *Zollernia*, all from America.

— Palmae (Arecaceae) (200–240/2 800–3 400) important in the understoreys of humid forests of Oceania, Asia, Madagascar and America, frequently on fresh or swampy soils or soils subject to inundation; their importance is much more limited in Africa, where the palms are not found in mountain forests. Several palms inhabit dry zones where, as in Madagascar and South America, they form large but frequently open stands of very peculiar character, *Borassus aethiopum* in South-East Asia and in Africa, *Orbignya* spp. in Brazil and Mexico.

Moore (in Meggers *et al.*, 1973) presents data on their distribution. There are 97 genera and 1 385 species in the eastern tropics; in Madagascar, the Mascarene Islands and the Seychelles these figures diminish to 29 and 132; in Africa there are no more than 16 genera and 117 species; numbers rise again to 64 and 83 in South America and to 48 and 339 in North America. The percentage of endemic genera is high everywhere and tends to exceed 80 per cent in the regions of the Far East (92 of 97) and in Madagascar (25 out of 29); certain groups occur only in the palaeotropical regions (Phoenicoideae, Borassoideae, Nipoideae, Caryotoideae, Podococcoideae), others are exclusively neotropical (Pseudophoenicoideae, Iriarteoideae, Geonomoideae, Phytelephantoideae).

— Rhamnaceae (60/900), poorly represented in intertropical forest environments; its chief representatives are lianas belonging to: *Gouania* (70), pantropical and *Ventilago* (40), Oceania and Asia (with some species in Madagascar and Africa).

— Rosaceae (100/2 000–3 000), particularly well represented in the northern hemisphere. Only the genus *Pygeum* (*Prunus, Laurocerasus*) (90) palaeotropical, chiefly Malesian, can claim some importance in intertropical forest environments.

— Rubiaceae (450–500/7 000). There are many large

genera in intertropical environments often in the form of understorey shrubs: *Canthium* (*Plectronia*) (200), palaeotropical; *Cephaëlis* (*Uragoga*) (200), pantropical, chiefly American; *Faramea* (150), American; *Grumilea* (100), palaeotropical; *Guettarda* (100), American (with a few species in New Caledonia); *Ixora* (400), pantropical, chiefly Malesian; *Lasianthus* (100), palaeotropical chiefly Malesian (with one species in America); *Mussaenda* (200), palaeotropical; *Palicourea* (200), American; *Pavetta* (400), palaeotropical; *Psychotria* (800), pantropical; *Randia* (a genus that often is split) (300), pantropical; *Rudgea* (150), American; *Sabicea* (100), from Madagascar, Africa and America; *Tarenna* (*Chomelia*) (400), palaeotropical; *Timonius* (150), from Oceania, Malesia (chiefly), Asia and Madagascar. *Hallea*, palaeotropical and *Calycophyllum*, American, are of economic interest.

— Rutaceae (and Flindersiaceae) (150/900–?2 000): *Chloroxylon* (*C. swietenia*), India and Sri Lanka; *Fagara* (250), pantropical; *Flindersia*, Oceania and Malesia.

— Thymeleaceae (50/650), chiefly palaeotropical and poorly represented in tropical forests.

— Ulmaceae (15/200): *Celtis* (80), pantropical.

— Violaceae (20/900): *Rinorea* (350), pantropical, mainly African.

Intertropical families

Families frequently of limited continental distribution and chiefly represented by trees and shrubs.

— Avicenniaceae (2/15), Rhizophoraceae (16/120), Sonneratiaceae (Blattiaceae) (2/7). These three families contain the essential constituents of mangroves. The Avicenniaceae, with one genus *Avicennia* (14), are distributed world-wide. The Rhizophoraceae include, apart from two secondary genera (*Ceriops* and *Kandelia*), *Brugeria* (6) which is palaeotropical but absent from the Atlantic coast of Africa and the pantropical *Rhizophora* (8, of which *R. conjugata* and *R. mucronata* occur chiefly in Asia and East Africa, *R. racemosa* chiefly in Africa and America on the Atlantic side and *R. mangle* chiefly on the Atlantic side of Africa, and the Atlantic and Pacific sides of America, but also in New Caledonia). The Sonneratiaceae, including *Sonneratia* (7), surround the Indian Ocean and spill over to the western shores of the Pacific. This distribution of mangrove species has been much discussed, especially by Van Steenis (1958a) and Schnell (1971).

— Caryocaraceae (2/25). An American family, represented especially by the genus *Caryocar* (20).

— Dichapetalaceae (Chailletiaceae) (4/250) from Africa and Madagascar; *Dichapetalum* (220) also has some species in South-East Asia and South America.

— Dipterocarpaceae (*ca.* 20/400–600) essentially South-East Asia and Malesia, except New Guinea. Gottwald (1968) states that the Malayan Peninsula, Borneo and the Phillipines contain 61–300 species, Sumatra and the eastern part of the peninsula of Indo-China 51–60, and Sri Lanka, Burma, Thailand, Laos and Cambodia 16–50 species. All these species belong to the sub-family of Dipterocarpoideae (which is also represented in the Seychelles) and are chiefly inhabi-

tants of humid forests. Some dipterocarps, belonging to the sub-family Monotoideae (2/52, *Marquesia* and *Monotes*), are only found in Madagascar and Africa especially in dry forests and their degraded forms. The Dipterocarpaceae are absent from America and Australia.

Some Asian and Malesian genera have numerous species: *Shorea* (and *Anthoshorea*) (*ca.* 180), *Dipterocarpus*, *Hopea* (and *Balanocarpus*), *Vatica* (60–90 each), while others are less important and play a more secondary role (*Anisoptera, Cotylelobium, Doona, Dryobalanops, Parashorea, Pentacme, Stemonoporus* (and *Monoporandra*), *Upuna, Vateria*).

Some figures may indicate the importance of this family in the forests of Asia and Malesia: Brown and Matthews (1911) estimated that 75 per cent of the forests of the Philippines were dipterocarp forests. Fox (1967) states that on *ca.* 35 ha enumerated in Borneo (Sabah) 750 among a total of 1 000 trees with a DBH $\geqslant$ 50 cm were dipterocarps; or 90 per cent if only trees with diameters $>$ 90 cm were taken into account; the number of species represented was 36 (of which 4 species each exceeded 8 per cent of the trees enumerated and together accounted for over 50 per cent of the total).

— Humiriaceae (8/50): American, having one species (*Sacoglottis gabonensis*) on the western African coast.

— Pandanaceae (3/900). A palaeotropical family of species which grow mostly on swampy ground or near the sea-shore: *Freycinetia* (250), from Sri Lanka and Taiwan to Australia, New Zealand and Polynesia, mainly climbing shrubs; *Pandanus* (630), palaeotropical, trees and shrubs.

— Sarcolaenaceae (Chlaenaceae) (8/40): Madagascar, where it is important in the dry forests of the western region.

— Vochysiaceae (6/200): American, with one genus (*Erismadelphus*) in western Africa; *Qualea* (60); *Vochysia* (100).

Pantropical families

Trees and shrubs

— Actinidiaceae (Saurauiaceae) (3/350): *Saurauia* (300), Malesia, Asia and America (and one species in Australia).

— Agavaceae (20/670): *Cordyline* (20), Oceania and Malesia; *Dracaena* (150), palaeotropical.

— Anacardiaceae (60–80/600): *Astronium*, America; *Dracontomelum*, Oceania and Malesia; *Mangifera* (40), Malesia and South-East Asia; *Melanorrhea*, Malesia and South-East Asia; *Schinopsis*, America.

— Annonaceae (120/2 100): *Artabotrys* (100), palaeotropical, chiefly lianas; *Annona* (120), America, some species in dry Africa; *Guatteria* (250), America; *Polyalthia* (150), palaeotropical, especially Malesia; *Popowia* (100), palaeotropical; *Uvaria* (180), palaeotropical, mostly lianas; *Xylopia* (160), pantropical, especially Africa.

— Apocynaceae (200/2 000): *Alstonia* (60), palaeotropical, especially Oceania, Malesia and Asia; *Alyxis* (130), Oceania, Asia and Madagascar; *Aspidosperma* (80), neotropical; *Ervatamia* (100), Oceania, Asia and Madagascar; *Landolphia* (60), Madagascar and Africa, mostly lianas; *Mandevillea* (130), neotropical, mostly lianas; *Parsonsia* (100),

Oceania and Asia; *Prestonia* (70), neotropical; *Rauvolfia* (100), pantropical (excluding Australia); *Strophantus* (60), Malesia, Asia, Madagascar and Africa, mostly lianas; *Tabernaemontana* (140), Africa and America.

— Araliaceae (50–70/700, chiefly in Oceania, Malesia and America): *Dendropanax* (*Gilibertia*) (80), pantropical; *Didymopanax* (40), neotropical; *Oreopanax* (120), neotropical; *Polyscias* (80), palaeotropical but rare in Africa; *Schefflera* (200), pantropical.

— Bignoniaceae (120/800): *Arrabidaea* (100), neotropical, mostly lianas; *Jacaranda* (50), neotropical; *Tabebuia* (100), neotropical.

— Bombacaceae (20–30/200, chiefly in America): *Bombax* (a genus often split) (60), chiefly American; *Ceiba* (20), pantropical; *Chorisia*, America; *Durio* (and *Boschia*) (30), Malesia and South-East Asia; *Ochroma*, America; *Quararibea* (50), America.

— Burseraceae (20/600): *Bursera* (100), chiefly in dry regions of America; *Canarium* (70), palaeotropical; *Commiphora* (180), chiefly in dry regions of India and Africa, in dry forests in Madagascar; *Protium* (90), Malesia, Asia and Madagascar, but chiefly in America; *Santiria* (50), Malesia. Genera with fewer species are: *Aucoumea* (*A. klaineana* from Central Africa), *Dacryodes* (*Pachylobus*), pantropical, chiefly African, and *Tetragastris*, American.

— Chrysobalanaceae (12/400, chiefly American): *Acioa* (30), Africa (with some species in America); *Couepia* (60), neotropical; *Hirtella* (90), America (and some species in Africa and Madagascar); *Licania* (and *Moquilea*) (130), tropical and warm-temperate America (and some species in New Caledonia and Malesia); *Magnistipula* (10), Africa; *Maranthes* (10, chiefly in Africa) and *Parinari* (60), pantropical.

— Combretaceae (20/600): *Anogeissus*, Asia and Africa; *Combretum* (250), pantropical (except Australia), with many lianas; *Terminalia* (250), pantropical.

— Dilleniaceae (10/400, chiefly in Australia, rare in Africa); *Dillenia* (60), Oceania, Asia and Mascarene Islands; *Tetracera* (60), pantropical.

— Ebenaceae: *Diospyros* (including *Maba*) (nearly 500), is the most important genus of this family occurring in all intertropical regions, chiefly in humid, but also in dry forests; also in hot temperate regions. White (1971) used this genus to demonstrate in an African context the close interrelations existing between taxonomy, chorology and ecology.

— Elaeocarpaceae (12/400, hot and warm-temperate regions): *Elaeocarpus* (250), Oceania and Asia; *Sloanea* (120), Oceania, Asia and America.

— Erythroxylaceae (4/250): only important genus *Erythroxylum*, pantropical, chiefly Madagascar and America.

— Flacourtiaceae (including Samydaceae) (90/1 300): *Casearia* (160), pantropical; *Homalium* (200), pantropical.

— Guttiferae (Clusiaceae) and Hypericaceae (40–50/1 000): *Calophyllum* (100), palaeotropical, chiefly Asia (and a few species in America); *Clusia* (200), epiphytes and stranglers in America, also in New Caledonia and Madagascar; *Garcinia* (200–400), palaeotropical, chiefly Asia; *Mammea* (*Ochrocarpos*) (50), pantropical, chiefly Oceania

and Malesia (25), also in Madagascar (20) (and a few species in Africa and America); *Rheedia* (40), America (and a few species in Madagascar); *Symphonia* (20), chiefly Madagascar (18) (a few species in Asia and America); *Tovomita* (60), America; *Vismia* (40), America (and some species in Africa). Mention should also be made of various species of *Caraipa*, and *Platonia insignis* in America and *Mesua ferrea* in Malesia, South-East Asia and India.

— Icacinaceae (45–60/400): *Stemonurus* (*Gomphandra*) (50), Malesia and South-East Asia.

— Lauraceae (30/2 500): *Aniba* (40), America; *Beilschmiedia* (200), pantropical (well represented in Madagascar and Africa); *Cinnamomum* (270), Oceania and Asia; *Cryptocarya* (250), not found in central Africa, only in South Africa and in Madagascar; *Endiandra* (80), Oceania and Malesia; *Licaria* (and *Acrodiclidium*) (50), America; *Litsea* (400), Australia, Malesia, Asia and America in hot and warm-temperate regions; *Neolitsea* (80), Australia, Malesia and Asia; *Ocotea* (and *Nectandra*) (400), Madagascar and America, rare in Africa; *Persea* (and *Machilus*) (150), Malesia and America, also warm-temperate region in Asia and America; *Phoebe* (70), Malesia, America (not in Africa); *Eusideroxylon* (*E. zwageri*) from Malesia should also be mentioned.

— Lecythidaceae (20/400, chiefly neotropical) and Barringtoniaceae (5/100, chiefly palaeotropical): *Barringtonia* (100), palaeotropical; *Petersianthus*, one species in the Philippines and one (*P. macrocarpus*) in Africa; *Couroupita* (20), neotropical; *Eschweilera* (120), neotropical (and one species in New Guinea); *Lecythis* (50), neotropical. The two smaller American genera *Cariniana* and *Couratari* deserve mention.

— Loganiaceae (and Strychnaceae) (18/500): *Strychnos* (200), pantropical, with very many lianas.

— Lythraceae (25/550): *Cuphea* (250), America; *Lagerstroemia* (50), Australia, Malesia and Asia.

— Malpighiaceae (60/800), chiefly South America: *Byrsonima* (120), neotropical; *Heteropteris* (and *Banisteria*) (160), neotropical (with one African species); *Tetrapteris* (80), neotropical.

— Melastomataceae (and Memecylaceae) (200–240/4 000), chiefly in America: *Memecylon* (300), palaeotropical; *Miconia* (900), neotropical.

— Meliaceae (50/1 400): *Aglaia* (and *Amoora*) (300), Oceania and Asia; *Dysoxylum* (200), Oceania and Malesia; *Guarea* (170), America (and some species in Africa); *Trichilia* (300), Madagascar, Africa and America (and one species in Malesia); this family contains also a number of genera with fewer species: *Melia*, palaeotropical; *Toona*, Australia and Asia; *Azadirachta*, Malesia and Asia; *Sandoricum*, Malesia and Mauritius; *Entandrophragma*, *Khaya* (found also in Madagascar), *Lovoa* and *Turraeanthus* in Africa; *Carapa* (with some African species), *Cedrela* and *Swietenia* in America.

— Moraceae (60/1 500): *Artocarpus* (50), Malesia and South-East Asia; *Dorstenia* (170), Africa and America; *Ficus* (1 000), a pantropical genus containing many epiphytes and stranglers, chiefly Oceania and Malesia. In addition to these three important genera there are *Antiaris*, palaeotropical; *Chlorophora*, Madagascar, Africa and America; *Brosimum*, neotropical. The genera *Cecropia* (100), in America, comprising some very common species, and *Musanga* in Africa, with only one very common species (*M. cecropioides*), are capable of forming virtually pure stands in secondary forests only.

— Myristicaceae (18/300), chiefly in Asia: *Myristica* (120), palaeotropical; *Pycnanthus* (*P. angolensis*), Africa; *Virola* (60), neotropical.

— Ochnaceae (30–40/600): *Lophira* (*L. alata*), Africa; *Ochna* (90), Asia and Africa; *Ouratea* (300), pantropical, chiefly Africa and America.

— Olacaceae (25/250): *Coula* (*C. edulis*), Africa; *Heisteria* (60), America (and some species in Africa); *Strombosia*, palaeotropical.

— Sapindaceae (150/2 000): *Allophylus* (190), pantropical; *Cupania* (50), neotropical; *Guioa* (70), Oceania and Malesia; *Matayba* (50), neotropical; *Paullinia* (180), neotropical, chiefly lianas; *Serjania* (220), neotropical, chiefly lianas.

— Sapotaceae: one of the most representative humid tropical forest families, has been rather neglected, but has recently been very thoroughly analysed by Aubréville (1965a) who recognized 126 genera and some 600–800 species (other authors recognize only 40–80 genera).

The distribution of genera and, to a large extent, of the number of endemic genera and tribes, over the continents is fairly even: 42 genera (of which 38 are endemic) in Oceania and Asia; 44 genera (39 endemic) in Madagascar and Africa; and 47 genera (43 endemic) in America. The distribution of one very large taxon, comprising four tribes (Chrysophylleae, Malacantheae, Plachonelleae, Pouterieae) as well as the Manilkareae, is pantropical, while the other nine tribes have a more regional distribution. *Manilkara* (*ca.* 80 species) is the only pantropical genus but is is nevertheless preponderantly African and American. The numerical importance of other genera varies greatly: the Pouterieae tribe (21 genera in South America) contains a great number of genera, so far very little known (36 in all have been defined to date). On the whole, there is a need for more detailed studies of this family, especially in Oceania and America.

Apart from *Manilkara* the following are among the principal genera: *Planchonella* (80) from Malesia and chiefly Oceania; *Lucuma* (*sensu lato* 100), from Oceania, Malesia and America; *Madhuca* (80) and *Palaquium* (120), centred in Malesia; *Bumelia* (30–60), from South-East Asia and America; and finally *Chrysophyllum* (150), *Micropholis* (40) and *Pouteria* (50), all from America. In addition to these genera there are others especially from Madagascar (*Faucherea*), Africa (*Aningueria*, *Autranella*, *Baillonella*, *Gambeya*, *Tieghemella*) and America (*Achras*), which deserve note.

— Simaroubaceae and Irvingiaceae (24/120): *Quassia* (*Simarouba*) (40), America (and some species in Africa). The Irvingiaceae include three African genera: *Desbordesia* (*D. glaucescens*), *Irvingia* (which also occurs in Malesia and South-East Asia) and *Klainedoxa*, but as regards number of species the importance of these three genera is small.

— Sterculiaceae (60–70/1 000): *Byttneria* (90), pantropical; *Cola* (120), Africa; *Sterculia* (300), pantropical. The genera *Heritiera* (and *Tarrietia*) (30), palaeotropical and *Mansonia*, Asian and African, as well as the genus *Triplochiton* (*T. scleroxylon*), exclusively African, are important.

— Symplocaceae (2/400): *Symplocos* (400), Oceania, Asia and America.

— Tiliaceae (50/450): *Grewia* (160), palaeotropical; *Pentace*, Malesia and South-East Asia.

— Verbenaceae (75–100/3 000): *Aegiphila* (160), neotropical; *Citharexylum* (110), neotropical; *Gmelina* (30), Oceania and Asia; *Premna* (200), palaeotropical; *Tectona*, Malesia and Asia; *Vitex* (270), pantropical.

Lianas

Several of the most characteristic genera mentioned above are partly or wholly represented by lianas: *Dalbergia* and *Machaerium* (Papilionaceae), *Gouania* and *Ventilago* (Rhamnaceae), *Artabotrys* and *Uvaria* (Annonaceae), *Landolphia*, *Mandevillea* and *Strophanthus* (Apocynaceae), *Arrabidaea* (Bignoniaceae), *Combretum* (Combretaceae), *Strychnos* (Loganiaceae), *Paullinia* and *Serjania* (Sapindaceae). To these the *Clusia* (Guttiferae) and *Ficus* (Moraceae) species which are stranglers, the *Freycinetia* (Pandanaceae) and also various Araceae of climbing habit might be added, but all these belong to families in which trees and shrubs predominate. There are certain families which are much richer in lianas than in trees or shrubs.

— Aristolochiaceae (7/600), pantropical but absent from Australia: *Aristolochia* (500).

— Connaraceae (16–24/400), pantropical but chiefly palaeotropical: *Connarus* (100), pantropical.

— Hippocrateaceae (18/300), pantropical, but chiefly South American: *Hippocratea* (120), pantropical; *Salacia* (a genus frequently split) (200), pantropical.

— Menispermaceae (65/420), pantropical.

— Passifloraceae (12/600), chiefly African and American: *Adenia* (100), palaeotropical but chiefly African; *Passiflora* (500), chiefly American (some species in Oceania and Asia).

— Piperaceae (5/800–?2 000), pantropical (except for herbaceous Peperomiaceae): *Piper* (700–?1 900), pantropical.

Herbs

Only the chief families will be mentioned.

— Acanthaceae (250/2 600), hot regions, especially Malesia and South America, rare in temperate regions.

Araceae (115/2 000), hot regions, especially in Malesia and South America, rare in temperate regions; terrestrial plants, often climbing, even liana-like, sometimes epiphytes. Many important genera: *Aglaonema* (45), Malesia; *Amorphophallus* (and *Hydrosme*) (100), palaeotropical; *Anthurium* (550), America; *Arisaema* (and *Alocasia*) (150), pantropical, except Australia and South America; *Dieffenbachia* (30), America; *Homalomena* (140), Malesia, Asia and South America; *Monstera* (50), America; *Philodendron* (270)

America; *Pothos* (70), Malesia (and one species in Madagascar); *Rhaphidophora* (100), Malesia (and one species in Africa); *Schismatoglottis* (100), Malesia (and one species in South America); *Spathiphyllum* (40), America (and some species in Malesia); *Xanthosoma* (40), America.

— Begoniaceae (5/900), hot regions, especially in South America; absent from Australia, rare in temperate regions: *Begonia* (900).

— Commelinaceae (40/600), hot regions, chiefly Africa and America, rare in temperate regions. Some genera are purely forest plants.

— Cyperaceae (70–90/4 000), cosmopolitan: in the tropics well represented in non-forest environments, but a limited number of species in forests, especially on swampy ground and, exceptionally, on firm ground (*Mapania* (80), pantropical).

— Gesneriaceae (120–140/2 000), hot regions, not very numerous in Africa and rare in temperate regions; frequently epiphytic.

— Maranthaceae (30/400), hot regions, especially Africa and America; often capable of spreading considerably in secondary forests: *Calathea* (150), neotropical; *Maranta* (30), neotropical.

— Peperomiaceae (4/1 000), pantropical (this family is often combined with the Piperaceae): *Peperomia* (1 000), pantropical, but chiefly American; often epiphytic.

— Zingiberaceae (and Costaceae) (50/1 500), hot regions, especially in the palaeotropics and more especially in Malesia; few in America; often capable of spreading considerably in secondary forests. Many large genera: *Aframomum* (50), Africa; *Alpinia* (250), Oceania and Asia; *Amomum* (100), Oceania and Asia; *Boesenbergia* (50), Malesia; *Costus* (150), pantropical, chiefly Africa and America; *Curcuma* (60), Oceania and Asia; *Globba* (100), Malesia and Asia; *Hedychium* (50), Malesia, Asia (and some species in Madagascar); *Kaempferia* (70), Malesia and Asia; *Renealmia* (70), Africa and America; *Riedelia* (50), Malesia; *Zingiber* (90), Australia, Malesia, South-East Asia.

To these should be added the terrestrial ferns, for they occupy an important position in the understorey of tropical forests, particularly humid ones. They often belong to pantropical families and genera the systematics and nomenclature of which are not always completely well-established so it is difficult to make general statements. The physiognomic importance of tree ferns, especially in mountain areas of both Oceania and Asia has to be mentioned; tree ferns are however of markedly less significance in Africa and America.

Epiphytes, parasites and saprophytes

Only the principal families in which these plants occur will be discussed. Of the families already dealt with, the Guttiferae and Moraceae, for instance, contain species of *Clusia* and *Ficus* that are epiphytes in their juvenile stages, but later develop into strangler-type lianas. The Araceae, Gesneriaceae, Melastomataceae (*Medinilla* (300), Oceania and Malesia) and the Peperomiaceae contain both terrestrial plants and epiphytes.

— Balanophoraceae (18/120), parasites of hot regions;

no genus is common to both the palaeotropics and the neotropics: *Balanophora* (80), palaeotropical.

— Bromeliaceae (46/1 700), terrestrial and epiphytic plants of dry or humid intertropical America including high mountains: *Pitcairnia* (260), chiefly terrestrial; *Tillandsia* (500) and *Vriesea* (200) chiefly epiphytic.

— Cyclanthaceae (11/180), shrubby or rhizomatous climbing plants, sometimes epiphytic, in tropical America; the most representative genera are *Asplundia*, *Carludovica*, *Cyclanthus*, *Dicranopygium* and *Sphaeradenia*.

— Loranthaceae (and Viscaceae) (40/1 400); a cosmopolitan family well represented within the tropics by parasites including *Loranthus* (600) and *Viscum* (500) (some authors prefer to split these two large palaeotropical genera).

— Marcgraviaceae (5/120), climbing shrubs or, frequently, epiphytes of intertropical America: *Marcgravia* (50); *Norantea* (40).

— Orchidaceae (700/17 000–20 000); a cosmopolitan family well represented in the intertropical zone by species which may be terrestrial and non-forest, but are mostly forest epiphytes, and in exceptional cases, terrestrial saprophytes. Among the genera richest in species there are: *Bulbophyllum* (1 500) chiefly palaeotropical; *Dendrobium* (1 500), Oceania and Asia; *Epidendrum* (800), *Oncidium* and *Pleurothallis* (1 000), neotropical.

This family shows considerable discontinuities in its distribution pattern over the continents at the sub-tribe and generic level. There are grave uncertainties about the composition of certain genera (*Bulbophyllum*, *Habenaria*, *Malaxis*, *Physurus*), but it is significant that Brieger (1971) names only 7 genera which are common to the tropical zones of America, Asia, and Africa (with Madagascar) and only 13 genera common to Asia and Africa, of which some are much better represented in Asia than in Africa and Madagascar (*Calanthe*, *Oberonia*, etc.) and others much less so (*Satyrium*, *Polystachya*, etc.).

Several sub-tribes essentially belonging to one or other of these three continents, contain isolated genera dispersed over one of the others, where they are occasionally represented by a single species: the American sub-tribes Spiranthineae and Sobraliineae each have one species in Africa (one species of *Manniella*, and one of *Diceratostele* respectively); the African sub-tribe Habenariineae is represented by one species of *Sylvorchis* in South-East Asia; and the Asian sub-tribes Gastrodiineae, Liparidineae and Phajineae by the monospecific genera *Uleiorchis*, *Androchilus* and *Ghiesbreghtia* in America.

Brieger has published a scheme of the distribution of American orchids in which he distinguishes 5 areas: the Greater Antilles, Mexico and northern central America; southern central America and Andes; the Lesser Antilles and upper Amazonia; lower Amazonia and southeastern Brazil. He also points out the considerable discontinuities of distribution between Africa (containing some 1 200 species) and Madagascar (containing some 700 known species but probably nearer 1 000). For India, South-East Asia, Malesia and Polynesia the available information is still too scanty to be considered. On the other hand, the paucity of orchids in Australia is commented upon: there is only one endemic tribe, the Diurideae (Neottioideae), to which must be added some isolated elements of other sub-families.

— Rafflesiaceae (9/50), parasites of hot regions: *Rafflesia*, Malesia.

To these must be added epiphytic ferns belonging to different families (especially Hymenophyllaceae) and to various genera; like the terrestrial ferns they represent a by no means negligible part of the epiphytic synusia of the tropical forests, especially of the humid types.

In the same environment, the mosses constitute epiphytes of equal importance. In mountain regions, especially at the zone of permanent clouds, they, and the lichens, become a significant physiognomic character of the vegetation. While in lowland dry forests, epiphytic ferns, mosses and lichens play a much more restricted role, these plants are still met with in considerable numbers in dry mountain forests.

Finally, mention should be made of the epiphyllic organisms, belonging to various groups of the Algae, Fungi (equally important as saprophytes), Lichens and Hepaticae. Studies of the lower plants in intertropical forest environments have been fragmentary and sporadic; there appears to be a need for an intensification of research in this field.

Flora and altitude

So far altitudinal distributions have been discussed more as side issues. These distributions are so bound up with local phenomena that it is difficult to make general statements, especially as the boundaries between lowland and mountain forests vary greatly from one part of the world to another in response to latitude and continentality of climate; they frequently vary within one floristic province from one mountain range to the next, and even between two slopes of the same mountain. It is however possible to trace the major outlines of such distributions.

Between 1 000 and 2 000 m there is a decline and later a disappearance of certain lowland floristic elements: the Dipterocarpaceae and Lythraceae in Malesia and Asia, the Caesalpiniaceae in Africa and America. Simultaneously there is the appearance and increase of the Lauraceae and various other families, differing between the continents:

— in Malesia and Asia, Araucariaceae and Podocarpaceae, Aceraceae, Cunoniaceae, Elaeocarpaceae, Ericaceae, Fagaceae, Hamamelidaceae, Juglandaceae, Magnoliaceae, Myrtaceae, Symplocaceae and Theaceae;

— in Madagascar and Africa the increase in altitude is only indicated by some Coniferae (*Podocarpus, Widdringtonia, Juniperus*) and Ericaceae; the presence of Cunoniaceae and Monimiaceae in the montane forest formations of Madagascar should be noted;

— in central and southern America there appear the Araucariaceae and Podocarpaceae, Actinidiaceae, Araliaceae, Cunoniaceae, Ericaceae, Fagaceae, Hamamelidaceae, Magnoliaceae, Monimiaceae, Myrsinaceae, Myrtaceae, Proteaceae, Symplocaceae, Theaceae and Winteraceae.

In addition to these families which are quite characteristic, and others of minor importance, there will be found a great number of genera which belong to the flora of the plains, but are represented in these altitudes by species which form part of the environment of mountain forests.

Floristic features of the continents

According to Francis (1951), Australia is extraordinarily rich in Cunoniaceae, Epacridaceae, Myrtaceae and Proteaceae, and in Lauraceae, Rutaceae, and Sapindaceae. It is also characterized by a paucity of Leguminosae and several other families (Bignoniaceae, Guttiferae, Myristicaceae). A noteworthy feature is the occurrence, even at low altitudes of two genera of Coniferae, *Araucaria* and *Agathis*. In general, there is a progressive reduction, from north to south, of tropical species and a progressive increase of species characteristic for warm-temperate zones.

The flora of Melanesia represents an impoverished Malesian flora. Malesia and Asia are the exclusive domain of the Dipterocarpaceae; the flora of Malesia is closer to the flora of Asia than to that of Australia, and this manifests itself largely by floristic elements found in the mountains of Malesia. Three floristic provinces can be distinguished in Malesia: a western province (Malayan Peninsula, Sumatra, Borneo, Philippines); a southern province (Java, the Lesser Sunda Islands); and an eastern province (Celebes, the Moluccas, New Guinea, Bismarck Archipelago, Solomon Islands). Whitmore (1975) also stresses the marked concentration and rich diversity of the Dipterocarpaceae (and of climbing palms) in the zone corresponding to the tectonic plate of the Sunda (Malayan and Indo-Chinese Peninsulas, Sumatra), in contrast to the relative poverty of the plate of Sulu (New Guinea). Some dry, degraded forests are encountered in Van Steenis' (1934, 1935, 1936) southern province and its environs. In central South-East Asia and in India such dry forests, also frequently degraded, are characterized by the presence of *Tectona grandis* (Verbenaceae) and special dipterocarps. Legris (in Anon., 1974), pointed out that the vegetation of India differed more clearly from that of Malesia than the flora of South-East Asia.

For Madagascar, Humbert (1965) enumerated the various families encountered in the different forest zones in the eastern part; he distinguishes, *inter alia*, a low altitude domain characterized by Myristicaceae which is independent of montane zones. He also mentions the dry forests characterized by Burseraceae and Sarcolaenaceae in the western floristic domain. It is noteworthy that the flora of Madagascar shows strongest affinities with Africa, but is enriched by Asian and Australian genera.

The flora of Africa is relatively poor by comparison with the other continents, a fact stressed by Richards (in Meggers *et al.*, 1973); according to him, this relative poverty has its roots in palaeogeography, in the relative aridity of the continent and finally in recent human influences. Africa represents one of the areas of abundant Leguminosae; there is an equatorial belt of dense humid forest largely composed of Caesalpiniaceae, fringed on the north by a forest border in which Sterculiaceae and Ulmaceae predominate, and in the south a belt of peripheral savannas and a zone of dry Caesalpiniaceae forests. The flora of Africa forms a homogeneous whole, but with generic and specific variations from west to east, towards the centre and the south.

America too is dominated by the Leguminosae. The inventories carried out by Heinsdijk (1960) give an idea of the importance of this and other families in the Amazonian forest. As regards numerical importance, the Leguminosae represent 14–31 per cent of the trees enumerated in the 15 forest types distinguished by the author, the Lecythidaceae (6–26 per cent)—this family even overtaking the Leguminosae in eastern Amazonia, the Sapotaceae, Burseraceae, Lauraceae and the Rosaceae (*sensu amplo*, 1–12 per cent); next come other families typical for the American part of the southern hemisphere, such as the Humiriaceae, Myristicaceae and Vochysiaceae. Heinsdijk also describes variations from east to west: these are marked by the increase in importance of the Annonaceae, Elaeocarpaceae, Lauraceae and Monimiaceae, and the corresponding diminution of that of the Sterculiaceae and Violaceae. Aubréville (1961) also notes that the species of palms are more numerous in the west than in the east of Amazonia, but he adds that the number of individuals is higher in the east, especially near the Amazon estuary. The same conclusions are reached by Hueck and Seibert (1972) who divide the Amazon-Orinoco basin into a number of sub-regions each characterized by various genera and species. Their study covers the whole of South America and concerns all its intertropical forest ecosystems. Like Aubréville (1961), they distinguish between the forests of the Atlantic coast of Brazil and the Amazonian forests, but no clear differentiation on the level of genera or families emerges; Schnell (1971) points out the important role played by epiphytes in the coastal forests of Brazil.

There are few dry forests in America; in Central America such forests are only found in patches, and they appear to be equally rare in South America, where, in Amazonia, they tend towards an open shrub-forest with special edaphic conditions. The *campos cerrados* of Brazil never exhibit the aspect of close dry forest, as the wooded savannas do in southern central America.

Forest types

Physiognomic and structural inventories

Analyses of physiognomic structure are often little more than qualitative descriptions. Often they are supported by tables or illustrations based on transects, profiles, quadrats or sample plots; Rollet (1974) quotes numerous such studies of humid lowland forest formations of different degrees of completeness; similar illustrated studies exist also for various dry forests and mountain forests. In a paper on the largerly forest vegetation of Surinam, Van Dillewijn (1957) demonstrated the usefulness of stereophotographic documents combining terrestrial and aerial aspects (see chapter 5).

Where the aim is the greatest possible completeness,

analyses are normally confined to small areas. On larger areas the analyses tend to be partial, usually restricted to trees and shrubs exceeding certain minimum dimensions. Detailed physiognomic inventories are frequently based on unsystematic sampling within a given forest so it may be harzardous to generalize from their results. However even small area physiognomic inventories are very valuable, provided they are as thorough as possible, and more are required. Unfortunately the felling of many trees and the time required will always form serious obstacles to such studies.

Terminology

All classifications of forest types are based on a similar hierarchy: largely disregarding topography, they classify in the first place by climatic factors since such factors are always applicable to whole regions at least, and secondly by edaphic factors as these are important at the local level. Any further subdivisions are then based on the floristic composition. Some authors follow Beard's classification for America (1955) in employing a hierarchy which distinguishes, on a regional scale, communities, associations, assemblages, facies, locations, faciations and consociations, but to others these terms appear to cause confusion.

Difficulties frequently arise from the use of local terms which, though they relate to well defined local forest types, are inacceptable for general application. Such terms include: *bush* or *scrub* to describe the high forests of the intertropical zone of Australia (where the term forest means essentially an *Eucalyptus* forest); *heath forest* used in Malesia to describe forests on sandy coastal terraces, though they have nothing to do with Ericaceae; *jungle* used in South-East Asia or India; *miombo* relating to certain dry forests of southern central Africa; *igapó* and *mata de várzea* used to describe swamp forests and periodically inundated forests along the river Amazon. Difficulties may also arise with names applied to shrubby high altitude formations—*elfin forest* for instance which describes vegetation rather different from what is normally called a forest. Other difficulties are experienced when terms are taken over into other languages or given a different sense by various authors. The following examples may suffice:

— *forêt équatoriale* and *tropical forest* (in the sense of intertropical);

— *rain forest* may be tropical, subtropical or temperate in Australia and other subtropical or temperate regions (Chile, the Pacific coast of the United States, Canada);

— *bosque pluvial tropical* and *bosque pluvial subtropical* are two badly defined terms;

— wet and moist are both translated by humide in French;

— *bosque húmedo, bosque muy húmedo, bosque pluvial* are used as corresponding to *moist forest, wet forest*, and *rain forest*, however this last-named term is frequently used for all intertropical forests, whereas the real meaning of *bosque pluvial* is *superhumid forest*;

— *semi-evergreen* and *semi-deciduous*, or sometimes

moist deciduous imply very subtle distinctions as to the percentage of trees in the upper stratum shedding their leaves;

— *evergreen rain forest* and *evergreen seasonal forest*, sometimes *half evergreen seasonal forest*, also involve very fine differentiations as to the length of the very short dry season these formations have to endure; the same is true of *dry facies of tropical wet evergreen forest* and *moist facies of tropical dry evergreen forest* in Sri Lanka;

— there is also the *semi-evergreen mesophyll vine forest* of Australia; though this may be taken to be similar to *semi-evergreen forest*, it is difficult to find strictly equivalent terms in the literature relating to other continents;

— the term *peat swamp forest* used in Malesia might be translated into French as *forêt sur tourbe de Sphagnum* but this would be wrong, as the term refers to other organic residues.

The interpretation of altitudinal terms used is equally difficult: a *forêt d'altitude* in Cambodia or the Malayan Peninsula would be a mere *forêt de colline* in Africa, just as a *bosque pluvial premontano* on certain slopes of the Andes is called a *bosque pluvial montano bajo* elsewhere. The altitudinal zonation of vegetation is complicated by the occasional intrusion of a mass elevation effect; this phenomenon has been observed by various authors and has been thoroughly analysed by Van Steenis (1936) for the mountains of Malesia, but so far it has not yet received a satisfactory general explanation. Because of this effect, a given species, a given limited grouping of vegetation or a particular formation tends to occur on small isolated mountains at a lower altitude than on higher compact mountain range; thus the *mountain moss forest* occurs at just above 500 m altitude in the Seychelles, above 1 000 m in the Philippines, at about 2 400 m on Mt. Kaindi and 3 100 m on Mt. Wilhelm in New Guinea, according to Jeffrey, Brown and McVean respectively (in Flenley, 1974).

It is therefore not surprising to find a multitude of more or less equivalent terms: Rollet (1974) lists some 50 terms to designate the closed lowland humid forests; Cain and De Oliveira Castro (1959) mention a dozen different terms, all used before 1950, to characterize a single montane forest type in Tanganyika.

Classifications

The number of authors who have tried during the last decades to establish classifications of intertropical forests applicable to whole continents or sometimes, the whole world, is comparatively small: Schimper and Von Faber (1935), Champion for Asia (1936), Burtt Davy (1938). These were reviewed by Puri (in Anon., 1958b), Baur (1962) (who distinguished 10 types of *rain forest* and provides useful schematic profiles and tables of structural characteristics), Ellenberg and Mueller Dombois (1967), and by Champion and Seth (1968); Richards (1952) was concerned with this problem, Beard for America (1955) and Webb (1959) for Australia—his classification is unusual by being, in the first place, physiognomic; further, Holdridge (in Espinal and Montenegro, 1963), Aubréville (1965b) and

Fosberg (1970) who stresses that such classifications should be based chiefly and equally on physiognomy, structure and physiology, and thereafter on floristics, dynamics, vegetation history and ecology—most other authors would probably put ecology first. Thus Brünig (1972, see chapter 2) attaches the greatest importance to the gradient of humidity which allows him indirectly to define what he calls macroecosystems on the basis of their physiognomic characters. His classification is a hierarchy of ecosystems (macro-, meso-, micro-, nano-, pico-ecosystems, and communities) and it founded on the same basic principles of ecology, physiognomy, structure and physiology.

Africa deserves a special mention in this context because of a conference held in Yangambi (Zaire) in 1956, where considerable agreement was reached (Anon., 1956) in respect of French and English terminology for major vegetation types in Africa south of the Sahara. It provides a useful framework for definitions and further studies. It has been the subject of commentaries by Trochain (1957) and others, and of various constructive criticisms especially those of Boughey (1957), Monod (1963) and Greenway (1973). No one classification is accepted as valid or sufficiently detailed; they are often modified and enriched by additions which frequently relate to transitional forms. This is another factor that helps to explain the difficulties involved in making comparisons from one region or continent to another. Three examples will illustrate the problem.

Richards (1952) tried to establish a correspondence between the *semi-evergreen seasonal forest* (i.e. a forest with a short dry season) of Trinidad, the *dry evergreen forest* and *mixed deciduous forest* of western Africa, and the *monsoon forest* and *moist teak forest* of southeastern Asia; now other authors regard the *dry evergreen forests* of Asia as a type of *forêt dense sèche* (a closed dry forest), while the west African *mixed deciduous forest* as described by Richards would correspond to the *forêt dense humide semi-décidue* (closed humid semi-deciduous forest) of other authors, and the *mixed deciduous forest* of Indian plant geographers would come close to the *forêts sèches, denses ou ouvertes* (open or closed dry forests) of Africa.

Vareschi (1968) has vividly demonstrated the different interpretations possible for Schimper's term *selva pluvial*, (which corresponds to Grisebach's *jungla tropical* and Humboldt's *hilea*) in Venezuela, by mapping the areas occupied by this formation according to six different authors, notwithstanding the lacunae and inaccuracies of cartographic material used.

Van Steenis (1958a) is of the opinion that *rain forests* of the Asiatic type are very rare in Africa and that the *monsoon forests* of Asia (admittedly in their *evergreen* facies) are the equivalents of *rain forests* of the African type, and Vareschi (1968) equates the neotropical *selvas alisias* or *selvas húmedas* with the *monsoon forests* of Asia and calls the rain forests of the Asiatic type *selvas eupluviales*. As regards America the supplementary concept of *selvas muy húmedas* is often used. Some authors reserve the expression *monsoon forests* for Asia, identifying them in their entirety with the *deciduous facies* (which Van Steenis also mentions), and talk of rain forests in Africa and America.

The explanation for these terminological anomalies appears to lie in the fact that Africa (with the possible exception of the Gulf of Benin and isolated other points) does not receive such high amounts of rainfall as certain regions in Asia (e.g. the Gulf of Siam, the southern part of the Malayan Peninsula or Borneo) or America (especially the upper basin of the Amazon, and the Pacific coast of Central and South America). In addition there seems to exist a general difference between the humidity scales applied in Asiatic and African classification, resulting in the latter classing as *dry formations* what for the former are *subhumid formations* though rainfall and seasonal dry conditions may be very close; this difference of humidity grading is particularly noticeable in connection with the closed semi-deciduous forests described in South-East Asia (Legris, in Anon., 1974).

All this is most disconcerting. For this reason a permanent committee for the classification and mapping of vegetation on a world-wide scale, sponsored by Unesco, published *A framework for the classification of world vegetation* (Anon., 1969). This classification, sometimes referred to as the Unesco classification, is largely modelled on that published by Ellenberg and Mueller Dombois (1967) and is more pragmatic than systematic. It is designed to make possible the coordinated preparation of maps at scales of 1/1 000 000 or less, but could be adapted for maps on a larger scale by adding subdivisions. It is largely based on physiognomic, structural and ecological distinctions, and is divided into classes, sub-classes and groups of formations which in turn may contain subformations and further subdivisions. Eventually the colours, textures and conventional symbols relating to the classification will contribute considerably towards a greater uniformity of vegetation maps. Though the classification is still fairly new, it has already been tested in the field in Costa Rica, a country very rich in different formations of tropical vegetation; the criticisms advanced were all minor and frequently of no more than local significance.

The Unesco classification represents an excellent basis for further work; the section on tropical formations is satisfactory, though it may perhaps require some further elaboration. Once correct equivalents of the various English terms proposed are published in other languages, an important first step will have been taken towards its wide practical application. Such a project does not appear to pose very serious problems, but the incorporation of local or regional synonyms for the various vegetation types is a far more difficult and subtle task. However, if those working on a local or regional scale make the effort of replacing the terms they have used more or less traditionally by the best-fitting equivalents of the new classification, the science of vegetation classification will have made significant progress. The chief remaining task will then be to refine the classification in detail.

Forest types and climates

Physiognomic descriptions of humide lowland forests on dry land have been published by many authors, including Burtt Davy (1938), Aubréville (1949a), Richards (1952),

Van Steenis (1958a and b), Cain and De Oliveira Castro (1959), Knapp (1965, 1973), Mildbraed and Domke (1966), Schnell (1971), Whitmore (1975). These deal chiefly with density and height of trees and shrubs, apparent stratification, buttresses and stilt roots, stem and crown forms, buds, size and shape of leaves, phenology of leaves and shoots, cauliflory, flowering, pollination, fruit and seeds, lianas and stranglers, epiphyllic, epiphytic and saprophytic plants, parasites, herbs, and plants of particular habits (tree ferns, conifers, *Pandanus* spp., palms, bamboos). Where climatic conditions are similar many common characteristics are found on all continents; however, it is important to note that the forests of America, and especially those of the Amazon region contain fewer large trees and fewer buttressed trees than the forests of Africa, that their understorey is less dense and that they contain many palms of all sizes especially on sites where soil moisture is relatively high. For instance, near Manaus in Amazonia, Takeuchi (1962) recorded 114 palms on 1 850 m² of *terra firme* forest and 167 palms on 2 200 m² in a terrain depression forest. According to Schnell (1962), similar differences exist between forests of South-East Asia and Africa.

On a world-wide scale three physiognomic forest types which closely correspond to three types of climate can be distinguished; all are characterized by a rainfall of not less than 1 500 mm, dry periods rarely exceeding three months.

— Hyperhumid climates: the Asiatic *rain forests* of Van Steenis, the American *forêts superhumides* of Aubréville and the *pluviales* of Spanish-speaking authors; they are rare in Africa but occur in Queensland (Australia).

— Very humid climates: evergreen types of American and African *rain forests*, evergreen types of *monsoon forests* (=*semi-evergreen rain forests* of Baur); also occur in Australia.

— Humid climates: African *rain forests* of the semi-deciduous type, deciduous types of *monsoon forests* in Asia; infrequent in America (Central America, Venezuela, forest of Goias and Mato Grosso in Brazil) and Australia.

When dry conditions become more severe, the following types can be distinguished: the closed dry forests of southern central Africa and western Madagascar which have been described by various authors and mapped by Keay (1959); the deciduous dry forests of India and South-East Asia. Dry forests seldom occur in America; in Australia, certain closed stands of *Eucalyptus* spp. appear to represent equivalent formations. They are often subdivided into *dry evergreen forests*, and *dry deciduous forests* but it is not possible to establish clear boundaries between these two types.

Forest types and soils

Several authors have stressed the importance of edaphic factors for dry land forests without in any way denying the even greater importance of regional climates. They include Ashton (in Anon., 1958a; Anon., 1965) for Borneo, Robbins and Wyatt-Smith (1964), for the Malayan Peninsula, Schmid (in Anon., 1958b) for South-East Asia and for New Caledonia, and Mangenot (1955) for the Ivory Coast. There are physiognomic and floristic differences between forests growing on predominantly clayey soils and others growing on predominantly sandy soils. The latter appear to be relatively less luxuriant and less rich in species; this is true of the *heath forests* of Asia, the low and sometimes even shrubby forests of the Amazon basin (the *pseudo-caatingas* and *carrascos* which have been shown by RADAM, the recent Brazilian survey by radar photography, to be far more extensive than had previously been believed, especially in the river basins of the Rio Branco and the Rio Negro where they were studied by Takeuchi (1962) under the ambiguous name of *campinas*. Forests on calcareous soils seem to be of rather limited extension; most are dry forest types; woody formations on ultra-basic parent rocks are also more often close to dry thickets than to the humid forest type. However pedological factors are by no means always preponderant on the physiognomic level, and Murça Pires (see annex) discusses the case of the *matas de cipó* (forests with climbers) which occur along the Transama-zonian road between Marabá and Altamira on both very poor and rich soils; no satisfactory explanation of this phenomenon has yet been given. It may be due to palaeo-phytogeographic causes, in as much as a final stable relationship with soil types may not yet have been reached as suggested by Letouzey (1968) for the *forêts clairsemées à strate inférieure de Marantacées* (open forests with an understorey of Marantaceae) in the western part of Zaire. Although superficially ecologically homogeneous, computer analyses of species frequencies demonstrate subtle influences of topography, edaphic and other factors. In a study in Sarawak, Ashton (see chapter 8) was able to show that there is often no correlation between floristic and structural variations, the former being connected with nutrient factors (P content), the latter with rooting depth and the moisture retention capacity of the soil; similar results were reported by Austin *et al.* (1972) from Brunei and by Peeris (unpublished) from Sri Lanka. Such analyses of floristic variation are undoubtedly of great value in deepening understanding of the relations between plant communities and their environment.

Mention should also be made of the tendency towards monophytism in forests growing under poor edaphic conditions and its effect on their physiognomy. It manifests itself on hydromorphic soils; the forests often exhibit floristic compositions and physiognomic appearances of a highly peculiar character. Many authors have tried to describe their characteristics and the main types are the following.

— Mangroves are the subject of a very large literature, though their extension is in fact very limited and their interest merely local (exploitation for fuel wood or clearing for rice cultivation).

— Swamp forests immediately behind the coast, found in various parts of the world consist largely of stunted trees but sometimes gradually merge into inland types of swamp forests. They were studied in Sarawak under the name of *peat swamp forests*. They may derive from old mangroves.

— Riverine forests, which form narrow fringes along the banks of rivers and large streams, such as the *restingas* along the Amazon.

— Periodically inundated forests, which cover appreciable surfaces, especially the *mata de várzea* of the Amazon region characterized by an abundance of medium-sized trees with stilt roots, many buttressed trees with pneumatophores, an abundance of lianas and a poorly developed understorey. Several well-known types have been distinguished in Asia and Africa related to the amount of inundation, but in America there is need for more differentiation. For Amazonia, Murça Pires (see annex) describes the considerable differences between the *várzeas of the estuary* (forests without herbaceous plants, but with many palms) and *várzeas of the lower Amazon* (associated always with tall herbaceous plants, *canaranas*, but devoid of palms).

— Swamp forests are relatively small in extension but occur on all continents. The *igapó* of Amazonia is particularly well developed; palms are abundant in the understorey; the swamp forests of southern Mexico are similar whereas in African swamp forests this group of plants is almots exclusively represented by the genus *Raphia* (though the spread of the Raphiales in these forests appears frequently to be a result of human influence). Swamp forests are also found in Asia but appear to be much less important in Australia.

Forest types and altitude

Mountain forests differ from lowland forests in the reduced tree height, less obvious crown stratification, reduced number of buttresses, stems often crooked, smaller and coriaceous leaves, persistent foliage, less cauliflory, large lianas not well developed, stranglers often present, epiphyllic and epiphytic plants abundant (at least in the higher parts of such forests), mosses and lichens abundant on the forest floor, and here and there tall herbaceous plants, occasional palms and bamboos (especially in Asia, Madagascar and America). A detailed physiognomic, structural and floristic comparison between lowland forests (at 400 m) and mountain forests (at 1 700 m) was carried out by Grubb *et al.* (1963) on plots of 400 m² in Ecuador (and by other authors elsewhere).

Generally speaking and allowing for the mass elevation effect already mentioned, the lowland forest looses its physiognomic and floristic characteristics at an altitude of *ca.* 1 000 m where it changes into a mountain forest. Above this altitude, many authors distinguish between sub-montane and montane forests but the limits between these two formations are rather fluid and comparisons between continents, or even between regions are difficult. All major mountain ranges have *cloud forest* or *moist forest, forêt de nuage, forêt néphéliphile, Nebelwald, selva nublada, ceja de la montaña* in the Andes, or *mossy forest, sylve à lichens* in Madagascar, etc. The high humidity favours the development of mosses, lichens and epiphytes; often they are exposed to strong winds which uproot trees and cause gaps. Above the layer of clouds may be dry montane forests with evergreen foliage, a type which also occurs at lower altitudes on very dry slopes. At the upper limits of dry and humid mountain forests there is a rapid transition to stunted, shrubby formations (*elfin forest, elfin woodland*) which mark the upper boundary of woody vegetation and the passage to the herbaceous formations of high altitudes.

It has long been believed that in the tropics the effect of altitude counteracted the effect of latitude and that differences in physiognomy between mountain and lowland forests could be attributed exclusively to the temperature factor. Today, this explanation has been relinquished and expressions like *montane temperate forests* or *serva pluvial templada* have been dropped. A variety of climatic and edaphic factors, peculiar to the tropics, are now taken into account to explain the characteristic physiognomy of these mountain forests. Flenley (1974) discussed the importance of these ideas by reference to the mountains of Malesia, and Whitmore (1975) deals with them in a more detailed and general manner in an attempt to find explanations for the physiognomic and structural characteristics of different forest formations in mountain areas.

Conclusions: research needs and priorities

The floristic composition of intertropical forests is characterized by richness and heterogeneity.

Floristic studies of forest ecosystems should largely be based on existing herbaria and floras, which are very incomplete for tropical areas. For America in particular a great deal of effort is required.

The following measures and research projects are necessary to further the floristic study of these ecosystems.

1. More substantial material aid ought to be given towards the formation and operation of herbaria and the preparation of floras on local, national and regional levels.
2. The theoretical and practical training of taxonomic botanists and field foresters of all grades should be intensified; in particular there is need for improvement as regards the identification of trees and lianas in the field and the collection of botanical material.
3. The publication of serious popular floras (such as the *Flora of Nigeria* and *Nigerian Trees*) should be encouraged.
4. Scientific names should replace vernacular ones in all inventories.
5. All inventories undertaken for economic purposes should be at least partly combined with properly conducted botanical inventories. However, the inclusion of such scientific inventories must not be used to conceal the aims of the forest inventories themselves which may involve a harsh and often completely destructive modification of the ecosystem.
6. Botanical inventories carried out with the aim of studying an ecosystem must take into account all plants present, since all such plants have affected, are affecting or may in future affect the life of the ecosystem from the points of view of ecology, biology, economy, etc. Floristic inventories of secondary forests

are equally important, for such forests are the object of greatly varying forms of utilization by the local populations.

It is difficult to apply to communities of plants in tropical forests the phytosociological methods employed in temperate regions where the flora is completely known and the structures are very much less complex; provisionally at any rate, the determination of leading species should be concentrated on.

The study of the continental distribution of the families and genera of plants is hampered by the gaps in taxonomic knowledge. In particular the study of certain large genera of importance in the understorey of forests has hardly begun. A brief survey of nearly 100 families provided a picture of the richness and complexity of the floristic composition of tropical forests. Studies on a very limited number of important families (Coniferae, Leguminosae, Palmae, Dipterocarpaceae, Sapotaceae, Orchidaceae) have given a more integrated insight into the place and role of these families on certain continents but the number of such studies is still regrettably small. Knowledge of the non-vascular plants is still very fragmentary indeed.

1. Monographs on certain large genera (*Psychotria, Ixora, Rinorea, Ourata, Drypetes, Symplocos, Pavetta, Garcinia, Memecylon, Elaeocarpus, Saurauia, Miconia, Cuphea, Alpinia, Anthurium, Begonia, Peperomia, Piper, Tillandsia*, etc.) should be encouraged and supported.
2. The study of the non-vascular plants of tropical forest ecosystems should be intensified.

The complexity of floristic compositions and the gaps in knowledge make it extremely difficult to draw a picture, however sketchy, of the floristic content of each continent at the level of families, sub-families or tribes of genera.

The terminology of forest types is extremely complicated and confused, even when leaving out the difficulties caused by local names, the problems inherent in translation and the arbitrary lines along which ecological, climatic and topographical factors are subdivided. Several general classifications of forest types have been published during the last 50 years and an enormous number have been proposed at a local or national scale; these have been combined by some authors on a regional or continental basis; usually their systems have included all types of vegetation. Recently attempts have been undertaken to construct classifications applicable to all terrestrial vegetation. The suggested universal classifications are based on physiognomic, structural, physiological and ecological considerations, but the priority accorded to these three or four factors varies; they all disregard floristic, dynamic and historical factors except possibly as elements justifying subdivisions. The so-called Unesco classification (Anon., 1969), a synthesis elaborated by a group of people working on a world scale, is a great achievement. Already some tests of application have been made and allow to assess its merits and defects.

1. The so-called Unesco classification should henceforth be followed for the definitions, distinctions, and

cartographic projects (which should agree as regards colours, textures and symbols).

2. Particular attention should be given to preparing an official translation of the terminology employed in that classification into a number of languages.

3. The synonymy of terms so far used with those used in the new classification should be worked out. Such an operation would make it possible to modify, complete and enlarge the Unesco classification if necessary.

Universal classifications are important; they are more than just intellectual exercises. In the same way as do the taxonomic names used in floristics, the names of vegetation types should correspond to certain characteristics which should be as precise as possible; such a name, forming part of a hierarchical classification, is a convenient means of scientific synthesis and enables valid comparisons to be made in the fields of physiognomy, ecology, biology, economy, etc.

Modern methods of numerical analysis have an im-

portant part to play in the refinement of knowledge of the relations between floristic elements and ecological factors. Notwithstanding certain inconsistencies on an intercontinental level, the existence of links between forest types and gradients of humidity or dryness appear to have been established. Similarly, the forest types associated with the major distinct soil types are well known. The altitudinal mass elevation effect has not been satisfactorily explained. While cloud forests refer to well defined units of vegetation on all continents, the great confusion still reigning above and below their zone can only be solved by means of detailed observations of the floristics and on various factors of soil and climate that characterize mountain areas.

a. Detailed studies of floristic variations using numerical analysis and based on adequate sampling, should be encouraged to increase ecological knowledge.

b. Special attention should be given to the various factors of climate and soil that characterize mountain areas.

Selective bibliography

AIRY SHAW, H. K. (Willis, J. C.) *A dictionary of the flowering plants and ferns*. 8th ed. Cambridge, Cambridge University Press, 1973, 1 245+66 p.

ANON. *C.S.A./C.C.T.A. Phytogéographie—Phytogeography, Yangambi (1956)*. London, Publ. n° 22, 1956; réimp. Publ. n° 53 (1961), 35 p.

——. *Proceedings of the Symposium on humid tropics vegetation. Tjiawi (Indonesia)*. Council for Sciences of Indonesia and Unesco Science Co-operation Office for South-East Asia, 1958a, 312 p.

——. *Study of tropical vegetation. Proceedings of the Kandy Symposium. Recherches sur la zone tropicale humide*. Paris, Unesco, 1958b, 226 p.

——. *Symposium on ecological research in humid tropics vegetation. Kuching, Sarawak (1963)*. Government of Sarawak and Unesco Science Co-operation Office for South-East Asia, 1965, 376 p.

——. *A framework for a classification of World vegetation*. Paris, Unesco, 1969, 26 p. multigr. *International classification and mapping of vegetation*. Ecology and Conservation no. 6. *Classification internationale et cartographie de la végétation*. Écologie et Conservation n° 6. Paris, Unesco, 1973, 93 p.

——. *Ressources naturelles de l'Asie tropicale humide*. Paris, Unesco, Recherches sur les ressources naturelles, vol. XII, 1974, 490 p.

ASHTON, P. S. *Ecological studies in the mixed dipterocarp forests of Brunei State*. Oxford Forestry Memoirs, 25, 1964, 110 p.

——. Some problems arising in the sampling of mixed rain forest communities for floristic studies. In: *Ecol. Res. Humid Tropics Veg.*, Proc. Kuching Symposium, p. 235–240. Paris, Unesco, 1965, 376 p.

—— and M. (eds.). *The Quaternary era in Malesia*. Transactions of the second Aberdeen-Hull Symposium on Malesian Ecology (Aberdeen, 1971). University of Hull, Department of Geography, Miscellaneous Series no. 13, 1972, 122 p.

AUBRÉVILLE, A. *Climats, forêts et désertification de l'Afrique tropicale*. Paris, Soc. Éd. géographiques, maritimes et coloniales, 1949a, 351 p.

AUBRÉVILLE A. *Étude écologique des principales formations végétales du Brésil et contribution à la connaissance des forêts de l'Amazonie brésilienne*. Nogent-sur-Marne, Centre technique forestier tropical (CTFT), 1961, 268 p.

——. La forêt dense de la Lobaye. *Cahiers de la Maboké* (Paris), 2, 1964, p. 5–9.

——. Sapotacées. *Adansonia*, Mém. n° 1, 1965a, 157 p.

——. Principes d'une systématique des formations végétales tropicales. *Adansonia*, 5, 1965b, p. 153–196.

——. Les Césalpinioïdées de la flore camerouno-congolaise. *Adansonia*, 8, 1968, p. 147–175.

AUSTIN, M. P.; GREIG-SMITH, P.; WHITMORE, T. C. The application of quantitative methods to vegetation survey. I. Association analysis and principle component ordination of rain forest. *J. Ecol.*, 55, 1967, p. 483–503.

——; GREIG-SMITH, P. The application of quantitative methods to vegetation survey. II. Some methodological problems of data from rain forest. *J. Ecol.*, 56, 1968, p. 827–844.

——; ASHTON, P. S.; GREIG-SMITH, P. The application of quantitative methods to vegetation survey. III. Reexamination of rain forest data from Brunei. *J. Ecol.*, 60, 1972, p. 305–324.

BAUR, G. N. *The ecological basis of rain forest management*. Forestry Commission New South Wales, 1961–62. Rome, FAO, André Meyer Fellowship Programme Report, 1962, 499 p.

BEARD, J. S. The classification of tropical American vegetation types. *Ecology*, 36, 1955, p. 89–100.

BOUGHEY, A. S. The physiognomic delimitation of West African vegetation types. *J. West Afr. Sc. Ass.*, 3, 1957, p. 148–165.

BRIEGER, F. G. Botanische Grundlagen der Orchideenforschung. In: Brieger, F. G.; Maatsch, R.; Senghas, K. (Schlechter, R.). *Die Orchideen*, p. 123–137. Berlin und Hamburg, Parey, 1971.

BURTT DAVY, J. *The classification of tropical woody vegetation types*. Oxford, Imp. For. Inst. Paper no. 13, 1938, 85 p.

CAIN, S. A.; DE OLIVEIRA CASTRO, G. M. *Manual of vegetation analysis*. New York, 1959, 325 p.

CHAMPION, H. G. A preliminary survey of the forest types of India and Burma. *Indian For. Rec.*, n.s. 1, 1936, p. 1–286.

CHAMPION, H. G.; SETH, S. K. *Forest types of India*. Dehra Dun, Forest Research Institute, 1965, revised ed. 1968, multigr.

COUSENS, J. E. Some notes on the composition of lowland tropical rain forest in Rengam Forest Reserve, Johore. *Malayan Forester*, 14, 1951, p. 131–139.

DILLEWIJN, F. N. Van. *Sleutel voor de interpretatie van begroeiingsvormen uit luchtfoto's 1:40 000 van het Noordelijk deel van Suriname*. Paramaribo, 1957, 45 p.

ELLENBERG, H.; MUELLER DOMBOIS, D. Tentative physiognomic-ecological classification of plant formations of the earth, based on a discussion draft of the Unesco working group on vegetation classification and mapping. *Ber. geobot. Inst. ETH, Stiftg. Rübel* (Zürich), 37, 1967, p. 21–55.

EMBERGER, L. Les végétaux vasculaires. In: Chadefaud M.; Emberger, L. *Traité de Botanique systématique*. 2 vol. Paris, Masson, 1960, 1 539 p.

ESPINAL, L. S.; MONTENEGRO, E. *Formaciones vegetales de Columbia. Memoria explicativa sobre el mapa ecológico (1/1 000 000)*. Bogotá, 1963, 201 p.

FLENLEY, J. R. (ed.). *Altitudinal zonation in Malesia*. Transactions of the third Aberdeen-Hull Symposium on Malesian ecology (Hull, 1973). University of Hull, Dept. of Geography, Misc. ser. no. 16, 1974, 119 p.

FOSBERG, F. R. A classification of vegetation for general purposes. In: Peterken, G. F. *Guide to the checksheet for IBP areas. IBP Handbook no. 4*, 2nd ed., p. 73–120. London, Oxford, Blackwell, 1970, 133 p.

FOX, J. E. D. An enumeration of lowland dipterocarp forest in Sabah. *Malayan Forester*, vol. 30, no. 4, 1967, p. 263–279.

FRANCIS, W. D. *Australian rain forest trees*. Sydney, London, 1951, 469 p.

GÉRARD, P. *Étude écologique de la forêt dense à Gilbertiodendron dewevrei dans la région de l'Uele*. Publ. INEAC (Bruxelles), Sér. sci., n° 87, 1960, 159 p.

GERMAIN, R.; ÉVRARD, C. *Étude écologique et phytosociologique de la forêt à Brachystegia laurentii*. Publ. INEAC (Bruxelles), Sér. sci., n° 67, 1956, 105 p.

GOTTWALD, H. P. J. L'identification et l'appellation des bois de 'Lauan' et de 'Meranti'. *Bois et Forêts des Tropiques*, 121, 1968, p. 35–45.

GREIG-SMITH, P. Notes on the quantitative description of humid tropics vegetation In: *Ecol. Res. Humid Tropics Veg.*, Proc. Kuching Symposium, p. 227–230. Paris, Unesco, 1965, 376 p.

GREENWAY, P. J. A classification of East African vegetation. *Kirkia*, 9, 1973.

GRUBB, P. J.; LLOYD, J. R.; PENNINGTON, T. D.; WHITMORE, T. C. A comparison of montane and lowland rain forest in Ecuador. *J. Ecol.*, 51, 1963, p. 567–601.

HALLÉ, F.; LE THOMAS, A.; GAZEL, M. Trois relevés botaniques dans les forêts de Bélinga. Nord-Est du Gabon. *Biologica Gabonica*, vol. 3, n° 3, 1967, p. 3–16.

HEINSDIJK, D. *Interim report to the Government of Brazil on the dry land forests on the Tertiary and Quaternary south of the Amazon River*. Rome, FAO Report no. 1284, 1960.

——; DE MIRANDA BASTOS, A. *Report to the Government of Brazil on forest inventories in the Amazon*. Rome, FAO Report no. 2080, 1965.

HUECK, K.; SEIBERT, P. Vegetationskarte von Südamerika. In: Walter, H. *Vegetationsmonographien der einzelnen Grossraüme*. Band IIa. Stuttgart, 1972.

HUMBERT, H.; COURS DARNE, G. *Notice de la Carte de Madagascar, au 1/1 000 000*. CNRS/ORSTOM. Carte Internationale du Tapis végétal, 1965.

JACOBS, M. (ed.). *Flora Malesiana Bulletin*. Leiden, Foundation Flora Malesiana and Rijksherbarium. Annual review providing information and contact between institutes and individual botanists of Malesia and the tropical parts of Asia, Australia and the Pacific, in botany, ecology, plant geography, exploration and bibliography.

JONES, E. W. Ecological studies in the rain forest of Southern Nigeria. IV. The plateau forest of the Okomu Forest Reserve. Part 1. The environment, the vegetation types of the forest and the horizontal distribution of species. *J. Ecology*, 43, 1955, p. 564–594.

KEAY, R. W. J. *Vegetation map of Africa south of the tropic of Cancer, 1/1 000 000*. Oxford, Unesco/AETFAT, 1959.

KNAPP, R. *Die Vegetation von Nord und Mittelamerika*. Stuttgart, 1965, 373 p.

——. *Die Vegetation von Afrika*. Stuttgart, 1973, 626 p.

LANJOUW, J. *Index Herbariorum (A guide to the location and contents of the World's public Herbaria)*. Lanjouw, J.; Stafleu, F. A. *The Herbaria of the World*. Part 1. IUBS/Unesco; Utrecht (106 Lange Nieuwstraat), The International Bureau of Plant Taxonomy and Nomenclature, 1964.

LEBRUN, J.; GILBERT, G. *Une classification écologique des forêts du Congo*. Publ. INEAC (Bruxelles), Sér. sci., n° 63, 1954, 89 p.

LETOUZEY, R. *Étude phytogéographique du Cameroun*. Paris, Lechevalier (Encyclopédie Biologique 69), 1968, 508 p.

——. Flore du Cameroun. *Boissiera* (Genève), 24, 1976, p. 571–573.

LOWE-MCCONNELL, R. H. Speciation in tropical environments. *Biol. J. Linn. Soc.*, 1, 1969, p. 97–133 (Van Steenis, C. G. G. J.), p. 135–148 (Hedberg, O.), p. 149–153 (Richards, P. W.).

MACARTHUR, R. H.; WILSON, E. O. *The theory of island biogeography*. Princeton, N. J., Princeton University Press, 1967, 203 p.

MANGENOT, G. Étude sur les forêts des plaines et plateaux de la Côte d'Ivoire. *Études éburnéennes*, 4, 1955, p. 5–81.

MEGGERS, B. J.; AYENSU, E. S.; DUCKWORTH, W. D. *Tropical forest ecosystems in Africa and South America: a comparative review*. Washington, D.C., Smithsonian Institution, 1973, 350 p.

MELCHIOR, H. *A. Engler's Syllabus der Pflanzenfamilien*, II. Berlin, 1964, 666 p.

MILDBRAED, J.; DOMKE, W. Grundzüge der Vegetation des tropischen Kontinental-Afrika. *Willdenowia*, 2, 1966, 253 p.

MONOD, Th. Après Yangambi (1956): notes de phytogéographie africaine. *Bull. IFAN*, vol. 25, sér. A, 2, 1963, p. 594–619.

PRANCE, G. T. Phytogeographic support for the theory of Pleistocene forest refuges in the Amazon basin, based on evidence from distribution patterns in *Caryocaraceae, Chrysobalanaceae, Dichapetalaceae* and *Lecythidaceae*. *Acta amazonica*, 3, 1973, p. 5–28.

RAVEN, P. H.; AXELROD, D. I. Angiosperm biogeography and past continental movements. *Annals Missouri Botanical Garden*, 61, 1974, p. 539–673.

REES, T. I. *Report to the Government of British Guiana*. Expanded Program of technical assistance. Rome, FAO, Report no. 1762, 1963, 68 p.

RICHARDS, P. W. *The tropical rain forest: an ecological study*. Cambridge, Cambridge University Press, 1952, 450 p. 4th reprint with corrections, 1972.

ROBBINS, R. G.; WYATT-SMITH, J. Dry land forest formations and forest types in the Malayan Peninsula. *Malayan Forester*, 27, 1964, p. 188–216.

ROLLET, B. *L'architecture des forêts denses humides semperviventes de plaine*. Nogent-sur-Marne, Centre technique forestier tropical (CTFT), 1974, 298 p.

SCHIMPER, A. F. W.; VON FABER, F. C. *Pflanzengeographie auf physiologischer Grundlage*. 3 ed. Iena, 1935, 2 vol., 1 613 p.

SCHMID, M. *Végétation du Viet-Nam. Le massif sud-annamitique et les régions limitrophes*. Paris, Mémoire ORSTOM n° 74, 1974, 243 p.

SCHNELL, R. Remarques préliminaires sur quelques problèmes phytogéographiques du Sud-Est asiatique. *Rev. gén. Bot.*, 69, 1962, p. 301–366.

——. *Introduction à la phytogéographie des pays tropicaux. Les problèmes généraux*. 2 vol. Paris, Gauthier-Villars, 1970 et 1971, 500 p., 452 p.

——. *Introduction à la phytogéographie des pays tropicaux. La flore et la végétation de l'Afrique tropicale*. Paris, Gauthier-Villars, 2 vol., 1977, 950 p.

SCHULZ, J. P. *Ecological studies on rain forest in northern Surinam*. Amsterdam, 1960, 267 p.

STEENIS, C. G. G. J. Van. On the origin of the Malesian mountain flora. *Bull. Jard. Bot. Buitenzorg*, ser. 3, 13 and 14, 1934, 1935 and 1936, p. 135–362, 289–417 and 56–72.

——. Tropical lowland vegetation: the characteristics of its types and their relation to climate. In: *Proceedings of the Ninth Pacific Science Congress* (1957), vol. 20 (Humid tropics), 1958a, p. 25–37.

——. *Commentary on the vegetation map of Malaysia (1/5 000 000)*. Unesco, Humid tropics research projects, 1958b.

TAKEUCHI, M. The structure of the Amazonian vegetation. *J. Fac. Sc. Univ. Tokyo*, Sect. III, Bot. 8, 1961, p. 1–26, 27–35; 1962, p. 279–288, 289–296, 297–304.

TROCHAIN, J. L. Accord interafricain sur la définition des types de végétation de l'Afrique tropicale. *Bull. Inst. Ét. Centrafr.*, 13–14, 1957, p. 56–93.

VAN DER HAMMEN, T. Climatic periodicity and evolution of South American Maestrichtian and Tertiary floras. *Boletín Geológico* (Bogotá), vol. 5, no. 2, 1957, p. 49–91.

VARESCHI, V. Comparación entre selvas neotropicales y paleotropicales en base a su espectro de biotipos. *Acta Botanica Venezuelica*, 3, 1968, p. 239–263.

WALTER, H. *Ecology of tropical and sub-tropical vegetation* (transl. Mueller Dombois, D.). Edinburgh, Oliver and Boyd, 1971, 539 p.

WEBB, L. J. A physiognomic classification of Australian rain forest. *J. Ecol.*, 47, 1959, p. 551–570.

——; TRACEY, J. G.; WILLIAMS, W. T.; LANCE, G. N. Studies in the numerical analysis of complex rain forest communities. I. A comparison of methods applicable to site/species data. *J. Ecol.*, 55, 1967, p. 171–191.

——; ——; ——; ——. Studies in the numerical analysis of complex rain forest communities. II. The problem of species sampling. *J. Ecol.*, 55, 1967, p. 525–538.

WHITE, F. The taxonomic and ecological basis of chorology. In: *Mitt. Bot. München 10, Proceedings 7th plenary Meeting AETFAT* (Munich, 1970), 1971.

WHITMORE, T. C. *Tropical rain forests of the Far East*. Oxford, Clarendon Press, 1975, 278 p., 550 references.

Organization

Definition

Closed tropical forests may be studied from the point of view of their organization, i.e. how they are built, their architecture and the kinds of structure hidden in the apparently disorderly mixture of trees and species.

Although the emphasis is on lowland evergreen tropical rain forests, which are the most complex, the following formations are also considered:
— semi-deciduous rain forests
— deciduous rain forests
— montane rain forests
— freshwater swamp forests
— heath forests
— mangroves.
Tree and shrub savannas and open deciduous forests such as dry dipterocarp or miombo forests are not considered.

It should be pointed out that there is a continuum of types between open and closed deciduous forests; all kinds of transitional forests exist between the two extremes and they have been given convenient local names (llanos, miombo, dry dipterocarp forest, etc.). An open canopy normally induces a herbaceous layer, the presence of which is a criterion for exclusion from the present study.

Much work is still required to reach acceptable definitions for the various closed tropical formations, and equivalence between nomenclatures of the various authors (see chapter 4). Even at the highest level of the tropical forest hierarchy, considerable effort is needed for the systematization of phenological and mapping studies.

Much success has been achieved in the study of the organization of closed tropical forests. The best known are the lowland rain forests, i.e. lowland wet evergreen and semi-deciduous rain forests, because of their commercial importance. The numerous available forest inventories make it possible to draw conclusions about their architecture and their structures. The phrase organization of a forest covers two concepts: the architecture of this forest and its internal structures, i.e. the geometry of the stands and the tree populations.

The architecture of a forest is reflected in an assemblage of relationships in the dimensions of the various parts of the trees (irrespective of species). Therefore, the study is primarily morphological. It seems useful to distinguish the principal biological types: trees and shrubs, climbers, epiphytes, etc. Their relative importance and their position in the ecosystem vary and can be studied separately. Trees are obviously the most important biological type. Study methods will be different for each biological type because of dimensional differences and varying degrees of difficulty in gathering and identifying specimens and in drawing individuals *in situ*. The architecture means at least two different things: the general architecture of the forest stand and the architecture of the various tree species.

The architecture of tropical tree species has been studied by a number of authors (e.g. Corner, 1952; Koriba, 1958). An interesting treatise on the subject is available in Hallé and Oldeman (1970), who distinguish a number of architectural types. Their effect on the morphology of riverain and dry-land rain forests has been described by Oldeman (1972a, b) who maintains that several levels of branching occur in the canopy. The habit of trees is unfortunately not sufficiently considered in floras and dendrologies; readers with little tropical experience can hardly visualize the infinite variety of types which are visible on photographs or on accurate profile diagrams (Hallé *et al.*, 1967). These individual shapes give distinct aspects to the forests, especially in the case of gregarious species which are particularly obvious on aerial photographs, e.g. stands of *Gilbertiodendron*, or riverain *Guibourtia* forests in Africa and, in general, most stands which contain a small number of species (deciduous *Lagerstroemia* forests, *Terminalia-Triplochiton* formations, freshwater swamp forests and mangroves). In the case of rain forests, the multiplicity of species often makes species differentiation difficult on aerial photographs; on small scales, e.g. 1/50 000, it is practically impossible to separate forest types except by a stratification according to crown density and size.

The word structure has been used in many ways to describe assemblages of trees or species which seem to follow particular mathematical laws, e.g. distributions of diameters at breast height, i.e. 1.30 m, of total heights, spatial distributions of trees and species, of floristic diversity, and of associations. It is therefore possible to speak of diameter structures, height structures, crown structures, canopy structures, spatial structures, etc. A more complete definition of structure will be examined later, but it is clear that the biological significance of the phenomena, as expressed by mathematical formulae, is the real objective of the investigations.

Architecture

Various general systems have been proposed to describe the architecture of plant communities: Tansley and Chipp (1926), Burtt Davy (1938), Richards *et al.* (1940), Dansereau (1951), van Dillewijn (1957), Addor *et al.* (1970). The number of published photographs which illustrate works on tropical forests is very large, but many authors are more sensitive to aesthetic than to scientific motivations. Van Dillewijn's (1957) method, supplemented by profiles and inventories, should be used.

Profiles

The architecture of forests has often been described by means of transects or profiles. The relevant literature on lowland evergreen rain forests has been reviewed by Rollet (1968, 1969, 1974), who describes the principal styles of the various authors.

Different dimensional parameters, such as width, length of profile, and the minimum height of shrubs or trees, are used by various authors. Similarly, there is considerable divergence in the amount of simplification which is used in the drawing of the vertical profile, in the horizontal projection in which the position of the trees is shown or in the drawing of buttresses, and the projection of crowns and fallen dead trees. On vertical profiles, deciduous trees can be distinguished from evergreens, and the stems and crowns of vines, and some groups, such as dipterocarps, conifers and epiphytes, can be distinguished.

In view of the above, the results are extremely varied, from the schematic profiles of Smit (1964) for Colombia, Cousens (1951) and Robbins (1964) for Malaysia, Fanshawe (1952) for Guyana, Stehlé (1945, 1946) for the West Indies, to more sophisticated profiles like those of Hallé *et al.* (1967) for Gabon, Lindeman and Moolenaar (1959) for Surinam, Oldeman (1972a) for French Guyana and Ashton (1964) for Brunei. Some authors have produced profiles without any defined width, only to show the general outlook of the stands (Aubréville, 1947, 1949). Obviously, a wide range of intermediates exists between the simplest and the most elaborate type.

British authors have produced numerous profiles in a number of tropical countries, following procedures recommended by Burtt Davy (1938) and Richards *et al.* (1940). Profile dimensions were: 25 feet width, 200 feet length; 15 feet for minimal height. Davis and Richards actually developed this method in Guyana in 1933.

The multiplicity of independently conceived methods has resulted in a range of representations; e.g. for widths: 25 feet (7.6 m), 66 feet (*ca.* 20 m), Cousens (1951); and 5 m, 10 m, 20 m or 30 m strips 5 or 10 m wide; for minimum heights 1 and 3 m (Aubréville, 1961), 1 and 5 m (Aubréville, 1947), 3 m (Takeuchi, 1961, 1962), 5 m, and sometimes 10 and 20 m (Lindeman and Moolenaar, 1957), 4 m (Rollet, 1969). Instead of minimum heights, minimum diameters have been also used: 10 cm (Lamprecht and Veillon, 1957), 6 inches girth or about 5 cm DBH and 2 feet girth or about 20 cm DBH (Holmes, 1958).

Authors very rarely accurately portray boughs and foliage which occur in the geometric profiles. This particularly applies to climbers, leaning stems and crown volumes.

Drawing a profile of, for example, 10×50 m, may take as much as 4–8 days or more, depending on the degree of accuracy, the detail shown in the crowns, and the need for specimens for species determination. In spite of this amount of effort, the value of these profiles should not be overestimated, in view of their small area and the variability of the forest; one should realize that two contiguous or neighbouring profiles can show very different crown structures and stem distributions. However, a profile can certainly portray important aspects of stand geometry, understorey, crowding and morphological characteristics of the forest, such as crown position, size distribution of stems, conspicuous biological types (e.g. tree ferns, palm trees, conifers, vines, epiphytes and the form of stem bases). The total area in the world surveyed in this way is probably not more than 30 ha, which is a minute fraction of the area occupied by tropical forests; this poses therefore a problem of representativeness. It is possible to imagine that this same area, conveniently distributed among all the tropical forests, e.g. 400–500 profiles 1/20 to 1/10 ha each, could give an excellent picture of reality.

If the morphology of the main tropical formations, or their architectural types, can be well characterized through profiles, it is doubtful whether forest types, as characterized by their floristic composition, could be distinguished in the same way. Excessive cost, as well as too much simplification, are extreme situations which should be avoided if profiles are designed to describe the morphology and architecture of the main tropical formations with acceptable accuracy. The ideas of Newman (1954) on idealized or reworked profiles, the simplifications of Dansereau (1951) appear to require an exaggerated manipulation of the observations and to be too arbitrary a representation of reality. Finally, with a very advanced interpretation of the data of one, or a very small number, of profiles (Takeuchi, 1961, 1962a, b) one may overestimate the quantitative significance of the conclusions, which necessarily can reflect only what the sample can show.

Many general handbooks on phytogeography show profiles from various sources, often simplified from the original. Aubréville (1965) used Hallé's oblique view of highly representative block-diagrams portraying the main tropical vegetation types: wet evergreen rain forest, semi-deciduous rain forest, montane rain forest, dry deciduous thicket, open deciduous forest, savanna woodland. Francke (1942) had similar ideas but for profiles. These representations have considerable value for summarizing architectural types and phenological behaviour.

Non-exhaustive lists of forest profiles have been established by formations, and by countries within each formation. The following formations are considered: lowland evergreen tropical rain forests, semi-deciduous rain forests, montane rain forests, freshwater swamp forests, mangroves, heath forests or kerangas. These profiles can be found in the publications of the following authors: Davis *et al.* (1933, 1934), Richards (1936, 1939, 1952, 1963), Brooks (1941), Beard (1942, 1944b, 1946, 1949), Stevenson (1942), Nelson-Smith (1945), Stehlé (1945, 1946), Aubréville (1947, 1961), Eggeling (1947), Foggie (1947), Louis (1947), Donis (1948), Cousens (1951), Fanshawe (1952), Asprey *et al.* (1953), Lamprecht (1954, 1956, 1958, 1964), Jones (1955), Germain *et al.* (1956), Lamprecht *et al.* (1957), Holmes (1958), Keay (1959), Lindeman *et al.* (1959), Robbins (1959, 1968), Anderson (1961), Burgess (1961), Rodrigues (1961), Takeuchi (1961, 1962), Thai-van-Trung (1962), Veillon (1962), Grubb *et al.* (1963), Ashton (1964), Robbins *et al.* (1964), Smit (1964), Voorhoeve (1964), Mayo Melendez (1965), Morales (1966), Vega *et al.* (1966), Whitmore (1966), Hallé *et al.*

(1967), Huttel (1967), Brünig (1968a), Vareschi (1968), Hozumi *et al.* (1969), Fox (1970), Smith (1970), Blasco (1971), Havel (1972), Oldeman (1972) and Rollet (1974).

In conclusion, the number of available forest profiles is still very low and is not sufficiently representative. The different representations and styles of the authors certainly provide useful information but the dimensions of the profiles and the rules for recording the data should be standardized through international recommendations. It is difficult to extrapolate results from a small number of profiles (their length does not exceed 50–60 m generally) and it is misleading to draw conclusions on the presence or absence of strata from the evidence of a single profile. Formations are easily represented by profiles and it could be interesting to use them to describe certain forest types.

Distribution of number of trees by diameter classes

Introduction

At least one measurement is considered in any kind of forest inventory, namely diameter at breast height (DBH) or girth at breast height (GBH). This is understandable as DBH is the easiest tree measurement, in spite of the difficulty caused by the presence of buttresses. All other measurements are more difficult: commercial height to the first defect, height to base of crown, total height, diameters at various other levels, crown diameters, etc. Moreover, DBH is the main parameter for gross volume prediction, as there is a strong correlation between this volume and the square of the diameter.

Many inventory data are available for trees already classified into diameter classes, and several million trees have been recorded and processed in this way. Measurement units, class intervals, and lower limits for diameter or girth, differ significantly from one country to another. Frequently, a limited number of species are taken into account. Hence, only complete inventories down to a certain diameter are considered here. Many inventories of this type are available down to 60 cm DBH or 6 feet girth breast height (GBH), or even down to 40 cm, and a much lesser number down to 20 cm. With smaller DBH, the number of species is so high and the difficulty of identification so great, that inventories down to 10 cm or 5 cm or less are extremely scarce and limited to small areas. Of course the understanding of rain forest composition and structure needs all the trees to be measured to the smallest possible DBH; due to the inverse relationship between DBH and stem number subsampling techniques must therefore be used, at the lower end of the diameter range.

It should be noted that, with a 10 cm-diameter class interval, the number of trees decreases almost geometrically from large to small DBH classes, with nearly twice as many trees in each class as in the one above. Nevertheless, a careful study of data shows that the ratio itself always increases from large to small diameter classes, and that the geometric progression model does not hold if small DBSs are included. Hence there is a need for improved models. For almost a century, many authors have drawn attention to this DBH distribution, and proposed various tentative mathematical models: de Liocourt (1898), Huffel (1919–26), Schaeffer *et al.* (1930), Meyer (1933, 1952), François (1938), Prodan (1949), Le Cacheux (1955), Dawkins (1958), Pierlot (1966), Loetsch *et al.* (1967), Zöhrer (1969), Rollet and Caussinus (1969), Caussinus and Rollet (1970).

The main conclusion is that authors have mostly been in search of a mathematical representation which would fit their data; they were satisfied when this fit was statistically acceptable, and they generally did not look for biological interpretations of the parameters. Le Cacheux (1955) was probably the first to give some theoretical reason for the exponential model.

On the other hand, the more abundant the data, the more sensitive the statistical tests. They consequently reject the model and ever more improved models are needed which, in their turn, are discarded when additional data become available. This search for perfection may well be considered as pure mathematical virtuosity. Hence, this approach cannot but lead to an impasse. Undoubtedly a better understanding of competition and growth, regeneration and mortality, and possibly mechanical properties of trees, will be required to produce a model which accurately depicts biological reality. Diameter distributions of all species are only one characteristic of the stand, reflecting an equilibrium which exists wherever rain forests are in a natural state. Distributions of other characteristics, such as total height, crown volume, basal area and biomass, will be considered later; DBH distributions for each species can help to clarify the concept of the balanced primitive forests. Data on total heights are too few to allow a similar interpretation by species.

Distribution of trees by diameter classes

For conciseness, but with the disadvantage of poor terminology, this distribution will be refered to as 'total structure'. A tentative explanation of this distribution is found in Rollet (1974), where the theoretical justification of the model, choice of a 1– or 2– parameter model, Pareto's model, goodness of fit, comparison of stands, and influence of slope and altitude are discussed.

The larger the area and the smaller the lower DBH limit, the higher is the chance that the model will be rejected for poorness of fit. This may lead to an excessive rigour concerning the relevance of models which describe biological phenomena. However, the quest for statistically and biologically sound models is worth while and should be continued.

Most investigations of 'total structure' show a high variability of tree number within each DBH class. With a plot size of 1 ha, it can be shown that the number of trees in the 10–19 cm class is distributed according to a flattened, bell-shaped histogram in a given region, and that the same shape of curve is found in other, distant regions; e.g. Venezuela (fig. 1), Brazil, Gabon, Kampuchea, etc.

When higher DBH classes are considered, this distribution becomes less and less flat, more and more asymetrical, then L shaped with many plots without trees. This

tendency seems quite general; quantitatively it shows, for a given plot size, the actual fluctuation of the means of the smoothed frequency curve in each DBH class.

Influence of slope and altitude

No reliable statistical study has ever been undertaken. Lamprecht (1954) thinks that total structure reflects the environmental conditions in the montane rain forests of Venezuela. In Brazilian Amazonia, Heinsdijk (1957) has distinguished the forests according to their topographic situation (upper slope, lower slope, upper and lower plateaux). The slopes seem to be richer in small stems and poorer in large stems than the plateaux. De Milde and Groot (1970) distinguished several forest types in Guyana: forests on rolling terrain or on hills do not seem to show very different total structures; on the contrary, riverain or lateritic terrace forests show lower stem densities, especially between 10 and 29 cm DBH.

Altitude differences of about 400 m in the Imataca range, in Venezuela, seem too small to induce noticeable differences in total structures. Steep slopes ($\geqslant$30 per cent) have fewer stems per hectare than slight or moderate slopes (Rollet, 1969). According to White (1963) the total number of stems per hectare increases from 600 to 900 m altitude in Puerto Rico.

Pierlot (1966) published forest inventories from Zaire at altitudes between 450 m and 2 200 m; apparently no well defined altitudinal gradient can be found for the total structures, but, in southwestern Nigeria, low stands without large diameters do occur only on narrow high ridges (ridge forests of Ray, 1971).

On the contrary Wyatt-Smith (1960) found a higher number of stems per unit area on ridges than on upper or lower slopes in Malaysia. The trend towards an increase in the number of stems with altitude—if there is any—is weak according to Arnot's (1934) data from Malaysia, but in Brunei, Ashton (1964), showed that slopes or low ridges were richer in large stems and poorer in small stems than valleys. Rollet (1974) observed high large stem densities in Sarawak and Sabah on tops of dissected plateaux 50–80 m above valley lowlands.

Veillon (1965) seems to be one of the few authors who studied systematically how the principal characteristics of undisturbed forest types vary with an increase in altitude (50 plots each 0.5 ha scattered between 70 and 3 250 m in the Venezuelan Andes). The number of trees with a DBH$\geqslant$20 cm per plot increased progressively from 52 to 138 between 70 m and 1 590 m, it then fluctuated somewhat with a relative maximum of 123 at 1 940 m and an absolute maximum of 168 at 2 960 m, it then fell to 71 stems at 3 250 m.

Influence of forest formations and of forest types

The various closed tropical forests (mangrove, swamp forest, lowland and montane evergeen rain forest, semideciduous rain forest, dry deciduous forest) can be easily

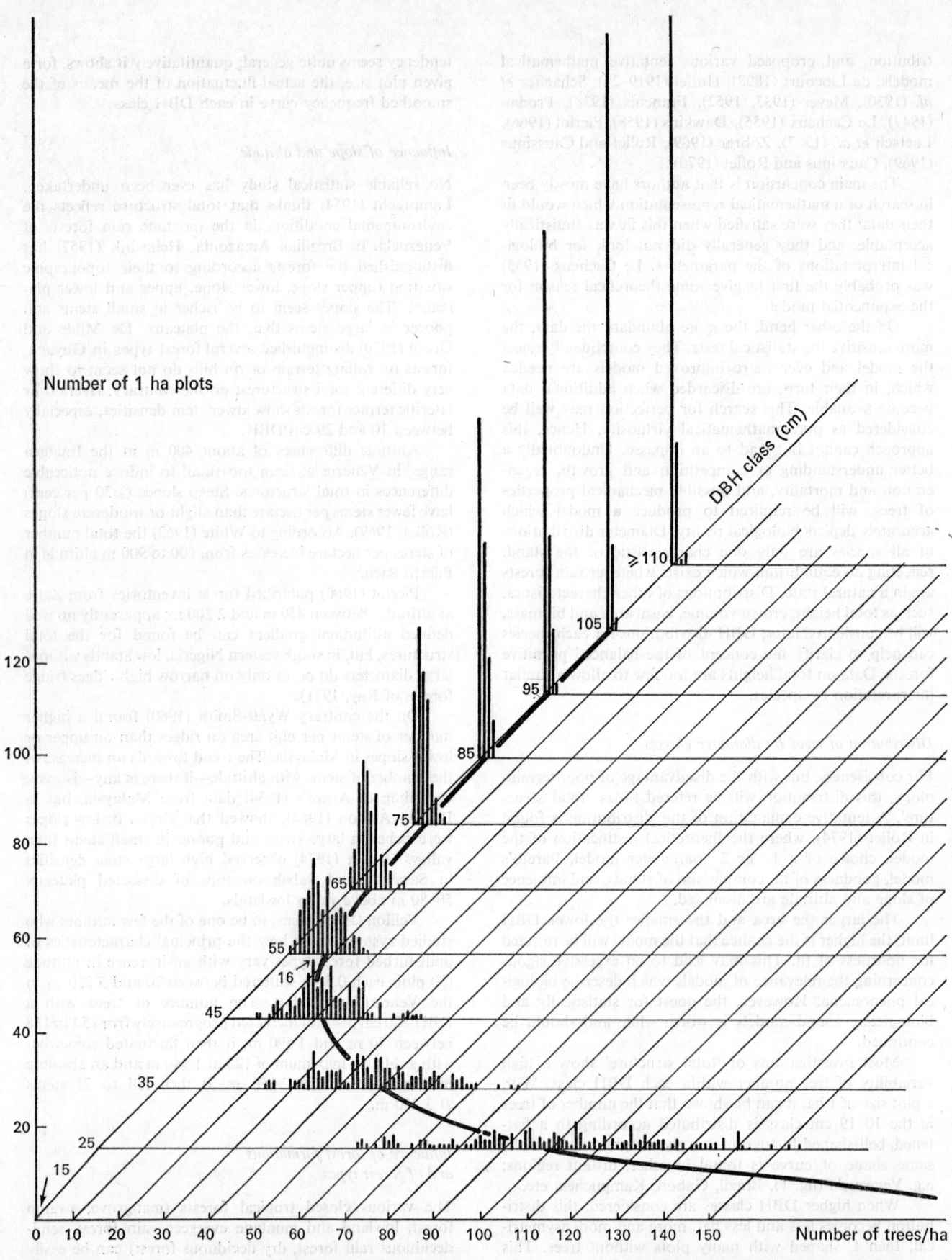

FIG. 1. Structure of lowland humid tropical forest at Imataca, Venezuelan Guyana, on a 151 ha stand. Three-dimensional graph showing the relationships between the number of 1 ha plots and tree numbers in 10 cm DBH classes. The bold line shows the medium points of the relation between diameter and tree number; it is not possible to represent the value for the 10–19 cm class on this scale.

distinguished by their total structures. The comparison is easier with semi-logarithmic graphs: DBH classes as abscissae and logarithms of number of stems as coordinates. Such a graph can be found in Rollet (1974). The lines showing the various formations on an equal basis of 100 ha are fairly distinct by their slopes and the number of large and small stems. This number varies in the proportion of one to three for small stems, in wider proportions for large stems (e.g. DBH $\geqslant$ 100 cm, although the latter dimension never exists in several formations).

Statistics on the total structures of the above-mentioned formations are found in Rollet (1974, annexes 4 and 5); for mangroves, see Salandy (1964) and Mayo Melendez (1965); for swamp forests, see Marshall (1939), Beard (1946b, c), Anderson (1961), Hegyi (1962, 1963a, b), Redhead (1964, 20 acres in Ebué); for semi-deciduous and deciduous forests, see Rollet (1952, 1962), Mooney (1961), Lamprecht (1961, 1962), Anon. (1964, 1966); for montane forests, see Dawkins (1958), Morales (1966), Aubréville (1967), Pereira (1970) and for tropical pines, see Rollet (1952).

Within each formation, some forest types show conspicuous differences in total structures but forest types are not in general well studied and difficult to distinguish according to their architecture.

Beard (1946b) performed four inventories, each of 40 ha, corresponding to four different facies or types of evergreen *Carapa-Eschweilera* rain forests in Trinidad where the numbers of small and large stems are very different. According to de Milde and Groot (1970), riverain forests, lowland forests on lateritic terraces, and *Dicymbe* forests on white sands contain only half the number of stems of other Guyana rain forests in the DBH interval 10 to 29 cm. In Zaire, *Brachystegia laurentii* forests seem richer in large stems (DBH $\geqslant$ 60 cm) than other types of rain forests (Germain and Évrard, 1956; Pierlot, 1966). In Kampuchea and Viet-Nam, and more generally in Indo-China *lato sensu*, some semi-deciduous *Lagerstroemia* forests show a strong accumulation of large diameters, up to 30 or 40 stems (DBH $\geqslant$ 60 cm) per hectare (Rollet, 1962). The same phenomenon is observed in Sabah (Nicholson, 1965; Fox, 1967) and in Sumatra and Kalimantan.

These concentrations of large diameters are linked to the behaviour of certain species, and probably also to the history of the stands. Burgess (1972) attributes the gregariousness of *Shorea curtisii* in Malaysia to drought resistance of the seeds being higher than in other species, while Lee Peng Choong (1967) considers the high number of large trees of *Dryobalanops aromatica* in the stands is due to more frequent and abundant mast-years. Of course such species must be very long-lived.

In the exponential model with 10 cm classes, the sum of stems larger than a given diameter is equal to the number of stems in the immediate lower class when the ratio r of the convergent progression is 1/2. When r is greater than 1/2, this sum is greater than the number of stems in the next lower class; it is twice when r=2/3, three times when r=3/4. When r increases this means that survival percentage from a class to the next one increases and becomes, instead of 1/2, 2/3, 3/4, etc. Wyatt-Smith (1963) published

a number of fairly detailed inventories for Malaysia from which one concludes that most of the large trees belong to a few species only. Such is the case for *Dryobalanops aromatica, Shorea curtisii, S. parviflora, S. leprosula, S. pauciflora, Dipterocarpus cornutus, D. kerrii, Koompassia* (Wyatt-Smith, 1963, part 2). While few large trees occur in the swamp forests in comparison with lowland rain forests, hill and lower montane rain forests seem to be richer in large trees than lowland ones.

It would be interesting to study systematically total structures in relation to soils, topographic position and altitude.

Conclusions

A great number of data are scattered in the forestry literature and especially in the archives of the forest services. A tentative review of forest inventories of 24 tropical countries was carried out for lowland evergreen rain forests and trees with a DBH $\geqslant$ 20 cm by Rollet (1974) and possibly to include DBH $\geqslant$ 10 cm. Many inventories could be quoted but their quality is very uneven: a number of them show obvious bias in number of stems among DBH classes, others did not record all the species or trees. The main difficulty in comparing forest inventories is the heterogeneity of their conception: different lower DBH limit and class intervals, and units of measurement.

Particularly, when a pantropical structure is tentatively computed, it is necessary to find something in common between all available inventories; many of them are therefore discarded. Dawkins (1958) proposed a pantropical structure for rain forests on the basis of 11 inventories among which two were rather small. Rollet (1974) gave the total structure for 30 1 ha plots surveyed in 24 different sites of America, Africa and Asia. The figures are similar to Dawkin's, but they cannot unfortunately be considered as representative. The pantropical structure computed in 1969 and reproduced in 1974 seems more satisfactory.

One can easily imagine how difficult is the calculation of representative means. The areas surveyed and the sampling intensities in various countries are very unequal; the consequence is sometimes a rather unexpected situation with small countries heavily surveyed and large countries poorly surveyed. Hence obviously erroneous conclusions occur if numbers of stems are merely summed up by DBH classes; therefore, the suggested pantropical structures must be accepted in a most critical way.

Above 20 cm DBH, total structures are obviously exponential. Any anomalous result, especially on a semi-logarithmic graph and showing a strongly broken line, should be regarded as issued from low quality, biased inventories, or from disturbed forests (by man or by natural causes); disturbance does not exclude the former hypothesis.

When DBHs smaller than 20 cm are considered, a simple exponential model is no longer convenient and a semi-logarithmic graph shows a concavity towards the upper side, which means that the number of stems increases more quickly than expected from a simple exponential model. This tendency is accentuated with smaller and smaller DBHs, e.g. on 0.162 ha in Venezuela with 2 cm DBH classes (2–4; 4–6; 6–8; 8–10 cm), the number of stems are respectively 477, 134, 50, 27; ratios of one number to the next vary from 2 (50/27) to over 3 (477/134). The same phenomenon is observed with 5 cm classes on 2 ha (Rio Grande, Venezuela) where ratios vary from 1.4 (157/110) to over 3 (1 401/448):

Diameter interval (cm)	5–9	10–14	15–19	20–24	25–29	30–34	35–39
Stem number	1 401	448	229	157	110	71	40

Diameter distributions by species will be considered later.

In short, the distribution of trees by diameter classes is basic to study tropical forests and is the most accessible and studied information. Very often no other data are available. It is rarely analysed in detail; in particular, the influence of soils, topography and altitude has not been studied systematically. It makes possible the assessment of the effects of the main environmental factors on forest architecture. An exponential law of distribution is unable to account for observed distributions of trees by diameter classes. Studies on competition should allow a better formulation of these theoretical laws governing such distributions.

Height of trees and strata

Besides the establishment of profiles, very few total heights (or heights up to the base of the crown) have been measured. Mildbraed (1933) made some measurements in 1 ha in Fernando Po but his measures are highly biased; also Davis and Richards (1934) on 0.19 ha in Guyana; Richards (1939) in a small collection of trees in Nigeria; Takeuchi (1961, 1962) for trees over 10 cm or 5 cm DBH in 0.16 ha in dry land rain forest and 0.22 ha in marsh forest (Brazilian Amazonia); Ashton (1964) in five profiles 25×200 feet in Brunei with 10 foot classes; Ogawa *et al.* (1965a, b) for trees of a DBH⩾4.5 cm in five plots 40×40 m in Thailand (two of them in rain forest); Kira and Ogawa (1971): two sites in rain forests of Sabah, two in Thailand, one in heath forest of Kuala Rompin (Malaysia), one in a montane forest in the Cameron Highland (Malaysia); Klinge (1973b) and Fittkau and Klinge (1973) on 0.2 ha, for trees⩾1.5 m high, in a rain forest near Manaus, Brazil (unequal classes below 5 m, equal classes over 5 m); Rollet (1974) in various rain forests of America, Africa and Asia for trees⩾4 m high.

Heinsdijk (1957–65) estimated ocularly heights up to the base of the crown (and total heights for emergents) over 1 200 ha in Amazonia for trees of DBH⩾25 cm.

Brünig (1968a, b) drew graphs of total height and DBH for 5 important species of heath forests in Sarawak. Liew

That Chim (unpublished) measured total heights in several mangrove plots in Sabah.

Gray (1975) studied the size and spatial distribution of trees on plots 50×50 m square in remnant forest in the Bulolo-Wau region, Papua-New Guinea. The two species concerned were *Araucaria cunninghamii* and *A. hunsteinii*. The first is found growing naturally between 600 m and 2 800 m altitude (mostly 900–2 200 m), usually associated with steep slopes and ridges. The second species is found between 550 m and 2 100 m (mostly 750–1 500 m) and is characteristic of less steep slopes. Natural stands of both species have been utilized for timber extraction; in addition they have been extensively employed in reforestation. The tallest *A. hunsteinii* was 78.7 m, which is rather exceptional, as most of the largest veterans appear to be 65–75 m high. In the case of *A. cunninghamii*, several of the largest trees slightly exceed 60 m but very few attain 70 m or more; the highest measured was 70.8 m. Richards (1952) reports two trees of *Koompassia excelsa* from Sarawak and Malaya 84 m and 81 m high respectively.

The interest of the measurement of total heights, or heights up to the base of the crown, lies in its relevance to the distribution of leaf-masses and crowns.

Many authors used to distinguish several strata in the crowns, following the traditional description of temperate forest (herb-, shrub-, tree-layers). Olberg (1952–53) proposed height classes to describe temperate forests: I emergents, II upper layer, III 75 per cent of mean height of upper layer, IV between 50 and 75 per cent, V between 25 and 50 per cent, VI below 25 per cent. Unfortunately, these considerations are supported by no or very few measurements, or by ocular observations. Thus, Richards (1936) recognizes 3 strata of trees at 34 m, 18 m and 8 m levels, plus a shrub layer. He thinks that profiles illustrate this stratification by themselves. Wyatt-Smith (1963) shares this opinion for Malaysia. There are emergents between 30 and 45 m, an upper layer between 21 and 30 m, an understorey and a shrub layer. Swamp, hill, and heath forests have also 3 strata, montane forests 2 and Ericaceae thicket one.

Gérard (1960) distinguishes 5 strata in the *Gilbertiodendron dewevrei* forests of Zaire but considers that only 2 strata could be recognized: a shrub layer and two tree strata combined; he noticed some discontinuity in a rather empty level at 10 m. Taylor (1960) recognizes 4 strata in the rain forests, which probably include also semi-deciduous forests, of Ghana: <6 feet, between 6 and 60 feet, up to 130 feet and emergents up to 200 feet.

Stevenson (1942) also considers that there are 4 strata in the rain forests of British Honduras. Stehlé (1945, 1946) mentioned 2 to 4 strata in the West Indies. In Surinam, Lindeman and Moolenaar (1959) speak of 3 to 5 strata while Schulz (1960) keeps the concept but thinks it is somewhat arbitrary. In Guyana, Fanshawe (1952) distinguishes 4 strata, or 3 if emergents are not considered. Takeuchi (1961, 1962) recognizes 3 strata in dry-land Amazonian forests but no strata in marsh forests.

According to Webb (1959), there are 3 strata in Australian rain forests and sometimes an emergent layer, but he thinks that whether or not to consider strata is a very personal and subjective judgment. The strata are reduced to 2 in montane forests. Robbins (1968) thinks of 3 concentrations of heights in lowland rain forests, 2 in lower montane, and one in montane rain forests.

Oldeman (1972a) considers that strata exist in the rain forests of French Guyana but he gives the word a special meaning: a densification of branching at certain levels; hence a shrub layer, a first assemblage of trees and a second assemblage of tall trees.

Ashton (1964 personal communication) asserts the existence of strata. However the five profiles and three histogrammes of total heights produced do not show the clear existence of strata. The histogrammes are based on trees over 15 feet total height, grouped in 10 foot classes with a logarithmic scale for frequency. The scale exaggerates the gaps for large heights where classes without trees are in fact neighbouring classes with one or two trees. With small plots (*ca.* 0.09 ha), it is not unusual to get discontinuities of this kind.

Some authors are more careful in their way of thinking; others, less in number, think that no strata exist in rain forests (some authors even changed their mind). Aubréville (1932) denies the existence of 3 strata in the rain forests of Ivory Coast.

Beard (1946b) considers that there is no visible stratification in the forests of Trinidad. Although two modes appear in the frequencies for the 76–85 and 86–95 foot height classes in rain forests of Guyana, Davis and Richards (1933, 1934) think that it is difficult to speak of a stratification: 'The stratification is very irregular and ill-marked'.

Walton, Barnard and Wyatt-Smith (1952), after writing 'Lowland and hill dipterocarp forest is arranged in several storeys' add 'It must be stressed that no rigid division exists between these five storeys or layers, trees growing gradually through the various storeys till they reach their own particular niche within the biological framework'.

Heinsdijk (1960, 1965), Schulz (1960), Murça Pires (unpublished) deny the presence of strata. Murça Pires gives three distributions of total heights with 1 m classes for trees over 2 m high on 0.2 ha of rain forest and 2 plots of 0.2 ha in swamp forest in Brazil. He concludes that there is no stratification of total heights.

Ogawa *et al.* (1965b) describe a method to detect the existence of strata, through a graph, with total height as coordinates and corresponding height below the crown as abscissae; more or less isolated groups of points indicate that the in between levels are virtually unoccupied by crowns.

In undisturbed lowland evergreen rain forest, Rollet (1969, 1974) measured all individuals ≥4 m high on 10 plots of 0.5 ha in Venezuelan Guyana and 20 plots of 0.25 ha in America, Africa and Asia. In the first case, 1 200 stems were measured, and over 15 000 in the second. Five conclusions can be drawn from these observations:

— Total heights

Their distributions are steep L-shaped; when the height-class moves from the 48–50 m to the 4–6 m interval, the increase in number of trees is greater than a simple

exponential model would predict. On the other hand, the distributions are not plurimodal. There is a tendency to get a plurimodal distribution with small areas, e.g. <0.25 ha, or with small height intervals, e.g. ≤2 m. For areas ≥0.50 ha and intervals of height ≥4 m, height-distributions are regularly decreasing and get smoother and smoother when areas increase.

— Heights below the crowns

Their distributions decrease regularly. Irregularities have the same origin as already mentioned for total heights; they disappear with larger areas or intervals. Hence it is impossible to conclude that the bases of crowns are concentrated at particular levels.

— Spatial distribution of crowns

If a crown position is represented by a vertical line from the base to the top of the crown on a profile, one can observe a progressive densification of those lines as one gets closer to the ground level; there is no evidence of gaps.

— Spatial distribution of foliage density

The volumes occupied by crowns, in layers 2 m thick, or quantities that are grossly proportional, can be calculated; these show bell shaped distributions with a maximum at *ca.* 18–20 m. There is some analogy with the fact that the DBH class 30 to 39 cm or 35 to 44 cm generally corresponds to the maximum value in the distribution of volumes or biomass with for example a 10 cm DBH interval.

— Distribution of total heights in the regeneration

The progressive decrease of tree number with increase in height classes, holds also for stems <4 m high. Thus, on 5 ha with 2 m classes, the following numbers were found for stems ≥4 m high:

Height class (m)	4–5.9	6–7.9	8–9.9	10–11.9	12–13.9
Stem number	6 315	3 162	1 710	925	687

and in 0.17 ha with 1 m classes, for stems <4 m high:

Height class (m)	<1	1–1.9	2–2.9	3–3.9
Stem number	17 113	1 494	582	315

This trend holds also in the layer <1 m high with 2 dm intervals:

Height class (m)	1–2.9	3–4.9	5–6.9	7–8.9
Stem number	10 505	3 831	1 617	862

All these observations seem to show that many lowland evergreen forests have no strata but rather a progressive decrease of stem number with height increase; for leaf masses there seems to be a maximum density at about half way between the tops of the highest trees and the ground.

It has been observed that the architecture is more simple when altitude increases, with total heights getting progressively shorter. It is often stated that, in the tropics, dwarf forests or montane thickets have respectively 2 and 1 strata. Because of lower mean temperatures, one cannot help but compare them with temperate forests but the architecture and dynamics of the latter seem to depend very much on the species and in all cases the number of the latter is limited. If some still undisturbed temperate forests show an architecture similar to uneven-aged stands with all kinds of ages and dimensions in any small area, and an equilibrium quite similar to tropical rain forests (cf. Tregubov, 1941), others look rather like even-aged stands, i.e. collections of trees of about the same dimensions with bell-shaped height—and diameter—distributions. Moreover, virgin temperate forests show cycles in their dynamics with alternate dominancy of shade species and light demanders (cf. Dobroć forest in Slovakia, Leibundgut, 1959). In lowland evergreen rain forests, light demanders are always present but scattered; they are a normal constituent of the forest but they are completely outnumbered by the bulk of shade tolerant species; they are casually found in a given spot and in Van Steenis' terminology (1958) they are nomads in the forest.

To return to the architecture of rain forests, it seems that the exponential model already mentioned is no longer adequate in peculiar soil conditions and therefore, also in the case of special floristic compositions. It has been already mentioned that Gérard (1960) found a fairly unoccupied level at 10 m above the ground in *Gilbertiodendron* forests in Zaire. Some periodically inundated marsh forests with *Triplaris surinamensis*, *Licania apetala* and *Macrolobium acaciaefolium* in Venezuela (Rollet, 1974) and with *Guibourtia demeusii* in the People's Republic of Congo (Rollet, 1963), are conspicuous for the absence of undergrowth. Crowns seem therefore to concentrate in one layer, giving an even canopy without emergents; likewise, certain deciduous formations in South-East Asia with a dominancy of *Lagerstroemia* (a light demanding species in a succession phase after burning for shifting cultivation) have practically no undergrowth, and show table-like canopies without emergents when inspected from the air. Young secondary forests have a similar aspect where low, short-lived, gregarious light demanders like *Musanga* and *Cecropia* occur practically in pure stands.

The same aspect of table canopies exists for most of the mangroves, and for peat swamp forests, with *Melaleuca leucadendron* in Kampuchea and Viet-Nam and *Shorea albida* in Sarawak. For the latter, Anderson (1961) gives aerial photographs and profiles, practically without any undergrowth.

On the contrary, some formations show an entirely different aspect of canopy with scattered emergents towering high above the main canopy: giant *Agathis* scattered and very emergent over a rain forest in Melanesia (Whitmore, 1966); tall emergent *Araucaria cunninghamii* and *A. hunsteinii* in New Guinea dense forest (Womersley, 1958); large *Pinus merkusii* in similar conditions and

without any regeneration near Kontum, Viet-Nam (Rollet, unpublished); tall emergent *Shorea platyclados* between 750 and 2 500 feet in the hill dipterocarp forest of Malaysia (Burgess, personal communication); high *Lumnitzera* emerging in even-canopied mangroves; large *Dinizia excelsa* in western Amazonian rain forests (Rollet, 1974). Scattered trees emerging from a low carpet rich in vines are characteristic of the *selva de bejucos* in Venezuela (Rollet, 1968), or the liane forest of Surinam (Lindeman and Moolenaar, 1959) or the *mata com cipoal* of Brazilian Amazonia, or *forêts denses ouvertes* in the Ivory Coast, or open woodlands of Congo (Rollet, 1963).

All these examples are almost always—if not always—linked to peculiar successions, i.e. to the history of the stands.

In conclusion, it is necessary to increase total height measurements (and heights below the crown) to clarify the concept of strata. Many authors refer to strata without producing measurements. Except in particular cases, forest stands do not seem to show strata in the spatial distribution of the crowns, i.e. of the foliage. From such height measurements growth and competition phenomena will be better known and through these measures rather than diameter records, architecture will become understood.

Crowns

Description

It is well known that some plant families show conspicuous branching, e.g. subhorizontally radiating, more or less numerous and thin branches of Myristicaceae, Guttiferae, Annonaceae; the crowns in some species have very peculiar shapes (Rollet, 1963): stratified (*Terminalia*, *Piptadeniastrum*, *Canthium*), disklike (*Austranella congolensis*), in loose little balls (*Erythrophloeum*, *Entandrophragma*, Irvingiaceae), thick dense crowns (*Lophira*, *Mangifera*), light crowns (some species of *Celtis*, *Brachystegia nigerica*, numerous Leguminosae); weeping (*Fagara*, *Chlorophora*, *Cleistopholis*, *Salix humboldtii*). Letouzey (1969) gives schematic drawings of crowns and branching for some African species. Brünig (1970) mentions various crown shapes: disk, umbrella, spherical, cauliflower, etc. Dawkins (1958) suggested describing crowns according to four criteria for sylvicultural uses: *crown position* (a score, 5 to 1 is given to emergents, to trees belonging to upper- or lower canopy, upper- or lower understorey); *crown form* (a score 5 to 1: perfect, good, tolerable, poor, very poor); *crown contact* and finally *crowding*. These ideas were applied and further developed in Malaysia by Wyatt-Smith (1963), and by Walker (1962) in eastern Nigeria to appreciate the regeneration of interesting species before logging.

Cover

The distribution of crown diameters is generally considered in connection with the DBH distribution (see below). Francis (1966) studied a crown diameter distribution on aerial photographs in Sabah, for 1 000 trees occurring on about 9 ha by 10 foot classes starting with 40 feet. Rollet (1969, 1974) studied about 1 200 trees on 10 profiles made in Venezuelan Guyana. In both cases, the distributions look exponential like the DBH or total height distributions.

The measurement of crowns is not easy because of irregular shapes; it is advisable to take an average of half of two perpendicular diameters; this measure is in all cases more accurate in the field than on aerial photographs where understorey crowns cannot be observed and where crown confluences are frequent. From these data, the cover may be determined, with some restrictions since compactness of foliage is not considered. Summing all the crowns as if there were circles, percentage cover for Venezuelan Guyana fluctuates around 4 times the area of the corresponding plot for stems ≥ 4 m high and around 2.5 times for trees with a DBH≥ 10 cm. The determination of cover is important since quality and quantity of light reaching the various levels inside the stands govern growth and competition to a large extent. Several techniques use hemispherical photographs: Anderson (1964); Johnson and Vogel (1967); Chartier *et al.* (1973). Johnson (1970) measured the quantity of transmitted light in the central part of the hemispherical photograph, corresponding to a 90° angle cone called canopy closure index to distinguish it from the per cent cover. Other authors determined the percentage of sky obscured by the leaves, without considering foliage overlap at various levels, e.g. Emlen (1967); others measured this global cover at various levels, e.g. between 5 and 13 m and above 13 m (Desmarais and Vazquez, 1970).

Competition

Competition is difficult to study even in the simplest cases: e.g. by weighing 2 or 3 annual plants growing together at various densities. In the case of a tropical forest, the large number of species, the impossibility to identify and separate the field of activity of the various root systems, the difficulty of adequate evaluation of biomass variations of a tree make the quantification of competition terribly complex.

Foresters used to appreciate competition qualitatively by taking into account boughs and crown relative positions, their shape and apparent luxuriance. Leibundgut (1956) recommends the use of crown depths (3 classes: >0.5 of the tree total height, between 0.25–0.50, <0.25) and tree luxuriance from the evolution of growth. The proposals of Krajicek *et al.* (1961) for North American forests must be mentioned but they are difficult to adapt to the conditions of tropical forests because of their floristic diversity; they obtained a very good correlation between DBH and crown diameter whenever trees grow isolated. The various broad-leaved species show a similar behaviour and can be considered together. Competition between crowns is determined by computing, for all the stems on one acre of closed stand, the area which would be occupied by the crowns, had the trees been isolated; if the result is 2 acres, the crown competition factor is 2. Taking into consideration known productivities, sylvicultural techniques are sought which will approximate the value of the factor which optimizes the production.

Disposition of foliage masses

Odum *et al.* (1963) think that the existence of strata in a rain forest of Puerto Rico below 600 m altitude is doubtful: by weighing foliage masses at various levels there is no evidence of concentration at certain levels. McArthur and Horn (1969) proposed building a foliage profile by noting optically the height of the lowest leaf on a large number of verticals at random. They relate mathematically the foliage density between two levels, to the probability to meet a leaf between those levels. Rollet (1974) showed that in Venezuelan Guyana leaf-masses have a roughly bell-shaped distribution with maximal density at 18–20 m, i.e. half the height of the tallest trees in the area.

Structure and roughness

Brünig (1970) studied the structure of canopies of different formations in Sarawak from the point of view of their influence on the reduction of wind velocity and increase of turbulence at various levels within the forest, which has a direct incidence on evapotranspiration. Wind velocity within the stands is a function of the logarithm of height above the ground, corrected by a factor which he calls canopy roughness, and which depends on crown dimensions and their height of emergence over the main canopy and, possibly, on leaf-size. Heat diffusion depends also on the shape, size and orientation of leaves.

It is believed that the study of canopy morphology on aerial photographs could improve the classic photo-interpretation of formations and forest types. Rollet suggests statistically analysing the measurements of a great number of distances from the canopy outer surface to a reference plane (i.e. the study of the replica of the detailed stereoscopic image of the canopy), to detect in the cover structures using the theory of regionalized variables (cf. Millier *et al.*, 1972) and to study aerial photographs in coherent light. Finally, the morphology of canopies and the nature of leaf surface determine selective absorption and reflection among the various wave-lengths; this field of remote sensing is still poorly explored and could be a key to detailed interpretation of tropical forest types and species determination.

In short, crowns have been little studied. However, practically only crowns are observed on aerial photographs; they reflect light selectively, slow down the wind, diffuse heat, re-evaporate part of the rainfall. Their architecture is often overlooked in dendrologies. The study of latent structures in crown canopies could possibly improve and even renovate photo-interpretation, although the latter has been considerably widened by remote sensing. The proportion of senile trees, detectable in 'false' colour, could be an important characteristic of tropical forests.

Relations between the main dimensions of trees

The liaisons between the most commonly measured dimensions of trees, DBH (D), total height (H) and horizontal crown diameter ($\varnothing$) are examined.

DBH and height

There are many proposed functions $H=f(D)$; usually H or log H are polynomial expressions with D or log D (Sandrasegaran, 1971); Ogawa *et al.* (1965b) proposed a generalized allometric equation: $1/H=1/AD^n+1/H^*$. Practically, n is shown to be very close to 1; thus H is considered as an hyperbolic function of D; H^* is the maximal height possible in the stand, and A is a constant. Examples are given from rain forests, semi-deciduous and dry dipterocarp forests. Graphs were drawn on double log. coordinates. Kira and Ogawa (1971) present similar graphs and use the above relation to avoid lengthy measures of all total heights and to compute quickly biomasses. The logarithms of biomasses are correlated linearly with the logarithms of D^2H.

The functions $H=f(D)$ proposed by Sandrasegaran (1971) are all empirical and without theoretical justification. It seems that the groups of points (D, H) look similar for all lowland evergreen rain forests (Rollet, 1974) and even for dry dipterocarp forests (Ogawa *et al.*, 1965b): it becomes curvilinear and funnel-like when D and H increase; but when D is high, e.g. $\geqslant 60$ cm, H increases more and more slowly—this is a good justification for the choice of Ogawa *et al.* Beyond a certain DBH, total heights fluctuate independently; this has important dendrometric consequences. On the scatter of points (D, H) when using ordinary coordinates, for a given diameter class, height distributions are L-shape (exponential) for small DBHs, then move progressively to bell-shaped for large DBHs. On the contrary, for a given height class, diameters are becoming more and more diversified when the height class increases. Generally speaking, for a given DBH class, a number of very different total heights are possible and vice versa; the higher is the class, the more varied are the heights.

A list of measurements of total heights (generally with corresponding diameters) has been given already. A large amount of data has been collected by Heinsdijk (1957–1965) but only for $D\geqslant 25$ cm.

DBH and crown diameter

Paijmans (1951) measured crown diameter ($\varnothing$) of emergents on 5 plots 200×300 m from aerial photographs at the scale of 1/10 000 and studied the relation with DBH; the ratio $\varnothing/D$ was about 30. In the Philippines, Macabeo (1957) studied 200 trees of *Pentacme contorta*, and also the relation between $\varnothing$ and commercial height, and between $\varnothing$ and commercial volume. In Guyana, Swellengrebel (1959) studied the relation ($\varnothing$, D) in small collections of trees (<200 individuals) corresponding to 3 formations (swamp forest with *Mora excelsa* and two rain forests, one with *Eperua*) in order to predict standing volumes from $\varnothing$ measurements on aerial photographs; the correlation was good, with a strong dispersion especially for *Mora*. Heinsdijk (1960b) studied the relation between D and $\varnothing$ for emergents only, for over 27 000 trees distributed on 1 200 ha south of the Amazon between Manaus and Belém. The ratio $\varnothing/D$ varied between 25 and 36 and was frequently near 33. It would have been

interesting to see if the most abundant species had a different behaviour. Graphs of $\varnothing$ and D, from the field data kept in Rio de Janeiro, look like the above mentioned graphs of H and D. Marginal distributions of $\varnothing$ are asymetrically bell-shaped because they only concern emergent trees. The familiar exponential distribution would have been obtained if all trees over a much smaller DBH had been considered.

Dawkins' (1963) overall study mentions the results of Heinsdijk for Amazonian rain forests, of Swellengrebel in Guyana and Paijmans in Celebes. Here regressions are computed between D and $\varnothing$. Dawkins also deals with certain species from rain forest and semi-deciduous forest, e.g. *Triplochiton* between 15 and 100 cm. The regression coefficient between D and $\varnothing$ varies considerably from one species to another, but the correlation is generally strong. He suggests that the regression is almost linear during the maturity of the tree; during youth $\varnothing$ increases faster than D, and slower and slower during senility; this gives a sigmoid relationship; if D and $\varnothing$ are in the same units the ratio $\varnothing/D$ fluctuates between 16 and 27 and is often near to 20.

For 197 trees of the eastern coast of Sabah (especially *Shorea*, *Parashorea*, *Dipterocarpus* and *Dryobalanops*) Francis (1966) found the relation somewhat linear but with a strong dispersion increasing with $\varnothing$. Rollet (1969) found ratios $\varnothing/D$ varying between 13 and 18 for 1 500 trees $\geqslant 4$ m high in Venezuelan Guyana; the regression between $\varnothing$ and D was somewhat curvilinear. Perez (1970) analysed the relation between $\varnothing$ and D for *ca.* 4 500 trees from different formations in Puerto Rico, Dominica and Thailand (lowland rain forest, lower and upper montane rain forest, semi-deciduous and dry deciduous forest). On 29 double logarithmic graphs the regression was almost linear. Rollet (1974) analysed the relation ($\varnothing$, H) on emergent trees using Heinsdijk's data from Brazil; on 13 ha of one area, $\varnothing$ varied from 10.7 to 20.0 m when H varied from 16 to 40 m; on 15 ha in another area, $\varnothing$ varied from 9 to 23 m when H varied from 16 to 52 m. On the first area, the largest $\varnothing$ reached 34–40 m, and 28 m in the second; the most frequent maximum is respectively equal to 11 m and 13 m.

It is probable that the ratio $\varnothing/D$ varies during the life of a tree, that it depends on the species and also on the forest type. The ratio is smaller for understoreys and codominants than for the emergents. The practical interest of a relation between $\varnothing$ and D among emergents is to allow some prediction on timber volumes (Heinsdijk, 1957, 1958). In fact, since it is difficult to recognize species on air photographs and impossible with panchromatic or infra-red film to detect infected or unsound trees, volumes predicted in this way cannot be but rough estimates.

In summary, the correlation between DBH and crown diameter is not always very strong; the ratio of these two variables varies during the life of the tree, and it is probably different from one species to another. The study of crown diameters is of little interest from the point of view of dendrometry, at least when commercial volumes are needed. The shape of their distribution is exponential. The relation between DBH and total height is curvilinear with a comet-like tail. The scatter of points and the regression line are a good synthetic representation of the architecture of tropical forests, for their marginal distributions are those of DBH and heights and the regression line is probably characteristic of the stand.

Root systems

Jeník (1971) gives 200 t/ha as the root system biomass of the rain forest of Ghana; Klinge (1973b) gives a comparative table of results for Ghana, the Ivory Coast, Trinidad and the Brazilian Amazon. Coster (1932, 1935), Wilkinson (1939), Louis (1947), Kerfoot (1963), Huttel (1967, 1969), Jeník and Mensah (1967), Mensah and Jeník (1968), Odum (1970), Jeník (1971), Kira and Ogawa (1971), Klinge (1973a, b), Fittkau and Klinge (1973) and Leroy-Deval (1973, 1974) studied the root systems of certain tropical trees.

Jeník (1971) and many other workers mention that soil types may influence the root distribution at various depth levels.

Louis (1947) gives drawings of buttresses and root systems of *Cynometra alexandri* and the tap-root of *Macrolobium (Gilbertiodendron) dewevrei*. Gérard (1960) gives photographs of root systems of large *Gilbertiodendron dewevrei* on deep soils with a primary and secondary tap-roots; on shallow gravelly soils the roots are superficial and tap-roots are rotten. In the same way *Julbernardia* shows different types of buttresses according to the soil: plank buttresses with 45°–60° slopes on deep soil, irregular, creeping and decurrent buttresses in shallow soils. Rollet (1969) observed root systems and buttress shapes on about 30 species in Venezuelan Guyana; tap-roots seemed to be absent but *Catostemma commune* had a large one, *Eschweilera chartacea*, *E. corrugata* and *Parinari rodolphii* had medium ones, while *Alexa imperatricis* and *Inga* sp. had several small tap-roots. Even when the soil is deep, most species seem to have superficial root systems. Tree falls are frequent in time of high winds; cyclones provoke huge gaps transforming undisturbed forests into vine or liane forests which return very slowly to the climax. Superficial root systems on rocky soils or lateritic pans were observed by Lindeman and Moolenaar (1959) in Surinam in the liane forests where fallen trees are abnormally numerous. Budowski (in Jeník, 1971) noticed that high rainfall (8 000 mm/a) in a lowland rain forest of Colombia, made practically all root systems superficial, while with 2 000 mm, lowland Colombian rain forests had deeper roots. Leroy-Deval (1973, 1974) observed anastomosis between roots of *Aucoumea klaineana* in Gabon.

Observations on root system morphology could be increased at low cost by taking advantage of road or railway construction, trees uprooted by wind especially at the end of the wet season or after intensive erosion; this would permit a better understanding of the influence of topography, of soils (depth, texture, waterlogged conditions) and rainfall on root systems.

Other morphological characteristics

Buttresses and aerial roots

Buttresses are an important morphological character of closed tropical forests and, though systematical studies are not available, buttresses become more frequent nearer the equator and less frequent with increasing altitude.

An interesting introduction to buttresses can be found in Richards (1952). Morphological (and anatomical) studies are fairly numerous (Chalk and Akpalu, 1963; Guéneau, 1973) but frequency studies by species, diameter classes, forest types, etc., are far less numerous. The relation between height and DBH (or ground level width) can be examined, and interesting morphological and quantitative characters for field determinations can be noted. Literature reviews have been made by Chipp (1922); Stehlé (1945, 1946) for the West Indies; Jimenez Saa (1967) for Costa Rica; Rollet (1969) for Venezuela; Taylor (1960) for Ghana; Letouzey (1969) for Africa; Rosayro (1954) for Sri Lanka and Wyatt-Smith (1954) for Malaysia. Some dendrologies illustrate buttresses, but taxonomists generally neglect this character because it is too variable during tree life or depends too much on soil conditions. Quantitative studies are scarce: Setten's (1954a, b) for Malaysia are the largest; but see also Lebrun (1936) for Zaire; Vincent (1960) for Malaysia; Takeuchi (1961) for Amazonia and Rollet (1969) for Venezuela.

Buttresses are important practically in dendrometry: volumes are evaluated above buttresses; also buttresses complicate growth measurements.

Aerial roots are also a conspicuous feature of tropical forests but are generally less frequent than buttresses although they are sometimes plentiful in swamp montane forests, e.g. in *Euterpe-Symphonia-Tovomita* facies (Beard, 1949), and also in the dwarf montane *Clusia* thickets of Stehlé (1945, 1946), and in mangroves. They occur among very different botanical genera (Rollet, 1969).

Biological types

Phanerophytes are obviously omnipresent; it is possible to study separately climbers and palms, which through their abundance and species diversity are highly characteristic of certain forest types. Epiphytes are the next most abundant type, whilst chamaephytes and geophytes are much less common. Therophytes are rather scarce. The Raunkiaer classification was revised and adapted to tropical conditions by Richards *et al.* (1940), Lebrun (1947, 1964), Aubréville (1963), Vareschi (1966, 1968) and Mangenot (1969).

The preparation of biological spectra helps the classification of tropical forests. A number of studies of rain forests used this concept: Lebrun (1947); Germain and Évrard (1956), Gérard (1960) and Évrard (1968) in Zaire; Mangenot (1955) and Guillaumet (1967) in the Ivory Coast; Vidal (1966) in Laos; Vareschi (1966) and Rollet (1969) in Venezuela. Anderson (1961) and Brünig (1968) gave biological spectra for swamp—and heath—forests in Sarawak whilst Hosokawa (quoted in Tixier) and Tixier (1966) studied epiphytes in considerable detail in Indo-China.

Climbers

Climbers are the second most important life-form in lowland rain forests. Their contribution to the herb and shrub layers is much greater than to the tree layers (Rollet, 1969). It seems that in certain cases and at certain stages, the proportion of climbers reaches a maximum among the seedlings (when climbers still retain an erect habit). As stems grow older it diminishes and finally gives way to the tree life-form.

Climber diameter distributions show that they do not necessarily behave as light-demanders. All types of diameter distribution already mentioned for trees can be found; climbers and trees do not behave very differently.

Many climbers show high potential growth rates. If the equilibrium of the formation is not disturbed they become more hindered than trees in the course of their life; competition among vines becomes more severe than among trees and the number of the former decreases rapidly when their diameter increases. On the contrary, in gaps there is an extraordinary development of vine thickets that live a long time and return slowly to a rain forest. It seems also that shallow soils are favourable to the development of climbers with scattered trees as in the liane forest of Surinam (Lindeman and Moolenaar, 1959).

The tropical shelterwood system and its variants (see chapter 20) progressively opens up the canopy by poisoning undesirable trees and climbers; too intense an opening may induce an excessive development of climbers.

Climbers have at least three behaviour types:
— light demanding species which do not regenerate in closed forest and which have a very sporadic distribution;
— relatively shade tolerant species with a weak regeneration but rather high survival;
— shade tolerant species with abundant regeneration but relatively low survival.

Some are locally very abundant, e.g. *Bauhinia guianensis* in the rain forest of Venezuelan Guyana. Numerous species occur on wet soils, especially along rivers. They rarely exceed 50 cm DBH.

In dry-land rain forest, their spatial distribution follows the Poissonian model (e.g. for vines with a diameter ⩾10 cm DBH on 1/8 ha plots). There is some evidence of gregariousness which can be tested by a systematic analysis of the influence of plot size. Before reaching the climber life-form, their height distribution seems to follow an exponential model; the same occurs later on for their diameters. The measurement of the height, length and crown diameter of climbers is impracticable. The importance of climbers in the canopy can be measured by their leaf weight: it is 10 per cent of the leaf total in a rain forest of Kampuchea (Hozumi *et al.*, 1969), one third in southern Thailand (Ogawa *et al.*, 1965), and more than 36 per cent in a lower montane forest in Gabon (Hladik, 1974). In an undisturbed forest, the proportion of climbers (by weight) is low. In Amazonia Fittkau and Klinge (1973) and Klinge and Rodrigues (1974) found 46 t/ha fresh weight of climbers in a total of 687 t/ha.

Competition among climbers in tree crowns has been studied by Kira and Ogawa (1971). It seems that there are no exclusion phenomenon between climbers and non-climbers among seedlings but rather a vague proportionality (Rollet, 1969). Among seedlings ⩽1 m high, there is about 1 climber seedling for 5 tree seedlings. This ratio is 1 to 100 for stems with a DBH⩾10 cm in Venezuelan Guyana.

Leaves

Leaves comprise only a small part of the total biomass. According to Ogawa *et al.* (1965a) it reaches 7.7–8.2 t/ha compared to a total biomass of 320–400 t/ha, i.e. about 2 per cent. However, their crowding is a characteristic feature of the forest and they play an important role by their area development (besides photosynthesis and light interception) in evapotranspiration as a screen against wind, rainfall, and night condensations. Therefore, it is useful to know the leaf area index. According to Ogawa *et al.* (1965a) it varies from 1.6 to 1.8 in dry dipterocarp forests of South-East Asia; it reaches 3.9 for monsoon forests and 10.8 for rain forests; undergrowth increases this index from 1.2 to 4.0; its total variation is from 3.0 (dry dipterocarp forest) to 12.3 (rain forest). Hozumi *et al.* (1969) quote still lower figures.

Raunkiaer (1934) described a size scale to describe leaves (lepto-, nano-, micro-, meso-, macro-, mega-phylls). Webb (1959) distinguished 2 subcategories among mesophylls. He described Australian forests using leaf size as a principal character. Vareschi (1968) regards the rain forests in Sabah as true or eupluvial rain forests, because their leaves have practically the same shape and size. He gives comparative drawings for leaf-assemblages found in various forest types of Venezuela: size and shape become increasingly diversified as drier types are considered. The proportion of driptips has always been considered as a character linked to high rainfall: Baker (1938) in Sri Lanka, Gessner (1956) in Guatemala and Rollet (1969) in Venezuela.

The area of individual leaves for a given species is very variable (coefficient of variation of 25–50 per cent); the same is true for the specific density (0.40–2.60 g/cm^2 in Venezuelan Guyana). For the estimation of leaf area index, these density differences should be taken into account in order to avoid bias. The distribution of areas of individual leaves in a forest is almost log-normal, without gaps. Any subdivision in classes, though convenient, is therefore somewhat arbitrary (Rollet, 1969).

Leaf sizes in rain forests were studied by Stehlé (1945, 1946) in the West Indies, Duvigneaud *et al.* (1951) in Zaire, Richards (1952) in Nigeria and the Philippines, Cain *et al.* (1956) in Brazil, Loveless and Asprey (1957) in Jamaica, Tasaico (1959) and Petit Betancourt (1964) in Costa Rica, Ashton (1964) in Brunei and Rollet (1969) in Venezuela. Leaf sizes were studied in other tropical forest types by Stehlé (1945, 1946) for the West Indies, Webb (1959) for Australia, Ashton (1964) for heath forests of Brunei and Brünig (1968) for heath forests of Sarawak (kerangas). Van Steenis (1958) doubts that statistics of leaf size can show significative differences to distinguish types among lowland tropical rain forests.

Health conditions

Health conditions are important for two reasons: they have a considerable effect on the commercial value of a forest and serve to characterize primitive forests. In forests inventories, shape, quality and visible defects are generally recorded for each tree. Often a small sample is felled to evaluate hidden defects. Also each tree may be given a score for a better evaluation of the forest from the point of view of utilization.

The more primitive a forest, the higher is its proportion of decay. Secondary forests are generally more sound although site has a considerable influence on this proportion (e.g. swamps) and some species are more susceptible than others to special defects.

Conclusion

Such a study of the architecture of closed forests is global, descriptive and static.

It is global because species are not distinguished and because their spatial distribution is not considered. It describes the incidence of the various constituents and shows that some tree dimensions have similar well defined distributions even on small areas. Finally, this description is static. Its value lies in the fact that it will not change in time since an equilibrium has been reached. This concept of equilibrium takes all its significance when a formation is studied species by species.

It has been noticed that in the exponential expression of the distribution of diameters, heights, crown diameters, the time factor does not appear. It was introduced for considerations of growth and mortality in the description of the number of stems within a given diameter- (or height-) class and is eliminated when it is discovered that the variation of this number is null, i.e. that the stand is in equilibrium. This is not the case with rapidly growing formations like secondary forests or mangroves at various ages of their succession.

Attention is drawn to the need for advanced studies on the two categories of secondary forests following shifting cultivation and gaps due to windfalls. Balanced rain forests are in fact a peculiar case and the end of the successional processes. Although internal modifications are always the rule in climax forests (tree-falls and mortality), the global appearance is unchanged qualitatively (number of species) and quantitatively (biomass, diameter or height-distribution, etc.). Hence the need to look for growth—and competition—models. Such models exist in simple cases of even-aged, mono- or paucispecific tree-populations (Turnbull, 1963). In the case of rain forests, the number of species is very high and their behaviour and performances are extremely varied. Therefore research should be orientated towards the dynamics of stands to understand fully their internal functioning by studying *competition, growth, regeneration* and *mortality*.

Structure

The word structure has become rather vague in all languages because of the various meanings it has acquired. Thus structure has been used in the sense of distribution of tree-numbers in diameter classes by Meyer *et al.* (1943), Meyer (1952), Turnbull (1963), White (1963), Rollet (1969) and Mervart (1971); for basal area distribution in diameter classes by Turnbull (1963); or for components of biomass by Golley *et al.* (1969). Richards (1939) used structure in the sense of distribution of individuals in life-forms or in strata, afterwards (1940) to designate architecture organization, and in the same way as Takeuchi (1961) and Grubb *et al.* (1963). Goldstein *et al.* (1972) use the word structure in the sense of species associations; Brünig (1970) uses stand structure of various formations in Sarawak to mean their canopy architecture; likewise, Jones (1945), for Nigeria, but adding the concept of age-groups. The same variety of meanings applies to the word pattern which roughly covers what we call here structure: any stable or evolving non-anarchic situation of a population or community in which some kind of organization can be detected, how even small, and which can be represented by a mathematical model, a statistical law of distribution, a classification or a characteristic parameter. Thus, it is possible to speak of diameter structure, total height structure, crown-, canopy-structure, global spatial structure (gregarious, homogeneous structure), spatial structure of a species, floristic richness structure (species-area curve), floristic diversity structure, association structure and balanced structure.

In all cases, there are obviously preferential groupings of species in tropical forest vegetation, e.g. dominancy of life-forms, species associations due to special edaphic conditions, clusters of large trees on certain topographic sites; there are also more subtle assemblages that require advanced data processing to be detected.

Diameter distributions for individual species

Examples of diameter distributions are shown in fig. 2.

Diameter structures for all species have already been discussed. Diameter structures for individual species show very different behaviours. If the logarithms of the number of trees are plotted in 10 cm diameter classes, species can be subdivided into 2 groups by a line that represents a theoretical species for which the number of trees would double from one diameter class to the immediate lower class.

Certain species show an erratic diameter distribution (type 1, on fig. 2) or a flattened bell-distribution (type 2), or small numbers of trees decreasing progressively when diameter increases, so that using normal coordinates a line of slight negative slope is obtained (type 3). Others follow the exponential model, i.e. using semi-logarithmic coordinates gives almost a straight line, but with very different slopes (type 4): high slopes correspond to species without large diameters, but with extremely numerous small diameters (type 5); others may show an upward (type 6) or a

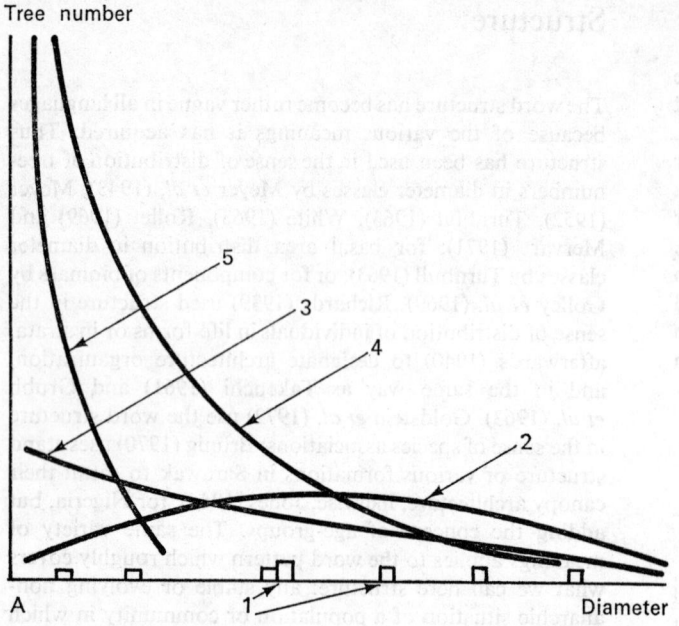

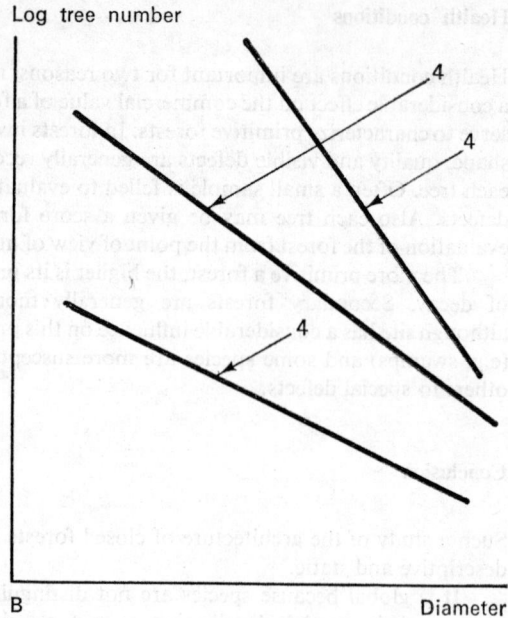

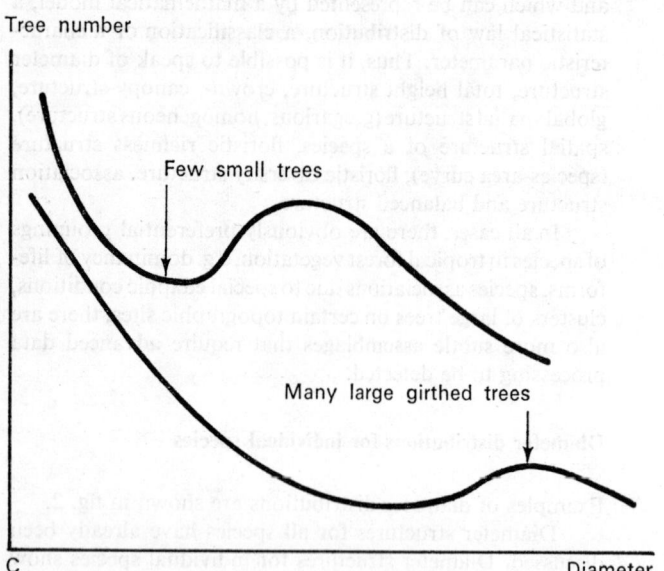

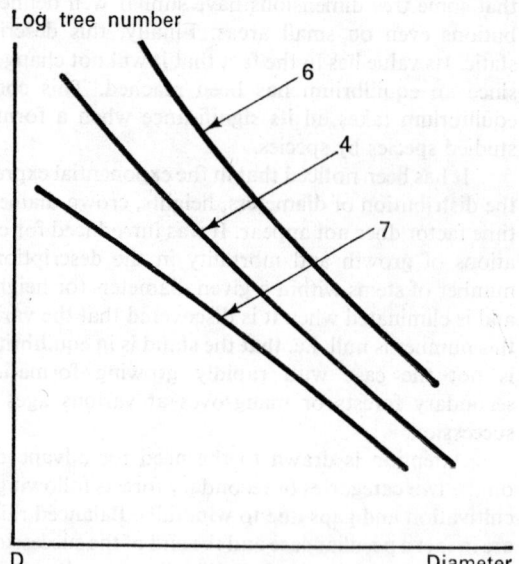

FIG. 2. The relations between tree number and diameters.

A. The different tree distributions reflect their behaviour types; 1, 2 and 3 are sun plants, 4 and 5 are shade plants.

B. As A, using a semi-log. scale; the species have a type 4 distribution with different slopes.

C. Irregularities due to forest history and the behaviour of certain species.

D. Using a semi-log. scale, the exponential model is a straight line (type 4); the actual distributions diverge from this because they have either too many (type 6) or too few (type 7) trees with small diameters.

downward concavity (type 7), which means that as diameter decreases, numbers of trees grow quicker (type 6), or more slowly (type 7) than in an exponential model. These various types of structure are easily related to different sylvicultural behaviours. Thus, species belonging to types 1, 2 and most of 3 are light demanders—nomad species of Van Steenis (1958). Species of types 4 and 5 are shade tolerant species.

There are species which cannot be represented by any of these 7 types, in particular those showing a relative gap for medium diameters and an apparent accumulation of large diameters. Caussinus *et al.* (1969, 1970) proposed a general analytical expression for the 7 types of diameter structures.

A stand may be characterized by a measure of its equilibrium; the ratio of number of species of types 4, 5, 6 and 7 over the total number of species. This ratio can be expressed either in number of species or corresponding number of trees, basal area, volumes or even biomass. In this way *coefficients of equilibrium* are obtained which make comparisons of stands possible. A coefficient of 0.75 expressed in basal area can be considered as typical of a balanced forest.

Examples of structures of diameters by species (Venezuelan Guyana, 155.5 ha). Number of trees by diameter classes, for the 5 types of diameter distributions (Rollet, 1974).

Type	Diameter class (cm)	10–19	20–29	30–39	40–49	50–59
1	*Tabebuia serratifolia*	3	1	3	3	2
2	*Inga alba*	84	90	105	73	27
3	*Sclerolobium* sp.	49	38	23	24	14
4	*Licania densiflora*	1 632	796	450	237	106
5	*Trichilia schomburgkii*	1 225	140	10	1	

Type	Diameter class (cm)	60–69	70–79	80–89	90–99	⩾100
1	*Tabebuia serratifolia*	1	1		3	1
2	*Inga alba*	20	3	5	2	
3	*Sclerolobium* sp.	10	9	4	1	2
4	*Licania densiflora*	33	4			
5	*Trichilia schomburgkii*					

Uses of diameter distributions

Basal area

The basal area of a stand is the sum of the trunk sections at DBH level, or above buttresses, of all trees on a given area and above a certain diameter. The basal area gives an idea of crowding in a forest; it is generally expressed for one hectare and various lower diameter limits, e.g. ⩾10 cm, ⩾20 cm, ⩾40 cm, ⩾60 cm.

Thus, in Venezuelan Guyana, the basal area of rain forests is for these limits respectively equal to 23.1, 18.7, 6.9, 3.7 m²/ha. About half of the basal area ⩾10 cm DBH refers to trees between 20 and 40 cm.

Examples of basal areas in m²/ha for trees ⩾10 cm DBH on various stands are as follows. They are arranged in increasing order of basal areas:
— swamp forest, Guyana, Hegyi (1962, 1963), 15.3;
— lowland evergreen rain forest, Venezuela, Rollet (1974), 23.1;

— mangrove, Venezuela, Salandy (1964), 23.6;
— lowland evergreen rain forest, Malaysia, Barnard (1954), 33–41;
— lowland evergreen rain forest, Ivory Coast, Bernhard-Reversat *et al.* (1975) 32 (⩾13 cm DBH).

Examples of basal area for trees ⩾20 cm DBH in various formations:
— swamp forest, Guyana, Hegyi (1962, 1963), 12.3;
— lowland evergreen rain forest, Venezuela, Rollet (1974), 18.7;
— montane rain forest (Archives, Facultad Ciencias Forestales, Mérida, Venezuela), 20.0;
— mangrove, Rio San Juan, Venezuela, Salandy (1964), 22.0;
— semi-deciduous forest, Kampuchea, Rollet (1962), 26.1;
— *Mora* swamp forest, Trinidad, Beard (1946b), 37.6.

Those averages based on large samples are more interesting than individual figures found on single plots. Lamprecht (1972) found that montane rain forests have larger basal areas than lowland rain forests: 40–60 m²/ha instead of 30–40 m²/ha. This statement needs substantiation by additional large sampling.

Basal area is easily computed from diameter distributions. If the midpoints of diameter classes are D_1, D_2, D_3, etc., with tree numbers n_1, n_2, n_3, etc., the basal area is $\frac{\pi}{4}(D_1^2 n_1 + D_2^2 n_2 + D_3^2 n_3 + \text{etc.})$. The basal area of climbers is negligible.

According to the pantropical structure established by Rollet (1974), the mean basal area of trees with a DBH⩾20 cm, in lowland evergreen rain forests is about 21 m²/ha (or 7 m²/ha for trees with a DBH⩾60 cm). Dawkins (1958) gives respectively 28.7 and 13.9 m²/ha; the latter figures seem to be rather high.

The basal areas show a strong variability, as a consequence of the fluctuation in distribution of trees by diameter classes (see fig. 1). The numerous data gathered in the annex to Rollet (1974) give an idea of that variation, either between plots of a region, or between regions of a given country, or between countries. For example, in Nigeria, the basal areas for trees with GBH⩾2 feet (*ca.* 20 cm DBH) vary from 11.6 to 24.1 m²/ha for different regions of lowland tropical rain forest, and between 4.4 and 12.8 m²/ha for trees with a DBH⩾60 cm. The variation is of course higher between small plots; see Huttel in Bernhard-Reversat *et al.* (1975) for the Ivory Coast and Brünig (1973) for Amazonia.

The concept of basal area in tropical forests is somewhat arbitrary since, because of buttresses, diameters are measured at variable heights; 40 per cent of trees with diameters ⩾40 cm have buttresses exceeding 1.50 m high in Venezuelan Guyana.

If it is accepted that diameter distributions follow exponential laws—which is generally true for trees with a DBH⩾20 cm—then the distribution of basal area follows an incomplete Γ function. Practically, the distribution is asymmetrically bell-shaped, with a long tail for large diameters.

Dendrometric observations

In stand architecture, both total heights and heights to the crown seem to vary as a function of DBH, according to a parabolic, or hyperbolic law (Ogawa *et al.*, 1965b). However, for large diameters, e.g. $\geqslant 60$ cm, there is a quasi-independence between DBH and total height. This tendency is even more marked for commercial heights (to the first defect) for DBH $\geqslant 40$ cm. Hence it is possible to estimate commercial volumes (uncorrected for defects) from the product of the basal area and the mean commercial height weighed by a form factor; this makes it unnecessary to measure commercial heights for all trees and saves considerable time. Thus, de Saint Aubin (1963) indicated that a good estimation of commercial volumes (in m^3) of *Aucoumea klaineana* in Gabon can be obtained by multiplying the basal area (in m^2/ha) by 10. Walker (1964) showed that for evergreen and semi-deciduous rain forests of Nigeria, the volumes without bark of six important commercial species are generally almost linearly related to basal areas, but the slopes of the regression lines vary greatly with species, from 21 units for *Afzelia* to 32 units for *Triplochiton*. Rollet (1967) indicated a well marked proportionality between basal areas and commercial volumes in Venezuelan Guyana; here the factor is slightly below 10.

Therefore the mean of the products of commercial height and the form factor for each tree within a diameter class is approximatively constant for DBH $\geqslant 40$ cm. Consequently unbiased estimates for uncorrected commercial volumes can be predicted on the basis of small representative samples, but it is wise to proceed with a certain number of local checks.

Uncorrected volumes and commercial values

Usable commercial volumes may be much smaller than uncorrected volumes because of defects. Grading is generally easily carried out during inventories but hidden defects and losses due to logging and sawmilling must be investigated in separate studies. FAO, forest research institutes and various consultants feel that it is absolutely necessary to carry out inventories to obtain more realistic figures of usable volumes than in the past and, in particular, more accurate than the uncorrected volumes.

The definition of commercial species is constantly changing, either because of fashion or following an increase of the general demand, or the marketing of new abundant species or through technical improvements in timber utilization (sawmilling, peeling, drying, impregnation, log protection). Owing to economical conditions, logging companies may extract only the top qualities or all inferior qualities and even large branches. In these conditions, it is easy to understand how difficult are short or medium-term predictions for the commercial value of a forest; that is why an inventory must always be detailed enough to allow high flexibility in the presentation of results.

A first estimation of the commercial value of a forest can be given by the number (or their basal area) of large stems per hectare; this may be detailed for important species. This mean number of large trees ($\geqslant 55$ or $\geqslant 60$ cm DBH) per hectare varies in a proportion of 1 to 3 in different countries of America (7.6–23.0); of 1 to more than 2 in Africa (12.1–26.1); in Asia it varies from 16 to 28 with the highest densities in Borneo and the Philippines. Basal areas follow the same variation: 5.7–10.7 m^2/ha in America; 6.3–13.9 m^2/ha in Africa; 8.2 to more than 20 m^2/ha in Asia. Uncorrected volumes in m^3/ha are more than 10 times the basal areas in m^2/ha (Rollet, 1974). A large quantity of information on uncorrected and commercial volumes can be found in inventory reports by FAO and bilateral aid bodies.

Biomass

Biomass measurements in closed tropical forests are still scarce and have been conducted on areas generally less than 0.25 ha. Due to the International Biological Programme (IBP) and some scientific institutes, some very accurate studies were done in Manaus, Amazonia (Klinge, 1973; Klinge and Rodrigues, 1974), in the Ivory Coast (Huttel, 1967), in Thailand, Kampuchea, and Malaysia (Ogawa *et al.*, 1965a; Hozumi *et al.*, 1969).

During most of this time, methodologies were built up but replications were rarely carried out so the variability of biomass is unknown; as it is strongly correlated with uncorrected volumes, it obviously fluctuates a great deal. Thus it is only possible to give biomass data for areas.

Nevertheless, the available studies show the relative importance of the principal components of biomass: bole wood, bark, leaves, flowers and fruit, litter, aerial roots and underground roots. Weight measurements of the latter are not always carried out; indirect estimates are used. Litter measurements should take into account the periodicities in leaf-fall. Various techniques have been proposed to evaluate the standing ligneous masses of the evergreen forests by Ogawa *et al.* (1965a), Hozumi *et al.* (1969) in Thailand, Kampuchea and Malaysia, or to estimate the leaf weight in a mangrove of Florida (Heald, 1969). Leaves represent a small portion of total biomass (flower and fruit contribution is also very small); their weight is much smaller than the sampling error on the total biomass estimation.

Dawkins (1959, 1961) concludes from data collected in Trinidad, Puerto Rico and Europe that there is a very small underestimation if the weight of branches under 4 cm girth is excluded and even under 4 cm diameter (0.1 per cent of the total volume). He suggests a simple rule to estimate the total volume of a tree including bark: the product of basal area, total height and 0.526. However, it would be useful to take other measures, to study the incidence of species and in particular of their timber densities which are very different from one another. It is better to compute basal area from actual diameters rather than from the midpoints of diameter classes because of the incidence of rare big trees.

The volume distribution with 5 cm diameter classes has an asymmetrical bell shape with a mode for the 35–39 cm class (30–39 cm class with 10 cm interval), recalling an incomplete Γ function. Considering the above mentioned relation between H and D, the probability density of the distribution law would be $K.H.D^2.e^{-\theta D}$, i.e. with an expression similar to $K'D^4.e^{-\theta D}$. Stands are very variable, not only in their basal areas, but also in the distribution of their total heights. On 20 0.25 ha plots surveyed in 11 tropical countries, and using Dawkins' suggestion, the mean total volume for all stems $\geqslant 4$ m high was found to be 570 m³/ha with an error of 10.6 per cent ($P=0.05$). In this example, stems between 4 m high and 5 cm DBH represent less than 0.5 per cent of the volume $\geqslant 10$ cm DBH, while trees between 10 and 19 cm DBH account for 9.2 per cent of the volume $\geqslant 10$ cm DBH. Total volume distribution per plot is bell-shaped (Rollet, 1974).

Results of biomass for lowland evergreen rain forests are available in Ogawa *et al.* (1965a), Hozumi *et al.* (1969) for South-East Asia; Veillon (1965) for Venezuela; Golley *et al.* (1969) for Panamá; Ovington *et al.* (1970) for Puerto Rico and in Fittkau *et al.* (1973) and Klinge *et al.* (1974) for Amazonia.

Golley *et al.* (1969) studied various forest types (lowland, hill, gallery, mangrove). Veillon studied the altitudinal variation of total volume (DBH$\geqslant 20$ cm) in 50 0.5 ha plots regularly distributed between 70 and 3 250 m of altitude in the Venezuelan Andes. The Japanese authors also studied semi-deciduous forests and dry dipterocarp forests in South-East Asia, but results are still fragmentary.

Spatial distributions

The distribution of trees of individual or all species on small areas will now be considered. It has already been shown that the number of individuals of all species in a particular diameter class follows a distribution law and shows a variability which depends on the plot size; the smaller the size, the nearer to a Poisson model are the distributions. This phenomenon is more marked if the analysis is carried out species by species; either by studying the total number of individuals of a species per plot or, the number for a particular diameter class. The principal objective is to evaluate to what extent and to what degree species are gregarious, how far are they from the extremes of random and regular distributions. The best example of the latter is a forest plantation.

These studies define precisely the concepts of homogeneity and gregariousness of vegetation. By definition, a species has a homogeneous spatial distribution if its plants are randomly distributed; if so, the probability of its presence in a plot is constant (concept of homogeneity) independent of its presence or absence in neighbouring plots (absence of gregariousness). Several approaches are possible: one considers either the presence or absence or the number of individuals of a species on area units.

Distributions of presences and absences by species: runs theory

This method was used in quantitative studies of grasslands in temperate countries (Dagnelie, 1968). Long continuous strips or large blocks must be available; for this reason it has been rarely used in tropical forests. The homogeneity of distribution of a species is quantified by its number of runs of presence, or absence. A run is a continuous succession of the same character in any number of contiguous plots. If presence in a plot is noted P and absence a, the following succession shows 8 runs: *PPP aaaaaa Paaa PP aaaa PPPPPP aa.*

For each species the number of presences m, the number of absences n and the number of runs U are noted. If the presence in a plot is independent of the presence in a neighbouring plot, the number of runs U must be near to $\dfrac{2\,mn}{m+n}+1$. U fluctuates around this mean with a variance $= \dfrac{2\,mn\,(2\,mn-n-m)}{(m+n)^2\,(m+n-1)}$. A tendency to gregariousness is shown when U is less than expected, while high values of U would indicate super-homogeneity.

The influence of plot size may be systematically studied by pooling neighbouring plots. In the only available study of this type in Venezuelan Guyana (Rollet, 1969) the number of randomly distributed species ($\geqslant 10$ cm DBH) increased gradually from 1/8 ha to 1 ha plots. For 1/8 ha plots, half of the species were randomly distributed whereas on 1 ha plots the number of random species was 2 or 3 times the number of non-random species. The same effect occurred with trees with a DBH$\geqslant 40$ cm. A provisional conclusion, for a given area of lowland evergreen rain forest, is that the proportion of random distributed species increases when plot size or diameter limit increase (Rollet, 1969). Similar studies in other regions should be done to test this conclusion. Moreover competition studies should be carried out, in order to increase knowledge of sylvicultural behaviour.

Spatial distribution of trees

Several distribution models may be studied: Poisson model and contagious models of Neyman (1939) and Thomas (1946) for various plot sizes and diameters. There are some difficulties in the interpretation of the results: conclusions vary irregularly with sizes and it may be arbitrary to prefer one model to another. Skellam (1952) showed that different models can give identical distributions.

In the expression of the law of a contagious distribution two parameters with simple meaning are used: the mean number of cluster per area and the mean number of individuals by cluster. Neyman as well as Thomas assume that the centers of clusters are randomly distributed and that the number of individuals per cluster is a random variable following Poisson's law.

The studies of spatial distributions are very scarce in tropical forests: Rollet (1952) in an evergreen forest of Viet-Nam for *Dipterocarpus* spp. $\geqslant 60$ cm; Jones (1955) in an evergreen forest of Nigeria; Jack (1961) in a semi-deciduous forest of Ghana only for trees with a DBH$\geqslant 90$ cm; Webb and Tracey (1965) in a rain forest of North Queensland; Faber (1967) in an evergreen forest of New Guinea for trees with a DBH$\geqslant 35$ cm; Rollet (1969), in an evergreen forest of Venezuelan Guyana for trees with a DBH$\geqslant$10 cm, concluded that distributions are seldom Poissonian : Thomas' distribution fits for 1/4–1/3 of the species according to plot size, Neyman's for 1/3–1/2 of the species, and Poisson's up to 7 per cent of the cases. Several thousand tests were done on more than 200 species for 1/16, 1/8, 1/4 and 1 ha plots with diameters $\geqslant$10, $\geqslant$20 and $\geqslant$40 cm.

Neyman's model indicates a mean number of clusters per plot much less correlated to plot size than Thomas', and this could be an argument in favour of Thomas' in spite of a lesser number of fits.

To conclude, the runs method and the test of goodness of fit to a Poisson series give satisfactory results accepting or rejecting the random hypothesis, although the Poisson method is the more severe and probably the more refined. A great number of species show the existence of some gregariousness which means it is unjustified to speak of the homogeneity of the evergreen rain forest. The fact that individuals of a species have a tendency to be gregarious on plots of a given size does not necessarily means that the species occurs in clumps in the sylvicultural sense.

In summary it is possible to analyse the spatial distribution of trees in the field using the presence or absence in plots by the method of runs, or the number of trees per plot. These studies are carried out for a certain plot size and a given lower limit of DBH or height. Since there are many theoretical distributions to which actual distributions can be compared, it is easy to imagine the large number of possible results, which are not necessarily consistent. Gregariousness is not expressed in the same way for different plot sizes; hence the necessity to synthesize those particular structures into a more general characteristic, using the theory of regionalized variables.

These studies on gregariousness (or non-homogeneity) bring into light the individual performances of species from the point of view of natural regeneration and interspecific competition. They are still very few. The high cost of the required inventories and mapping are partly responsible for this situation. However a number of existing inventories could be processed for partial studies, at least for important commercial species. But the reluctance of the organizations in charge of inventories towards this kind of analysis which is considered too academic largely explains those gaps of knowledge. However it is perfectly obvious that regeneration of tropical forests and species behaviour cannot be studied without such accurate methods.

Quantitative expressions of floristic richness and diversity

Only tree richness and diversity will be examined here.

It can easily be shown that the floristic composition of a rain forest changes gradually when larger and larger areas and smaller and smaller diameters (or heights) are considered. The species present in 1 m² of the forest-floor have very little in common with trees $\geqslant$40 cm DBH on the surrounding hectare. Therefore, it is necessary to specify areas and the size of individuals and to compare areas of about the same size.

Floristic richness is the total number of species of any size present on a given area (see chapter 4). In fact, even a high quality forest inventory underestimates the floristic richness.

Floristic diversity is the way species are distributed among the individuals. Trees and shrubs differ from herbs and epiphytes in that they are generally well defined as individuals—although suckers of climbers and of some tree species are sometimes fairly frequent. One difficulty is the very different sizes among individuals, each of them counting for one unit in the computation. This difficulty has been felt in many fields of the living world. To express plankton diversity Dickman (1968) showed that certain species of very small size can be represented by an enormous number of individuals; on the contrary, the highly dominant volume of some individuals of a few species makes negligible the contribution of other species. It has always been difficult to adequately represent the place occupied by a species in nature, hence various systems of weighing: number of individuals, basal area, volume, biomass, Curtis' importance value index, or other combinations such as the complexity index of Holdridge (1967), the *indice biotaxonómico* of Vareschi (1968), and the *erweiterter bedeutungswert-index* of Finol (1971).

Richness

The floristic richness of a formation may be expressed in several ways and raises interesting theoretical problems.

Number of species in a given area. Species-area curve

One of the simplest way to express the floristic richness of a tropical forest is to count the number of species occurring among the first *n* enumerated individuals, e.g. among the first 100 individuals: Brünig (1968) for the heath forests of Sarawak, or among a lesser number: Rollet (1969) in the rain forest of Venezuelan Guyana. This method is somewhat artificial, as the *n* individuals occupy a variable area and for this reason it is preferable to consider floristic richness in a fixed area. A large number of examples are available for small areas but difficulties quickly increase when diameter or height limits decrease: e.g. undetermined species become very numerous.

The following results can be quoted: 246 species of trees $\geqslant$10 cm DBH on 1 ha in a lowland rain forest of Malaysia (Rollet, 1974); 230 species of all plants on 1/100 ha in Cameroon (see chapter 4); 502 species $\geqslant$1.5 m high in 0.2 ha near Manaus, Amazonia (Fittkau and Klinge, 1973). To compare floristic richnesses, equal areas and equal lower limits of tree size are required; hence the idea of plotting a species-area curve. The results of various authors may be shown on the same graph if the lower limit of size is the same; thus one can get an idea of the relative floristic richness of different regions (Ashton, 1964; Rollet, 1969) and within a region for various limits of size of individuals. In this manner, the greatest possible number of results may be compared on a common basis. If a collection of identified trees, e.g. DBH$\geqslant$10 cm are available, a family of species-area curves $\geqslant$20 cm, $\geqslant$30 cm, $\geqslant$40 cm, etc., can be drawn.

Relatively few species-area curves have been established for tropical forests and, except for rare cases, they are only for trees on small areas, 1–3 ha in general. The following table shows the mean number of species of trees and climbers with a DBH ⩾ 10 cm as a function of the area in Venezuelan Guyana:

Area (ha)	1/8	1/4	1/2	1	2	4
Mean number of species	27	42	62	87	117	151

Area (ha)	8	16	32	64	128
Mean number of species	194	236	284	355	442

Species-area curves for lowland rain forests are very different from temperate formations; the former are in Emberger's (1950) phrase progressively saturated, while the latter show a quick increase in the number of species followed by a leveling off or at least a very slow increase of species. The phenomenon is particularly typical of bogs where almost all species are present on a very small area: this vegetation is quickly saturated, while the increase in number of species is indefinitely sustained in rain forests. For poor floras like bogs, it is attractive to try to define the smallest area which contains all or almost all the species, hence the concept of minimal area, which produced an abundant literature reviewed by Goodall (1952) and criticized by Gounot (1961). The minimal area concept is probably less useful in the tropics than in the temperate zone, at least for the more complex formations. *A priori*, saturation processes are different among seedlings and big trees; therefore, it is necessary to fix a minimal diameter or height. It will be of interest to consider separately the lower or so-called regeneration strata because of the considerable difference in size. Since species-area curves for seedlings do not show any tendency to flattening, the figures of minimal area suggested for the lowland evergreen rain forests must be accepted with utmost care: 100–150 m² in the Ivory Coast (Emberger, 1950; Mangenot, 1955); 100–800 m² for Zaire (Germain and Évrard, 1956); 1 100 m² in the *Gilbertiodendron* forests of Zaire (Gérard, 1960); 2 500 m² in Bangka, eastern coast of Sumatra (Meijer Drees, 1954). It is essential to study species-area curves for large samples, but such samples are very scarce. In conclusion the concept of minimal area has a lesser interest than in the temperate zone.

Various mathematical expressions of the species-area curves have been proposed. Fisher, Corbet and Williams (1943) think that the number of species is proportional to the logarithm of the number of individuals, i.e. approximately to the logarithm of the area. Williams (1964) tried to generalize this hypothesis to a number of populations, in particular to animal populations. Preston (1948, 1962) proposed a log-normal law for the distribution of species: number of species as a function of the number of individuals grouped in classes and showing a geometric progression. From this hypothesis, it follows that the logarithm of species number increases as the logarithm of area. But it seems that the number of species increases faster than the logarithm of area, but slower than a power function of area. In other words, a model for species-area curves would lie in between Williams' and Preston's. Goodall (1952) and Tüxen (1970) reviewed the abundant literature on species-area curves. A probabilistic formulation has been tried by Godron (1970, 1971) and Poissonet (1971) with the hypothesis of a random distribution of species on the area studied. Looking at a family of species-area curves for different DBH limits it is conspicuous that the ⩾ 90 cm and above curves are very flat and quite close to the ideal model where each area unit would contain one additional species. This model in which species numbers increase proportionally to areas is of course a straight line.

From the family of species-area curves it can be deduced how the number of species of a given area increases when the lower DBH limit decreases (see table below). The numbers seem to vary according to a power function of DBH.

Number of species (trees and climbers) as a function of a lower limit of DBH, in a lowland rain forest in Venezuelan Guyana, 128 ha (strip 25 m × 51.2 km). From Rollet (1969).

DBH (cm)	⩾ 100	⩾ 90	⩾ 80	⩾ 70	⩾ 60
Number of species (trees and climbers)	24	45	60	98	137

DBH (cm)	⩾ 50	⩾ 40	⩾ 30	⩾ 20	⩾ 10
Number of species (trees and climbers)	167	246	277	343	442

These considerations make it possible to forecast by extrapolation the number of species with a DBH ⩾ 10 cm out of an inventory where only species with a DBH ⩾ 20 cm are available.

Such families of curves could be drawn for any region where the detailed inventories are available.

Sampling problems

Nothing has been said about the spatial distribution of plots used to establish a species-area curve. The number of species encountered will be greater when plots are far apart instead of being contiguous. Thus, the sampling system (block, long strip, grid with more or less wide intervals) introduces modifications in the theoretical curve. The study of these aspects has been neglected (Rollet, 1969).

It should be stressed again that very few relatively important areas have been enumerated with care. Ashton (1964) gives an example in Brunei where a hilly area of 50 acres (about 20 ha) has 420 tree species with a DBH ⩾ 10 cm and a lowland 50 acre area 472 species. Altogether on 100 acres 760 species with a DBH ⩾ 10 cm were found, which reveals a very high floristic richness. In his strip of 128 ha in Venezuelan Guyana, Rollet (1969) certainly underestimated the number of trees and climbers with a DBH ⩾ 10 cm (442 species). Dubois (personal communication) found

322 tree species with a DBH$\geqslant$45 cm on 77 ha between Santarém and the Transamazónica highway. Heinsdijk's (1957, 1958, 1960) inventories for trees with a DBH$\geqslant$25 cm on over 1 200 ha scattered along a belt more than 1 500 km long between Belém and Manaus, considerably underestimated the number of species (400).

Diversity

Several attempts have been made to relate mathematically the number of stems and the corresponding number of species. One of the first and certainly the simplest is the *Mischungsquotient* of Jentsch (1911) quoted by Mildbraed (1933). This mixture coefficient is the ratio of the number of stems to the corresponding number of species. In itself this ratio does not mean anything if the area surveyed or the DBH lower limit are not mentioned, and since this ratio is steadily increasing with area, it is not of much interest. Mildbraed used it for one hectare survey in Isla de Macías Nguema Biyogo (formerly Fernando Po).

Much more interesting are the attempts of Fisher *et al.* (1943), of Preston (1948, 1962), of Williams (1964) and the use of the information theory. Earlier Pidgeon and Ashby (1942) had suggested that the number of species n increased with the area a according to the law $n=m \log a+b$. They called m the floristic variation. Following the theoretical studies of Fisher *et al.* (1943), Williams (1964) thought he could generalize the results obtained from a number of populations where the number of species with n individuals is $\dfrac{\alpha x^n}{n}$, x being between 0 and 1. Thus, the number of species with 1, 2, 3 individuals is αx, $\dfrac{\alpha x^2}{2}$, $\dfrac{\alpha x^3}{3}$, etc. The total number of species is the sum of a convergent series and equals $\alpha \log\left(1+\dfrac{N}{\alpha}\right)$, where N is the total number of individuals and α a constant called *floristic diversity coefficient*. N/α being large, the number of species is proportional to the logarithm of the number of individuals, and, therefore, practically proportional to the surveyed area.

Several authors computed values of α for various tropical forests in Zaire and the Ivory Coast. William's hypothesis underestimates the number of species in tropical forest communities. Moreover, there are practical drawbacks in the utilization of this series where often many terms are missing for field data. On the other hand, when the size of the sample increases, species with only one individual may become scarce. Preston's hypothesis (1948) has no such drawbacks; this author proposed not to determinate the number of species with 1, 2, 3 ... individuals, but the number of species present, between intervals in geometric progression that he calls octaves: 1 to 2, 2 to 4, 4 to 8, 8 to 16, etc. The number of species in such intervals has a Gaussian distribution; in the Rth octave away from the mode n_0, the number of species would be $n_0 \cdot e^{-(aR)^2}$; a is a coefficient. The breaking up of the variation interval into octaves and especially the corrections proposed by Preston for the limits of small octaves are somewhat arbitrary. This can be avoided by using cumulative numbers of species with 1, 2, 4, 8,

16 stems, etc., and the bell-shaped distribution becomes sigmoid. In fact, the sigmoid curves obtained are asymetrical, their upper half showing lesser curvature than expected, which indicates a deficit of species in the third quartile of the variation interval and an excess in the fourth (Rollet, 1969).

Black *et al.* (1950), Murça Pires *et al.* (1953) applied the Preston method to Amazonian forests in order to deduct the total number of species of a forest from a sample.

Later, Preston (1962) proposed a formula to predict the floristic richness N_{a+b} of two regions a and b, from their respective floristic richnesses N_a and N_b:

$$(N_{a+b})^{1/z}=(N_a)^{1/z}+(N_b)^{1/z};$$

with z near to 0.27. The formula is additive and if p equal areas are considered with the number of species N_{a1}, $N_{a^2} \ldots N_{ap}$ fluctuating around a mean $N_{\bar{a}}$, then follows:

$$(N_{pa})^{1/z}=p \cdot (N_{\bar{a}})^{1/z}, \text{ hence } N_{pa}=p^z \cdot N_{\bar{a}},$$

which means that p areas contain a number of species roughly proportional to a power function of p.

For large areas, this model seems to overestimate the total number of species, especially when extrapolating the results of a small area to a large area. However, it is more satisfactory than Williams' model, since the number of species is generally underestimated in the field, even in very good inventories. Therefore, the predictions of Preston's model are probably nearer to reality provided extrapolation to too large areas is avoided.

These two models are tentative interpretations of observed floristic diversities; the underlying hypotheses do not implicate any biological considerations. Without any hypothesis floristic diversity can be expressed by applying information theory and using the concept of mean quantity of information, H, per individual, which measures the degree of uncertainty concerning the identification of any individual. Following the interpretation of Brillouin (1959), a number of biologists applied these concepts in various fields: plankton, birds, etc.

Pielou (1966) calls N the total number of individuals of a population of s species, N_1 the number of individuals of species 1, N_2 the number of individuals of species 2, etc., and calculates the quantity

$$H = \frac{1}{N} \log \frac{N!}{N_1! \, N_2! \ldots N_s!}$$

She suggests quantifying the relative disorder of a population by the ratio H/H_{max} where H_{max} measures the maximal theoretical disorder occurring when the s species have the same number of individuals N/s. Hence $H_{max}=\text{Log } s$. The ratio H/H_{max} seems almost independent of area whereas H slowly increases as the area increases. In Venezuelan Guyana, on areas varying from 4 to 128 ha, the relative disorder H/H_{max} fluctuates between 76.3 and 78.6 per cent for the tree species with a DBH$\geqslant$10 cm. More systematical studies might demonstrate that H/H_{max} is a characteristic coefficient for each of the main tropical forest types.

Godron (1971) thinks that the diversity indices of Margalef (1957) and Pielou (1966) are but a first approach;

he suggests that several indices of structural heterogeneity should be calculated to take into account the topographical position of vegetation masses belonging to the various species. The general heterogeneity for each species will be expressed by the quantity of information brought by the knowledge of the number of presences E of a particular species in S plots, i.e. $\log_2 \frac{2^S-1}{C_S^E}$ which equals practically to S minus the logarithm (base 2) of the number of combinations of S plots E at a time. Using binary distances, regions with maximal heterogeneity are determined. The analysis of biological heterogeneity is completed by the study of gregariousness by means of the above mentioned runs and of interspecific liaisons. Thus, he determined ecological sub-assemblages by detecting groups of coexisting indicator species or of species with opposite ecological requirements.

In conclusion, floristic richness and diversity are two very important characteristics of tropical forests. They lead to fruitful theoretical developments, particularly concerning the species-area curve, different indices of diversity, comparison problems and the prediction of the floristic richness of an area from a sample.

These studies require an advanced floristic knowledge of forests. Therefore they are slow and costly. The inventories that are necessary need to be of high quality. They are very few, especially for DBH of 10 or 5 cm.

Species communities

Gregarious structures, both global and by species, have been already mentioned. Their effect is a decrease of local complexity in floristic mixtures. The fact that certain species are living preferentially together accentuates this tendency and contributes to structurize the forests, but for different reasons it seems that the hierarchy of associations is premature. When more detailed inventories are available, the interpretation of large floristic tables becomes practically impossible. Although the association concept cannot be questioned *a priori*—it is obvious in certain niches or in special biotopes—it seems far less obvious in lowland tropical rain forests where no noticeable variations of the environment occur to create marked floristic differences.

To overcome these difficulties, various methods of multivariable analysis and of classification were tried. The dilemma of ordination or classification and the preferable strategies will not be discussed here but obviously important methodological work is still necessary: data processing with species presence or species abundance; quantification of the abundance of a species; elimination of rare species; influence of minimal size of species and of plot size; consideration of ecological factors; criteria of similarity between plots.

Computation methods are many and can be simple or very sophisticated, from Bray and Curtis' ordination to Benzecri's factorial analysis of correspondences, with in between the classic principal component analysis, various versions of association analysis developed by Williams, the nearest neighbour method of Webb and Tracey (1965), etc.

It seems that close cooperation is always desirable between mathematicians and ecologists to clarify the present situation. Here are some examples for the application of numerical analysis to tropical forests: for Brunei, use of Bray and Curtis' ordination by Ashton (1964); in Australia, use of the association analysis by Webb *et al.* (1967, see chapter 4); for the Solomons association analysis and principal components analysis; for the Central African Empire, principal component analysis; for Venezuelan Guyana, association analysis and factorial analysis of correspondences by Rollet (1969).

Conclusions

The laws of organization of tropical forests are difficult to apprehend. What is striking is the multiplicity of species and strange morphologies (buttresses, climbers); there is a general and vague feeling of anarchy and confused mixture of species. To determine an organization requires an advanced knowledge of the species, large numbers of measurements and a certain effort of abstraction.

As for the architecture, structures can be considered globally without taking species into consideration, or studying species individually.

A structure exists whenever a phenomenon is not erratic, i.e. when it seems to follow a law and is therefore predictable and liable to extrapolation. Total heights of trees in a stand, their diameters, are not erratically distributed; their global arrangement is the result of competition and of internal dynamic processes. It is translated by a statistical equilibrium, either in relative frequences of trees in diameter classes, or in the proportions of species groups having different behaviours and which allow the assessment of the equilibrium status of the stands. Special ecological conditions and historical factors may locally modify structures.

Analysis demarcate new structures, linked to the behaviour of each species and which are reflected in specific spatial distributions. The assemblage of these behaviours is translated by several characteristics such as floristic richness and diversity. It is worth while to fit observed spatial distributions with theoretical models and deduct the properties of the tropical forest. Thus, the concepts of homogeneity and gregariousness can be examined critically.

The concept of structure is applicable to communities. Although soils can have a decisive influence on floristic compositions, communities seem more ill-defined than in temperate countries. Phytosociological studies are only just starting.

Uses of structures *lato sensu* are numerous. Global structures lead naturally to practical considerations of basal area, standing volumes, and biomass. The existence of specific spatial distributions leads to the study of the behaviour of species: phenology, dissemination, germination, survival with all their sylvicultural and management consequences.

Thermodynamic considerations lead to the hypothesis that nutrient-depleted humid tropical climax soils with a high degree of entropy should carry stands of low entropy, that is a high diversity value, to maintain efficiently a stable state of the whole system. Relatively rich and extremely poor soils should carry stands of lower diversity. This hypothesis agrees with observations in Sarawak, Guyana and the Brazilian Amazon (Fraenzle, personal communication).

The structures of lowland tropical rain forests are relatively unstable on small areas and always liable to modifications following the falling or death of trees. These are natural micro-catastrophies. But when larger and larger areas of forests are considered, those local disturbances have a more and more buffered modifying effect, and equilibrium of structures can be defined (Rollet, 1969). However, when submitted to the destructive action of man, particularly to his shifting cultivation, the tropical rain forests are fragile in a sense that they are replaced abruptly by new stages which regain their original architecture and structures rather slowly. Sometimes this destructive action seems irreversible. The resistance of these forests to man's impact varies according to soils, topography and climate. The tropical forests seem to shrink quicker and quicker following the progress of permanent and shifting agriculture. But if they are easily destroyed, they can also be put under management for a sustained production.

General conclusion: research needs

For 50 years, the studies of the tropical rain forests have moved more and more from a qualitative knowledge (Schnell, 1950a,b; Richards, 1952) to quantitative, thanks to the great development of commercial inventories which yielded general informations, and also to detailed studies on small areas carried out by forest research institutes, universities, the IBP and isolated researchers.

In fact, advanced quantitative studies can develop only after a thinking and conception phase during which topics and methodologies are defined, and after some experimental research.

Small samples are not sufficient to understand the organization of tropical forests, e.g. to elucidate spatial distributions, species communities, and regeneration processes. When a large amount of data are available in a forest (e.g. several million tree dimensions, spatial position and the botanical identity of trees), it is desirable to reduce them into something simpler and to find out the laws of architecture and organization, to build explicative models (simple models explaining particular aspects and more complex ones covering larger and larger numbers of those aspects). It is necessary to evaluate the costs of the various operations (data recording, punching, programming, computer time, interpretation, publication, data conservation) in order to optimize the results.

One realizes the importance of mathematical tools and the necessity of cooperation with statisticians, computer scientists and eventually mathematicians. It is obvious that it is not mathematical laws which we wish to extract from forest structures *per se*, but to interpret them biologically and

to use them to elucidate biological phenomena. Taking into account the supplement of interpretation which can be carried out on classic forest inventories and the infrastructure offered by the forest inventory projects of FAO, bilateral technical assistances, a scientific cooperation between these organizations and also with interested researchers, is highly desirable. Since these projects often last several years, association agreements should be found to ensure the scientific exploitation of certain results and publication.

Forest inventories must be considered as the first step towards the management of natural forests. The statistical appraisal of the relative importance of the species, in number of stems of various dimensions, their distribution in the field, the adequate interpretation of the inventory of small stems—the so-called regeneration inventories—the conclusions on the behaviour of the various species, the study of the evolution of natural clearings or gaps, of growth (natural and stimulated), of mortality and competition, of the incidence of climbers, are information that make decisions for taking certain selected parts of a forest into production possible. The knowledge of architecture and structures is therefore indispensable for forest management. Gaps existing in knowledge are guide-lines for useful research. These must be conducted with rigorous methodologies in a multi-disciplinary spirit, in view of a rational long-term management of tropical forest ecosystems. This management should be given the highest priority.

Research needs

Priority topics are indicated with an asterisk.

Description (see chapter 4)

— *Standardization of forest profiles (width, length, minimum height, subdivision into strips for canopy trees, undergrowth, regeneration layers; cf. Louis, 1947). Need for international agreement.
— High quality monographs of forests. Inventories of large areas, in particular for montane, swamp forests and mangroves.

Architecture

— Architecture of trees: branching, crown morphologies, aspects seen from the air and on large-scale aerial photographs.
— Structures of canopies on aerial photographs.
— Problems of automatic cartography.
— Climbers: their importance in regeneration layers and in crowns; regeneration, dynamics; populations.
— Epiphytes: populations; epiphytism and forest types.
— Leaves and physiognomic definition of forest types. Webb's system. Leaf area index; densities, consistency, biomass.
— Root systems and buttresses (morphology, dimensions, frequency, mechanical and anatomical problems).

Structure

— Structures of diameters: influence of rainfall, altitude, slope, soils. Variability, abnormalities (high numbers of large diameters). Types of structure by species; balanced structures; pantropical structures.
— Biomass: need for more measurements; sampling techniques; variability; errors.
— *Spatial distributions (global and by species). Ecological determinisms; research methods (runs, Poisson's and contagious laws; regionalized variables theory).
— *Dynamic models. Structures of diameters, heights, biomass. Competition models, growth and mortality models. Regeneration. Growth of secondary forests after shifting cultivation. Growth in natural clearings.
— Floristic richness and diversity. Species-area curves. Prediction of floristic richness from a sample. Associations.

Methodology

Some standardization in the recording of field data seems desirable in order to compare easily results from different sources. The metric system (or SI units) is almost universally accepted; international agreement is highly desirable on the adoption of simple rules for the measurement of tree dimensions (flexible enough to allow any grouping in classes, e.g. a 43 cm diameter would mean any diameter between

43.0 and 43.9 cm, a 16 m height, any height between 16.0 and 16.9 m) and for profile strip-widths (multiples of 5 m could be considered). Plots should be small, square or rectangular, in blocks, strips or grids. Some subsampling could be used.

It is difficult to describe in detail desirable processing, since new methods are always discovered, and new measurements taken into consideration. Commercial forest inventory data can very often be processed in more sophisticated ways. Therefore, these data could be reworked for scientific purposes.

For data processing, facilities in an international computing centre should be sought. A minimal processing for the architecture and structures of tropical forests could be recommended: distribution of DBH by classes (all and individual species); spatial distributions and competition; establishment of floristic tables, investigations on species associations.

There is little chance that much money will be used to save old archives of very uneven value; however, this would be worth while for some high quality data. Inventory data are rarely fully reproduced; for those of high quality, conservation measures (e.g. microfilms) could be contemplated. The most convenient way to keep data is on magnetic tapes. Any tape should be accompanied by an explanatory leaflet. The centralization in a data bank, near to a computing centre, could be contemplated, or more simply the wider exchange of data and programmes between specialists.

Bibliography

ADAMSON, R. S. The classification of life forms of plants. *The Botanical Review*, 5, 1939, p. 546–562.

ADDOR, E. E.; RUSHING, W. N.; GRABAU, W. E. A procedure for describing the geometry of plant assemblage. In: Odum, H. T.; Pigeon, R. F. (eds.). *A tropical rain forest. A study of irradiation and ecology at El Verde, Puerto Rico*, p. B 151–B 167. Division of Technical Information, U.S. Atomic Energy Commission (USAEC, Oakridge, Tenn.), 1970, 1 678 p.

ANDERSON, J. A. R. The enumeration of 235 acres of dipterocarp forest in Brunei. *Malayan Forester*, 20, 1957, p. 144–150.

*——. *The ecology and forest types of the peat swamp forests of Sarawak and Brunei in relation to their silviculture*. Edinburgh, unpublished Univ. thesis, 1961, 191 p., annexes, 53 fig., 50 plates, 85 ref.

——. *Revised checklist of trees of peat swamp forests of Sarawak*. Sarawak, Forest Department, 1971, 33 p.+10 p. multigr.

ANDERSON, M. C. Studies of the woodland light climate. 1. The photographic computation of light conditions. *J. Ecology*, 52, 1964, p. 27–41.

*ANDREWS, J. R. T. *A forest inventory of Ceylon*. Hunting Survey Corp. (Canada-Ceylon Colombo Plan Project), 1961, 116 p.

ANON. *Inventaires forestiers du bassin de la M'Baere et en Haute Sangha. République Centrafricaine*. Nogent-sur-Marne (France), Archives du Centre technique forestier tropical (CTFT), 1963 (M'Baere), 4 fasc., 14, 14, 15, 79 p. multigr.; 1964 (Haute Sangha), 8 fasc., 15, 29, 16, 15, 16, 16, 16, 87 p. multigr.

——. *Inventaire forestier dans le Lom et Kadei, Cameroun.*

Nogent-sur-Marne, Archives CTFT, 1966, 6 fasc., 28, 55, 55, 52, 54, 101 p. multigr.

*ASHTON, P. S. *Ecological studies in the mixed dipterocarp forests of Brunei State*. Oxford Forestry Memoirs, 25, 1964, 110 p.

ASPREY, G. F.; ROBBINS, R. G. The vegetation of Jamaica. *Ecological Monographs*, vol. 23, no. 4, 1953, p. 359–412.

ASSMANN, E. *Waldertragskunde: organische Produktion, Struktur, Zuwachs und Ertrag von Waldbestanden*. Munich, 1961, 490 p.

*AUBRÉVILLE, A. La forêt de la Côte d'Ivoire. Essai de géobotanique forestière. *Bull. Comité d'études historiques scientifiques de l'AOF*, vol. 15, n° 2–3, 1932, p. 205–249.

*——. Les brousses secondaires en Afrique équatoriale. Côte d'Ivoire, Cameroun, AEF. *Bois et Forêts des Tropiques* (Nogent-sur-Marne), 2, 1947, p. 24–52.

*——. *Climats, forêts et désertification de l'Afrique tropicale*. Paris, Soc. Éd. géographiques, maritimes et coloniales, 1949, 351 p.

*——. *Étude écologique des principales formations végétales du Brésil et contribution à la connaissance des forêts de l'Amazonie brésilienne*. Nogent-sur-Marne, CTFT, 1961, 268 p.

*——. Classification des formes biologiques des plantes vasculaires en milieu tropical. *Adansonia*, vol. 3, n° 2, 1963, p. 226–231.

*——. Principes d'une systématique des formations végétales tropicales. *Adansonia*, vol. 5, n° 2, 1965, p. 153–196.

* Major reference.

AUBRÉVILLE, A. La forêt primaire des montagnes de Bélinga. *Biologica Gabonica*, fasc. 2, 1967, p. 95–112.

*BAKER, J. R. Rain forest in Ceylon. *Kew Bull.*, 1938, p. 9–16.

BARNARD, R. C. A manual of Malayan silviculture for inland lowland forest. *Res. Pamph. For. Res. Inst.* (Kepong), 14, 1954, 199 p.

*BAUR, G. N. *The ecological basis of rain forest management*. Forestry Commission of New South Wales, 1961–62. Rome, FAO, André Meyer Fellowship Programme Report, 1962, 499 p.

*BEARD, J. S. Montane vegetation in the Antilles. *Caribbean Forester*, vol. 3, no. 2, 1942a, p. 61–74.

*——. The use of the term deciduous as applied to forest types in Trinidad (British West Indies). *Empire For. J.*, 21, 1942b, p. 12–17.

*——. The natural vegetation of the island of Tobago (British West Indies). *Ecol. Monogr.*, 14, 1944a, p. 135–163.

*——. Climax vegetation in tropical America. *Ecology*, 25, 1944b, p. 127–158.

——. Los climax de vegetación en la América tropical. *Revista Fac. Nac. de Agronomía de Medellín* (Colombia), vol. 6, nº 23, 1946a, p. 225–293.

——. The *Mora* forest of Trinidad (British West Indies). *J. Ecology*, vol. 33, no. 1, 1946b, p. 173–192.

*——. *The natural vegetation of Trinidad*. Oxford Forestry Memoirs, 20, 1946c, 152 p.

*——. *The natural vegetation of the Windward and Leeward Islands*. Oxford Forestry Memoirs, 21, 1949, 192 p.

BERGEROO-CAMPAGNE, B. Évolution des méthodes d'enrichissement de la forêt dense de la Côte d'Ivoire. *Bois et Forêts des Tropiques* (Nogent-sur-Marne), 1958, 58, p. 17–32; 59, p. 19–35.

BERNHARD-REVERSAT, F.; HUTTEL, C.; LEMÉE, G. Recherches sur l'écosystème de la forêt subéquatoriale de Basse Côte d'Ivoire. *La Terre et la Vie* (Paris), vol. 29, nº 2, 1975, p. 169–264.

BITTERLICH, W. Das Relaskop. *Allgemeine Forst- und Holzwirtsch. Zeitung*, 60, 1949, p. 5–6.

*BLACK, G. A.; DOBZHANSKY, T.; PAVAN, C. Some attempts to estimate species diversity and population density of trees in Amazonian forests. *The Botanical Gazette* (Chicago), vol. 3, no. 4, 1950, p. 413–425.

BLASCO, F. *Montagnes du sud de l'Inde, forêts, savanes, écologie*. Trav. section scientifique et technique (Institut français de Pondichéry), vol. 10, nº 1, 1971, 436 p., 1 carte, 19 fig., 43 phot. Thèse Université de Toulouse.

BRILLOUIN, L. *La science et la théorie de l'information*. Paris, Masson, 1959, 302 p.

BRISCOE, C. B.; WADSWORTH, F. H. Stand structure and yield in the Tabanuco forest of Puerto Rico. In: Odum, H. T.; Pigeon, R. F. (eds.). *A tropical rain forest. A study of irradiation and ecology at El Verde, Puerto Rico*, p. B 79–B 89. Division of Technical Information, U.S. Atomic Energy Commission (USAEC, Oakridge, Tenn.), 1970, 1 678 p.

BROOKS, R. L. The regeneration of mixed rain forest in Trinidad. *Caribbean Forester*, vol. 2, no. 4, 1941, p. 164–173.

BROWN, W. H.; MATTHEWS, D. M. Philippines dipterocarp forests. *Philip. J. Sci.*, 9, Sect. A, 1914, p. 413–561.

*BRÜNIG, E. F. Der Heidewald von Sarawak und Brunei. *Mitteilungen Bundesf. für Forst- u. Holzwirtschaft* (Reinbek), no. 68, 2 vol., 1968a, 431 p., 14 fig., 16 tab., 28 phot.

——. Some observations on the status of heath forests in Sarawak and Brunei. In: Misra, R.; Gopal, B. (eds.). *Proc. Symp. Recent Adv. Trop. Ecol.*, 1968b, p. 451–457.

*——. Stand structure physiognomy and environmental factors in some lowland forests in Sarawak. *Tropical Ecology*, vol. 2, no. 1, 1970, p. 26–43.

BRÜNIG, E. F. Species richness and stand diversity in relation to site and succession of forests in Sarawak and Brunei (Borneo). *Amazoniana*, vol. 4, no. 3, 1973, p. 293–320.

——. *Ecological studies in the kerangas forests of Sarawak and Brunei*. Borneo Lit. Bureau for Sarawak, For. Dpt. (Kuching), 1974, 237 p.

——; KLINGE, H. Comparison of the phytomass structure of equatorial rain forest in central Amazonas, Brazil and in Sarawak, Borneo. In preparation.

*BURGESS, P. F. The structure and composition of lowland tropical rain forest in North Borneo. *Malayan Forester*, vol. 24, no. 1, 1961, p. 66–80.

——. Studies on the regeneration of the hill forests of the Malay Peninsula: the phenology of dipterocarps. *Malayan Forester*, vol. 35, no. 2, 1972, p. 103–123.

BURTT DAVY, J. *The classification of tropical woody vegetation types*. Imp. For. Inst. Paper no. 13, 1938, 85 p.

*CAIN, S. A.; CASTRO, G. M. O.; PIRES, J. M.; DA SILVA, N. T. Application of some phytosociological techniques to Brazilian rain forest. *Amer. J. Bot.*, vol. 43, no. 10, 1956, p. 911–941.

*CAUSSINUS, H.; LAMBERT, E.; ROLLET, B. Sur l'utilisation d'un nouveau modèle mathématique pour l'étude des structures des forêts denses humides sempervirentes de plaine. *C.R. Acad. Sci. Paris*, vol. 269, 1969, p. 2547–2549.

*——; ROLLET, B. Sur l'analyse, au moyen d'un modèle mathématique, des structures par espèces des forêts denses humides sempervirentes de plaine. *C.R. Acad. Sci. Paris*, 270, 1970, p. 1341–1344.

CHALK, L.; AKPALU, J. P. Possible relation between the anatomy of the wood and buttressing. *Commonwealth Forestry Review*, vol. 42, no. 1, 1963, p. 53–68.

*CHAMPION, H. G. A preliminary survey of the forest types of India and Burma. *Indian For. Rec. (new series)*, vol. 1, no. 1, 1936, 8+286 p.

CHARTIER, P.; BECKER, M.; BONHOMME, R.; BONY, J. P. Effets physiologiques et caractérisation du rayonnement solaire dans le cadre d'une méthode d'aménagement sylvicole en forêt dense africaine. *Bois et Forêts des Tropiques* (Nogent-sur-Marne), nº 152, 1973, p. 19–35.

CHIPP, T. F. Buttresses as an assistance to identification. *Kew Bull.*, 1922, p. 265–268.

*CORNER, E. H. J. *Wayside trees of Malaya*. 2nd edition. Singapore, vol. 1, 1952, 772 p.; vol. 2, 228 pl.+5 p., 359 fig.

COSTER, C. Wortelstüdien in de Tropen. *Tectona*, vol. 25, 1932, p. 828–872; vol. 28, 1935, p. 861–878. Reedition in 1957.

*COUSENS, J. E. Some notes on the composition of lowland tropical rain forest in Rengam Forest Reserve, Johore. *Malayan Forester*, vol. 14, no. 3, 1951, p. 131–139.

*——. A study of 155 acres of tropical rain forest by complete enumeration of all large trees. *Malayan Forester*, 21, 1958, p. 155–164.

——. Some reflections on the nature of Malayan lowland rain forest. *Malayan Forester*, 28, 1965, p. 122–128.

DAGNELIE, P. Quelques méthodes statistiques d'étude de l'homogénéité et de caractérisation de la végétation. In: *Fonctionnement des écosystèmes terrestres au niveau de la production primaire* (Actes du Colloque de Copenhague, 1965), p. 481–486. Paris, Unesco, 1968, 516 p.

*DANSEREAU, P. Description and recording of vegetation upon a structural basis. *Ecology*, 32, 1951, p. 172–229.

*DANSEREAU, P.; BUELL, P. F.; DAGON, R. A universal system for recording vegetation. 2. A methodological critique and an experiment. *Sarracenia*, 10, 1966, p. 1–64.

*DAVIS, T. A. W.; RICHARDS, P. W. The vegetation of Moraballi Creek, British Guiana. An ecological study of a limited area of tropical rain forest. *J. Ecology*, vol. 21, 1933, p. 350–384; vol. 22, 1934, p. 106–155.

*DAWKINS, H. C. *The management of natural tropical high forest with special reference to Uganda*. Imperial Forestry Institute (Oxford), Institute Paper no. 34, 1958, 155 p.

——. The volume increment of natural tropical high forest and limitations on its improvement. *Empire For. Rev.*, 38, 1959, p. 175–180.

——. Estimating total volume of some Caribbean trees. *Caribbean Forester*, vol. 22, no. 3–4, 1961, p. 62–63.

*——. Crown diameters: their relation to bole diameters in tropical forest trees. *Commonwealth Forestry Review*, vol. 42, no. 114, 1963, p. 318–333.

DESMARAIS, A. P.; VAZQUEZ, A. Upper canopy crown closure at El Verde. In: Odum, H. T.; Pigeon, R. F. (eds.). *A tropical rain forest. A study of irradiation and ecology at El Verde, Puerto Rico*. Division of Technical Information, U.S. Atomic Energy Commission (USAEC, Oakridge, Tenn.), 1970, 1 678 p.

DICKMAN, M. Some indices of diversity. *Ecology*, vol. 49, no. 6, 1968, p. 1191–1193.

*DILLEWIJN, F. J. van. *Sleutel voor de interpretatie van begroei-ingsvormen uit luchtfoto's 1/40 000 van het Noordelijk deel van Suriname*. Paramaribo, 1957, 45 p.

DONIS, C. *Essai d'économie forestière au Mayumbe*. Publ. INEAC (Bruxelles), Sér. sci., n° 37, 1948, 92 p.

DULAU, L. *Le milieu physique et les aspects actuels de la végétation de la Guadeloupe*. Thèse Toulouse, 1955, 486 p.

*DUVIGNEAUD, P. (éd.). *Productivité des écosystèmes forestiers* (Actes du Colloque de Bruxelles, 1969). Paris, Unesco, 1971, 707 p.

——; SMET, S.; KIWAK, A.; MESOTTEN, G. Écomorphologie de la feuille chez quelques espèces de la 'laurisilve' du Congo méridional. *Bull. Soc. Roy. Bot. Belgique*, 84, 1951, p. 91–95.

EGGELING, W. J. Observations on the ecology of the Budongo rain forest, Uganda. *J. Ecology*, 34, 1947, p. 20–87.

EMBERGER, L. Observations phytosociologiques de la forêt dense équatoriale. *Arch. Inst. Grand-ducal Luxemb.*, 19, 1950, p. 119–123.

*——; MANGENOT, G.; MIÈGE, J. Existence d'associations végétales typiques dans la forêt dense équatoriale. *C.R. Acad. Sci. Paris*, 231, 1950a, p. 640–642.

*——; ——; ——. Caractères analytiques et synthétiques des associations de la forêt équatoriale de Côte d'Ivoire. *C.R. Acad. Sci. Paris*, 231, 1950b, p. 812–814.

EMLEN, J. T. A rapid method for measuring arboreal canopy cover. *Ecology*, vol. 48, no. 1, 1967, p. 158–160.

ÉVRARD, C. *Recherches écologiques sur le peuplement forestier des sols hydromorphes de la cuvette centrale congolaise*. Publ. INEAC (Bruxelles), Sér. sci., n° 110, 1968, 295 p., 33 phot.

FABER, P. J. An application of discontinuous distributions in the analysis of tropical rain forests. In: *Proc. 14th IUFRO Congress*, sect. 25, 1967, p. 115–128.

——; LUITJES, J.; VERSTEEGH, C. *Bosinventarisatie in de Sekoliviakte*, 1959, 18 p., 1 map+annex, multigr.

——; ——; VINK, W. *Verslag over de Bosinventarisatie in de Warsamson*, 1960–1961.

*FANSHAWE, D. B. *The vegetation of British Guiana, a preliminary review*. Imp. For. Inst. (Oxford), Inst. Paper no. 29, 1952, 96 p.

FINOL, H. Berücksichtigung neuer Parameter in der Strukturanalyse von tropischen Urwaldern. *Mededelingen Fakulteit Landbouw-Wetenschappen* (Gent), vol. 36, n° 2, 1971, p. 701–709.

*FISHER, R. A.; CORBET, A. S.; WILLIAMS, C. B. The relation between the number of species and the number of individuals in a random sample of an animal population. *J. Animal Ecology*, 12, 1943, p. 42–58.

*FITTKAU, E. J.; KLINGE, H. On biomass and trophic structure of the Central Amazonian rain forest ecosystem. *Biotropica*, vol. 5, no. 1, 1973, p. 2–14.

FLEMMICH, C. O. *Marketing of wood and wood products with particular reference to the export of timber*. British Guiana, FAO report no. 1737, 1963, 83 p.

FOGGIE, A. Some ecological observations on a tropical forest type in the Gold Coast. *J. Ecology*, 34, 1947, p. 88–106.

*FOX, J. E. D. An enumeration of lowland dipterocarp forest in Sabah. *Malayan Forester*, vol. 30, no. 4, 1967, p. 263–279.

——. Ultra basic forest North East of Ranau. *Archives Forest Service Sandakan*, 1970, 29 p. multigr.

*FRANCIS, E. C. Crowns, boles and timber volumes from aerial photographs and field surveys. *Commonwealth Forestry Review*, 45, 1966, p. 32–66.

*FRANCKE, A. Zur Gliederung der forstlich wichtigeren Vegetationsformationen des tropischen Afrikas. *Kolonialforstl. Mitteilungen*, 1, 1942, 44 p.

FRANÇOIS, T. La composition théorique normale des futaies jardinées de Savoie. *Rev. Eaux et Forêts*, vol. 76, n° 1–18, 1938, p. 101–115.

GÉRARD, P. *Étude écologique de la forêt de Gilbertiodendron dans l'Uele*. Gembloux, Inst. Agron., thèse de doctorat, 1959, non publié.

*——. *Étude écologique de la forêt dense à Gilbertiodendron dewevrei dans la région de l'Uele*. Publ. INEAC (Bruxelles), Sér. sci., n° 87, 1960, 159 p.

GERMAIN, R.; ÉVRARD, C. Caractères structurels du groupement à *Brachystegia laurentii* (D. Wild) Louis ex Hoyle dans la région de Yangambi (Congo belge). In: *Proc. 8th Int. Bot. Congress*, Sect. 7–8, 1954, p. 148–151.

*——; ——. *Étude écologique et phytosociologique de la forêt à Brachystegia laurentii*. Publ. INEAC (Bruxelles), Sér. sci., n° 67, 1956, 105 p.

GESSNER, F. Die Wasseraufnahme durch Blätter und Samen. In: *Handbuch der Pflanzenphysiologie*, Bd. 3, p. 222–225. Berlin, Springer Verlag, 1956.

GLERUM, B. *Forest inventory in the Amazon Valley*. FAO report, part 7, no. 1492, 1962, 8 p.

——; SMIT, G. *Forest inventory in the Amazon Valley*. FAO report, part 6, no. 1271, 1960, 14 p.

——; ——. *Combined forestry-soil survey along the road BR 14 from São Miguel do Guamá to Imperatriz*. FAO report no. 1483, 1962, 139 p.

*GLORIOD, G.; LANLY, J. P. *Inventaire de 100 000 hectares de forêt dense dans la région de Kango (République gabonaise)*. Nogent-sur-Marne, CTFT, 1964, 189 p.

*GODRON, M. Un modèle pour la courbe aire-espèces. *Le naturaliste canadien*, 97, 1970, p. 491–492.

*——. Comparaison d'une courbe aire-espèces et de son modèle. *Oecologia Plantarum*, vol. 6, n° 2, 1971, p. 189–195.

——. Trois problèmes posés par l'extension des observations relatives à la production d'un écosystème forestier. In: *Productivité des écosystèmes forestiers* (Actes du Colloque de Bruxelles, 1969), p. 579–581. Paris, Unesco, 1971, 707 p.

GOLDSTEIN, R. A.; GRIGAL, D. F. Definition of vegetation struc- ture by canonical analysis. *J. Ecology*, vol. 62, no. 2, 1972, p. 277–284.

*GOLLEY, F. B.; McGINNIS, J. T.; CLEMENTS, R. G.; CHILD, G. Y.; DUEVER, M. J. The structure of tropical forests in Panama and Colombia. *Bio-Science*, vol. 19, no. 8, 1969, p. 693–696.

*GOODALL, D. W. Quantitative aspects of plant distribution. *Biol. Rev.*, 27, 1952, p. 194–245.

*GOUNOT, M. Les méthodes d'inventaire de la végétation. *Bull. Serv. carte phytogéogr.*, Sér. B, vol. 6, n° 1, 1961, p. 7–73.

GRAY, B. Size, composition and regeneration of *Araucaria* stands in New Guinea. *J. Ecology*, vol. 63, no. 1, 1975, p. 273–289.

*GRUBB, P. J.; LLOYD, J. R.; PENNINGTON, T. D.; WHIT- MORE, T. C. A comparison of montane and lowland rain forest in Ecuador. I. The forest structure physiognomy and floristics. *J. Ecology*, vol. 51, no. 3, 1963, p. 567–599.

GUÉNEAU, P. Contraintes de croissance. *Cahiers scientifiques n° 3, Bois et Forêts des Tropiques* (Nogent-sur-Marne), 1973, 52 p.

*GUILLAUMET, J. L. Recherches sur la végétation et la flore de la région du Bas Cavally (Côte d'Ivoire). Mém. Office de la recherche scientifique et technique outre-mer (ORSTOM), n° 20, 1967, 247 p., 30 phot.

HALEY, C. Forestry in tropical rain forests of Australia. In: *4th World Forestry Congress*, 3, 1954, p. 219–234.

*HALLÉ, F.; LE THOMAS, A.; GAZEL, M. Trois relevés botaniques dans les forêts de Bélinga. Nord-Est du Gabon. *Biologica Gabonica*, vol. 3, n° 3, 1967, p. 3–16.

*HALLÉ, F.; OLDEMAN, R. A. A. *Essai sur l'architecture et la dynamique de croissance des arbres tropicaux*. Paris, Masson, 1970, 178 p.

HAVEL, J. J. New Guinea forests. Structure, composition and management. *Australian Forestry*, vol. 36, no. 1, 1972, p. 24–37.

*HEALD, E. J. *The production of organic detritus in a south Florida estuary*. Univ. of Miami, Coral Gables, Fla., Ph. D. thesis, 1969, VIII+110 p.

HEGYI, F. *Forest valuation survey on the right bank Takutu River, left bank Mazaruni River*. Arch. Forestry Dpt. Guiana (Georgetown), 1962, 13 p.

——. *Forest valuation survey of the Amakura-Koriabo watershed, south of Wauna Creek, N.W. District*. Forestry Dpt. (Georgetown), 1963a, no. 38–63.

——. *Forest valuation survey of the Waini-Manawarin watershed, N.W. District, Kwabanna Survey Area*. Forestry Dpt. (Georgetown), 1963b, no. 39–63.

HEINSDIJK, D. Begroeiing en luchtfotografie in Surinam. Centr. Bur. Luchtkaart. Paramaribo, Publ. 13, 1953.

*——. *Forest inventory in the Amazon Valley*. FAO report no. 601, part 1, Region between Rio Tapajós and Rio Xingú, 1957, 135 p.

*——. The upper story of tropical forests. *Tropical Woods*, no. 107, 1957, p. 66–84; no. 108, 1958, p. 31–45.

*——. *Region between Rio Xingú and Rio Tocantíns*. FAO report no. 949, part 2, 1958a, 94 p.

*——. *Region between Rio Tapajós and Rio Madeira*. FAO report no. 969, part 3, 1958b, 83+17 p.

*——. *Region between Rio Tocantíns and Rios Guamá and Capim*. FAO report no. 992, part 4, 1958c, 72+17 p.

*——. *Region between Rio Caete and Rio Maracassume*. FAO report no. 1250, part 5, 1960a, 677 p.

——. *Dryland forest on the Tertiary and Quaternary south of the Amazon River*. FAO report no. 1284, 6 parts, 1960b, 2+28+15+16+24+25 p.

*HEINSDIJK, D.; MIRANDA BASTOS, A. de. *Forest inventories in the Amazon*. FAO report no. 2080, 1965, 78 p.

*HLADIK, A. Importance des lianes dans la production foliaire de la forêt équatoriale du Nord-Est du Gabon. *C. R. Acad. Sci. Paris*, Série D, 8 avril 1974, p. 1–4.

*HOLMES, C. H. The broad pattern of climate and vegetational distribution in Ceylon. In: *Proc. Kandy Symposium*, p. 99–114. Paris, Unesco, 1958, 226 p.

*HOZUMI, K.; YODA, K.; KOKAWA, S.; KIRA, T. Production ecology of tropical rain forest in Southwestern Cambodia. 1. Plant biomass. *Nature and Life in South-East Asia*, 6, 1969, p. 1–51.

HUFFEL, G. *Économie forestière*. Paris, 1919–1926, 3 vol.

HUTTEL, C. *Écologie forestière en Basse Côte d'Ivoire*. ORSTOM, Centre d'Adiopodoumé, 1967, 33 p. multigr., 2 profils, 8 graph.

——. *Rapport d'activité pour l'année 1961*. ORSTOM, Côte d'Ivoire, Centre d'Adiopodoumé, 1969, 37 p. multigr.

*JACK, W. H. The spatial distribution of tree stems in a tropical high forest. *Empire For. Rev.*, 40, 1961, p. 234–241.

*JENÍK, J. Root structure and underground biomass in equatorial forests. In: *Productivity of forest ecosystems* (Proc. Brussels Symposium, 1969), p. 323–331. Paris, Unesco, 1971, 707 p.

*——; MENSAH, K.O.A. Root systems of tropical trees. *Preslia*, 39, 1967, p. 59–65.

JIMENEZ SAA, J. H. *La identificación de los árboles tropicales por medio de características del tronco y la corteza*. IICA (Turrialba, Costa Rica), 1967, 136 p., 71 fig., 2 maps, multigr.

JOHNSON, P. L. Hemispherical photographs at El Verde. In: Odum, H. T.; Pigeon, R. F. (eds.). *A tropical rain forest. A study of irradiation and ecology at El Verde, Puerto Rico*, p. 309–311. Division of Technical Information, U.S. Atomic Energy Commission (USAEC, Oakridge, Tenn.), 1970, 1 678 p.

——; VOGEL, T. C. *Evaluating forest canopies by a photographic method*. US Army Cold Regions Research and Engineering Laboratory, Research Report 253, 1967.

JONES, E. W. The structure and reproduction of the virgin forest of the north temperate zone. *New Phytol.*, 44, 1945, p. 130–148.

*——. Ecological studies in the rain forest of Southern Nigeria. IV. The plateau forest of the Okomu forest Reserve. Part 1. The environment, the vegetation types of the forest and the horizontal distribution of species. *J. Ecology*, 43, 1955, p. 564–594; Part 2. The reproduction and history of the forest. *J. Ecology*, 44, 1956, p. 83–117.

KEAY, R. W. J. *An outline of Nigerian vegetation*, 3rd ed. 1959, 46 p.

——. Increment in the Okomu Forest Reserve, Benin. *Nigerian Forestry Information Bulletin (new series)* no. 11, 1961, 33 p.

*KERFOOT, C. The root systems of tropical forest trees. *Common- wealth Forestry Review*, 42, 1963, p. 19–26.

*KIRA, T.; OGAWA, H. Assessment of primary production in tropical and equatorial forests. In: *Productivity of forest ecosystems* (Proc. Brussels Symposium, 1969), p. 309–321. Paris, Unesco, 1971, 707 p.

*KLINGE, H. Struktur und Artenreichtum des zentralamazo- nischen Regenwaldes. *Amazoniana*, vol. 4, no. 3, 1973a, p. 283–292.

*——. Root mass estimation in lowland tropical rain forests of Central Amazonia, Brazil. I. Fine root masses of a pale yellow latosol and a giant humus podzol. *Tropical Ecology*, vol. 14, no. 1, 1973b, p. 29–38. II. *Anais Acad. Bras. Ciencias* (Rio de Janeiro), 45, 1973, p. 595–609.

*KLINGE, H.; RODRIGUES, W. A. *Phytomass estimation in a Central Amazonian rain forest*. In: Young, H. E. (ed.), *IUFRO biomass studies*, p. 339–350. Orono, Univ. of Maine, 1974.

*KORIBA, K. On the periodicity of tree growth in the tropics, with reference to the mode of branching, the leaf-fall and the formation of the resting bud. *Gardens Bull. Sing.*, vol. 17, no. 1, 1958, p. 11–81.

*KRAJICEK, J. E.; BRINKMAN, K. E.; GINGRICH, S. F. Crown competition. A measure of density. *Forest Science*, vol. 7, no. 1, 1961, p. 35–42.

*LAMPRECHT, H. Über Strukturuntersuchungen im Tropenwald. *Zeitschr. für Weltforstw.*, vol. 17, n° 5, 1954, p. 162–168.

——. Über Profilaufnahmen im Tropenwald. In: *12th Congr. IUFRO* (Oxford), Sect. 23/5, 1956, p. 35–43.

*——. Der Gebirgs-Nebelwald des venezolanischen Anden. *Schweiz. Zeitschr. f. Forstw.*, 2, 1958, p. 1–27.

——. Tropenwälder und tropische Waldwirtschaft. *Soc. For. Suiss.*, Suppl. n° 32, 1961, 110 p.

——. Ensayo sobre unos métodos para el analisis estructural de los bosques tropicales. *Acta Científica Venezolana*, vol. 13, n° 1, 1962, p. 57–65.

*——. Ensayo sobre la estructura florística de la parte sur-oriental del bosque universitario 'El Caimital'. Estado Barinas. *Revista For. Venezolana*, vol. 6, n° 11–12, 1964, p. 77–119.

——. Einige Strukturmerkmale natürlicher Tropenwaldtypen und ihre waldbauliche Bedeutung. *Forstwissenschaftliches Zentralblatt*, vol. 91, no. 4, 1972, p. 270–277.

——; VEILLON, J. P. La Carbonera. *El Farol*, 168, 1957, p. 17–24.

LANDON, F. H. Malayan tropical rain forest. In: *Proc. 4th World Forestry Congress*, vol. 3, 1954, p. 107–118.

——. Malayan tropical rain forest. *Malayan Forester*, 13, 1955, p. 30–38.

LEBRUN, J. Observations sur la morphologie et l'écologie des contreforts du *Cynometra alexandri* au Congo belge. *Bull. Inst. Roy. Col. belge*. vol. 7, n° 3, 1936, p. 573–584, 1 pl., 1 phot.

——. *La végétation de la plaine alluviale au sud du lac Édouard*. Inst. Parcs Nat. Congo belge, 2 vol., 1947, 800 p., 104 phot., 2 cartes.

*——. A propos des formes biologiques des végétaux en régions tropicales. *Bull. Acad. Roy. Sc. Outre-Mer, Belgique*, 1964, p. 926–937.

*——; GILBERT, G. *Une classification écologique des forêts du Congo*. Publ. INEAC (Bruxelles), Sér. sci., n° 63, 1954, 89 p.

*LE CACHEUX, P. Applications des méthodes statistiques à l'étude des forêts équatoriales. In: *Proc. 4th World For. Congress*, vol. 3, 1954, p. 698–709.

——. Analyse statistique de la forêt tropicale en vue de son utilisation pour la production de la cellulose. *J. Agric. Trop. Bot. Appl.* (Paris), vol. 2, n° 1–2, 1955, p. 1–17.

——; TUDDO, A. de; HUGUET, L. *Proyecto para la fabricación de celulosa y papél en centro América*. Rome, FAO/CAIS/59/1, 3 fasc., 1959, 378 p., 51 p.+maps, 232 p.

*LEE PENG CHOONG, *Ecological studies on Dryobalanops aromatica Gaertn. f.* University of Malaya, Ph. D. thesis, 1967, 145 p.+XIX, 10 pl.

LEIBUNDGUT, H. Beispiel einer Bestandesanalyse nach neuer Baumklassen. (Empfehlungen für die Baumklassenbildung und Methodik bei Versuchen über die Wirkung von Waldpflegemassnahmen. In: *12th Congr. IUFRO* (Oxford), vol. 2, Sect. 23, 1956, p. 92–103.

——. Über Zweck und Methodik der Struktur und Zuwachsanalyse von Urwäldern. *Schweiz Zeitschr. f. Forstwesen*, vol. 110, n° 3, 1959, p. 111–124.

LÉONARD, J. Les divers types de forêts du Congo belge. In: *Symposium Bruxelles* (Association pour l'étude taxonomique de la flore de l'Afrique tropicale, AETFAT), 1950, p. 81–93.

LEONIDAS VEGA, C. Observaciones ecológicas sobre los bosques de roble de la Sierra Boyacá, Colombia. *Turrialba*, vol. 16, n° 3, 1966, p. 286–296.

LEROY-DEVAL, J. Les liaisons et anastomoses racinaires. *Bois et Forêts des Tropiques* (Nogent-sur-Marne), n° 152, 1973, p. 37–49, 5 fig., 7 phot.

——. *Structure dynamique de la rhizosphère de l'okoumé dans ses rapports avec la sylviculture*. Nogent-sur-Marne, CTFT, thèse Ingénieur Docteur, 1974, 113 p.

*LETOUZEY, R. *Étude phytogéographique du Cameroun*. Paris, Lechevalier, 1968, 508 p.

*——. *Manuel de botanique forestière*. Nogent-sur-Marne, CTFT, 2 vol. en 3 fasc., 1969, 1970, 1972, 189+461 p.

*LINDEMAN, J. C.; MOOLENAAR, S. P. *Preliminary survey of the vegetation types of northern Surinam*. The vegetation of Surinam, 2. Utrecht, 1959, 45 p.

LIOCOURT, F. de. De l'aménagement des sapinières. *Bull. Soc. For. de Franche-Comté et Belfort* (Besançon), vol. 4, n° 6, 1898, p. 396–409.

*LOETSCH, F; HALLER, E.; HENNING, N. Beitrag zur mathematischen Formulierung abnehmender Stammzahlverteilung. In: *14th Congr. IUFRO* (Munich), Sect. 25, 1967, p. 168–181.

LOUIS, J. Contribution à l'étude des forêts équatoriales congolaises. In: *C.R. Semaine agr. Yangambi*, Publ. INEAC (Bruxelles), hors série, 2, 1947, p. 902–923.

LOVELESS, A. R.; ASPREY, G. F. The dry evergreen formations of Jamaica. 1. The limestone hills of the South Coast. *J. Ecology*, vol. 45, no. 3, 1957, p. 799–822.

*MACABEO, M. E. Correlation of crown diameter with stump diameter, merchantable length and volume of white Lauan, *Pentacme contorta* (Vid.) Merr et Rolfe in Tagkawayan forests, Quezon Province. *Philip. J. For.*, vol. 13, no. 1–2, 1957, p. 99–117.

*MANGENOT, G. Étude sur les forêts des plaines et plateaux de la Côte d'Ivoire. *Études éburnéennes*, 4, 1955, p. 5–81.

*——. Réflexions sur les types biologiques des plantes vasculaires. *Candollea*, vol. 24, n° 2, 1969, p. 279–294.

*MARSHALL, R. C. *Silviculture of the trees of Trinidad and Tobago*. Oxford Univ. Press, 1939, 247 p.

*MAYO MELENDEZ, E. *Algunas caracteristicas ecológicas de los bosques inundables de Darien, Panamá*. Turrialba, M. Sci. thesis, 1965, 158 p.

*McARTHUR, R. H.; HORN, H. S. Foliage profile by vertical measurements. *J. Ecology*, vol. 50, no. 5, 1969, p. 802–804.

*MENSAH, K. O. A.; JENÍK, J. Root system of tropical trees. *Preslia*, 40, 1968, p. 21–27.

MERVART, J. *Frequency curves of the growing stock in the Nigerian high forests*. Federal Department of Forestry (Ibadan), 1971, 12 p.

——. Growth and mortality rates in the natural high forest of Western Nigeria. *Nigeria Forestry Information Bull.* (new series, Ibadan), 22, 1972, 28 p., 14 diagr., 50 tabl.

MEYER, H. A. Eine mathematisch-statistische Untersuchung über den Aufbau des Plenterwaldes. *Schweiz Zeitschr. f. Forstwesen*, 84, 1933, p. 33–46; 88, p. 124–131.

——. Die rechnerischen Grundlagen des Kontrollmethoden. *Beiheft zur Zeitschr. des Schweiz Forstvereins*, 13, 1934, 122 p.

*——. Structure, growth and drain in balanced uneven-aged forests. *J. For.*, vol. 50, no. 2, 1952, p. 85–92.

MEYER, H. A.; STEVENSON, D. D. The structure and growth of virgin Beech-Birch-Maple-Hemlock forests in Northern

Pennsylvania. *Journ. Agric. Res.*, vol. 67, no. 12, 1943, p. 465–484.

*MILDBRAED, J. Ein Hektar Regenwald auf Fernando Po. *Notizbl. Bot. Gart. Berlin*, vol. 11, n° 109, 1933, p. 946–949.

*MILDE, R. de; GROOT, D. *Inventory of the Ebini-Itaki area*. Techn. Rep. no. 9, UNDP Forest Industry Development Survey, 1970, 106 p.

*MILLIER, C.; POISSONNET, M.; SERRA, J. Morphologie mathématique et sylviculture. In: *3e Conférence du Groupe consultatif des statisticiens forestiers IUFRO* (1970), p. 287–307. INRA, 1972, 332 p.

MOONEY, J. W. C. Classification of the vegetation of the high forest zone of Ghana. In: *Tropical soils and vegetation*, p. 85–86. Paris, Unesco, 1961, 115 p.

MORALES, U. J. A. *Estudios estructurales en el Rodal 3 del Bosque Universitario San Eusebio. Edo Mérida*. Mérida, ULA, Fac. Ciencias forestales, tesis de grado, 1966, 84 p.

*MURÇA PIRES, J.; DOBZHANSKY, T.; BLACK, G. A. An estimate of the number of species of trees in an Amazonian forest community. *The Botanical Gazette* (Chicago), 114, 1953, p. 467–477.

NEAL, D. G. *Statistical description of the forests of Thailand*. Bangkok, Military Research and Development Center, 67–019, May 1967, 346 p., map.

NELSON-SMITH, J. H. Forest associations of British Honduras. *Caribbean Forester*, vol. 6, no. 1–2, 1945, p. 45–61; no. 3, p. 131–147.

*NEWMAN, I. V. Locating strata in tropical rain forest. *J. Ecology*, 42, 1954, p. 218–219.

*NEYMAN, J. On a new class of 'contagious' distribution applicable in entomology and bacteriology. *Ann. Math. Statis.*, 10, 1939, p. 35–57.

*NICHOLSON, D. I. A study of virgin forest near Sandakan, North Borneo. In: *Symposium on ecological research in humid tropics vegetation* (Kuching, 1963), p. 67–87. Unesco Science Co-operation Office for South-East Asia, 1965, 376 p.

ODUM, H. T. Rain forest structure and mineral cycling homeostasis. In: Odum, H. T.; Pigeon, R. F. (eds.). *A tropical rain forest. A study of irradiation and ecology at El Verde, Puerto Rico*, p. H 3–H 52. Division of Technical Information, U.S. Atomic Energy Commission (USAEC, Oakridge, Tenn.), 1970, 1 678 p.

*——; COPELAND, B. J.; BROWN, R. Z. Direct and optical assay of leaf mass of the lower montane rain forest of Puerto Rico. *Proc. Nat. Acad. Sci.* (Washington), 49, 1963, p. 429–434.

*OGAWA, H.; YODA, K.; OGINO, K.; KIRA, T. Comparative ecological studies on three main types of forest vegetation in Thailand. 2. Biomass. *Nature and Life in South-East Asia*, 4, 1965a, p. 49–80.

*——; ——; KIRA, T.; OGINO, K.; SHIDEI, T.; RATANA-WONGSE, D.; APASUTAYA, C. Comparative ecological studies on three main types of forest vegetation in Thailand. 1. Structure and floristic composition. *Nature and Life in South-East Asia*, 4, 1965b, p. 13–48.

OLBERG, A. Über die Kennzeichnung der Bestandesstruktur. *Allgem. Forst- Jagd Zeitung*, vol. 124, no. 8; vol. 125, 1952–1953.

*OLDEMAN, R. A. A. *L'architecture de la forêt guyanaise*. Thèse Fac. Sci. Montpellier, 1972a, 247 p., 113 fig.

——. L'architecture de la végétation ripicole forestière des fleuves et criques guyanais. *Adansonia*, Sér. 2, vol. 12, n° 2, 1972b, p. 253–265, 31 fig.

OLDHAM, P. D. On estimating the arithmetic means of log normally-distributed populations. *Biometrics*, vol. 21, no. 1, 1965, p. 235–239.

OVINGTON, J. D.; OLSON, J. S. Biomass and chemical content of El Verde lower montane rain forest plants. In: Odum, H. T.; Pigeon, R. F. (eds.). *A tropical rain forest. A study of irradiation and ecology at El Verde, Puerto Rico*, p. H 53–H 77. Division of Technical Information, U.S. Atomic Energy Commission (USAEC, Oakridge, Tenn.), 1970, 1 678 p.

*PAIJMANS, K. Een voorbeeld van interpretatie van Luchtfoto's van oerwoud : Het Malilicomplex op Celebes (Interpretation of aerial photographs in a virgin forest complex: Malili, Celebes, Indonesia). *Tectona*, 41, 1951, p. 111–135.

PEREIRA, J. A. *Silvicultura y manejo de bosques tropicales*. Informe técnico 6, PNUD/FAO, 1970, 160 p.

*PEREZ, J. W. Relation of crown diameter to stem diameter in forests of Puerto Rico, Dominica and Thailand. In: Odum, H. T.; Pigeon, R. F. (eds.). *A tropical rain forest. A study of irradiation and ecology at El Verde, Puerto Rico*, p. B 105–B 122. Division of Technical Information, U.S. Atomic Energy Commission (USAEC, Oakridge, Tenn.), 1970, 1 678 p.

PETIT BETANCOURT, P. *Relaciones fisionómicas de las hojas de 2 asociaciones vegetales*. Turrialba, tesis, 1964, 60 p.

PIDGEON, I. M.; ASHBY, E. A new quantitative method of analysing of plant communities. *Austr. J. Sci.*, 5, 1942, p. 19–21.

*PIELOU, E. C. The measurement of diversity in different types of biological collections. *J. Theoret. Biol.*, 13, 1966, p. 131–144.

*PIERLOT, R. *Structure et composition des forêts denses d'Afrique centrale, spécialement celles du Kivu*. Bruxelles, Mémoires Acad. Roy. Sci. Outre-Mer, nouvelle série, vol. 16, n° 4, 1966a, 367 p.

——. La relation entre le nombre de tiges à l'hectare et leur diamètre. In: *6e Congrès forestier mondial* (Madrid), 1966b.

——. Une technique d'étude de la forêt dense en vue de son aménagement : la distribution hyperbolique des grosseurs. *Bull. Soc. Roy. For. Belgique*, 2, 1968, p. 122–130.

*POISSONET, P. Relation entre le nombre d'espèces par échantillon et la taille de l'échantillon dans une phytocénose. *Oecologia Plantarum*, vol. 6, n° 3, 1971, p. 289–296.

*PRESTON, F. W. The commonness and rarity of species. *Ecology*, 29, 1948, p. 254–283.

*——. The canonical distribution of commonness and rarity. *Ecology*, vol. 43, no. 2, 1962, p. 183–215; no. 3, p. 410–432.

PRODAN, M. Die theoretische Bestimmung des Gleichgewichtszustandes im Plenterwald. *Schweiz Zeitschr. f. Forstwesen*, 1949, p. 81–99.

*RAUNKIAER, C. *The life forms of plants and statistical plant geography*. Oxford (being the collected papers of Raunkiaer), 1934, 14+632 p.

RAY, R. G. *Six forest inventories in the tropics. Report no. 5. Nigeria. Summary of the final report*. Directorate of Program coordination, Department of Fisheries and Forestry (Ottawa), 1971, 18 p.

REDHEAD, J. F. *Stand tables of Nigerian Forest Reserves*. Department of Forestry, University of Ibadan, 1964 (unpublished).

*RICHARDS, P. W. Ecological observations on the rain forest of Mount Dulit, Sarawak. *J. Ecology*, vol. 24, no. 1, 1936, p. 1–37; p. 340–360.

*——. Ecological studies on the rain forest of Southern Nigeria. The structure and floristic composition of the primary forest. *J. Ecology*, 27, 1939, p. 1–61.

*——. *The tropical rain forest: an ecological study*. Cambridge Univ. Press, 1952, 450 p. 4th reprint with corrections, 1972.

——. The upland forests of Cameroons Mountain. *J. Ecology*, vol. 51, no. 3, 1963, p. 529–554.

*RICHARDS, P. W.; TANSLEY, A. G.; WATT, A. S. The recording of structure, life forms and flora of tropical forest communities as a basis for their classification. *J. Ecology*, 28, 1940, p. 224–239.

*ROBBINS, R. G. The use of the profile diagram in rain forest ecology. *The Journal of Biological Sciences*, vol. 2, no. 2, 1959, p. 53–63.

——. The biogeography of tropical rain forest in South-East Asia. In: *Proc. Symposium Recent Adv. Trop. Ecology*, part 2, 1968, p. 521–535.

*ROBBINS, R. G.; WYATT-SMITH, J. Dryland forest formations and forest types in the Malayan Peninsula. *Malayan Forester*, vol. 27, no. 3, 1964, p. 188–216.

*RODRIGUES, W. A. Aspects phytosociologiques des pseudo-caatingas et forêts de várzea du Rio Negro. In: Aubréville, A., 1961, p. 209–265. See above.

——. Estudo de 2,6 hectares de mata de terra firme da terra do Navio, Território do Amapá. *Boletim do Museu Paraense E. Goeldi, Nov. ser. Botânica*, n° 19, 1963, 22 p.

*ROLLET, B. *Études sur les forêts claires du Sud indochinois*. Recherches forestières (Saigon), 1952, 250 p.

*——. *Inventaire forestier de l'Est Mékong*. Rapport FAO n° 1500, 1962, 184 p.

*——. *Introduction à l'inventaire forestier du Nord Congo*. Rapport FAO, 2 vol., n° 1782, 1963, 142+111 p.

*——. *Inventario forestal de la Guyana venezolana*. Informe 3. Caracas, Ministerio de Agricultura y Cría, 1967, 352 p.

*——. Étude quantitative de profils structuraux de forêts denses vénézuéliennes. Comparaison avec d'autres profils de forêts denses tropicales de plaine. *Adansonia*, vol. 8, n° 4, 1968, p. 523–549.

——. La régénération naturelle en forêt dense humide sempervirente de plaine de la Guyane vénézuélienne. *Bois et Forêts des Tropiques* (Nogent-sur-Marne), 124, 1969, p. 19–38.

*——. *Études quantitatives d'une forêt dense humide sempervirente de plaine de la Guyane vénézuélienne*. Thèse doctorat, Faculté des Sciences, Toulouse, 1969, 473 p.+annexe (173 p.), 94 fig., 9 pl., 239 tabl.

——. *L'architecture des forêts denses humides sempervirentes de plaine*. Nogent-sur-Marne, CTFT, 1974, 298 p., 8 pl., 155 tabl.+annexes.

*——; CAUSSINUS, H. Sur l'utilisation d'un modèle mathématique pour l'étude des structures des forêts denses humides sempervirentes de plaine. *C.R. Acad. Sci. Paris*, vol. 268, sér. D, 14, 1969, p. 1853–1855.

ROSAYRO, R. A. A reconnaissance of Sinharaja rain forest. *Ceylon For. N.S.*, vol. 1, no. 3, 1954a, p. 68–74.

——. Field characters in the identification of tropical forest trees. *Empire For. Rev.*, 33, 1954, p. 124–141, 8 phot.

ROSEVEAR, D. R.; LANCASTER, P. C. Historique et aspect actuel de la sylviculture en Nigéria. *Bois et Forêts des Tropiques* (Nogent-sur-Marne), 28, 1953, p. 3–12.

*SAINT-AUBIN, G. de. Aperçu sur la forêt du Gabon. *Bois et Forêts des Tropiques* (Nogent-sur-Marne), 78, 1961, p. 3–17.

——. *La forêt du Gabon*. Publ. CTFT, 21, 1963, 208 p.

*SALANDY GUEVARA, C. *Inventario sobre los manglares comprendidos en la región oriental de Venezuela. Plan de manejo y aprovechamiento*. Caracas, Ministerio de Agricultura y Cría, 1964, 42 p.

SANDRASEGARAN, K. Height-diameter-age multiple regression models for *Rhizophora apiculata* Bl. (syn. *Rhizophora conjugata* Linn.) in Matang mangroves, Taiping, West Malaysia. *Malayan Forester*, vol. 34, no. 4, 1971, p. 260–275, 8 tabl.

SCHAEFFER, A.; GAZIN, A.; D'ALVERNY. *Sapinières. Le jardinage par contenance*. Paris, Presses universitaires de France, 1930.

SCHNELL, R. Remarques préliminaires sur les groupements végétaux de la forêt dense ouest-africaine. *Bull. IFAN*, vol. 12, n° 2, 1950a, p. 297–314.

——. *La forêt dense*. Paris, Lechevalier, 1950b, 323 p.

*SCHULZ, J. P. *Ecological studies on rain forest in Northern Suriname*. Amsterdam, Van Eedenfonds, 1960, 267 p.

*SETTEN, G. G. K. The height of buttress structure on trees of Meranti tembaga, *Shorea leprosula* Miq. *For. Res. Inst., Res. Pamphlet* no. 7, 1954a, 5 p. multigr.

——. The effect of situation on the height of buttress structure on trees of *Shorea parvifolia* Dyer, *Shorea leprosula* Miq., *Shorea curtisii* Dyer ex King, *Dryobalanops aromatica* Gaertn. f. and *Koompassia malaccensis* Maing ex Benth. *For. Res. Inst., Res. Pamphlet* no. 12, 1954b, 9 p. multigr.

SKELLAM, J. G. Studies in statistical ecology. 1. Spatial pattern. *Biometrika*, 39, 1952, p. 346–362.

SMIT, I. G. *Inventario de bosques con fotografías aereas de la región Río Carare, Río Opón, Santander, Colombia*. 1964, 48 p.

SMITH, R. F. The vegetation structure of a Puertorican rain forest before and after short-term gamma irradiation. In: Odum, H. T.; Pigeon, R. F. (eds.). *A tropical rain forest. A study of irradiation and ecology at El Verde, Puerto Rico*, p. D 103–D 140. Division of Technical Information, U.S. Atomic Energy Commission (USAEC, Oakridge, Tenn.), 1970, 1 678 p.

*STEENIS, C. G. G. J. Van. Tropical lowland vegetation: the characteristics of its types and their relation to climate. In: *Proc. 9th Pacific Science Congress* (Bangkok, 1957), 20, 1958, p. 25–37.

*——. Rejuvenation as a factor for judging the status of vegetation types: the biological nomad theory. In: *Proc. Kandy Symposium*, p. 212–218. Paris, Unesco, 1958, 226 p.

*STEHLÉ, H. Les types forestiers des Iles Caraïbes. *Caribbean Forester*, 6, suppl. 1re partie, 1945, p. 273–468.

*——. Les types forestiers des Iles Caraïbes. *Caribbean Forester*, 7, suppl. 2e partie, 1946, p. 337–709.

STEUP, F. K. H. Bijdragen tot de kennis der bosschen van Noord Celebes. Celebes (Tjempaka-Hoetan complex, Minahassa). *Tectona*, 25, 1932, p. 119–147.

STEVENSON, N. S. Forest associations of British Honduras. *Caribbean Forester*, vol. 3, no. 4, 1942, p. 164–172.

*SWELLENGREBEL, E. T. G. On the value of large scale aerial photographs in British Guiana forestry. *Empire For. Rev.*, 38, 1959, p. 54–61.

*TAKEUCHI, M. The structure of the Amazonian vegetation. Tropical rain forest. II. *J. Fac. Sci. Univ. Tokyo*, Sect. 3:8, 1961, p. 1–26.

*——. Idem. IV. High Campina forest of the upper Rio Negro. Sect. 3:8, 1962a, p. 279–288.

*——. Idem. V. Tropical rain forest near Uaupès. Sect. 3:8, 1962b, p. 289–296.

*TANSLEY, A. G.; CHIPP, T. F. *Aims and methods in the study of vegetation*. London, 1926, 383 p.

TASAICO, R. *La fisionomía de las hojas de árboles en algunas formaciones tropicales*. Interamerican Institute of Agricultural Sciences, IICA (Turrialba, Costa Rica), tesis de grado, 1959, 86 p.

TAYLOR, C. J. La régénération de la forêt tropicale dense dans l'Ouest africain. *Bois et Forêts des Tropiques* (Nogent-sur-Marne), 37, 1954, p. 19–26.

*——. *Synecology and silviculture in Ghana*. Univ. College of Ghana, 1960, 417 p.

*Thai-van-Trung, *Écologie et classification de la végétation fo-restière du Vietnam*. Inst. Bot. Leningrad, 1–41, 1962, 1 carte (en russe).

*Thomas, M. A theory for analysing contagiously distributed populations. *Ecology*, 27, 1946, p. 329–341.

Tixier, P. *Flore et végétation orophile de l'Asie tropicale. Les épiphytes du flanc méridional du massif sud-annamitique*. Paris, SEDES, 1966, 240 p.

Tregubov, S. S. *Les forêts vierges montagnardes des Alpes dinariques. Massif de Klakovatcha-Guermetch*. Montpellier, thèse, 1941, 118 p.

*Trochain, J. L. Nomenclature et classification des types de végétation en Afrique noire occidentale et centrale. *Ann. Univ. Montpellier, suppl. scientif., sér. Botanique*, 2, 1946, p. 35–41.

——. Nomenclature et classification des types de végétation en Afrique noire française. *Bull. Inst. Étud. Centr.*, nouvelle série, 2, 1951, p. 9–18.

*Turnbull, K. J. *Population dynamics in mixed forest stands. A system of mathematical models of mixed stand growth and structure*. Univ. Washington, Ph. D. thesis, 1963, 196 p.

*Tüxen, R. Bibliographie zum Problem des Minimi-Areals und der Art-Areal Kurve. *Excerpta Botanica*, B, vol. 10, n° 4, 1970, p. 291–314.

*Vareschi, V. Sobre las formas biológicas de la vegetación tropical. *Bol. Soc. Venezolana Ciencias Naturales*, vol. 26, no. 110, 1966, p. 504–518.

*——. Comparación entre selvas neotropicales y paleotropicales en base a su espectro de biotipos. *Acta Botanica Venezuelica*, 3, 1968, p. 239–263.

Vega, L.; Gomez, G. *Muestro lineal de la regeneración natural en el bosque hidrofítico de Carare*. Inst. Invest. Proyectos Forestales y Madereros, Univ. Dist. Francisco J. de Caldas, Bogotá, 1966.

Veillon, J. P. *Coníferas autóctonas de Venezuela. Los Podocarpus*. Universidad de Los Andes (Mérida), 1962, 156 p.

*——. Variación altitudinal de la masa forestal de los bosques primarios en la vertiente nor occidental de la cordillera de los Andes, Venezuela. *Turrialba*, vol. 15, no. 3, 1965, p. 216–224.

*Vidal, J. E. Types biologiques dans la végétation forestière du Laos. *Bull. Soc. Bot. France, Mémoires*, 1966, p. 197–203.

*Vincent, A. J. A quantitative analysis of buttress dimensions of *Shorea leprosula* Miq. (*Meranti tembaga*). *Malayan Forester*, 23, 1960, p. 288–313.

Voorhoeve, A. G. Somes notes on the tropical rain forest of the Yoma-Gola national forest near Bomi Hills, Liberia. *Commonwealth Forestry Review*, vol. 43, no. 1, 1964, p. 17–24.

Walker, F. S. Diagnostic sampling in Eastern Nigeria. *Malayan Forester*, 25, 1962, p. 123–139.

——. Volume table for *Triplochiton scleroxylon, Chlorophora excelsa, Pycnanthus angolense, Khaya* spp., *Terminalia ivorensis* and *Afzelia* sp. *Bulletin of Nigerian Forest Department*, April 1964, p. 9–10, 6 graphs.

Walton, A. B. La sylviculture des forêts à *Dipterocarpus* des basses terres en Malaisie. *Unasylva*, 7, 1953, p. 22–27.

Walton, A. B. The regeneration of dipterocarp forests after high load logging. *Empire For. Rev.*, vol. 33, no. 4, 1954, p. 338–344.

——. Forests and forestry in North Borneo. *Malayan Forester*, 18, 1955, p. 20–23.

*——; Barnard, R. C.; Wyatt-Smith, J. The silviculture of lowland dipterocarp forest in Malaya. *Malayan Forester*, 15, 1952, p. 181–197.

Watson, J. G. The regeneration of tropical rain forest. *Malayan Forester*, 5, 1928, p. 20–23.

*Webb, L. J. A physiognomic classification of Australian rain forest. *J. Ecology*, vol. 47, no. 3, 1959, p. 551–570.

*——. A new attempt to classify Australian rain forests. *Silva Fennica*, 105, 1960, p. 98–105.

——; Tracey, J. G. Current quantitative floristic studies in Queensland tropical rain forest. In: *Symposium on ecological research in humid tropics vegetation* (Kuching, Sarawak, 1963), p. 257–261. Unesco Science Co-operation Office for South-East Asia, 1965, 376 p.

*White, H. H. Jr. Variation of stand structure correlated with altitude in the Luquillo Mountains. *Caribbean Forester*, vol. 24, no. 1, 1963, p. 46–52.

*Whitmore, T. C. The social status of *Agathis*, in a rain forest in Melanesia. *J. Ecology*, vol. 54, no. 2, 1966, p. 285–301.

Wilkinson, G. Root competition and silviculture. *Malayan Forester*, 8, 1939, p. 11–15.

*Williams, C. B. *Patterns in the balance of nature and related problems in quantitative ecology*. London, N.Y., Academic Press, 1964, 324 p.

*Womersley, J. S. The *Araucaria* forest of New Guinea. A unique vegetation type in Malaysia. In: *Proc. Symposium on humid tropics vegetation* (Tjiawi, Indonesia), p. 252–257. Unesco Science Co-operation Office for South-East Asia, 1958, 312 p.

*Wyatt-Smith, J. *An ecological study of the structure of natural tropical lowland evergreen rain forest in Malaya*. Imperial Forestry Institute (Oxford), thesis, 1948, 67+137 p. (unpublished).

——. Storm forest in Kelantan. *Malayan Forester*, 17, 1954, p. 5–11.

——. Suggested definitions of field characters (for use in the identification of tropical forest trees in Malaya). *Malayan Forester*, 17, 1954, p. 170–183.

*——. Stems per acre and topography. *Malayan Forester*, 23, 1960, p. 57–58.

*——. Manual of Malayan silviculture for inland forests. History (part 1). Environmental factors (part 2). Silviculture and forest management (part 3). *Malayan For. Rec.*, 23, 1963, 400 p.

*Zöhrer, F. Ausgleich von Häufig keitsverteilungen mit Hilfe der Beta-Funktion. *Forstarchiv.*, vol. 40, n° 3, 1969, p. 37–42, 6 fig.

*——. The Beta-distribution for best fit of stem-diameter distributions. In: *3ᵉ Conférence du groupe consultatif des statisticiens forestiers* (IUFRO, 1970), p. 91–106, 4 fig. INRA, 1972, 332 p.

Animal palaeogeography and autecology

Palaeogeography

The geological history of the tropics and its significance for the present day faunal distributions (see also chapter 3) are considered in this section.

Current views of global plate tectonics (Dickinson, 1971; Tarling and Tarling, 1971; Bird and Isaacs, 1972; Dewey, 1972) are that the outer shell of the earth's mantle (the lithosphere) is divided into numerous plates separated by three major kinds of boundaries: mid-ocean ridges, where the ocean floor is formed by the injection of hot magma from below; deep trenches, where the ocean floor moves back down into the mantle; and faults, where the plates move laterally with respect to one another. All three are regions of intense seismic activity.

At the end of the Palaeozoic, *ca.* 225 million years B.P., all the continents were united into a supercontinent, Pangaea. Pangaea began to rift apart by the middle Triassic. The first movement separated a northern land mass, Laurasia, from a southern one, Gondwanaland. During the Cretaceous a Laurasian faunal element apparently evolved which probably included ancient fishes, salamanders, primitive frogs, various turtles, lizards, booid snakes, some prototherians, marsupials and early placental mammals. At the same time, evolved a Gondwanaland faunal element including fishes, caecilians, frogs, lizards, boid snakes, avian groups ancestral to ratites and galliformes, prototherians, and metatherians. Dinosaurs were characteristic of both continents.

During the late Cretaceous, South America and Africa drifted apart, but South America apparently remained connected with Antarctica, whilst Madagascar separated from Africa and later from India. Laurasia was still in existence, but the North Atlantic was opening up and Europe was isolated from eastern Asia by shallow seas at least part of the time. Western North America was also temporarily isolated from eastern North America by interior seas extending from the Gulf of Mexico to Arctic Ocean. These late Cretaceous land movements effectively isolated South America from Africa, and Madagascar from Africa and India.

During the early Cenozoic, the continents were probably more isolated than at any other time. South and North America were still separated, India had not yet joined Asia, and connections between Africa and Europe still had not been established. India collided with Asia during the Eocene, and Europe and Africa were apparently united for a period in the Palaeocene and again in the Oligocene. North America was probably joined to Asia during much of the

Cenozoic and there was extensive floral and faunal interchange at this time. The differentiation of floras and faunas continued to take place on the isolated southern continents during the early and mid-Cenozoic when these continents apparently had tropical climates which favoured the evolution of species in most taxa.

During the late Cenozoic period, southern Central America was formed and for the first time there was floral and faunal exchange between North and South America. At this time major mountain building in the Himalayas, Andes and Rockies occurred and there were major coolings and extensive glaciations in the northern hemisphere. There was little exchange of fishes along the Central American corridor, but there was a substantial net northward movement of amphibians and reptiles, whilst the net movement among birds and mammals seems to have been from north to south, though the pattern is less clear for birds.

Except for oscine passerine birds and most mammals, the fauna of South America is primarily derived from the ancient Gondwanaland faunal element. In Africa, there is a much greater mixture of Gondwanaland and Laurasian elements because of the regular connections between Africa and Eurasia. Some of the dominant faunal elements in Africa, such as scincid, elapid and viperid reptiles, oscine passerine birds and placental mammals are derived from the Laurasian faunal element.

Tropical vertebrate faunas

Geological histories are most clear for freshwater fishes, amphibians and reptiles. Freshwater fishes differentiated primarily in the two large land masses, Laurasia and Gondwanaland. The richest fauna evolved in Gondwanaland, especially in the tropics of South America and Africa. This Gondwanaland element includes orders and families common to Africa and South America, families endemic to each continent (presumably having evolved since their separation), and families which apparently evolved in Gondwanaland but now also occur in Laurasia (table 1). Amphibians can be divided into the same two major faunal elements (table 2). Reptiles are similar, but there are a number of uncertain taxa (table 3).

A comparable analysis is not possible for birds.

Most extant mammal families arose in the Tertiary; consequently their present day distributions are not strongly influenced by the large-scale continental movements. There are, however, important Gondwanaland placental mammal faunas in South America and Africa. The South American forms include edentates, condylarths, notoungulates and probably bats, caviomorph rodents and platyrrhine primates. They probably had a northern origin. The ancient African groups include cetaceans, proboscidians, hyracoids, embrithopods and, probably, insectivores, primates, carni-

TABLE 1. Postulated geographical origins of freshwater fishes (Cracraft, 1974).

Gondwanaland faunal element		Laurasian faunal element
Dipnoi	Gymnotoidei	Acipenseroidea
Lepidosirenidae	Rhamphichthyidae[4]	Polyodontidae
Ceratodontidae[1]	Apteronotidae[4]	Acipenseridae
Polypterini	Gymnotidae[4]	Holostei
Polypteridae[2]	Electrophoridae[4]	Lepisosteidae
Osteoglossiformes	Siluriformes	Amioidea
Osteoglossidae	25 families	Amiidae
Notopteridae	Atheriniformes[1]	Cypriniformes
Hiodontidae[3]	Cyprinodontidae	Cyprinoidei
Mormyriformes	Poeciliidae	Catostomidae
Mormyridae[2]	Goodeidae	Cyprinidae
Gymnarchidae[2]	Adrianichthyidae	Cobitidae[6]
Gonorhynchiformes	Perciformes	Homalopteridae[5]
Kneriidae[2]	Cichlidae	Salmoniformes
Phractolaemidae[2]	Nandidae	Umbridae
Cypriniformes	Anabantidae[6]	Esocidae
Characoidei	Luciocephalidae	Salmonidae
Characidae	Mastacembelidae[6]	Percopsiformes
Citharinidae[2]	Chaudhuriidae	Percopsidae[3]
Hemiodontidae[4]	Channiformes	Aphredoderidae[3]
Anostomidae[4]	Channidae[6]	Amblyopsidae[3]
Ctenoluciidae[4]	Clupeiformes	Perciformes
Gasteropelecidae[4]	Denticipitidae[2]	Centrarchidae[3]
		Percidae

1. Tentatively assigned to Gondwanaland faunal element.
2. Restricted to Africa.
3. Restricted to North America.

4. Restricted to South America and southern Central America.
5. Restricted to Eurasia.
6. Common to Africa and Eurasia.

vores and rodents. A northern origin is also probable for these groups. The primates and rodents, which evolved in Africa, probably dispersed to South America when the two continents were much closer together.

The Australian region, including all of the islands lying to the east of Wallace's line, is faunistically very distinct from the other tropical regions. The transition from the Asian floras and faunas characteristic of Java, Borneo and islands to the west, to the Australian ones is not sharp and the line dividing the two regions might be drawn anywhere from Wallace's line to almost as far east as New Guinea according to the taxon considered (Darlington, 1957).

In the Australian region there are very few freshwater fishes (but an endemic family of lungfishes); a few frogs (Leptodactylidae, Hylidae, Ranidae and Brevicipitidae); many reptiles (several families of turtles, crocodiles, geckos, agamids, skinks, varanids, pythons, colubrids and many elapids); a distinctive avifauna, and a distinctive but species-poor mammalian fauna.

TABLE 2. Postulated geographical origins of amphibians (Cracraft, 1974).

Gondwanaland faunal element	Laurasian faunal element
APODA	CAUDATA
Caeciliidae	Hynobiidae
ANURA	Cryptobranchidae
Pipidae	Ambystomidae
Leiopelmatidae	Salamandridae
Microhylidae	Amphiumidae
Hylidae	Plethodontidae
Bufonidae	Proteidae
Myobatrachidae	Sirenidae
Rhinodermatidae	ANURA
Leptodactylidae	Ascaphidae
Dendrobatidae	Rhinophrynidae
Pseudidae	Discoglossidae
Centrolenidae	Pelobatidae
Ranidae (part)	Ranidae (in part)
Sooglossidae	
Hyperoliidae	

TABLE 3. Postulated geographical origins of reptiles (Cracraft, 1974).

Gondwanaland faunal element	Laurasian faunal element	Undefined element
CHELONIA	CHELONIA	CHELONIA
Chelyidae	Dermatemydidae	Testudinidae
LACERTILIA	Chelydridae	Cheloniidae
IGUANIA	Carettochelyidae	Dermochelyidae
Iguanidae	Trionychidae	Pelomedusidae
Agamidae	LACERTILIA	RHYNCHOCEPHALIA
Chamaeleontidae	SCINCOMORPHA	Sphenodontidae
GEKKOTA	Xantusiidae	LACERTILIA
Gekkonidae	Dibamidae	SCINCOMORPHA
Pygopodidae	Scincidae	Amphisbaenidae
SCINCOMORPHA	ANGUOIDEA	OPHIDIA
Teiidae	Anguidae	TYPHLOPOIDEA
Lacertidae	Anniellidae	Typhlopidae
Gerrhosauridae	Xenosauridae	Leptotyphlopidae
OPHIDIA	VARANOIDEA	COLUBROIDEA
BOIDEA	Helodermatidae	Colubridae
Boidae (part)	Varanidae	Hydrophiidae
	Lanthanotidae	CROCODILIA
	OPHIDIA	Crocodylidae
	BOIDEA	
	Boidae (part)	
	Aniliidae	
	COLUBROIDEA	
	Elapidae	
	Viperidae	

Birds

The birds of the tropical forests of the world are quite well-known. In Africa, the lifetime work of Chapin resulted in a series of four volumes which appeared in 1954. More recently, the avifauna of the entire continent of Africa has been treated exhaustively by Moreau (1966), supplemented by Hall and Moreau (1970). For India and South-East Asia,

important reference works are those of Delacour and Jobouville (1931), Mayr (1941), Delacour (1947), Ali (1955, 1968) and Ripley (1961). The birds of the Neotropics have been less intensively studied but complete check lists and distributional guides are available (Eisenmann, 1955; de Schauensee, 1966, 1970; Olrog, 1968). Regional studies are available for Central America (Skutch, 1954, 1960, 1969). In South America, regional contributions to the study of avian natural history have been made by Haverschmidt in Surinam (1968), Pinto (1938, 1944) in Brazil and Sick and Pabst (1968), Olrog (1959) in Argentina and Johnston (1965) in Chile. Distribution patterns of South American birds have been analysed in the light of recent climatic changes (Haffer, 1969).

Most of the families of birds found in the Old World tropics were at one time much more widely spread in Eurasia, suggesting a northern origin for them. For example, families of birds now restricted to the tropics, but known from fossils in temperate areas of Europe or Asia include the Struthio-nidae, Anhingidae, Phoenicopteridae, Sagittariidae, Numi-didae, Rostratulidae, Otididae, Pteroclididae, Psittacidae, Musophagidae, Coliidae, Trogonidae, Phoeniculidae, Buce-rotidae, Capitonidae, and Eurylaimidae. Other families, such as Turnicidae, Heliornithidae, Jacanidae, Rhyncho-pidae, Indicatoridae, Pittidae, Campephagidae, Nectari-niidae, Zosteropidae, Estrildidae, and Dicruridae, which are currently restricted to the tropics, but have disjunct dis-tributions, some of them including the New World, were doubtless widely distributed at higher latitudes when climate were more moderate there. Thus, much of the avifauna of the Old World Tropics seems to be relict in its present distribution.

Despite its large size and general climatic stability during the Cretaceous and Cenozoic, Africa has very few endemic avian taxa above the species level (Moreau, 1966; Cracraft, 1974) and even species richness is fairly low (Keast, 1972). Within restricted areas of tropical wet forest, a patient observer is likely to encounter over twice as many species in South America as in Africa (Amadon, 1973). There are apparently two main sources of the African avifauna. Some groups, such as the Struthionidae, Musophagidae, Coliidae, Pteroclidae, Columbidae, Psittacidae, and Eury-laimidae, have their closest living relatives in South America and may represent elements of an ancient Gondwanaland avifauna (Cracraft, 1974). The other source is families that are wide-spread in tropical and subtropical Eurasia. Some of these have radiated in Africa, but there is no good evidence that any song-bird family originated in Africa and then spread to Asia and subsequently to the New World.

The avifauna of the Indian subcontinent is composed almost entirely of taxa with strong Eurasian affinities. There is only one endemic family, the Irenidae, and only 10 endemic genera (Ripley, 1961). This argues against the suggestion of Darlington (1957) and others that tropical Asia was a major center for avian evolution. Ripley's (1961) analysis of the 175 endemic species from the Indian subcontinent sugges-ted that 17 per cent have Palaearctic affinities, 17 per cent African affinities and 62 per cent Indo-Chinese affinities.

In contrast, the avifauna of South America is extremely rich in species and endemic families. Families of birds restricted to South America or which have only recently spread northward in Central America include the Tina-midae (45 species), Anhimidae (3 species), Cracidae (38 species), Psophiidae (3 species), Eurypigidae (1 species), Cariamidae (2 species), Thinocoridae (4 species), Nyc-tibiidae (5 species), Steatornithidae (1 species), Trochi-lidae (319 species), Galbulidae (15 species), Bucconidae

(30 species), Ramphastidae (37 species), *Dendrocolaptidae (48 species), *Furnariidae (215 species), *Formicariidae (222 species), *Conopophagidae (11 species), *Rhinocryp-tidae (26 species), *Cotingidae (90 species), *Pipridae (59 species), *Tyrannidae (365 species), *Oxyruncidae (1 species), Phytotomidae (3 species), Troglodytidae (63 species), Coerebidae (39 species), Tersinidae (1 species) and Thraupidae (222 species). The greatest richness of these endemic groups occurs in the sub-oscine passerines (indicated by an asterisk above), which dominate the South American tropical avifaunas, but which are poorly represented in the Old World tropics. The oscines are generally considered to be more 'advanced' and many observers have concluded that they are in the process of replacing the sub-oscines (Mayr and Amadon, 1951; Amadon, 1957). Their prevalence in South America is explained by these authors by the extreme iso-lation of South America during most of geological history and during the main evolution and spread of the oscines. This view has been strongly challenged by Slud (1960) and the matter will probably not be resolved satisfactorily in the near future.

The birds of the Australian region include representa-tives of about 63 families of land and freshwater birds. Of these, 44 families are widely distributed, two (the frogmouths and wood swallows) are shared with the Indian subcontinent, and the remainder are strictly Australian. They are the Casuariidae (3 species), Dromiceiidae (2 species), Aptery-gidae (3 species), Megapodidae (10 species), Pedionomidae (1 species), Rhynochetidae (1 species), Aegothelidae (8 species), Acanthisittidae (4 species), Menuridae (2 species), Atrichornithidae (2 species), Cractidae (10 species), Gralli-nidae (4 species), Ptilorhynchidae (18 species), Paradisaeidae (43 species), Neosittidae (5 species), Callaeidae (3 species), and Meliphagidae (160 species). These families total 279 species and the proportion of species in exclusive fam-ilies is larger in the Australian region than in any other.

Mammals

As indicated previously, the dispersal of mammals into the Old World Tropics seems easy to interpret. All of recent families appear to have had a Laurasian origin though some have radiated in the tropics. The break-up of Gondwana-land primarily affected opportunities for movements within the tropical regions by forms radiating there from Laurasian ancestors. There are about 13 orders and 52 families of mammals currently living in the Old World Tropics. Of these, the Insectivora (6 families), Dermoptera (1 family), Primates (6 families), Lagomorpha (1 family), Rodentia (13 families), Carnivora (3 families), Perissodactyla (3 fam-ilies), and Artiodactyla (6 families) are almost universally considered by paleontologists to have had their origin in Laurasia (McKenna, 1969). Bats (Chiroptera) are presently of world-wide distribution, but they are closely related to orders believed to have Laurasian origins (Insectivora, Der-moptera). The pangolins (Pholidota), currently restricted to tropical Africa and Asia, are known from fossils in the early Tertiary of North America and Europe. If the aardvark (Tubulidentata) is closely related to the Condylartha, as suggested by McKenna (1969), then its origin is also Laurasian. For the same reason, a Laurasian origin is likely for the elephants (Proboscidea) and conies (Hyracoidea).

Even though most mammalian taxa seem to have entered Africa from the North, many of them have radiated in Africa and then reinvaded Asia and Europe. Included in this group are the Crocidurine Soricidae, Lorisidae, Cercopithecidae, Oreopithecidae, Pongidae, Hominidae, Cricetidae, Dendromuridae, Thryonomyidae, Viverridae, Hyaenidae, Felidae, Rhinocerotidae, Orycteropodidae, Deinotheriidae, Gomphotheriidae, Elephantidae, Procaviidae, Hippopotamidae, Suidae, Tragulidae, Giraffidae, and Bovidae (Cooke, 1968). Some of these families are included in this list primarily because fossils are known earlier from Africa than from Asia, but the fossil record is in general poor and Asian origins for the Crocidurine Soricidae and Pongidae are quite possible.

Unlike the situation with birds, the mammalian faunas of Africa and South America are very similar in richness. Africa has 51 families, 240 genera and about 756 species while the Neotropical area has 50 families, 278 genera and about 810 species (Keast, 1972). The slightly higher richness of the neotropics is due to more species of bats and rodents in South and Central America than in Africa. The greater richness of bats in the neotropics is also characteristic of small-sized areas. For example, the Cameroon has 76 species of bats, while Colombia has 104. Zaire has only 115 known species, while 154 are recorded for Mexico. The greater richness of rodents in the neotropics is due, however, to geographical replacements of species, since Zaire has 92 species and Brazil 95. Forest-inhabiting primates are similar in the two continents with 5 families, 14 genera and 44 species in Africa and 3 families, 16 genera and 42 species in tropical America. However, in Africa, five species have adapted to savanna environments, while none of the South American monkeys has.

The most striking difference among the two mammalian faunas is the preeminence of ungulates in Africa. Twenty-seven species are known from the wet forests of Africa and 68 species from the grasslands, while in South America,

TABLE 4. Australian region mammals.

	Australia		
	Families	Genera	Species
Monotremes	2	2	2
Marsupials	6	52	119
Rodents (Muridae)	1	13	67
Bats	7	21	41+
	New Guinea		
	Families	Genera	Species
Monotremes	1	2	3
Marsupials	4	24	47
Rodents (Muridae)	1	20	56
Bats	6	21	45+

there are only 9 forest and 6 grassland species. Nevertheless, ecological equivalents are easily recognizeable, with the deer and large caviomorph rodents of South America being the equivalents of the Tragulidae and Cephalophinae of West Africa (Dubost, 1968).

The native land mammals of the Australian region are only a few monotremes, many marsupials, a diverse fauna of rodents (in the family Muridae), fruit bats, and members of six wide-spread families of insectivorous bats (table 4).

Fish

There is strong evidence for a union of South America and Africa during the Mesozoic when freshwater fishes evolved over the entire land mass and after which they developed independently in the two continents (see Takeuchi *et al.*, 1967, for an excellent summary of the geological evidence). The biological evidence includes the affinity of the most primitive fish groups of these two continents. Even Darlington (1965) has abandoned his earlier belief (1957) in the origin of the South American fishes from Central America; the passage of fishes over this land bridge without leaving a trace is highly unlikely given the success of the cypriniform ancestors (Myers, 1960).

Central America

The fish fauna of Central America (including Mexico from the isthmus of Tehuantepec to the border of Colombia) is distinctly different in origin and composition from that of South America (and Africa) and can appropriately be called a Middle America fauna (Miller, 1966). The history of Middle America has been considerably different from that of South America. There is geological evidence for most of what today is Honduras, El Salvador, and large adjacent sections of Nicaragua and Guatemala, having remained above sea level since the Palaeozoic. However, it has been a large island with the sea at the isthmus of Panamá and in the Tehuantepec region. There was also an oceanic connection across southern Nicaragua at some time in the Palaeozoic. These ocean barriers effectively prevented the invasion into Middle America of primary fishes (all but a very few species of this category are restricted to fresh water, making them important indicator animals for zoogeographers) from both South and North America. It has been only since the Panamá land bridge arose during the late Pliocene that representatives of the extremely diverse South American fauna have been able to migrate up into Middle America. Thus, while primary fish families predominate in the fish fauna of Africa and South America, Middle America has only 104 such species and two thirds of the 18 families and 74 of the species are found in Costa Rica and Panamá only. Further north, where only 5 families and 27 species of primary fishes occur, it is the secondary freshwater fishes, which have some tolerance for sea-water, which have radiated and dominated the Middle America fauna. Of the 165 secondary freshwater fishes, there is no concentration; *ca.* 75 per cent occur north of Costa Rica as might be expected for this considerably greater area. For its area Middle America has fewer primary freshwater fish species, and many more secondary or peripheral fresh-water species than elsewhere in the world. Besides the primary and secondary species (represented by 34 genera and 6 families) there are 187 peripheral species represented by 73 genera in 30 families; nearly half of these belong to five families (Ariidae, Atherinidae, Gerridae, Eleotridae and Gobiidae); also, 31 peripheral species are resident. The predominant families of Middle America are the Poeciliidae (14 genera, 57 species) and the Cichlidae (6 genera, 82 species), including 76 or more species of *Cichlasoma*, which has undergone great radiation in this region, paralleling on a small scale the radiations of cichlids in the African lakes. These two families constitute 84 per cent of the secondary fishes and more than 50 per cent of the primary and secondary groups combined. Also important are the Cyprinodontidae and the pimelodid cat-fishes of the genus *Rhamdia*. The characins, the dominant group of South America primary fishes, have 20 genera and 51 species in Middle America, but fewer than 10 species are north of Costa Rica. The other important primary family, Gymnotidae, is represented by 6 genera and 7 species.

The only families of primary or secondary freshwater fishes to reach Middle America from North America are the Lepisosteidae (one species in the Costa Rican tributaries of Lake Nicaragua), the Catostomidae (two species in southern Mexico and Guatemala), and the Ictaluridae (one species in Guatemala and Belize). Apart from these, the cyprinodonts, cichlids, and marine derivatives, all the fish fauna of Middle America north of Costa Rica is derived from South America. Only two South American species have reached the United States. Five of the South American cat-fish families are present in Middle America but only the Loricariidae (8 genera, 15 species) have more than 2 species.

Miller (1966) has distinguished 5 provinces for the Middle America fish fauna.

The Usumacinta Province includes the area of south-eastern Mexico from the Río Papaloapam eastward and southward to the Río Polochic of eastern Guatemala, including the Yucatan Peninsula and Lake Izabal. There are over 200 fish species, with 108 species, 29 genera and 9 families of primary and secondary fishes. There is a large number of marine derivatives, which have permanent freshwater populations. The secondary fishes include the highly diversified poeciliids and the cichlids, represented by *ca.* 44 species. There are only 5 species of characins. There is moderate to strong endemism.

The Chiapas-Nicaraguan Province extends along the Pacific slope from the Río Tehuantepec basin to the Nicoya Peninsula of extreme western Costa Rica. Water-bodies here are very limited and the fauna is generally impoverished, with only 30 species, 14 genera and 8 families of primary and secondary fishes. There is one gymnotid, 2 species of characins and cyprinodontids, only 8 poeciliids, and 10 species of cichlids of which 7 are endemic.

The San Juan Province includes the Great Lakes of Nicaragua and their outlet the Río San Juan, including a portion of Costa Rica. There are *ca.* 34 species in 16 genera for the cichlids, characins, and poeciliids. Many of the former have undergone a radiation in the Great Lakes, as seen in the genus *Cichlasoma*. There is also a shark, two saw-fishes and a tarpon.

The Isthmian Province includes the wet tropical lowlands south and east of Nicaragua, including the isthmus of Costa Rica and Panamá. This is the richest province, with 136 species in 55 genera and 17 families of primary and secondary fishes due to the influence of the South American fish fauna, especially the cat-fishes, characins, gymnotids, and cichlids which together comprise over 100 species. Also, the poeciliids indicate that this area has been a center of diversification. There are less than half the freshwater fish of marine origin than occur in the Usumacinta Province. In the Panamanian extension of this province, many of the Atlantic slope species have their closest relatives on the Pacific slope, emphasizing the importance of the land mass rise in speciation. There is a great similarity between the faunas of the Pacific slope of Panamá and that of the Caribbean slope of Colombia.

The Chipas-Guatemalan Highlands Province includes the highlands of Chipas, Mexico eastward to beyond Honduras. With elevations above 1 200 m, its only indigenous fish is the cyprinodontid genus *Profundulus* which presumably evolved within this province.

The general pattern for Middle America can be equated to an island being invaded by peripheral species (with marine affinities) and especially by freshwater species capable of crossing marine barriers (secondary fishes). These, especially the families Poeciliidae and Cichlidae, radiated and comprised the bulk of the freshwater fishes prior to the establishment of a connection between this island and the two diverse sources of fishes in North and especially South America. Subsequently, few North American species have become incorporated into the Middle American fauna, perhaps because of the physiological limitation on northern invaders. The greatest proportion of the South American components are concentrated in Costa Rica and Panamá indicating that the invasion is recent.

South America

The basic stock of the ostariophysans, which constitute the major group of primary freshwater fishes, appear to have diverged when Gondwanaland broke up. Thus both the siluroids and characoids were present in South America, the latter presumably giving rise to the gymnotids, which are restricted to the Neotropics. These groups have radiated to produce the bulk of the South American fauna of more than 2 500 species—the world's richest region. The predominant group of primary fishes is the characiformes, of 13 families with 200–220 genera and over 1 000 species. The South American cat-fishes, the siluriformes, also contain more than 1 000 species in 14 families. The Osteoglossidae is a primitive primary freshwater family which has strong affinities with African and Australian species. Of non-primary fishes, the family Cichlidae again is important, with 20 genera and *ca.* 125 species. The final important group is the cyprinodonts, *ca.* 60 species, whose closed affinities are with African, not North American, cyprinodonts. Géry (1969) gives a complete list of families and

species numbers. The numbers of species in each country, from Géry (1969) as modified by Fowler (1954), are:
— Brazil (mostly Amazonian) 1 334;
— Peru (mostly Amazonian) 503;
— Paraguay (including many of the primitive forms) 447;
— Colombia 397;
— the Guyanas 364;
— Venezuela 325;
— Bolivia (mostly Amazonian) 277;
— Ecuador 118;
— Argentina 110;
— Uruguay 105;
— Chile 23.

In general, those countries without Amazonian areas, such as Argentina, have mostly an Andean fauna, which is rather sparse in species.

According to Darlington (1957), the main pattern is central richness, greatest in the Amazon, somewhat less in the adjacent river systems north and south, and still less in the most isolated Magdalena; moderate poverty in western drainage of Colombia and Ecuador; progressive poverty southward from southeastern Brazil, but interrupted by the south-flowing Paraguay-Paraná system; poverty and specialization of torrent fishes in the Andes; and radiation of certain cyprinodonts, mostly in Lake Titicaca; and transition in the far south to a completely different fauna of antarctic peripheral fishes.

Géry (1969), following Eigenmann and his collaborators, has distinguished eight major faunistic regions.

The Orinoco-Venezuelan Region contains 325 species distributed into four provinces: the Maracaibo basin, the fauna of which is related to that of the Magdalena as well as the Orinoco; the poorly known Coastal Caribbean; the Orinoco, showing affinities with the Guyanas and the upper course of the Amazon; and Trinidad, with half its species primary, many apparently of recent origin.

The Magdalenean Region is small with *ca.* 150 species derived mostly from the upper Orinoco. A number of taxa, chiefly in the Cauca Basin, are isolated and several species are endemic. There are no ancestral non-ostariophysan taxa.

The Trans-Andean Region of *ca.* 390 species from the Pacific slope of Ecuador and Peru, Colombia west of the Cordillera de Bogotá, and eastern Panamá. About 26 per cent of the species are endemic and 60 per cent similar to Atlantic slope species suggesting that this fauna was separated from the South American fauna by the Andes some time ago. There is a marked impoverishment from north to south into the increasingly dry coastal climate.

The Andean Region includes elevations from 1 000 to 4 000 m where the fishes are found in the fast streams and torrents and the altiplano areas where Lake Titicaca has had an explosive radiation of cyprinodonts (many of which may have been reduced by recent trout introductions).

The Paranean Region includes the La Plata-Uruguay-Paraná-Paraguay area of north and northwest Argentina, eastern Bolivia, and southern Brazil. Most of its fauna appears to be derived from the Amazon.

The Patagonian Region includes northern Chile and Patagonia to the southern tip of Argentina. The fauna is rather depauperate due to climate, altitude, and irregular river flow of the few rivers. There are *ca.* 20 species related to fishes with marine affinities found in Australia, New Zealand and South Africa.

The Guyanean-Amazonian Region, including the Amazon basin, has 1 300 species with representatives from almost all taxa. Initially, it was thought that the Amazon basin fauna spread in all directions, but it is more likely that there was a peripheral fauna around the basin and that the fauna of the Amazon diversified later.

The East-Brazilian Region is small and includes: northeastern Brazil with a poor fauna, possibly due to climate, similar to the lower Amazon; the isolated Río São Francisco, with a fauna only distantly related to the Amazon; the Río Ribeiro and other small coastal rivers in southeastern Brazil. Half of the fauna is endemic, showing strong resemblances in characids with the most remote region, the Trans-Andean, Colombia, and Ecuador fauna, almost down to the species level; there is no known explanation.

The major trend in South America has resulted from the split from Africa and evolution in isolation from the rest of the world for a long time in an area with intertwining river systems, vast seasonally flooded areas, and bays.

Africa

Although the continent of Africa was connected to South America, its freshwater fish fauna indicates a substantially different history. The richest fauna in Africa is found near the equator; over 400 species in 24 families in the river Zaire (Congo). The upper Niger has at least 134 species and the Zambezi has 155, although less than one hundred species occur south of the Zambezi and more than 50 per cent of these are cyprinids. According to Roberts (1972), in the Congo Basin 15 per cent of species are characoids, 23 per cent siluroids, and 16 per cent are cyprinoids, giving a total of 54 per cent ostariophysians. For comparison, in the Amazon, 43 per cent of the fishes are characoids, 39 per cent siluroids, and 3 per cent gymnotids, giving a total of 85 per cent ostariophysians. Africa has a much more complex fish fauna. Besides the groups in common with South America (ostariophysians, Lepidosirenidae, Osteoglossidae, Nandidae, Cichlidae, Cyprinodontidae, Poeciliidae, Galaxiidae, and Percichthyidae), there are many common to Asia including the Notopteridae, Cyprinidae, Mastacembelidae, Anabantidae, and Ophiocephalidae or Channidae. Also, there are primitive primary freshwater forms known only from Africa (the Polypteridae, Denticipitidae, Pantodontidae, Phractolaemidae, Kneriidae, Mormyridae, and Gymnarchidae). Thus, the taxonomic diversity in Africa is considerably higher, although there are less species than in South America.

According to Darlington (1957), the pattern of distribution in Africa has great richness and a diversity of ancient stocks equaled nowhere else from the west-tropical area to the Nile and with the richest of all in the Zaire; there is moderate poverty in the east, partly compensated for by cyprinids of Asiatic origin; local radiation, especially of cichlids in the great lakes; progressive poverty in the Sahara, but a few fishes still present in isolated waters; and, with tenuous transition, isolation in the northwestern corner of the continent of a very different limited freshwater fish fauna closely related to that of Europe. This pattern is obviously determined by climates, the size of drainage systems and their histories.

One of the unanswered questions is why the characoids have never assumed the predominance they have in South America, where they seem to have radiated into an amazing array of ecological niches, as seen most dramatically in the great diversity of dentition and feeding types within this group. There are several possible answers. First, Africa may have had as diverse a characoid fauna as South America in the past, but many species were displaced and became extinct by the recent influence of the cyprinids. Secondly, the cyprinids may have arrived in Africa and assumed these niches before the characoids were able to radiate. Thirdly, the characoids may never have been able to speciate in Africa.

In Africa, there are many large central eastern lakes, in which the cichlids radiated to become a numerically dominant group of fishes. South America has never had a great system of lakes. Thus, while each continent has had an important fish radiation, they are of two different groups.

The African fauna is obviously a product of the original groups present. However, the presence of several primitive groups found in Africa only suggests that either these were extant but did not occur in South America, or were developed after the split. Africa's fauna also shows Asian influence; the acquisition of the cyprinids was probably quite recent.

Asia

Unfortunately, there is no good, single study of fish distribution of Asia, although it is covered by several sources. The main part of tropical Asia including the main islands of the Sunda shelf (Sumatra, Java, Borneo), has a large freshwater fish fauna, extremely rich in cyprinids and their allies, moderately rich in spiny-rayed fishes, but lacking all the archaic African groups and, in fact, showing almost no additions from the latter's fauna. The Asian fauna was distributed fairly evenly through the eastern Oriental region during recent land connections. A great variety of lowland fishes are included in this main Oriental fauna, but some families specialized in mountain torrents. It is presumed from present distributions that the dominant cyprinids and their allies evolved in tropical Asia and spread north and west, with the older north-temperate forms withdrawing before them (Darlington, 1957). In the eastern region there is a steady but complex transition from a tropical to a temperate fauna, from tropical Asia northward through China. There is also apparently an east-west interdigitation of Chinese fishes with those of the interior Asia highlands. As an example of the high diversity in tropical Asia, Smith (1945) lists 546 species in 49 families in Thailand, of which 38 per cent are cyprinids and 18 per cent siluriform cat-fishes. The cyprinids are always the predominant group whether in Lake Lanao, where there are 13 species flock as well as 5 other cyprinids in 4 endemic genera (Myers, 1960), or the

depauperate fauna on islands, such as Sri Lanka, where of the only 54 indigenous species, about half are cyprinids. Other families in these islands include Anabantidae, Channidae, Anguillidae, and one indigenous Cichlidae species (Fernando and Indrasena, 1969). Also, other than the cyprinodonts, this is the only true oriental secondary freshwater fish. According to Darlington (1957), the pattern of distribution is: richness (in spite of absence of archaic groups) in the main part of tropical Asia and on the recent continental islands of the Sunda shelf; poverty and specialization in the central highlands; transition involving many groups over a great area through eastern Asia from the tropics north to the river Amur; and wide east-west distribution of special northern groups.

The fauna of tropical Asia seems to have been isolated from Africa and the tropical cyprinids especially have spread outward and become a part of the African fauna.

Invertebrates

Knowledge of the distributions of orders, families and species of invertebrates is so meager that it is not possible to assemble comprehensive data on faunal richness in the various tropical regions. A few groups of insects, however, are well enough known to permit some preliminary comparisons. Tropical butterflies are moderately well known and are richest in species in South America (table 5), which has a distinctive butterfly fauna of mostly endemic genera (table 6). Termites are richest in both species and genera in Africa (table 7), while South America has more genera, but fewer species than the Oriental region.

The role of movements of continents on patterns of distribution of terrestrial invertebrates is little understood and syntheses of current knowledge are yet to be written.

TABLE 5. Butterfly fauna of various world regions (data compiled from Lewis, 1973).

	Europe	North America	Asia	Indo-Australasia	Africa	South America
Families	7	10	10	11	9	13
Genera	113	116	177	350	250	618
Species	241	344	409	1 353	1 199	1 912

TABLE 6. Percentage of butterfly genera that are endemic or occur in other tropical regions. Since genera that co-occur in temperate regions adjacent to tropical regions are not considered in this table, the columns do not sum to 100 per cent (data compiled from Lewis, 1973).

	South America	Indo-Australasia	Africa
Endemic	90.8	66.6	75.8
South America	—	1.5	0.4
Indo-Australasia	0.8	—	17.6
Africa	0.2	12.5	—
All three	1.0	1.7	2.5

TABLE 7. Termite fauna of various world regions.

	Neotropical[1]	Ethiopian[2]	Oriental[3]
Genera	60	89	38
Species	408	570	(525)*

*Includes species and sub-species.

1. Araujo, R. L. Termites of the neotropical region. In: Krishna, K.; Weesner, F. M. (eds.). *Biology of Termites*, vol. II. New York, Academic Press, 1970, 643 p.
2. Buillon, A. Termites of the Ethiopian region. In: as above.
3. Roonwal, M. L. Termites of the Oriental region. In: as above.

Autecology

Animal ecology may be directed toward any one of or a combination of goals: annual cycles, reproductive rates, resource utilization, habitat selection, population dynamics, relationships with competitors and predators. Some of these types of information require a relatively short-term study, but others require many years of work with marked individuals. Long-term studies are particularly infrequent for all tropical forest animals and information from shorter studies is scarce.

Mammals

There are now enough studies on rodents and primates to permit generalizations concerning some of their basic ecological traits.

Rodents

The most complete studies of rodents in tropical forests have been made by Fleming (1970a, b, 1971, 1974) in Panamá, and Harrison (1951, 1952, 1955, 1956, 1958) in Malaysia. Tropical and temperate rodent reproductive rates are summarized in table 8. In general, temperate species produce more litters per month while breeding. Litters are only slightly smaller in the tropical species, and therefore the total number of young produced per year is not very different in the species so far studied. This contrasts with the situation in birds where clutch sizes are conspicuously smaller in tropical species than in temperate species.

It has been generally assumed that populations of tropical rodents fluctuate less than temperate ones but this may be due to the paucity of information on the tropical species. After reviewing the evidence available on outbreaks of tropical rodents, Herschkovitz (1962) concluded that fluctuations might be as drastic in tropical regions as in the Arctic and north-temperate areas. In Panamá, Fleming (1971) found that rodent populations declined during the wet and early part of the dry season when fruits were relatively unavailable and increased in the latter part of the dry season and early part of the wet season when fruits were more abundant. It was not clear whether this was due to the food supply or a seasonal change in diet by several important predators.

Rodents appear to be about as common in tropical forests as they are in temperate forests (Fleming, 1971) and there are no conspicuous differences in home ranges and survivorship patterns between them.

TABLE 8. Comparison of several reproductive parameters of temperate and tropical rodents (after Fleming, 1971).

Characteristic	Temperate species		
	Murids (24 spp.)	Heteromyids (11 spp.)	Zapodids (3 spp.)
Number of litters per month	0.63 ± 0.12	0.40 ± 0.15	0.36
Number of litters per year (season)	4.06 ± 0.85	2.00 ± 0.49	1.50
Average litter size	4.06 ± 0.62	3.76 ± 0.84	5.27

Characteristic	Tropical species	
	Murids (14 spp.)	Heteromyids (1 sp.)
Number of litters per month	0.26 ± 0.02	0.24
Number of litters per year (season)	3.19 ± 0.80	1.44
Average litter size	3.71 ± 0.51	3.20

Primates

Tropical forests support high densities of primates on all continents but they are less dense in South America than in either Africa or Asia (Bernstein, 1967; Chivers, 1973). Troop size and structure are highly varied in primates, but forest-dwelling species show a strong tendency to live in small, single male groups. In particular, leaf-eating monkeys, which are a predominant part of the forest primate community, live in small groups at high densities (Eisenberg *et al.*, 1972). All African rain forest *Cercopithecus* species live in small, one-male groups but *Mandrillus leucophaeus*, *Cercocebus* spp., and *Miopithecus talapoin* live in larger groups with more than one male (Gartlan and Struhsaker, 1972). The significance of this difference is unkown. Regardless of habitat, most primates give birth to a single young at a time. There are apparently two strong selective pressures that have favoured the minimal litter size. One is the difficulty of movement by a female with more than one young (Goss-Custard, Dunbar and Aldrich-Blake, 1972). In those species with litters of two or three, the young are maintained in a nest most of the time and no females habitually carry more than one young at a time. A second may be the long dependency period coupled with seasonal fluctuations in food supply. There is ample evidence that reproductive rates and survival of young primates are related to fluctuations in food supply (Goss-Custard, Dunbar and Aldrich-Blake, 1972; Chivers, 1974) but much more work needs to be done on the relationships between population dynamics and resource availability.

Earlier studies of tropical primates indicated a strong vertical segregation of species within forests (Booth, 1956).

Subsequent work has not substantiated this amount of segregation, the foraging heights of species actually overlapping a great deal, but there are differences in the means (Gartlan and Struhsaker, 1972; Chivers, 1973; Clutton-Brock, 1973). There are strong habitat segregations among ecologically similar species in Africa (Gartlan and Struhsaker, 1972) and South-East Asia (Chivers, 1973). Species also change with altitude, mostly by the loss of species characteristic of the lowland forests. The strongest habitat segregations seem to occur between similar-sized species that appear to be ecologically equivalent, suggesting that competition may be important in favouring the habitat-use patterns observed. Relatively little is known about patterns of community structure in neotropical forest primates since most of the work has been done in Central America where there are relatively few species present (Carpenter, 1965).

Birds

Though much more purely descriptive work is needed, current knowledge of the basic biology of tropical birds and their adaptations is not likely to be seriously modified by new discoveries.

In only a few cases have careful studies of population dynamics been made. These include a small frugivore, *Manacus manacus*, of South America (Snow, 1962), and a small obligate insectivore, *Hylophylax naeviodes*, in Panamá (Willis, 1974); species very different in their population characteristics reflecting their different environments and food.

The black-and-white manakin (*Manacus manacus*) lives in secondary evergreen forests that are rich in fruit-bearing trees and shrubs, especially of the families Melastomataceae and Rubiaceae. The males gather on communal display grounds and a given male occupies his display court most of the year, except when molting. Nest building, incubation and care of the young are performed entirely by the dull females. Clutch size is usually two, occasionally one; incubation is 18–19 days, long for a bird of that size; the young fledge at 13–15 weeks, but are dependent on the female for another 3–4 weeks. The females regurgitate a mixture of insects and fruit to the young.

Only 40 per cent of clutches hatched and only 19 per cent fledging young. At least 86 per cent of the losses were due to predation mainly by snakes. Lost broods were replaced a few days to over 6 weeks later and successful fledging was followed by another clutch 3–6 weeks later. Correlated with the low reproductive success was a high adult survival. Snow's estimate of 11 per cent annual mortality is much lower than reported for any small passerine birds so far studied.

Population dynamics of the spotted ant-bird (*Hylophylax naeviodes*) studied by Willis were similar to those reported from the manakins with a long breeding season, pairs attempting to rear a number of broods, very high nest predation and adult survival. However, males and females share in the care of the young, compared with only the female in the manakin.

Biology

Though details are available for only a few species, enough is known about the biology of tropical birds to generalize about a number of their ecologically significant characteristics.

Breeding seasons. Nearly all tropical birds have distinct breeding seasons though in more humid regions these are on the average longer than for temperate species. Within the tropics the longest breeding seasons are found in species of humid forests and the shortest breeding seasons, usually associated with the wet season (though some of the larger raptors and aquatics breed during the dry season), are found in the drier areas (Moreau, 1950; Lack, 1968; Ricklefs, 1969).

Clutch size. The clutches of tropical birds are significantly smaller than their temperate zone counterparts. In wet tropical regions clutches of two are the mode while in the more seasonal areas, clutches of three are prevalent; temperate species lay 4–5 eggs. Nevertheless, because breeding seasons average longer in the tropics, the number of eggs laid per year may be roughly similar.

Survivorship. In humid tropical regions all observers have reported very high nest predation rates and relatively little nestling starvation compared to temperate zone conditions. The few studies available suggest higher adult survival rates for small tropical birds than for their temperate counterparts; high survivorship may be inferred from the low rates of breeding success in tropical species.

Dispersal. Tropical forest birds are highly sedentary and are readily stopped by very narrow habitat barriers. Species of disturbed habitats, on the other hand, are adapted to moving about and locating newly disturbed sites. These differences are critical for conservation practices.

Social organization. Tropical frugivorous birds, such as manakins, cotingas and cocks-of-the-rock in South America and birds-of-paradise in New Guinea, are well known for their unusual breeding systems involving communal displays of the males associated with brilliant and often exaggerated male plumages. There is also increasing evidence that the incidence of communal breeding systems in tropical birds is much higher than in temperate species (Fry, 1972). The reason for this is not clear. The extent of territorial behaviour is difficult to determine since territories may be occupied all the year, so all neighbours have a high familiarity with one another. Thus the level of territorial activity is much less than is the case with migratory species that must establish territories and breed during a relatively short time. For this reason, knowledge of spacing patterns in most small forest birds is relatively poor.

Habitat. There is considerable evidence that birds in tropical forests are restricted to narrower vertical zones of the vegetation than they are in temperate forests and that this is associated with the greater species richness of tropical forests. In addition, minor changes in vegetation characteristics cause greater changes in the species composition of tropical bird communities than similar changes in temperate vegetation.

Community structure

The structure of tropical bird communities has been studied in some detail, especially in the neotropics (MacArthur, Recher and Cody, 1966; Diamond and Terborgh, 1967; Orians, 1969; Terborgh, 1970; Karr, 1971; Diamond, 1975) but some data are also available for New Guinea and Borneo (Pearson, D. L., personal communication).

Species richness. Tropical habitats support more species of birds than temperate habitats of similar vegetative structure (MacArthur, Recher and Cody, 1966) and this does not appear to be dependent on the richness of plant species in tropical forests (Orians, 1969). More species may be supported because there are more resources above threshold levels all the year, but patterns of resource utilization by tropical birds are still very poorly known and the contributions to richness of greater range of resources versus more specialized utilization of resources cannot be assessed.

Species richness of birds decreases enormously at altitudes above *ca.* 1 000 m (Terborgh, 1970; Orians, unpublished). Large insectivorous species are the first to disappear, but there are substantial declines in all feeding types of the community (Orians, 1969).

The total density of birds in tropical habitats is often similar to summer densities in the temperate zone, yet the number of species is much greater. Not surprisingly, the most conspicuous difference between the two latitudes is that tropical habitats support a large number of rare species of birds whereas temperate habitats have relatively few rare species.

Reptiles and amphibians

In nearly all taxa, there are more species of amphibians and reptiles in tropical regions than in temperate ones. All of the approximately 73 species of caecilians are tropical. The *ca.* 2 000 species of anurans occur at nearly all latitudes but species richness is much higher in the tropics. The neotropical anuran fauna is dominated by the families Bufonidae, Leptodactylidae (over 300 species), Atelopodidae, and Hylidae. The dominant families in the Old World tropics are the Pelobatidae, Ranidae, Rhacophoridae, and Brevicipitidae. The Australian region also has a rich anuran fauna dominated by the families Leptodactylidae, Ranidae, and Brevicipitidae. In contrast, salamanders are primarily a temperate group and have penetrated the tropics only in Central America where there are nearly 100 species, mostly characteristic of montane wet forests (Wake, 1970).

Turtles are abundant in all tropical regions but predominant families differ on the various continents. In the Neotropics most turtles belong to the families Chelydridae (15 spp.), Pelomedusidae (7 spp.) and Chelydidae (13 species). In Africa and tropical Asia the dominant families are Testudinidae (40 spp.) and Trionychidae (18 species). Tropical lizard faunas are also very distinct. In the Neotropics there are many species of Iguanidae (*ca.* 300 spp.) and Teiidae (175 species); the African fauna is dominated by the Chamaeleontidae (80 spp.), Lacertidae (130 species) and Cordylidae (46 spp.); and the tropical Asian fauna is dominated by the Agamidae (250 spp.). In contrast, most families of snakes are shared by all tropical regions.

Relatively few studies of the ecology of reptiles and amphibians have been done in the wet tropics. For a good review of the status of these investigations, see Pianka (1973). The lizards of the islands of the West Indies, where abundance and relative habitat simplicity have made the systems attractive, have been studied intensively. Some of these studies have included forested habitats, but island patterns are quite different than in similar habitats on the mainland. Long-term population studies of tropical reptiles and amphibians are non-existent.

Nevertheless, we are able to tentatively identify the most important aspects of lizard communities (Schoener, T. W., personal communication). First, all reptile communities show some marked species differences in habitat use and in food taxon or size. Secondly, separation of species by daily activity period is common in diurnal lizards and known for turtles, but has not yet been observed for nocturnal lizards or snakes. Thirdly, separation of species by breeding season is relatively rare in reptiles even though most species are seasonal breeders even in wet tropical forests. Such non-segregation of breeding seasons is typical of animals whose time to maturity is long relative to the yearly cycle. Fourthly, micro-habitat differences between species increase as the number of species living together increases. Fifthly, separation with respect to food type is more important for animals feeding on large foods (relative to their own body size) than for those feeding on relatively small foods.

Fish

The freshwater fish faunas of Africa are much better known than those of Central and South America, but Miller (1966), Myers (1966) and Géry (1969) summarize the information for the Neotropics.

Most studies in Africa concern fishery biology, an outgrowth of which was the work on radiation of cichlid species in the great lakes. Other commercially important species, such as cyprinid *Labeo* have also been studied. Fryer and Iles (1972) summarize the biological and fisheries information for the great lakes of Africa and give an extensive bibliography.

There is little information about New World or Asian fishes. Taxonomic works constitute the bulk of the literature, but there are also economically oriented studies. This is exemplified by Boucault (1973) on Amazon fishes, Mago's (1973) description of ornamental species from the Venezuelan Amazon, Fontenele (1950) on production of one of the major food fishes, the cichlid *Cichla*, Cordiviola de Yuan (1971) and Solano (1973) on another commercially important fish, *Prochilodus*, and by Machado (1971) and Eyzaguirre *et al.* (1973) on basic production and growth studies of some economically important species. There are also many studies on fish-rearing in Asia, although most concern *Tilapia* spp., important food fish introduced from Africa (see references in Fernando and Indrasena, 1969). There are almost no ecological studies in Central America (Miller, 1966), South

America or Asia; only four papers deal with the ecology of freshwater fishes of the Neotropics (Lowe-McConnell, 1964, 1969b; Knöppel, 1970; Zaret and Rand, 1971); a few present useful natural history notes (Géry, 1969; Roberts, 1972).

Taxonomy. The knowledge of taxonomy of freshwater fishes is well-developed. The best references many of which have extensive bibliographies, for the four regions covered in this chapter, are presented in table 9.

TABLE 9. Key references on the taxonomy and distribution of tropical fishes.

Region	References
Central America	Miller, 1966
Costa Rica	Bussing, 1970
Honduras	Miller and Carr, 1974
Mexico	del Villar, 1970
Panamá	Meek and Hildebrand, 1916; Hildebrand, 1938
South America	Ellis, 1913; Géry, 1969
Argentina	Ringulet and Arámbury, 1967
Bolivia	Eigenmann and Allen, 1942
Brazil	Fowler, 1954; Britski, 1969
Chile	Steindachner, 1905; Eigenmann and Allen, 1942; Mann, 1954
Colombia	Fowler, 1945; Dahl, 1971
Guyana	Lowe-McConnell, 1964, 1969
Peru	Eigenmann and Allen, 1942
Surinam	Boeseman, 1952
Uruguay	Devincenzi, 1924, 1933
Venezuela	de Beaufort, 1940; Mago, 1970
Africa	Boulenger, 1909–1916
Asia	Mori, 1936
China	Nichols, 1943
India	Hora, 1944
Sunda Islands	Weber and Beaufort, 1911–1951
Thailand	Smith, 1945.

Ethology. There is a vast literature on behavioural studies of cichlids beginning with Baerends and Baerends-Van-Roon (1950). Most concern African species and virtually all have been done in the laboratory.

Ecology. The overwhelming majority of ecological studies concern the African lakes. It is unknown how fish communities are organized, what they eat, what effects they have on food resources, what effect have the seasons on their breeding, and how adaptable they are.

Resource management. Most studies have been carried out in Africa, where introduced species, such as the Canadian brood trout, *Salvelinus fontinalis*, or the indigenous but widely introduced *Tilapia* species account for most of the work. In the majority of cases, introductions have had little success; either they exterminated commercially important or even endemic fishes (see Zaret and Paine, 1973), or required extensive inputs for meager gains. In a few cases there have been excellent results (see Fernando and Indrasena, 1969), but the ability to predict these is lacking. The best achievements have been in Africa, with indigenous fish.

Invertebrates

Population studies of tropical insects are mainly restricted to species of medical, agricultural or sylvicultural importance. The emphasis on pest species (see summaries of Clark *et al.*, 1967; and Boer and Gradwell, 1971) probably means that models of insect population are not sufficiently general to make strong statements or predictions about randomly selected insects (Gilbert and Singer, 1973). In the tropics where studies are few and diversities of species and of life-histories are high, predictions will be especially difficult.

There are few adequate studies of tropical insects. Recent long-term marking studies on certain butterflies indicate differences between temperate and tropical insect populations as well as between disturbed and natural sites in the tropics. *Acraea encedon*, which as a larva feeds on the ruderal *Commelina*, is present throughout the year around and is not unlike ecologically similar temperate species (*Euphydryas, Ghlosyne*) in terms of reproductive value curves, egg sizes, larval biology, adult longevity and population size and stability (Owen, 1971).

However, the *Passiflora*-feeding genus *Heliconius* in Trinidad is very different (Ehrlich and Gilbert, 1973). Adults live up to 6 months with no inactive period. Egg production is evenly spaced and adults collect most of the resources for the eggs. *Heliconius* collects pollen, extracts amino-acids and uses them as a major means of producing eggs (Gilbert, 1972). The plants provide pollen at a regular rate throughout the year so that *H. ethilla* populations remain constant even during the dry season. In fact, a constant population size was maintained for at least 27 generations. Mortality due to egg parasitoids alone was about 90 per cent and probably changes seasonally. However, two factors act to decrease these changes: parasitoids and predators (ants) on the later stages of the juveniles increase their take if egg parasites decline; pollen for egg production is a limited resource. *Passiflora* possesses extra-floral nectaries, which attract and help maintain both ants and hymenopterous parasitoids. Also, a strong mutualism exists between *Heliconius* and its pollen plant *Anguira* (Gilbert, 1972).

Other tropical forest insects have similar life-history and population dynamic patterns. For example, euglossine bees live up to one year (Janzen, 1971), have relatively constant populations levels, and depend upon plants which provide pollen and nectar in low amounts over long periods of time. Long adult insect life in the humid tropics is possible because of whole year resource availability and warm temperatures.

Most that is known about broad habitat distributional patterns of tropical insect faunas comes from Janzen's (1973) extensive sweep net samples of foliage insects in Central America. For the two herbivore guilds, adult sucking bugs and chewing beetles, the major patterns are as follows.

During the growing season, the number of individuals and total biomass is much greater in secondary vegetation than in adjacent primary forest understorey; the primary forest understorey contains the greatest species richness of insects. In areas with a slight dry season, secondary and primary understorey vegetation have only a small fraction of bug and beetle species in common. This fraction increases in areas with a more severe dry season. Some taxa, notably ants, are exceptionally wide-spread with many species occurring in diverse habitats (Carroll, 1974). Greater habitat overlap is likely to be more common in groups which are largely comprised of generalized predatory and scavenger species.

The greatest species richness and biomass in forest understorey samples occurs at sites between *ca*. 1 000 and 2 000 m. It has been postulated that the lower night temperatures result in a greater net plant productivity. Insect abundance and species richness declines drastically above 3 000 m and groups that are limited to the cool, wet substrate, such as ground-nesting ants and wood-burrowing beetles and termites, are virtually absent from wet, tropical mountain-tops.

Extensive areas of nutrient-poor white sand soils occur in the Amazon basin along the drainage of the Río Negro and in parts of the lowland Asian tropics. These areas are notoriously poor in species and abundance of animals. Janzen (1974) proposed that the low abundance of animals is largely due to the exceptionally high concentrations of toxic secondary chemicals in the plants coupled with the generally low turnover rates of their vegetative parts. Tropical islands also appear to be poor in both numbers of species and individuals. As Allen *et al.* (1973) point out, one set of 800 sweeps from the Osa Peninsula of Costa Rica produces more morpho-species of beetles and bugs than Drewry (1970) found in five years of collecting in Puerto Rico's Luquillo Forest. Allen *et al.* (1973) and Janzen (1973) noted that the proportion of predators in Puerto Rico arthropod samples were much higher than comparable mainland samples. The successful island colonization by predators (and the relatively polyphagous homopterans) may be due simply to the greater array of food available to them as contrasted with the more food-specialized beetles and bugs.

In regions with a severe dry season, insect abundance and species richness decline greatly during the dry months when moist refugia, such as fringing forests, support large numbers. Ant foraging appears to become spatially more patchy during the dry season as the ground surface temperature increases (Carroll, 1974). Some species of adult Macro-Lepidoptera are active in the dry season, but a high percentage are in reproductive diapause. The majority of the trees flower during the dry season in the deciduous forests of Central America and this is also the time of greatest activity by the pollinating bee community. In the more seasonally arid woodlands of Ethiopia, Burger (1974) found little dry season flowering.

Long-term evolutionary processes are undoubtedly important for understanding the present faunal composition, but there has been very little work on this. Pleistocene events appear to have influenced the present distribution of some butterflies in the Amazon basin (Turner, J. R. G., personal communication) and be partly responsible for the poor stem-nesting ant fauna in western Africa compared to Central America (Carroll, 1974).

Research needs and priorities

The knowledge about most aspects of most tropical species is still very meager. Even the most important crop pests are poorly known and economically important species are difficult to manage because the knowledge about their basic biology is lacking. There is clearly a continuing need for basic ecological studies of tropical animals and communities, whether or not there is an immediately apparent application of the knowledge (see Farnworth and Golley, 1974). It is impossible to know which pieces of information will be important for problems we do not yet anticipate. For example, the equilibrium theory of island biogeography was developed without consideration for conservation measures and yet it is cited several times as providing evidence that tropical forest reserves need to be larger than temperate ones and that they should be provided with habitat corridors whenever possible.

On the other hand, tropical forests are rapidly becoming reduced in area and scattered. They are becoming islands and their faunal composition may well be regulated by those same forces regulating species composition on typical islands in the sea (see the following chapter).

Studies of individual species

The properties of ecosystems depend on the characteristics of their component species but little is known of the biology of the majority of tropical forest animals. Therefore, high priority needs to be given to comparative studies of their morphology, physiology and behaviour. If ecologically equivalent species from different continents were selected it would enable the generality of the results obtained to be assessed.

Comparative studies of time and energy budgets of tropical animals, including intraspecific comparisons in different habitats, should illuminate the ways in which these animals are related to their environments and how they are likely to be affected by various alterations.

It is very important that the species to be studied and the measurements to be made in expensive long-term studies are carefully selected. The following criteria are useful in the design of population studies:
— intensive studies involving marked individuals that can be followed long enough to obtain estimates of their movements, survival and reproduction, are needed for much of the basic information;
— populations should be studied in a variety of habitats with known histories of disturbance; if possible, several ecological islands of differing sizes should be included in the design;
— representative taxonomic groups of different ecological characteristics and species in disturbed and primary forest habitats should be selected;
— manipulations should be initiated *after* the populations have been studied intensively enough so that their responses can be predicted and the disturbance can therefore represent a test of hypotheses.

While studies are urgently needed of native species of animals, the increasing rate at which exotic species are being introduced necessitates an understanding of their impacts on native animals. Species of different successional stages need study to increase our ability to generalize about interactions between native and introduced species.

Tropical forestry and agriculture are especially sensitive to attacks from a range of pest species and temperate zone experience with these problems may be of limited use and even misleading if applied uncritically to tropical regions. Much better information is needed on the rate of loss of agricultural and timber products to pests, the influences of monocultures on these losses, and the potential for utilization of knowledge of the ecologies of natural predators on crop pests as a component of pest control systems not dependent on massive use of toxic chemicals. Many tropical forests are still relatively free of pesticides and provide opportunities to study factors responsible for the natural population regulation of the important insect pests. Such studies of the 'life systems' of these insects must include their predators, competitors, host plant phenology and defenses, the physical environment and the life-histories of the insects themselves.

Studies of important disease agents (see chapter 17) should be patterned after the life systems concept introduced above. It should then be possible to develop complex systems of disease control not dependent simply on the use of chemicals to which the organism inevitably acquire resistance.

Priorities for research on fish populations

There is considerable justification for emphasis on a continuation of basic scientific research into all areas of fish studies, including taxonomic, bio-geographical and ecological. This is because, especially in the New World, scientific studies have barely scratched the surface of fish biology.

Because of the limited concentrations of freshwater fish and aquatic resources in Central America, scientific studies are likely to remain poorly funded. Thus, it is likely that most fisheries management will involve impoundments, such as the successful programme with African *Tilapia*. Inadvertent introductions can lead to undesirable consequences (see Zaret and Paine, 1973). Knowledge of indigenous fish, their habits, habitats, and communities is the necessary basis for successful fish culture.

In South America, only taxonomy has been undertaken with any thoroughness, and this is far from complete. Knowledge of fish densities, population distributions, food habits, and the effects of seasonality is required. The best approach would involve regional cooperation. South American rivers can provide sport fishing and this constitutes one of the best uses of the indigenous fishes. High mountain lakes provide a certain amount of fish and government agencies have enhanced these. Some effort has been made at rearing the more important commercial fish species (such as *Cichla ocellaris*, *Astronotus ocellatus*, and *Arapaima gigas*) in artificial ponds (Brazil).

One of the obstacles to the success of fisheries pro-grammes has been the absence of national large permanent ponds from the South American lowlands. As a result, there is no real understanding of how to manipulate these water-bodies; siltation occurs rapidly or disease vectors become a problem. For example, in the llanos of Venezuela, large artificial ponds were built in the Mantecal region to hold drinking-water for cattle during the dry season. These became silted within one or two years and also developed large populations of the snail vector of schistosomiasis (see chapter 17). It is clear that a great deal of basic ecological work on natural ponds will have to be conducted before stable artificial communities can be managed successfully.

The natural permanent water-bodies of Africa have been the focus of considerable research, and while there are great problems from disease the co-evolution of these systems for long periods of time has allowed the development of natural population controls, but there is a great need for adequate understanding of their management to ensure a continuous supply. There is evidence that the harvest is declining in certain areas. For this reason, it is increasingly important to direct future research into areas of resource management as well as understanding the ecological principles involved.

A second area of concern is potential pollution, and especially eutrophication, of some of these lakes. An early warning system should be developed now.

Tropical Asia has a diverse fish biota. The river systems have been greatly augmented by artificial reservoirs, ponds, and drainage ditches. Thus, the water system is complex and efforts should be made to integrate the indigenous fishes to complement man's efforts as happens in rice paddies which utilize grass carp to remove undesirable aquatic species.

In summary, it seems that the major concentration of biological research in the tropics should be ecological. In the New World, much more information about artificial waters is needed, including the determination of which species can be cultured, what problems are most likely to occur and how they might be solved. The Amazon fish fauna is almost unknown ecologically and a regional multidisciplinary effort will be most useful. Another potentially important area is river fish, since certain populations (e.g. cat-fishes) probably can be managed effectively. This will depend on research to determine species abundances, distributions and production. A final area for study is the effects of seasonal changes such as the relation of wet season and high river periods to breeding, habits, and production.

In Africa, the resources must be protected from the potentially damaging effects of man. Over-exploitation, pollution and eutrophication can result in detrimental effects; adequate preparations to safeguard and to correct negative changes are required. These ecological investigations of species populations and communities should, where possible, involve the use of natural or man-made perturbations.

There is a great need for basic research into the biology and ecology of the indigenous fish fauna of Asia. There is also a need for one comprehensive review of this field.

Bibliography

ALI, S. A. *The book of Indian birds*. Bombay, India, Bombay Nat. Hist. Soc., 1955.

——; RIPLEY, S. D. *Handbook of birds of India and Pakistan*. Oxford, Oxford University Press, 1968.

ALLEN, J. D.; BARNTHOUSE, L. W.; PRESTBYE, R. A.; STRONG, D. R. On foliage arthropod communities of Puerto Rican second growth vegetation. *Ecology* (Durham, N.C.), vol. 54, no. 3, 1973, p. 628–632.

AMADON, D. Remarks on the classification of the perching birds (Order Passeriformes). *Proc. Zool. Soc. Calcutta* (Calcutta), Mookerjee Mem. Vol., 1957, p. 259–268.

——. Birds of the Congo and Amazon forests: a comparison. In: Meggers, B. J.; Ayensu, E. S.; Duckworth, W. D. (eds.). *Tropical forest ecosystems in Africa and South America: a comparative review*, p. 267–277. Washington, D.C., Smithsonian Institution Press, 1973, 350 p.

*ARAUJO, R. L. Termites of the Neotropical region. In: Krishna, K.; Weesner, F. M. (eds.). *Biology of termites*. New York, Academic Press, 1970, vol. II, 643 p.

BAERENDS, G. P.; BAERENDS-VAN-ROON, J. M. *An introduction to the study of the ethology of cichlid fishes*. Leyden, E. J. Brill, 1950, 235 p.

BEAUFORT, L. F. de. Freshwater fishes from the Leeward group, Venezuela, and eastern Colombia. *Stud. Fauna Curaçao, Aruna, Bonaire and Venezuela Islands*, vol. 2, 1940, p. 109–114.

BERNSTEIN, I. S. Intertaxa interactions in a Malayan primate community. *Folia Primatologica* (Basel), vol. 7, no. 1–2, 1967, p. 198–207.

BIRD, J. M.; ISAACS, B. (eds.). *Plate tectonics*. Washington, Am. Geophys. Union, 1972, 563 p.

BOER, P. J. den; GRADWELL, G. R. (eds.). *The dynamics of populations*. Proc. Adv. Study Inst. on Dynamics of Numbers in Populations. Oosterbeek, Netherlands, Wageningen, 1971, 611 p.

BOESEMAN, M. A preliminary list of Surinam fishes not included in Eigenmann's enumeration of 1912. *Zool. Meded.*, 21, 1952, p. 179–200.

BOOTH, A. H. The distribution of primates in the Gold Coast. *J. West African Sci. Ass.* (London), 2, 1956, p. 122–133.

BOUCAULT, F. R. Algunas espécies de interés económico de fauna ictiológica amazonica. In: *Simposio internacional sobre fauna silvestre e pesca fluvial e lacustre Amazonica. Programa cooperativo para o desenvolvimento do trópico Americano*, IICA-Trópicos, vol. II IVE, 1973, p. 1–18.

BOULENGER, G. A. *Catalogue of the fresh-water fishes of Africa in the British Museum*. London, British Museum, 1909–1916, vol. I–IV.

BRITSKI, H. A. Lista do tipos de peixes das colecoes do Departmento de Zoologia de Secretaria da Agricultura de São Paulo. *Papéis Avisos Zool. S. Paulo* (São Paulo), vol. 22, 1969, p. 197–215.

BUILLON, A. Termites of the Ethiopian region. In: Krishna, K.; Weesner, F. M. (eds.). *Biology of termites*. New York, Academic Press, 1970, vol. II, 643 p.

BURGER, W. C. Flowering periodicity at four altitudinal levels in eastern Ethiopia. *Biotropica* (Pullman, Washington), vol. 6, no. 1, 1974, p. 38–42.

BUSSING, W. A. New species and new records of Costa Rican freshwater fishes with a tentative list of species. *Rev. Trop. Biol.*, 14, 1970, p. 205–249.

CARPENTER, C. R. The howlers of Barro Colorado Island. In:

De Vore, I. (ed.). *Primate behavior: field studies of monkeys and apes*, p. 250–291. New York, Holt, Rinehart, and Winston, 1965, 654 p.

CARROLL, C. R. *The structure of arboreal ant communities*. University of Chicago, Ph. D. thesis, 1974.

CHAPIN, J. P. *Birds of the Belgian Congo. Bull. Amer. Mus. Nat. Hist.*, vol. 65, 1933 (and also 1935, 1939, 1954).

CHIVERS, D. J. An introduction to the socio-ecology of Malayan forest primates. In: Michael, R. P.; Crook, J. H. (eds.). *Comparative ecology and behavior of primates*, p. 101–146. New York, Academic Press, 1973, 847 p.

——. The Siamang in Malaya. *Contr. to Primatology* (Basel), 4, 1974, p. 1–335.

CLARK, L. R.; GEIER, P. W.; HUGHES, R. H.; MORRIS, R. F. *The ecology of insect populations in theory and practice*. London, Methuen, 1967, 232 p.

CLUTTON-BROCK, T. H. Feeding levels and feeding sites of Red Colobus (*Colobus badius tephrosedes*) in the Gombe National Park. *Folia Primatologica* (Basel), vol. 19, no. 5, 1973, p. 368–379.

CONWAY, G. R. Ecological aspects of pest control in Malaysia. In: Farvar, M. T.; Milton, J. P. (eds.). *The careless technology: ecology and international development*. Garden City, N.Y., The Natural History Press, 1972, 1 030 p.

*COOKE, H. B. S. Evolution of mammals on southern continents. II. The fossil mammal fauna of Africa. *Quart. Rev. Biol.* (Baltimore, Md.), vol. 43, no. 3, 1968, p. 234–264.

CORDIVIOLA DE YUAN, E. Crecimiento de peces de Paraná Medio. I. Sabalo (*Prochilodus platensis* Holmberg) (Pisces, Tetragonopteridae). *Physis* (Buenos Aires), 30, 1971, p. 438–504.

*CRACRAFT, J. Continental drift and vertebrate distribution. *Ann. Rev. Ecology and Systematics* (Palo Alto, Calif.), 5, 1974, p. 215–261.

DAHL, G. *Los peces del Norte de Colombia*. Bogotá, Ministerio de Agricultura, INDERENA, 1971, 391 p.

*DARLINGTON, P. J. *Zoogeography: the geographical distribution of animals*. New York, Wiley, 1957, 675 p.

——. *Biogeography of the southern end of the world*. Cambridge, Harvard Univ. Press, 1965, 236 p.

DELACOUR, J. *Birds of Malaysia*. New York, Macmillan, 1947, 382 p.

——; JOBOUVILLE, P. *Les Oiseaux de l'Indochine française*. Paris, Exposition coloniale internationale, 1931, 4 volumes.

DE SCHAUENSEE, R. M. *The species of birds of South America*. Wynnewood, Pa., Livingston Publ. Co., 1966.

——. *A guide to the birds of South America*. Wynnewood, Pa., Livingston Publ. Co., 1970, 470 p.

DEVINCENZI, G. J. Peces del Uruguay. *Anales del Museo Nacional de Montevideo. Act. Museo de Historia Natural*, serie 11, Entrega 5, 1924, p. 139–293.

——. Peces del Uruguay. Notas complementarias, II. *Anales del Museo de Historia Natural de Montevideo*, serie 4, no. 3, 1933, p. 1–11; serie 4, no. 4, 1933, p. 1–5.

DEWEY, J. F. Plate tectonics. *Scientific American* (New York), 226, 1972, p. 56–68.

*DIAMOND, J. M. Distributional ecology of New Guinea birds. *Science* (Washington, D.C.), vol. 179, no. 4075, 23 Feb. 1973, p. 759–769.

* Major reference.

DIAMOND, J. M. Assembly of species communities. In: Cody, M. L.; Diamond, J. M. (eds.). *Ecology and evolution of communities*, p. 342–444. Harvard Univ. Press, 1975, 545 p.

——; TERBORGH, J. Observations on bird distribution and feeding assemblages along the Río Callaroa, Department of Loreto, Peru. *Wilson Bull.* (Lawrence, Ka.), vol. 79, no. 3, 1967, p. 273–282.

DICKINSON, W. R. Plate tectonic models of geosynclines. *Earth Planet. Sci. Lett.*, 10, 1971, p. 165–174.

*DOBBEN, W. H. van; LOWE-McCONNELL, R. H. (eds.). *Unifying concepts in ecology*. Report of the plenary sessions of the first international congress of ecology (The Hague, September 8–14, 1974). The Hague, W. Junk B. V. Publishers; Wageningen, Centre for agricultural publishing and documentation, 1975, 302 p. Chapters on 'Flow of energy and matter between trophic levels'; 'Comparative productivity in ecosystems'; 'Diversity, stability and maturity in natural ecosystems'; 'Diversity, stability and maturity in ecosystems influenced by human activities'; 'Strategies for management of natural and man-made ecosystems'.

DOBZHANSKY, T.; PAVAN, C. Local and seasonal variations in relative frequencies of species of *Drosophila* in Brazil. *J. Anim. Ecol.* (Cambridge), vol. 19, no. 1, May 1950, p. 1–14.

DREWRY, B. A list of insects from El Verde, Puerto Rico. In: Odum, H. T.; Pigeon, R. F. (eds.). *A tropical rain forest. A study of irradiation and ecology at El Verde, Puerto Rico.* Division of Technical Information, U.S. Atomic Energy Commission (USAEC, Washington), 1970, 1 678 p.

DUBOST, G. Les niches écologiques de forêts tropicales sud-américaines et africaines, sources de convergences remarquables entre Rongeurs et Artiodactyles. *La Terre et la Vie*, (Paris), vol. 22, nº 1, 1968, p. 3–28.

*EHRLICH, P. R.; GILBERT, E. Population structure and dynamics of the tropical butterfly *Heliconius ethilla*. *Biotropica* (Pullman, Wa.), vol. 5, no. 2, 1973, p. 69–82.

EIGENMANN, C. H.; ALLEN, W. R. *Fishes of western South America. I. The intercordilleran and Amazonian lowlands of Peru. II. The high pampas of Peru, Bolivia, and northern Chile.* Lexington, Univ. of Kentucky Press, 1942, 494 p.

EISENBERG, J. F.; MUCKENHIRN, N. A.; RUDRAN, R. The relation between ecology and social structure in primates. *Science* (Washington, D.C.), vol. 176, no. 4037, 1972, p. 863–874.

EISENMANN, E. The species of Middle American birds. *Trans. Linn. Soc. N.Y.* (New York), 7, 1955, p. 1–128.

ELLIS, M. M. The gymnotid eels of tropical America. *Mem. Carnegie Mus.* (Pittsburgh), 6, 1913, p. 109–195.

EYZAGUIRRE, H. F.; CÓRDOVA, R. V.; HURTADO, F. E.; RAMUS, V. C. Aspectos básicos de la producción piscicola de caracidos tropicales. In: *Simp. Int. Sobre Fauna Silv. Pesca Fluv. Lacus. Amaz. Prog. Coop. Desen. Trop. Amer.*, IICA-Trópicos, vol. II, IXA, 1973, p. 1–11.

*FARNWORTH, E. G.; GOLLEY, F. B. (eds.). *Fragile ecosystems. Evaluation of research and applications in the Neotropics.* Berlin and New York, Springer Verlag, 1974, 258 p.

FERNANDO, C. H.; INDRASENA, H. H. A. The freshwater fishes of Ceylon. *Bull. Fish. Res. Stn.* (Ceylon), 20, 1969, p. 101–134.

FLEMING, T. H. Notes on the rodent faunas of two Panamanian forests. *J. Mammalogy* (Baltimore, Md.), vol. 51, no. 3, 28 August 1970a, p. 473–490.

*——. Comparative biology of two temperate-tropical rodent counterparts. *Amer. Midl. Natur.* (Notre Dame, Ind.), vol. 83, no. 2, 1970b, p. 462–471.

*FLEMING, T. H. Population ecology of three species of neotropical rodents. *Misc. Publ. Mus. Zool., Univ. Mich.* (Ann Arbor), 143, 31 Aug. 1971, p. 1–77.

*——. The population ecology of two species of Costa Rican heteromyid rodents. *Ecology* (Durham), vol. 55, no. 3, 1974, p. 493–510.

FONTENELE, O. Contribucaô para o conhecimento da biologia dos Tucunarés (Actinopterygii, Cichlidae) em captiveiro. *Rev. Brasil. Biol.*, 10, 1950, p. 503–519.

FOWLER, H. W. Colombian zoological survey. Part I. The freshwater fishes obtained in 1945. *Proc. Acad. Nat. Sci. Phila.* (Philadelphia), 97, 1945, p. 93–135.

——. *Os peixes de agua doce do Brasil.* Vol. I, Entrega 1–4. São Paulo, Departamento de Secretaria da Agricultura, 1954, 625 p.

FRY, C. H. The social organization of bee-eaters (Meropidae) and cooperative breeding in hot climate birds. *Ibis* (London), vol. 114, no. 1, 1972, p. 1–14.

*FRYER, G.; ILES, T. D. *The cichlid fishes of the Great Lakes of Africa.* New Jersey, R.F.H. Pub., 1972, 641 p.

FUTUYMA, D. J. Community structure and stability in constant environments. *Amer. Nat.* (Chicago), vol. 107, no. 955, May–June 1973, p. 443–446.

GARTLAN, J. S.; STRUHSAKER, T. T. Polyspecific associations and niche separation of rain forest anthropoids in Cameroon, West Africa. *J. Zool. Soc. Lond.* (London), vol. 168, pt. 2, 1972, p. 221–266.

*GÉRY, J. The freshwater fishes of South America. In: Fittkau, E. J.; Illies, J.; Klinge, H.; Schwabe, G. H.; Sioli, H. (eds.). *Biogeography and ecology in South America*, vol. 2, p. 828–848. The Hague, Junk, Monographiae Biologicae no. 18, 1968–69, 2 vol., 946 p.

GILBERT, L. E. Pollen feeding and reproductive biology of *Heliconius* butterflies. *Proc. Nat. Acad. Sci.* (Washington, D.C.), vol. 69, no. 6, 1972, p. 1403–1407.

——; SINGER, M. C. Dispersal and gene flow in a butterfly species. *Amer. Nat.* (Chicago), vol. 107, no. 953, 1973, p. 58–72.

GOSS-CUSTARD, J. D.; DUNBAR, R. I. M.; ALDRICH-BLAKE, F. P. G. Survival, mating, and rearing strategies in the evolution of primate social structure. *Folia Primatologica* (Basel), vol. 17, no. 1, 1972, p. 1–19.

*HAFFER, J. Speciation in Amazonian forest birds. *Science* (Washington, D.C.), vol. 165, no. 3889, 1969, p. 131–137.

HALL, B. P.; MOREAU, R. E. *An atlas of speciation in African passerine birds.* London, British Museum (Nat. Hist.), 1970.

HARRISON, J. L. Reproduction in rats of the subgenus *Rattus. Proc. Zool. Soc. Lond.* (London), vol. 121, pt. 3, 1951, p. 673–694.

——. Breeding rhythms of Selangor rodents. *Bull. Raffles Mus.* (Singapore), 25 August 1952, p. 109–131.

*——. Data on the reproduction of some Malayan mammals. *Proc. Zool. Soc. Lond.* (London), vol. 125, pt. 2, 1955, p. 445–460.

*——. Survival rates of Malayan rats. *Bull. Raffles Mus.* (Singapore), 27 Oct. 1956, p. 5–26.

*——. Range of movement of some Malayan rats. *J. Mammalogy* (Baltimore, Md.), vol. 39, no. 2, 1958, p. 190–206.

HAVERSCHMIDT, F. *Birds of Surinam.* London, Oliver and Boyd, 1968, 445 p.

HERSCHKOVITZ, P. Evolution of neotropical cricetine rodents (Muridae). *Fieldiana* (Chicago), vol. 46, no. 1, 1962, p. 1–524.

HILDEBRAND, S. F. Freshwater fishes of Panama. *Field Mus. Nat. Hist. Zool.* (Chicago), ser. 22, 1938, p. 215–359.

HORA, S. L. On the Malayan affinities of the freshwater fish fauna of peninsular India. *Proc. Nat. Inst. Sci. India* (New Delhi), 10, 1944, p. 423–439.

JAGO, N. D. The genesis and nature of tropical forest and savanna grasshopper faunas, with special references to Africa. In: Meggers, B. J.; Ayensu, E. S.; Duckworth, W. D. (eds.). *Tropical forest ecosystems in Africa and South America: a comparative review*, p. 187–196. Washington, D.C., Smithsonian Inst. Press, 1973, 350 p.

JANZEN, D. H. Euglossine bees as long-distance pollinators of tropical plants. *Science* (Washington, D.C.), vol. 171, no. 3967, 1971, p. 203–205.

*——. Sweep samples of tropical foliage insects: effects of seasons, vegetation types, elevation, times of day, and insularity. *Ecology* (Durham) vol. 54, no. 3, 1973, p. 687–700.

*——. Tropical blackwater rivers, animals, and mass fruiting by the Dipterocarpaceae. *Biotropica* (Pullman, Wa.), vol. 6, no. 2, 1974, p. 69–103.

JOHNSTON, A. W. *The birds of Chile and adjacent regions of Argentina, Bolivia, and Peru*. Buenos Aires, Platt Establecimientos Gráficos, S.A. vol. I, 1965, 398 p.; vol. II, 1967, 447 p.

KARR, J. R. Structure of avian communities in selected Panama and Illinois habitats. *Ecol. Monogr.* (Durham), 41, 1971, p. 207–233.

KEAST, A. Continental drift and the evolution of the biota on southern continents. In: Keast, A.; Erck, F. C.; Glass, B. (eds.). *Evolution, mammals, and southern continents*. Albany, State Univ. N.Y. Press, 1972, 543 p.

KNÖPPEL, H. A. Food of central Amazonian fishes. *Amazonia II* (Sebunday, Colombia), 3, 1970, p. 257–359.

*LACK, D. L. *Ecological adaptations for breeding in birds*. London, Methuen, 1968, 409 p.

LEWIS, H. I. *Butterflies of the world*. Chicago, Follett Publ. Co., 1973, 312 p.

LOWE-MCCONNEL, R. The fishes of the Rupunini savanna district of British Guyana, South America. *Zool. J. Linn. Soc.* (London), vol. 45, no. 1, 1964, p. 103–144.

——. Speciation in tropical freshwater fishes. *Biol. J. Linn. Soc.* (London), vol. 1, no. 1 and 2, 1969a, p. 51–75.

——. The cichlid fishes of Guyana, South America, with notes on their ecology and breeding behaviour. *Zool. J. Linn. Soc.* (London), vol. 48, no. 2, 1969b, p. 255–302.

——. *Fish communities in tropical freshwaters*. London and New York, Longman, 1975, 337 p.

MACARTHUR, R. H.; RECHER, H.; CODY, M. On the relation between habitat selection and species diversity. *Amer. Nat.* (Chicago), vol. 100, no. 913, 1966, p. 319–332.

*——; WILSON, E. O. *The theory of island biogeography*. Princeton, Princeton Univ. Press, 1967, 203 p.

MACHADO, A. A. Contribución al conocimiento de la taxonomía del genera *Cichla* (Perciformes, Cichlidae) en Venezuela. Part I. *Acta Biol. Venez.*, 7, 1971, p. 459–497.

MAGO, L. F. *Lista de los peces de Venezuela*. Caracas, Ministerio de Agricultura y Cría, Oficina Nacional de Pesca, 1970, 241 p.

——. Estudio de los peces ornamentales de acuario de la Amazonia venezolana. In: *Simp. Int. Sobre Fauna Silv. Pesca Fluv. e Lacus. Amaz. Prog. Coop. Desen. Trop. Amer.*, IICA-Trópicos, vol. II IXE, 1973, p. 1–7.

MANN, G. F. *Vida de los peces en aguas chilenas*. Santiago, Ministerio de Agricultura, Universidad de Chile, 1954, 342 p.

MANN, G. Die Ökosysteme Sudamerikas. In: Fittkau, E. J.;

Illies, J.; Klinge, H.; Schwabe, G. H.; Sioli, H. (eds.). *Biogeography and ecology in South America*, vol. 1, p. 171–229. The Hague, Junk, Monographiae Biologicae no. 18, 1968, 446 p.

MAYR, E. *List of New Guinea birds*. New York, Amer. Mus. Nat. Hist., 1941, 260 p.

——; AMADON, D. A classification of recent birds. *Amer. Mus. Novit.* (New York), 1496, 1951, p. 1–42.

MCKENNA, M. C. The origin and early differentiation of therian mammals. *Ann. N.Y. Acad. Sci.* (New York), vol. 167, art. 1, 1969, p. 217–240.

MEEK, S. E.; HILDEBRAND, F. The fishes of the fresh waters of Panama. *Field Mus. Nat. Hist.* (Chicago), Pub. 191, Zool. Series; vol. 10, no. 15, 1916, p. 217–374.

*MILLER, R. R. Geographical distribution of Central American freshwater fishes. *Copeia* (New York), 4, 1966, p. 773–802.

——; CARR, A. Systematics and distribution of some freshwater fishes from Honduras and Nicaragua. *Copeia* (New York), 1, 1974, p. 120–125.

MOREAU, R. E. The breeding season of African birds. I. Land birds. *Ibis* (London), vol. 92, no. 2, 1950, p. 223–267.

*——. *The bird faunas of Africa and its islands*. New York, Academic Press, 1966, 429 p.

MORI, T. *Studies on the geographical distribution of freshwater fishes in eastern Asia*. Chosen, Japan, privately published, 1936.

MYERS, C. W.; RAND, A. S. Checklist of amphibians and reptiles of Barro Colorado Island, Panama, with comments on faunal change and sampling. *Smithsonian Contr. to Zool.* (Washington, D.C.), no. 10, 1969, p. 1–11.

MYERS, G. S. The endemic fauna of L. Lanao and the evolution of higher taxonomic categories. *Evolution* (Lawrence, Ka.), vol. 14, no. 3, 1960, p. 323–333.

——. Derivation of the freshwater fish fauna of Central America. *Copeia* (New York), 4, 1966, p. 766–773.

NICHOLS, J. The freshwater fishes of China. In: Tyler, R. (ed.). *Natural history of central Asia*. New York, Amer. Mus. Nat. Hist., vol. 9, 1943, 322 p.

OLROG, C. C. *Las aves argentinas. Una guia del campo*. Tucumán, Argentina, Instituto Miguel Lillo, 1959, 343 p.

——. *Las aves sudamericanas*. Tomo primero, fundación. Tucumán, Argentina, Instituto Miguel Lillo, 1968, 506 p.

ORIANS, G. H. The number of bird species in some tropical forests. *Ecology* (Durham), vol. 50, no. 5, 1969, p. 783–801.

OWEN, D. F. *Tropical butterflies*. Oxford, Clarendon Press, 1971, 214 p.

PANT, N. C. Important entomological problems in humid tropical Asia. In: *Natural resources of humid tropical Asia*, p. 307–329. Paris, Unesco, 1974, 456 p.

*PIANKA, E. R. The structure of lizard communities. *Ann. Rev. Ecol. and System.* (Palo Alto, Calif.), 4, 1973, p. 53–74.

PINTO, O. M. Catalogo des aves do Brasil. Pt. I. *Revista do Museu Paulista*, 22, 1938, 566 p.

——. Catalogo des aves do Brasil. Pt. II. *Revista do Museu Paulista*, 1944, 700 p.

RICKLEFS, R. E. The nesting cycle of songbirds in tropical and temperate regions. *Living Bird* (Ithaca, N.Y.), 8, 1969, p. 165–175.

RINGULET, R. A.; ARAMBURY, R.; ARAMBURY, A. de. *Los peces argentinos de agua dulce*. La Plata, Com. Invest. Cient. Prov. Buenos Aires, 1967, 602 p.

RIPLEY, S. D. *A synopsis of the birds of India and Pakistan*. Bombay, Bombay Nat. Hist. Soc., 1961, 703 p.

*ROBERTS, T. R. Ecology of fishes in the Amazon and Congo basins. *Bull. Mus. Comp. Zool.* (Cambridge, Mass.), vol. 143, no. 2, 1972, p. 117–142.

ROONWAL, M. L. Termites of the Oriental region. In: Krishna, K.; Weesner, F. M. (eds.). *Biology of termites*, vol. II. New York, Academic Press, 1970, 643 p.

SICK, H.; PABST, L. F. Las aves do Rio de Janeiro (Guanabara) (lista sistemática anotada). *Separato de Arquivos do Museu Nacional*, 53, 1968, p. 99–160.

SKUTCH, A. F. *Life histories of Central American birds*. Berkeley, Cooper Ornithological Soc., Cooper Ornithological Society Pacific Coast Avifauna. Vol. I, Miller, A. H.; Pitelka, F. A. (eds.); no. 31, 1954, 448 p. Vol. II, Miller, A. H.; Davis, J.; Pitelka, F. A. (eds.); no. 34, 1960, 573 p. Vol. III, Davis, J.; Christman, G. M. (eds.); no. 35, 1969, 580 p.

SLUD, P. The birds of 'Finca La Selva', Costa Rica. *Bull. Amer. Mus. Nat. Hist.* (New York), vol. 121, art. 2, 1960, p. 49–148.

SMITH, H. M. The freshwater fishes of Siam or Thailand. *Bull. U.S. Nat. Mus.* (Washington, D.C.), 188, 1945, p. 1–622.

*SNOW, D. W. Field study of the black–and–white manakin *Manacus manacus* in Trinidad. *Zoologica* (New York), vol. 47, pt. 2, 1962, p. 65–104.

SOLANO, M. J. M. Reproducción inducida del bocachico, *Prochilodus reticulatus* Valenciennes. In: *Simp. Int. Sobre Fauna Silv. Pesca Fluv. Lacus. Amaz. Prog. Coop. Desen. Trop. Amer.*, IICA-Trópicos, vol. II IXB, 1973, p. 1–34.

STEINDACHNER, F. Die fische de sammlung Plate. In: *Fauna Chilensis*, vol. 6, 1905, p. 200–214.

STRONG, D. R., Jr. Rapid asymptotic species accumulation in phytophagous insect communities: the pests of cacao.

Science (Washington, D.C.), vol. 185, no. 4156, 1974, p. 1064–1066.

TAKEUCHI, H. S. U.; KANAMORI, H. *Debate about the Earth*. San Francisco, Freeman Cooper, 1967, 281 p.

TARLING, D. H.; TARLING, M. *Continental drift*. Garden City, Doubleday, 1971, 140 p.

TERBORGH, J. Distribution on environmental gradients: theory and a preliminary interpretation of distributional patterns in the avifauna of the Cordillera Vilacabamba, Peru. *Ecology* (Durham), vol. 52, no. 1, 1970, p. 23–40.

*——. Preservation of natural diversity: the problem of extinction-prone species. *Bio-Science* (Arlington, Va.), vol. 24, no. 12, 1974, p. 715–722.

VILLAR, J. A. del. *Peces mexicanos (claves)*. Mexico, Comisión nacional consultativa de pesca, 1970, 166 p.

WAKE, D. B. The abundance and diversity of tropical salamanders. *Amer. Nat.* (Chicago), vol. 104, no. 936, 1970, p. 211–213.

WEBER, M.; BEAUFORT, L. F. de. *The fishes of the Indo-Australian Archipelago*. Leiden, E. J. Brill, vols. 1–9, 1911–1951 (vol. 8 by Beaufort, L. F. de, only; vol. 9 by Beaufort, L. F. de; Chapman, W. M.).

*WILLIS, E. O. Populations and local extinctions of birds on Barro Colorado Island, Panama. *Ecol. Monogr.* (Durham), 44, 1974, p. 153–169.

ZARET, T. M.; RAND, A. S. Competition in tropical stream fishes: support for the competitive exclusion principle. *Ecology* (Durham), vol. 52, no. 2, 1971, p. 336–337.

*ZARET, T. M.; PAINE, R. T. Species introduction in a tropical lake. *Science* (Washington, D.C.), vol. 182, no. 4111, 1973, p. 449–455.

Animal populations

Introduction

This chapter surveys the recent literature showing some animal interactions within tropical forests and indicating the problems which are likely to be encountered in manipulating the system. The cited papers provide extensive references to earlier work. Three useful advanced texts, all of which are theoretically oriented and treat much tropical material, are those by MacArthur and Wilson (1967), Levins (1968), and MacArthur (1972). Among the texts which stress the use of models, statistics and other mathematical approaches, those of Pielou (1969), Poole (1974) and J. M. Smith (1974) are particularly useful. Southwood (1966) and Poole (1974) provide useful summaries of ecological methods.

A number of areas related to the subject matter of this chapter are included in other chapters of this volume. Of particular interest are chapters 3, 6, 11, 14 and 17.

Species richness

Tropical ecosystems are characterized by high plant and animal species richness. This central fact is not easy to understand.

The following terminology is adopted. The term *species richness* is used for the number of species in some taxon or taxa in an area; one area would be twice as rich as another if it had twice as many species of the taxa under consideration. The term *species diversity* is used for an index that weights species by their contribution to biomass, energy flow, cover, or other measure. The latter term is often used for both meanings, but there is sufficient distinction to make terminological precision desirable.

Regardless of the various measures used, the increase of richness and diversity in terrestrial systems from the higher latitudes towards the equatorial region is well documented for many of the larger taxa. The increase in richness does not necessarily have a constant slope nor achieve peak values exactly in the equatorial zone [e.g. for salamanders see Wake (1970), for rodents Fleming (1970) and for birds Tramer (1974)]. Data illustrating this general trend are presented in Simpson (1964) for mammals in nearctic and neotropical North America, MacArthur (1972) for birds in the same region and for ants in South America, Fischer (1960) for ants and snakes within the New World, Lloyd,

Inger and King (1968) for reptiles, and Greenwalt (1960) for New World humming-birds.

Other faunal changes have been noted along the temperate-tropical gradient. Schoener (1971) observed that there was a greater diversity of bill lengths among tropical species of insectivorous birds. The major difference was the increase in numbers of tropical species having longer bills. Schoener and Janzen (1968) examined the size of insects collected by sweep net sampling done in Massachusetts, U.S.A., and in Costa Rica. They found most tropical samples contained significantly larger insects than temperate samples and concluded that insect size increases with desiccation resistance and is probably related to the length of the growing season. Lindsey (1966) also looked for latitudinal gradients in morphological categories among poikilothermic vertebrates. Large adult amphibians tended to be proportionally more frequent among temperate species; a similar but weaker trend was observed for snakes.

Dobzhansky (1950), Fischer (1960) and Pianka (1966) discussed the cause or causes of species richness. Pianka clarified the various hypotheses which had been proposed to account for the increase in richness by placing them in one of six general categories: time-based, spatial heterogeneity, competition, predation, climatic stability, and productivity. He stressed that these hypotheses should be considered separately so that the factor(s) regulating richness could be identified. Concrete examples of the various hypotheses are presented below.

Spatial heterogeneity hypothesis

MacArthur and MacArthur (1961) showed that there was a positive linear relationship between bird species richness and foliage height diversity in temperate deciduous forests. MacArthur, Recher and Cody (1966), Karr and Roth (1971) and many others extended this to the tropics. MacArthur *et al.* (1966) showed that, with certain modifications, the relationship held for birds in Puerto Rico and Panamá.

Total diversity or richness may be divided into two major components: within-habitat and between-habitat (MacArthur, 1965). The former consists of those species, 'which co-exist in spite of regular overlap in their place of feeding, or which have made other adjustments to co-exist'. MacArthur maintained that, for birds, there was an upper limit of within-habitat species richness so that habitats soon become saturated with species. Since within-habitat levels are relatively fixed, total species richness must increase through the addition of new habitats. This addition, the between-habitat component, accounts for the difference between total richness and within-habitat richness.

Predation hypothesis

If a generalized predator feeds on various prey species in proportion to their abundance, no prey species can predominate (Paine, 1966). Spight (1967) showed that the major way predators might affect prey diversity is by regulating the distribution of space utilizable by the various prey species. Janzen (1970a, b) argued that herbivores, particularly those consuming seeds and seedlings, could depress

the numbers of individual offspring of their food plants. Their ability to do so was a function of the density of the seeds or seedlings, which was, in turn, a function of the distance from the parent tree. The probability of a tree maturing increases with its distance from its parent. Hence the co-existence of many tree species.

Janzen (1971a) found that squirrels and bruchid beetles, which fed on the seeds, and several moth larvae, which fed on the seedlings of *Dioclea megacarpa*, damaged seedling vines more heavily close to the parent. Similarly, attack by the bug, *Dysdercus fasciatus*, was a function of distance from the parent tree (Janzen, 1972a). The rate of seed discovery was high beneath the parent tree but declined to zero at distances of 30–60 m. However, Wilson and Janzen (1972) exposed *Scheelea* palm seeds to attack by ovipositing bruchid beetles, but the attacks were not consistently related to distance from parent trees or nut density.

Climatic stability and productivity hypotheses

Many workers have been impressed by the purported amelioration, predictability and stability of abiotic factors in the tropics. Connell and Orias (1964) presented a model for the production and regulation of species richness. With few violent fluctuations in physical factors, less energy was necessary for responses meeting environmental stresses. More energy was available for growth and reproduction and caused larger populations. These larger populations, living in a relatively energy-rich area, would be genetically more variable and less vagile, resulting in semi-isolated populations and speciation. This continuous rapid speciation would eventually culminate in the formation of large numbers of specialized species, which would introduce instability into the system, providing a check upon further increases in richness. This model is yet to be tested.

The capability of tropical vegetation to produce material throughout the year in a wide variety of forms (fruit, flowers, etc.), allows species dependent upon these to survive without migration. This, for example, accounts for the large tropical diversities of mammals (Fleming, 1973) and bats (Wilson, 1973).

Any definite statement concerning the causes of tropical species richness is premature. The need for both between- and within-habitat studies of species richness for taxa in addition to conspicuous vertebrates is required. Such studies should focus upon distinguishing between the possible explanatory hypotheses summarized by Pianka (1966) and in developing alternative ones. Emphasis should be placed upon mechanistic explanations which can account for the observed distributions.

Population dynamics

The richness of tropical biota has certain implications which may or may not be valid. Foremost among these implications are those dealing with the size and stability of single species populations and with the relative role of abiotic and biotic factors in population dynamics.

Population density

It is often assumed that tropical species consist of fewer individuals than temperate species; some studies [e.g. Elton (1973) for some insects] support this. The numbers of individuals per species varies: few species are abundant, a few extremely rare, but most have intermediate values (Preston, 1948). Furthermore, as Preston (1960) points out for birds, there is a higher total of individuals in tropical areas than in temperate ones. The abundant species may play key roles within their particular habitats in the utilization, storage and transport of energy and nutrients between trophic levels, in the domination of resources such as space, etc.

The iguanid lizard, *Anolis limifrons*, is an example of a species whose local population densities may sometimes attain values as high as comparable lizard species in temperate zones (Heatwole and Sexton, 1966). This species ranges from Vera Cruz, Mexico, southward at least into Colombia. There it is probably replaced by a closely related species, *A. fuscoauratus* (Peters and Donoso-Barros, 1970). It has been extensively studied in eastern and central Panamá. Body size varies from 17 mm at hatching to *ca.* 52 mm. It is restricted to lowland forested areas where it is a member of the forest-floor and lower vegetation stratum (0–2 m). Both sexes perch facing downward on vertical stems during the day; males slightly higher than females (Sexton *et al.*, 1972). The type of perch varies geographically. In eastern Panamá there is a tendency to occupy trees with large trunk diameters, especially those trees having buttresses (Sexton *et al.*, 1964); in central Panamá much smaller diameters are utilized. Territorial and feeding activities occur on both the vegetation and ground, but most such activities seem to be launched from the vertical perch sites.

The species is largely insectivorous with at least 60 per cent of prey having a body length of 4–10 mm (Sexton *et al.*, 1972), although the prey size of adults varies from *ca.* 2 to 32 mm. There was no evidence that one population was food-limited. *A. limifrons* is eaten by a wide variety of especially vertebrate predators.

This ecologically annual species breeds seasonally (Sexton *et al.*, 1963; Sexton *et al.*, 1971) with breeding activity increasing from dry season to wet season. Within the dry season the reproductive activity varies positively as a function of dry season precipitation.

Hatched individuals and eggs are sensitive to moisture conditions. Adults lose water rapidly and cannot survive in large outdoor screened cages if not supplied with a canopy (Sexton and Heatwole, 1968). The single egg is usually laid beneath the leaf litter and on top of the mineral soil from which it must absorb water. Dry season water availability appears to be the ultimate control of the reproductive cycle. During the dry season eggs are produced in low numbers if at all; concomitant with this is the increase of lipid reserves, presumably for the rapid production of eggs at the beginning of the wet season (Sexton *et al.*, 1971).

A. limifrons can be classified as a relatively generalized species, particularly regarding its reproductive system. Within years of normal wet and dry season precipitation

there is a burst of egg production in early wet season (Sexton *et al.*, 1963) which continues at fairly high levels. During unusual years, especially with heavy precipitation during the dry season, reproduction apparently can proceed at levels higher than normal, building population size up far above usual levels. Thus, population levels on Barro Colorado Island dropped from 0.117 anoles per m² in the unusually moist dry season of 1962–1963 to 0.039 anoles per m² during the unusually dry dry season of 1964–1965. These seasonal changes in population increase within years and changes in population density between years suggest that pulses of new individuals in common species must have an effect on community organization and function.

There are other densities estimates available for other common species. Fleming (1971) studied three species of rodents in seasonal lowland forest on the more seasonal Pacific coast and on the wetter Caribbean coast of Panamá. At the Pacific site *Liomys adspersus* averaged 10 per ha from June to October (middle of the wet season) and 5.5 per ha in February and March (middle of dry season). *Oryzomys capita* peaked at 3.2 individuals per ha on the Pacific site in October and 4.3 per ha in November at the Caribbean area. The third species, *Proechimys semispinosus*, was at its lowest values at the Pacific in June (1.1 per ha) and its highest from October until January (3.8 per ha). On the opposite side of the Isthmus of Panamá this species varied annually from 2.3 individuals per ha to 5.0–5.6 per ha. He feels that these species have densities at the same level as temperate forest rodents. In another paper contrasting latitudinal changes in the diversity of mammalian communities, Fleming (1973) concluded 'the relative abundances of species of small rodents inhabiting temperate and tropical forests are quite similar. Likewise, the relative abundances of temperate and tropical bats are similar except that there are apparently many more rare (or seldom mistnetted) species in tropical communities'.

A. J. Berry (1966) counted the numbers of five species of terrestrial snails on bare and moss covered limestone in a Malaysian forest. The estimates for three species were in excess of 400/m².

Biotic versus abiotic factors

It is thought (e.g. Dobzhansky, 1950) that the alleged reduction of abiotic stresses in tropical situations is less a factor in the evolution and maintenance of tropical communities than are biological interactions (e.g. mutualism, competition, etc.). Hence, tropical communities are often said to be under biological control, while higher latitude systems are under physical controls. Sanders (1968, 1969) has proposed the stability-time hypothesis; 'Where physiological stresses have been historically low, biologically accommodated communities have evolved. As the gradient of physiological stress increases, resulting from increasing physical fluctuations or by increasingly unfavorable physical conditions *regardless* of fluctuations, the nature of the community gradually changes from a predominately biologically accommodated to a predominately physically controlled community. Finally, when the stress conditions become greater than the adaptive abilities of the organisms, an abiotic condition is reached. The numbers of species present diminish continuously along the stress gradient.' MacArthur (1972), after briefly discussing yearly fluctuations in populations of tropical animals in terms of seasonal factors, stated: 'In view of the foregoing evidence of reduced seasonality in the tropics, we tentatively infer that any great tropical population fluctuations are the consequences of closely packed competitors.'

These views represent the consensus of many biologists. Nevertheless, the role of abiotic factors in the regulation of tropical ecosystems has been neglected.

r *and* K *selection*

These views assume little if any coupling between the abiotic environment and tropical organism. This section discusses the relations of populations with their physical environment. Later sections will deal with the very important multiple species interactions. If tropical communities are regulated by their interspecific interactions, each species must be tightly coupled to its abiotic environment so as to extract all the resources it can to compete effectively.

Dobzhansky's ideas of the type of selective forces acting on temperate versus tropical populations were made more precise by MacArthur and Wilson (1967) who developed the concept of *r* and *K* selection. These two terms are parameters from the logistic equation describing population growth:

$$\frac{dN}{dt} = rN\frac{K-N}{K}$$

where N=number of individuals, r=rate of increase, and K=the carrying capacity of the environment.

Those populations occupying habitats with unexploited resources, such as open islands, early successional stations, areas subject to strong and unpredictable abiotic events, etc., are selected for genotypes which produce a large number of offspring; that is, they are selected for a high *r*. Populations existing under the opposite set of conditions, as members of complex ecosystems in equilibrium, are in severe intraspecific and interspecific competition, are at or near the maximal possible (K). Selection under these conditions would be for genotypes which would utilize the limited resources with maximal efficiency. Species undergoing *r*-selection will tend to have short generations, high productivity and great vagility. Those which are *K*-selected would exhibit greater longevity, fewer offspring and would be more sedentary. Thus, *r*-selected species are more likely to occur at the higher latitudes and *K*-selected species at the lower ones. Similarly, *r*-selected species would predominate at the pioneer stages of a sere and *K*-selected ones at the climax.

The reproductive biology of various tropical species have been examined to see if they conform to these predictions. Moreau (1944) has shown that tropical birds produce fewer eggs than do their temperate counterparts. Fogden (1972), after discussing the strong seasonal breeding patterns exhibited by forest birds of Sarawak, Borneo, contrasts the availability of food for temperate versus tropical species of

birds. Those of temperate latitudes produce large and repeated broods during a season when food abundance may be several thousandfold that obtained in non-breeding seasons; differences in insect abundance in the tropics tend to be only a fewfold. Consequently, young of tropical species are limited in number and receive more parental attention. Cody (1966) has constructed a three-dimensional model which predicts bird clutch size in various habitats. The underlying theme is that the time and/or energy resources of the adult are limited, and that these must be allocated to maximize the bird's contribution to the next generation. Energy is allocated between clutch size (first axis) and meeting the demands of competition (second axis) and predation (third axis). In those environments in which competition and predation are high (i.e. the tropics), energy available for egg production will be low.

Inger and Greenberg (1966b) showed that clutch size in lizards from Bornean rain forests was smaller than that of lizards from the more seasonal forests of tropical Asia.

Among the more direct studies are those concerned with insects. Landahl and Root (1969) compared the life-tables of the temperate lygaeid, *Oncopeltus fasciatus*, with that of the tropical *O. unifasciatellus*. Eggs were raised under similar conditions except that photoperiods corresponded to those of the area of origin of the insects. For *O. fasciatus* $r=0.044$ and for *O. unifasciatellus* $r=0.034$. The former species began to oviposit earlier than the latter, but the duration and survivorship of the egg stage were nearly identical. Both species attained adulthood at the same age. *O. fasciatus* lays more egg clutches.

Young (1972) determined *r* for seven species of Costa Rican butterflies inhabiting primary forests which had a long life-span and low but long-term fecundity. Their *r* values varied from 0.115 to 0.163 per day in contrast to $r=0.316$ for a butterfly of secondary forests. Young and Muyshondt (1973) compared differences in average size and developmental time in two sub-species of *Morpho peleides*: *M. p. hyacinthus* in El Salvador, a country with little or no virgin tropical wet forest, and *M. p. limipida* in Costa Rica. They concluded: 'From the differences in average size and developmental time between *peleides limipida* in Costa Rica and *peleides hyacinthus* in El Salvador, we postulate that the latter subspecies has experienced a longer evolutionary history in unstable environments. Smaller size associated with faster development would seem to be further adaptations to existing in unstable communities since there is a selective response to reproduce faster under such conditions.'

Cook *et al.* (1971) and Turner (1971) both worked on Trinidad butterflies, the former with two species of *Parides* in lower montane rain forest and the latter with *Heliconius erato* in southeastern Trinidad. *H. erato* lives either in secondary forest or on the fringes of primary forest. In spite of expectations, Cook *et al.* found that *Parides* spp. had an observed survivorship in the wild much less than expected on the basis of longevity in captivity. They believed that the difference might have been caused by the frequent heavy rains. In contrast, Turner found that his species had a much longer life-span in nature, close to that observed in captivity.

Fleming (1974) compared demographic features of two

rodents, *Liomys salvini* from a more seasonal environment in Central America and *Heteromys desmarestianus* from less seasonal habitats. *L. salvini* reached sexual maturity in 3–4 months, *H. desmarestianus* in eight months. The former species had a slightly higher mean litter size and slightly lower juvenile survivorship.

Seasonal population stability

The concepts of *r* and *K* selection are very important if the selective forces producing them are capable of being shifted easily as a result of human perturbations. Disturbance may either select *r*-selected species in human controlled ecosystems or *K*-selected species may alter to *r*-selected ones by changes in selective pressures. Faunal studies of the reproductive response to fluctuations in biotic and abiotic factors are very important.

Populations under *K* selection would not fluctuate much. Some tropical species in complex ecosystems show seasonal and annual fluctuations in numbers. Other populations, however, seem to be relatively constant in number.

There have been few long-term studies of population size for tropical forest species. Ehrlich and Gilbert (1973) followed a population of the butterfly *Heliconius ethilla*, in the northern hills of Trinidad for 27 generations during which the population level was very constant over. *H. ethilla* had many characteristics of a *K*-selected species (a long adult life, iteroparity, delayed reproduction and limited resources), but it did not exhibit the expected low reproductive effort expected of *K*-selected species. In the moist evergreen Mayanja Forest, Uganda, Delany (1971) observed no large-scale fluctuations of rodent numbers, although reproductive cycles of some species varied with the seasons.

Most studies have utilized indirect techniques for estimating numbers of individuals. A common method has been to follow the course of the reproductive cycle throughout a year. If the frequency of reproductive activity is similar at all seasons, it is assumed that the population level is relatively constant. P. Y. Berry (1964) found that the seven anuran species breed throughout the year in Singapore and animals of all reproductive stages were found each month. One species did show some evidence of change in breeding condition relative to climate, and others increased their reproductive activity at the onset of rains following brief dry spells. Inger and Greenberg (1963) found that some adults of the frog, *Rana erythraea*, at Kuching, Sarawak, were in breeding condition in all months. The proportion of females ready to breed varied monthly, however, without any obvious relation to climatic factors. Inger and Bacon (1968) could find no evidence of seasonal trends in spermatocyte counts, egg size, or proportion of ripe females in six species of Bornean rain forest frogs. In a study of lizards from Borneo, Inger and Greenberg (1966b) concluded that four species appeared to breed continuously throughout the year and that year-round reproduction probably buffered population fluctuations. Campbell (1973) showed that an iguanid lizard, *Anolis poecilopus*, which lived along the sides of large shaded streams in Panamá, did not exhibit marked monthly differences in reproduction. Males and females of the teid lizard, *Ameiva bifrontata*, in Estado Sucre, Venezuela, appear to have no evident seasonal reproductive component (Leon and Ruiz, 1971). The same conclusion was reached by Alcala and Brown (1967) for the scincoid lizard, *Emoia atrocostata* in mangrove forests of southern Negros, Philippines. In equatorial Colombia, Tamsitt and Valdivieso (1965) found no change in the male reproductive system of the frugivorous bat, *Artibeus lituratus*. Mares and Wilson (1971) found that frugivorous bats were reproductively active during the dry season in Costa Rica while the insectivorous ones were not.

Owen (1964) showed that some individuals of the pulmonate snail, *Limicolaria martensiana* contained eggs at all seasons in Uganda but that there were two breeding peaks, when about one third of the individuals contained eggs, which coincided with the two annual dry seasons. (But see discussion of Owiny (1974) later). Benson and Emmel (1973) showed that the nympholine butterfly, *Marpesia berania*, was at equilibrium conditions for the dry season and the first month of the wet season in Costa Rica. Thus, a wide variety of tropical species are known or suspected to vary little in population numbers.

In contrast, many species show seasonal and annual differences in numbers. Again, the evidence is both direct and indirect. N. G. Smith (1972) observed that a diurnal moth, *Urania*, of Central and South America has undergone population fluctuations and/or migrations at frequent intervals since 1868. Sexton (1967) demonstrated that a population level of the lizard, *Anolis limifrons*, differed by a factor of three when measured at a comparable season two years apart. Struhsaker (1973) noted a decline in numbers of vervet monkeys in Kenya which he attributed to a decline in the availability of food. Owen *et al.* (1972) collected butterflies in a garden in Sierra Leone over 27 months. Ten of the 24 species of Acraeidae were common enough to indicate monthly variations in relative abundance.

Most evidence, from studies of the change in reproductive condition, is for seasonal changes in population levels. A number of bats have been observed to be seasonal breeders, e.g. *Eidolon helvum*, *Epomops franqueti*, and *Micropterus pusillus* in Rio Muni, West Africa (Jones, 1972), *Eidolon helvum* in Uganda (Mutere, 1967) and *Miniopterus australis* in northern Borneo (Medway, 1971). Seasonal patterns of reproduction in tropical rodents have been documented by Delany (1964) and Okia (1973) in Uganda, Rood and Test (1968) in Venezuela, and Fleming (1970) in Panamá. The reproductive cycle of the African elephant was studied by Smith and Buss (1973) in Uganda and Zambia. They found that, although this species may breed at any time, there were more young borne during the wet season than the dry one.

Moreau (1950) indicated that the birds of Africa tended to fall into one of five seasonal breeding classes: the big raptors and scavengers laid in the middle of the dry season (perhaps because prey was more abundant at this period of reduced vegetational cover); ground nesters in wet season in areas with infrequent ground fires but in dry season after the likely fire period in other areas; birds dependent upon tall grasses for nests and/or food nested after the start of the rains; water-birds reproduced when their watery habitat was at its maximal extent; and most other birds, nearly all of which were dependent upon insects, began laying with the pre-rain flush of new foliage and insects.

Voous (1950) summarized information about the breeding patterns of birds in Sumatra, Borneo and Java. Most species, even those inhabiting evergreen forests, showed distinctly seasonal patterns; breeding generally started at the end of the wet season and reached a peak before the driest month. He discounted the importance of photoperiod and thermal changes. The few species of birds which breed throughout the year inhabited cultivated areas. Fogden (1972) found that the forest birds of Sarawak, Borneo, were also highly seasonal breeders.

Sexton *et al.* (1963, 1971), Jenssen (1970), Licht and Gorman (1970), Ruibal *et al.* (1972), Campbell (1973), Fitch (1973), and Fleming and Hooker (in press) have all shown that various tropical and subtropical species of the large iguanid genus *Anolis* are seasonal breeders.

Studies of terrestrial invertebrates have shown that many taxa vary seasonally in numbers and in reproductive activity. Dobzhansky and Pavan (1950) discovered changes in relative frequency in species composition among the populations of *Drosophila* spp. trapped at many localities throughout Brazil. Pipkin (1965) observed changes in numbers of *Drosophila* spp. in Panamá. An oviviviparous African land snail, *Limicolaria martensiana*, breeds continuously but has bimodal peaks during the dry season (Owiny, 1974). The reproductive production of various species of terrestrial snails was affected by rainfall at Bukit Chintamini, Pahang, Malaysia (A. J. Berry, 1966). Frequent and heavy rains promoted simultaneous reproductive activity and survival of all age classes; dry periods had the opposite effect.

Factors affecting seasonality

Most changes in population size or structure are periodic and usually expressed in terms of the seasonal changes in precipitation. The coincidence of the periodicity of the reproductive cycle with some regularly occurring environmental phenomenon is not proof of causal relationship between them. If the relationship is not coincidental, it can be operating at one or both of two levels: ultimate and proximate. Ultimate factors are those which act at the level of natural selection on the phenotype, while proximate factors are those which act as physiological cues to produce an adaptive response to a changing environment. Both can be either biotic or abiotic.

For many species of animals, the ultimate factors responsible for seasonality in population dynamics are biotic, with food availability for the young being an obvious example. Thus, Ward (1969) found that the bird *Pycnonotus goiavier*, in the very uniform climate of Singapore, was a seasonal breeder with the breeding period coinciding with maximum in insect abundance.

Smythe (1970a) described a study of the seasonality of fruit-fall for two types of fruit on Barro Colorado Island, Panamá. Type 1 had seeds 1.5 cm, and type 2 was 1.5 cm in diameter and was not wind-dispersed. The peak for total fruit fall was May–June, early in the wet season. This was due to the increase in number of species fruiting, in average fruit size, and in number of fruits. However, type 1 fruit-fall was more seasonal and type 2 more uniform throughout the year. Kaufman (1962) had studied the social organization of a medium-sized omnivorous carnivore, the coati (*Nasua narica*), on Barro Colorado, and Smythe (1970b) reviewed this organization as it was influenced by fruit-fall. He reported that all classes of coati feed on fruit as a major part of their diet from February to August. As fruit becomes relatively scarce in the late wet season and early dry season, the adult males hunt solitarily while the bands of female and young forage for small animals in the leaf litter.

The biological factors, such as flower or fruit availability, are under their own selective forces and animals dependent upon such seasonal plants often become locked into a mutually dependent relationship with them. For example, Janzen (1967a) was interested in the synchrony of flowering and fruiting of trees observed during the dry season in Central America. He postulated that reproductive activities of the plants occurred at that time because all available energy was allocated to intra- and interspecific competition (i.e. growth) during the wet season. During the relatively dormant dry season energy reserves built up during the wet season could be channeled into reproduction. Furthermore, during the time of leaf-fall flowers would be more visible to pollinators and fruit to animal dispersing agents. A further advantage to the animals was that the nectar and fruit served as auxillary water sources during a period of low water availability. Thus pollinating and dispersal agents became locked into the flowering regime of the plants.

The phenotypes selected must be coupled with their ultimate selective forces through some environmental factor. This coupling mechanism involves some factor which acts as a cue predicting change in the environmental status of the animal and the physiological response adjusted to meet the new conditions. These cues, then, would be proximate factors.

As a possible example of a biotic proximate factor regulating a reproductive cycle, Ward (1969) suggested that changes in the internal protein levels of *Pycnonotus goiavier* could institute change in its reproductive cycle. Precipitation has received considerable attention as a possible abiotic proximate factor. Most studies of the reproductive cycle of anoline lizards have indicated that breeding is concentrated in or restricted to the wet season. Sexton *et al.* (1971) have shown that the extent of breeding activity in *Anolis limifrons* in Panamá varies with the severity of the dry season. Brown and Sexton (1973) showed that the reproductive condition of females of *Anolis sagrei* from Belize is a function of relative humidity. It remains to be shown how direct is the relationship between moisture and their reproductive activity. Recently, Gorman and Licht (1974) suggested that there is a great deal of plasticity in the reproductive response of male anolines to varying environments, and that photothermal cues may be more important than humidity.

Epple *et al.* (1972) also discussed the role of photoperiod in regulating the testicular cycle of the Costa Rican bird, *Zonotrichia capensis*. They confirmed earlier work of Miller (1965) on a Colombian population that photoperiod was a factor initiating testicular recrudescence, but concluded that photoperiodicity played little, if any, role in reproductive periodicity of the Costa Rican birds. Detailed experimental studies of the factors regulating reproductive cycles, coupled with comparative field studies, are necessary to determine the complex set of interactions which must govern the cycles of tropical species.

Certain species exhibit periodicity in other related phenomena. The bat *Eidolon helvum*, is a seasonal breeder in Uganda and it exhibits delayed implantation (Mutere, 1967). There are two precipitation peaks per year, October–November and April–May (the higher peak). Mating occurs in April–June, but implantation is observed in October–November. Birth follows in four months. The young are weaned at one month of age, at the time of greatest insect abundance.

Owiny (1974), also working in Uganda, stated that the land snail, *Limicolaria martensiana*, was induced to breed when precipitation decreased so that there was an inverse relationship between precipitation and breeding intensity. Ovoviviparity had evolved in this species with the eggs and/or young retained *in utero* during aestivation in the dry season. The young and/or unhatched eggs were deposited early in the wet season.

The identification of ultimate and proximate factors and their impact upon reproduction is very important for an understanding of the population dynamics of both *K* and *r*-selected species. Such studies are particularly important in areas subject to human intervention where evolutionary forces might change direction and where *r*-selected species might prevail. The study of the control of reproductive systems of tropical species will be given great emphasis in the near future.

Weather

Since population changes seem related to meteorological events, mention should be made of weather periodicity.

Brookfield and Hart (1966) have commented upon the use of several statistics other than mean or modal values to describe rainfall pattern. Their comments stress the need for developing wide-spread and long-term stations. They use not only means and standard deviations, but also coefficients of variation, rainfall intensity and spells of dry or wet weather. The concept of spells of weather seems particularly useful to agronomists and ecologists because one can arbitrarily adjust the definition of a dry or wet spell to fit levels of some meteorological factor into the requirements of a crop, species or ecosystem under study. Of particular importance is the need to relate such spells of weather, however defined, to the generation time of particular species.

Brünig (1969) has commented upon the seasonality of droughts in Sarawak, the lowland forest area of Borneo often used as an example of an equable climate. The annual temperature variation is 2° C, with diurnal ranges of 5–7° C. Wind speeds are low, and high winds are localized. The

annual precipitation exceeds 3 000 mm with a small seasonal variation. There is, however, a rainfall peak in December–February. Brünig used running 30 day totals. A dry period (less than 100 mm in 30 days) induced drought responses in certain plants. He concluded 'We can further conclude that the distribution of these dry or arid periods exhibits a seasonal pattern in all stations, even in those stations with an apparently extreme evenness of rainfall throughout the year'.

Interactions among species

Interactions among organisms fall into three basic categories: competition; predation or parasitism; and mutualism.

Competition

Competition occurs whenever species overlap in their utilization of resources so that an increase in the density of one adversely affects the others. Competition is difficult to observe, and most of the evidence for it is indirect. A good review of the theory and its consequences is provided by MacArthur (1972).

Laboratory studies of competition in a uniform environment providing a single resource usually result in the elimination of all but one species. Therefore, the persistence of many species in nature must be due to environmental heterogeneity. More species could be accommodated if the resource range were greater and/or the resources were more finely subdividable among the species.

In an unvarying environment there is no theoretical limit to the similarity of competing species, but fluctuations in resource availability make very similar species vulnerable to competitive exclusion. Diamond (1973) found that among co-existing New Guinea birds the ratio between their weights averaged 1.90 and was never less than 1.33 or greater than 2.73.

Bird distribution patterns appear to be largely explicable in terms of competitive interactions which may explain some of the gaps in the ranges between closely related pairs of species (Wilson, 1958; Terborgh, 1970; Diamond, 1972; MacArthur, 1972). There is less evidence that competitive interactions are important in determining community patterns of other groups, for example, tropical bats (McNab, 1971).

In order to elucidate the effects of different factors, field ecologists have generally used either removal studies or the patterns of resource partitioning.

Inger and Greenberg (1966a) examined the effects of competition among three sympatric species of ranid frogs (*Rana blythi*, *R. ibanorum* and *R. macrodon*) inhabiting stream-sides in Sarawak by removing all the individuals of one species and observing the effect on the abundance of the remaining species. While there was extensive niche overlap among the three species, *Rana blythi* tended to be found farther from the stream bank than either *R. ibanorum* or *R. macrodon*. On one area, which had only *R. blythi* and *R. ibanorum* present originally, the latter species was removed and the population of *R. blythi* increased. The elimination of *R. blythi* from a second stream where all three species were initially present resulted in a *ca.* fourfold increase in *R. macrodon*.

Division of food resources has been the area most frequently studied. Moreau (1948) studied 172 bird species on 7 770 km² of Africa. Each species was categorized according to the 31 habitat types in which it foraged. Excluding the weaver finches, he found that within genera only 16 per cent of the theoretically possible cases of overlap occurred and that in only one third of these was there overlap in diet. Ninety-four per cent of the congeneric species and 98 per cent of the non-congeneric ones were distinct either in habitat or diet or both. Moreau concluded that most of the cases of habitat overlap were not due to diet overlap because the species had differences in body or beak size, or foraged in different manners or at different ranges. Later workers stressed the importance of habitat structure in influencing the number of bird species present.

Following the earlier work of Oliver (1948), Collette (1961), Ruibal (1961) and Corn (1971) on the allocation of resources, especially of perch characters, among sympatric species of the lizard *Anolis*, Rand (1964) concluded that 'Differences in structural and climatic habitats among sympatric species of *Anolis* produce a spatial separation which appears to reduce interspecific competition and increase the efficiency with which the available habitat is exploited'. Later, Rand and his co-workers discussed the structural habitats of several other lizard species (Rand and Rand, 1966, 1967; Rand, 1967a, 1967b; Rand and Humphrey, 1968; Rand and Williams, 1969). Rand's idea of structural and climatic habitats formed a base for Schoener's work on species packing and resource partitioning in sympatric species of island anoles (Schoener, 1968; Schoener and Gorman, 1968; Schoener, 1970; Schoener and Schoener, 1971). However, Andrews (1971) in a study of the structural habitat of a mainland anole in Costa Rica obtained different results with considerable intraspecific variation. Gorman *et al.* (1971) found that two anoles introduced to Trinidad occupied similar perch types and hybridized to some extent supporting the view that structural differences may also be partially responsible for reproductive isolation. Crump (1971) concluded that one component influencing the distribution of a large number of reptiles in Brazil was vegetation density. Lazelle (1972) described the habitats of the 16 species of anoles inhabiting the Lesser Antilles.

Schoener (1974) analysed resource allocations among groups of competing species in studies in which the ecological differences were identified. He concluded that habitat dimensions were more important than food-type dimensions and both were more important than time; predators were more often active at different times than members of other groups; terrestrial poikilotherms subdivide food resources through temporal segregation; vertebrates are less seasonal than are invertebrates and, subdivision of the food resources is more important for animals feeding on larger food items than for animals feeding on smaller prey.

In spite of the abundant evidence that potential competitors partition resources through a variety of mechanisms the degree of overlap can change with the availability of the resources. Moreau (1948) found that the species of weaver finches showed much more overlap in feeding than did the members of the other avian groups he studied. The finches fed on superabundant grain resources and, as supplies became depleted, migrated to another source. From observations of large numbers of bird species feeding on the berries of a single plant species, Willis (1966c) suggested that 'biological or physical fluctuations create niches which can temporarily be exploited by more than one species, since full exploitation of a niche always lags behind the appearance of the niche'. From similar observations Terborgh and Diamond (1970) concluded that the extensive niche overlap between species was often reduced by interspecific aggression and utilization of different feeding levels.

Leck (1972) investigated seasonal changes in the feeding habits of frugivorous birds on Barro Colorado Island, Panamá. Most species fruited at the end of the dry season and beginning of the wet season. During the dry season obligatorily frugivorous bird species exploited fruit produced by the continuously fruiting species of plants whilst the facultatively frugivorous ones switched to alternate food sources. Intraspecific aggression of tanagers and fly-catchers declined during the period of fruit abundance.

High quality resources can generate both intra- and interspecific interactions which will limit their utilization. Johnson and Hubbel (1974) reported that agonistic behaviour between rival colonies of stingless bees at Turrialba, Costa Rica, increased as the quality (concentration and amount of nectar and pollen as well as spacing) of the food increased. Furthermore, non-aggressive species of bees were excluded from high density food by aggressive species.

To elucidate the complex communities of Malaysian and Australian rain forests, Harrison (1962) devised a simplified classification of the feeding habits of mammals and birds relative to the gross structural organization of the forests. The subdivisions were:
— the upper air community: the birds and bats hunting above the canopy; most are insectivorous, but there is a large proportion of carnivores;
— the canopy community: birds, bats and other mammals confined to this zone; most species feed on leaves, fruit or nectar, but there are some insectivores and mixed feeders;
— middle-zone flying animals: birds and insectivorous bats; most species are insectivorous, but there are a few carnivores;
— middle-zone scansorial animals: mammals ranging up and down the tree trunks, entering both the canopy and ground zones; these animals are predominantly mixed feeders with some carnivores;
— large ground animals: mostly mammals, largely plant feeders (some, such as the elephant, can reach up and feed in the canopy);
— small ground animals: birds and mammals that can do some climbing getting into the lower portions of tree trunks, predominantly insectivorous or mixed feeders with some herbivores and carnivores.

The establishment of such a classification is the first step to an understanding of the organization and functioning of complex systems. Such schemata enable an investigator to assess the importance of various groups of animals in the passage of energy and nutrients throughout the entire community (e.g. Eisenberg and Thorington, 1973).

Colwell (1973) studied the interspecific dynamics of four flowering plants (*Centropogon valerii*, *C. talamancensis*, *Macleania glabra* and *Cavendishia smithii*), four species of foraging birds (three humming-birds of which *Colibri thalassinus* and *Eugenes fulgens* were seasonally nomadic and *Panterpe insignis* was resident, and a resident flower piercer, *Diglossa plumbea*) and two species of nectarivorous mites (*Rhinoseius colwelli* and *R. richardsoni*) which lived within the flowers and were transported between them by the humming-birds. *Panterpe insignis* was an opportunistic. *Colibri thalassinus* was a sequential specialist which relied upon a limited number of resources at any one time and switched from one to another as they became available, or migrated to another region. Both *Eugenes fulgens* and *Diglossa plumbea* were interstitial species dependent on low-density energy sources such as widely dispersed plants. In addition, *D. plumbea* utilized heavy vegetative cover to minimize attacks by humming-birds.

The mites were incidentally picked up by feeding humming-birds and transported within their nasal cavities. They were restricted to the flowers of certain species: *Rhinoseius richardsoni* to the Ericaceous *Cavendishia smithii* and *Macleania glabra* and *Rhinoseius colwelli* to the Lobeliaceous *Centropogon valerii* and *C. talamancensis*. The distribution of the mites was essentially controlled, not by interspecific killings, but by the selection of plants by feeding humming-birds. Colwell believes that interstitial, sequential, specialist and hypercontingent species are favoured by environments with a high degree of constancy and predictability.

Predation

The results of predator-prey interactions depend whether the predator has a threshold of prey abundance below which it cannot meet its metabolic requirements or whether it is capable of exterminating the prey. In the latter case oscillations in predator-prey abundance tend to be amplified, and the predator-prey system is characterized by local extinctions and emigrations (Huffaker, 1970). In the former case the fluctuations tend to be damped. The importance of spatial distributions for predation rates on tropical tree seedlings has been demonstrated by Connell (1970).

In general, the oscillations inherent in predator-prey interactions are reduced when the predator has a high threshold below which it cannot exploit the prey; the predator has traits, such as territoriality, which prevent build-up of dense populations and when the environment is heterogeneous, making it more difficult for the predators to find and exploit its prey.

Knowledge of predator-prey interactions in tropical forests is meager, and because of their economic importance plant-herbivore interactions are concentrated on. Parasite-host interactions are treated in chapter 14.

Plant-herbivore interactions

The nature of plant defenses against herbivores is partly known, and in a few cases the selective activity of herbivores has actually been measured (Jones, 1962, 1966; Feeny, 1973; Cates, 1975). Tropical shrubs tend to be high in chemical defenses and are the source of most of the world's spices (Baker, 1964). Early successional tropical species appear to have fewer defenses than later successional and climax species. The growth of these early successional species in monocultures, i.e. typical modern agriculture, is therefore an especially vulnerable system to decimation by herbivores (Janzen, 1973b; Tahvanainen and Root, 1972).

Tropical herbivorous insects are generally thought to be much more host specific in comparison with temperate zone ones, though the evidence for this is meager. The rationale offered for this is the increased heterogeneity of the plant community and predictability of the resource base (e.g. see Janzen, 1970b). It is difficult to sort out cause and effect relationships; to know whether high cost specificity of herbivores generates (or is a product of) high tropical plant diversity. Janzen (1970a) has argued that density-dependent grazing by herbivores in large measure prevents domination of tropical forests by competitively superior tree species. Burdon and Chilvers (1974) showed how higher host specificity by fungal and insect herbivores contributes to niche differentiation in mixed species of young *Eucalyptus* in southeastern Australia. Janzen (1970a) proposed a probability model based on the distribution from the parent tree of seed-fall and the response of seed predators to explain the numbers of tree species in tropical forests. The experimental results obtained provide mixed support for the model (e.g. Janzen, 1971a, 1972a). For lygaeid bugs feeding on fig seeds, Slater (1972) suggested, and, for bruchids feeding on legume seeds Janzen (1971a) has shown that almost all the seeds may be destroyed.

The number of herbivorous insects found on an individual tropical tree seem too small for interspecific competition, but the resource base used by any species may be much smaller than is generally realized. There may be considerable competition among shoot-tip feeding cossid moth larvae and cerambycid beetles. Competitive interactions must be considered within the context of the small fraction of the plant that is actually utilized. Also, shoot-tip feeders may reduce the amount of resource available in flushing leaves, and these could influence the available carbohydrate levels in the roots.

Increased distances between conspecific trees in tropical forests as compared to temperate forests undoubtedly reduces the frequency of outbreaks of particular insect pests, but such outbreaks do occur. There are periodic outbreaks of lepidopterans whose larvae defoliate many individuals of their highly dispersed host trees in Costa Rican deciduous forests (Carroll, unpublished). If selection has not provided tropical trees with the tolerance to survive major defoliation, the rare insect outbreak could have a profound effect on the abundance and distribution of host trees.

Janzen (1973a) and Feeny (in press) have pointed out that a plant's evolutionary response to one species of herbivore might greatly reduce the suitability of the plant for another herbivore species. Therefore, higher order inter-actions from density responsive predators and parasitoids might influence host plant selection. In time highly host specific predators and parasitoids could sufficiently lower the herbivore population on a given host plant to increase the probability of successful invasion by a new herbivore. The assemblage of predators and parasitoids that are attracted to a host plant are part of that plant's defense mechanism and might influence the allocation of the plant's defense budget. The evolution of extrafloral nectaries (Bentley, 1975a) and, in some instances, the plant's associ-ation with sugar-secreting homopterans are generalized defense mechanisms that function by attracting predators, such as ants, and parasitoids, such as tachinids and small wasps, to the vicinity of the plant. Vines are important pathways for arboreal ants and frequently have more extra-floral nectary glands than other plant life-forms (Bentley, 1975b). From a survey of vines and herbs, Berlin and Henry (1973) found that vines frequently had epidermal fissures that suggested secretory tissue which supported large yeast populations; vines might be adapted for the selection of more organized extrafloral nectaries.

Mutualism

Mutualistic activities include pollen transfer, seed dispersal and defense of vegetative tissues; most terrestrial examples involve plants and animals.

Mutualistic interactions are perhaps commonest in the tropics. For example, there are no obligate ant-plant mutu-alisms or orchid bees north of 24° N and no nectarivorous or frugivorous bats north of 33° N. Within the tropics mutu-alistic interactions are more prevalent in the warm, wet evergreen forests than in cooler and more seasonal habitats. This may be due to the higher all year impact of insect herbivores on plants under moist tropical conditions, which increases selective pressure on plants to evolve anti-herbivore devices. Theoretically responses to herbivores are faster for outbreeding than inbreeding plant species (Ehrlich and Raven, 1965; Gilbert, 1972). If potential pollinators exist animal pollination should be favoured over wind polli-nation, because the stronger specificity between animal and plant increases the efficiency of pollen transfer.

Pollination

Pollination biology is an enormously complex discipline (Van der Pilj and Dodson, 1966; Faegri and Van der Pilj, 1971; Proctor and Yeo, 1973). Coupling between the plant and its pollinator may be very specialized. *Ficus* has an obligatory relationship with its minute agaenid wasps polli-nators (Ramirez, 1969). In *F. sycomorus*, a monoecious species, the wasp *Ceratosolen arabicus* is a mutualistic symbiont. Females enter the syconium and deposit a single egg in ovules having short styles. These ovaries gradually swell into galls. The developing male and female wasps emerge and mate in the closed syconium. As the males bore their way out the anthers are knocked into the cavity where

they break open. The pollen adheres to the female wasps who exit through the holes made by the males (Galil and Eisikowitch, 1968).

In contrast are the observations of Baker *et al.* (1971): 'But very few plants have only one kind of visitor capable of effective pollination (species of *Yucca* and *Ficus* are probable exceptions) and practically none have only one kind of visitor capable of gaining sustenance from the floral parts of the nectars.' These authors classify the visitors to the flowers of *Ceiba acuminata* into major pollinators, e.g. the bat, *Leptonycteris sanborni*; minor pollinators, such as humming-birds; irrelevant species (those visiting the intact flowers but unlikely to effect pollination), such as bumble bees (*Bombus*), carpenter bees (*Xylocopa*), honey bees (*Apis mellifera*), wasps and a skipper butterfly; and species which destroy the intact or fallen flowers.

The most conspicuous difference between insect polli-nation in the tropics and in the temperate zone is the greater elaboration of flowering patterns in the tropics. Gentry (1974) suggests that a major part of these patterns (at least for the Bignoniaceae) has resulted from competition for pollinators among the relatively depauperate bee fauna in developing a long flowering period. In the deciduous forest the greatest peak of flowering occurs during the dry season (Janzen, 1967a). As the dry season becomes increasingly severe the flowering peaks may be shifted into the wet season (Burger, 1974), and in regions with a very mild dry season flowering pulses are weak and dispersed throughout the year.

It is possible to distinguish several phenologies of insect pollination (Janzen, 1967a, 1971c; Gentry, 1974). Night blooming flowers are often pollinated by sphingid moths especially if the flowers are pale and have long tubular corollas. Highly synchronized mass flowering occurs among several genera of trees in the dry season, e.g. *Tabebuia* spp. Another pattern is the production of large numbers of flowers over an extended flowering period, e.g. *Arrabidaea* vines. Gentry (1974) found that species of *Arrabidaea* had largely non-overlapping flowering peaks and tended to share many of the pollinating bees, especially species of *Eulaema* and *Euglossa*. By lengthening the seasonal flight period of many bee species such plants create a greater temporally divisible pool of pollinators. In deciduous forests during the wet season or throughout the year in more evergreen forests, traplining (Janzen, 1971c; Gentry, 1974), in which a plant produces a few high quality flowers each day over a long flowering period, is common. Presumably, this maximizes long distance outcrossing. Traplining pollinators may fly many kilometers in their flower circuit (Janzen, 1971c); one of many reasons why tropical nature reserves need to be large. Furthermore, the removal of one plant or even the removal of a few flowers over a period of several days could prove disastrous to the pollinator and greatly affect the fitness of other plants on the circuit.

Heinrich and Raven (1972) discussed the energetic relationships between flowers and their pollinators: 'a balance must exist between the energy expended by foragers and the caloric reward of the flowers if cross-pollination is to be maximal. A flower must provide sufficient reward to attract foragers, but it must limit this reward so that the animals will go on to visit other plants of the same species'. In humid lowland tropics where individuals of a tree species may be widely separated they predict that the flowers should offer increased energy rewards for the larger pollinators necessary for cross-pollination.

Heithaus *et al.* (1974) examined the relationship of *Bauhinia pauletia* to two bat pollinators, a large species (*Phyllostomus discolor*) which visited the flowers in groups and a smaller species (*Glossophaga soricina*) which did so singly. The activities of both species promote outcrossing. *Phyllostomus discolor* drains the nectar and probably depletes it so that visits must be spaced far enough apart in time to permit renewal. The social movement of this species from area to area would clearly aid in outcrossing. The smaller species probably flies from flower to flower in a trapline (Janzen, 1971c), laps nectar but does not drain it, so that shorter returns are energetically reasonable.

Ants and plants

Ants are involved in many mutualistic associations with flowering plants. Janzen studied the interactions of swollen-thorn acacia trees with pseudomyrmecine ants (Janzen, 1967b and c, 1969b). One taxon of acacias has enlarged stipular thorns which serve as domiciles for the ants. The trees also have well-developed foliar nectaries and Beltian bodies at the tips of the leaflets. The former provide carbohydrates and the latter proteins and lipids to the ants. The trees are rapidly growing inhabitants of disturbed areas, but they are susceptible to both attack by phytophagous insects and to shading by competing trees, shrubs and vines. Shoots of the acacias not occupied by ants have much greater numbers of phytophagous insects than do occupied ones and suffer greater damage. The ants also adversely affect surrounding vegetation by killing their leading shoots which contact the host tree. This enables the acacia to grow in a relatively open cylinder, free of other vegetation. Rehr *et al.* (1973) made a study of the defenses of non-ant-acacias which constitute about 90 per cent of the acacia species found within Central America. They have better chemical defense systems against phytophagous insects than the ant-acacias.

Another example of obligatory mutualism involves attine ants, a fungus-culturing group inhabiting tropical and subtropical areas of the New World (Weber, 1969, 1972). The three sub-groups of attines are:
— primitive genera which utilize insect feces and carcasses of insects as a culture medium for fungi;
— intermediate genera which use both plant debris and freshly cut leaves and flowers for a culture medium;
— the most complex genera (*Atta* and *Acromyrmex*) which culture the fungi on leaves and flowers cut from living plants. Most fungi involved have not been identified since only mycelia and no sporophores are found in the gardens.

Field and laboratory experimental studies explain the attine-fungal interaction. In the garden of the attines, the fungi live as pure cultures on an artificial medium; elsewhere the same fungi are slow growing and are displaced by contaminating species. Martin (1970) found that the attine fungi lacked certain proteolytic enzymes necessary for obtaining nitrogen from the polypeptide-rich substrate of fresh leaves. The ant fecal material contains various nitrogenous compounds which supplement the nitrogen of the culture medium, but the effect is not long-lasting. There are also proteolytic enzymes in the fecal liquid. The activities of the ant also help the fungus become a dominant competitor within the garden. The mechanical effect of chewing in rupturing cell walls enables the fungi to get at the plant protoplasm easier. In addition, proteolytic enzymes which originated within the plant cell are also available to digest the protoplasm of the disrupted cells. The planting of mycelia within the garden by the ants enables the fungi to expand spatially very rapidly.

Cherrett (1968, 1972) found that *Atta cephalotes* foraged more at 31–47 m distant from the colony and did not necessarily utilize the individual plants (of the 36 used species of the 72 which occurred) nearest the nest. The overall effect is that grazing was spread over a large area. Other research on attines had dealt with pheromones (Moser and Silverstein, 1967; Blum *et al.*, 1968; Moser *et al.*, 1968), stridulation (Markl, 1965) and local distribution (Rockwood, 1973).

Army ants

E. O. Wilson (1971) has reviewed much of the literature dealing with army ants and of social insects in general. Rettenmeyer (1962a) divided arthropod associates of army ants into three categories: flies which accompany the moving front of ants; species utilizing their refuse heaps and species living within their bivouac or nest. In Panamá, Rettenmeyer (1961) showed that these dipterans at the advancing front are parasites and deposit either eggs or larvae on insects fleeing from the ants. In combination with the ants they kill as many as 90 per cent of the prey. There are also species of millipedes (Rettenmeyer, 1962b) and thysanurans (Rettenmeyer, 1963) associated with the nests or columns of Panamanian army ants. The relationship between ants and millipedes is probably mutualistic but not obligatory; the thysanurans are most likely obligatory myrmecophiles. The former probably keep the nest free of attack by fungi by scavenging the organic debris which accumulate around a nest. The latter feed on liquid oozing from booty and on surface secretions or other materials from the ants.

Many species of migrant birds wintering in the neotropics follow ant swarms (Willis, 1966a), but they are either restricted to zones peripheral to the swarm while resident species feed in zones more central to the swarm or they accompany ant species whose swarms are smaller and less frequent. Similarly, the position of a species relative to the ant swarm may be determined by the presence or absence of other resident species of ant-associated birds (Willis, 1966b). Willis believes that superabundant food sources are frequently obtainable in tropical forests and that there is a lag between the onset of a particular superabundant food source and its utilization. The more efficient resident species of ant-following birds would exclude migrant species and the less efficient (or smaller) resident species of ant-followers when food is not superabundant. Other treatments of the biology of ant-following birds are Willis (1972) and Willis and Oniki (1972).

Man and animal populations

The three preceding sections have stressed the organization and functioning of animal populations in tropical ecosystems as though human beings had had no effect upon them. Human interactions with the forests of tropical areas are ancient, and older human cultures have had a substantial impact upon forest animals which is still evident. For example, Bennett (1968) has ascribed curious gaps in the present distribution of certain species of Panamanian vertebrates to precolonial agricultural clearing. These distributional gaps are best explained as the result of forest recolonization in the cleared areas.

The utilization of forest by earlier cultures is quite different from today (Watters, 1971). Formerly, with shifting agriculture, there was a regular temporal pattern of use within relatively restricted areas providing a series of farming and successional stages and so preserving a relatively high species richness and a source of colonizing species. Current practices radically divert tropical systems to types which differ in their abiotic and biotic composition and their terminal stage. The implications of such severe changes are the extinction of numerous species and of species associations, erosion, declining soil fertility and increasing agricultural and human pests.

The most pressing research needs are to exploring the practical and theoretical effects, in both spatial and temporal terms, of these new patterns of land use. Especially useful will be studies directed towards developing some concept of the balance necessary between forested and deforested areas which would minimize ecological and economic disasters.

Most of the zoological work relevant to these concerns has been performed by scientists interested in health or pest problems and will be treated in chapters 14 and 17. The relative lack of research by other zoologists indicates the magnitude of the effort needed to understand the ecology of perturbed tropical ecosystems.

Stability

The concept that mature tropical forests represent stable ecosystems is undergoing a rapid change, and much of the future research in tropical systems should be directed towards determining the mechanics by which these systems revert to mature communities after single and multiple perturbations.

Much less is known of the zoological aspects than of the botanical ones of successional changes following disturbance. A recent study by Opler, Baker and Frankie (1976) on secondary plant succession can be used as an instructive model for both its *modus operandi* as well as for its findings. They studied the recovery of vegetation on three plots in lowland wet forest and on one lowland dry forest in Costa Rica; changes in the gross structural organization, species

richness, dispersal types, reproductive strategies, etc., were measured. The results indicate that, initially, secondary succession may be quite rapid and that forests may soon acquire a superficial resemblance to their condition before disturbance. The factors influencing this rate of recovery include: the persistence of plants or of their seeds in the area after disturbance; the size of the area disturbed; the distance from the area to other sources of recolonizing species. One overall conclusion is that early secondary successional stages are quite resistant to perturbation and may make rapid recovery. The authors do not extend this conclusion to mature communities.

May (1975) has reviewed the stability of mature tropical moist forests; members of complex systems living in relatively predictable environments are less liable to resist gross environmental change because of their post evolutionary commitments to certain reproductive and trophic strategies (e.g. *K*-selected species, storage of nutrients in woody tissue, etc.).

There may be more turnover in species; or more importantly in species assemblages, than previously thought in mature tropical ecosystems. For example, Wilson (1958), in discussing the unevenness of and distribution within forests of New Guinea, stated 'Patchiness exists at both a local, clearly ecological level, and a broader geographic level not easily correlated with environmental influences. The combined irregularities in the distribution of multiple species result in distinct shifts of faunal composition and relative abundance over distances of only a few kilometers even in relatively continuous, homogeneous rain forest'.

Dobzhansky and Pavan (1950) found the same type of patchy distribution for species of *Drosophila* in Brazil. Samples taken a short distance apart often contained some species tending to form discrete nuclei with others tending to be more uniformly distributed.

Futuyma (1973) expects that sets of tightly coupled species to show demarcations between them which would vary from place to place. Environmental change may cause changes in the composition and existence of various species sets, particularly if less adaptable species are eliminated because of the mutual dependency of members of species sets. Thus there may be considerable temporal change in species sets in habitats untouched by man. The effects of human disturbance would exacerbate the elimination of species sets, probably selecting those more generalized species adapted to secondary communities.

Effects of human intervention

The preceding section indicated that there may be a rapid recovery during the early stages of a secondary sere but the species compositions of the terminal stages involve a great deal of substitutability between various species or species assemblages. Thus, recovery from disturbance may well depend upon the initial species composition of the disturbed area. The substitutability has important consequences in two areas.

Direct effects

The species present in a disturbed area immediately after the disturbance are obviously dependent upon the original species composition. Often, both species richness and density decline precipitously after human intervention. Lasebikan (1975) studied the change in species composition and density of soil micro-arthropods in a Nigerian forest over a short period. One site remained undisturbed; on another one the underbrush and the small and medium sized trees were removed. The upper 10 cm of soil were sampled initially and 3, 8 and 10 months after treatment. Initially there was no difference between the two sites, but the soil fauna of that area declined both in species richness and in population density on the cleared site.

Howden and Nealis (1975) trapped scarabaeid beetles in undisturbed forests, in cleared areas within formerly forested sites, and in forests subject to annual flooding near Leticia, Colombia. The fauna of the forested areas were much more diverse than those of the other two sites with the fauna of the clearings being particularly depauperate in both species richness and density. Only six species were taken in the clearing, four of which were not found in the adjacent forest.

This decline in species diversity is significant for economic management, particularly in agriculture and in tourism. The use of native species to regulate the levels of pests of agricultural products is a prime example. Strong (1974) found that the number of insect pests found on cacao (*Theobroma cacao*) in regions where it was cultivated quickly rose to an asymptotic value which was a function of the area in cultivation or the annual production. Most of these pests were local in origins.

Most agricultural pests, particularly of cacao, in Malaysia, are members of the secondary communities (Conway, 1972). Cacao plantations simulate the characteristics of local secondary communities and insects living on plants of this stage of the successional sere invaded the cacao trees. Chemical control programmes were ineffective in permanently reducing the level of the pests, presumably because the local predators were able to regulate pest numbers.

Island effects

The long-term effect of forest clearance is to produce a mosaic of forested and open areas. Thus, the remaining forested areas may well be viewed as islands (Simberloff, 1974). MacArthur and Wilson (1967) provided a useful and influential model of island communities which may have great importance for understanding the functioning of terrestrial islands. In their model the number of species present is the result of the interaction of the rate at which the island is colonized by new species and the rate at which species disappear from it. These rates are affected by other factors such as area, distance from a colonizing source and topographic heterogeneity.

Diamond (1973) studied the bird fauna of New Guinea and its off-shore islands and made some suggestions that are directly relevant to wooded terrestrial areas on islands. Many of the islands he studied had more than the predicted number of species, and these were islands that experienced recent change in area due to a wide variety of causes (decrease in area of island because of volcanic eruptions, rise in sea level, etc.). Diamond states 'On such islands S (the number of bird species) must be gradually returning to equilibrium as a result of a temporary imbalance between immigration and extinction; this process may be termed relaxation'. When areas are subdivided, a decline in species abundance will occur until the species abundance is in accord with immigration and extinction rates for islands of that size.

Isolated, complex forest communities show a decline in the numbers of species present over a period of several decades. Myers and Rand (1969) and Willis (1974) have shown that the number of species of reptiles and amphibians and of birds, respectively, has declined on Barro Colorado Island since it was set aside as a nature preserve. This 15.6 km² hill-top was isolated from the surrounding forests when Gatun Lake filled with water in 1910–1914. For about ten years thereafter, part of it was farmed before it became a preserve in 1923. Many species which inhabited second growth successional stages have disappeared as those stages were replaced by maturing forests. In addition, invasion of the island by species from the mainland has been minimal, even though the distance between them is slight. Thus, even undisturbed areas may decline in species abundance. The disappearance of one species, if it is a member of some interacting complex such as a mutualistic one, also means the disappearance of associated species.

Suggested research

The effects of disturbance briefly enumerated above suggest several lines of research necessary for understanding the management of tropical systems.

Species assemblages. There is little concrete information about the reality of interacting assemblages of species in mature systems. There is even less information available about the substitutability of such assemblages. Their investigation should provide information about the likely composition of derived faunas in damaged ecosystems.

Investigation of the direction and rate of recovery of disturbed forest ecosystems will provide information necessary to estimate the impact of wide-spread deforestation. Information to be gathered at all seral stages should include species composition and demography, dispersal and trophic characteristics of the different species. The design of these studies should include not only single manipulations of the ecosystems but also successive manipulations with varying intervals between them. Such studies would duplicate the serial attacks humans make upon forest systems. The manipulations themselves should be varied and should include selective and clear cutting, fire, grazing, hunting, mineral exploration, pest introduction, food gathering, recreation and chemical control of vegetation, and animals.

Finally, the effect of forest size on maintaining diversity

must be fully investigated. Diamond (1973) has stressed the idea that forest reserves should be as large as possible to maximize species richness. A factor which has apparently not yet been considered in such studies is what might prove to be a distinction between apparent area and biological area. The former would be the physical area as measured in hectares. The latter would be the area within that physical area which is really utilized by a particular species or species assemblage. For example, forest preserves set aside for large vertebrates may actually represent smaller preserves than indicated by the preserve boundaries because of incursions by man from the outside. Thus, hunting, utilization of preserves by cattle, pigs, etc., or similar human activities effectively reduce the area of the preserve by eliminating certain protected species (e.g. the Wilpattu National Park in Sri Lanka, Eisenberg and Lockhart, 1972). See chapter 21.

Bibliography

ALCALA, Angel, C.; BROWN, Walter, C. Population ecology of the tropical scincoid lizard, *Emoia atrocostata*, in the Philippines. *Copeia* (Washington, D.C.), 3, September 1967, p. 596–604.

ANDREWS, Robin, M. Structural habitat and time budget of a tropical *Anolis* lizard. *Ecology* (Durham), vol. 52, no. 2, 1971, p. 262–270.

BAKER, H. G. *Plants and civilization.* Belmont, California, Woodsworth Pub. Co., 1964.

——; BAKER, I.; CRUDEN, R. W. Minor parasitism in pollination biology and its community function: the case of *Ceiba acuminata*. *Bio-Science* (Washington, D.C.), vol. 21, no. 22, 15 November 1971, p. 1127–1129.

BENNETT, Charles, F. Human influences on the zoogeography of Panama. *Ibero-Americana*, 51, 1968, p. 1–112.

BENSON, Woodruff, W.; EMMEL, Thomas, C. Demography of gregariously roosting populations of the nympholine butterfly *Marpesia berania* in Costa Rica. *Ecology* (Durham), vol. 54, no. 2, 1973, p. 326–335.

BENTLEY, B. The protective function of ants visiting the extrafloral nectaries of *Bixa orellana* L. (Bixaceae). Submitted to *J. Ecol.*, 1975a.

——. Extrafloral nectaries and plant life form. *J. Ecol.*, 1975b.

BERLIN, J. M.; HENRY, P. Répartition des levures à la surface de la tige de vigne. *C.R. Acad. Sci. Paris*, 277, 1973, p. 1885–1887.

BERRY, A. J. Population structure and fluctuations in the snail fauna of a Malayan limestone hill. *Journal of Zoology, Proceedings of the Zoological Society of London* (London), vol. 150, part 1, September 1966, p. 11–27.

BERRY, P. Y. The breeding patterns of seven species of Singapore Anura. *Journal of Animal Ecology* (Oxford), vol. 33, no. 2, June 1964, p. 227–243.

BLUM, Murray, S.; PADOVANI, Flavio; AMANTE, Elpidio. Alkanones and terpenes in the mandibular glands of *Atta* species (Hymenoptera: Formicidae). *Comparative Biochemistry and Physiology* (London), vol. 26, no. 1, July 1968, p. 291–299.

BROOKFIELD, H. C.; HART, D. *Rainfall in the tropical southwest Pacific.* Canberra, The Australian National University, Research School of Pacific Studies, Department of Geography, Publication G/3, 1966.

BROWN, Kathren, M.; SEXTON, Owen J. Stimulation of reproductive activity of female *Anolis sagrei* by moisture. *Physiological Zoology* (Chicago), vol. 46, no. 2, April 1973, p. 168–172.

*BRÜNIG, Eberhard, F. On the seasonality of droughts in the lowlands of Sarawak (Borneo). *Erdkunde* (Bonn), vol. 23, no. 2, June 1969, p. 127–133.

BURDON, J. J.; CHILVERS, G. A. Fungal and insect parasites contributing to niche differentiation in mixed species stands of eucalypt saplings. *Austr. J. Bot.*, 22, 1974, p. 103–114.

BURGER, W. C. Flowering periodicity at four altitudinal levels in eastern Ethiopia. *Biotropica*, 6, 1974, p. 38–42.

CAMPBELL, Howard, W. Ecological observations on *Anolis lionatus* and *Anolis poecilopus* (Reptilia, Sauria) in Panama. *American Museum Novitates* (New York), no. 2516, 4 April 1973, p. 1–29.

CARROLL, C. R. *Beetles, parasites and morning glories.* 1975, multigr.

CATES, R. G. The interface between slugs and wild ginger: some evolutionary aspects. *Ecology*, 1975, in press.

CHERRETT, J. M. The foraging behaviour of *Atta cephalotes* L. (Hymenoptera: Formicidae). I. Foraging pattern and plant species attacked in tropical rain forest. *Journal of Animal Ecology* (Oxford), vol. 37, no. 2, June 1968, p. 387–403.

——. Some factors involved in the selection of vegetable substrate by *Atta cephalotes* L. (Hymenoptera: Formicidae) in tropical rain forest. *Journal of Animal Ecology* (Oxford), vol. 41, no. 3, October 1972, p. 647–660.

CODY, M. L. A general theory of clutch size. *Evolution* (Lawrence), vol. 20, no. 2, June 1966, p. 174–184.

COLLETTE, B. B. Correlations between ecology and morphology in anoline lizards from Havana, Cuba and southern Florida. *Bulletin of the Museum of Comparative Zoology* (Cambridge, Mass.), 125, 1961, p. 137–162.

*COLWELL, Robert, K. Competition and coexistence in a simple tropical community. *American Naturalist* (Chicago), 107, November–December 1973, p. 737–760.

CONNELL, J. H. A predator-prey system in the marine intertidal region. I. *Balanus glandula* and several predatory species of *Thais. Ecol. Monogr.*, 40, 1970, p. 49–78.

——; ORIAS, Eduardo. The ecological regulation of species diversity. *American Naturalist* (Chicago), vol. XCVIII, no. 903, November–December 1964, p. 399–414.

CONWAY, Gordon, R. Ecological aspects of pest control in Malaysia. In: Farvar, M., Taghi; Milton, John, P. (eds.). *The careless technology.* Garden City, N.Y., The Natural History Press, 1972, 1 030 p.

COOK, Laurence, M.; FRANK, Kenneth; BROWER, Lincoln, P. Experiments on the demography of tropical butterflies. I. Survival rate and density in two species of *Parides. Biotropica* (Pullman, Wash.), vol. 3, no. 1, June 1971, p. 17–20.

CORN, J. J. Upper thermal limits and thermal preferenda for three sympatric species of *Anolis. Journal of Herpetology* (USA), vol. 5, no. 1, 1971, p. 17–21.

* Major reference.

CRUMP, Martha, L. Quantitative analysis of the ecological distribution of a tropical herpetofauna. *Occasional Papers of the Museum of Natural History* (Lawrence), 3, 1971, p. 1–62.

DELANY, M. J. A study of the ecology and breeding of small mammals in Uganda. *Journal of Zoology, Proceedings of the Zoological Society of London* (London), vol. 142, part 2, March 1964, p. 347–370.

——. The biology of small rodents in Mayanja Forest, Uganda. *Journal of Zoology, Proceedings of the Zoological Society of London* (London), vol. 165, part 2, September 1971, p. 85–129.

*DIAMOND, J. M. Comparison of faunal equilibrium turnover rates on a tropical island and a temperate island. *Proc. Nat. Acad. Sci.* (USA), 68, 1972, p. 2742–2745.

*——. Distributional ecology of New Guinea birds. *Science* (Washington, D.C.), vol. 179, no. 4075, 23 February 1973, p. 759–769.

*DOBBEN, W. H. van; LOWE-MCCONNELL, R. H. (eds.). *Unifying concepts in ecology*. Report of the plenary sessions of the first international congress of ecology (The Hague, September 8–14, 1974). The Hague, W. Junk B.V. Publishers; Wageningen, Centre for agricultural publishing and documentation, 1975, 302 p.

Chapters on 'Flow of energy and matter between trophic levels'; 'Comparative productivity in ecosystems'; 'Diversity, stability and maturity in natural ecosystems'; 'Diversity, stability and maturity in ecosystems influenced by human activities'; 'Strategies for management of natural and man-made ecosystems'.

*DOBZHANSKY, T. Evolution in the tropics. *American Scientist* (USA), 38, April 1950, p. 209–221.

——; PAVAN, C. Local and seasonal variations in relative frequencies of species of *Drosophila* in Brazil. *Journal of Animal Ecology* (Cambridge), vol. 19, no. 1, May 1950, p. 1–14.

*EHRLICH, P. R.; RAVEN, P. H. Butterflies and plants: a study in coevolution. *Evolution*, 18, 1965, p. 586–608.

——; GILBERT, Laurence. Population structure and dynamics of the tropical butterfly *Heliconius ethilla*. *Biotropica* (Pullman, Wash.), vol. 5, no. 2, September 1973, p. 69–82.

EISENBERG, John, F.; LOCKHART, Melvyn. An ecological reconnaissance of Wilpattu National Park, Ceylon. *Smithsonian Contributions to Zoology* (Washington, D.C.), no. 101, 1972.

——; THORINGTON, Richard, W. Jr. A preliminary analysis of a neotropical mammal fauna. *Biotropica* (Pullman, Wash.), vol. 5, no. 3, December 1973, p. 150–161.

ELTON, Charles, S. The structure of invertebrate populations inside neotropical rain forest. *Journal of Animal Ecology* (Oxford), vol. 42, no. 1, February 1973, p. 55–104.

EPPLE, August; ORIANS, Gordon, H.; FARNER, Donald, S.; LEWIS, R. A. The photoperiodic testicular response of a tropical finch, *Zonatrichia capensis costaricensis*. *The Condor* (Lawrence), vol. 74, no. 1, 1972, p. 1–4.

*FAEGRI, Knut; PILJ, L. van der. In: *The principles of pollination ecology*, p. 1–29. Second edition. Oxford, Pergamon Press, 1971, 248 p.

FEENY, P. Biochemical coevolution between plants and their insect herbivores. In: Gilbert, L. E.; Raven, P. H. (eds.). *Coevolution of animals and plants* (Symposium at First Int. Congr. Syst. and Evol. Biol., 1973, Boulder, Colorado). Austin, University of Texas Press.

FISCHER, A. G. Latitudinal variations in organic diversity. *Evolution* (Lancaster), vol. 14, no. 1, March 1960, p. 64–81.

FITCH, H. S. Observations on the population ecology of the Central American iguanid lizard *Anolis cupreus*. *Caribbean Journal of Science*, vol. 13, no. 3–4, 1973, p. 215–229.

FLEMING, Theodore, H. Notes on the rodent faunas of two Panamanian forests. *Journal of Mammalogy* (USA), vol. 51, no. 3, 28 August 1970, p. 473–490.

*——. Population ecology of three species of neotropical rodents. *Miscellaneous publications of the Museum of Zoology, University of Michigan* (Ann Arbor), no. 143, 31 August 1971, p. 1–77.

*——. Numbers of mammal species in North and Central American forest communities. *Ecology* (Durham), vol. 54, no. 3, 1973, p. 555–563.

——. The population ecology of two species of Costa Rican heteromyid rodents. *Ecology* (Durham), vol. 55, no. 3, 1974, p. 493–510.

——; HOOKER, Roderick, S. *Anolis cupreus*: the response of a lizard to tropical seasonality. *Ecology*, in press.

*FOGDEN, M. P. L. The seasonality and population dynamics of equatorial forest birds in Sarawak. *Ibis* (London), vol. 114, no. 3, 1972, p. 307–343.

FUTUYMA, Douglas, J. Community structure and stability in constant environments. *American Naturalist* (Chicago), vol. 107, no. 955, May–June 1973, p. 443–446.

GALIL, J.; EISIKOWITCH, D. On the pollination ecology of *Ficus sycomorus* in East Africa. *Ecology* (Durham), vol. 49, no. 2, 1968, p. 259–269.

GENTRY, A. H. Flowering phenology and diversity in tropical Bignoniaceae. *Biotropica*, vol. 6, 1974, p. 64–68.

GILBERT, L. E. Pollen feeding and reproductive biology of *Heliconius* butterflies. *Proc. Nat. Acad. Sci.* (USA), 69, 1972, p. 1403–1407.

GORMAN, G. C.; LICHT, P.; DESSAUER, H.; BOOS, J. O. Reproductive failure among the hybridizing *Anolis* lizards of Trinidad. *Systematic Zoology* (Lawrence), vol. 20, no. 1, March 1971, p. 1–18.

——; LICHT, P. Seasonality in ovarian cycles among tropical *Anolis* lizards. *Ecology* (Durham), 55, 1974, p. 360–369.

GREENWALT, Crawford, H. *Hummingbirds*. Garden City, N.Y., Doubleday, 1960, 250 p.

*HARRISON, J. L. The distribution of feeding habits among animals in a tropical rain forest. *Journal of Animal Ecology* (Oxford), vol. 31, no. 1, February 1962, p. 53–63.

HEATWOLE, Harold, F.; SEXTON, Owen, J. Herpetofaunal comparisons between two climatic zones in Panama. *American Midland Naturalist* (Notre Dame), vol. 75, no. 1, January 1966, p. 45–60.

*HEINRICH, Bernd; RAVEN, Peter, H. Energetics and pollination ecology. *Science* (Washington, D.C.), vol. 176, no. 4035, 12 May 1972, p. 597–602.

HEITHAUS, E., Raymond; OPLER, Paul, A.; BAKER, Herbert, G. Bat activity and pollination of *Bauhinia pauletia*: plant-pollinator coevolution. *Ecology* (Durham), vol. 55, no. 2, 1974, p. 412–419.

HILL, M. O. Diversity and evenness: a unifying notation and its consequences. *Ecology* (Durham), vol. 54, no. 2, 1973, p. 427–432.

HOWDEN, H. F.; NEALIS, V. G. Effects of clearing on the composition of the coprophagous scarab beetle fauna (Coleoptera). *Biotropica*, 7, 1975, p. 77–83.

HUFFAKER, C. B. (ed.). *Biological control*. Proc. Amer. Assoc. Adv. Sci. Symp. Biol. Control. New York, Plenum Press, 1970, 511 p.

HURLBERT, Stuart, H. The nonconcept of species diversity: a critique and alternative parameters. *Ecology* (Durham), vol. 52, no. 4, 1971, p. 577–586.

INGER, Robert, F.; GREENBERG, Bernard. The annual reproductive pattern of the frog *Rana erythraea* in Sarawak. *Physiological Zoology* (Chicago), vol. 36, no. 1, January 1963, p. 21–33.

——; ——. Ecological and competitive relations among three species of frogs (genus *Rana*). *Ecology* (Durham), vol. 47, no. 5, 1966a, p. 746–759.

*——; ——. Annual reproductive patterns of lizards from a Bornean rain forest. *Ecology* (Durham), vol. 47, no. 6, 1966b, p. 1007–1021.

——; BACON, James, P. Jr. Annual reproduction and clutch size in rain forest frogs from Sarawak. *Copeia* (Washington, D.C.), 3, 31 August 1968, p. 602–606.

*JANZEN, Daniel, H. Synchronization of sexual reproduction of trees within the dry season in Central America. *Evolution* (Lawrence), vol. 21, no. 3, 27 September 1967a, p. 620–637.

*——. Interaction of the bull's-horn acacia (*Acacia cornigera* L.) with an ant inhabitant (*Pseudomyrmex ferruginea* F. Smith) in eastern Mexico. *University of Kansas Science Bulletin* (Lawrence), vol. 47, no. 6, October 1967b, p. 315–558.

——. Fire, vegetation structure, and the ant *x* acacia interaction in Central America. *Ecology* (Durham), vol. 48, no. 1, 1967c, p. 26–35.

——. Allelopathy by myrmecophytes: the ant *Azteca* as an allelopathic agent of *Cecropia*. *Ecology* (Durham), vol. 50, no. 1, 1969a, p. 147–153.

——. Birds and the ant *x* acacia interaction in Central America, with notes on birds and other myrmecophytes. *The Condor* (Lawrence), vol. 71, no. 3, July 1969b, p. 240–256.

*——. Herbivores and the number of tree species in tropical forests. *American Naturalist* (Chicago), vol. 104, no. 940, November–December 1970a, p. 501–528.

——. Comments on host-specificity of tropical herbivores and its relevance to species richness. In: Heywood, V. H. (ed.). *Taxonomy and ecology*. London, Academic Press, 1970b.

*——. Escape of juvenile *Dioclea megacarpa* (Leguminosae) vines from predators in a deciduous tropical forest. *American Naturalist* (Chicago), vol. 105, no. 942, March–April 1971a, p. 97–112.

*——. Seed predation by animals. In: Johnston, R. F.; Frank, P. W.; Michener, C. D. (eds.). *Annual review of ecology and systematics*, vol. 2. Palo Alto, Annual Reviews Inc., 1971b, 510 p.

——. Euglossine bees as long-distance pollinators of tropical plants. *Science* (Washington, D.C.), vol. 171, no. 3967, 15 January 1971c, p. 203–205.

——. Escape in space by *Sterculia apetala* seeds from the bug *Dysdercus fasciatus* in a Costa Rican deciduous forest. *Ecology* (Durham), 53, 1972a, p. 350–361.

——. Sweep samples of tropical foliage insects: effects of seasons, vegetation types, elevation, times of day, and insularity. *Ecology*, 54, 1973, p. 687–700.

——. Host plants as islands. II. Competition in evolutionary and contemporary time. *American Naturalist* (Chicago), 107, 1973a, p. 786–788.

*——. Tropical agroecosystems. *Science*, 182, 1973b, p. 1212–1219.

JENSSEN, Thomas, A. The ethoecology of *Anolis nebulosus* (Sauria, Iguanidae). *Journal of Herpetology* (USA), vol. 4, no. 1, 1970, p. 1–38.

JOHNSON, Leslie, K.; HUBBELL, Stephen, P. Aggression and competition among stingless bees: field studies. *Ecology* (Durham), 55, 1974, p. 120–127.

JONES, Clyde. Comparative ecology of three pteropid bats in Rio Muni, West Africa. *Journal of Zoology, Proceedings of the Zoological Society of London* (London), vol. 167, part 3, July 1972, p. 353–370.

JONES, D. A. Selective eating of the acyanogenic form of the plant *Lotus corniculatus* L. by various animals. *Nature* (London), 193, 1962, p. 1109–1110.

——. On the polymorphism of cyanogenesis in *Lotus corniculatus*. I. Selection by animals. *Canad. J. Genet. Cytol.*, 8, 1966, p. 556–567.

KARR, James, R.; ROTH, Roland, R. Vegetation structure and avian diversity in several New World areas. *American Naturalist* (Chicago), vol. 105, no. 495, September–October 1971, p. 423–435.

KAUFMAN, J. H. The ecology and social behavior of the coati, *Nasua narica*, on Barro Colorado Island, Panama. *University of California Publications in Zoology* (Berkeley), 60, 1962, p. 95–222.

LANDAHL, John, T.; ROOT, Richard, B. Differences in the life tables of tropical and temperate milkweed bugs, genus *Oncopeltus* (Hemiptera: Lygaeidae). *Ecology* (Durham), vol. 50, no. 4, 1969, p. 734–737.

LASEBIKAN, B. A. The effect of clearing on the soil arthropods of a Nigerian rain forest. *Biotropica*, 7, 1975, p. 84–89.

LAZELLE, J. D., Jr. The anoles (Sauria, Iguanidae) of the Lesser Antilles. *Bulletin of the Museum of Comparative Zoology* (Cambridge, Mass.), no. 143, 1972, p. 1–115.

LECK, Charles, F. Seasonal changes in feeding pressures of fruit- and nectar-eating birds in Panama. *The Condor* (Lawrence), vol. 74, no. 1, 1972, p. 54–60.

LEON, Juan, R.; RUIZ, Lila, J. Reproducción de la lagartija, *Ameiva bifrontata* (Sauria: Teiidae). *Caribbean Journal of Science*, vol. 11, no. 3–4, September–December 1971, p. 195–201.

*LEVINS, Richard. *Evolution in changing environments; some theoretical explorations*. Princeton, Princeton University Press, 1968, 120 p.

*LICHT, P.; GORMAN, G. C. Reproductive and fat cycles in Caribbean *Anolis* lizards. *University of California Publications in Zoology* (Berkeley), 95, 23 October 1970, p. 1–52.

LINDSEY, C. C. Body sizes of poikilotherm vertebrates at different latitudes. *Evolution* (Lawrence), vol. 20, no. 4, December 1966, p. 456–465.

LLOYD, Monte; INGER, Robert, F.; KING, F., Wayne. On the diversity of reptile and amphibian species in a Bornean rain forest. *American Naturalist* (Chicago), vol. 102, no. 928, November–December 1968, p. 497–516.

MACARTHUR, Robert, H. Patterns of species diversity. *Biological Reviews* (USA), 40, 1965, p. 510–533.

*——. *Geographical Ecology*. New York, Harper and Row, 1972, 269 p.

*——; MACARTHUR, J. W. On bird species diversity. *Ecology* (Durham), vol. 42, no. 3, July 1961, p. 594–598.

——; RECHER, H.; CODY, M. On the relation between habitat selection and species diversity. *American Naturalist* (Chicago), vol. 100, no. 913, July–August 1966, p. 319–332.

*——; WILSON, Edward, O. *The theory of island biogeography*. Princeton, Princeton University Press, 1967, 203 p.

MARES, M. A.; WILSON, D. E. Bat reproduction during the Costa Rican dry season. *Bio-Science* (Washington, D.C.), vol. 21, no. 10, 15 May 1971, p. 471–477.

MARKL, Hubert. Stridulation in leaf-cutting ants. *Science* (Washington, D.C.), vol. 149, no. 3690, 17 September 1965, p. 1392–1393.

MARTIN, Michael, M. The biochemical basis of the fungus-attine ant symbiosis. *Science* (Washington, D.C.), vol. 169, no. 3940, 3 July 1970, p. 16–20.

*MAY, Robert, M. *Diversity, stability and maturity in natural ecosystems, with particular reference to the tropical moist forests.* Rome, FAO, 1975, 9 p. multigr.

MEDWAY, Lord. Observations of social and reproductive biology of the bent-winged bat *Miniopterus australis* in northern Borneo. *Journal of Zoology, Proceedings of the Zoological Society of London* (London), vol. 165, part 2, October 1971, p. 261–273.

MILLER, A. H. Capacity for photoperiodic response and endogenous factors in the reproductive cycles of an equatorial sparrow. *Proceedings of the National Academy of Science* (Washington, D.C.), 54, 1965, p. 97–101.

MOREAU, R. E. Clutch size: a comparative study, with special reference to African birds. *Ibis* (London), 86, 1944, p. 286–347.

——. Ecological isolation in a rich tropical avifauna. *Journal of Animal Ecology* (Cambridge), 17, 1948, p. 113–126.

*——. The breeding season of African birds. I. Land birds. *Ibis* (London), 92, 1950, p. 223–267.

MOSER, J. C.; SILVERSTEIN, R. M. Volatility of trail marking substances of the town ant. *Nature* (London), vol. 215, no. 5097, 8 July 1967, p. 206–207.

——; BROWNLEE, R. C.; SILVERSTEIN, R. M. Alarm pheromones of the ant *Atta texana. Journal of Insect Physiology* (Oxford), vol. 14, no. 4, April 1968, p. 529–535.

MUTERE, Festo, A. The breeding biology of equatorial vertebrates: reproduction in the fruit bat, *Eidolon helvum* at latitude 0°20′N. *Journal of Zoology, Proceedings of the Zoological Society of London* (London), vol. 153, part 2, October 1967, p. 153–161.

MYERS, C. W.; RAND, A. S. Checklist of amphibians and reptiles of Barro Colorado Island, Panama, with comments on faunal change and sampling. *Smithsonian Contributions to Zoology* (Washington, D.C.), no. 10, 1969, p. 1–11.

OKIA, N. O. The breeding pattern of the soft-furred rat, *Praomys morio* in an evergreen forest in southern Uganda. *Journal of Zoology, Proceedings of the Zoological Society of London* (London), vol. 170, part 4, August 1973, p. 501–504.

OLIVER, J. A. The anoline lizards of Bimini, Bahamas. *American Museum Novitates* (New York), no. 1383, 24 September 1948, p. 1–36.

*OPLER, Paul, A.; BAKER, Herbert, G.; FRANKIE, Gordon, W. *Recovery of tropical lowland forest ecosystems.* Blacksburg, Virginia, USA Center for Environmental Studies, 1976.

OWEN, D. F. Bimodal occurrence of breeding in an equatorial land snail. *Ecology* (Durham), vol. 45, no. 4, 1964, p. 862.

——; OWEN, J.; CHANTER, D. O. Seasonal changes in relative abundance and estimates of species in a family of tropical butterflies. *Oikos* (Copenhagen), vol. 23, no. 2, 1972, p. 200–205.

OWINY, A. M. Some aspects of the breeding biology of the equatorial land snail, *Limicolaria martensiana* (Achatinidae: Pulmonata). *Journal of Zoology, Proceedings of the Zoological Society of London* (London), vol. 172, part 2, February 1974, p. 191–206.

*PAINE, Robert, T. Food web complexity and species diversity. *American Naturalist* (Chicago), vol. 100, no. 910, January–February 1966, p. 65–75.

PETERS, James, A.; DONOSO-BARROS, Roberto. *Catalogue of the neotropical squamata, Parts I and II. U.S. National Museum Bulletin*, 292, 1970.

*PIANKA, Eric, R. Latitudinal gradients in species diversity: a review of concepts. *American Naturalist* (Chicago), vol. 100, no. 910, January–February 1966, p. 33–46.

PIELOU, E. C. *An introduction to mathematical ecology.* New York, Wiley-Interscience, 1969, 286 p.

PIPKIN, Sarah, Bedichek. The influence of adult and larval food habits on population size of neotropical ground-feeding *Drosophila. American Midland Naturalist* (Notre Dame), vol. 74, no. 1, July 1965, p. 1–27.

POOLE, Robert, W. *An introduction to quantitative ecology.* New York, McGraw-Hill, 1974, 532 p.

*PRESTON, F. W. The commonness, and rarity of species. *Ecology* (Durham), vol. 29, no. 3, July 1948, p. 254–283.

——. Time and space and the variation of species. *Ecology* (Durham), vol. 41, no. 4, October 1960, p. 611–627.

PROCTOR, Michael, C.; YEO, Peter. *The pollination of flowers.* New York, Taplinger Publishing Co., 1973, 418 p.

RAMIREZ, William, B. Fig wasps: mechanism of pollen transfer. *Science* (Washington, D.C.), vol. 163, no. 867, 7 February 1969, p. 580–581.

*RAND, A. S. Ecological distribution in anoline lizards of Puerto Rico. *Ecology* (Durham), vol. 45, no. 4, 1964, p. 745–752.

——. Ecology and social organization in the iguanid lizard *Anolis lineatopus. Proceedings U.S. National Museum* (Washington, D.C.), vol. 122, no. 3959, 1967a, p. 1–77.

——. The ecological distribution of the anoline lizards around Kingston, Jamaica. *Breviora* (Cambridge, Mass.), 272, 17 November 1967b, p. 1–18.

——; RAND, Patricia, J. Aspects of the ecology of the iguanid lizard *Tropidurus torquatus* at Belém, Pará. *Smithsonian Miscellaneous Collection* (Baltimore), vol. 151, no. 2, 8 July 1966, p. 1–16.

——; ——. Field notes on *Anolis lineatus* in Curaçao. *Fauna of Curaçao and other Caribbean islands* (Netherlands), 24, 1967, p. 112–117.

——; HUMPHREY, Steven, S. Interspecific competition in the tropical rain forest: ecological distribution among lizards at Belém, Pará. *Proceedings U.S. National Museum* (Washington, D.C.), vol. 125, no. 3658, 1968, p. 1–17.

——; WILLIAMS, Ernest, E. The anoles of La Palma: aspects of their ecological relationships. *Breviora* (Cambridge, Mass.), 327, 1969, p. 1–19.

REHR, S. S.; FEENEY, P. P.; JANZEN, D. H. Chemical defense in Central American non-ant-acacias. *Journal of Animal Ecology* (Oxford), vol. 42, no. 2, June 1973, p. 405–416.

RETTENMEYER, Carl, W. Observations on the biology and taxonomy of flies found over swarm raids of army ants (Diptera: Tachinidae, Conopidae). *University of Kansas Science Bulletin* (Lawrence), vol. 42, no. 1, 29 December 1961, p. 993–1066.

——. The diversity of arthropods found with neotropical army ants and observations on the behavior of representative species. *Proceedings North Central Branch of the Ecological Society of America* (USA), 17, 1962a, p. 14–15.

——. The behavior of millipedes found with neotropical army ants. *Journal of the Kansas Entomological Society* (USA), vol. 35, no. 4, October 1962b, p. 377–384.

——. The behavior of Thysanura found with army ants. *Annals of the Entomological Society of America* (Lawrence), vol. 56, no. 2, March 1963, p. 170–174.

ROCKWOOD, Larry, L. Distribution, density, and dispersion of two species of *Atta* (Hymenoptera: Formicidae) in Guanacaste Province, Costa Rica. *Journal of Animal Ecology* (Oxford), vol. 42, no. 3, October 1973, p. 802–817.

Rood, J. P.; Test, F. H. Ecology of the spiny ray *Heteromys anomalus*, at Rancho Grande, Venezuela. *American Midland Naturalist* (Notre Dame), vol. 79, no. 2, January 1968, p. 89–102.

Ruibal, Rodolfo. Thermal relations of five species of tropical lizards. *Evolution* (Lawrence), vol. 15, no. 1, March 1961, p. 98–110.

——; Philibosian, Richard; Adkins, Janet, L. Reproductive cycle and growth in the lizard *Anolis acutus*. *Copeia* (Washington, D.C.), 3, 8 September 1972, p. 509–518.

Sanders, H. L. Marine benthic diversity: a comparative study. *American Naturalist* (Chicago), vol. 102, no. 125, May–June 1968, p. 243–282.

——. Benthic marine diversity and the stability-time hypothesis. In: *Diversity and stability in ecological systems*, p. 71–81. Report of a symposium held May 26–28 1969, Upton, N.Y., Biol. Dept., Brookhaven National Laboratory. Brookhaven Symposia in Biology, no. 22.

*Schoener, Thomas, W. The *Anolis* lizards of Bimini: resource partitioning in a complex fauna. *Ecology* (Durham), vol. 49, no. 4, 1968, p. 704–726.

——. Size patterns in West Indian *Anolis* lizards. II. Correlations with the sizes of particular sympatric species—displacement and convergence. *American Naturalist* (Chicago), vol. 104, no. 936, March–April 1970, p. 155–174.

——. Large-billed insectivorous birds: a precipitous diversity gradient. *The Condor* (Lawrence), vol. 73, no. 2, 1971, p. 154–161.

*——. Resource partitioning in ecological communities. *Science* (Washington, D.C.), vol. 185, no. 4145, 5 July 1974, p. 27–39.

——; Gorman, G. C. Some niches differences in three Lesser Antillean lizards of the genus *Anolis*. *Ecology* (Durham), vol. 49, no. 5, 1968, p. 819–830.

*——; Janzen, Daniel, H. Notes on environmental determinants of tropical versus temperate insect size patterns. *American Naturalist* (Chicago), vol. 102, no. 295, May–June 1968, p. 207–224.

——; Schoener, Amy. Structural habitats of West Indian *Anolis* lizards. I. Lowland Jamaica. *Breviora* (Cambridge, Mass.), vol. 368, no. 1, 1971, p. 1–53.

Sexton, Owen, J. Population changes in a tropical lizard *Anolis limifrons* on Barro Colorado Island, Panama Canal Zone. *Copeia* (Washington, D.C.), 1, 20 March 1967, p. 219–222.

——; Heatwole, H. F.; Meseth, Earl, H. Seasonal population changes in the lizard, *Anolis limifroms*, in Panama. *American Midland Naturalist* (Notre Dame), vol. 69, no. 2, April 1963, p. 482–491.

——; ——; Knight, Dennis. Correlation of microdistribution of some Panamanian reptiles and amphibians with structural organization of the habitat. *Caribbean Journal of Science*, vol. 4, no. 1, March 1964, p. 261–295.

——; Heatwole, Harold, F. An experimental investigation of habitat selection and water loss in some anoline lizards. *Ecology* (Durham), vol. 49, no. 4, 1968, p. 762–767.

*——; Ortleb, Edward, P.; Hathaway, Loline, M.; Ballinger, R. E.; Licht, Paul. Reproductive cycles of three species of anoline lizards from the isthmus of Panama. *Ecology* (Durham), vol. 52, no. 2, 1971, p. 201–215.

——; Bauman, Joan; Ortleb, Edward. Seasonal food habits of *Anolis limifrons*. *Ecology* (Durham), vol. 53, no. 1, 1972, p. 182–186.

*Simberloff, Daniel, S. Equilibrium theory of island bioge- ography and ecology. *Annual Review of Ecology and Systematics* (Palo Alto, California), 1974, p. 161–182.

Simpson, George, G. Species density of North American recent mammals. *Systematic Zoology* (Lawrence), vol. 13, no. 2, June 1964, p. 57–73.

Slater, J. Lygaeid bugs (Hemiptera: Lygaeidae) as seed predators of figs. *Biotropica*, 4, 1972, p. 145–151.

Smith, J. Maynard. *Models in ecology*. Cambridge, Cambridge University Press, 1974, 146 p.

Smith, Neal, Griffith. Migrations of the day-flying moth *Urania* in Central and South America. *Caribbean Journal of Science*, vol. 12, no. 1–2, June 1972, p. 45–51.

Smith, Norman, S.; Buss, Irven, O. Reproductive ecology of the female African elephant. *Journal of Wildlife Management* (Washington, D.C.), vol. 37, no. 4, October 1973, p. 524–534.

*Smythe, Nicholas. Relationships between fruiting seasons and seed dispersal methods in a neotropical forest. *American Naturalist* (Chicago), vol. 104, no. 935, January–February 1970a, p. 25–35.

——. The adaptive value of the social organization of the coati (*Nasua narica*). *Journal of Mammalogy* (Lawrence), vol. 51, no. 4, November 1970b, p. 818–820.

Southwood, T. R. E. *Ecological methods*. London, Methuen, 1966, 391 p.

Spight, Tom, M. Species diversity: a comment on the role of the predator. *American Naturalist* (Chicago), vol. 101, no. 992, November–December 1967, p. 467–474.

Strong, Donald, R., Jr. Rapid asymptotic species accumulation in phytophagous insect communities: the pests of cacao. *Science* (Washington, D.C.), vol. 185, no. 4156, 20 September 1974, p. 1064–1066.

Struhsaker, Thomas, T. A recensus of vervet monkeys in the Masai-Amboseli Game Reserve, Kenya. *Ecology* (Durham), vol. 54, no. 4, 1973, p. 930–932.

Tahvanainen, Jorma, O.; Root, Richard, B. The influence of vegetational diversity on the population ecology of a specialized herbivore, *Phyllotreta cruciferae* (Coleoptera; Chrysomelidae). *Oecologia* (Berlin), vol. 10, no. 4, 1972, p. 321–346.

Tamsitt, J. R.; Valdivieso, Dario. The male reproductive cycle of the bat, *Artibeus lituratus*. *American Midland Naturalist* (Notre Dame), vol. 73, no. 1, January 1965, p. 150–160.

*Terborgh, John. Distribution on environmental gradients: theory and a preliminary interpretation of distributional patterns in the avifauna of the Cordillera Vilacabamba, Peru. *Ecology* (Durham), vol. 52, no. 1, 1970, p. 23–40.

*——; Diamond, J. M. Niche overlap in feeding assemblages of New Guinea birds. *The Wilson Bulletin* (USA), vol. 82, no. 1, March 1970, p. 29–52.

Tramer, Elliott, J. On latitudinal gradients in avian diversity. *The Condor*, 75, 1974, p. 123–130.

Turner, John, R. G. Experiments on the demography of tropical butterflies. II. Longevity and home-range behaviour in *Heliconius erato*. *Biotropica* (Pullman, Wash.), vol. 3, no. 1, June 1971, p. 21–31.

*Van der Pilj, L.; Dodson, C. H. *Orchid flowers, their pollination and evolution*. Coral Gables, Florida, Univ. of Miami Press, 1966.

Voous, K. H. The breeding seasons of birds in Indonesia. *Ibis* (London), 92, 1950, p. 279–287.

Wake, David, B. The abundance and diversity of tropical salamanders. *American Naturalist* (Chicago), vol. 104, no. 936, March–April 1970, p. 211–213.

WARD, Peter. The annual cycle of the yellow-vented bulbul *Pycnonotus goiavier* in a humid equatorial environment. *Journal of Zoology, Proceedings of the Zoological Society of London* (London), vol. 157, part 2, January 1969, p. 25–45.

WATTERS, R. F. *Shifting cultivation in Latin America*. Rome, FAO, 1971, 305 p.

WEBER, Neal, A. Ecological relations of three *Atta* species in Panama. *Ecology* (Durham), vol. 50, no. 1, 1969, p. 141–147.

——. The attines: the fungus-culturing ants. *American Scientist* (USA), vol. 60, no. 4, July–August 1972, p. 448–456.

WILLIS, Edwin, O. The role of migrant birds at swarms of army ants. *The Living Bird* (Ithaca, N.Y.), 5, October 1966a, p. 187–231.

——. Interspecific competition and the foraging behavior of plain-brown wood creepers. *Ecology* (Durham), vol. 47, no. 4, 1966b, p. 667–672.

——. Competitive exclusion and birds at fruiting trees in western Colombia. *Auk* (Lawrence), 83, 1966c, p. 479–480.

——. Taxonomy, ecology and behavior of the sooty ant-tánager (*Habia gutturalis*) and other ant-tanagers (Aves). *American Museum Novitates* (New York), no. 2480, 3 February 1972, p. 1–38.

*WILLIS, Edwin, O. Populations and local extinctions of birds on Barro Colorado Island, Panama. *Ecological Monographs* (Durham), 44, 1974, p. 153–169.

——; ONIKI, Yoshika. Ecology and nesting behavior of the chestnut-backed antbird (*Myrmeciza exsul*). *The Condor* (Lawrence), vol. 74, no. 1, 1972, p. 87–98.

WILSON, D. E. Bat faunas: a trophic comparison. *Systematic Zoology* (Lawrence), 22, 1973, p. 14–29.

——; JANZEN, Daniel, H. Predation on *Scheelea* palm seeds by bruchid beetles: seed density and distance from the parent palm. *Ecology* (Durham), vol. 53, no. 5, 1972, p. 954–959.

*WILSON, Edward, O. Patchy distributions of ant species in New Guinea rain forests. *Psyche* (USA), 65, March 1958, p. 26–38.

*——. *The insect societies*. Cambridge, Mass., Belknap Press, 1971, 548 p.

YOUNG, Allen, M. Breeding success and survivorship in some tropical butterflies. *Oikos* (Copenhagen), vol. 23, no. 3, 1972, p. 318–326.

——; MUYSHONDT, Alberto. Notes on the biology of *Morpho peleides* in Central America. *Caribbean Journal of Science*, vol. 13, no. 1–2, June 1973, p. 1–50.

8 The natural forest: plant biology, regeneration and tree growth

Introduction

Scope; relevance to land development

This chapter reviews present knowledge of the processes of change in both time and space in rain forests. This includes short-term changes with seasons, of leaf production and fall, flowering and fruiting, and their variation from year to year; the changes in composition brought about by the life-cycles of the trees and by competition; changes wrought by such natural catastrophes as typhoons and volcanoes; and briefly, geographical variation in composition and its relation to evolutionary history and the functional significance of the diversity of composition which distinguishes rain forest from all other ecosystems.

An understanding of the internal dynamics forms the scientific basis for sylvicultural management: studies of the growth of timber species in natural forests and plantations, and of whole forests in dynamic equilibrium and after timber extraction, have already been made in several countries. Their purpose has been to calculate the rotation period in order to estimate future yields of commercial timber, or to estimate the effects of sylvicultural treatment. A critical review of this work and of methods of measuring growth is included. The still largely unsolved problem of ensuring adequate natural regeneration following felling requires detailed knowledge of flowering and fruiting phenology, of the processes and conditions of germination, establishment and subsequent competition and growth of the preferred species. In the foreseeable future, as the regenerated and refined forests reach maturity, problems of nutrient conservation, of pest epidemics, and of tree improvement will arise; all these require knowledge of forest dynamics.

The interests of commercial forestry are incompatible with the conservation of all aspects of primary tropical rain forest, as the process of sylvicultural refinement to enhance future timber yield inevitably leads to the impoverishment or extinction of non-timber species, and to major changes in animal population densities also. With the growing realization that the diversity of primary tropical rain forest provides the most promising remaining resource for agricultural, as well as forestry, tree crop diversification in the humid tropics, research into the breeding systems, phenology, autecology and physiology of rain forest trees has assumed a new importance. It is becoming apparent too that the hill land of the humid, and especially aseasonal, tropics cannot generally support permanent intensive herbaceous agriculture. Nevertheless traditional tropical hill farming systems have been successful for centuries in the past. These always involve periods of fallow during which a succession reconstitutes soil fertility; this is the subject of chapter 9. Nevertheless, the species of this fallow period originate from the gap phase of the primary forest cycle, and an understanding of their biology and of forest dynamics is essential for research into this important field. Scientific management and conservation, not only of managed forests and plantations as a timber resource but also inviolate primary forests as a genetic resource, is now seen to be essential to man's very survival in the tropical hill lands and hence an integral part of land-use policy. Knowledge of the internal dynamics of these forests, and particularly of their population structure and the breeding systems of their species is fundamental to these objectives.

Change in tropical forests

Rain forest is usually of such stature and complexity, that it appears permanent and static. But this is deceptive, for it is continuously changing; a mass equivalent to that of the whole forest dies and is renewed every 40–100 years.

The pattern of change is extremely complex, for it is brought about by many different processes which operate at different scales; some affect individual trees, others species populations, and the scale of this may vary between species; others may affect part or all of a forest, or yet the forests of a region. On the smallest time scale are phenological changes, that is seasonal or periodic changes in leaf, flower and fruit, distinctive for each species. This, through the timing and form of seed production and dispersal, will affect the spatial arrangement and composition of the regeneration and future forest. In part this will also be determined by the inherited competitive characteristics of the species including their intrinsic pattern of change in leaf, habit and phenology from seedling to maturity (ontogenesis) and by opportunities for establishment created by the chance death of an emergent tree. In the long term these processes will interact through natural selection to determine future evolution. Indeed, the whole forest is gradually changing through interaction of the living ecosystem and the environment. What is observed now is the product of a continuous and often ancient sequence of events, some fortuitous and some themselves predetermined, and this will largely predetermine the composition and pattern of change in both the near and distant future. The distribution of species and forests will also, of course, vary in space and time according to limits imposed by environmental factors.

The management and modification of tropical rain forest environments, though aimed at controlling dynamic processes within a human life-span must take account of the dangers of long-term regression or face ultimate catastrophe. In an increasing number of regions that catastrophe is already upon us. The ability to predict and interpret the response of forest to various alternatives depends on our capacity to define and interpret the existing pattern of variation. The difficulty lies in determining the relative importance of the various factors.

Man induced changes are discussed in chapter 9. This chapter reviews knowledge of the dynamic aspects of tropical rain forest structure and species composition under the prevailing environmental conditions first in the humid tropics and later where edaphic or climatic conditions are more limiting or unpredictable; finally, knowledge of forests in the seasonal tropics is reviewed.

Biotic environment

The dynamics of forests must be based on the concept of the canopy as a continuously changing entity; see Whitmore (1975) for a comprehensive review. The climate above the canopy varies diurnally, but is seasonally relatively uniform in temperature; seasonality may or may not be manifested in rainfall.

The canopy structure and density vary with its age and according to the architecture of its species components, thus the transmission reflection, absorption and conversion to heat of solar radiation reaching the canopy will vary; the proportions will change in cloudy and sunny weather, and the patches of shade and sunfleck will move during the day. The path of light, its intensity and spectral composition, through the understorey to the ground will also vary according to the structure of the canopy (see chapter 2). The arrangement of leaves below the canopy will be largely determined by the available light in the current season and in the arrangement of the understorey trees. This arrangement depends on light conditions and root competition, from their time of establishment; so all is ultimately dependent on the canopy trees.

The first problem is to define the units of spatial change in the canopy. The species might appear to be the obvious choice, but to follow and interpret the spatial changes of such a highly definitive element is over-ambitious and impractical with present knowledge. Natural cyclical change, initiated by gap formation, is accompanied by a sequence of species which at each stage share an array of biological characteristics. Most ecologists who have studied this have intuitively classified the species into groups on the basis of various biological characteristics. In the simplest terms, such classification has followed the empirical dichotomy of shade tolerant and light demanding species (cf. Jones, 1950, 1955, 1956). Other workers have utilized other characteristics such as longevity, growth rates, wood density, dispersal mechanisms, fruiting regularity, seed production, seed viability, and frequency of occurrence in various seral stages (see Richards, 1952; Ross, 1954; Budowski, 1965).

A further set of species characteristics undoubtedly important in understanding the biological processes underlying natural cyclical change, and which frequently distinguish the species of each stage, are those associated with phenology, mode of shoot extension and hence architecture; these have been the object of elegant studies by morphologists (e.g. Corner, 1949; Scaronne, 1957; Koriba, 1958; Roux, 1964–65; Prévost, 1966; and the comprehensive review of Hallé and Oldeman, 1970) but their ecological significance has yet to be fully appreciated. The size and arrangement

of leaves in influencing the leaf area index is a principal determinant of rates of production and tend to change at different stages in the forest cycle; the work of Horn (1972) in temperate broad-leaved forest needs to be extended to the tropics.

Any space may at one time be occupied by any of a number of pioneer, succession, or mature canopy species. Intrinsic spatial change will be reviewed by considering units based on the degree of canopy opening, following Whitmore (1975), for this is a principal determinant of the mature canopy. The characteristics of the mature canopy phase and its components will be treated first, and then succession in canopy openings of successively greater scale will then be considered. In closed canopy tropical forests three phases have been defined by Whitmore: the gap phase, following opening up of the canopy (in forestry convention that is when the regenerating trees are under 2.7 m tall); the building phase, during which the species grow logarithmically and the increments of height and girth of trees are almost linearly related (conventionally, when the largest trees are 0.3–0.9 m GBH); and the mature phase, when height increment declines until a maximum height is reached but girth increment, often at a declining level, continues.

Owing to its longevity, the mature phase occupies the major area. Poore (1968), for example, considered that only 10 per cent of the Jengka Forest Reserve, Malaysia, consisted of gaps. The total area covered by gaps, and their size depend on features such as the maturity of the canopy, the structure and wind profile of the canopy, and the susceptibility of the forest to wind and storm damage (see Wyatt-Smith, 1954; Jones, 1956; Webb, 1958). In forests not subject to periodic catastrophe, the mature canopy imposes a certain order and uniformity on structure, often including a demonstrable stratification (Ashton, 1964; unpublished), which has sometimes been disputed (see chapter 5).

Phenology, particularly of the canopy component, is a major determinant of forest dynamics especially in the seasonal zone; leaf-shed and flushing affecting the climate below the canopy, flowering phenology the breeding systems, and fruiting phenology affecting the means of dispersal (hence, indirectly, breeding systems also), conditions of germination and establishment.

This intrinsic canopy heterogeneity varies with climate in the three major tropical regions; in a similar way, it also varies with soil conditions. The concept of a monolithic and uniform climatically determined tropical rain forest is facile and hinders detailed understanding of the factors underlying its dynamics (Ashton, 1964).

Species diversity is so great, especially in aseasonal climates, that many species appear to be spatially complementary within one phase of the forest cycle. This change of species with time, and the mechanisms which maintain low but apparently stable population numbers, constitute a separate aspect of dynamics to be reviewed.

Plant biology and regeneration in aseasonal forests

Mature phase

Canopy characteristics

On well watered freely drained soils of average fertility the canopy possesses the following characteristics (see also chapter 5).

The structure is heterogeneous. The sympodially branched hemispherical crowns of the emergents (mainly Dipterocarpaceae in South-East Asia, Bombacaceae and Vochysiaceae in South America, Leguminosae throughout the tropics) are more or less free standing and arise at the height of the less easily defined main canopy; this is of variable height and density, and composed mainly of more or less narrow-crowned trees, of varying architecture though predominantly sympodial. Climbers and epiphytes are sparse. The consequent unevenness of the top of the canopy may enhance local air turbulance and thus, through increased transpiration rates, reduces leaf temperatures. The predominantly dome-shaped emergent crowns on the other hand shade the main canopy and are themselves partially in sun and partially in shade at all times of day; this should lead to reduced transpiration during the middle of the day and increase it—and lengthens the period of photosynthesis—in the emergent crowns during the morning and evening (Brünig, 1970, 1971).

Leaf size classes are discussed in chapter 5. Evidence suggests that the stomatal sensitivity of emergent tree leaves is usually weak after they mature, the trees transpiring like giant wicks (Kenworthy, 1971). They nevertheless possess adaptations which reduce rates of transpiration. Most are sclerophyllous, with thick cuticles and stomata more or less confined to the lower face. They are also strongly sclerenchymatous, the vascular bundles often having stone cells above and below; this provides both rigidity and fragments the mesophyll. Most have high reflectance owing to shiny, waxy surfaces and, on their underfaces, pale scales or hairs which also reduce diffusion rates in moving air. The leaves are generally held at an angle to the mean direction of solar radiation thus reducing the quantity received per unit area of leaf (see Brünig, 1970). As a consequence the canopy as a whole has high reflectance (albedo), but still much light is transmitted and such forests are often highly structured, with a high leaf area density (mean area of leaf per unit volume of space) beneath the canopy.

The canopy, as well as individual crowns, permit moderate light penetration; moderate shade, and moderate distribution of sunflecks on the forest floor during the four hours around noon (see e.g. Wong and Whitmore, 1970).

Under well watered, but not swampy, conditions the density of emergents increases, e.g. *Mora* forests of Guyana, dipterocarp forests of the Philippines and East Sabah, Sarawak basalt (Ashton, unpublished). There are also higher densities of crown epiphytes and canopy climbers. This appears to be associated with an increase in LAI in the canopy and an observable decline in shade light intensity and density of sunflecks below; there is a related lower tree density and LAI in the understorey, where there is a significant increase in macrophylls. The dense canopy of emergents may absorb more radiant energy and thus become warmer during the day and have higher transpiration rates than the previous type.

Conditions of low soil nutrient availability, especially in physiologically shallow soils and on excessively drained siliceous soils are associated with a decline in stature and density of emergents which often become absent, leading to an even, aerodynamically smooth, canopy. Conifers are often present however and, where large (especially *Agathis* in South-East Asia), remain monopodial and often prominently emergent (Brünig, 1970; Ashton, unpublished). The exceptional emergent dipterocarp *Shorea albida* of the Bornean peat swamps forms a continuous smooth emergent canopy high above the diffuse main canopy and dense understorey (Anderson, 1961). There is a decrease in leaf size among trees on these infertile soils with notophylls predominating and microphylls greatly increasing; sclerophylly, albedo, and epidermal indumenta of various kinds all greatly increase. These canopy and leaf features are associated with a clearly observable increase in light intensity and sunfleck density on the forest floor, especially around noon, and a great increase in the density and LAI of the understorey. Shade epiphytes and myrmecophytes may become common. The leaf characteristics will lead also to reduction of transpiration rates yet the even canopy structure is likely to act as a trap of humid air within the understorey. These features thus probably have a profound effect on dynamics within and below the mature phase canopy.

Phenology

Mature phase species such as dipterocarps generally only begin flowering after they attain the canopy; however, in culture, many will flower within five years though rarely set fruit (Ng, 1966).

The only systematic observations in mature phase rain forest are restricted to canopy species in submesic sites (Holmes, 1942; McClure, 1966; Medway, 1972; Yap, unpublished). Much of this type of information remains hidden in forest records and needs collating and publishing. Observations on the phenology and flower biology of the Meliaceae have been made by Styles (1972). Earlier observations on mature phase species in cultivation provide unreliable information, probably owing to the different micro-climatic conditions prevailing. Though leaf change and reproductive phases of the forest as a whole are often only weakly seasonal they conform to various patterns and most individuals of a population flower together. Few canopy species frequently flower out of phase (Wood, 1956; Ashton, 1969; Start, unpublished). Canopy species do not loose and produce leaves continuously. Leaf abscission is usually associated with drought (Wycherley, 1973) though there is a correlation with day-length in the understorey tree cocoa (Alvim, 1964; Greenwood and Posnette, 1969) and bud break has been found by experiment to be similarly affected among species of the seasonal tropics (Longman and Jeník, 1974). The young flush usually emerges at more or less the same time or before leaf-fall; the proportion of even shortly deciduous species is rather small, less than 3 per cent of the individuals in on example from Borneo (Ashton, unpublished). Many species flush more than once a year, only a very few flush branch by branch (see, e.g. Koriba, 1958). Young leaves are often pendant, bright anthocyanin-red and distinctly warmer than the mature leaves in direct sunlight (Smith, 1909). The function of the colour is unknown, also whether the higher temperature is a result of the colour or greater stomatal sensitivity. It appears likely that the life of most canopy leaves is less than 18 months.

The canopy as a whole has regular rather than aperiodic flowering, but with great between year variation in intensity; in the Far East, years when the dominant Dipterocarpaceae flower gregariously also have heavier than average flowering of many other families. Beneath this canopy a small but significant number of species grow and flush continuously; none are deciduous. A few understorey genera appear to flower more or less continuously. The causative factors behind initiation of inflorescence primordia remain little understood, intrinsic mechanisms (biological clocks) and extrinsic influence probably dominating in different species and different environments. Clear examples of the intrinsic mechanism causing flowering have been found in *Ficus* spp. (McClure, 1966) and *Homalium* (Holttum, 1940). Extrinsic mechanisms that have been identified include drought followed by light rain, in the Dipterocarpaceae for instance, though some exceptional droughts were not associated (Webb, 1958; Schulz, 1960; Whitehead, 1969; Baillie, 1972; Burgess, 1972; Fox, 1973). Wycherley (1973) has identified high insolation as the individual extrinsic factor most likely to be responsible for flowering years, with periods of high diurnal temperature range playing some secondary and usually correlated role. Heavy rain is well known to promote flowering in the epiphytic orchid *Dendrobium crumenatum* (Coster, 1937) and the herbs *Zephyranthes* and *Pancratium*. Koopman and Verhoef (1938) observed that *Eusideroxylon zwageri* in Sumatra and Kalimantan, flowered irregularly from one place to another, but usually after dry periods.

No detailed comparative phenological observations exist for forests on different soils. Contrary to initial expectations there is a significant increase in deciduous species on well watered fertile sites (Ashton, unpublished), the shortly deciduous *Firmiana*, *Pterocymbium*, *Alstonia scholaris*, *Octomeles* spp. being confined to them in Malesian perhumid climates. There is some evidence to suggest that leaf longetivity is shorter on the mesic sites; no differences in flowering phenology are known.

Pollination biology

Research into the dynamics of these important aspects in natural forest of the perhumid tropics has only recently started. The following patterns are suggested for the mature phase (Ashton, 1969).

Among the emergents flowers are copiously produced,

and are usually small and inconspicuous though scented and commonly with prominent nectaries. They are usually hermaphrodite, with less than 10 per cent having morphological modifications favouring outcrossing. Bearing in mind the irregular or periodic but synchronous heavy flowering it would appear that the pollinators (rarely identified in any truly mature phase species) are mostly small unspecialized insects—flies, thrips, beetles, polylectic bees. Self-pollination appears to be the rule among the emergents. This is contrary to the opinion, that specialized often conspicuous flowers, with vectors are the case, held by Faegri and Van der Pilj (1971); most of the species they considered were of secondary forests, forest fringes and the seasonal tropics. The South American mature phase emergent *Bertholletia excelsa* is, however, pollinated, by large insects such as *Bombus*, which have sufficient strength to raise the androphore which covers the stigma of the hermaphrodite flower. This tree though, as most others in the canopy, only produces flowers at a certain time of year; thus, the pollinator is polylectic and must depend on other species to provide food when this species is not in flower. The disappearance of some species of animals or plants will therefore affect the life of other species and can interfere with the natural ecosystem. Many other examples could be cited, demonstrating complexity of mixed tropical forests. In most species petals fall directly after anthesis.

In the understorey over 40 per cent of the trees are either dioecious or possess other morphological adaptations favouring outcrossing; the main canopy species are intermediate in this respect, but in others conform with the emergents. Floral morphology and size is more variable: many families have large flowers; often few or very few; sometimes highly coloured (especially red and white), sometimes dingy purplish, brown or green; and smelling of offal or rotting fruit. This suggests a greater diversity of pollinators and, in the significant number of species which flower at frequent intervals or continuously, that vectors are more specialized (though moths, beetles, flies and other insects have also been mainly identified), and that panmixis prevails.

Though the flower biology of woody climbers conforms to that of the emergent trees, that of crown epiphytes and parasites is conspicuously different. Their bright highly diversified large flowers, more or less continuously produced, suggest a high vector specificity. For instance, in many Amazon orchids, the conspicuous and colourful labellum is adapted as a platform for alighting insects. In the terrestrial genus *Costus* (Zingiberaceae) a very similar structure derives from a modification of a stamen.

No patterns of variation related to soils have been observed. Amongst canopy species a few of the deciduous species of mesic* soils bear brilliant tubular flowers during the deciduous phase and are often bird (e.g. *Firmiana*) and

---

* Mesic is used in the American soil taxonomy to characterize soils or micro-sites where the mean annual temperature is between 8 and 15 °C. See: *Soil taxonomy: basic system of soil classification for making and interpreting soil surveys*. Washington, D.C., USDA, Soil Conservation Service, Agricultural Handbook no. 436, 1975, 754 p.

bee (e.g. *Wightia*) pollinated, and the largely wind pollinated conifers and Casuarinaceae are frequently gregarious and confined to skeletal and podsolized lowland soils.

Fruit biology

Though emergent species may produce high quantities of fruit at infrequent intervals, these are produced by a minute proportion of the original flowers. This has been variously attributed to failed pollination, rain during flowering, intrinsic physiological reasons and predation. Understorey species frequently produce a few, very large fruit. Very few main canopy or emergent tree and climber genera produce light winged fruit or seeds (e.g. *Koompassia*, *Engelhardtia* in South-East Asia, but compare the more seasonal West African forests). It is not known how effectively they are dispersed in a usually windless environment, or how their seedlings, with such small food resources, become established.

It is not known to what extent agamospermy occurs among rain forest trees although the seeds of some dipterocarps (Maury, 1968, 1970) and *Eugenia* are polyembryonic. This has to be taken into account in theories of speciation.

More or less heavy regular fruiting occurs each year in the forest as a whole and there are occasional heavy fruiting years in which many other taxa which fruit at longer intervals are involved. Related dipterocarps with staggered flowering times appear to have differential rates of fruit development leading to such synchronous fruiting (Holmes, 1942; Wood, 1956; McClure, 1966; Ashton, 1969; Medway, 1972). Such periodicity must sustain a low carrying capacity of fruit predators, however generalized, and thus increase the proportion of seed that survive. At the genetic level biochemical specialization may maintain high predator-host specificity and further lower predator carrying capacities (Janzen, 1974).

Dispersal

Many mature phase species possess no known means of fruit dispersal with the fruit falling beneath or near the perimeter of the parent crown. For instance, the winged fruit of the South-East Asian dipterocarps gyrates into the main canopy and then falls randomly, but more or less vertically, below. This can lead to aggregated clumping of seedlings and possibly repetition of floristic composition in time; by favouring temporary inbreeding among small numbers of individuals sharing a common parentage, it may also be conducive to rapid ecotypic diversification (Ashton, 1969). However, in some species, differential predation has been found to favour survival of the seedlings furthest from the parent (Janzen, 1970).

Many propagules are dispersed by animals, particularly monkeys, other arboreal mammals and birds. Such fruit are usually conspicuously coloured when ripe—usually red, but also black, white, yellow or blue. They mostly have freshy pericarps, mesocarps, endocarps or with arils or arillodes, the latter often with a dry protective pericarp that dehisces to reveal a colourful juicy interior (Corner, 1949, 1952, 1956). The seeds themselves are frequently toxic

and are either rejected or pass unharmed through the gut of the vector; see Ridley (1930), for a general account. Such wider dispersal may be associated with lower rates of speciation (Ashton, 1969) and to changing local pattern of floristic composition of a forest type (e.g. Rollet, 1974). This does not imply a regular alternation of repeating species associates as implied by Aubréville (1938) for which there is not convincing evidence in natural undisturbed forest.

Dormancy

In general the seeds of mature phase tree species have no dormancy period, although Koopman and Verhoef (1938) found *Eusideroxylon zwageri* seeds retained a good germination capacity for a year. However, there are some conspicuous exceptions in Malaysia, notably in *Anisophyllea* (Rhizophoraceae) and various Leguminosae (Ng, 1973, 1974), and some of the small seeds of the wind-dispersed mature phase *Koompassia malaccensis* and *Dyera costulata* germinated 56 and 130 days following sowing respectively (Wyatt-Smith, 1963), whilst dipterocarp seed dormancy could only be extended for four weeks by desiccation or by combined desiccation and cooling (Tang, 1971; Tang and Tamari, 1973); the African *Terminalia ivorensis* and *Triplochiton scleroxylon* seeds loose viability if stored for any length of time, and the dormancy of 350 woody species from a variety of habitats in the Ivory Coast, varied from less than two weeks to 3 years (De La Mensbruge, 1966).

Germination and establishment

The highest mortality in the life cycle occurs between flowering and seedling establishment and its appreciation is crucial in the development of suitable methods of natural regeneration of logged forest (e.g. Wyatt-Smith, 1963); nevertheless it is not known to what extent mortality is probabilistic rather than assignable to specific selective factors. The conditions for germination and establishment of mature phase tree species are nearly always highly specialized and this is the major factor preventing their return after forest clearance (Gómez-Pompa *et al.*, 1972). There is often considerable interspecific variation in moisture requirement for successful germination; this may help to explain both intrinsic spatial pattern and extrinsic spatial variation in forest floristics. Brunck (in De La Mensbruge, 1966) found that seeds of the West African okoumé (*Aucoumea klaineana*) lost viability if relative humidity fluctuated more than 8 per cent though they may be stored for several years if both temperature and humidity are lowered (see under seasonal forests). Burgess (unpublished) was able to show that the Malayan ridge-top dominant *Shorea curtisii* required moister conditions for germination than did *S. leprosula* and *S. parvifolia* on hill-side. It was more tolerant of low light intensities following establishment, and was thus well adapted to regenerate under the dense ground-cover of the stemless palm *Eugeissona tristis* which abounds in the same habitat and which provides moisture and light conditions in which *S. curtisii* alone among its congeners can survive and become established. For successful establishment, mature phase species on submesic sites require constantly moist conditions; low light intensity is not limiting, indeed shade is usually obligatory to prevent drying out.

Predation is a major selective factor. Studies have been carried out by Synnott (1973) in the Budongo forests of Uganda, under a mature forest canopy of *Cynometra alexandri*, *Entandrophragma* spp. and *Khaya* spp., on the regeneration of *Entandrophragma utile*. Under simulated natural seed-fall, *ca.* 40 per cent of the seeds were eaten by rodents before and immediately after germination and nearly 30 per cent were killed within 2 years by antelope browsing. Other losses were caused by seed-rot, insect and fungal attacks and drought. The survival rate after 2 years from seed-fall was 2 per cent. Chan (unpublished) recorded 83 per cent abortion in *Shorea ovalis* Korth., possibly owing to failed pollination; over 90 per cent of the survivors were killed by a single insect predator before falling, and 16 per cent of these were killed by three others on the ground; predation decreased with distance from the tree. These predators are not host specific, but in other cases they are and their importance as agents maintaining diversity is reviewed later.

No comparative information on germination and establishment in different forest types is available. No significant differences in fruit biology are known between different forests on average and well watered fertile sites. On infertile, and particularly xeric sites (especially ridge crests and podzols) there is an increase in mature species with dry, dehiscent, small-seeded, and wind-dispersed propagules, e.g. *Cratoxylon*, *Austrobuxus*, *Ctenolophon*, *Allantospermum*, *Axinandra*, *Agathis*, *Casuarina* in South-East Asia. This might be expected from environments which are the frequently windier, and have light conditions on the forest-floor which allow for the rapid initiation of photosynthesis.

Another important but ill-studied aspect is the development of the rhizosphere flora. It is now fairly well established (e.g. Edmiston in Odum and Pigeon, 1970) that many rain forest families are mycorrhizal (e.g. Dipterocarpaceae, some Tiliaceae, Sterculiaceae with ectotrophic and Myrtaceae with vesicular arbuscular mycorrhizae). It also appears that, as in the temperate zone, mycorrhizae become more wide-spread and physiologically important in infertile soils. They may prove to be a principal determinant of species composition, acting at the stage of seedling establishment.

Seed production, dispersal, dormancy and establishment problems appear to be similar for woody climbers and trees, but are obviously entirely different for epiphytes, which generally produce small seeds, with little food store, which are dispersed by convection currents as well as animals.

Physiology of regenerating trees

As a result of the generally plentiful seed resources, and the periodic heavy fruit production, perhumid rain forests are characterized by a dense though fluctuating and often clumped seedling and young sapling distribution on the ground. This sylviculturally important fact has been demonstrated in West Malaysian mixed dipterocarp forest (De Leeuw, 1936; Soepono and Ardiwinata, 1957; Wyatt-Smith,

1963) and also mixed lowland forest in Irian (Loekito and Hardjono, 1965), Venezuela (Rollet, 1974), Surinam (Schulz, 1960), Trinidad (Beard, 1946), Sumatran mixed swamp forest (Rakoen, 1955), New Guinea *Araucaria* forest (Gray, 1975) and Kalimantan heath forest (Soebidja, 1955), though this is by no means always the case for all species, and especially in edaphic subclimax forests dominated by single species, e.g. *Hopea mengarawan* Miq. in Sumatra (Thorenaer, 1924) and *Shorea albida* Sym. in Sarawak (Anderson, 1961). The Malayan uniform system of regenerating lowland dipterocarp forest relies on such an assumption. Nevertheless, the vast majority of seedlings are destined to die in the first few years following fruit fall. The survivors (known to sylviculturists as recruits) may in part be those which through chance or genotype were able to survive initial root competition. Increase in light intensity through partial canopy opening is known to lead to rapid height increment among seedlings, though many turn yellow and stagnate if complete canopy clearance is undertaken. The effects of light intensity on the growth of seedlings of some tropical forest tree species were investigated by Wadsworth and Lawton (1968) who concluded that *Khaya grandifoliola* was the only species that put on a useful growth rate at 1 per cent daylight, and *Pinus caribaea* showed the least adaptability to shade. Sunderland (unpublished) compared the rates of photosynthesis and respiration in the slow-growing shade tolerant Malaysian *Shorea maxwelliana* with the faster-growing light demanding, yet essentially mature phase, *Shorea leprosula*; he demonstrated that:

— neither were able to assimilate more than they respired in the absence of sunflecks;

— stomatal response in both was immediate, remaining closed in shade and opening within seconds of the arrival of a sunfleck;

— the shade tolerant species had the lower compensation point, but was unable to photosynthesise at as high a rate as the light demander in direct sunlight.

Some authors recognize a suppression period (Brown and Matthews, 1914; Richards, 1952) in the relatively young phase of certain Philippine species, whereby light demanding species are able to persist without significant growth in deep shade until a gap appears. This suppression period does not seem to exist under Amazonian conditions; it is probable that the data obtained elsewhere are being interpreted incorrectly, through the mixing of observations from clearings and mature forests, and from light demanders and slow growing shade tolerant species.

The crown architecture, and leaf disposition and anatomy, of understorey trees and saplings are as would be expected from Sunderland's experiments. The crowns are diffuse, taller than broad, and thus allowing a high leaf area index. The leaves are mostly horizontal, thin, larger than in the canopy, and with long narrow acumina and petioles. Leaf change is gradual and leaves apparently live longer than in the canopy; continuous shoot extension is infrequent but commoner than in the canopy.

Subsequent to the sapling stage the growing trees eventually cease to respond to changes in the canopy above and around their crowns; this has been observed both in western Africa (Lowe, 1968) and the Solomon Islands (Whitmore, 1974) and explains why selective felling systems and improvement thinnings have never been demonstrably successful.

Species diversity

For the definitions and generalizations about species and genetic diversity, see table 1. See also chapters 4 and 5.

The intrinsic species diversity, particularly of trees poses a problem for the interpretation of natural selection in tropical forests, and hence also for internal dynamics and genetics. Janzen (1970, 1971) has emphasized the biochemical (phytotoxin) diversity of the tree components in rain forest synusiae, and interpreted it as a consequence of predator (particularly of seeds, seedlings and juveniles) selection pressures leading to coevolution of highly specific predator-host relationships (see also Connell, 1970; Wilson and Janzen, 1972; Rockwood, 1973; Center and Johnson, 1974). This provides a density-dependent mechanism which may explain the ecological complementarity of rain forest tree species in space, and a mechanism for natural selection which is not manifested in recurrent spatial

patterns and species associations. The time of attack and the feeding behaviour of the predator can, however, affect the spatial distribution of survivors and hence the pattern of change within tree populations and, in the long term, their genecology. The work of Janzen and others has principally involved trees in different genera in the seasonal tropics; biochemical differences and host specificity between sympatric species in large genera appear not to occur; e.g. the lack of host specificity among beetles in dipterocarp fruits (Daljeet Singh, 1974). This suggests that the initial process of speciation among trees is usually, perhaps always (Ashton, 1969), allopatric in response to more conventional selection pressures, but that biochemical differences at generic and family level maintain floristic diversity in these forests. Much of the evidence supporting these hypotheses, however, is indirect and is inferred from the behaviour of the particular species in monoculture (e.g. Harper, 1969; Janzen, 1971) or in areas isolated from natural predators. Janzen (1974) has also observed that plant toxins are most prevalent on infertile soils and in the mature phase of the forest. Species diversity however is maximal on average to somewhat infertile soils (Ashton, unpublished).

TABLE 1. Species and genetic diversity.

Definitions

Species diversity: the number of species per unit area.
Genetic diversity: the number of gene alleles within a population.

Generalizations

1. Both species and genetic diversity are high in geologically diverse areas.
2. With increasing latitude species diversity decreases and genetic diversity increases.
3. In spatially heterogeneous environments there is more opportunity for disruptive selection and thus speciation; both species and genetic diversity are high.
4. Both species and genetic diversity increase with increasing proximity to ecotones and physical interfaces.
5. Continental species tend to have higher genetic diversities than island species.
6. Geological Time Theory: the older an area is geologically the greater its species diversity.
7. Ecological Time Theory: species diversity increases with time since an environmental disturbance.
8. During successions, species diversity first increases, then decreases as the climax is approached.

The ecological instability of the single species dominance which can occur on limiting soils is demonstrated by the epidemic increases of a lethal defoliator in the pure stands of the Bornean peat swamp dipterocarp *Shorea albida* (Anderson, 1961). Large patches up to several square kilometres in area of *Nothofagus* in the highlands near Onim, south-east slopes of Mt. Giluwe, Papua-New Guinea, were recently observed to have many dead or dying trees, and pest attack seems a plausible explanation (R. Hynes, personal communication). One of the best documented examples of the same effect is provided by *Hevea brasiliensis* which can be monocultured successfully in Malaysia, yet, in its native Brazil, it can only be grown in mixed communities because of pest epidemics (Purseglove, 1968; Harper, 1969). Similarly, in Queensland the monoculture of certain rain forest species has been abortive due to seedling damage by rats, insects, and fungal disease (Webb, 1968).

Webb *et al.* (1967) investigated the failure of the rain forest species, *Grevillea robusta*, in monoculture in eastern Australia and reported that the actively growing roots inhibited the growth of its own seedling. Moreover, some factors in the leaf drip produced the same result. In Brazil monospecific plantations of *Dinizzia excelsa* produced dark black, very acid humus in ten years which can only support very specialized understorey plants; also litter from *Simaruba amara* and *Bertholetia excelsa* may contain autotoxins as these species appear not to regenerate under their own canopy. In western Africa Mangenot (1958) has noted the absence of regeneration beneath *Scaphopetalum amoenum*. Other rain forest species, the so-called non-gregarious species, show similar behaviours both in monoculture and in natural forests. The general confirmation of such findings in the wild would provide a manifestation of long-term cyclical change in conformity with Aubréville's (1938) theory.

Possible long-term degraded change in primary forest has been observed in Brunei by Ashton (1964) who noticed collars of bleached soil around the principal roots of the dipterocarp *Dryobalanops aromatica*, the leaves of which are rich in phenolic compounds. Such long-term change may be universal on stable surfaces in the everwet tropics, but defies careful investigation.

Altitudinal variation (see chapter 4)

The dynamics of montane forests have been little studied and no comparative analysis has been made; though H. T. Odum's *A tropical rain forest* (1970) is a rather specialized study of hill and submontane forests in the hurricane belt of the West Indies. Descriptive ecological and physiological studies in South-East Asian rain forest are summarized by Flenley (1974) and Whitmore (1975). On the moist acid soils above *ca.* 1 000 m the forest assumes the increasingly smooth canopy structure and also the sclerophylly of lowland forest on infertile soils. Many species are common, and the increased light on the forest-floor provides similar conditions for apparently essentially similar forest dynamics and reproduction biology. Though Ashton recorded a decline in increment with altitude in Sarawak, the reverse is generally assumed, based in particular on work in Africa; in the latter however the increased increment at higher altitudes would seem to be in part correlated with increased humidity and lower rainfall seasonality in the mountains of the seasonal tropics (but see below).

In the cloud zone under truly perhumid conditions the low stature montane forest trees become enshrouded in hepatics and other epiphytes. These hepatics form a spongy mass round the bases of the larger trees, often enlarging and coalescing so that the ground and canopy become ill-defined from one another. Terrestrial animals maintain gulley-like pathways especially along ridges in which litter collects and prevents moss growth. Two totally different substrates therefore exist for seedling establishment and there is some evidence that seedling survival on the two differs among species.

No rain forest trees are frost hardy when young. At altitudes where frosts become frequent forest gives way to alpine grasslands in which highly specialized non-forest woody pachycauls form diffuse populations (*Cyathea* in New Guinea, *Rhododendron zeylanicum* in Sri Lanka, Bromeliaceae in the Andes, Compositae, *Lobelia* in Africa and South America, etc.). The interphase between forest and grassland is sharp and often accentuated by fires. At its altitudinal limits the forest canopy is even but dense and apparently excludes frost from the understorey. In parts of montane Sri Lanka increasing frost, associated with increasing aridity in recent years, has apparently penetrated to the understorey, and is leading to an irreversible change from forest to grassland as the canopy trees die without being replaced.

Wind forests

In some regions (notably southwestern Sri Lanka, southwestern coastal Indo-China, eastern coastal Luzon and Samar, southeastern New Guinea, the South Pacific and West Indian islands) the perhumid climate is modified by rain-bearing trade-winds for several months of the year; these are characterized by their constancy and are sometimes rather strong. This climate is associated with a dense low stature forest, in which emergents are either absent or barely raised above the main canopy, and by sclerophylly. Crown sizes are generally small, and gaps between them large owing to frequent wind-movement; the understorey and regeneration of canopy species is consequently usually dense. In Luzon the canopy species, though so different phenotypically, are the very same species which compose the mature phase of the tall and structurally heterogeneous mixed dipterocarp forest on the leeward slopes. These forests must not be confused with storm forests.

Gap and building phases

Without the death or partial death of some of the mature phase canopy individuals the forest has no intrinsic mechanism to initiate anything other than phenological change; the canopy would be complete and only the shade tolerant seedlings of the mature phase would maintain themselves beneath it. Such a static situation, however, obviously never arises because the continuity of the forest canopy is continually eroded by disturbance of two main kinds. One involves the immediate environment of a single tree and is caused by the collapse of a large branch or a senile tree as the result of age or disease. The other kind results from some natural physical force, such as a storm, landslip, or even large animals. The consequent rupture of the canopy produces the opportunities for change. Owing to the intrinsic heterogeneity of the mature phase canopy gaps are more or less indeterminate, but in forests where a single species dominates the canopy their full extent becomes appreciated, thus gaps caused by lightning and wind in *Shorea albida* dominated peat swamp forests of Sarawak have been the subject of studies by Anderson (1961) and Brünig (1964, 1974).

Nevertheless, there have been few quantitative assessments of the natural regeneration in canopy gaps and there is therefore little agreement as to the factors which influence the outcome of the process. Van Steenis (1958) feels that chance plays a large part in many cases, and aggregates of one species result solely from the proximity of a seeding parent to a newly-created canopy gap. Poore (1968) suggested that the regeneration in these gaps depends primarily on the established seedlings and saplings that are present when the gap is formed; the existence of particular seedlings and saplings, of course, is dependent on many factors such as nearness of parents, efficiency of dispersal, dormancy, periodicity of flowering and fruiting, and degree of shade tolerance. Whitmore (1974) demonstrated that the reproductive biology and growth rates of the 12 dominant species in a Solomon Islands forest are adapted differentially to the degree of canopy disturbance by cyclones. Webb *et al.* (1972) concluded that many factors affect the final composition of the gap and that the outcome is primarily probabilistic. It would appear that many further studies are required before comparative importance of the many factors which could possibly affect the final species composition of the gaps can be determined.

The gaps in the canopy of the forest vary in size from a few square metres to many hectares. As will be seen, the type of regeneration, and the degree and direction of change which result from this disturbance depend essentially on the size of the canopy gaps produced.

Small gaps

These are the commonest form in forest rarely influenced by strong winds, land-slides or other unpredictable catastrophes. They are found mainly through two causes.

a. Natural death of main canopy or emergent trees. As a rule the crown dies back as the trunk rots from the inside and finally falls creating a more or less narrow track through the canopy.

b. Lightning strike. Emergents are generally struck and killed instantly, but the young saplings and understorey are killed, and some canopy trees may be partially or even wholly killed for a radius of some 20 m from the base of the bole, presumably through transmission of the shock through the roots. They all usually fall in the same manner as other dead individual trees. This is therefore more devastating than (a) but both provide canopy gaps of similar dimensions. The pattern of regeneration is rather different in each case.

a. The saplings are not killed and they rapidly respond to the increased light by height increment; apical dominance is maintained until the leader emerges from the main canopy when sympodial branching, usually associated with onset of flowering (but see Hallé and Oldeman, 1970) leads to the development of the mature crown shape. Meanwhile the number of individuals growing in the gap is increasingly reduced by competitive self-thinning. Though strictly mature phase species occupy these small gaps, the faster growing species (such as *Shorea leprosula*) have the competitive advantage and the slow growing shade tolerant species make their way up to the main canopy under its shade. These small gaps are consequently unlikely to initiate any immediate change in the total floristic composition of the canopy. However, since there are no data which suggest that each canopy individual is replaced by individuals of the same species, regeneration in these small gaps will probably effect a change in the horizontal distribution of the canopy species array. Over a long period regeneration in these small gaps results in a canopy whose composition is a flux of species of the mature phase. This continuous regeneration of the more shade tolerant species in small canopy gaps has been described by Van Steenis (1958) as diffuse regeneration.

b. Lightning strikes eliminate the available regeneration. They frequently remain vacant therefore for prolonged periods, and may be semi-permanent at sites which for physiographic reasons are particularly lightning-prone. Generally they are first colonized by herbs, including *Pandanus* spp., forest grasses and ferns, then by the germination and establishment of more light demanding mature phase species with sometimes a few, usually transient, pioneer tree species.

Large gaps

These may be caused by a number of natural factors including, in increasing order of magnitude, localized windthrow, land-slides, volcanism and typhoons (for the latter two, see catastrophes, below). The first and the last differ from the others in that the regeneration is not eliminated. Fundamentally, large gaps differ from small gaps in two respects: their longevity and the quantity of light which penetrates to the lower strata. These larger gaps can only be filled by individuals from a lower storey and the possibility exists for an individual to grow from the seedling stage to the canopy with its crown always receiving uninterrupted direct sunlight. These conditions are therefore simulated by sylviculturists, who recognize in them the optimal conditions of growth for the seedlings and saplings. By opening the forest canopy, Philip (1968) increased the seedling regeneration of light demanding species including the valuable Meliaceae family in Uganda rain forest whereas in untouched forest the seedlings were mainly of very shade tolerant species. The Malayan uniform system similarly aims to enhance regeneration of merantis (*Shorea*). The observation that meliaceous regeneration occurred in young mixed communities on the forest edge supports Eggeling's (1947) view, and the view held by many workers in western Africa, that forests rich in mahogany are secondary. Under those circumstances where the understorey is not destroyed the ground is quickly occupied and the surviving mature phase saplings are often overtopped by the rapid growth of tree and woody climbers representing a distinct floristic component of the forest, often known as pioneers or biological nomads (Van Steenis, 1958). These form the first stage in true seral successions within the forest and dominate early successions especially where the soil has been scarified.

It appears, from Symington's (1933) experiments in Malaysia and Keay's (1960) in Nigeria that seeds of many pioneer species are photoresponsive and may lie dormant in the soil for long periods in the shade. This simple yet important experiment requires repetition and extension for the results, if correct, imply greater predictability of species composition in the early building phase.

Pioneers are characterized by the following properties.

— Light seeds (<1 mg e.g. in *Musanga cecropioides* and *Anthocephalus chinensis*) with minimal food storage, efficiently dispersed usually by wind or birds.

— Seed dormancy; but more needs to be known; it is certainly true of the plantation species *Anthocephalus chinensis* and *Albizia falcataria*.

— Germination in response to light and/or temperature; requiring further experimental confirmation, but see Olatoye (1968).

— Immediate photosynthesis following germination.

— Early onset of a logarithmic phase of growth and its continuation until more or less sudden death.

— A growth strategy involving the formation of many vertical or inclined pithy stout stems bearing large leaves in a dense single layered canopy which excludes initial competitors and maximizes temperature during the heat of noon and consequently also rates of transpiration and photosynthesis; this deserves greater physiological study, e.g. see Coombe and Hadfield (1962).

— Leaves which change orientation daily so that the degree of insolation is reduced.
— High growth rates.
— In some species the accumulation of nutrients from subsoil into leaves and eventually litter; e.g. K in *Musa* and *Manihot* and P in *Mallotus resinoides* (Sajise, personal communication).
— Frequent association with ants, which may augment nutrients through excreta, protect the frequently non-toxic leaves from herbivores, and even themselves consume invasive climbers that might otherwise smother their crowns (Janzen, 1969).
— Early onset of more or less continuous flowering and copious fruit production.
— Intolerance of shade.
— Short life-span.

Examples: trees: *Macaranga* spp., *Dillenia suffruticosa*, *Mallotus paniculatus*, from South-East Asia; *Musanga* from Africa; *Cecropia* from South America; climbers: *Uncaria*, *Merremia*, *Ipomoea* from South-East Asia.

Through this tangle a significant number of the mature phase survivors persist. In due course, in as little as 15 years in *Homalanthus* (Kellman, 1970), these species begin to die, often preceded by the opening up of the pithy ascending branches under their own weight. The mature phase species may now take over but are often preceded by a further more or less distinct group of tree species which attain prominence in the late seral succession and are characterized by:

— apical dominance leading to rapid height increment (some eventually become emergents);
— horizontal branches of large obtrullate densely arranged leaves in more or less widely separated superimposed layers;
— high growth rates;
— deep root systems and often prominent tap-roots;
— high nitrogen fixation activity in some (e.g. *Anthocephalus chinensis* in leaves; *Albizia* in roots); more investigation is required here;
— intolerance of shade;
— moderate life-span, sometimes exceeding 100 years.

In other respects they share most characters of the pioneers, though they are frequently more shade tolerant. Many, such as species of *Octomeles*, *Terminalia*, *Bombax*, *Ceiba*, *Alstonia* and *Sterculia*, are shortly deciduous. They thus combine the leaf size and disposition which favours maximum rates of assimilation within the crown causing rapid height increment. When mature they form hemispherical rather diffuse crowns which produce a micro-climate conducive to the mature phase species. There is some evidence that trees of this synusia in particular, with their deep feeder roots and possibly a high nitrogen fixation activity, play an important part in restoring the soil nutrient status and re-establishing the forest mineral cycle. *Albizia* and *Samanea* (Leguminosae) are examples in which the leader eventually becomes horizontal, the apical dominance being transferred to an axillary branch; their leaf-

lets move diurnally so that incident radiation is minimized.

The categories pioneer, late seral succession, mature phase light demander and mature phase shade tolerant are thus seen to be based on a range of biological characteristics. They are not, of course, discrete in nature, and the different permutations of characteristics provide an endless array of growth strategies in relation to light and competition that adds a further subtle dimension to those processes of natural selection that lead to the intrinsic pattern of forest variation. Thus *Homalanthus*, *Trema*, *Muntingia* and *Anthocephalus* combine the longevity of pioneers with the crown architecture of late seral succession species; while the distinction between the latter and mature phase light demanders is blurred by many examples, some species of *Terminalia* and *Alstonia*, as well as *Dyera* and *Pterocymbium*. On the other hand, species of the mature phase continue to regenerate with only slight natural disturbances, and this regeneration is cyclical within the group of mature species which behave vicariously, so that the composition of the mature canopy is in this sense determinate. As they cannot persist without moderate disturbance the distribution of late secondary species is, by contrast, probabilistic among those species lacking seed dormancy and small clumps of the same species may occur. In Australian wet sclerophyll forest dominants such as *Eucalyptus torelliana*, *E. grandis*, *Tristania conferta*, and *Casuarina torulosa* may be regarded as belonging to marginal late seral succession—mature phase types.

The frequency and extent of landslips and slumps causing large gaps in which the soil and vegetation are entirely removed are correlated with factors such as rainfall intensity, slope, angle of bedding (if the parent rocks are sedimentary), physical properties of the soil notably structure, and the type of vegetation. If earthquakes are common, landslips may occur even on moderate slopes, and the area of soil mantle removed is much more extensive than that caused by torrential rains. Viewed from the air, landslips on hill and mountain-sides following earthquakes on the north coast of Papua-New Guinea appear to cover areas ranging from tens to hundreds of hectares, but no measurements have been made on the ground. Although the boundaries of the landslip may be visible for many years, the area of partly bare soil is rapidly covered by a layer of vegetation. In North Queensland landslips are relatively minor in occurrence and <10 ha. Large shallow-rooted trees have been observed to collapse on slopes above *ca.* 15° on soils derived from granites or schists, especially when the soils have been saturated by heavy rains and there is strong wind. Trees are more vulnerable and soils more liable to instability near a sudden increase in slope, so that a dimpling pattern of old root craters tends to be found in special topographic positions on mountain-sides (Williams *et al.*, 1969).

In large gaps, where the mature phase component has been entirely removed, the late seral succession is eventually invaded by an association of species mainly of mature phase origin and showing their slow growth rates and range of crown architecture, yet characterized by effective propagule dispersal, especially by animals. In Asia the extremely slow rate of return of the emergent dipterocarps distinguishes such extensive forest for hundreds of years (e.g. aeras formerly cultivated by tribes now extinct in Central Borneo). The rate of spread of dipterocarps is estimated to be *ca.* 2 m/a (Ashton, 1969). Though Dipterocarpaceae are a conspicuous example many other rain forest trees, including palms (Whitmore, 1973) are effectively permanently eliminated when large areas are completely cleared. Relatively small obstacles such as rivers can act as semi-permanent barriers to reinvasion of extensive areas by many mature phase species; see Ashton (1969); also, for Central America, Gómez-Pompa *et al.* (1972).

What are the effects of the plant species? Patterns in the distribution of the individuals within regeneration appear early. Some are due to environmental influences, in particular micro-site differences, others to non-environmental factors such as seed accessibility (Greig-Smith, 1952; Ross, 1954; Richards, 1955; Williams *et al.*, 1969). There is little evidence with which to correlate such patterns with pre-existing patterns in the original vegetation or with discontinuities produced by the patterns of disturbance. To what degree are these floristic patterns reflected in succeeding phases? This problem was examined in the first 12 years of regeneration of complex notophyll vine forest in southeastern Queensland. Williams *et al.* (1969) showed that the earliest phase was characterized by rapid species changes in time. All the regeneration changed in unison even though the distribution of these initial colonizers was not random. The pioneer phase mirrored micro-site differences which were produced during the original clearing process. These patterns changed in time. The patchiness in the distribution of the establishing primary forest species was unrelated to these environmental patterns (Webb *et al.*, 1972). Thus the species of a phase do not appear to affect the floristic composition of successive phases. However, Williams *et al.* did show that the introduced species *Lantana camara* was capable of completely arresting the succession, for which it is well-known in regenerating forests elsewhere in the tropics. The long-term effects of the occupation of a site by such a species are unknown.

Occasional catastrophes

The principal forms of natural catastrophe which modify forest on a regional scale are typhoons and volcanic explosions. Cyclones with wind velocities over 200 km/h, increased locally depending on topography and wind-funnelling effects, occur every few years in various localities in the tropical cyclone belt along the western fringes of the major oceans at latitudes greater than 5° north and south of the climatic equator, in particular in north-eastern Philippines, Guam, Queensland, Melanesia, and the West Indies. This area partly coincides with that of wind forests.

The effects of cyclones are strinkingly more devastating in areas which only experience them at long intervals. In North Queensland rain forest areas of major disturbance caused by cyclones have been mapped from air photos and surveyed by helicopter (Webb and Tracey, unpublished). They are mostly restricted to altitudes below 700 m and to ridges and spurs of the ranges in the coastal corridor which favours acceleration of wind velocities. The areas are irregularly shaped and vary from less than a hectare to several hectares to irregular strips and swathes up to several hundred metres wide and several kilometres long. Cyclonic damage has a very patchy distribution throughout a forest; it affects the large trees by defoliation, the shearing-off of large branches, and uprooting, especially of shallow-rooted and plank-buttressed species. The penetration of light to lower layers of the forest liberates the suppressed seedlings and promotes a dense growth of vines in the understorey and over the crowns of standing trees to form climber-towers. The humus floor and the seed sources of the forest remain intact, and the debris decomposes rapidly. The canopy of North Queensland rain forests intermittently damaged by cyclones is characteristically very broken, and the associated rejuvenation of *Calamus* spp. and other light demanding lianas in the understorey renders them almost impenetrable; hence, the term cyclone scrubs.

Although there are references in the literature to the effects of cyclones or hurricanes in tropical rain forest areas (see Webb, 1958), detailed descriptions of the biological effects and measurements of the scale of disturbance are few (though see Whitmore, 1974). Again the understorey and regeneration are not lost, though the forest presents an even-aged appearance for many years and possibly centuries afterwards, with the light demanding species predominating (Wyatt-Smith, 1963; Whitmore, 1974, 1975). It is suggested that localized windthrow, as well as landslips, are necessary for the successful regeneration and hence survival of *Araucaria* spp. stands in New Guinea (Gray, 1975).

Volcanic catastrophes affect tropical forests in the Philippines, West Java, certain areas of Papua-New Guinea and other islands in the andesitic belt of the south Pacific. The destructive effects of volcanism and the reaction of the vegetation are best known from the studies of Krakatau (Drs. van Leeuwen, 1936). The devastated areas may be extensive, occupying many square kilometres, but volcanic episodes are exceptional agents in the dynamics of rain forest vegetation. In the extreme but well documented case of Krakatau, after a long initial colonization phase by bryophytes, ferns, and eventually grasses had built up humus, woody seral succession proceeded slowly but normally, the species depending on the availability of a seed source. Where volcanism occurs in mainland areas, and the soils are young such as those derived from pyroclastic material in Papua-New Guinea, the rain forests exhibit local dominance of tree species generally belonging to late seral succession. The local abundance of species such as *Anisoptera thurifera* (e.g. Mt. Victory, Hydrographers Range, Ioma-Popondetta and Morobe coast) and *Pometia pinnata* (e.g. New Britain) may be correlated with past episodes of volcanic ash showers.

Forest fringes

Natural forest fringes occur by water. Like gaps there is a high light intensity on the forest-floor but in this case it is semi-permanent. This, particularly along river-banks, provides the original habitat for many species which have subsequently invaded secondary succession and for many others also associated with gaps.

Many species have water, or even fish (some *Ficus*, *Dysoxylum* spp., *Sandoricum borneense*) dispersed propagules, and a significant number are pollinated and dispersed by wind, e.g. *Octomeles*, *Casuarina* spp. The exposed oblong crowns of many others bear conspicuous large flowers; at least in South-East Asia bat pollination is an important feature of river fringes (Start, unpublished) and several bat-pollinated tree species (*Duabanga*, *Oroxylum* spp.) are confined to it.

Variations with soils

The present description of gap phase dynamics reaches its clearest manifestation on mesic soils, where macrophyll forms may greatly predominate, even in relatively small gaps, and where light demanding mature phase species find soil conditions suitable for reinvasion. Owing to the canopy density there is a tendency for the gradual domination of the mature phase by slow-growing shade tolerant trees which are able to survive and grow under the shade of the early mature phase. This accentuates the contrast between gap and mature phase in the forest.

On infertile skeletal and/or podzolized soils and peats the soil surface dries. This is exacerbated in large gaps and especially land-slides by oxidation or removal of the humus as well as the forest in which most of the available nutrients are stored. Though a few specialized macrophylls exist, usually with waxy or hairy leaves (e.g. *Ficus grossularioides*, *Macaranga pruinosa* in South-East Asia) most gap species occur in the mature phase on these soils. The main difference between mature and gap phase is therefore a quantitative one, the gap phase favouring species with relatively high growth rates, efficient dispersal and plentiful seed production. The return of mature phase forest in large gaps on podzolized soils is very slow indeed; there is reason to suppose that soils are frequently irreversibly impoverished and further podzolized (see Ashton, 1964; Brünig, 1970); on skeletal podzols succession may even be deflected more or less permanently to scrub.

Altitudinal variation

With the exception of the notable experiments of Kramer (1926, 1933) the gap phase on mountains has not been observed in any detail. The pattern varies with altitude, possibly because water is less limiting on the mountains in spite of increased soil acidity. Regeneration of the upper montane forest, and stunted forest on crests at lower altitudes, is essentially similar to that on infertile lowland soils, but the canopy always retains an impoverished array of macrophylls, some of which are unknown in the lowlands (e.g. *Symingtonia populnea* in South-East Asia). The work of Kramer on the regeneration in artificial openings in the mountain rain forest of Mt. Gede, western Java, showed a regeneration similar to that in natural openings. He found that in artificial openings <0.1 ha in extent, the primary forest dominants survived and made good growth. In

openings of 0.2–0.3 ha, the regeneration and growth of primary forest dominants were suppressed by the dense growth of secondary species. Further development of regeneration depended on the tree species and the degree of enlargement of the openings. After five years the tops of these species reached the lower layer of the crown of the surviving adjacent canopy trees.

Species diversity

Species diversity is universally less in the gap than in the mature phase. The leaves of many species of pioneer and later seral stages (e.g. *Anthocephalus*, *Octomeles* spp., *Shorea leprosula*) on submesic and mesic sites are rapidly reduced to skeletons by herbivorous insects. It appears that the products of photosynthesis are channelled to maximize growth, and a low premium is put on leaf longevity; gap phase species apparently divert less energy into toxin production. This is supported by the well-known fact that browsing mammals preferentially feed on gap phase species.

As with the mature phase, the gap phase flora becomes impoverished on isolated land surfaces of limited area. This allowed Whitmore (1974) to monitor the growth of the 12 commonest species over 5 years in the floristically poor forests of Kolombangara, Solomon Islands. Adequate populations for study could be identified on limited areas, and thus the characteristics of their reproduction biology and growth, which allowed each to adopt a specific role in the growth cycle of the forest, were identified.

Plant biology and regeneration in seasonal forests

Categories and their delimitation

Only aspects of seasonal forest dynamics which differ from those of evergreen forests are reviewed. In addition to leaf-fall, the two major differences which distinguish seasonal forests from evergreen are their phenology, particularly of flowering and fruiting, and their diversity; both are causally related with fundamental differences in their dynamics.

Within areas occupied by perhumid forest limestone hills support seasonal forests; while in areas normally considered as within the seasonal forest zone perhumid forests can occur on basalts and fertile alluvium. In both cases the forests may largely conform in dynamics and leaf phenology with the normal perhumid types, but are floristically poorer and frequently with a tendency to single species dominance; flowering and fruiting, however, is a more predictable annual event.

Fire

The presence or absence of fire has such fundamental effects, particularly on the structure, germination, early establishment and floristic composition and hence the dynamics of tropical forest ecosystems, that two categories of seasonal forest are considered separately under these headings: closed seasonal forests, and savannas. Savannas are outside the scope of this Report, but they have greatly spread into closed seasonal forest areas through the agency of man especially on free-draining and shallow soils.

The influence of wildfires within tropical and subtropical humid forests is somewhat controversial and appears always to be associated with human activity. In seasonal humid forest types with a prolonged dry season, such as in the Gogol Valley near Madang, Papua, there is evidence from historical records and the presence of charcoal in top soils that wildfires burnt extensive areas of rain forest in the 1890s and in 1914. The accompanying dry seasons were so severe that the rivers in the Gogol Valley were said to have dried up. The dwila (*Intsia palembanica, I. bijuga*) is readily regenerated by fire and is estimated to provide about a quarter of the log volume of the Gogol forests. This is regarded by some foresters as further evidence for catastrophic fires throughout the region in earlier times. In tropical and subtropical humid coastal Queensland, wildfires have not been observed inside unlogged rain forest areas, but often invade rain forest margins where there is inflammable weed growth or slash from logging. However, Gardner and Ridley (1961) claimed that wildfires may penetrate subhumid rain forests in South Queensland. They quoted the destruction of 80 ha in 1957, and noted that thickets of the introduced *Lantana camara* are commonly established on such areas. Attrition of the edges of rain forest by frequent fires from adjacent savanna and open woodland is documented throughout moist tropical and subtropical areas. For example, in North Queensland, annual burning of rain forest margins has resulted in the replacement of rain forest by grassland, especially on relatively shallow and infertile soils derived from acid schists and granite. The occurrence of fire-tolerant sclerophyll forest (characterized by species of *Eucalyptus, Tristania, Acacia, Casuarina, Melaleuca*, etc.) on the latter soils within the humid tropical rain forest region of North Queensland is also correlated with wildfires before European settlement (Webb, 1968). In the wettest seasonal and perhumid climates repeated burning by man leads to species poor grassland dominated by coarse perennial species (e.g. *Imperata* in Asia and Africa, see chapters 9 and 20).

Mature phase

Characteristics of canopy structure

In weakly seasonal forests the heterogeneous canopy structure of perhumid forest is retained, with more or less isolated emergents, a dense main canopy and structured understorey. Such forests are broadly confined to areas having a mean monthly rainfall of at least 75 mm, and include the tropical evergreen rain forests of Indo-Burma, Assam, and the Western Ghats, all those of Africa and the majority in South America. In regions with increasingly long dry seasons the forest stature declines; emergents become scattered and eventually disappear, and the canopy becomes even; epiphytes, but not climbers, also decline.

Phenology

The region of deciduous forests is distinguished by a single more or less prolonged dry season. For miombo in Zaire, Malaisse (1974) recognizes five seasons of different phenological activity, a hot dry season following the cold dry one—the southern analogues of those of South-East Asia; also early and main wet seasons followed by a less dependable late wet season.

Leaves

Though the deciduous habit becomes prominent in the seasonal tropics it is not universal and is primarily related to soil nutrient status and not climate as generally supposed. If nutrient status is moderate or high there is a greater tendency towards deciduousness, initially among the emergents, then the main canopy, and eventually in the whole forest in regions with increasing dry season.

Examples

Forest	Degree of deciduousness	Climate
Dipterocarp forests of Luzon	Low, a few emergent species	4 months dry season with intermittent rain throughout.
Rain forest of Gorakpur, Assam	Moderate, mainly among emergents	3 months dry season, usually no rain.
Semi-deciduous forest, Laos	Over 55 per cent of main canopy; 17 per cent in understorey; 35 per cent in herb layer	3–6 continuous months without rain.
Teak forests of Gir, Kathiawar	Completely deciduous canopy; 90 per cent deciduous understorey	1 000 mm of very unreliable rain during 4 months; no rain in dry season.

As has been noted in Indo-China (e.g. Vidal, 1966) leaf break generally precedes the onset of the monsoon, both in terrestrial plants and epiphytes, and Ho (unpublished) suggests this must be in response to the surge in atmospheric humidity that is known to occur then; the possibility of a capillary upward movement of soil water at that time was suggested, but no explanatory mechanism appears available. Different species are leafless for different periods and one species remains leafless for different periods in different climates; frequently individuals of a species in a single forest differ in their leafless period (Legris and Blasco, 1972). For example Vidal (1966) records that *Hevea* loses its leaves for less than a month, while *Delonix regia* (also an exotic in Asia) does so for longer in Laos; in Sarawak both are evergreen. Where soil moisture is higher, trees in deciduous forest may not be totally defoliated. For example near the living quarters of Sakaerat Experiment Station in northeastern Thailand where seepage from sewage occurs the deciduous trees still remain green in February, wehreas trees of the same species growing farther away have already shed their leaves. In the Thai dry dipterocarp forests there are two peaks of defoliation, a major peak in the middle of December and a minor in February which is associated with an increase in wind speed.

However, evergreen forests occur on excessively drained infertile soils in all these climates.

Examples: the seasonal dipterocarp forests of Arakan (Burma) and Cardamomes (Kampuchea) on yellow podzolic soils are almost entirely evergreen; and *forêt sempervirente basse* at low altitudes on humic podzols in the Kampuchea lacks deciduous species, yet the seasonal dipterocarp forests on latosols in the same region have a high proportion of deciduous species in the canopy.

The dry evergreen forests of southeastern peninsular India occur on coarse dry gravels, in the same climate region as totally deciduous seasonally flooded *Acacia* woodlands on montmorillonitic alluvial soils, and about 80 per cent dry deciduous forests on the semi-arid slopes of the Eastern Ghats.

In South America evergreen savanna is the rule in many regions, often adjacent to mesic valley sites bearing semi-deciduous closed forest.

A large amount of information on the phenology of African trees has been collected by Taylor (1960).

Leaf morphology and anatomy. Leaves of evergreens broadly conform to the perhumid forest canopy norm. Ferri (1963) has confirmed that these evergreen leaves continue to transpire freely throughout the dry season in Brazilian *cerrado*, and evidence suggests that even here stomatal control is weak after the leaf matures.

Leaves of deciduous species, by contrast, are diverse, many species adopting the growth strategy of perhumid gap phase species during their leafy wet season phase though owing to the resultant seasonal fluctuation in rates of xylem formation, and hence its density, pithy soft wooded tree species are a minority in the drier climates. Many are nevertheless characterized by large thin leaves on long slender petioles, usually borne in irregularly trullate or domed crowns rather than the flat dense crown-shapes seen in *Bombax*, *Terminalia*, *Alstonia* spp. and young *Ceiba* spp., which are prominent in the gap phase of weakly seasonal semi-evergreen forests. Physiological observations in these deciduous species, especially of stomatal response, appear not to exist but high stomatal sensitivity may be predicted.

Flowering

There are few detailed phenological studies of flowering from the seasonal tropics, though much casual observation is fragmented, or hidden in floristic or forestry accounts such as those of Capon (1947) in Zaire, De La Mensbruge (1966) in Ivory Coast and Champion and Seth (1968) in India. The great majority of trees in seasonal forests flower annually; mostly following leaf-fall and either before (most deciduous species) or at the same time or following (most evergreen species) the opening of the next flush. Flowering mainly takes place at the beginning or towards the end of the dry season (see e.g. Vidal, 1966; Longman and Jeník, 1974). Thus in Amazonia peaks of flowering occur at the beginning

of the wet and dry seasons; nevertheless some flowers are present throughout the year. Vidal recorded the proportion of species with different flowering times in Laos as follows.

	Period of flowering	Closed semi-evergreeen forest
Dry season	while deciduous	6 per cent
	while in leaf	65 per cent
Wet season		29 per cent

	Period of flowering	Mixed deciduous forest	Savanna
Dry season	while deciduous	19 per cent	5 per cent
	while in leaf	52 per cent	29 per cent
Wet season		29 per cent	64 per cent

Pongumphai (1970) recorded the approximate time for flowering and fruit ripening of some species in Thai semi-evergreen forests. Tree species in Dipterocarpaceae and *Parkia streptocarpa* (Leguminosae) flowered from December to February and produced ripe fruits form March to May, while *Dialium cochinchinensis* (Leguminosae) flowered from April to June and produced ripe fruit from December to February. *Irvingia malayana* (Simarubaceae), *Diospyros* spp. (Ebenaceae) and *Hydnocarpus ilicifolius* (Flacourtiaceae) flowered from February to April and produced ripe fruits from July to August, July to September and August to December respectively for each species.

By contrast with the aseasonal humid tropics many species, especially but not exclusively those which flower when leafless, bear large and brightly coloured flowers. For miombo in Zaire Malaisse (1974) observed that different woody synusiae possess differing flowering seasonality in semi-deciduous *Brachystegia* woodland. The understorey has a well marked flowering peak during the hot dry season, though some shrubs flowered towards the end of the rains; the herbaceous field synusia has two maxima the most important of which, mainly comprising hemicryptophytes, occurred at the height of the wet season, while another, clearly defined and mainly involving the geophytes, occurred at the end of the hot dry season. As Malaisse observes, the flowering may be postponed or telescoped according to the time of burning. Many evergreen irregular (though synchronous) flowering species in the perhumid zone flower annually in seasonal climates; e.g. *Hevea brasiliensis* (Euphorbiaceae), *Dipterocarpus gracilis* (Dipterocarpaceae) and *Terminalia catappa* (Combretaceae). Though so far as the species is concerned, flowering is annual, only a part of the populations in many weakly seasonal forests flower each year, individuals frequently failing to flower in two successive years. Florence (1964) compared flowering and seed production in six tropical Australian *Eucalyptus pilularis* stands and found that within an apparently homogeneous area there was a wide variation in the time of flowering. Where flowering was light no clearly defined seeding peak occurred, but there was a well defined peak of seed fall in a summer after heavy flowering. Florence suggests that variation in timing and intensity of flowering may reflect the considerable genetic diversity within species which results from environmental selection.

Triplochiton scleroxylon, a west African tree, has periodic flowering and fruiting habits. Abundant flowering is often associated with a low July-August rainfall, but defoliation by insects may also play a part and insect population dynamics may of course be affected by the climate (Jones, 1974, 1976). This species flowers annually but the intensity varies considerably throughout its range. Flowering as well as fruit set varies in intensity from year to year in all seasonal forests (see e.g. Champion and Seth, 1968). Evidence for flowering and leaf flushing in response to day-length and temperature variation in the seasonal tropics is summarized by Longman and Jeník (1974).

In the Asian monsoon tropics two major groups of gregarious monocarpic plants also occur (besides certain palms, e.g. *Corypha* spp.): various bamboos, see Janzen (1976) for a speculative discussion on causation and *Strobilanthes* spp. (Acanthaceae) mostly between 1 000–2 500 m. The gregarious flowering, fruiting and death of their dense understorey cover every 7–20 years, often on a regional scale has a profound effect on forest dynamics, especially germination and establishment of forest tree seedlings as well as on granivore populations and those of their predators.

Flower biology

Bawa (1974) has shown that the great majority of trees in Costa Rican seasonal semi-deciduous forests are outbreeders, often with self-incompatibility mechanisms. The climatically versatile teak (*Tectona grandis*), a gap phase species at the humid end of its distribution, is an outcrosser (Bryndum and Hedegart, 1969; Hedegart, 1976). It is noteworthy that hybridization in the wild and continuous morphological variation are not infrequent in seasonal closed forests and are characteristic of many closed seasonal forest and savanna genera (e.g. *Brachystegia*, *Monotes* spp. in Africa, savanna dipterocarps in Indo-Burma).

It would appear that a far greater proportion of trees in the seasonal tropics are pollinated by large, far-flying and systematic pollinators than in the everwet zone—these include large bees (e.g. for *Cassia* spp. and many other Leguminosae, *Tabebuia*, *Lagerstroemia* spp.) and birds (for *Bombax*, *Butea*, *Erythrina*, *Firmiana* spp.).

More is undoubtedly known of the flowering biology of tropical trees in cultivation than in the wild, and in particular the Caribbean and Asian pines which are wind-pollinated (Burley, 1976).

Fruit production

Though there is great diversity, many species share the characteristics of mature phase perhumid forest species; increasing length of dry season is associated with increase in fruit production (often by reduction of percentage mortality following anthesis) and decrease in fruit size. As a rule fruiting takes place towards the end of the dry season, but there are many exceptions. De La Mensbruge (1966) recorded

fruiting times among 350 species in semi-evergreen seasonal forest in the Ivory Coast; he found there that species producing many small fruit were generally more abundant than those producing a few large fruit, but this may be in part a reflection of the late secondary state of the more humid forests. Fruit production varies greatly from year to year: Malaisse (unpublished) recorded 160 kg/ha dry weight in Zaire miombo woodland in 1968, yet 2 t/ha in 1972 in a very dry sunny year following a wet year. In Thailand heavy seed years occur at intervals of 3–5 years. Fruiting phenology in Viet-Nam has been discussed by Hôi (1972). No examples of irregular embryo development are known from seasonal deciduous forests, though adventive polyembryony was demonstrated in the evergreen dipterocarp *Hopea odorata* of Indo-Chinese riparian forests (Maury, 1970).

Dispersal

Fleshy fruit decline in regions of increasing dry seasons. Dry, especially winged or minute wind-dispersed, diaspores commensurately increase. The proportion of wind-dispersed canopy species in West African (Longman and Jeník, 1974) and South American (e.g. Rollet, 1974) rain forests is much higher than those of oceanic South-East Asia. Long distance dispersal is therefore probably of regular occurrence (Gómez-Pompa *et al.*, 1972) in many families, local endemism limited, except in certain highly specialized species, especially of dry evergreen forests, e.g. *Diospyros* spp., and many species are extremely wide-spread—several from Africa to Moluccas and even pantropical. Such species appear to be ecologically tolerant.

Dormancy

Whereas most species of semi-evergreen forest lack seed dormancy, see e.g. Hôi (1972) for Viet-Nam, in areas with a marked dry season and an increasingly unreliable wet season there is an increasing proportion of species whose seeds possess dormancy (e.g. Troup, 1921; Dent, 1942) while many increase in germination capacity for many months following fruit-fall, e.g. *Adina cordifolia*, *Albizia lebbeck*, *Cassia fistula*, *Terminalia chebula*, and *Trewia nudiflora* in India. This is most marked among species which fruit early in the dry season, e.g. *Mesua ferra* in South-East Asia (Richards, 1952). In Africa it has been found that dormancy can be prolonged if relative humidity is controlled; according to Brunck (in De La Mensbruge, 1966) okoumé seeds retain their viability for a year if kept between 6–8 per cent relative humidity at normal temperature, and for several years if at 2 per cent (Gauchette, 1958).

Germination and establishment

Closed forests. Increasing seasonality of rainfall imposes a season of increasing length unfavourable to seedling germination and establishment, counteracted in deciduous forests by the particularly favourable light climate near the ground during the dry season.

Germination generally follows fruit-fall early in the wet

season; rapid increment leads to self-thinning during the deciduous phase in moist deciduous forest where many of the saplings remain evergreen. In dry forests the saplings too are frequently deciduous, and growth becomes seasonal and confined to the wet season. Mortality may be principally caused by:

— unreliability of rainfall, especially at the time of commencement of the ensuing wet season (particularly in drier forests—see Rollet, 1962);
— intense drought in the first dry season following establishment;
— herbivore predation;
— self-thinning.

As a consequence seedlings are relatively ephemeral in most kinds of seasonal forests and sylvicultural methods involving seed trees are usually necessary if planting is to be avoided; seasonal evergreen rain forests are the most frequent, but not universal exception.

For a review of the phenology and seed properties of a wide range of tree species of the seasonal tropics, see FAO (1955). For a detailed account of dormancy, germination and establishment in West African seasonal evergreen and semi-deciduous forests, see Gilbert (1938), Jones (1950, 1956) and De La Mensbruge (1966); special studies include those on *Brachystegia laurentii* (Germain and Evrard, 1956), *Gilbertiodendron dewevrei* (Gérard, 1960) and various other commercial species (CTFT monographs).

A special case is provided by Asian forests, of which the Burmese teak forests form an example, in which the understorey consists of a dense brake of monocarpic bamboos. Light intensity beneath the brake is inadequate for seedling survival, and regeneration in nature occurs as a result of, and in managed forests is planned around, the successful establishment of seed falling in the year the bamboos gregariously fruit and die.

Species diversity

Species diversity of forest declines abruptly at the margin of the aseasonal tropics. 10 ha of perhumid forest on soils of average fertility in Malaysia may contain 550 spp., Thailand seasonal evergreen forests on similar soils a maximum of 100. This (besides Pleistocene history) may largely explain the lower floristic diversity of African and most South American forests compared with those of West Malaysia. Diversity at generic level is retained in seasonal evergreen forest however, and so apparently is its biochemical diversity (Janzen, 1971). Both appear to decline most rapidly in deciduous forest.

Altitudinal variation

The cloud layer which often persists on seasonal tropical mountains and marks a sudden transition to vegetation essentially similar to that in similar habitats in the perhumid zone, probably explains why girth increments on African mountains are said to exceed those of the seasonally arid adjacent lowlands.

Gap phase

The distinction between mature and gap phase is less clear in seasonal forests as few have never been felled. This is particularly true in western and central Africa were it is frequently observed that the principal canopy species fail to regenerate under their own shade (Jones, 1956; Longman and Jeník, 1974); this led Aubréville (1938) to propose his mosaic or cyclical theory of regeneration. The most likely explanation is that these forests are in a very late stage of seral succession in which shade tolerant species are still replacing their more light demanding predecessors. Where primary evergreen forests exist the gap phase conforms to the same principles as in perhumid forests, though the deciduous element on comparable soils increases. Gap phase species of moist seasonal semi-evergreen forests in Asia (e.g. *Gmelina arborea*, *Tectona grandis*) and Africa (Taylor, 1960; Longman and Jeník, 1974) come to occupy mature phase niches in dry deciduous forests, the deciduous element being particularly represented in late succession in the gap phase of semi-evergreen forests. In larger gaps an initial phase of fast-growing herbaceous species is recognized in West African forests, which may persist and hinder regeneration of economic tree species.

Mangrove forests

Mangrove forests are distributed throughout the tropics where suitable accumulating mud banks or sheltered sandy bays occur, their latitudinal limits being defined by frost. On the west coast from the River Senegal to the River Longa in Angola, including the islands of the Gulf of Guinea, there are patches of mangrove dominated by *Rhizophora racemosa*, *R. harrisonii* and *R. mangle*; *Avicennia nitida* sometimes occurs behind the *Rhizophora* zones. The most extensive mangrove forests are in the Niger delta. The West African mangrove species also occur in the Atlantic shores of tropical America and the West Indies. See chapter 4.

The mangroves of Madagascar and the eastern shores of Africa have a strong Asian affinity. Patches are found at intervals from the Red Sea at about latitude 15° N, southwards to about latitude 32° S. The important species are *Rhizophora mucronata*, *Sonneratia caseolaris*, *Ceriops tagal* and *Bruguiera gymnorrhiza*—all members of the Rhizophoraceae—and *Avicennia marina*.

All mangrove species are evergreen. Their physiology is still inadequately known but apparently depends on constant uptake and leaching or secretion of salt in such a way that the osmotic concentration within the plant cells always exceeds the salt concentration in the water in which they stand. Arrest of uptake (e.g. by frost or defoliation) is immediately lethal.

Leaf change, flowering and fruiting take place in most species in frequent cycles at more or less regular intervals throughout the year. Fruits are generally water-borne, and several have special adaptations (e.g. precocious elongation of photosynthetic hypocotyl in many Rhizophoraceae) which facilitate establishment in soft mud. Few species regenerate by sucker shoots if damaged, which provides a problem in the regeneration of clear-out forest (see Watson, 1928).

Tree growth rates

Introduction

The popular conception of tropical rain forest is of an immensely fecund and fast-growing type of vegetation, luxuriant and with great potential. This view has been held by travellers from Hanno onwards, and is caused by familiarity with the margins of the rain forest, the edges of clearings and water-sides. Even today the 'great bulk, luxuriance and potential of the tropical forest' is referred to by scientists; this recent quotation comes from Longman and Jeník (1974). To be sure, tropical high forest still covers a substantial area of land; see chapter 1. Multiplying these values by others for standing biomass and net biomass production does indeed produce some very large figures. However, much careful qualification is needed before one can claim that the values indicate that tropical rain forest has great potential.

The invasive, pioneering species which regenerate in gaps and clearings and on abandoned land—the biological nomads in the terminology of Van Steenis (1958)—certainly are capable of large volume increments. But generally the components of forest 40 or more years old do not grow exceptionally fast. There are numerous tales of foresters being sent back to check measurements 'since such slow growth could obviously not be correct in tropical rain forest' (Wadsworth, quoted in Baur, 1962). Nevertheless tropical forests are notable for the large biomass that they are able to build up even on inherently infertile soils (see chapter 10).

There is demonstrably great variation in growth rates between species (e.g. Schulz, 1960, for Surinam). A 14 m girth *Bertholletia excelsa* in the Jari River region, Brazil, was considered to exceed 1 400 years (Murça-Pires, unpublished). The following ranges of life-span have been predicted.

Life-span (years)	Nigeria (Redhead, 1968)	Malaysia (Cousens, 1965)
20–40	*Musanga, Macaranga* spp.	
50–100	*Albizia, Ceiba, Pycnanthus, Ricinodendron* spp.	
100–300	*Anonidium, Celtia, Myrianthus, Pentaclethra* spp.	
300–400	*Gossweilerodendron, Guarea, Chlorophora* spp.	*Shorea leprosula*
1 400		*Balanocarpus heimii*

This adds a further dimension to our understanding of temporal change in an intrinsic spatial pattern. Normal maximum ages have been estimated as 200–450 years in Malaysian dipterocarps (Wyatt-Smith, 1968) and 300–350 years in *Lophira alata* and *Guarea cedrata* in western Africa (Jones, 1956).

Heinsdijk (1963) analysed the stand structure of 120 000 trees in Amazon rain forest and predicted that the average tree of diameter class 25–35 cm, would have a diameter increase in the order of 8 mm/a; between 25 and 155 cm, the increment would be between 8 and 3.7 mm/a; and the average age of such ideal trees would range between 27 for trees between 25–35 cm diameter and 418 years for those between 145–155 cm diameter. It is dangerous to infer increment rates from stand structure, for the latter may in part be a product of differentials in growth rates with age and in part long-term fluctuations in recruitment either on a local (in which case it can often be detected) or regional scale; for a detailed discussion see Schulz (1960) and Rollet (1974). The frequent existence of dense stands of large trees of a particular species with little or no regeneration beneath may be an indication of past disturbance, as in *Campnosperma brevipetiolata* forests in the Solomons (Whitmore, 1974), *Goupia glabra* in Surinam (Schulz, 1960), or in many West African forests; or long-term phasic succession as in *Shorea albida* peat swamps in Sarawak (Anderson, 1961); or yet again competitive exclusion, in dense canopied forests in dynamic equilibrium, by species with slow increment and great longevity in the canopy. These latter forests are common on well watered fertile sites, as *Mora* forests in Guyana, dipterocarp forests on basalt in Sarawak (Ashton, unpublished), or *Eusideroxylon zwageri* forests in Sumatra (Koopman and Verhoef, 1938).

Tables 2 and 3 show how much greater is the potential for growth under plantation conditions, at more or less controlled stand density, and especially on the rich volcanic soils of Indonesia, than in primary forest. Some plantation species produce up to 28 t/ha/a. On more common, less fertile soils production has reached 9 t/ha/a (Dawkins, 1963), but more usually the range is 3–6 t.

The highest yields in plantations come from rich soils, high latitudes or altitudes and narrow-crowned species. The greater productivity of the outer tropical lowlands or inner tropical highlands is explicable in terms of less strongly leached soils and by the cool nights theory. The very substantial losses from gross production because of continual respiration, 75 per cent in Müller and Nielsen (1965), 55 per cent in Weck (1960), are thought to be reduced under conditions where there is a substantial difference between daytime and night-time temperatures. In primary forest, mean girth increments for individual girth classes in samples of Sarawak mixed dipterocarp forests (Ashton, unpublished) mainly varied in relation to altitude, obscuring the little variation in relation to soils. Trees in forest at 500 m altitude had approximately half the mean girth increment (in all size classes) of those near sea level (16 and 28 mm at 10–20 cm diameter; 25 and 56 mm at 40–50 cm diameter) in contrast with observations in plantations. It is obvious that more stems of narrow-crowned species can stand on a unit area than those of wide-crowned species. Dawkins (1963) found that, over the range 4–36 m²/ha, increased basal area density was generally associated with higher yields while individual tree diameter increments were reduced. This could explain why Ashton also found that volume increment of natural stands varied more in relation to soils than altitude.

TABLE 2. Estimates of above-ground biomass.

Locality	Forest type	Altitude (m)	Volume (m³/ha)	Oven dry weight (t/ha)	Author
Siak/Mandai, Riau, Sumatra	Lowland dipterocarp	0–150	198	143	Dilmy, 1971
Dumai, Riau, Sumatra	»		219	164	»
Duri, Riau, Sumatra	»		239	183	»
Semangus, Bianchi, Sumatra	»		188	128	»
Nunukan, NE Kalimantan	»		434	293	»
Tarakan, NE Kalimantan	»		320	214	»
Sangkulirang, NE Kalimantan	»		246	170	»
Sampit, Central Kalimantan	Peat swamp forest		193	132	»
Sampit, Central Kalimantan	*Agathis borneensis*		429	233	»
Gede-Pangerango West Java	*Altingia excelsa* montane rain forest	1 100–1 500	796	549	»
Ghana	Secondary forest	lowland		370	Greenland and Kowal, 1960
Anguédédou Ivory Coast	Almost natural *Turraeanthus* sp.	lowland	421	240	Müller and Nielsen, 1965
Malaysia	Budded rubber plantation (7 years)	lowland		138	Wycherley, 1969
Malaysia	Oil palm plantation (7 years)	lowland		43	»
Malaysia	Budded rubber plantation (30 years)	lowland		170	Wycherley and Templeton, 1969
Malaysia	Mature tropical high forest	lowland		400	»
Malaysia	Lowland dipterocarp	lowland	490–700		Wong, 1967
Estimates for tropics	Tropical forest			450 (range 60–800)	Longman and Jeník, 1974
Malesian region	Mixed dipterocarp			960	Brünig, 1967

TABLE 3. Estimates of net above-ground biomass production.

Locality	Forest type	Altitude (m)	Volume (m³/ha/a)	Weight (t/ha/a)	Author
Indonesia	*Michelia velutina* plantation (20 years)		28	13	Ardikosoema *et al.* 1955
Anguédédou Ivory Coast	Almost natural *Turraeanthus* sp.	lowland	13.1	7.5	Müller and Nielsen, 1965
Malaysia	Budded rubber plantation (7 years) (4 years)	lowland		20.4 35.5	Wycherley, 1969
»	Oil palm plantation (7 years)	lowland		6.4	Wycherley, 1969
»	*Gmelina arborea* plantation	lowland		13.9±1.0	Wycherley, 1969
»	*Albizia falcata* plantation	lowland		26.6	Wycherley, 1969
»	*Eucalyptus robusta* plantation	lowland		10.2	Wycherley, 1969

Locality	Forest type	Altitude (m)	Volume (m³/ha/a)	Weight (t/ha/a)	Author
Malaysia	*Pinus merkusii*	lowland		13.7	Wycherley, 1969
»	*Pinus caribaea*			6.3	Wycherley, 1969
»	*Pinus kesiya*			9.6	Wycherley, 1969
»	*Dryobalanops aromatica* dipterocarp forest	lowland		8.2±1.1	Wycherley, 1969
»	*Dipterocarpus* spp. plantation			10.1±0.8	Wycherley, 1969
Indonesia	*Agathis loranthifolia* plantation (35 years)	> 300	42	20	Sudarmo, 1956
»	*Agathis loranthifolia* plantation (25 years)	> 300	29	14	Sudarmo, 1956
»	*Altingia excelsa* plantation (15 years)	> 300	24	16	Wulfing, 1949
»	*Altingia excelsa* plantation (55 years)	> 300	9	6	Wulfing, 1949
»	*Anthocephalus chinensis* plantation (6 years)		24	11	Sudarmo, 1957
»	*Anthocephalus chinensis* plantation (24 years)		16	7	Sudarmo, 1957
»	*Dalbergia latifolia* plantation (20 years)	lowland	37	28	Wulfing, 1949
»	*Dalbergia latifolia* plantation (45 years)	lowland	19	15	Wulfing, 1949
»	*Ochroma bicolor* plantation (2 years)		132	16	Wulfing, 1949
»	*Ochroma bicolor* plantation (4 years)		61	7	Wulfing, 1949
»	*Pinus merkusii* plantation (20 years)	> 300	46	25	Wulfing, 1949
»	*Pinus merkusii* plantation (25 years)	> 300	24	13	Wulfing, 1949
Malaysia	*Shorea leprosula* (yield table: 64 years)	lowland	24	12	Noakes, 1937
Indonesia	*Swietenia macrophylla* plantation (15 years)	> 300	46	21	Wulfing, 1949
»	*Swietenia macrophylla* plantation (35 years)	> 300	16	7	Wulfing, 1949
»	*Tectona grandis* plantation		23	15	Wulfing, 1949
»	*Tectona grandis* plantation		14	9	Wulfing, 1949
»	*Tectona grandis* plantation		9	6	Wulfing, 1949
Mean estimate for tropics	Tropical high forest			3.0–3.5	Weck, 1961
Sumatra and Kalimantan	Dipterocarp forest	lowland		5.8	Soerianegara, 1965
Gede-Pangerango, West Java	Montane rain forest	1 100–1 500		12.4	Soerianegara, 1965
Ghana	Secondary forest	lowland		8	Greenland and Kowal, 1960
Malaysia	Oil palm plantation	lowland	maximum rate	30.3	Wycherley and Templeton, 1969
Estimates for tropics	Tropical forest			20 (range 10–50)	Longman and Jeník, 1974

Where both volumes and weights are given, they have been taken from Dawkins (1963). The volumes are maximum rates taken from the yield tables.

The largest size classes are only represented on average and more fertile sites and are densest in the latter. On infertile and xeric sites the standing volume of wood over bark (including twigs) is of the order of only half those on fertile sites; however, their annual volume increment may be similar for they have a much denser understorey. Forests on sites of average fertility have higher standing volumes and increments than both. The trade-off point in forest management would be associated with decreased crown exposure and increased crown contact. It would be difficult to find such a point for any one combination of species, site and age; in highly variable tropical forest a more pragmatic approach is needed.

Brünig (1967) points out that the wide-spreading crowns have a low ratio of assimilating crown surface to respiring crown volume; narrow-crowned trees would have a higher production.

Chapter 10 considers the measurement of biomass and production. A number of formulae have been proposed by workers in the temperate zone to estimate world-wide productivity from climatic data. While these attempts may have some local comparative value within a region (Theron, 1973; Kingston, 1974), most workers in tropical forests prefer the more direct approach (Jones, 1959; Dawkins, 1963; Brünig, 1967).

Standing biomass and production data comprise only a fraction of the data on growth rates. Most of the figures are held by government forest services in unpublished records collected for two main purposes:
— to estimate the effects of sylvicultural treatment;
— to estimate yields of commercial timber.
Most of the data were intended for the second purpose or, more specifically, for the calculation of a definitive rotation age. Rotation length is a difficult concept to apply to tropical forest. What managers actually need is a measure of felling cycle length; the rate at which the forest can be cut while maintaining sustained yields (Wyatt-Smith, 1968).

Forest managers are principally interested in standing volumes and volume increments per unit area for large areas of forest. It is impracticable to measure these quantities directly; it is necessary to select and to measure parameters more readily accessible but having some close relation with the two parameters required by the managers. The principal parameters selected are diameter (DBH) or girth (GBH), at usually 1.3 m above ground level or above buttresses and, to a lesser extent, height. Other parameters which may be measured include diameters at various stem heights, crown depth and diameter. A number of other characteristics which are not susceptible to simple measurement have been scored, with varying degrees of success.

Growth rings

The autoregressive effects of past growth on future increment have been known for many years in Europe and North America (Gevorkiantz and Duerr, 1938). The study of past growth by examination of annual growth rings is a routine operation in most temperate forests. If similar work could be carried out in tropical forest, the forest manager could be supplied with growth estimates a great deal more rapidly than is the case at present. Unfortunately many of the desirable timber trees do not form consistently and easily recognizable annual growth rings. Hummell (1946) thought he could distinguish concentric parenchyma bands formed at the beginning of each of two growing seasons in a single tree of *Entandrophragma angolense* from Ghana; the rings in a *Khaya grandifoliola* were indistinct.

Lowe (1961, 1968) showed that iodine staining would show up the starch in the probable annual rings in *Triplochiton scleroxylon* in Nigeria, but only in the sap-wood so it was possible to count back for about twenty years. Of six 8-year-old trees in Gambari Forest Reserve, the four larger showed 8 rings and the two smaller 6 rings. Lowe thought that making up for mortality in the original crop by replanting could have caused the discrepancy. Rings which were not sharply delimited, or which did not show a concentration of starch in the late wood, were regarded as false and were not counted. Similar criteria were used in his study of three 21-year-old trees in Sapoba Forest Reserve. Two trees showed 19 rings plus 3 heart-wood rings, the other showed 15 rings plus 9 heart-wood rings. The Sapoba area has a rainfall of *ca.* 2 030 mm/a, the Gambari area *ca.* 1 125 mm/a; and both have a single dry season. Onochie (1947, quoted in Lowe, 1961, 1968), also working in southern Nigeria, found evidence of annual growth rings in *Triplochiton* in plantations and in natural forest. Roberts (1961) examined monthly cores from 30 *Triplochiton* trees in the Ofram Headwaters Forest Reserve, northwest of Kumasi in Ghana. He noted that growth rings are not always defined, but his technique was more coarse than that used by Lowe, and possible anatomical differentiation was not sought. Evidence supporting non-seasonal growth in Ghana is given by Lowe (1968) who quotes four years of study girth increment for two *Triplochiton* trees in the Bobiri arboretum near Kumasi.

Other West African species have been studied by Amobi (1973) in forest at Ibadan, Nigeria. Cores were take at weekly intervals for 9–17 months from one tree of *Bombax buonopozense, Bosquesia angolensis, Daniellia ogea, Hildegardia barteri, Monodora tenuifolia* and *Ricinodendron heudelotii*. He found that growth rings were periodic, not always distinct, and generally annual; there were interspecific differences. The staff of the Centre technique forestier tropical have re-examined a number of West African woods: *Afzelia* sp., *Entandrophragma* sp., and *Isoberlinia doka* have parenchymal bands limiting growth periods. These bands are, as in *Tectona grandis*, more pronounced in strongly seasonal climates. A change in the size and spacing of the bands is noticeable in *Acacia albida, Cassia siamea* and *Pterocarpus erinaceus*, while the late-wood fibres are flattened in *Aucoumea klaineana* and *Khaya ivorensis*. Mariaux (1967, 1969), reporting this work, mentioned several techniques besides anatomical examination—chemical treatment with phloroglucin and iodine, ultra-violet light, beta-ray densitometry—but none have proved completely successful. Even when trees appear to be circular the rings are frequently wavy, discontinuous and indistinct.

In west-central Malaysia, Menon (1947) noted that

Tectona was diffuse-porous and produced two rings a year; it is ring-porous in northeastern Malaysia, under a seasonal climate, with one ring produced apparently annually (Wyatt-Smith, personal communication). The partial association between a deciduous habit and ring formation was noted in Java by Coster (1927, quoted in Lowe, 1968) and confirmed by Lowe (1968). Coster studied 85 species over three years and found that sharply defined growth rings were formed only in species that were deciduous, though not all deciduous species formed rings.

Tschinkel (1966) worked on *Cordia alliodora* in Costa Rica. After examining sanded discs from various heights in trees of more or less known age, he concluded that annual growth rings could be distinguished. Like Hummell (1946), Tschinkel emphasized the need to examine whole discs in order to differentiate between whole annual rings and incomplete other rings. Two or more very narrow rings could appear to be only one, but as the rings normally become more distinct towards the top of the tree it is possible to separate the rings by reference to upper discs. Tschinkel noted the importance of the pattern of wide and narrow rings, rather than their absolute size, in helping to trace indistinct rings.

Brünig (1971) noted that clearly visible bands of late-wood occur commonly in heath forest trees in Sarawak and quoted *Agathis borneensis* as an example although it has not been shown that the irregular bands are formed annually.

Even tropical pines cannot be relied upon to produce annual rings. Using beta-ray densitometry, Harris (1973) found a small zone of core-wood with poorly developed late-wood was succeeded by a rapid change to multiple bands of very dense wood within each annual layer in *Pinus caribaea*, *P. merkusii* and *P. oocarpa* from Malaysia, *P. caribaea* from Bukoba on the western, wetter, side of Lake Victoria is being examined by X-ray densitometry at Oxford; it has shown almost no changes in density from pith to bark after 10 years' growth (Plumptre, personal communication).

It is noteworthy that all the species for which annual rings have been demonstrated or are claimed, are deciduous. Evergreen trees do not seem to form easily recognizable growth rings. Even in the deciduous species, it seems necessary to adopt destructive sampling in order to trace whole rings. For this reason, ring marking by wounding techniques, such as the micro-needle used by Wolter (1968) and McKenzie (1972), is likely to be useful only in species where the rings are normally clear.

Wounding, by micro-needle or bark windows, as a mean of determining growth between known dates, suffers from several disadvantages: the extraction and measurement of wedges of wood is a skilled task; it is very damaging to the tree; and growth along any one radius may be unrepresentative of growth over the whole disc. However one can discount the likelihood of exaggerated growth due to stimulation of the cambium; for with careful wounding this is barely detectable. Wounded trees are unsuitable as representative of the surrounding stand or potential crop and the method cannot be recommended for large-scale growth studies.

Accuracy of growth measurements

For the majority of tropical forest trees, recurrent measurement of girth or diameter is preferable to disc or wedge studies. Most areas standardize the measurement height at that recommended by IUFRO—1.30 m above ground level on the uphill side of a tree standing on a slope although some countries measure at 1.37 m. For buttressed trees there are many conventions. It would seem sensible to allow for further height growth of buttresses, and therefore to measure either well above most buttresses, e.g., at a standard height of 3 m, or at a certain distance above the top of the buttresses as they are when the tree is first assessed. The former method is preferable.

The smallest division of the measurement scale should be large enough to be read unambiguously and to be unaffected by a reasonable range of tensions in the measuring tape. For intervals or measurement of a year or more, a smallest division of 1 mm diameter is frequently used, with instructions to read either to the completed division or to the nearest division. Most research work is now carried out with unstretchable steel or virtually unstretchable fibre-glass tapes.

1 mm diameter is the smallest generally useful division on a measurement scale, but *ca.* 5–10 per cent of the annual growth. Mensurational error may thus be rather high. For periods of less than a year smaller divisions must be used. Hand-held tapes then raise problems of error due to differences in tension and, on large trees, to the difficulty of positioning a tape directly on a marked point or paint ring. Yacom's (1970) vernier scale, to read tenths of the smallest tape division, cannot avoid these problems.

In Uganda, Dawkins (1956) conducted a test with a reliable forest ranger and a steel tape with gradations of one sixteenth of an inch (1.6 mm) for girth measurements. Agreement between two girth and three height measurements (by Haga hypsometer) within one week on 316 trees was given as:

girth

difference between 2 measurements	0 –	67 %
difference between 2 measurements 1 scale division		31 %
difference between 2 measurements 2 scale divisions		2 %

height
difference between most divergent
of 3 measurements

< 3 feet	76 %
4–6 feet	22 %
> 6 feet	2 %

Dawkins does not say how these height differences relate to divisions on the hypsometer scale. He went on to develop his ten-ring method (Dawkins, 1956). Ten narrow horizontal rings were painted on a tree, with the inter-ring distance such that there was less than one inch (25.4 mm) difference in girth between the largest ring and the smallest. The mean of the ten rings was expressed in hundredths of an inch. Discrepancies between measures on the same day were:

0	– 78 %
1 scale division (2.5 mm)	– 20 %
2 scale divisions (5.1 mm)	– 2.5 %

The ten-ring method was used by Schulz (1960) in Surinam and by Nicholson (1958) in Sabah. It is time consuming and expensive, and would not normally be used except in areas where fixed dendrometer bands and dial gauges are likely to be stolen.

For very accurate work there are a variety of electrical instruments, mainly based on displacement transducers (Impens and Schalck, 1965; Dobbs, 1969; La Point and Cleve, 1971; Kinerson, 1973). The Japanese workers seem to favour electrical strain gauges (Kuroiwa, 1959; Ninokata and Miyazato, 1959). A review of instruments is given by Geissler (1970).

Absolute growth rates

In trees height increment with time follows a sigmoid curve; after an initial phase while the seedling becomes established it passes through a logarithmic phase of growth, finally declining into the mature phase. Among rain forest trees the height increment curve varies with crown architecture, but in mature phase species there appears to be no clearly defined logarithmic phase of height increment, growth initiating relatively rapidly owing to the large seed food reserves, and proceeding rather erratically and relatively slowly (even when conditions are optimal) until the final height is approached. The girth increment continues throughout the life of the tree, although it declines with senility.

Nevertheless, there are several records of apparently healthy tropical forest trees making no measurable diameter growth over long periods: 17.5 years in Trinidad (Bell, 1971); 12 years in New South Wales (Baur, 1962). 28 years in Nigeria (Keay, 1961). Non-growth is known in many other places, but it has frequently been ascribed to poor mensurational technique (Dawkins, 1956). Very slow growth is also well known. At the other extreme, rapid growth in length of shoots, roots and in diameter have also been recorded, mainly in nurseries and in young plantations. The following are a few representative data.

Annual increment

Species	Height increment (m/a)	Locality	Author
Albizia moluccana	3	Andaman Islands	Bradley, 1922
Albizia moluccana	10	Mindanao, Philippines	Revilla, personal communication
Terminalia superba	2.8	Zaire	Lebrun and
Musanga cecropioides	3.8	Zaire	Gilbert, 1954
Ochroma lagopus	5.5	Stann Creek, Belize	Anon., 1960
Cedrela odorata	3.1	Sapoba, Nigeria	Lamb, 1968
Cecropia peltata	5.0	Costa Rica	Davis, 1970

Species	Diameter increment (cm/a)	Locality	Author
Ochroma lagopus	9.0	Stann Creek, Belize	Anon., 1960

	Root increment		
Mean rate for fast-growing seedlings	2.0 cm/day	Java	Coster, 1932
Cissus sicyoides (aerial roots)	9.6 cm/day	Java	Coster, 1927
Theobroma cacao (tap-roots)	1.5 cm/month	Ghana	McKelvie, 1954

Effects of hydration and dehydration

Dendrometer measurements cannot distinguish between the various causes of tree expansion and contraction: xylem production, phloem production, periderm production, bark shedding, shrinkage and swelling due to dehydration and hydration. It is generally observed that tree size is at a maximum when humidity is high and at a minimum when humidity is low and transpiration high (Iyamabo, 1971). The difference between any two successive troughs or crests in a dendrograph plot is a measure of apparent growth on a daily basis.

Schulz (1960) in Surinam used the Dawkins ten-ring method of tree measurement on 250–300 trees of seven economically important species and thought the mean girth per tree was likely to be within 0.4 mm of the true value. During the dry season a tree of 100 cm GBH may show a contraction of more than 0.2 mm between 08.00 and 16.00 h.

The only attempt to differentiate between swelling and wood production is recorded from outside the tropical forest zone. In the northern guinea savanna zone of Nigeria, McComb and Ogigirigi (1970) took sections of a *Eucalyptus citriodora* stem, dried to equilibrium with air at 21° C and 20 per cent relative humidity, sealed the ends of the 25 cm long sections and placed them in a chamber with 50 per cent relative humidity. Equilibrium was reached after 4–5 days during which the radius was measured daily with a micro-dendrometer to 0.01 mm. They adjusted the swelling values of each of the six stem sections, which ranged from 76 to 180 mm in diameter, to a single value relative to the largest section. This latter value was used in an examination of the apparent radial growth of a live *E. citriodora*, 18 cm DBH, during a 14-day period during which ambient relative humidity rose from 32 to 49 per cent. They concluded that 51 per cent of the apparent radial growth of 0.3 mm was due to swelling rather than to real growth. They also noted the marked swelling when rain fell directly on the bark of the tree. In the rain forest zone, relative humidity is generally considerably higher and more uniform than in the northern guinea zone. See chapters 2 and 12.

In subtropical areas such as the dry sclerophyll forests

of Australia greater water stress develops. Hopkins (1968) points out that if evaporation rates are sufficiently high, even though the soil moisture matrix potentials differ little from maximum values, water may not be able to flow through the soil fast enough to supply the tree's needs. Then a rapid increase in tension could be related directly to stem shrinkages. Stewart *et al.* (1973) suggest that living cambial tissues in *Eucalyptus regnans* may be under moderate to severe water stress sufficiently great to cause diurnal contraction and expansion of the developing cells almost daily in the growing season. Hopkins (1968), in a markedly seasonal climate with temperatures falling to below 4° C, found that the bark of *E. obliqua*, *E. radiata* and *E. regnans* did not respond to the rain directly falling on it. The bark types range from smooth gum (*E. regnans*) to slightly flaky and fibrous (*E. radiata*) to rough and stringy (*E. obliqua*).

Kenworthy (1971), referring to hydrological work in the Sungai Gombak catchment, Selangor, peninsular Malaysia, suggests that contractions might be of the order of 0.01 mm diameter by the evening.

Temperate trees appear to fluctuate less violently than tropical forest trees. A summary of the effects, showing marked within-tree variation is given by Kozlowski and Winget (1964). There is evidence for variation between different sides and at different heights of the same tree.

The hydration/dehydration cycle may produce fluctuations of a magnitude comparable with the annual increment of slowly growing trees. It is therefore normal practice in intensive research to measure trees while turgor is high, that is, in the early morning, but this is impracticable for many forest services.

Short-term fluctuations

For most studies, daily or more than once-daily measurement of trees is impracticable or unnecessary. A number of tropical workers report the results of weekly or fortnightly assessments. Dawkins (1956), using his ten-ring method, noted that a tree of 48 cm diameter could contract by more than 0.8 mm in diameter in a week and 3 mm in ten weeks at Mpanga in Uganda. Boaler, working on *Pterocarpus angolensis* in miombo woodland in Tanzania and using Liming (1957) vernier dendrometer bands on two trees at each of four sites, found shrinkages of 0.3 mm diameter over five days and maximum growth in the range 0.3–2.2 mm (Jeffers and Boaler, 1966). In the less seasonal climate of Sandakan, Sabah, Nicholson (1958) used Dawkins' ten-ring method on five trees of *Shorea smithiana*. Fortnightly measurements showed shrinkages of 0.08 mm diameter and growth of up to 1 mm. Iyamabo (1971) at Ibadan in Nigeria used a steel tape with a vernier to measure weekly increment on three small (8.4–9.4 cm diameter) trees of *Triplochiton scleroxylon*. He found decreases of up to 0.5 mm and increases of 2.0 mm per week. As in most studies, there is a general similarity in behaviour between different trees, but individuals can show marked deviations from the norm. Trees which can increase fast are frequently also those which shrink fast. During low rainfall periods there is a general damping down of fluctuations.

Bell (1969) studied 30 *Mora excelsa* trees of 20–62 cm DBH, using Liming (1957) type steel dendrometers in the Matura Forest Reserve in Trinidad over one year. Maximum monthly shrinkage was 1.3 mm and maximum monthly increase 3.0 mm, though the majority of the values were in the range —0.8 to +1.6 mm.

Murphy (1970) at El Verde, Puerto Rico, also used Liming (1957) bands on 50 trees of each of five species. After twelve months, second bands were fitted and showed clearly that there may be a two or three months period during which readings will be low. Unlike most authors, Murphy found that only *Manilkara bidentata*, and only about a third of its trees, showed shrinkage during the two years of study.

Seasonal fluctuations

A relatively large number of studies have attempted to link periodic growth and shrinkage to seasonal climatic changes and to correlate them with phenology. The coarser studies, perhaps limited by the availability of meteorological data and/or by calculating facilities, have attempted to show by graphs and histograms relations between monthly growth and rainfall in the same month.

In any one study period, neither the rainfall nor the rates of growth may be described easily in terms of good fit to a mathematical function. Consequently the reader of these reports must use his own judgment in deciding whether to agree or to disagree with an author who claims good correlation. Much depends on how the graphs and histograms are drawn. At this subjective level one may say that correlations look better in the more seasonal climates.

More sophisticated analysis involves the comparison of five-day running mean growth rates against rainfall. Dawkins (1956) found good agreement for *Lovoa brownii* and *Entandrophragma angolense* provided that the trees were in full leaf and not flowering.

In the rather aseasonal climates of the Far East, the behaviour of trees is especially erratic. Moving thirty-day totals of rainfall, used in Sarawak by Brünig (1969, 1971) and Baillie (1972), show periods of possible water stress which cannot be detected from monthly rainfall totals. It is unlikely that rainfall itself is a determining factor in growth; the availability of soil moisture is more probably a directly influencing factor. Baillie's (1972) studies, using relatively crude models, show that moisture stress is much more frequent on shallower soils. Nevertheless Ashton (unpublished) found no correlation between mean girth increment and soil types in Sarawak. Tropical rain forest studies on growth rates have not examined environmental factors in much detail; most information is available from North America. The tropical studies cannot, therefore, draw more precise conclusions than, for example, 'These observations suggest that drought is the chief factor determining the periodicity of this species' (Schulz, 1960).

So far as there is any seasonal rhythm, there is a tendency for tropical trees to behave in a similar manner to temperate trees. Bud break may take place well before the beginning of the rains, and leaf flush is often accompanied by a measurable bole shrinkage (Boaler, 1963). Radial growth usually does not begin before the crown is at least half covered by new foliage. Stem growth sometimes slows coincidentally with subsequent minor flushes; growth during flowering and fruiting, if they take place in the main growing season, is often erratic. Resting buds frequently form in the early part of the latter half of the wet season but stem growth may persist until after the rains cease. In areas where the dry season is especially marked there may be stem swelling before new growth at the beginning of the next wet season.

Growth correlations

The strongest correlation is frequently between tree size and growth. It is obvious that in a collection of trees of the same age, the larger trees must have grown faster than the smaller ones, even if at the time of examination the relationship is less simple.

Dawkins (1963), in the most thorough study so far, used diameter increment (i.d.), and increment in basal area relative to ground resource expressed per hectare of ground per year (i.g.b.), calculated from:

$$(2d \times i.d. - (i.d.)^2) \times G/d^2,$$

where d=diameter at the last measurement in cm,
$\quad$ $i.d.$=current annual diameter increment in cm/a, and
$\quad$ G=estimate of stand basal area (at breast height and over bark) in the locality of the subject tree in m²/ha.

Wycherley (1965) preferred to use a relative rate of girth increment, calculated as:

$$\frac{\log_e Gf - \log_e Gi}{t_f - t_i},$$

where G=girth, t=date, f=final and i=initial.

Wycherley devised this measure for the assessment of *Hevea brasiliensis* plantations in Malaysia as an attempt to take into account different initial tree sizes. In this particular case, the use of initial tree size as a covariate in the analysis would have been preferable, and probably absolute or relative basal area increment as a measure of growth.

Most foresters have been unduly casual in their consideration of suitable expressions of growth. A short discussion is given in Palmer (1975) together with an illustrated comparison of different expressions.

It seems that the more careful studies have been less successful in finding correlations between increment and other tree parameters than have the smaller scale and less careful studies. Lowe (1971), working mainly with tree plantations in southern Nigeria, has emphasized the effect of an early established competitive hierarchy in the crops. Initial advantage, through more favourable micro-site, differential nursery treatment, or genetical superiority or any other factor, is very important as soon as competition occurs between trees. Lowe's work suggests that this begins rather earlier than had previously been expected. It further underlines the strong correlation between tree increment and initial tree size. The findings are in agreement with those from seedling studies, such as those summarized by Harper

(1961), Sweet and Wareing (1966) and White and Harper (1970).

Other factors correlated with individual tree increment are various crown parameters such as diameter, form and position relative to neighbouring crowns, but, apparently equivalent trees may show great differences. Dawkins (1956) quotes Hummell as finding differences of the order of 1:5 in measurements covering 10–14 years. Computer analyses of large number of forest trees for many tropical countries, undertaken at the Commonwealth Forestry Institute, Oxford suggest that differentials can be as much as 1:20; this excludes the trees which remain stagnant but apparently healthy for many years.

Further, crown characteristics change with time, though the process is known only in outline and for some species. Young and vigorous dipterocarps frequently have narrow deep rather conical crowns but, after reaching the canopy, they spread out. This spreading does not appear to lead to a decreased increment, though Brünig (1967) points out that a narrow crown has a more favourable ratio of assimilating surface to respiring crown volume.

Conclusions

Individual tree growth may be highly irregular in the short and in the long term. The natural life-span of these trees is long enough for climatic change to be significant (Dale, 1954). Mervart (1972) finds indication of a secular change in growth rates, besides other changes. In Uganda Dawkins (1963) noted that annual fluctuations of diameter increment of the order of 1:2 are common, even over whole plantations, but that over a period of 4–5 years averages are more steady. Subsequent work at Oxford indicates that, even for Uganda, this is only partly true, and that for less seasonal climatic areas there is little good evidence for stability in growth rates. This finding is not evident in publications from foresters, who are often required to find one value for a rotation applicable to all forest sites and species in any country. Mervart (1969) points out how unsuitable mean values are; this has been recognized by some foresters (Nicholson, 1965) while others affirm that with a biological resource of great variability average predictive parameters must be relied on to a considerable extent (Fox, 1973). A further, more critical, view is that the latter will lead to further obscuring of a complex situation, and that it is inconceivable that simplistic measures and explanations of demonstrably complicated behaviour have any scientific value. The problem is to devise techniques for handling these complex situations, to allow mathematical modelling, simulation studies and predictions. Clearly this calls for a massive effort; even the great work at El Verde, Puerto Rico (Odum and Pigeon, 1970) barely scratches the surface.

Recommendations for future research depend on the objectives. From the foresters' point of view there is a strong need for systems which allow analysis and prediction of tree crop growth. The work of Dawkins (1963) and Lowe (1971) in particular show what may be expected from individual tree studies; the work of Mervart (1972) and the Commonwealth Forestry Institute indicates the direction of crop work. To a very large extent these studies are empirical. There is clearly a need for seedling studies on the lines of those begun in Ghana (Longman and Jeník, 1974), with more emphasis on the competitive effects demonstrated by temperate zone workers. It must be realized that it will be very difficult to synthesize models of tropical forest growth from studies of individual species. On the other hand, autecological studies can be illuminating when one species forms an important part of the crop, as is shown by Burgess (1968, 1969, 1972) on *Shorea curtisii* in Malaysian hill dipterocarp forest.

For foresters the chief lack is of twenty years and more long records of growth from individual trees. There is still a lamentable lack of interest in properly establishing and assessing growth plots. For example, the yield plots started by Fox (1970) have been measured more often than they have been analysed. There is also a no less lamentable lack of interest in preserving the older plots, which have sometimes been felled just as they are producing data of unique value. Simply following growth by measurements in naturally developing crops is not enough. Well designed, established, assessed and analysed experimental plots are also required for a great deal of past effort has gone into treatment plots whose results cannot be analysed statistically.

Tropical forest is the most complex vegetation on earth, and one of the most complicated of all ecosystems. This means that co-operative efforts through interdisciplinary studies are more likely to be productive, as shown by the El Verde study, than the rather traditional idea of having one research sylviculturalist in a forestry department.

Summary and conclusions

The subjects reviewed in this chapter are fundamental to an understanding of the consequences of man-induced changes in tropical forest lands; these are described in chapter 9.

The opportunity for niche specialization in complex tropical forests is immense. In such a community, the environment varies in space and time. Some of this variation may be at least partly determinate and depend, for example, on the specific character of the existing or pre-existing vegetation in any particular spot. The chemical and physical character of the soil at any particular micro-site may be determined in such a fashion. Similarly, the spatial variation of physical resources such as light under a complete canopy may be partially determined by the floristic variation in the canopy. The existence of such biologically-induced variations in the availability of resources allows almost unlimited possibilities for biological specializations. Some ecologists have consequently speculated on the existence of mutual compatibility or even specific interdependence between the species in such a system (e.g. Webb, 1968). However, the largest and probably the most significant temporal variations in environmental factors are spatially indeterminate and depend on the largely fortuitous occurrence of canopy gaps. This temporal and spatial variability in the existence of regenerative opportunities, and the largely unpredictable environmental nature of the micro-sites so produced, empha-

sizes the importance of interference as a facet of competition. Theoretically species can coexist not only because of variations in tolerances to particular micro-site conditions and because of differing strategies of competition (large seeds versus small seeds, rapid growth versus slow growth, shade tolerance versus light exigence, etc.), but also because of chance temporal and spatial avoidance in their utilization of the available regeneration opportunities.

Hopkins (1970) holds the view therefore that only four biological conditions are essential to prevent more than one species from occupying the same space. First, the species must have regular, occasional flowering and fruiting times which do not coincide. Secondly, the seeds must have limited viability. Thirdly, there should be a virtually uninhibited climatic period for establishment and growth. Fourthly, it should maintain adequate seed production despite competition.

Though all these conditions are met by many species of the mature phase, the first two are probably not as a rule characteristics of those of the gap and building phases. Nevertheless the floristic diversity of rain forest appears to be maintained in part by the specialized nature of the species' ecological range and biological characteristics and also in part, by the probabilistic nature of establishment and consequently regeneration, and of the fortuitous interchangeability of species which possess irregular flowering times and short-lived seeds.

It is well known, that these safeguards can fail; secondary successions may be deflected and certain species may gain dominance under special conditions (Richards, 1952). There are four practical implications of our present knowledge.

First, the dynamic balance which maintains species diversity with low population densities in the mature phase of tropical rain forest is very easy to destroy, and difficult to recreate owing to:

— the predominantly inefficient propagule dispersal and hence slow rates of reinvasion;
— the specialized requirements for successful germination and establishment;
— the more or less irreversible effects (on free draining soils) of removing the living forest and humus in which the available nutrients, as well as the infinitely complex macro- and micro-fauna and flora, are stored.

For discussion see Gómez-Pompa *et al.* (1972) and May (1973, 1975).

Secondly, it is often assumed that a natural forest in equilibrium will manifest maximum sustainable biomass production for its particular habitat. The evidence suggests, however, that natural selection ultimately favours only those species that can, through their tall stature and dense crowns, shade out their competitors, allowing only shade tolerant species to survive (Horn, 1972) and not necessarily those of highest production. On very freely draining and infertile soils intermittent water stress, even in the perhumid tropics, prevents the development of such a stand of tall canopy trees with a dense single-layered canopy and consequently high transpiration rates. A canopy consisting of small inclined shiny leaves, so placed that the canopy as a whole has high reflectance, replaces it. Here, water stress is the dominant factor in natural selection but the resultant canopy structure, as it were by accident, allows more light to penetrate. Consequently, a multilayered understorey has high leaf area index and allows such high rates of volume increment (and presumably biomass increment and mortality) that the total gross volume increment of forests on infertile sites can equal that on the mesic sites. The mature phase forest on intermediate sites is generally not only floristically and structurally the most diverse, but also produces the highest gross volume increment.

Thirdly, if the mature phase is selectively or clear-felled, the nutrient and water resources in the mineral soil allows sustained rapid growth of the gap species which are so adapted only on fertile soils. On other sites nutrients are mainly stored in the vegetation. These become limiting and succession is consequently retarded; it may be deflected into a semi-permanent regressive state with further impoverished soil. Without fertilizer input repeated cropping of fertile soils will also lead to impoverishment.

Fourthly, it will be the gap phase and perhaps the faster growing mature phase light demanders which will be selected for wood production. *Albizia falcataria*, a late pioneer in the Moluccas, is utilized for both pulp and construction in the Philippines, where it has been known to reach 20 m tall within 2 years of planting and is cropped on an 8-year rotation in small holdings. Again, some of the most valuable stands exploited in the humid tropics, e.g. *Terminalia superba* in Mayumbe, Africa, and *Tectona grandis* in the Far East, are the product of seral succession following abandonment after shifting cultivation.

Further, the forester has to compete with the agriculturist for fertile land. Dry infertile soils may never sustain a land-use system involving periodic forest removal, but could support crops producing fruit, latex, oils, or pharmaceuticals. The mature phase, which occupies 90 per cent of the rain forest area, has no place in forest managed for wood production. However, it has a vital place in resource development:
— in watershed management, where its multistoreyed structure provides the most effective cover against soil erosion and its mat of roots and micro-climate a soil environment which maximizes water storage and ameliorates local flooding; in hilly districts the peculiar features of the mature forest canopy allow it to obtain water from moving clouds, providing permanent streams to adjacent lowlands in seasonal climates (see Mueller-Dombois, 1972); other vegetation cover compares lamentably in this respect;
— as the richest pool of genetic resources in existence, not only for timber tree improvement, but for new or improved crops yielding edible oils, fruits, pharmaceuticals and other chemicals and to maintain the options needed for continuously changing world and national markets;
— as the control or base-line ecosystem against which the long-term sustainability of man-made ecosystems must be judged, and from which lessons must be learned when the latter fail, as they often inevitably will.

Forests of the seasonal tropics

It is broadly true that, though evergreen forests in the seasonal tropics are slow to regenerate, and frequently undergo deflected succession, as a consequence of human interference, forests on more fertile land have greater powers of regeneration after damage than have those in the perhumid zone. This is for the following reasons:
— prevalent ability to regenerate by suckers as in sal (*Shorea robusta*) and *Tectona grandis* in India;
— more efficient seed dispersal;
— seed dormancy;
— less demanding requirements for seed germination;
— higher proportion of the total nutrients stored in the soil;
— higher proportion of nutrients returned annually to the soil.

All these factors become increasingly apposite with increasing length of the dry and the unpredictability of the wet season.

This is already known to traditional herdsmen and agriculturalists, who are exploiting them. The remarkable capacity for recovery even of the mutilated forests of Gir, at the driest end of the range of *Tectona grandis* in India, has been demonstrated in temporary exclosures (Hodd, unpublished) which provides evidence that rotation of cattle in the wet season and cutting and storage of the excess grass as hay could maintain high cattle numbers in good condition during the dry season and without having to top trees for forage.

Research needs and priorities

Methods

The study of many dynamic aspects of tropical forest is technically simple and cheap yet very few have been undertaken methodically on a large enough scale over an adequate period of time. The principal requirements are adequate areas of inviolate unmodified natural forest and enthusiastic people willing to spend long periods in the field and to continue studies over several years. The criteria for choice of adequate representative areas for research are many-faceted and their scientific basis yet to be established. Areas of less than 1 000 ha may generally be considered inadequate and 2 000 ha with a surrounding buffer zone of commercially managed forest of comparable area should be aimed for. The same sample plots and sample populations laid out and labelled may be used for many of the studies recommended (see also chapter 21).

Recommended research

Topics are first listed, and their relevance to problems of management and exploitation cited at the end.

Synecology

1. Studies of pattern and process in different forest types. Periods of at least 5 years are required, with annual measurement of all individuals. Studies should be designed for continued measurement on a permanent basis.
a. Studies of rates of cyclical change.
— Periodic measurement of girth and height increments for all trees in representative samples of several hundred numbered and mapped individuals.
— Periodic measurement of seedling recruitment and mortality in representative samples on the same basis.
— Continuous recording of leaf change, flowering and fruiting of all numbered individuals under study (see 4 below).
b. Studies of indirectional change.
— Collection of evidence from soil and vegetation of enhanced leaching, nitrification, cation accumulation, etc.
c. Studies of understorey dynamics relative to canopy structure in both natural forests and forest plots subjected to experimental manipulation.
— Development of rapid methods for characterizing the light climate.
— The influence of canopy structure on understorey composition, canopy species regeneration, understorey dynamics.
— Effect of gap size on gap phase succession, with and without soil scarification.
— Effect of gap size on soils.
— Studies on the degree of determinism in gap succession: seed dormancy, germination, seedling competition, etc. (see 6 and 7 below).
— Distribution in space and time of canopy gaps.
2. Dynamics within the soil.
Studies of changes in root systems are technically difficult but establishment of, for example, glass-sided trenches filled in between observations, in combination with destructive samplings, should be attempted.
a. Dynamic changes in root systems.
b. Species interactions in the rhizosphere.
c. Development of root systems during gap and building phases.
d. Re-establishment of the nutrient cycle following catastrophe.
3. Spatial variation in dynamics.
a. Variation in forests on differing soils.
b. Variation with climate.
c. Variation with altitude; particularly trends in mean increment rates of all species in mixed forests.

Autecology

4. Phenological observations of numbered individuals in population samples over at least 5 years.
a. Recording dates and frequency of leaf change, flowering, fruiting and their synchrony.
b. Tagging of leaves to establish their duration.
c. Tagging of inflorescences to establish rates and causes of mortality among developing flowers and fruits.

d. Physiological studies of causes of phenological change.
5. Reproductive biology.
a. Studies of floral mechanisms favouring outcrossing, e.g. dioecism and its variation within populations.
b. Studies of pollination and pollen vectors.
c. Studies of genetical variation in populations.
d. Studies of embryogenesis in relation to reproduction methods.
6. Fruiting biology.
a. Methods of dispersal.
b. Pre- and post-dispersal predation and mortality; ecology of predation; predator-host relationships.
c. Seed dormancy within and outside the forest.
d. Physiological aspects of germination.
7. Seedling ecology.
a. Studies of selective mortality with individuals of wild populations in mature and gap phases.
b. Comparative physiology of seedlings and saplings of selected species; especially water relations, photosynthesis and respiration, nutrient cycling and growth.
c. Studies of establishment, especially of root systems.
8. Studies of allelopathy in mixed forests.
9. Comparative studies of tagged individuals of mature and gap phase species on a variety of sites and conditions, both in natural unfelled forest, carefully managed regenerating forest and plantations.
a. Comparisons of growth rates.
b. Relationship of girth increment with that of height and phenological changes.
c. Comparative physiology.
d. Estimates of length of life of trees:
— ^{14}C dating on basal cores of old trees;
— deductions of age from studies in increment and mortality.
10. Surveillance of cyclical changes in numbers of animal and plant populations and their interrelationships.
11. Ecology of exotic invaders into tropical forests.

Collation

It is essential that workers in different regions actively collaborate and communicate; much fragmented work has already been accomplished and is lost in old files; it needs to be searched on and made available.

Application of research

(The numbers refer to the numbered paragraphs in the above recommended research topics.)

Sylvicultural needs (see also chapter 9)

To allow analysis and prediction of crop growth:
1a, 1c, 2a, 2b, 3a-c, 7a-c, 9a-c, 11.
To minimize nutrient loss from forest systems:
1b, 1c, 2a-d, 3a-c, 4b.

To maximize the phase of rapid growth:
1a, 1c, 2a, 2c, 3a-c, 4b, 7a-b, 9a-c.
To ensure adequate regeneration at all times:
1a, 1c, 3a-c, 4a, 4c, 4d, 5b, 6a-d, 7a-c, 8, 10, 11.
To reduce pest epidemics:
6a-c, 8, 10, 11.

Exploitation of genetic resources

Problems of ascertaining conditions for establishment and growth:
1a, 2a, 2c, 3a-c, 6a-d, 7a-c, 8.

Problems of ascertaining genetic systems for crop improvement:
1a, 3a-c, 4a, 4c, 4d, 5a-d, 10.

Management and conservation (see also chapter 21)

The assessment of minimum size of populations to ensure their permanent maintenance:
1c, 3a-c, 4a, 4c, 5a-d, 6a-d, 7a, 8, 9d, 10.
The problem of ensuring the maintenance of environmental stability (see chapter 9):
1b, 1c, 2b, 2d, 3a-c, 5b, 6b, 8, 9b, 9d, 10, 11.

Bibliography

Major general references

ASHTON, P. S. Speciation among tropical forest trees: some deductions in the light of recent evidence. *Biol. J. Linn. Soc.*, 1, 1969, p. 155–196.

——. *Report on research undertaken during the years 1963–1967 on the ecology of mixed dipterocarp forests in Sarawak.* Unpub. ms. for Sarawak Government.

AUBRÉVILLE, A. La forêt coloniale : les forêts de l'Afrique occidentale française. *Ann. Acad. Sci. Colon.* (Paris), 9, 1938, p. 1–245.

BAKER, H. G. Evolution in the tropics. *Biotropica*, 2, 1970, p. 101–111.

BAWA, K. S. Breeding systems of tree species of a lowland tropical community. *Evolution*, 28, 1974, p. 85–92.

BEARD, J. S. *The natural vegetation of Trinidad.* Oxford Forestry Memoirs, 20, 1946, 152 p.

BROWN, W. H.; MATTHEWS, D. M. Philippines dipterocarp forests. *Philip. J. Sci.*, 9, A, 1914, p. 413–561.

BRYNDUM, D.; HEDEGART, T. Pollination of teak (*Tectona grandis* L.). *Silv. Genet.*, 18, 1969, 57 p.

BUDOWSKI, G. Distribution of tropical American rain forest species in the light of successional processes. *Turrialba* (Costa Rica), 15, 1965, p. 40–42.

BURLEY, J. Breeding systems, variation and genetic improvement of tropical conifers. In: Burley, J.; Styles, B. T. (eds.). *Tropical trees: variation, breeding and conservation* (Proceedings of an international symposium, Oxford, April 1975; Linnean Society Symposium Series no. 2). London, New York, Academic Press, 1976, 244 p.

——; NICKLES, D. G. *Selection and breeding to improve some tropical conifers.* 2 vol. Oxford, Commonwealth Forestry Institute, 1973.

CATINOT, R. Sylviculture tropicale en forêt dense africaine. *Bois et Forêts des Tropiques* (Nogent-sur-Marne), 1965, 100, p. 5–18; 101, p. 3–16; 102, p. 3–16; 103, p. 3–16; 104, p. 17–30.

——. Premières réflexions sur une possibilité d'explication physiologique des rythmes annuels d'accroissement chez les arbres de la forêt tropicale africaine. *Bois et Forêts des Tropiques*, 131, 1970, p. 3–36.

CHAMPION, H. G.; SETH, S. K. *General sylviculture for India.* Delhi, Manager of Publications, 1968, 511 p.

CONNELL, J. H. On the role of natural enemies in preventing competitive exclusion in some marine animals and in rain forest trees. In: *Proc. Adv. Study Inst. Dynamics Numbers Popul.* (Oosterbeek, 1970), p. 298–312.

DAUBENMIRE, R. F. *Plants and environment: a textbook of plant autecology.* 3rd ed. New York, Wiley, 1974, 422 p.

DAWKINS, H. C. *The management of natural tropical high forest with special reference to Uganda.* Oxford Imperial Forestry Institute, paper no. 34, 1958, 155 p.

——. The volume increment of natural tropical high forest and limitations on its improvement. *Empire For. Rev.*, 38, 1959, p. 175–180.

DE LA MENSBRUGE, C. *La germination et les plantules des essences arborées de la forêt dense humide de la Côte d'Ivoire.* Nogent-sur-Marne, Centre technique forestier tropical (CTFT), 1966, 389 p.

DOBBEN, W. H. van; LOWE-MCCONNELL, R. H. (eds.). *Unifying concepts in ecology.* Report of the plenary sessions of the first international congress of ecology (The Hague, September 8–14, 1974). The Hague, W. Junk B.V. Publishers; Wageningen, Centre for agricultural publishing and documentation, 1975, 302 p.
Chapters on 'Flow of energy and matter between trophic levels'; 'Comparative productivity in ecosystems'; 'Diversity, stability and maturity in natural ecosystems'; 'Diversity, stability and maturity in ecosystems influenced by human activities'; 'Strategies for management of natural and man-made ecosystems'.

FAEGRI, K.; PILJ, L. van der. In: *The principles of pollination ecology*, p. 1–29. Oxford, Pergamon Press, 1971.

FAO. *Tree seed notes.* Rome, FAO, 1955, 354 p.

GÓMEZ-POMPA, A.; VÁZQUEZ-YANES, C.; GUEVARA, S. The tropical rain forest: a non-renewable resource. *Science*, 177, 1972, p. 762–765.

GREIG-SMITH, P. Ecological observations on degraded and secondary forest in Trinidad, British West Indies. II. Structure of the communities. *J. Ecol.*, 40, 1952, p. 316–330.

HALLÉ, F.; OLDEMAN, R. A. A. *Essai sur l'architecture et la dynamique de croissance des arbres tropicaux.* Paris, Masson, 1970, 178 p.

HARPER, J. L. The role of predation in vegetational diversity. In: *Brookhaven Symp. Biol.*, 22, 1969, p. 48–62.

HEDEGART, T. Breeding systems, variation and genetic improvement of teak, *Tectona grandis.* In: Burley, J.; Styles, B. T. (eds.). *Tropical trees: variation, breeding and conservation* (Proceedings of an international symposium, Oxford, April 1975; Linnean Society Symposium Series no. 2). London, New York, Academic Press, 1976, 244 p.

HOLMES, C. H. Flowering and fruiting of forest trees of Ceylon. *Indian Forester*, 68, 1942, p. 411–420, 488–499, 580–585.

HOPKINS, B. Vegetation of the Olokemeji Forest Reserve,

FANSHAWE, D. B. Regeneration of forests in the tropics with special reference to British Guiana. In: *Brit. Comm. For. Conf.* (Ottawa), 1952, 8 p.

——. *The vegetation of British Guiana, a preliminary review.* Oxford, Imperial Forestry Institute, paper no. 29, 1952, 96 p.

FAO. *Forest genetic resources.* Rome, FAO Forestry occasional papers, 1973, 1975.

FERRI, M. G. Contribution to the knowledge of the ecology of the Rio Negro caatinga (Amazon). *Bull. Res. Council Israel* (Jerusalem), 80 (3/4), 1960, p. 195–207.

——. *Simposio sobre a cerrado.* Univ. São Paulo, vol. 1, 1963, 375 p.; vol. 3, 1971, 239 p.

FLENLEY, J. R. (ed.). *Transactions of the third Aberdeen-Hull Symposium on Malesian ecology: altitudinal zonation in Malesia* (Hull, 1973). Univ. Hull, Dept. Geography, miscellaneous series no. 16, 1974, 119 p.

FLORENCE, R. G. A comparative study of flowering and seed production in six blackbutt (*Eucalyptus pilularis* Sm.) forest stands. *Aust. For.*, 1964, p. 28–33.

FOX, J. E. D. Dipterocarp seedling behaviour in Sabah. *Malayan Forester*, 36, 1973, p. 205–214.

FRANKIE, G. W. Tropical forest phenology and pollinator plant co-evolution. In: *Co-evolution of animals and plants* (1st Int. Congr. Syst. and Ecol. Biol., Boulder, Colorado), 1973, p. 192–209.

——; BAKER, H. G.; OPLER, P. A. Tropical plant phenology: applications for studies in community ecology. In: Lieth, H. (ed.). *Phenology and seasonality modelling.* Berlin and New York, Springer Verlag, Ecological Studies no. 8, 1974, 444 p.

——; ——; ——. Comparative phenological studies of trees in tropical wet and dry forest in the lowlands of Costa Rica. *J. Ecol.*, 62, 1974, p. 881–919.

GARDNER, A.; RIDLEY, W. F. Fires in rain forest. *Aust. J. Sci.*, 23, 1961, p. 226.

GAUCHETTE, J. Essais de prolongation du pouvoir germinatif des graines d'okoumé. In: *Conf. Forest. de Pointe-Noire*, 1958.

GÉRARD, P. *Étude écologique de la forêt dense à Gilbertiodendron dewevrei dans la région de l'Uele.* Publ. INEAC, Sér. sci., n° 87, 1960, 159 p.

GERMAIN, R.; ÉVRARD, C. *Étude écologique et phytosociologique de la forêt à Brachystegia laurentii.* Publ. INEAC, Sér. sci., n° 67, 1956, 105 p.

GILBERT, G. *Observations préliminaires sur la morphologie des plantules forestières au Congo belge.* Publ. INEAC, Sér. sci., n° 17, 1938, 28 p.

GOPAL, M.; PATTANATH, P. G.; KUMAR, A. A comparative study of germination behaviour of *Tectona grandis* of some Indian provenances. In: *Proc. and Tech. Papers Symp. man-made forests in India* (Dehra Dun, Society of Indian Foresters), 1972, p. III B2–III B 27.

GRANT, V. and K. *Hummingbirds and their flowers.* New York, Columbia University Press, 1968, 115 p.

GRAY, B. Size, composition and regeneration of *Araucaria* stands in New Guinea. *J. Ecol.*, 63, 1975, p. 273–289.

GREENWOOD, M.; POSNETTE, A. F. The growth flushes of cocoa. *J. Hort. Sci.*, 25, 1969, p. 164–174.

HEINSDIJK, D.; MIRANDA BASTOS, A. de. *Forest inventories in the Amazon.* Rome, FAO report no. 1284, 1963, 6 parts: 2+28+15+16+24+25 p.

HÔI, L. V. *Forest calendar and forestry seed.* Saigon, Inst. Rech. For., 1972, p. 1–100 (in Vietnamese).

HOLTTUM, R. E. Periodic leaf change and flowering of trees in Singapore. *Gard. Bull. Sing.*, 11, 1940, p. 119–175.

HORN, H. *The adaptive geometry of trees.* Princeton, Monographs in population biology, 3, 1972.

JAIN, N. K. Physical characters and output of sal seeds. *Tropical Ecology*, vol. 3, no. 1 and 2, 1962, p. 133–138.

JANZEN, D. H. Synchronization of sexual reproduction of trees with the dry season in Central America. *Evolution*, 21, 1967, p. 620–637.

——. Allelopathy by myrmecophytes: the ant *Azteca* as an allelopathic agent of *Cecropia. Ecology*, 50, 1969, p. 147–153.

——. Seed predation by animals. *Ann. Rev. Ecol. and Syst.*, 2, 1971, p. 465–492.

——. Tropical blackwater rivers, animals and mast fruiting by the Dipterocarpaceae. *Biotropica*, 6, 1974, p. 69–103.

——. Interactions of seeds and their insect predators/parasitoids in a tropical deciduous forest. In: Price, P. W. (ed.). *Evolutionary strategies of parasitic insects and mites.* London, Plenum, 1975.

——. Two patterns of pre-dispersal seed predation by insects in Central American deciduous forest trees. In: Burley, J.; Styles, B. T. (eds.). *Tropical trees: variation, breeding and conservation* (Proceedings of an international symposium, Oxford, April 1975; Linnean Society Symposium Series no. 2). London, New York, Academic Press, 1976, 244 p.

——. Why do bamboos wait so long to flower? In: Burley, J.; Styles, B. T. (eds.). *Tropical trees: variation, breeding and conservation* (Proceedings of an international symposium, Oxford, April 1975; Linnean Society Symposium Series no. 2). London, New York, Academic Press, 1976, 244 p.

JONES, N. Records and comments regarding the flowering of *Triplochiton scleroxylon* K. Schun. *Commonwealth For. Rev.*, vol. 53, no. 1, 1974, p. 155.

——. Some biological factors influencing seed setting in *Triplochiton scleroxylon* K. Schum. In: Burley, J.; Styles, B. T. (eds.). *Tropical trees: variation, breeding and conservation* (Proceedings of an international symposium, Oxford, April 1975; Linnean Society Symposium Series no. 2). London, New York, Academic Press, 1976, 244 p.

KADAMBI, K. Observations on natural regeneration of teak. *Indian Forester*, vol. 85, no. 11, 1959.

KARTAWINATA, K. *A provisional bibliography on tropical forest ecosystems.* BIOTROP, Bogor, Indonesia, 1972, 34 p.

——; ATMAWIDJAJA, R. (eds.). *Co-ordinated study of lowland forests in Indonesia* (symposium volume). BIOTROP, Bogor, Indonesia, 1974, 183 p.

KEAY, R. W. J. Wind dispersed species in a Nigerian forest. *J. Ecol.*, 45, 1957, p. 471–478.

——. Seeds in forest soils. *Nigeria Forestry Information Bulletin* (new series), 4, 1960, p. 1–12.

——. Increment in the Okomu Forest Reserve, Benin. *Nigeria Forestry Information Bulletin* (new series), 11, 1961, 33 p.

KENWORTHY, J. B. Water and nutrient cycling in a tropical rain forest. In: Flenley, J. R. (ed.). *The water relations of Malesian forests. Transactions of the first Aberdeen-Hull symposium on Malesian ecology*, p. 49–65. Univ. Hull, Dept. of Geography, miscellaneous series no. 11, 1971, 97 p.

KERSHAW, A. P. A pollen diagram from Lake Euramoo, north-east Queensland, Australia. *New Phytol.*, 69, 1970, p. 785–805.

KESLER, W. Note sur la multiplication du parasolier. *Bull. Agric. Congo belge*, 1950.

KOELMEYER, K. O. The periodicity of leaf change and flowering in the principal forest communities of Ceylon. *Ceylon Forester* (new series), vol. 4, no. 2, 1959, p. 157–189.

KOOPMAN, M. J. F.; VERHOEF, L. *Eusideroxylon zwageri*

T. & B. het ijzervoet van Borneo en Sumatra. *Tectona*, 31, 1938, p. 381–399.

KRISHNASWAMY, V. S.; MATHAUDA, G. S. Phenological behaviour of a few forest species at New Forest, Dehra Dun. *Indian For. Rec.* (*new series*), *Silviculture*, vol. 9, no. 2, 1954, p. 89–134.

LEROY-DEVAL, J. Vie et mort des parasoliers (*Musanga cecropioides*). *Bois et Forêts des Tropiques*, 112, 1967, p. 265–274.

LOEKITO; HARDJONO. Survey regenerasi hutan bekas tebangan di Orasbari Utama. *Rimba Indonesia*, 10, 1965, p. 265–274.

LOWE, R. G. The effect of competition on tree growth. In: *2nd Nigeria For. Conf.* (Enugu, 1966), 1968.

MALAISSE, F. Phenology of the Zambezian woodland area, with emphasis on the miombo ecosystem. In: Lieth, H. (ed.). *Phenology and seasonality modelling*, p. 269–286. Berlin and New York, Springer Verlag, Ecological Studies no. 8, 1974, 444 p.

MANDOUX, E. La régénération naturelle dans les forêts remaniées du Mayombe. *Bull. Agric. Congo belge*, 45, 1947, p. 403–421.

MANGENOT, G. Les recherches sur la végétation dans les régions tropicales humides de l'Afrique occidentale. In: *Recherches sur la zone tropicale humide. Actes du Colloque de Kandy*, p. 115–125. Paris, Unesco, 1958, 226 p.

MAURY, G. Germination anormale chez les Diptérocarpacées de Malaisie. *Bull. Soc. d'Hist. Nat. Toulouse*, 104, 1968, p. 187–202.

——. Différents types de polyembryonnie chez quelques Diptérocarpacées asiatiques. *Bull. Soc. d'Hist. Nat. Toulouse*, 106, 1970, p. 282–288.

MAY, R. M. *Diversity, stability and maturity in natural ecosystems, with particular reference to the tropical moist forests*. Rome, FAO, 1975, 9 p. multigr.

MCCLURE, H. E. Flowering, fruiting and animals in the canopy of a tropical rain forest. *Malayan Forester*, 29, 1966, p. 182–203.

MILLER, R. S. Competition and species diversity. In: *Brookhaven Symp. Biol.*, 22, 1969, p. 63–70.

MOHAN LAL, K. S. Manipulation of undergrowth for aiding natural regeneration of evergreen and semi-evergreen species. *Indian Forester*, vol. 86, n° 7, 1960.

MUELLER-DOMBOIS, D. Natural vegetation and agricultural development in the hill country of Ceylon. *Loris*, 12, 5, 19, 1972, p. 262–263.

NG, F. S. P. Age at first flowering in dipterocarps. *Malayan Forester*, 29, 1966, p. 290–295.

——. Germination of fresh seeds of Malaysian trees. *Malayan Forester*, 36, 1973, p. 127–132.

——. Seed for reforestation: A strategy for sustained supply of indigenous species. *Malayan Forester*, 37, 1974, p. 271–277.

——; LOH, H. S. Flowering to fruiting periods of Malaysian trees. *Malayan Forester*, 37, 1974, p. 127–132.

ODUM, H. T.; PIGEON, R. F. (eds.). *A tropical rain forest. A study of irradiation and ecology at El Verde, Puerto Rico.* Oak Ridge, Division of Technical Information, U.S. Atomic Energy Commission, 1970, 1 678 p.

OLATOYE, S. T. Seed storage problems in Nigeria. In: *Proc. 9th Brit. Comm. For. Conf.* (New Delhi), 1968.

PANDE, D. C. A short note on seeding of some species in Uttar Pradesh. In: *Proc. Ninth Silvicultural Conf.* (Dehra Dun, Forestry Research Institute), 1956.

——. Progress of the sal natural regeneration in Uttar Pradesh under the normal prescriptions of the working plans. In: *Proc. Ninth Silvicultural Conf.* (Dehra Dun, Forestry Research Institute), 1956a.

PERCIVAL, M.; WOMERSLEY, J. S. *Floristics and ecology of the mangrove vegetation of Papua-New Guinea*. Lae, Papua-New Guinea, Dept. of Forest, Division of Botany, Botany Bulletin no. 8, 1975, 96 p.

PHILIP, M. S. The dynamics of seedling populations in a moist semi-deciduous tropical forest in Uganda. In: *Proc. 9th Brit. Comm. For. Conference* (New Delhi), 1968.

POINSIER, J. L. Le parasolier, essence de reboisement pour la forêt secondaire. *Bois et Forêts des Tropiques*, 3, 1947, p. 31–34.

PONGUMPHAI, S. *Introduction to forest dendrology*. Bangkok, Forest Faculty, Kasetsart Univ., 1970, 127 p.

POORE, M. E. D. Studies in Malaysian rain forest. I. The forest on Triassic sediments in Jengka Forest Reserve. *J. Ecol.*, 56, 1968, p. 143–196.

PRÉVOST, M. F. Architecture de quelques Apocynacées ligneuses. In: *Colloque sur la physiologie de l'arbre. Mém. Soc. Bot. Fr.* (Paris), 914, 1966, p. 23–36.

RAJKHOWA, S. Regeneration of Upper Assam *Dipterocarpus-Mesua* forests. *Indian Forester*, vol. 87, no. 1, 1961.

RAKOEN, M. P. Beberapa pendapat mengenai daerah hutan Bangkalis Kampar. *Rimba Indonesia*, 4, 1955, p. 223–248.

RICHARDS, P. W. What the tropics can contribute to ecology? *J. Ecol.*, 51, 1953, p. 231–241.

——. The secondary succession in the primary rain forest. *Sci. Prog. Lond.*, 43, 1955, p. 45–57.

ROCKWOOD, L. L. The effect of defoliation on seed production of six Costa Rican tree species. *Ecology*, 54, 1973, p. 1363–1369.

ROLLET, B. *Inventaire forestier de l'Est Mékong*. Rome, FAO, Report no. 1500, 1962, 184 p.

——. *Études quantitatives d'une forêt dense humide sempervirente de plaine de la Guyane vénézuélienne*. Thèse de doctorat, Faculté des Sciences, Toulouse, 1969, 473 p.+173 p. (annexe).

ROSS, R. Ecological studies on the rain forest of southern Nigeria. III. Secondary succession in the Shasha Forest Reserve. *J. Ecol.*, 42, 1954, p. 259–282.

ROTHE, P. L. Régénération naturelle en forêt tropicale. *Bois et Forêts des Tropiques*, 8, 1948, p. 368.

ROUX, J. Espèces à rameaux végétatifs dimorphes. I. L'appareil aérien de *Notobuxus acuminata* (Gilgs) Hutch. (*Buxaceae, Buxoideae*). *Nat. Monsp. S. Bot.*, 16, 1964–1965, p. 177–193.

——. Sur le comportement des axes aériens chez quelques plantes à rameaux végétatifs polymorphes. Le concept de rameau plagiotrope. *Ann. Sc. Nat. Bot.* (Paris), 12ᵉ sér., 9, 1968, p. 109–256.

SCARONNE, F. *Contribution à l'étude des dormances en milieu équatorial (Ouest Cameroun)*. Rapport inédit, 1957, 68 p. multigr.

SEN GUPTA, J. N. Seed weights, plant per cent, etc., for forest plants in India. *Indian For. Rec.* (*new series*), *Silviculture*, vol. 2, no. 5, 1937, p. 175–221.

——. *Dipterocarpus* (Garjan) forests in India and their regeneration. *Indian For. Rec.* (*new series*), *Silviculture*, vol. 3, no. 4, 1939.

SETH, S. K. The problem of sal natural regeneration in Uttar Pradesh. In: *Proceedings All India Sal study Tour and Symposium*. Dehra Dun, Forestry Research Institute, 1953.

SMITH, A. M. On the internal temperature of leaves in tropical insolation, with special reference to the effect of their colour on their temperature; also observations on the periodicity of the appearance of young coloured leaves growing in Peradeniya Gardens. *Ann. R. Bot. Gardens Peradeniya*, 4, 1909, p. 229–298.

SOEBIDJA, R. S. Persoalam silvikultur di Sampit. *Rimba Indonesia*, 4, 1955, p. 83–89.

SOEPONO, R.; ARDIWINATA, B. Beberapa segi silvikultur dari perkerjaan mekanisasi dalam hutan *Dipterocarpaceae* di Mentwir (Balikpapan). *Pengumuman Balai Besar Penjelikikan Kehutanan Indonesia*, 58, 1957.

START, A. N.; MARSHALL, A. G. Nectarivorous bats as pollinators of trees in West Malaysia. In: Burley, J.; Styles, B. T. (eds.). *Tropical trees: variation, breeding and conservation* (Proceedings of an international symposium, Oxford, April 1975; Linnean Society Symposium Series no. 2). London, New York, Academic Press, 1976, 244 p.

STEENIS, C. G. G. J. Van. Basic principles of rain forest sociology. In: *Study of tropical vegetation. Proceedings of the Kandy Symposium*, p. 159–163. Paris, Unesco, 1958, 226 p.

SYNNOTT, T. J. Seed problems. In: *IUFRO Int. Symp. on seed processing* (Bergen), 1973.

TANG, H. T. Preliminary tests on the storage and collection of some *Shorea* spp. seeds. *Malayan Forester*, 34, 1971, p. 84–98.

——; TAMARI, C. Seed description and storage tests of some dipterocarps. *Malayan Forester*, 36, 1973, p. 38–53.

WADSWORTH, R. M.; LAWTON, J. R. S. The effects of light intensity on the growth of seedlings of some tropical tree species. *J. West African Sci. Ass.*, vol. 13, no. 2, 1968.

WATSON, J. G. *The mangrove swamps of the Malay Peninsula.* Malayan Forestry Rec., 6, 1928, 275 p.

WEBB, L. J. Biological aspects of forest management. *Proc. Ecol. Soc. Aust.*, 3, 1968, p. 91–95.

——; TRACEY, J. G.; HAYDOCK, K. P. A factor toxic to seedlings of the same species associated with living roots of the nongregarious subtropical rain forest tree *Grevillea robusta*. *J. Appl. Ecol.*, 4, 1967, p. 13–25.

WHITEHEAD, D. R. Wind pollination of angiosperms: evolutionary and environmental considerations. *Evolution*, 23, 1969, p. 28–35.

WHITMORE, T. C. (ed.). *Tree Flora of Malaya*, vol. 1. Kuala Lumpur, Longman, 1972, 471 p.

——. *Palms of Malaya*. Kuala Lumpur, Oxford University Press, 1973, 129 p.

WILLIAMS, W. T.; LANCE, G. N.; WEBB, L. J.; TRACEY, J. G.; DALE, M. B. Studies in the numerical analysis of complex rain forest communities. III. The analysis of successional data. *J. Ecol.*, 57, 1969, p. 515–535.

——; ——; ——; ——; CONNELL, J. H. Studies in the numerical analysis of complex rain forest communities. IV. A method for the elucidation of small scale forest pattern. *J. Ecol.*, 57, 1969, p. 635–654.

WILSON, D. E.; JANZEN, D. H. Predation of *Scheelea* palm seeds by bruchid beetles: seed density and distance from the parent palm. *Ecology* (Durham), 53, 1972, p. 954–959.

WONG, Y. K.; WHITMORE, T. C. On the influence of soil properties on species distribution in a Malayan lowland dipterocarp rain forest. *Malayan Forester*, 33, 1970, p. 42–54.

WOOD, G. H. S. The dipterocarp flowering season in North Borneo. *Malayan Forester*, 19, 1956, p. 193–201.

WYATT-SMITH, J. Storm forest in Kelantan. *Malayan Forester*, 17, 1954, p. 5–11.

References exclusive to section on growth

AMOBI, C. C. Periodicity of wood formation in some trees of lowland rain forest in Nigeria. *Annals of Botany* (London), vol. 37, no. 149, 1973, p. 211–218.

ANON. *British Commonwealth Forest Terminology*. Part 1. Empire Forestry Association (London), 1953, 163 p.

——. *Increment data of forest plantations in Mexico, West Indies and Central and South America*. Second annual report of the Section on Planting, FAO Latin American Forestry Commission, Regional Committee on Forest Research. Compiled at the Tropical Forest Research Centre, USDA Forest Service, Rio Pedras, Puerto Rico, 1960.

ARDIKOSOEMA, R. I.; KAMIL, R. N. *Musanga cecropioides* as an exotic tree in Indonesia. *Rimba Indonesia*, 4, 1955, p. 10–28.

BAKER, J. R.; BAKER, I. The seasons in a tropical rain forest (New Hebrides). Part 2. Botany. *Journal of the Linnean Society, Zoology* (London), 39, 1936, p. 507–519.

BELL, T. I. W. *An investigation into some aspects of management in the Mora (Mora excelsa Benth.) forests of Trinidad with special reference to the Matura Forest Reserve*. Univ. of the West Indies, Ph. D. thesis, 1969, 180 p. multigr.

——. *Management of the Trinidad Mora forests with special reference to the Matura Forest Reserve*. Trinidad & Tobago, Forestry Division, 1971, 9+70 p.

BLUM, B. M.; SOLOMON, D. S. *The accuracy of mean-growth estimates made with dial-gage dendrometers*. USDA Forest Service, Northeastern Forest Experiment Station Research, Note no. NE–54, 1966, 4 p.

BOALER, S. B. The annual cycle of stem girth increment in trees of *Pterocarpus angolensis* D.C., at Kabungu, Tanganyika. *Commonwealth Forestry Review* (London), vol. 42, no. 3, 1963, p. 232–236.

BORMANN, F. H.; KOZLOWSKI, T. T. Measurement of tree growth with dial-gage and vernier tree ring bands. *Ecology*, 43, 1962, p. 289–294.

BOWER, D. R.; BLOCKER, W. W. Accuracy of bands and tape for measuring diameter increments. *Journal of Forestry*, vol. 64, no. 1, 1966, p. 21–22.

BRADLEY, J. W. A plantation of remarkable growth. *Indian Forester*, 48, 1922, p. 637–640.

BRÜNIG, E. F. On the limits of vegetable productivity in the tropical rain forest and the boreal coniferous forest. *Journal of the Indian Botanical Society*, vol. 46, no. 4, 1967, p. 314–322.

BURGESS, P. F. An approach towards a silvicultural system for the hill forests of the Malay peninsula. *Malayan Forester* (Kuala Lumpur), vol. 33, no. 2, 1970, p. 126–134.

CANNELL, M. G. R.; LAST, F. T. (eds.). *Tree physiology and yield improvement*. London and New York, Academic Press, 1976, 568 p.

COSTER, C. Zur Anatomie und Physiologie der Zuwachszonen und Jahresringbildung in den Tropen. *Annals Jardin Bot.* (Buitenzorg, Java), 37, 1927, p. 49–160.

——. Wortelstudien in de Tropen. I. De jeugdontwikkeling van het wortelstelsel van een zeventigal boomen en groenbemesters. *Tectona* (Buitenzorg), 25, 1932, p. 828–872.

DALE, I. R. Forest spread and climatic change in Uganda during the Christian era. *Empire Forestry Review* (London), vol. 33, no. 1, 1954, p. 23–29.

DAVIS, R. B. Seasonal differences in internodal lengths in *Cecropia* trees; a suggested method for measurement of past growth in height. *Turrialba* (Costa Rica), vol. 20, no. 1, 1970, p. 100–104.

DAWKINS, H. C. *Interim results from research plots—Mpanga RP 13—measurement accuracy*. Forest Department, Uganda, Technical Note no. 8/56, 1956, 1 p.

——. *The productivity of tropical high forest trees and their reaction to controllable environment*. Univ. of Oxford, Ph. D. thesis, 1963, 5+4+111 p. multigr.

DOBBS, R. C. An electrical device for recording small fluctuations and accumulated increment of tree stem circumference. *Forestry Chronicle*, 45, 1969, p. 187–189.

FLENLEY, J. R. (ed.). *The water relations of Malesian forests. Transactions of the first Aberdeen-Hull symposium on Malesian ecology*. University of Hull, Department of Geography, miscellaneous series no. 11, 1971, 97 p.

FOX, J. E. D. Yield plots in regenerating forest. *Malayan Forester*, vol. 33, no. 1, 1970, p. 7–41.

——. *Natural vegetation of Sabah and natural regeneration of the dipterocarp forests*. University College of North Wales, Bangor, Ph. D. thesis, 1972, 477 p. multigr.

FRANSON, C. G. B. The course of growth of *Pinus caribaea* var. *hondurensis* throughout the year in West Malaysia. *Malayan Forester*, vol. 33, no. 3, 1970, p. 240–242.

FRITTS, H. C. *Tree rings and climate*. London and New York, Academic Press, 1977, 562 p.

GEISSLER, H. Instruments for measuring the seasonal course of diameter increment in forest trees. *Wiss. Z. Tech. University of Dresden* (GDR), vol. 19, no. 6, 1970, p. 1589–1596.

GEVORKIANTZ, S. R.; DUERR, W. A. *Methods of predicting growth of forest stands in the forest survey of the Lake States*. USDA Forest Service, Lake States Forest Experiment Station, Economic Note no. 9, 1938, 59 p.

HALL, R. C. A vernier tree-growth band. *Journal of Forestry*, 42, 1944, p. 742–743.

HARPER, J. L. Approaches to the study of plant competition. In: Milthorpe, F. L. (ed.). *Mechanisms of biological competition*, p. 1–39. Symposium Society of Experimental Biology, no. 15, 1961.

HARRIS, J. M. The use of beta rays to examine wood density of tropical pines grown in Malaya. In: Burley, J.; Nickles, D. G. (eds.). *Selection and breeding to improve some tropical conifers*, p. 86–94. Symposium of IUFRO section 22 working group on breeding tropical and subtropical species (Gainesville, Florida, USA, 1971). Oxford, Commonwealth Forestry Institute, vol. 2, 1973.

HOPKINS, E. R. Fluctuations in the girth of regrowth eucalypt stems. *Australian Forestry*, vol. 32, no. 2, 1968, p. 95–110.

HUMMELL, F. C. The formation of growth rings in *Entandrophragma macrophyllum* A. Chev. and *Khaya grandifoliola* C. DC. *Empire Forestry Review* (London), vol. 25, no. 1, 1946, p. 103–107.

IMPENS, I. I.; SCHALCK, J. M. A very sensitive electric dendrograph for recording radial changes of a tree. *Ecology*, vol. 46, no. 1/2, 1965, p. 183–184.

IYAMABO, D. E. *Some aspects of girth and radial growth patterns of Triplochiton scleroxylon K. Schum*. Department of Forest Research (Ibadan, Nigeria), Research Paper (Forest Series), no. 1, 1971, 7 p.

JEFFERS, J. N. R.; BOALER, S. B. Ecology of a miombo site, Lupa North Forest Reserve, Tanzania. I. Weather and plant growth, 1962–1964. *Journal of Ecology* (Oxford), vol. 54, no. 2, 1966, p. 447–463.

JONES, E. W. Review of Regenwalder, eine vergleichende studie forstlichen produktionspotentials, by J. Weck. *Empire Forestry Review* (London), vol. 38, no. 4, 1959, p. 429.

KEAY, R. W. J. Increment in the Okomu Forest Reserve, Benin. *Nigeria Forestry Information Bulletin (Ibadan), new series*, 11, 1961, 34 p.

KINERSON, R. S. A transducer for investigation of diameter growth. *Forest Science*, vol. 19, no. 3, 1973, p. 230–232.

KINGSTON, B. A climatic classification for forest management in Uganda. In: *10th Commonwealth Forestry Conference* (Oxford), 1974, 22 p.

KOZLOWSKI, T. T.; WINGET, C. H. Diurnal and seasonal variation in radii of tree stems. *Ecology*, vol. 45, no. 1, 1964, p. 149–155.

KUROIWA, K. Measurement of radial change of stems by strain gauge. *Journal of the Japan Forestry Society*, vol. 41, no. 9, 1959, p. 331–333.

LAMB, A. F. A. *Fast growing timber trees of the lowland tropics. No. 2. Cedrela odorata*. Commonwealth Forestry Institute (Oxford), 1968, 46 p.

LA POINT, G.; CLEVE, K. van. A portable electronic multi-channel dendrograph and environmental factor recording system. *Canadian Journal of Forest Research* (Ottawa), vol. 1, no. 4, 1971, p. 273–277.

LEBRUN, J.; GILBERT, G. *Une classification écologique des forêts du Congo*. Publication INEAC (Bruxelles), Sér. sci., n° 63, 1954, 89 p.

LIMING, F. G. Homemade dendrometers. *Journal of Forestry*, vol. 55, no. 8, 1957, p. 575–577.

LOWE, R. G. *Periodic growth in Triplochiton scleroxylon* K. Schum. Federal Department of Forest Research (Ibadan, Nigeria), Technical Note, no. 13, 1961.

——. Periodicity of a tropical rain forest tree: *Triplochiton scleroxylon* K. Schum. *Commonwealth Forestry Review* (London), vol. 47, no. 2, 1968, p. 150–163.

——. *Some effects of stand density on the growth of individual trees of several plantation species in Nigeria*. University of Ibadan (Nigeria), Ph. D. thesis, 1971, 10+239 p. multigr.

MARIAUX, A. Les cernes dans les bois tropicaux africains, nature et périodicité. *Bois et Forêts des Tropiques* (Nogent-sur-Marne, France), n° 113, 1967, p. 3–14; n° 114, 1967, p. 23–37.

——. La périodicité des cernes dans le bois de limba. *Bois et Forêts des Tropiques* (Nogent-sur-Marne, France), n° 128, 1969, p. 39–54.

McCOMB, A. L.; OGIGIRIGI, M. *Features of the growth of Eucalyptus citriodora and Isoberlinia doka in the northern guinea savanna zone of Nigeria*. FAO, Federal Department of Forest Research, Savanna Forestry Research Station (Samaru, Nigeria), research paper no. 3, 1970, 6 p.

McKELVIE, A. D. Root studies on seedlings. In: *Annual Report 1953–1954*, p. 24–25. West African Cocoa Research Station (Tafo, Ghana), 1954.

McKENZIE, T. A. Observations on growth and a technique for estimating annual growth of cativo (*Prioria copaifera*). *Turrialba* (Costa Rica), vol. 22, no. 3, 1972, p. 353–354.

MENON, P. K. B. Growth rings of locally-grown teak. *Malayan Forester*, vol. 11, no. 1, 1947, p. 26–27.

MERVART, J. Growth studies in the natural THF for forest management purposes. *Obeche* (Ibadan), vol. 1, no. 5, 1969, p. 48–59.

——. Growth and mortality rates in the natural high forest of western Nigeria. *Nigeria Forestry Information Bulletin*, (Ibadan), *new series*, 22, 1972, 28 p.

MESAVAGE, C.; SMITH, W. S. Timesavers for installing dendrometer bands. *Journal of Forestry*, vol. 58, no. 5, 1960, p. 396.

MÜLLER, D.; NIELSEN, J. Production brute, pertes par respiration et production nette dans la forêt ombrophile tropicale. *Det Forstlige Forsavaesen i Danmark*, vol. 29, n° 2, 1965, p. 69–160.

MURPHY, P. G. Tree growth at El Verde and the effects of ionizing radiation. In: Odum, H. T.; Pigeon, R. F. (eds.). *A tropical rain forest. A study of irradiation and ecology at El Verde, Puerto Rico*, p. D–141–D–171. Division of Technical Information, U.S. Atomic Energy Commission, 1970, 1 678 p.

NICHOLSON, D. I. One year's growth of *Shorea smithiana* in North Borneo. *Malayan Forester*, vol. 21, no. 3, 1958, p. 193–196.

——. A study of virgin forest near Sandakan, North Borneo. In: *Symposium on ecological research in humid tropics vegetation* (Kuching, 1963), p. 67–87. Government of Sarawak and Unesco Science Co-operation Office for South-East Asia, 1965, 376 p.

NINOKATA, K.; MIYAZATO, M. Measurement of daily variation of trunks by electrical strain gauge. *Bulletin Faculty of Agriculture* (University of Kogoshima), no. 8, 1959, p. 76–99.

NOAKES, D. S. P. A yield table for meranti. *Malayan Forester*, 6, 1937, p. 204–208.

ONOCHIE, C. F. A. *A preliminary study on girth, increment and age in Triplochiton*. Federal Department of Forest Research (Ibadan, Nigeria), unpublished record, 1947.

PALMER, J. R. Towards more reasonable objectives in tropical high forest management for timber production. *Commonwealth Forestry Review* (London), 1975.

REDHEAD, J. F. Historical evidence and speculation on some aspects of the forest kingdom of Benin. *Obeche* (Ibadan), vol. 1, no. 4, 1968, p. 49–54.

ROBERTS, H. Seasonal variation in the starch content of the sapwood of *Triplochiton scleroxylon* K. Schum. *Sterculiaceae* (trade name—wawa/obeche) in Ghana, West Africa. *Empire Forestry Review* (London), vol. 40, no. 1, 1961, p. 61–65.

STEWART, C. M.; THAM, S. H.; ROLFE, D. L. Diurnal variation of water in developing secondary stem tissues of eucalypt trees. *Nature* (London), vol. 242, no. 5398, 1973, p. 479–480.

SWEET, G. B.; WAREING, P. F. *The relative growth rates of large and small seedlings in forest tree species*. Supplement to Forestry, Oxford University Press, 1966, p. 110–117.

THERON, J. *Report on initial climatic classification of East Africa*. Yield studies project R 2533 report. Commonwealth Forestry Institute (Oxford), 1973.

TSCHINKEL, H. M. Annual growth rings in *Cordia alliodora*. *Turrialba* (Costa Rica), vol. 16, no. 1, 1966, p. 73–80.

WECK, J. The importance of the tropical forest types in world wood production. *Holz Roh- u. Werkstoff*, 18, 1960, p. 273–281.

——. The yield potential of the productive forest of the world. In: *Papers commemorating 150 years of university forestry education in Hungary* (University of British Columbia), 1961, p. 113–118.

WHITE, J.; HARPER, J. L. Correlated changes in plant size and number in plant populations. *Journal of Ecology* (Oxford), 58, 1970, p. 467–485.

WOLTER, K. E. A new method for making xylem growth. *Forest Science*, vol. 14, no. 1, 1968, p. 102–104.

WYATT-SMITH, J. Determination of a cutting period for THF. *Obeche* (Ibadan), vol. 1, no. 4, 1968, p. 36–48.

YACOM, H. A. Vernier scales for diameter tapes. *Journal of Forestry*, vol. 68, no. 11, 1970, p. 725.

9 Secondary successions

Introduction

As a preamble to this chapter on the disturbed forest ecosystem, it is particularly relevant to quote a paragraph from Richards' (1952, 1972) *The tropical rain forest*: 'Though many types of secondary vegetation derived from tropical rain forest have been described, they have seldom been closely studied and very few systematic observations have been made on the successions of which they are stages. It is unfortunate that this should be so, because no aspect of the ecology of tropical rain forest is of greater value or promises results of more theoretical importance. From the fragmentary observations available, it is evident that these secondary and deflected successions are complex and vary greatly from place to place, depending on differences in the habitat and in its previous history. The details can only be fully elucidated by careful quantitative observations continued over a long period of years. Such observations would help to fill the most serious gaps in our present knowledge of the rain forest.' Such a statement, which reflects the extent of the knowledge acquired, the gap and the action to be taken to remedy the situation, remains valid. This is confirmed by Whitmore's statement (*Tropical rain forests of the Far East*, 1975): 'There has been remarkably little general advance knowledge of secondary successions in the two decades which have elapsed since Richards' book was published. There have been several additional detailed studies, but still enormous gaps remain in our understanding.'

Several authors have referred to the rain forest as being a stable community. This statement requires clarification. When ecologists say the rain forest is stable, they are not thinking of a static situation but one in which the community, over a sufficiently large area, is continually adjusting so that it maintains essentially the same overall structure, constitution and functioning. For instance, if it were possible for an ecologist to observe the same area at intervals of 50 years over the last 500 years, he would still recognize it as the same community. This means that because of its resiliency the community can withstand minor perturbations of climate and animal impact with only small changes in specific constitution or structural form. To do this, the community must be sufficiently close to a state of dynamic equilibrium in which the overall changes during the period of historic observation are so small as not to affect the essential nature of the community. The area under consideration must be sufficiently large to take account of such local variations as shifting cultivation, windfall damage, land-slides, lightning strikes, biotic effects, etc. Within such limitations the system is self-maintaining and can be recognized as a biological entity which has persisted for a considerable time, although it is clear from evidence given in chapter 3 that over periods of thousands of years major changes have occurred.

The stability of the system depends on two important provisions. First, the forces acting on the system do not exceed certain threshold values. Secondly, that the environmental conditions are themselves relatively constant. In some instances a small disturbance can lead to a progressive alteration of the dynamic balance. This relative stability therefore, is different from the massive (in magnitude, scale or duration) changes induced by man in very recent times where the community has been completely destroyed and replaced by some other land-use system or where the disturbance has exceeded the capacity of self-maintaining processes and the dynamic equilibrium is destroyed.

Any kind of apparently stable system (in which flora and fauna represent a climatic climax) has therefore, as a prerequisite, that the system is left essentially undisturbed, or that disturbances are light, such as, for example, the death and fall of individual large trees. Heavier disturbances of the system will lead to other changes and this in turn will result in a succession which may, or may not, ultimately build up a system similar to the original. Such successions are generally called secondary, a term which includes forests resulting from repeated major disturbances.

This chapter deals with the secondary forests and the disturbances originating the successions. It leans heavily on the description of secondary and deflected successions by Richards (1952, 1972), and on recent publications especially by Webb *et al.* (1972), Gómez-Pompa *et al.* (1974) and Golley *et al.* (1975).

Depleted and secondary rain forests are rapidly increasing in extent as a result of human activities all over the tropics. If present day trends continue, within a few decades, these will be the principal kinds of forest in the tropical area. The following major questions are of interest to the land manager. First, how are successions recognized? Do they form a continuum so that a given community can be placed in the sequence even if a long time series of observations is not available? Secondly, what are the rates of change in structural and functional parameters? Thirdly, how are these rates influenced by the environment and by the disturbances? And, fourthly, are there several end points of recovery?

There are two main approaches in the observations of succession. First, a site might be examined sequentially leading to a description of the changes on one plot after a major disturbance. Secondly, a number of plots of different known ages may be examined at the same time and, assuming that the initial ecological conditions were similar on all the plots, the series can be interpreted as revealing the process of change. Both methods have their limitations. The first requires a long time; there are more data on the earliest stages of recovery and very few on the later stages. In the second it is difficult to obtain reliable data on successions older than 30 years. A good example of the first method is the continual sampling of permanent plots subjected to different cultivation regimes, initiated in Sarawak and mentioned by Whitmore (1975) who also suggested the study of forests along logging roads, which can be easily dated and which show a clear succession along their edges.

It is necessary to use both methods for an analysis of the entire sequence of succession in any tropical forest type. However, the impact of initial conditions is most important in the earliest stages, which can be observed directly, and may be less serious in the later stages of succession, the age of which must be inferred from historical records or the memories.

The species should be the obvious choice of unit, but to follow and interpret their change would be over ambitious and impractical in many instances. The problem is to define broader units which will facilitate the comparative description of vegetation changes in geographically and floristically distinct forest types. Some ecologists (e.g. Jones, 1955, 1956) have classified the species on the basis of various biological characteristics into groups such as shade tolerant and light demanding. Others (Richards, 1952; Ross, 1954; Budowski, 1965) have utilized characteristics such as longevity, speed of growth, wood density, dispersal mechanisms, regularity of fruiting, seed production, seed variability, and frequency of occurrence in various seral stages. Finally, Webb, Tracey and Williams (1972) have used forest canopies: the pioneer canopy; the early secondary canopy; the late or advanced secondary canopy; and the mature or climax canopy.

The modifications

By far the greatest amount of disturbance of the tropical forest ecosystem has been caused by man. The degree of disturbance 'varies widely between the three great world areas in which tropical forest occurs, the American forests being relatively undisturbed, while in Asia and particularly in Africa the changes and destruction caused by man in clearing, cutting and burning, are very conspicuous' (FAO, 1958); and this is confirmed by recent investigations. In Africa, the human impact on the natural vegetation has been more severe and for a longer period than in either the Amazon or in large areas of the Indo-Malaysian rain forest. Tropical western Africa has for a long time had a relatively high population of agricultural people who have influenced the most remote areas. Even in the depths of the so-called primary forest, there is often evidence of former human occupation in the form of pottery and charcoal fragments (see chapter 19). In Amazonia there may still be large areas of forest which up to now have been very little affected by man. In Asia, largely as a result of human occupation from early times and shifting cultivation, most forests have been degraded and have reverted to what has been called a peniclimax, pedoclimax or plesioclimax. There is no primary forest left in India, Bangladesh or Sri Lanka, except in the Andaman Islands (see also chapter 3).

The disturbances, when not originated by fire, most frequently commence with the creation of gaps in the upper canopy. Quite small gaps (such as referred to in chapter 8) are not disturbances in the sense used here, but are considered a normal feature of natural regeneration of the primary forest, even though the fall of a living, large tree may create a gap of appreciable extent (Lebrun and Gilbert, 1954; Baur, 1962). The size of the gap has a decisive influence on the development. Richards (1952) mentions some observations (by Kramer) from upland rain forest in Java: 'In artificial openings less than 10 ares in extent the existing regeneration of the primary forest dominants survived and made good growth. If the openings were 20–30 ares in extent, however, the regeneration was completely suppressed by the luxuriant growth of secondary forest species.'

The influence of soil types and condition on the vegetation and the early successive stages is a much disputed subject, and some authors postulate that the vegetation nutrient level is not directly related to the soil nutrient level (cf. Golley et al., 1975). However, soil type and condition play a role in seedling and plant establishment and early succession, and this would explain the success of taungya system in South-East Asia where forest exists not only on poor soils (ferralsols), but also on soils which have a definite capability for agriculture such as cambisols, acrisols and nitosols. In Africa and tropical America, soils are generally nutrient poor and forests have subsisted mainly on the poorer soils (ferralsols), and the use of taungya system to restore forest has been questioned.

Shifting cultivation

Most of the types of primitive agriculture practised on forest-land are referred to as shifting or swidden cultivation, which is described in more detail in chapters 19 and 20. The disturbance of the forest has three phases: fellings of all or most of the trees on a patch of forest, burning, and cultivation of one or more crops. The soil is soon exhausted, the farm abandoned, and a new patch of forest attacked. Secondary succession begins on the abandoned plot.

Plantation agriculture may often take a parallel course. 'In most tropical countries there are large areas of abandoned plantations showing various stages of secondary successions' (Richards, 1952). An example is given by Rhus taitensis forest in the Solomon archipelago growing under coco-nut trees which were abandoned at the onset of the second world war (Whitmore, 1975).

Burning

Burning in connection with primitive agriculture has been mentioned above. Fire is also widely used for the renewal of pastures and hunting. Others are due to abandoned camp-fires, and some are lit for no apparent purpose. Because of all these and other causes, fire is termed a biotic factor though a small number[*] of fires may be attributed to natural causes.

Tropical evergreen rain forest is generally held to be fairly immune to fire. There are, however, exceptions (e.g. Guyana), and local fires may develop during dry spells,

or where spots of semi-xerophilous vegetation occur. Towards the more seasonal forests, when the dry season exceeds 3 to 4 months, fires become increasingly frequent. Disturbed rain forest is more prone to fire damage; when as a result of disturbances, grasses invade, it is more easily burnt and a cycle of fire and further grass invasion may commence which results in the substitution of a savanna. A detailed account of burning habits and fire patterns in the three major tropical zones is given by Batchelder (1967), who summarizes the trends as follows: 'Man's use of fire in the tropical world is no longer in the stage of ecological climax, wherein a stable, harmonious relationship to the environment exists. The impact of the Western culture has set in train a process that has led to rapid changes in the environment, and destroyed the old balance between man and nature (Watters, 1960). Thus the specific conformation of the interrelations of man, land and fire in any particular area can be seen as stages along a continuum of change. At one end are the areas just emerging from ecological climax, with relatively little soil erosion and forest or pasture degradation. The highlands of New Guinea and the interior rain forests of the Amazon Basin are representative of this stage. Near the middle of the continuum are areas in ecological disequilibrium, in which population growth, changes in value systems, soil erosion, and forest and pasture degradation have created a critical situation. Shortages of food in the rural areas have led to widespread undernourishment, accelerated deterioration of the land, and migration of rural people to cities. Many areas of tropical Africa and Latin America are representative of this stage. At the other end of the continuum, in the stage of ecological rehabilitation, are the areas in which the application of scientific knowledge and techniques has resulted in the use of fire as an instrument of forest and grassland management. Representative of this stage in the continuum are those areas in East and South Africa, parts of India, Burma, Thailand and Australia where control burning is used with adequate safeguards against escape of fire and a tested year-by-year plan of fire use.' The author adds that 'the use of fire and its effects appear to have declined in Oceania and parts of South and South-East Asia, notably India. Here, agricultural populations have become sedentary, while primitive cultivators represent minorities located in remote geographic areas. On the other hand, cultural use of fire is increasing in much of Africa and parts of tropical and subtropical Latin America. The rate of population increase is high, particularly among rural peoples, and new lands are being sought for clearing by expanding populations. It is clear that for much of the tropical world, the effects of fire on the natural and cultural environment are increasing at an accelerating rate'.

Fires occurring at the beginning of the dry season generally have a lower impact, while fires towards the end of the dry season, have a stronger regressive effect (e.g. Freson et al., 1974).

---

[*] In those countries that maintain fire records, less than two per cent of the total fires reported are attributed to natural phenomena (Batchelder, 1967).

Exploitation

The disturbance of the forest by exploitation consists of opening the upper canopy and in the influence on the remaining vegetation and on the soil. Selective logging, limited to one or a few species and to a minimum diameter, may produce relatively small and sometimes only few gaps per hectare. Such logging will deplete the forest of certain species and dimensions of trees, but ecologically it will not constitute a major disturbance of the system. Between this and the extreme of complete removal of the ligneous vegetation are all degrees of exploitation. Felling large trees causes damage to the remaining smaller trees. Baur (1962) quotes an example from Nigeria where logging 6 trees (40 m³) per hectare damaged 32 per cent of the remaining trees; similar figures for North Borneo were 12/ha (90 m³) and 53 per cent. The heavy logging equipment causes much damage to trees and seedlings as well as through compaction of the soil (see chapter 20). The effects of the technologically-advanced methods of harvesting tropical forests are potentially catastrophic because they are irreversible. In many cases, clear-felling operations may be sufficiently extensive to cause the extinction of at least some species of the original forest.

Erosion

Erosion is an important factor where the forest cover has been destroyed, such as in shifting cultivation, and where the terrain is sloping. 'When the slope of the ground is anything but gentle, and some authors note that the process starts with 3 per cent slope, surface erosion of a more serious type sets in, the top soil is removed, silting up the valleys, soil fertility is gone and grass savanna installed. Even in the absence of a new cultivation cycle erosion is resumed after every annual grass fire. The basin is subject to flooding. Apart from spectacular gullies and hill slides, common in lighter soils, erosion may go as far as to uncover lithosols, and stone and concretion layers. The site is lost for agriculture and will certainly be given up to foresters to try their trade and restoration' (Donis, 1965). Tricart and Cailleux (1965) give an account of soil degradation and erosion following removal of the forest cover, including the particular patterns of the *lavakas* in Madagascar (*voçorocas* in Brazil).

Conclusion

The effects of man as an agent of disturbances, and hence of change, are many. On a small scale, forests are dissected by roads, clearings and dwellings. Animals are hunted and seed dispersal mechanisms may be affected. Even the most innocuous recreational pursuits increase trampling and soil compaction. On a larger scale, man extracts timber, manages the forest in a number of ways to ensure adequate natural or artificial regeneration of commercial timber species, and in the extreme case, clear-fells the forests. It is sometimes difficult to make a clear difference between changes induced by man and changes induced by natural phenomena (human versus wildfires, natural flooding versus flooding resulting from dam building, etc.), but there are a number of obvious reasons to consider separately the changes caused by natural disturbances and those caused by man. First, the areas which can be cleared by man occur on a scale far greater than those which occur naturally. Secondly, man-induced disturbances usually result in the removal of most of the vegetation from the affected areas. This, of course, constitutes a major nutrient drain from the system. Thirdly, in many cases at least, natural regeneration of the cleared area is actively discouraged for varying periods of time after the initial clearing.

In view of the lack of systematic information on the causes and nature of modifications resulting from man's activities in various ecological zones, there is a clear need for establishing in any large development scheme an operational research unit to analyse the causes and to monitor the changes.

Regeneration

Environmental factors

Environmental factors are of the utmost importance in controlling the direction of regeneration. 'The removal of the forest cover at once changes the illumination at ground-level from a small fraction to full daylight. The temperature range greatly increases and the average and minimum humidity of the air become much lower. There is a change from the complicated system of microclimates characteristic of high forest to conditions closely approximating to the standard climate of the locality. Exposure to sun and rain very quickly alters the properties of the soil. Where the slope is sufficient, erosion will begin to remove the surface layers or their finer fractions. The rise in soil temperature leads to a rapid disappearance of humus' (Richards, 1952). He mentions the loss of nitrogen and other nutrients and emphasizes the importance of the time factor: 'If the secondary succession is allowed to start immediately after the forest has been cleared, there is little opportunity for soil erosion and a closed cover of vegetation is formed in a few weeks; the process of soil impoverishment is arrested before it can go very far and soil and microclimate begin at once to return to their original condition. If on the other hand there is a long period of cultivation after clearing, particularly if the crop does not provide an adequate ground-cover, the long exposure of the soil, together with the losses of plant nutrients in the harvested crops, leads to such large changes in the structure, humus-content and nutrient status of the soil that the time needed to restore the equilibrium between soil and vegetation, even where there is no large-scale erosion, becomes much greater. The importance of the length of the cultivation period and the extent of humus destruction to the course and length of the subsequent succession are thus easily understood.' Time is a selective factor in evolution (Gómez-Pompa *et al.*, 1974), but it is generally difficult to 'ascertain the past history of an area so as to build up a picture of succession after different treatments' (Whitmore, 1975). Considering the end point of recovery, or possibly the several end points, the time factor may appear less important: 'it is not the timing of the appearance of the life form stages that is significant but the sequence itself' (Golley *et al.*, 1975). These authors, like several others, also point out that the impact of the initial conditions on the succession is most important in the earliest stages and of lesser influence in the later stages.

Fire affects the regeneration both directly through the burning of seeds, seedlings and trees and indirectly through its action on the soil. The latter is often superficial and of brief duration, though surface temperatures may be high, reducing the amount of organic matter, modifying soil texture, facilitating erosion (Schnell, 1971) and reducing nitrogen fixation. 'There is often a deterioration of structure due in part to the dismembering of soil aggregates resulting in clogging of the porous space' (Budowski, 1966).

Many seeds, and most seedlings, will be killed, and trees are damaged by fire, but organs located more than a few centimeters under the soil surface will not normally be affected. There are great differences in fire tolerance between species, and fire is thus a selective factor, tending towards a simplification of the floristic composition and a complication of the successional pattern. In the moist deciduous tropical forests fire may be of use in arresting the normal successional process and maintaining vegetation in a more economically useful seral stage. There seems little doubt that, in the past, annual or almost annual burning, plus other disturbance such as grazing and shifting cultivation, has perpetuated aggressive fire tolerant species such as teak (*Tectona grandis*) and sal (*Shorea robusta*) over large areas in the Asian region. For example, gregarious flowering of bamboo, at intervals of 30–50 years, followed by fire and natural seedling, has produced much natural teak in Asian forests (FAO, 1958). Fires could stimulate seed germination of some secondary species. Vázquez-Yanes (1974) demonstrated this for *Ochroma lagopus* and underlined the importance of light and temperature for the germination of secondary species (among the genus *Piper* there are photoblastic species, the seed dormancy of which is regulated by red-far red illumination periods; there are other species, including some legumes and *Heliocarpus*, the seed dormancy of which is due to the seed envelopes and can be broken by high temperatures or scarification).

Repeated fires will often deflect the succession and lead to a grassland fire-climax. Many so-called savanna areas are not a climatic climax but essentially the results of repeated (often annual) fires (Budowski, 1966; Schnell, 1971).

Biotic factors

The presence of propagules will determine the composition of the early successional communities. The role sometimes played by suckers, showing more vigorous growth than seedlings, is underlined by Webb *et al.* (1972). The great extent of forest clearing has favoured the immigration of species belonging to less hygrophilous communities (Lebrun et Gilbert, 1954). Years with a longer than usual dry period may have considerable ecological significance (Schulz, 1960).

Guevara and Gómez-Pompa (1972) reported that primary forest in Veracruz, Mexico, contained from *ca.* 200–900 seeds per m² comprising 13-26 species; they have studied the presence of seeds in samples of primary and secondary vegetation soils and have demonstrated that the seeds of secondary species such as *Ageratum conyzoides*, *Bidens pilosa*, *Clibadium arboreum*, *Eupatorium* sp., *Heliocarpus* sp., *Irosine celosia*, *Phytolaca decandra* and *Robinsella mirandae* were dominant. They have also demonstrated that dormancy of these seeds plays an important role in succession. These results are similar to those obtained by Keay (1960) in Nigeria, who, by studying the seeds of the forest soils, has concluded that very few species are important from a forestry standpoint, while there is a great preponderance of seeds of secondary *Musanga*.

Regeneration is also controlled by biotic interactions. Anaya and Rovalo (1975) have shown that plant competition may be regulated by secondary substances produced by some species of *Piper* and *Croton* which inhibit germination and growth of other species. The whole problem of allelochemistry and allelopathy is discussed in chapters 8 and 20.

The study of seedlings is of the utmost importance. Liew and Wong (1973) demonstrated the drastic effects of clearing on the density, mortality and growth of dipterocarp seedlings in virgin and logged forests in Sadak (Malaysia) where dipterocarps can be considered as primary species; in some cases 98 per cent of the seedlings disappeared.

Recovery may also occur from root stocks, from coppicing and by other vegetative means. The relative importance of each depends partly on the extent and type of disturbance and the community. Del Amo and Gómez-Pompa (in Gómez-Pompa *et al.*, 1974) have observed that young trees, within the forest, maintain a meristematic dormancy which may continue for very long periods; this is used for forest management and improvement in dipterocarp forests. It has also been demonstrated (Janzen, 1970), that predators have an impact on species distribution and this can be altered by man's intervention. Vázquez-Yanes (1974) has demonstrated that the presence of peculiar species is facilitated by seed dispersal due to bats and birds (see chapter 7).

Conclusion

Some factors influencing forest regeneration can be easily affected by human activities. Further experiments will have to be carried out to obtain a clear and comprehensive understanding of the regeneration process and in order to know how to influence it by manipulations. A controversy remains about the probability and predictability of the regeneration process. Some authors (Kellman, 1970; Gómez-Pompa *et al.*, 1974) consider that regeneration is not random, and predictable, while others (Webb, Tracey and Williams, 1972) consider that the pattern of distribution of species in the first stage may be affected by environmental factors, but is mainly probabilistic. This question will be discussed in the section dealing with successional stages (see also chapters 8 and 21).

Another important issue which needs to be studied is the impact of environmental disturbances on specific richness, since genera like *Euphorbia* have 2 000 species, *Solanum* 1 700 and *Piper* 2 to 3 000. The destruction of reproductive and ecological barriers between isolated populations has facilitated the free genetic exchange between them (Gómez-Pompa *et al.*, 1974).

Secondary forest characteristics

Secondary communities on rain forest sites are extremely varied, but it is useful for descriptive purposes to single out a 'typical secondary forest, i.e. the earlier seral stages found on areas which have been cultivated or exploited for timber, but not subsequently grazed or burnt . . . secondary forest in this sense has well-marked features and is as a rule easily recognized' (Richards, 1952). These features are the smaller average dimensions of the trees, though scattered large trees may remain from the original stand; the often regular and uniform structure of the very young stages and the very irregular structure of somewhat older stages with many lianas. Floristically, the secondary forest is much poorer in tree species than the primary and there are few epiphytes (Lebrun and Gilbert, 1954).

'Old secondary forest is difficult or impossible to distinguish from undisturbed virgin forest, but since secondary forest consists of stages in a succession leading ultimately to the re-establishment of the climatic climax, this is only to be expected. Biotic climaxes resulting from deflected successions may closely resemble other climatic climaxes, though not the rain forest from which they are derived' (Richards, 1952). In certain cases, the weak representation of small trees as compared to the number of large trees of given species may be explained by the forest being an old secondary community.

The trees constituting the earlier stages have a number of common characteristics which differ from those of the predominant trees of the primary forest. The typical secondary forest trees are light demanding, requiring a light intensity of at least 75 per cent of full daylight. Many are unable to regenerate under their own shade. They show rapid growth, 1–4 m/a in height, and 2–4 cm/a in diameter. They have early flowering and efficient propagule dispersal, usually by wind or animals. They are short-lived, soon reaching their maximum dimensions, often dying at *ca.* 15 years of age and only exceptionally living beyond 25–30 years. They are frequently gregarious, and their timber is light, soft and not very durable (Richards, 1952; Lebrun et Gilbert, 1954; CTFT, 1974). There is also a secondary fauna of predators, pollinators and dispersal agents. A description of the life-cycle of typical secondary species, with examples from tropical America, is given in fig. 1 from Gómez-Pompa and Vázquez-Yanes (1974). Hopkins, Webb and Williams (1975, unpublished) describe the characteristics of the forest canopy types (pioneer, early secondary, advanced secondary and mature).

Secondary species appear to be somewhat more selective with regard to the soil than primary species (Richards, 1952; CTFT, 1974). Kellman (1969) suggests that secondary species may cycle elements in greater amount than primary species, and that specific plants (*Trema orientalis* and *Melastoma polyanthum*) restore phosphorus and potassium to

the upper soil layers. This explains also why they can establish themselves on poor soils. Nye and Greenland (1960, see chapter 20) have recorded the changes in nutrients with succession. The accumulation of soil nutrients with succession is largely associated with an increase in humus derived from litter fall. The rate of humus increase depends upon the maximum level of humus characteristics of the mature system and the degree the soil falls short of this level. They also suggest that the ratio of carbon to nitrogen does not change greatly during the succession, the increase in nitrogen being thought to be mainly derived from non symbiotic nitrogen fixation. Increases in phosphorus, potassium, calcium and other elements are also closely tied to the vegetation, although the total amounts in the vegetation are much greater than that in the soil.

The build-up of plant biomass and, consequently, nutrient storage, may be rapid. Leaf cover and leaf area may in certain cases reach that of mature forest in about 6 years (Golley *et al.*, 1975). Thus, although the initial occupation of the site by a given set of species may appear random or highly variable and the full recovery of the ecosystem require hundreds of years, recovery follows a repeatable, predictable sequence of life-forms and the essential functional characteristics of the ecosystem are re-established relatively rapidly (Golley *et al.*, 1975). The secondary forest life-form stage may be 'that point when mass and production are maximum in the tropical forest sere' (Kira and Shidei, 1967, see chapter 10). In Puerto Rico, Jordan (1971, see chapter 10) has shown that annual biomass increase is greater in secondary vegetation than in primary.

'. . . It is noteworthy to consider the floristic and physiognomic similarities between the young secondary formations of the various tropical regions. Some genera are common, which have a comparable aspect and play a similar role in the landscapes: *Trema* (Africa, America, Asia), *Vismia* (Africa, America), climbing *Scleria* (Africa, America), *Mussaenda* (Africa, Asia), *Leea* (Africa, Asia), etc. The *Cecropia* play, in the young secondary vegetation of the tropical humid America, the same role as *Musanga cecropioides* in that of Africa; morphology and stand are identical; both *Musanga* and *Cecropia* originate from some heliophilic riverain associations, where they live naturally' (Schnell, 1971). There are several exceptions to this; some of these are found in the examples of successions in the following section.

Successional stages

In general, succession in rain forests shows similar features in all three regions of the humid tropics and unless deflected by grazing, or repeated burning and cultivation, tends to re-establish the climatic climax in a series of communities, beginning with herbaceous weeds, shrubs, small, rapidgrowing, short-lived trees, larger trees typical of second growth forest, usually light demanding and rapid-growing, back to primary forest species. The secondary forest stage is often marked in America by species of *Cecropia* and in Africa and Malaysia by *Musanga cecropioides* and species of

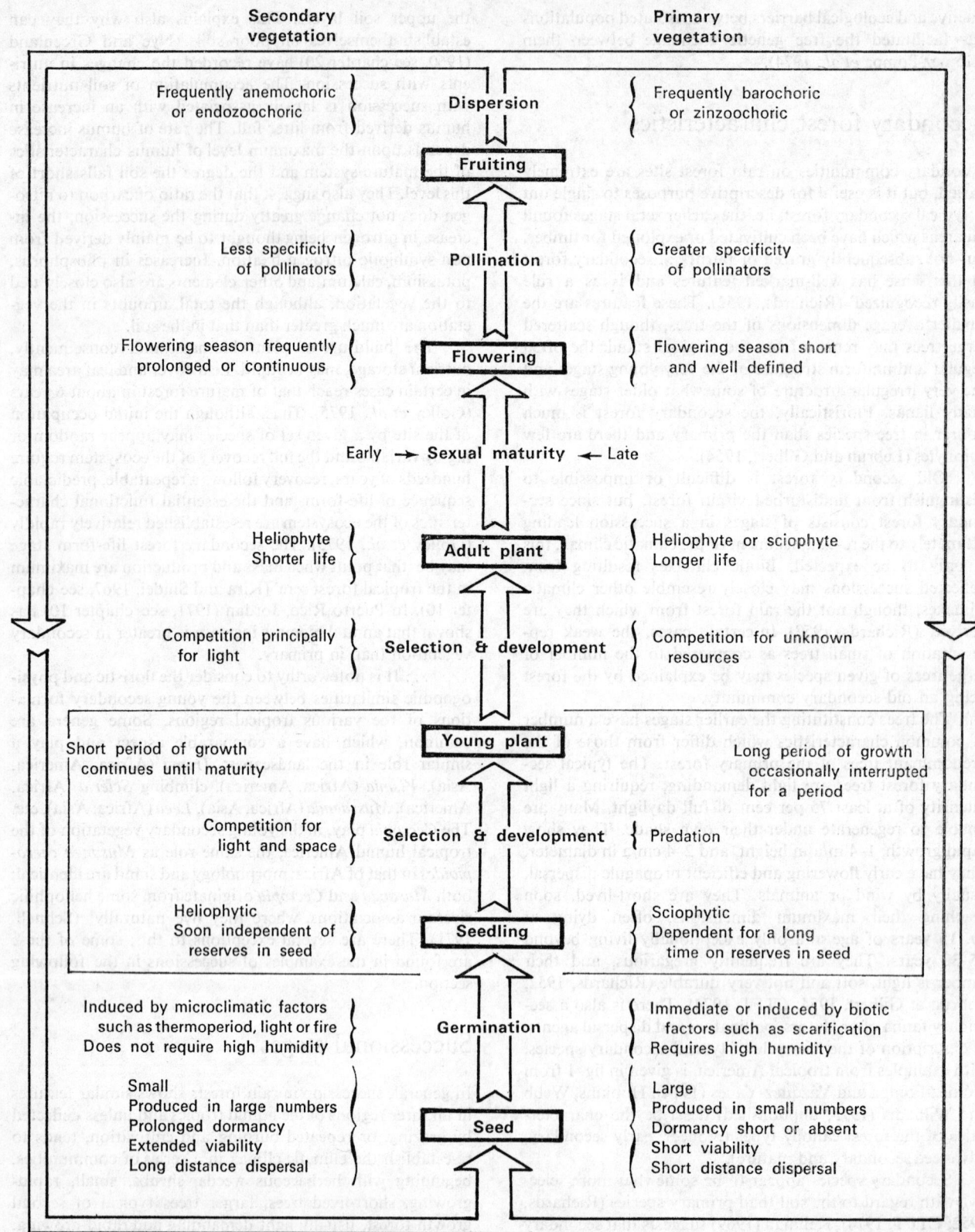

**Secondary
vegetation**   **Primary
vegetation**

Fig. 1. Life-cycles of plant species. Each step can be isolated
and a series of research projects may be planned that will contribute to understanding the whole (Gómez-Pompa and Vázquez-
Yanes, 1974).

Macaranga respectively, all rapid-growing forest species with good means of seed dispersal, thus facilitating a temporary foothold. *Trema* species are found in both Africa and Malaysia. Secondary forests, if undisturbed by fire, grazing or felling, are gradually invaded by primary forest species and will tend toward the climatic climax though probably over a very long period (FAO, 1958). The weed and grass stage may be of only some weeks' or months' duration.

Repeated cultivation and burning and other factors 'alter the course of the succession by giving an advantage to certain species over others. This often leads to the replacement of forest by grassland closely simulating an edaphic or climatic climax' (Richards, 1952). It may also lead to the formation of a scrub with scattered medium-sized trees, as over large areas in Africa. On the related question of the forest-savanna boundary, most authors concur that most savannas have an anthropogenic origin (Budowski, 1965; Schnell, 1971).

In moist deciduous forests, the successional trend is often toward semi-evergreen types, with deciduous light demanding species gradually crowded out by shade-bearing evergreen species, under which the former cannot regenerate. This trend is accelerated by fire protection, while annual fires tend to arrest this change (FAO, 1958).

General characteristics

Budowski (1965) compiled a list of general characteristics of the life-form sequence for use in land management. Table 1 is a summary of his large number of observations of the sequence of successional stages in tropical forest communities. There are several features which require emphasis. First, the number of species per unit area generally increases as the succession proceeds. Not only does one group of species replace another, but the number of species and the distribution of their individuals change. Secondly, the vegetation biomass increases as the succession proceeds, and so does nutrient storage, leaf cover and leaf area. In secondary stands in Panamá these statistics reached those of the mature forest in about six years (Golley *et al.*, 1975). Thus, the photosynthetic and transpiration capacities of the ecosystem are quickly re-established. These changes also rapidly re-establish the micro-climatic conditions; Ross (1954) found that the environmental conditions at ground level in 14-year-old secondary forest closely resembled those in mature forest. Thus, although the initial reoccupation of the site by a given set of species may appear random or highly variable and full recovery of the ecosystem may require hundred of years, recovery follows a repeatable, predictable sequence of life-forms and the essential functional characteristics of the ecosystem are re-established relatively rapidly.

Rate

There are few studies of the rate of successions. These studies should be carried out using replicated plots with careful descriptions of the environmental conditions. Kellman (1970) studied plots of the same age with different environments, and plots of different ages having similar environments in Mindanao, Philippines. A summary of some of his data (table 2) clearly shows the variability between plots of the same age. While there is an overall trend of increasing biomass and number of species, the variation between plots may be fivefold. Snedaker (1970) carried out a similar study at Lake Izabel, Guatemala. The range of biomass (table 3) between plots is also very large. These two studies illustrate that when an adequate sample is collected the seral communities appear to become more similar. It is clear that we do not have a set of samples sufficient to describe the rate processes of changes in species composition, biomass, production or nutrient storage in tropical forest successions. From their study of production and biomass of forest ecosystems in the western Pacific, Kira and Shidei (1967, see chapter 10) suggest the shape of the expected rate curves for tropical forests. There is an initial period of rapid increase in biomass and production which reaches a maximum level late in the sere and declines slightly to the mature stage. The secondary forest life-form stage, mentioned earlier, is probably the point where biomass and production are maximal. Assuming that the Kira and Shidei curve represents a reasonable description of rate processes in tropical forests and knowing the characteristics of the mature forest, the average rate during the succession could be predicted. These would provide a basis for observational tests in actual stands. These tests would require adequate samples in different ages of forests, using correct plot designs, in a forest area large enough to provide undisturbed areas and permanently protected plots which can be experimentally manipulated. It would be extremely

TABLE. 1 Characteristics of forest succession in tropical lowlands (after Budowski, 1965).

	Pioneer stage	Older stages
Floristic composition	few species of wide distribution	many species
Strata	few	several
Large stem diameters	none	present
Undergrowth	dense	less dense
Shape of upper crowns	uniform, thin, light green	varied, dark green
Age composition	even-aged	uneven-aged
Seeds	small	large
Regeneration of dominants	absent	common
Growth in diameter and height	rapid	slow
Life-span of dominants	short	long
Size and shape of leaves	many macrophylls	mostly microphyll
Hardness and weight of wood	soft, light	hard, heavy
Climbers	few species many individuals mostly herbaceous	many species few individuals large, woody
Epiphytes	few species	many species

valuable to have a number of sets of these experiments in different environments and regions. In this way, it would be possible to build predictions of the course of succession under different disturbances and calculate environmental and economic costs of making errors in land management.

TABLE 2. Above-ground biomass and plant species in plots of secondary vegetation in Mindanao, Philippines (after Kellman, 1970).

Age (years)	Fresh weight biomass (kg/ha × 10⁻³)	Number of plant species
1.0	31	48
1.0	23	104
2.5	39	45
3.0	27	21
6.5	119	66
7.0	117	77
7.0	83	46
7.0	27	26
19.0	136	50
19.0	116	61
19.0	103	61
19.0	81	55
21.0	134	93
23.0	84	23
27.0	419	81
27.0	370	78
Mature	11 205	85

TABLE 3. Variation in biomass of successional vegetation at Lake Izabel, Guatemala (after Snedaker, 1970).

Age (years)	Number of plots	Biomass (kg/ha)	
		Mean	Range
0.83	10	9 415	4 430– 14 140
1	17	8 327	3 870– 15 690
2	16	14 199	5 600– 27 560
3	7	22 820	9 480– 46 240
4	16	26 908	12 920– 48 180
5	16	36 666	12 210– 84 420
6	16	44 669	14 730– 70 160
7	16	46 658	14 400– 96 830
8	4	65 823	29 070– 97 930
9	4	72 403	26 090–114 240
10	32	53 303	10 000–210 320

End point

Richards (1952) and Cousens (1965) question the existence of a stable mature state in tropical rain forest. They suggest that natural forest always consists of a mosaic of seral stages of different ages where early, middle and late seral components occur in varying proportions. This concept of a dynamic equilibrium of the tropical forest is probably much closer to reality than earlier concepts of climax forest communities. These successions lead toward a set of possibilities which are a function of the history and availability of the biota, and the environmental conditions.

Where disturbance is very wide-spread and continuous, the succession may be retarded and deflected towards a biota that can tolerate these conditions. The pine forest in the Cordillera Central of Luzon is an example of a wide-spread fire climax in what would ordinarily be broad-leaved tropical forest. It is extremely important that accurate predictions of the end points of succession are developed so that landscapes are not dealt with *post facto* but disturbances are regulated so that if management errs, recovery can proceed to the prior ecosystem and the process can recommence. See chapter 8.

Regional patterns

Successional patterns following disturbance in tropical forests will be described for various regions. These are based on the following sources:

Asia-Pacific

Indonesia (Sastrapradja and Kartawinata, 1975, unpublished)
Philippines (Mabesa, 1958)
Malaysia (Landon, 1958)
Thailand (Sanga Sabhasri, 1975, unpublished)
Viet-Nam (Pham-Hoang-Ho, 1975, unpublished)
Teak (Kermode, 1958)
Sal (FAO, 1958; Hewetson, 1958; Seth, Kaul and Srivastava, 1975, unpublished)
Queensland (Hopkins, Webb and Williams, 1975, unpublished)

Africa

Zaire (Lebrun and Gilbert, 1954)
Nigeria (Ross, 1954)
Western Africa (Schnell, 1971)

America

West Indies (Wadsworth, 1958)
Panamá (Richards, 1952, summarizing Kenoyer, 1929)
Surinam (Boerboom, 1974)
Brazil (Richards, 1952, summarizing Freise, 1938)

Pantropical formations

Mangroves (FAO, 1958; Noakes, 1958)
Conifers (FAO, 1958)
Bamboos (FAO, 1958; Seth, 1958)
Mountain forests (Schnell, 1971).

Asia-Pacific region

Indonesia

The extent of secondary forest and *Imperata cylindrica* grassland in Indonesia is *ca.* 23 and 16 million ha respectively, and is mainly the result of shifting cultivation, plantation agriculture and fire. The recent selective exploitation of timber from primary rain forest has been very rapid, creating millions of hectares of depleted forest which is a mixture of primary and secondary types.

In the last decade, some investigations have focused on

the regeneration of economic species (*Hopea mengarawan, Shorea* spp., *Pinus merkusii, Agathis borneensis, Pometia* spp., etc.) but the most detailed investigation on regeneration of primary forest species after canopy opening is still the work of Kramer (1926, 1933, see Annex by Whitmore) in montane forests of Mt. Gedeh, West Java.

There are only a few systematic investigations of successions. Jochem (1928) described the early stages at Deli, Sumatra, and recognized three stages: herbaceous—a floristically rich community consisting of short-lived herbs and young trees; *Trema-Blumea*—a community where *Trema* spp., *Blumea balsamifera* and *Abroma angustum*, were the main small species; and tree—a secondary growth in which one or more of *Trema* spp., *Macaranga tanarius, M. denticulata, Callicarpa tomentosa, Melochia umbellata* predominate. The abandoned cultivated land in South Banten, West Java, is usually invaded first by *Imperata*, and later by shrubs such as *Melastoma malabathricum* and *Lantana camara*, and then by tree species of secondary forest such as *Vitex, Grewia* and *Dillenia*. Similar observations were made by Eussen and Wijahardja (1973) in Lampung, southern Sumatra, who stated that the established *Imperata* grassland was invaded by *Eupatorium odoratum* and then by secondary forest species, such as *Melanolepis multiglandulosa, Piper aduncum, Pterocymbium javanicum, Grewia glabra*, etc. They contended that the competitive ability of the *Imperata* was partly attributed to allelopathy.

Frey-Wyssling (1931) compared heath-like scrub community (consisting of *Leptospermum flavescens, Melastoma decemfidum, Rhodomyrtus tomentosa*, etc.) on Habinsaran plateau and Lake Toba region (northern Sumatra) with grasslands found under similar ecological conditions on the Karo plateau (northern Sumatra). He concluded that both communities arose through human disturbance of rain forest; the scrub developed where fire and grazing were less and grassland when human influence was more intense.

Kostermans (1960) considered that the floristic composition of successional stages depends on various factors including the extent of the vegetation destroyed, the flora of the immediate surroundings, the season in relation to the seed-bearing period of the neighbouring vegetation, the dispersal agents, the climate and in particular, the direction and velocity of winds as seed dispenser and rain necessary for germination, the soil, the way the original forest was cleared, the duration of man's interference, and the characters of the plant species available to invade the destroyed area. The interaction among these factors defines the type of secondary successions. Kostermans listed species typically found in secondary vegetation. Secondary forest species share several common characteristics; they are: generally light demanding and intolerant of shade, have efficient means of propagule dispersal adapted for wind or animal transport, fast-growing, short-lived, fruiting abundantly and independently of climate, flower when young, have high germination capacity, are drought and fire-resistant, highly tolerant to climate, temperature and soils, and usually have underground organs (Van Steenis, 1958).

Philippines

The logged dipterocarp forests show various patterns of succession, where the area is not taken over by shifting cultivation.

In northern Negros, wild bananas, *Musa* spp., occur abundantly with patches of the grass *Panicum sarmentosum*; fast-growing species (genera *Trema, Melanolepis, Homolanthus, Macaranga, Piptorus*) follow. Trees spared by the logging may develop epicormic branching and may supply seed, and less injured pole trees may continue growth.

In Bataan, herbaceous weeds (e.g. *Blumea balsamifera*) and grasses are followed by the pioneer species mentioned above, plus *Ficus variegata*. Dipterocarp species will gradually take over provided seed sources are present. However, where the area is invaded by climbing bamboos (*Schizostachyum* spp.) which form a mat over the ground, there is little chance for the winged seed of the dipterocarps to reach the soil and the erect bamboo *S. lumampao* is likely to invade. Few saplings of tree species are found here.

On Mt. Makiling where the dipterocarp forest had been degraded to savanna through shifting cultivation, the scattered trees and shrubs are mainly followed by trees of the genera *Melochia, Columbia, Litsea, Macaranga, Ficus, Mallotus, Premna, Trema, Canarium* and *Artocarpus*. Thereafter seedlings of the dipterocarps *Parashorea plicata* and *Pentacme contorta* may appear when a mother tree is nearby. The degradation of the original forest in this area may well take place in other areas if uncontrolled logging and shifting cultivation continue

In Surigao, a thick growth of the vine *Merremia nymphaefolia*, and wild bananas, covers the ground in the cut-over forest and delays the regeneration of the pioneer species of trees (*Trichospermum, Alphitomia* and other genera). Twelve years after a tornado, the dipterocarps *Shorea almon, S. polysperma* and *Pentacme contorta* were predominant where no clearing and burning had taken place.

At elevations above 1 000 m, *Pinus insularis* becomes dominant if fires are infrequent.

Malaysia

When forest is cleared and cultivated for a short time, coppice shoots and pioneers, such as *Trema* spp., *Macaranga* spp., and *Mallotus* spp., soon establish a dense secondary forest. If the area is small or if seed-bearers of *Shorea* spp. or *Dipterocarpus* spp. have been left, high forest conditions may be restored within a lifetime, and probably much of the so-called virgin jungle is really secondary forest of this type. When disturbance is prolonged for 2–3 years and three or more crops are harvested, there is usually an invasion of *Imperata cylindrica* grass and the succession is seriously delayed. *Imperata* will sooner or later be invaded and suppressed by *Melastoma malabathricum, Eupatorium odoratum*, and other pioneers. However, it is very easily burnt and, because of its underground rhizomes, recovers quickly after a fire while its competitors are weakened or destroyed. Such areas are often deliberately burnt for grazing and extensive fire sub-climaxes are established.

The soil becomes so impoverished that rehabilitation is very difficult and would, in the course of nature, take several centuries.

Thailand

In addition to logging, hunting, overgrazing and over-browsing by domesticated livestock, shifting cultivation associated with fire is the main cause of the disturbances, both in the lowlands and in the mountains. Shifting cultivation, which was mainly practised in evergreen forests and rarely in deciduous forests, especially the dry dipterocarp forest, is now extensively done in this type at lower elevations. No study has been made on the effect of fire in dry dipterocarp forest and little is known about its general effects.

Viet-Nam

The forests in Viet-Nam have been modified by shifting cultivation for a long time. The considerable extent of steppe grasslands on the basalt plateaux of Pleiku, of the pine formations on the slopes of cristalline or shale-sandstone areas near Dalat, is probably due to man. The effects on the forest of the use of defoliating chemicals and the impact of bombing during the recent war have been studied and they are obviously very different from those of fire, shifting cultivation or selective logging; they are reviewed in chapter 20.

Teak (*Tectona grandis*)

Selective logging of teak, with a minimum diameter, may involve as few as 1–3 trees per hectare. The ecological effect is small, and the species is not depleted. In the more accessible areas, logging operations comprise other hardwoods as well, and the disturbance of the forest is greater. The immediate effects may only be apparent in gaps where *Eupatorium odoratum* invades. Later the effect becomes apparent when the bamboo flowers, following which there is often a dense growth of young teak saplings. These are mostly derived from plants which survived the removal of the canopy and were prevented from developing by the bamboo shade.

A considerable part of the moister teak forest is kept in a pre-climax state by the annual fires. These conditions appear to favour teak. Young teak can tolerate annual burning. On removal of the cover the additional light enables it, in one growing season, to produce a tall vigorous coppice shoot that it is no longer in any danger from light surface fires. Recent investigations have shown that a considerable number of other species in the mixed deciduous forest can also tolerate burning.

There is also the effect of complete felling and burning of the forest for taungya cultivation. In the semi-evergreen forest this often results in a pre-climax with pure patches of early colonists such as *Trema amboinensis* and teak usually does not reappear. In the mixed deciduous forests the results of taungya vary. In some of the moister lower areas a pre-climax may result with patches of soft-wooded colonists, but in the drier lower moist areas, the forest usually regenerates after taungya to its original climax or pre-climax. The effect of complete burning of the forest for shifting cultivation often, in the drier areas, favours the teak at the expense of other species and the young teak shoots are so vigorous that they keep ahead of the weeds.

Sal (*Shorea robusta*)

The distribution of sal in India is primarily due to man. Fire and grazing exert a powerful influence, and it is difficult to determine whether the sal forest is a true climatic climax or a pre-climax to a moister evergreen forest type, for there are examples of development to the latter, following protection from fire. However, it seems that the moist deciduous form of sal forest, as in central India, is a true climatic climax. Even when burnt annually, it does not regress to a drier mixed deciduous type, nor are there signs here of progress to a moister type of vegetation. The highly gregarious character of sal, its dense canopy and very short leafless period, together with fire tolerance, appear to support succession of the species and generally prevent invasion of light demanding trees. Natural regeneration is often present under the more or less mature sal.

In many parts where the original sal forest has been destroyed, *Shorea robusta* seedlings gradually colonize the burnt grassland which has taken its place. Protection from fire may result in shading out the grass sufficiently to enable the trees to withstand further fires. Sal forest which originates in this way, tends to be purer than those which have originated from successions.

In other areas the effect of fire is different. In wet areas, the sal forest may be able to maintain itself only when frequent fires restrain the evergreen species, while in drier areas exclusion of fire is necessary to permit sal to survive.

Deterioration of climate or sinking of the water-table may produce the same results as grazing, burning and felling, as appears to be happening in some forests in Uttar Pradesh.

Queensland

An area of 1 600 m^2 of virgin subtropical rain forest on Mt. Glorious was cleared. The first year the floristic composition varied and pioneer species whose long-lived seed was already present in the top soil of the mature forest predominated. Later four groups of species developed; these were related to micro-site differences caused by the clearing. Group 1 was composed of the aggressive introduced shrub *Lantana camara* and the native pioneer shrubs *Solanum mauritianum* and *Trema aspera*. Earlier these were scattered over the entire area, but had been eliminated by the intense competition (especially shading) of the later seral species except on areas where bulldozer compaction of the soil had been heavy. Group 2 was ubiquitous; it consisted of two species (*Daphnandra micrantha* and *Eupomatia laurina*) which sucker freely from residual stumps and roots. Group 3 consisted mainly of mature forest species, but it also contained the long-lived pioneer trees *Homalanthus populi-*

folius and *Tieghemopanax elegans*. Their seeds are long-lived and were probably in the soil before clearance. Group 4 consisted entirely of species of the mature forest genera *Actephila, Alangium, Dysoxylum, Eugenia, Pseudocarapa* and *Wilkiea*. Some originated from suckers, some from seed already present, and some from seed entering the plot from adjacent trees.

A patchy structure apparently unrelated to the microsite differences was established 12 years after clearing. The patches only differed quantitatively and thus represented facies of a single association. It is unlikely that any succession will reproduce the original species-community.

Africa region

Zaire

Following clearing and a short cultural cycle, the succession follows the typical scheme: post-cultural ubiquitous nitrophytes, pre-forest phase, young secondary forest, old secondary forest, reconstitution of original forest, with each of these phases being of longer duration than the foregoing. However, only the practically uninhabited areas are still covered by primary forest; elsewhere the shifting cultivator attacks stages of secondary growth mixed with remnants of primary forest. This leads to an extremely intricate pattern.

At the end of a cultural cycle, or after the cessation of fires, the area is occupied by a nitrophilous association named after *Caloncoba welwitschii* and *Trema orientalis* (when a secondary savanna is protected from fire, it is not the pyrophilous and xerophilous shrubs of the savanna, but those belonging to the afore-mentioned association that occupy the area). This phase is soon followed by a young secondary *Musanga cecropioides* forest stage which reaches its optimum in 8–10 years when the former phase has practically disappeared. The *Musanga* reaches a height of 15–20 m, is an obligate light-demander which renews its foliage at a rapid rate, thus actively contributing to a reconstitution of the humus supply in the soil. Its umbrella-like canopy allows through sufficient light for facultative heliophytes and other secondary tree species which penetrate the canopy and the *Musanga* declines. The next stages are dominated by tolerant heliophytes, fairly rapid-growing and reaching a height of 35 m whilst understoreys of the umbrella *Musanga* phase persist for a long time. The gradually developing species of the old secondary forest represent the stage leading to a reconstitution of the initial forest type. Most of the predominant species of the old secondary forest are deciduous.

Uncontrolled felling for charcoal production has in recent years assumed considerable proportions in certain areas of Zaire.

Nigeria

In the Shasha (now Omo) Forest Reserve observations were made on plots which had been left uncultivated for $5\frac{1}{2}$, $14\frac{1}{2}$ and $17\frac{1}{2}$ years. Herbaceous weeds of cultivated land are

the first, but transient, invaders. The next phase consists of typical secondary forest trees such as *Musanga cecropioides, Trema guineense, Vernonia conferta, V. frondosa* and *Fagara macrophylla*. These species are tall enough to suppress the herbaceous weeds; they assume and maintain dominance for 15–20 years. Which species first attains dominance may to some extent be a matter of chance, of presence on or near the area of propagules, at the right time. The chance factor may become less important, and the competition factor more so, when a closed community has formed.

Everywhere *Musanga cecropioides* had become dominant after about 3 years. Apparently it does not colonize bare ground, but requires a vegetational cover. Its subdominant is *Macaranga barteri*. *Musanga* does not regenerate in its own shade; it dies of senescence at 15–20 years of age. At this stage, various other species have reached a height of 20–25 m (species of *Albizia, Anthocleista, Diospyros, Discoglypremna, Funtumia, Sarcocephalus*, etc.). Also young trees of dominant high forest species are present. The trend towards re-establishment of the climax is clear, but the young high forest still contains a high proportion of light demanding species. Further changes over a long period of years are likely to take place, until a more stable condition is reached.

Western Africa

Various species which are normally present in the primary forest and which possess a fairly broad ecological amplitude may participate in the young successional stages (e.g. *Heisteria parvifolia*, certain *Ficus, Palisota hirsuta*), as also light-demanders such as *Pycnanthus, Triplochiton* and *Terminalia*. But the low secondary growth is particularly characterized by the occurrence of a number of shrubs and small trees which do not normally belong to the primary forest of the area. Some of these belong to riverain communities (e.g. *Alchornea cordifolia, Macaranga, Musanga cecropioides, Costus*), others to relatively low and dry edaphic formations such as rocky ridges (e.g. *Holarrhena, Newbouldia*). On the forest-savanna borders, species from the savanna (e.g. *Piliostigma thonningii, Parkia biglobosa*) may participate or dominate. Light demanding species from the semi-deciduous forest may install themselves in openings in the evergreen rain forest, and some of these are long-lived and grow into large trees, e.g. *Triplochiton scleroxylon* in the Ivory Coast.

The successions no doubt continue for centuries. Large 80–100-year-old *Ceiba pentandra* trees on the sites of former villages indicate this forest is still far from the climax. Such successions are mostly interrupted as the shifting cultivator attacks the younger secondary stages. This often results in dense shrub communities with scattered medium-sized or small trees, and often large oil palms (*Elaeis guineensis*) which were spared when the clearings were made. If human interference ceases, the progressive succession will recommence towards reconstitution of the forest; otherwise the regressive succession may lead to a savanna. Fire is a predominant factor in the development, and regressive successions will more easily take place in the semi-deciduous forests than in the true evergreen rain forest. The anthropogenic nature of the forest-savanna border is supported by its clear-cut appearance, the presence of groves of dense forest in the savanna, the patchy mixture of forest and savanna in certain regions, and the presence in the savanna of large clean-boled trees belonging to the closed forest.

America

West Indies

The most important forest type on several of the West Indies islands is the *Dacryodes-Sloanea* association, somewhat less complex in structure and composition than optimum tropical forests. The most common tree is *Dacryodes excelsa*, associated with species of *Sloanea*, especially *S. berteriana*. Disturbance of these forests has been widespread. Even where man has not modified the forest, periodic hurricanes have. The most striking evidence of succession is on the extensive areas cleared for agriculture and subsequently abandoned. Pioneer grasses and herbs usually cover the soil within a month or two and trees predominate within 5 years. This first tree stand is replaced by species which are of value for posts, poles or fuel-wood as soon as the protection afforded by the first trees makes possible a major improvement in the physical condition of the soil. Subsequent succession is less marked and may require several tree rotations for substitutions of species. Some species from this stage persist indefinitely retaining a position in the middle storey of climax forests.

Timber cutting, which is usually highly selective, tends to weaken the top storey and to reduce the representation of *ca.* 6 species. The residual stands, however, usually retain about the same structure and species as the virgin stands. Cutting for posts and fuel-wood has nearly the same effect as clearing and cultivation, with the exception that certain of the second generation species reappear almost immediately from sprouts.

No factors other than hurricanes, timber cutting, and land clearing are important in the modification of the *Dacryodes-Sloanea* association. The forest does not burn, and the fauna is almost entirely avian.

Panamá

In the first stage of colonization of clearings on Barro Colorado Island the chief plants are grasses and sedges (species of *Cyperus* and *Scleria*), but dicotyledonous herbs (Amaranthaceae, Euphorbiaceae, Compositae, Solanaceae, Mimosaceae, etc.) are also abundant. Most of these pioneers are short-lived. After a year large-leaved monocotyledons, especially *Heliconia* spp. and the palm *Carludovica palmata* appear together with numerous seedlings of the trees which dominate the next stage (species of *Trema*, *Cecropia*, *Apeiba*, *Ochroma*, *Cordia*, etc.) and various herbaceous plants not present in the first year. Lianas make this *ca.* 2 m tall vegetation an impenetrable tangle.

After 2 years a young secondary forest is formed of

which the most conspicuous trees are *Cecropia mexicana* and *C. longipes*. Other trees found at this stage are *Ochroma limonense* (balsa) and the palm *Attalea gomphococca*. After 15 years this community still persists, but has become much richer in species. *Ficus* spp., 8–10 species of *Inga*, many small melastomaceous trees and shrubs, *Protium* spp. and very many others are now present. Lianas are numerous, including the conspicuous ribbon-like *Bauhinia excisa*. There is an abundant shrubby and herbaceous undergrowth. By this time a number of primary forest species have established themselves, but the community is still much denser and more difficult to penetrate than primary forest. The indications are that the succession would eventually lead to the development of climax forest.

Surinam

At Blakawatra, in the eastern part of the country, about 70 km from the coast, a forest area was cleared. The debris was piled in rows about 10 m wide and burnt. There were three sites: 1) a nearly flat well-drained sandy loam with the debris removed; 2) a slightly concave sticky loam with the debris removed; and 3) as 1) but with ashes and piles of trunks and stumps covering the surface.

On site 1, the vegetation cover was less than 0.1 per cent three months after debris removal. Eight months after debris removal, weeds, shrubs and tree seedlings covered 10–30 per cent of the surface. At eighteen months, the vegetation consisted of an open, up to 4–5 m high upper layer of *Cecropia obtusa*, and an impenetrable 1.5 m high layer of *Solanum* spp. and *Scleria secans* interspersed with young secondary trees. At 3 years there was a stratified community with an open, 8–11 m high, *Cecropia* layer, a dense 3–7 m high tree layer of the genera *Palicourea*, *Maprounea*, *Goupia*, *Isertia*, *Laetia*, *Inga*, *Vismia*, etc., and the *Solanum-Scleria* layer. Later changes were less marked. In five years *Cecropia* reached its maximum height of 11–14 m, and at seven years *Palicourea* and *Isertia* were penetrating the upper layer and *Cecropia* was declining. The undergrowth now contained representatives of all strata of the original forest, such as *Dicorynia guianensis*, *Paypayrola guianensis* and *Ischnosiphon gracilis*. *Virola surinamensis* was established in the fourth year; this is a species of hydromorphic sites with seeds the size of nutmegs and the nearest seed trees were some hundreds of metres away.

On site 2, heavy rains led to a temporary inundation, resulting in a savanna-like vegetation which only 7 years after was slowly invaded by trees from the small elevations on this site.

On site 3, where the debris was burnt, the ashes became covered with algae. This delayed the succession, and, even after seven years, part of the site still bears only grasses and some ruderal plants.

On the area as a whole, the basal area of the secondary forest after 7 years amounted to 40–70 per cent of that of the original forest. The effect of small differences of terrain and soil (site 2) and of concentrated burning (site 3) are notable.

Coastal Brazil[1]

The secondary succession on abandoned cultivated ground in the area between Pernambuco and S. Catharina seems to be different in some respects from that in the rain forest region of central and northern South America. The first phase is the colonization of the ground by grasses (e.g. *Andropogon* spp., *Chloris bahiensis, Brachiaria reptans*) and herbs such as *Thalia* and *Alternanthera* spp. The pioneers are soon joined by very rapidly growing climbers (*Canavalia, Pachyrrhizus* and *Phaseolus* spp.). Some shrubs survive from the previous cultivation and these together with invading light demanding trees form islands which tend to spread and shade out the grasses. In the young secondary forest which now develops several species of *Cecropia,* each characteristic of different soil conditions, occur; these live for 8–12 years. Of the remaining secondary forest trees, some (e.g. Euphorbiaceae, Sterculiaceae, etc.) reach an age of 25–30 years; others live longer; most are soft-wooded. Some of the secondary forest species are said never to occur in primary forest and some primary forest species never in secondary. In Freise's view the succession would ultimately lead to a climax of different composition to the original primary forest. If this is so, it may be because the primary forest has been so extensively destroyed that few suitable seed-parents now remain. The greatest age of the secondary forest is estimated to be 150–200 years; in this time there would be five to eight generations of the shorter-lived, and three to five generations of the longer-lived, tree species.

The secondary forest of this region is often interfered with by felling, burning, etc. In this way deflected successions are started, leading to the development of grassland or *Pteridium* communities or to the denudation to bare rock.

Pantropical formations

Mangroves

In Malaysia, the only normal disturbance is felling. Mangrove species fruit annually and prolifically, but the closeness of the normal canopy inhibits the survival of seedlings of the desired species. Opening of the canopy by thinnings is therefore beneficial. Felling tends to cause crop deterioration by either retarding the early succession stages, or hastening the later stages, or increasing the proportion of inferior species in the climax. Any attempt to hasten the succession by felling the pioneer *Avicennia* species results in prolific regeneration and coppicing. In drier areas felling encourages *Bruguiera gymnorhiza* into *Rhizophora* forest; and in the driest area, felling is almost invariably followed by the dense growth of *Acrostichum* fern that marks the end of the mangrove succession. In the climax wet *Rhizophora* forest, regeneration is generally fairly successful, but even here felling undoubtedly increases the proportion of the very inferior *Bruguiera parviflora.*

Succession, with some differences in pioneer species, is essentially the same in the eastern and western mangrove zones.

Conifers

Although very little research has been done on conifer succession, it is ordinarily towards mixed hardwood forest. Only where conditions are unfavourable for hardwoods, such as low soil fertility or moisture content, or high acidity, do tropical conifers develop in pure stands. On the warmer, wetter, more fertile sites, they are unable to compete successfully against the hardwoods. Repeated fires, however, make it possible for conifers to take over hardwood sites. In addition, many tropical coniferous species can establish themselves and grow well in a wide range of climate and soil, and the chance occurrence of open sites and available seed have given rise to many oddities of distribution. Grazing (e.g. India), extensive fires (e.g. Central America) and insect epidemics (e.g. Honduras) may threaten these forests.

Bamboos

The main disturbance in bamboo forests are clear-cutting, fire and grazing. Clearcutting tends to decrease the production of new culms. Fire may be very intense, especially after gregarious flowering, and kill the seedlings, and grazing (India), combined with fire or alone, may be severe enough to exterminate the bamboo over large areas. In contrast to what happens in the case of fire, seedlings do survive under heavily grazing inside the dead bamboo clumps and, if given a chance, eventually grow into new clumps.

Bamboos are generally markedly gregarious. Thus, *Melocanna bambusoides* and *Dendrocalamus strictus* sometimes form the entire secondary regrowth over abandoned clearings in India; they may remain for a long time as a biotic subclimax, as fires do not destroy the subterranean rhizomes. Without fire, trees of the original forest may gradually become established.

Mountain forests

The destructive action of man is felt at altitudes up to 2 000 m in the Malayan region and in East Africa and up to 4 000 or even 5 000 m in the Andes; fires may extend to 4 000 m in East Africa. The fires have a profound effect on the mountain forests. At middle altitudes in Africa the forest is replaced by grassland, with species from the nearby savanna. In South-East Asia, pine forests have extended their area into destroyed primary hardwoods. The mountain grasslands above 2 000 m in Cameroon are subjected to fires which appear to be responsible for the regression of the forest borders. There are many examples of forests destroyed by fire at middle altitudes in the western African mountains.

---

1. Descriptions of early successions following disturbance of the primary forest in the Amazon appear to have been made mainly in connection with soil studies. The action of man—which has only been felt to some appreciable extent in recent years—has consisted of the removal of forest for the construction of highways, urban nuclei, mining, railways, and agricultural settlements. Forest utilization for timber extraction has until recently been of little impact on the greater part of these forests.

Conclusions

The disturbances caused by natural phenomena and forest animals, though sometimes of importance locally, are of minor impact from a global point of view. Fire, one of the most important agencies influencing these ecosystems, is essentially a biotic factor. Soil erosion, causing losses which can only be compensated over long periods of time, is almost always caused by man. Consequently, man is the factor causing by far the greatest disturbance and having had the strongest effect. In thinly populated forest regions, such as the Amazon and parts of New Guinea, disturbance has been light and vast areas of primary forest are still found. Where the population is dense, changes have occurred in the extent, composition and character of the forests. These are most wide-spread in Africa, where in places the climax forest has disappeared completely or is represented by vestiges only. Similar situations are found in many other parts of the tropics.

The forest-savanna border is generally anthropogenic, though certain savannas are very old. Without fire, forest invades the savanna in most places. Most savanna is a fire-climax.

Disturbances of the primary forest ecosystem lead to secondary successions which, in spite of the manifold variations, tend to show a similar pattern in different regions. The early stages present considerable homology physiognomically and floristically. The typical secondary forest trees display a number of common characteristics. If left undisturbed, secondary forest will gradually be invaded by tree climax. However, this development will take many years, and more often than not the succession will be interrupted at a young or intermediate stage. When the disturbance ceases, succession recommences. Repeated disturbances, including fires, often lead to a regressive succession which may result in grassland or shrub formations, or open forests. Disturbances by sylviculture also change, or even completely transform, the ecosystem (e.g. forest plantations for industrial purposes).

Knowledge of what happens after disturbance of the moist tropical forest ecosystem is partial. Transfer of experience from one place to another is debatable unless we know exactly the conditions under which experience has been built; and transfer from one generation to the next is not easy. A better knowledge of cause and effects of man-induced disturbances is a pre-requisite to tropical forest ecosystem manipulation and sylviculture.

Research needs and priorities

Research in moist tropical forests should therefore concentrate on the following areas.

— Study of the faunistic, floristic and structural changes which accompany man-induced disturbances in predominantly intact humid tropical forests. Such studies will make it possible to clarify: where and in what form the propagules occur in canopy gaps (seed in the soil and released from dormancy, seed entering the areas after the creation of the gaps, seedlings and saplings present on the forest floor prior to the creation of the gaps); the environmental effects of the size and type of gap; how the critical colonizers survive or die; and how subsequent recruits appear within the gaps.

— Study of the faunistic, floristic and structural changes which occur in secondary vegetation on sites previously occupied by humid tropical forests including: effects of duration, size, original floristic and faunistic composition, and the former land use on the process of succession; effects of previous vegetation and adjacent vegetation's composition and proximity on the direction, speed and composition of succession; specific effects of the floristic composition and structure of the phases of succession on the later phases; the role of soil type and condition on the establishment and constitution of the early successions; the productivity of the various seral stages as compared with the mature forest.

The studies could be carried out through ecosystem analysis of a set of replicated and carefully designed experiments, on permanent, guarded and protected study areas for a variety of treatments and undisturbed control areas, to show the rates of succession under different treatments, the accumulation of essential nutrients and the interaction of soil, micro-organisms, plants and animals.

— Autecological and ecophysiological studies of key species. Such studies will assist in understanding the role of each species within the ecosystem and in identifying the features of the species which influence their success in regeneration. They will include: phenology of flowering and fruiting; methods, efficiency, reliability of dispersal mechanisms, conditions required for germination and early establishment; seed longevity and viability under various conditions; degree of shade tolerance at various stages of growth and under various conditions of stress from other limiting factors; length of time that the species can maintain themselves and revert to active growth under increased illumination; rates of growth under specified environmental conditions.

— Investigations on groups of organisms or on ecosystem components as they relate to succession.

These groups might be consumers or decomposers, mycorrhiza, reptiles, seed predators, etc. Some priority should be given to seed predators. The role of such organisms in the successional system should be explored so that patterns of influence throughout the system can be established.

— Other specific studies could be carried out on the physical factors which affect the micro-climate during the succession; on mineral cycling during and after disturbance; on the role of tropical conifers, their regression or expansion in relation to felling, fire and grazing; on the forest-savanna boundaries, through fire protected and grazing protected plots; etc.

— Finally, studies should be initiated on ways and means for manipulating recovery. If the process can be speeded up by introducing key organisms or nutrients, or other action, the negative impact of habitat modification might be reduced (see chapter 20).

This research is mainly autecological and will aid the sylvicultural management of some forest types. If enough information is collected about individual species, the properties of the forest as a whole may be synthesized, but this may not be true for complex tropical forests where the overall properties cannot be defined from isolated events. Thus Hopkins, Webb and Williams (1975, unpublished) assume that a complex tropical forest is a stochastic, not a determinate system, with a self-healing process regeneration, comparable with that of a living organism. The overall properties of the stochastic system cannot be deduced from a detailed study of the properties of a few of its individuals. There are many measurable quantities of the habit, growth and life-cycle of some forest tree species; but it is not the quantities themselves, but their distribution and the parameters which define these distributions which are needed. They propose collecting information on the more immediately relevant distributions in time and space which are those of flowering, seed viability and canopy gaps of various sizes. In this work on distributions, the identification of the trees is quite unnecessary; the only interest in identification would be to ascertain those species on the limits of the distributions, and therefore in a particularly favourable or unfavourable situation as regards regeneration. Once these distributions are known, it might be possible to devise a computer simulation of the regeneration process, and thereby investigate the boundary conditions under which the system becomes unstable and ultimately fails.

Selective bibliography

ANAYA, A. L.; ROVALO, Y. M. Alelopatía en plantas superiores: diferencias entre el efecto de la presión osmótica y los alelopáticos sobre la germinación y crecimiento de algunas especies de la vegetación secundaria de una zona cálido húmeda de México. In: Gómez-Pompa, A.; Vázquez-Yanes, C. (eds.). *Investigaciones sobre la regeneración de selvas altas en Veracruz.* México, Inst. Biol. UNAM, 1975, in press.

AUBRÉVILLE, A. La forêt coloniale : les forêts de l'Afrique occidentale française. *Annales Acad. Sci. Colon.* (Paris), 9, 1938, p. 1–245.

BATCHELDER, R. B. Spatial and temporal patterns of fire in the tropical world. In: *Proc. Sixth Annual Conf. Tall Timbers Fire Ecology* (Tallahassee, Florida), 1967, p. 171–190.

BAUR, G. N. *The ecological basis of rain forest management.* Forestry Commission of New South Wales, Australia, 1961–62, 499 p. Rome, FAO, André Meyer Fellowship Programme Report, 1962, 499 p.

BOERBOOM, J. H. A. Succession studies in the humid tropical lowlands of Surinam. In: *Proc. First Intern. Congress of Ecology* (The Hague), 1974, p. 349–357.

BOURLIÈRE, F. The comparative ecology of rain forest mammals in Africa and tropical America: some introductory remarks. In: Meggers, B. J.; Ayensu, E. S.; Duckworth, W. D. *Tropical forest ecosystems in Africa and South America: a comparative review,* p. 279–292. Washington, D.C., Smithsonian Institution Press, 1973, 350 p.

BRETSKY, P. W.; BRETSKY, S. S.; LEVINTON, J.; LORENZ, D. M. Fragile ecosystems. *Science,* 177, 1973, p. 1147.

BRÜNIG, E. F. Species richness and stand diversity in relation to site and succession of forests in Sarawak and Brunei (Borneo). 3rd Symp. Biogeogr. and Landscape Res. in South America (Max Planck Inst., Plön, 3.5.1972). *Amazoniana,* vol. 4, no. 3, 1973, p. 293–320.

BUDOWSKI, G. Distribution of tropical American rain forest species in the light of successional processes. *Turrialba* (Costa Rica), 15, 1965, p. 40–42.

——. Fire in tropical American lowland areas. In: *Proc. Fifth Annual Conf. Tall Timbers Fire Ecology* (Tallahassee, Florida), 1966, p. 5–22.

CATINOT, R. Sylviculture en forêt dense africaine. *Bois et Forêts des Tropiques,* n° 100 à 104, 1965.

CENTRE TECHNIQUE FORESTIER TROPICAL (CTFT). *Caractères des espèces forestières secondaires.* Nogent-sur-Marne, Publ. multigr., 1974, unpublished.

COUSENS, J. E. Some reflections on the nature of Malayan lowland rain forest. *Malayan Forester,* vol. 28, no. 2, 1965, p. 122–128.

DOBBEN, W. H. van; LOWE-McCONNELL, R. H. (eds.). *Unifying concepts in ecology.* Report of the plenary sessions of the first international congress of ecology (The Hague, September 8–14, 1974). The Hague, W. Junk B. V. Publishers; Wageningen, Centre for agricultural publishing and documentation, 1975, 302 p.

Chapters on 'Flow of energy and matter between trophic levels'; 'Comparative productivity in ecosystems'; 'Diversity, stability and maturity in natural ecosystems'; 'Diversity, stability and maturity in ecosystems influenced by human activities'; 'Strategies for management of natural and man-made ecosystems'.

DONIS, C. A. Shifting agriculture (Shag). In: *Proc. Duke Univ. Tropical Forestry Symp.* (Durham, USA), 1965, p. 30–43.

EUSSEN, J. H. H.; WIJAHARDJA, S. Studies of alang-alang (*Imperata cylindrica* (L.) Beanv.) vegetation. *BIOTROP Bulletin,* 6, 1973, p. 3–24.

FAO. *Tropical silviculture,* vol. I, II. Rome, 1958, 190 p., 415 p.

FARNWORTH, E. G.; GOLLEY, F. B. (eds.). *Fragile ecosystems. Evaluation of research and applications in the Neotropics.* Berlin and New York, Springer Verlag, 1974, 258 p.

FOSBERG, F. R. Nature and detection of plant communities resulting from activities of early man. In: *Proc. of the Symposium on the impact of man on humid tropics vegetation* (Goroka), p. 251–262. Unesco Science Co-operation Office for South-East Asia (Djakarta), 1960, 402 p.

FRASER DARLING, F. Impact of man on the biosphere. L'impact de l'homme sur la biosphère. *Unasylva* (Rome, FAO), vol. 22 (2), no. 89, 1968, p. 3–13.

FRESON, R.; GOFFINET, G.; MALAISSE, F. Ecological effects of the regressive succession muhulu—miombo—savanna in Upper-Shaba (Zaire). In: *Proc. First Intern. Congr. of Ecology* (The Hague), 1974, p. 365–371.

FREY-WYSSLING, A. *Over de struikwildernis van Habinsaran. Trop. Natuur.,* 20, 1931, p. 194–198.

GOLLEY, F. B.; McGINNIS, J. T.; CLEMENTS, R. G.; CHILD, G. I.; DUEVER, M. J. *Mineral cycling in a tropical moist forest ecosystem.* Athens, Georgia, Univ. of Georgia Press, 1975, 248 p.

GÓMEZ-POMPA, A.; VÁZQUEZ-YANES, C.; GUEVARA, S. The tropical rain forest: a non-renewable resource. *Science,* 177, 1972, p. 762–765.

GÓMEZ-POMPA, A.; VÁZQUEZ-YANES, C. Studies on the secondary succession of tropical lowlands: the life cycle of secondary species. In: *Proc. First Intern. Congr. of Ecology* (The Hague), 1974, p. 336–342.

GUEVARA, S.; GÓMEZ-POMPA, A. Seeds from surface soils in a tropical region of Veracruz, Mexico. *J. Arnold Arboretum*, 53, 1972, p. 312–335.

HEWETSON, C. E. Other species (sal). In: *Tropical silviculture*, vol. II, p. 367–377. Rome, FAO, 1958, 415 p.

JANZEN, D. H. Herbivores and the number of tree species in tropical forests. *American Naturalist* (Chicago), 104, no. 940, 1970, p. 501–528.

JONES, E. W. Ecological studies on the rain forest of southern Nigeria. IV. The plateau forest of the Okomu Forest Reserve. Part I. The environment, the vegetation types of the forest and the horizontal distribution of species. *J. Ecol.*, 43, 1955, p. 564–594.

——. Ecological studies on the rain forest of southern Nigeria. IV (continued). The plateau forest of the Okomu Forest Reserve. Part II. The reproduction and history of the forest. *J. Ecol.*, 44, 1956, p. 83–117.

KEAY, R. W. J. Seeds in forest soils. *Nigeria Forestry Inform. Bull.* (new series), 4, 1960, p. 1–12.

KELLMAN, M. C. Some environmental components of shifting cultivation in upland Mindanao. *J. Trop. Geogr.*, 28, 1969, p. 40–56.

——. *Secondary plant succession in tropical montane Mindanao*. Canberra, Australia, Australian National University, Dept. Biogeogr. Geomorph., Publ. BG/2, 1970, 174 p.

——. The viable weed seed content of some tropical agricultural soils. *J. Appl. Ecol.*, 11, 1974, p. 669–678.

KENOYER, L. A. General and successional ecology of the lower tropical rain forest of Barro Colorado Island, Panamá. *Ecology*, 10, 1929, p. 201–222.

KERMODE, C. W. D. Teak. In: *Tropical silviculture*, vol. II, p. 168–178. Rome, FAO, 1958, 415 p.

KLINE, J. R. *Terrestrial ecosystems*. Contribution to mineral cycling in south-eastern ecosystems, May 1974. Augusta, Ga., 1975.

KOSTERMANS, A. J. G. H. The influence of man on the vegetation of the humid tropics. In: *Symposium on the impact of man on humid tropics vegetation* (Goroka, New Guinea), p. 332–338. Unesco Science Co-operation Office for South-East Asia (Djakarta), 1960, 402 p.

LANDON, F. H. Malayan tropical rain forest. In: *Tropical silviculture*, vol. II, p. 1–12. Rome, FAO, 1958, 415 p.

LEBRUN, J.; GILBERT, G. *Une classification écologique des forêts du Congo*. Publ. INEAC (Bruxelles), Sér. scient., n° 63, 1954, 89 p.

LIEW, T. C.; WONG, F. O. Density recruitment mortality and growth of dipterocarp seedlings in virgin and fogged-over forests in Sadak. *Malayan Forester*, vol. 36, no. 1, 1973, p. 3–15.

MABESA, C. The Philippine forests. In: *Tropical silviculture*, vol. II, p. 57–87. Rome, FAO, 1958, 415 p.

MEGGERS, B. J.; AYENSU, E. S.; DUCKWORTH, W. D. *Tropical forest ecosystems in Africa and South America: a comparative review*. Washington, D.C., Smithsonian Institution Press, 1973, 350 p.

MICHELSON, A. *Considérations sur la forêt spontanée africaine et son exploitation. Statistiques relatives à la régénération naturelle de cette forêt*. Étud. Forest., Comité national du Kivu, Nouv. Sér., n° 5, 1953, 91 p.

NOAKES, D. S. P. Mangrove. In: *Tropical silviculture*, vol. II, p. 309–318. Rome, FAO, 1958, 415 p.

NYE, H. Some soil processes in the humid tropics. I. A field study of a catena in the West African forest. *J. Soil Sci.*, 5, 1954, p. 7–21.

RICHARDS, P. W. *The tropical rain forest: an ecological study*. Cambridge Univ. Press, 1952, 450 p., 4th ed. with corrections, 1972.

ROLLET, B. *L'architecture des forêts denses humides sempervirentes de plaine*. Nogent-sur-Marne, Centre technique forestier tropical (CTFT), 1974, 298 p.

ROSS, R. Ecological studies on the rain forest of southern Nigeria. III. Secondary succession in the Shasha Forest Reserve. *J. Ecol.*, 42, 1954, p. 259–282.

SCHNELL, R. *Introduction à la phytogéographie comparée des pays tropicaux*. 2 vol. Paris, Gauthier-Villars, 1970, 500 p.; 1971, 452 p.

SCHULZ, J. P. *Ecological studies on rain forest in northern Surinam*. Verhand. Koninkl. Nederl. Akad. van Wetensch. 2/LIII, no. 1, 1960.

SETH, S. K. Natural regeneration and management of bamboos. In: *Tropical silviculture*, vol. II, p. 298–303. Rome, FAO, 1958, 415 p.

SNEDAKER, S. C. *Ecological studies on tropical moist forest succession in eastern lowland Guatemala*. Unpublished Ph. D. thesis, University of Florida (Gainesville), 1970, 131 p.

SOEDIARTO, R. Some points on planning and management of the tropical rain forest in Indonesia. In: Kartawinata, K.; Atmawidjaja, R. (eds.). *Co-ordinated study of lowland forests in Indonesia*, p. 67–70. Bogor, BIOTROP and IBP, 1974, 183 p.

STEENIS, C. G. G. J. Van. Basic principles of rain forest sociology. In: *Study of tropical vegetation. Proceedings of the Kandy Symposium*, p. 159–163. Paris, Unesco, 1958, 226 p.

——. Plant speciation in Malesia, with special reference to the theory of nonadaptive saltatory evolution. *Biol. J. Linn. Soc.*, 1, 1969, p. 97–133.

SYMINGTON, C. F. The study of secondary growth on rain forest sites in Malaya. *Malayan Forester*, 2, 1933, p. 107–117.

TRICART, J.; CAILLEUX, A. *Le modelé des régions chaudes — forêts et savanes. Traité de géomorphologie*, vol. V. Paris, Société d'édition de l'enseignement supérieur, 1965, 322 p.

VÁZQUEZ-YANES, C. Studies on the germination of seeds of *Ochroma lagopus* Swartz. Turrialba (Costa Rica), vol. 24, no. 2, 1974, p. 176–179.

WADSWORTH, F. H. Tropical rain forest. In: *Tropical silviculture*, vol. II, p. 13–23. Rome, FAO, 1958, 415 p.

WEBB, L. J.; TRACEY, J. G.; WILLIAMS, W. T. Regeneration and pattern in the sub-tropical rain forest. *J. Ecol.*, vol. 60, no. 3, 1972, p. 675–695.

WHITMORE, T. C. *Tropical rain forests of the Far East*. Oxford, Clarendon Press, 1975, 278 p., 550 ref.

WYATT-SMITH, J. A note on tropical lowland evergreen rain forest in Malaya. *Malayan Forester*, 12, 1949, p. 16.

——. Natural plant succession. *Malayan Forester*, vol. 12, no. 3, 1949, p. 148–152.

Gross and net primary production and growth parameters

Introduction

Plant organic material is produced largely through the process of photosynthesis. Some of the production is observed as an increase in biomass which is then available for harvest by man or other consumers. Another portion of the production is expended by the vegetation in its own maintenance processes. Photosynthesis takes place in green leaves, which are distributed over the branches of individual plants, which in turn are organized in populations making up communities or forest stands. The communities in turn are arranged in a mosaic of landscape units. One can deal with production at any level of this hierarchy. However, to fully understand production at any one level it is necessary also to examine the levels above and below it. For example, the study of production of a tree species requires the study of how environmental factors influence individual trees of different ages, sizes and locations. The conclusions about the population also are interpreted in the context of all species, both plant and animal, making up the forest, and of forests generally. For this reason, ecologists and foresters have studied production of the entire forest occurring in the tropics as an abstract entity, as vegetation formations, as specific forest communities, as separate species of trees and as individual plants. All these studies are necessary for a full understanding of production of tropical forests; unfortunately, the emphasis has not been equal or adequate at any level.

Definitions

The conventional definitions of forest production, as used in the International Biological Programme (IBP), are quoted directly from Newbould (1967):

"The assimilation of organic matter by a plant community during a certain specified period (e.g. one year), including the amount used up by plant respiration, is called the *gross primary production*. Gross production minus respiration or the formation of plant tissues and reserve substances during the period is the *net primary production*, NPP, which may be known simply as primary production. When production is measured as dry weight it includes some mineral salts incorporated into the products of photosynthesis. If ash content is estimated and excluded, or some method is used which estimates only the formation of organic compounds, then *organic production* should be specified.

"A general starting point for comparing photosynthetic primary production is the cumulative course of 'net assimilation' (net dry matter production of green parts) over the year(s) or vegetation period(s). This can be determined by the sum total of the following features determined periodically through the year:

— biomass change of photosynthetic plants;
— plant losses by death and the shedding of parts above and below ground;
— man's harvest (in some cases);
— consumption of photosynthesizing plants by animals (botanical and zoological methods will be used to estimate amount lost).

"The unit of study will commonly be a whole biological system, i.e., the sum total of standing crops, which are the populations of living organisms under consideration in a defined area at a defined time. Biomass is the total amount of living matter present at a given moment in a biological system (in this case the photosynthetic plants making up a woodland stand). It is taken to include heart-wood and bark (which may no longer be alive) but no dead roots and branches (with no viable buds). In the present context it would be expressed in terms of dry weight, or ash-free dry weight (=organic weight). Biomass may be estimated directly by weighing or indirectly from measurements of the volume and density of the various components concerned."

As stated above, these conventional definitions refer to the collection of plants which comprise the first or the primary component of forest ecosystem. However, the plants or primary producers are only one component in ecosystems and each component may have production. Further, the ecosystem, considered as a whole, may also have production (ecosystem production to distinguish it from primary production of the vegetation). For a climax ecosystem in equilibrium the net production of the ecosystem, the net primary production minus the total respiration of all the heterotrophs, approximates to zero; this value will of course fluctuate considerably on small areas.

Determinants of production

Production is fundamentally a physiological process involving the synthesis of organic matter through photosynthesis. Higher plants can be divided into two photosynthetic types, C_3 plants where phosphoglyceric acid is the primary fixation product and C_4 plants where malic or aspartic acids are the primary products (Hatch and Slack, 1966). C_3 plants have photorespiration which reduces the efficiency of photosynthesis; C_4 plants are apparently without photorespiration. C_4 plants appear to be especially adapted to hot dry environments although they do appear elsewhere. Many tropical grasses are C_4 plants. The study of the ecological significance of C_4 plants is just beginning and is a topic of considerable interest.

In the C_3 plants, which are more characteristic of tropical forests, the photosynthetic process is limited by the chemical process itself and by the physiology and architecture of the leaves. For example, Bonner (1962) points out that the

quantum efficiency of photosynthesis is theoretically about 20 per cent. Actually, the leaf does not increase production as light energy increases above the light saturation point which is reached at about 1/20 or 1/10 full sunlight. In contrast, photorespiration increases with increase in light energy and, further, the leaves shade one another as they are arranged through the canopy. Wadsworth (1941–47) describes the effect of this shading in a lower montane rain forest (Tabanuco forest) in Puerto Rico. Over three years, the average annual trunk diameter growth of four crown classes (table 1) was highest in the dominants which were exposed to full sunlight. Trees growing in the forest shade had about one third the diameter growth of the dominants. These data are supported by the direct measures of photosynthesis in sun and shade leaves by Stephens and Waggoner (1970). These various phenomena lead to a rather low efficiency of the utilization of solar energy by vegetation, which is often about 2–3 per cent.

TABLE 1. Annual diameter growth of the trunk of trees in four crown classes in lower montane rain forest, Puerto Rico (after Wadsworth, 1941–47). These data were obtained on a 1.8 acre plot, with all trees having a diameter over 1.5 inches remeasured annually. The data represent remeasurement over three years of 931 trees of 57 different species.

Crown class	Number of trees	Average annual diameter growth	
		inches	mm
Dominant	67	0.25	6.35
Codominant	100	0.23	5.84
Intermediate	509	0.14	3.55
Suppressed	255	0.08	2.03
Total	931	Average 0.14	3.55

In addition, production is controlled by a variety of other environmental factors. Lieth and Box (1972) have summarized data on net primary production, temperature, and precipitation in their development of a model of primary production on a world basis. Like most ecological processes, production reaches a maximum at some median level of the environmental factor. That is, as temperature increases production increases, but at a certain temperature further increase causes a decrease in production. Lieth and Box have described the global patterns of primary production for change in energy, temperature, water and nutrient supply. The primary production observed in the field is an integration of all these patterns for a given locality and for the species of plants being considered.

Temperature and rainfall elements may be considered primary factors in regulating overall ecosystem structure and function while soil characteristics may be considered as secondary (or modifying) factors, often, but not always, exerting their strongest effects during early succession. The absence of an extended cold season allows growth to continue throughout the year in tropical areas as long as moisture conditions also remain favourable. Only at high elevations do temperatures decline to values that may be inhibitory. Of course, there are factors other than temperature which reduce plant productivity at high elevations; including high wind velocities and inadequate or unstable substrates and nutrient conditions (in the Blue Mountains, Jamaica, see Grubb, 1974). There has been some speculation in the literature that the high average annual temperatures of tropical regions may promote high rates of plant respiration and thus reduce NPP, but according to Walter (1971) conclusive evidence to support this proposition appears to be lacking. In the tropics, temperature seems to have its most important effect on NPP by influencing the water relations of plant communities as discussed in the following section.

Because of high rates of actual evapotranspiration in tropical latitudes, the amount of precipitation necessary to sustain a specific type of vegetation is greater than at higher latitudes. Annual precipitation of 760 mm, for example, supports well developed mesic forest at 45° north latitude in Michigan, United States, but only scrub vegetation in southwestern Puerto Rico, where rates of evapotranspiration are much higher. Thus rainfall and annual actual evapotranspiration are equally important in determining the type of vegetation that will occupy a particular site. It follows that these are also important determinants of NPP.

The effect of rainfall on NPP is most clearly reflected by vegetation types that are water limited. In many tropical forest areas the periodicity and seasonality of rainfall is as important as the total annual amount received in regulating productivity. Some tropical areas have slight seasonal trends in the abundance of rainfall which probably have only a slight effect on NPP but in other regions, especially the monsoonal areas of South and South-East Asia, the seasonality of rainfall has as distinct an effect on the productivity of vegetation as does the seasonality of temperature at higher latitudes. In India, for example, a prolonged dry season of up to nine months duration greatly restricts the growing season and consequently the total annual NPP. In forests that occur in areas subject to only moderate drought, seasonal responses may be restricted to only a fraction of the tree flora. The lower montane rain forest of Puerto Rico, for example, is subject to a slight decline in rainfall during February and March and this is expressed in a heavier leaf-fall during those months but only a few tree species actually produce identifiable growth rings. In Panamá and the Canal Zone there is a rainfall gradient from the Caribbean coast to the Pacific coast. The Caribbean coast, which is down wind of the prevailing northeast trade-winds, is subject to a rainfall of 3 300 mm/a with a monthly range of 38–570 mm. It is vegetated by evergreen forest. The Pacific coast, less than 65 km away, has an annual rainfall of 1 780 mm and receives less than 14 mm during some months. It is vegetated by deciduous forest.

Whereas NPP is most frequently limited by low levels of rainfall, in some situations very high levels of rainfall in combination with high atmospheric humidity may inhibit NPP. Small stunted forests, called elfin forests, are found at high elevations in many tropical areas (see chapter 4): the epiphyte-covered trees in some elfin forests, such as the one at about 1 000 m in northeastern Puerto Rico, where the productivity is thought to be relatively low. The reason for this stunting is not clearly understood. Temperatures at these elevations are not low enough to retard growth directly. A more likely explanation is the excessively high rainfall (5 000 mm in northeastern Puerto Rico), the low rates of transpiration and mineral transport and the low levels of oxygen in the soil because of the waterlogged conditions. The low transpiration rates are attributable to the prevailing cool temperatures and the high atmospheric humidity. Strong winds at high elevations may inhibit some plants.

The physical and chemical properties of the soil on which an ecosystem occurs will influence its production. Porous sandy soils allow rapid percolation of rain-water to the water-table where it is largely unavailable to plants. Consequently, under similar levels of rainfall, vegetation on sandy sites commonly displays a different aspect than vegetation on less porous soils. In the case of forest vegetation, the trees on sandy sites are often relatively small and rates of NPP are assumed to be relatively low although the data to conclusively demonstrate this are lacking (the increment studies of Ashton, in Sarawak, have actually demonstrated the reverse: growth rates appeared correlated to leaf area index and LAI was greatest when there was water stress in the canopy).

Many of the nutrients in a mature tropical ecosystem are in the living vegetation rather than in the soil. The removal of vegetation usually leaves a site with very limited potential to sustain high rates of productivity, agricultural or otherwise. Sandy soils in particular are usually nutrient poor.

Finally, Best (1962) in analysing agricultural production in tropical and temperate environments calls attention to the differences in daily solar radiation at different latitudes during the growing season. The long periods of growth in the tropics reduce total production in annuals, while perennials may benefit from the year long growing season. Thus, production of tropical annuals may be less than that of temperate annuals. As examples, Best cites rice grain yields of 2 t/ha in the tropics compared with 4–5 t/ha in temperate regions. In contrast, perennial crops such as sugar cane, annually produce 260 t/ha shoot fresh weight under the continuous growth conditions in the tropics versus 120 t/ha from sugar-beet under a 7 month growing season in temperate areas.

Jordan (1971a, b) analysing trends in world organic production concluded that on the gradient of decreasing solar energy from equator to the poles wood production is constant but leaf production decreases. The explanation for these relationships is not completely clear. While Jordan suggests that fast wood growth might be of advantage to the vegetation where solar radiation is less available, there is seldom a case where the vegetation utilizes more than a few per cent of the solar input.

This brief consideration of the determinants of production suggests that solar energy, moisture availability and soil nutrients are key factors in the development of predictive statements about tropical forest production. Many more data are required to show exactly how these factors are interacting with each other and the vegetation to produce the patterns which have been observed. To fill this gap data on solar energy and precipitation must be measured at the same site and time as production is measured. Data from meteorological stations are seldom very useful for this purpose.

Sampling problems

Gross primary production

Forest gross primary production has been defined as the assimilation of organic matter by a plant community during a specified time interval including the amount used by plant respiration. This parameter can be estimated by measuring the uptake of gaseous oxygen or loss of carbon dioxide over 24 hours. In the simplest case, the carbon dioxide uptake during the daylight hours represents the production minus the photosynthate used during the day in respiration. In the past, it has been assumed that the day-time respiration proceeds at the same rate as night-time respiration and therefore, day respiration can be estimated from the night rate. Actually, day-time respiration probably operates at a higher rate than at night because of the temperature differential between these periods. This problem is technically difficult and requires further study.

Assuming that these measurement problems can be overcome, then it is necessary to estimate the gaseous metabolism of the forest component sample or a species population sample and expand the estimate to the entire forest. Ideally, this procedure would involve samples of each component, calculation of the variation and estimation of the number of samples required for a given level of confidence in the estimate. In actual practice, forests are so tall, contain such a large mass and are so diverse, that no one has developed methods of estimating the variance of forest gross production. Techniques range from the whole cylinder method (Odum and Jordan, 1970), to aerodynamic methods (Lemon *et al.*, 1970) and to the summation of subcomponent method (Yoda, 1967). In all cases, these give, at best, a crude approximation of the production parameter.

Net primary production

Net primary production refers to the increase in plant material which can often be observed as plant growth and which can be measured by weight. For this reason it is more amenable to estimation since it is only necessary to measure the increase in dry weight over time. These time intervals are crucial in the measurements since the plant system is dynamic with some tissue increasing in mass while other tissue shrinks or dies. Further, herbivores feed on the tissue during the period and this loss also represents production.

The principal approach to estimation of net primary production is the harvest method (Kira *et al.*, 1967), where sequential samples of the vegetation are removed and weighed. The difference in weight between two periods, when corrected for death between periods of measurement and that eaten by herbivores, represents net production. The accuracy of the estimate depends upon the sampling design since forest production varies over space and time. In an esti-

mation technique where the forest is considered to be made up of a complex of stems of differing size and composition and production is the increase in volume of these stems and the leaves and other materials which are caught as they fall to the ground, then the samples must be of sufficient size and number to cover the spatial variation in the type. Conversely, if the technique focuses on individual tree species, then these species must be sampled adequately so that each species' contribution to the total forest production can be estimated. The latter technique is used most effectively where the forest is composed of few species.

Plot size

The plot size for the estimation of production of tropical forest is not really known and is a subject for more research. Poore (1968) made an extensive analysis of the composition of the canopy trees in undisturbed lowland dipterocarp forest in Malaysia. His study showed that the differences even within small areas were such that a plot of 2–4 ha was necessary in fairly uniform forest to obtain a sample representative of species. Specifically he found that a plot of a least 2 ha was necessary to obtain a coefficient of similarity of 50 per cent for canopy trees. However, these conclusions may well be biased toward larger trees, since Drees (1954) found a plot of 0.25 ha was adequate to determine species numbers in Indonesia heath forest if all sizes of plants were tallied. Obviously, restriction of the sample to mature plants eliminates young individuals and seedlings, as well as seeds, and is a strongly biased sample of the vegetation on a plot.

In addition, Brünig (1973) has shown that there is an upper limit to plot size since large plots include environmental diversity and no longer reflect the forest type being studied. He states that plots of 0.2–0.5 ha are necessary if homogeneity is desired. Expansion of plot area beyond this size introduces excessive heterogeneity by including different site and stand conditions. Klinge and Rodrigues (1973) also found this to be true when testing transect length. Very large transects underestimated forest biomass. They also tested plot size and found that reduction of plot dimensions from 40×50 m to 40×40 m made no difference to biomass measurement, but reduction to 30×30 m or less increased the variability of the estimate of the mass and hence required an increased number of samples.

Brünig (1973) concludes that in sampling biomass the investigator must use relatively small plots, stratify the sampling according to site and species composition and supplement the plot data by intensive survey of the area surrounding the plots to detect local differences. These suggestions provide a reliable guide to biomass or harvest production sampling at this stage in our knowledge of tropical forests.

Sample number

The number of plots required to sample forest biomass is another problem. Klinge and Rodrigues (1973) found that 900 m² plots or 100 m² plots were inadequate to estimate the mass on 2 000 m² within 5 per cent accuracy. Hozumi *et al.*

(1969) used 20 plots with an area of 1 m² to estimate ground vegetation in Cambodian rain forest and discovered that over 300 plots would be required for estimation at the 95 per cent confidence level. Similarly, for sampling small tress (<4.5 cm diameter), 20 times the 15(2×2 m) plots used would be required for similar confidence levels. Obviously, this sampling problem must be resolved by the investigator using standard statistical techniques. No general rule for tropical forests can be given.

Relation between weight and stem diameter

In most studies of net forest production, the entire vegetation on a plot is not cut and weighed. Rather, for the larger trees the frequency distribution of diameters is determined and biomass samples of each diameter class is made. Then, the average biomass per class can be multiplied by the number of stems in that class and the mass of the forest can be reconstructed. Hozumi *et al.* (1969) sampled 130 stems divided among the sizes and species present to determine the allometric relations between mass and tree diameter and height. Sabhasri *et al.* (1968), studying a dry evergreen forest at Sakaerat, Thailand, selected 214 stems distributed among species and over the range of diameter-height groups. Golley *et al.* (1975), in Panamá tropical moist forest, took samples of 10 per cent of each diameter class so that the most abundant stems were most heavily sampled. Other important similar studies have been made.

The Japanese workers have shown (Ogawa *et al.*, 1965) that in Thailand the most significant correlations were between weight of a tree and d^2h, where d is the diameter of the tree at 1.3 m from the ground (DBH) and h is the tree height. The actual allometric relationship must be established for each forest because the relationship between d, h, volume and weight are influenced by site conditions. This method has so much promise that it is worth while reproducing the detailed methodology of these Japanese ecologists:

"On square sample plot, 40 m × 40 m in size, was set up in each stand. All trees and woody climbers over 4.5 cm in DBH contained in each plot were then measured of their stem diameter at 1.3 m height above the ground (DBH), total tree height, and height of the lowest living branch. In trees that have plank buttresses surpassing 1.3 m in height, the stem diameter at the level just above the upper end of buttress was substituted for DBH. Stem diameter of lianas was measured at 1.3 m height where the stems rose up toward the canopy, irrespectively of the distance from the stem base. Diameter tape and caliper were used for diameter measurements. Height of felled trees was directly measured with a tape, while that of standing trees was estimated by the use of Weise's hypsometer. A scaled metal pole (7 m long) was also employed for the height measurement of lower trees.

"After the census of all trees, sample trees of various sizes and of different species were felled from inside and adjacent areas of the plot. Fresh weights of stem, branches and leaves were separately recorded for all sample trees. Oven-dry weights were later determined in the laboratory with air-dried samples after they had been brought to Japan. Leaf area/leaf fresh weight relations were also determined in the field.

"Because of the trunk of tropical broadleaf trees is very often divided into boughs of nearly equal thickness, the distinction between stem and branch is inevitably arbitrary. In practice, the one reaching the highest level was artificially treated as the main stem or the extension of bole. As for lianas, the separation of individuals was almost impossible, because they often tied two or more trees together passing from one to another high up in the canopy or crept for a long distance on the ground. Their biomass was therefore determined taking the host tree as a unit whenever it was felled. It was usually necessary to cut off the stems of lianas climbing up a host tree before we cut it down, when the leaves of the lianas had already been withered up and were easily scattered and lost with the falling of the host. About one-third of the amount of leaves of woody climbers were estimated to have been lost in this way.

"The procedure used for the estimation of plant biomass utilizes the allometric regressions derived from sample tree measurements between the weights of stem, leaves, etc. per tree and such linear quantities as DBH and tree height, for the purpose of estimating the former amounts from the latters which are relatively easy to be measured. The details of the method are, however, considerably improved, especially as applied to the leaf biomass estimation.

"Estimating stem weight. The dry weight of stem (W_s) of sample trees is most closely correlated with the square of their DBH (D) multiplied by their height (H) as shown in fig. 1. No significant difference is found among different forest types with respect to the $W_s \sim D^2H$ relations. The regression can be written as

$$W_s = 0.0396(D^2H)^{0.9326}$$
$$[W_s]\text{:kg.} \quad [D]\text{:cm.} \quad [H]\text{:m.} \tag{1}$$

D^2H is expected to be proportional to the volume or weight of stem, if the stem is approximately cone-shaped. In fact, the allometry constant (0.9326) in Eq. (1) is close to unity. Considering the large number of species involved, it is rather surprising that the observed values so well fit a single regression. There is a wide variety in the apparent specific gravity of woods among the species. Such difference may cause considerable error in estimating the stem weight of individual trees by Eq. (1), but the observed values in fig. 1 are more or less randomly distributed on both sides of the regression line, showing that the errors for individual stems may offset each other in the calculation of total stem biomass of a stand.

"Estimating branch weight. Similar allometric relation exists between the dry weight of branches per tree (W_B) and D^2H, but the errors for individual trees are much larger than in stem weight, as indicated by the distribution of observed values in a very wide zone on the log W_B. log D^2H diagram (fig. 2). The regression line thus represents only the approximate trend. In fig. 2, trees heavily loaded with lianas are omitted, since the amount of branches in such trees is generally unusually small, often being less than one-tenth of the amount expected from the regression. It is therefore unreasonable to estimate their branch weight by Eq. (2).

"Estimating root weight. The weight of underground organs was actually measured with only three individuals.

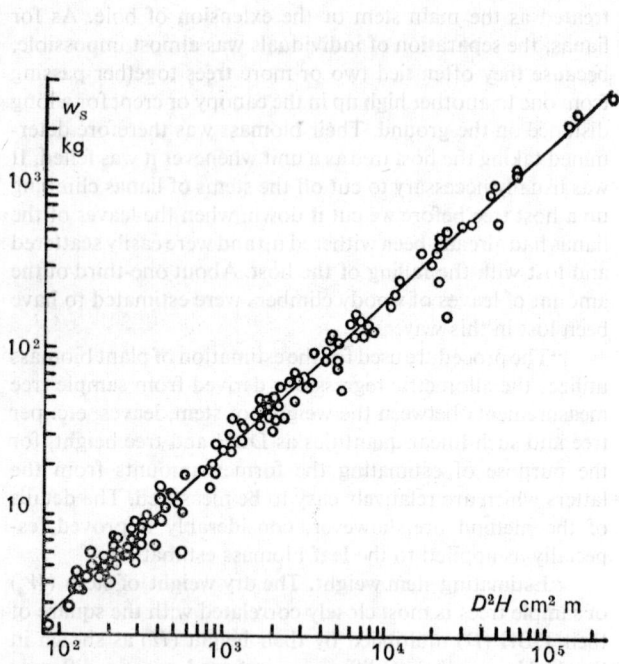

"FIG. 1. Simple allometry between the oven-dry weight of stem per tree (W_s) and D^2H (D: Diameter of stem at breast-height, H: height of tree), expressed by the linearity between the logarithms of the two amounts.

No significant differences are found between different forest types with respect to the $W_s \sim D^2H$ regression.

The biggest of them was only 15.1 cm in its DBH and any bigger trees could not be dug up because of the enormous amount of labor needed. As far as the three sample trees are concerned, however, the following allometric equation very closely fits the result of measurement.

$$[W_R] = 0.0264(D^2H)^{0.775} \tag{3}$$

$[W_R]$:kg.　$[D]$:cm.　$[H]$:m.

The allometric constant (0.775) in Eq. (3) is significantly smaller than unity whereas it is nearly equal to unity in stems and branches. This means that the weight ratio of root to aerial shoot becomes smaller with increasing tree size (fig. 3).

"Estimating leaf weight. The amount of leaves borne by a tree is so sensitive to such factors as incident light intensity reaching its crown, stand density, age of the tree, etc., that its estimation is liable to greater error. The fitness of the relations between the leaf amount and D or D^2H to the allometric regression is even less satisfactory as compared with the above-mentioned case of branch weight. What is most remarkable about the leaf amount is the fact that it tends to approach a certain asymptotic value with the increase of tree size. The regression between the

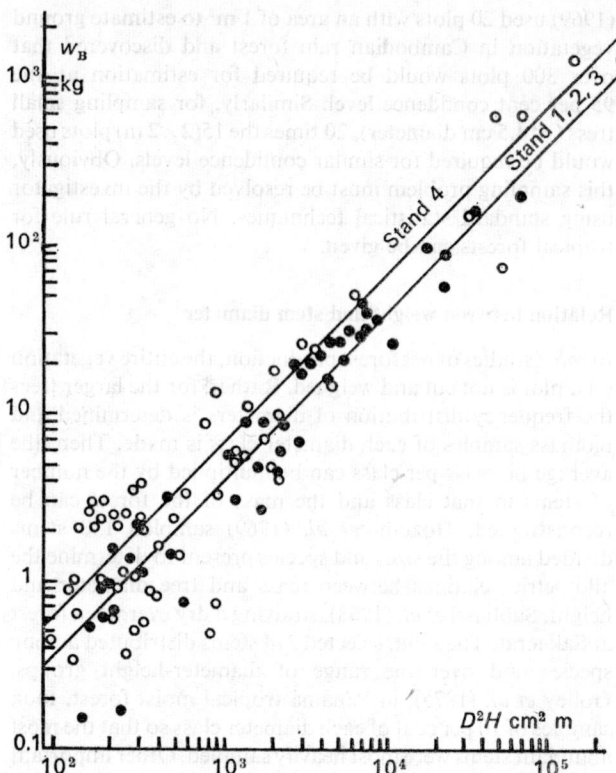

'FIG. 2. Simple allometry between the oven-dry weight of branches per tree (W_B) and D^2H.

Although the relations in individual trees vary over a considerably wide range, different regressions may be recognized in the deciduous forest types (stands 1, 2, 3, solid circles) and the rain forest (stand 4, open circles) respectively.

leaf dry weight per tree (W_L) and the size or amount of other organs is not linear on the log-log coordinates unlike the cases of ordinary allometric regression, but it is a kind of ceiling curve as illustrated in fig. 4.

"After various trials, it was found that the leaf weight-stem weight relation was the simplest and most stable, and can properly be approximated by a hyperbola. The equations show that the leaf amount born by a tree cannot exceed 40 (=1/0.025) kg, however large the tree size may be. This may well be expected in trees growing as aggregated stand. The trial to estimate the leaf amount by the $W_L \sim D$ or $W_L \sim D^2H$ allometry, which has frequently been made by recent workers, is therefore likely to make serious overestimation especially for big trees over 40 cm DBH.

"The effect of liana on the leaf amount of its host tree should be noted. The reduction of branch amount of host tree due to the lianas has already been noticed above. In the similar way, the weight of leaves of a tree loaded with lianas is reduced to a considerable extent, often under one-tenth of the normal amount expected. If, however, leaves of the lianas climbing a host tree and those of the host itself are put together, the total sum tends to agree in amount with the expected value. In other words, the climbers reduce the amount of the host's leaves by covering the surface of

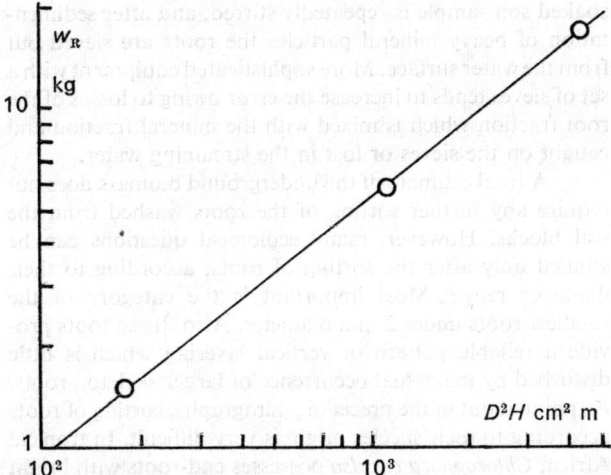

"FIG. 3. Simple allometry between the oven-dry weight of roots per tree (W_R) and D^2H in three sample trees from stand 4.

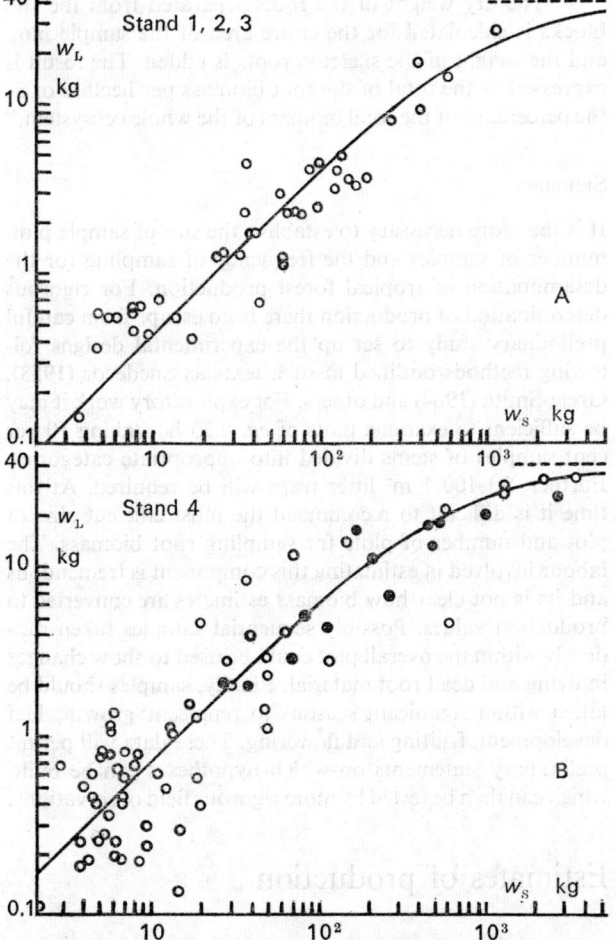

"FIG. 4. Hyperbolic relations between the oven-dry weight of leaves per tree (W_L) and that of stem (W_S).
Broken lines represent the asymptote ($W_L = 40$ kg/tree) beyond which the leaf amount of a tree cannot increase, however large the tree may be. Solid circles indicate the sum of the amount of tree leaves and that of lianas borne by the tree.

the latter's crown, but the reduced amount is almost filled up by the climbers' own leaves. This fact is very important since it enables us to estimate the total leaf biomass of trees and climbers from the stem size of trees alone. The leaf biomass is thus considered to represent the value for the total forest canopy, regardless of the presence or absence of lianas." (Ogawa *et al.*, 1965.)

Time of sampling

A further problem in sampling forest production involves the timing of the sample collection. Tropical forests almost always display a seasonality which reflects periodicity in water input, light and photoperiod, and other environmental factors. There have been numerous studies of forest phenology and these are summarized by Lieth (1974); see also chapter 8. Five features of the forest are studied: leaf emergence, leaf-fall, flowering, fruiting, and growth. In the most moist of tropical rain forests there appears to be continuous activity of leaf production, flower and fruiting throughout the year, although most species have a peak of activity which usually falls in the drier season (Richards, 1952). However, where the dry season is longer, seasonality in production is more evident. Leaf flush may occur in the dry season, as may flowering of a major number of species. Leaf-fall is attuned to dryness, reaching a peak in the dry period. Daubenmire (1972) points out that maximum growth rates are higher in trees where the unfavourable season is cold in comparison to where it is dry—evergreen species also have a lower capacity for rapid growth. These observations suggest that production estimates must be frequent enough to include the variation in leaf and fruit production over the year. Further, Hopkins (1970) observed that between year differences in phenology were smaller than between individual plants of the same, suggesting that genetic or micro-site factors were more important than climatic factors. This is a further reason for considering the forest as a whole over an area sufficient to include the range of genetic or micro-site effects present in the system.

The phenological observations suggest that production studies should be based on a knowledge of the rainfall and temperature pattern for the site following the scheme of Walter and Lieth (1960). If the site has two wet and dry seasons then each season must be adequately sampled. Further, between year variation also must be considered.

The effects of variation of forest mass over time and space are especially significant when the net production is measured directly as in the collection of litter fall or in tree diameter-height growth. Litter fall is conventionally measured using 6 to 10 traps of 1 m² placed at random in the forest plot, with litter removal occurring every two weeks. Klinge and Rodrigues (1968) studied litter fall in an Amazonian terra firme forest near Manaus, Brazil, using ten 0.25 m² traps. They found that the number of collectors was too small to estimate fall of wood and fruit if less than a 10 per cent standard error is required. Reducing the standard error to 10 per cent of the mean would require about 100 traps for these two litter components. However, leaf litter was successfully sampled by this method.

Growth of wood is conventionally measured with tapes or dendrometers which can be fixed to appropriate trees and sampled at sequential periods. It is essential that an adequate sample over space, time, and species be taken (see chapter 8 and especially the section on *Growth rates of rain forest trees*).

Estimate of the underground biomass

Jeník (1971) states: "Unlike the single-dominant forests and plantations in mixed tropical forests the identification of underground roots according to species is nearly impossible. Only exceptionally do thin end-roots and smaller skeleton roots possess morphological and/or anatomical features which could be practically used in the tedious sorting of roots. Moreover, the number of species and wide range of sizes of trees add to the difficulties in quantitative assessment of the underground biomass. Hitherto, only rough estimates were carried out.

"Five successive steps are recommended in order to obtain a rough estimate of the underground biomass in the tropical forest.
1. Estimate of the biomass of larger skeleton roots.
2. Excavation of the soil blocks.
3. Separation of the roots from the soil.
4. Sorting of the roots and dry-weight determination.
5. Calculation and interpretation of results.

"Within a sample plot, possibly an area of 50 m by 50 m, all larger horizontal roots creeping near the ground can be directly measured, and their volume readily calculated. The portion protruding above the ground must be subtracted from the portion situated under the ground. The size and volume of the tap-roots, sinkers, 'heart' roots and deeply situated horizontal roots is estimated in wedge-shaped soil pits excavated at the foot of representative trees. These trees are selected mainly according to the shape of root spurs, buttresses, stilt roots and surface horizontal roots, suggesting one of the above-mentioned five types of root systems. The wedge-shaped soil transect should be situated between two larger horizontal roots, the depth being greater near the tree base. All larger skeleton roots within a certain volume of soil are removed, dried and weighed. Alternatively, one can calculate the approximate biomass by using the volume and average specific gravity of wood in tropical tree roots. According to our experience 0.6 specific gravity would be a reasonable value for such a calculation.

"Soil blocks for the estimate of the biomass of smaller skeleton roots and end-roots are situated either at random over the given sample plot or along a transect line connecting two larger trees selected at random. Tree trunks and larger skeleton roots must be avoided. In ferrallitic soils, soil blocks were sampled on an area of 25 cm by 25 cm. In the top-soil, two rectangulars of 5 cm thickness were removed ... In ferrallitic soils, 50 cm depth seems to include the majority of roots which could play a role in the rough estimate of the biomass ... These soil blocks appear to be a minimum, ten blocks an optimum for the rough estimate of the underground biomass.

"The separation of roots from soil samples is best done by the method of successive flotation in wash basins. The soaked soil sample is repeatedly stirred, and after sedimentation of heavy mineral particles the roots are sieved out from the water surface. More sophisticated equipment with a set of sieves tends to increase the error owing to losses of the root fraction which is mixed with the mineral fraction and caught on the sieves or lost in the streaming water.

"A total estimate of the underground biomass does not require any further sorting of the roots washed from the soil blocks. However, many ecological questions can be studied only after the sorting of roots, according to their diameter range. Most important is the category of the smallest roots under 2 mm diameter. Also, these roots provide a reliable pattern of vertical layering which is little disturbed by the casual occurrence of larger skeleton roots. As pointed out in the preceding paragraphs, sorting of roots according to their species origin is very difficult. In tropical Africa, *Chlorophora excelsa* possesses end-roots with bright yellow coloration, and older skeleton roots with bright yellow lenticells; despite this, the difficulties encountered in sorting these roots appeared to be insurmountable.

"The dry weight of the roots separated from the soil blocks is calculated for the entire area of the sample plot, and the weight of the skeleton roots is added. The result is expressed as the total of the root biomass per hectare or as the percentage of the total biomass of the whole ecosystem."

Summary

It is therefore necessary to establish the size of sample plot, number of samples and the frequency of sampling for the determination of tropical forest production. For rigorous determination of production there is no escape from careful preliminary study to set up the experimental designs following methods outlined in such texts as Snedecor (1953), Greig-Smith (1964) and others. For exploratory work it may be sufficient to examine plots of *ca.* 0.25 ha, taking 10 per cent samples of stems divided into appropriate categories. Further, 20–100 1 m² litter traps will be required. At this time it is difficult to recommend the most efficient size of plot and number of plots for sampling root biomass. The labour involved in estimating this component is tremendous and its is not clear how biomass estimates are converted to production values. Possibly sequential samples taken randomly within the overall plot could be used to show changes in living and dead root material. Finally, samples should be taken within significant seasons to represent growth, leaf development, fruiting and flowering. These data will permit preliminary statements on which hypotheses can be built, which can then be tested by more rigorous field observations.

Estimates of production

Gross primary production

There are relatively few estimates of the gross primary production of forest ecosystems in any region of the earth including the tropics. Of nine different estimates available from the literature, five were made in Thailand and

Democratic Kampuchea (Cambodia) by the Japanese team from Osaka City University. One other is from Malaysia, one from Ivory Coast and two from Puerto Rico (table 2). Considering all the forests, the average gross primary production is 86 t/ha/a. This average is quite similar to the estimates of Brünig (1968) and Weck (1960, quoted by Brünig) of 56–89 t/ha/a and 82–95 t/ha/a respectively based on theoretical considerations. Even though these estimates are quite limited in number they span a rather wide range of tropical forest types and, therefore, provide a basis for hypothesis building. The rainfall received over the forests in table 2 ranges from *ca.* 1 600–3 700 mm/a and the distribution of precipitation from a distinct and intense dry season to adequate moisture in all months. From table 2, it can be predicted that gross primary production will range from above 140 t/ha/a in the most equitable evergreen forest type to as little as 30 t/ha/a in dry savanna.

In each case gross primary production has been calculated from net primary production plus respiration. In mature tropical forest respiration is typically a large part of gross primary production (as much as 70 per cent; Golley, 1972) and, therefore, the accuracy of the production estimate depends upon the accuracy of the respiration measurement. In the judgment of Yoda *et al.* the highest estimate of GPP in table 2 may be in error because of an overestimate of respiration. In the Khao Chong forest samples of trunks and branches required considerable handling before respiration of the tissue could be measured. The workers feel that injury of the plant material and the long preparation time may have caused a stimulation of respiration. In contrast, at Pasoh Forest (Malaysia) the leaf, branch, stem and root samples were taken from 11 representative species of various sizes and the time between felling a sample tree and starting the measurement of respiration in the respiration containers where KOH absorbed CO_2 was minimized. Respiration of the vegetation at 25 °C was 54.5 t/ha/a in dry matter consumed versus 118.5 t/ha/a in Khao Chong forest. Since the biomass of these forests was 475 t/ha/a in Pasoh and 324 t/ha/a in Khao Chong, it may be true that the respiration was overestimated in the latter stand. These data illustrate the problems associated with the estimates of gross primary production.

It is worth mentioning that some attempts have been made to calculate the photosynthetic efficiency.

TABLE 2. Estimates of gross primary production of tropical forests. Data marked with an asterisk indicate values calculated from several of the authors' papers.

Forest type	Location	Production	Source
Equatorial rain	Khao Chong, Thailand	144*	Kira *et al.* (1964) and Yoda (1967)
Lower montane	El Verde, Puerto Rico	119	Odum (1970)
Evergreen seasonal	Southwestern Democratic Kampuchea (Cambodia)	117	Hozumi *et al.* (1969)
Lowland dipterocarp	Pasoh, Malaysia	85	Yoda *et al.* (1975)
Dry monsoon	Chieng Mai, Thailand	70*	Ogawa *et al.* (1961)
Red mangrove	Puerto Rico	58	Golley *et al.* (1962)
Savanna-monsoon forest ecotone	Chieng Mai, Thailand	53*	Ogawa *et al.* (1961)
Seasonal rain	Coastal Ivory Coast (Anguédédou)	53	Müller and Nielsen (1965)
Dipterocarp savanna	Chieng Mai, Thailand	32*	Ogawa *et al.* (1961)

Net primary production

There are not many more records of tropical forest net primary production than there are measures of gross primary production. The many tropical forest inventories only enumerate that part of the ecosystem which is of interest for immediate exploitation of timber—the merchantable volume of commercial or potentially commercial tree species. They do not contain much useful information on non-timber tree vegetation or on all components of the forest and so cannot be used for determining the net primary production.

The average of fourteen values found in the literature is about 20 t/ha/a and ranged from 9 to 32 t/ha/a (table 3). Again, there is a tendency for the highest values to be in regions with the most equitable moisture conditions and the lowest to be in areas with an extended dry season or other limiting environment.

None of the studies cited in table 3 are of young growing stands. As noted in fig. 1, the age of the stand has an important influence on the net production and the relative proportions of the production in stems, leaves and roots. The net primary production in mature forests would be expected to be less than in young forests since most of the gross primary production would be employed in maintenance. Actually this proportion is near 70 per cent for mature forests as contrasted to 40 per cent for young stands (Golley, 1972a). Unfortunately, there are no examples of the trend in net primary production in a single tropical forest type to illustrate these relationships. Yet forest plantations are often expected to equal or exceed the unmanaged rates of production. Net production of forest plantations of *Elaeis guineensis* in Yangambi (Zaire) to 14 years age are reported to have an annual production of 37 t/ha (Westlake, 1963) and 37–44-year-old *Shorea robusta* plantations at Dehra

Dun, India, may almost reach this level (Subba Rao *et al.*, 1972). These production values are slightly above those cited in table 3 for natural tropical forests, which are older and more complex, but comparisons are difficult to make (see chapter 8 for more data on plantations).

There are, in addition to these values, a large number of reports on the growth of individual tropical forest trees under managed conditions. Often these values are very high but they should not be used to indicate realistic rates of net primary production of forest stands. This is because individual trees measured for growth are often less than 20 years of age and the growth plots are often first generation plots after natural forest or secondary shrub and their performance is unlikely to represent sustained timber production. At best these data show the capacity of the species to survive and grow under the given conditions and the general pattern of growth rhythms.

TABLE 3. Estimates of net primary production (t/ha/a) in tropical forests. An asterisk means that the estimate does not include roots.

Forest type	Location	Net production	Source
Equatorial	Yangambi, Zaire	32	Bartholomew *et al.* (1953)
Equatorial	Khao Chong, Thailand	29	Kira *et al.* (1964)
Secondary forest 40 years old	Kade, Ghana	24	Nye (1961)
Lowland dipterocarp	Pasoh, Malaysia	22	Bullock, in Gist (1973)
Bamboo in monsoon forest	Burma	20*	Rozanov and Rozanova (1964)
Subequatorial (Banco plateau)	Ivory Coast	17	Lemée *et al.* (1975)
Bamboo in rain forest	Burma	16*	Rozanov and Rozanova (1964)
Dry deciduous	Varanasi, India	16	Misra (1972)
Lower montane	El Verde, Puerto Rico	16	Odum (1970)
Subequatorial (Yapo plateau)	Ivory Coast	15	Lemée *et al.* (1975)
Seasonal rain	Anguédédou, coastal Ivory Coast	13	Müller and Nielsen (1965)
Mangrove	Puerto Rico	9	Golley *et al.* (1962)

Litter production

Both the estimates of gross primary production and net primary production require rather extensive inputs of labour and time to obtain a sample representative of the site. In contrast, it is easier to measure the loss of organic material from the vegetation in the form of dead plant parts. This material is termed litter, and litter fall has been determined for over 40 tropical forests (table 4). Actually, litter measurements only encompass the fall of leaves, fruits, flowers and small twigs. Dead stems and roots are not included in litter fall measurements, and for this reason, litter fall is not a true measure of net production. However, litter fall does estimate the production dynamics of the rapid growing and highly dynamic part of the vegetation.

Fall of fruits, flowers, twigs and leaves averaged 10 t/ha/a in the 42 different forests or about one-half of the average net primary production estimate in table 3. The frequency distribution of these litter data suggest that there may be four groups of data representing very dry forests, dry forests, semi-deciduous forests and wet forests. The median of the distribution when graphed with 2 t/ha/a intervals is 7–8 t/ha/a. The highest litter falls were 23 and 25 t/ha/a and represented equatorial rain forest and evergreen gallery forest.

Leaf litter fall alone has been measured in over twenty forests and averages 5.5 t/ha/a (table 4). The two measures of litter are not strictly comparable because relatively few forests (12 in table 4) have both leaf and total litter estimates. In the forests with both leaf and total litter estimates, leaf litter was 54 per cent of the total. This proportion is higher (70 per cent) in a tropical humid montane forest (2 100 m) of southern India, where the average total litter over three years is 5.6 t/ha/a, of which leaf litter represents 3.9 t/ha/a, i.e. 70 per cent, and 1.3 t/ha/a (23 per cent) are contributed by twigs and bark (Blasco and Tassy, 1975).

Several investigators have measured litter fall over consecutive years. In most instances, the measurements represent two years; however, Malaisse *et al.* (1972) and Subba Rao *et al.* (1972) have records over three and four years respectively. The litter fall over two years shows that the measurement for one year is usually within 88 per cent of the second value (range of 75 to 100 per cent). For those cases where more than two years data are available, the lowest reading is usually within 80 per cent of the highest yearly value. Blasco and Tassy (1975) measured litter production over three years and found that it varied from 3.8 to 7.3 t/ha/a, the average being 5.5 t/ha/a. In this tropical humid montane forest ecosystem of southern India, these variations do not seem to be directly related to rainfall, temperature or dryness; the contribution to leaf litter production by few species of the canopy (8 species produce 70 per cent of the average annual leaf production) may be the important factor, as their massive defoliation could explain the high increase in leaf litter production which is noticed some years.

Two studies also measure litter fall in variants of the same forest. Bernhard (1970) compared a valley with a plateau and Cornforth (1970) compared two soil types. Again the litter values for the comparison sites are usually within 88 per cent of one another (range of the 5 cases is 67 to 98 per cent).

TABLE 4. Estimates of litter fall in tropical forests (t/ha/a). Leaf litter is presented separately from total litter which also includes fruit and flower parts and twigs.

Formation type	Location	Leaf litter	Total litter	Source
Evergreen	Banco, Ivory Coast	8.2	11.9	Huttel and Bernhard-Reversat (1975)
Evergreen	Yapo, Ivory Coast	7.1	9.6	Huttel and Bernhard-Reversat (1975)
Evergreen	Olokemeji, Nigeria	—	7.2	Hopkins (1966)
Moist semi-deciduous	Omo, Nigeria	—	4.6	Hopkins (1966)
Secondary moist semi-deciduous	Nigeria	—	5.6	Madge (1965)
Secondary forest (40 years)	Kade, Ghana	6.9	10.5	Nye (1961)
Semi-deciduous	Tafo, Ghana	—	20.9	Cunningham (1963)
Mixed forest	Yangambi, Zaire	—	12.4	Laudelout and Meyer (1954)
Brachystegia	Yangambi, Zaire	—	12.3	Laudelout and Meyer (1954)
Macrolobium	Yangambi, Zaire	—	15.3	Laudelout and Meyer (1954)
Musanga	Yangambi, Zaire	—	14.9	Laudelout and Meyer (1954)
Dry evergreen	Lubumbashi, Zaire	4.7	9.2	Malaisse *et al.* (1975)
Miombo forest	Lubumbashi, Zaire	3.0	4.0	Malaisse *et al.* (1972)
Miombo	Lubumbashi, Zaire	2.9	7.5	Malaisse *et al.* (1975)
Riparian	Lubumbashi, Zaire	4.5	5.9	Malaisse *et al.* (1975)
Tropical moist	Darien, Panamá	—	11.3	Golley *et al.* (1975)
Premature wet	Darien, Panamá	—	10.5	Golley *et al.* (1975)
Gallery	Darien, Panamá	—	11.6	Golley *et al.* (1975)
Second growth	Canal Zone, Panamá	—	6.0	Tropical Test Center (1966)
Rain forest	Colombia	—	8.5	Jenny *et al.* (1949)
Rain forest	Colombia	—	10.1	Jenny *et al.* (1949)
Amazonian upland	Manaus, Brazil	5.6	11.3	Klinge and Rodrigues (1968)
Evergreen seasonal (*Mora*)	Trinidad	7.0	—	Cornforth (1970)
Lower montane	El Verde, Puerto Rico	4.8	11.4	Wiegert (1970) and Odum (1970)
Montane	Rancho Grande, Venezuela	—	7.8	Medina and Zelver (1972)
Dry forest	Calabozo, Venezuela	—	8.2	Medina and Zelver (1972)
Lowland dipterocarp	Malaysia	—	7.2	Mitchell, in Bray and Gorham (1964)
Lowland dipterocarp	Malaysia	—	5.5	Mitchell, in Bray and Gorham (1964)
Upland dipterocarp	Malaysia	—	6.3	Mitchell, in Bray and Gorham (1964)
Secondary forest	Malaysia	—	8.3	Mitchell, in Bray and Gorham (1964)
Secondary forest	Malaysia	—	10.5	Mitchell, in Bray and Gorham (1964)
Secondary forest	Malaysia	—	14.4	Mitchell, in Bray and Gorham (1964)
Lowland dipterocarp	Pasoh, Malaysia	8.3	12.6	Bullock, in Gist (1973)
Evergreen gallery	Thailand	—	25.3	Ogawa *et al.* (1961)
Temperate evergreen	Thailand	—	18.9	Ogawa *et al.* (1961)
Dipterocarp savanna	Thailand	—	7.8	Ogawa *et al.* (1961)
Mixed savanna	Thailand	—	8.0	Ogawa *et al.* (1961)
Equatorial rain	Khao Chong, Thailand	11.9	23.2	Kira *et al.* (1964)
Terminalia-Shorea	Chakia, India	6.2	—	Singh (1968)
Tectona	Chakia, India	5.0	—	Singh (1968)
Diospyros-Anogeissus	Chakia, India	4.2	—	Singh (1968)
Shorea-Buchanania	Chakia, India	3.1	—	Singh (1968)
Butea	Chakia, India	1.0	—	Singh (1968)
Dry deciduous	Varanasi, India	—	7.7	Misra (1972)
Montane (2 100 m)	Gundar, India	3.9	5.5	Blasco and Tassy (1975)
Chir plantation	Dehra Dun, India	—	7.8	Subba Rao *et al.* (1972)
Teak plantation	Dehra Dun, India	—	7.8	Subba Rao *et al.* (1972)
Sal plantation	Dehra Dun, India	—	10.9	Subba Rao *et al.* (1972)

Production of forest components

Forest components have been treated separately by some investigators and the productions of leaves, stems, branches and roots are presented in table 5.

In forests other than those studied by Lemée *et al.* (1975) in the Ivory Coast, most of the net production is by stems and branches even in what are presumably mature forests.

Further, leaf production appears to be considerably greater than root production. These results are rather troubling. One might expect larger rates of production for roots which are very dynamic but possibly the dynamism involves only the root hairs and small rootlets which have a small mass. Similarly, one might expect that mature forests would exhibit low rates of turnover of wood. In Panamá, Golley *et al.* (1975) found a few large trees weighing a ton or more dry weight

TABLE 5. Production of forest components in selected tropical forests (t/ha/a).

Forest type	Leaves	Stems and branches	Roots	Source
Equatorial	9.5	19.2	2.8	Bartholomew et al. (1953)
Subequatorial (Banco)	11.7*	4.6	0.6	Lemée et al. (1975)
Subequatorial (Yapo)	10.2*	4.6	0.7	Lemée et al. (1975)
Seasonal rain (Anguédédou)	2.1	9.1	1.5	Müller and Nielsen (1965)
Secondary forest	6.9	13.6	2.3	Nye (1961)
Equatorial	12.0	16.2	0.4	Kira et al. (1964)
Dry deciduous	6.2	9.3	—	Misra (1972)

* including flowers and fruits.

which had fallen to the forest floor. Nevertheless, it would be necessary to have as many as 20 such trees falling on a hectare each year if the loss of dead trees is to equal the wood production on some sites listed in table 5. In contrast, Yoda *et al.* (1975) found that fall of dead wood over 10 cm diameter was 3.3 t/ha/a at Pasoh, Malaysia. In addition, they suggest that about 10 per cent of the wood of large trees is lost by decomposition before the trees fall to the ground. The fall of dead wood was also variable; it ranged from 3.1 to 20.5 t/ha/a over five 2 ha plots. Clearly, there are questions about the production of specific forest components that need to be resolved by further studies. Root production appears to be *ca.* 12 per cent, and leaf production *ca.* 63 per cent of stem production.

Generalizations

A number of authors have used data on tropical forest production to speculate about the relationship between the various terrestrial and aquatic systems in the biosphere. For example, Golley (1972) showed that the tropical forests contributed a larger proportion of the world gross production than any other system including the oceans. Golley and Lieth (in Golley and Golley, 1972) used these data to build maps of primary production on a world scale, which illustrate the high production rates in certain tropical locations. These conclusions show the importance of tropical forests to the biosphere's dynamic processes. However, while tropical forests have a greater gross production than temperate or polar forests, they do not necessarily have a greater rate of production of harvestable wood (see chapter 21). Bray and Gorham (1964) examining litter fall data from a more limited data set than available today, concluded that equatorial forests produced less wood per leaf mass than temperate forests. And, they concluded further that use of leaf litter fall as an index to forest productivity would overestimate equatorial forest production and underestimate production of evergreen gymnosperm forests. Jordan (1971a, b) more recently, but still reasoning from a relatively limited data set, concluded than on a gradient of decreasing solar energy wood production is constant, but litter production decreases—thus supporting Bray and Gorham's conclusions. However, Jordan also suggests that where the gradient is decreasing precipitation litter production is constant and wood production decreases (see also *Litter production* by

Pugh and Nicholson, 2 vol., Academic Press, 1975). These various conclusions do not include the hidden cost of loss of wood from rot of standing trees. If decomposition of standing trees is very high, this could shift the estimates of litter.

Brünig and Klinge (personal communication) also note that the Amazon Guianese lowland forest appears to be less tall and of smaller biomass than Malaysian mixed dipterocarp forest on comparable sites. The former forest almost looks like a mixed dipterocarp forest without the tallest individuals of dipterocarp trees. In addition, Amazonian forest trees seem to more frequently be characterized by heavy, silica-rich and defective (hollow) timber. Wood decomposition in living trees is also wide-spread in the dipterocarp forests of South-East Asia and probably an important ecological factor (soil substrate factor in Sarawak, Ashton) and production element. Regional comparisons of production and other functional characters of tropical forest would be extremely valuable.

Aside from ecological considerations, it is important to know if the production of a forest which has evolved without human influence is a measure of the site's productivity for timber or agricultural production, as suggested by Webb *et al.* (1971) for rain forest sites in Australia. Even though agriculture and, increasingly forestry, is subject to considerable manipulation of environmental limiting factors, agricultural and forest production is mainly based on the solar energy input to the site and is controlled largely by precipitation and soil nutrients. For this reason, natural forest production may be a useful measure against which to compare the yields from managed production systems. Such comparisons show that managed systems often only crudely simulate the unmanaged system that was selected for the site by evolution and history of the biota and landscape so that the yield is much less than expected and the side-effects of soil erosion, decimation of the flora and fauna and other degradation factors are severe. The challenge of management is to build systems which are ecologically sound and require minimum maintenance by man. To do this we need to understand the internal dynamics of these forest systems (see chapters 20 and 21). Therefore, for production studies permanent plots are needed in the variety of forest types found within the three tropical forest areas of the globe. On these sites, growth, species population dynamics and production physiology could all be measured so that the production process can be understood at the level necessary for management.

Summary

The data describing production in tropical forests are extremely limited and, therefore, it is not possible to discuss regional differences in production. As discussed earlier, the relationship of net production and such environmental factors as precipitation and solar energy have been described by Lieth and Box (1972) as well as by Rosenzweig (1968) and others and, therefore, theoretically production can be predicted for any broad location if environmental data are available. It is an important gap in our knowledge not to have the observational data with which to test these predictions. Finally it should be pointed out that estimates of standing crop biomass are not estimates of productivity. Biomass provides the baseline upon which production is measured but productivity requires sequential measures of the vegetation.

Conclusions

Production is a dynamic process and as such requires a number of measurements over appropriate time intervals for rigorous description. Nowhere on the earth does there exist published data of forest gross or net primary production at one site adequate to determine variability of production under yearly variation in the environment. Such data are required to judge the significance of the data for a single stand collected over one year. Permanent study plots should be established and protected for such measures of production.

Although a few sets of data on gross and net tropical forest primary production have been published, these are insufficient to show regional differences in production to serve as a baseline against which to judge management. Most studies have been carried out in South-East Asia and the papers of Kira, Ogawa, Hozumi and associates are the best summary of these production studies. Very few studies represent the Amazonian forest or the Congo forest of Africa. Since these latter forests represent some of the major areas of tropical forest on the earth there is a major gap in our knowledge of forest production. It is disheartening to hear numerous plans for increasing production of forestlands in the tropical regions of the earth when the rate of natural production of the indigenous vegetation is not known. Tropical countries should insist on knowing natural rates of production before they allow such schemes to be organized,

otherwise they may lose a precious heritage for short-term gain which will not pay the cost of recovery of the landscape if the scheme fails (see chapter 21).

Research needs and priorities

For management data on forest ecosystems, including the soil, climate, micro-organisms, animals, lower plants, as well as the trees, are required. The production must be examined in the context of the structure and function of all ecosystem components. Hopefully, these studies could be carried out in each region of tropical forests, on a variety of types of forest, with adequate replication over space and time. Obviously this effort would require large inputs of manpower and financing, yet a phased series of studies could be initiated and lead toward the larger goal as long as the overall project was adequately designed at the onset. Further, the overall project could incorporate on-going studies of productivity in several parts of the world.

In the design of forest production ecology, the following priorities should be considered.

1. Estimates of year to year and site to site variations in gross and net primary production are regularly needed.
2. Many more data are needed on the production of underground parts.
3. Ecosystem components such as epiphytes, lianas, algae, lichens and mosses are seldom examined in production studies; their roles should be evaluated.
4. Much more attention should be paid to the dynamics of dead vegetation, including litter fall, branch fall, trunk or stem fall, and decomposition rates of each of these components, as well as decomposition of standing dead trees (standing dead wood).
5. Finally, it is extremely important that production processes be determined under the influence of constraints from the environment (such as soil fertility and moisture availability) and the biota (such as intensity of herbivory).

For long-term planning, it is necessary to have a comparison of the productivity and, therefore, profitability, of the two main systems of management in vogue today, namely selective working of the tropical forest and the establishment of plantations of fast-growing, relative shorter rotation species (see chapters 20 and 21). In either case, knowledge of the net primary production and the factors controlling production will be helpful in such planning decisions.

Bibliography

BARTHOLOMEW, W. V.; MEYER, J.; LAUDELOUT, H. *Mineral nutrient immobilization under forest and grass fallow in Yangambi (Belgian Congo) region, with some preliminary results on the decomposition of plant material on the forest floor.* Publ. INEAC (Bruxelles), Sér. sci., n° 57, 1953, 27 p.

BERNHARD, F. Étude de la litière et de sa contribution au cycle des éléments minéraux en forêt ombrophile de Côte-d'Ivoire. *Oecol. Plant.*, 5, 1970, p. 247–266.

BERNHARD-REVERSAT, F.; HUTTEL, C.; LEMÉE, G. Quelques aspects de la périodicité écologique et de l'activité végétale saisonnière en forêt ombrophile sempervirente de Côte-d'Ivoire. In: Golley, P. M.; Golley, F. B. (eds.). *Tropical ecology with an emphasis on organic production*, p. 217–234. Athens, Univ. of Georgia, 1972, 418 p.

BLASCO, F.; TASSY, B. Étude d'un écosystème forestier montagnard du sud de l'Inde. *Bull. Écol.*, 6, 1975, p. 525–539.

BEST, R. Production factors in the tropics. *Netherlands J. Agric. Sci.*, vol. 10, no. 5, 1962, p. 347–353.

BONNER, J. The upper limit of crop yield. *Science*, 137, 1962, p. 11–15.

BRAY, J. R.; GORHAM, E. Litter production in forests of the world. In: Cragg, J. (ed.). *Advances in ecological research*, 2, p. 101–157. Academic Press, 1964.

BRÜNIG, E. F. On the limits of vegetable productivity in the tropical rain forest and the boreal coniferous forest. *J. Indian Bot. Sci.*, vol. XLI, no. 4, 1968, p. 314–322.

——. Biomass diversity and biomass sampling in tropical rain forest. In: Young, H. E. (ed.). *IUFRO biomass studies*, p. 269–293. Orono, Univ. of Maine, 1973.

CORNFORTH, I. S. Litter fall in tropical rain forest. *J. Appl. Ecol.*, 7, 1970, p. 603–608.

COSTES, C. (ed.). *Photosynthèse et production végétale*. Paris, Gauthier-Villars, 1975, 284 p.

CUNNINGHAM, R. K. The effect of clearing a tropical forest soil. *J. Soil Sci.*, 14, 1963, p. 334–344.

DAUBENMIRE, R. Phenology and other characteristics of tropical semi-deciduous forest in North-Western Costa Rica. *J. Ecol.*, 60, 1972, p. 147–170.

DOBBEN, W. H. van; LOWE-MCCONNELL, R. H. (eds.). *Unifying concepts in ecology*. Report of the plenary sessions of the first international congress of ecology (The Hague, 8–14 September 1974). The Hague, W. Junk B. V. publishers; Wageningen, Centre for agricultural publishing and documentation, 1975, 302 p.

Chapters on "Flow of energy and matter between trophic levels"; "Comparative productivity in ecosystems"; "Diversity, stability and maturity in natural ecosystems"; "Diversity, stability and maturity in ecosystems influenced by human activities"; "Strategies for management of natural and man-made ecosystems".

DREES, E. M. The minimum area in tropical rain forest with special reference to some types of banka (Indonesia). *Vegetatio*, vol. 5, no. 6, 1954, p. 147–170.

GIST, C. S. *Some tropical modelling efforts*. Ecology Center, Utah State University (Logan, Utah), 1973, multigr.

GOLLEY, F. B. Energy flux in ecosystems. In: Wiens, J. A. (ed.). *Ecosystem structure and function*, p. 69–90. Corvallis, Oregon State Univ. Press, 1972.

——. Summary. In: Golley, P. M.; Golley, F. B. (eds.). *Tropical ecology with an emphasis on organic production*, p. 407–413. Athens, Univ. of Georgia, 1972a, 418 p.

——; ODUM, H. T.; WILSON, R. F. The structure and metabolism of a Puerto Rican red mangrove forest in May. *Ecology*, 43, 1962, p. 9–19.

——; LIETH, H. Bases of organic production in the tropics. In: Golley, P. M.; Golley, F. B. (eds.). *Tropical ecology with an emphasis on organic production*, p. 1–26. Athens, Univ. of Georgia, 1972, 418 p.

——; MCGINNIS, J. T.; CLEMENTS, R. G.; CHILD, G. I.; DUEVER, M. J. *Mineral cycling in a tropical moist forest ecosystem*. Athens, Georgia, Univ. of Georgia Press, 1975, 248 p.

GREIG-SMITH, P. *Quantitative plant ecology*. 2nd ed. London, Butterworth, 1964, 207 p.

GRUBB, P. J. Factors controlling the distribution of forest types on tropical mountains: new facts and a new perspective. In:*Transactions of the third Aberdeen-Hull Symposium on Malesian ecology: altitudinal zonation in Malesia*, p. 13–46. Univ. of Hull, Dept. of Geography, misc. ser. no. 16, 1974, 119 p.

HATCH, M. D.; SLACK, C. R. Photosynthesis by sugar cane leaves. A new carboxylation reaction in the pathway of sugar formation. *J. Biochem.*, 101, 1966, p. 103–111.

HOPKINS, B. Biological productivity in Nigeria. *Sci. Assoc. Nigeria Proc.*, vol. 1, no. 3, 1962, p. 20–28.

——. Vegetation of the Olokemeji Forest Reserve, Nigeria. IV. The litter and soil with special reference to their seasonal changes. *J. Ecol.*, 54, 1966, p. 687–703.

——. Vegetation of the Olokemeji Forest Reserve, Nigeria. VI. The plants on the forest site with special reference to their seasonal growth. *J. Ecol.*, 58, 1970, p. 765–793.

HOZUMI, K.; YODA, K.; KIRA, T. Production ecology of tropical rain forest in southwestern Cambodia. II. Photosynthetic production in an evergreen seasonal forest. *Nature and Life in S.E. Asia*, 6, 1969, p. 57–81.

HUTTEL, C.; BERNHARD-REVERSAT, F. Biomasse végétale et productivité primaire, cycle de la matière organique. In : Recherches sur l'écosystème de la forêt sub-équatoriale de basse Côte-d'Ivoire. *La Terre et la Vie* (Paris), 29, 1975, p. 169–264.

JENÍK, J. Root structure and underground biomass in equatorial forests. In: *Productivity of forest ecosystems*, p. 323–330. Paris, Unesco, 1971, 707 p.

JENNY, H.; GESSEL, S. P.; BINGHAM, F. T. Comparative study of decomposition rates of organic matter in temperate and tropical regions. *Soil Sci.*, 68, 1949, p. 419–432.

JORDAN, C. F. A world pattern in plant energetics. *Amer. Sci.*, 59, 1971a, p. 425–433.

——. Productivity of a tropical forest and its relation to a world pattern of energy storage. *J. Ecol.*, 59, 1971b, p. 127–242.

KIRA, T.; OGAWA, H.; YODA, K.; OGINO, K. Primary production by a tropical rain forest of southern Thailand. *Bot. Mag. Tokyo*, 77, 1964, p. 428–429.

——; ——; OGINO, K. Comparative ecological studies on three main types of forest vegetation in Thailand. IV. Dry matter production with special reference to the Khao Chong rain forest. In: Kira, T.; Iwata, K. (eds.). *Nature and Life in S.E. Asia*, 6, 1967, p. 149–174. Fauna and Flora Research Society (Kyoto), 1967.

——; SHIDEI, T. Primary production and turnover of organic matter in different forest ecosystems of the western Pacific. *Japanese Journal of Ecology*, vol. 17, no. 2, 1967, p. 70–87.

KLINGE, H. Root mass estimation in lowland tropical rain forests of Central Amazonia, Brazil. I. Fine root masses of a pale yellow latosol and a giant humus podzol. *Tropical Ecology*, vol. 14, no. 1, 1973, p. 29–38.

II. "Coarse root mass" of trees and palms in different height classes. *An. Acad. Brasil. Ciênc.*, vol. 45, no. 3/4, 1973, p. 595–609.

III. Nutrients in fine roots from giant humus podsols. *Tropical Ecology*, vol. 16, no. 1, 1975, p. 28–38.

——; RODRIGUES, W. A. Litter production in an area of Amazonian terra firme forest. Part I. Litter fall, organic carbon and total nitrogen contents of litter. *Amazoniana*, vol. 1, no. 4, 1968, p. 287–302. Part II. Mineral nutrient content of the litter. *Amazoniana*, vol. 1, no. 4, 1968, p. 303–310.

——; ——. Biomass estimation in a Central Amazonian rain forest. *Act. Cient. Venez.*, 24, 1973, p. 225–237.

——; ——. Phytomass estimation in a Central Amazonian rain forest. In: Young, H. E. (ed.). *IUFRO biomass studies*, p. 339–350. Orono, Univ. of Maine, 1974.

——; ——; BRÜNIG, E. F.; FITTKAU, E. J. Biomass and structure in a Central Amazonian rain forest. In: Golley, F. B.; Medina, E. (eds.). *Tropical ecological systems: trends in terrestrial and aquatic research*, p. 115–122. Berlin and New York, Springer Verlag, Ecological Studies no. 11, 1975, 398 p.

LAUDELOUT, H.; MEYER, J. Les cycles d'éléments minéraux et de matière organique en forêt équatoriale congolaise. In: *Trans. 5th Int. Cong. Soil Sci.*, 2, 1954, p. 267–272.

LEMÉE, G.; HUTTEL, C.; BERNHARD-REVERSAT, F. Recherches sur l'écosystème de la forêt sub-équatoriale de basse Côte-d'Ivoire. *La Terre et la Vie* (Paris), 29, 1975, p. 169–264.

LEMON, E. R.; ALLEN, L. H.; MULLER, L. Carbon dioxide exchange of a tropical rain forest. Part II. *Bio-Science*, vol. 20, no. 19, 1970, p. 1054–1059.

LIETH, H. (ed.). *Phenology and seasonality modelling*. Berlin and New York, Springer Verlag, Ecological Studies no. 8, 1974, 444 p.

——; BOX, E. Evapotranspiration and primary productivity: C. W. Thornthwaite memorial model. In: Mather, J. R. (ed.). *Papers on selected topics in climatology*, vol. 2, p. 37–46. New York, Elmer, Thornthwaite memorial volume 2, 1972.

——; WHITTAKER, R. H. *Primary productivity of the biosphere*. Berlin and New York, Springer Verlag, Ecological Studies no. 14, 1975, 340 p.

MALAISSE, F.; ALEXANDRE, J.; FRESON, R.; GOFFINET, G.; MALAISSE-MOUSSET, M. The miombo ecosystem: a preliminary study. In: Golley, P. M.; Golley, F. B. (eds.). *Tropical ecology with an emphasis on organic production*, p. 363–405. Athens, Univ. of Georgia, 1972, 418 p.

——; FRESON, R.; GOFFINET, G.; MALAISSE-MOUSSET, M. Litter fall and litter breakdown in miombo. In: Golley, F. B.; Medina, E. (eds.). *Tropical ecological systems: trends in terrestrial and aquatic research*, p. 137–152. Berlin and New York, Springer Verlag, Ecological Studies no. 11, 1975, 398 p.

MEDINA, E.; ZELVER, M. Soil respiration in tropical plant communities. In: Golley, P. M.; Golley, F. B. (eds.). *Tropical ecology with an emphasis on organic production*, p. 245–269. Athens, Univ. of Georgia, 1972, 418 p.

MISRA, R. A comparative study of net primary productivity of dry, deciduous forest and grassland of Varanasi, India. In: Golley, P. M.; Golley, F. B. (eds.). *Tropical ecology with an emphasis on organic production*, p. 279–293. Athens, Univ. of Georgia, 1972, 418 p.

MITCHELL, H. L. Trends in the nitrogen, phosphorus, potassium and calcium content of leaves of some forest trees during the growing season. *Black Rock For. Pap.*, 1, 1936, p. 30–44.

MÜLLER, D.; NIELSEN, J. Production brute, pertes par respiration et production nette dans la forêt ombrophile tropicale. *Det Forstlige Forssvaesen i Danmark*, 29, 1965, p. 69–160.

NEWBOULD, P. J. *Methods for estimating the primary production of forests*. London, Blackwell Scientific Publications, IBP Handbook no. 2, 1967.

NYE, P. H. Organic matter and nutrient cycles under moist tropical forest. *Plant and Soil*, vol. 13, no. 4, 1961, p. 333–346.

ODUM, H. T. Summary: an emerging view of the ecological system at El Verde. In: Odum, H. T.; Pigeon, R. F. (eds.). *A tropical rain forest. A study of irradiation and ecology at El Verde, Puerto Rico*, p. I–191 to I–289. Division of Technical Information, U.S. Atomic Energy Commission (USAEC), 1970, 1678 p.

——; JORDAN, C. F. Metabolism and evapotranspiration of the lower forest in a giant plastic cylinder. In: Odum, H. T.; Pigeon, R. F. (eds.). *A tropical rain forest. A study of irradiation and ecology at El Verde, Puerto Rico*, p. I–165 to I–189. Division of Technical Information, U.S. Atomic Energy Commission (USAEC), 1970, 1678 p.

OGAWA, H.; YODA, K.; KIRA, T. A preliminary survey of the vegetation of Thailand. *Nature and Life in S.E. Asia*, 1961, p. 49–80.

——; ——; ——; OGINO, K.; SHIDEI, T.; RATANAWONGSE, Du-

ongkeo; APASUTY, Charn. Comparative ecological studies on three main types of forest vegetation in Thailand. I. Structure and floristic composition. In: Kira, T.; Iwata, K. (eds.). *Nature and Life in S.E. Asia*, 4, 1965, p. 13–48.

PESSON, P. (ed.). *Écologie forestière. La forêt : son climat, son sol, ses arbres, sa faune*. Paris, Gauthier-Villars, 1974, 282 p.

POORE, M. E. D. Studies in Malaysian rain forest. I. The forest on Triassic sediments in Jengka Forest Reserve. *J. Ecol.*, 56, 1968, p. 143–196.

REES, A. R.; TINKER, P. B. H. Dry matter production and nutrient content of plantation oil palms in Nigeria. *Plant and Soil*, 19, 1963, p. 19–32.

REICHLE, D. E.; FRANKLIN, J. F.; GOODALL, D. W. (eds.). *Productivity of world ecosystems*. Washington, D.C., National Academy of Sciences (NSF), 1975, 166 p.

RICHARDS, P. W. *The tropical rain forest: an ecological study*. Cambridge University Press, 1952, 450 p. 4th reprint with corrections, 1972.

ROSENZWEIG, M. L. Net primary productivity of terrestrial communities: prediction from climatological data. *American Naturalist*, 102, 1968, p. 67–74.

ROZANOV, B. G.; ROZANOVA, I. M. Biological cycle of bamboo (*Bambusa* spp.) nutrients in the tropical forests of Burma. *Bot. Zhur.* (Moscow), 49, 1964, p. 348–357.

SABHASRI, S.; KHEMNARK, C.; AKSORNKOAE, S.; RATISOONTHRON, P. Rept. I. *Primary production in dry-evergreen forest at Sakaerat Amphoe Park Thong, Chai Changwat Nakhou Ratchasima. I. Estimation of biomass and distribution among various organs*. Bangkok, ASRCT, 1968.

SINGH, K. P. Litter production and nutrient turnover in deciduous forests of Varanasi. In: *Proc. Symp. Recent Adv. Trop. Ecol.*, 2, 1968, p. 655–665.

SNEDECOR, G. W. *Statistical methods*. The Iowa State College Press (Ames, Iowa), 1953, 485 p.

STEPHENS, G. R.; WAGGONER, P. E. Carbon dioxide exchange of a tropical rain forest. Part I. *Bio-Science*, vol. 20, no. 19, 1970, p. 1050–1053.

SUBBA RAO, B. K.; DABRAL, B. G.; PANDE, S. K. Litter production in forest plantations of Chir (*Pinus roxburghii*), Teak (*Tectona grandis*) and Sal (*Shorea robusta*) at New Forest, Dehra Dun. In: Golley, P. M.; Golley, F. B. (eds.). *Tropical ecology with an emphasis on organic production*, p. 235–243. Athens, Univ. of Georgia, 1972, 418 p.

TROPICAL TEST CENTER. *Environment data base for regional studies in the humid tropics*. Semi-annual Report no. 1 and 2. USATECOM Project no. 9–4–9913–01, 1966, p. 71–76.

WADSWORTH, F. H. Growth in the lower montane rain forest of Puerto Rico. *Caribbean Forester*, 8, 1941–47, p. 27-43.

WALTER, H. Le facteur eau dans les régions arides et sa signification pour l'organisation de la végétation dans les contrées sub-tropicales. In: Colloques internationaux du Centre national de la recherche scientifique, vol. 59, *Les divisions écologiques du globe*. Paris, Centre national de la recherche scientifique, 1955, 236 p.

——. *Ecology of tropical and sub-tropical vegetation*. Van Nostrand Reinhold, 1971, 539 p.

——; LIETH, H. *Klimadiagramm-Weltatlas*. Jena, Gustav Fischer, 1960–1967, 245 p.

WEBB, L. J.; TRACEY, J. G.; WILLIAMS, W. T.; LANCE, G. N. Prediction of agricultural potential from intact forest vegetation. *J. Appl. Ecol.*, 8, 1971, p. 99–121.

WECK, J. Klimaindex und forsliches produktions potential. Ein weiterer beitrag zum problem ihrer korrelation. *Forstarchiv.*, vol. 31, no. 7, 1960, p. 101–104.

WESTLAKE, D. F. Comparisons of plant productivity. *Biological Review*, 38, 1963, p. 385–425.

WHITTAKER, R. H., LIKENS, G. E. The primary production of the biosphere. *Hum. Ecol.*, 1, 1973, p. 301–369.

WIEGERT, R. G. Effects of ionizing radiations on leaf fall, decomposition and litter micro-arthropods of a montane rain forest. In: Odum, H. T.; Pigeon, R. F. (eds.). *A tropical rain forest. A study of irradiation and ecology at El Verde, Puerto Rico*, p. H–89 to H–100. Division of Technical Information, U.S. Atomic Energy Commission (USAEC), 1970, 1678 p.

YODA, K. Comparative ecological studies on three main types of forest vegetation in Thailand. III. Community respiration. *Nature and Life in S.E. Asia*, 5, 1967, p. 83–148.

——; OGAWA, H.; KIRA, T. Structure and productivity of a tropical rain forest in West Malaysia. Paper presented to the *12th Int. Cong. Botany* (Leningrad, 1975), 23 p. multigr.

ZAVITKOVSKI J.; ISEBRANDS, J. G.; CROW, T. R. Application of growth analysis in forest biomass studies. In: *Proceedings 3rd North American Forest Biology Workshop* (Colorado State University, 9–12 September 1974), p. 196–226.

11 Secondary production

Introduction

Secondary production suffers from the same problems of definition as primary production. Many ecologists define secondary production as the elaboration of new tissue which is added to the standing crop biomass of animals (McCullough, 1970). This definition is technically correct but it is insufficient since secondary production may be discussed at any level of ecological organization and the concept is not identical at each of these levels. For example, the growth of an individual animal is secondary production of that individual. Secondary production of the population is the increase in biomass of the individuals making up the population; it is the sum of the growth of individuals minus their loss of weight together with the population's natality, mortality, immigration and emigration. Equally, all the animals in an ecosystem can be considered together as the heterotrophs. The secondary production of this fauna is not the sum of the productions of the populations since one population may feed on another transferring organic material throughout the heterotroph component. Rather, secondary production at this level of organization is the increase in the biomass of the total fauna. Obviously the time interval is exceedingly important in the calculation of each of these productivities. Usually the period of time used for individual production is the life-span of the individual, for the population one year or one season, and for the heterotrophic component of the ecosystem one or more years.

In discussions of secondary production it is therefore necessary to specify the level of organization and the time interval. Information on secondary production of individuals, populations and heterotrophic compartments of ecosystems is relatively well known in temperate faunas but very poorly known in tropical forests. For instance, the pattern of growth of individuals is common knowledge to biologists and detailed explanation of this process can be found in elementary animal physiology or animal science textbooks. A variety of environmental and genetic factors influence growth of individuals. Nutrition and temperature are especially significant factors. There is no evidence that tropical animals have a fundamentally different growth pattern than temperate animals (see chapter 7).

There is much less known about production at the population level and general laws of population secondary production have not yet been accepted. Population secondary production is a function of the inputs and outputs; an excess of inputs over outputs results in storage, or increase in

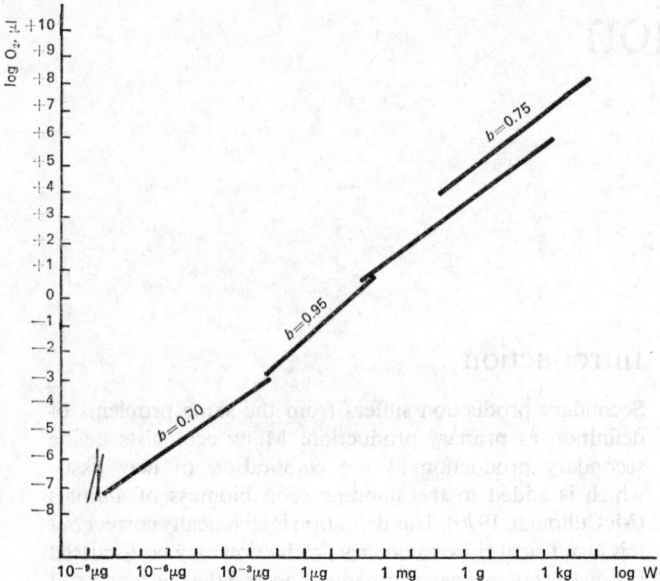

FIG. 1. Relationship between basal metabolism (log O_2 in μl) and body weight (log W, in μg, mg, g and kg) in various animals.

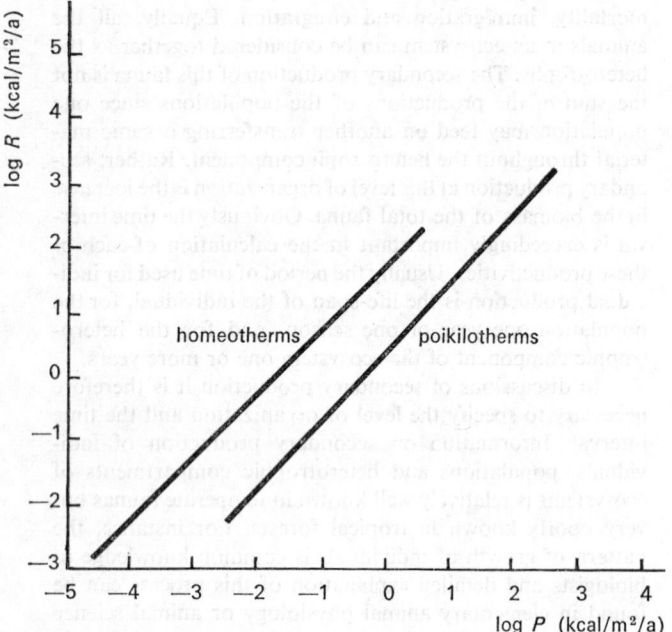

FIG. 2. Relationship between production (log production in $kcal/m^2/a$) and metabolism (log respiration in $kcal/m^2/a$) in field populations.

biomass, and, therefore, is by definition secondary production. In animal populations input consists of food consumption, natality and immigration to the population. Output is faecal, urinary and ruminary gas loss, mortality due to death and emigration. Obviously the calculation of secondary production of a population is complicated since there is seldom a possibility of weighing the entire population at two separate times and determining its weight gain or loss.

There are very few studies of production of an animal species over a full year in tropical forests. Gliwicz's (1973) study of the forest rodent, *Proechimys semispinosus*, in Panamá is the closest to these requirements. There are many studies of the systematics, distribution, and natural history of tropical forest animals but these observations seldom can be directly used for estimating secondary production; they are mainly useful in interpreting differences in the production ecology or trophic dynamics of individuals, populations, or the faunas of various types of tropical forest (see chapters 6 and 7).

Another approach to the determination of secondary production involves the collection of animals of a variety of species, the determination of the production ecology features of one species and extrapolating from this to the entire collection. For example, birds characteristic of the forest undergrowth can be captured by mist nets. The feed intake, assimilation, growth, natality and mortality of the commonest species are determined and under the assumption that these common species are representative of the entire set of species, the production ecology of the bird fauna is estimated. Very seldom, if ever, is the assumption of applicability tested. The variation between species may be greater in tropical forests since the number of species is larger and dominance is often less extreme than in temperate faunas. Even so, the production ecology of groups of animals such as granivores, arboreal folivores, soil fauna, litter fauna, small mammals and birds are occasionally studied as a group.

A large body of information is available on the metabolic requirements and the secondary production of selected populations and individuals. If the diversity of individual heterotrophs and their body size or weight are known, the basal metabolic rate can be determined from standard graphs (fig. 1). While the basal rate must be corrected for the energy cost of feeding and activity as well as thermoregulation, these corrections can also be made for many types of populations (Golley, 1968).

If the annual metabolic expense of a population is known, then it is possible to estimate its production since McNeil and Lawton (1970) have shown that production and metabolism are related in field populations (fig. 2). They found that the relationship for homeotherms (10 different species of birds and mammals) have, for the same metabolic level, a lower production rate than poikilotherms (42 species of invertebrates and fish).

Production and metabolism together represent the material assimilated by the population. If the percentage of consumed food which is assimilated is known, the consumption or input to the population can be estimated from pro-

Secondary production 251

duction plus metabolism. Unfortunately, assimilation percentage varies with the type of food, physiological conditions as well as type of digestive tract. Some vertebrates can assimilate 80–90 per cent of the food consumed, while some detritus-feeding animals such as isopods and millipedes only assimilate 10–30 per cent. However, data on the ecological energetics of specific organisms probably can be extrapolated to the population within a habitat and season. More data of this type are needed.

Secondary production data

The basic information required for estimation of heterotrophic production is density and biomass. There have been few attempts to determine the density or biomass of the fauna for any tropical forest. Data exist only for red mangrove forest (Golley et al., 1962), lower montane rain forest (Odum et al., 1970), Amazonian terra firme forest (Fittkau and Klinge, 1973), and miombo ecosystem (Malaisse et al., 1972). In addition, an estimate of faunal dynamics has been presented for tropical moist forest (Golley et al., 1975).

The animal biomass in these tropical systems (in kg/ha) are:

Amazonian terra firme forest	210
Lower montane rain forest	118
Miombo woodland	76
Mangrove forest	64
Tropical moist forest	73.

Unfortunately, none of these estimates represent adequate samples for all groups of heterotrophs over all seasons. Indeed, the authors, in most cases, present them as incomplete and preliminary estimates. Nevertheless, it is doubtful if the total biomass is more than twice the preliminary estimates. If this is correct, then these data suggest that in the forest the animal biomass is much less than the vertebrate animal biomass, including large ungulates, in African grassland and savanna. For example, eastern African grassland may have 100–300 kg/ha of large herbivores alone (Bourlière, 1963; Lamprey, 1964). However, in an eastern African thornbush savanna (Hendrichs, 1970) there are

50 kg of ungulates, 4 kg of small mammals and 250 kg of soil animals per hectare, which is approximately equal to the forest animal biomass.

Montgomery and Sunquist (1975) suggest that while the mammal herbivore biomass of neotropical forests and savannas are roughly similar, the number of species is greater in savannas. Few large-sized species have developed an ability to feed in an arboreal habitat. For example, on Barro Colorado Island, Panama Canal Zone, the nonvolant mammal biomass was 53 kg/ha, of which sloths made up 60 per cent and howler monkeys 12 per cent.

There are a few studies of density, seasonal patterns, feeding and production ecology of small mammals in tropical forests. Fleming (1975) has reviewed these data and tentatively concludes that while tropical small mammal faunas have more species than temperate faunas (for example, the continental United States contain about 267 small mammals while Panamá, which has an area one hundredth that of the United States, contains 166 species), dominance of a few small mammal species is probably equally true in tropical or in temperate forests. Tropical small mammals respond to the periodicity in rainfall and while their numbers and reproduction may vary seasonally, their overall production of young appears to be less than that of temperate small mammals. Biomass of small mammals is known for a few locations (table 1) but the data are too few for generalization. Seasonal variation in biomass is observed. For example, biomass of three Central American dry forest species Liomys adspersus, Oryzomys capito, and Proechimys semispinosus varied from 208 to 629 g/ha, 0 to 159 g/ha, and 125 to 1099 g/ha respectively. On Barro Colorado Island, Panama Canal Zone, Gliwicz (1973) said the average biomass of Proechimys semispinosus was 2900 g/ha; the population on a 16 ha island had a secondary production of 97×10^3 kcal/a and a cost of maintenance of 1973×10^3 kcal/a.

Karr (1975) examined certain features of the production ecology of tropical forest bird faunas in comparison with temperate forest faunas. His data were derived from studies in Panamá, Liberia and Illinois. The numbers of species were 75 per cent higher and the number of individuals were 49 per cent higher in tropical forest habitats. However, the existence energy required to maintain the birds at constant weight was

TABLE 1. Fresh weight biomass estimates for populations or communities of tropical forest small mammals, from Fleming (1975).

Location and habitat	Number of species	Average density (no./ha)	Average biomass (wet weight, g/ha)	Source
Panamá: dry tropical forest	11 (rodents)	18.9	4 025	Fleming (1970, 1971, 1972)
	5 (marsupials)	3.5	1 538	
	16 (total)	22.4	5 562	Fleming (1970, 1971, 1972)
Panamá: moist tropical forest	9 (rodents)	11.3	6 304	
	4 (marsupials)	2.1	1 293	
	13 (total)	13.4	7 597	
Malaysia: primary forest	11 (rodents)	4.9	840	Harrison (1969)
secondary forest	4 (rodents) 1 (tree shrew)	7.1	810	Harrison (1969)
scrub	5 (rodents)	3.2	282	Harrison (1969)
grassland	4 (rodents)	5.5	324	Harrison (1969)

TABLE 2. Biomass and energy relations of resident forest avifaunas, from Karr (1975).

	Biomass (g/ha)		Existence energy[e] (kcal/ ha/day)	Mean weight (g)
	Standing crop[a]	Con- suming[b]		
Illinois	1.21	0.25	0.41	49.5
Liberia	1.24	0.30	0.47	34.9
Panamá	1.32	0.34	0.42	36.2
Puercos Island	0.95	0.30	0.44	22.1

[a]Standing crop biomass $= \sum\limits_{i=1}^{s} (Ai)(Wi)$

[b]Consuming biomass $= \sum\limits_{i=1}^{s} (Ai)(Wi^b)$

[e]Existence energy $= \sum\limits_{i=1}^{s} (a)\ (Ai)(Wi^b)$,

where Ai = abundance of species i; Wi = individual live weight of species i; b = slope of weight metabolism regression; $a = y$ intercept of weight metabolism regression.

the same in both locations: 420 kcal/ha/day in Panamá and 410 kcal/ha/day in Illinois. Further, the mean weight of the tropical birds was less than the temperate ones. Thus, biomass, energy consumption, and existence energy costs of the bird populations were relatively similar (table 2). Finally, the avifauna in both environments appears to use about 0.2 per cent of the annual net primary production of the forest. Karr concludes that the increase in bird species diversity in tropical forests is due to their variety of additional food resources (such as fruits).

The greatest part of the heterotrophic biomass is in the soil and litter. The mass of the saprophages is 5 to 6 times that of herbivores and carnivores, as shown below:

	Biomass (kg/ha)		
	Soil fauna	Herbivores	Carnivores
Amazon lowland rain forest	165	30	15
Lower montane rain forest	80	25	10
Red mangrove forest	52	8	4

(Note that these data do not coincide exactly with those given earlier because they have been calculated from the range of values given in the lists of the various authors.) This emphasis on decomposition is probably due to both the refractory nature of the wood and leaves in forests as compared with grasslands and the arboreal problem for herbivores which must feed on leaves, flowers and fruits located sometimes as high as 90 metres or more above the soil surface; also to differences of phenology and seasonal availability.

Secondary production of tropical animals has been studied at the community level only in the palm savanna at Lamto, Ivory Coast (Lamotte, 1975), although there are other studies focusing on a species population or set of populations at other tropical research sites; the results will not be reported here as this Report concerns tropical forest ecosystems.

Golley (1972) pointed out that the relationship between biomass of a system compartment and the flux through the compartment is a clue to the role of the compartment in control of ecosystem processes. Compartments with small mass, rapid turn-over time, and large fluxes are most likely to be important in control of processes. There is little information on the per cent of the leaf mass or leaf production directly consumed by herbivores in tropical forests. Odum and Ruiz-Reyes (1970) estimated 7 per cent of the leaf area in lower montane rain forest was consumed by animals, de la Cruz (1964) estimated 6 per cent consumption in Costa Rica, and Golley et al. (1975) estimated 10 per cent in Panamá. Bray and Gorham (1964) examining a variety of mainly temperate forests, concluded that 6–10 per cent of the total leaf surface was consumed by insects annually. In contrast, Klinge and Rodrigues (1973) quote Eidmann (1942, 1943) and Buchler (in Mann, 1968), as well as Hopkins (1967), that 25 to over 60 per cent of the leaf mass may be consumed by herbivores. These rates seem unlikely given the biomass of herbivores found in forests, but they may represent special, local situations.

Thus, in forests there is a large mass of leaf matter of which only a small percentage is consumed. Golley (1973) and others have suggested that certain of the heterotrophs, especially insects, may have characteristics of cybernetic regulators of ecosystem processes, including primary production. For example, phytophagous insects occupy the strategic position of a potential regulator through consumption of foliage which is the plants, primary site of energy capture and biochemical synthesis. Also, species of phytophagous insects are usually coextensive with every plant species so that plants seldom occur without their consumers. Further, these plants and consumers have a long history of association and coevolution. And finally, plants and insects react and respond physiologically, ecologically and evolutionarily to variations in the state of each other. Thus, plants, and other consumers and insects must be viewed as two coevolving, competing, interdependent, ecological subsystems (see chapter 8).

Insects and other consumers respond to a variety of host conditions, such as the nutritional value of the plant, the presence or absence of substances with growth regulator effects on insects and the specificity of volatile components of plant metabolism. There are many examples of temperate insects outbreaks on vegetation under stress of low nutrient or moisture conditions. It is also well known that old, weakened or stressed trees are more susceptible to insect attack than vigorous ones. It seems clear that as the state of the host plants changes due to ageing, climatic or other stresses, the insect consumers can sense the variations and react to them. One of these causal mechanisms might be alterations in the normal sugar or nitrogen levels of the foliage (Schwenke, 1968). See also chapter 14.

After insects respond with increased reproduction to the changed host conditions, a feedback message can be transmitted to the host through the insect activity. For example, severe defoliation disrupts the hosts' physiological status and can cause increased litter fall, resulting in increased nutrient input to the soil-litter system. Weakened and old plants often die after defoliation, causing changes in the physical environment. The net result can be more available nutrients, light and moisture for the surviving plants and probably greater growth. Thus, consumers can help restore primary production and nutrient cycling to more optimal rates. These rate changes can come about through alteration of the efficiency of the photosynthetic process, the mobilization and utilization of stored reserves, interaction and differentiation of bud primordia, and change in the vertical distribution of biomass production. Grazers can interact directly with producer components of tropical forests and significantly affect the other components and thus the small rate of consumption should be considered as representing highly efficient activity of components which have as their main function the control of system processes.

In contrast, the litter in the forest ecosystem must be decomposed rapidly so that the essential elements tied up in the biomass can be released for uptake by the plants. In tropical forests the total litter fall may exceed 20 t/ha/a. This represents a large decomposition task for the system. Fungi are of more importance than in temperate forests and may be the first step in the decomposition process (Went and Stark, 1968; Beck, 1970, 1971). However, many authors have

also demonstrated that termites are important. For example, Maldague (1970) studying evergreen forest in Zaire showed that there were 870 termite colonies per hectare, representing a mass of 16 t/ha. He estimated that 2.6 t per hectare of organic matter was immobilized in these colonies, with 0.7 t of calcium and 61 kg of nitrogen. These termites may have consumed *ca.* 6 t/ha/a. Huttel and Bernhard-Reversat (1975) report similar densities of termites in subequatorial forest in the Ivory Coast. Matsumoto (1974) also estimates that at Pasoh Forest, Malaysia, termites consume 38.8 kg leaf litter per hectare per week or the equivalent of 32 per cent of the leaf-fall rate.

Besides the termites, earthworms and micro-arthropods may also be important in tropical forests. In a subequatorial forest in Ivory Coast located on a thalweg, Huttel and Bernhard-Reversat (1975) report a large number of earthworm casts. These casts represent mechanical movement of soil and mineralization of nitrogen, both of which enrich the surface substantially (table 3). The exact role of micro-arthropods in the surface soil is poorly known and the large numbers extracted from soil and litter suggest that even larger numbers are present (table 4).

TABLE 3. Comparison of element content of surface soil and worm casts (Huttel and Bernhard-Reversat, 1975).

| | C ‰ | N total ‰ | P total ‰ | Base exch., m. equiv./100 g soil | | | | pH |
				Exch. cap.	Mg	K	Ca	
worm casts	30.0	1.88	0.69	7.5	0.61	0.16	0.79	4.3 ± 0.2
	24.5	1.33	0.83	6.5	0.37	0.12	0.34	
soil	13.2*	0.96*	0.45	4.6	0.33	0.08	0.30	4.3 ± 0.2
0–10 cm			0.37	4.8	0.14	0.19	0.39	

* annual mean

TABLE 4. Number of micro-arthropods per m² in tropical forest in the surface soil.

Banco, plateau } (Ivory Coast; Huttel	54 000
Banco, thalweg } and Bernhard-Reversat, 1975)	17 500
Yapo }	26 000
Zaire (Maldague, 1970)	64 000–72 000
Nigeria (soil and litter) (Madge, 1969)	65 200
Puerto Rico (Wiegert, 1970)	1 890– 7 620
Amazonia (Beck, 1971)	14 700

Research needs and priorities

The entire subject of heterotrophic ecology needs greater attention. While the study of so-called secondary production is of little value to understanding the ecological system, measures of the flow of energy and materials through the food web are very important. Ideally research would proceed in the following steps:

— sampling to demonstrate the numbers of different species of heterotrophs;
— identification of all heterotrophs from these samples;

— determination of the feeding habits of heterotrophs;
— construction of a food web showing spatial and temporal changes in feeding;
— measurement of the amount of food consumed and, thus, the flows in the food web;
— determination of energy and mineral contents of foods and expression of the flows in these terms;
— integration of functional data with those concerning vegetation to give the entire system dynamics.

It is a measure of the task, as nowhere in the tropics there has ever been a truly adequate food web description. The problem is due partly to the fact that few animal ecologists or systematists are trained to study all taxa. Rather, teams of researchers are needed but when these are developed it is seldom possible to prevent each specialist from making his studies in a traditional and separate way, with the result that the overall study ends up with a great number of individual projects in various stages of maturity, which cannot be integrated together. Possibly the solution to specialization is to focus on the flows in the food web rather than the species; this does not avoid the need to know thoroughly the biology of the species (see chapters 6 and 7).

Finally, it should be emphasized that there is another whole sequence of important functions that heterotrophs perform which are associated with the feeding activities but which do not figure in energy and material flow calculations. These activities include pollination, seed dispersal, and seed and fruit predation, and appear to be exceedingly important in tropical forests. Very little is known about their biology and nothing about their quantitative significance to ecosystems. This is a major gap and should be an area of active research in the future (see chapter 8).

Summarizing, there is an immense literature available on features of the biology of individual heterotrophs in tropical forests—these data have been accumulating as long as scientists and naturalists have visited the tropics. However, a complete study of the kinds of heterotrophs in one forest does not exist. Research should, therefore, be focused on a few sites where complete studies can be made. Any but the most general speculation about heterotroph function is not warranted until these studies can be completed.

The entire area of heterotrophic activity needs expanded attention by research workers. Studies need to be carried out in a system context focusing on food chains and other interactions as well as the dynamics of the component of interest. For example, a study of the metabolism of a population should include an evaluation of the source of energy to the population. Aside from increased research on standing crops and input and output from forest populations, it is also essential to consider other functions of animals in the system. The input-output of energy and materials for herbivores is relatively unimportant compared to that of soil-litter fauna, but the functional roles of the herbivores may be exceedingly important. Very little is known about the quantitative character of these functional relationships.

Bibliography

BECK, L. *Zur Ökologie der Bodenarthropoden im Regenwaldgebiet des Amazonasbeckens.* Bochum, Ruhr-Universität, Habilitationsschrift, 1970.

——. Bodenzoologische Gliederung und Charakterisierung des amazonischen Regenwaldes. *Amazoniana,* 3, 1971, p. 69–132.

BOURLIÈRE, F. Observations on the ecology of some large African mammals. *African Ecol. Human Evol.,* 36, 1963, p. 43–54.

BRAY, J. R.; GORHAM, E. Litter production in forests of the world. In: Cragg, J. B. (ed.). *Advances in ecological research,* vol. 2, p. 101–157. London, Academic Press, 1964.

CRUZ, A. de la. A preliminary study of organic detritus in a tropical forest ecosystem. *Rev. Biol. Trop.,* 12, 1964, p. 175–185.

EIDMANN, H. Der tropische regenwald als Lebensraum. *Kolonial forstl. Mitt.,* 5, 1942, p. 91–147.

——. Zur Ökologie der Tierwelt. *Beitr. Kolonial forstl.,* 2, 1943, p. 25–45.

FITTKAU, E. J.; KLINGE, H. On biomass and trophic structure of the Central Amazonian rain forest ecosystem. *Biotropica,* vol. 5, no. 1, 1973, p. 2–14.

FLEMING, T. H. Notes on the rodent fauna of two Panamanian forests. *J. Mamm.,* 51, 1970, p. 473–490.

——. Population ecology of three species of neotropical rodents. *Misc. Publ. Mus. Zool.* (Univ. Mich.), 143, 1971, p. 1–77.

——. Aspects of the population dynamics of three species of opossums in the Panama Canal Zone. *J. Mamm.,* 53, 1972, p. 619–623.

——. The role of small mammals in tropical ecosystems. In: Golley, F. B.; Petrusewicz, K.; Ryszkowski, L. (eds.). *Small mammals, their productivity and population dynamics,* p. 269–298. Cambridge, Cambridge Univ. Press, 1975.

GIST, C. S. *Some tropical modelling efforts.* Utah State University (Logan, Utah), Ecology Center, 1973, multigr.

GLIWICZ, J. A short characteristic of a population of *Proechimys semispinosus* (Towers, 1860), a rodent species of the tropical rain forest. *Bull. Acad. Pol. Sci., Biol. Sci.,* 21, 1973, p. 413–418.

GOLLEY, F. B. Secondary productivity in terrestrial communities. *Am. Zoologist,* 8, 1968, p. 53–59.

——. Summary. In: Golley, P. M.; Golley, F. B. (eds.). *Tropical ecology with an emphasis on organic production,* p. 407–413. Athens, Univ. Georgia, 1972, 418 p.

GOLLEY, F. B. Impact of small mammals on primary production. In: Gessaman, J. A. (ed.). *Ecological energetics of homeotherms, a view compatible with ecological modelling,* p. 142–174. Utah State University (Logan), Mono. Series, vol. 20, 1973.

——; ODUM, H. T.; WILSON, R. F. The structure and metabolism of a Puerto Rican red mangrove forest in May. *Ecology,* 43, 1962, p. 9–19.

——; McGINNIS, J. T.; CLEMENTS, R. G.; CHILD, G. I.; DUEVER, M. J. *Mineral cycling in a tropical moist forest ecosystem.* Athens, Georgia, Univ. of Georgia Press, 1975, 248 p.

HARRISON, J. L. The abundance and population density of mammals in Malayan lowland forests. *Malayan Naturalist Journal,* 22, 1969, p. 174–178.

HENDRICHS, H. Schätzungen der Huftierbiomasse in der Dornbuschsavanne nördlich und westlich der Serengeti steppe in Ostafrika nach einem neuen Verfahren und Bemerkungen zur Biomasse der anderen pflanzenfressenden Tierarten. *Säugetierkund. Mitt.,* 18, 1970, p. 237–255.

HOPKINS, B. A comparison between productivity in forest and savanna in Africa. *J. Ecol.,* 55, 1967, p. 19–20.

HUTTEL, C.; BERNHARD-REVERSAT, F. Biomasse végétale et productivité primaire, cycle de la matière organique. In: Recherches sur l'écosystème de la forêt sub-équatoriale de basse Côte-d'Ivoire, p. 203–228. *La Terre et la Vie* (Paris), 29, 1975, p. 169–264.

KARR, J. Production, energy pathways and community diversity in forest birds. In: Golley, F. B.; Medina, E. (eds.). *Tropical ecological systems: trends in terrestrial and aquatic research,* p. 161–176. Berlin and New York, Springer Verlag, Ecological Studies no. 11, 1975, 398 p.

KLINGE, H.; RODRIGUES, W. A. Phytomass estimation in a Central Amazonian rain forest. In: Young, H. E. (ed.). *IUFRO Biomass Studies,* p. 337–350. Orono, Univ. Maine, 1973.

LAMOTTE, M. The structure and function of a tropical savanna ecosystem. In: Golley, F. B.; Medina, E. (eds.). *Tropical ecological systems: trends in terrestrial and aquatic research,* p. 179–222. Berlin and New York, Springer Verlag, Ecological Studies no. 11, 1975, 398 p.

LAMPREY, H. F. Estimation of the large mammal densities, biomass and energy exchange in the Tarangire game reserve and the Masai Steppe in Tanganyika. *East African Wildlife J.*, 2, 1964, p. 1–46.

MADGE, D. S. Field and laboratory studies on the activity of two species of tropical earthworms. *Pedobiologia*, 9, 1969, p. 188–214.

MALAISSE, F.; ALEXANDRE, J.; FRESON, R.; GOFFINET, G.; MALAISSE-MOUSSET, M. The miombo ecosystem: a preliminary study. In: Golley, P. M.; Golley, F. B. (eds.). *Tropical ecology with an emphasis on organic production*, p. 363–403. Athens, Univ. Georgia, 1972, 418 p.

MALDAGUE, M. E. *Rôle des animaux édaphiques dans la fertilité des sols forestiers*. Publ. INEAC, Sér. sci., n° 112, 1970, 245 p.

MANN, G. Die Okosyteme Sudamerikas. In: Fittkau, E. J.; Illies, J.; Klinge, H.; Schwabe, G. H.; Sioli, H. (eds.). *Biogeography and ecology in South America*, vol. 1, p. 171–229. The Hague, Junk, Monographiae Biologicae no. 18, 1968, 446 p.

MATSUMOTO, T. The role of termites in the decomposition of leaf litter on the forest floor of Pasoh Study Area. In: *IBP synthesis meeting* (Kuala Lumpur, August 1974), 7 p. multigr.

McCULLOUGH, D. R. Secondary production of birds and mammals. In: Reichle, D. E. (ed.). *Analysis of temperate forest ecosystems*, p. 107–130. Berlin and New York, Springer Verlag, Ecological Studies no. 1, 1970.

McNEIL, S.; LAWTON, J. H. Annual production and respiration in animal populations. *Nature*, 225, 1970, p. 472–474.

MEDINA, E.; ZELVER, M. Soil respiration in tropical plant communities. In: Golley, P. M.; Golley, F. B. (eds.). *Tropical ecology with an emphasis on organic production*, p. 245–269. Athens, Univ. Georgia, 1972, 418 p.

MONTGOMERY, G. G.; SUNQUIST, M. E. Impact of sloths on neotropical forest energy flow and nutrient cycling. In: Golley, F. B.; Medina, E. (eds.). *Tropical ecological systems: trends in terrestrial and aquatic research*, p. 69–98. Berlin and New York, Springer Verlag, Ecological Studies no. 11, 1975, 398 p.

ODUM, H. T.; ABBOTT, W.; SLEANDER, R. K.; GOLLEY, F. B.; WILSON, R. F. Estimates of chlorophyll and biomass of the Tabanuco forest of Puerto Rico. In: Odum, H. T.; Pigeon, R. F. (eds.). *A tropical rain forest*, p. I–3 to I–19. Division of Technical Inf., United States Atomic Energy Commission (USAEC), 1970, 1 678 p.

——; RUIZ-REYES, J. Holes in leaves and the grazing control mechanism. In: Odum, H. T.; Pigeon, R. F. (eds.). *A tropical rain forest*, p. I–69 to I–80. Division of Technical Inf., USAEC, 1970, 1 678 p.

PESSON, P. (ed.). *Ecologie forestière. La forêt : son climat, son sol, ses arbres, sa faune*. Paris, Gauthier-Villars, 1974, 282 p.

SCHWENKE, W. New indicators of the dependence of population increases of leaf and needle-feeding forest insects on the sugar content of their diet. *Z. fur Angew. Entomol.*, 61, 1968, p. 365–369.

WANNER, H. Soil respiration, litter fall and productivity of tropical rain forest. *J. Ecol.*, 58, 1970, p. 543–547.

WENT, F. W.; STARK, N. The biological and mechanical role of soil fungi. *Proc. National Acad. Sci.* (Washington), 60, 1968, p. 497–504.

WIEGERT, D. F. Effects of ionizing radiation on leaf fall, decomposition and the litter micro-arthropods of a montane forest. In: Odum, H. T.; Pigeon, R. F. (eds.). *A tropical rain forest*, p. H–89 to H–100. Division of Technical Inf., USAEC, 1970, 1 678 p.

——; EVANS, F. C. Investigations of secondary productivity in grasslands. In: Petrusewicz, K. (ed.). *Secondary productivity of terrestrial ecosystems*, p. 499–518. Warsawa, Polish Academy of Sciences, 1967, 2 vol., 879 p.

12 Water balance and soils

Water balance

The water balance of an area covered with forest can be expressed in terms of the two generally applicable formulae:

$$(1) \qquad P = ETR + R + D \pm \Delta H$$

in which P, precipitation above the forest
ETR, actual evapotranspiration
R, run-off
D, infiltration
ΔH, variations in available soil water;

$$(2) \qquad P = P_{sol} + E_t + I$$

in which P, precipitation above the forest
P_{sol}, throughfall
E_t, stemflow
I, interception.

Rainfall

The rainfall regimes of the humid tropical zones and their characteristically large between year variations are well known.

A study undertaken in Indonesia by Schmidt and Ferguson (1951) is a good illustration of the use of this knowledge. They made a climatic map of the area on the basis of the relationship between the number of dry and wet months during the year. This gave a measure of the water deficit using the simplest type of meteorological station. The areas thus identified showed that evergreen forest formations are associated with climates having only a small number of dry months. Whitmore (1975) has used the same index in Malaysia. Gaussen, Legris and Blasco (1967) have studied Indochina, Thailand and the Malay peninsula using slightly different criteria: they counted the number of dry months in which the rainfall (in mm) was less than twice the temperature (in degrees Celsius). Studies of the same kind were carried out in Africa by Aubréville (1949) who considered all months receiving less than 30 mm of rain as dry and those which received more than 100 mm as very moist. There are few such studies in Latin America. However, very little is known of the precipitation which is received within tropical forests due to the very small number of permanent measuring points. It seems that the whole of Africa and Madagascar only have one properly equipped station (Périnet forest in Madagascar) and a few other partially equipped ones (in the Ivory Coast and Nigeria). Similar stations in South-East Asia are equally rare (in the Philippines and the Pasoh station, Malaysia), especially in the very moist dense evergreen forests. Three forest stations have functioned in French Guyana: two run by ORSTOM (Crique Virgile and Grégoire-forêt) and one by the National Meteorological Service at Maripasoula-forest; at Crique Virgile, an observation tower which reached 30 m, i.e. the average level of the tree crowns, was used, and measurements could be performed simultaneously near the soil and in the highest part of the canopy. An effort should be made to perfect long-term automatic data recording apparatus capable of withstanding the hot and humid climate, including radio or satellite transmitting systems.

Sustained attention should also be given to problems of the number, siting and representativeness of stations (whether in clearings, beneath forest, etc.). One can only surmise that precipitation has been grossly underestimated when run-off measurements attain values of 70 and even 90 per cent of rainfall. There is a methodological problem in this field which will not be solved until the design of meteorological stations within the humid tropical forest is carefully detailed.

Throughfall

Throughfall is derived more from calculation than from direct measurement because of methodological inadequacies. Estimates of the amount of water reaching the ground by leaf drainage are in the order of 70–80 per cent of annual precipitation. This was inferred for the semi-deciduous forest of Indonesia by Mohr and Van Baren (1954). It is confirmed by calculations made in Africa, where Malaisse (1973) shows that penetration by leaf drainage in the Zaire miombo (woodland with *Brachystegia* and *Isoberlinia*) amounts to 890 mm and 1 148 mm in the years 1973 and 1974, in which the respective rainfall was 1 105 mm and 1 463 mm. This fact is also confirmed in India at Dehra Dun where respective throughfall of 69.7 and 64.1 per cent of annual rainfall was measured in plantations of *Pinus roxburghii* and *Tectona grandis* (Dabral and Rao, 1968). Finally, in dense evergreen forest at Pasoh in Malaysia, 91 per cent of the annual rainfall of 2 000–2 400 mm is throughfall. It is probable that these percentages give a close enough approximation to reality.

It would be of considerable interest to have direct measurements by a method recently perfected at Banco forest in the Ivory Coast (Huttel, in Bernhard-Reversat, Huttel and Lemée, 1975). Precipitation is collected in rectangular troughs with a collecting surface of 500 cm² and a cross-section designed to avoid loss by splashing. These are placed 50 cm above the ground and water from them is collected in 25 litre containers. Precision in weekly measurements is such that it approaches 0.1 mm of rainfall. The number of such pluviometers located at random within the Banco forest was raised from 3 to 12 and then 24 since it is only above 12 that a precision of 10 per cent of the mean is reached, when precipitation is greater than 5 mm.

Teoh (in Anon., 1974) has shown that in the case of a 6.5-year-old rubber plantation at Pasoh the mean percentage of throughfall of monthly rainfall decreases with increasing density of trees whereas the mean monthly stemflow increases.

Stemflow

Stemflow was measured for over a year in the dense evergreen forest of Banco (Huttel, 1962) on a plot of 300 m² bearing 16 trees, each provided with a spiral plastic gutter at the base of the trunk. Stemflow was less than 1 per cent of throughfall. Sollings and Drewry (1970) obtained the same result in Puerto Rico by comparing the leaf drainage from isolated trees with consequent stemflow. Nye (1961) in a semi-deciduous forest in Ghana and Malaisse (1973) in a woodland (miombo) in Zaire respectively found values of 1 per cent and 0.4–1.4 per cent. Lastly, Chunkao (in Anon., 1974) showed a value of 0.5 per cent for the dense evergreen forest of Pasoh. In contrast, higher values have been obtained: 28 per cent in Brazil (Freise, 1936) and 18 per cent on a group of 27 trees in Puerto Rico (Kline, Jordan and Drewry, 1968). Intermediate values of 3.3 and 7.1 per cent were found in India by Dabral and Rao (1968) under *Pinus roxburghii* and *Tectona grandis* and 7.2 per cent under *Shorea robusta* (1969).

The contradictions disclosed by these result lead to consideration of the need to follow up such measurements using standard methods for it is probable that the principal reason for the variation shown lies in the diversity of methods used (in particular the number and spacing of trees). It would also be advisable to take into account the magnitude and intensity of rainfall and the effect of wind on the angle of fall of precipitation, etc. Dabral (1967) shows, in a very preliminary study, that for the same amount of water, stemflow decreases with increasing deviation of rainfall under a plantation of *Tectona grandis*, this showing the influence of rainfall intensity.

Interception

Interception is the quantity of water held by the vegetation, that which does not reach the ground.

The proportion of precipitation intercepted by tropical and subtropical forests is very variable: 3 per cent in Nigeria (Hopkins, 1965), 5 per cent in Costa Rica (McColl, 1970); 8 per cent in Malaysia (Chunkao, in Anon., 1974); 12 per cent in the Ivory Coast (Huttel, 1962); 12–26 per cent in Puerto Rico (Kline, Jordan and Drewry, 1968); 18–20 per cent in Zaire (Malaisse, 1973); 31 per cent in Mauritius (Vaugham and Wiehe, 1947); 35 per cent in Uganda (Hopkins, 1960); 38 per cent in India in a 37-year-old plantation of *Shorea robusta* (Dabral and Rao, 1969) and even 65–70 per cent in subtropical forest in Brazil (Freise, 1936). These represent means calculated over different durations and are derived from comparison between throughfall and rainfall measured at points close to the area of study. This variation is not surprising since rainfall interception is, among other things, a function of the stand characteristics as Czarnowski and Olzewski (1968) have shown in the following formula:

$$I—f(H\sqrt{SN}),$$

in which I, interception, is a function of
H, the height of the stand
S, the basal area of the stand
N, the stem density.

It seems, nevertheless, that research in this field has not been advanced enough. The variability of rainfall within an area should be the object of more attention when the comparative methods outlined above are applied. More precise studies, such as those on the effect of the length and intensity of rainfall on interception (Malaisse, 1973), and of the interception by the understorey and the herb stratum need to be carried out.

Evapotranspiration

Direct measurement of evapotranspiration (ETR) is very difficult. Because of this, potential evapotranspiration is often used during periods when $P>ETP$ since ETP and ETR are then very similar. In dry periods the available soil water or the difference between precipitation (P) and the soil moisture content during a given period are used to calculate ETR. The actual annual evapotranspiration of a number of tropical and subtropical forests was thus evaluated.

— Evergreen dense forest of Banco (Ivory Coast): *ca.* 1 150 mm (Huttel, 1962).
— Evergreen semi-deciduous forest at Yapo (Ivory Coast): 1 168 mm (Huttel, 1962).
— Woodland in Zaire (miombo): 1 050 mm (Malaisse, 1973).
— Woodland in northeastern Thailand: 948 mm (Sabhasri *et al.*, 1970).
— Plantation of *Eucalyptus* (Bengal): 1 136 mm (Banerjee, 1972).

These values agree with a direct measurement of transpiration and an estimation of soil evaporation of a tropical montane forest in Java (Coster, 1937): 870 mm and 200 mm respectively, giving a total of 1 070 mm annual evapotranspiration.

Formulae have been frequently used to evaluate potential evapotranspiration (Thornthwaite, 1948; Mohr and Van Baren, 1954; Turc, 1961). Even if their degree of accuracy can be questioned, at least they demonstrate variation of the phenomenon at a regional level although adjustment from one region to another is required.

Very few tropical countries have undertaken evapotranspiration studies within forests: these include Indonesia, Thailand and Malaysia in South-East Asia and the Ivory Coast and Zaire in Africa. This shows just how much a geographical extension of these studies is necessary, especially as additional studies should be undertaken on different types of forest.

Research on the study of vegetation transpiration need considerable methodological perfection (type of equipment, measuring techniques, interpretation of results). A network of evaporation measuring stations should always be linked to meteorological stations.

Run-off and drainage

This is probably one of the elements in the water balance of tropical forests for which information is the least well developed and the least precise. Studies such as those of Ruangpanit (1971) in Thailand, which have shown the fall-off in run-off as a function of increase in area covered by vegetation, are very scarce. This showed the importance of 60–70 per cent vegetation cover as well as the role of the intensity and duration of the precipitation on the volume of run-off (table 1).

TABLE 1. Relationship between the percentage of leaf cover and run-off recorded for three classes of rainfall intensity and duration.

Percentage of leaf cover	Run-off (m³/ha) for 3 classes of rainfall intensity		
	0–10 mm/h	10–20 mm/h	>20 mm/h
20–30	3.1	6.5	8.0
50–60	2.7	6.5	7.8
60–70	1.7	4.1	5.1
80–90	1.4	3.6	5.1

Percentage of leaf cover	Run-off (m³/ha) for 3 classes of rainfall duration		
	0–60 min.	60–180 min.	>180 min.
20–30	2.6	4.7	7.4
50–60	2.6	4.6	6.4
60–70	1.4	2.7	4.4
80–90	1.4	2.3	4.3

Run-off measurements beneath forest have been evaluated from watercourses. These estimates would lead one to think that run-off is high in tropical humid areas. This is particularly true in the very wet areas where annual precipitation is over 2 500 mm because the coefficient of flow of watercourses then oscillates at around 50 to 70 per cent. In India, Rao, Dabral and Ramola (1973) measured a run-off of 41 per cent of rainfall under *Shorea robusta* at Dehra Dun where the mean annual precipitation is 2 427 mm.

However, in most lowland forests which receive less rainfall (1 600–2 200 mm), annual run-off is rarely above 50 per cent of the rainfall and may be less than 10–30 per cent, as indicated by the Hydrological Service of ORSTOM for Africa. In Madagascar, on basins of *ca.* 15 ha, at the Forest Station of Périnet (1 800 mm/a), Bailly *et al.* (1974) have measured run-off rates of *ca.* 36 per cent under natural forest, 20 per cent under *Eucalyptus* plantation and 56 per cent under a secondary forest. On basins of *ca.* 1 km², these annual run-off rates were of 43 per cent and 58 per cent under natural and secondary forest respectively.

But on the other hand, studies of run-off immediately beneath a forest cover on various slopes undertaken by

Roose (1967) at Adiopodoumé (Ivory Coast) using plots of a new type, showed that this never exceeds 1 per cent. Thus water probably reaches the rivers mainly through vertical or oblique internal drainage. The importance of these factors is stressed by Douglas (1971) for Malaysia. On very small basins (1.5–2 ha), the scientists of the Centre technique forestier tropical (Bailly *et al.*, 1974) have shown that the secondary forest in Madagascar had only 3 per cent run-off of the annual rainfall over 9 years, a considerably smaller value than for cleared areas.

The investigation of the water regime of soils beneath forest in the intertropical zone was generally limited to that of the rhizosphere. Excluding a few rare exceptions (such as that of Roose in the Ivory Coast), little study has been made of the total soil profile. The perfection of neutron soil moisture probes (neutron scattering method) has not changed this. This is not due either to an underestimation of the role of water as a principal agent in soil formation since it is, for example, given adequate consideration in the study of both soil solutions and hydrolysis. In reality, it seems that it has been preferred to infer the role of soil water from the transport of material resulting from the movement of water.

The behaviour of the soil-forest complex remains poorly understood. The composite role of soil should be taken into account in remedying this gap. The absence or scarcity of a litter layer means that the upper layers of the soil play a fundamental role in the division between run-off and infiltration. Structural characteristics are more important than grain size, because of the kaolinitic nature of the clay particles and their weak capacity for water retention. This is well understood but has not yet been the subject of any studies in depth. On identical experimental plots, Roose (1967) in the Ivory Coast found that run-off was 1 per cent of annual rainfall whereas Blancanneaux (1974) in French Guyana found it to be 56 per cent due to a massive, compact, impervious B horizon in the forest soils there. A detailed analysis of the internal structure of material should throw some light on this question. A joint approach of hydrological specialists and soil physicists is recommended here.

Variations in available soil water

The study of soil water variations is the field of the soil physicist and geographer who wishes to compare the dynamics of water in soils of savannas and of forests. One of the most recent studies in this subject was completed in the Ivory Coast (the Banco and Yapo forests) with the help of neutron probes. Huttel (1962) stresses the effect of the size of soil aggregates on this phenomenon. Where there is a homogenous texture, the range of variation in the water content of the profile decreases regularly with depth; it is higher in horizons where root activity is present. If the lower horizons are clayey this decrease is less evident. If the upper horizons become very coarse (gravel for example) they can release much more water than the lower horizons.

Avenard (1971) stated that the processes of soil moisture change are slower beneath forest than beneath savanna and progress downwards, because the root systems slow down infiltration and water does not penetrate in the deep horizons until after the retention capacity of the surface ones has been exceeded.

The gaps in knowledge are much the same as those indicated for run-off and drainage. It is to be hoped that soil physicists will devote more attention to this matter, for which information does exist, but in a very dispersed form, and is often obtained from more detailed studies on cultivated land.

Conclusions: research needs and priorities

Complete and coherent studies of the hydrological dynamics beneath different types of forest and under different climatic regimes are lacking. This information is necessary so as to provide those studying cultivated lands with the references to the original environments, to enable comparison to be made between the present state of such environments and their initial dynamic equilibrium.

In order to achieve this, a program of research with the following priorities is recommended.

1. An increase in the number and density of stations giving hydropluviometric measurements within the tropical forest ecosystems. For this considerable methodological efforts need to be carried out:
   — perfection of automatic data recording and long distance transmission apparatus for rainfall measurement, capable of withstanding the climate;
   — design of methods and equipment for direct measurement of throughfall, stemflow and interception. This presupposes a detailed study of the siting of such equipment.
2. New studies on evapotranspiration through increasing:
   — stations measuring soil evaporation;
   — measurement of vegetation transpiration, which requires methodological improvement. In this field it seems that Africa and South-East Asia have already made some progress which is lacking in America.
3. Studies of run-off and internal soil drainage beneath forests using experimental plots of the type perfected by Roose at the ORSTOM station at Adiopodoumé (Ivory Coast). These do not appear to exist outside French-speaking western Africa and this is a large gap to be filled.
4. Additional studies of available soil water, using the neutron soil moisture probe.

Soils

A first important property of tropical soils is the extent of alteration of their parent rock (Aubert, 1965). This largely arises from the rapid rate of rock decomposition due to high temperature and humidity. These soils have developed over a great depth due to the absence of Quarternary glaciations and the protection of the forest.

They have a low mineral content, with kaolinite predominant within their clay fraction, giving a weak capacity for cation adsorption. They are rich in silica and free iron oxides and the majority contain equally large quantities of free alumina which sometimes is dominant. They are generally acid and poor in macro-nutrients. Nevertheless the presence of the forest modifies this; the upper horizon is enriched with organic matter and bases through the decomposition of plant litter. Finally, these soils are generally well drained and their structure involves good aeration, but once the forest disappears they are very susceptible to leaching.

These soils have been given various names, including latosols, ferrallitic soils, kaolisols, oxisols and inceptisols.

All tropical soils did not develop on old lanscapes; low-lying areas and mountains are sources of differentiation. A more or less lengthy dry season changes the conditions of soil development and evolution; unfavourable structures may occur during the dry season and erosion can start more easily. Iron oxides dominate largely; these have been called tropical ferruginous soils, leached ferrallitic soils, red-yellow podzolics, ultisols or grey podzolics. They are given less consideration in this chapter which concerns primarily forest soils.

Physical properties

Soil temperature

The temperature of the soil beneath humid tropical forest decreases during the hot season and increases during the less hot season with regard to air temperature. The range of thermal fluctuation as well as seasonal fluctuations are reduced. This is most clearly shown in the upper 30 cm of the soil as indicated by Kittredge (1962) and confirmed by observations made beneath humid deciduous forest at Dehra Dun, India (Dabral *et al.*, 1969).

Soil structure

Soil structure beneath forest is generally good and maintains a good infiltration level. It is only when the vegetation is removed that problems arise due to their exposure to high temperatures and intense rainfall.

French and English-speaking Africa (Pereira, 1954, 1956; Hénin *et al.*, 1958; Monnier, 1965, among others) and the Pacific islands, especially Hawaii (Uehara *et al.*, 1962; Cagauan and Uehara, 1965) give the basic sources for description of the soil structure in the humid tropics. Uehara and his colleagues have linked the formation of stable aggregates of oxisols, ultisols and inceptisols to the nature and structure of their minerals: presence of clays of a 1:1 type, mainly of kaolinite and halloysite, hydrated and non-hydrated iron oxides and alumina, concentrated in a mineral fraction less than 2 μ and present in the form of coatings. Combeau and Quantin (1963) stress the beneficial role of organic matter in the stability of soil structure. This results in a crumb structure with strong cohesion in the upper horizons of soils beneath forest and in other parts of the tropics with similar conditions (no explanation has been given for the existence of such a structure).

French and English soil scientists, in the course of their studies in the African continent, have supplied the main methods used for defining structure. Measurements of water-stable aggregates have been shown to be insufficient and Pereira perfected a test of reaction to water-drops to evaluate structure. French pedologists use a coefficient of instability developed by Hénin *et al.* (1958) which is expressed as:

$$S = \frac{A+L}{Ag_{air}+Ag_{alcohol}+Ag_{benzene}}$$

in which $A+L$, percentage of clay and silt

Ag_{air}, percentage of stable aggregates in water without pre-treatment

$Ag_{alcohol}$, percentage of stable aggregates after pre-treatment with alcohol

$Ag_{benzene}$, percentage of stable aggregates after pre-treatment with benzene.

S varies form 0.1 to 100 and the stability is large for values less than unity. Values of 0.3–0.7 are commonly recorded for the upper horizons of forest soils.

Porosity

One of the very first measurements of the porosity of a tropical soil which had not been subjected to major anthropogenic modifications was carried on a lateritic soil now called oxisol (Roberts, 1933). Its total porosity equalled 57 per cent in the first 12 cm and 47 per cent between 75 cm and 150 cm depth; hence a satisfactory permeability. This judgment was borne out and confirmed in detail by recent measurements in Hawaii (Baver and Trouse, 1965) and by others in Thailand (Ruangpanit, 1971); see table 2.

Oxisols have a high permeability but that of tropical inceptisols (Tropepts) is higher. The dense horizon present at depth in ultisols and in some ferrallitic soils suppresses

TABLE 2. Bulk density and total porosity of forest soils within the humid tropics.

Soils	Bulk density (g/cm³)		Total porosity (%)	
	Area	Depth	Area	Depth
Oxisols (Low humic latosols) (Hawaii)	1.1	1.2	62	59
Inceptisols (Humic lotosols) (Hawaii)	0.8	0.8	70	70
Inceptisols (Hydrohumic latosols) (Hawaii)	0.5	0.2	82	93
Ultisols (Humic ferruginous latosols) (Hawaii)	1.2	2	60	
Soil beneath humid forest (Thailand)	0.48	0.67		

this favourable characteristic in the soils of humid tropical forests and restricts root development.

Under the drier forests of Thailand or those of a subtropical type in Japan (Okinawa island), the porosity decreases: the soils of Thailand (Chunkao, in Anon., 1974) have bulk densities which vary between 1.37 and 1.80 g/cm³ and those of Japan a total porosity of *ca.* 50–55 per cent across the profile.

This high permeability is due to a large number of micropores which do not close when the profile is moistened. This is shown by Roberts (1933), for Puerto Rican soils by Smith and Cernuda (1952) and for those of Brazil (Grohmann and Conagin, 1960).

Finally, analyses made in Puerto Rico, South America and Africa show the great stability of aggregates in the surface horizons of the soil. At Puerto Rico this is 10–500 times higher than that of aggregates of the lower horizons (Smith and Cernuda, 1952; Lugo-Lopez and Juarez, 1959). In South America it is attributed to the presence of organic colloids originating from aerobic decomposition of cellulose (Molina and de Giuffre, 1961). The stability of aggregates of the forest soils of Africa is equally dependent on organic matter (Boyer and Combeau, 1960) through creation of a clay-humus complex in which the two elements provide mutual protection.

Hydrodynamic properties

These have already been discussed with reference to the water balance. The high rate of percolation resulting from a favourable structure should be stressed. The mean permeability for ferrallitic soils in Africa rises to 14 cm/h at the surface and remains 6.5 cm/h at depths. Bonnet (1968), in Puerto Rico, observed infiltration rates of 13.3, 15.4 and 23.7 cm/h for the upper horizons of inceptisols, oxisols and ultisols respectively. The disappearance of the forest and bringing the soils into cultivation decreased these values within two or three years due to the dispersion of the clay fraction and the diminution of porosity, especially the macropores (Boyer and Combeau, 1960); cases have been cited where infiltration fell to below 3 to 4 cm/h.

The role of hydromorphy and texture (as a factor of water retention) needs to be underlined as regards the distribution of humid tropical forests.

In fact, dense evergreen forest and evergreen semi-deciduous forest show, often for geomorphological reasons, a particular physiognomy in very wet areas or waterlogged soils. Lebrun and Gilbert (1954) have given evidence of this fact in Zaire where they distinguish: (1) swamp forests on water-saturated soils, such as forests of *Symphonia globulifera* and *Mitragyna ciliata*; (2) riverine forests in areas where the water-table has considerable variation; (3) periodically flooded forests and (4) valley forests withstanding occasional floods of short duration. The *Mora excelsa* forest, described in Guyana by Davis and Richards (in Lemée, 1961), belongs to group 3 and that of *Eugenia heyneana* and *Terminalia glabra*, described by Misra (in Lemée, 1961) from the Ganges valley, would appear to belong to groups 4 and 2.

Table 3 summarizes the relations between soil texture and forest type. Although such relations are not always obvious in well drained areas, it is clear that the available water for plant growth will be lower in sandy than in clay soils.

TABLE 3. Distribution of some forest types in relation to soil texture (in Lemée, 1961).

	Pure white sand	Coloured sands	Sandy-loam soils	Clay soils
Guyana (Davis and Richards)	*Eperua falcata* forests (Wallaba forest)	*Ocotea radiaei* forest	*Eschweilera-Licania* mixed forest	*Mora gonggrijpii* forest
Amazonia, Amapa area (Aubréville)	*Humiria* low forest	—		*Dinizia excelsa* forest
Ivory Coast (Mangenot)	—	Turraeantho-Heisterietum	Diospyro-Mapanietum	—
Borneo (Kostermans)	*Shorea longifolia* forest	*Koompassia malaccensis* forest		*Koompassia excelsa* forest
Sarawak (Richards)	*Agathis* heath forest	—		Mixed dipterocarp forest
Central Viet-Nam (Schmid)	—	—	Dipterocarp forest	Forest with Meliaceae and Sapindaceae

Chemical properties

Organic matter and nitrogen

All authors are unanimous in recognizing that forest soils of the intertropical zone have a high organic matter content. Bartholomew (1972) has indicated this in *Soils of the humid tropics* as have Charreau and Fauck (1970) and Babalola and Cheheda (1972) for Africa; Hardon (1936) for Indonesia; Khemnark et al. (1972) for Thailand; as well as Baver (1970) for Hawaii. The organic content increases, as Khemnark et al. (1972) have shown, from dry and semi-deciduous forests to evergreen forests (table 4).

TABLE 4. Organic matter content of the upper (15 cm) soil horizon under different forests in Thailand (Khemnark *et al.*, 1972).

Forest type	Organic matter (per cent dry weight)
Montane evergreen forest	11.0
Lowland evergreen forest	9.5
Semi-deciduous teak forest	7.3
Dipterocarpus dry forest	5.7
Dipterocarpus and pine dry forest	3.7

This organic matter is generally well decomposed and has a C/N ratio of *ca.* 12.

In some conditions, mainly those associated with waterlogging, the accumulation of undecomposed organic matter leads to the formation of peat. Where there is good drainage, strong leaching and acid parent rocks, the formation of much less decomposed organic matter leads to the appearance of podzols, as in Latin America (Amazonia and Guyana) and the Far East.

The high organic content decreases considerably after clearing and development of agriculture. Charreau and Fauck (1970) indicate losses of 20–35 per cent during the first year of exploitation in Africa. Babalola and Cheheda (1972) give values as high as 63 per cent loss after several years of cultivation.

Hardon (1936) indicated two other factors which are now better understood. The first is the fall in the C/N ratio in the lower levels of the soil and also its indifference, at this depth, to the type of vegetation present. The second is the increase of C/N both in the upper and lower soil horizons in relation to altitude, especially over 500 m; this is due principally to the rise in the percentage of carbon (from 9.4 to 21.7 per cent under primary forest in Indonesia).

Changes in the organic matter content are always accompanied by parallel changes in organic nitrogen, which constitutes the dominant form of soil nitrogen. Bartholomew (1972) estimates that some tropical soils beneath primary forest (or old secondary forest) accumulate 10 000 kg/ha of nitrogen in their humus horizon. This becomes available at the rate of 400 kg/ha/a (Schreiner and Brown, 1938) under exploitation. Ahn (1959) confirms this fact in demonstrating nitrogen contents of 9 000 and 8 000 kg/ha respectively under semi-deciduous and evergreen forest in Ghana, in the upper 90 cm of soil. Thus the organic nitrogen of soil is no longer an available source of nitrogen after a long period of cultivation and in terms of the natural processes one must rely on atmospheric nitrogen and its non-biological fixation.

Other aspects

Tropical soils generally have an inadequate supply of sulphur, but Dabin (1970) has shown in Africa the organic nature of sulphur present and the scarcity of sulphur deficiency under forests.

Studies in depth in the temperate and tropical zones

have shown a fall in soil phosphorus content with increasing intensity of alteration. In cool subhumid climates, this content is 3 000 ppm whereas under hot humid climates of the temperate zone (Pierre and Norman, 1953) it falls to 500 ppm, and less than 200 ppm in very old soils of the humid tropics (Nye and Bertheux, 1957; Bouyer and Damour, 1964; Enwezor and Moore, 1966). This phenomenon is known from Africa where Ahn (1959) has measured a total phosphorus content of 139 ppm in the upper 5 cm of soil beneath dense evergreen forest and 290 ppm in that beneath semi-deciduous evergreen forest. This is also found in South-East Asia, where only 43 ppm of assimilatable phosphorus were found in Thailand beneath dense evergreen forest and 36 ppm beneath semi-deciduous evergreen forest (Khemnark *et al.*, 1972). This decrease is accompanied by a relative increase in organic phosphorus in some soils (latosols, oxisols and podzols in particular).

The recycling of bases is an important characteristic of humid tropical forest soils. With an annual precipitation of less than 1 700 mm the upper horizons are relatively base-rich because of biogeochemical cycles (Aubert and Tavernier, 1972). With annual rainfalls of more than 1 700 mm this enrichment becomes less pronounced.

After deforestation and agricultural development, soil is more easily leached and its base content diminishes, sometimes rapidly. Hence the low content generally indicated by agronomists; and thereafter the development of a large number of studies which have led to relatively good information on minimal, and sometimes optimal, content in exchangeable potassium for crops, on the conditions for application of potassium fertilizers and on magnesium deficiency thresholds. These studies are still incomplete in so far as calcium and the balance of potassium-calcium-magnesium in the soil is concerned.

Soils beneath humid tropical forests are often acid at depth (pH of *ca.* 5.5 falling sometimes to 4 in very old soils) but have a little higher pH (*ca.* 6) at the surface because of the liberation of bases during decomposition of organic matter. Table 5 gives some examples of pH at the surface and at depth in forest soils of Africa and South-East Asia.

There is an abundant literature on manganese, boron, copper, zinc, molybdenum, chlorine, etc., in all continents, but this only refers to cultivated soils. It seems that almost nothing is known about strictly forest soils.

TABLE 5. pH of soils beneath humid tropical forests.

	pH at surface	pH at depth
Thailand (Khemnark *et al.*, 1972)		
Dense evergreen forest	5.7	5.4 (30 cm)
Semi-deciduous teak forest	6.7	5.8 (30 cm)
Woodland	6.2	5.7 (30 cm)
Ivory Coast (Aubert, 1959)		
Dense evergreen forest	6.1	5.5 (50 cm)
Madagascar (Aubert, 1959)		
Evergreen semi-deciduous forest	6.1	5.7 (40 cm)

Conclusions: research needs and priorities

In the great majority of humid tropical regions pedological studies refer to cultivated soils and almost totally neglect soils beneath forest from which they derive. One of the most fundamental problems of tropical pedology is the marked change in the physical and chemical characteristics when forest soils are cleared and brought into cultivation. Measurements of the rate of soil development and the making of decisions to maintain the initial favourable conditions presuppose knowledge of these conditions.

The physical characteristics, stability of structure, porosity and permeability as well as the hydrodynamic properties, which form a fundamental element in the study of soil-water-plant relationships, do not require the perfection of new techniques of analysis; it is more a matter of choice between existing methods so as to facilitate comprehension and the sharing of knowledge.

On the chemical side, a first priority should be accorded to the following:

1. The constitution and nature of organic matter, especially a knowledge of the role of more or less decomposed fractions in soil structure. With reference to fertility, three points deserve more detailed attention:
   — the acceptable organic content to give a satisfactory yield in ecosystems transformed by agriculture;
   — the dynamics of nitrogen, especially organic nitrogen;
   — the dynamics of sulphur beneath forest so as to give a better understanding of the deficiencies of this element in crop lands.
2. The evolution of phosphorus in soils rich in iron and aluminium hydroxides and oxides capable of immobilizing this element.
3. Soil phosphorus directly assimilatable by plants.
4. Available potassium for plants. For this an analytical method should be perfected for determination of assimilatable forms of potassium in soils.
5. Study of leaching of potassium in order to better understand the leaching of potash in agroecosystems.
6. The content of important trace elements in forest soil profiles in order to predict deficiencies or toxicities when they are cultivated. A methodological problem will arise in the undertaking of this study since the soil characteristics due to trace element deficiencies must first be identified in a more precise manner.
7. Soil acidity in so far as it affects cation exchange.

Erosion

With the exception of desert areas where wind erosion dominates, and the polar regions of permanent frost, land is subjected to erosion by water. In tropical humid regions this is now much better understood since the equation of Wischmeier and Smith (1960) has been applied to the prediction of erosion. The equation gives a better definition of the role of different factors in erosion. It states that:

$$E = R.K.SL.C.P$$

and means that the loss of soil (E) is a function of five factors: climate (R), the resistance of the soil (K), topography (SL), the vegetation cover (C) and anti-erosive measures (P).

In humid tropical regions, climatic weathering is very high: the R index is over 200 in the semi-arid zones of Africa and up to 1 500 and even 2 000 on the coasts of the Gulf of Guinea, from Guinea to Cameroon (Roose, in Anon., 1975).

The resistance of the soil to attack by water appears, in general, more satisfactory than that of a good number of leached soils of temperate regions. Roose (in Anon., 1975) surmises this from the fact that K is from 0.30 to 0.60 in the United States for soils which are more susceptible to erosion, while for ferrallitic soils in the Ivory Coast, K is between 0.05 and 0.18. This conclusion should not be surprising if one looks at the fact that soils formed on basic rock are generally very well structured and more resistant and also that the more acid soils always contain sufficient alumina to ensure flocculation of their clay fraction (Greenland, in Anon., 1975). El Swaify (in Anon., 1975) has completed a very detailed analysis of this K factor in Hawaii using the 7th Approximation classification; it shows differences in behaviour arising from seasonal drought or humidity.

As elsewhere, the influence of the length of slope (L) is neither constant nor very high in the humid tropics. In contrast, that of slope angle (S) is a determining factor: loss of soil increases in an exponential manner with the percentage of slope or according to a second degree equation of a similar nature (Wischmeier and Smith, 1960).

The vegetation cover of the soil dominates all factors in the control of erosion. This is especially true under very high rainfall, the impact of which is so strong that it reduces and minimizes the effect of soil erodability. The forest ecosystem therefore plays an important role.

Where the forests still exist, erosion remains at a minimum level because the ground protection which it gives counteracts rainfall and the conditions which regulate its intensity. Table 6 drawn up by Roose (in Anon., 1975) for western Africa is revealing on this subject.

The role of the forest is clear and it is not surprising that land deprived of vegetation is the most eroded. This small loss of soil is seen clearly in tropical America: 2.55 t/ha/a beneath natural forest in Trinidad (Bell, 1973); 0.41 t/ha/a in French Guyana, at Crique Grégoire (ORSTOM, 1975); and in South-East Asia: less than 1 t/ha/a at Bandoeng (Java) on volcanic soil beneath forest (Gonggrijp, 1941). At Périnet, in Madagascar, the measurements made on small watersheds of *ca.* 1.5 ha have shown that during 9 years of study the soil losses were nil under secondary forest, but increased to 9 t/ha/a after clearing and cultivation (Bailly *et al.*, 1974).

Two additional points should be underlined. First, dense evergreen forest less reduces erosion than temperate forest because its litter is separate from the soil and thus

TABLE 6. Erosion (t/ha/a) and run-off (percentage of annual precipitation) beneath various types of vegetation cover in western Africa (Roose, in Anon., 1975).

Station, vegetation type and mean annual precipitation	Slope %	Erosion (t/ha/a)			Run-off (% of annual rainfall)		
		natural environment	bare ground	crop land	natural environment	bare ground	crop land
Adiopodoumé (1954–1973) (Ivory Coast) Evergreen secondary forest 2 100 mm: 4 seasons	4.5	—	60		—	35	
	7	0.03	138	0.1–90	0.14	33	0.5–30
	20	0.2	570		0.7	24	
	65	1.0	—		0.7	—	
Divo (1967–1970) (Ivory Coast) Semi-deciduous forest 1 750 mm: 4 seasons	9	0.5	—				
Bouaké (1960–1970) (Ivory Coast) Dense shrubby savanna 1 200 mm: 4 seasons	4	* b. 0.20 n.b. 0.01	18–30	0.1–26	b. 0.3 n.b. 0.03	15–30	0.1–26
Cotonou (1964–1968) (People's Republic of Benin) Dense thicket 1 300 mm: 4 seasons	4	0.3–1.2	17–27.5 after clearing	10–85	0.1–0.9	17	20–35

*The signs b. and n.b. mean "burnt" or "not burnt".

allows run-off beneath this layer in the upper soil horizon. Bennett (1939) frequently cites losses of soil of less than 0.01 t/ha/a for erosion beneath forest in the United States. Secondly, the intensity of erosion beneath forest is influenced by the intensity of rainfall and the variation in density of the vegetation cover. This has been clearly shown by Ruangpanit (1971); see table 7.

TABLE 7. Relationship between the loss of soil and rainfall intensity in northern Thailand beneath different vegetation densities on more or less forested experimental plots (Ruangpanit, 1971).

Rainfall intensity (mm/h)	Soil losses (kg/ha) Density of vegetation cover (% of the area which is covered)			
	20 to 30	50 to 60	60 to 70	80 to 90
0–10	6.1	4	2.9	2.6
10–20	19.1	19.2	9.8	10.9
> 20	43.6	25.2	28.1	16.9

It is evident that the removal of soil particles to be carried away by water and run-off is considerably greater when raindrops reach the ground with a high kinetic energy, i.e. when they have not been intercepted by dense vegetation.

The main form of soil erosion in forests is soil slumping, where the slope angle and the presence of impermeable soil horizons create a sliding surface; this is especially prevalent in cases of waterlogging. A good illustration of this phenomenon is found in northern Sarawak, where slumping is often caused by the presence of a sand layer overlying a clay horizon on steep slopes.

Soil creep can also occur beneath forest, as shown in the soil profile, e.g. in the planing of quartz veins at a certain level (a type of stone line) (Fournier, 1958).

It is only when the forest ecosystem is destroyed that erosion intensifies and can reach catastrophic values. This increase in erosion remains relatively low if the forest is replaced by a woody ecosystem. For example, at Lake Alaotra (Madagascar) a low level of erosion beneath a forest was increased to only 0.025 t/ha/a beneath afforestation with *Eucalyptus* but reached more than 59 t/ha/a under various crops (measurements made in 1959–1960 at the Agricultural Research Station) (Bailly *et al.*, 1974).

The management of woody plantations requires the greatest care as pointed out by Bell (1973) with reference to teak plantations in Trinidad. Fire must be suppressed since it results in the exposure of bare ground through destruction of the shrubby species which form a protective understorey in the plantation. Spacing of trees for optimal growth as well as tree location along contours, both favour soil conservation.

In Java, Coster (1938) has shown the important role of a dense herbaceous layer over which water may run-off but which retains the soil. This is well known in Africa where, for example, an *Aristida* grassland covering 60–100 per cent of the ground will lose only 4 t/ha/a and 0.025 t/ha/a respectively; this loss increases to 12 t/ha/a when only 20 per cent of the area is protected. In Madagascar, on the High Plateaux, soil losses are nil under thick grass cover whilst they increase to 25 t/ha/a when the plant cover is degraded (Bailly *et al.*, 1967).

Finally, it is mainly beneath crop land that conditions for a dangerous increase in erosion occur, and require soil conservation measures. Erosion values are intermediate between the very low ones in the forest and the very high ones on bare ground. They vary in relation to the type of crop, the rate at which it covers the ground and the agricultural techniques used: density, stage of development of the crop, tillage, fertilization and eventually mulching, all play an important role. Roose (in Anon., 1975), studying the C factor of Wischmeier and Smith's equation, drew up table 8 which gives coefficients of erosion for different crops and vegetation covers in relation to bare ground which has a coefficient of 1 (results of work undertaken on experimental plots).

TABLE 8. Erosion as a function of vegetation cover (Roose, in Anon., 1975).

	mean annual C
Bare ground	1
Dense forest or cultivation with copious mulch	0.001
Savanna and grassland in good condition	0.01
Cover plants which develop rapidly or early plantation	0.1
Intensive rice cultivation	0.1 to 0.2
Palms, coffee, cocoa, with cover plants	0.1 to 0.3
Cotton, tobacco (second cycle)	0.5
Yams, cassava	0.2 to 0.8
Ground-nuts (as a function of date of planting)	0.4 to 0.8
Maize, sorghum and millet (as a function of yield)	0.3 to 0.9

It is not surprising that a study undertaken in the framework of the International Hydrological Decade (UNESCO/WMO, 1974) gave estimates for the transport of sediments in the watercourses of forested areas in humid tropical regions which are as low as 18–37 t/km²/a (Congo) or 67–87 t/km²/a (Amazon), whilst the intensely cultivated basins of South-East Asia yield as high as 500–600 t/km²/a (Mahanadi in India or Si-Kiang in China) or more than 1 500 t/km²/a (Damodar in India).

A seminar held at the International Institute of Tropical Agriculture in Ibadan, Nigeria, in 1975 (Anon., 1975) summarized the state of erosion in Latin America (Pla), in Africa (Ahn; Okigbo; Roose), in the Caribbean region (Ahmad) and in South-East Asia (Panabokke).

Conclusions: research needs and priorities

Studies of erosion need to be carried out on cultivated lands and to provide data for soil conservation policies. Measurements made beneath forests serve only to supply base-line values. It is from this angle that measurements of soil losses in the humid tropical regions should be considered. The main approaches which should guide these studies are the following:

— The factorial analysis of the erosion process undertaken on experimental plots under natural rainfall, is far too long term for developing countries; it is preferable to substitute analysis by use of rainfall simulators now technically adequate. One of these is operating in the Ivory Coast (Roose, in Anon., 1975); research could be undertaken with the same method in South-East Asia and Latin America. Control plots such as those of Wischmeier and Smith (1960) could be used for making comparable studies.

— The use of representative and experimental basins for the evaluation of soil erosion under unaltered vegetation and under man-made ecosystems, by measuring the flow of solid material (sedimentation) in watercourses.

— The study of soil conservation through biological techniques must be considerably advanced. In other words, rational agricultural conservation measures must be used rather than very expensive mechanical means, which should only be used in case of failure of the former.

Selective bibliography

Water balance

ANDRIEUX, C.; BUSCARLET, L.; GUITTON, J.; MERITE, B. Mesure en profondeur de la teneur en eau des sols par ralentissement des neutrons rapides. In: *Radioisotopes in soil-plant nutrition studies*, p. 187–219. Vienne, AIEA, 1962, 461 p.

*ANON. *IBP synthesis meeting* (Kuala Lumpur, 12–18 August 1974). Several papers by Ashton, P. S.; Chunkao, K.; Leigh, C. H.; Teoh, T. S. Kuala Lumpur, Pasoh Forest Reserve, 1974, multigr.

AVENARD, J. H. *La répartition des formations végétales en relation* avec l'eau du sol dans la région de Man-Touba. Paris, ORSTOM, Travaux et Documents n° 12, 1971, 159 p.

*AUBRÉVILLE, A. *Climats, forêts et désertification de l'Afrique tropicale*. Paris, Soc. Ed. géographiques, maritimes et coloniales, 1949, 351 p.

BANERJEE, A. K. Evapotranspiration from a young *Eucalyptus* hybrid plantation of West Bengal. In: *Proc. and Tech. Papers Symp. man-made forests in India* (Society of Indian Foresters, Dehra Dun), 1972.

* Major reference.

BERNHARD-REVERSAT, F.; HUTTEL, C.; LEMÉE, G. Some aspects of the seasonal ecologic periodicity and plant activity in an evergreen rain forest of the Ivory Coast. Quelques aspects de la périodicité écologique et saisonnière en forêt ombrophile sempervirente de Côte-d'Ivoire. In: Golley, P. M.; Golley, F. B. (eds.). *Tropical ecology with an emphasis on organic production*, p. 217–234. Athens, Univ. of Georgia, 1972, 418 p.

*——; ——; ——. Recherches sur l'écosystème de la forêt sub-équatoriale de basse Côte-d'Ivoire. *La Terre et la Vie* (Paris), 29, 1975, p. 169–264.

BONZON, B.; PICARD, D. Matériels et méthodes pour l'étude de la croissance et du développement en pleine terre des systèmes racinaires. *Cah. ORSTOM*, *Sér. Biol.*, 9, 1969, p. 3–18.

COSTER, C. De verdamping van verschillende vegetatie vormen of Java. *Tectona*, 30, 1937, p. 1–102.

*CZARNOWSKI, M. S.; OLZEWSKI, J. L. Rainfall interception by a forest canopy. *Oikos*, 19, 1968, p. 345–350.

DABRAL, B. G.; RAO, B. K. S. Interception studies in chir and teak plantations, New Forest. *Indian Forester*, 94, 1968, p. 541–551.

*——; ——. Interception studies in sal (*Shorea robusta*) and khair (*Acacia catechu*) plantations, New Forest. *Indian Forester*, 95, 1969, p. 314–323.

DELVIGNE, J. *Pédogenèse en zone tropicale. La formation des minéraux en milieu ferrallitique.* Paris, Dunod, 1965, 117 p.

FLENLEY, J. R. (ed.). *The water relations of Malesian forests. Transactions of the first Aberdeen-Hull Symposium on Malesian ecology.* Univ. Hull, Dept. of Geography, miscellaneous series no. 11, 1971, 97 p.

*FREISE, F. Das Binneklimma von Urwäldern im subtropischen Brasilien. *Petermanns Geogr. Mitteilungen*, 82, 1936, p. 301–307.

GAUSSEN, H.; LEGRIS, P.; BLASCO, F. *Bioclimats du Sud-Est asiatique.* Inst. français de Pondichéry, Trav. Sect. scientifique et technique, 3, 1967.

GEIGER, R. *Das Klima der bodennahen Luftschicht.* 4 edition. Braunschweig, Friedr. Vieweg und Sohn, 1961, 646 p.

GORNAT, B.; GOLDBERG, D. The relation between moisture measurements with a neutron probe and soil texture. *Soil Sci.*, 114, 1972, p. 254–258.

GOSSE, G. *Calcul des paramètres agronomiques utilisés dans la formule de Turc pour différentes localités de Côte-d'Ivoire.* Adiopodoumé, ORSTOM, 1973, 24 p. multigr.

HEWLETT, J. D.; DOUGLAS, J. E.; CLUTTER, J. L. Instrumental and soil moisture variance using the neutron scattering method. *Soil Sci.*, 97, 1964, p. 19–24.

HOPKINS, B. Vegetation of the Olokemeji Forest Reserve, Nigeria. III. The microclimate, with special reference to their seasonal changes. *J. Ecol.*, 53, 1965, p. 125–138.

*HUTTEL, C. Estimation du bilan hydrique dans une forêt sempervirente de basse Côte-d'Ivoire. In: *Radioisotopes in soil-plant nutrition studies.* Vienne, AIEA, 1962, 461 p.

KLINE, J. R.; JORDAN, C. F.; DREWRY, G. Tritium movement in soil of a tropical rain forest (Puerto Rico). *Science*, 160, 1968, p. 550–551.

*MALAISSE, F. Contribution à l'étude de l'écosystème forêt claire (miombo). Note 8. Le projet miombo. *Ann. Univ. Abidjan*, E, vol. 6, n° 2, 1973, p. 227–250.

*McCOLL, J. G. Properties of some natural waters in a tropical wet forest of Costa Rica. *Bio-Science*, 20, 1970, p. 1096–1100.

*McGINNIS, J. T.; GOLLEY, F. B.; CLEMENTS, R. G.; CHILD, G. I.; DUEVER, M. J. Elemental and hydrology budgets of the Panamian tropical moist forest. *Bio-Science*, 19, 1969, p. 697–700.

MOUTONNET, P.; BUSCARLET, L.; MARCESSE, J. Emploi d'un humidimètre à neutrons de profondeur associé à un réflecteur pour la mesure de la teneur en eau des sols au voisinage de la surface. *Ann. I.T.B.T.P.*, 233, 1967, p. 1–5.

*NYE, P. H. Organic matter and nutrient cycles under moist tropical forest. *Plant and Soil*, 13, 1961, p. 333–346.

*ODUM, H. T.; MORE, A. M.; BURNS, L. A. Hydrogen budget and compartments in the rain forest. In: Odum, H. T.; Pigeon, R. F. (eds.). *A tropical rain forest*, H105–H122. Division of Technical Information, USAEC, 1970, 1678 p.

*OVINGTON, J. D. A comparison of rainfall in different woodlands. *Forestry*, 27, 1954, p. 41–53.

RAO, B. K. S.; DABRAL, B. G.; RAMOLA, B. C. Quality of water from forested watersheds. *Indian Forester*, vol. 99, no. 12, 1973.

*ROOSE, E. J. Quelques exemples des effets de l'érosion hydrique sur les cultures. *C. R. Coll. sur la fertilité des sols tropicaux* (Tananarive), II, 1967, p. 1385.

——; HENRY DES TURREAUX, P. Deux méthodes de mesure du drainage vertical dans un sol en place. *Agron. Trop.*, 25, 1970, p. 1079–1087.

*RUANGPANIT, N. Crown cover of hill-evergreen trees as affected to soil and water losses. Faculty of Forestry, Kasetsart University (Bangkok), *Kog-Ma Watershed Research Bulletin*, vol. 7, no. 7, 1971, 25 p.

SABHASRI, S.; CHUNKAO, K.; NGAMPONGSAI, C. *The estimation of evapotranspiration of the old clearing and the dry evergreen forest, Sakaerat, Nakorn Rachasima.* Bangkok, Faculty of Forestry, Kasetsart University, 1970, 6 p. multigr.

SCHMIDT, F. H.; FERGUSON, J. H. A. Rainfall types based on wet and dry period ratios for Indonesia with western New Guinea. *Verhan. Djawatan Meteorologi dan Geofisik* (Djakarta), 42, 1951.

SOLLINGS, P.; DREWRY, G. Electrical conductivity and flow rate of water through the forest canopy. In: Odum, H. T.; Pigeon, R. F. (eds.). *A tropical rain forest*, H137–H153. USAEC, 1970, 1678 p.

THORNTHWAITE, C. W. An approach toward a rational classification of climate. *Geographical Review*, 38, 1948, p. 85–94.

*TURC, L. Evaluation des besoins en eau d'irrigation, évapotranspiration potentielle. *Annales agri. INRA*, 12, 1961, p. 13–49.

ULLAH, W.; JAISWAL, S. D.; RAWAT, U. S. Accuracy of rainfall sampling in forest clearings. *Indian Forester*, 96, 1970, p. 195–202.

*WHITMORE, T. C. *Tropical rain forests of the Far East.* Oxford, Clarendon Press, 1975, 278 p., 550 references.

General soil science and classification

*AHN, P. M. *West African soils.* London, Oxford University Press, 1970, 332 p.

*AUBERT, G. Influence des divers types de végétation sur les caractères et l'évolution des sols en régions équatoriales et subéquatoriales ainsi que leurs bordures tropicales semi-humides. In: *Sols et végétation des régions tropicales* (Travaux du Colloque d'Abidjan, 1959), p. 41–48. Paris, Unesco, 1961, 115 p.

*——. Classification des sols utilisée par la section de Pédologie de l'ORSTOM. *Cahiers ORSTOM, Série Pédologie*, 3, 1965, p. 269–288.

BAVER, L. D. *Summary of Hawaiian contributions to the study of tropical soils.* Washington, D.C., Committee on Tropical Soils, U.S. National Research Council, 1970, 17 p. multigr.

BLANCANNEAUX, Ph. *Essai de synthèse pédo-géomorphologique et sédimentologique de la Guyane francaise*. Paris, ORSTOM, 1974, 141 p. multigr.

*BOYER, J. *Essai de synthèse des connaissances acquises sur les facteurs de fertilité des sols en Afrique intertropicale francophone*. Washington, D.C., Committee on Tropical Soils, U.S. National Research Council, 1970, 175 p. multigr.

*BURINGH, P. *Introduction to the study of soil in tropical and subtropical regions*. 2nd ed. Wageningen, Centre for agricultural publishing and documentation, 1970, 99 p.

*Commonwealth Bureau of Pastures and Field Crops. *A review of nitrogen in the tropics with particular reference to pastures*. Farnham Royal, Bucks. (England), Commonwealth Agricultural Bureaux, 1962, Bulletin 46, 185 p.

*COULTER, J. K. *Soils of Malaysia. A review of investigations on their fertility and management*. Washington, D.C., Committee on Tropical Soils, U.S. National Research Council, 1970, 54 p. multigr.

*DUDAL, R. *Dark clay soils of tropical and subtropical regions*. Rome, FAO Agricultural Development, Paper no. 83, 1965, 161 p.

FAO and UNESCO. *Approaches to soil classification*. Rome, FAO, World Soil Resources Report no. 32, 1968, 143 p.

——. *Definitions of soil units*. Rome, FAO, World Soil Resources Report no. 33, 1968, 72 p.

——. *Soil map of South America*. Rome, FAO, World Soil Resources Report no. 34, 1968.

——. *Supplement to report no. 33* (see above). Rome, FAO, World Soil Resources Report no. 37, 1969, 72 p.

*FINCK, A. *Tropische Boden*. Berlin, Verlag Paul Parey, 1963, 188 p.

Interamerican Institute of Agricultural Sciences. *Volcanic ash soils in Latin America*. Turrialba (Costa Rica), Training and Research Center of Interamerican Institute of Agricultural Sciences, 1969, 341 p.

LE MARE, P. H. *A review of soil research in Tanzania*. Washington, D.C., Committee on Tropical Soils, U.S. National Research Council, 1970, 25 p. multigr.

LEMOS, R.; COSTA, D. The main tropical soils of Brazil. Rome, FAO, *World Soil Resources Report* no. 32, 1968, p. 95–106.

*MOHR, E. J. C.; VAN BAREN, F. A. *Tropical soils*. London and New York, Interscience, 1954, 498 p.

*MOSS, R. P. (ed.). *The soil resources of tropical Africa*. A symposium of the African Studies Association of the United Kingdom. London, Cambridge University Press, 1968, 226 p.

*NYE, P. H.; GREENLAND, D. J. *The soil under shifting cultivation*. Farnham Royal, Harpenden, Bucks. (England), Commonwealth Bureau of Soils, Commonwealth Agricultural Bureau, 1960, Technical Communication no. 51, 156 p.

*PHILLIPS, J. F. V. *The development of agriculture and forestry in the tropics*. 2nd ed. London, Faber and Faber, 1966, 212 p.

SIEFFERMANN, G.; MILLOT, G. Equatorial and tropical weathering of recent basalts from Cameroon. In: *Proc. Int. Conf. Tokyo*, 1969, p. 417–430.

*SOMBROEK, W. G. *Amazon soils*. Wageningen, Centre for agricultural publishing and documentation, 1966, 292 p.

*STACE, H. C. T.; HUBBLE, G. D.; BREWER, R.; NORTHCOTE, K. H.; SLEEMAN, J. R.; MULCAHY, M. J.; HALLSWORTH, E. G. *Handbook of Australian soils*. Glenside (South Australia), Rellim Technical Publications, 1968.

*STEPHENS, C. G. *A manual of Australian soils*. Melbourne, CSIRO, 1962, 62 p.

——. Soils of Uganda. In: Jameson, U. D. (ed.). *Agriculture in Uganda*, 2nd ed. London, Oxford University Press, 1970, 414 p.

*SYS, C.; VAN WAMBEKE, A.; FRANKART, R.; GILSON, P.; JONGEN, P.; PÉCROT, A.; BERCE, J. M.; JAMAGNE, M. *La cartographie au Congo*. Bruxelles, INEAC, Série technique n° 66, 1961, 149 p.

TAVERNIER, R.; SYS, C. *Classification of the soils of the Republic of the Congo*. Kinshasa, Pedologie, Special no. 3, 1965, multigr.

USDA. *Soil classification; a comprehensive system, 7th Approximation*. Washington, D.C., U.S. Government Printing Office, 1960, 265 p.

VAN WAMBEKE, A. *Republic of the Congo (Kinshasa): status of soil studies*. Washington, D.C., Committee on Tropical Soils, U.S. National Research Council, 1970, 13 p. multigr.

Soil physical properties

*BAVER, L. D. *Soil physics*. Third edition. London, Chapman and Hall, 1956, 398 p.

——; TROUSE, A. C., Jr. *Tillage problems in the Hawaiian sugar industry. I. Basic principles of compaction in relation to Hawaiian soils*. Exp. Sta. Hawaiian Sugar Planters Association, Techn. Suppl. Soils Rep. no. 9, 1965, multigr.

BONNET, J. A. Relative infiltration rates of Puerto Rican soils. *J. Agr. Univ. Puerto Rico*, 52, 1968, p. 233–240.

BOYER, J.; COMBEAU, A. Etude de la stabilité structurale de quelques sols ferrallitiques de la République Centrafricaine. *Sols Africains*, 5, 1960, p. 6–42.

CAGAUAN, B.; UEHARA, G. Soil anisotropy and its relation to aggregate stability. *Soil Science Soc. Am. Proc.*, 29, 1965, p. 198–200.

COMBEAU, A.; QUANTIN, P. Observations sur les variations dans le temps de la stabilité structurale des sols en région tropicale. *Cahiers ORSTOM, Sér. Pédologie*, 3, 1963, p. 17–26.

EL SWAIFY, S. A.; AHMED, S.; SWINDALE, L. D. Effects of adsorbed cations on physical properties of tropical red and tropical black earths. II. Liquid limit, degree of dispersion, and moisture retention. *J. Soil Sci.*, 21, 1970, p. 188–198.

GROHMANN, F.; CONAGIN, A. Técnica para o estudo da estabilidade de agregados do solo. *Bragantia*, 19, 1960, p. 329–343.

*HÉNIN, S.; MONNIER, G.; COMBEAU, A. Méthode pour l'étude de la stabilité structurale des sols. *Ann. Agron.*, 9, 1958, p. 73–92.

KITTREDGE, J. *The influence of the forest on the weather and other environmental factors. Forest influences*. Rome, FAO Forests and Forest Production Studies no. 15, 1962.

LEBRUN, J.; GILBERT, G. *Une classification écologique des forêts du Congo*. Bruxelles, Publ. INEAC, Sér. sci., n° 63, 1954, 89 p.

LEMÉE, G. Effets des caractères du sol sur la localisation de la végétation en zones équatoriale et tropicale humide. In: *Sols et végétation des régions tropicales (Tropical soils and vegetation)*, p. 25–39. Paris, Unesco, 1961, 115 p.

LUGO-LOPEZ, M. A.; JUAREZ, J. Jr. Evaluation of the effects of organic matter and other soil characteristics upon the aggregate stability of some tropical soils. *J. Agric. Univ. Puerto Rico*, 43, 1959, p. 268–272.

MOLINA, J. S.; DE GIUFFRE, L. S. Colloid production in the aerobic decomposition of cellulose and their influence upon the structure of different soil types. In: *Arq. 5th Cong. Int. Microbiol.* (Rio de Janeiro), 2, 1961, p. 594–601.

*MONNIER, G. Action des matières organiques sur la stabilité structurale des sols. *Ann. Agron.*, 16, 1965, p. 327–400.

*PEREIRA, H. C. Soil structure criteria for tropical crops. *Trans. 5th. Int. Cong. Soil Sci.*, 2, 1954, p. 59–64.

——. The assessment of structure in tropical soils. *J. Agric. Sci.*, 45, 1954, p. 401–410.

——. A rainfall test for structure of tropical soils. *J. Soil Sci.*, 7, 1956, p. 68–74.

ROBERTS, R. C. Structural relationships in a lateritic profile. *Am. Soil Surv. Assoc. Bull.*, 14, 1933, p. 88–90.

SHARMA, M. L.; UEHARA, G. Influence of soil structure on water relations in low humic latosols. I. Water retention. *Soil Sci. Soc. Am. Proc.*, 32, 1968, p. 770–774.

SMITH, R. M.; CERNUDA, C. F. Some characteristics in the macrostructure of tropical soils in Puerto Rico. *Soil Sci.*, 73, 1952, p. 183–192.

UEHARA, G.; FLACK, K. W.; SHERMAN, G. D. Genesis and micromorphology of certain soil structural types in Hawaiian latosols and their significance to agricultural practices. *Int. Soc. Soil Sci. New Zealand, Trans. Joint Meet. Comm. IV and V*, 1962, p. 264–294.

Soil chemical properties

ACQUAYE, D. K.; MACLEAN, A. J.; RICE, H. M. Potential and capacity of potassium in some representative soils of Ghana. *Soil Sci.*, 103, 1967, p. 79–89.

AHN, P. M. The principal areas of remaining original forest in Western Ghana and their agricultural potential. *Journal of the West African Science Association*, vol. 5, no. 2, 1959, p. 91–167.

ALLOS, H. F.; BARTHOLOMEW, W. V. Replacement of symbiotic fixation by available nitrogen. *Soil Sci.*, 87, 1959, p. 61–66.

ATTOE, O. J. Potassium fixation and release in soils occurring under moist and drying conditions. *Soil Sci. Soc. Am. Proc.*, 11, 1946, p. 145–149.

AUBERT, G.; TAVERNIER, R. Soil survey. In: *Soils of the humid tropics*, p. 17–44. Washington, D.C., National Academy of Sciences, 1972, 219 p.

AYRES, A. S.; HAGIBARA, H. H. Available phosphorus in Hawaiian soil profiles. *Hawaiian Planter's Rec.*, 54, 1952, p. 81–99.

——; ——. Effect of the anion on the sorption of potassium by some humic and hydrol-humic latosols. *Soil Sci.*, 75, 1953, p. 1–17.

BABALOLA, O.; CHEHEDA, H. R. Effects of crops and soil management systems on soil structure in Western Nigerian soils. *Nigerian Journal of Science* (Ibadan Univ. Press), vol. 6, no. 1, 1972.

BARBIER, G. La dynamique du potassium dans le sol. In: *Potassium symposium 1962*, p. 231–258. Bern (Switzerland), International Potash Institute, 1962, 632 p.

BARROW, N. J. A comparison of the mineralization of nitrogen and of sulfur drom decomposing organic materials. *Aust. J. Agr. Res.*, 11, 1960, p. 960–969.

BARTHOLOMEW, W. V. Nitrogen loss process—a recapitulation. *Soil Sci. Soc. N.C. Proc.*, 7, 1964, p. 78–81.

*——. Mineralization and immobilization of nitrogen in the decomposition of plant and animal residues. In: Bartholomew, W. V.; Clark, F. E. (eds.). *Soil nitrogen. Agronomy Monograph no. 10*, p. 285–306. Madison, Wis., American Society of Agronomy, 1965, 615 p.

*——. Soil nitrogen and organic matter. In: *Soils of the humid tropics*, p. 63–81. Washington, D.C., National Academy of Sciences, 1972, 219 p.

BLOOMFIELD, C.; COULTER, J. K.; KANARIS-SOTIRIOU, R. Oil

palms on acid sulphate soils in Malaya. *Trop. Agr.*, 45, 1968, p. 289–300.

*BOUYER, S.; DAMOUR, M. Les formes du phosphore dans quelques types de sols tropicaux. *Transactions 8th Int. Cong. Soil Sci.*, 4, 1964, p. 551–561.

CHANG, S. C.; JACKSON, M. L. Soil phosphorus fractions in some representative soils. *J. Soil Sci.*, 9, 1958, p. 109–119.

CHARREAU, C.; FAUCK, R. Mise au point sur l'utilisation agricole des sols de la région de Sefa (Casamance). *Agron. Trop.*, vol. 25, n° 2, 1970, p. 151–191.

DABIN, B. Méthode d'étude de la fixation du phosphore sur les sols tropicaux. *Coton Fibres Trop.*, 25, 1970, p. 213–234.

DÖBEREINER, J. Non-symbiotic nitrogen fixation in tropical soils. *Pesqui. Agropecu. Bras.*, 3, 1968, p. 1–6.

ENWEZOR, W. O.; MOORE, A. W. Phosphorus status of some Nigerian soils. *Soil Sci.*, 102, 1966, p. 322–328.

FOX, R. L.; OLSON, R. A.; RHODES, J. F. Evaluating the sulfur status of soils by plant and soil tests. *Soil Sci. Soc. Am. Proc.*, 28, 1964, p. 243–246.

GASSER, J. K. R.; BLOOMFIELD, C. The mobilization of phosphate in water-logged soils. *J. Soil Sci.*, 6, 1955, p. 219–232.

HARDON, H. J. Factoren, die de organische stof en stickstof-gehalten van tropische gronden beheerschen. *Korte Med. Algem. Proefsta Landb.*, no. 18, 1936.

HESSE, P. R. Sulphur and nitrogen changes in forest soils of East Africa. *Plant and Soil*, 9, 1957, p. 86–96.

KHEMNARK, C.; WACHARAKITTI; AKSORNKOAE, S.; KAEWLA-IAD, T. Forest production and soil fertility at Nikhom Doi Chiangdao, Chiangmai Province. *Kasetsart Univ. For. Res. Bull.* (Faculty of Forestry), 1972, p. 22–24.

LE MARE, P. H. Soil fertility studies in three areas of Tanganyika. *Emp. J. Exp. Agr.*, 27, 1959, p. 197–222.

McCLUNG, A. D.; DE FREITAS, L. M. M. Sulphur deficiency in soils from Brazilian campos. *Ecology*, 40, 1959, p. 315–317.

MIDDELBURG, H. A. Potassium in tropical soils: Indonesian archipelago. In: *Potassium symposium*, p. 221–257. Bern (Switzerland), International Potash Institute, 1955, 613 p.

MOSS, P.; COULTER, J. K. The potassium status of West Indies volcanic soils. *J. Soil Sci.*, 15, 1965, p. 284–298.

NG, S. K. Potassium status of some Malayan soils. *Malayan Agric. J.*, 45, 1965, p. 143–161.

NYE, P. H.; BERTHEUX, M. H. The distribution of phosphorus in forest and savannah soils of the Gold Coast and its agricultural significance. *J. Agri. Sci.*, 49, 1957, p. 141–159.

*PIERRE, W. H.; NORMAN, A. G. *Soil and fertilizer phosphorus in crop nutrition*. Agron. Monogr. no. 4, New York, Academic Press, 1953, 492 p.

ROOSE, E.; GODEFROY, J. Lessivage des éléments fertilisants en bananeraie. *Fruits*, 23, 1968, p. 580–584.

——; ——; MULLER, M. Estimation des pertes par lixiviation des éléments fertilisants dans un sol de bananeraie de basse Côte-d'Ivoire. *Fruits*, 25, 1970, p. 403–420.

SCHREINER, O.; BROWN, B. E. Soil nitrogen. In: *Soils and men*, p. 361–376. Washington, D.C., U.S. Department of Agriculture, U.S. Government Printing Office, 1938, 1232 p.

WIKLANDER, L. Forms of potassium in the soil. In: *Potassium symposium*, p. 109–121. Bern (Switzerland), International Potash Institute, 1954, 445 p.

WOODRUFF, C. M. Estimating the nitrogen delivery of soil from the organic matter determination as reflected by Sanborn field. *Soil Sci. Soc. Am. Proc.*, 14, 1950, p. 208–212.

YOUNGE, O. R.; PLUCKNETT, D. L. Quenching the high phosphorus fixation of Hawaiian latosols. *Soil Sci. Soc. Am. Proc.*, 30, 1966, p. 653–655.

Soil erosion

*Anon. Workshop on *Soil conservation and management in the humid tropics* (Ibadan, 30 June–4 July 1975). Several papers by Ahmad, N.; El Swaify, S. A.; Greenland, D. J.; Okigbo, B. N.; Panabokke, C. R.; Pla, I.; Roose, E. Ibadan, International Institute of Tropical Agriculture, 1975, multigr. Proceedings to be published in 1977.

*Aubréville, A. Érosion sous forêts et érosion en pays déforesté dans la zone tropicale humide. *Bois et Forêts des Tropiques* (Nogent-sur-Marne), 68, 1959, p. 3–14.

Baillie, I. C. *An occurrence of charcoal in soil under primary forest.* Unpublished report, Forest Department, Kuching, Sarawak, 1971, 10 p.

Bailly, C.; de Vergnette, J.; Benoit de Coignac, G.; Velly, J.; Celton, J. Essai de mise en valeur d'une zone des Hauts Plateaux malgaches (Manankazo) par l'aménagement rationnel. Effet de cet aménagement sur les pertes en terre et le ruissellement. In: *C.R. Coll. sur la fertilité des sols tropicaux* (Tananarive), II, 1967, p. 1362–1383.

*——; Benoit de Coignac, G.; Malvos, C.; Ningre, J.M.; Sarrailh, J. M. Étude de l'influence du couvert naturel et de ses modifications à Madagascar. Expérimentations en bassins versants élémentaires. *Cahiers Scientifiques, Suppl. Bois et Forêts des Tropiques* (Centre technique forestier tropical, Nogent-sur-Marne), n° 4, 1974, 114 p.

Bell, T. I. W. Erosion in the Trinidad teak plantations. *Commonwealth Forestry Review* (London), vol. 52, no. 3, 1973, p. 223–233.

Bell, W. V. *Protecting plantations of long-fibre tree species from moss by insects and disease.* Rome, FAO, Technical Report, UNDP/SF project MAL/12, no. 4, 1971, 24 p.

*Bennett, H. H. *Soil conservation.* New York, MacGraw-Hill, 1939, 1 vol., 993 p.

Coster, C. Bovengrondsche afstromingen en erosie op Java. *Med. Boschb. Proefsta*, no. 64, 1938.

*FAO. *Forest influences.* Rome, 1962, 307 p.

*——. *Contribution à l'étude de la conservation du sol en Afrique occidentale française.* Paris, ORSTOM, 1958, 134 p. multigr.

*Fournier, F. Research on soil erosion and soil conservation in Africa. *Sols Africains* (Paris), 1965, p. 53–96.

Freeman, J. D. *Report on the Iban of Sarawak.* Kuching, Sarawak, Government Printing Office, 1955, 54+85 p.

Gonggrijp, L. Het erosie-onderzoek. *Tectona* (Buitenzorg, Java), no. 34, 1941, p. 200–220.

*Goujon, P.; Bailly, C.; de Vergnette, J.; Benoit de Coignac, G.; Roche, P.; Velly, J.; Celton, J. Conservation des sols en Afrique et à Madagascar. *Bois et Forêts des Tropiques*, n°s 118, 119, 120, 121, 1962.

UNESCO/WMO. *Records of the International Conference on the results of the International Hydrological Decade and on future programmes in hydrology.* Paris, Unesco, 5 vol., 1974, 112 p., 130 p., 91 p., 108 p., 74 p.

*Wischmeier, W. H.; Smith, D. D. A universal soil-loss estimating equation to guide conservation farm planning. In: *7th Int. Cong. Soil Science*, vol. 1, 1960, p. 418–425.

Decomposition and biogeochemical cycles

Introduction

The first part of this chapter examines the processes of plant litter decomposition. This consideration of one of the major nutrient pathways from the vegetation leads on to an examination of the cycling of chemicals in the second part.

Decomposition

Methods

For a forest in equilibrium, the annual litter fall is usually assumed to equal its decomposition. Whilst this is true, it does not follow that the rates of the two are equal. The average amount of litter on the ground is also required in order to determine the rate of decay.

    The ratio of the mean weight of litter on the ground (litter standing crop) to the annual litter fall gives the average time (in years) for the litter to decompose, although, as pointed out by Olson (1963), this is not strictly true where the litter fall is periodic. Alternatively the decay may be calculated periodically: decay $(t_{1,2})$ = litter on ground (t_1) + litter fall $(t_{1,2})$ — litter on ground (t_2), (1) where t_1 and t_2 are the consecutive observations (preferably not more than two weeks apart and certainly not longer than one month apart because of possible decay in the litter traps) and $t_{1,2}$ is the change between these. From such amounts of decay data the time for decomposition can be calculated for any season.

    An alternative method is to observe the same materials at intervals. The usual technique is to place a known area or weight of leaves or leaf discs in plastic mesh bags pegged in the forest litter. Periodically bags are collected, the amount of remaining material determined and the rate of disappearance calculated. Madge (1969) found that ants appeared reluctant to enter bags where the rate of disappearance was correspondingly less than from open cages. Wood and branch samples may be used in a similar manner.

    The results may be expressed in one of four ways:
— weight of decay/area/time;
— weight of decay/weight of litter/time;
— coefficient of decomposition (k), *decay* in formula (1) divided by the mean litter on the ground $((t_1 + t_2)/2)$ (Jenny *et al.*, 1949; Nye, 1961; Olson, 1963);
— mean time for complete (or a percentage) disappearance
These measurements are not comparable and each has its disadvantages: the calculation of k assumes that decompo-

sition is logarithmic, and the amount of litter on the ground is difficult to measure accurately. The values resulting from direct measurement of weight loss are more precise, but the variety of experimental techniques makes comparisons difficult.

Results

There is little information on the rates at which dead plant material decomposes in tropical forests; there is none at all on animal material. The data on litter fall presented in table 4 of chapter 10 give no more than an indication of possible decay rates, for there are very few data on litter on the ground.

The rates of leaf decomposition are presented in table 1 which show that most tropical forest leaves decay in *ca.* 6 months, the range between the different forests being 2.5–11 months (or, if the one value for subtropical forest is included, up to 19 months). These very rapid rates are due to the high temperatures and humidities within tropical forests causing rapid biological activities.

The disappearance of woody material is obviously much slower than that of leaves and, as demonstrated by IBP work at Pasoh, the smaller pieces decay fastest. It is impossible to compare adequately the few available results because of their differing measurements.

Hopkins (1966) found that $25 \times 25 \times 75$ mm pieces of wood placed in Olokemeji Forest lost 50 per cent of their dry weight in *ca.* 7 months whilst in the somewhat wetter Omo Forest the rate was twice as slow. At Banco, 1–2 cm diameter branches lost 32 per cent of their weight in 6.5 months (Bernhard-Reversat, personal communication). IBP studies at Pasoh showed that the loss of weight of 1.5–6.5 cm diameter branches was 19–39 per cent in *ca.* 13 months and, in another study, 3–6 cm diameter branches lost 37–81 per cent of their weight in *ca.* 18 months.

The Pasoh workers determined that *ca.* 330 g/m²/a of dry material was returned to the forest floor from branches > 10 cm diameter. At El Verde, Odum and Pigeon (1970) determined the following rates of return:

small branches	400 g/m²/a
logs	55 g/m²/a
stumps	145 g/m²/a
total	600 g/m²/a.

These figures probably give no more than the order of magnitude, for it was estimated that *ca.* 10 per cent of dead tree material at Pasoh decayed before the tree fell.

TABLE 1. Rates of decomposition of fresh leaf-fall in tropical and subtropical forests.

Locality	Forest type	Remarks	Rate of decomposition			Source
			mg/g/d	k	months to complete disappearance	
Kade, Ghana	moist evergreen	40-year-old		4.7	2.5	Nye, 1961
Zaire			3.3			Bartholomew, Meyer and Laudelout, 1953
Yangambi, Zaire	moist evergreen				4	Laudelout and Meyer, 1954
Olokemeji, Nigeria	moist semi-deciduous				4	Hopkins, 1966
Omo, Nigeria	moist evergreen				5	Hopkins, 1966
Ibadan, Nigeria	moist semi-deciduous	15–20-year-old	2.2*		5*	Madge, 1965
Banco, Ivory Coast	moist evergreen	valley plateau	13.1* 7.1*	4.2 3.3	5* 9*	Bernhard, 1970 Bernhard-Reversat, 1972
Yapo, Ivory Coast	moist evergreen		3.6*		11*	
El Verde, Puerto Rico			1.4–2.9			Odum and Pigeon, 1970
Santa Fe, Panamá	pre-montane wet				5.5	Golley *et al.*, 1975
	tropical moist				6.5	Golley *et al.*, 1975
	riverine				13	Golley *et al.*, 1975
Panamá	secondary				17	Tropical Test Center, 1966
Murcielagos, Guatemala	moist deciduous	0–14-year-old secondary			2.5*/**	Ewel, 1976
Calima, Colombia	perhumid				7	Jenny *et al.*, 1949
Chinchiná, Colombia	subtropical				19	Jenny *et al.*, 1949

* using plastic mesh bags; all other results from collecting leaves on the ground.
** to 50 per cent disappearance.

Factors

As pointed out in chapter 10, leaf-fall for one year is usually within 88 per cent of the adjacent year. The variation for decay is much greater; in two consecutive years at Olokemeji, Hopkins (1966) found leaves took averages of 2.5 and 6 months to disappear. The considerable variation amongst the Ivory Coast forests and at El Verde (see table 1) suggests that local site factors are at least equally important as regional differences.

Different species decompose at different rates as found by different workers: Bernhard-Reversat (1972) in the Banco forest; Ewel (1976) in Guatemala, where the leaves of a palm took one fifth longer than four other tree species; Madge (1965, 1969) in Nigeria; Odum and Pigeon (1970) at El Verde, where the leaves of some species took 2.5 times as long as others; Singh (1969) for a pot experiment in India; and by the IBP workers on branch decay at Pasoh, where one species took 2.1–2.5 times as long as another.

Odum and Pigeon (1970) found significant differences of leaf litter decay rates under different trees. By carrying out litter exchanges between sites in the Banco forest, Bernhard-Reversat (1972) found that the site of decay (with the valley site having the fastest rate) was more important than leaf species which was more important than the site of origin. She considered this was due to top soil characters. By contrast, in Guatemala, Ewel (1976) found similar rates on upland and alluvial sites and between the different stages of a secondary succession with the exception of the initial bare ground which had a slower rate.

The rate of decay decreases with leaf age as demonstrated by Bernhard-Reversat (1972) in Banco forest and

by Singh (1969) by CO_2 evolution from a pot experiment in India. Several workers have studied the relation between rate of decomposition and leaf mineral content with somewhat diverse results (Singh, 1969; Bernhard-Reversat, 1972; Ewel, 1976).

Another major factor is seasonal change. At Olokemeji, Hopkins (1966) found leaves disappeared in one month during the wet season and eight months during the dry, whilst pieces of wood took *ca.* 5 and 7–8 months respectively for 50 per cent disappearance. At Omo, he found leaves disappeared in 3–4 months during the wet season and 6–7 months during the dry. Madge (1965, 1969) found very similar values for a secondary forest at Ibadan and at Olokemeji. Odum and Pigeon (1970) also found considerable differences depending on the season at which leaves fell at El Verde.

These seasonal results strongly suggest a relation between precipitation and decomposition as has been commented upon by most workers. This probably also explains the between year variation.

Soil and litter organisms

However, rainfall does not directly control decomposition; it acts through the litter and soil fauna and micro-organisms. Unfortunately little critical work has been carried out in this difficult field. Some counts have been made of organisms in forest soils, litter and in litter placed out in bags. In forests at Santa Fe in Panamá, Golley *et al.* (1975) found the biomass of micro-arthropods to be almost twice that of the macro-arthropods. Table 2 presents the

TABLE 2. Micro-arthropod populations of tropical forest soils and litters.

Locality	Forest type	Remarks*	Population density (number/m²)	biomass (g/m²)	Source
Ibadan, Nigeria	moist semi-deciduous secondary	wet season L	38		Madge, 1965
		dry season L	0.4		
Olokemeji, Nigeria	moist semi-deciduous	wet season S and L	65		Madge, 1969
		dry season S and L	6		
Banco, Ivory Coast	moist evergreen	wet season L	*ca.* 13		Bernhard-Reversat, 1972
		dry season L	2.5		
Banco, Ivory Coast	moist evergreen	plateau S	54		Huttel and
		valley S	18		Bernhard-Reversat, 1975
Yapo, Ivory Coast	moist evergreen	S	26		
Zaire	moist evergreen	S	64–72		Maldague, 1970
El Verde, Puerto Rico	moist evergreen	L	22–33		Odum and Pigeon, 1970
		S	1.9–7.6		
Santa Fe, Panamá	tropical moist	mean of S and L one forest in wet season and another in dry season		3.36	Golley *et al.*, 1975
Amazon	lowland rain	S		17	Fittkau and Klinge, 1973
		S	15		Beck, 1971
	lower montane	S		8	Odum and Pigeon, 1970
	mangrove	S		5	Golley and Lieth, 1972

* L in litter; S in soil.

Decomposition and biogeochemical cycles 273

available data on the numbers and biomass of micro-arthropods in the soil and litter. Many workers have demonstrated a relation between faunal abundance and precipitation: Bernhard-Reversat (1972) found this was strongest after *ca.* 8 weeks of rainfall and Madge (1965) found 'a clear parallel between the number of soil fauna and the rate of disappearance' of leaf litter.

Of the micro-arthropods, usually defined as those passing through a *ca.* 2.5 mm mesh, the mites and Collembola are the most abundant and, presumably, the most important in litter decomposition. Two other invertebrate taxa, the termites (Insecta: Isopoda) and earthworms (Annelida: Oligochaetae) have also received some attention because of their apparent importance. In secondary forest at Ibadan, earthworms had a density of 34/m², a biomass of 10 g/m², and produced *ca.* 3.6 kg/m²/a of casts (Madge, 1965). The texture of worm casts usually differs from that of the forest soil in general; work in the Ivory Coast has shown that their rate of CO_2 production is about the same, but that their rate of nitrogen mineralization is greater than that of the soil in general (see chapter 11 and annex on the rain forest ecosystems of Ivory Coast).

In the IBP plots at Pasoh, 24 species of termites have been identified. Different species were present in different numbers and proportions at different times and on different plots. Matsumoto (1974) found they consumed 202 g/m²/a, the equivalent of one third of the leaf-fall. At El Verde, Odum and Pigeon (1970) found a termite biomass of 0.6 g/m² and that the most abundant species respired at a rate of 9 kJ/m²/a. In Zaire, Maldague (1970), found 260 g/m² of organic material in termite colonies; this included 70 g/m² of carbon and 6 g/m² of nitrogen. He estimated that termites consumed three quarters of the litter fall.

A lot more research is necessary before the role of these organisms becomes clear and the somewhat variable results obtained by most workers are adequately explained. However, it must be remembered that ultimate decomposition is by micro-organisms, not invertebrates. Fungi appear to be more important in tropical than temperate forests, but very little is known about microbial processes in tropical forest soils. The importance of the microflora is shown by the significant rate of weight loss of litter when the microfauna is suppressed. When the litter is fresh its biological activity is greatest; later, as the easily degradable substances are utilized, the biological activity decreases (Singh, 1969; Bernhard-Reversat, 1972). Madge (1965) considers that litter is first attacked by bacteria and fungi; arthropods intervene only in the second phase.

Litter decomposition is closely related to biogeochemical cycles. The rate of litter decomposition determines the stores of inorganic elements remaining in the litter component. This store is generally smaller than the others (see tables 4 to 8) except in some cases as a Surinam forest where the litter is 1 m deep (Stark, 1970). Seasonal changes in decomposition rate are reflected in the amounts of minerals lost by leaching. Also, one of the causes of the differences between the biogeochemical cycles of the various elements is their relative rates of liberation as the litter decomposes.

Biogeochemical cycles

Introduction

Terminology

Chemical elements are stable (apart from some radio-active isotopes). Their amount on the earth, and often in stable ecosystems, is constant. Elements do move between one part of the earth and another and between the different components of an ecosystem, but such movements are part of an overall cycle. This may be simply soil→plant →litter→decomposers→soil, or much more complex and including, for example, herbivores, carnivores and parasites within the ecosystem, stream flow out from it and precipitation into it. Hence the use of the term *cycle*. Since these movements take place between living organisms and between various parts of the earth's crust the cycles are called *biogeo*chemical.

Chemicals consist of one or more elements and many react within the bodies of living organisms and elsewhere in the biosphere. Hence it is best to consider the cycling of individual elements. However, some chemicals are inactive and cycle as compounds; it is useful to treat these as such. This was done in chapter 12 where the hydrological cycle was described; although water does react to form many essential compounds for living things, the vast bulk of it cycles unchanged. Persistent pesticides are another example. Hence the use of the term biogeo*chemical* rather than element cycles. Other terms, especially mineral and nutrient cycles, are in the literature; these are older and shorter and hence preferred by some authors. Clearly not all cycling are nutrients or indeed minerals. Generally all these terms are used indiscriminately to mean biogeochemical cycles.

The model

Fig. 1 shows a generalized scheme of biogeochemical cycles in tropical forests. Only the main components and pathways are shown. The major route for most elements is, as in most terrestrial ecosystems, soil→vegetation→litter→decomposers→soil. There is some direct flow from the vegetation to the soil by rain leaching, and some via the litter, and a diversion from the vegetation via the browsing food chains. Ecosystem gains are from the atmosphere and subsoil and losses go to the subsoil and streams. For clarity the various trophic levels within the browsing and decomposer components, and the organs (especially the roots, stems and leaves) within the vegetation, are not shown on the diagram. Also omitted are several other pathways including gains and losses by wind and animal movements, evaporation (of substances such as ammonia as well as water), direct transfers from litter to plant roots by mycorrhiza and nitrogen fixation.

It is a fairly easy if somewhat labourious task to determine the chemical composition of the various components and a considerable amount of data are available for tropical soils and vegetation. Much of these are on a proportion (per cent or ppm) basis but can be converted to an area basis

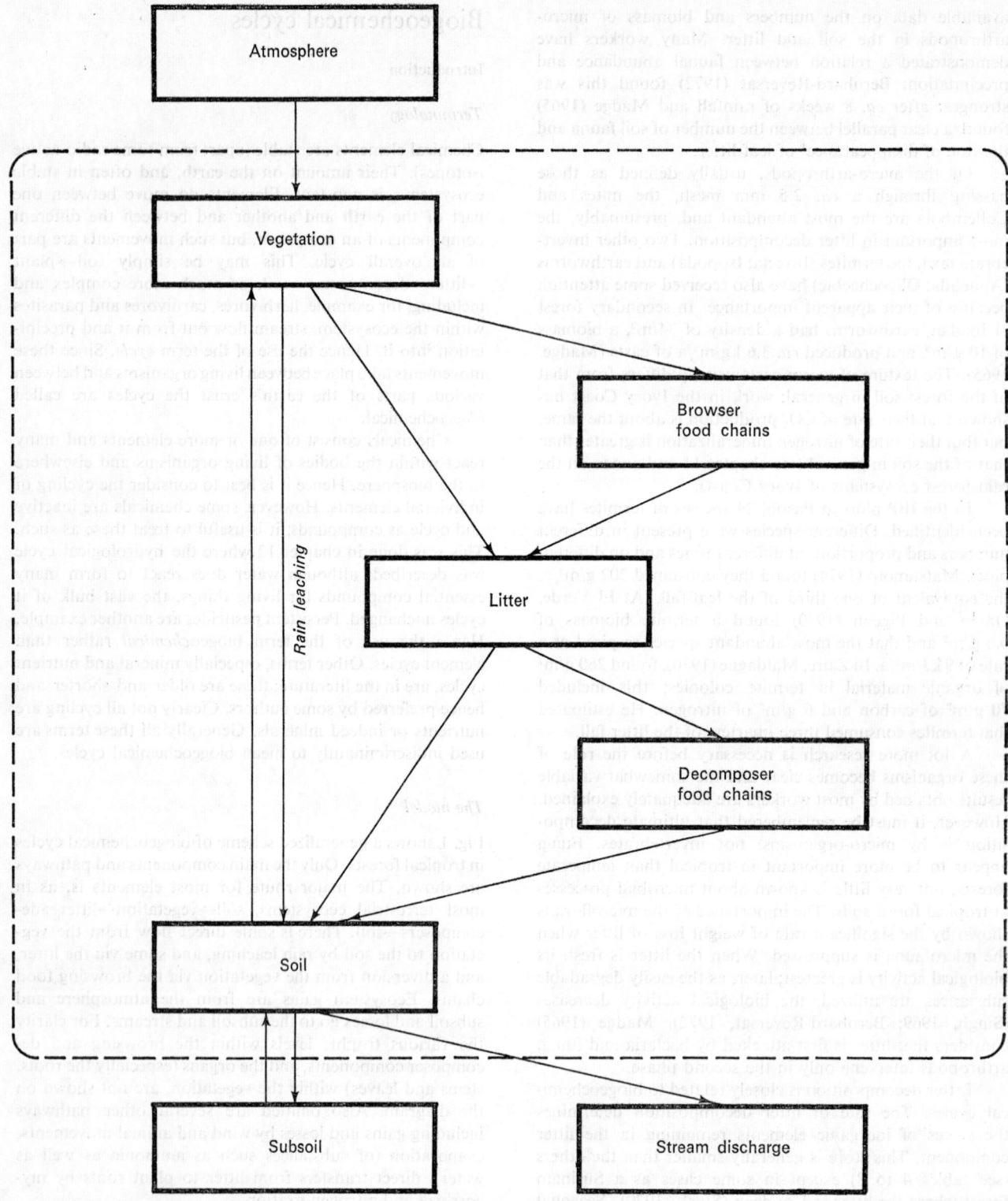

FIG. 1. A generalized scheme of biogeochemical cycles in tropical forests. The interrupted line forming the large rectangle indicates the limits of the forest ecosystem. Only the major stores and pathways are shown.

where the biomass is known. The determination of rates of flow between components is much more difficult and consequently little information is available except on litter fall. In no study have all the pathways been determined.

Variation

The chemical composition of the air is virtually constant, although the carbon dioxide concentration is higher within a few centimeters of the soil and there is also variation due to industrial pollutants. In contrast the chemical composition of the other components is very variable. A certain amount is known about spatial variation—between different forests, species and plant organs—but very little about variation in time for few studies have continued for more than one year.

The main soil variation is either with depth (as pointed out in chapter 12) due to leaching and surface litter fall or topography, which can cause even greater differences within a single forest.

In the vegetation there are considerable differences between species and between plant organs. For example, Bernhard-Reversat (1975) found the following interspecific ranges of variation in the per cent dry weight composition of the woody shoots of 19 species from two forests in the southern Ivory Coast:

N	P	K	Ca	Mg
0.09–0.38	0.016–0.50	0.02–0.28	0.13–0.75	0.01–0.25.

It is unfortunate that in the most detailed study of all mineral cycling in tropical forests, Golley *et al.* (1975) had, because of inaccessibility, to analyse different sites in the wet and dry seasons. Woody biomass differences accounted for over half the between site variation. Thus it is unwise to compare them for seasonal differences. Nevertheless, comparisons may be made within each site. Amongst the macronutrients analysed, significant differences were found between the amounts of potassium, calcium and sodium in the overstorey and understorey leaves in the dry season and between the amount of phosphorus in the overstorey and understorey stems in the wet season. No such significant differences were found for magnesium, the only other macronutrient analysed. The micronutrients showed more significant differences; for example, between overstorey and understorey leaves in both seasons for boron and cobalt.

Epiphyllae generally have high nutrient contents compared to the leaves on which they grow (Odum and Pigeon, 1970; Jordan and Kline, 1972).

There are also differences in rates of movement. In round figures, *ca.* 80 per cent of the elements in plant litter are in the leaf component and *ca.* 10 per cent in both the wood component and in the flower and fruit component. However, there is considerable variation in these figures. In general the leaf component accounts for 70–90 per cent of the litter elements. The variation in the other two components is considerably greater, usually being in the range of 3–20 per cent of the total.

Klinge and Rodrigues (1968) presented figures for two consecutive years for an Amazonian *terra firme* forest. The between year variation in the percentages of the various

elements in the litter component were considerable as shown in table 3. Again the leaf litter was the least variable with each of the two years differing by *ca.* 10 per cent of their mean, but the values for the wood and for fruits averaged *ca.* 25 and 50 per cent respectively, with those for the wood being particularly variable.

A major source of variation is in the value for calcium which is the most abundant element in wood. The variation in the proportions of calcium amongst the litter components may well be largely due to the variation in, and the consequent difficulty of measuring, the woody litter.

Despite these sources of variation, it is possible to determine the mean rates of movements of elements and these are the data of most relevance to the ecosystem as a whole. Nevertheless, all the results presented in this chapter should be regarded as showing an order of magnitude only.

TABLE 3. Between year variation in the chemical composition of plant litter components for an Amazonian *terra firme* forest. Data are percentage variation of the mean: 100 (year 1—mean)/ mean, where the amounts are the percentages of the elements which occur in the three components of the litter. Calculated from Klinge and Rodrigues (1968).

Litter component	P	K	Na	Ca	Mg
leaves	9	10	8	14	8
wood	7	6	50	26	23
fruits	62	61	50	47	49

Choice of denominator

One further problem in studies of element cycling is the choice of a denominator. Weight per area, or per area per time, are probably the most useful. However, as the amounts of the various elements differ considerably, one often wishes to know the proportions contained in the different components of the ecosystem or the proportions of flow along the various routes. The total store is often unobtainable or irrelevant as, for example, the gases of the atmosphere, the depth to which a soil should be analysed, and the amount of calcium in a limestone soil. The total amount cycling per annum is seldom known and variable so that the proportion of this can rarely be given.

Results

In no case have all the pathways shown on fig. 1 been studied. In many cases only one or two pathways have been investigated. Tables 4–8 summarize the available data from the more comprehensive studies. These show for each of the major nutrients the amounts contained (in g/m²) in the main ecosystem components and the measured (or calculated by difference) flows between them (in g/m²/a). The three most complete studies are all in moist evergreen forests: at El Verde, Puerto Rico (Odum and Pigeon, 1970), at Santa Fe, Panamá (Golley *et al.*, 1975) and the still incompletely analysed and reported IBP study of the Pasoh Forest Reserve, Malaysia.

Stores

Store is probably not the best term to use for the element content of ecosystem components, for changes are constantly occurring. Nevertheless it is very widely used; content, inventory and standing crop are alternative terms.

The element composition of the main ecosystem components are shown in the upper part of tables 4–8. The large amounts in the vegetation compared to the soil contrast with most terrestrial types of vegetation where the soil is the major store. The exceptions are due to particularly high concentrations in the soil parent material as, for example, calcium and magnesium in Panamá.

The amounts in the roots are in the order of 5–10 per cent of the element in the total vegetation. The litter generally contains less than 10 per cent, and usually less than 5 per cent, of the elements in the vegetation. The Surinam forest studied by Stark (1970) is exceptional because of its 1 m deep litter. The content of the animals is almost negligible; less than 0.01 per cent of the total in the ecosystem for all cases so far investigated.

Most workers have found that the nutrient concentrations of leaves, flowers and fruits are higher than those of the woody organs. Golley *et al.* (1975) carried out detailed analyses in Panamá and discovered the following sequence of decreasing concentrations: understorey flowers and fruits, understorey leaves, overstorey leaves, litter, overstorey flowers and fruits, roots, understorey stems, and overstorey stems.

There are considerable differences between the various forests. In Panamá, Golley *et al.* (1975) found that tropical moist and secondary forests generally contained higher concentrations than premontane wet forest. Mangrove forests were very different with higher concentrations of magnesium, sodium and iron and lower ones of nitrogen, phosphorus, potassium and calcium. Considerably more research is required before the ecological implications of these differences are fully understood. The two sets of results on stores at El Verde (Odum and Pigeon, 1970) are not identical but are of the same order of magnitude except for the soils where, surprisingly, one set is about two or more times the other. The inconsistencies in the El Verde data are much greater in the flow results. Only one set of leaf-fall results is presented and it appears to be of the same order as for other tropical forests. For other flows their set 'H' is *ca.* 50 times set 'I' suggesting that the units stated as per week in set 'H' should read per year. However, even if this is the case, there are still considerable differences between the two sets of data. Set 'I' is of the same order as those for other tropical forests; set 'H' is clearly erroneous and has been placed in brackets in tables 4–8. Apart from its leaf-fall data it will not be considered further.

TABLE 4. Ecosystem component stores and flows of nitrogen in different tropical forests.

Locality*	Puerto Rico El Verde	Ivory Coast			Ghana Kade	Malaysia Pasoh
		Banco plateau	Banco valley	Yapo		
ECOSYSTEM STORES (g/m²)						
above-ground vegetation	120	140	140	100	180	
below-ground vegetation	3				21	
total vegetation	123				201	
litter	13				3.5	
soil		170;650	120;580	160;260	460	500–1 000
(depth analysed in cm)		(10;50)	(10;50)	(10;50)	(25)	(Ao horizon)
Total store					665	
ECOSYSTEM GAINS (g/m²/a)						
precipitation	1	2	2	2	1.5	
ECOSYSTEM LOSSES (g/m²/a)						
stream flow	3					
FLOWS WITHIN ECOSYSTEM (g/m²/a)						
throughfall leaching	7	6	6	1.3	1.2	
leaf-fall		14	14	9	20	
flower and fruit fall		1	0.7	1.6		
branch fall		2	0.9	0.8	4	
total plant litter fall	16	17	16	12	24	10
total vegetation to litter flow	23	23	22	13	25	

* Panamá: near Santa Fe; tropical moist forest data are the means of the Río Lara site in the wet season and the Río Sabana site in the dry season (Golley *et al.*, 1975).
Puerto Rico: El Verde; moist evergreen (Odum and Pigeon, 1970).
Ivory Coast: Banco Forest plateau and valley sites are 400 m apart; Yapo forest is 45 km away; evergreen forests on various soil types (Bernhard, 1970; Bernhard-Reversat, 1972, 1975).
Ghana: Kade on the ecotone between moist evergreen and moist semi-deciduous forest; 40 year mature secondary forest (Greenland and Kowal, 1960; Nye and Greenland, 1960; Nye, 1961).
Malaysia: Ulu Gombak Forest Reserve, moist evergreen (Kenworthy, 1971); Pasoh Forest Reserve (IBP site), moist evergreen (Lim, 1974).

TABLE 5. Ecosystem component stores and flows of phosphorus in different tropical forests (see table 4 for details of localities and sources).

	Panamá		Puerto Rico El Verde	Ivory Coast			Ghana Kade	Malaysia Pasoh
	tropical moist	premontane wet		Banco plateau	Banco valley	Yapo		
ECOSYSTEM STORES (g/m²)								
above-ground vegetation	14		(7)	10	10	7	12	
below-ground vegetation	0.6		11				1.1	
total vegetation	15	2.6					13	
animals	0.005							
litter	1.4	0.08	0.1	5*	33*	2.5*	0.1	
soil	2.2		56				1.2*	3–6
(depth analysed in cm)	(30)		(25)	(50)	(50)	(50)	(25)	(100)
Total store	19		67				14	
ECOSYSTEM GAINS (g/m²/a)								
precipitation	0.1		(941)	0.05	0.05	0.05	0.04	
subsoil weathering	−0.02		(369)					
ECOSYSTEM LOSSES (g/m²/a)								
stream flow	0.07		(1 310)				0.4	
FLOWS WITHIN ECOSYSTEMS (g/m²/a)								
throughfall leaching	0.06		(1 009)	0.15	0.9	0.6		
stemflow leaching			(406)					
total leaching from vegetation			(1 415)					
leaf-fall	0.8			0.6	1.3	0.3	0.7	
flower and fruit fall	0.04			0.08	0.06	0.08		
branch fall	0.08	0.01		0.1	0.05	0.02	0.3	
total plant litter fall	0.9	1.1		0.8	1.4	0.4	1.0	
total vegetation to litter flow	1.0		0.1	1.2	2	1.1	1.4	0.3
herbivore consumption	0.2							

* available phosphorus

TABLE 6. Ecosystem component stores and flows of potassium in different tropical forests (see table 4 for details of localities and sources).

	Panamá		Puerto Rico	Ivory Coast			Ghana	Malaysia
	tropical moist	premontane wet	El Verde	Banco plateau	Banco valley	Yapo	Kade	Ulu Gombak
ECOSYSTEM STORES (g/m²)								
above-ground vegetation	298		(102)	60	60	35	81	
below-ground vegetation	8						9	
total vegetation	306	171	92				90	
animals	0.02		0.2 (0.1)					
litter	4		9 (5)				1	
soil	35			8*	16*	12*	65*	
(depth analysed in cm)	(30)		(25) (25)	(50)	(50)	(50)	(25)	
Total store	345	201	101 (107)					
ECOSYSTEM GAINS (g/m²/a)								
precipitation	0.9		(182)	0.6	0.6	0.6	1.8	1.3
subsoil weathering	−0.02		(26)					
ECOSYSTEM LOSSES (g/m²/a)								
stream flow	0.9		(208)					1.1
FLOWS WITHIN ECOSYSTEM (g/m²/a)								
throughfall leaching	5		(801)	6	17	8	22	0.4
stemflow leaching			(718)					
total leaching from vegetation			(1 519)					
leaf-fall	11		(0.2)	2	7	2	7	
flower and fruit fall	0.3			0.5	0.6	0.8		
branch fall	2	0.8		0.3	0.2	0.1	0.6	
total plant litter fall	13	9		3	8	3	8	
total vegetation to litter flow	18			9	25	11	29	3.2
herbivore consumption	1.5							

* exchangeable potassium

TABLE 7. Ecosystem component stores and flows of calcium in different tropical forests (see table 4 for details of localities and sources).

| | Panamá | | Puerto Rico | Ivory Coast | | | Ghana | Malaysia |
	tropical moist	premontane wet	El Verde	Banco plateau	Banco valley	Yapo	Kade	Ulu Gombak
ECOSYSTEM STORES (g/m²)								
above-ground vegetation	358			120	120	190	248	
below-ground vegetation	21						15	
total vegetation	379	189	124 (129)				263	
animals	0.04							
litter	32		4.5 (1.7)				4.5	
soil	2 217		114 (35)	10*	20*	22*	258*	
(depth analysed in cm)	(30)		(25) (25)	(50)	(50)	(50)	(25)	
Total store	2 627	394	(242) (166)					
ECOSYSTEM GAINS (g/m²/a)								
precipitation	2.9		3.4 (218)	1.6	1.6	1.6	1.2	1.4
subsoil weathering	13		(213)					
ECOSYSTEM LOSSES (g/m²/a)								
stream flow	16		4.4 (432)					0.2
FLOWS WITHIN ECOSYSTEM (g/m²/a)								
throughfall leaching	4		4.0 (239)	2.3	3.1	1.9	2.9	1.5
stemflow leaching			0.6 (73)					
total leaching from vegetation			4.6 (312)					
leaf-fall	22	10	(4)	5	7.5	8.5	21	
flower and fruit fall	0.1			0.3	0.1	0.6		
branch fall	1.7	0.8		0.9	0.9	1.5	8.2	
total plant litter fall	24	10	4.2	6	8	11	29	
total vegetation to litter flow	28		9	7	11	13	32	
herbivore consumption	2							—7

* exchangeable calcium

Description, functioning and evolution of tropical forest ecosystems

TABLE 8. Ecosystem component stores and flows of magnesium in different tropical forests (see table 4 for details of localities and sources).

	Panamá		Puerto Rico El Verde	Ivory Coast			Ghana Kade	Malaysia Ulu Gombak
	tropical moist	premontane wet		Banco plateau	Banco valley	Yapo		
ECOSYSTEM STORES (g/m²)								
above-ground vegetation	38		34 (35)	53	53	18	34	
below-ground vegetation	3						4.4	
total vegetation	41	37					37	
animals	0.02							
litter	0.2		1 (1)				0.6	
soil	226		79 (29)	8*	11*	8*	37*	
(depth analysed in cm)	(30)		(25)	(50)	(50)	(50)	(25)	
Total store	268	170	114 (64)					
ECOSYSTEM GAINS (g/m²/a)								
precipitation	0.5		2.6 (47)	0.7	0.7	0.7	1.1	0.3
subsoil weathering	4		(104)					
ECOSYSTEM LOSSES (g/m²/a)								
stream flow	4		1.8 (151)					0.2
FLOWS WITHIN ECOSYSTEM (g/m²/a)								
throughfall leaching	1		0.4 (68)	3.4	4.1	1.6	1.8	0.2
stemflow leaching			0.1 (16)					
total leaching from vegetation			0.5 (83)					
leaf-fall	2	3	(1)	4	3	1.8	4.5	
flower and fruit fall	0.04			0.3	0.1	0.2		
branch fall	0.15	0.15		0.8	0.2	0.2	0.8	
total plant litter fall	2	3	0.9	5	3.6	2.3	5	
total vegetation to litter flow	3		1.4	8	7.5	4	7	
herbivore consumption	0.3							1.9

* exchangeable magnesium

Inputs and outputs

Where all major inputs and outputs have been studied the central parts of tables 4–8 show that these are approximately equal. However, it will be observed that all these values are only available for forests where the stream flow losses have been determined and where impervious rocks prevent any loss to a deep water-table; the rates of subsoil weathering were obtained by subtraction. In some cases this results in negative values which are usually interpreted as net gains to the ecosystem. The total ecosystem inventory divided by the annual loss gives the turnover rate; 100 or more years were required for potassium, phosphorus and calcium in Panamá (Golley *et al.*, 1975) whereas at El Verde the longest turn-overs were *ca.* 30 years for calcium and magnesium and much less than one year for the other macronutrients (Odum and Pigeon, 1970). However, these calculations are of little significance because of the imprecision of both the loss measurements and the delimitation of the ecosystem.

Flows within ecosystems

Fig. 1 shows that the major flow within the ecosystem in from the soil to the vegetation; rates for this have been calculated, but not measured, in tropical forests. From the vegetation there are three possible pathways: via the decomposer or browser food chains, or by direct leaching of the foliage by rainfall. The only data on browsing food chains is that of Golley *et al.* (1975) from Panamá who found that it accounted for *ca.* 10 per cent of the total from vegetation flow. The bulk of the flow, as in all woody ecosystems, is thus via the plant litter and decomposers.

However, flow from the vegetation to the soil may be as litter fall or due to the action of rainfall (in practice there is direct flow from the vegetation to the soil by throughfall and stemflow not encountering or not inter-acting with the surface litter at times of year when this is sparse and indirect flow in which there is an intermediate stage with the elements incorporated into the surface litter). The proportions of this vegetation to soil flow as litter fall varies with the elements: generally it is over three quarters for nitrogen, phosphorus and calcium, about half for magnesium, and less than half for potassium and sodium. Here there is clearly some relationship with the chemical mobility of the ions. Within the litter flow, leaves generally account for 75–85 per cent, wood for 5–20 per cent, and flowers and fruits for 5–15 per cent of the annual mineral flows (Klinge and Rodrigues, 1968; Bernhard-Reversat, 1972).

They are differences in the chemical composition of rainfall, throughfall and stemflow depending on whether the rain-water has leached elements from the leaves and stems or whether these organs have absorbed elements from the rain-water. Stemflow has rarely been measured in tropical forests; it is normally a small percentage of the throughfall (see chapter 12). Throughfall generally leaches potassium, magnesium, nitrogen and calcium from the leaves; other elements being either unchanged or absorbed by the leaves. However, work at Pasoh has shown that on some days nitrogen, potassium, sodium, calcium and magnesium are leached from the canopy by rainfall whilst, on other days, they are absorbed by the leaves from the rainfall. Sodium and calcium were often absorbed by the canopy rather than leached from it, whereas magnesium and especially potassium were seldom absorbed. Nitrate and ammoniacal-nitrogen varied considerably but formed only a small part of the nitrogen content of the leachate which was mainly albuminoid-nitrogen which, of course, was all leached from the canopy.

Thus the process is one of considerable complexity. More research is needed. In the meantime the great majority of tropical forest flow data, which are litter or leaf-fall analyses, cannot be used to suggest even the order of magnitude of the total vegetation to soil litter flow.

Nutrients in the litter layer may pass directly to the soil with rain-water penetration or go via the decomposer fauna and micro-organisms. Golley *et al.* (1975) found that, in Panamá, *ca.* 70–90 per cent of the major nutrients passed through the decomposers but less than 10 per cent of sodium did; this element was noted as being particularly abundant in the throughfall because of the proximity of the sea.

Bartholomew *et al.* (1953) and Bernhard-Reversat (1972) measured the liberation of elements during litter decomposition. Potassium is liberated very rapidly and is able to be recycled quickly, whereas calcium appears to be liberated only on complete mineralization and thus principally during the wet season when it is easily leached. Phosphorus and magnesium are intermediate. Part of the litter elements can be absorbed directly by the dense mass of fine roots in evergreen forests through the agency of mycorrhiza. This is particularly important in some Amazonian forests (Went and Stark, 1968) where the upper soil horizon has a high root density allowing efficient absorption of elements by the vegetation and relatively small losses to streams.

As previously pointed out, the only study of browsers is that of Golley *et al.* (1975) who calculated and estimated their food intakes in Panamá. Their results, which are presented in the lowest line of tables 5–8 show that the herbivores consume *ca.* 10 per cent of the standing crop of leaves and fruits per annum. The amount passing to the carnivores is *ca.* 4 per cent of this for phosphorus, potassium and calcium, 11 per cent for magnesium, and *ca.* 300 per cent for sodium. About 90 per cent of these flows were through mammals. These results need to be treated with caution until others are available for comparison. Their value for the sodium flow from herbivores to carnivores shows either mineral sources are being used or their assumptions or estimates require revision. Most herbivores probably feed to a large extent on young leaves and shoots; the consumption of which is almost certainly under recorded and their mineral content is normally higher than that of mature organs. Also no account has been taken of sap sucking insects. Thus the herbivore consumption may be somewhat greater than usually envisaged. Unpublished work in western Africa, which showed that driver ants were probably the major forest carnivores, supports this possibility.

The amount of an element in an ecosystem component divided by its annual output from that component gives its mean turnover time (in years). Most tropical forests show little variation in the turnover times of different elements in the same component except for the soil where the store largely depends on the parent material. Average turnover times are of *ca.* 15 years for the vegetation and somewhat less than one year for the litter. Golley *et al.* (1975) calculated the following range or turnover times for phosphorus, potassium, calcium and magnesium in different organs and ecosystem components in Panamá:

leaves	0.8– 1.6 year
stems	11.7–15.2 years
roots	0.4– 0.9 year
fruits	0.3 year
litter	0.2– 1.2 year
detritivores	0.4– 2.2 days
herbivores	0.4– 2.6 days
carnivores	7.3–18.3 days.

Vegetation and litter data from Puerto Rico and Ghana are of the same order of magnitude (Nye, 1961; Jordan and Kline, 1972).

Clearly element cycling is very rapid in tropical forests whether it is measured as a turnover rate or in terms of weight per unit area. It is due to high temperatures and humidities, and the absence of a resting period, allowing high productivity (except for nitrogen, mineral levels in the vegetation are generally not greater than in temperate forests). The apparent high fertility of tropical forests is due largely to this rapid cycling and not to high mineral contents in the ecosystem and certainly not to high contents in the soil.

Tables 4–8 also demonstrate considerable differences between the various forests. The two sites in the Banco Forest emphasize this for the annual flows of canopy leaching and of litter fall on the valley site were generally 2–3 times those on the plateau site. With such a large variation over a small area (the two sites were only 400 m apart) it is pointless to speculate about possible causes of the differences between the various forest types.

Other aspects

Micronutrients have been little studied. In Panamá Golley *et al.* (1975) analysed ecosystem components for thirteen of these and determined flow rates for cobalt, copper, iron, manganese, lead, strontium and zinc, and at El Verde, Odum and Pigeon (1970) analysed the flow rates of chloride, copper, iron, manganese and sulphur. There are no other results with which to compare these data.

Seasonal changes have been investigated in some forests. Their magnitude depends on the seasonal differences in rainfall which affect the degree of seasonality of leaf-fall—a major pathway. In Panamá, Golley *et al.* (1975) found that whilst the amounts of elements transported to the litter by branch fall were approximately equal in the two seasons, the wet season leaf-fall contained about one and a half times the amount of the dry season leaf-fall. Perhaps

this is due to their study of different forest sites in the two seasons; most work has shown much higher mineral movement by leaf-fall during the dry season as would be expected in the more seasonal tropical forests. Not surprisingly, they also found element flows in throughfall in the wet season to be about twenty times those in the dry season and changes in the vegetation/soil elements store ratios due to leaching. Jordan and Kline (1972) pointed out that some seasonal changes will depend on site aspect and wind direction during the wet season. They also pointed out that elements with a more or less constant concentration in stream flow (with variable flow volumes) are probably not deficient in the ecosystems from which the streams flow, whereas variation in stream flow concentration suggests inadequate soil exchange capacity and probably impoverished soil.

Successions result in an increase in vegetation biomass as has been demonstrated many times. The rates at which elements accumulate during successions are shown in table 9. The rates clearly decrease with time and, assuming the 40-year-old forest at Kade, in Ghana, is in equilibrium, probably attain equilibrium rates in *ca.* 20 years. In Guatemala, Ewel (1976) measured the rate of return of nutrients to the soil from litter at different successional stages. His data (table 10) demonstrate the gradual increase in the amounts cycling during the succession. The high rates of uptake of phosphorus and potassium from the soil by the very early secondary vegetation are particularly noteworthy as was mentioned in chapter 9.

TABLE 9. Accumulation rates of elements in secondary successions (in g/m²/a).

Age (years)	N	P	K	Ca	Mg
Sante Fe, Panamá (Golley et al., 1975)					
2		2.5	15	16	2
4		3	16	13	2
6		2	9	12	1.5
Yangambi, Zaire (Bartholomew et al., 1953)					
5	11	0.6	9	8	
18	4	0.6	3	5	
Kade, Ghana (Nye, 1961)					
40	5	0.3	2	7	1

TABLE 10. Rates (g/m²/a) of flow of elements from litter to the soil in secondary successions in Guatemala (from Ewel, 1976).

Age (years)	N	P	K	Ca	Mg
1	7	0.3	1.1	7	4
3	10	0.4	1.7	8	5
4	10	0.4	1.1	4	5
5	9	0.3	1.2	6	6
6	14	0.6	2.0	15	4
9	14	0.7	2.4	21	4
14	14	0.7	2.4	21	4
mature	17	0.6	2.0	88	6

Nitrogen is a somewhat exceptional element in that it is one of the most important to living things, is rather more difficult than most other major nutrients to analyse, in the soil is associated only with the organic matter, has volatile compounds, and is fixed from the atmosphere.

Many tropical members of the Papilionaceae possess root nodules containing *Rhizobium* or similar micro-organisms, although there seem to be few amongst forest species. Nitrogen fixation has been critically demonstrated for very few species and its rate under field conditions is unknown. Some tropical non-legumes have nodules on their roots, stems or leaves. These contain micro-organisms and may also fix nitrogen symbiotically. These include species of *Cycas* and *Dioscorea* and some number of the Rubiaceae.

Non-symbiotic fixation also occurs by the free-living bacteria *Azotobacter* spp., *Beijerinckia indica* and *Clostridium* spp. Moore (1966) quotes rates of 4–24 g/m²/a and considers that it may be an important input of nitrogen.

The total nitrogen exchange with the air may be considerable. At El Verde it was estimated to be over 10 g/m²/a; a rate over ten times that of the input from rainfall (Odum and Pigeon, 1970). These estimates are by difference. Direct determinations using the acetylene reduction technique are necessary for indisputable results (Balandreau *et al.*, 1973).

Few studies have been carried out on soil nitrogen in tropical forests. Singh (1969) in pot experiments in India demonstrated that it was very rapidly lost from decomposing leaves. In Ivory Coast forests, Bernhard-Reversat (1974) determined the following rates of nitrogen mineralization in g/m²/a:

Forest	Banco (plateau)	Banco (valley)	Yapo
litter	1	1	0.3
soil (0–10 cm)	17	16	13
total	18	17	13.3

It was not clear if the differences between Banco and Yapo forests were due to soil type or the type of leaf nitrogen. De Rham's (1971) results are of the same order.

Mycorrhiza are very common on the roots of tropical forest trees and it has been suggested that most of the nutrient flow from the litter to the roots takes place through these without mineralization (Went and Stark, 1968). This may be a major reason for the relatively small stream flow losses observed.

Root exudates have been demonstrated but it is unknown if they form a significant element pathway.

Conclusions

With the present paucity of data it is dangerous to try and draw too many conclusions. Certain points do require emphasis.

Tropical forests differ from most ecosystems in three ways: their very high vegetation biomass; the very high element stores in this biomass; and their rapid rates of cycling.

Soil/vegetation element store ratios are very variable because of differences in parent material and selective uptake of the required elements by the vegetation. Hence, comparisons of cycling rates using the soil, vegetation, and total ecosystem stores as a basis give different results.

Golley *et al.* (1975) distinguished four categories of elements:
1. small amounts in the soil and rapid cycling;
2. small amounts in the soil and slow cycling;
3. large amounts in the soil and rapid cycling; and
4. large amounts in the soil and slow cycling.

They considered elements in category 1 to be limiting and category 3 to be possibly limiting. Category 4 elements were considered possibly not and category 2 ones probably not limiting. For the forests of Panamá phosphorus was limiting and potassium possibly limiting; calcium, magnesium and sodium were possibly not limiting and all the other (micro-nutrient) elements analysed probably not limiting. They point out that these forests occurred under optimal conditions and that non-optimal conditions as in the more seasonal tropics may produce different patterns of cycling. The pattern at El Verde, in Puerto Rico, is not dissimilar to that observed in Panamá, but the western African forests show several differences. These appear to be mainly due to the lower amounts of elements in the soils which probably cause some elements, particularly potassium, to be limiting.

These patterns and possible limiting elements are important considerations when perturbations are considered. They are associated with two major factors of the biogeochemical cycle in tropical forest ecosystems. First, a large proportion of the elements are contained in the vegetation and, secondly, the rates of cycling are very rapid—especially when compared with the rates of inputs and outputs. These need to be remembered when the effects of major perturbations are considered. Thus a process like deforestation removes much of both the nutrient stores and the main nutrient pathways. Tree roots will be much less abundant so that nutrients made available by litter decomposition may be lost to stream flow. If mycorrhiza do permit considerable direct flow of elements from the litter to tree roots, then the mycorrhizal destruction which follows felling may cause further element outputs.

Whatever is planted on the cleared area will require nutrients. There are three possible sources: the soil, precipitation and man (for nitrogen there is also fixation). The data in tables 4–8 show that the input from the rainfall is negligible compared to the stores in the soil and, of these, phosphorus and potassium are generally the most likely elements to be limiting.

Under shifting cultivation considerable amounts of nutrients are removed in harvests and then, as discussed above, the fertility gradually increases under the ensuing secondary succession fallow. The nutrient aspects of these processes were discussed by Nye and Greenland (1960) (see chapter 20). With tree crops, particularly of fibres, the nutrient losses on harvest are much less. However, such crops are normally permanent and it is possible that the second and later harvests may yield considerably less because of nutrient deficiencies (Bernhard-Reversat, 1976). Monitoring can easily check on potential limiting elements; it is essential that this is carried out. Deficiencies can easily be overcome by artificially adding fertilizers and such inputs will probably have to be carried out in all permanent cropping systems to balance the harvest outputs.

Research needs and priorities

The general paucity of data is obvious and the major research needs are to fill the gaps in the tables presented and to greatly expand the range of forests studied.

The processes discussed are complex and their adequate study requires an integrated team of research workers from many disciplines. Also, as the work at Pasoh has demonstrated, even day to day variations may be significant and it is essential to study a single site in considerable detail for a few years. No doubt the final results from the IBP studies in the Pasoh Forest will be a most valuable addition to knowledge of biogeochemical cycling. Similar detailed investigations are necessary from drier types of forest and from other continents.

Such research into the natural forest ecosystems is complimentary to and should link up with parallel studies on plantations and food crop agroecosystems (including shifting cultivation and the ensuing fallows) on adjacent areas so that the various land uses can be adequately compared (see chapter 20).

Clearly there cannot be more than a few such detailed studies. It is therefore vital that many other forest types and areas be studied less intensively so that the major patterns of element cycling can be elucidated. These studies should concentrate on the main pathways: leaf-fall and canopy leaching within ecosystems, and gains to and losses from ecosystems.

The following aspects are in particular need of research.
— Investigations on litter and soil organic matter decomposition including:
  • factors controlling the rate of decomposition;
  • the rate of decomposition of woody material and the impact of dead trunks;
  • the role of the microflora and microfauna, and the importance of termites in consuming dead organic matter.
— Investigations on mineral flows into and from ecosystems including:
  • the input by rainfall;
  • quantitative measures of nitrogen fixation;
  • losses by leaching, stream flow and deep drainage. Precise data on these should enable the determination of the degree to which biogeochemical cycles are closed in both forests and derived agroecosystems.
— Investigations on flows within ecosystems including:
  • the processes of liberation of elements from litter and soil organic matter;
  • the chemical state of elements in the soil with reference to their availability to absorption by plants and including the role of soil organic material;
  • the role of mycorrhiza in organic compounds assimilation, including its quantitative importance;
  • secondary food chains and especially the role of insects.

Bibliography

BALANDREAU, J.; TOUTAIN, F.; HUTTEL, C.; REVERSAT, F. Mesure de la fixation de l'azote moléculaire en forêt. In: *Nouveaux documents pour une étude intégrée en écologie du sol*, p. 31–33. Paris, Recherche coopérative sur programme du Centre national de la recherche scientifique (CNRS) n° 40, écologie du sol, vol. 3, 1973, 289 p.

BARTHOLOMEW, W. V.; MEYER, J.; LAUDELOUT, H. *Mineral nutrient immobilization under forest and grass fallow in the Yangambi (Belgian Congo) region, with some preliminary results on decomposition of plant material on the forest floor*. Bruxelles, Publ. INEAC, Sér. sci., n° 57, 1953, 27 p.

BECK, L. Bodenzoologische Gliederung und Charakterisierung des amazonischen Regenwaldes. *Amazoniana*, 3, 1971, p. 69–132.

BERNHARD, F. Étude de la litière et de sa contribution au cycle des éléments minéraux en forêt ombrophile de Côte-d'Ivoire. *Oecol. Plant.*, 5, 1970, p. 247–266.

BERNHARD-REVERSAT, F. Décomposition de la litière de feuilles en forêt ombrophile de basse Côte-d'Ivoire. *Oecol. Plant.*, 7, 1972, p. 279–300.

——. L'azote du sol et sa participation au cycle biogéochimique en forêt ombrophile de Côte-d'Ivoire. *Rev. Ecol. Biol. Sol* (Paris), 11, 1974, p. 263–282.

——. Recherches sur l'écosystème de la forêt sub-équatoriale de basse Côte-d'Ivoire. VI. Les cycles des macroéléments. *La Terre et la Vie* (Paris), 29, 1975, p. 229–254.

BERNHARD-REVERSAT, F. Essai de comparaison des cycles d'éléments minéraux en plantation de framiré (*Terminalia ivorensis*) et en forêt naturelle de Côte d'Ivoire. *Bois et Forêts des Tropiques* (Nogent-sur-Marne), no. 167, 1976, p. 25–38.

DICKINSON, C. H.; PUGH, G. J. F. (eds.). *Biology of plant litter decomposition*. London and New York, Academic Press, 1974, vol. 1, 288 p.; vol. 2, 610 p.

EWEL, J. J. Litter fall and leaf decomposition in a tropical forest succession in eastern Guatemala. *J. Ecol.*, 64, 1976, p. 293–308.

FITTKAU, E. J.; KLINGE, H. On biomass and trophic structure of the Central Amazonian rain forest ecosystem. *Biotropica*, 5, 1973, p. 2–14.

GOLLEY, F.; ODUM, H. T.; WILSON, R. F. The structure and metabolism of a Puerto Rican red mangrove forest in May. *Ecology*, 43, 1962, p. 9–19.

GOLLEY, F. B.; LIETH, H. Bases of organic production in the tropics. In: Golley, P. M.; Golley, F. B. (eds.). *Tropical ecology with an emphasis on organic production*, p. 1–26. Athens, Univ. of Georgia, 1972, 418 p.

——; MCGINNIS, J. T.; CLEMENTS, R. G.; CHILD, G. I.; DUEVER, M. J. *Mineral cycling in a tropical moist forest ecosystem*. Athens, Univ. of Georgia Press, 1975, 248 p.

GREENLAND, D. J.; KOWAL, J. M. L. Nutrient content of a moist tropical forest of Ghana. *Plant and Soil*, 12, 1960, p. 154–174.

HOPKINS, B. Vegetation of the Olokemeji Forest Reserve, Nigeria. IV. The litter and soil with special reference to their seasonal changes. *J. Ecol.*, 54, 1966, p. 687–703.

HUTTEL, C.; BERNHARD-REVERSAT, F. Recherches sur l'écosystème de la forêt sub-équatoriale de basse Côte-d'Ivoire. V. Biomasse végétale et productivité primaire, cycle de la matière organique. *La Terre et la Vie* (Paris), 29, 1975, p. 203–228.

JENNY, H.; GESSEL, S. P.; BINGHAM, F. T. Comparative study of decomposition rates of organic matter in temperate and tropical regions. *Soil Science*, 68, 1949, p. 419–432.

JORDAN, C. F.; KLINE, J. R. Mineral cycling: some basic concepts and their application in a tropical rain forest. *Ann. Rev. Ecology and Systematics*, 3, 1972, p. 33–50.

KENWORTHY, J. B. Water and nutrient cycling in a tropical rain forest. In: Flenley, J. R. (ed.). *The water relations of Malesian forests. Transactions of the first Aberdeen-Hull Symposium on Malesian ecology*, p. 49–65. University of Hull, Dept. of Geography, miscellaneous series no. 11, 1971, 97 p.

KLINGE, H. Biomasa y materia orgánica del suelo en el ecosistema de la pluviselva centro-amazónica. *Acta Cient. Venez.*, 24, 1973, p. 174–181.

——. Root mass estimation in lowland tropical rain forests of Central Amazonia, Brazil. III. Nutrients in fine roots from giant humus podsols. *Tropical Ecology*, vol. 16, no. 1, 1975, p. 28–38.

——. Nährstoffe, Wasser und Durchwurzelung von Podsolen und Latosolen unter tropischem Regenwald bei Manaus/Amazonien. *Biogeographica* (The Hague), vol. 7, 1976, p. 45–58.

——. Bilanzierung von Hauptnährstoffen im Ökosystem tropischer Regenwald (Manaus) vorläufige Daten. *Biogeographica* (The Hague), vol. 7, 1976, p. 59–77.

——; RODRIGUES, W. A. Litter production in an area of Amazonian terra firme forest. Part. I. Litter fall, organic carbon and total nitrogen contents of litter. *Amazoniana*, vol. 1, no. 4, 1968, p. 287–302. Part II. Mineral nutrient content of the litter. *Amazoniana*, vol. 1, no. 4, 1968, p. 303–310.

——; FITTKAU, E. J. Filterfunktionen im okosystem des zentralamazonischen Regenwaldes. *Mitteilgn. Dtsch. Bodenkundl. Gesellsch.*, 16, 1972, p. 130–135.

LAUDELOUT, H.; MEYER, J. Les cycles d'éléments minéraux et de matière organique en forêt équatoriale congolaise. In: *Trans. 5th Int. Cong. Soil Sci.*, 2, 1954, p. 267–272.

LIM, M. T. Litter fall and mineral nutrient content of litter in Pasoh Forest Reserve. In: *IBP Synthesis Meeting* (Kuala Lumpur, August 1974), multigr.

MADGE, D. S. Leaf fall and litter disappearance in a tropical forest. *Pedobiologia*, 5, 1965, p. 273–288.

——. Field and laboratory studies on the activity of two species of tropical earthworms. *Pedobiologia*, 9, 1969, p. 188–214.

MALDAGUE, M. E. *Rôle des animaux édaphiques dans la fertilité des sols forestiers.* Bruxelles, Publ. INEAC, Sér. sci., n° 112, 1970, 245 p.

NYE, P. H. Organic matter and nutrient cycles under moist tropical forest. *Plant and Soil*, vol. 13, no. 4, 1961, p. 333–346.

——; GREENLAND, D. J. *The soil under shifting cultivation.* Harpenden, Commonwealth Bureau of Soils, Technical Communication, no. 51, 1960, 156 p.

ODUM, H. T.; PIGEON, R. F. (eds.). *A tropical rain forest. A study of irradiation and ecology at El Verde, Puerto Rico.* Division of Technical Information, U.S. Atomic Energy Commission (USAEC, Oakridge, Tenn.), 1970, 1678 p.

OLSON, J. S. Energy storage and the balance of producers and decomposers in ecological systems. *Ecology*, 44, 1963, p. 322–331.

RAMAM, S. S. Primary production and nutrient cycling in tropical deciduous forest ecosystems. *Tropical Ecology*, vol. 16, no. 2, 1975, p. 140–146.

RHAM, P. DE. *L'azote dans quelques forêts, savanes et terrains de culture d'Afrique tropicale humide (Côte-d'Ivoire).* Université de Lausanne, thèse; Zurich, Buchdruckerei Berichthaus, 1971, 124 p.

RODIN, L. E.; BAZILEVICH, N. I. *Production and mineral cycling in terrestrial vegetation* (transl. ed. G. E. Fogg). Edinburgh and London, Oliver and Boyd, 1967, 288 p.

SINGH, J.; SINGH, U. R. An ecological study of soil Micro-Arthropods from soil and litter of tropical deciduous forest of Varanasi, India. *Tropical Ecology*, vol. 16, no. 2, 1975, p. 81–85.

SINGH, K. P. Studies on decomposition of leaf litter of important trees of deciduous forests at Varanasi. *Tropical Ecology*, vol. 10, no. 2, 1969, p. 292–311.

STARK, N. The nutrient content of plants and soils from Brazil and Surinam. *Biotropica*, 2, 1970, p. 51–60.

STARK, N. Direct nutrient cycling in the Amazon basin. In: Idrobo, J. M. (ed.). *II Simp. Foro Biol. Trop. Amazon.* (Florencia, Caqueta y Leticia, Amazonas, 1969), p. 172–177. Bogotá, 1970.

——. Nutrient cycling. I, II. *Tropical Ecology*, 12, 1971, p. 24–50, 177–201.

——. Nutrient cycling pathways and litter fungi. *Bio-Science*, 22, 1972, p. 355–360.

Tropical Test Center. *Environment data base for regional studies in the humid tropics.* Semi-annual Report no. 1 and 2. USATECOM Project no. 9–4–9913–01, 1966, p. 71–76.

TURVEY, N. D. *The nutrient cycling under tropical rain forest in Central Papua.* Univ. of Papua-New Guinea, Dept. of Geography, Occasional Papers no. 10, 1974, 96 p.

WENT, F. W.; STARK, N. Mycorrhiza. *Bio-Science*, 18, 1968, p. 1035–1039.

14 Pests and diseases in forests and plantations

Introduction

Many papers on tropical forests pests and diseases were presented at a meeting organized jointly by IUFRO and FAO in 1964 at Oxford and were published in two volumes in 1965 entitled, *IUFRO/FAO symposium on internationally dangerous forest pests and diseases*. Consequently only the major problems in the past decade, with emphasis on the period 1969–1974, are reviewed in this chapter. Spaulding's (1961) annotated list of viral, bacterial and fungal diseases of forest trees should also be consulted.

A pest or pathogen is defined as a destructive biological organism, while the term disease refers to the physiological and structural disturbances caused by a pest or lack of a needed symbiont or nutrient. Damage is the social and economic value of the destruction caused. These organisms play an essential role in ecosystems, particularly in forests where most participate in the vital functions of degradation, decomposition and recycling of nutrients.

Pest organisms and damage

Parasitic flowering plants are a major problem in some areas, although weeds and dodders are excluded from this chapter. Mistletoes (Loranthaceae) appear to cause more loss than any other family of parasitic phanerogams (Greenham and Hawksworth, 1965) and many species attack a wide range of hosts. Seed dispersal may be by explosive ejection in dwarf mistletoes and by animals, chiefly birds, in leafy mistletoes. Dwarf mistletoes are a highly specialized genus (*Arceuthobium*) of epiphytic dicotyledons parasitic on the Cupressaceae and Pinaceae (Hawksworth and Wiens, 1972). Their infection induces the production of profuse dense masses of distorted branches called witches' brooms. The infection reduces the tree's growth rate when the upper half of the crown is affected, and a heavy attack eventually kills the tree.

Viruses are ultramicroscopic infectious particles consisting of a single or double stranded central helix of nucleic acid with or without a protein coat (Wellman, 1972; Carter, 1973). Symptoms vary: those that destroy chlorophyll irregularly in leaf tissues or prevent its uniform formation are commonly called mosaics or mottles; others cause dwarfing, leaf enations (outgrowths), curling and necrotic spotting of leaves, yellowing and rosettes, witches' brooms, leaf galls, big bud and floral gigantism, adventitious buds and roots, root swellings and tumours (Seliskar, 1965; Carter, 1973). While symptoms are typical for host species, they are extremely variable when different species are infected. Viruses may be transmitted by many vectors and means: insects (notably species of Aphididae, Aleyrodidae, Miridae and Coccoidea), mites, nematodes, fungi, seeds, pollen, parasitic plants (dodder), grafts, budding and sap (Carter, 1973). Wellman (1972) considers viruses next to fungi as the most destructive agents of crops in the neotropics, but their role in forestry is certainly less than that of fungi and insects. Seliskar (1965) lists few viruses as tropical forest pests.

Mycoplasma-like organisms (MLO) have gained prominence in the past few years. Diseases grouped together as yellow diseases were attributed to viruses until 1967, but several of these have now been shown to be caused by MLO. Characteristic symptoms include proliferation of axillary buds and shortening of internodes giving rise to witches' brooms, a reduction in leaf size, brittleness and chlorosis of the leaves, and replacement of floral parts by leaves (Hull, 1972). Transmission occurs mainly by insect vectors, especially by Cicadellidae.

Bacteria, generally about 1 μ in diameter, require a court of entry; lenticels, stomata and water pores are natural courts, while mechanical wounds are another means of entry (Carter, 1973). Transmission is accomplished by a variety of agencies—rain, wind, insects, etc. In the warm, moist tropics, bacteria are abundant on plants and in the soil. Few studies have been made of their ecology in tropical forests (Wellman, 1972). Their mode of attack is more restricted and prolonged. Bacteria are found mostly in the parenchyma cells of the sap-wood. In newly felled trees they establish themselves at the wood-bark interface and spread rapidly. In living trees symptoms include blight, leaf spots, rot, hyperplasia and wilting; wetwood or watermark symptoms are also caused by bacteria (Rossell *et al.*, 1973).

Boyce's (1961) excellent introduction to forest pathology mentions few tropical problems. A number of accounts of tropical fungal diseases and associated pathogens have been published since 1965 covering a specific geographical region or a particular host or pest. Perhaps the most general and notable contribution has been Wellman's (1972) book on diseases in the neotropics which includes several forestry diseases and many observations on the ecology of different pests and diseases. Areas covered in other contributions include southern Brazil (May, 1964); Cuba (Morellet, 1969); Trinidad, Tobago and part of Jamaica (Pawsey, 1970); Dominican Republic (Etheridge, 1971); Fiji (Firman, 1972); India (Bakshi *et al.*, 1972); Africa (Gibson, 1965; Gibson, 1967); and Malawi (Peregrine and Siddiqi, 1972). With respect to tree species: Natawiria (1972/73) reviewed the diseases (rots, nematodes and insects) of *Albizia falcata* (= *falcataria*) in Indonesia; Etheridge (1968) made preliminary observations on fungal and other diseases of *Pinus caribaea* in Belize; Gibson (1970) reviewed fungal diseases of *Pinus patula*; Procter (1965) surveyed the diseases of *Pinus* in southern Tanzania; Bakshi (1966) listed diseases of man-made forests, including several from the tropics; and Ivory (1974) described the fungal problems of exotic conifers in West Malaysia; Levy (1968) reviewed sapstain in stored

wood products; Bakshi (1971) published on the Polyporaceae of India; Schönhar (1974) reviewed the literature published during 1969–1973 on *Heterobasidium annosum* (Fr.) Brev. (= *Fomes annosus* (Fn) Cooke, Polyporaceae).

The more common diseases caused by fungi (and other agents) include the following. Damping-off or mortality among seedlings in nurseries has been frequent and often alarmingly high. According to Reddy (1969) it may occur at three stages: pre-emergence blight; post-emergence mortality at the cotyledon stage soon after germination; and nursery root rot in older seedlings which is seldom lethal but frequently reduces growth. Damping-off is normally caused by a number of soil inhabiting facultative pathogenic fungi. Their activity as pathogens may be favoured by certain environmental conditions unfavourable to the plant but favourable to the fungi such as soil texture, extreme pH and excess moisture or drought.

Diseases caused by heart rot fungi may be classified into trunk, butt and root rots (Bakshi, 1965). In trunk rot the fungi may enter via two types of infection courts: wounds due to fires, pruning and thinning, wind, cankers and animal damage or dead branches, stubs and knots which develop normally on the stem. Fungi causing butt and root rot enter via healthy or injured roots and progress into the butt; butt rot may also be initiated via wounds at the base of the stem. Root rot is a more insidious disease because it is extremely difficult to determine the extent of the infection without seriously harming or killing the host (Foster, 1965). The incidence of heart rot is high in older trees of certain species and there is much wastage during logging. Bakshi (1965), Foster (1965) and Bakshi and Singh (1970) provide excellent accounts of these diseases.

The incidence of bark lesions commonly known as cankers is relatively high in several species although economic damage is seldom conspicuous. Krstic (1965) divided cankers into four categories: open perennial cankers; covered perennial cankers; covered ephemeral cankers; and generalized bark necrosis characterized by the absence of callus tissue. Bacteria and fungi are the principal pathogens of cankers, and frequently two or more species are responsible. Insects and nematodes are also often present. Environmental and subcultural factors (poor site, drought, age, high humidity and adverse tree vigour) are instrumental in creating and assisting the spread of cankers. Krstic (1965) lists few canker problems in the tropics.

Rust fungi (Protobasidiomycetes: Uredinales) are extremely numerous in the tropics. Wellman (1972) reports that tropical rusts are commonly autoecius (having a single host) whereas those in the temperate zone are heteroecius. In forest trees, they are generally heteroecius and associated with conifers. Wellman also states rusts are more scattered in mixed tropical forests due to intermingling with many nonsusceptible hosts and 'most of the time a tropical tree host does not have even a good chance to build up an excessive abundance of spores, and to become a serious menace in the wild'. A reddish discolouration of foliage or seed is a widely recognized symptom of rust diseases; they also form cankers and galls on branches and stems, deform shoots, and occasionally produce witches' brooms (Gremmen, 1965).

The fungi that form mycorrhizae with plant roots play an important ecological role in forestry; their absence gives rise to disease symptoms, especially severe nutrient deficiency. Mycorrhizal are more efficient than non-mycorrhizal roots in absorbing minerals in short supply and they expose a much greater absorbing surface (Bakshi and Kumar, 1968). Ectomycorrhizae, in which the root system is heterozoic, occur in the Fagaceae, Myrtaceae, Pinaceae and a few other families and vesicular-arbuscular mycorrhizae, in which hyphae penetrate host cells, occur in most others including the genera *Araucaria, Cupressus, Podocarpus, Swietenia* and *Tectona*. In the tropics mycorrhizae are essential in the successful establishment of many exotic trees and must be considered in the afforestation of former treeless areas; the initial failure of *Pinus kesiya* in the Philippines and *P. merkusii* in Indonesia was attributed to lack of mycorrhizae (Bakshi and Kumar, 1968). A comprehensive review of mycorrhizae in afforestation has been published by Mikola (1970) who refers to several examples in the tropics. More recently, Smith (1974) has reviewed the physiology and associations of mycorrhizal fungi.

Although insects have received most attention it would be misleading to attribute most damage to them rather than to fungi. Nearly all tropical forest insect pests belong to one of five orders: Coleoptera, Hemiptera, Isoptera, Lepidoptera and Orthoptera. Forest insect damage may be classified into four broad categories: defoliation (direct or indirect) caused largely by larvae of Lepidoptera and by a few Coleoptera and Orthoptera; boring or mining inside seeds, bark, wood and shoots, mainly due to species of Coleoptera and Isoptera and to a few Lepidoptera; chewing of bark and wood largely by Coleoptera; and necrosis and wilting due to chemicals secreted by some Hemiptera. In addition, insects are also vectors of many major diseases. A marked increase in research on insect pests in the tropics, especially in plantations, has been evident in the past five years.

A comprehensive account of major insect problems in tropical forestry, including forest products, has been published by Gray (1972a) and Lamb (1974) makes brief reference to a few forest pests. Other accounts pertaining to specific geographical areas,[1] insect groups or tree species published recently or not in Gray's (1972a) paper include: neotropics (Orians, 1974), Colombia (Bustillo and Lara, 1971; Bustillo, 1973), Salvador (Berry, 1959), Cuba (Morellet, 1969), Dominican Republic (Etheridge, 1971), Fiji (Roberts, 1973a), Southern Pacific Islands (Gray, 1974a), Papua-New Guinea (Gray and Wylie, 1974), Indonesia (Natawiria and Tarumingkeng, 1971), Thailand (Chaiglom, 1974), Burma (Hanson, 1963), Bangladesh and Pakistan (Zethner, 1973), and Malawi (Lee, 1971), *Pinus caribaea* in Nicaragua (Yates, 1971); Isoptera by Araujo (1970) on the termites of the Neotropical Region, Bess (1970) on Hawaii and the Oceanic Islands, Paulian (1970) on Madagascar, Roonwal (1970) on the Oriental Region; Gay and Calaby (1970) and Weesner (1970) on the termites of the Australian and Nearctic Regions respectively, Gay (1970) on introduced termites, Kudler (1967) on insect defoliators, largely Lepidoptera, in Ghana, Austara and Jones (1971) list 63 species of Lepidop-

tera, their host plants and pest status of exotic *Cupressus* and *Pinus* species in Africa.

Ruehle (1973) states that 'nematode disease of forest trees remain virtually unknown. Disregard of rhizosphere ecology by forest scientists probably accounts for past faitures to recognize nematodes as important soil and site factors in forestry. Root losses due to nematodes on forest trees have generally gone unnoticed'.

Other animals can also be very destructive in tropical forests and plantations. Most of these are relatively large and indigenous, for example, Rosevear (1969) notes that several rodents in West African forests are vegetarian and live on fruits, flowers, leaves and soft inner bark of trees. However, cattle, goats and deer are among the more prominent introductions and can be serious nuisances if left unchecked. Kulkarni (1968) classified the types of damage experienced in tropical forests due to wildlife as: compacting the soil; trampling; digging for roots; browsing; debarking; and damage to seeds. Although widlife damage is not rare, there are few reports published in the literature in recent years.

Seed mortality is often heavy due to pests. According to Janzen (1971) seed predators strongly influence plant composition by maintaining species diversity. Bacteria, fungi, insects, birds and mammals are the dominant seed pests. Noble and Richardson (1968) have prepared a list of seed-borne diseases. Seed production or ripening occurs at a certain time of the year, either annually or at longer intervals, and during this period several predators, notably birds and insects, appear in large numbers and take a heavy toll. Many organisms may attack seed of a single host (Gray and Wylie, 1974).

Pest and disease problems of important timber trees in the tropics have been collated for *Gmelina arborea* (Lamb, 1968a), *Cedrela odorata* (Lamb, 1968b), *Araucaria* spp. (Ntima, 1968), *Pinus merkusii* (Cooling, 1968), *P. caribaea* (Lamb, 1971) and *Terminalia ivorensis* (Lamb and Ntima, 1971).

---

1. There are about 1 500 species of insect pests known to damage forests and forest products in the Indian subcontinent and the earliest references to forest insect pests in the Indian region date back to 1810–1840. All the information available up to 1898 was compiled by Stebbing in a pamphlet entitled *Injurious insects of Indian forests* (in Beeson, 1941). Regular pest diseases research and surveys in India, however, began with the inauguration of the present Forest Research Institute in 1906–1907. Insect pest surveys were conducted as a result of which a treatise on *Ecology and control of the forest insects of India and the neighbouring countries* was published by Beeson (1941). During this period a reference collection of insects was also started at the Forest Research Institute, Dehra Dun, which now contains 200 000 insects of 20 000 species. The publication of a list of insect pests of forest plants in India and the neighbouring countries in nine parts (Bashin and Roonwal, 1954; Roonwal, 1954; Bashin, Roonwal and Singh, 1956; Mathur and Singh, 1956–1960) may be considered as a supplement to Beeson's (1941) publication.

Major pest and disease problems

Rarely are pests and diseases detected during the incipient stages of damage or infection. Tree destruction or mortality is usually first noticed when very obvious and in some instances the pest may have already disappeared. Several pests may occur at the site, and it is difficult to identify the primary pathogen(s) (Viennot-Bourgin, 1974). Another problem is identifying the pest since many species are undescribed (Gray, 1974b).

The examples given below provide a good insight into the problems confronting tropical forestry. It is impossible to delineate the pests strictly according to type of damage, or to specific hosts or to a single subheading; for example, the well-known shoot-tip borer *Hypsipyla grandella* (Zeller) (Lepidoptera : Pyralidae) is a pest in plantations, regeneration and forest, and of seed, seedlings and trees. Of necessity, the mention of major pest and disease problems was confined in this chapter largely to bionomics: identity of pest and type of associated disease, significant parasites and predators, extent of damage, age of principal hosts and brief history of occurrence in locality.

The account of insects is limited to more recent papers and to those unavailable to Gray (1972a). With the exception of insects, the amount of space in this chapter is a fair reflection of the literature published (or possibly the number of specialists) on each category of pest rather than of their numbers, damage or economic significance. Major pest or disease status has been arbitrarily accorded where severe damage to seedlings or trees or commercial importance has been reported, or where authors consider a potentially serious problem has arisen. The increase in the number of reports on pest and disease damage in the neotropics during the past decade is noteworthy.

It is difficult to give accurate data on the losses due to diseases and pests, but a survey carried out in India during the late 1950s showed that combined losses due to all causes was *ca.* 13 per cent (Ministry of Overseas Development, 1958). This estimate appears to be rather conservative because of the difficulty in carrying out a total surveillance of the forested land. Also, the estimate relates mostly to natural forests where disease and insects generally do not normally become epidemic.

In India, decay in the standing tree is the most important single factor causing loss in timber. In a survey of 704 *Shorea robusta* trees in western Uttar Pradesh, over 70 per cent were decayed entailing a loss of about 10 per cent of timber (Bakshi, Rehill and Chowdhury, 1963). Random sample surveys conducted in central India showed that 33 per cent of *Shorea robusta* and 27 per cent of *Tectona grandis* trees are useless due to diseases and insects. *Shorea robusta* is also subject to serious periodic borer epidemics. In one instance 60 000 ha in two divisions of Madhya Pradesh were affected. The value of timber saved by trap control operation in this area was enormous as compared to the expenditure involved in the operation.

Nursery

The nursery is somewhat ambivalent because it offers pests a highly nutritious, frequently even-aged food source in a confined space; on the other hand, control is often relatively easy. Several of the pest and disease problems in the nursery have arisen as a consequence of poor siting and management and failure to appreciate or study the nutrient requirements of the young plants. With few exceptions (e.g. heart rot), seedling diseases include all types of diseases found on larger plants. Hodges (1965) has given an excellent description of the important diseases occurring in forest nurseries.

Fungi

Cylindrocladium scoparium Morgan (Moniliales : Moniliaceae), is the most important pathogen of *Eucalyptus* in nurseries in Brazil (Hodges and May, 1972). Ko *et al.* (1973) reported wilting of *Flindersia brayleyana* seedlings and browning of the stems near the soil by *Rhizoctonia solani* Kuhn (Tubasnellales : Thanatephoraceae). More than 50 per cent of the seedlings were killed in parts of the nursery beds in Hawaii. Susceptibility was found to be inversely related to seedling age, with up to complete mortality at one month and zero at four months in a trial after inoculation with *R. solani.*

Damping-off of young seedlings of *Araucaria hunsteinii* was an obstacle to establishing plantations prior to 1959 in Papua-New Guinea, but the problem has been largely overcome through using a better soil medium (Havel, 1965).

Damping-off in *Pinus merkusii* nurseries in Java and North Sumatra in Indonesia was recorded by Suharti (1971), who isolated eight fungi from the diseased seedlings. *Fusarium* sp. (Moniliales : Hyphomycetaceae), *Pythium* sp. (Peronosporales : Pythiaceae), *Rhizoctonia solani* and other species were most virulent, especially under heavy shade.

A seedling blight of *Pinus*, especially of *P. caribaea*, was observed first in the nursery in Malaysia in 1963. By 1967 it had reached epidemic proportions and the production of seedlings was halted in nurseries in West Malaysia (Freezaillah and Low, 1968). Initially the disorder was attributed to nutritional imbalance and a fungus, *Pestalotia* sp. (Melanconiales : Melanconiaceae) which was always found associated was considered as secondary. Later investigation revealed *Colletotrichum gloeosporioides* Penzig (Melanconiales : Melanconiaceae) (perfect state *Glomerella cingulata* (Stonem). Spauld. and Schrenk) as the pathogen of the disease which was most infectious during the wetter months of the year (Lim, 1970). However, Ivory (1972, 1974) considers *Cercospora pini-densiflorae* Hori and Nambu (Moniliales : Dematiaceae) as a more likely candidate of the blight, commonly called brown needle disease, rather than *C. gloeosporioides.*

In India, damping-off has been high in several states (Reddy, 1969). All *Pinus* species are susceptible to varying degrees; for example, complete mortality was recorded in *Pinus kesiya* (= *insularis*), whereas *Pinus roxburghii* was unaffected. Though *Fusarium*, *Phytophthora* (Peronosporales : Pythiaceae) and *Pythium* species may cause the disease in

India, Reddy (1969) observed *Rhizoctonia solani* to be most prevalent in isolation tests of the diseased seedlings.

Losses in *Pinus* nurseries in Kenya and Tanzania were up to 59 per cent in four-month-old *P. caribaea* and 56 per cent in five-month-old *P. patula*. Hocking and Jaffer (1969) suspected several fungi including *Botryodiplodia theobromae* Pat. (Sphaeropsidales : Sphaerioidaceae), *Corticium solani* (Prill. and Delacr.) Bourd. and Galz. (Agaricales : Corticiaceae), and *Fusarium solani* (Mart.) Sacc. The soil-borne fungi *Fusarium* spp., *Pythium* spp. (mainly *P. ultimum* Trow), and *Rhizoctonia solani* were primarily responsible for severe damping-off in *Pinus* nurseries at periodic intervals in eastern Africa (Gibson and Hudson, 1969).

Lesions on leaves of young plants of *Vitex keniensis* were very evident during the colder wet season in nurseries in Kenya. *Phoma viticis* Celotti (Sphaeropsidales : Sphaerioidaceae) and *Phyllosticta ragatensis* Ivory (Sphaeropsidales : Sphaerioidaceae) were responsible for the blight. Damage was inflicted primarily by the former which caused death of the leaves and shoot tips (Ivory, 1967a). Many graft failures of *Cupressus lusitanica* recorded in Kenya appeared to be largely due to fungal infection of the union, particularly by *Monochaetia unicornis* (Cooke and Ellis) Sacc. (perfect state *Rhynchosphaeria cupressi* Nattrass, Booth and Sutton, Melanconiaceae) (Gibson and Howland, 1969; Howland and Gibson, 1969). *Colletotrichum acutatum* Simmonds f. sp. *pinea* Dingley and Gilmour is the pathogen of the terminal crook disease of *Pinus*, especially *P. radiata*, in three or four forest nurseries between 1 800 and 2 500 m elevation in Kenya (Gibson and Munga, 1969). It is spread over long distances by the transfer of diseased plants. Abnormal thickening and rigidity of the stem of seedlings 3–30 cm high results from necrosis of the shoot-tips. Violet root rot caused by *Helicobasidium compactum* Boedijn (Auriculariales : Auriculariaceae) was first noticed in teak seedlings in 1963 in Tanzania and severe patch mortality (an estimated 100 000 plants were killed) occurred during the wet months of 1964 (Hocking and Jaffer, 1967). No further recurrence has been observed following removal of the nursery to a better site.

Insects

White grubs (Coleoptera : Scarabaeidae) are a problem in *Pinus merkusii* nurseries in Indonesia (Intari and Natawiria, 1973).

Psyllid attack is very severe in young seedlings and transplants of *Triplochiton scleroxylon* in nurseries and transplant beds in Nigeria (see Gray, 1972a). Two species, *Diclidophlebia eastopi* Vondracek and *D. harrisoni* Osisanya (Hemiptera : Psyllidae) are responsible for the damage—the psyllids feed and complete their life-cycle on the leaves, which are shed prematurely, and they also kill the apical buds. This results in stunting and copious branching (Osisanya, 1970). Osisanya (1969) concluded that seedlings less than 6 cm girth should not be used for planting in the field.

Mites

A serious infestation of the spider mite *Oligonychus milleri* (McGregor) (Acariiformes : Tetranychidae) on nursery seedlings of *Pinus caribaea* in Jamaica was reported by Muma and Adeji (1970), especially following drought periods. Moderate to severe yellowing of the needles occurred.

Nematodes

In Brazil a root-lesion nematode *Pratylenchus brachyurus* Godfrey (Tylenchida : Pratylenchidae) was found in seedlings and young *Eucalyptus alba* and *E. saligna* trees. Stunting, yellowing, a reduced root system and necrosis of the other roots were the main symptoms and many plants died (Lordello, 1967). *E. citriodora* was resistant to the disease.

In the Philippines the root-knot nematode *Meloidogyne incognita* (Kofoid and White) Chitwood (Tylenchida : Heteroderidae) attacked seedlings of *Anthocephalus chinensis* in nurseries and was responsible for severe stunting (Postrado and Glori, 1968).

Plantations

Fungi

A new pathogen was isolated in 1969 from roots of dying 10–15-year-old *Araucaria angustifolia* growing in plantations in Parana State, Brazil. It has been subsequently isolated from seedlings of *Eucalyptus saligna* and roots of several planted *Pinus* spp. (notably *P. caribaea*, *P. elliottii*, *P. kesiya* (= *insularis*), *P. oocarpa* and *P. taeda*) in the southern States of Minas Gerais, Sao Paulo and Espirito Santo. The fungus appears to be highly virulent to all pine species and is considered a serious threat to plantations of susceptible tree species in Brazil by Hodges and May (1972), who named it *Cylindrocladium clavatum*. The disease is characterized by discrete circular infection centers of up to 25 trees. Copious resin encrustation of the roots with infiltration of the inner root tissues is common on *Pinus*. The practice of using central nurseries for the distribution of seedlings has apparently hastened the dispersal of the pathogen (Hodges and May, 1972).

A serious disease was observed in 1965 on the 1962 plantings of *Eucalyptus grandis* and *E. saligna* in Surinam; by 1966 nearly all trees were affected with more than 50 per cent mortality (Boerboom and Maas, 1970). Boerboom and Maas showed the pathogen to be *Endothia havanensis* Brener (Sphaeriales : Hyalodidymae). First symptoms appear on the trees 14–18 months after planting when they average *ca.* 5 cm diameter; the base of the stem swells and longitudinal cracks appear in which reddish-brown to black pycnidia form and a ruby-coloured gum exudes. The tree reacts by forming callus around the site of infection. After 2–3 years as many as five cankers may appear along the stem and these coalesce. A survey in 1967 by Boerboom and Maas (1970) indicated the disease was present in all localities and they suggested *Eucalyptus* was unsuitable for growing in the tropical lowlands in Surinam.

Corticium salmonicolor Berk. and Br. was responsible for the death of young *Eucalyptus kirtoniana* and attack on a hybrid Eucalyptus in Turrialba, Costa Rica, while *Pellicularia koleroga* Cke (Agaricales : Thelephoraceae) infected young *Eucalyptus saligna* (Segura, 1970a).

Botryodiplodia theobromae (= *Diplodia natalensis* Pole Evans) causes stem canker and die-back in the exotic *Casuarina equisetifolia* during drought years in Puerto Rico (Liu and Martorell, 1973).

Mortality of 50–100 per cent of *Eucalyptus* plantations reported from Karnataka, Kerala and Goa States has been attributed to the indigenous pink disease caused by *Corticium salmonicolor* which has killed or badly damaged nearly 40 000 ha of plantations in high rainfall and low altitude areas. The disease manifests during the second year when the main stem and branches are attacked and girdled. It assumes epidemic level from the third year onwards. While the seedlings are killed outright in *Eucalyptus citriodora* and *E. grandis*, it is not so with *Eucalyptus* hybrid which produces epicormic shoots below the girdling. The infection progresses downwards killing the affected parts. The incidence of the disease varies from place to place; it may be up to 100 per cent and is generally more severe in high rainfall areas. In certain plantations (Uttar Pradesh) about 5 per cent of the trees were windblown due to a root disease. Also, in some locations yellowing followed by mortality occurred over an area of *ca*. 10 000 ha. *Eucalyptus* plantations are raised on a short rotation of about ten years and the juvenile decay will result in serious loss in pulpwood yield.

The fungus *Corticium scoparium* was also found pathogenic to *Eucalyptus*, causing defoliation the severity of which varied according to species: a hybrid from India was very susceptible, *E. alba* and *E. saligna* moderately susceptible, *E. grandis* and *E. maculata* only slightly susceptible, while *E. deglupta* exhibited high resistance (Segura, 1970b).

Up to 14 per cent mortality was recorded on 23 ha *Acacia catechu* plantations in Uttar Pradesh State, India (Bakshi *et al.*, 1968). Of this, 12 per cent was due to the root rot fungus *Ganoderma lucidum* ((W. Curt.) Fr.) Karst. (Agaricales : Polyporaceae); the other 2 per cent was attributed to rats. The young seedlings had apparently become infected via the soil through diseased roots of stumps left after logging. *Ailanthus excelsa* and *Bombax ceiba* raised on the same sites were found to be resistant to the world-wide fungus.

Three fungi, *Botryodiplodia theobromae*, *Colletotrichum gloeosporioides* and *Pestalotiopsis adjusta* (Ell. and Ev.) Stey. (Melanconiales : Melanconiaceae) cause leaf spot lesions on *Swietenia mahogani* and *S. macrophylla* in Bihar State, India (Verma, 1972b); up to 55 per cent of the foliage may be affected with premature leaf-fall.

A very high incidence of *Peniophora rhizomorphasulphurea* Bakshi and Singh (Aphyllophorales : Corticiaceae) and *Trichaptum zonale* (Berk.) G. H. Cunn. (= *Polyporus zonalis* Berk.) (Aphyllophorales : Polyporaceae) was found in wind-thrown teak in New Forest in Uttar Pradesh State in India (Bakshi *et al.*, 1966). The former heavily colonizes coppiced stems and stumps and other dead wood and it spreads mainly through rhizomorphs in the soil. *Trichaptum*

zonale spores enter through fresh injuries usually in the butt region and spread into the roots.

Fuscoporia livida (Kalchbr.) G. H. Cunn. (= *Fomes lividus* (Kalchbr.) Sacc.) and *Trichaptum zonale* followed *Phialophora* sp. (Moniliales : Dematiaceae) in wounds resulting from coppicing dry teak forests flush with the ground in India (Singh *et al.*, 1973). The latter is a nondecay fungus which may accelerate decay fungi reaching the heart-wood as well as making the trees more susceptible (Singh and Tewari, 1970). After establishment in the stools, the decay migrates into the callous and high side shoots, but the low side shoots remain free of infection because the heart-wood is not connected with that of the stool. The incidence of decay was as high as 50 per cent on trees at the end of the either 30 or 48 year rotation and extended for up to 4 m up the stem. Singh *et al.* (1973) recommended selection of the low side shoots and a stool height 10–15 cm above the ground to restrict the incidence of heart-wood decay in coppice crops.

Root rot in plantations of *Tectona grandis* caused by *Trichaptum zonale* has also been recorded from West Bengal. The decay is caused in the root and the butt region resulting in the decay of both heart-wood and sap-wood and makes trees liable to windthrow.

Wilt of *Casuarina equisetifolia* caused by *Trichosporium vesiculorum* Butler (Hyphales : Delatiaceae) has been reported in India from the States of Orissa, Andhra Pradesh, Tamil Nadu, West Bengal and Gujarat (Bakshi *et al.*, 1972). The pathogen infects the trees through injury caused during pruning and lopping. The fungus eventually spreads into roots. Secondary spread of the disease then occurs through root to root contacts and grafts resulting in diseased trees appearing in groups. Infected trees are killed in 6 to 8 months, and the mortality may be as high as 75 per cent. The stem blister rust of *Pinus roxburghii* is caused by *Cronartium himalayense* Bagchee (Uredinales : Melampsoraceae) and the disease resulted in serious mortality in the early 1920s and 1930s but subsequently declined (Bakshi *et al.*, 1972). It is, however, still present and the alternate stages of the rust are common on *Swietenia* spp. It causes mortality particularly in young crops, the attack occurring mainly on branches which are killed by girdling.

Leaf rust of *Tectona grandis* caused by *Chaconia tectonae* T. S. and K. Ramakr. (*Olivea tectonae*; Uredinales : Pucciniaceae) is a microcyclic rust, attacking leaves only and is wide-spread (Bakshi *et al.*, 1972). Damage is severe in young plantations, especially on nurseries where young plants have a retarded growth due to premature defoliation.

The root rot *Poria vincta* var. *cinerea* (Bres.) Setliff (Aphyllophorales : Polyporaceae) was pathogenic on 10 exotic tree species and on logs and stumps of 21 indigenous species in the highland forest areas of eastern Africa (Ivory, 1973), but it does not attack *Pinus*. The first sign of disease was chlorosis of the foliage at the base of the crown; it spread rapidly up the crown and the tree died. The root system became covered with white mycelial fans which developed into a black fungal sheath bearing white flecks. In Kenya, Ivory (1973) recorded 5.5 and 9.9 per cent average mortality in eight-year-old plantations of *Acacia angustifolia* and *Cupressus lusitanica* respectively in January 1970, and in

Uganda the mortality was up to 10 per cent in line samples and 20 per cent in square plot samples of *C. lusitanica*. Negligible losses were evident in trees older than 10 years, whereas there was considerable loss of volume and in quality in younger trees. The pathogen spreads by root contact and by air-borne basidiospores.

Plantings of *Cupressus macrocarpa* in Kenya were abandoned mainly because of serious infection by *Monochaetia unicornis* in the early 1950s (Olembo, 1969). The cypress canker occurs in the lower half of the tree and is mostly associated with branch crotches or injury due to pruning. Olembo (1969) isolated the virulent Strain A from diseased *Cupressus lusitanica*, which is less susceptible. In a survey of 6 300 trees in eastern Africa, he found 4 per cent with cypress canker and only 0.1 per cent mortality; the different provenances of *C. lusitanica* displayed variability in susceptibility. This decline in the severity of the canker has been due to the gradual removal of more susceptible plants and selection of more resistant provenances (Gibson, personal communication).

Armillariella mellea (Vahl. ex Fr.) Kummer (= *Armillaria mellea*) (Agaricales : Agariceae) occasionally causes severe losses in Kenya and Tanzania where plantations of *Pinus patula* occur. Olembo (1972) has suggested the presence of a heat-labile inhibitor in the soil which prevents damage by *A. mellea* at other sites free of the pathogen.

Dothistroma pini Hulbary (Sphaeropsidales : Sphaerioidaceae) (perfect state *Scirrhia pini* Funk and Parker) invades and kills the foliage, notably of *Pinus radiata*, resulting in a reduction in growth rate and death in extreme cases. The fungus has become a major problem in plantations in subtropical Africa (Gibson, 1972). In the case of *P. radiata* the susceptibility of the trees decreases with age, with those less than about one year being severely infected, but there is little difference between young and old foliage (Gibson, 1974).

Fusicoccum tingens Goidanish, conidial state of *Botryosphaeria ribis* (Grossenb. and Dugg.) (Sphaeriales : Gnomoniaceae), is a facultative wound pathogen causing die-back of pines in eastern Africa in association with debilitating factors such as drought, insect damage and poor site. Ivory (1967a) regards the disease as potentially serious and has demonstrated pathogenicity of the fungus to *Pinus elliottii* and *P. radiata*.

Deaths in teak plantations in Tanzania observed in 1966 were due to root rot, mostly by *Helicobasidium compactum* Boedijn (Auriculariales : Auriculariaceae), which had spread through the soil. None was recorded in the 1906, 1922 and 1962 plantations but it was present in the 1961, 1964 and 1965/1966 plantations (Hocking and Jaffer, 1967). *Fusarium solani* was isolated from collar and stem cankers and from pink-stained wood of 40-year-old teak in Tanzania by Hocking (1968), who suggested the trees had previously been subjected to severe water stress which resulted in bark fissures that provided infection courts for the pathogen.

Insects

Acanthoderes jaspidea Germar (Coleoptera : Cerambycidae) is reported as a serious pest in plantations of *Schizolobium*

parahybum in Sao Paulo State, Brazil (Vila, 1965–66). The beetle attacks trees mostly 7–15 cm girth and kills them.

Amante (1967) described damage to 5–10-year-old plantations of *Eucalyptus alba* in Sao Paulo State, Brazil, caused by leaf-eating ants, *Atta laevigata* F. Smith and A. *sexdens rubropilosa* Forel (Hymenoptera : Formicidae). Up to 15 per cent of the trees were defoliated. The same ants have also damaged *Pinus elliottii* plantings in Sao Paulo State, Brazil (Amante, 1967).

In May, 1970, nearly 2 000 ha of a *Eucalyptus* plantation in Sao Paulo State, Brazil, were severely infested by larvae of *Eupseudosoma involutum* (Sapp.) (Lepidoptera : Arctiidae) (Balut and Amante, 1971). Premature leaf-fall and withering of the branches were experienced. Large numbers of parasitic Diptera and microhymenoptera emerged from Lepidoptera pupae collected in the field.

Serious outbreaks by *Euselasia eucerus* Hewiston (Lepidoptera : Riodinidae) and *Thyrinteina arnobia* Cramer (Lepidoptera : Geometridae) occurred in *Eucalyptus* plantations in Minas Gerais State, Brazil (Briquelot and Osse, 1969). Altogether 336 ha of *E. citriodora* and 6 332 of *E. paniculata* out of a total of 18 120 ha were affected in 1967. Most damage was caused by the larvae of *Thyrinteina arnobia*.

Gryllus assimilis Fabricius (Orthoptera : Gryllidae), a serious pest that bites off leaves and debark small branches, has caused deaths in a 1972 *Eucalyptus saligna* plantation in Fazenda Monte Alegre, Brazil (Grodzki, 1973).

Pteroplata ?adustus Burmeister (Coleoptera : Cerambycidae) infests *Eucalyptus* in Bolivia, where damage was first observed in 1971 (Squire, 1972). The pest attacked trees in a plantation, where groups of up to 50 trees died. Trees as high as 9 m are attacked along the entire bole by as many as 20 larvae.

In 1968–69 a moth, *Glena bisulca* Rindge (Lepidoptera : Geometridae) killed many trees in plantations of the *Cupressus lusitanica* in Colombia (Drooz and Bustillo, 1972). The insect has three overlapping generations per year. The larvae are wasteful feeders and they may completely defoliate the trees which die as a consequence. The severest infestation was observed on the most favourable sites for tree growth at elevations of *ca.* 2 300 m. Drooz and Bustillo (1972) recorded numerous parasites, predators and diseases of which a fungus *Cordyceps* sp. (Hypocreales : Clavicipitaceae) and tachinid *Siphoniomyia melas* Bigot (Diptera : Tachinidae) were predominant, but were not apparently responsible for the heavy toll of the pest population.

The shoot-tip borers *Hypsipyla grandella* and *H. robusta* (Moore) are primarily responsible for the failure to establish monocultures of *Cedrela*, *Khaya*, *Swietenia* and *Toona* species in many tropical areas. Since the reviews by Gray (1972a) and Orians (1974) were prepared a substantial number of publications have appeared, especially on *Hypsipyla grandella* as a result of concerted activity in Costa Rica and Venezuela. Of note is the paper by Roovers (1971) describing the development, behaviour and population dynamics of *H. grandella* in plantings of *Cedrela angustifolia*, *C. odorata* and *Swietenia macrophylla* at Barinitas in Venezuela. In these studies the life-history of the pest was completed during 26–64 (mean 44) days. Most oviposition was recorded on *Cedrela angustifolia* and least on *Swietenia macrophylla*; preferences with respect to host species were exhibited by the insect. In another important development in Venezuela, Carruyo (1973) extracted and separated the essential oils of the terminal buds and leaves of *Cedrela angustifolia*, *C. odorata*, *Guarea guara* and *Swietenia macrophylla*. He then compared their attractiveness to *Hypsypila grandella*. Further work is required to identify and isolate the fraction(s) involved and determine a possible control agent. The moth showed a preference equivalent to the relative quantities present in the species; most was obtained from *Cedrela odorata*. Inoue (1973) compared the growth of four different sources of *Cedrela* (one of *C. ?odorata*, and three provenances of *C. fissilis*) planted in the open and in shade beneath a *Eucalyptus* species. He observed differences in growth, resistance to frost and to attacks by *Hypsipyla grandella*, with *Cedrela odorata* showing superior resistance to attack. Styles (1972) noted that *Cedrela* was not severely infested by *Hypsipyla robusta* in Africa.

In Central America, notably Costa Rica, intensive research has been proceeding over the past six years on *H. grandella* and many papers have been published (Anon., 1973a, 1973b). In these studies, emergence and maturing of adults (Sliwa and Becker, 1973), the flight and host selection behaviour have been investigated (Gara *et al.*, 1972–73), the presence of a female sex attractant confirmed (Holsten and Gara, 1974), and the efficacy of biological control agents considered such as introduced parasites (Bennett and Yaseen, 1972), egg parasites (Grijpma, 1972), and bacteria *Bacillus thuringiensis* (Hidalgo-Salvatierra and Palm, 1972). Geary *et al.* (1973) compared many different provenances of *Swietenia humilis*, *S. macrophylla* and *S. maghoni* grown from seed collected from Mexico to Panamá and in the Virgin Islands and planted at 13 locations in Puerto Rico and the Virgin Islands. Only *S. maghoni* showed immunity to attack from *Hypsipyla grandella* while a greater percentage of *Swietenia macrophylla* trees were attacked than *S. humilis*. There appeared to be little or no variation in resistance within *S. macrophylla* or *S. humilis* but two provenances showed greater susceptibility.

Morellet (1969) mentions *Hypsipyla grandella* as a pest of *Swietenia macrophylla* in Cuba.

Little research has been carried out on *Hypsipyla robusta*, an equally if not more serious problem because of its more wide-spread distribution (Gray, 1972a), due mainly to disinterest in planting susceptible tree species while no effective control measures are yet available against the shoot-tip borer.

In India, *Hypsipyla robusta* is the most important pest of Meliaceae (*Swietenia macrophylla* and *Toona ciliata*) within its range. It can cause considerable loss of seed, particularly of *Toona ciliata*, but its importance lies mainly in its activity as a shoot borer; infestation being heaviest on young vigorous trees growing in full sun. Heavy infestation with repeated destruction of terminal buds results in forking, development of crooked stems and often in permanent stunting. *Hypsipyla robusta* is also a major pest of *Swietenia* and combined with the attack of a collar borer, *Pagiophloeus longiclavis* Marshall (Coleoptera : Curculionidae), causes a

lot of damage (Bakshi *et al.*, 1972). These two pests are most injurious in the evergreen areas of India (Wynaad and South Coimbatore) where entire crops have been destroyed. The collar borer larva tunnels in the cambium and sap-wood, causing formation of a cankerous collar which may extend for some distance below the ground, and young trees are commonly girdled and killed. The abundance of pests is chiefly determined by the breeding facilities provided by fellings of *Toona ciliata* nearby. Data show that liability to attack is greater in new plantations on open sites than in those with overhead shade given by the canopy or by undergrowth or by cover plants.

Zethner (1973) briefly refers to the insects as pests in Bangladesh, Burma and India.

Gray and Wylie (1974) confirmed the presence of *Hypsipyla robusta* in Papua-New Guinea by rearing adults from infested shoots of *Toona sureni* (= *australis*) and they recorded the number of new infestations and mortality in a 0.3 ha plot over three years. Prior to a fire which destroyed many of the trees, they had been heavily attacked by *Hypsipyla robusta* and *Xylotrupes gideon* (Linnaeus) (Coleoptera : Dynastidae) and as a consequence the trees were severely scarred, stunted and deformed. Gray (1974a) gives brief notes on the occurrence of *Hypsipyla robusta* in the Solomon Islands and Papua-New Guinea.

In Fiji there was a high incidence of ambrosia beetle attacks in 1971 on *Swietenia macrophylla* logs (Gray, 1974a). Two species of Platypodidae and two Scolytidae were identified, of which the former were suspected as being responsible for much of the damage. Further planting was suspended in 1972 (Williams, unpublished). Gray (1974a) suggested that the beetles were breeding successfully elsewhere, possibly in the dead or dying plantation trees poisoned for thinning. Roberts (1973a) found two platypodid species, *Crossotassus externedentatus* Fairmaire and *Platypus gerstaeckeri* Chapuis (Coleoptera : Platypodidae), were primarily responsible for the damage. His studies show several behavioural differences between these species in time of flight, pattern of attack and nest structure. Breeding was observed in poisoned, wind-blown and fallen trees, and especially in thinnings (Roberts, 1973a). Most attacks occurred in trees 43 cm girth and over as were common in plantings aged seven or more years; no attacks were evident in trees aged less than five years or in the large indigenous trees left standing in the plots. The Department of Forestry has recommended planting mahogany on a small scale with a greater mixture of hardwoods with emphasis on several sylvicultural treatments prescribed by Roberts (Williams, unpublished).

The over 15–20-year-old *Araucaria cunninghamii* plantations at Bulolo and Wau in Papua-New Guinea have been less susceptible than the younger ones to insect attack by *Hylurdrectonus araucariae* Schedl (Coleoptera : Scotytidae). This invaded 2 632 ha of 4 438 ha of plantations by December 1971 compared with 629 ha of 3 345 ha by December 1966 (Gray, 1975). A large proportion of the infestation was in plantations under 15 years old, which were severely infested with little increment and considerable mortality, especially among the less vigorously growing trees (Gray, 1975). Although still infested, few of the trees die as a consequence of attack by the termite *Coptotermes elisae* (Desneux) (Isoptera : Rhinotermitidae) whereas an attack on younger trees is often fatal. The older and more vigorously growing trees also show a lower incidence of infestation by *Vanapa oberthuri* Pouillaude (Coleoptera : Curculionidae) (Gray and Barber, 1974; Gray and Wylie, 1974). No serious pest problems have been observed in *Araucaria hunsteinii* the planting of which was increased following cessation of *A. cunninghamii* in 1968.

Larvae of *Euchtora viridis* (Fabricius), *Holotrichia constricta* (Burmeister), *Holotrichia helleri* Brenske, *Lepidiota stigma* (Fabricius) and *Leucopholis rorida* (Fabricius) (Coleoptera : Scarabaeidae) are damaging the roots of one to two-year-old trees (*Albizia falcata, Anthocephalus cadamba, Leucaena glauca, Swietenia* spp. and *Tectona grandis*) in taungya plantations in East Java, Indonesia (Intari and Natawiria, 1973).

Trial plantings of *Araucaria cunninghamii* in Sabah have been infested by *Coptotermes curvignathus* Holmgren (Isoptera : Rhinotermitidae) with high mortality (Thapa and Shim, 1971). Root rot infection was considered to be an important predisposing factor, but this is doubtful in view of the small area (0.3 ha) of plantations and experience elsewhere (Gray, 1972a; Gray and Wylie, 1974), where the incidence of attack is strongly related to the presence nearby of parent colonies living in remnant stumps or logs after logging operations.

In three-year-old taungya plantations in Malaysia, Cheah (1971) noted increasing mortality due to an unidentified insect (possibly a gryllid as observed by Gray (1974a) in a similar case, rather than a cicada as suggested by Cheah). Up to 14 per cent kill was recorded in all tree species including *Dryobalanops aromatica, Swietenia* sp., *Shorea leprosula, S. ovalis* and in the exotic *Pinus caribaea*. Many punctures were present in the lateral and terminal shoots of the young trees which changed colour and die-back was evident before they fell off about three weeks after attack. A chalcid parasite was later found to be common in the infested areas.

The most notorius pest of *Acacia nilotica* plantations in India is the *Celosterna scabrator* Fabricius (Coleoptera : Cerambycidae) borer. The adults of the pest eat irregular patches of bark, sometimes girdling and killing young shoots. Only young trees of basal diameter 7 or 8 cm are subject to infestation. Though the pest is important for *Acacia nilotica*, it has also been recorded as injurious to *Prosopis cineraria*. The eggs are laid on young living plants and the larva bores in the stem and roots which are hollowed out. *Celosterna scabrator* is also a serious pest in plantations of *Casuarina equisetifolia, Prosopis spicigera*, teak and other trees in India (Gray, 1972a). More recently the beetle has become a serious threat to *Eucalyptus* plantings (*E. citriodora, E. grandis* and *E. tereticornis*) in Andhra Pradesh, Bihar, Madhya Pradesh and Tamil Nadu. The adults feed on green bark, sometimes completely girdling and finally killing the growing shoots. Oviposition takes place on the bark of 1–4-year-old trees. The eggs hatch, the larva enters the stem and eats its way down into the root thus killing the young sapling. The older plantations are resistant to attack as they are not suitable

for oviposition. *Sahyadrassus malabaricus* Moore (Lepidoptera : Hepialidae) occurs in southern India and is injurious to young plantations of *Eucalyptus* and *Tectona grandis*. Stems of *ca.* 1 cm diameter are attacked, the tunnels in larger host plants being relatively short and restricted to sap-wood. Saplings usually break at the canker point but on larger trees the wound may eventually heal depending on the size and vigour of the tree.

Urostylis punctigera Westwood (Hemiptera : Urostylidae) is a very destructive pest of pure plantations of *Michelia champaca* in West Bengal which have been destroyed by this sap-sucker. It was first noticed in 1927 and in subsequent years it was discovered in 1–14-year-old plantations; the highest incidence being in Darjeeling and Kurseong Forest Divisions. The damage caused by the pest is very localized and the infested area does not spread rapidly.

Atteva fabriciella Swederus (Lepidoptera : Yponomeutidae) and another lepidopterous pest, *Eligma narcissus* Cramer (Lepidoptera : Sarrothripinae), have been serious defoliators of 1–5-year-old monocultures of *Ailanthus excelsa* in Madhya Pradesh State in India (Gray, 1972a). Mathur *et al.* (1970) gave details on the species' biology and control. They found little defoliation in mixed plantations. Flowering trees are heavily defoliated by the sixth or final instar larvae in January-February and later in mid-year the first instar larvae of the next generation defoliate the trees at the beginning of the monsoon rains. *Atteva fabriciella* is a serious pest in *Ailanthus* plantations in northern and central India. It is rather scarce in summer on account of its large-scale mortality due to unfavourable weather and in winter large numbers of the pupae are destroyed by high degree of parasitism by *Brachymeria hime-attevae* Joseph *et al.* (Hymenoptera : Chalcididae). Repeated defoliation occurs over large areas and growth is seriously affected; one to two-year-old seedlings are sometimes killed, while older seedlings are badly weakened by killing of their laterals and the leaders. The defoliator may also attack the inflorescence, seriously affecting seed production. Trees growing on better sites are less susceptible to attack. In the absence of leaves the larvae bore inside young twigs. Besides the pupal parasite *Brachymeria hime-attevae*, a tachinid parasite (*Ptychomia remota* Aldrich or *Bessia remota*, Diptera : Tachinidae) attacks third instar larvae. Though parasitism by *Ptychomia remota* is quite common, the parasitism percentage is not high. Another tachinid larval parasite (*Carcelia* spp., Diptera : Tachinidae) and a nematode parasite (*Mermes* sp.) has been recorded to attack larvae of *Atteva fabriciella*. *Eligma narcissus* primarily attacks *Ailanthus* trees in southern India which occurs abundantly from September to February. The natural enemy complex of this pest is rather inadequately known. A fungal disease caused by *Beauveria bassiana* and two larval tachinid parasites *Eutachina civiloides* and *Sturmia inconspicuella* have been recorded for this pest.

Tectona grandis is periodically defoliated by *Hyblaea puera* Cramer (Lepidoptera : Noctuidae) and *Pyrausta machaeralis* Walker (Lepidoptera : Pyralidae), popularly known as teak skeletonizer, in Bangladesh, Burma, India and South-East Asian countries (Gray, 1972a). In India, losses from outbreaks of these pests have been estimated to be *ca.* 13 per cent of the annual increment (Bakshi, 1967). Complete defoliation of young plants may cause die-back and change the feeding habit of the larvae to boring of tender growing leaders which results in forking of the main stem. Repeated annual defoliation is believed to stimulate production of epicormic branching. A large volume of data have been collected on the parasite-predator complexes of these defoliators and on the value of miscellaneous undergrowth supporting caterpillars which are collateral hosts of their parasites. One of the most important parasites recorded for these two defoliators is *Cedria paradoxa* which can be produced on a mass scale.

Among other tree defoliators in India, that of *Gmelina arborea* by *Calopela leayana* Latreille (Coleoptera : Chrysomelidae), *Dalbergia sissoo* by *Plecoptera reflexa* Guenee (Lepidoptera : Noctuidae) and *Anthocephalus chinensis* (*Anthocephalus cadamba*) by *Arthroschista* (*Margaronia*) *hilaralis* Walker (Lepidoptera : Lymantriidae) need mention. *Gmelina arborea* defoliator is a serious pest of plantations in the eastern region and occurs wherever the tree is grown. In heavy infestations the leading shoots of young trees dry up and remain leafless for about four months of the growing season and are eventually killed. The defoliator of *Dalbergia sissoo* is serious in irrigated plantations and can cause substantial losses. The *Anthocephalus chinensis* defoliator is of importance in West Bengal where entire plantations may be defoliated. Terminal buds may also be destroyed leading to stunted growth and epicormic branching.

A shoot borer, *Tonica niviferana* Walker (Lepidoptera : Oecophoridae), severely attacked one and two-year-old transplanted *Bombax ceiba* grown in alternate rows with teak in Kerala State, India (Sebastian, 1969). No infestation was observed in the nursery. One pupal parasite, *Xanthopimpla tonicae* Townes and Shui-Chen Chiu (Hymenoptera : Ichneumonoidea) has been recorded for this pest.

Fabre and Brunck (1974) reported infestation of *Khaya ivorensis* by *Hypsipyla robusta* in the Ivory Coast.

Terminalia ivorensis is a promising major plantation species in western Africa, but insect attack in the fruiting stage could impede orchard management as many fruits fall off before they mature (Jones and Kudler, 1971). In Ghana two weevils were most likely responsible for the damage: *Auletobius kuntzeni* Voss (Coleoptera : Attelabidae) greatly reduced the quantity of fruits likely to mature by infesting early the developing fruits and, later, *Nanophyes* spp. (Coleoptera : Curculionidae) attacked the ripening fruits, thus affecting both the quantity and quality of the mature fruits. Fabre and Brunck (1974) report attack by *Tridesmodes ramiculata* Warren (Lepidoptera : Thyrididae) in the shoots of *Terminalia ivorensis* and *T. superba* in young plantations in the Ivory Coast.

The pine woolly aphid, *Pineus pini* Linnaeus (Homoptera : Adelgidae) not *Pineus ?sylvestris* Annand (Gray, 1972a), has become a major danger to many *Pinus* species in eastern Africa. Odera (1974) concludes that it was introduced from Australia on pine scions, probably in 1964. The pest has since spread into Kenya, Tanzania and Uganda, and Odera believes it will continue to disperse, mainly via transportation of infested nursery stock and by wind,

throughout the entire range of *Pinus* in Africa. Of the 39 host species of *Pinus* recorded in eastern Africa by Odera (1974), *P. elliottii* and *P. radiata* are among the more susceptible while *P. patula* is less susceptible, and *P. strobus* var. *chiapensis* appears to be resistant to attack.

In nine-year-old *P. patula* plantations at Muko in Uganda, *Gonometa podocarpi* Aurivillius (Lepidoptera : Lasiocampidae) heavily defoliated 16 ha in 1965 and 7 ha in 1971. Austara (1971) observed two generations per year, the presence of a noninclusion type virus which caused high larval mortality and development of the larvae primarily during the dry season. He found that 1–2 larvae per metre of branch length was necessary to result in complete defoliation.

Incidence of attacks by the pine defoliator *Orgyia hopkinsi* Collenette (Lepidoptera : Lymantriidae) have increased with the expansion of the *Pinus* plantations in Kenya (Odera, 1972a). Wide-spread outbreaks of *Orgyia hopkinsi* occurred in plantations of *Acacia mearnsii* and *Pinus* in Turbo in 1969, in which *P. patula* and *P. radiata* were severely defoliated. Odera (1972a) reports studies on the biology (description of stages, life-history and habits) and bionomics (hosts, epidemiology, chemical control) of the species. A virus appeared to be associated with high larval mortality in the field.

A native tussock moth, *Orgyia mixta* Snellen (Lepidoptera : Lymantriidae), has invaded exotic softwoods planted in Kenya. Austara and Migunda (1971) observed the life-cycle to take 48–85 days and 20–25 larvae per metre of branch resulted in total defoliation; a nematode *Charops* sp., in the larval stage, appeared to be the main agent of the natural control. Younger plantations appeared more vulnerable to attack because the early larval instars prefer feeding on the underside at the base of growing needles which were more prevalent on the younger trees.

Lee (1972) has given a detailed preliminary account of the biology and bionomics of *Plagiotriptus* spp. (Orthoptera : Eumastacidae) which persistently defoliate *Pinus*, especially *P. patula* in southern Malawi. Three species of grasshoppers have been identified, formerly determined as *Manowia* (Gray, 1972a; MacCuaig and Davies, 1972), which were endemic in the evergreen forest. Outbreaks initially occur on poor sites and later spread to better sites.

The termites causing maximum damage to plantations belong to the genera *Odontotermes* and *Microtermes*, which are fungus-growing species. They commence their attack in the upper 15–25 cm of the soil. The tap-root is ring barked often extending for some distance above the soil which usually results in the death of the plant.

Mites

The spider mite *Oligonychus milleri* has infested three-year-old *Pinus caribaea* in Jamaica in 1969; there was some mortality (Muma and Adeji, 1970).

Mammals

Monkey damage to exotic softwoods is wide-spread in Kenya. Studies by Omar and De Vos (1970) between July 1967 and July 1968 showed that most damage was caused by

Cercopithecus mitus kolbi Neuman (Primates : Cercopithe-cidae) during the dry season (July-October) and along the boundary zone of the natural forest and plantation. Little or no damage was observed in enrichment planted areas where grassland areas covered the boundary zone. The monkeys rip the bark off the stems and the trees die. The affected areas provide suitable oviposition sites for another serious pest of *Cupressus* spp., the beetle *Omedia gahani* Distant (Coleoptera : Cerambycidae). Incidence of monkey damage in *Cupressus lusitanica* plantations was 30–60 per cent of the trees, 26.0 per cent of 424 *Pinus patula* trees, and only 6 per cent of 92 *Pinus radiata* trees (Omar and De Vos, 1970).

Sharma (1967) mentions that cattle are a major problem in India and cause much damage to forest trees. Elephants pull out seedlings and young trees in *Pinus* plantations at Devikulam in India (Nair, 1971).

Natural forests

Parasitic plants

The dwarf mistletoe, *Arceuthobium bicarinatum* Urban, is a major pathogen of *Pinus occidentalis* in the Dominican Republic where more than half the forests are affected (Hawksworth and Wiens, 1972). Another dwarf mistletoe, together with a canker is a major pest of *P. rudis* and *P. tenuifolia* at higher elevations in Guatemala (Clark, 1973).

Viruses and mycoplasmas

Santalum album is a very valuable timber, mainly because of its scented oil extractives. Sandal spike disease has for 70 years, taken a heavy toll of the tree in India and future supplies are in danger. New leaves become progressively reduced in size, pale and stand stiff and crowded on shortened internodes giving a spike-like appearance which eventually spreads over the entire tree and eventually kills it. The disease was considered to be caused by a virus and the virus nature of the spike disease had been fairly well estab-lished by graft transmission, observation of inclusion bodies in infected tissues, and the general similarity of the symptoms of this disease with those of other well known virus infections of plants, though sap transmission of the disease by artificial inoculation had failed (Venkatesh and Kedharnath, 1964). Studies by Iyengar (1965) indicate that the sandal trees are both an autophyte and root parasite, which may be of some significance to infection. Nayar (1967) considered insect vec-tors or element deficiencies as unlikely explanations, but intimated that virus transmission via pollen grains could be responsible. Following a description of the infected anatomy of the trees, Nayar and Padmanabha (1969) suggested mass-ing of mycoplasma could account for the observed disor-ganization of the phloem. This proposal has been strength-ened by more recent findings (Hull, Horne and Nayer, 1969; Varma, Chenulu, Raychaudhuri, Prakash and Rao, 1969). For example, Padmanabha *et al.* (1973) detected MLO in the xylem and parenchymatous tissues as well as in the phloem of infected plants, and Dijkstra and Lee (1972) were able to transmit both the sandal spike disease and MLO to *Vinca rosea*.

Fungi

Botero (1972) described die-back of young *Cecropia tessmanii* trees in Colombia due to *Botryodiplodia theobromae* (= *Di-plodia theobromae* Nowell) (perfect state *Physalospora tho-dina* (Bak. and Curk) Cooke).

Schieber (1967) and Wellman (1972) noted the striking occurrence of the rust *Cronartium conigenum* Hedge. and Hunt (Uredinales : Melampsoraceae) on the cones of *Pinus caribaea*, *P. leiophylla*, *P. montezumae*, *P. oocarpa* and *P. pseudostrobus* in southern Mexico, Guatemala and in the West Indies. The alternate host are tropical oaks, including *Quercus corrugata* and *Q. tomentosa*. The rust presents a barrier to natural reforestation of the pines. The cones are infected shortly after pollination and they increase up to four times in size and are very bright orange.

Approximately 27 000 ha of the most prevalent *Metro-sideros collina* subsp. *polymorpha* on Hawaii have declined (Kliejunas and Ko, 1973). *Phytophthora cinnamomi* was sus-pected, but later excluded as the primary causative agent. More recent investigations indicate that the trees are de-clining because of an inorganic nutrient deficiency, but in certain areas root pathogens may be involved (Kliejunas and Ko, 1974).

Heart rots account for major losses of timber in natural forests of India. *Fomes caryophylli* (Racib.) Bres. (Aphyllo-phorales : Hymenochaetaceae), *Fomes fastuosus* (Lev.) Cooke and *Hymenochaete rubiginosa* (Fr.) Lev. (Aphyllo-phorales : Hymenochaetaceae) cause considerable loss of timber in high and coppice *Shorea robusta* forests and ap-proximately 10 per cent of timber is lost due to heart rot in felling coupes in high forests. The infection occurs mostly through branch stubs, broken branches, knots and fire wounds and the decay in heart-wood is white fibrous in *Fomes caryophylli* and white pocket in other two fungi. These fungi attack young and mature trees; up to 70 per cent of the trees may be infected. Associates of *Shorea robusta* and other hardwoods are also attacked. *Fomes badius* (Berk.) Cooke a perennial heart rot fungus almost exclusively on *Acacia catechu*, attacks the heart-wood of this species causing considerable damage both in natural and artificial stands. The heart-wood becomes decayed, the rot being white and spongy. The incidence of attack and the damage caused is rather heavy and at places nearly all trees may be attacked. *A. catechu* is subject to various injuries through which the infection occurs and generally trees above 10 cm DBH when heart-wood develops, are attacked.

Corticium coprosmae G. H. Cunn. (Aphyllophorales : Corticiaceae) has been observed causing heart rot in *Ery-thrina suberosa* in India (Bakshi, Reddy, Puri and Singh, 1972). Butt rot in *Pinus patula* due to *Poria monticola* Murrill. (Aphyllophorales : Corticiaceae) has been reported in West Bengal; this is a new host record for the pathogen. Mature and over mature trees of *Xylia xylocarpa* have been observed to be attacked by butt rot fungus (*Fomes fastuosus*) in natural forests in Goa, where the infection may be as high as 80 per cent.

Among the root diseases of forest species is wilt disease of *Dalbergia sissoo* caused by *Fusarium solani* (Mart.) Sacc.

(Hyphales : Tuberculariaceae). The disease is systemic and the diseased trees die in 4 to 6 months. *F. solani* is a soil inhabiting fungus and the infection occurs through the roots. Trees of all ages are attacked though seedlings and saplings are less susceptible to infection. The disease is absent or rare on loose sandy soils; incidence is high in both natural forests and plantations where the soil is stiff and clayey with poor drainage.

Polyporus shoreae, a common fungus in *Shorea robusta* forests of northern India causes root rot. The disease is of economic significance where rainfall exceeds 2 000 mm/a and is encountered on good sites. The fungus attacks healthy uninjured roots and the affected trees show symptoms of top dying after about 3/4 of the root system is attacked. Dying may extend downwards till all the tree is dead or is wind-blown. The disease is serious in wetter *Shorea robusta* forests of West Bengal and Assam, and healthy, vigorous trees of advanced age are attacked with 20 per cent or more mortality. High moisture content of the soil favours the activity of the fungus.

A fatal disease affecting all tree sizes of *Pterocarpus angolensis*, a valuable furniture wood, has spread through central Africa (Geary, 1972). The affected trees develop chlorotic and wilting leaves which fall off resulting in stag heads; epicormic shoots are also common. One forest reserve in Zambia had 14 per cent of the trees dead and another 36 per cent dying. Other tree species are also affected. A mistletoe *Loranthus* sp., though common, was not confirmed as a causal agent and a white fungus (*?Armillariella mellea*) was present at the root collars of a few trees, but otherwise no causative agent was apparent.

Insects

In a survey of termites in rain forests near the Magdalena river in Colombia, Becker (1972) observed *Coptotermes niger* Synder (Isoptera : Rhinotermitidae) in the bole of many trees (*Goupia glabra*, *Humiriastrum coloumbiarium* and *Xylopia* species); the heart-wood had been destroyed up to heights of 10 m.

Hypsipyla ferrealis (Hampson) and *H. grandella* account for the destruction of over 90 per cent of the seeds of *Carapa guianensis* in Trinidad (Bennett and Yaseen, 1972). Becker (1973) described the larval and pupal stages of *Hypsipyla ferrealis* in Brazil, and Becker (1973) studied important aspects of the biology of *H. ferrealis* in Costa Rica, in which the life-cycle took an average of 35 days, with three and four larval instars. Attack was observed in 36 per cent of the 4 328 seeds of *Carapa guianensis* collected for his studies. Biological control of the seed pests was investigated by Bennett and Yaseen (1972) using parasites of *Hypsipyla robusta* introduced from India; however, only one of the four parasites, *Trichogrammatoidea nana* (Zehntnr.) (Hymenoptera : Trichogrammatidae) was later recovered.

Juniperus bermudiana was the dominant tree on Bermuda but high mortality occurred following a hurricane in 1950 and the introduction of scale insects (*Carulaspis visci* Schrank in 1944 and *Lepidosaphes newsteadi* Sulc, Hemiptera : Diaspididae, in 1943) (Gray, 1972a). However,

Carulaspis minima (Targ.) is apparently the main pest responsible for the damage according to Challinor and Wingate (1971), who observed a few partly resistant remnant patches; resistance declines with age. Closely related species of *Juniperus* on Bermuda have not been severely attacked by the scale insects.

In early 1973 a reoccurrence of *Dendroctonus frontalis* Zimmerman (Coleoptera : Scolytidae) at epidemic levels was reported in Central America and Mexico (Clark, 1973; Vite *et al.*, 1974). Studies by Vite *et al.* (1974) indicate differences in biology, host preferences and in pheromone production between beetles from different countries which suggests more than one species is involved.

Thapa (1970a) observed outbreaks of *Margaronia hilaralis* Walker (Lepidoptera : Pyralidae) on gregarious *Anthocephalus chinensis* growing in recently logged forests in Sabah. Trees less than one year old were severely defoliated resulting in die-back of the leaders and subsequent deformation into branching boles, but little mortality. Natural control by parasites, especially *Hilomastrix* sp. (Hymenoptera : Encyritidae), appeared to be effective in reducing the damage.

An interesting account of up to 97 per cent mortality on 0.4 ha sample plots of *Anogeissus latifolia* and *A. pendula* in Rajasthan State in India is given by Verma (1972a). These trees dominate 77 per cent of the forests in the State. Mortality was mainly experienced during 1966–69 when a severe drought was recorded. Many of the trees were attacked by the borers *Olenecamptus anogeissii* Gardner and *Olenecamptus idianus* Thomson (Coleoptera : Cerambycidae) which contributed to increased mortality. Attacks were confined largely to the upper stems and none was observed below 0.5 m above the ground. A marked reduction in mortality and insect damage was noted in 1970 and 1971 when heavy monsoonal rains fell, although seedlings of *Anogeissus latifolia* were still attacked.

Roonwal (1972) has described the larval galleries, pupation and emergence of the borer, *Atractocenus reversus* Walker (Coleoptera : Lymexylonidae), a serious pest of standing trees, felled logs and sown timber of *Boswellia serrata* and *Lannea coromandelica* in India. Roonwal (1971) observed *Carphoborus boswelliae* (Stebbing) (Coleoptera : Scolytidae) infestation in branches and stems of dying and dead *Boswellia serrata* in Sillari Forest in Maharashtra State. However, the trees had been killed by poison-girdling but the intense boring of the scolytid rendered the wood useless except for fire-wood.

Hoplocerambyx spinicornis borer is the most important pest of *Shorea robusta* and in very heavy attacks it can kill the trees though more often it causes severe degradation of timber and also exposes the host to infection by other wound parasites notably *Polyporus weberianus*. Epidemics have been reported from central India, Uttar Pradesh and West Bengal in which all sizes of trees were attacked causing heavy losses of timber. The attacks can be detected by accumulation of dust at the base of the tree.

The borer *Pammene theristis* Meyrick (Lepidoptera : Eucosmidae) had periodically killed many one-year-old *Shorea robusta* seedlings in regeneration areas (Chatterjee and Thapa, 1970). Infestation is up to 80 per cent in areas devoid of undergrowth. Though high percentages of the seed are also damaged by the pest and other insects, it has little adverse effect on regeneration due to the excellent viability and germination of the remaining seed.

Mammals

The African elephant *Loxodonta africana* Blumenbach (Proboscidea : Elephantidae) poses a major conservation issue in Africa and has been a serious problem to forestry (Laws, 1970). Elephants have selectively suppressed the regeneration of economically desirable species such as *Chrysophyllum albidium* and *Khaya anthotheca* in Uganda; up to 36 per cent of the trees were damaged in a survey in Bugondo Forest (Laws, 1970). On the other hand, the ecological role of elephant in forests is very important, particularly in maintaining high productivity, according to Wing and Buss (1970) who show that with proper management more intensive use can be made of the forest products due to the selective feeding of the animals.

Biodeterioration of newly felled bark and wood

Biodeterioration of sound wood in tropical forests is generally much faster than in temperate forests. A complex of organisms is involved and these may differ considerably at various stages and on different wood species. Shigo and Hillis (1973) have reviewed biodeterioration in living trees caused by micro-organisms. The incidence of heart rot in several commercial species is very high in certain instances or areas and often leads to rejection of large segments of a felled tree, as in the case of *Araucaria hunsteinii* in Papua-New Guinea. The role of insects has been more intensively studied following the felling of trees (Gray, 1972a) whereas other organisms such as bacteria and fungi have been seldom studied, though they play a prominent role. The fungus *Botryodiplodia theobromae*, which is a common staining organism in freshly fallen trees in tropical Africa, enters through the vessels of the cut wood soon after felling while other sap staining fungi (*Phialophora* sp.) enter through the ray parenchyma cells (Olofinboba, 1974).

Of the insects, ambrosia beetles, notably Platypodidae and Scolytidae, are the major pests because they rapidly infest commercial lumber in large numbers causing degradation and even rejection of the logs. This problem has been very serious where export logs and shipping delays are involved, since the longer the lumber is left in the forest or on the wharf, the more likely the intensity of infestation will increase (Gray, 1974a). In a few instances ambrosia beetles attack living trees (see Gray, 1972a) and considerable staining may be caused by associated fungi with resultant degradation (Thapa, 1971). Three species provide important and difficult problems: *Dendroplatypus impar*, which habitually infests living *Shorea* spp. in Malaysia; *Trachyostus ghanaensis*, which habitually infests living *Triplochiton scleroxylon* in Africa; and *Doliopygus dubius* which commonly infests living *Terminalia superba* in Africa. Schedl (1972) has published a detailed account of the taxonomy, biology and

bionomics of the Platypodidae. Roberts (1971, 1973b) published detailed accounts of the distribution of the Platypodidae of Cameroon and Nigeria, noting that weather conditions are a salient factor affecting distribution and behaviour. Beaver (1972) studied the Platypodidae and Scolytidae in Brazil; he observed a sharp progressive increase in larval mortality in nests according to presence or absence of parent adults. Gray and Wylie (1974) have given detailed collection records of major ambrosia beetles in Papua-New Guinea and Gray (1974a) has mentioned them as serious problems in several Pacific Islands and as a quarantine hazard.

Despite an urgent need for investigations on the ecology of ambrosia beetles, little recent progress has been made (the work of the West African Timber Borer Research Unit, with various reports on biology and control issued in 1958–1962, is well known), except mainly in Papua-New Guinea where intensive observations on flight activity using sticky traps (Gray, 1974d; Wylie, in preparation) and attack patterns on two different lumber species (Gray *et al.*, in preparation) were carried out over the period 1972 to early 1974. This research has shown the necessity for caution in depicting the behaviour of individual species as being similar. Thapa (1970b) studied the periodicity and intensity of attack by insects over 12 months in freshly felled billets of *Parashorea tomentella* in Sabah, while Karstedt (1968) has described the general attack by insects and fungi over a period of three months in freshly felled billets of five tree species in Venezuela.

Conclusions

There has been a marked increase in publications on pests and diseases in tropical forests and plantations in the decade 1965–74, especially on those in Central and South America on which few reports had been published previously. Several serious problems recorded earlier have remained and a number of new ones experienced. A comparative assessment of the major pests and diseases in terms of impact and potential threat or hazard is not feasible. However, table 1 lists several of the major pests and diseases according to their principal hosts, habitat and distribution.

Few reliable assessments of impact have been carried out. In many instances, monetary losses have been relatively small despite intensive damage or high mortality, because the total number of trees affected was not great. This has been particularly evident in species trials. Thus there have been few outbreaks approaching those experienced in temperate coniferous forests. This may also be a consequence of the greater areas planted in temperate regions, as well as reflecting the greater homogeneity of temperate forests.

All but a few of the major pests and diseases have occurred in situations developed by man (forest nurseries, plantations, felling sites, etc.). These areas have comparatively large sources of food. In several instances the pest and disease problems have developed following the implementation of sylvicultural measures such as pruning and thinning.

No part of the plant is free from pests and diseases. However, there are differences in the type of pest and disease

TABLE 1. Ecological and geographical distribution of selected major pests and diseases in tropical forests.

Regions	Forest ecosystems and plantations	Major pests and diseases
Central and South America	*Cedrella* plantations	*Hypsipyla grandella* (mahogany shoot borer)
	Pinus forests	*Dendroctonus frontalis* (southern pine beetle)
	Pinus and *Quercus* forests	*Cronartium conigenum* (southern cone rust)
	Eucalyptus nurseries	*Cylindrocladium scoparium* (fungus)
Oceania	*Araucaria cunninghamii* plantations	*Hylurdrectonus araucariae* (needle miner)
	Swietenia and *Toona* plantations	*Hypsipyla robusta* (shoot borer)
Asia	*Swietenia* and *Toona* plantations	*Hypsipyla robusta*
	Dipterocarp forests	*Hoplocerambyx spinicornis* (heart-wood borer)
	Tectona grandis forests and plantations	*Xyleutes ceramica* (bee-hole borer)
Africa	*Khaya* and *Swietenia* plantations	*Hypsipyla robusta*
	Savanna woodlands, semi-deciduous forests and plantations	*Loxodonta africana* (elephant)
	Pinus nurseries and plantations	*Dothistroma pini* (blight)
	Pinus plantations	*Pineus pini* (woolly aphid)
	Chlorophora plantations	*Phytolyma* spp. (gall bugs)
	Cupressus plantations	*Omedia gahani*
	Rain forests (especially *Terminalia superba*)	*Doliopygus dubius* (borer)

and resultant damage found associated with these parts. Insects have been the most commonly recorded pests and are comparatively restricted in geographical and host distribution. On the other hand, the next most commonly recorded pests, fungi, appear to possess a greater geographic range and they are frequently associated with a larger number of host species. Further, fungi have been dispersed much more in recent years by man due to an increased international exchange of plant material and to inadequate quarantine.

Control and prevention

Control of tropical forest pests and diseases has presented more problems than in temperate forests for several reasons: greater difficulty in identifying the pest and disease (many are undescribed; several may occur simultaneously; etc.); most have many life-cycles per year; many forest operations are carried out on a relatively small scale and in the case of plantations on a trial and error basis. When a serious problem arises a substitution is usually made. An investigation into the causative agent may be neglected, or the cause is diagnosed wrongly in many instances—poor site being a common excuse (Gray, 1974b), because a specialist is not available or the costs do not warrant the effort, etc.; few, if any, regular systematic surveys of pests and diseases are carried out. Consequently, the gravity of the problem may not be realized or appreciated until many trees are destroyed or damaged. Often no prescribed treatments are readily available. Several additional reasons could also be given: adequate knowledge (identification of causal species, epidemiology and population trends, etc.) of a pest and disease should be gathered before implementing large-scale control measures (Gray, 1974b; Viennot-Bourgin, 1974) unless the gravity of the situation dictates emergency measures; a change in the basic philosophy of managers towards pest and disease problems (Gray, 1974b) is required before the protection of man-made forests can be undertaken in a more orderly basis; foresters and managers must take greater cognizance of these problems and be more aware of the difficulties of such research.

Methods

Chemical

Most control measures reported recently have involved biocides or pesticides (insecticides, fungicides, nematocides, etc.). The availability and effectiveness of these chemicals in the tropics varies enormously between pests, regions and trees (Gray, 1972a). There are many general accounts on control (Franz, 1965; Kuntz, 1965; McNabb, 1965; Lamb, 1974). Although this literature is heavily orientated towards the control of temperate pests and diseases, the general principles and techniques described are generally applicable, with a few important changes or allowances, to the tropics. For example, high temperatures and humidity make the wearing of protective clothing very uncomfortable and often unbearable, and the high frequency of rain may rapidly reduce the effectiveness of slow acting biocides applied to the surface of the trees (Gray, 1972b).

Though figures are not available, reports in literature suggest an increasing use of biocides in tropical forestry. This may be unavoidable where a valuable resource requires urgent protection, but the trend is a cause of concern. Scientific evidence has shown many side-effects from different biocides. Few investigations have been undertaken in the tropics, which parallel or even approximate those carried in North America, for example, into the ecological impact (effect on decomposers and wildlife, drift patterns, stream contamination, etc.) of biocides sprayed over large forest areas. Similar investigations are certainly warranted in the tropics if large-scale spraying operations are contemplated.

The use of growth regulating, hormonal, physiologically active compounds and pheromones (chemicals released by insects, etc.) and other aggregative, attractant or anti-attractant substances could be investigated more vigorously.

This particularly applies to those problems which are the same in tropical and temperate areas in which much basic research has already been carried out. For example, the use of population aggregating and anti-aggregating pheromones (e.g. Vite, 1970; Rudinsky *et al.*, 1974) to regulate bark beetle control measures and as a means of reducing the intensity of *Ambrosia* beetle attack on newly felled trees at logging sites in tropical forests.

Biological

Huffaker (1974) suggests biological control of pests in the tropics is more likely than in temperate habitats. He corroborates this thesis with the data of DeBack (1972) who found biological control projects in the tropics were 60 per cent successful, compared with 45 per cent or less elsewhere. On checking DeBack's (1972) data, it is noticed that 23 of the 25 completely successful projects in the tropics and subtropics were conducted on small and often remote islands which are noted for their highly endemic faunas; furthermore, most of the projects were against pests that had been introduced. Therefore, it would appear that Huffaker's (1974) assumption does not apply to the large tropical landmasses which constitute 98 per cent or more of the tropical terrestrial habitat, where successful projects have been few. This may of course simply reflect the little effort expended on biological control on these landmasses.

In recent years there has been a growing awareness of the possible dangers of the wide-spread and repetitive use of biocides. Several people have urged that greater emphasis be given to other methods, notably biological control (e.g. Simmonds, 1968). However, it is apparent from recent literature, that chemical control is still most frequently resorted to. Gray (1974a) has stressed the need for studies on biological control of forest insect pests in the tropics as a matter of high priority. In the case of bacteria, fungi and viruses, there may be less possibility of biological control.

There are few reports of adequate or successful biological control of pests and disease in forest plantations, regeneration and enrichment plantings in the tropics, but several authors (Thapa, 1970a; Austara, 1971; Chatterjee and Misra, 1974; Gray and Wylie, 1974) refer to the presence of many parasites and predators. Whether this suggests biological control is a doubtful proposition or one that is of low probability is unknown. Because of limited resources and time and the usually quicker impact of biocides, research on biological control is often accorded low priority. Subsequently, little is achieved other than the identification of potential or known parasites and predators and a few observations on their biology. This useful, but preliminary work is frequently abandoned when the outbreak subsides or another more urgent problem arises. Consequently, there are few instances in which biological control has been vigorously investigated in tropical forestry.

As emphasized by Bennett (1974), caution is required in undertaking a biological control especially in little known ecosystems. Advice from specialist organizations (for example, the Commonwealth Institute of Biological Control) should be sought, at least in the preliminary stages. These organizations often have excellent facilities and access to potential control agents known to be active against similar pests in other situations or countries. There are too many instances of biological control agents being released in the tropics with little follow-up research, other than a cursory inspection some months later wherein the demise of the released populations is noted since no individuals can be found.

Sylvicultural

Sylvicultural or management control is usually the more efficacious and practical solution to a serious pest or disease problem. There are many choices available: substitution of a non-susceptible or more resistant tree species (Omar and De Vos, 1970; Gray, 1972a); change of site for a given tree species (Ivory, 1967a); modification of planting, thinning or pruning regimes (Bakshi *et al.*, 1966; Roberts, 1973a); addition of materials (fertilizers, hormones, etc.) to increase tree vigour. In some cases the pathogen has arisen directly or indirectly as a consequence of a sylvicultural technique. Where large areas of plantations or forests are affected, the implementation of sylvicultural measures to control the problem may be costly, though necessary, if the damage is to be restricted; for example, Clark (1973) has recommended the destruction of trees infested by *Dendroctonus frontalis* in Guatemala.

Prevention of spread

Quarantine was a salient topic at the joint IUFRO/FAO Symposium in 1964 (see Francke-Grossmann, 1965; Granhall, 1965; Nordin, 1965). While the need for quarantine may appear obvious, attention must again be drawn to it because current measures appear inadequate in forestry. The introduction of the pine woolly aphid into Africa (Odera, 1974) is probably a consequence of this. Quarantine standards in many tropical countries are low to virtually non-existent, especially with regard to micro-organisms (viruses, bacteria and fungi).[1] It is largely these introductions plus those less visible, such as termites (Gray, 1970), which have been reported in the past decade.

The risk of new introductions has increased due to a greater exchange of plant material and to increased travel across borders by potential vectors. Other hazards include the planting of trees near borders and the natural movements of animals. This exchange is acceptable, but more serious attention must be given to the type of material because of the risk of introducing pests and diseases differs with each. Herrera Autter (1968) has suggested that, among other measures, trees, shrubs and asexual parts be excluded from South America, which is relatively isolated. Further, May (1967) has recommended that only exotic trees, which are

---

1. Instances in India of inadvertently introduced parasites are twig blight of Mexican pines caused by *Diplodia pinea*; *Dothistroma* and *Cercospora* blight of *Pinus radiata*; *Lophodermium* (Helotiales : Phacidiaceae) needle cast of *Pinus elliottii* and *P. patula* among others.

superior to native species and which can be successfully established, be introduced into Brazil.

In quarantine, the onus for detecting pests and diseases has been largely placed on the country of entry or the consignee. This allows the pathogen access, if it has survived the trip; unless identified, it becomes a threat. In view of the large number of entries reported, there would appear to be a need for additional precautions and provision of appropriate specialists and facilities to identify and isolate dangerous pathogens. One precaution is stricter measures or more onus placed on consignor to ensure that the plant material is not infected. Another suggestion is the provision of special, centralized, quarantine stations in isolated areas to service several countries at lower cost.

Control of major pests and diseases

Only control measures recorded in the literature as successful or promising will be mentioned.

Nursery

Control or prevention of pests and diseases in the nursery has been achieved frequently. Many countries have evaluated a large number of tree species, especially the exotics *Araucaria*, *Cupressus*, *Eucalyptus* and *Pinus*, on different ecological sites. Major disease problems appear to have arisen largely through a lack of knowledge of the species requirements and consequently a high rate of failure has occurred due to poor siting, adverse soil, etc. The small scale, trial and error basis, and relative ease of adjustment or applying biocides generally enables an economic solution to be found.

The majority of successful control measures have involved the use of biocides (table 2); additional measures have also been taken such as reducing soil acidity or moisture to make conditions less favourable to fungi. Other successful measures include: use of a better soil media for reducing root rot in *Araucaria hunsteinii* in Papua-New Guinea (Havel, 1965); transfer to a drier and warmer site to virtually eliminate seedling blight and improve the growth of *Vitex keniensis* in Kenya (Ivory, 1967b); trenching and removal of infected or suspect trees has been recommended in controlling root rot in teak seedlings in Tanzania (Hocking and Jaffer, 1967).

Plantations

Control of serious pests and diseases of planted trees, particularly in large plantations, is much more difficult than in nurseries. Most plantations are being established on less favourable sites usually hilly or more rugged terrain. Effective control using biocides has been rarely attained because of inaccessibility of the plantations, undergrowth, and other reasons noted above. With the wide range of biocides now available, excellent control has been frequently obtained in topical application or field trials, but when the most promising biocide is tried over a large area control is usually ineffective.

Bakshi *et al.* (1966) recommended trenching of diseased stumps, disposal of wood debris and treatment of cut ends of thinned stems with creosote as preventive measures in controlling root rot in teak plantations in India.

Control of *Dothistroma pini* blight on *Pinus radiata* trees in Kenya by copper fungicides is related to the pathogen's high sensitivity to copper translocated by the host to germinating conidia (Gibson and Howland, 1970). In field trials light dosages of copper fungicides gave high levels of protection against *Dothistroma pini* (Gibson, 1971) but large-scale spraying has proved impracticable in Africa due to elevation and topography (Gibson, 1974).

Aerial application of a ULV spray mixture of 47 per cent toxaphene (chlorinated camphene) and 23 per cent DDT in oil at 3.8 1/ha gave good control of the *Eupseudosoma involuta* infestation in a *Eucalyptus* plantation in Brazil (Balut and Amante, 1971). Where feasible ground dusting with Malatol gave satisfactory control of the moths *Euselasia eucerus* and *Thyrinteina arnobia* in Eucalyptus plantations in Brazil (Briquelot and Osse, 1969).

Gamma-BHC was found the most effective of three insecticides tried against the gryllid *Gryllus assimilis* in *Eucalyptus saligna* plantations in Brazil (Grodzki, 1973).

A promising technique for controlling the depredations of leaf-eating *Atta* species in the neotropics has been described by Lewis (1972). It involves aerial applications of an attractive toxic bait (4 per cent aldrin, 5 per cent soya-bean oil and 91 per cent dried citrus meal) which appears to be collected in preference to fresh leaves. An application rate of 2.2 kg/ha on an island in Trinidad killed 91 per cent of the ants; higher dosages gave lower kills. The bait, which remains attractive for 2–4 weeks in the dry season, is carried to the nest where the colony dies by contact. During the wet season the bait is waterproofed with 2 per cent methyltrichlorosilane. The amount of insecticide use is considerably less than applied by other techniques.

Gamma-BHC, DDT and endrin were found very effective in trials against the moth *Atteva fabriciella* in India. Mathur *et al.* (1970) recommended large-scale spraying of *Ailanthus excelsa* plantations at the beginning of the wet season when the first instar larvae were present.

Brown (1970) found 0.05 per cent gamma-BHC with 1 per cent teepol the most effective of five insecticides against the pine woolly aphid in field trials on a plot of young *Pinus thumbergiana* at Mugugu in Kenya. Later, this mixture was observed to be only partially effective and additional testing of eight insecticides in the laboratory and field was carried out (Odera, 1972a, b). Baygon (propoxur) and Thiodan (endosulphan) were most effective of which the latter is cheaper (Odera, 1972a, b).

Austara (1971) found gamma-BHC most effective of three insecticides in topical application trials against *Gonometa podocarpi* in Uganda. He recommended a dosage rate 0.5 per cent weight/volume in protecting the *Pinus patula* plantations.

Repeated applications of 1 per cent dieldrin were effective in controlling the cerambycid *Analeptes trifasciata* in *Anacardium occidentale* plantations in the Ivory Coast (Brunck and Fabre, 1970).

Sumithion (fenitrothion) was the most effective of three

TABLE 2. Chemical control of seedlings and graft diseases.

Plant species	Country	Chemical	Pathogen and disease	Reference
Flindersia brayleyana	Hawaii, U.S.A.	Pentachloronitrobenzene	*Rhizoctonia solani* - damping-off of seedlings	Ko et al. (1973)
Anthocephalus chinensis *Pinus caribaea*	Philippines Malaysia	Telone (1, 3-dichloropropene) Daconil 2787 (tetrachloroisophthalonitrile)	*Meloidogyne incognita* *Colletotrichum gloeosporioides* - damping-off of seedlings	Glori and Postrado (1969) Lim and Anthony (1970)
Pinus caribaea, P. merkusii, P. oocarpa	Malaysia	Benlate ((Methyl 1-butyl-carbamoyl)-2-benzimidazole carbamate) and Difolatan 4F ((N-tetrachloroethylthio) cyclohexene dicarboximide) for *P. merkusii*	?*Cercospora pinidensiflorae* - brown needle disease in seedlings	Ivory (1972)
Pinus kesiya (= *insularis*), *P. patula, P. radiata*	India	Blitox (50% dust of copper oxychloride), zineb (65% dust zinc ethylene-bis-dithiocarbamate)	Mainly *Rhizoctonia solani* - damping-off of seedlings	Reddy and Misra (1970)
Cupressus lusitanica	Kenya	Benlate (50% (methyl 1-butyl-carbamoyl) 2-benzimidazole carbamate)	Mainly *Monochaetia unicornis*. Disease of grafts	Gibson and Howland (1969)
Pinus spp.	Kenya	Fundilan (2, 4, 5, 6, tetrachloroisophthalonitrile diluted in C7-7002)	*Colletotrichum acutatum* - terminal crook disease of seedlings	Sang and Munga (1973)
Pinus patula	Kenya, Tanzania, Uganda	Rhizoctol (10% methyl arsenic sulfide) and Rhizoctol Combi (Rhizoctol plus 5% 4-benzoquinone-n'-benzoylhydrazone oxine) in seed pellets	*Fusarium* spp., *Pythium* spp. and *Rhizoctonia solani*. Damping-off of seedlings	Gibson and Hudson (1969); Hocking and Jaffer (1969)

insecticides against the tussock moth *Orgyia mixta* in topical application trials in Kenya (Austara and Migunda, 1971). It was recommended for use in the field because of its low toxicity to fish and mammals. Spraying with methidathion has shown promise in controlling the shoot borers *Hypsipyla robusta* in *Khaya ivorensis* and *Tridesmodes ramiculata* in *Terminalia ivorensis* and *T. superba* in the Ivory Coast (Fabre and Brunck, 1974).

MacCuaig and Davies (1972) conducted a series of tests in the laboratory and field of several insecticides for use against the wingless grasshoppers, *Plagiotriptus* spp., in Malawi. They recommended gamma-BHC and suggested aerial spraying where large areas had to be treated. Lee (1972) has indicated that the aerial spraying with this chemical has given adequate control.

Protection of man-made forests from wildlife has been reviewed by Holloway (1966) who divided the methods into two main types: empirical, comprising population control (shooting, poisoning and trapping), barriers and repellents; and ecological, using management through adjusting the animal populations and the forest habitat aimed at preventing serious forest damage (see chapter 20).

Omar and De Vos (1970) recommended growing of less palatable tree species (*Pinus patula* and *P. radiata*), patrolling, and consolidation of planted and grassland areas as preventive measures against monkey damage in Kenya. Formerly, many monkeys were shot.

Natural forests

Benlate was the most effective of five fungicides tested *in vitro* against *Diplodia natalensis* on *Casuarina equisetifolia* in Puerto Rico (Liu and Martorell, 1973).

Grodzki (1973) used CH_3Br and CS_2 in trials to control *Laspeyrasia araucariae* Pastrana (Lepidoptera : Eucosmidae) in *Araucaria angustifolia* seeds.

Biodeterioration of newly felled bark and wood

Little progress has been made in protecting newly felled bark and wood and although several measures designed to reduce the losses from insects have been promulgated (Gray, 1972a), reports of serious damage still appear regularly (Gray, 1974a). Biocides offer immediate protection, but their efficacy varies according to the chemical, its concentration, the frequency of rain and several other factors. If applied promptly, an effective biocide will usually give good results over a prolonged period; for example, Karstedt (1968) obtained protection in Venezuela against insects with gamma-BHC and dieldrin (but the latter insecticide may not be recommended because of its toxicity; Gray, 1972b) and against fungi with pentachlorophenol, organic mercury compounds and sodium phenylphenate; in another trial in Sabah, Thapa (1970b) found that 1 and 2 per cent gamma-BHC gave protection for six months against insect attack. In Africa protection against sap staining fungi was obtained for up to four weeks using antitranspirants, whereas *Antiaris africana* was stained blue within four days of felling if left untreated (Olofinboba, 1974).

A major problem is the tendency of the newly felled logs to crack, especially if stored in the open as is frequently practised. Bacteria, fungi and insects rapidly inhabit the fissures, particularly on the more shaded side, and commence biodeterioration even though the logs may have been sprayed with biocides. Unless pressure treated, kept moist or repeated applications or biocides are made, it is difficult to overcome this problem.

Conclusions

Implementation of research on the control of pests and diseases in tropical forestry has lagged behind that in temperate regions. This has been partly due to reticence in employing specialists in protection until a serious problem arises, to unavailability of specialists, and to the small scale of most forestry operations. The exceptions have generally been relatively short periods of concerted research by groups of scientists; for example, the West African Timber Borer Research Unit (1953–1962), the Entomology Section in Papua-New Guinea (1966–1975) and current research on *Hypsipyla* in Central and South America. These have been very productive; though seldom providing ideal solutions they have provided a much better understanding and knowledge of the pest or disease.

Control measures used vary according to the biology and ecology of the pests and diseases, the extent of damage, the preferences of the specialists and/or administrator, and the resources available.

Control has often been achieved with biocides, but seldom in larger established plantations. Because of cost and probably lower efficiency in the tropics (due to greater frequency of rainfall areas and higher temperatures), repeated dosages are a doubtful proposition. The particular chemical and method may be dictated by local supply and efficiency of the equipment. Aircrafts have seldom been utilized except in recent years.

Control methods other than the direct application of biocides have been largely concentrated on *Hypsipyla* the shoot-tip borer of *Cedrela* and *Swietenia*. These include of controlled release of biocides and isolation of pheromones, attractants and anti-attractants, as well as more conventional biological control. Mention should also be made of the elucidation of resistance in certain species or provenances of *Cedrela* and *Swietenia* to certain *Hypsipyla*. The research has greatly advanced knowledge and it could well result in practical control. Similar research should be encouraged on other multi-national forest pests and diseases.

Unless projected rotational short-term plantings and massive harvesting of forests in Brazil and in Sarawak are altered an increase in diseases and pests is expected. This has certainly been evident in the past decade and a consequence of the large number of species trials and established plantations. The increasing use of biocides is not encouraged where blanket spraying of large areas is carried out. Although not offering such a short-term solution to pest and disease problems, research on resistance, biological control and other more desirable control methods is preferable. Certainly in the case of biodeterioration of freshly felled timber, research is urgently required into methods of reducing infestation by ambrosia beetles and fungi. Certain chemicals have proved effective but have not been widely utilized and they are of doubtful value when the timber is stored for long periods.

Comparison of plantations and forests

Plantations and mixed tropical rain forests offer many contrasts. Plantations are characterized by trees of similar age, size and structure usually growing in a pattern prescribed by man. Ecologically they provide phytophagous and other organisms with few choices of food (host and type), few habitats and micro-environments and an abundance of a relatively homogeneous resource. They may exhibit variation with the smaller and often stressed trees on the poorer sites, which may increase their susceptibility to pests and diseases. Regeneration and enrichment planting may be categorized with plantations, although they are generally less homogeneous and may offer more resistance to infection.

In comparison, mixed tropical rain forests have a very complicated structure and composition of many tree species of different ages, sizes and structure growing adjacent in seemingly myriads of patterns (Harrison, 1962; Richards, 1973). Other types or stages of tropical forests may be less complex and less diversified; for example, in early stages of succession, at higher elevations, under extreme climatic or edaphic conditions, and in forests of conifers (Orians, 1974). These provide phytophagous organisms with many choices of food, numerous habitats and micro-environments; they represent a very heterogeneous resource which may be abundant, depauperate (Richards, 1973).

Tropical forests are highly dynamic. At a local level (a tree or patch of vegetation) the forest is one of fast growth and transition, whereas over a large area the system is seemingly relatively stable in species composition. Most if not all organisms in tropical forests exhibit strong selectivity of one form or another (host or site preference, behaviour, reproduction, etc.) that is highly developed at all levels. Competition of differing degrees and even competitive exclusion is another characteristic of many organisms which has seldom been investigated. Thus, although the forests appear abundant in food resources, there may not be an abundance at the species level due to selectivity. In addition to harbouring injurious forest pests and diseases, tropical forests are a reservoir for many agricultural and human ones.

Numbers and composition of organisms inhabiting forests and plantations differ markedly, but most species found in plantations originated from the forest. The presence of nearby or adjacent grasslands or transitions may result in a plantation containing species of mixed origin, but with most being found on natural host plant. There are scant quantitative data available on relative numbers and composition of organisms inhabiting the forest. Whitmore (1975) provides some comparative information on larger animals in different forest types of the Far East. There is the lack of information on the zonation of phytophagous organisms

and other animal life in tropical forests; in one of the few studies published, Harrison (1962) classified the birds and mammals of tropical rain forests in Australia and Malaysia into six communities by their horizontal stratification and range of foodstuffs. Many species frequently have few individuals (Wellman, 1972; Elton, 1973). On Barro Colorado Island in the neotropical zone Elton (1975) found numbers per species of invertebrates (largely insects and spiders) to be very low and most species were very small and scarce by day. Total numbers of phytophagous or herbivorous organisms are generally low (Richards, 1973). Calculations of biomass by Fittkau and Klinge (1973) in central Amazonian rain forest in Brazil showed 940 t/ha for living plants and only *ca.* 200 kg/ha for animals, of which 50 per cent were soil fauna and only 7 per cent fed on living plant matter and 19 per cent on living and dead wood. Organisms living in dead wood and in the soil may exhibit less selectivity, but a richness of species is nearly always present. Populations in similar, as well as different, forests may also differ considerably in numbers and composition for reasons as yet unexplained (Gray, 1972a), although the numbers may be correlated with other factors, such as humidity in the case of bacteria and fungi. See also chapters 6, 7 and 11.

Besides diversity of species and numbers of organisms, complex structure of the forest, and abundant habitats, the hosts of many phytophagous and wood-feeding organisms also have attraction and defence mechanisms. These are highly evolved in the tropics (Levin, 1973). These may be a decisive factor in several species, and responsible for the low proportion of animals feeding on living tissue compared with those feeding on wood and decaying matter. Complex organic profiles in chromatograms of volatiles taken from air samples adjacent to trees noted for high resistance to biological attack in Panamá, suggested that vapours from the wood may control decomposition and resistance (Hutton and Rasmussen, 1970). Their hypothesis is worth further investigation, since other studies (Gray, 1972a) show the presence of specific chemical compounds responsible for the resistance of certain tropical trees. It is generally agreed that more vigorously growing or healthy trees exhibit greater resistance to pests. The ability of the trees to give off volatiles may be a function of the physiological health of the tree. The specificity of these odours may partly explain the selectivity of the organisms.

Certain defensive mechanisms are well known, but others are not. Trichomes (hair-like appendages growing from the epidermis) play a significant role in protecting plants from phytophagous insects (Levin, 1973). Ethylene production in response to infection (Shain and Hillis, 1972) is another mechanism of defence about which little information exists for tropical timbers.

The tropical climate is fairly equitable with respect to temperature throughout the year, but daily variations may be considerable according to elevation, insolation and other factors. Rainfall is usually >2 000 mm/a and with humidity not so equitable, with many areas experiencing wet and dry seasons while in other parts droughts or wet periods of varying intensity may be experienced irregularly.

Wind and light intensity also have an important influence on organisms although their effects are probably less inside the forest. There is little data on the stratification of climate in different forests; however, beneath the upper canopy, temperatures are considerably lower and the majority of organisms are found in these cooler micro-environments and in the less exposed situations. In general, severe weather fluctuations appear to cause fluctuations in populations of organisms, with drought conditions exercising a notable effect (Gray, 1972a); for example, numbers and species of insects decreased during dry season in Costa Rica (Janzen, 1973), while several insect outbreaks have occurred in dry periods (Gray, 1972a). There appear to be differences in numbers and composition with elevation; Janzen (1973) observed highest insect density at intermediate elevations. The *ca.* 700–1 500 m zone is a transitional one between lowland and montane forest, it often has a rich flora and a milder climate. It is interesting that the temperature preferences of most insects (and possibly other organisms) in temperate and tropical regions is between 20 and 27° C; most activity occurs in the temperate zone during late spring and summer, and in the tropics in the morning (06.00–10.00 h) and late afternoon (17.00–21.00 h).

Plantations appear to be inhabited by fewer species often represented by more individuals. A surprising abundance may be collected in plantations, particularly in areas subjected to winds or near forest or transitional zones. A large proportion of these are transitory or migratory (Gray, 1974c). There may also be an interchange of organisms either daily or periodically between the plantations and adjacent vegetation; Janzen (1973) observed a strong movement of insects into moister areas during the dry season in Costa Rica. The climate in tropical plantations differs from that in forests, with temperatures and evaporation being generally higher and with greater fluctuation in humidity. Organisms occuring in the plantations will largely comprise those found in the adjacent forest, if the same tree species are present, or in the case of an exotic, those occupying similar niches which feed on the tree without deleterious effects. A substantial population of organisms may exist on other plants growing in the plantation and in the soil. The overall composition of organisms in plantations may change noticeably as the trees grow large and the foliage changes; the most dramatic changes follow canopy closure.

Another important contrast between forests and plantations is the relative frequency of occurrence of outbreaks in each. The term outbreak has been commonly used to describe a population increase of great magnitude, of pests causing extensive damage. Such outbreaks have been frequently recorded in temperate forests. A number of infestations have also been reported in tropical forests, but these have all been associated with gregarious hosts (Gray, 1972a; 1974a). No comparable outbreaks have been recorded for mixed rain forests in the tropics. This has aroused much interest and speculation. Localized outbreaks, in which only a few trees or a small patch are affected are relatively uncommon and poorly documented. They are brief (less than six months) and due to rapid recovery or to canopy closure, little visual evidence remains, unless several adjacent trees are killed. Forest surveys have recorded old patches of dead or dying trees covering a small area with apparently no evidence of fire, wind or human damage, although lightning could be responsible; pests and diseases are usually considered the most likely cause. In other cases, only a few partially defoliated trees are evident or large numbers of an organism are found in a patch within the forest (Gray, 1972a). These outbreaks are probably limited by the great richness of host plants and by the selectivity of the organism wherein only a few trees of a certain age class of a particular host provide acceptable food. These localized outbreaks could be triggered-off by climatic extremes or stress.

Little is known of the efficacy of these population regulatory factors and their variation. Much has been published on diversity and stability in agro- and forest-ecosystems in which tropical forests have often been alluded to. It is evident that any simple ecosystem consisting of a single host species is more likely to be severely affected by pests and diseases, whereas a complex ecosystem is rarely, if ever, affected. This has been expressed in the dogma that diversity creates stability. Several hypotheses have been proposed to explain it. Van Emden and Williams (1974) show that the usage of the words diversity and stability has been confounded. They accept the general association between diversity and stability in natural systems. The structural variation of the ecosystems as well as their patterns must also be important factors (Murdoch *et al.*, 1972). Tropical forests have evolved over a long time and their distribution and abundance, as well as composition and structures, have changed considerably (Fittkau and Klinge, 1973; Richards, 1973; see also chapter 3). They are very responsive to incessant environmental changes and competition of major magnitudes. In the past they have been relatively stable, but the recent activities of man have created largely insettled ecosystems.

Regulation of populations in tropical forests would appear crucial in view of the more equitable climate and greater reproductive potential. That few large outbreaks have been observed suggests that the factors regulating populations are effective. This is poorly understood and various hypotheses have been proposed. In mixed rain forests where the density of individual tree species is very low, the problem of an organism locating its host has been suggested as a major regulating factor (Gray, 1972a; Wellman, 1972). Host age composition may also be important; for example, older trees possess many more crevices of loose bark and epipytes, and they harbour many organisms not found on younger trees. Further evidence is the rapid increase of certain organisms during logging operations.

A conspicuous feature is the high proportion of parasites or predators which has led to the hypothesis that these are the primary regulating factor (Janzen, 1971; Evans, 1974). Evans (1974) has stated 'it is now suggested that, because environmental conditions are optimal for fungal development, those fungi pathogenic to arthropods have evolved to the extent where they play an important role in regulating host populations'. He found fewer diseased arthropods in secondary or depleted forests than in undisturbed forests in Ghana. However, he provides little evidence as to whether the regulation is direct or indirect, active or passive (for example, he lists several arthropods recognized as predators or parasites which were also diseased) and no account is given of the species' relative abundance in the different situations. In many instances, predation or parasitism especially by fungi, is secondary (Gray, 1972a). In other cases where an organism is hidden or able to move rapidly, predation or parasitism may be low, while in other cases it is high. More quantitative information is required on the extent and nature of predation and parasitism at the different levels in the food web. Food webs and interactions between organisms also require study as little is known about them in the major sub-ecosystems (canopy, subcanopy, ... forest floor and soil strata) before their role can be properly determined.

Another very important, not regulatory, influence is climate (Gray, 1972a). Little emphasis has been given to the effects of climate (particularly micro-climate) on populations, since the belief that climate varies little throughout the year or day has been commonly accepted, as seemingly often borne out by meteorological data. However the data are often scanty or at best monthly or annual averages which may give a biased result; for example, one week of extremely wet weather may cancel out three weeks of dry in the rainfall average, but the effects of the dry weather on the forest may be considerable. Because of the irregularity of severe dry periods, the absence of suitable recording instruments and difficulty of sampling, information on these effects is limited, except for the above-mentioned study by Janzen (1973) in Costa Rica and on a Caribbean Island. There, seasonal differences were clearly evident in different localities. Tropical organisms appear to be more fragile with respect to weather fluctuations whereas temperate organisms can survive a wider range. The former have adapted to a relatively narrow range (15–30° C), and throughout most of the year the organisms experience no hardship or adverse weather conditions. However, at irregular or regular intervals (where marked wet and dry seasons are experienced) adverse climatic conditions may occur. A change in the numbers and composition of the organisms may occur, with the more fragile ones showing greater stress. Documentation of these conditions and their effects is necessary for an understanding of population regulation.

In contrast, the frequency of outbreaks in plantations has been high, though not entirely unexpected. While the reasons are varied and complex, they have been primarily attributed to a breakdown in those factors responsible for stability (Webb, 1968), or because natural forests have a better defence against pests and disease (Lamprecht, 1966). Whether any one explanation is accurate or sufficient is doubtful; it is necessary to study the ecology of the pest and host in both the plantation and forest to verify the reason and this has been rarely done. Certainly, the homogeneity of plantations is an important factor because less attack has been reported when a mixture of species are planted or the structure of the plantation is altered (Mathur *et al.*, 1970; Gray, 1974a). Indigenous species may suffer heavily from indigenous diseases and insects when raised in monocultures. *Evodia lunu-ankenda* (*E. roxburghiana*), a component of the

natural forest in India, was unsuccessful when it was raised in plantations, possibly due to a root disease. Sample surveys conducted in felling coupes of dry coppice *Tectona grandis* in Gujarat State showed that 50 per cent of trees were decayed rendering the basal logs useless, up to 5 m. Similarly insect pests developed into serious epidemics when indigenous species like *Ailanthus, Swietenia, Tectona grandis, Toona ciliata*, were planted in pure stands. In *Tectona grandis* plantations, increment losses due to defoliation by insects may be up to 30 per cent annually. High mortality due to repeated defoliation by an insect resulted in abandonment of *Gmelina* plantations in India and Burma. This should be examined more closely. Webb (1968) has suggested that a limit to the reduction in the number of host species should be observed in sylvicultural operations to avoid pest invasions. The selectivity of organisms for a particular host species and in turn for a particular age class, is highly developed. This could explain why certain species occur in low numbers in the forest, but in epidemic numbers in monocultures of a specific age class. In other cases, the failure of predators or parasites to establish themselves (due to absence of alternate host, adverse climate, etc.) is probably the primary reason for outbreaks, and in others, the adverse effects of climatic changes. In nearly all cases a number of factors are jointly operative. Of note is the greater frequency of outbreaks in younger plantations in the tropics, whereas the older trees appear more susceptible to pests and diseases (Gray and Wylie, 1974), except where interference (wounding) has been responsible. There is little information on the reasons for this, but different nutrient status (for example, a higher nitrogen content) or sensitivity to stress and changes in attraction and defence mechanisms are possibles. Whether tropical plantations are more likely to be invaded by pests and diseases than those in temperate regions has not been investigated, though the greater number of candidate organisms in tropical forests and the high frequency of infestation tend to support the observation that the former are more prone to injury.

Conclusions

Tropical forest ecosystems and those created or disturbed by man contrast strongly. These contrasts have seldom been enumerated or identified and studied in detail, especially with respect to protection and to their associated fauna and flora.

From the protection viewpoint, a notable difference has been the low incidence of major pests and disease problems in relatively undisturbed mixed forests, and the large number of problems reported from the man-made situations. There are almost certainly many reasons for these outbreaks: climatic stresses, absence of predators and parasites or their ineffectiveness, and the spatial distribution of the host and its evenagedness are three of the more pominent explanations, but there are scant quantitative data available confirming their role and significance. There is also a serious lack of comparative studies on pest organisms in both the epidemic and natural environment localities. A list of host species and cursory observations on abundance are generally the only data available. Another serious deficiency is the study of traditional patterns of use of the tropical forests, especially shifting or swidden agriculture: are pests and diseases a major problem and how do the people contend with or compensate for these? Although yields are low, these patterns of use are often accompanied by a better plant health situation than in plantations. More information is needed for exploited forest areas, as the results of traditional methods of pest control could have an impact on the methods employed in modern plantations.

Research needs and priorities

1. Creation of two or more research units to study the ecology of organisms in tropical forests and plantations, including studies on the population dynamics and the effect of climate, predators and parasites, selectivity and interchange of organisms between plants, species richness and numbers, and roles of organisms and interactions.

With the gradual or possibly rapid depletion of tropical forests in the future due to the demand for timber and an increase in plantations, it is imperative that the present void in knowledge on these basic ecological aspects and others be bridged. It is highly preferable that a multidisciplinary approach be undertaken, with emphasis on micro-organisms (viruses, bacteria, fungi, and nematodes) in view of the little quantitative information available on them. The research units should be established for a period of more than five years.

2. Establishment of a central register of pests and diseases in tropical forestry.

This register could possibly be maintained by the Forestry Department of FAO. Hepting (1965) proposed a system of literature retrieval for the field of forest pathology on a world basis. However, a more complete system is needed for several reasons: the entire field of forest protection should be covered; several tropical pests and diseases are not mentioned in the literature; many reports give scant details of important particulars—causal organisms, their description, epidemiology, extent of damage, etc.—, but the information may be available from a technician or specialist (Gray, 1974b); many species' trials, especially with exotics, have been carried out in the past decade and there has been a marked increase in plantations (Gray, 1972a) (there has also been a concomitant increase in pest and disease problems, but there has been relatively little interchange of data on these pests between tropical countries). The register would enable more effective quarantine by identifying the more dangerous pests likely to be introduced. It could be updated annually or more often through direct liaison with each national organization responsible for forestry.

3. Evaluation of pest control methods and study of population regulation.

Orians (1974) has recommended evaluation of temperate pest control methods and their applicability in the tropics, and a comparable evaluation of theoretical insect regulation concepts. Insect regulation concepts and those concerning other pests in temperate regions are still incomplete and subject to polemics. Current control methods in temperate forests are being subjected to criticism (for example, the efficacy or necessity for large-scale application of biocides over forests in North America) and change due to the development of alternate methods of control such as the possible usage of aggregating pheromones and anti-attractants and use of helicopters to extract fallen or damaged lumber containing potential or breeding populations of pests. Furthermore, there is a serious dearth of information on the regulation of pest organisms in tropical forestry which would make evaluation inopportune. See Bowers *et al.* (1976) for new prospects in biological control.

Direct application of temperate pest control methods to the tropics may be inadvisable (Gray, 1974b; Lamb, 1974), because modification or adjustment is often necessary; for example, the forest may be used as an important food source and be inhabited by the local people, thus there would be a need to examine carefully the effects of applying biocides. Many major tropical pests and diseases require evaluation on their own merits, because it is difficult to compare the problem with those experienced elsewhere, and especially with those in the temperate region, due to their peculiarity (for example, *Hylurdrectonus araucariae* in hoop pine plantings in Papua-New Guinea). On the other hand, notably among the micro-organisms, the same species may occur in the temperate and tropical zone, thus enabling more direct evaluation.

4. Adoption of sylvicultural practices which reduce the amount of wounding and infection of planted trees and regeneration, and initiation of research into the defence mechanisms of tropical trees when wounded. A programme of research into the feasibility and limitations of rotational planting in the tropics, rather than a trial and error procedure.

Two major sources of concern loom large in protecting plantations in the future, that have been inconspicuous in the past.

a. Most plantations are young and routine sylvicultural practices, pruning and thinning, aimed at enhancing form and growth, have yet to be carried out on a large scale. The current techniques used are largely based on those used in temperate regions or on other tree species, and they will require modification before becoming standardized. Because the trees are wounded in pruning or thinning operations, the implementation of large-scale routine measures will therefore greatly increase the number of potential infection courts. Under tropical conditions the risk of entry of harmful viruses, bacteria, fungi and insects is also greatly increased.

b. Few second rotation plantings, other than replanting to replace losses have been established. This could change dramatically in the future following the recent technological breakthrough in pulping tropical woods and emphasis on fast-growing species. The harvesting of plantations and replanting at 10-year intervals or less is an appealing proposition likely to be widely adopted in the humid lowland tropics. It will therefore resemble an agricultural cropping system, where experience has shown the necessity for rotational cropping, resting of the soil and its replenishment. The adoption of similar practices in plantations appears obvious in the tropics.

A greater risk of pests and diseases is likely in these plantings for two primary reasons:
— The growth and health of the second crop of trees is likely to be less than the first crop. These trees would also be more sensitive to stress and stressed trees are more susceptible to damage by destructive agencies.
— An increase in pest and disease problems could be expected if there is little time lapse between harvesting and replanting the crop. Many pathogens persist for a time after felling, particularly root rots (Bakshi, 1965; Bakshi and Singh, 1970) and termites (Gray, 1972a), which infest the new trees via the soil. Consideration should also be given to changing the current practice of planting sites soon after logging, as for example recommended by Roberts (1973a) for reducing the incidence of ambrosia beetle attack.

5. More emphasis on breeding of resistance in major tree species.

Breeding trees resistant to pests and diseases is an exercise to which tropical sylviculturalists and biologists have paid little attention (Gerhold, 1973). A better knowledge of associated mycorrhizal fungi could also help in obtaining higher yields and enhanced disease resistance (Smith, 1974). Possibly, the neglect has been a consequence of the long-term studies and expense necessary in such research and the often devastating nature of many pests and diseases and temporary status of many of the researchers.

In view of the increasing number of trials of favoured tree species (for example, *Araucaria*, *Eucalyptus* and *Pinus*) and the all too frequent reports of damage, the feasibility of breeding resistant trees should be earnestly investigated. There are few reports of active searching for resistance; for example, in Africa, the selection, on the appearance of the phenotype and propagation, of individuals showing resistance to *Dothistroma* blight is being carried out (Ivory and Patterson, 1970; Gibson, 1974). A number of reports have already been published indicating differences in the susceptibility of closely related tree species in the tropics; for example: *Araucaria* (Gray, 1972a), *Cupressus* (Olembo, 1969), *Eucalyptus* (Segura, 1970b), *Pinus* (Odera, 1974), *Swietenia* (Geary et al., 1973). However, these give little or no indication as to the overall performance of the more resistant tree species.

Introduction of new provenances or selection of more resistant varieties may not provide a final solution (Levin, 1973; Viennot-Bourgin, 1974) because the wood quality or some other desirable quality may be less or another pest may attack the new introduction. These considerations should not deter research on breeding for resistance, but rather extend the objectives and encourage caution.

Bibliography

AMANTE, E. Prejuizos causados pela formiga sauva em plantaçoes de *Eucalyptus* e *Pinus* no estado de Sao Paulo. *Silv. S. Paulo*, 6, 1967, p. 355–363.

ANON. World situation with regard to forest diseases. *Unasylva* (Rome, FAO), vol. 19, no. 3, 1965a, p. 107–112.

——. World situation with regard to forest insects. *Unasylva*, vol. 19, no. 3, 1965b, p. 113–120.

——. *Proceedings of the first symposium on integrated control of Hypsipyla*. Costa Rica, Turrialba (5–12 March 1973), IICA, 1973a.

——. *Studies on the shoot borer Hypsipyla grandella (Zeller) Lep. Pyralidae*. Vol. 1 (Grijpma, P., ed.). Turrialba, IICA, Miscellaneous Publications, no. 101, 1973b, 91 p.

ARAUJO, R. L. Termites of the Neotropical Region. In: Krishna, K.; Weesner, F. M. (eds.). *Biology of Termites*, p. 527–576. New York, Academic Press, vol. II, 1970, 643 p.

AUSTARA, O. *Gonometa podocarpi* Aur. (Lepidoptera : Lasiocampidae) a defoliator of exotic softwoods in East Africa. The biology and life cycle at Muko, Kigezi District in Uganda. *East African Agric. For. J.*, vol. 36, no. 3, 1971, p. 275–289.

——; MIGUNDA, J. *Orgyia mixta* Snell. (Lepidoptera : Lymantriidae) a defoliator of exotic softwoods in Kenya. *East African Agric. For. J.*, vol. 36, no. 3, 1971, p. 298–307.

——; JONES, T. Host list and distribution of lepidopterous defoliators of exotic softwoods in East Africa. *East African Agric. For. J.*, vol. 36, no. 4, 1971, p. 401–413.

BAKSHI, B. K. Accomplishments: Asia. In: *Internationally dangerous forest tree diseases*. Washington, D.C., USDA, Forest Service, Misc. Pub., no. 939, 1963.

*BAKSHI, B. K. Known and potential hazards from stem diseases, heart rots. In: *FAO/IUFRO Symp. on int. dangerous forest diseases and insects* (Oxford, 1964), vol. 1, 1965, 8 p.

——. Diseases of man-made forests. In: *World Symp. on man-made forests and their industrial importance* (Canberra), 1966, p. 639–661.

——. Quantification of forest disease losses; report for Asia. In: *Proc. 14th IUFRO Congress* (Munich), 1967.

——. *Indian Polyporaceae (on trees and timber)*. New Delhi, Indian Counc. of Agric. Res., 1971, 246 p.

——; REHILL, P. S.; CHOWDHURY, T. G. Field studies on heart-rot in sal (*Shorea robusta* Gaertn.). *Indian Forester*, vol. 89, no. 1, 1963.

——; SINGH, S.; SINGH, U. A new root rot disease complex in teak. *Indian Forester*, vol. 92, no. 9, 1966, p. 566–569.

*——; KUMAR, D. Forest tree mycorrhiza. *Indian Forester*, vol. 94, no. 1, 1968, p. 79–84.

——; REDDY, M. A. R.; SINGH, S.; PANDEY, P. C.; MUKHERJEE, S. N. Khair seedling mortality in plantations. *Indian Forester*, vol. 94, no. 9, 1968, p. 659–661.

——; DOBRIYAL, N. D. Effect of fungicides to control damping-off on development of mycorrhiza. *Indian Forester*, vol. 96, no. 9, 1970, p. 701–703.

*——; SINGH, S. Heart rot in trees. In: Romberger, J. A.; Mikola, P. (eds.). *International Review of Forestry Research*, vol. III, 1970, p. 197–251.

——; REDDY, M. A. R.; PURI, Y. N.; SINGH, S. *Forest disease survey (Final technical report), 1967–1972. Survey of the diseases of important nature and exotic forest trees in India*.

Dehra Dun (India), Forest Research Institute and Colleges, 1972, 117 p., 20 pl.

Balut, F. F.; Amante, E. Nota sobre *Eupseudosoma involuta* (Sepp., 1952). Lepidoptera, Arctiidae, plaga de *Eucalyptus* spp. *Biologico*, vol. 37, n° 1, 1971, p. 13–16.

Bashin, G. D.; Roonwal, M. L. A list of insect pests of forest plants in India and their adjacent countries. Part 2. *Indian For. Bull.*, vol. 171, no. 1, 1954.

——; ——; Singh, B. A list of insect pests of forest plants in India and their adjacent countries. Part 3. *Indian For. Bull.*, vol. 171, no. 2, 1956.

Beaver, R. A. Biological studies of Brazilian Scolytidae and Platypodidae (Coleoptera). I. *Camptocerus* Dejean. *Bull. Ent. Res.*, 62, 1972, p. 247–256.

Becker, G. Termiten in regenwald des Magdalenenstromals in Kolumbien. *Z. Angew. Ent.*, vol. 71, no. 4, 1972, p. 431–441.

Becker, V. O. Estudios sobre el barrenador *Hypsipyla grandella* (Zeller) (Lep., Pyralidae). XVI. Observaciones sobre la biología de *H. ferrealis* (Hampson), una especie afín. *Turrialba*, vol. 23, no. 2, 1973, p. 154–161.

Beeson, C. F. C. *The ecology and control of the forest insects of India and the neighbouring countries*. Dehra Dun (India), Forest Research Institute, 1941.

Bennett, F. D. Criteria for determination of candidate hosts and for selection of biotic agents. In: Maxwell, F. G.; Harris, F. A. (eds.). *Proceedings of the summer institute of biological control of plant insects and diseases*, p. 87–96. Jackson, University Press of Mississipi, 1974, 647 p.

——; Yaseen, M. Parasite introductions for the biological control of three insect pests in the lesser Antilles and British Honduras. *Pest and News Summaries*, vol. 18, no. 4, 1972, p. 468–474.

Berry, P. A. Entomología económica de El Salvador. Ministerio de Agricultura y Ganaderia (Santa Tecla, El Salvador), *Bol. técnico*, n° 24, 1959, 255 p.

Bess, H. A. Termites of Hawaii and the Oceanic Islands. In: Krishna, K.; Weesner, F. M. (eds.). *Biology of Termites*, p. 449–476. New York, Academic Press, vol. II, 1970, 643 p.

Boerboom, J. H. A.; Maas, P. W. Th. Canker of *Eucalyptus grandis* and *E. saligna* in Surinam caused by *Endothia havanensis*. *Turrialba*, vol. 20, no. 1, 1970, p. 94–99.

Botero, R. O. Secamiento descendente del yarumo (*Cecropia tessmanii*). *Revista Facultad Nac. Agr.* (Medellín), vol. 27, n° 3, 1972, p. 59–62.

*Bowers, W. S.; Ohta, T.; Cleere, J. S.; Marsella, P. A. Discovery of insect anti-juvenile hormones in plants. *Science*, vol. 193, no. 4253, 1976, p. 542–547.

Boyce, J. S. *Forest pathology*. New York, McGraw-Hill, 1961, 572 p.

Briquelot, A.; Osse, L. Presencia de insectos en las plantaciones de Eucaliptos de la companía siderúrgica 'Belgo-Mineira' y su control experimental por diversos medios. *Instituto Forestal Latino-Americano de Invest. y Capacitación* (Mérida, Venezuela), 29, 1969, p. 35–41.

Brown, K. W. Tests of insecticides against *Pineus* spp. in East Africa. *East African Agric. For. J.*, vol. 34, no. 2, 1970, p. 200–201.

Browne, F. G. *Pests and diseases of forest plantation trees*. Oxford, Oxford University Press, 1968, 1 330 p.

Brunck, F.; Fabre, J. P. Note sur *Analeptes trifasciata* Fabricius, Coléoptère Cérambycide, grave ravageur d'*Anacardium occidentale* en Côte-d'Ivoire. *Bois et Forêts des Tropiques* (Nogent-sur-Marne), 134, 1970, p. 15–19.

Bustillo, A. E. Lista preliminar de insectos que atacan los cultivos forestales en Antioquia. *Revista Instit. Colombiano Agropecuario*, vol. 8, no. 1, 1973, p. 81–86.

Bustillo A. E.; Lara, L. Plagas forestales. *Boletin de divulgación* (Instituto Colombiano Agropecuario), no. 3, 1971, 31 p.

Carruyo, L. J. Estudio preliminar de extractivos de las Meliaceas que atraen a *Hypsipyla grandella* Zeller. *Instituto Forestal Latino-Americano de Invest. y Capacitación* (Mérida, Venezuela), 43, 1973, p. 13–19.

*Carter, W. *Insects in relation to plant disease*. 2nd ed. New York, Wiley, 1973, 759 p.

Chaiglom, D. *Dangerous insect pests of forest plantation in Thailand*. Bangkok, Royal Forest Dept., 1974, 35 p. (published in Thai language).

Challinor, D.; Wingate, D. B. The struggle for survival of the Bermuda Cedar. *Biological Conservation*, vol. 3, no. 3, 1971, p. 220–222.

Chatterjee, P. N.; Thapa, R. S. Sal regeneration: the problem of sal seed and seedling borer *Pammene theristis* Meyrick in Dehra Dun sal forests (Lepidoptera : Eucosmidae). *Indian Forester*, vol. 96, no. 8, 1970.

——; Misra, M. P. Natural insect enemy and plant host complex of forest insect pests of Indian region. *Indian For. Bull.*, 265, 1974.

Cheah, L. C. A note on taungya in Negeri Sembilan with particular reference to the incidence of damage by oviposition of insects in plantations in Kenaboi Forest Reserve. *Malayan Forester*, vol. 34, no. 2, 1971, p. 133–147.

Clark, E. W. Informe al gobierno de Guatemala sobre infestaciones de Dendroctonus en los pinares de Guatemala. Rome, FAO, Tech. Rep. n° 3164, 1973, 27 p.

Cooling, E. N. G. *Fast growing timber trees of the lowland tropics. Pinus merkusii*. Comm. For. Institute, Dept. Forestry (Oxford), no. 4, 1968, 169 p.

DeBack, P. The use of imported natural enemies in insect pest management ecology. In: *Proc. Tall Timbers Conf. Ecol. Anim. Control Habitat Manage.* (Tallahassee, 1971), 3, 1972, p. 211–233.

Dijkstra, J.; Lee, P. E. Transmission by dodder of sandal spike disease and the accompanying mycoplasma-like organisms via *Vinca rosea*. *Neth. J. Pl. Path.*, vol. 78, no. 5, 1972, p. 218–224.

Drooz, A. T.; Bustillo, A. E. *Glena bisulca*, a serious defoliator of *Cupressus lusitanica* in Columbia. *J. Econ. Ent.*, vol. 65, no. 1, 1972, p. 89–93.

Elton, C. S. The structure of invertebrate populations inside neotropical rain forest. *J. Anim. Ecol.*, vol. 42, no. 1, 1973, p. 55–104.

——. Conservation and the low population density of invertebrates inside neotropical rain forest. *Biological Conservation*, vol. 7, 1975, p. 3–15.

*Emden, H. F. Van (ed.). *Insect/plant relationships*. Oxford, Blackwell, for the Royal Entomological Society, 1973, 215 p.

——; Williams, G. F. Insect stability and diversity in agroecosystems. *Ann. Rev. Ent.*, 19, 1974, p. 455–475.

Etheridge, D. E. Preliminary observations on the pathology of *Pinus caribaea* Morelet in British Honduras. *Comm. For. Rev.*, vol. 47, no. 1, 1968, p. 72–80.

——. *Inventario y fomento de los recursos forestales (República Dominicana); patología y entomología forestales*. Rome, FAO, Tech. Rept. 1, FD: SF/DOM 8, 1971, 31 p.

Evans, H. C. Natural control of arthropods, with special reference to ants (Formicidae), by fungi in the tropical high forest of Ghana. *J. Appl. Ecol.*, 11, 1974, p. 37–49.

Fabre, J. P.; Brunck, F. Action du methidathion sur les

chenilles mineuses des pousses des jeunes plants de Framire en Côte-d'Ivoire. *Bois et Forêts des Tropiques* (Nogent-sur-Marne), 155, 1974, p. 58–60.

FIRMAN, I. D. A list of fungi and plant parasitic bacteria, viruses and nematodes in Fiji. *Comm. Mycological Institute, Phyto-pathological Pap.*, no. 15, 1972, 36 p.

FITTKAU, E. J.; KLINGE, H. On biomass and trophic structure of the Central Amazonian rain forest ecosystem. *Biotropica*, vol. 5, no. 1, 1973, p. 2–14.

*FOSTER, R. E. Known and potential hazards from root rot. In: *FAO/IUFRO Symp. on int. dangerous diseases and insects* (Oxford, 1964), vol. 1, 1965, 11 p.

FRANCKE-GROSMANN, H. Some investigations on the hazard of intercontinental spread of forest and timber insects. In: *FAO/IUFRO Symp. on int. dangerous diseases and insects* (Oxford, 1964), vol. 1, 1965, 11 p.

FRANZ, J. M. Forest insect control by biological measures. In: *FAO/IUFRO Symp. on int. dangerous diseases and insects* (Oxford, 1964), vol. 2, 1965, 24 p.

FREEZAILLAH, B. C. Y.; LOW, C. M. Some preliminary observations on the attack and control of a fungal (*Pestalotia* sp.) disease on seedlings of *Pinus caribaea*. *Malayan Forester*, vol. 31, no. 1, 1968, p. 15–19.

FRIEND, J.; THRELFALL, D. R. *Biochemical aspects of plant-parasite relationships*. Phytochemical Society Symposia Series no. 13. London and New York, Academic Press, 1977, 354 p.

GARA, R. I.; ALLAN, G. C.; WILKINS, R. M.; WHITMORE, J. L. Flight and host selection behaviour of the mahogany shoot borer, *Hypsipyla grandella* Zeller (Lepid., Phycitidae). *Z. Angew. Ent.*, vol. 72, no. 3, 1972–73, p. 259–266.

GAY, F. J. Species introduced by man. In: Krishna, K.; Weesner, F. M. (eds.). *Biology of Termites*, p. 459–494. New York, Academic Press, vol. I, 1970.

——; CALABY, J. H. Termites of the Australian Region. In: Krishna, K.; Weesner, F. M. (eds.). *Biology of Termites*, p. 393–448. New York, Academic Press, vol. II, 1970, 643 p.

GEARY, T. F. Mukwa blight in Central Africa. *Plant Disease Recorder*, vol. 56, no. 9, 1972, p. 820–821.

——; BARRES, H.; YBARRA-CORONADO, R. *Seed source variation in Puerto Rico and virgin islands grown mahoganies*. USDA, Forest Service, Institute Tropical Forestry (Puerto Rico), Res. Paper ITF, 17, 1973, 24 p.

GERHOLD, H. D. Forest trees. In: Nelson, R. R. (ed.). *Breeding plants for disease resistance: concepts and applications*, p. 375–386. University Park, Pennsylvania State University Press, 1973.

GIBSON, I. A. S. The impact of disease on forest production in Africa. In: *FAO/IUFRO Symp. on int. dangerous forest diseases and insects* (Oxford, 1964), vol. 1, 1965a, 14 p.

——. Forest pathology in East Africa. *East African Agric. For. J.*, vol. 31, no. 2, 1965b, p. 194–198.

——. The influence of disease factors on forest production in Africa. In: *14th IUFRO-Congress* (München), vol. 5, 1967, p. 327–360.

——. Diseases of *Pinus patula*. A review. *Comm. For. Rev.*, vol. 49, no. 3, 1970, p. 267–274.

——. Field control of *Dothistroma* blight of *Pinus radiata* using copper fungicide sprays. *East African Agric. For. J.*, vol. 36, no. 3, 1971, p. 247–274.

*——. *Dothistroma* blight of *Pinus radiata*. *Ann. Rev. Phyto-pathology*, 10, 1972, p. 51–72.

*——. Impact and control of *Dothistroma* blight of pines. *Eur. J. Forest Path.*, vol. 4, no. 2, 1974, p. 89–100.

——; HOWLAND, P. Graft failure in young *Cupressus lusitanica*.

East African Agric. For. J., vol. 35, no. 1, 1969, p. 52–54.

GIBSON, I. A. S.; HUDSON, J. C. Pelleting of pine seeds with rhizoctol and other fungicides for control of damping off in Kenya highland nurseries. *East African Agric. For. J.*, vol. 35, no. 1, 1969, p. 98–102.

——; MUNGA, F. M. A note on terminal crook disease of pines in Kenya. *East African Agric. For. J.*, vol. 35, no. 2, 1969, p. 135–140.

——; HOWLAND, A. K. The action of copper fungicides in the control of *Dothistroma* blight of pines, a pilot study. *East African Agric. For. J.*, vol. 36, no. 1, 1970, p. 139–153.

GLORI, A. V.; POSTRADO, B. T. *Control of root-knot nematode* (*Meloidogyne incognita* (Koifoid and White) Chitwood) *on Kaatoan Bangkal* (*Anthocephalus chinensis* (Lamk.) Rich. ex Walp.) *by soil fumigation*. Philippines, Dept. of Forestry, Res. Note Reforest., no. 3, 1969, 6 p.

GRANHALL, I. Quarantine measures against forest diseases and pests. In: *FAO/IUFRO Symp. on int. dangerous diseases and insects* (Oxford, 1964), vol. 2, 1965, 8 p.

*GRAY, B. Economic tropical forest entomology. *Ann. Rev. Ent.*, 17, 1972a, p. 313–354.

——. Some aspects of insecticide usage in tropical Asia. *Asian J. of Med.*, 8, 1972b, p. 299–301.

——. The economics and planning of research into tropical forest insect pests. *Pest Articles and News Summaries*, vol. 20, no. 1, 1974a, p. 1–10.

——. Forest insect problems in the South Pacific Islands. *Comm. For. Rev.*, vol. 53, no. 1, 1974b, p. 39–48.

——. Observations on insect flight in a tropical forest plantation. III. Flight activity of Platypodidae (Coleoptera). *Z. Angew. Ent.*, vol. 75, no. 1, 1974c, p. 72–78.

——. Distribution of *Hylurdrectonus araucariae* Schedl (Coleopera : Scolytidae) and progress of outbreak in major hoop pine plantations in Papua-New Guinea. *Pacific Insects*, vol. 16, no. 4, 1975, p. 383–394.

——; BARBER, I. Studies on *Vanapa oberthuri* Pouillaude (Coleoptera : Curculionidae), a pest of hoop pine plantations in Papua-New Guinea. *Z. Angew. Ent.*, vol. 76, no. 4, 1974, p. 394–405.

——; WYLIE, F. R. Forest tree and timber insect pests in Papua-New Guinea. II. *Pacific Insects*, vol. 16, no. 1, 1974, p. 67–115.

GREENHAM, C. G.; HAWKSWORTH, F. G. Known and potential hazards to forest production by the mistletoes and dwarf mistletoes. In: *FAO/IUFRO Symp. on int. dangerous forest diseases and insects* (Oxford, 1964), vol. 1, 1965, 11 p.

*GREMMEN, J. Stem diseases of conifers caused by rust fungi. In: *FAO/IUFRO Symp. on int. dangerous forest diseases and insects* (Oxford, 1964), vol. 1, 1965, 15 p.

GRIJPMA, P. Studies on the shoot borer *Hypsipyla grandella* (Zeller) (Lep., Pyralidae). X. Observations on the egg parasite *Trichogramma semifumatum* (Perkins) (Hym. : Trichogrammatidae). *Turrialba*, vol. 22, no. 4, 1972, p. 398–402.

GRODZKI, R. M. *Gryllus assimilis*: daños causados e metodos de combate. *Revista Florestal*, vol. 4, nº 2, 1973, p. 34–37.

HANSON, H. C. *Diseases and pests of economic plants of Burma*. Washington, D.C., Amer. Inst. Crop Ecology, 1963, 68 p.

HARRISON, J. L. The distribution of feeding habits among animals in a tropical rain forest. *J. Anim. Ecol.*, vol. 31, no. 1, 1962, p. 53–63.

HAVEL, J. Plantation establishment of klinkii pine (*Araucaria hunsteinii*) in New Guinea. *Comm. For. Rev.*, vol. 44, no. 3, 1965, p. 172–187.

*HAWKSWORTH, F. G.; WIENS, D. *Biology and classification of* dwarf mistletoes (*Arceuthobium*). Washington, D.C., USDA, Agric. Handbook, no. 401, 1972, p. 1–234.

HEPTING, G. H. Appraisal and prediction of international forest disease hazards. In: *FAO/IUFRO Symp. on int. dangerous forest diseases and insects* (Oxford, 1964), vol. 1, 1965, 12 p.

HERRERA AUTTER, S. Protección forestal continental suramericana. In: *Proc. 6th. World Forestry Cong.* (Madrid, 1966), vol. 2, 1968, p. 1995–1997.

HIDALGO-SALVATIERRA, O.; PALM, J. D. Studies on the shoot borer *Hypsipyla grandella* Zeller. (Lep., Pyralidae). XIV. Susceptibility of the first instar larvae to *Bacillus thuringiensis*. *Turrialba*, vol. 22, no. 4, 1972, p. 467–468.

HOCKING, D. Stem canker and pink stain of teak in Tanzania associated with *Fusarium solani*. *Plant Disease Reporter*, vol. 52, no. 8, 1968, p. 628–629.

——; JAFFER, A. A. Field observations on root rot of teak in Tanzania. *FAO Plant Prot. Bull.*, vol. 15, no. 1, 1967, p. 10–14.

——; ——. Damping-off in pine nurseries: fungicidal control by seed pelleting. *Comm. For. Rev.*, vol. 48, no. 4, 1969, p. 355–363.

*HODGES, Jr., C. S. Seed and seedling diseases of forest trees of the world. In: *FAO/IUFRO Symp. on int. dangerous forest diseases and insects* (Oxford, 1964), vol. 1, 1965, 8 p.

HODGES, C. S.; MAY, L. C. A root disease of pine, *Araucaria* and *Eucalyptus* in Brazil caused by a new species of *Cylindrocladium*. *Phytopathology*, vol. 62, no. 8, 1972, p. 898–901.

HOLLOWAY, C. W. The protection of man-made forests from wildlife. In: *World Symp. on man-made forests and their industrial importance* (Canberra), 1966, p. 697–715.

HOLTSTEN, E. H.; GARA, R. I. Studies on attractants of the mahogany shoot borer, *Hypsipyla grandella* Zeller (Lepidoptera : Phycitidae) in Costa Rica. *Z. Angew. Ent.*, vol. 76, no. 1, 1974, p. 77–86.

HOWLAND, A. K.; GIBSON, I. A. S. A note on *Diplodia* spp. on pines in East Africa. *East African Agric. For. J.*, vol. 35, no. 1, 1969, p. 45–48.

HUFFAKER, C. B. Some implications of plant-arthropod and higher-level arthropod-arthropod food links. *Environmental Ent.*, vol. 3, no. 1, 1974, p. 1–9.

*HULL, R. Mycoplasma and plant diseases. *Pests and News Summaries*, vol. 18, no. 2, 1972, p. 154–164.

——; HORNE, R. W.; NAYER, R. M. Mycoplasma-like bodies associated with sandal spike disease. *Nature* (London), 224, 1969, p. 1121–1122.

HUTTON, R. S.; RASMUSSEN, R. A. Microbiological and chemical observations in a tropical forest. In: Odum, H. T.; Pigeon, R. F. (eds.). *A tropical rain forest. A study of irradiation and ecology at El Verde, Puerto Rico*, p. F-43–F-56. U.S. Atomic Energy Commission (USAEC), Div. of Technical Information, 1970, 1 678 p.

INOUE, M. T. Ensaio de procedencia de *Cedrela* em Santo Antomio de Platina, Pr. *Revista Florestal*, vol. 4, n° 2, 1973, p. 49–57.

INTARI, S. E.; NATAWIRIA, D. White grubs in forest tree nurseries and young plantations. *Laporan, Lembaga Penelitian Hutan*, no. 167, 1973, 22 p.

IVORY, M. H. *Fusicoccum tingens* Goid.: a wound pathogen of pines in East Africa. *East African Agric. For. J.*, vol. 32, no. 3, 1967a, p. 341–343.

——. A seedling blight of *Vitex keniensis* Turrill. *East African Agric. For. J.*, vol. 32, no. 4, 1976b, p. 393–398.

——. A technique for root cuttings of *Pinus radiata* in Kenya. *East African Agric. For. J.*, vol. 36, no. 4, 1971, p. 356–360.

——. Pilot plantations of quick-growing industrial tree species

in Malaysia. *Pathological problems of fast-growing exotic conifers in West Malaysia*. UNDP, FAO, Report prepared for the Govt. of Malaysia, Tech. Report no. 6, 1972, 40 p.

IVORY, M. H. *Poria* root disease of exotic forest trees in East Africa. *East African Agric. For. J.*, vol. 39, no. 2, 1973, p. 180–188.

——. Pathological problems of fast-growing exotic conifers in West Malaysia. *Malayan Forester*, vol. 35, no. 4, 1974, p. 299–308.

——; PATERSON, D. N. Progress in breeding *Pinus radiata* resistant to *Dothistroma* needle blight in East Africa. *Silvae Genetica*, vol. 19, no. 1, 1970, p. 38–42.

IYENGAR, A. V. V. The physiology of root-parasitism in sandal (*Santalum album* Linn.). *Indian Forester*, vol. 91, no. 4, 1965, p. 246–258 (part I); vol. 91, no. 5, 1965, p. 341–355 (part II); vol. 91, no. 6, 1965, p. 423–437 (part III).

*JANZEN, D. H. Seed predations by animals. *Ann. Rev. Ecology Systematics*, 2, 1971, p. 465–492.

——. Sweep samples of tropical foliage insects: effects of seasons, vegetation types, elevation, time of day and insularity. *Ecology*, vol. 54, no. 3, 1973, p. 687–708.

JONES, N. Records and comments regarding flowering of *Triplochiton scleroxylon* K. Schum. *Comm. For. Rev.*, vol. 53, no. 1, 1974, p. 52–56.

——; KUDLER, J. Fruit development and insect pests of *Terminalia ivorensis* A. Chev. *Comm. For. Rev.*, vol. 50, no. 3, 1971, p. 254–262.

KARSTEDT, P. Ataque de los insectos y hongos a las hojas recien tumbadas en el bosque tropical y su prevención. *Instituto Forestal Latino-Americano de Invest. y Capacitación* (Mérida, Venezuela), vol. 27–28, 1968, p. 37–57.

KLIEJUNAS, J. T.; KO, W. H. Root rot of ohia (*Metrosideros collina* subsp. *polymorpha*) caused by *Phytophthora cinnamomi*. *Plant Disease Reporter*, vol. 57, no. 4, 1973, p. 383–384.

——; ——. Deficiency of inorganic nutrients as a contributing factor to ohia decline. *Phytopathology*, vol. 64, no. 6, 1974, p. 891–896.

KO, W. H.; HUNTER, J. E.; KUNIMOTO, R. K. *Rhizoctonia* disease of Queensland maple seedlings. *Plant Disease Reporter*, vol. 57, no. 11, 1973, p. 907–909.

KRSTIC, M. Cankers of forest trees. In: *FAO/IUFRO Symp. on int. dangerous forest diseases and insects* (Oxford, 1964), vol. 1, 1965, 18 p.

KUDLER, J. Problem of forest insect defoliation in Ghana. In: *14th IUFRO-Congress* (München, 1967), vol. V, 1967, p. 618–628.

KULKARNI, D. H. Protection of forests against wildlife in the tropics. In: *Proc. 6th World Forestry Cong.* (Madrid, 1966), vol. II, 1968, p. 1924–1927.

KUNTZ, J. E. Forest disease control: indirect measures. In: *FAO/IUFRO Symp. on int. dangerous forest diseases and insects* (Oxford, 1964), vol. 2, 1965, 10 p.

LAMB, A. F. A. *Fast growing timber trees of the lowland tropics. Gmelina arborea*. Comm. For. Institute, Dept. Forestry (Oxford), no. 1, 1968a, 31 p.

——. *Fast growing timber trees of the lowland tropics. Cedrela odorata*. Comm. For. Institute, Dept. Forestry (Oxford), no. 2, 1968b, 46 p.

——. *Fast growing timber trees of the lowland tropics. Pinus caribaea*. Comm. For. Institute, Dept. Forestry (Oxford), no. 5, 1971, 254 p.+appendix.

——; NTIMA, O. O. *Fast growing timber trees of the lowland tropics. Terminalia ivorensis*. Comm. For. Institute, Dept. Forestry (Oxford), no. 5, 1971, 72 p.

LAMB, K. P. *Economic entomology in the tropics*. London, Academic Press, 1974, 195 p.

LAMPRECHT, H. La silvicultura tropical en relación con el establecimiento de plantaciones forestales y el manejo de los bosques naturales. *Instituto Forestal Latino-Americano de Invest. y Capacitación* (Mérida, Venezuela), 22, 1966, p. 18–32.

LAWS, R. M. Elephants as agents of habitat and landscape change in East Africa. *Oikos*, 21, 1970, p. 1–15.

LEE, R. F. *A preliminary annotated list of Malawi forest insects*. Malawi, Forest Research Institute, Research Record, no. 40, 1971, 132 p.

——. *A preliminary account of the biology and ecology of Plagiotriptus spp. (Orthoptera : Eumastacidae)*. Malawi, Forest Research Institute, Research Record, no. 48, 1972, 100 p.

LEVIN, D. A. The role of trichomes in plant defence. *Quart. Rev. Biology*, vol. 48, no. 1, 1973, p. 3–15.

*LEVY, C. *Sapstain of stored wood products and its control. A review of the literature*. Port Moresby, Forest Products Research Centre, 1968, 37 p. multigr.

LEWIS, T. Aerial baiting to control leaf-cutting ants. *Pests and News Summaries*, vol. 18, no. 1, 1972, p. 71–74.

LIM, T. M. Seedling blight of pine. *FAO Plant Prot. Bull.*, 18, 1970, p. 119–120.

——; ANTHONY, J. Control of seedling blight of *Pinus caribaea* Mr. caused by *Colletotrichum gloeosporioides* Penz. *Malayan Forester*, vol. 33, no. 2, 1970, p. 144–148.

LIU, L. J.; MARTORELL, L. F. *Diplodia* stem canker and dieback of *Casuarina equisetifolia* in Puerto Rico. *J. Agr. Uni.* (Puerto Rico), vol. 57, no. 3, 1973, p. 255–261.

LORDELLO, L. G. E. A root-lesion nematode found infesting *Eucalyptus* trees in Brazil. *Plant Disease Reporter*, vol. 51, no. 9, 1967, 791 p.

MACCUAIG, R. D.; DAVIES, R. L. The toxicity of some insecticides against *Manowia* sp. (Acridoidea : Eumastacidae), a pest of *Pinus* plantations in Malawi. *East African Agric. For. J.*, vol. 37, no. 4, 1972, p. 272–278.

MATHUR, R. N.; SINGH, B. A list of insect pests of forest plants in India and their adjacent countries. Parts 4–10. *Indian For. Bull.*, vol. 171, nos. 3, 4, 5, 6, 7, 8, 9, 1956–1960.

——; CHATTERJEE, P. N.; SEN-SARMA, P. K. Biology, ecology and control of *Ailanthus* defoliator *Atteva fabriciella* Swed. (Lepidoptera : Yponomeutidae), the defoliator of *Ailanthus excelsa. Indian Forester*, vol. 96, no. 7, 1970, p. 538–552.

MAY, L. C. Molestias de coniferas ocorrentes no estado de Sao Paulo. *Silv. S. Paulo*, 3, 1964, p. 221–245.

——. O perigo da propagacao de molestias com a introducao de essencias florestais. *Silv. S. Paulo*, 6, 1967, p. 183–187.

McNABB, Jr., H. A 'new' concept of forest tree disease control: physiological suppression. In: *FAO/IUFRO Symp. on int. dangerous diseases and insects* (Oxford, 1964), vol. 2, 1965, 2 p.

*MIKOLA, P. Mycorrhizal inoculation in afforestation. In: Romberger, J. A.; Mikola, P. (eds.). *International Review of Forestry Research*, vol. III, 1970, p. 123–196.

Ministry of Overseas Development (Ministry of Agriculture). *FAO/ECAFE timber trend study for the Far East. Country report for India*. New Delhi, Government of India, 1958.

MORELLET, J. Problèmes forestiers à Cuba. *Bois et Forêts des Tropiques* (Nogent-sur-Marne), 123, 1969, p. 3–17.

MUMA, H. H.; ADEJI, S. A. *Oligonychus milleri* on *Pinus caribaea* in Jamaica. *Florida Ent.*, vol. 53, no. 4, 1970, 241 p.

MURDOCH, W. W.; EVANS, F. C.; PETERSON, C. H. Diversity and pattern in plants and insects. *Ecology*, vol. 53, no. 5, 1972, p. 819–829.

NAIR, P. N. Preliminary trials with tropical conifers in Kerala State. *Indian Forester*, vol. 99, no. 5, 1971, p. 233–242.

NATAWIRIA, D. Pests and diseases of *Albizia falcataria. Rimba Indonesia*, vol. 17, no. 1/2, 1972/73, p. 58–69.

——; TARUMINGKENG, R. C. Some important pests of forest trees in Indonesia. *Rimba Indonesia*, vol. 16, no. 3/4, 1971, p. 151–165.

NAYAR, R. M. Some noteworthy features in Sandal Spike disease. *Indian Forester*, vol. 93, no. 1, 1967, p. 78–79.

——; PADMANABHA, H. S. A. Pathological anatomy of sandal trees affected by Spike disease. *Van Vigyan* (Dehra Dun), vol. 7, no. 3, 1969, p. 68–72.

*NOBLE, M.; RICHARDSON, M. J. An annotated list of seed-borne diseases. *Comm. Mycological Institute, Phytopathological Pap.*, no. 8, 1968, 191 p.

NORDIN, V. J. The intercontinental spread of forest pathogens. In: *FAO/IUFRO Symp. on int. dangerous forest diseases and insects* (Oxford, 1964), vol. 1, 1965, 14 p.

NTIMA, O. O. *Fast growing timber trees of the lowland tropics. The Araucarias*. Oxford, Comm. For. Institute, Dept. Forestry, no. 3, 1968, 139 p.

ODERA, J. A. Insecticidal control of *Pineus* sp. (Homoptera : Adelgidae) in East Africa. *Pests and News Summaries*, vol. 17, no. 4, 1971, p. 464–467.

——. A defoliator of pines, *Orgyia hopkinsi* (Lepidoptera : Lymantriidae) in Turbo, Kenya. *Canadian Ent.*, vol. 104, no. 3, 1972a, p. 355–360.

——. Insecticidal control of *Pineus* sp. (Homoptera : Adelgidae) in East Africa. *East African Agric. For. J.*, vol. 37, no. 4, 1972b, p. 308–312.

——. The incidence and host trees of the pine woolly aphid, *Pineus pini* (L.), in East Africa. *Comm. For. Rev.*, vol. 53, no. 2, 1974, p. 128–136.

ODUM, H. T. PIGEON, R. F. (eds.). *A tropical rain forest. A study of irradiation and ecology at El Verde, Puerto Rico*. Div. of Techn. Information, U.S. Atomic Energy Commission (Oak Ridge), 1970.

OLEMBO, T. W. The incidence of Cypress canker disease in East Africa. *East African Agric. For. J.*, vol. 35, no. 2, 1969, p. 166–173.

——. Studies on *Armillaria mellea* in East Africa. Effect of soil chelates on penetration and colonization of *Pinus patula* and *Cupressus lusitanica* wood cylinders by *Armillaria mellea* (Vahl. ex Fr.) Kummar. *Eur. J. Forest Path.*, vol. 2, no. 3, 1972, p. 134–140.

OLOFINBOBA, M. O. Sap stain in *Antiaris africana*, an economically important tropical white wood. *Nature*, 249, 1974, p. 860.

OMAR, A.; DE VOS, A. Damage to exotic softwoods by Sykes monkeys (*Cercopithecus mitus kolbi* Neuman). *East African Agric. For. J.*, vol. 35, no. 4, 1970, p. 323–330.

ORIANS, G. Tropical population ecology. In: Farnworth, E. G.; Golley, F. B. (eds.). *Fragile ecosystems. Evaluation of research and applications in the neotropics*, p. 5–65. Berlin and New York, Springer Verlag, 1974, 258 p.

OSISANYA, E. O. The effect of attack of *Diclidophlebia eastopi* (Vond.) (Homoptera : Psyllidae) on the survival of *Triplochiton scleroxylon* (K. Schum.). *Nigerian Ent. Mag.*, vol. 2, no. 1, 1969, p. 19–25.

——. Effect of shade on the rate of infestation of *Triplochiton scleroxylon* by *Diclidophlebia* species. *Ent. Exp. & Appl.*, vol. 13, no. 2, 1970, p. 125–132.

PADMANABHA, H. S. A.; BISEN, S. P.; NAYAR, R. Mycoplasma-like organisms in histological sections of infected sandal spike (*Santalum album* L.). *Experientia*, vol. 29, no. 12, 1973, p. 1571–1572.

PANT, N. C. Important entomological problems in humid tropical Asia. In: *Natural resources of humid tropical Asia*, p. 307–329. Paris, Unesco, 1974, 456 p.

PAULIAN, R. The termites of Madagascar. In: Krishna, K.; Weesner, F. M. (eds.). *Biology of Termites*, p. 281–294. New York, Academic Press, vol. II, 1970, 643 p.

PAWSEY, R. G. Forest diseases on Trinidad and Tobago, with some observations in Jamaica. *Comm. For. Rev.*, vol. 49, no. 1, 1970, p. 64–77.

PEREGRINE, W. T. H.; SIDDIQI, M. A. A revised and annotated list of plant diseases in Malawi. *Comm. Mycological Institute, Phytopathological Pap.*, no. 16, 1972, 51 p.

POSTRADO, B. T.; GLORI, A. W. *Root-knot nematode: potential threat to Kaatoan Bangkal* (*Anthocephalus chinensis* (Roxb.) Miq.). Philippines, Dept. of Forestry, Res. Note Reforest., no. 2, 1968, 5 p.

PROCTER, J. E. A. Diseases of pines in the Southern Highlands Province, Tanganyika. *East African Agric. For. J.*, vol. 31, no. 2, 1965, p. 203–209.

REDDY, M. A. R. Damping-off in conifer nurseries in India. *Indian Forester*, vol. 95, no. 7, 1969, p. 475–479.

——; MISRA, B. M. Fungicidal soil treatments to control damping-off of diseases in pines. *Indian Forester*, vol. 96, no. 3, 1970, p. 270–275.

RICHARDS, P. W. The tropical rain forest. *Scientific American*, vol. 229, no. 6, 1973, p. 58–67.

ROBERTS, H. The Platypodidae of Nigeria (Coleoptera). I. The mountains. *Rev. Zool. Bot. Afr.*, vol. 83, no. 3–4, 1971, p. 243–301.

——. *Forest entomology. Fiji.* 1973a, 33 p. multigr.

——. The Platypodidae of Nigeria (Coleoptera). III. The low altitude rain forest. *Revue Zool. Botanique Africaines*, vol. 87, no. 2, 1973b, p. 344–378.

ROONWAL, M. L. A list of insect pests of forest plants in India and their adjacent countries. Part 1. *Indian For. Bull.*, vol. 171, no. 1, 1954.

——. Termites of the Oriental Region. In: Krishna, K.; Weesner, F. M. (eds.). *Biology of Termites*, p. 315–391. New York, Academic Press, vol. II, 1970, 643 p.

——. Taxonomical and biological observations on bark beetles of genus *Carphoborus* (Coleoptera : Scolytidae) from West Pakistan, Western Himalayas and Central India. *Z. Angew. Ent.*, vol. 67, no. 3, 1971, p. 305–316.

——. Field observations on biology of salai borer, *Atractocerus reverus* Walk. (Coleoptera : Lymexylonidae) in India. *Z. Angew. Ent.*, vol. 72, no. 1, 1972, p. 92–97.

ROOVERS, M. Observaciones sobre el ciclo de vida de *Hypsipyla grandella* Zeller en Barinitas, Venezuela. *Instituto Forestal Latino-Americano de Invest. y Capacitación* (Mérida, Venezuela), 38, 1971, p. 3–46.

ROSEVEAR, D. R. *The rodents of West Africa*. London, British Museum (Natural History), Publ. no. 677, 1969, 604 p.

*ROSSELL, S. E.; ABBOT, E. G. M.; LEVY, J. F. Bacteria and wood. A review of the literature relating to the presence, action and interaction of bacteria in wood. *J. Inst. Wood Science*, vol. 6, no. 2, 1973, p. 28–35.

RUDINSKY, J. A.; SARTWELL, C. Jr.; GRAVES, T. M.; MORGAN, M. E. Granular formulation of methylcydokexenone: an antiaggregative pheromone of the Douglas fir and spruce bark beetles (Coleoptera : Scolytidae). *Z. Ang. Ent.*, vol. 75, no. 3, 1974, p. 254–263.

RUEHLE, J. L. Nematodes of forest trees. In: Webster, J. M. (ed.). *Economic Nematology*, p. 312–334. New York, Academic Press, 1972, 563 p.

——. Nematodes and forest trees, types of damage to tree roots. *Ann. Rev. Phytopathology*, 11, 1973, p. 99–118.

SANG, F. K. A.; MUNGA, F. M. Trials of fungicides for control of terminal crook in *Pinus radiata* in Kenya. *East African Agric. For. J.*, vol. 39, no. 1, 1973, p. 41–45.

SCHEDL, K. E. *Monographie der Familie Platypodidae, Coleoptera*. The Hague, Junk, 1972, 322 p.

SCHIEBER, E. Pine cone rust in the highlands of Guatemala. *Plant Disease Reporter*, vol. 51, no. 1, 1967, p. 44–46.

*SCHÖNHAR, S. *Fomes annosus* (Fr.) Cooke in nadleholz-beständen und möglichkeiten zur seiner bekampfung. Sammelbreferat über die in den jahren 1969–1973 erschienene literatur. *Z. Pflkrankh. Pfl. Schutz.*, vol. 81, n° 4, 1974, p. 52–64.

SEBASTIAN, U. O. Observations on the biology of *Tonica niviferana*, the shoot borer of *Bombax ceiba*. *Indian Forester*, vol. 95, no. 7, 1969, 487 p.

SEGURA, C. B. DE. La enfermedad rosada (*Corticium salmonicolor*) y el mal el hilachas (*Pellicularia koleroga*) sobre varias especies de *Eucalyptus* en Turrialba, Costa Rica. *Turrialba*, vol. 20, no. 2, 1970a, p. 254–255.

——. Manchas foliares causadas por el hongo *Cylindrocladium scoparium* Morg. en *Eucalyptus* spp. en Turrialba, Costa Rica. *Turrialba*, vol. 20, no. 3, 1970b, p. 365–366.

SELISKAR, C. E. Virus and virus like disorders of forest trees. In: *FAO/IUFRO Symp. on int. dangerous forest diseases and insects* (Oxford, 1964), vol. 1, 1965, 44 p.

SHAIN, L.; HILLIS, W. E. Ethylene production in *Pinus radiata* in response to *Sirex-Amylostereum* attack. *Phytopathology*, vol. 62, no. 12, 1972, p. 1407–1409.

SHARMA, D. G. The problems of forest production in India with special reference to Madhya Pradesh. *Indian Forester*, vol. 93, no. 6, 1967, p. 407–410.

SHIGO, A. L.; HILLIS, W. E. Heartwood, discolored wood, and micro-organisms in living trees. *Ann. Rev. Phytopathology*, 11, 1973, p. 197–222.

SIMMONDS, F. J. Economics of biological control. *Pests and News Summaries*, vol. 14, no. 3, 1968, p. 207–215.

SINGH, S.; TEWARI, R. K. Role of precursor fungus in decay in standing teak. *Indian Forester*, vol. 96, no. 12, 1970, p. 874–876.

——; PURI, Y. N.; BAKSHI, B. K. Decay in relation to management of dry coppice teak forests. *Indian Forester*, vol. 99, no. 7, 1973, p. 421–430.

SLIWA, D.; BECKER, V. O. Studies on the shoot borer *Hypsipyla grandella* (Zeller) (Lep., Pyralidae). XX. Observations on emergence and maturing of adults in captivity. *Turrialba*, vol. 23, no. 3, 1973, p. 352–356.

SMITH, S. E. Mycorrhizal fungi. *Critical Rev. Microbiology*, vol. 3, no. 3, 1974, p. 275–313.

SPAULDING, P. *Foreign diseases of forest trees of the world*. Washington, D.C., USDA, Agric. Handbook no. 197, 1961, 361 p.

SQUIRE, F. A. Entomological problems in Bolivia. *Pests and News Summaries*, vol. 18, no. 3, 1972, p. 249–268.

STYLES, B. T. The flower biology of the Meliaceae and its bearing on tree breeding. *Silvae Genetica*, vol. 21, no. 5, 1972, p. 175–182.

SUHARTI, M. Causes of the damping-off disease in *Pinus merkusii* seedlings and environmental influence on its development. *Laporan, Lembaga Penilitian Hutan*, no. 162, 1971, 35 p.

THAPA, R. S. Bionomics and control of larval defoliator, *Margaronia hilaralis* Wkr. (Lepidoptera : Pyralidae). *Malayan Forester*, vol. 33, no. 1, 1970a, p. 55–62.

——. Borers of freshly felled timbers of *Parashorea tomentella* and their control. *Malayan Forester*, vol. 33, no. 3, 1970b, p. 230–242.

——. Results of preliminary investigation on black stains in commercial trees in dipterocarp forest of Sabah. *Malayan Forester*, vol. 34, no. 1, 1971, p. 53–64.

——; SHIM, P. S. Termite damage in plantation hoop pine, *Araucaria cunninghamii* D. Don, in Sabah and its control. *Malayan Forester*, vol. 34, no. 1, 1971, p. 47–52.

VARMA, A.; CHENULU, V. V.; RAYCHAUDHURI, S. P.; PRAKASH, N.; RAO, P. S. Mycoplasma-like bodies in tissue infected with sandal spike and brinjal little leaf. *Indian Phytopathology*, 22, 1969, p. 289–291.

VENKATESH, C. S.; KEDHARNATH, S. Breeding sandal for resistance to the spike disease. *Indian For. Bull.*, no. 243, 1964.

VERMA, S. K. Observations on the mortality in the forests of *Anogeissus* in Rajasthan. *Indian Forester*, 3, 1972a, p. 199–205.

VERMA, R. A. B. Leaf spot diseases of mahogany. *Indian Phytopathology*, vol. 25, no. 1, 1972b, p. 33–35.

VIENNOT-BOURGIN, G. The role of phytopathological research in developing countries. *Phytopathology*, vol. 64, no. 7, 1974, p. 912–917.

VILA, M. M. Uma broca do guapuruvu (*Acanthoderes jaspidea* Germ.). *Silv. S. Paulo*, 41, 1965–66, p. 305–309.

VITE, J. P. Pest management systems using synthetic pheromones. *Contrib. Boyce Thompson Institute*, vol. 24, 1970, p. 343–350.

VITE, J. P.; FEDERICO ISLAS, S.; RENWICK, J. A. A.; HUGHES, P. R.; KLIEFOTH, R. A. Biochemical and biological variation of southern pine beetle populations in North and Central America. *Z. Angew. Ent.*, vol. 75, no. 4, 1974, p. 422–435.

WAHEED KHAN, M. A. Root-rot and patch-mortality disease in equatorial teak of the Sudan. In: *FAO/IUFRO Symp. on int. dangerous forest diseases and insects* (Oxford, 1964), vol. 1, 1965, 3 p.

WEBB, L. J. Biological aspects of forest management. *Proc. Ecol. Soc. Aust.*, 3, 1968, p. 91–95.

WEESNER, F. M. Termites of the Nearctic Region. In: Krishna, K.; Weesner, F. M. (eds.). *Biology of Termites*, p. 477–525. New York, Academic Press, vol. II, 1970, 643 p.

*WELLMAN, F. L. *Tropical American plant disease.* (*Neotropical phytopathology problems.*) The Scarecrow Press (Metuchen, N.J.), 1972, 989 p.

*WHITMORE, T. C. *Tropical rain forests of the Far East*. Oxford, Clarendon Press, 1975, 278 p., 550 references.

WING, L. D.; BUSS, I. O. Elephants and forests. *Wildlife Monographs*, vol. 19, no. 2, 1970, p. 1–92.

YATES, H. O. *Investigación sobre el fomento de la producción de los bosques del noreste de Nicaragua*. Rome, FAO, Tech. Rep. 1, 1971, 13 p.

ZETHNER, O. Forstentomologische problemen i det indiske silkontinent belyst ved eksempler fra Pakistan og Bangladesh. *Ent. Meddr.*, vol. 41, n° 3, 1973, p. 129–143.

Part II

Man and patterns of use
of tropical forest ecosystems

Introduction

Biocoenoses are communities associating various species. Biotopes are the physical environments in which such communities live. An 'extraordinarily complex series of interactions come into play between the occupants of the same biotope and the object of ecology is to sort out their principal characteristics and relations between them and the abiotic factors. It is this network of multiple interactions that . . . permits the ecosystem to be defined completely'.[1] Each species in an ecosystem has its place and fulfils an appropriate role. Ecosystems are stable in the absence of modification of the physical environment, but this does not exclude minor evolutionary changes. European authors have often used terms such as equilibrium, climax or homeostasis to describe this state but in the United States there is more insistence on both an endogenous and exogenous dynamics which is capable of inducing a certain degree of reorganization between the different elements within upper and lower limits. This equilibrium is not static as ecosystems are capable of withstanding modifications of the surrounding environment and marked variations in population densities.

It is within this framework that man is considered an integral part of the ecosystem. Such an approach leads to a sound biological basis for a new type of human geography that analyses man's place in the environments where he lives. This new type of human geography is a return to fundamental sources and more precisely to the time when Vidal de la Blache (1921)[2] wrote: 'from the geographical point of view, the fact of cohabitation, i.e. the common use of a certain area, is the foundation for everything'. This 'interdependence of all the co-inhabitants of the same area' justifies an ecological approach to man and the way in which he occupies and uses the land.

However, the concepts and techniques of ecology should not be applied to the study of the relationships between man and the natural environment without some precautions. Environmental conditions affect man like all wild animals. Chapters 15, 16, 17 and 18 will deal successively with the demographic factors, nutrition, health and the impact of the tropical forest environment on the human organism. This concerns the autecology of the human species and it should be possible to assess the limits, constraints and stresses that man has to withstand within the tropical forest ecosystems: his reactions; how and to what extent he has been able to resist and adapt or has become weakened. The presence of man in forest environments implies his adaptation. It is only the extent of this adaptation and the way in which it is expressed, either individually during the course of a lifetime or by the effect of selective pressure on a population, that is in question. The responses to the environment should not be reduced to purely biological mechanisms. They are largely conditioned by the customs, attitudes and interrelationships rooted within the culture and the social system.

The exploitation of forest ecosystems by human groups shows at a much higher degree the autonomy of the latter towards the genetic determinism of the species. It is true that systems of production are more or less effectively adapted to their environment in the same way as the human organism is; societies would not otherwise survive. It is equally true that the general form of such adaptations takes the genetic characteristics of humanity into account. A selective genetic drift may be produced under the pressure of particular environments (such as sickle cell anaemia or the blood characteristics of some high altitude Andean populations). But no general or particular genome feature can rationally be linked to the type of tools, the choice of plants used or the fertilizing techniques; nor to the combination of factors that determine the way of life of a particular population as compared to others. The variety in this field is considerable. It results from inventions or ideas borrowed from other human groups that it would be wrong to derive from one or another modification of the genetic make-up. Human groups facing the same natural constraints or advantages use very many different adaptations. Distance or geographical obstacles have been the cause of a certain degree of genetic differentiation. Hereditary features have been developed and today would appear to be characteristic of some areas. But even when associated with well-defined morphological characteristics, such features only have minor importance. No-one has been able to prove that they have any effect on the way or the nature of the efforts by which man draws his livelihood from the environment. The diversity of systems of production, and of each of their components, is the result of cultural and especially agricultural differentiation, during the very short time span from the Neolithic to the present day. This is a major point that stresses the gap between natural and human ecology. The response of an animal species towards the environment is not always rigorously determined by heredity; there is a possibility of learning by experience and thus a minimum

1. Lamotte, M. Ecologie. In: *Encyclopaedia universalis*, vol. 5, p. 923–933. Paris, 1969, 1106 p.
2. Vidal de La Blache, P. *Principes de géographie humaine*. Paris, Armand Colin, 1921, 327 p.

of indetermination with regard the place and role of a species in the ecosystem into which it is integrated. The studies of the last twenty years have produced evidence of a certain degree of continuity from the animal to man as regards this and other concepts. Among animals, the acquired behaviour partly belongs to a social heritage. But even among the most highly evolved species, this remains a minor part and the resulting variation in behaviour is infinitely less than in the case of man.

The variety of cultural responses by man to the problem of survival in a given environment thus implies a certain freedom of action as compared to the other components of ecosystems. Man the animal does not only come under the effects of the environment, having an assigned place within narrow limits, instead he manages to extract himself from this biological position and manipulates ecosystems to his advantage according to his organizational and technological capacities. In other words, he ceases to remain part of the equilibrium and breaks it. It must be stressed that in nature nothing is really stable and changes occur and have been produced long before man. Here again it is more a question of extent and of relative speed. Transformations accelerate from the process of food-gathering to plantations and to towns. Each of these transformation levels represents only a transitory period of equilibrium. The mastering of ecosystems which is closely linked to demographic growth and to an enlargement of the communication network gives a striking uniqueness to human ecology. Nature is not only the place and the framework in which human activities take place, it has become an object of exploitation. The concern for the conservation and protection of the forest that is threatened by an excess of such exploitation reintroduces the idea of a relative equilibrium at the highest level of the relationship between nature and man. But, since the first stages of agriculture, man has no longer been the gatherer that at all times had to come to terms with natural forces.

If he always commands nature, while obeying it, according to Bacon's famous phrase, it is henceforth him that calls the tune. This is why a simple classification is not adequate in describing man-made ecosystems. Chapters 19, 20 and 21 in considering the relationships between man and the forest will emphasize the dynamic changes occurring from one to another more productive form of exploitation. Two factors that are linked together promote such changes: technological innovation and demographic growth. The second acts in two ways: the progressive occupation of virgin lands, i.e. intact forest which will be involved in the cropping and fallow cycles; and the shortening of fallows and the development of permanent cropping.

The concept of scale in terms of both time and space plays an important role in considering relationships between human societies and ecosystems. Chapter 19, which approaches these problems from the aspect of human geography, will give considerable attention to this. Geographers can no longer simply take into consideration the relationship between the natural environment and man. This objective has become inseparable from the consideration given to all kinds of mediations which exist between natural areas and their occupants. These mediations refer to: the history, the number of generations responsible for the perfection and transmission, at the level of each group, of a particular range of production techniques; the society as functioning at present and its particular features such as the division of the tasks of production and the distribution of resources; the ways by which areas are linked together and ranked on the basis of exchange of goods and information, and of political organization, so that each portion of the world may be decisively influenced by other areas and peoples; lastly, mediation through population densities that are the result of a complex series of interactions between demographic growth, the progress of production techniques and the effectiveness of political and territorial control.

15 Demography

Introduction

Human groups in tropical forest ecosystems include both those whose activities are confined to that ecosystem and those whose activities are linked with societies outside it. Relatively intact tropical forest survives only in a few areas, as do the very small human populations whose lives are not directly affected by human activities based on economic and social systems with essential portions or extensions outside the tropical forest zone.

Demographic features of human populations include definition of population and subgroup boundaries, age and sex specific birth, death and migration rates, and the resultant patterns of population composition, growth, change and distribution in time and space.

Anthropology can be criticized for presenting functional analyses of the idealized descriptions of social structure of small numbers of obscure peoples, rarely showing how the structure operates in the face of environmental, demographic or historical variation, and without systematic reference to interaction with the rest of the world (Hackenberg, 1974). Leach (1954) broke with the traditions of staticism and homogeneity, and attempted to demonstrate variations in social structure within a single ethnic group which were associated systematically with differences in land-use patterns (swiddening versus irrigated agriculture) but he failed to connect the changes or variations with demographic patterns.

Demographers usually choose national samples as units of analysis and description. The data are pooled despite major differences in the ways in which people behave in organized groups with common interests. Attributes and indicators may be considered with respect to individuals, but social structure and its relation to demography and environment are not. Attempts have been made to summarize the knowledge of demographic conditions at the continental or subcontinental level in tropical areas (e.g. Davis, 1951; Caldwell and Okonjo, 1968) but these have been more concerned with description and historical reconstruction of demographic patterns than with analysis of relationships between environment and demographic variables. Furthermore, most demographic methods assume populations are closed to migration, and many of the more powerful techniques assume that birth rates and death rates are not changing—assumptions which are not valid in most of the populations considered here.

The range of error in predictions of growth rate and population size is often greater than acceptable, since small errors have large economic consequences. Also, the ability of nations to control their population sizes by deliberate efforts have frequently failed to meet their targets. This suggests that the application of the theory of populations is not well developed, despite the prodigious data. Population theorists have generally depended on physiological-biological causes of death or capacity to reproduce, or human nature, as explanatory principles, and have been relatively unsuccessful or disinterested in using environment as a determinant of population characteristics. Nor have they been particularly interested in looking at the consequences of demography for environment. There is no generally accepted systematic theory which relates the tropical forest ecosystem to demography.

Most, if not all of the outstanding theoretical issues which face demography world-wide could be or have been considered in studies of human populations in the tropics, but no sweeping generalizations have emerged which bear the test of comparative studies. Societies may modify (consciously or unconsciously) the relationship between their population size and their environment in a number of ways. These include modifications of the size of the resource base by changing territorial control or through social mechanisms such as trade, or through technological innovations modifying the number of entrants into the population by influencing fertility and migration, or by modifying the number of departures from the population by influencing mortality or migration (see chapter 19, part 1). These strategies usually operate simultaneously. They do not ordinarily operate homogeneously for all segments of the population, so that distributions within the population may change.

Several rationalistic demographic theories assume that people are trying to maximize or optimize some particular thing (sexual pleasure, social prestige, control of production, economic well-being in old age, perpetuation of lineage, etc.). Generally these theorists have not considered the simultaneous operation of more than one motivation or even the extent to which long-term versus short run goals may be balanced, mutually supporting, or contradictory. Presumably all people have a need for prestige, sexual pleasure, control of resources, old age security, desires to avoid domination and desires to maximize their economic well-being, etc. How do all these factors interact?

Non-rational or a-rational theories seek to explain differences or changes in vital rates through considering biological factors affecting births, deaths and migration, without requiring an analysis of motivation. The mathematical determination of population characteristics is deceptively simple. Population size is the sum of births, minus deaths, plus net migration; population age and sex composition is the result of the application of age and sex specific birth, death and migration rates. Thus factors which increase probabilities of survival tend to increase population growth, while factors decreasing fertility decrease population growth, but each of these also affects the age structure and thus has a feedback on future growth.

As has often been noted, the usual initial effects of

population contact have been to increase deaths through the introduction of new diseases, vectors, reservoirs, or disease transmitting human contact. The secondary effect has usually been the decline in mortality due to increased biological or behavioural immunity—shunning malarial areas, wearing protective clothing, suppressing vectors, preventing or curing illness and avoiding famines.

The biological determinants of fertility are at least as complicated as those affecting mortality. Calhoun (1962), on the basis of rodent studies, points to the empirical generalization of a direct interaction between high population density and control of fertility, showing for example change of social structure and failure to mate among rats crowded in cages. Analogous mechanisms have been suggested as causes of human fertility decline in cities. Romaniuk (1968) studied the relatively low fertility in tropical Africa and concluded 'the sterility is physiological, not voluntary. This view is supported by the study of various cultural and biological factors related to fertility... the physiological sterility observed in low fertility areas is caused by venereal disease'. The clearest biologically mediated regulatory mechanism on fertility is the association between lactation (nursing) and post-partum amenorrhoea (Potter *et al.*, 1965; Jain and Sun, 1972; Simpson-Hebert, 1975). According to this argument, post-partum return of fecundability is inhibited by the suppression of menstruation by lactation. Thus the risk of a nursing mother conceiving is reduced in comparison with a woman of similar age and parity who is not nursing. Factors inhibiting lactation (e.g. early substitution of other foods for mother's milk, cessation of nursing due to cultural preferences, post-partum employment of woman physically removed from her infant, early infant death, etc.) should hasten the return of ovulation and decrease the time between conceptions. Factors in modernization (bottled milk, industrial employment of women) thus should decrease post-partum infertility, possibly upsetting unconscious fertility regulating mechanisms which were inherent in traditional agricultural societies. Substitution of processed foods for mother's milk might then have a synergistic effect on raising fertility by exposing the infant to greater risk of malnutrition, infection and mortality and exposing the mother to greater risk of closely spaced pregnancies.

Davis and Blake (1956) are correct in pointing to social factors which inhibit or promote the risk of fertilization, development and delivery of a live-born infant, often irrespective of conscious motivation on the part of the parents. Bulmer (1971) has described a series of such measures operating in New Guinea, which have been effective in reducing birth rates. Nag (1967) investigated the influence of family structure on coital frequency (all other things being equal, high coital frequency should be associated with high fertility). He has found evidence that fertility in India is higher in nuclear families than in extended families, and argues this is the result of greater coital frequency associated with increased privacy in the nuclear families. On the other hand industrialization, which has usually been associated with nucleation, has also been generally associated with a decline in fertility. Evidently the privacy theory will not explain all cases. The Davis and Blake scheme does not treat socio-biological factors of fecundability (the ability of a female to become pregnant). Recent research in the relationship between reproductive physiology and nutrition (Frisch and Revelle, 1969; Frisch, 1974; Frisch and McArthur, 1974) imply that the period of fecundability is delayed by malnutrition. Malnutrition also contributes to mortality through the malnutrition—infection syndrome (Scrimshaw, Taylor and Gordon, 1968; National Academy of Sciences, 1970, etc.), and, it is argued, may contribute both to high fertility (through attempting to make up for infant mortality), and high maternal mortality (through the maternal depletion syndrome in which malnourished women are progressively weakened due to frequent, closely spaced pregnancies) (Mata, 1975).

The demographic transition: a unifying concept

Demographic transition theory does not specify causal mechanisms in detail. It is a set of descriptive generalizations developed initially from experience in Europe and North America (Davis, 1945; Notestein, 1945). In these areas fertility and mortality rates are assumed to have been originally high ($\geqslant 40$ per thousand) and in approximate balance, so that the population did not grow. In Europe, mortality rates fell slowly as a result of improved public health, production and distribution of food, medical technology, etc., and this was followed by a slow decline in fertility more or less associated with industrialization and urbanization, the need for education, the availability of material goods, and the possibility of social security in the form of state or other non-familial rather than family institutions. Modern fertility and mortality rates are now *ca.* 10 per thousand. During the transition the excess of fertility over mortality in Europe averaged only *ca.* 0.5 per cent per year.

The history of the demographic transition in the tropical world has been different from that of Europe. Tropical populations have entered the transition more recently, as mortality rates have fallen very rapidly, especially in the past 30 years, to the point where they are now approaching modern European levels. The characteristics of contemporary tropical populations can be summarized in a strong consistent negative association between latitude and demographic rates:

	World mean (%)	Correlation equation	Correlation coefficient (r)
Crude birth rate	35.3	$=47.6-0.511$ lat.	—0.65
Crude death rate	14.2	$=18.0-0.157$ lat.	—0.40
Natural increase (%)	2.1	$=2.87-0.031$ lat.	—0.51

In most tropical areas there is now a large gap between birth and death rates, as birth rates have tended to remain at high levels, with consequent rapid and massive population growth, sometimes at rates of 3 per cent or more. Thus population growth (with the implied characteristics of large population size and high child dependency ratio) has become

characteristic of contemporary tropical populations, whose cultural traditions and economies are different from and whose socio-economic modernization has not reached European levels, and who do not have the options of moving to less densely settled regions.

All this implies that case studies should relate to the interaction between environment and the determinants and consequences of fertility, mortality and migration, specifically as they may be associated with the processes of recent socio-economic and technological modernization. Particularly important questions are: what were the demographic conditions prior to modernization? what were the pre-modern patterns of mortality and how have these changed? what have been the social and environmental influences on fertility and migration associated with modernization? what have been the environmental effects and social structural adaptations to rapid population growth and large population size and concentration?

Methodological problems

Ideally, in order to make demographic projections for a defined area, the present population size and composition should be known, along with the fertility, mortality and migration rates. The size and age and sex composition of the population is usually known from a census. Vital rates are normally calculated from knowledge of the population size and vital events during a defined time. Where vital events are not recorded with sufficient completeness or accuracy, rates may be reconstructed by retrospective surveys, collection of reproductive histories, or by comparing two successive and relatively closely spaced censuses (assuming there has been no net migration). Where these types of data are lacking, vital rates may be estimated by comparing the observed age distribution in a single census with model life-tables which have been constructed by assuming various combinations of age specific mortality and fertility (Bourgeois-Pichat, 1957; Collver, 1965; Keyfitz and Flieger, 1971; Weiss, 1973; Brass, 1975) and by an extension of Brass's techniques fertility rates may be estimated from census data on numbers of children born and surviving and ages of their mothers (Cho, 1973).

Methods and standards for demographic analysis have generally been developed for dealing with national populations for which repeated censuses exist, where vital events have been reliably recorded for the majority of the population, where ages can be ascertained with accuracy, where net international migration is negligible, and where calculation of demographic characteristics of small size subgroups is not a primary goal. Further development, testing and application of techniques of incomplete data on anthropological populations is essential for greater understanding of population-ecosystem interaction.

Because good quality census and registration information is lacking for many parts of the world, the classical demographic techniques may be neither applicable nor appropriate—appropriateness hinges on the nature of the populations from which models have been drawn which relate mortality and fertility to population age structure.

Demographers distinguish between stationary populations which result from a balance between births and deaths and in which there is no net migration, and stable populations in which there is no net migration and age and sex composition remains constant as a result of constant vital rates, but in which an imbalance between vital rates may result in a change of population size. If there is no net migration and age-sex-specific birth and death rates remain constant over a few generations, the population will achieve and maintain a structure dependent on those rates, regardless of the age and sex structure with which the population started. The assumption that age and sex specific vital rates remain constant greatly simplifies demographic analysis by allowing demographers to calculate the vital rates necessary to produce an observed age and sex distribution. A closely allied method is that of model life-tables which are based on the assumption that the general shape of the distributions of mortality by age are similar, regardless of the total mortality (i.e., that mortality shortly after birth is relatively high, then declines in early childhood and then increases gradually with age; Coale and Demeny (1966). This suggests that there is a limited number of families of distributions of mortality by age, within which the crude death rate predicts the age and sex specific levels of mortality. This allows the fitting of age and sex specific mortality patterns which could have produced that population structure. Because the overall pattern of association between age and mortality has generally been found to be highly correlated with the mortality experienced by infants and young children, demographic analysts sometimes feel that the assumption that the rates are stable over several generations need not be rigidly adhered to in this type of analysis. The fertility and mortality experience of the few years prior to the census will be represented by the age distribution of the young children and their presumptive mothers, and the age distribution can be fitted to the lowest ages of a model life-table which in turn will yield the mortality schedule for the entire population. Since the general pattern of variation of fertility by age has been found to be similar in a number of different populations, the age specific fertility rates can be calculated by fitting the number of children born (calculated from numbers of children surviving to time of census) and the number of women in the reproductive ages to a model fertility schedule. Thus model life-table analysis has been applied to populations where vital rates are unknown but believed to be unchanging or slowly changing, and where two successive censuses are not available. In such circumstances the analysis yields a calculation of the average vital rates of the population over the past few years and allows a prediction of the future size and composition of the population assuming the calculated rates persist.

If the general assumption on which model life-table and related analyses are based is accepted, anomalies in the mortality distribution of modern tropical populations should be looked for as indicators of interactions with the environment. One such anomaly apparently exists in sub-Saharan Africa, where the mortality of children between ages one and five is higher relative to mortality at other ages than is

found in other populations (Cantrelle, 1974; Page, 1974). This excess mortality is due to infections, parasitic diseases and the interaction of malnutrition. The differences between African and Asian mortality patterns have resulted from the isolation of Africa from technological developments (Cantrelle, 1974). In discussing reasons for the apparent divergence of tropical African populations from the model life-tables of Coale and Demeny (1966), Page (1974, citing Jelliffe, 1968) also refers to nutritional factors, especially post-weaning malnutrition. The pattern of excess mortality among young African children seems likely to be a recent phenomenon, rather than an inherent characteristic. The malnutrition-infection may have resulted from rapid population growth associated with changes in traditional child feeding practices (e.g., early weaning) which may in turn be associated with socio-economic changes (e.g., increased participation of women in the non-family labour force), unaccompanied by socio-economic development which would lead to adequate restructuring of infant and child diets (Welbourn, 1955, 1958) or adequate substitution for traditional forms of birth limitation (e.g., lactational amenorrhoea, taboos on post-partum intercourse, etc.).

Tropical forest populations living under non-modern conditions generally meet few or none of the ideal requirements for stable model life-table analysis. The extent to which these techniques yield reasonable approximations despite the failure of the assumptions has not been systematically tested.

The fact that the local groups are liable to be small creates special problems of analysis and probably has a direct effect on the levels of the rates themselves. Analysis assumes that demographic events occur with a given probability associated with age and sex. Life-tables are simply statements of the probability of a person of given age and sex surviving (or dying) in a given time interval. Because demographic analysis is usually applied to large populations the probabilities are applied directly to large numbers of people within any category, and it is assumed that any variations will be proportionally small, random and self-cancelling. Thus the probabilities are given as averages without any statement of variance. These assumptions are not appropriate for small populations where random variations may have large proportional effects on the composition of the total population. Over a series of years the errors might be self-cancelling or self-reinforcing. An imbalance of one sex at reproductive age suggests the importance of the possibility of exchange of population between subunits of populations, as is typical of tropical forest hunters and gatherers. Thus migration must be considered as an important factor in the demographic analysis of small populations, especially as it applies to management of population/resource relationships.

The difficulties are compounded in the analysis of small non-modern populations by the lack of reliable information on age. The errors may be systematic and life-table estimates may be seriously biased causing large errors in the estimation of vital rates without good means of verification.

Intergroup migration, which seems characteristic of anthropological populations also has serious effects on the applicability of conventional demographic techniques. Because migration is an important means of regulating population/resource interaction, a method must be devised which will combine analysis of meaningful social-behavioural units with meaningful demographic units. The boundaries of these two types of units may be quite different.

The characteristic distribution of mortality over time in the relevant socio-demographic units is not known. If mortality tends to be constant, stable population assumpttions may be met, and the results of their analysis may be meaningful. If, on the contrary, the long run pattern is of oscillations between low and high rates of mortality, the average may be misleading. This problem can be addressed through a comparative series of historical studies, but the necessary information seems to be lacking. Another approach might be computer simulation.

Demographic interactions with the ecosystem

Details of the interaction of human population variables with the ecosystem must be sought within each of the major use-patterns in the tropical forest zone.

Features of the ecosystem which might affect fertility probably do so primarily through disease and nutrition interactions. For example, delayed onset of menarche associated with malnutrition is characteristic of many tropical populations, with a consequent theoretical increase in time between generations and decrease in length of the fecundable period for women. This condition is not unique to the tropics, but has been characteristic of undernourished populations in general. Another factor is the reduction of fertility by venereal disease, again not unique to the tropics. Small population size (which restricts access to marital partners and thus may delay age of marriage and increase time between marriages in the case of widowhood) is a feature of size and marriage rules of the social group; it is not unique to the tropical forest. Thus there is in general no unique tropical condition affecting fertility.

There are few physical barriers to migration within and between tropical forest zones except the oceans (which may actually be channels rather than barriers); again these are not unique to the tropics. The barriers are either sociocultural (which are not unique to the tropics), or biomedical (especially vector-borne diseases, only a few of which are tropical).

Apparently then tropicality is not an independent variable with a major deterministic effect on demography. Linkages of the tropical forest environment with demographic variables tend to be loose, indirect or trivial. For this reason, demographic conditions in the tropics are more likely to reflect socio-economic-cultural conditions and modes of adaptation than the environment *per se*. Most of the tropics is characterized by primary dependence on agriculture and/or extractive industries, low level of secondary or tertiary industrialization, frequent national emphasis on a single crop for subsistence and often for export and the consequent vulnerability of the national economy to climatic and world market variations, low income, and relatively poor nutrition. Demographically these areas are characterized by relatively high birth rates, high but rapidly falling death rates, rapid population growth, rapid growth of cities and a lag in the development of basic social and educational services. Large-scale international migration which was often encouraged to fill the under-exploited areas is no longer possible or encouraged; large-scale internal migration has become important as has small-scale and often temporary international migration for the deliberate purpose of introducing technical and social innovations. Almost all tropical populations have been touched and most have been substantially altered by these processes.

Demographic, cultural, social and economic interchange between populations within the ecosystem and interchange between populations from different ecological zones is an important, perhaps essential feature of the total socio-demographic system. Pygmies, for example, do not live in isolation, but have essential relationships with Bantu groups (see chapter 19, part 2); Yanomamo marriage, genetic structure and trade have essential links with the more sedentary Makiritare (Chagnon *et al.*, 1970); Iban men regularly spend many years travelling away from home (*bejalai*) to collect prestige goods or to supplement agricultural income (Freeman, 1955); regular oscillation occurs between social structural and land-use patterns of swiddeners/irrigated rice farmers in highland Burma (Leach, 1954); and there is demographic, social and economic flow between upland and lowland in northern Thailand (Kunstadter, 1972). Such interchanges are parts of a wide-spread continuous series of processes with a very long history.

To sum up, the tropical ecosystem characteristics most likely to have demographic effects operate primarily in the form of vector-borne diseases. Demographic effects on the tropical forest ecosystem however are wide and varied, depending primarily on social, economic and cultural circumstances. Population size and density, for example, affect the ability of the system to maintain itself under swidden agriculture or hunting and gathering. Increase in population size, perhaps resulting from social or economic changes outside of the tropics, may lead to a decrease in the productivity of the environment, which may lead in turn to migration or increased mortality (thus reducing population growth) or to technological change (thus increasing productivity through further modification of the environment, allowing a higher population to be maintained). See chapter 19, part 1.

Patterns of human use and associated demographic characteristics

Hunting, gathering and non-commercial fisherfolk make relatively little destructive modification of their ecosystems. They operate as food and other commodity harvesters, occasionally with small cultivated or domesticated supplements. Their communities are small, and population densities are low. Long-term population growth is slow or zero, unless factors determining mortality are disturbed from the outside. The initial reaction of these isolated communities to outside contact has often been a decline in population size due to increased mortality from newly introduced causes. Fertility is maintained at a moderate level, high enough to recoup any local disaster, but not so high as to lead to sustained population growth. Migration may take place within a defined seasonal cycle, but migration out of or into the use-system is rare or absent.

Such populations, often ethnically distinct from the majority populations, remain only in areas which are marginal to exploitation at a higher level, usually internal mountainous or heavily forested or swampy zones not suited for agriculture. Populations occupying these areas may have been pushed into them from other more favourable areas, and they do not necessarily represent the cultural, ecological or demographic characteristics of pristine hunters and gatherers. At present these groups are under increasing pressure to stop their nomadic way of life.

Commercial lumbering without replacement of the removed species, may be considered a variant of hunting and gathering, but control is from outside of the tropical forests; there is no intention of permanent settlement and often no intent to maintain sustained yields. Population density is low, largely adult male, non-self reproducing, and highly mobile. Commercial logging is often in direct competition for resources with hunters and gatherers or swiddeners. Because roads are built, commercially logged areas are often opened for settlement by more intensive land-users, and the areas are often made biologically and socially unsuitable for further hunting and gathering. Short range effects include changing species composition, eliminating rare or commercially valuable species, unless the exploitation is rigidly controlled (which is rare).

Swidden farmers, cultivating root or grain crops as their major source of subsistence, modify the forest by clearing, burning, selective cultivation and weeding as well as by some hunting and gathering. The length, and thus the ecological characteristics of fallow depends on population crowding and competition for land, and on customs regarding use of fire (see chapter 19, part 1).

Local, semi-isolated swidden communities are relatively small (probably <1 000 on the average), and population density is relatively low (probably *ca.* 25–50 persons/km²) when the essential areas used for fallow, hunting and gathering are included in the calculation. Mortality in the semi-isolated village communities has probably been characterized by sharp peaks caused by epidemic diseases from external contact (see chapter 17), or by occasional local disasters (failure of rainfall, plagues of pests, etc.), or by local warfare. Fertility probably remains relatively high, but short range variations in birth rate interact with marriage regulations and local fluctuations in availability of mates due to the effects of epidemics on the age structure (Kunstadter, 1966). Migration may help readjust local population resource imbalances, and may be associated with marriage or fission of the community and colonization of ecologically similar but less densely populated areas, or by temporary or permanent movement to areas of more intense economic activity.

Modification of the basic swidden system has taken place in several directions as a result of contact with cash economies and more highly organized socio-political systems. One example is the generally ecologically and economically successful subsistence-plus-cash crop small-holder, characteristic of some parts of Indonesia and elsewhere. The farmers retain most or all of their traditional swiddening practices with large numbers of crops, and produce most or all of their subsistence commodities, while intercropping with commercial species increases the intensity of production but not the basic ecological effects. They participate in a cash economy, but are not totally dependent on it. Population characteristics may be modified from those of integral swiddening, with some increase in density, and probably related changes in mortality patterns. Participation in a cash economy gives value to production of non-subsistence surpluses; thus there may be a motive for and profit in expanding family size.

Another modification of swiddening is its use as a supplement to economies based on permanent fields for growing subsistence or cash crops in areas of land shortage, or as a preliminary stage in preparing land for permanent fields, or as a means of establishing a claim to land which will subsequently be used for commercial cropping. These are probably the cause of the most rapid wide-spread destruction or modification of tropical forests at present. The major activity in such areas is temporary production of crops, usually on the margins of more densely settled zones of permanent cultivation, either in the foot-hills or heavily forested areas not previously regularly cleared. Repeated migration and then sale of the cleared land has been an important aspect of this type of land use, e.g., in Thailand. Local communities are usually small, and settlements are often dispersed but their degree of isolation, population density and other demographic characteristics vary. Residence in the forest is often temporary and the system is usually a stage of transformation to permanent field cultivation.

The large-scale plantations or commercial farms make extensive modification of the environment. They employ a population which is often highly skewed in age and sex distribution (emphasizing young adult males) and thus depend often on migrant labour rather than on reproduction. The demographic effects thus pertain both to the area directly affected, and to the area from which labour is recruited. The individuals moving into the cultivated area often come from a different environment and may be carriers of new diseases, and susceptible to the diseases of the plantation area and may transmit these diseases to their homes when they return (see chapter 17). In addition they often act as agents of social and cultural change in connecting their home communities and cash economy (see chapter 19, part 2). Diet of the plantation-dwellers, based on a cash economy, will vary from that of the subsistence agricultural areas from which the plantation workers come, and may result in a decline in nutritional status and increased morbidity (see chapter 16).

The population density on the plantations tends to be relatively low since the objective is to maximize output, rather than to optimize support of the human population. Where plantations are established in previously settled areas they may result in a decrease in human population density, and act as a force to induce migration to other areas, through consolidation of small holdings. The net effects on the tropical forest ecosystem and the demographic implications depend on the nature of technology and the linkages with external socio-economic systems.

A variant of the plantation system, the taungya system of planting commercial lumber trees after allowing farmers to clear and plant their own crops for a year or two (see chapter 20), supports a much lower population density than a normal swiddening system since the period of rotation for forest lumber is *ca.* 50–100 years. Where use of the commercial forest is controlled in this fashion, the farming population is either forced into new territory or into different occupations, or suffers increased mortality. The increase in commercial value of the system goes generally to the forest owners and managers, not to the farmers, although in some areas forest workers are the taungya farmers in their spare time.

Another variant is cash cropping by swiddeners whose subsistence economy is incomplete, and who are dependent on these sales. A number of opium producers, e.g., Hmong or Meo (see Geddes, 1976) function in this manner. Apparently they attempt to maximize cash income, rather than to stabilize their use of a particular site, by concentrating their population for intensive farming where and when it is possible. They maximize the productive members of the household through high fertility, plural marriage, adoption and incorporation of hired labour.

Increased production brings increased income and may lead successful families to settle outside the forest and engage in other occupations. Such outflow probably does not balance the increase of population due to natural increase and immigration into the system. The population/resource relationship is maintained by community fission, movement to new land resources, and temporary community fusion where adequate land resources allow.

Permanent cultivation makes use of extensive management of vegetation, soil and moisture characteristics (irrigated field farming). Forest is completely cleared, land is levelled, increasingly large-scale water control projects are developed, with consequent effects on freshwater (and sometimes marine) biota, number of uncultivated plant species is reduced and productive effort concentrates on a few species. Such populations are characteristic of much of South-East Asia and parts of Africa, Oceania and South America.

Most of the readily irrigable land has already been levelled but modern technology allows increased production through water control; this allows multiple cropping using fertilizers and pesticides.

The human population is large and dense. Though most people may live in dispersed settlements, these societies are almost always organized with local and regional markets and a national economy supporting a large urban settlement. Remote settlements are increasingly linked by trade and administrative services to other settlements and ultimately to the nation, linking human settlements of diverse types.

Their mortality has generally fallen to 10–20/1 000, while fertility has generally been maintained at $\geqslant$30–40/1 000, resulting in a young population ($\geqslant$40 per cent under age 15) with a consequently high dependency ratio and an annual growth rate of at least 1–3 per cent. Fertility is subject to some conscious control. Migration feeds urban growth and also includes rural to rural movement in response to population pressure and economic opportunities.

Coupled with the general fall in mortality, there have been changes in causes of death. The increasing size and concentration of population increases the dangers of intestinal and respiratory diseases. At the same time as public health measures reduce the threat of major traditional causes of death, extensive environmental modification often favours the spread of vector-borne diseases and results in major increases in these causes of morbidity and mortality (see chapter 17). Increasing modernization of agriculture increases the risk of accidents. The balance favours a lower mortality than before. Another cause of mortality is inadequate nutrition especially of children, even if total calorie production is adequate (see chapter 16). The synergistic effects of malnutrition and infection make such populations more prone to infectious disease mortality than are adequately nourished populations.

Industrial and urban communities range in size up to several million people and they are the locus of political, economic and cultural systems which dominate the rest of the country; for this reason they are important in determining the future of the tropical forests (McGee, 1967).

Population in Asian cities may be extremely dense by European and American standards, but the density of the total population of the social system which they represent and from which they draw their resources must be calculated on the basis of a much larger area. These cities are growing very rapidly due both to immigration and to natural increase. Their age structures often show surpluses of young adults and children, and they usually have a sex ratio favouring males especially in the older working ages.

In general, mortality due to the major infectious diseases and most of the major vector-borne diseases have been controlled, but they exhibit major social class differences in mortality rates and causes due to extreme socio-economic inequalities and maldistribution of social benefits. The lower classes often have more contact with disease agents and suffer more from malnutrition and the associated consequences from infectious and vector-borne diseases, while the upper classes are coming to resemble the disease picture of the urbanized western societies, with very low infant and child mortality and an increasing importance of degenerative diseases in older ages.

Apparently, major fertility differentials parallel social differences. Fertility has generally declined among the upper classes and professionals, while remaining relatively high among the lower socio-economic groups. Fertility control by deliberate family planning is now available and widely used in most tropical cities, augmenting these fertility differentials.

Migration, a major factor in the growth of cities, can be

attributed both to limited economic opportunity and crowding in rural areas, and perceived economic, social, cultural and political opportunities in cities. Thus rural to urban migration favours professionals and working age individuals (more rarely families), and urban to rural movement has favoured individuals who were only temporarily in the urban wage labour market. The recent suburbanization trend in tropical cities parallels that in the temperate zone; the people in the suburbs are still generally urban in terms of occupation and cultural orientation, although the rapidly growing cities sometimes encapsulate truly rural villages on their outskirts.

Wars have been taking place in some tropical forest zones for a generation or more. Large numbers of people have been involved, destruction may have been very great and the forest over wide areas has been altered in major and various ways.

An understanding of the future trends of the human population and its interactions with tropical forest ecosystems will require an understanding of the larger-scale social and economic developments in the tropics, at least as much as a knowledge of the details of interaction between man and the tropical forests on a local level.

Determinants and consequences of demographic characteristics

Introduction

The determinants and consequences of demographic charac teristics in the tropical forest areas lie both within and outside of the tropics. Tropical populations with high fertility and high mortality contained a potential for rapid growth; this has occurred as a result of biomedical and socio-economic changes introduced from outside. The characteristics of tropical populations have changed radically within the past 200 years, and especially within the past 30 years. Prior to the colonial period, populations in the tropics were not necessarily stationary; various language groups, sometimes representing coherent socio-political groups, expanded and contracted. The arrival of the colonial powers resulted in the drawing of borders often for ethnic, socio-political or ecological distributions (Leach, 1960; Kunstadter, 1967). The political realities which these borders represented were associated with changes in patterns of migrations. Thus, for example, people from the Indian subcontinent moved throughout the British Empire, Vietnamese throughout the French colonies, and in more recent years, Filipinos in areas linked to the United States.

The establishment of political order, the introduction of larger-scale socio-economic structures, and the provision of medical services lowered mortality, but colonialism also created conditions which tended to increase mortality (at least temporarily) or change its causes and distribution. New diseases were introduced during the earliest periods of contact with semi-isolated tropical populations. The result was occasionally devastating. Another series of effects on mortality was related to modifications of the environment,

technological change, etc. Less direct was the effect of socio-environmental change resulting from the population growth.

Population growth has been sustained by a maintenance or even increase in fertility. Colonial regimes may have induced motives for increased fertility or sustained high fertility as a result of a change from primarily subsistence local economies to an emphasis on labour intensive or cash economies in which surplus production could be stored or converted to money and invested in devices for increasing productivity.

Colonial regimes also led to the establishment of urban centres, previously absent from many tropical areas, and induced migration into them. In recent years the birth control devices and the cultural patterns associated with their use also have been introduced and have been associated with reduced fertility and slowing the population growth rate in some countries, as cultural traditions which maintained high fertility, have changed.

In the post-colonial era, new demographic characteristics have emerged. Death rates in general have continued to decline and causes of death began to resemble those found in industrialized countries of the temperate zones. Nonetheless, socially perpetuated inequalities within and between nations are associated with important differentials in mortality in many third world nations of the tropics by age, sex and causes. Patterns of fertility and mortality in the small upper classes tend to resemble that in the western world, whereas high fertility and mortality may remain the lot of the poor. Undernutrition undoubtedly contributes to their high mortality and is in many cases a result of changed socio-economic conditions and ecological adaptation from largely self-contained subsistence to largely cash oriented occupations and from less to more dense settlement and the elimination of most uncultivated or undomesticated products from the diet.

The range, total amounts and rates of migration have increased. Age and sex distribution has been modified with an emphasis on movement by individuals rather than families. In initial stages there has usually been a surplus of young males moving to work in extractive industries and plantations. With the development of wage work, young adult women also moved as individuals. National political considerations changed the migrations across previously loosely administered borders. Migration across ecological boundaries has apparently become much more common within the nations than it was before they were established and national economic system developed.

A proper knowledge of population dynamics among and within local groups requires an understanding of the social and cultural patterns, the constraints on population for the interaction between their technologies and the natural environment.

Regretably this has only occasionally been approximated to through interdisciplinary studies with a truly ecological perspective; most studies tend to be focused on a single problem using the tools of a single discipline. This means that the following case studies are necessarily incomplete.

Amazonia: the Yanomamo

The Amazon basin forest can be divided into two major types with differing potential for human settlement: the slightly elevated *terra firme* above the flood plain, and the smaller, but much more productive *várzea*, lining the periodically flooded main channels (see chapter 4 and the annex on the forest ecosystems of the Brazilian Amazon). Because of this, there are two basic strategies of socio-demographic adaptation relating population to resources.

Aboriginal *terra firme* groups are represented by the Camayura, Jivaro, Kayapo, Sirionó, Yanomamo and Wai-wai. Land is plentiful but relatively fragile and subject to overutilization due to limited soil fertility and, where large mammals are sparse, population size is maintained at a low level and villages move frequently. Common mechanisms for control of population include such behaviour patterns as post-partum sexual taboos, infanticide and warfare. Cultural patterns acting to disperse populations include intergroup warfare, blood revenge, fear of sorcery and desire to be independent (Meggers, 1971). Capture, mostly of women and young children and their incorporation, as well as reproduction is used to increase the local group size.

In the *várzea*, where land and fish resources are relatively concentrated and are much more rapidly renewed than in the *terra firme*, but where productivity is highly seasonal, population density and residential stability are greater. Nonetheless, control of population growth is essential, because although average productivity is much higher than on the *terra firme*, variations are large due to floods which restrict the growing season, and occasional very high water and shifting river courses which force the villages to move and which may destroy stored food. Among *várzea* groups such as the Omagua and Tapajós, mechanisms for controlling population include infanticide and warfare although *várzea* warfare is primarily with non-*várzea* groups, and captives from the *terra firme* are incorporated not as kin, but as property for expanding the labour force, to be disposed of during times of food scarcity. This apparently allows the *várzea* groups to persist at a higher level of living, and with a higher population, while avoiding the socially disruptive effects of periodic high mortality within their own social group (Meggers, 1971).

If population and resources are systematically linked, the forces which tend to cause or sustain population growth and concentration must be examined more closely.

The Yanomamo, a *terra firme* group, have been studied intensively by physical and social anthropologists (Arends *et al.*, 1967; Chagnon, 1968a, 1968b, 1974; Neel and Chagnon, 1968; Chagnon *et al.*, 1970; Lizot, 1971; MacCluer, Neel and Chagnon, 1971; Spielman *et al.*, 1972; Ward, 1972; Neel and Weiss, 1975; for demographic information on other nearby groups see Salzano, 1961, 1964; Salzano *et al.*, 1967) and a large amount of demographic data have been collected. Two aspects of the questions outlined above will be emphasized: is the Yanomamo population in balance, and what are the demographic effects of warfare in this group in controlling population size, density and distribution?

The Yanomamo Indian population has apparently been expanding for the past 100 years in northern Brazil and southern Venezuela, to cover a 250 000 km² area between 0° to 5° N, and 62° to 66° W, around the headwaters of the Rio Negro and Orinoco. Most of the 125 to 200 villages, with a total of about 15 000 inhabitants, are in forests at 120–220 m altitude; a few are at 600 m.

The basic economy of the Yanomamo is slash-and-burn gardening of *ca.* 50 species, especially bananas and plantains which supply up to 75 per cent of their food. They also hunt and gather, especially for nuts, palm-fruits, tubers, wild bananas, wild honey, etc. Although many species are taken large mammals and birds are not abundant. Technology is simple, depending mostly on wooden tools (bow and arrow, club, blow-gun) supplemented only recently by metal blades received in trade (Chagnon, 1968a; Chagnon, 1968b; Chagnon *et al.*, 1970).

Their demography has been studied extensively, and the techniques developed for this study (Chagnon, 1974) in an extremely difficult social and natural environment, amongst people who live dispersed in temporary groupings characterized by inter-village warfare, are a monument to anthropological enterprise. Interpreting the facts is almost as difficult as gathering them. One basic question is what do the Yanomamo represent—untouched pristine forest people unaffected by civilization (Ward, 1972), or remnants of an older group, driven into environmentally marginal and undesirable area, affected in basic ways by indirect (and more recently by direct) contact with the rest of the world: introduction of the cooking banana as their major food crop; introduction of falciparum malaria as a major source of morbidity and mortality; trade supplying metal tools; and, within the past generation, missionaries who may have reduced mortality by restraining warfare and infanticide and by providing antibiotics, or outsiders who may have increased mortality through contagious diseases such as measles (Neel and Weiss, 1975). Because of the lack of historical demographic data and the difficulty of establishing their ages, it seems appropriate to consider them as they are today without making strong evolutionary or historical arguments.

Neel and Chagnon (1968) argue that the Yanomamo and the nearby Xavante Indians represent the stage of primitive populations with intermediate birth and death rates, as contrasted with the high death rates of settled agricultural societies in developing countries, and the low rates in contemporary industrialized societies. However, the reported vital rates are not consistent. The definition of the demographic unit of analysis, under socio-demographic conditions such as those of the Yanomamo, is difficult and leads to a discussion of group structure and boundaries.

Villages are clusters of families centered around a headman. Dominance apparently depends on personality and warlike ability, and secondarily on kinship ties. Thus village composition is inherently unstable. Intervillage warfare, associated with changing patterns of alliances, village fusion and fission and raiding for wives are also important as a cause of exchange of personnel between villages. Thus although villages are (at least temporarily) territorially distinct units of up to 250 people, they do not form the appropriate unit for demographic analysis which depends on assumptions of populations in which there is no net migration. Their size and age-sex composition at any moment, and thus the vital rates which might be determined on the basis of reproductive histories of women currently living in them may be more a function of recent migration than of the balance between births and deaths.

Nonetheless, villages obviously are important units in understanding the interaction between population and environment. This depends on knowledge of patterns of structure and interactions between villages. Village size and composition varies systematically between the centre of Yanomamoland (villages of 40–250 people) and the periphery (25–100 people). Chagnon believes this is a result of the requirements and results of warfare. Villages in the centre have more warfare, more inter-village alliances and trading, and when they split or move, the distance of the moves is shorter than among peripheral villages. The minimum size of *ca.* 40 is the size necessary to have a large enough number of men to form a raiding party and to leave a few to protect the village. The maximum of 250 is the most that can be sustained before competition for leadership between two or more adult men, in the absence of adequate social restraint of conflict and the threat of violence, causes this village to split. Because of the danger of war, small groups in the centre are also constrained to fuse for self protection; moving out of this field of combat is not attractive because of the presence of more hostile surrounding groups. Villages on the periphery have less warfare and are more widely dispersed (Chagnon, 1968a, 1968b). The settlement pattern and use of natural resources thus cannot be understood solely on the basis of knowledge of a single village, but must take into account the dynamics of the population and society at the supra-village level as well as the patterns of social relationships between villages. Among the Yanomamo it is apparent that the composition of the villages is a result of general population increase and warfare between villages.

The sex distribution of the Yanomamo is unbalanced in favour of males (Chagnon, 1974); this has been attributed to female infanticide. The sex-ratio is more unbalanced in the central (57 per cent males) than in the peripheral villages (53 per cent), and in the 0–14 year age group (55–61 per cent) (Chagnon, 1968b). The balance is restored at later ages apparently by excess adult male deaths in violence and warfare (table 1). Disease plays a much more important role in deaths of females than of males; reported deaths associated with childbirth account for less than one per cent of all female deaths.

The population is young, suggesting that both birth and death rates are high. Customs maintaining high birth rate include early marriage—girls are often married many years before their first menses and are remarried quickly after being widowed or separated. Although men marry at a later age, because of polygyny and the unbalanced sex-ratio, there are still too few women for all adult males to be married simultaneously. This is one of the chief sources of tension that leads to warfare, especially in the central villages. There is a high incidence of polyandry in peripheral

TABLE 1. Causes of death (as percentages) reported in genealogies from two Yanomamo groups (from Chagnon, 1974).

	Male	Female
Warfare and other violence	32.5	6.1
Disease	41.1	71.8
Supernatural	19.4	16.7
Accident, injury	2.2	2.0
Old age	4.8	2.9
Childbirth	—	0.4
Total number of deaths	314	245

Most of the reported deaths are of adults. The figures should not be considered as *rates* because the source from which they were drawn does not represent a population of known size during a given time interval.

villages, which, along with the less unbalanced sex-ratio probably reduces sexual tension as a source of fighting among males.

Chagnon *et al.* (1970) estimate that Yanomamo women begin to reproduce at about age 15, and state that women reaching age 40 report having *ca.* 3.8 live births, with an average birth interval of 3–4 years. Neel and Chagnon (1968) believe that due to under-reporting of live births and infanticide, the total should actually be *ca.* 4.5 live-born children per woman (the inconsistency may result from failure to report infanticide victims as live-born). According to reproductive histories, the number of children surviving to maturity for each woman reaching age 40 is *ca.* 3.2 (Neel and Chagnon, 1968). Almost all adult women are reportedly fertile (Neel and Weiss, 1975). Customs acting to reduce fertility include sexual abstinence for 12–18 months following childbirth. This should yield a minimum birth interval of 21–27 months, but birth intervals are much greater than this.

Neel and Chagnon (1968) believe the Yanomamo are characterized by moderate birth and death rates, and have pointed to some of the apparent contradictions in the published data on Yanomamo reproduction. The problem in determining vital rates of the Yanomamo is that the data required are not available. In their early attempt, Neel and Chagnon compare Yanomamo reproduction with that of the Xavante, another South American group who apparently had much higher reproduction. They also argue that the health of the Yanomamo appears good, and death frequencies, where they have been recorded, appear to be low. In a later study Neel and Weiss (1975) refer to measurements of uterus size and arrive at a 'gross pregnancy rate' of 0.25 per woman per year. If all of these apparent pregnancies resulted in live births, it would correspond to a four-year interval between births; a urine test for pregnancy resulted in an estimate of a three-year interval between births. Because of the observed sex-ratio of 140 males per 100 females during the first two years of life plus anecdotal data they estimate *ca.* 15–20 per cent of all pregnancies are followed by infanticide, yielding an effective live birth interval of about four years. Using Weiss's technique (1973) for life-table estimation they conclude that the population had a growth rate of *ca.* 0.85 per cent/a, a male birth rate of 58.6/1 000 (female 56.0/1 000 and crude death rate of 50.1/1 000 (female 47.5/1 000). These are high. This conclusion seems more congruent with the facts than the earlier one and with the historical evidence (Chagnon, 1974) of rapid population growth. Chagnon's data suggest the growth rate estimated by Neel and Weiss is too low.

Using the Bourgeois-Pichat (1957) life-table technique (Keyfitz and Flieger, 1971) Kunstadter has analysed the age distribution data presented by Neel and Weiss (1975) to determine possible vital rate and natural increase values. The ranges are: crude birth rates 52.1–81.4/1 000; death rates 10.6–62.5/1 000; natural increase 1.89–4.15 per cent. The mean estimated birth rate would thus be *ca.* 60/1 000 (close to the Neel and Weiss estimate) but the mean estimated crude death rate is *ca.* 27/1 000 (lower than Neel and Weiss) with a growth rate of *ca.* 3.3 per cent. This is closer to the estimate of 2.5 per cent and derived from Chagnon's historical evidence for fission of his faster growing subgroup of villagers.

There are major differences in the genetic and social contributions of different individuals in the population. Politically dominant men (who may gain support from their relatives) are liable to contribute far more than the average number of descendants through the practice of polygyny. Analysis of the genealogies shows how large the contribution of a single man can be: one man was reported as fathering 42 children, and one of his sons fathered 33 as compared with the average of deceased males of *ca.* 5–7 (Chagnon *et al.*, 1970; MacCluer, Neel and Chagnon, 1971; Chagnon, 1974). Obviously the maximum and average number of children a woman can have is much smaller. Given the association between reproductive performance and political power, breeding is not random; the contribution of a few males may be disproportionately very large and genetic change may take place in small groups much more rapidly than in larger more nearly randomly breeding populations. The bio-evolutionary significance of this in the Yanomamo population is not clear, but it may help to account for the apparent reproductive advantage of one subgroup over another reported by Chagnon (1974). He states that after splitting from a single village 100 years previously, one subgroup, Namoweiteri, has fissioned into five villages with 700 people, while the other, Shamateri, has split into 12 villages with about 2 000 people and a much larger territory. Comparison of the genealogies of the two subgroups shows that at present the more rapidly growing one has a net reproductive performance about one-third higher than the more slowly growing one.

It is not possible to give the long range implications of this pattern which apparently is taking place in a situation restrained socially (by warfare from neighbouring groups) but not environmentally (no evident shortage of land or other resources). The effects on the Yanomamo of environmental constraints and the reasons for their population explosion of the past 100 years cannot be determined. It is evident, however, that the contemporary Yanomamo population, as described by Chagnon and others, cannot be thought of as being in long-term equilibrium with the environment. Sustained growth at the present rate would rapidly outstrip the resources. With changes in their growth rate and density some changes both in their use of resources and in their social structure would be expected. On the other hand, the lack of information does not allow the interpretation of present Yanomamo population characteristics as representative of a portion of a cycle of population increase and decline which might ultimately be thought of as equilibrium.

Warfare does not seem explainable purely on economic terms and Chagnon (1974) attributes it primarily to political motives. He recognizes scarcity of women as a precipitating factor, but as has already been indicated politics are closely related to access to women. Despite the conclusion that pressure on resources does not cause warfare (0.06–0.35 person/km^2 and, as Chagnon states, there was ample space between the villages, unused for cultivation and abounding with large mammals), the effect of war may still have important implications for use of resources. Fear of war evidently helps maintain unoccupied buffers between villages especially in the central area, and may result in expansion on the periphery. Thus warfare might prevent any permanent large local concentration of population which could easily damage the local environment beyond the point where it could easily recover.

Clearly there is much that is not yet known about the Yanomamo which would be helpful in understanding their relationship with their ecosystem. Neel and Weiss (1975) correctly point to the importance of knowledge of mortality. Their present rapid expansion is more likely the result of a sustained decline in mortality than of some change in fertility. In addition it would be useful to know much more about the environment—specifically the limits of soil, plant and animal regeneration given varying intensities of use by Yanomamo techniques. What would be the effects of larger or more permanent concentrations of population on the transmission of diseases, especially vector-borne diseases? Is there any evidence that the area into which the Yanomamo are expanding was previously occupied and, if so, what techniques were used, and what density of population was sustained for what length of time? Until such questions are answered the present characteristics of the Yanomamo population and their use of the environment cannot be placed into proper historical or ecological perspective.

Borneo: the Iban and Land Dayak

Two major indigenous ethnic groups of Borneo, the Iban and Land Dayak (Noakes, 1948; Jensen, 1965, 1966–67, 1974; Lee, 1966, 1970; Miles, 1970) are located primarily in Sarawak, as well as adjacent parts of Kalimantan and Brunei (except for the work of Miles (1970) on the Ngadju Iban ethnographic descriptions or population figures are not available for the Kalimantan segments of these groups; it is known that there was substantial migration of Ibans and Land Dayaks from Kalimantan to Sarawak in the early 1960s, but prior to that migration seems to have been relatively small since 1945; the 1971 Indonesian census did not distinguish these groups from other segments of the

Kalimantan population). They have been chosen as representatives of rice swiddeners in tropical areas because the ethnographic studies of them have been used frequently in investigating ecological hypotheses (Vayda, 1961; Allen, 1970; Appell, 1971; Hallpike, 1973) and because there is fairly detailed demographic data on them. The ethnographic descriptions of Freeman (1955, 1970) among the Ulu Ai Iban of the Baleh river region, Kapit District, Third Division of Sarawak, and Geddes (1954, 1957) on the Land Dayak of Mentu Tapuh, Serian District, First Division, emphasize traditional patterns, and suggest the directions which modernization was taking in *ca*. 1950. Wrights, Morrison and Wong (1972) depict recent changes among the Ibans, but provide no detailed ethnographic or statistical information. Demographic data are derived primarily from official census sources (Noakes, 1948; Jones, 1962a, 1962b; Chander, 1972).

The Ibans have been moving in historic times out of the Kapuas river basin of interior Kalimantan towards and along the Sarawak coast, and towards the northeastern part of Sarawak (Sandin, 1956). Thus they are now living in hilly and mountainous country, mostly along secondary tributaries of major rivers (especially the Rajang and Lupar), and in coastal plains and swamps, ranging in altitude from sea level to above 500 m (Freeman, 1955; Lee, 1970). The Land Dayaks are concentrated in the broken hill country under 500 m in southeastern Sarawak and adjoining parts of Kalimantan from which some of them moved to Sarawak in the early 1960s.

Rainfall in the Iban and Land Dayak areas is generally 3 000–4 000 mm/a; the highest rainfall is from November to February (usually 250–300 mm/month) and in the driest months (June-August) there is usually at least 200 mm/month (Lee, 1970; Appell, 1971). Mean daily temperature is *ca*. 25° C but slightly cooler at higher altitudes. Relative humidity is over 80 per cent most of the time (Lee, 1970). Soils are generally nutrient poor; localized pockets of rich soils are associated with recent volcanism and with siltation by flood-waters. This has restricted agricultural development in the interior. More intensive agriculture is found in coastal and alluvial areas; upland soils with a volcanic base are cultivated more intensively by shifting cultivation than are nearby soils based on sedimentary rock (Lee, 1970). The characteristic vegetation of the area is the inland dipterocarp forest, with montane forests at higher altitudes, and peat swamp forests and mangroves on the coasts (Lee, 1970).

Both Ibans and Land Dayaks are of Proto-Malay ethnic stock, they are primarily agricultural and mostly use swidden techniques for growing upland rice. The Ibans prefer to use the land only once or twice before moving to previously uncut or slightly used forests. They have less permanent settlements than do the Land Dayak. According to results of the 1960 census, 86 per cent of the Ibans of 15 years and older are economically active; this is associated with the important economic role of women (Freeman, 1970; Jensen, 1974) and the relatively low child dependency ratio as compared with other Sarawak populations. 98 per cent are engaged in agriculture, of which 91 per cent is dry rice, and 8 per cent is rubber. Some are becoming more sedentary

and growing wet rice, and since at least the early 1950s a few men have been taking wage labour jobs (Lee, 1970). The Iban men, in their search for prestige and profit, also continue to carry on the traditional *bejalai* (journeying) pattern, leaving their home villages for months or years at a time to cut trees or collect forest products to sell to Chinese merchants, or voyaging to remote parts of South-East Asia for trade or employment (Freeman, 1955, 1970; Lee, 1970).

The Land Dayaks sow dry rice in the hills and wet rice in the low-lying swamps, using swidden techniques for both. They tend to confine their swiddening to a regular cultivation and fallow cycle, and use the same parcels repeatedly. 77 per cent of the Land Dayaks are economically active, of whom 95 per cent are agricultural, with *ca*. 10 per cent in rubber (Jones, 1962a). For several generations they have been in close contact with Chinese and Malays, and generally have been in an inferior socio-economic position. Although there is considerable population redistribution within Land Dayak territory, they have not expanded their territorial control recently as have the Ibans.

Both Iban and Land Dayak groups were traditionally organized at three levels which have been maintained under modern administrative conditions: the household or family (Iban: *bilek*) which is the basic unit of economic production and consumption, and which occupies a single apartment in the longhouse; the longhouse, which is a basic dwelling unit composed of several distinct households; and the village, composed of one or several longhouses which may or may not be in close proximity to one another, but which farm within what is recognized to be a common territory. Each village has a headman (Iban: *tuai rumah*) and a village augur (Iban: *tuai burong*). At present the *tuai rumah* headman is recognized by the government as the community representative, empowered to settle disputes by reference to customary law within his longhouse or village community (Freeman, 1970). In neither Iban nor Land Dayak communities is the headman elevated in wealth, control of property or resources, as compared with other villagers. Though he enjoys superior prestige, he is essentially an arbiter, not a chief with powers of redistribution (Geddes, 1954; Allen, 1970; Freeman, 1970). An area chief (Iban: *penghulu*) is elected by members of 15–30 longhouse villages, and he is officially appointed and paid by the government. His position allows him to impose small fines in the settlement of disputes between longhouses (Jensen, 1965); there seems to have been no comparable office in traditional times. The village augur enjoys no civil power, other than as a respected elder; his function is to tell omens at important village agricultural rites (Jensen, 1965). In neither group is there any class or caste stratification, although headmanship tends to be inherited.

The Ibans of Baleh, as studied by Freeman in 1949–51 (Freeman, 1955, 1970) were pioneering in virgin forest, a situation similar to that of 50–60 per cent of the Iban population at that time (Freeman, 1970). Among these people, the original clearing is made by the *bilek* families of a large longhouse group. After a few years, as more forest is cleared, and distance to the farms increases, small, usually closely related groups of *bilek* families split off from the large longhouse, to move closer to the primary forest which they are cutting for swiddens. Each of these groups builds a subsidiary small longhouse (*dampa*) in which the *bilek* families stay together for 2–6 years, until there is no nearby primary forest to clear. The *bilek* families then dismantle the *dampa* and rebuild it at another location more favourable. After 15–25 years or more, the *dampa* groups may return to their original longhouse community and each *bilek* (or its successor household) farms for a second time its old swidden (which by now has grown up into secondary vegetation). The *bilek* families retain their rights to recultivate their old fields as long as they remain within the general region of the original big longhouse, with which they maintain social ties and where they may store property and valuables which they do not move to the *dampa*. Before government control, Iban communities treated land as if there were an unrestricted supply. When all the primary forest had been felled in one area, and the land had been recultivated once, the large longhouse community moved to another area, if necessary clearing the new area of its previous inhabitants by war or threat of war. Iban groups ordinarily did not fight among themselves, but rather looked outwards for territory and human heads. Under government control, warfare, head-hunting and territorial expansion were forbidden, and territories within a given administrative district were alloted to longhouse villages existing within that district. This has been interpreted by the villagers as meaning that each *bilek* family has equal use-rights to the land within the communal allotment. As long as the *bilek* family remains in the district, it retains rights to the land it or its ancestors first cleared. The use-rights may be borrowed or exchanged between *bilek* families, but the original clearers, or their descendents, do not thereby lose their future rights unless they move away from the district. Use-rights to swidden land are inherited in the same way as other family valuables. All those born into or marrying into, and remaining in the *bilek* family retain their rights to the land. Those moving out or marrying out lose these rights. When the *bilek* household splits, the use-rights are retained by the descendents of the original claimants, and any disputes in use are decided in a conference between competing claimants. The first of the claimants to put a parcel into secondary cultivation thereafter will have exclusive use-rights to that parcel.

In contrast with the Iban, who traditionally viewed virgin unowned land as continuously available for the clearing, who viewed land occupied by other ethnic groups as available for the taking, and who were expanding their territory rapidly in recent times, the Land Dayaks described by Geddes (1954) believe that all land is owned, and have a concept of stability of territory associated with a given village. There is considerable movement of individuals, households, longhouse groups and villages within Land Dayak territory. Villages, especially those with growing populations, sometimes split into two segments which take on independent identity within the same territory. Villages or village segments may change location, but only within areas already occupied by other Land Dayaks. They may

try to lay claim to uncleared or long unused land at some distance from an existing village. Their claims will usually be resisted by the prior residents. In such cases, the old residents will often attempt to incorporate the new into their village, thus retaining control over more distant lands which may be more productive for crops or hunting than the more frequently worked land close to the settlement. Incorporation of outsiders has the advantage of adding to the size, prestige and potential strength of the village in resisting other incursions, but it has the disadvantage of increasing population pressure on the land.

The differences in the approach of the two groups to land tenure are associated with different practices as regards migration and warfare during the traditional period. The Ibans apparently used warfare and head-hunting deliberately as means of terrorizing non-Ibans and expanding into their territory (Vayda, 1961), while the Land Dayaks took heads only when defending their own villages. Ibans actively engage in extensive trade with non-Ibans, actively seek new economic opportunities including acquisition of both coastal and inland farming areas, and even joined ocean-going pirates. Land Dayaks traditionally were confined to the inland foot-hill regions by preference and by Iban and coastal Malay raiders (Lee, 1970). They believe their souls may get lost and their bodies may be exposed to illness-causing demons if they leave home, and even in the late 1940s most Land Dayaks had never even seen the ocean (Geddes, 1954).

Among the Land Dayaks, swidden land belongs to the one who first clears the primary forest, and use-rights are passed by inheritance to all descendents, males and females. This set of people constitutes a bilateral descent group with respect to a particular parcel of land. In practice not all members of this descent group are equally recognized or equally likely to claim or use their rights. Claims are more likely to be remembered if the land was used recently by a close relative. Because everyone has a large number of ancestors, everyone theoretically has claims on large numbers of land parcels. Although a claim may theoretically be traced through individuals who marry outside of the household in which the individual heir was born, in practice individuals who marry and move into different households acquire land rights in the households into which they have moved and give up the rights in the household into which they were born. Thus, the size of the land-using descent groups does not expand indefinitely. Post-marital residence may be arranged to assure access to land. When a child is born to a new couple, that couple normally establishes a new household, retaining its share of land rights derived from ancestors, or it may stay within the household in which it was formed, taking care of the older generation, and acquiring thereby their land rights. Unlike the Iban, among the Land Dayaks, the strongest claim to land is held by a member of the descent group who has not previously used the land, or whose ancestors have used it least recently. If two claimants have similar claims due to previous uses, the elder of the two has priority over the younger. The cultivators enjoy the fruits of the field they cultivate, regardless of previous or subsequent users. This system applies to land used for shifting cultivation and in recent years has been applied to land planted with trees, even though this removes the land from circulation among members of the descent group for an extended period (Geddes, 1954). According to Geddes, land disputes within villages are rare, usually result from genuine misunderstandings and are always settled within the village. Disputes between villages, often resulting from the attempt of one village to extend its territory at the expense of another, were traditionally settled by ordeal, threat of force, or warfare. Warfare is now banned, and such disputes are settled by the government on the basis of available historical information.

Differences of Iban and Land Dayak land tenure systems are consistent with two principles suggested by Geddes. The Land Dayaks are attempting to maintain equality of opportunity and social status by maximizing everyone's possibilities of claims to land allowing (through the ambilateral post-marital residence system) reallocation of population among descent groups, and assuring that land tenure will be neither permanent nor directly heritable, thus precluding the development of a few descent groups with greater control of productive land resources and assuring a fairly even distribution of population over the available land. The second principle is that force of numbers allows claimants to exert their will on less populous groups, and allows them to gain access to land even if the historical or genealogical basis for the claim is weak. Among the Iban, however, apparently, there was no need for such a redistribution system in the traditional system, as land was viewed as plentiful and unbounded. Anyone not having access to secondary forest could and did cut new fields in primary forest.

Both Ibans and Land Dayaks had mechanisms which allowed the reallocation of land resources between households, groups of households and even villages. Both groups also had important means of reallocating individuals among household which had similar effects of maintaining a relatively even distribution of population (at the household level) towards the resources. Two major devices were used: adoption and post-marital residence. Among the Ibans, for example, adoption is common and gives children to childless adult siblings within the same *bilek* family into which the child had been born, and between *bilek* families, in which the adoptee takes on all the rights of an individual born into the adopting family. In Freeman's (1970) study, *ca.* 36 per cent of the families gained members by extra-*bilek* adoption. Slightly more males (52 per cent) than females were adopted. Freeman explains this high rate of adoption on the basis of the relatively high rate of childlessness. 20 per cent of couples married more than 10 years were childless, and it was these who adopted children to avoid the extinction of their *bilek* household. Post-marital residence is another device by which people may be moved from one household to another. In societies with strongly unilocal post-marital residence, imbalances are bound to occur between households as a result of random variation in sex-ratio of the offspring. The ambilocal system followed by the Ibans allows the new couple to establish itself in the

bilek family where it is needed for labour or where land resources permit. The result was that there were an approximately equal number of males moving in with their wife's family as there were women moving in with their husband's family. Geddes (1954) describes a similar situation among the Land Dayaks.

Both Freeman and Geddes suggested that the traditional land tenure system was in danger of failure in the late 1940s, due to two major changes: imposition of strong external governmental authority which restricted population movement, attempted to control land use, prevented the use of force to settle land disputes and attempted to stabilize the claims of village communities to bounded land resources; and the development of opportunities for participation in a cash economy associated with more or less permanent cultivation of rubber, irrigated rice and other crops, changing the nature of land use from temporary to more or less permanent. Both Geddes and Freeman seemed to see the possibilities of involvement in wage labour as a threat to the integrity of the villages, as it might draw off the men, leading them to abandon their traditional agricultural pursuits. In the years since their work it has also become obvious, as Freeman (1955) hinted, that rapid population increase will cause land scarcity. No recent ethnographic work has been done to show how the land-use and land tenure systems have been modified in response to population pressure due to this rapid population growth.

Iban and Land Dayak age distributions (table 2) indicate that both populations are relatively young. The Land Dayaks have higher child dependency and child/woman ratios than do the Ibans, suggesting higher fertility and perhaps higher mortality and a heavier burden of economically non-productive children. Comparison of the figures from the 1947 and 1960 censuses (table 3) show that both populations are increasing their proportions at young ages, and thus increasing their dependency burdens.

TABLE 2. Analysis of Iban (Sea Dayak) and Land Dayak age structures.

	Iban		Land Dayak	
	1947	1960	1947	1960
Mean age	23.7	21.1	18.4	16.8
Child/woman ratio	0.55	0.72	0.76	0.88
Child dependency ratio	0.62	0.74	0.80	0.94

The age distributions show some unexplained anomalies, for example, the deficit in males in ages 15–34, and the smaller number of infants and children under age 2 as compared with those age 2 and older in both censuses and both groups. This suggests a consistent pattern of age misreporting, and problems of analysis requiring either detailed rechecking of the originals, or adjustment of available figures. The age distribution of older adult Land Dayaks suggests that men live longer than women; a fact which Geddes (1954) remarked.

TABLE 3. Iban (Sea Dayak) and Land Dayak age distributions from the 1947 and 1960 censuses of Sarawak.

Age group (years)	Iban—1947			Land Dayak—1947		
	Male	Female	Total per cent	Male	Female	Total per cent
0–4	12 530	12 598	13.2	3 334	3 285	15.7
5–9	13 375	12 636	13.7	3 480	3 412	16.3
10–14	9 251	7 977	9.1	2 330	2 228	10.8
15–19	8 084	8 848	8.9	2 066	2 314	10.4
20–24	5 942	7 272	6.9	1 408	1 670	7.3
25–29	7 425	9 015	8.6	1 553	2 030	8.5
30–34	7 229	7 346	7.7	1 532	1 560	7.3
35–39	7 357	7 298	7.7	1 514	1 356	6.8
40–44	6 157	6 004	6.4	1 197	1 035	5.3
45–49	4 963	4 786	5.1	793	651	3.4
50–54	4 184	4 101	4.3	683	584	3.0
55–59	2 243	2 176	2.3	302	288	1.4
60–69	3 400	3 716	3.7	513	475	2.3
70–79	1 468	1 648	1.6	206	190	0.9
>80	624	764	0.7	121	85	0.5
Total	94 232	96 094	190 326	21 032	21 163	42 195

Age group (years)	Iban—1960			Land Dayak—1960		
	Male	Female	Total per cent	Male	Female	Total per cent
0–4	19 664	19 556	16.5	5 411	5 650	19.2
5–9	18 434	17 194	15.0	4 942	4 875	17.0
10–14	10 688	8 868	8.2	3 132	2 874	10.4
15–19	9 333	11 236	8.7	2 479	2 982	9.5
20–24	7 439	9 475	7.1	1 844	2 246	7.1
25–29	8 291	10 753	8.0	2 098	2 598	8.2
30–34	8 615	8 696	7.3	1 870	1 833	6.4
35–39	8 209	7 771	6.7	1 784	1 639	5.9
40–44	6 985	6 654	5.7	1 459	1 310	4.8
45–49	5 145	4 711	4.1	1 045	898	3.4
50–54	4 801	4 665	4.0	889	845	3.0
55–59	2 655	2 523	2.2	471	400	1.5
60–64	2 972	3 311	2.6	445	539	1.7
65–69	1 352	1 530	1.2	206	178	0.7
70–74	1 454	1 627	1.3	198	163	0.6
>75	1 441	1 693	1.3	178	138	0.5
Total	117 478	120 263	237 741	28 451	29 168	57 619

Source: 1947 census figures from Noakes (1948), 1960 census figures from Jones (1962a).

Jones (1966) has calculated the vital rates of the two groups from these censuses:

	Iban	Land Dayak
Crude birth rate (per thousand)	48	57
Crude death rate (per thousand)	30	32
Expectation of life at birth (years)	50	46
Total fertility rate (average total births per woman of completed fertility)	5.8	7.3

He used the reverse survival method to calculate birth rates, and estimated death rates by using the difference between birth and growth rates. His estimated birth rates for 1942–46 were 43.6 for Iban and 52.0 per thousand for Land Dayak, indicating that birth rates had increased following the

second world war. His estimates of birth rates for Land Dayak are extremely high, approaching the level if natural fertility was achieved; moreover his estimated birth, death and life expectancy figures are inconsistent with estimates derived from model life-tables and the stable population theory. If the vital rates are correct, the Coale-Demeny (1966) model life-tables yield an estimate of about 32 (not 46) years life expectancy at birth for the Land Dayaks. Hypotheses to be investigated in a reanalysis of these data include: Land Dayak birth rates are at the extreme upper end of those reported for human populations; mortality declined very rapidly during the 1950s when there was an increase in fertility of such magnitude as to seriously distort model life-table assumptions; inconsistencies or distortions in age reporting and serious underreporting of infants in the censuses require age-adjustments other than those made by Jones (1966). These are important for understanding the interaction between the human population and the ecosystem because of the suggestion made later that population growth among Iban and Land Dayak seems to have been positively associated with the length of residence and density of population.

Despite these unresolved questions about the vital rates, the figures do suggest that the fertility, mortality and natural increase of the Land Dayaks are higher than those of the Ibans, and that the major difference is substantially higher fertility among the Land Dayaks. This is supported by a comparison of age-specific fertility among the two groups (see table 4). Both start child bearing relatively early, but Land Dayak women have more children at all ages from 20–24 onwards, and continue bearing children at higher rates and greater maternal age than do the Ibans. This pattern has apparently existed at least since 1927. The data suggest that total fertility increased between the two censuses, but the lower total number of children born to the women over age 40–44 in the 1960 census may be due to memory lapse, rather than a difference in cohort fertility.

TABLE 4. Mean numbers of children born alive to all Iban and Land Dayak women of selected ages (Jones, 1966).

Women's age (years)	Iban (Sea Dayak)		Land Dayak	
	1947	1960	1947	1960
15–19	0.2	0.4	0.2	0.3
20–24	1.0	1.0	1.3	1.8
30–34	2.6	3.3	3.7	4.8
40–44	3.6	4.0	4.9	5.6
50–54	n.a.	3.9	n.a.	5.3
60–64	n.a.	3.6	n.a.	5.3
>70	n.a.	3.3	n.a.	5.0
Total 15–44	2.0	2.4	2.6	3.1
Total >45	n.a.	3.8	n.a.	5.4

Noakes (1948) analysed the reproductive performance of ever-married women of the two groups and found that a higher proportion of ever-married Land Dayak women age 10–44 (88 per cent) than of Iban (Sea Dayak) women (75 per cent) had borne children. Also, higher proportions of Land Dayak mothers had borne larger numbers of children, for example, 25 per cent of Land Dayak mothers had borne over 5 children as compared with 16 per cent of Iban mothers. Jones' analysis of reproductive performance by age of mother confirms the impression of higher fertility among the Land Dayaks. In addition he shows a higher child mortality among the Land Dayaks, as illustrated in table 5.

The explanation of the demographic differences between the two groups is not clear and must await further quantitative description and analysis, e.g., of marriage and other behaviour. Contrary to what might be expected from the difference in fertility, the proportions of females married

TABLE 5. Numbers of Iban and Land Dayak children born, surviving and dying before the 1947 census, per 100 mothers of given age (Jones, 1948).

Mother's age group (years):	15–19		
Child's vital status:	born	survived	died
Iban (Sea Dayak)	113	111	22
Land Dayak	150	130	20

Mother's age group (years):	20–24		
Child's vital status:	born	survived	died
Iban (Sea Dayak)	117	149	28
Land Dayak	227	185	42

Mother's age group (years):	25–29		
Child's vital status:	born	survived	died
Iban (Sea Dayak)	256	225	31
Land Dayak	322	251	71

is generally similar in both groups and marriage seems to be slightly earlier and more stable for the Ibans than for the Land Dayaks, as shown in table 6.

The ethnographic sources do not allow detailed quantitative comparisons of the populations of the villages. It is therefore difficult to judge exactly how well the ethnographic descriptions fit the demographic conditions of those villages, or how representative the village demographies (and thus the relationship of demography and behaviour) might be with regard to the total ethnic group. There are some demographic statements in the ethnographies which can be checked against census results, and there are tabulations of the census materials broken down by district which allow the demographic situation in the study area to be compared with the population of the ethnic group as a whole.

Geddes (1954) does not give figures for age at marriage or proportion marrying but indicates that although most marriages are stable, divorce is allowed. These generalizations are borne out in the census results already described.

TABLE 6. Proportions of Iban and Land Dayak females married by age, 1960 (Jones, 1962a).

Age group (years)	Iban (%)	Land Dayak (%)
10–14	0.4	0.4
15–19	44.7	33.9
20–24	78.9	79.6
25–29	88.9	91.1
30–34	89.2	87.5
35–39	88.7	88.7
40–44	80.9	75.0
45–49	75.2	70.0
50–54	60.7	52.2

Iban men normally marry women 6–7 years younger than themselves and usually marry for the first time in their middle twenties (Freeman, 1970). These statements are borne out by examination of the census data for proportion married by age and sex (Noakes, 1948). Out of a group of Ibans known to Freeman (1970), only about 5 per cent of those over 31 years old were either never married or permanently divorced. This seems consistent with census results from the division where Freeman's villages were located: 8 per cent of 25 385 males and females 30 years and above (Noakes, 1948). Geddes (1954) gives no comparable figures for the Land Dayak but, as already indicated, census figures give the Ibans slightly higher proportions married.

Ibans would be expected to have higher fertility than the Land Dayaks, because of the greater proportion of women who are married. Despite their higher rate of marriage, contraception and migration might reduce their fertility as compared with the Land Dayak. Jensen (1966–67) states that among the Iban 'for economic reasons the need to limit children is as important as the need to have them. For as long as they can remember, the Iban have practised various kinds of birth-control (references occur even in legends and myths)'. He mentions continued breast feeding of a child as the commonest means of restricting births through lactational suppression of ovulation. Jensen also mentions self-induced abortions and the existence of abortion specialists. Geddes (1954, 1957) is silent on these matters, suggesting that birth-control may have been traditionally less important for the Land Dayaks.

The young Iban men are well known for their custom of journeying which is not interrupted by marriage (Freeman, 1970). Young men leave their farms, wife and children in the charge of their father or father-in-law and stay away from home for several months or even years at a time, and may continue journeying on and off until they are well over 40 if they have not yet taken over responsibility for farm management. This custom clearly has major effects on the composition of the community, with *ca.* 20 per cent of the adult males absent even during time of maximum need for farm labour. Much adult male labour is thus converted into prestige goods, tools or cash collected during the journeying, while female labour is devoted to subsistence production. Land Dayak men do not journey as frequently as do Ibans, but Geddes (1954) gives no figures. Some idea of the demographic reality of the ethnographic descriptions can be gathered from the Ibans having *ca.* 28 times more international movements per caput than the Land Dayaks (State of Sarawak, 1967).

If there are real differences in vital rates and migration, what are their economic effects? Both Geddes (1954) and Freeman (1970) give figures on labour input and production for four sample farms, but the variability is so large as to make comparisons of limited value, even if the samples were larger and the methods identical.

Vayda (1961) has argued that Iban journeying is an important part of their subsistence economic system as it helps them locate new land resources as well as trading for tools. Freeman (1955) states that prestige goods gathered on these trips are stores of value, directly relevant to the

subsistence system in times of fluctuation in production. It is difficult to judge the economic effectiveness of journeying as a part of the Iban economic system as compared with the apparently more intensive cultivation system (including use of wet rice in swamp swiddens, lower value of prestige goods and greater employment of males in subsistence agriculture) of the Land Dayaks.

The few data available, together with the descriptions of land-use and land tenure systems, suggest that a detailed demographic and economic study comparing Iban and Land Dayak populations and economics would be rewarding in terms of understanding the interaction between populations of swiddening groups and the tropical forest ecosystem in addition to population growth and over-cultivation. A consideration of the available information may imply lines which such an inquiry should follow. Geddes and Freeman give only limited clues as to the history and characteristics of the communities they studied.

Total Iban and Land Dayak population size and growth are summarized in table 7. The Land Dayak popu-

TABLE 7. Intercensal population increase of Sarawak Iban (Sea Dayak) and Land Dayak (Jones, 1962a; Chander, 1972).

Year	1939	1947	1960	1970
Iban number	167 700	190 326	237 741	302 984
Average annual increase (%)		1.58	1.71	2.42
Land Dayak number	36 936	42 195	57 612	83 276
Average annual increase (%)		1.65	3.40	3.68

lation was growing slightly faster than Iban population after 1939 and considerably faster after 1947. In Kapit and Serian districts, where Freeman and Geddes worked, the results are very similar, although the growth rates are lower (table 8). These figures support the picture given by Freeman and Geddes, that the populations had relatively low rates of increase in the period before their research. Iban settlement in the area studied by Freeman was controlled by the

TABLE 8. Intercensal population increase of Iban and Land Dayak in study districts (Jones, 1962a; Chander, 1972).

Year	1947	1960	1970
Iban (Kapit District)			
Number	25 020	30 822	38 658
Average annual increase (%)		1.60	2.26
Land Dayak (Serian District)			
Number	17 674	23 102	32 369
Average annual increase (%)		2.06	3.37

government. The Baleh River area was one of the zones into which the Ibans were moving at the time the government extended its control to the interior (Sandin, 1956) and the government forced the Iban settlers out of the Baleh basin in the 1870s and in the early 1900s. They were granted permission to settle only in 1922, about 25 years before Freeman began his research. The rate of increase was slow during the early settlement, as compared with more recent times, and with the more rapid growth of the Land Dayaks in the area studied by Geddes, which had apparently been settled for a long time. This suggests that surplus land area does not *per se* lead to rapid population growth; on the contrary, rapid population growth occurred only after the area had been settled.

Whether the length of settlement, or some threshold level of population density, or change in economic system, or external factors, or combination of these led to the increase cannot be determined, but the available historical and demographic data give some clues as to what might have happened. Tribal warfare had been suppressed early in the century, but the effects of reduction of mortality and social disruption are not known quantitatively. Clearly there was a substantial lag between the suppression of warfare and the onset of rapid population growth. Smallpox epidemics are mentioned occasionally in the period before the second world war (Sandin, 1956), but not thereafter; there were also apparently successful attempts to control malaria in the post 1945 period. Thus some important causes of morbidity and mortality were apparently reduced, although Jones (1966) believes malaria control could not have occurred soon enough to have affected the population growth by the time of the 1960 census. In any case, in 1953 the Iban population seems to have been relatively healthy and unaffected by venereal disease which might have limited fertility (Griffith, 1955). Migration was not in sufficient quantities in the 1947–1960 period to account for much of the increase, as shown in table 9.

As already suggested, fertility seems to have increased markedly. In 1947 the proportion of females married was lower at all ages in the Third Division (where Freeman worked) than among the Sarawak Ibans as a whole; so were the age-specific numbers of children born to women. The proportions married and fertility were not tabulated by divisions in the 1960 census, but figures for Kapit District show that the Iban population grew by 5 802 between 1947 and 1960 (Jones, 1962a), but only 835 of the residents of the district had been born in other districts (Jones, 1962a). Thus more than 85 per cent of the growth was due to natural increase. By 1960 the proportion of Sarawak Iban

TABLE 9. Birth-place of Sarawak Iban and Land Dayak in the 1960 census (Jones, 1962a).

	Sarawak	Elsewhere*
Iban	236 752 (99.6)	989 (0.4)
Land Dayak	57 543 (99.9)	76 (0.1)

* Most were males born in Indonesia.

TABLE 10. Iban nuptiality and fertility by age (1947 and 1960).

Age group (years)	Proportion of Iban women married		
	Third Division 1947[1]	Sarawak 1947[1]	Sarawak 1960[2]
10–14	0.2	0.3	0.4
15–19	26.8	28.2	44.7
20–24	65.7	69.3	78.9
25–29	82.1	83.5	88.9
30–34	82.0	84.7	89.2
35–39	84.0	85.3	88.7
40–44	71.5	74.5	80.9

Age group (years)	Mean number of live-born children*		
	Third Division 1947[1]	Sarawak 1947[1]	Sarawak 1960[2]
10–14	0.06	0.03	0.03
15–19	0.16	0.17	0.36
20–24	0.87	1.00	1.45
25–29	1.69	1.90	2.58
30–34	2.34	2.64	3.27
35–39	3.03	3.29	3.88
40–44	3.27	3.58	4.05

1. Calculated from Noakes (1948).
2. Jones (1962a).

* The denominator for these means is total number of women in the age group, regardless of marital status and proven fecundity.

women married at all ages increased, the age of marriage declined, and fertility at all ages increased as shown in table 10. These are consistent with the hypothesis of changed demographic behaviour with the socio-economic system as they became more permanently settled.

For both Iban and Land Dayak land tenure and land-use systems, as much land was available as could be used by the populations given the limits of their technologies and labour. Primary forest still existed, there is no mention of land tenancy or landlord problems, and population densities were below the level at which Freeman (1955) calculated soil fertility could be maintained. He calculated, on the basis of amounts used per *bilek* family, the theoretical carrying capacity of the land for an adequate 15 year fallow as 34 per square mile (1955). This is far above the population densities of 2 and 3 persons/square mile in 1947 and 1960 in Kapit District (Jones, 1962a) and about equal to the 33 persons/square mile in Serian and Sadong Districts in 1947 (Jones, 1962a), where Geddes made his study. The density in Serian District reached 46/square mile in 1960, by which time the population had increased *ca.* 31 per cent over the 1947 level. These figures lend support to the idea that in neither Iban nor Land Dayak areas could land shortage *per se* explain the economic shortfall, nor could unequal access to the land, resulting from the land tenure system, be the cause. Nevertheless, rice production was inadequate and the agricultural subsistence economy was deteriorating. Among the Land Dayaks, Geddes (1954) reports that the interest rate on unmilled rice increased from 50 per cent before the second world war to 75–100 per cent in 1949–50, the value of prestige goods relative to productive goods and staples was declining and about half the households had less than enough rice to last them through the year. Geddes concludes that the shortage of food did not indicate shortage of land or soil infertility. He attributed the variation in the ability of families to meet their needs to the organizational ability of the household heads, the existence of supplemental income from something other than rice (e.g., rubber) which could be exchanged for purchased goods, and the ability to conserve resources for necessities. Freeman (1970) is more explicit regarding the inherent variability of production as related, among other things, to variations in rainfall. He concludes that in the very poor season 1949–50, only 32 per cent of his sample *bilek* households produced more rice than they required, and estimates that in a normal year 70–80 per cent would meet their ordinary requirements.

Some of the possible ecological effects of the traditional Iban and Land Dayak land-use systems may be considered in attempting to understand why the Ibans cut primary forests, where labour requirements were higher than in secondary forests. The Ibans were in a situation where they could and did expand into primary forest whereas Land Dayaks were confined to traditionally held territories by hostile neighbours and their own customs. A direct least effort economic interpretation does not explain the Iban preference for primary forest. If Freeman's and Geddes' figures are correct, Ibans swiddening in primary forest produce on the average *ca.* 30 per cent less rice per unit of labour than do Land Dayaks on hill swiddens. Freeman (1970) labels Iban land use as prodigal, since the immediate recultivation after the first year of use 'inflicts permanent injury on the land'. In making this judgment he seems to disregard the fact that Ibans do return to secondary forest some years after they have cut and farmed the primary forest, suggesting that the land has not been permanently damaged (though the fallow period may thereby be lengthened). He also suggests that men prefer to cut primary forest because women can clear the regrowth in the second year, thus freeing the man to go on journeys. He gives no figures on the relative amounts of labour involved or the actual amount of journeying following the cutting of primary and secondary forest areas. Aside from preference, are there other explanations why the Ibans continue to cut primary forests when they are aware of, and sometimes use secondary forests? Land shortage and loss of soil fertility alone will not explain this, as Ibans are far below the theoretical carrying capacity. One possible explanation is that although cutting primary forest requires more labour, it results in a more adequate yield per unit of labour and reduces variability in the yield. Freeman's yield figures show less variability than do Geddes'. The reason for this may not be soil fertility. Burning the large logs which contain most of the biomass and nutrients may be more difficult than burning the slash from secondary forests, which has a larger proportion held in more readily combustible form (Sabhasri, 1976). Soil moisture variability may be a more likely explanation. If soil moisture in the early growing season is

a limiting factor of rice production in Borneo, the larger amount of soil moisture available under primary (as compared with secondary) forests would reduce the variability and threat of crop loss due to water shortage early in the growing season when rainfall is lowest and most uncertain. Additional measurements of soil fertility, soil moisture, labour inputs and variations in productivity would be required to test this hypothesis.

Freeman (1970) believes that the traditional Iban economy was one in which scarcity might always occur for a sizeable portion of the population. In times of shortage Ibans borrowed from more fortunate relatives, sold or exchanged prestige goods or supplemented their diet with sago palm. It appears that in Borneo, as in northwestern Thailand, the limiting factor is labour for weeding throughout the growing season (Kunstadter, 1976).

At the time of Freeman's study the Ibans were changing from being pure subsistence swiddeners to a cash economy by growing rubber. This was stimulated by Chinese traders and rapidly accepted when rubber growing was three times as profitable as rice. At about the same time the value of rice as cash rather than as barter was becoming recognized. By 1950, most of the families had small rubber plantations. A few had begun to employ labourers in the rubber groves and some of the *bilek* families had completely abandoned rice growing for rubber. The Iban arguments against relying completely on rubber were because: it was risky as a sole source of income because it depended on an external market with notorious price fluctuations; rice was not always readily available for purchase; they preferred the taste of their own rice; and rubber growing modified the traditional division of labour, men would have to turn over the management of the rice to women and devoted themselves entirely to rubber tapping (Freeman, 1970). This had changed by 1960, when a higher number engaged in rubber growing and when females almost equaled males as rubber growers or employees (Noakes, 1948; Jones, 1962a). Parallel trends occurred among the Land Dayaks, who apparently started growing rubber somewhat earlier than the Ibans, as shown in table 11.

The demographic and social structural implications of this change had become apparent at the time of Freeman's and Geddes' research, but the greater diversity of economic resources and the possibility of a cash profit, an increase in the proportions of both sexes (including a very large increase in females) involved in rubber growing, was associated with increases in fertility at all ages, in child survival, and proportion married, and decrease in age at marriage (Jones, 1962a) among both groups between 1947 and 1960.

Besides the possible causes of population growth some of the environmental consequences may be also considered. The problem of swidden agriculture is the decrease in soil fertility and general environmental degradation which results from recultivation after too brief a fallow. Environmental degradation was occurring in Freeman's study area, so it should not be seen as a result of population growth and land scarcity, but as a result of the field rotation system employed in the presence of ample land. Comparison with

TABLE 11. Changes in the percentage of Iban and Land Dayak labour force employed in various occupations, 1947–60 (Noakes, 1948; Jones, 1962a).

Occupation	Iban			
	1947		1960	
	Male	Female	Male	Female
Rice farming	97.0	91.3	85.8	92.0
Rubber growing and processing	0.6	0.1	9.4	7.1
Other agriculture, fishing, forestry	0.5	0.2	0.9	0.4
Non-agricultural	1.8	8.3*	4.0	0.6
Number of persons active in the labour force for all occupations	62 175	65 532	64 083	58 618

Occupation	Land Dayak			
	1947		1960	
	Male	Female	Male	Female
Rice farming	94.4	82.9	80.2	91.1
Rubber growing and processing	2.9	0.6	11.5	6.5
Other agriculture, fishing, forestry	0.5	0.7	1.2	1.0
Non-agricultural	2.2	15.8*	7.1	1.5
Number of persons active in the labour force for all occupations	12 209	13 325	13 933	9 565

* Mostly housewives.

the Land Dayaks, where there was over 10 times as dense a population and where Geddes maintained there was no problem of loss of soil fertility, suggests that carrying capacity must be related to local and temporary concentrations of population, as well as of density averages.

Ibans and Land Dayaks are traditionally swidden rice farmers undergoing modernization through monetization of their agricultural systems, especially in the form of rubber growing. Under the traditional system there was considerable individual and annual fluctuation in rice production. Traditional land tenure and social systems resulted in relatively equal access to land, both through reallocation of land-use rights and by reallocation of individuals between families, movements of sections of longhouses, and even whole villages. Ibans did this largely by expanding into underpopulated forested areas. Land Dayaks rearranged populations and landholdings within land which had already been cleared and claimed. Both groups traditionally had concepts of shared property similar to those reported among other peasant groups, allowing redistribution of food during times of local shortage, and in neither group were social classes developed in association with unequal access to production or distribution of essential goods. Both had the means of storing value, which could be traded for food, and alternative sources of food which could be used during

times of rice shortage. These are all means of reducing fluctuations between population and resources.

Population growth through natural increase seems to have been relatively slow among both groups before 1939. In both groups the potential for rapid population growth seems to have been present in the form of early age and high rate of marriage for females, and relatively easy reassortment of mates in the event of divorce or widowhood. Controls on population growth may have been in the form of mortality due to disease rather than famine; the Ibans are also known to have practised birth-control consciously (in the form of abortion) and perhaps unconsciously (in the form of *bejalai* journeying which separated adult males from their families for months or even years). No good quantitative data exist to explain the slow rate of population growth prior to 1939, nor the rapid growth thereafter. Reduced disease mortality and increase in security and social stability as a result of suppression of warfare may have been preconditions for increased rate of growth, but demographic figures also suggest an increase in nuptiality and fertility contributing in a major way to the growth. Jones (1966) claims the acceleration came before malaria control and Griffith (1955) indicates the Ibans were already quite healthy in 1953. The acceleration was associated with a modification of the economic systems of both groups, with the addition of rubber growing, a cash crop which could absorb additional labour throughout the year, and which could smooth out the fluctuations in the subsistence economy.

These are examples in which the sedentarization and agricultural intensification is taking place before the carrying capacity is reached, and in the case of the Ibans, before it is even approached. This suggests that either Boserup's (1966) hypothesis of population pressure as the cause of agricultural intensification does not apply here (see chapter 19, part 1) or Freeman's estimate of carrying capacity is in error by a factor of 10. Kunstadter suggested that the motivations for agricultural intensification may include a desire to reduce fluctuations in production, as well as a desire to balance population and production, to maximize production or to maximize productivity.

Bangladesh

Bangladesh is one of the most densely settled parts of the world, with *ca.* 550 persons/km². With an average of 1 600 kcal/caput/day and an average gross national product of $70/person/a, it is one of the world's poorest countries. Over 90 per cent of the population is agricultural and there has been little industrial development.

The country lies astride the Tropic of Cancer but, due to the sheltering effects of the Himalayan mountains, its climate is relatively warm throughout the year and is tropical or subtropical, with a mean maximum monthly temperature *ca.* 24° C, and a mean monthly minimum of *ca.* 5° C less. Rainfall is moonsonal with most falling from June to October.

The population is concentrated in the deltaic plain and flood-plain of the Ganges and Brahmaputra rivers, a low-lying area, reticulated with natural watercourses and canals. Except for the mountainous northern and eastern borders, the advancing front of the delta, and a small slightly elevated area north of Dacca, there is little remaining natural vegetation. 80 per cent of the cultivable land is used for wet rice. The main cash crop is jute, supplemented by sugarcane and tea.

Bangladesh is a good case study in which to examine the interaction of the human population with the tropical forest ecosystem because: it has the 'most serious population problem in the world' (UNROD, 1972); the size of its population; the almost complete modification of the original tropical forest, and the relative excellence and quantity of the demographic data.

What follows is primarily of the population of Matlab Bazaar thana, a *ca.* 259 km² area, intensively studied since the early 1960s by the Cholera Research Laboratory in Dacca, which has maintained almost continuous surveillance over vital events. As a result, this is the outstanding population laboratory in the tropical world (Mosley, Chowdhury and Aziz, 1970). The area is centered about 40 km southeast of Dacca, right on the Tropic of Cancer and 91° E. It is in the Brahmaputra-Meghna flood-plain, and much of the land is subject to annual flooding. Average annual rainfall is *ca.* 2 100 mm, mainly from June to November, with less than 20 mm falling in December and January. There is high humidity throughout the year. The original vegetation has long been cleared for agriculture and gardening, but may have resembled that of the Dacca forest now growing in a better drained area 20–30 m above the flood plain, *ca.* 85 km to the north. This is a secondary *Shorea robusta* forest with trees growing to a height of 15–25 m and a diameter of 30–100 (–130 cm) in 20 years.

The people are ethnic Bengalis, 80 per cent are Muslim and most of the rest are Hindu. Population density is >770/km². The people are almost all rice and jute farmers, supplementing their diet with a variety of garden crops and fishing, and sending a steady stream of temporary and permanent migrants to seek wage work in Dacca and other towns.

Between 1966 and 1971 the crude birth rate in Matlab thana averaged *ca.* 45.6/1 000, while the crude death rate was *ca.* 15.3/1 000, giving an average annual natural increase of *ca.* 3.03 per cent. This is higher than the estimated national average annual growth rate of *ca.* 2.5 per cent (Hossain, 1972), based on a crude birth rate of *ca.* 45.7 and crude death rate of 17.2/1 000 in 1975. The death rate at Matlab thana may have been lower than the national average because of the presence of a modern prevention and treatment centre for diarrhea and dysentary (Mosley, Chowdhury and Aziz, 1970). This growth rate was reduced locally by net annual outmigration of about 1.3 per cent, bringing the local population growth down to about 2 per cent/a. The age distribution in 1966 is shown in table 12. This is a young population, with a median age of 16.3, and with an unbalanced sex-ratio in favour of males at all ages except 20–34 (when many men are away from their home villages seeking wage work in the cities). The age distribution

TABLE 12. Age distribution in Matlab thana, Bangladesh (April 1966).

Age (years)	Numbers		Males
	Males	Females	100 females
0– 4	9 913	9 669	102.5
5– 9	10 581	10 007	105.7
10–14	7 456	6 236	119.6
15–19	4 042	3 831	105.5
20–24	2 889	4 259	67.8
25–29	3 456	4 643	74.4
30–34	3 343	3 579	93.4
35–39	3 424	2 872	119.2
40–44	2 757	2 262	121.9
45–49	2 228	1 752	127.2
50–54	2 057	1 761	116.5
55–59	1 322	977	135.3
60–64	1 383	1 229	112.5
65+	2 268	1 477	153.5
Unknown	30	45	—
Total	57 149	54 599	Mean 104.6

Source: Mosley, Chowdhury and Aziz (1970).

is consistent with the following generalizations which are also supported by other evidence:

— sex-ratio at birth is within the normal human range;
— except for ages less than one year, death rates for females are generally higher than for males;
— net outmigration rates are higher for males than for females;
— most male migration is associated with the search for work, especially in cities; most female migration is associated with marriage, or as a dependent (Stoeckel *et al.*, 1972);
— fertility is high, with an average of over 6 children born to women of completed fertility, and 2.2 female children surviving to age 15;
— high fertility is maintained by marriage of almost all women by 13–16 years of age, and most remain married throughout their reproductive years;
— there is a strong preference for male offspring backed by Islamic religious sanctions (Yankauer, 1959; Jahan, 1974), which apparently affects adversely the life chances of females at all ages;
— absence or death of male children may determine the desire for and the birth of additional children (Welch, 1974).

Robinson (1967) in his discussion of the recent history of mortality in Bangladesh distinguishes four stages: first, elimination of famines and many epidemics by improved transportation of food and elementary public health measures during the first several decades of the twentieth century; secondly, control of the chief epidemic diseases including smallpox, plague and cholera by the end of the 1930s through public health programme; thirdly, absolute reduction in all causes of death except maternal mortality and tuberculosis from the 1950s; and fourthly,

Table 13. Mortality changes in India and Bangladesh.

Cause of death	India[1] ca. 1900 Death rate (per 1 000)	India[1] ca. 1900 Percentage of total deaths	Indian sub-continent[1] 1930s Death rate (per 1 000)	Indian sub-continent[1] 1930s Percentage of total deaths	Bangladesh[1] ca. 1960 Death rate (per 1 000)	Bangladesh[1] ca. 1960 Percentage of total deaths	Matlab Bazaar[2] thana (1966–71) Death rate (per 1 000)	Matlab Bazaar[2] thana (1966–71) Percentage of total deaths
Cholera	3.7	9.5	1.8	5.5	0.5	2.5		
Smallpox	0.4	1.0	0.4	1.2	0.2	1.0	Smallpox and measles 0.6	3.9
Plague	0.3	0.8	0.4	1.2	—	—	—	—
Malaria	Fevers 22.8	58.6	8.8	27.0	3.0	15.0	Fevers	
Typhoid			5.0	15.4	3.0	15.0	2.5	16.3
Complications of childbirth	?	?	3.0	19.2	2.0	10.0	0.3[3]	2.0[3]
Tuberculosis	?	?	5.0	15.4	2.5	12.0		
Pneumonia and other respiratory diseases	?	?	0.7	2.2	1.0	5.0	1.4*	9.1
Diarrhea and dysentery	2.5	6.4	1.0	3.1	3.0	15.0	2.0	13.1
All other	8.8	22.6	6.7	20.6	5.0	25.0	8.8**	57.5
Total	38.9	100.0	32.8	100.0	20.2	100.0	15.3	

1. Robinson (1967).
2. Curlin, Chen and Hussain (1975).
3. Maternal deaths were 5.7–7.7/1 000 live births (Chen, Gesche, Ahmed, Chowdhury and Mosley, 1974). This figure is not included as a separate cause of death in Curlin *et al.* (1975).
* All respiratory causes.
** Including accidents 0.8/1 000, unknown 4.1/1 000 and other 3.9/1 000.

TABLE 14. Sex-age specific death rates
(Matlab thana, 1966–71).

Age	Male	Female
1	150.0	134.8
1– 4	21.4	30.2
5– 9	3.4	4.0
10–14	1.3	1.8
15–44	3.4	4.1
45–64	15.1	17.0
65+	67.7	81.4
Total	15.0	15.5

Source: Curlin, Chen and Hussain (1975).

decline in the remaining causes. These changes are illustrated in table 13. The figures suggest the great changes in numbers and distribution of deaths among causes, with a movement away from particular infectious or vector-borne epidemic diseases, and towards a variety of other causes. The mortality transition is incomplete because the total death rate is still 50–75 per cent higher than European standards, infectious diseases, especially gastro-intestinal ones, are still important, and because the age distribution of deaths still includes a very large proportion of infants and young children—some 10 times the rate in European populations (table 14).

There have been episodes of high mortality due to a variety of causes—epidemics, famines, warfare, tidal waves and cyclones. Curlin, Chen and Hussain (1975) studied the demographic effects of the April 1971-March 1972 war.

There was no detectable change in birth rate (45.6/1 000). This declined to 41.8 in the following year, but in 1973–74 rose to a higher level than that of the four pre-war years. The crude death rate which had been *ca.* 15.0/1 000 rose to 21.4 during the war year, fell to 16.2 in 1972–73, and returned to its pre-war level the next year. The decline in the birth rate during the year following the war represented reduced conceptions during the war year. A slight increase in births during November 1971–January 1972 followed the return of males from jobs in Dacca during the civil disturbances of January-April 1971 which preceded the outbreak of the war. After the war, the birth rate went up temporarily beyond previous normal levels, as delayed marriages were consummated, and previously married couples who had separated as refugees were reunited. Prior to these years a slight decline in birth rate had been observed, apparently correlated with an increase in female age at marriage; this was reversed in the initial stages of recovery from the war but may have resumed in succeeding years.

Mortality rates also changed. Prior to the war, neonatal (infants age 0–29 days) deaths represented 68 per cent of all infant (under one year) deaths and the total infant mortality rate was *ca.* 141.5/1 000 live births, while the crude death rate (for the total population) was about 15.3/1 000. Infant mortality rate rose to 174.7/1 000 during the war due entirely to an increase in post-neonatal deaths. Apparently neonatal deaths did not increase from the high

rate of 86.8/1 000 because conditions could not get any worse. Meanwhile, post-neonatal deaths and deaths of children aged 1–9 increased considerably in association with malnutrition and infectious diseases for which they no longer had maternal antibodies and to which they were exposed by the mass movement of people. The excess mortality in these age groups continued in the immediate post-war years, indicating the sensitivity of these age groups to social and environmental disruptions. The causes of death which increased in association with the war included especially cholera, *Shigella* infections, and other gastro-intestinal ailments probably associated with lower resistance due to malnutrition. In Bangladesh as a whole, it is estimated that over 500 000 lives were lost, while *ca.* 260 000 births were either averted or postponed.

The overall demographic effect, though major in terms of numbers of lives lost or births averted, was minor in terms of the growth of the population, because of the large surplus of births over deaths in the normal years, the brevity of the disruption and the speed of the demographic recovery. The increase in birth rates to a level above the pre-war years does not seem to be as a result of conscious motivation, but to the availability of large numbers of fecundible women whose exposure to risk of pregnancy had been temporarily restricted. This interpretation is consistent with the results of a study on relationship between birth interval and infant mortality (Stoeckel and Chowdhury, 1974).

There is nothing characteristically tropical about the Bangladesh population. The demographic history does suggest the potential for rapid population growth in tropical areas under marginal economic conditions once the constraints of mortality due to infectious diseases are reduced, if food shortages can be averted and if there is a cultural or religious support for high fertility.

Despite the recent slight increase in average age at marriage and the slight drop in fertility (Afzal, 1967; Stoeckel and Chowdhury, 1973; Sirageldin *et al.*, 1975), the population of Bangladesh will continue to grow rapidly for at least two generations even if there is a radical and immediate drop in average number of children born per woman (Frejka, 1974). Apparently this is the course of the future even if there are short periods of abnormally high mortality. Neither biological nor socio-economic homeostatic mechanisms seem to have been called into play at this stage.

Surprisingly, the low level of income and nutrition in Bangladesh have not been sufficient to reduce fertility below the highest world levels. Better nutrition might speed the onset of menarche and increase the number of surviving children by improving the survival rates especially of neonates. One study suggests that increased infant survival in Bangladesh has been associated with increased spacing between births, due to prolongation of lactational amenorrhoea in mothers who continue to nurse their surviving infants. The net effect of this increase in child spacing is to accelerate the population growth (Schultz, 1972). The radical decline in overall mortality (from *ca.* 40/1 000 in the 1900s to *ca.* 15/1 000 in the early 1970s) including an increase in infant and child survival has not been sufficient to induce people to lower their net fertility. Only *ca*. 4 per cent of the eligible women were practising family planning in the early 1970s in special demonstration areas (Stoeckel and Chowdhury, 1973). The data from a study of pregnancy termination intervals in Matlab thana do not support the position that 'women with higher child mortality will tend to make up for their losses by producing a shorter birth interval than produced by women with lower child mortality' (Stoeckel and Chowdhury, 1974). This suggests that high fertility is being maintained at a level near the physiological maximum by forces of society rather than by individual socio-economic or psychological needs. These data are in direct contradiction to Chandrasekhar's (1972) predictions concerning the relationships between infant mortality and fertility.

There are no published community studies which show the connections at the local level between population pressure and socio-economic or ecological change. One therefore has to take into account the cultural and social descriptions of Bangladesh society as a whole. Jahan (1974) indicates the cultural support in Bangladesh for high fertility. Women's status is kept low by *purdah* and patrilocal marriage. Because of the limitations on female economic and social roles, 'the only way (a woman) can get security in her husband's household is by producing a male heir. The more male children she has, the greater is her security and acceptance' (Jahan, 1974). Although there is no Islamic law against family planning, custom and the lack of socially acceptable and economically viable alternatives, and the social importance of the role of women as reproducers are strong enough forces to maintain high fertility in spite of the economic pressures it creates. These customs are only weakly counteracted by low prevalence of widow remarriage in spite of the high prevalence of widowhood.

Economic development strategies which have been followed at the national level have emphasized increased rice production through use of farm machinery, fertilizers, irrigation, improved seed and mechanical processing. Lindenbaum (1974) argues this has the effect of further depressing the status of women by reducing their economic contributions to the family. Social development has emphasized more equitable distribution of land and other important resources among families, which might improve the distribution of food, but again is unlikely to benefit the status of women, since they have little or no control over property. The economic value of women is declining, as shown by the increasing importance of dowry payments from bride's to groom's family as a requirement for marriage, in contrast with the traditional brideprice payment from groom's to bride's family (Lindenbaum, 1974). Thus despite the higher mortality of females as compared with males, marriageable women have evidently become surplus. The demographic consequences of this further depression of women's status are unknown. The future of women in urban areas appears bleak since they have an even lower status there than in rural areas (Jahan, 1974) suggesting that the urban labour force is unlikely to absorb the surplus females where there is direct competition for jobs with

males. They still have no alternative to family roles, even though their apparent underutilization in the labour force may contribute to the lag of economic development behind population growth. If low social status of women is causally associated with high fertility, as has frequently been argued, there is no suggestion that social change for women in Bangladesh will lead to higher status and lowered fertility.

Over the past 15 years economic growth has fallen behind population growth. A larger number and proportion of the population has become poorer as measured by the level of 2 100 kcal/caput/day and 45 g of protein/caput/day (Chen and Chowdhury, 1975). The average in the 1970s fell to 1 600 kcal and 40 g (with protein increasingly of vegetable origin), equal to the minimum requirement (FAO/WHO, 1973). Meanwhile the proportion of landless farmers increased from 15 per cent in the 1950s and 20 per cent in the 1960s to 40 per cent in the 1970s. By the 1970s wages constituted over half the income of agricultural workers, indicating the extent of landlessness and the vulnerability of the majority of the population to monetary as well as crop fluctuations (Chen and Chowdhury, 1975).

The nation is being kept alive by increasing food imports (0.6 million t/a in 1960–65, 1.4 million t/a in 1970–75). Mortality continues to decrease more rapidly than fertility, and population growth proceeds more rapidly than increase in agricultural production. 80 per cent of the increase in farm output has come from increasing the amount of land under cultivation, and only 20 per cent from technological improvements. Rice production is *ca*. 800 kg/ha—a low figure even by Asian standards (Asian Development Bank, 1969). Increased intensity of cultivation by traditional means has not increased production per unit land area or per person. Low population density districts have a slightly lower average number of crops per year and significantly higher productivity per caput and a slightly higher production per unit area than high density districts (Chen and Chowdhury, 1975). The low population density areas are found in the hilly regions away from the delta, suggesting that hill agriculture is more productive than lowland farming. This is apparently true even though cultivation/fallow cycles have been shortened to *ca*. three years with a resulting decline in productivity (Ahmad, 1968).

Studies of the patterns of change in population distribution between the 1961 and 1974 censuses show that urban centres grew 138 per cent, high density rural districts grew 28 per cent and low density rural areas grew 42 per cent (Chen and Chowdhury, 1975). A similar pattern of growth occurred between the 1951 and 1961 censuses. Studies of migration patterns in Matlab thana (Stoeckel, Chowdhury and Aziz, 1972) show that the densely settled rural area contributed population to Dacca and other urban centres. Urban areas also grew as a result of surplus of births over deaths, as urban fertility rates are still relatively high.

To sum up, contrary to what Wynne-Edwards (1962) might predict, the Bangladesh population seems to have grown far larger than the optimum for its environment, with little sign of the effective operation of homeostatic mechanisms. The response of the Bangladesh population to what appears by western, African and even many Asian standards to be Calhounian population pressure (Calhoun, 1962) has not been to innovate technologically or socially to any large extent (*contra* Boserup, 1966), nor to involute successfully (*contra* Geertz), nor to readily accept family planning devices (*contra* Ravenholt, 1975), nor even to be very successfully limited by Malthusian checks such as war or famine. Instead they continue to rely primarily on a strategy of redistribution: conserving their resources from shipment to United Kingdom, India and Pakistan by revolutions, redistributing their population in directions which for the moment appear to offer individual economic advantages in the form of wage labour or more productive land,[1] far beyond the point where people in other countries have chosen to restrict population growth voluntarily. They have simultaneously lowered their standard of living and death rate and maintained high fertility on a diet which has declined below physiologically established minimum standards.

Conclusions

Human population in tropical forest areas has grown with increasing speed in the past few generations as a result of declining death rates unmatched by declining birth rates. Although some new causes of morbidity and mortality have been introduced, they have not compensated for the decline in old causes such as malaria and smallpox. Most of the growth is from natural increase, with only a small proportion due to migration from other zones. The populations have grown at unprecedented rates. In general this increase has been accompanied by an increase in the size and scope of socio-economic systems, either spontaneously or as a result of outside contact and intervention.

As the human population expanded, the forest has been rapidly and extensively modified by human activity (see chapter 20). Several trends are evident in the patterns of human use. There has been a move from a high to a lower position in the food chain (from hunting and gathering to the controlled production of foodstuffs). The rate of turnover of essential nutrients has increased from cycles in the order of 100 years to single years. This implies an increasing efficiency of energy use, or, more often, an increasingly unbalanced dependence on fossil fuels or other resources not renewed at the rate at which they are used. There has been an increasing geographic, social and demographic scale of economic systems, including the development of large extractive industries and destructive environmental modification for the benefit of societies geographically or culturally based largely outside the tropical forest. As a result, the total amount of true tropical forest has declined rapidly.

These conditions have changed the population—environment relationships especially in use-systems which are sensitive to population density, or hunting and gathering. The productivity of these systems cannot be permanently raised by intensification of effort without environmental deterioration and decline in productivity (see chapter 19).

Migration to similar, but unoccupied niches, is increasingly difficult or impossible. Temporary intensification or use of marginal lands leads to declining productivity per caput, and possibly to increasing mortality (a Malthusian balance of resources, productivity and population) or to a transformation of the system of production through technological innovation leading to an increase in the ability of the agricultural system to sustain the population (Boserup, 1966). Neither warfare nor agricultural intensification seems directly correlated with population density among the Yanomamo or Borneo swiddeners, and mortality seems to be declining in Bangladesh despite extreme population pressure and lagging agricultural development. Other solutions to the increase in density include migration to and participation in other, more intense, socio-economic systems.

The increased migration which has occurred in some places has increased cultural heterogeneity and the chances of contacts with new diseases. In recent years such developments have been coupled with improved food distribution, public health and medical practices and increased social control. These have allowed the development of large cities which grow by natural increase as well as immigration despite the increased risks of illness and of intercommunal battles associated with greater contacts of different groups competing for limited resources.

Demographic, environmental and socio-economic changes have not occurred in isolation but there is no agreement that they follow a consistent sequence. Such changes are clearly interrelated, but the nature of this is incompletely understood. Theorists of the demographic transition have attempted to relate socio-economic and demographic changes, but rarely consider environmental variables either as determinants, consequences or correlates. The case studies suggest the value of analysing these changes at the level of functioning socio-cultural units. The natural boundaries of the tropical forest ecosystem do not correspond to the socio-cultural and demographic boundaries of the people now occupying these areas. Determinants of the patterns of interaction increasingly transcend the natural boundaries in demands for resources, influences on fertility, mortality and migration, and patterns of socio-cultural organization.

The number and variety of publications relevant to the interaction of population and environment in the tropical forest ecosystems is growing rapidly. Vayda (1969) has brought together a series of previously published papers offering theoretical models of population—environment interaction and case studies of particular areas. Harrison and Boyce (1972) have edited a volume dealing with the demographic, genetic, social and cultural adaptations of populations to a variety of environments. Hackenberg (1974) discusses the development of anthropological methods referring to a number of studies in tropical Africa

1. Elahi (1973) correctly states that such redistribution of the population cannot solve the problem of population pressure, implying that it will only succeed in spreading the problem to areas of lower population density.

and Asia. Kocher (1973) reviewing contemporary literature, develops a model relating rural development to decline of fertility.

Biological adaptation of human populations to the environment is described in several studies. One particularly relevant example is the study of the association between sex ratio at birth and infection with vector-borne disease (Hesser, Blumberg and Drew, 1976), suggesting a mechanism other than infanticide which might account for an abnormal sex ratio in a tropical population.

Social adaptations to population growth in South-East Asia are discussed by several authors in Kantner and McCaffrey's (1975) volume. Stevenson has attempted to relate the development of political systems to population density and cropping systems in several tropical African groups.

Relevant reviews of particular geographical areas include Megger's (1971) discussion of mechanisms for control of population size and density in Amazonia, and a special issue of *Études Rurales* (1974) devoted to agriculture and societies in South-East Asia. Summaries and collections of articles on the population of Africa (Brass, 1968; Caldwell and Okonjo, 1968; Condé, 1971; Ominde and Ejiogu, 1972; Caldwell *et al.*, 1973; Cantrelle, 1974), Papua-New Guinea and southern Asia (Davis, 1951) also provide useful information to begin comparative studies.

Gaps in knowledge of the demographic parameters of interactions between human populations and the tropical forest ecosystems exist for a number of reasons. There has been inadequate analysis of available incomplete data (population registration or census information) through available techniques developed by Brass, Collver and others which would allow estimation of vital rates, and thus form the basis for comparative studies of vital rates and environmental use systems.

Insufficient data have been gathered from relevant population-ecosystem combinations to allow the construction of detailed ecological-demographic models and the testing of Malthusian, Boserupian or other theories concerning consequences and limits of population growth.

There has been inadequate observation on the population units which are in close interaction with the environment (i.e., family, household and communities). Most statistics are gathered at the national level and are not susceptible to analysis according to ecological type. Because census data are now stored on computer tape it may be possible to make such analyses by detailed regional investigations of census information. Such studies need to be coupled with detailed ecological research on the population units in order to get adequate information on the distribution of people and resources in the economically functional units or socio-demographic communities where demographic decisions are made.

Mathematical models which have been developed for use with national population statistics may be inappropriate for prediction of demographic behaviour at the lower units. Laws of probability in stochastic processes are likely to be different at the lower levels, so strategies applicable for the

perpetuation of families, lineages or small communities are unlikely to be identical to those for nations.

Inadequate attention has been paid to understanding the processes of perception of the need to take action, the choice of action and the evaluation of actions affecting the relationships between population and environment, either by modifying the environment or by modifying demographic rates. The communities on which such perceptions occur have not been defined and are not the same for all types of societies or environments, or for all demographic strategies.

Research needs and priorities

I. Derivation of vital rates from incomplete census and vital registration data through use of modern methods for analysis of incomplete data and indirect methods of estimation of vital rates, including analysis of portions of national censuses selected on the basis of known regional, district ethnic group or village variations in environment, technology and social organization.

II. Research to measure the basic demographic variables, their interaction with each other and with environment and social structure.

A. Studies of mortality.

1. Cross-sectional non-specific indices of morbidity and mortality.
   — Census for age distribution to estimate mortality rates by stable population model methods.
   — Reproductive histories of samples of women of different birth cohorts to measure reproductive performance including foetal wastage, pregnancy spacing and child survival, and to reconstruct age-specific mortality especially at ages 0–4 years, as well as cohort differences in these variables.
   — Age of menarche as an index of population health status and potential fertility.
   — Blood surveys of haemoglobin or haematocrit and serum protein as a measure of nutritional status.
   — Blood surveys of immunoglobulins, as a non-specific measure of recent infection.

2. Cross-sectional specific indices of morbidity which will allow estimates of some major causes, more adequate estimates of effects on the population, and estimation of the effects of specific environment-related diseases.
   — Blood surveys of age-stratified samples for probable leading causes of morbidity which generate persistant antibodies, or for parasitemia (dengue fever, Japanese B encephalitis, malaria, filariasis, etc.).
   — Stool surveys for ova and parasites.
   — Health history by specific complaint to identify site and probable cause of morbidity (skin, respiratory, intestinal infections, etc.).

3. Longitudinal measurement of morbidity and mortality which will allow estimates of changing patterns of impact of specific environment-related diseases on the population.
— Diagnosis of changes in the cause of death or illness by age, through use of medical records or by continuous or intermittent monitoring of the population.

4. Ecological-epidemiological studies on distribution of specific diseases (e.g., malaria, schistosomiasis, cholera) including means of transmission, intermediate hosts, reservoirs, vectors, persistence in environment and non-human populations, human behavioural barriers and channels for spread, and human behavioural responses as a means of defining the population-environment interactions with regard to major causes of death.

B. Studies of fertility.

1. Cross-sectional non-specific indices of fertility.
— Census for age distribution to estimate fertility rates by stable population model or own-children methods.
— Estimation of fertility rates from birth registration records of parity by maternal age, etc.
— Reproductive histories.

2. Specific interactions of fertility and environmental factors.
— Genetic, disease and behavioural factors limiting fertility or inducing foetal or infant mortality, e.g., genetic incompatibility, venereal diseases, malnutrition, abortion, contraception, infanticide.
— Social structural and behavioural factors influencing the desire for children as related to patterns of exploitation of the environment, e.g., value of children, studies in groups with different socio-economic systems and different demands for labour in similar environments; effects of major socio-economic changes on these patterns.

C. Studies of migration.

1. Cross-sectional measures of migration in relation to environment through census, to map patterns of migration between different systems of environmental use, e.g., rural to urban, swidden to plantation, etc., allowing mapping of point of origin and point of destination with regard to socio-ecological variables in addition to moves between administrative units.

2. Retrospective study of migration patterns through interviews to determine paths of migration, and correlates and motivations for migration within and between different systems of environmental utilization.

D. Studies of socio-economic systems.

1. Cross-sectional description and analysis of the portions of the socio-economic system most directly involved in interaction with the environment, including technology of environmental exploitation, social units involved in production, distribution and consumption of environmental resources, allocation and reallocation of resources within and between population sub-units, allocation and reallocation of individuals and groups within the resource space through mechanisms such as division of labour, inheritance systems, taxation, land ownership and land tenure practices, marriage, adoption, trade, warfare and other means of modifying the boundaries of socio-economic units.

2. Longitudinal or retrospective studies of causes and consequences of environmental change, population change, or change in socio-economic systems or changes in the interactions between these variables, e.g., studies of environmental and socio-economic correlates of rapid population growth, studies of transformation from subsistence to cash cropping, effects of the availability of wage work in previously non-monetized economies, etc.

3. Studies of interaction among demographic variables and patterns of environmental use, directed at specific problems, e.g., the disease implications of changes in environmental use patterns and associated movements or changes in concentration and location of populations, especially the effects of persistant wide-spread clearing of forest through lumbering, through change in swidden fallow cover from forest to grass, or change from swidden to irrigated farming.

E. Study of individual, family, community and national perception of environment-population interaction.

1. Definition of communities with respect to environment-population interaction.

2. Study of cues used by communities in ensuring the state of the environment-population interaction.

3. Study of the choices among alternative actions believed by community members to affect environment-population interaction through modification of the environment or modification of demographic rates.

4. Study of community perception of the effects of these actions.
In addressing these tasks, it would be useful first to attempt to estimate total population size in each of the use-types, and secondly to develop a rational scheme of sampling from within the population on the basis of:
— representing the types in proportion to their population size;
— representing crucial variability within the use-systems, because they are liable to be lost in the near future, to become predominant in the future, because they represent scientifically important types of interaction between population and the tropical forest ecosystem, or because they represent important regional variability.

Bibliography

AFZAL, M. The fertility of East Pakistan married women. In: Robinson, W. C. (ed.). *Studies in the demography of Pakistan*, p. 50–91. Karachi, Pakistan Institute of Development Economics, 1967, 225 p.

AHMAD, N. *An economic geography of East Pakistan.* 2nd ed. London, Karachi, New York, Oxford University Press, 1968, 401 p.

ALLEN, M. R. A comparative note on Iban and Land Dayak social structure. *Mankind*, 7, 1970, p. 191–198.

AMIN, S. (ed.). *Modern migrations in Western Africa.* London, International African Institute, 1975, 426 p.

ANON. *Projection and estimate of population of Bangladesh.* Dacca, Census Organization, Ministry of Home Affairs, Government of the People's Republic of Bangladesh, Census 1974, Pub. no. 7, Bulletin 1, 1973, 107 p.

——. *Bangladesh population census 1974.* Dacca, Census Commission, Ministry of Home Affairs, Government of the People's Republic of Bangladesh, Census Publication no. 26, Bulletin 2, 1975, 218 p.

APPELL, G. N. Systems of land tenure in Borneo: a problem in ecological determinism. *Borneo Research Bulletin*, vol. 3, no. 1, 1971, p. 17–20.

ARENDS, T.; BREWER, G.; CHAGNON, N.; GALLANGO, M. L.; GERSHOWITZ, H.; LAYRISSE, M.; NEEL, J.; SHREFFLER, D.; TASHIAN, R.; WEITKAMP, L. R. Intratribal genetic differentiation among the Yanomama of southern Venezuela. *Proc. Nat. Acad. Sci.*, vol. 57, no. 5, 1967, p. 1252–1259.

Asian Development Bank. *Asian agricultural survey.* Seattle, University of Washington Press, 1969, 787 p.

BARTLETT, M. S. The critical community size for measles in the United States. *Journal of the Royal Statistical Society, Series A*, 48, 1960, p. 37–44.

BENDER, D. R. Population and productivity in tropical forest bush fallow agriculture. In: Polgar, S. (ed.). *Culture and population: a collection of current studies*, p. 32–45. Cambridge, Mass., and London, Schenkman; and Chapel Hill, N.C., Carolina Population Center, University of North Carolina, 1971, 195 p.

BHATTACHARYA, A. K. Income inequality and fertility: a comparative view. *Population Studies*, vol. 29, no. 1, 1975, p. 5–19.

BOSERUP, E. *The conditions of agricultural growth; the economics of agrarian change under population pressure.* Chicago, Aldine, 1966, 124 p.

BOURGEOIS-PICHAT, J. Utilisation de la notion de population stable pour mesurer la mortalité et la fécondité des pays sous-développés. *Bulletin de l'Institut International de Statistique* (Actes de la 30e Session), 1957, vol. 36, n° 2, p. 94–121.

——. Social and biological determinants of human fertility in non-industrial societies. *Proceedings of the American Philosophical Society*, vol. 111, no. 3, 1967, p. 133–193.

BRASS, W. (ed.). *The demography of tropical Africa.* Princeton, N.J., Princeton University Press, 1968, 539 p.

BRASS, W. *Methods for examining fertility and mortality from limited and defective data.* Chapel Hill, N.C., International Program of Laboratories for Population Statistics, Department of Biostatistics, Carolina Population Center, University of North Carolina, 1975, 159 p.

BREESE, G. (ed.). *Urban South-East Asia: a selected bibliography of accessible research, reports and related materials on urbanism and urbanization in Hong Kong, Indonesia, Malaysia, the Philippines, Singapore, Thailand, Vietnam.* New York,

South-East Asia Development Advisory Group of the Asia Society, 1973, 165 p.

BRUCE-CHWATT, L. J. Endemic diseases, demography and socio-economic development in tropical Africa. *Canadian Journal of Public Health*, 66, 1975, p. 31–36.

BULMER, R. Traditional forms of family limitation in New Guinea. *New Guinea Research Unit Bulletin*, 42, 1971, p. 137–162.

CALDWELL, J. C.; OKONJO, C. *The population of tropical Africa.* London, Longmans, Green and Co., 1968, 457 p.

——; ADDO, N. O.; GAISIE, S. K.; IGUN, A.; OLUSANYA, P. O. *Population growth and economic change in West Africa.* New York, Population Council, 1973, 763 p.

CALHOUN, J. B. Population density and social pathology. *Scientific American*, vol. 206, no. 2, 1962, p. 139–148.

CANTRELLE, P. La mortalité du jeune enfant en Afrique tropicale. *Carnets de l'Enfance* (UNICEF), n° 15, juillet-septembre 1971.

——. Is there a standard pattern of tropical mortality? In: Cantrelle, P. (ed.). *Population in African development*, p. 33–34. Liège, International Union for the Scientific Study of Population, vol. 1, 1974, 347 p.

—— (ed.). *Population in African development.* Liège, Belgium, International Union for the Scientific Study of Population, 2 vol., 1974, 550 p.

CARNEIRO, R. L. Slash and burn agriculture: a closer look at its implications for settlement patterns. In: Wallace, A. F. C. (ed.). *Men and cultures*, p. 229–234. Philadelphia, University of Pennsylvania Press, 1961, 810 p.

CHAGNON, N. A. *Yanomamö: the fierce people.* New York, Holt, Rinehart and Winston, 1968a, 142 p.

——. The culture-ecology of shifting (pioneering) cultivation among the Yanomamö Indians. In: *Proceedings 8th International Congress of Anthropological and Ethnological Sciences* (Tokyo and Kyoto, 1968), 2, p. 249–255. Tokyo, Science Council of Japan, 1968b, 3 vol. 832 p.

——. *Studying the Yanomamö.* New York, Holt, Rinehart and Winston, 1974, 270 p.

——; NEEL, J. V.; WEITKAMP, L. R.; GERSHOWITZ, H.; AYRES, M. The influence of cultural factors on the demography and pattern of gene flow from the Makiritare to the Yanomamö Indians. *American Journal of Physical Anthropology*, vol. 32, no. 3, 1970, p. 339–349.

CHANDER, R. *Banchi pendudok dan perumahan Malaysia 1970 (1970 population and housing census of Malaysia); gulongan masharakat (community groups).* Kuala Lumpur, Jabatan Perangkoan Malaysia, 1972, 292 p., map.

CHANDRASEKHAR, S. *Infant mortality, population growth and family planning in India.* Chapel Hill, N.C., University of North Carolina Press, 1972, 399 p.

CHEN, L. C.; GESCHE, M. C.; AHMED, S.; CHOWDHURY, A. I.; MOSLEY, W. H. Maternal mortality in rural Bangladesh. *Studies in Family Planning*, vol. 5, no. 11, 1974, p. 334–341.

——; CHOWDHURY, R. H. *Demographic change and trends of food production and availabilities in Bangladesh (1960–1974).* Dacca, The Ford Foundation, 1975, 56 p.

CHO, L. J. The own-children approach to fertility estimation: an elaboration. In: *International Population Conference*, p. 263–279. Liège, International Union for the Scientific Study of Population, vol. 2, 1973, 416 p.

COALE, A. J.; DEMENY, P. *Regional model life tables and stable populations.* Princeton, N.J., Princeton University Press, 1966, 871 p.

COLLVER, O. A. *Birth rates in Latin America: new estimates of historical trends and fluctuations*. Berkeley, Institute of International Studies; University of California. Institute of International Relations, Research series no. 7, 1965, 187 p.

Committee for International Coordination of National Research in Demography. *A repertory of research projects in priority areas of demographic study*. 1st edition. Paris, CICRED, 1974a, 135 p.

——. *Directory of demographic research centres*. Paris, CICRED, 1974b, 719 p.

CONDÉ, J. *The demographic transition as applied to tropical Africa, with particular reference to health, education and economic factors*. Paris, OECD Development Centre, 1971, 207 p.

COOK, R. C. *1975 world population estimates*. Washington, D.C., The Environmental Fund, 1975, 2 p.

CUMPSTON, I. M. A survey of Indian immigration to British tropical colonies to 1910. *Population Studies*, vol. 10, no. 2, 1956, p. 158–165.

CURLIN, G. T.; CHEN, L. C.; HUSSAIN, S. B. *Demographic crisis: the impact of the Bangladesh civil war (1971) on the births and deaths in a rural area of Bangladesh*. Dacca, The Ford Foundation, 1975, 55 p.

DAVIS, K. The world demographic transition. *The Annals of the American Academy of Political and Social Science*, 237, 1945, p. 1–11; reprinted in Bobbs-Merrill reprint series in the social sciences, S–370, 257 p.

——. *The population of India and Pakistan*. Princeton, N.J., Princeton University Press, 1951, 263 p.

——. The amazing decline of mortality in underdeveloped areas. *American Economic Review*, 66, 1956, p. 305–318.

——; BLAKE, J. Social structure and fertility; an analytic framework. *Economic Development and Cultural Change*, vol. 4, no. 3, 1956, p. 211–235.

Délégation générale à la recherche scientifique et technique (DGRST). *Afrique Noire, Madagascar, Comores : Démographie comparée*. Paris, 1967.

DOUGLAS, M. Population control groups. *British Journal of Sociology*, vol. 17, no. 3, 1966, p. 263–273. Reprinted in: Kannmeyer, K. C. W. (ed.). *Population studies: selected essays and research*. Chicago, Rand McNally, 1969, 481 p.

ELAHI, K. M. Geodemographic regions of Bangladesh. *Oriental Geographer* (Dacca, Bangladesh Geographical Society), vol. 17, no. 1, 1973, p. 7–25.

FAO/WHO. *Energy and protein requirements*. Geneva, WHO, Report of a joint FAO/WHO Ad Hoc Expert Committee, Technical Report Series no. 522, 1973, 118 p.

FAWCETT, J. T.; ARNOLD, F.; BULATAO, R. A.; BURIPAKDI, C.; CHUNG, B. J.; IRITANI, T.; LEE, S. J.; TSONG-SHIEN, Wu. *The value of children in Asia and the United States: comparative perspectives*. Honolulu, East-West Center, Papers of the East-West Population Institute, no. 32, 1974, 69 p.

FOGLE, C.; GLEITER, K. J.; MCINTYRE, M. *International directory of population information and library resources*. Chapel Hill, Carolina Population Center, University of North Carolina, 1972, 324 p.

FREDERIKSEN, H. Malaria control and population pressure in Ceylon. *Public Health Reports*, vol. 75, no. 10, 1960, p. 865–868.

——. Determinants and consequences of mortality trends in Ceylon. *Public Health Reports*, vol. 76, no. 8, 1961, p. 659–663.

FREEMAN, J. D. *Iban agriculture; a report on the shifting cultivation of hill-rice by the Iban of Sarawak*. London, Her Majesty's Stationery Office (HMSO), 1955, 148 p.

——. *Report on the Iban*. London and New York, The Athlone Press, University of London and Humanities Press, London School of Economics, Monographs on Social Anthropology no. 41, 1970, 317 p.

FREJKA, T. *Bangladesh country prospects*. New York, Population Council, 1974, 31 p.

FRISCH, R. E. *Demographic implications of the biological determinants of female fecundity*. Cambridge, Mass., Harvard University Center for Population Studies, Research Papers Series, no. 6, 1974, 7 p.

——; REVELLE, R. Variations in body weights and the age of the adolescent growth spurt among Latin American and Asian populations in relation to calorie supplies. *Human Biology*, 41, 1969, p. 185–212.

——; MCARTHUR, J. W. Menstrual cycles: fatness as a determinant of minimum weight for height necessary for their maintenance or onset. *Science*, vol. 185, no. 4155, 1974, p. 949–951.

GEDDES, W. R. *The Land Dayaks of Sarawak*. London, Her Majesty's Stationery Office for the Colonial Office, Colonial Research Studies no. 14, 1954, 113 p.

——. *Nine Dayak nights*. London, Oxford University Press, 1957, 144 p.

——. *Migrants of the mountains: the cultural ecology of the Blue Miao (Hmong Njua) of Thailand*. Oxford, Clarendon Press, 1976, 274 p.

GLEITER, K. J. *International directory of population information and library resources*. Chapel Hill, Carolina Population Center, University of North Carolina, 1975, 331 p.

GOODE, S. H. (ed.). *Population and the population explosion; a bibliography for 1973*. New York, The Whitston Publishing Company, 1975, 172 p.

GRAY, R. H. The decline of mortality in Ceylon and the demographic effects of malaria control. *Population Studies*, vol. 28, no. 2, 1974, p. 205–229.

GRIFFITH, G. T. Health and disease in young Sea Dayak men. *Sarawak Museum Journal*, 6, 1955, p. 322–327.

HACKENBERG, R. A. Genealogical method in social anthropology: the foundations of structural demography. In: Honigmann, J. J. (ed.). *Handbook of social and cultural anthropology*, p. 289–325. Chicago, Rand McNally College Publishing Company, 1974, 1 295 p.

HALLPIKE, C. R. Functionalist interpretations of primitive warfare. *Man*, vol. 8, no. 3, 1973, p. 451–470.

HANCE, W. A. *Population, migration and urbanization in Africa*. New York and London, Columbia University Press, 1970, 450 p.

HANKINSON, R. *Population and development; a summary information guide*. Paris, OECD Development Centre, 1973, 56 p.

HARRISON, G. A.; BOYCE, A. J. *The structure of human populations*. Oxford, Clarendon Press, 1972, 447 p.

HARRISSON, T. Classifying the people. Appendix X. In: Noakes, J. L. *A report on the 1947 population census, Sarawak and Brunei (the Colony of Sarawak and the British Protected State of Brunei)*, p. 271–279. Kuching, Sarawak, Government Printer, and London, The Crown Agents for the Colonies, 1948, 262 p.

HESSER, J. E.; BLUMBERG, B. S.; DREW, J. S. Hepatitis B surface antigen, fertility and sex ratio: implications for health planning. In: Kaplan, B. A. (ed.). *Anthropological studies of human fertility*, p. 73–81. Detroit, Wayne State University Press, 1976, 146 p.

HOSSAIN, M. *Bangladesh: population projections, 1961–2000*. Washington, D.C., Population and Human Resources Division, International Bank for Reconstruction and Development, 1972, 18 p.

HOUDAILLE, J. Une enquête rétrospective sur la population du Bangladesh. *Population*, vol. 28, n° 3, 1973, p. 688–690.

Indonesia, Biro Pusat Statistik. *Sensus Penduduk 1971 (1971 population census)*, E, no. 17–20, *Penduduk* (population of) *Kalimantan Barat, Kalimantan Tengah, Kalimantan Selatan, Kalimantan Timur*. Jakarta, Biro Pusat Statistik, 1974, 230 p.

INED, INSEE, ORSTOM, SEAEC. *Sources et analyse des données démographiques. Application à l'Afrique d'expression française et à Madagascar. 1ʳᵉ partie : sources des données. 2ᵉ partie : ajustements des données imparfaites*. Paris, 1973.

INGERSOLL, J.; JABBRA, N. W.; LENKERD, B. *Resettlement and settlement; an annotated bibliography*. New York, South-East Asia Development Advisory Group of the Asia Society, 1976, 30 p.

International Union for the Scientific Study of Population (IUSSP). Union internationale pour l'étude scientifique de la population. *Directory of member's scientific activities 1975*. Liège, Belgium, UIESP, 1975, 441 p.

JAHAN, R. *Women in Bangladesh*. Dacca, The Ford Foundation; reprint of paper written for the 9th International Congress of Anthropological and Ethnological Sciences (Chicago, August 1973), 1974, 45 p.

JAIN, A. K.; SUN, T. H. Inter-relationship between socio-demographic factors, lactation and postpartum amenorrhea. *Demography* (India), vol. 1, no. 1, 1972, p. 1–15. Reprinted by University of Michigan Population Studies Center, no. 101.

JELLIFFE, D. B. *Infant nutrition in the subtropics and tropics*. Geneva, WHO, 1968, 271 p.

JENSEN, E. Hill rice: an introduction to the hill padi cult of the Sarawak Iban. *Folk. Dansk etnografisk tidsskrift*, 7, 1965, p. 43–88.

——. Iban birth. *Folk. Dansk etnografisk tidsskrift*, 8–9, 1966–67, p. 165–178.

——. *The Iban and their religion*. Oxford, Clarendon Press, 1974, 242 p.

JONES, L. W. Sarawak. *Report on the census of population taken on 15th June 1960*. Kuching, Government Printing Office, 1962a, 337 p., maps.

——. North Borneo. *Report on the census of population taken on 10th August, 1960*. Kuching, Government Printing Office, 1962b, 305 p., maps.

——. *The population of Borneo; a study of the peoples of Sarawak, Sabah and Brunei*. London, University of London, the Athlone Press, 1966, 213 p.

KANTNER, J. F.; MCCAFFREY, L. *Population and development in South-East Asia*. Lexington, Massachusetts, Lexington Books, 1975, 323 p.

KEEN, F. G. B. Economic relationships in a Hmong economy. In: Kunstadter, P.; Chapman, E. C.; Sabhasri, S. (eds.). *Farmers in the forest*. Honolulu, University Press of Hawaii, 1976, in press.

KEYFITZ, N.; FLIEGER, W. *Population; facts and methods of demography*. San Francisco, Freeman, 1971, 613 p.

KOCHER, J. E. *Rural development, income distribution and fertility decline*. New York, The Population Council, 1973, 105 p.

KUNSTADTER, P. Residential and social organization among the Lawa of northern Thailand. *Southwestern Journal of Anthropology*, 22, 1966, p. 61–84.

—— (ed.). *South-East Asian tribes, minorities, and nations*, Princeton, N.J., Princeton University Press, 1967, 2 vol., 902 p.

——. Natality, mortality and migrations in upland and lowland populations in northwestern Thailand. In: Polgar, S. (ed.).

Culture and population: a collection of current studies, p. 46–60. Cambridge, Mass., and London, Schenkman; and Chapel Hill, N.C., Carolina Population Center, University of North Carolina, 1971, 195 p.

KUNSTADTER, P. Demography, ecology, social structure, and settlement patterns. In: Harrison, G. A.; Boyce, A. J. (eds.). *Biological and social structure of human populations*, p. 313–351. Oxford, Clarendon Press, 1972, 447 p.

——. Subsistence agricultural economics of Lua' and Karen hill farmers. In: Kunstadter, P.; Chapman, E. C.; Sabhasri, S. (eds.). *Farmers in the forest*. Honolulu, University Press of Hawaii, 1976, in press.

KURISU, K. Multivariate statistical analysis on the physical interrelationship of native tribes in Sarawak, Malaysia. *American Journal of Physical Anthropology*, 33, 1970, p. 229–234.

LEACH, E. R. *Political systems of highland Burma*. Cambridge, Mass., Harvard University Press, 1954, 324 p.

——. The 'frontiers' of Burma. *Comparative Studies in Society and History*, 3, 1960, p. 49–73.

LEE, Y. L. The population of Sarawak. *Geographical Journal*, vol. 131, no. 3, 1965, p. 344–356.

——. The Dayaks of Sarawak. *Journal of Tropical Geography*, 23, 1966, p. 28–39.

——. *Populations and settlement in Sarawak*. Singapore, Donald Moore for Asia Pacific Press, 1970, 257 p.

LINDENBAUM, S. *The social and economic status of women in Bangladesh*. Dacca, The Ford Foundation, 1974, 32 p.

LIVINGSTON, F. B. Population genetics and population ecology. *American Anthropologist*, 64, 1962, p. 44–53.

LIZOT, J. Aspects économiques et sociaux du changement cultural chez les Yanomami. *L'Homme*, vol. 11, n° 1, 1971, p. 32–51.

MACCLUER, J. W.; NEEL, J. V.; CHAGNON, N. A. Demographic structure of a primitive population: a simulation. *American Journal of Physical Anthropology*, vol. 35, no. 2, 1971, p. 193–207.

MATA, L. J. Malnutrition-infection interactions in the tropics. *American Journal of Tropical Medicine and Hygiene*, vol. 24, no. 4, 1975, p. 564–574.

MCGEE, T. G. *The southeast Asian city; a social geography of the primate cities of South-East Asia*. New York, Praeger, 1967, 204 p.

MEEGAMA, S. A. Malaria eradication and its effects on mortality levels. *Population Studies*, vol. 21, no. 3, 1967, p. 207–237.

——. The decline in maternal and infant mortality and its relation to malaria eradication. *Population Studies*, vol. 23, no. 2, 1969a, p. 289–302.

——. A reply. *Population Studies*, vol. 23, no. 2, 1969b, p. 305–306.

MEGGERS, B. J. *Amazonia: man and culture in a counterfeit paradise*. Chicago, New York, Aldine Atherton, 1971, 182 p.

MILES, D. The Ngadju Dayaks of central Kalimantan, with special reference to the Upper Mentaya. *Behavior Science Notes*, vol. 5, no. 4, 1970, p. 291–317.

MOSLEY, W. H.; CHOWDHURY, A. K. M. A.; AZIZ, K. M. A. *Demographic characteristics of a population laboratory in rural East Pakistan*. Washington, D.C., Center for Population Research, National Institutes of Health, 1970, 8 p.

MYERS, G. C.; MACISCO, J. J. Jr. *Selective bibliography on migration and fertility*. Durham, N. C., Duke University Center for Demographic Studies, working paper no. 6, 1972, 8 p.

MYERS, P. F.; BOUVIER, L. F.; ECHOLS, J. R. *1975 World population data sheet*. Washington, D.C., Population Reference Bureau, 1975.

NAG, M. *Factors affecting human fertility in non-industrial societies: a cross-cultural study*. Yale University, Department of Anthropology, Ph. D. thesis, 1962, 227 p.

——. Family type and fertility. In: *Proceedings of the World Population Conference* (Belgrade, 1965), 2, p. 160–163. New York, United Nations, 1967, vol. 2, 510 p.

National Academy of Sciences. National Research Council. Food and Nutrition Board, Committee on Maternal Nutrition. *Maternal nutrition and the course of pregnancy*. Washington, D.C., U.S. Government Printing Office, 1970, 241 p.

——. Committee on the Effects of Herbicides in Vietnam. *The effects of herbicides in Vietnam. Part A. Summary and conclusions*. Washington, D.C., National Academy of Sciences, 1974, 372 p.

NEEL, J. V.; CHAGNON, N. A. The demography of two tribes of primitive, relatively unacculturated American Indians. *Proceedings of the National Academy of Sciences*, vol. 59, no. 3, 1968, p. 680–689.

——; WEISS, K. M. The genetic structure of a tribal population, the Yanomama Indians. XII. Biodemographic studies. *American Journal of Physical Anthropology*, vol. 42, no. 1, 1975, p. 25–52.

NEWMAN, P. *Malaria eradication and population growth; with special reference to Ceylon and British Guiana*. Ann Arbor, Michigan, The University of Michigan, School of Public Health, Bureau of Public Health Economics, Research Series no. 10, 1965, 259 p.

——. Malaria control and population growth. *Journal of Development Studies*, vol. 6, no. 2, 1970, p. 133–158.

NOAKES, J. L. *A report on the 1947 population census, Sarawak and Brunei (the Colony of Sarawak and the British Protected State of Brunei)*. Kuching, Sarawak Government Printer, and London, The Crown Agents for the Colonies, 1948, 262 p., maps.

NOTESTEIN, F. W. Population: the long view. In: Schultz, T. (ed.). *Food for the world*, p. 36–57. Chicago, University of Chicago Press, 1945, 352 p.

OCDE. *La transition démographique en Afrique tropicale*. Paris, OCDE, 1971.

OECD. *Information on the OECD Development Centre's Programme in social development and demography*. Paris, OECD, Development Centre, 1973, 17 p.

OMINDE, S. H.; EJIOGU, C. N. (eds.). *Population growth and economic development in Africa*. London, Nairobi, Ibadan, Heinemann, in association with Population Council (New York), 1972, 421 p.

ORSTOM, INSEE, INED. Colloque de démographie africaine. In: *Cahiers ORSTOM* (Paris), sér. Sci. humaines, vol. 3, n° 1, 1970–1971.

PAGE, H. J. Infant and child mortality. In: Cantrelle, P. (ed.). *Population in African development*, p. 85–100. Liège, International Union for the Scientific Study of Population, vol. 1, 1974, 347 p.

PAKRASI, K.; MALAKER, C. The relationship between family type and fertility. *Milbank Memorial Fund Quarterly*, vol. 45, no. 4, 1967, p. 451–460.

PELZER, K. The swidden cultivator as a producer of agricultural exports. In: Kunstadter, P.; Chapman, E. C.; Sabhasri, S. (eds.). *Farmers in the forest*. Honolulu, University Press of Hawaii, 1976, in press.

PODLEWSKI, A. M. Bilan de l'état des connaissances démographiques concernant les écosystèmes pâturés et forestiers des régions tropicales (Afrique). *Cahiers ORSTOM* (Paris), sér. Sci. humaines, vol. 12, n° 4, 1975, p. 379–400.

Population Council. *A survey of institutional development needs and capabilities in developing countries. Vol. 1. Demography and related social sciences, fertility regulation and related health sciences*. New York. The Population Council, 1975, 345 p.

Population Reference Bureau. *Population education sources and resources*. Washington, D.C., Population Reference Bureau, Inc., 1975, 23 p.

POTTER, R. G.; NEW, M. L.; WYON, J. B.; GORDON, J. E. Applications of field studies to research on the physiology of human reproduction. *Journal of Chronic Diseases*, 18, 1965, p. 1125–1140. Reprinted by Harvard University Center for Population Studies, contribution no. 9.

RASHID, H. E. R. *East Pakistan; a systematic regional geography and its development planning aspects*. 2nd ed. Lahore, Peshawar, Hyderabad, Karachi, Sh. Chulam Ali and Sons, 1967, 387 p.

RAVENHOLT, R. T. *World population crisis and action toward solution*. Washington, D.C., U.S. Agency for International Development, 1975, 17 p.

REVELLE, R. Possible futures for Bangladesh. *Asia*, 29, 1973, p. 34–54.

ROBINSON, W. C. (ed.). *Studies in the demography of Pakistan*. Karachi, Pakistan Institute of Development Economics, 1967, 225 p.

ROMANIUK, A. Infertility in tropical Africa. In: Caldwell, J. C.; Okonjo, C. (eds.). *The population of tropical Africa*, p. 214–224. London, Longmans, Green and Co., 1968, 457 p.

SABHASRI, S. Effects of forest fallow cultivation on forest production and soil. In: Kunstadter, P.; Chapman, E. C.; Sabhasri, S. (eds.). *Farmers in the forest*. Honolulu, University Press of Hawaii, 1976, in press.

SALZANO, F. M. Studies on the Caingang Indians. I. Demography. *Human Biology*, vol. 33, no. 2, 1961, p. 110–130.

——. Demographic studies on Indians from Santa Catarina, Brazil. *Acta Geneticae Medicae et Gemellologiae*, vol. 13, no. 3, 1964, p. 278–304.

——; NEEL, J. V.; MAYBURY-LEWIS, D. Further studies on the Xavante Indians. I. Demographic data on two additional villages: genetic structure of the tribe. *American Journal of Human Genetics*, vol. 19, no. 4, 1967, p. 463–489.

SANDIN, B. The westward migration of the Sea Dayak. *Sarawak Museum Journal*, 7, 1956, p. 54–81.

SCHULTZ, T. P. Retrospective evidence of a decline of fertility and child mortality in Bangladesh. *Demography*, vol. 9, no. 3, 1972, p. 415–430.

SCRIMSHAW, N. S.; TAYLOR, C. E.; GORDON, J. E. *Interactions of nutrition and infection*. Geneva, World Health Organization, WHO Monograph Series, no. 57, 1968, 329 p.

SIMPSON-HEBERT, M. *Breastfeeding and human infertility*. Chapel Hill, N.C., Carolina Population Center, University of North Carolina. Technical Information Service, Bibliography Series, no. 9, 1975, 9 p.

SIRAGELDIN, I.; NORRIS, D.; AHMAD, M. Fertility in Bangladesh: facts and fancies. *Population Studies*, vol. 29, no. 2, 1975, p. 207–215.

SPIELMAN, R. S.; ROCHA, F. J. da; WEITKAMP, L. R.; WARD, R. H.; NEEL, J. V.; CHAGNON, N. A. The genetic structure of a tribal population, the Yanomama Indians. VII. Anthropometric differences among Yanomama villages. *American Journal of Physical Anthropology*, vol. 37, no. 3, 1972, p. 345–356.

State of Sarawak, Department of Statistics. *Annual Bulletin of Statistics, State of Sarawak, 1966*. Kuching, 1967, 128 p.

Stoeckel, J.; Chowdhury, A. K. M. A.; Aziz, K. M. A. Outmigration from a rural area of Bangladesh. *Rural Sociology*, vol. 37, no. 2, 1972, p. 236–245.

——; Chowdhury, M. A. *Fertility, infant mortality and family planning in rural Bangladesh*. Dacca, Oxford University Press, 1973, 154 p.

-——; Chowdhury, A. K. M. A. Pregnancy termination intervals in a rural area of Bangladesh. *Demography*, vol. 11, no. 2, 1974, p. 207–214.

Stott, D. H. Cultural and natural checks on population growth. In: Ashley-Montague, M. F. (ed.). *Culture and the evolution of man*, p. 355–376. New York, Oxford University Press, 376 p. Reprinted in Vayda, A.P. (ed.). *Environment and cultural behavior; ecological studies in cultural anthropology*, p. 90–120. Garden City, New York, American Museum of Natural History, Natural History Press, 1969, 485 p.

United Nations. *Proceedings of the world population conference* (Belgrade, 30 August–10 September 1965). New York, United Nations, Department of Economic and Social Affairs, 1966, vol. 1, 349 p.; vol. 2, 509 p.; vol. 3, 435 p.; vol. 4, 557 p.

——. Economic Commission for Asia and the Far East (ECAFE). *Directory of key personnel and periodicals in the field of population in the ECAFE region*. New York, United Nations, 1970, 124 p.

——. ——. *Research, teaching and training in demography; a directory of institutions in the ECAFE region*. New York, United Nations, 1972, 444 p. Supplement no. 1 and no. 2, 1974.

——. ESCAP. *Population periodicals. A directory of serial population publications in the ESCAP region*. New York, United Nations, Economic and Social Commission for Asia and the Pacific (Bangkok, Thailand), Asian Population Studies Series no. 17, 1974, 102 p.

——. Fund for Population Activities (UNFPA). *Inventory of population projects in developing countries around the world 1973/74*. New York, United Nations Fund for Population Activities, 1973–74, 397 p.

——. Relief Operation Dacca (UNROD). *Some social aspects of development planning in Bangladesh. Vol. 1, Population planning*. Dacca, UNROD, 1972, 30 p.

Vayda, A. P. Expansion and warfare among swidden agriculturalists. *American Anthropologist*, 63, 1961, p. 346–358.

——. (ed.). *Environment and cultural behavior; ecological studies in cultural anthropology*. Garden City, New York, American Museum of Natural History, Natural History Press, 1969, 485 p.

Ward, R. H. The genetic structure of a tribal population, the Yanomama Indians. *Annals of Human Genetics* (London), vol. 36, no. 1, 1972, p. 21–43.

Watts, E. S.; Johnston, F. E.; Lasker, G. (eds.). *Biosocial interrelations in population adaptation*. The Hague, Paris, Mouton, 1975, 412 p.

Weiss, K. M. *Demographic models for anthropology*. Memoirs of the Society for American Archeology, no. 27, 1973, 186 p.; issued as *American Antiquity*, vol. 38, no. 2, part 2, 1973.

Welbourn, H. F. The danger period during weaning. *Journal of Tropical Paediatrics and African Child Health*, vol. 1, no. 1, 1955, p. 34–36.

——. Bottle feeding: a problem of modern civilization. *Journal of Tropical Paediatrics and African Child Health*, vol. 3, no. 4, 1958, p. 157–166.

Welch, F. *Sex of children: prior uncertainty and subsequent fertility behavior*. Santa Monica, The Rand Corporation, R–1510–RF, 1974, 48 p.

White, B. Demand for labor and population growth in colonial Java. *Human Ecology*, vol. 1, no. 3, 1973, p. 217–236.

Wolf, E. Closed corporate communities in Meso-America and central Java. *Southwestern Journal of Anthropology*, vol. 13, no. 1, 1957, p. 1–18.

World Health Organization (WHO). *Bibliography on human reproduction, family planning and population dynamics. Annotated articles and unpublished work in the South-East Asia region (October 1974-January 1975)*. New Delhi, Regional Centre for Documentation on Human Reproduction, Family Planning and Population Dynamics, World Health Organization Regional Office for South-East Asia, 1975, 119 p.

Wrights, L.; Morrison, H.; Wong, K. F. *Vanishing world: the Ibans of Borneo*. New York, Tokyo, Hong Kong, Weatherhill/Serasia, 1972, 152 p.

Wynne-Edwards, V. C. *Animal dispersion in relation to social behavior*. New York, Hafner, 1962, 653 p.

Yankauer, A. An approach to the cultural base of infant mortality in India. *Population Review*, vol. 3, no. 2, 1959, p. 39–51.

Zinke, P. J.; Sabhasri, S.; Kunstadter, P. Soil fertility aspects of the Lua' forest fallow system of cultivation. In: Kunstadter, P.; Chapman, E. C.; Sabhasri, S. (eds.). *Farmers in the forest*. Honolulu, University Press of Hawaii, 1976, in press.

Appendix

The most comprehensive and thoroughly classified listing of published demographic studies is *Population Index*, published quarterly by the Office of Population Research, Princeton University. This contains an indexing by country or region as well as by subject matter and author. It annually lists current periodical demographic bibliographies and demographic periodicals. L'Institut national d'études démographiques (France) publishes *Population*, which includes summaries and reviews of world-wide literature as well as original articles. Specialized bibliographies and resource lists are published periodically by the Committee for International Coordination of National Research in Demography (CICRED, 1974a, 1974b), Carolina Population Center (Fogle, Gleiter and McIntyre, 1972; Gleiter, 1975), special or regional United Nations agencies (ECAFE, 1970,

1972; ESCAP, 1974; UNFPA, 1973–74; WHO, 1975) and OECD (Hankinson, 1973; OECD, 1973). These also contain guides to individuals and institutions working on demographic questions. General bibliographies on demography are now supplanted by annual bibliographies indexed under special topics such as environment, ecology and population (Goode, 1975), plus specialized bibliographies on such topics as population crowding, population education, migration, resettlement and urbanization in South-East Asia.

The largest international professional association of population scientists is the International Union for the Scientific Study of Population, whose directory lists members and their scientific activities and institutions (IUSSP, 1975).

16 Nutrition

Introduction

There are too insufficient sound scientific data to present an adequate account on the diet and nutritional status of most of the populations living in tropical forest ecosystems. It is not that nutrition has been deliberately ignored, but simply that it has been scientifically ignored for so long that the necessary information has never been collected and published. Therefore, there are serious deficiencies in the literature.

The objective of this chapter is to provide descriptions of the diets of peoples living in tropical forest ecosystems and, where possible, to assess their likely nutritional status. Important gaps in knowledge will be emphasized and the means of acquiring this necessary information suggested.

Information on individuals

Nutritional data which have been obtained for families or groups is of limited value. They only give some very general notions about the adequacy of diet. For example, the mean energy intake of a family in northern Thailand may be 1 700 kcal/head/day whereas in an area of the Ivory Coast it may be 2 400 kcal/head/day. The distribution of this intake and physical activity throughout family are unknown; the mothers or children might be malnourished. Assuming that the methodology is reasonable, the only conclusions to be drawn are that the mean intake in the Ivory Coast is greater than in northern Thailand and *probably* implies an adequate energy intake for the individual family members.

Without reliable nutritional data on individuals very little accurate scientific interpretation is possible. It is of little use knowing the different foods available in any region and their chemical composition without also knowing their relative consumption by individual people in the various communities. If careful selection is made from a population, it is much more valuable to have details of the diet of 20 adult men, or women, or 3-year-old children, than to have a general account of the diet of 500 families. A few such studies will allow a reasonably valid appraisal of the diet and, possibly, nutritional status of a region to be made.

The changing concepts of physiologically acceptable nutritional standards, particularly in the case of energy and protein intakes makes this more important. Nutritional threshold situations may be relatively common for a large proportion of people living in tropical forest regions;

ill-health or temporary food shortage may easily result in nutritional deficiencies. Therefore, considerable emphasis will be placed on such information when it is available. Accounts of social and cultural factors and of agricultural or food-gathering customs, will also be examined. The primary consideration is the energy and nutrient intake relative to the way of life of the individuals in their environment.

There is now a convincing amount of medical evidence showing that overeating is potentially dangerous; indeed the ideal diet may supply only the minimal necessary quantities of energy and fat, and possibly even of protein. With minerals and vitamins, the situation is less clear but there is little evidence for advantages of intakes larger than the minimal recommended levels. There is, of course, little likelihood that most people living in tropical and subtropical forest ecosystems will obtain anything other than the minimal recommended levels; increasing population, and the desire to enhance living standards ensure a greater demand for food. Therefore, unless there are clear-cut and persuasive arguments in favour, nutritionists should not suggest that more than the minimum will be better. Otherwise, the situation appears worse than it is and the chance of improvement becomes less likely. This is not to argue that nutritionists should not specify clearly those situations where the diet is inadequate. However, diet providing seemingly low intakes of energy and nutrients should not necessarily be assumed to be unsatisfactory unless accompanied by signs of inadequate nutritional status or of significantly and unequivocally improved physical health and performance as a result of dietary supplements.

Nutritional requirements

The use of tables of energy and nutrient requirements as standards of reference is fraught with problems. The standards, local or international, are open to various interpretations and much of the final assessment depends on the knowledge and prejudices of the assessor. The levels suggested are often far above the minimal. In the absence of clinical or biochemical indications or of diminished physical working capacity, it ought to be assumed that the intakes of energy and nutrients represent a state of balance, and that even peculiarly low (or high) values are possible in physiological circumstances. Such an interpretation, of course, puts a considerable onus on the reliability of the techniques which have been used in the investigation.

Energy and protein intakes and growth

One of the most difficult decisions on nutritional status, as it concerns energy-protein intakes, is whether the growth rate of the children is reasonable and, as a corollary, whether the heights and weights of the adults are also satisfactory. There are serious objections to the application of standards, such as the Harvard-Iowa growth charts, to populations with different environmental and genetic backgrounds; they may not be valid for all North American populations. Although FAO/WHO experts (FAO, 1975) and the Expert

Committee on Medical Assessment of Nutritional Status (WHO, 1963) both 'strongly suggest that whenever possible the standards should be assembled from statistically acceptable anthropometric data on children from healthy population groups within the country, whose growth has not been limited by environmental, socio-economic or nutritional constraints', the Harvard-Iowa standards are used because these data are seldom available. The inevitable result is that the nutritional status of the children is frequently diagnosed as poor, since their 'weights-for-age' will almost always be appreciably less than North American standards.

The problem is exceedingly difficult and complex. There is little doubt that almost any population of children if fed large quantities of a high energy and high protein from infancy will grow taller and heavier. The implication of extreme importance and debatable validity is that simply because children grow taller and heavier on high energy and high protein diets, this is a good thing. At the present stage of knowledge, it does not seem at all explicit that bigger is better. The dividing line between adequate nutrition and over-nutrition is not clear-cut. It is not certain that the increases in height and weight of children and, presumably, the increased height and weight of the adult population provides only benefits to the recipient. Metabolic and degenerative diseases seem to be much more prolific in wellnourished communities. There are few inherent disadvantages in being small in the physiological sense. Physical working capacity is little altered in absolute terms and, proportionally, is often higher in healthy populations of comparatively low body weight. Life expectation is higher and the incidence of most diseases is lower in thin healthy individuals.

Therefore, unambiguous concepts of adequate nutrition need to be emphasized. It ought to be repeatedly stated that health and the normal capacity to perform physical work are the main physical requirements of an adequate nutritional intake, and that these are largely independent of height and weight. There should be no acknowledgement of superior status conferred by extra height or weight. There might then eventually be a more ready acceptance that differences in body size are largely the result of genetic, environmental, and perhaps adaptational factors.

Methodology

In presenting data on nutritional surveys attempts will be made to assess the methodological accuracy, the possibility of interpretation for the individuals of the community, the representativeness of the sample surveyed, and the implications for the nutritional status of the people. The technique by which the data were collected is of considerable importance. Also, the calculation of the energy and nutrient value of the diet commonly makes use of tables of composition of foods. Much is often made of the fact that varying samples of any particular food have differing energy, fat, protein, mineral and vitamin contents, which will be influenced by the soil on which it was grown, its treatment before and during cooking, the season of the year, and the accuracy of the original chemical analyses. These are mostly undefinite factors which have not often been assessed. The errors involved in the use of food composition tables is sometimes quoted as ± 10–15 per cent, but studies have been reported (Consolazio *et al.*, 1955, 1956; Southgate and Durnin, 1970) where the calculations of the energy intake using tables of food composition were only about 1 per cent different from the net energy availability of the diet, measured by bomb calorimetry of replicate samples of the diet and the collection of energy losses in urine and faeces.

Common tropical foods

Table 1 summarizes the nutritional value of the common foods eaten in tropical regions.

Africa

In 1967, The Regional Food and Nutrition Commission for Africa of the Joint FAO/WHO/OAU-STRC reported that there was 'an overall shortage of protein' in the forest zone of western Africa and that 'distribution within the family is such that the adults often get just enough, whilst the children suffer from a severe deficiency. Children usually suffer also a calorie deficiency'. It suggested improving

protein availability by growing more protein-rich foods, such as beans and cereals like sorghum, maize or rice. However, it recognized the real problem was the storage of pulses, which are usually quickly attacked by storage pests. Increasing the number of crops per year, thus reducing storage time, may be one possibility. For cereals storage and continuous cultivation in forest areas are difficult.

Greater availability of animal protein would be a significant improvement to the diet but its implementation is problematical. Goats, to supply both milk and meat, and poultry, for meat and eggs, are theoretically capable of increases in production. The difficulties are that there may be taboos about the drinking of milk and eating of eggs and also that poultry compete with man for cereals. However, at the time of these recommendations, there was much more emphasis on increasing the supply of protein, in contrast to the current concept that the energy sources of the diet should be augmented.

The considerable range in protein-energy ratios in western African diets was examined by Annegers (1973). The protein-calorie percentage of starchy staples was at its lowest—2 to 5 per cent—throughout almost the whole of the tropical forest regions. The low protein staples were cassava, plantain, sweet potato, cocoyams (*Colocasia* and *Xanthosoma*) and yams, in contrast to the relatively high protein-calorie percentage of sorghum and millet, the staples of the

TABLE 1. Common foods and the nutritional value of major food types.

Group	Common name	Botanical name	General nutritional properties of the group
CEREALS AND GRAINS	barley	*Hordeum vulgare*	Most supply *ca.* 350 kcal and 11–12 g protein/100 g dry weight. Barley and Guinea corn have most protein, rice and some millets least. Ratio of protein: energy is sufficient to support child growth. Also supply thiamine, riboflavine, niacin (maize is low in niacin). Degree of milling affects vitamin content: polished rice has one quarter the thiamine of brown rice. Excellent staple foods.
	bulrush millet	*Pennisetum americanum* (= *P. typhoideum*)	
	cañihua	*Chenopodium pallidicaule*	
	findi, hungry rice	*Digitaria exilis*	
	finger millet, ragi	*Eleusine coracana*	
	glutinous rice	*Oryza glutinosa*	
	Guinea corn, jowar	*Sorghum* spp.	
	kodon millet	*Panicum miliere*	
	maize	*Zea mays*	
	quiñoa	*Chenopodium quiñoa*	
	rice	*Oryza sativa*	
	wheat	*Triticum vulgare*	
STEMS, ROOTS, TUBERS AND STARCHY FRUITS	banana	*Musa* spp.	Similar energy content to cereals on dry weight basis. Protein content of potatoes and yams just adequate for child growth; cassava and sago very low. Also supply ascorbic acid. Yellow yams, sweet potatoes, bananas, supply vitamin A. Low in B vitamins and minerals. Tendency for children to be underfed when weaned on to these foods unless fat- and protein-supplying relishes used.
	bread-fruit	*Artocarpus communis*	
	cassava, manioc	*Manihot utilissima*	
	cocoyams	*Colocasia* spp.	
		Xanthosoma spp.	
	oca	*Oxalis crenata*	
	plantain	*Musa* spp.	
	potato: Irish	*Solanum tuberosum*	
	Andean	*Solanum curtilobaum*	
		Solanum andigenum	
	sago palm	*Metroxylon sago*	
	sweet potato	*Ipomoea batatas*	
	ulluca	*Ullucus tuberosa*	
	yam: white	*Dioscorea rotundata*	
	yellow	*Dioscorea cayenensis*	

Group	Common name	Botanical name	General nutritional properties of the group
LEGUMES AND OIL SEEDS	bambara ground-nut	*Voandzeia subterranea*	
	black gram	*Phaseolus mungo*	
	broad bean	*Vicia faba*	
	chick-pea, Bengal gram	*Cicer arietinum*	
	coco-nut	*Cocos nucifera*	
	cow-pea	*Vigna unguiculata*	
	green gram, mung bean	*Phaseolus aureus*	Dried legumes contain about 320 kcal and 20–30 g protein per 100 g, plus iron, the B vitamins and in some cases (green, yellow or red coloured varieties), vitamin A. They are ideal supplements to a cereal- or root-based diet. Mixed legume and cereal proteins have as high a biological value as meat.
	ground-nut	*Arachis hypogaea*	
	haricot bean	*Phaseolus vulgaris*	
	jack bean	*Canavalia ensiformis*	
	lablab bean	*Dolichos lablab*	
	lentil	*Lens esculenta*	
	locust bean	*Parkia* spp.	
	oil palm	*Elaeis guineensis*	
	pea	*Pisum sativum*	
	pigeon-pea	*Cajanus cajan*	
	sesame	*Sesamum indicum*	
	soya-bean	*Glycine max*	
VEGETABLES	aubergine	*Solanum melongena*	
	cabbage	*Brassica oleracea*	
	carrots	*Daucus carota*	
	chili	*Capsicum* spp.	
	cucurbits (marrow, pumpkin, squash)	*Cucurbita* spp., etc.	This list is only a selection of vegetables.
	green leaves (cassava, pumpkin, spinach, water leaf, etc.)	Species of *Amaranthus, Beta, Bidens, Brassica, Luffa, Spinacia,* etc.	Besides providing much-needed variety and flavour, vegetables supply vitamin C, vitamin A (dark green, yellow and red varieties), iron (dark green varieties), folic acid and other B vitamins. Leafy vegetables contribute protein in small but sometimes important amounts. Their composition is variable, affected much by weather and soil conditions, etc.
	okra, gombo	*Hibiscus esculentus*	
	onions	*Allium* spp.	
	gourds	*Lagenaria* spp. and *Benincasa* spp.	
	peppers (green and red)	*Capsicum* spp.	
	sugar-cane	*Saccharum officinarum*	
	tomatoes	*Lycopersicum esculentum*	
	numerous herbs, spices and flavouring plants.		
FRUITS	baobab	*Andansonia digitata*	
	citrus (lemon, lime, orange, tangerine, etc.)	*Citrus* spp.	
	durian	*Durio zibethinus*	This list is also brief. Fruits contribute vitamin A (reddish-orange fleshed varieties), ascorbic acid (especially baobab, citrus and guavas), and variable amounts of sugar. Useful sources of water during field labour in hot climates.
	guava	*Psidium guajava*	
	mango	*Mangifera indica*	
	mangosteen	*Garcinia mangostana*	
	melon	*Cucumis* spp. and *Citrullus* spp.	
	pawpaw, papaya	*Carica papaya*	
	pineapple	*Ananas comosus*	
ANIMAL	milk		Source of energy, protein, vitamins A and D, calcium. Low in iron and B vitamins.
	eggs		Protein, energy, vitamins A, D and B group, iron (not well absorbed).
	meat, fish, shell-fish, insects, etc.		Sources of protein, iron (especially red meat), vitamins A, D and B group (especially offal) and calcium (fish eaten whole with bones). Energy content varies with fat content. Little carbohydrate or vitamin C.

TABLE 2. Protein-calorie percentage
of western African starchy staples.

Wheat flour (70 to 80 per cent extraction)	12.1
Sorghum (Guinea corn)	10.6
Millet	10.2
Fonio (*Digitaria exilis*)	7.4
Maize	7.3
Rice	7.2
Yams	6.0
Cocoyams (*Xanthosoma*)	4.5
(*Colocasia*)	4.4
Sweet potato	4.2
Plantain	3.1
Cassava	1.3

savanna (table 2). With the replacement of millet by cassava, as is happening in many areas of western Africa, the protein-calorie percentage is likely to fall further.

The supplementation of these staple starchy diets with the rich protein sources of pulses, and nuts and seeds, is very variable in different regions and seasons. In general, their consumption is less in western Africa than in other parts of the developing world, probably mostly because they do not have high yields in the humid forest zones. Throughout most of the forest area the total daily per capita protein consumption from pulses and other protein-rich seeds is less than 5 g. It is much higher in other tropical areas where it may reach 13–17 g/head/day (FAO, 1971). Protein from animal sources is, in general, very low—probably less than 10 g/head/day. In individual cases, the contribution of rodents, snails and insects to the diet may be significant but usually is only 1–2 g/head/day.

Throughout most of the region, the protein-calorie ratio of the total diet was between 6.5 to 10 per cent. Although Annegers (1973) puts some emphasis on the fact that this is very much lower than is supposed to be the case for most western diets and diets from the non-western world, in fact it is no less than has been found in generally adequately nourished populations in New Guinea (Norgan, Ferro-Luzzi and Durnin, 1974) and in Ethiopia. However, Annegers is convinced that protein-calorie malnutrition (kwashiorkor) is more common in regions with low protein- calorie ratios in their diet—although there is only scanty evidence for his statement in true forest populations.

Seasonal variations in food availability may be very important in leading to temporary nutritional deficiencies. In western Africa there are normally no shortages of food in the humid forest zones, although there may be seasonal food restrictions (Annegers, 1973). The staple food crops allow harvesting throughout most of the year and palm oil is often available. The development of cash crops such as cocoa, coffee and palm oil has caused some changes in staple food crops, but plantains, cocoyams, and cassava can be cultivated with tree crops (see chapter 20).

Périssé (1966) and Dupin (1968) have also drawn attention to the low protein-calorie ratio of the staple foods (cassava, plantain, etc.), and to the infrequent and low consumption of animal foods in French-speaking forest

areas. Périssé (1966) also emphasizes the generally low consumption of fats, even though palm oil, ground-nut oil and coco-nut oil are all produced in these areas. Green vegetables and fruit are capable of forming a larger part of the diet and, because of their mineral and vitamin content, would be nutritionally useful. However, in some cases, their increased consumption would increase the bulky diet, which may itself be limiting the food eaten particularly by children (Nicol, 1971). The low energy content and the low protein-calorie ratio of the bulky diet especially affect children at weaning (Dupin, 1959), and kwashiorkor is most virulent at this stage.

Money, derived from cash crops to purchase imported foods such as condensed milk, sugar, white rice, bread, tins of sardines, etc., plays a variable but increasing role in the subsistence economy. These foods may increase the total energy and protein content of the diet, but often they are not nutritionally superior to the traditional foods and simply enhance the carbohydrate consumption. The importance of alcoholic drinks in forest regions has rarely been studied in depth. Although palm wine is supposed to be consumed in moderate quantities (e.g. Nicol, 1959a) as much as 10 1/person/day may be taken (Périssé, 1966).

Endemic goitre, due to a toxic glucoside with antithyroid properties present in certain varieties of cassava, may be found in the forest regions (Masseyeff, 1955), although some forms of food preparation destroy these toxic products.

Périssé (1966) makes the very pertinent comment that the reliability of questionnaire methods (either coupled with food-weighing or used alone) depends more on the quality of the interviewer and on the questionnaire than on the ability of the consumer to reply to the questions. Table 3 is an abbreviation of some of his results for populations living in humid forest regions. Much useful information about cultural, social and economic influences on the food consumption is also included. Despite their general interest, little can be accurately deduced for individuals from them. Some of the results appear low. The energy intake in Cameroon, Nigeria and Ghana may represent inadequate energy intakes but much more information is needed to be sure that this is really the case. Many of the protein intakes are also low, but are potentially compatible with adequate intake. Fat intakes are uniformly low but this may not represent any health hazard. Calcium intakes are all low, by most standards, but calcium deficiency is usually uncommon in such populations. Iron intake is uniformly high enough to be capable of providing adequate intakes to everyone, but it may well be that certain individuals, such as pregnant women, are obtaining insufficient quantities. With the exception of Central African Empire, the intakes of vitamin A seem satisfactory whilst those of vitamins B_1 and B_2 appear to be low.

In studies repeated several times at different seasons on the same families living in forest areas in Cameroon, there were low energy intakes (1 400–1 750 kcal/head/day) with small seasonal variation ($< \pm 200$ kcal) and low protein intakes (31–41 g/head/day) with relatively greater variation (± 7–14 g) (Masseyeff, Cambon and Bergeret, 1958; Masseyeff, Pierme and Bergeret, 1958; Eyidi, Pierme and

TABLE 3. Family studies (intakes/head/day)

	Energy (kcal)	Protein (g)	Protein (%)**	Fat (g)	Ca (mg)	Fe (mg)	Vit. A (I.U.)	Vit. B₁ (mg)	Vit. B₂ (mg)
Ivory Coast	2 236	49	8.9	21	459	15	6 140	1.5	0.7
Ghana	1 833	45	9.7	36	434*	14	10 320	0.7	0.6
Togo	2 031	41	8.1	24	531*	13	4 850	1.4	0.6
Nigeria	1 785	46	10.3	19	420	17	5 000*	1.5	0.7
Cameroon	1 655	42	10.1	42*	455	10	7 900*	0.7	0.6
Gabon	1 892	73	15.4	—	—	—	—	—	—
Central African Empire	2 224	42*	7.6*	36	414*	11*	2 840	0.5	0.4*
Congo	2 043	46	9.0	32	361	12	7 750	0.8	0.6

* Large variability between results.
** Percentage of the total calories.

Masseyeff, 1961). These results indicate energy and protein deficiency in at least some individuals. These studies contain a great deal of useful background and dietary information.

A study of some Nigerian villages by Collis, Dema and Omololu (1962a, b) is also exceptionally informative. There is a useful account of the environment, land use, food sources, cropping practices, climate, etc. Food eaten by individuals was recorded at each meal time for seven consecutive days at quarterly intervals throughout one year. No precise information is given about the exact method of recording, who carried it out, and, most unfortunately, average consumption per-head basis were used to evaluate the nutritional qualities of the diets. Thus most of their valuable individual field information is not presented—even as means and standard deviations—so that their papers are of very much less value than they could have been. The general results of this otherwise interesting survey seem to correspond with previous conclusions about western African forest diets. The mean energy intake seems to be between 1 700 and 2 000 kcal/head/day; the protein, especially animal protein, intake is low; tubers contribute more than 50 per cent of the energy intake; and kwashiorkor occurs in the children, especially at *ca.* 2–3 years of age.

Nicol (1959a, b) presented the results of the weighed food consumption of people living in villages in seven different parts of Nigeria including one in the rain forest. In many ways, this was a most useful and interesting survey. The food consumption of individuals of different ages was measured at three different seasons. However, the precise methodology is not stated so that it is difficult to ascertain the accuracy of the food weights. The numbers of individuals were very small. Yams provided most of the diet. Palm wine was drunk by both men and women (670 ml/day and 280 ml/day respectively). A small amount of animal food, leafy vegetables, fruits, plantains and nuts completed the variety of food. Overall, the intake of energy and nutrients seemed moderately satisfactory. The men ('over 12 years') received 2 400 kcal/day, women 1 950, children, 4–6 years, 1 210 and, 7–9 years, 1 520 kcal/day; protein, minerals, vitamins A and of the B group were all present in apparently adequate amounts. No signs were seen of nutritional deficiency, other than a very small prevalence of 'protein malnutrition', the women apparently maintained

normal body weights, heights and skinfold thicknesses were compatible with reasonable nutritional status.

These results on individuals reinforce the deduction from the family surveys in other tropical forest areas of Africa that, in general, poor nutrition may not be a widespread and serious problem, but firm conclusions are not feasible on the limited data.

Asia

South Asia

Little detailed information is available on the peoples living within these forest regions. In general, the proportion of protein in the diet is low and that of carbohydrate is high, but there is considerable ethnic variation. Deforestation and degradation of the forests resulting from the incursions of the rapidly-expanding population of the majority groups of non-forest-dwellers causes variable degrees of stress on these peoples. Thus the sources of food available to the forest peoples, including wildlife, are being steadily reduced. Millets, coarse-grain crops, and—in some areas—rice, are the staples, but roots and tubers are also important items of the diet. Leaves, honey, fish, and the rare domestic animal, provide other sources of food. Beer, made from rice or millets, may be drunk in considerable quantities by both sexes. Distilled spirits have also become popular items for purchase.

Some most interesting studies were carried out on aboriginal tribes of the Abor Hills by Sen Gupta (1952, 1953, 1955). The raw foodstuffs of families from several villages were weighed before cooking. This five month study showed the economy was mostly a subsistence one, and the small amounts of money earned from selling commodities was spent on salt, cigarettes, beads, etc. Rice was the staple in the lower regions, but *annyat* (*Coix lachryma*) and *namdung* (*Perilla ocimoides*) were also eaten, especially in the hilly regions. Maize was cultivated but usually used as pig food, except in times of shortage. Millet was also grown, particularly to provide beer. Sweet potatoes and *Colocasia antiquorum* were extensively grown in the hills. Pumpkins, soya-beans, plantain flower, mushrooms and bamboo shoots were also eaten. Meat and fish were occasionally eaten but not milk or milk products. Fats, oils and sugar were almost never consumed. Oranges were the only fruits.

Food was usually prepared by boiling. Two meals a day were eaten, one before day-break and the other in the evening. Some food was usually carried to the fields and eaten around noon. *Apong* was drunk by both men and women, young and old, in place of water, as an alcohol, and for food; its importance was reflected in their saying 'Apong is not only drink but our food, our strength, our health and life: without food we can live but without Apong we cannot'. Its alcoholic content varied from 2 to 5 per cent by weight, and it contained small amounts of protein, carbohydrate, calcium, iron, thiamine and nicotinic acid.

TABLE 4. Maximum, minimum and average nutritive value per consumption unit per day of two family groups (I and II) belonging to aboriginal tribes of the Abor Hills, India (Sen Gupta, 1952).

| Family group | | Energy (kcal) | Protein | | Fat (g) | Calcium (g) | Iron (mg) | Vit. A and carotene (I.U.) | Thia-mine (mg) | Ribo-flavin (mg) | Niacin (mg) | Vit. C (mg) |
			Total (g)	Animal (g)								
I	Minimum	2 551	67.9	2.1	13.7	0.7	23.1	2 080	1.1	0.2	24.8	62.3
	Maximum	3 534	99.6	16.5	22.0	1.4	44.6	10 221	1.8	0.3	42.0	134.3
	Average	2 945	82.0	—	16.9	1.1	33.4	6 651	1.4	0.2	33.2	68.7
II	Minimum	2 112	62.1	Nil	10.9	0.3	31.6	1 623	0.9	0.1	20.3	16.3
	Maximum	3 348	88.7	9.5	18.6	1.2	36.4	9 188	1.9	0.3	40.7	70.2
	Average	2 726	73.0	—	14.8	0.9	29.1	4 943	1.4	0.2	32.2	48.0

The tribes were divided into two groups according to their degree of self-sufficiency. Table 4 shows the results from the poorest tribes who had 'food grains sufficient throughout the whole year' (I), and who 'had not sufficient food grains' (II). Even the minimum intakes of energy and protein of the poorer group of tribes (2 112 kcal and 62.1 g/head/day) do not appear very low. Indeed, the only nutrients which appear low enough to be potentially harmful are calcium, riboflavin and ascorbic acid. However, no mention is made by the author of clinical or other signs of deficiency. Body weight of the Abor children between 5 and 14 was almost 2 kg heavier than that of other Indian children.

Other tribes in the Abor hills give a wider but essentially similar picture. Cereals provided 97–99 per cent of the total energy of the diet, which reached mean levels of 2 748 kcal/head/day for group I villages and 2 516 kcal/head/day for group II villages, the lowest one of which had 2 140 kcal/head/day. On the other hand, 50 per cent of one village had less than 2 000 kcal/head/day—although 34 per cent of another village consumed more than 3 500 kcal/head/day, and 13 per cent more than 4 000 kcal/head/day. Intakes of the nutrients showed almost no low calcium intakes, but low vitamin A, riboflavin and vitamin C.

There is a great deal of useful information on foods and dietary customs, especially on the universality of the basic alcoholic beverage (Sen Gupta, 1956). These difficult and laboriously-effected studies are limited by the usual omissions: first, there is insufficient information about the environment and living conditions; secondly, there is very little information on their nutritional status; and thirdly, there is virtually no information on individuals. In one of Sen Gupta's papers on the Galong tribe (1955), the statement is made that 16 per cent were suffering from vitamin deficiencies, but these signs of deficiency have not been related in any way to the nutritional findings.

Further studies on forest peoples have been carried out by Sen Gupta *et al.* (1956a, b, 1960) in Travancore (Kerala State) and in the North-East Frontier, but little extra knowledge of the nutritional status of the people can be gathered from these family studies. There are indications that intakes of protein, iron, calcium, and vitamins A, C and several of the B group, might be deficient in some individuals. Sen Gupta and his colleagues (Sen Gupta, Rao and Biswas, 1961) are also responsible for a useful tabulation of the nutritive value of many foodstuffs eaten by the various tribes.

One standard of comparison which allows some assessment of a multiplicity of factors affecting nutritional status is the heights and weights of well-nourished children of the same or similar genetic stock—although, again, to be efficacious this requires more detail than just simply the mean height and weight of all children in a community of a certain age. Such a standard has been produced for Indian children (Raghavan, Singh and Swaminathan, 1971), although the method of selection of the children might have been more stringently controlled. By this standard, most of the Indian children measured by Sen Gupta and his colleagues were not well nourished.

Tribes living in the forest regions of central India have also been investigated; the young children by Rao and Satyanarayana (1974) and the adults by Pingle (1975a). The villages were scattered throughout Andhra Pradesh. Most are subsistence agriculturalists, who cultivate millets, rice and, to a small extent, pulses, roots and tubers; their diet is supplemented by fishing and collecting wild greens, fruits and nuts. A useful account of the relative importance of the cultivated and wild foods is given by Pingle (1975b). Heights, weights and age were recorded for 706 children (Rao and Satyanarayana, 1974), and a small number of adults (Pingle, 1975a). Clinical assessment for signs of nutritional deficiency were made on adults and children. The anthropometric measurements showed low levels of height and weight for age, and weight for height for the children and low heights and weights for the adults. Signs of vitamin and mineral deficiency were restricted to vitamin A in the adults and the vitamin B complex in a small percentage of children.

Unfortunately the methodology of the diet survey must be severely criticized, and the data are extremely imprecise. In general, it seems that there might be some protein, calcium, ascorbic acid and riboflavin deficiencies in some of these tribes.

Two studies have been published which incorporate some nutritional data on forest aboriginal peoples living on the Andaman and Nicobar Islands (Swaminathan, Krishnamurthi, Iyengar and Rao, 1971; Satyanarayana, Rao and Susheela, 1974). The sources of foods available are described; they include a considerable range of animal and sea-food, together with fruit, pandanus, roots and tubers, but no quantitative details are given of food intake, even on a family basis. Anaemia was said to be common, although haemoglobin levels usually averages 10 g or more/100 ml in females and about 13 g/100 ml in males. Little other signs of deficiency were present, apart from some related to vitamin A, but again weights and heights of the adult population were low.

A study which reports results which must be almost unique on tribal populations was also carried out on some Nicobarese on the Nicobar Islands (Roy and Roy, 1969). The people lived in coco-nut forests near the sea, and had a varied diet of coco-nuts, fish and other sea-food, bananas, taro, yams, papaya and fruits. Raw and cooked foods of families were weighed during 3 or 4 days, but the results are quoted per adult. The intakes averaged the high levels of 3 050 kcal, 130 g protein (of which 103 g was animal protein), 1 425 mg calcium and 70 mg iron.

It seems that there is no published work on nutritional status of forest peoples living in Sri Lanka.

The nutritional findings on several of the Indian forest peoples are summarized in tables 5 and 6. It must be reem-

TABLE 5. Protein and energy intakes of Indian forest peoples.

State	Tribe	Protein (g/day)	Energy (kcal/day)
Assam	Abors	69.5–84.7	2 430–2 550
Bihar	Santals	50.6	1 900
	Mal Paharia	51.3	1 850
	Sauria Paharia	74.2	2 210
Gujarat	Dublas	35.1	1 210
Kerala	Uralis and		
	Kanikkar	30.1–36.7	1 830–2 228
Maharashtra	Warlis	36.1	1 350
	Dublas	40.6	1 540
Tamil Nadu	Todas	48.4	2 410
	Kotas	50.3	1 870
	Irulas	26.5	1 310
	Kurumbas	36.5	1 690
Andaman and			
Nicobar Islands	Nicobarese	130.0	3 050
	Onges	136.5	2 620
Arunachal Pradesh	Mompas	53.3	2 530

TABLE 6. Common food-stuffs among Indian forest peoples.

State	Tribe	Main foods	Subsidiary foods
Assam	Abors	rice	animal foods, leafy vegetables, alcoholic drinks, fruit
Bihar	Santals	rice	millets, vegetables, pulses
	Mal Paharia	rice, millets	pulses, leafy vegetables
	Sauria Paharia	millet, rice, root vegetables	pulses
Gujarat	Dublas	millets	rice, fats and oils, fruit
Kerala	Uralis and Kanikkar	rice	pulses, root vegetables, fruit
Maharashtra	Warlis	rice	root vegetables, millets, sugar, pulses, animal foods
	Dublas	rice, root vegetables	millets
Arunachal Pradesh	Mompas	millets, rice, leafy vegetables	milk, pulses, alcoholic drinks

phasized that such tables do not show whether or not the people are adequately nourished and where the nourishment comes from. More detailed information should be provided about seasonal variation in food supply, the effects of cash cropping and the availability of money and the foods that may be purchased with it; and the many different types of food that may be consumed. The available evidence does not indicate if undernutrition exists, or to what extent it exists, which particular sex and age groups are mostly affected, whether the limitations in the types of food available are mainly responsible or whether it is simply a matter of quantity, what foods could best improve the diet, etc. The precise, useful, nutritional data are very sparse indeed, and it behoves nutritional investigators in the future to be very clear about the objectives of what is often a disruptive experience for the village.

An interesting investigation of the effect of nutritional supplementation of the diet on working efficiency has been carried out by Satyanarayana, Rao, Rao and Swaminathan (1972). 500 extra calories and 11 g of protein were provided daily to a group of miners and their coal production was compared with a control group. The supplemented group increased their weight but did not significantly alter their output. Such studies are notoriously difficult to control in relation to the many variables which may exert an influence, but an enhanced nutritional status may not produce immediately apparent benefits in working efficiency.

Another problem related to providing extra foods, particularly to children, concerns the intolerance to lactose which may be very common in India and throughout much of South-East Asia. Reddy and Pershad (1972) have shown that while lactose deficiency was present in all the adults and about 40 per cent of the children whom they studied, this did not imply an intolerance to milk, and that withholding skim milk supplements because of supposed lactose intolerance is not necessarily sensible.

South-East Asia

Although the South-East Asian/Melanesian forest ecosystem has an overall general nutritional style, it is composed of many distinctive fragments. It is possible to simplify a discussion of the region by considering Melanesia separately. The biogeographical line dividing South-East Asia from Melanesia is useful for the description of human biology. The Melanesian region contains principally Melanesian populations whose staple item of diet is usually some root crop or sago. They have not yet been exposed to a number of important parasitic and infectious diseases which are widely diffused in continental Asia. Grazing animals are either unknown or very recent introductions; the pig constitutes the principal domestic animal and can be regarded as a substitute for the grazing animals used by mankind elsewhere. In South-East Asia *sensu stricto*, cereal crops, particularly rice, are the basis of nutrition. Grazing animals are common and many wild mammals, which do not exist in Melanesia, are potential reservoir of pathogens.

In Indonesia, a swidden shifting agriculture within the tropical forest is dominant in the outer islands, and paddy rice in the inner islands (Geertz, 1969). To a substantial degree these two agricultural systems extend to the north and comprise most of South-East Asia. To the south and east paddy is not represented. The burgeoning population growth of the rich alluvial plains and valleys of South-East Asia appears to be associated with high soil fertility, an abundant supply of water and the existence of a storable cereal. Melanesia has not developed any comparable phenomenon, nor in fact have other extensive areas of South-East Asia which more closely emulate the Melanesian pattern.

Burma

Almost none of the nutritional reports refers to people living in the forest, but the study of *Energy expenditure in agricultural activities in Burma* by Nyunt-Khin, Hla-Win and Tin-May-Than (1968) might be applied to such people. The energy expended in carrying out different agricultural activities by *ca.* 6–13 individual men and women was measured by indirect calorimetry. The results seem roughly similar to those on healthy populations in other countries.

Between 5 and 10 per cent of a village adult population and 27 per cent of the children living in the Tenessarim coastal strip in the south of Burma were suffering from iron-deficiency anaemia, defined as a Hb level of less than 11 g/100 ml (Aung-Than-Batu, Hla-Pe, Thein-Than and Khin-Kyi-Nyunt, 1972). However, the iron intake appeared reasonable, and factors other than nutrition may be involved although the intake of iron was estimated from studies on a few nurses who may not be typical of the village population.

A nutritional project has been started in Burma (Durnin) and the first part of the project consists of collecting anthropometric and dietary data on a wide selection of populations, including forest people. In two or three years, a large amount of information on contrasting populations in Burma would be available and the comparative nutritional status of people living in the forest zones could be assessed.

Thailand

Many nutritional studies have been carried out and there is an excellent bibliography of 340 papers for 1950–1970 (Chapman, 1972). Nevertheless, the amount which is relevant to forest zones is somewhat small. Anaemia, vitamin B-group deficiencies, goitre, and urinary calculi seem to occur in various areas of Thailand (Halstead, Valyasevi and Umpaivit, 1967; Tanphaichitr, Vimokesant, Dhanamitta and Valyasevi, 1970; Pongpanich, Srikrikkrich, Dhanamitta and Valyasevi, 1974; Valyasevi, Benchakarn and Dhanamitta, 1974), but it is uncertain whether these nutritional disorders are found in forest peoples.

Malaysia

No precise nutritional data are known for Malaysian forest people. The diet of the forest-dwellers along the mountain ranges in the middle and northern parts of western Malaysia has cassava as the staple, supplemented by hill rice and maize and hunting. Lowland forest groups live in the mangrove fringe of the rivers of southern Malaysia, where the people are more settled and skilled cultivators using shifting agriculture and a little hunting. These aboriginal peoples are known collectively as the Orang Asli.

Seventy-five children aged from 5 to 10 years, from 6 villages in the deep forest of the main mountain ranges and 86 similar aged children from 6 villages along the Endau river in southern Malaysia were measured for height, weight and triceps skinfold (Robson, Bolton and Dugdale, 1973). The authors conclude that the 'Orang Asli children are at least as well nourished as town children', and that in those villages on the fringes of the forest where the traditional mode of living has been lost, there is a higher level of nutritional, parasitic and other diseases. There is little convincing supporting evidence for these judgments.

Anaemia has been reported in children (some of whom were aborigines), which was due to nutritional iron deficiency (Luan Eng and Virik, 1966). It is not known if this is a common deficiency.

Indonesia

Java contains *ca.* 2/3 of Indonesia's population, has 2/3 of its area as agricultural land and less than one quarter as forest. In the other islands, more than 3/4 is forest and just over 2 per cent is cultivated. The investigation of nutritional disorders has been actively undertaken since the 1880s. The first director of the Netherland's East Indian Medical Laboratory (1888–1938), Eijkman, was awarded a Nobel Prize for his discovery of the connection between beri-beri and a deficiency of what was later found to be thiamine.

From 1934 to 1945 an Institute of Nutrition Research carried out many extensive nutritional studies in Java, South Sumatra and West Ceram (Postmus and Van Veen, 1949). The main foods consumed were cassava, rice, corn, sweet potatoes, ground-nuts and soya-beans. The mean body weight of the male adult of West Java was 50.5 kg and 43.5 kg was the equivalent for the women. Nutritional deficiency diseases were similar to those of the present day, i.e. xerophthalmia due to vitamin A deficiency—particularly in preschoolchildren; kwashiorkor and marasmus because of 'protein-calorie' malnutrition; iron-deficiency anaemia and goitre. Most of the recent information comes from surveys in West Java and in the Gunung Kidul by Bailey (1961a, b, c; 1962a, b). Excellent reviews which put these and other studies into perspective have been compiled by Van Veen (1971) and Van Veen, Hong and Nio (1971). Hunger oedema in 1957–1959 was estimated to occur in *ca.* 2–5 per cent of the population and, in the cassava region, in >50 per cent lactating women.

Soekirman (1974) gives a useful summary of present nutritional knowledge with suggestions for improvement. Oomen (1969) has drawn attention to the complex nature of the reasons why vitamin A deficiency (xerophthalmia) occurs only in some children in families living in similar conditions. Soekirman (1974) mentions the study on anaemia in plantation workers by Basta and Churchil (1974), which purported to show a marked increase in work output when ferrous sulphate was given for 60 days to the men. However a control non-anaemic group receiving iron and a non-anaemic group receiving only a placebo also increased work output by almost 25 per cent. Many studies have been carried out in the past to try and relate the degree of mild anaemia to work capacity and to work output, without clear-cut conclusions. Anaemia may be a serious and common disability of mainly nutritional origin in Indonesia, but the evidence is not yet conclusive.

62–89 per cent young school village children in North Sumatra, West Sumatra, East Java and Bali have goitre (Nain, Sastroamidjojo, Sujardi, Halim and Maspaitella, 1972).

Preschoolchildren in Central Java were considered to be chronically undernourished because they were smaller in height and weight than well-nourished Jakarta children (Tie, Lian, Liong-Ong and Rose, 1967).

There is evidence that nutritional inadequacy may be wide-spread in certain parts of Indonesia. Mean intakes of energy as found by the rural surveys of Bailey (1961b) are between 1 320 and 1 390 kcal/head/day, whereas the results quoted by Postmus and Van Veen (1949) of surveys in 1938–39 show intakes of 2 000 kcal/head/day. If this represents a real decline in energy intake, then, together with the nutritional deficiency diseases mentioned above, it indicates a serious wide-spread problem. However, precise nutritional data acquired in the last 15–20 years are sparse. The surveys are household studies and data on individuals, especially children, are almost non-existent. It may be justifiable to consider almost all of Indonesia as coming from forest ecosystems, but full descriptions of the environment are seldom given and it is probable that often the data are not really relevant to these ecosystems. In northern Borneo, Wadsworth (1960) considered that deficiencies of vitamin A and of protein were 'potential dangers' but no nutritional studies had been done in that country.

Useful tables of the nutritive value of various foodstuffs in Indonesia have been compiled by Goan-Hong, Kam-Nio, Prawiranegara, Herlinda, Sihombing and Jus'at (1974a, b).

New Guinea

The island of New Guinea is perhaps uniquely relevant to an approach where the patterns of utilization of the forest may be observed within a remarkably limited geographical area. The nature of the human communities in many ways also typifies the situation over much of South-East Asia. In the interior, at altitudes below 1 000 m, there are societies which live within the confines of a completely forested environment—small scattered groups with a low population density who are dependent upon wild animals for food and use only such simple agricultural practices as planting wild sago. The bulk of the New Guinean population dwell on the slopes of the upland valleys of the central cordillera. They are rotational gardeners living in grasslands bordered by extensive rain forests. They have recourse to the forest as a source of fuel, housing material, clothing and wild produce, but their basic nutrition is derived from, and most of their lives are spent in, a region of degraded forest.

There are also very large areas in the lowlands, in proximity to the coast, the main river valleys and extending on to the lower slopes of the ranges, where life is governed by a complex system of rotational gardening. Land is covered by a patchwork of gardens, primary forest and various stages of forest regeneration. These are swidden agriculturalists whose general mode of life is found throughout large areas of South-East Asia.

At present, relatively few people live in areas where the forest has been replaced by coco-nut, cocoa and coffee plantations or in urban centres. It is these people who are at present exposed to the maximum pressures of change, particularly in regard to nutrition, psycho-social factors and exposure to unfamiliar pathogens.

Partly because New Guinea has a significance which spreads far beyond its own shores and partly because its mountains, rivers, forests, peoples and fauna provide such fascinating areas of study, it is a country which has attracted many different types of investigators whose statements have not always reflected an impartial scientific judgment. Thus a general review of human biology states that 'The dense population of the New Guinea Highlands exhibits to a marked degree the adaptive processes which man may need to effect in order to survive in a region where low protein diet prevails, and at the same time they exhibit all the phenomena which result from a failure of these adaptive processes. The diet is low in protein, providing 6.5% and 7.2% of the energy value of the diet of highland men, fat providing only 10% of the energy requirements, so that a caloric deficiency is also present to some degree. The people have adjusted to these circumstances by a complex series of physiological and socio-cultural processes which have ensured the preservation of physical fitness and the capacity to reproduce in what would otherwise be inhospitable terrain. Energy is also conserved by behavioural adaptations as well. Norgan *et al.* (1974) found that approximately 70% of the total day was spent in lying, sitting and standing, and that this accounted for 60% of the total energy expenditure. Walking occupied up to 10% of the 24 hours and between 20% and 27% of the total energy output. The people therefore were less active and consequently less demanding on their energy intake. Although these figures are evident to a marked degree among the highlanders, similar results were found among coastal Melanesians living on a root crop staple. It has been suggested that along with the adaptive processes many of the groups living on a staple root crop such as sweet potato may be adapted to their diet through unusual processes of intestinal function, including the fixation of nitrogen from free nitrogen into amino acids in the intestine. This work, which has important, far-reaching significance, has never been satisfactorily investigated'.

The impression given is very misleading. The body build of the average New Guinean is admirable—well muscled people of high average physical fitness (Cotes, Anderson and Patrick, 1974). Their energy intake, especially in the coastal areas, appears low but their protein intake is reasonable and their physical activity is moderately high.

The comprehensive monograph of Harrison and Walsh (1974) contains chapters on the geographical, historical and social background, demography, genetics, physical anthropology, epidemiology and health, nutrition, lung function and physical working capacity, and temperature regulation, of a coastal (Kaul) and a highland (Lufa) community. It is, in many ways, a model of how such an all-embracing investigation can be mounted.

The findings provide some controversial and interesting results of very wide nutritional significance (Norgan, Ferro-Luzzi and Durnin, 1974; Ferro-Luzzi, Norgan and Durnin, 1975). There has been considerable interest in New Guinean populations since the original report by Hipsley and Clements (1950) showed low intakes of energy, protein and other nutrients in adults who appeared physically well developed. Indeed, the intake of protein seemed so low that suggestions have been made that the New Guineans have adapted to this by developing nitrogen-fixing intestinal flora to supplement food nitrogen (Oomen, 1970; Oomen and

Corden, 1970). More than 200 adults and 500 children each had all the food eaten during 5–7 consecutive days periodically for almost a year weighed and recorded. The methodology was probably as accurate and meticulous as has ever been accomplished in a field study on such comparatively large numbers. All of the main items of food in the diet were analysed to determine the energy, protein, fat and carbohydrate composition. Energy expenditure was assessed by indirect calorimetric measurements on all of the daily activities of the adults combined with detailed recording of the timed pattern of activities throughout the day.

No evidence was found for any nutritional deficiency in the adults. Yet the intake of both energy and protein of the coastal people was low—1 940 kcal and 37 g protein per day for the men and 1 420 kcal and 25 g protein per day for the women (protein represented only 6–7 per cent of the total energy). Intakes were much higher for the highland men and women (tables 7 and 8), caused by the increased physical activity of the walking up and down hills necessitated by their way of life. The daily existence of even the coastal people was one of moderate activity. There therefore seems no doubt that these adults, the men weighing 56–57 kg and the women 47–48.5 kg, and of good physique, were able to maintain energy and protein balance on these apparently low but obviously adequate intakes.

The almost equal intakes of energy of the non-pregnant, the pregnant and the lactating women is also remarkable. However, this appears to be explicable on purely physiological grounds, mainly by reason of reduced physical activity during pregnancy and lactation; again there was no evidence of malnutrition, even after repeated pregnancies.

The New Guinean adult has therefore adapted in an acceptable physiological manner, with little or no undesirable consequences, to his environment. Whether he will adapt equally well to new town environments, with changed food habits and exposure to new diseases, is questionable.

Other sources of considerable information are the excellent critique of Oomen (1971) which has wide implications for all such societies, and the account by Luyken, Luyken-Koning and Pikaar (1964) of nitrogen balance studies on village children living in the central highlands of Irian Jaya. Even on the low protein intakes of the sweet potato diet of the highlands these children were in nitrogen balance, although when given 10 g daily protein supplement

TABLE 7. The mean daily energy and nutrient intake of the Kaul and Lufa men (Norgan, Ferro-Luzzi and Durnin, 1974).

| Community and age group (years) | Number | Energy | | Protein | | Animal protein | Carbo-hydrate | Fat |
		kcal	kcal/kg	g	g/kg	g	g	g
Kaul								
all	51	1 944	34.6	36.9	0.66	9.1	366	39
18–29	19	2 130	37.0	40.3	0.70	8.8	402	43
>30	32	1 833	33.2	34.9	0.63	9.2	344	37
Lufa								
all	43	2 523	44.2	47.1	0.82	9.5	529	29
18–29	28	2 478	42.6	48.6	0.83	11.1	507	33
>30	15	2 609	46.9	44.3	0.78	6.7	570	23

TABLE 8. The mean daily energy and nutrient intake of the Kaul and Lufa women (Norgan, Ferro-Luzzi and Durnin, 1974).

Community and age group	Number	Energy		Protein		Animal protein g	Carbo-hydrate g	Fat g
		kcal	kcal/kg	g	g/kg			
Kaul								
all	69	1 424	30.5	24.5	0.53	3.7	274	29
n.p.n.l.*	34	1 402	30.8	23.1	0.51	3.1	275	27
lactating, 0–1 year	13	1 412	28.9	24.1	0.51	3.0	262	33
lactating, >1 year	19	1 491	31.9	27.7	0.58	4.4	286	31
pregnant	9	1 414	27.5	25.4	0.49	3.6	261	34
18–29 non pregnant	29	1 424	29.4	23.8	0.50	3.5	277	28
>30 non pregnant	31	1 427	32.4	24.9	0.57	4.0	274	29
Lufa								
all	41	2 105	41.8	43.2	0.85	9.9	444	23
n.p.n.l.	14	2 068	39.8	43.6	0.83	10.4	445	17
lactating, 0–1 year	14	2 133	43.9	43.0	0.88	9.2	449	23
lactating, >1 year	6	2 247	46.8	39.6	0.81	8.2	470	28
pregnant	7	2 001	37.4	45.9	0.85	10.8	407	25
18–29 non pregnant	28	2 158	42.5	43.8	0.85	11.4	455	23
>30 non pregnant	6	1 977	48.8	37.3	0.83	3.4	433	18

* n.p.n.l.: non pregnant, non-lactating.

for 3 months they showed a gain in weight of 1–2 kg (no matched control group was able to be obtained so the exact significance of the weight gain is slightly uncertain).

Ferro-Luzzi, Norgan and Durnin (1975) also found a few signs of occasional deficiencies in energy and protein intake in children. The diet is thus probably marginally inadequate at some crucial periods of childhood and adolescence.

An interesting paper by Dornstreich and Morren (1974) has suggested that cannibalism in New Guinea has some nutritional value to the community, although the extent of the necessary consumption would seem to involve a some-what high rate of killing (5–10 adults being eaten per annum per group of 100 people).

The Philippines

There is no detailed information about the nutrition of people in the forest regions of the Philippines, but Schlegel and Guthrie (1973) describe important field studies on two Tiruray communities, which contrasted in exactly the manner which most needs investigation. The Figel still lived in the traditional manner in the forested mountains, engaging in swidden farming, hunting, fishing, and gathering, while the Kabakaba farmed in an area from which the forests had disappeared and where they were in contact with the cash and credit market economy typical of Filipino peasant life. Dietary data were collected from only 'one typical and randomly selected male from each community'. All the food eaten, meal by meal, was weighed and recorded for a year

and the authors seem reasonably certain that 'any foods overlooked would represent a very small portion of the total intake'. This type of study is one which is typical of exactly the sort of nutritional information that is so scarce in the scientific literature.

The forest community is typical of many such peoples who subsist by a mixture of slash-and-burn agriculture and hunting, fishing and gathering. The 'typical' male was Silu, aged 35 years, 168 cm tall and weighing 59 kg. He had access to a wide selection of plant foods and table 9 gives descriptions and uses of the foods, together with average-size portions and the number of times these were eaten during the year. The table also contains similar data for the typical Tiruray male Bekey, aged 54 years, 170 cm tall and weighing 64 kg. He worked his own assigned land, and grew a limited range of crops—rice, corn, tomatoes, and onions, the last three being largely sold in the market. His rice crop was consumed entirely by his own family. With the income, additional rice and other foods were bought. Bekey did virtually no hunting or fishing and very little gathering and his variety of diet was much more limited than Silu's.

The shift from swidden to plough farming has clearly resulted in major changes in diet pattern. However, not only was the pattern very different between the two men, the quantities of energy, of minerals and of vitamins were also markedly in contrast, and it is here that major criticisms must be made about this paper. The daily energy intake of Silu, living in the forest, is given as 1 176 kcal/day, and that of Bekey as 2 236 kcal/day. The latter result is low

TABLE 9. Foods of Tiruray (from Schlegel and Guthrie, 1973).

Name	Sources: Wild (W) Domesticated (D) or Purchased (P)		Use	Average size of single portion (g)	Portions throughout year	
	Figel (Traditional)	Kabakaba (Peasant)			Silu (Traditional)	Bekey (Peasant)
PLANT FOODS						
a bamboo	W	W	young shoots are eaten	200	2	8
urai	W	W	young leaves are eaten	30	22	34
squash	D	D	fruit is eaten	240	10	87
squash	D	D	young leaves are eaten	70	—	3
jackfruit	D	D	fruit is eaten	225	2	30
fishtail palm	W	W	palm heart is eaten	210	22	—
long beans, anapai, cow-peas	D	P	pods and seeds are eaten	80	17	39
a small palm tree	W	W	palm heart is eaten	200	7	—
garlic	D	P	used for flavouring only	trace	12	15
soursop	D	D	fruit is eaten	80	12	—
guava	D	D	fruit is eaten	50	9	—
lemon grass	D	D	used for flavouring only	trace	41	—
mung bean	D	P	beans (not pods) are eaten	100	22	122
pigeon-pea	D	D	beans (not pods) are eaten	100	1	—
common bean (white and red)	D	D	beans are eaten	100	4	—
coco-nut	D	D	soft white flesh from young nuts is eaten	45	26	28
coco-nut milk	D	D	flesh is grated and squeezed with water to yield an oily white milk which is used as a broth to cook vegetables	120	81	547
coconut oil	D	D	the milk (see previous entry) is boiled until it becomes oil and is used as a cooking oil	trace	26	14
pomelo	D	D	similar to, and is eaten like an orange	25	10	—
a fungus	W	W	flesh is boiled and eaten	20	11	—
a fern	W	W	fronds are eaten	15	4	—
rice	D	D/P	grain is boiled and eaten	425	321	730
snake gourd, ampalaya	D	D	fruit and leaves are eaten	10	—	4
wild ampalaya	W	W	fruit and leaves are eaten	10	2	—
a plant similar to taro	W	W	young leaves are eaten	10	3	—
ginger	D	P	used for flavouring only	trace	3	14
papaya	D	D	if mature, fruit is eaten; if young, is cooked and eaten as a vegetable	150	17	43
coffee	D	P	bean is boiled and drunk	200 ml	68	376
calamansi	D	D	juice is used as a spice	trace	8	—
a tree	D	D	the young leaves are eaten	15	4	—
corn	D	D	kernels are eaten as grits or 'on the cob'	250	320	68
tomato	D	D	fruit is eaten with other vegetables	5	12	646
wild aquatic morning glory	W	W	young leaves are eaten	70	8	54
a tree	W	W	fruit (very sour and green in-colour) is eaten	10	1	—
turmeric	W	D	used for flavouring only	trace	2	1
a small palm	W	W	the bud is eaten	25	5	—
jute	W	W	the young leaves are eaten	15	7	—
okra	D	D	fruit is eaten	60	—	4
common gourd	D	D	flesh is eaten	230	—	13
onion, leek, a legume	D	D	stalk is eaten	5	39	641
similar mung bean	W	W	pods and seeds are eaten	100	1	—
taro	D	P	young leaves are eaten	50	9	20
taro	D	P	tubers are eaten	70	6	2

Name	Sources: Wild (W) Domesticated (D) or Purchased (P)		Use	Average size of single portion (g)	Portions throughout year	
	Figel (Traditional)	Kabakaba (Peasant)			Silu (Traditional)	Bekey (Peasant)
sesame	D	D	seeds are toasted, powdered and eaten with rice	20	3	—
a herbaceous vine	D	D	the leaves are eaten	50	4	—
sugar-cane	D	D	peeled and chewed for the juices	40	113	—
peanut	D	D	nuts are shelled and eaten	60	1	—
mango	D	D	fruit is eaten	50	5	—
a tree	W	W	brown fruit (somewhat similar to an onion) is eaten	30	1	—
a rattan	W	W	the heart is eaten	30	15	—
a mushroom	W	W	flesh is boiled and eaten	50	4	15
pechay, mustard	D	P	the leaves are eaten	70	—	31
banana	D	D	the blossom is eaten	40	3	32
banana	D	D	the fruit is eaten	90	141	148
banana	D	D	the heart of the stalk is eaten	40	1	—
songe gourd, patola	D	D	the fruit and leaves are eaten	80	9	22
egg-plant	D	D	the fruit is eaten	85	127	115
chili pepper	D	P	the fruit and leaves are eaten	40	—	4
a tree	W	W	fruit (very similar to jackfruit) is eaten	50	1	—
cucumber	D	D	fruit is eaten	40	11	—
a tree	W	W	green coloured fruit is eaten	35	1	—
sweet potato	D	P	leaves are eaten	70	30	19
sweet potato	D	P	tubers are eaten	100	159	107
manioc	D	P	tubers are eaten	100	93	92
tugi yam	W	W	tubers are eaten	60	1	—
a tree	W	W	fruit (similar to jackfruit) is eaten	50	1	—
ANIMAL FOODS						
domestic pig	D	D	flesh is eaten	80	1	58
wild pig	W	W	flesh is eaten	80	29	—
a fish	W	W	flesh (lean) is eaten	15	1	—
python	W	W	flesh is eaten	40	1	—
goby fish	W	W	flesh (fatty) is eaten	30	12	—
mullet	W	W	flesh (fatty) is eaten	50	11	—
eel	W	W	flesh is eaten	50	123	—
sleeper fish	W	W	flesh (lean) is eaten	30	75	—
a bird	W	W	flesh is eaten	25	1	—
barb (fish)	W	W	flesh (fatty) is eaten	20	66	—
frog	W	W	legs are eaten	20	1	—
mountain bass	W	W	flesh (very fatty) is eaten	30	6	—
sailtail lizard	W	W	flesh is eaten	25	1	—
giant fruit bat	W	W	flesh is eaten	20	1	—
flathead goby fish	W	W	flesh (fatty) is eaten	25	2	—
freshwater cat-fish	W	P	flesh (fatty) is eaten	30	1	62
small fruit bat	W	W	flesh is eaten	25	3	—
owl	W	W	flesh is eaten	20	1	—
crab	W	W	flesh is eaten	15	4	—
shrimp	W	W	flesh is eaten	20	119	—
beetle larvae	W	W	larvae are boiled and eaten	15	2	—
goby (fish)	W	W	flesh (lean) is eaten	15	16	—
goby (fish)	W	W	flesh (lean) is eaten	15	2	—
a fish	W	W	flesh (fatty) is eaten	30	1	—
chicken (eggs)	D	D	eggs are boiled or fried and eaten	1 egg	18	7
chicken	D	D	flesh and viscera are eaten boiled or fried	60	27	42
wild chicken	W	W	flesh and viscera are eaten boiled or fried	40	21	—
sleeper (fish)	W	W	flesh (fatty) is eaten	50	75	—
honey-bee	W	W	honey is eaten	30	2	—
spotted pomadasid (fish)	W	W	flesh (fatty) is eaten	30	1	—

Name	Sources: Wild (W) Domesticated (D) or Purchased (P)		Use	Average size of single portion (g)	Portions throughout year	
	Figel (Traditional)	Kabakaba (Peasant)			Silu (Traditional)	Bekey (Peasant)
murrel (fish)	W	W	flesh (lean) is eaten	50	4	—
deer	W	W	flesh is eaten	80	3	—
a bird	W	W	flesh is eaten	5	1	—
water monitor (lizard)	W	W	flesh is eaten (mostly the tail)	60	4	—
a fish	W	W	flesh (lean) is eaten	20	11	—
honey-bee	W	W	honey is eaten	20	1	—
a bird	W	W	flesh is eaten	60	1	—
leather jacket (fish)	W	P	flesh (fatty) is eaten	50	32	32
a fish	W	W	flesh (lean) is eaten	15	1	—
a fish	W	W	flesh (lean) is eaten	20	18	—
monkey	W	W	flesh is eaten	60	12	—
a fish	W	W	flesh (lean) is eaten	50	4	—
a fish	W	W	flesh (fatty) is eaten	40	14	—
PROCESSED FOODS						
sugar	P	P		10	19	367
bagoong	P	P	very salty sauce of tiny fishes	10	—	23
vinegar	P	P		5	—	5
bread	P	P		20	3	10
cooking oil (canned)	P	P		60	—	18
dried fish	P	P		25	8	401
faniyalam	P	P	a kind of rice cake made of rice flour mixed with sugar and fried in coco-nut oil	60	4	1
milk	P	P	used in tea or coffee	30	6	1
corned beef (canned)	P	P		20	—	4
rice noodles	P	P		70	—	27
tea	P	P		250	2	1
sardines (canned)	P	P		15	2	62
squid (canned)	P	P		15	—	1
beer	P	P		350	—	1
suman	P	P	a glutinous rice cake mixed with sugar and salt	90	—	3
salt	P	P		5	648	788
tuba	P	P	a palm wine	1 500	1	1
dried shrimp	P	P		10	—	5

but possible and has certainly been matched in many comparable studies. The intake of 1 176 kcal/day seems impossibly low. If it were even remotely similar to the real intake, poor nutritional status, physical apathy and a varied assortment of other effects on the individual should have been obvious—and none of these seem to have been present.

The implications of this study are of some importance. From the purely nutritional point of view, it seems probable that the changed diet of the previously forest people has not necessarily deteriorated in its ability to supply the various nutrients in adequate quantity even though the variety of the diet has become much restricted. Indeed, the total intake may have considerably increased. Firm conclusions are difficult because of the unreliability of the assessed dietary intake of the forest people. Nutritional intake seems larger and adequate in the forest people which has changed to the peasant way of life compared to the people remaining in the forest.

Measurements of energy expenditure in the basic activities of lying, sitting, standing and walking of male and female Filipino adults were related to body size; they did not differ significantly from people of other ethnic groups (Florentino, 1966; Florentino, Guzman and Garcia, 1966).

Taiwan

Blackwell, Chow, Chinn, Blackwell and Hsu (1973) gave a daily supplement of 800 kcal and 40 g of milk protein, plus vitamins and minerals to half of a group of apparently undernourished village women whose original diet was estimated to provide about 2 000 kcal and 40 g of vegetable protein daily. The trial was a random double-blind design and results are reported on 216 women who had consumed the supplement and had produced 111 male and 105 female infants. The investigation was mainly concerned with assessing the effects of nutrition on birth weights and lengths. There was a significant but very small increase in weight and length of the male babies but not of the females. The authors conclude that their data 'do not demonstrate any remarkable benefit derived from the very substantial supplementation of the maternal diet". If diets are inadequate, they require improvement by supplementation, changed cultivation, or by economic means. These three approaches are very difficult, so the results may sometimes appear illogical or obscure.

Central and South America

Central America

The peoples who live in parts of central and southern American forests are likely to be subjected to appreciable stresses in the near future due to large-scale deforestation.

There is only patchy and inadequate nutritional information on these peoples. Nevertheless, certain areas have been among the most effectively and intensively investigated by the Institute of Nutrition of Central America and Panamá (INCAP). This large and extremely active institute has undertaken a great variety of studies involving biochemical and clinical measurements, anthropometry, fitness testing, dietary assessments and socio-economic surveys on well over 20 000 people. *Nutritional evaluation of the population of Central America and Panamá* (INCAP, 1971) summarizes its findings. Dietary studies were made on families and not on individuals by either a 24 h recall or a 3-day record, both of which can lead to considerable errors. However, some attempt was made to obtain information on the distribution of foods within the family by weighing individual portions on one day.

TABLE 10. Consumption of foods/person/day in rural areas of Central America and Panamá (1965–67), in grams or edible portion (INCAP, 1971).

	Guatemala	El Salvador	Honduras	Nicaragua	Costa Rica	Panamá
Number of families	203	293	331	355	456	361
Persons per family	6.5	6.0	7.1	6.3	6.9	6.3
Milk products (fluid equivalent)	84	190	194	243	193	73
Eggs	13	10	13	12	15	11
Meat, poultry, fish	44	37	41	58	40	90
Beans	54	59	56	72	57	20
Vegetables	66	53	51	27	66	25
Fruit	14	17	40	41	7	50
Bananas and plantains	20	16	43	72	47	99
Starchy roots and tubers	14	13	22	33	46	82
Cereal products (as grain, meal)	412	411	276	240	199	261
maize (subtotal as grain)	359	352	224	139	41	32
tortillas and tamales	544	533	340	190	62	6
degerminated maize	0	0	0	0	0	29
toasted meal (pinol)	0	0	0	14	0	0
rice	16	27	29	54	100	186
wheat bread	36	26	12	28	54	37
wheat flour and pastes	4	0	8	7	12	10
other cereals	2	6	5	16	0	0
Sugar	52	41	39	58	89	51
Fats and oil	4	15	16	19	19	26

Nutrition, as it is influenced by agriculture, availability of foods, socio-economic status and intestinal infections, as it affects growth, physical fitness, working capacity, and as it leads to different types of deficiency diseases, is dealt with in a manner so exhaustive and so well-coordinated as to be unparalleled anywhere in the world. This fact is the more remarkable in that it required constant co-operation between Costa Rica, El Salvador, Guatemala, Honduras, Nicaragua and Panamá. But, it is not possible to differentiate the types of rural communities and to determine whether these results are relevant to forest ecosystems. The significance of the data for young children is often stressed in the report and since this is where precise knowledge is required, it is regrettable that more reliable techniques were not used.

In relation to the principal foods consumed, there is some variation between the six countries (table 10). While cereals are the mainstay of all the diets, they vary in kind and amount from Guatemala in the north to Panamá in the south. Maize is the principal cereal in rural El Salvador, Guatemala, Honduras and Nicaragua, but decreases progressively to small amounts in Costa Rica and Panamá. It is more commonly used by rural than by urban families in all countries. In Guatemala, where the Indian proportion in the population is high, all the rural families in the survey consumed maize tortillas. As maize consumption declines, rice increases, and is the staple cereal in Costa Rica and Panamá. In all the countries, it is consumed more by urban than by rural families. Only in Panamá, do the rural families eat more rice than the urban. In rural El Salvador, Guatemala and Honduras, rice is commonly used as a thickener in soups but combined with beans or some other food, it becomes a main dish in Costa Rica and Panamá.

In Panamá it is the custom to wash rice vigorously before cooking to remove all adhering starch. This also removes a significant amount of the water-soluble vitamins of the grain, a fact reflected in the nutritive content of the diet.

Wheat is not a staple food in any Central American country.

In many coastal or hot lowland areas, cassava and other starchy roots form an important part of the diet, supplementing or sometimes replacing cereals. Bananas and plantains also often take the place of cereals. The plantains are cooked or fried green and are common in areas where they are easily produced.

The greatest sugar consumer is Costa Rica: 89 g of sugar/person/day (mostly panela) are used in the rural area, compared to 77 g (mostly white) in the urban. The flavour of the brown panela is highly esteemed and sugar water is drunk at meals and in between.

Beans (*Phaseolus vulgaris*) are as characteristic of Central America as maize tortillas. Their consumption is somewhat greater in rural than in urban areas of all countries, and is lowest in Panamá where maize tortillas are also scarce. Black beans are preferred in Guatemala but red ones in Honduras and the red beans, are, in general, more popular in the countries to the south. Although of no known significance, the preferences shown for the different colours are strong. Beans are cooked with onion, epazote (goosefoot) or coriander leaf for flavour, and lard is often added. Because of the long cooking time, enough for two or more days are sometimes cooked at one time. In some areas it is customary to strain the cooked beans to remove the hulls, especially as food for small children. In rural Guatemala, beans and sometimes uncultivated greens form the main dish of the meal, except on occasions when meat is available. In Costa Rica, beans with rice are a common breakfast dish, a combination also common in Panamá, but not in countries to the north.

Both vegetables and fruit are used in larger amounts by urban than by rural families. Vegetables are usually expensive. Many rural families grow them to sell in towns. The vegetables most commonly sold and less often eaten by rural families include lettuce, peas, cabbage, carrots, spinach, beets, etc. The Indians of Guatemala eat the green leaves of a variety of wild plants no longer eaten in most other parts of Central America. Most cultivated vegetables

TABLE 11. Average intake of calories and nutrients/person/day by families in rural areas of Central America and Panamá (INCAP, 1971).

		Guatemala	El Salvador	Honduras	Nicaragua	Costa Rica	Panamá
Energy	kcal	2 117	2 146	1 832	1 986	1 894	2 089
Total protein	g	68.0	67.9	58.0	64.4	53.6	60.1
Animal protein	g	15.4	17.3	18.5	23.6	18.5	26.6
Fat	g	31.4	39.3	44.1	47.5	43.9	49.8
Carbohydrate	g	411	396	315	338	332	357
Calcium	mg	1 100	1 092	883	763	580	301
Iron	mg	17.9	11.6	15.5	18.2	15.4	14.3
Vitamin A	I.U.	2 420	893	1 280	1 693	1 796	1 826
Thiamine	mg	1.3	1.06	0.89	0.86	0.76	0.92
Riboflavin	mg	0.8	0.78	0.79	0.93	0.84	0.69
Niacin	mg	16.6	12.0	10.3	10.7	10.7	14.3
Ascorbic acid	mg	38	36	59	66	52	87

eaten by rural families are indigenous and include tomatoes, peppers, squash, chayote (*Secchium edulis*), avocado. Tomatoes and onions, even though in minute amounts, are used by all classes in all areas to flavour stews and other dishes.

Oranges and bananas are the most common fruit consumed and are available for much of the year; mangoes, pineapple and papaya have shorter seasons. Urban families in almost all the countries consume more than twice as many vegetables and fruit as rural families.

Most rural families raise chickens but seldom eat them. Fish is eaten mainly on the coasts and islands, although dried, smoked and salted fish finds its way to interior markets to some extent, especially at Easter. Other marine products, for example, large sea-turtles, are seen in some coastal markets, in Costa Rica and elsewhere. Despite the long coast-line, fish is not an important food in most of Central America. In many countries, eggs are the most expensive source of protein and they are not widely eaten especially in the rural areas where they are produced. Many rural housewives trade them. They may become cheaper as a result of the poultry improvement projects.

Like the other expensive foods, fats are less consumed in rural than urban areas. Fat consumption increases from 4 g/person/day in rural Guatemala to 41 g in urban Costa Rica. The most common fat in rural areas is lard and in the cities, vegetable oil (80 per cent in Panamá City). Some Indian families cook without fat, believing that it is harmful.

These foods do not appear representative of the diets of most other forest communities and it seems clear that they are not typical of forest-living peoples. They may, none the less, be similar to the diets of such people when they live in villages or small towns, still within reach of the forest, but with access to a peasant market economy.

The results are based on surveys such as that on three Maya Indian communities living in highlands of Guatemala and described by Flores, Garcia, Flores and Lara (1964). These communities were studied annually over four years and their food consumption remained remarkably constant during this time. Intakes of energy were about 2 000 kcal/head/day and protein between 50–60 g/head/day. In general, all the requirements for vitamins and minerals seemed adequately met.

The INCAP (1971) report tabulates the dietary intakes found in the rural areas (table 11). The energy and protein intakes, on average, are adequate and indeed clinical signs of protein-energy deficiency had a relatively low prevalence. However, calcium intake in Panamá was low, probably due to the low consumption of maize. Iron intake was uniformly adequate although iron-deficiency anaemia was common in some areas. Of the vitamins, dietary (and biochemical) indications of marked deficiency in vitamin A were prevalent but clinical signs of deficiency were rare. Riboflavin intake was inadequate; biochemical and clinical assessments indicated serious riboflavin deficiency. Iodine deficiency, with resultant goitre, is present in all the Central American countries except Guatemala since 1956—due to the introduction of salt iodization; the prevalence of goitre in this country dropped from 38 per cent of the population in the late 1960s to 5 per cent.

Thus, while the nutritional state of rural people appears to be reasonable, deficiencies may be present in a considerable proportion of the population, but without knowledge of individuals within the family, it is difficult to be sure of the most effective remedy.

Viteri and Torun (1975) gave a group of 18 male agricultural workers a protein-energy supplement over three years. Their intake was 3 555 kcal/day and 107 g/day of protein. They were compared to 18 workers from a poor community whose intakes were 2 693 kcal/day and 82 g/day of protein. The supplemented group had a higher fat-free mass in the body, a higher fat mass, a higher maximal oxygen consumption (though not when assessed as ml/kg body weight), and apparently worked harder and accomplished given tasks with less fatigue in a shorter time. If these findings are valid, they are of wide significance and they are worthy of being repeated by other investigators, but it is difficult to see any physiological reason why supplementing the food intake of men already receiving 3 555 kcal/day and 107 g/day of protein should have any effect.

South America

Although several nutritional studies have been reported on populations living within the Amazonian forests, the data are usually difficult to assimilate in the specific context of this chapter. In Peru, Huenemann and Collozos (1954) described studies on young children, some of whom lived in a river town in the Amazonian forest. They found low growth rates and deficiencies of protein and vitamin A. Subsequent surveys have been undertaken by the Nutrition Institute of Lima. While gross nutritional deficiency seemed uncommon, some degree of malnutrition was fairly prevalent and many of the rural populations must have had diets which were only marginally adequate.

Boza and Baumgartner (1962) present a great deal of information on foods, dietary habits, social customs and living conditions of five groups of people inhabiting the Orinoco basin in southern Venezuela. Data are given of height and weight, but not on nutritional intake, although

the authors state that 'a clinical examination revealed that some 24% of these Indians show some signs which may be related to nutritional deficiencies'.

Barron, Carbajal and Garcia (1969) also described an investigation in the Amazonian region of Colombia, but only some of the data apply to forest peoples. Energy intake was 1 977 kcal/day, protein 53 g/day, and mineral and vitamin intake was, on average, adequate.

Conclusions. Research needs and priorities

A critical examination of the nutritional literature relevant to tropical forest ecosystems first reveals the enormous gaps in knowledge—whole countries, with virtually nothing precise known about the nutritional state of the rural inhabitants, and secondly, the unsatisfactory methodology and the inadequate nature of the results obtained because they are quoted in a manner which does not allow the reader to form his own conclusion—for example, dietary intakes being quoted as percentages of requirement only. There may be no indication of the variability in the results, either in absolute terms or standard deviations, or as related to body size. Anthropometric information is seldom adequate. A satisfactory description of the environment and the socioeconomic and working conditions of the population is uncommon.

The method for measuring food intake needs critical consideration. Presumably it was thought that if some families showed, for example, energy intakes of 2 000 kcal/head/day when the recommended level was 3 000 kcal/head/day, this would indicate nutritional deficiency. The position is seldom so simple. Intakes of energy and nutrients are often near the border of adequacy, and the border is not a sharp demarcation but a wide area which may vary with season, work situations, environmental temperatures, infections and infestations, and other smaller factors. The methodology needs to be chosen with care and the ultimate aim must be constantly in mind. The food intake of households has a limited interest and use. The food intake of individuals in a family is a little more useful. The nutritional state of the population—or any section of it, such as pregnant or lactating women, children, adult male agricultural workers—can only be determined by data on individuals, including their height, weight and body fat, physique and preferably simple measurements of their excercise capacity, clinical signs of nutritional deficiency, and background environmental and social information. With this sort of knowledge, a report on 50 representative statistically chosen individuals would provide more useful data than has often been obtained from 500 families.

Malnutrition undoubtedly exists in populations living in relation to the tropical and subtropical forest ecosystems. Even when it is not superficially obvious, it may be ready to appear under stress. It is difficult to detect and Lörstad (1974) has written a thoughtful and original account of the statistical problems of its determination. Specific deficiencies of wide importance seem to be restricted largely to the energy-protein content of the diet, vitamin A and iodine. Iron deficiency also may be prevalent, but the significance to health and working capacity of the resultant anaemia is unclear, unless the anaemia is severe. However, the existence of suspected nutritional deficiency in a population needs to be investigated and this should be treated as a matter of urgency. Situations which can be remedied should be remedied, though this is often difficult even if the remedies are to hand. But resources are limited in the nutritional field, remedies are scarce, and there is no point in wasting time, facilities, effort and money unless these are really required.

The most immediate items to assess in the diet of a community is whether the amount of energy and protein are adequate. The assessment, especially for energy, is complicated because the signs of insufficient dietary energy may not be obvious and may involve some subjective judgments. Growth rates of children may give some clues; but they should be used with care, the standards for comparison should be sensible ones, and some clear evidence as to disadvantages associated with slower growth rates should be produced. Too many unwarranted assumptions are made about the desirability of rapid growth in children without the precise short and long-term advantages being assessed. Judgment of the sufficiency of dietary energy can also be made by measuring exercise capacity, or by introducing dietary supplements and critically assess their benefits.

Protein intake should almost always be studied in conjunction with energy. Calloway (1975) has shown how the energy content of the diet influences nitrogen balance to a greater extent than does protein intake in the marginally adequate ranges of intake. Subjective impressions about the desirability of animal protein, etc., should be avoided. People can be healthy on a diet with no animal protein and where protein constitutes only 6–8 per cent of the energy.

Nutritional surveys should include a careful search for signs of nutritional deficiency.

Many people living in or near forest regions, or in areas recently deforested, may have their diet modified or even radically altered, their living conditions changed with either improvement or worsening hygiene, and be exposed to different infections, etc. These factors have been mentioned in several reports but they are still very inadequately understood and they require considerably extended investigation.

Food taboos have not been mentioned, mainly because they rarely obtrude in nutritional reports. However, their presence may be important, although the evidence for this is scanty. They should always be looked for; the whole question of taboos and dietary likes and dislikes receives insufficient attention. Changing life-style when it affects the amount and type of physical work, and patterns of activity in leisure, may have nutritional importance. Medically, in relation to the incidence and prevalence of metabolic and degenerative disease, these are interesting problems. However, they are probably peripheral to the main requirement of obtaining a greater volume of basic information on the primary communities in or near the forest ecosystem.

For most countries which have peoples living in tropical or subtropical forest ecosystems, the urgent overriding necessity is to collect more data on their nutritional state. Surveys require little apparatus and are not expensive. Their results are a prime requirement for the proper formulation of development programmes. The investigations require careful planning, clear aims and interdisciplinary consultation. If communities or sections of them are suffering from nutritional deficiencies, action can be planned. This may involve agricultural improvements, extra employment, new crops, food supplementation, subsidized foods, etc. These assistance programmes are likely to succeed only with the most careful planning and accurate nutritional information. *Such studies form probably the most important and immediately useful field of nutritional research at the present day.*

Bibliography

ANNEGERS, J. F. The protein-calorie ratio of West African diets and their relationship to protein calorie malnutrition. *Ecol. Food Nutr.*, 2, 1973, p. 225–235.

——. Seasonal food shortages in West Africa. *Ecol. Food Nutr.*, 2, 1973, p. 251–257.

AUNG-THAN-BATU; HLA-PE; THEIN-THAN; KHIN-KYI-NYUNT. Iron deficiency in Burmese population groups. *Amer. J. Clin. Nutr.*, 25, 1972, p. 210–218.

BAILEY, K. V. Rural nutrition studies in Indonesia: background to nutritional problems in the cassava areas. *Trop. Geogr. Med.*, 13, 1961a, p. 216–233.

——. Rural nutrition studies in Indonesia: clinical studies of hunger oedema in the cassava areas. *Trop. Geogr. Med.*, 13, 1961b, p. 234–288.

——. Rural nutrition studies in Indonesia: epidemiology of hunger oedema in the cassava areas. *Trop. Geogr. Med.*, 13, 1961c, p. 289–302.

——. Rural nutrition studies in Indonesia: oedema in lactating women in the cassava areas. *Trop. Geogr. Med.*, 14, 1962a, p. 11–19.

BAILEY, K. V. Rural nutrition studies in Indonesia: the Gunung Kidul problem in perspective. *Trop. Geogr. Med.*, 14, 1962b, p. 238–258.

BARRON, A. G.; CARBAJAL, C. P.; GARCIA, J. D. Evaluación nutricional de la zona de influencia de la carretera marginal. *An. del Prog. Acad. de Med.*, 52, 1969, p. 101–120.

BASTA, S. S.; CHURCHIL, A. *Iron deficiency and the productivity of adult males in Indonesia.* IBRD/World Bank Staff Working Paper no. 175, 1974.

BERGERET, B. Note préliminaire sur l'étude du vin de palme au Cameroun. *Médecine tropicale*, vol. 17, n° 6, 1957, p. 901–904.

BLACKWELL, R. Q.; CHOW, B. F.; CHINN, K. S. K.; BLACKWELL, B. N.; HSU, S. C. Prospective maternal nutrition study in Taiwan: rationale, study design, feasibility, and preliminary findings. *Nutrition Reports International*, 7, 1973, p. 517–532.

BOZA, F. V.; BAUMGARTNER, J. Estudio general, clínico y nutricional en tribus indígenas del Territorio Federal Amazonas de Venezuela. *Arch. Ven. de Nutr.*, 12, 1962, p. 143–225.

CALLOWAY, D. H. Nitrogen balance of men with marginal intakes of protein and energy. *J. Nutr.*, 105, 1975, p. 914–923.

CHAPMAN, V. *Food and nutrition in Thailand: medical, social and technological aspects.* WHO, New Delhi Regional Office, 1972.

COLLIS, W. R. F.; DEMA, J.; OMOLOLU, A. On the ecology of child health and nutrition in Nigerian villages. I. Environment, population and resources. *Trop. Geogr. Med.*, 14, 1962a, p. 140–163.

——; ——. On the ecology of child nutrition and health in Nigerian villages. II. Dietary and medical surveys. *Trop. Geogr. Med.*, 14, 1962b, p. 201–229.

CONSOLAZIO, C. F.; HAWKINS, J.; BERGER, F.; JOHNSON, O.; KATZENEK, B.; SKALA, J. *Nutrition surveys of two consecutive training cycles of the airborne training Bn. Company 'G' Fort Benning, GA, Oct.-Nov. 1953.* U.S. Army Medical Nutrition Laboratory, Report no. 166, 1955, 47 p.

——; ——; JOHNSON, O. C.; RYER, R. III; FARLEY, J. E.; SAUER, F.; FRIEDEMANN, T. E. *Nutrition surveys at five army camps in various areas of the United States.* U.S. Army Medical Nutrition Laboratory, Report no. 187, 1956, 64 p.

COTES, J. E.; ANDERSON, H. R.; PATRICK, J. M. Lung function and the response to exercise in New Guineans: role of genetic and environmental factors. *Phil. Trans. Roy. Soc. London*, B, 268, 1974, p. 349–361.

DORNSTREICH, M. D.; MORREN, G. E. B. Does New Guinean cannibalism have nutritional value? *Human Ecology*, 2, 1974, p. 1–12.

DUPIN, H. L'alimentation traditionnelle du jeune enfant dans l'Ouest africain. Déficiences, possibilités de supplémentation. *Diététique et Nutrition*, 1, 1959, p. 33–40.

——. L'alimentation, l'état de nutrition et les tendances actuelles de la consommation alimentaire en Afrique intertropicale. *Développement et Civilisation* (Paris), 35, 1968, p. 21–30.

——. Les enquêtes nutritionnelles dans les pays en voie de développement. Intérêt, difficultés, limites. *Rev. Hyg. et Méd. soc.*, vol. 17, n° 3, 1969, p. 223–238.

——; KITAN, Y. Nutrition et travail. Alimentation des travailleurs dans les pays tropicaux. *Afr. méd.*, vol. 11, n° 97, 1972, p. 121–132.

DURNIN, J. V. G. A. *Protein requirements and physical activity.* Warsaw, Sport Wyczynowy, 1976, in press.

EYIDI, B.; PIERME, M. L.; MASSEYEFF, R. Une enquête sur l'alimentation à Douala. *Recherches et Études camerounaises* (Paris), 5, 1961, p. 3–45.

FAO/WHO/OAU-STRC. *Food and nutrition in Africa.* News Bulletin of the Joint Regional Food and Nutrition Commission for Africa, 1967.

FERRO-LUZZI, A.; NORGAN, N. G.; DURNIN, J. V. G. A. Food intake, its relationship to body weight and age, and its apparent nutritional adequacy in New Guinean children. *Amer. J. Clin. Nutr.*, 28, 1975, p. 1443–1453.

FLORENTINO, R. F. Energy requirements of Filipinos. *Philipp. J. Nutr.*, vol. 19, no. 1, 1966, p. 50–71.

——; GUZMAN, P. E. de; GARCIA, L. P. The energy cost of basic activities in some Filipinos. *Philipp. J. Nutr.*, vol. 19, no. 4, 1966, p. 258–271.

FLORES, M.; GARCIA, B; FLORES, Z.; LARA, M. Y. Annual patterns of family and children's diet in three Guatemalan Indian communities. *Br. J. Nutr.*, 18, 1964, p. 281–293.

Food and Agriculture Organization. *Agricultural production yearbook.* Rome, FAO, 1971.

——. Energy and protein requirements. *Food and Nutrition*, vol. 1, no. 2, 1975, p. 11–19.

FYOT, R. La valorisation industrielle d'une boisson traditionnelle: le vin de palme pasteurisé. *Technique et Développement*, 8, 1973, p. 27–29.

GEERTZ, C. Two types of ecosystems. In: VAYDA, A. P. (ed.). *Environment and cultural behavior. Ecological studies in cultural anthropology*, p. 3–28. New York, Natural History Press, 1969, 485 p.

GOAN-HONG, L.; KAM-NIO, O.; PRAWIRANEGARA, D. D.; HERLINDA, J.; SIHOMBING, G.; JUS'AT, I. Nutritive value of various legumes used in the Indonesian diet. In: *First Asian Workshop on Grain Legumes* (Bogor, Indonesia), 1974a.

——; ——; ——; ——; ——. Available sources of food in Indonesia. In: *Third National Pediatric Congress* (Surabaya, Indonesia), 1974b.

GRANDE, F.; ANDERSON, J. T.; KEYS, A. Changes of basal metabolic rate in man in semi-starvation and refeeding. *J. Appl. Physiol.*, 12, 1958, p. 230–238.

HALSTEAD, S. B.; VALYASEVI, A.; UMPAIVIT, P. Studies of bladder stone disease in Thailand. V. Dietary habits and disease prevalence. *Amer. J. Clin. Nutr.*, 20, 1967, p. 1352–1361.

HARRISON, G. A.; WALSH, R. J. A discussion on human adaptability in a tropical ecosystem: an IBP human biological investigation of two New Guinean communities. *Phil. Trans. Roy. Soc. London*, B, 268, 1974, p. 221–400.

HIPSLEY, E. H.; CLEMENTS, F. W. *Report of the New Guinea Nutrition Expedition 1947.* Canberra, Department of External Territories, 1950.

HOLMES, S. A qualitative study on family meals in Western Samoa with special reference to child nutrition. *Br. J. Nutr.*, 8, 1954, p. 223–239.

HUENEMANN, R. L.; COLLOZOS, C. Nutrition and care of young children in Peru. I. Purpose, methods and procedures of study. *J. Amer. Dietet. Ass.*, 30, 1954, p. 554–558.

——; ——. Nutrition and care of young children in Peru. III. Yurimaguas, a jungle town. *J. Amer. Dietet. Ass.*, 30, 1954, p. 1101–1109.

INCAP. *Nutritional evaluation of the population of Central America and Panamá.* Guatemala, INCAP, 1971.

LÖRSTAD, M. H. On estimating incidence of undernutrition. *Nutrition Newsletter* (Rome, FAO), vol. 12, no. 1, 1974, p. 1–11.

LUAN ENG, L. I.; VIRIK, H. K. Anaemias in children in Malaya. *Trans. Roy. Soc. Trop. Med. Hyg.*, 60, 1966, p. 53–63.

LUYKEN, R.; LUYKEN-KONING, F. W. M.; PIKAAR, N. A. Nutrition studies in New Guinea. *Amer. J. Clin. Nutr.*, 14, 1964, p. 13–27.

MASSEYEFF, R. Le goître endémique dans l'Est Cameroun. *Bull. Soc. Path. exotique*, 48, 1955, p. 269–290.

——; CAMBON, A.; BERGERET, B. *Le groupement d'Evodoula: étude de l'alimentation.* Paris, ORSTOM, 1958, 66 p.

——; PIERME, M. L.; BERGERET, B. *Enquêtes sur l'alimentation au Cameroun. II. Subdivision de Batouri.* Paris, ORSTOM, Rapport n° 4173, 1958, 183 p. multigr.

MONDOT-BERNARD, J. *Essai d'analyse de la situation alimentaire en Afrique.* Paris, OCDE, Centre de développement, 1974, 52 p. multigr.

NAIN, D. A.; SASTROAMIDJOJO, S.; SUJARDI, A.; HALIM, A.; MASPAITELLA, F. J. The prevalence of endemic goitre among school children in some parts of Sumatra, Java and Bali, Indonesia. *Penelitian Gizidan Makanan*, Jilid 2, 1972.

NICOL, B. M. The nutrition of Nigerian peasants, with special reference to the effects of deficiencies of the vitamin B complex, vitamin A and animal protein. *Br. J. Nutr.*, 6, 1952, p. 34–55.

NICOL, B. M. Tribal nutrition and health in Nigeria: a comparative clinical study of primitive and urban nutrition. *Am. J. Clin. Nutr.*, 1, 1953, p. 364–371.

——. The calorie requirements of Nigerian peasant farmers. *Br. J. Nutr.*, 13, 1959a, p. 293–306.

——. The protein requirements of Nigerian peasant farmers. *Br. J. Nutr.*, 13, 1959b, p. 307–320.

——. Protein and calorie concentration. *Nutr. Rev.*, 29, 1971, p. 83–88.

NORGAN, N. G.; FERRO-LUZZI, A.; DURNIN, J. V. G. A. The energy and nutrient intake and the energy expenditure of 204 New Guinean adults. *Phil. Trans. Roy. Soc. London*, B, 268, 1974, p. 309–348.

NYUNT-KHIN; HLA-WIN; TIN-MAY-THAN. Energy expenditure in agricultural activities in Burma. *Union Burma J. Life Sci.*, 1, 1968, p. 359–363.

OOMEN, H. A. P. C. Clinical epidemiology of xerophthalmia in man. *Amer. J. Clin. Nutr.*, 22, 1969, p. 1098–1105.

——. Interrelationship of the human intestinal flora and protein utilization. *Proc. Nutr. Soc.*, 29, 1970, p. 197–206.

——. Ecology of human nutrition in New Guinea: evaluation of subsistence patterns. *Ecol. Food Nutr.*, 1, 1971, p. 3–18.

——; CORDEN, M. W. *Metabolic studies in New Guineans*. Nouméa, South Pacific Commission, Tech. paper no. 118, 1970.

PERISSÉ, J. *L'alimentation en Afrique intertropicale, étude critique à partir des enquêtes de consommation 1950–1965*. Paris, thèse faculté pharmacie, 1966, 131 p. multigr.

PINGLE, U. Some studies in two tribal groups of central India. Part. I: dietary intake and nutritional status. *Plant Foods for Man*, 1, 1975a, p. 185–194.

——. Some studies in two tribal groups of central India. Part II: nutritive importance of foods consumed in two different seasons. *Plant Foods for Man*, 1, 1975b, p. 195–208.

PONGPANICH, B.; SRIKRIKKRICH, N.; DHANAMITTA, S.; VALYASEVI, A. Biochemical detection of thiamine deficiency in infants and children in Thailand. *Amer. J. Clin. Nutr.*, 27, 1974, p. 1399–1402.

POSTMUS, S.; VAN VEEN, A. G. Dietary surveys in Java and East Indonesia. *Chron. Nat.*, 105, 1949, p. 229A.

RABARY, R. Quelques aspects actuels des problèmes alimentaires et nutritionnels à Madagascar. *Bulletin de Madagascar*, 173, 1960, p. 907–915.

RAGHAVAN, V. K.; SINGH, D.; SWAMINATHAN, M. C. Heights and weights of well-nourished Indian school children. *Ind. J. Med. Res.*, 59, 1971, p. 648–654.

RAO, D. H.; SATYANARAYANA, K. Nutritional status of tribal preschool children of Andhra Pradesh. *Ind. J. Nutr. Dietet.*, 11, 1974, p. 328–334.

REDDY, V.; PERSHAD, J. Lactase deficiency in Indians. *Amer. J. Clin. Nutr.*, 25, 1972, p. 114–119.

ROBSON, P.; BOLTON, J. M.; DUGDALE, A. E. The nutrition of Malaysian aboriginal children. *Amer. J. Clin. Nutr.*, 26, 1973, p. 95–100.

ROY, J. K.; ROY, B. C. Food sources, dietary habits and nutrient intake of the Nicobarese of Great Nicobar. *Ind. J. Med. Res.*, 57, 1969, p. 958–964.

SATYANARAYANA, K.; RAO, D. H.; RAO, D. V.; SWAMINATHAN, M. C. Nutrition and working efficiency in coalminers. *Ind. J. Med. Res.*, 60, 1972, p. 1800–1806.

——; ——; SUSHEELA, T. P. Nutritional status of people of Andaman and Nicobar Islands. *Ind. J. Med. Res.*, 62, 1974, p. 662–671.

SCHLEGEL, S. A.; GUTHRIE, H. A. Diet and the Tiruray shift from swidden to plow farming. *Ecol. Food Nutr.*, 2, 1973, p. 181–191.

SEN GUPTA, P. N. Investigations into the dietary habits of the aboriginal tribes of the Abor hills (northesatern frontier). Part I. Padam areas. *Ind. J. Med. Res.*, 40, 1952, p. 203–218.

——. Investigations into the dietary habits of the aboriginal tribes of Abor hills (northeastern frontier). Part II. Minyong and Pangi. *Bull. Dept. Anthropol. Ind.*, vol. 3, no. 2, 1953, p. 155–173.

——. Investigations into the dietary habits of the aboriginal tribes of Abor hills (northeastern frontier). Part III. Galong tribe. *Bull. Dept. Anthropol. Ind.*, vol. 4, no. 2, 1955, p. 69–95.

——. Studies on the nutritive values of tribal alcoholic beverages. *Bull. Anthropol. Survey Ind.*, 5, 1956, p. 67–80.

——. Investigations into the diet and nutrition of the Nokte tribe of Tirap frontier division, N.E.F.A. *Bull. Anthropol. Survey Ind.*, vol. 9, no. 2, 1960, p. 25–37.

——; BISWAS, S. K. Studies on the diet and nutritional status of the Kanikkar and Urali tribes of Travancore. *Bull. Dept. Anthropol. Ind.*, vol. 5, no. 1, 1956a, p. 43–52.

——; ——. Studies on the diet and nutritional status of the Malapantaram, Muthuvan and Ullatan tribes of Travancore. *Bull. Dept. Anthropol. Ind.*, vol. 5, no. 2, 1956b, p. 9–19.

——; RAO, R. K.; BISWAS, S. K. Investigation on the nutritive values of foodstuffs of different tribes of India. *Ind. J. Appl. Chem.*, vol. 24, no. 1, 1961, p. 46–52.

SOEKIRMAN. *Priorities in dealing with nutrition problems in Indonesia*. Cornell University, Cornell International Nutrition Monograph Series no. 1, 1974.

SOUTHGATE, D. A. T.; DURNIN, J. V. G. A. Calorie conversion factors. An experimental reassessment of the factors used in the calculation of the energy value of human diet. *Br. J. Nutr.*, 24, 1970, p. 517–535.

SWAMINATHAN, M. C.; KRISHNAMURTHI, D.; IYENGAR, L.; RAO, D. H. Health survey of the Onge tribe of Little Andamans. *Ind. J. Med. Res.*, 59, 1971, p. 1136–1147.

TANPHAICHITR, V.; VIMOKESANT, S. L.; DHANAMITTA, S.; VALYASEVI, A. Clinical and biochemical studies of adult beri-beri. *Amer. J. Clin. Nutr.*, 23, 1970, p. 1017–1026.

TEDDER, M. M. Les régimes alimentaires de base aux Iles Salomon britanniques, *Bulletin du Pacifique Sud*, 3, 1973, p. 12–16; p. 28.

TIE, L. T.; LIAN, O. K.; LIONG-ONG, T. W.; ROSE, C. S. Health, development, and nutritional survey of preschool children in Central Java. *Amer. J. Clin. Nutr.*, 20, 1967, p. 1260–1266.

VALYASEVI, A.; BENCHAKARN, V.; DHANAMITTA, S. Anaemia in pregnant women, infants and preschool children in Thailand. *J. Med. Ass. Thailand*, 57, 1974, p. 301–306.

VAN VEEN, A. G. Some ecological considerations of nutrition problems in Java. *Ecol. Food Nutr.*, 1, 1971, p. 25–38.

——; HONG, L. G.; NIO, O. K. Some nutritional and economic considerations of Javanese dietary patterns. *Ecol. Food Nutr.*, 1, 1971, p. 39–43.

VITERI, F. E.; TORUN, B. Ingestión calórica y trabajo físico de obreros agrícolas en Guatemala. *Boletín de la Oficina Sanitaria Panamericana*, Suppl. alimentario, 1975, p. 58–74.

WATERLOW, J. C. Nutrition and infection. *Bull. of Br. Nutr. Found.*, 14, 1975, p. 98–104.

World Health Organization. *Expert Committee on Medical Assessment of Nutritional Status*. Geneva, WHO, Technical Report Series, vol. 258, 1963.

Health and epidemiology

Introduction

Thoughtful scholars of epidemiology have discerned that the proper appreciation of disease can only be obtained by relating it to the total environment. Pavlovsky (1939, 1963) developed the concept of landscape epidemiology and focal localization he termed nidality of disease. Audy (1958), as a result of his studies on scrub-typhus in Malaysia, referred to the assemblage of plant and animal communities as a biocenose with disease (particularly infectious diseases) as being dependent on the type of biocenose. The ecological principle of animal dispersion in a habitat was extended to parasites by Mantner (1967). He speculates that distribution of parasites might be due to ecological conditions or conditions of the ancient past. Ecological-epidemiologists have almost always developed the principles of their disciplines from the study of a particular disease, often a vector-borne infection, and from these observations they have proceeded to formulate a general theory for epidemiological and environmental interactions. The alternate approach of selecting a global ecosystem and trying to discern the conditions of human life, particularly in respect to health and disease, within that overall ecosystem has been neglected. The most recent research on the last human isolates of the world stresses the necessity, in order to understand the way of life of a population and its inter-actions with the environment, of considering all flows of energy, matter and information through the ecosystem, the social unit or the individual; one should not therefore forget the genetic potential existing in a population at each moment of its evolution, and to follow through time and environmental transformations the future of such a genome under the influence of migrations, random drift, mutations and selection pressures. The study of population genetics is inseparable from the other research on human biology and ecology.

The tropical forest is peculiar in that it possesses a dimension absent in other ecosystems -height. The height is such that it must be considered as a series of horizontal strata (see chapters 5 and 8). The canopy, exposed to sunlight has very different physical conditions and plant and animal life than the relatively dark, humid forest floor. Between these two extremes other biocenoses have estab-lished themselves. There is also bionomic isolation—species in ecological niches tending to stay confined to that habitat. These two related phenomena, the stratification and the bionomic isolation have an important effect on disease transmission.

An illustration can be given for vector-borne infections. Obviously the nature of man-vector contact greatly influences the kind and intensity of disease present in a community. An important factor in this relationship is the life-cycle of vectors in interaction with the environment. Each vector species possesses characteristic biological and behavioural traits that influence its interaction with its hosts. Thus the selection for breeding water, host preferences and resting behaviour are mostly genetically controlled characteristics that may or may not place a particular vector species in proximity to man. The same principles apply to zoonotic or potentially zoonotic diseases where the reservoir hosts occupy selected ecological niches in the forest. The transmission of disease to man is further complicated in the case of vector-borne zoonoses where there must be communication between the reservoir, vector and man. Some vectors and reservoirs have the ability to penetrate to the forest fringe and surrounding savanna and it is at these ecotones that transmission of disease to man may take place.

In succeeding sections it will be shown how complicated this cycle may be for such infections as yellow fever and some forms of leishmaniasis. The effect of bionomic isolation is usually protective for the health of forest-dwelling man. Moreover, it appears that for many infectious organisms, continuous cycling in the wild animal host may result in strains for which man is insusceptible or poorly susceptible. However when man alters the forest ecosystem, particularly by felling portions for agriculture, these relationships change and the risk to health is usually increased. The nature of these changes will also be considered.

Human associations with forests: epidemiological implications

There is a wide variation in the degree of intimacy of man with tropical forests from nomadic hunter-gatherers living entirely within the forest ecosystem to individuals from communities far away who make occasional entry. Between these two extremes are the semi-settled hunter-gatherers, shifting agriculturalists, settled agriculturalists in clearings, and communities situated at the edge of the forest whose members enter the forest for forage, hunting or other forest products. The nature of the human association with the forest is of great epidemiological significance. An association of great relevance is that of the nomadic hunter-gatherers of the forest and they will be considered in some detail. These people are in a real sense an integral part of the forest ecosystem. They can serve as sentinels for the identification of the health hazards, inherent in the forest.

Man's residence on earth has been largely as a hunter-gatherer. The development of agriculture some 10 000 years ago resulted in settlement into communities that led to agricultural-urban societies. There are very few remaining hunter-gatherer groups not in contact with peoples from outside the forest. Those with little or no contact from beyond their territory almost invariably exist in such remote area as to be virtually inaccessible for study. Where medical-anthropological studies have been carried out, usually they do not allow the sophisticated technical examination necessary to provide the necessary data. Some studies have been made on archaeological remains, but due to the destruction of soft tissues, except in the case of mummies, the dry bones provide little information.

All hunter-gatherers share at least two characteristics: small social groups with limited contact between them; and nomadism—at the most they practise shifting agriculture. These have an important impact on their pattern of disease. The epidemiological concepts governing their health and disease are largely based on the work of three outstanding students: Black, Dunn and Fenner. Their collective epidemiological principles can be summarized as follows (Dunn, 1968; Fenner, 1970; Black, 1975).

1. Hunter-gatherer populations remain stable if not affected by outside contact. Dunn gives the formula (after Wynne-Edwards) for population stability as:

$$\text{Recruitment arising from reproduction}$$
$$+ \text{ immigration}$$
$$= \text{ uncontrollable losses}$$
$$+ \text{ emigration}$$
$$+ \text{ social mortality.}$$

Immigration and emigration are usually thought to be negligible in these situations. Uncontrollable losses are deaths due to predatism, parasitism, accidents, starvation and chronic diseases associated with old age. Social mortality comprises elements such as war, homicide, infanticide and abortion. The hazards of the hunter-gatherer's life are obvious although most observers consider the natural environment hazards are insignificant; disease appears to be a major cause of mortality—the uncontrollable loss in the equation.

2. Infectious disease is the main cause of morbidity and mortality but the kinds of disease are governed by their ecosystem and their population size. Their infections are either anthropozoonotic or of a kind that can be continuously cycled within their social group. 'Only those organisms which, like varicella, can exist in small groups, and those which like yellow fever virus, have a non-human reservoir, are likely to have played a role in the development of mankind through his much longer human history' (Black *et al.*, 1974).

The role of anthropozoonotic diseases may seem self-evident but, in an undisturbed ecosystem, they may not be so important owing to the bionomic isolation of the reservoir and, where there is an intermediate host, the vector.

The concept that each communicable disease requires a critical mass of hosts affects the pattern of disease. Diseases with a short infectious period such as measles cannot persist in isolated small communities, while chronic infections such as tuberculosis or viruses such as herpes simplex and cytomegalovirus that survive in the body for a long time with periods of reactivation can. It has been estimated that a community size of 200 000 is necessary to ensure the persistence of measles but only 2 000 people are necessary to sustain varicella (Black *et al.*, 1974) and less than 1 000 for chickenpox. Fenner has written an excellent account of the theoretical projections of critical mass for infectious disease under different ecological and social conditions.

An inventory of hunter-gatherers of the tropical forests

In 1964 and 1967 the World Health Organization, alarmed that the existence of many peoples was endangered, convened a scientific group to consider the issue and make recommendations for further study and action. This group's reports (*WHO Technical Report Series*, no. 279 and 387, 1968) listed the existing hunter-gatherers of the tropical forests. To this may be added the compendium of hunting and gathering peoples prepared by Murdock (1968). The collective list is as follows.

Africa
Pygmies of the Ituri forest (Zaire) and forests of the Cameroon, Central African Empire (Babinga, Bi-Aka), Gabon and People's Republic of the Congo.

Asia
Negrito and proto-malay tribes of Malaysia and southern Thailand
Yumbris of northern Thailand
Negritos of the Philippines and Sri Lanka
Negritos (Onges) of Little Andamans
Jarain of South and Middle Andamans
Punan and Penan groups of Sarawak and South Kalimantan
Veddahs of Sri Lanka
Chenchu, Kadar, Warli and Katkari of southern and central India.

South America
Guayaki and Moro of eastern Paraguay
Siriono of Bolivia
Ge families, especially the Cayapo of eastern and central Brazil
Yanomamo of Brazil and Venezuela
Guahibo of the Venezuela-Colombia border
Nambikuara of the Mato Grosso
Mortoko of the northern Chaco
Pakasnovas of the Brazil-Bolivian border.

This list is undoubtedly incomplete and there are remote groups of which virtually nothing is known. Nor does this list include the many settled or semi-settled peoples of the tropical forests in Africa, Asia, South America and Melanesia. Medical studies have been carried out for relatively few of these peoples.

Health and disease

Where possible a general demographic-epidemiologic regional picture is presented followed by accounts of the health of groups intimately associated with the forest, and finally a briefer account of populations more removed from the forest, so that the epidemiological characters of true forest-dwellers serve as a base-line from which the effects of other associations and dislocations could be discerned.

Africa

Demographic aspects

The eastern forest zone, from Cameroon to Zaire is sparsely inhabited with *ca.* 0.5–7 persons/km² whereas the western forests have densities ranging from 30–60/km² (although there are areas of low population density e.g. Liberia and the southwestern Ivory Coast; see chapter 19). In the eastern zone females appear slightly to outnumber males while the opposite seems to pertain to the western zone. It is very difficult to obtain reliable vital statistics that can be used for demographic characterization in respect of age pyramids, birth and death rates, etc. From the data available it is estimated that slightly less than 40 per cent of the population living in the eastern forests are under 15 years old whereas slightly more than 40 per cent are in the western zone. Those over 60 years constitute 6 per cent of the total population.

In the western zone the boundary between the forest and urban centre is not well marked while in the eastern zone it is clearer. This is important epidemiologically. The consequences for non-immunes from urban centres migrating permanently or temporarily into a forest environment through forced or voluntary resettlement schemes, recruitment as forest plantation labourers, or merely as foragers to supplement their diet, are largely unknown.

Despite the caveats regarding the paucity and unreliability of the statistical data, it appears that the birth rate in the western zone (45 births per 1 000 inhabitants per year) is higher than in the eastern zone (36/1 000/a). If the procreative age of women is taken as 14–49 years then in the west the average live births per woman is 6 and in the eastern zone 4. Mortality rates are even more unreliable. Nevertheless, in the eastern zone there is an astounding variation in mortality: 20 per 1 000 in Cameroon, 24 in the Congo, 30 in Gabon and 59 in Haut-Ogoué. On average, the death rates are lower than those of the grazing lands. The trend however is toward population growth: in Gabon (1960 estimate) 5 per cent/a while in southern Nigeria it is 2 per cent/a. See also chapter 15.

Ituri forest pygmies

There are approximately 176 000 pygmies living in the Ituri forest of Zaire (Murdock, 1968). The pygmies are of three groups, Aka, Efe and Mbutu which in turn are further subdivided into small family groups or clans of 5 to 34 individuals. They lead a semi-nomadic hunter-gatherer existence in the tropical rain forest of Zaire. Other pygmoid peoples exist in the forests of the Cameroon and Gabon but virtually nothing is known regarding their health or habits. While they are true hunter-gatherers living deep within the forest, pygmies cannot be considered a completely isolated people. They have a dependent relationship with Bantus living in villages at the forest fringe and emerge from time to time to barter wild meat for the agricultural products of the Bantu. It should also be noted that, in the Cameroon, the pygmies are being settled along the roads and some started school before 1958. However, in the Central African Empire and the People's Republic of the Congo, most of the pygmies continue to live within the forests (Bi-Aka) whilst some tend to settle in villages (Babinga).

The results of surveys for parasitic and infectious diseases are summarized in table 1. The prevalence of *Entamoeba histolytica*, 36 per cent, is surprisingly high for a nomadic people. How many of these infections cause clinical disorder is not known. Brumpt *et al.* (1972) have found a similar percentage of *E. histolytica* cysts among the Sara N'dindjo of the Central African Empire who are neither pygmies, nor forest-dwellers (they live in the Sudan zone of this country); but they noticed 20 per cent of *E. hartmanni* which could be confused with *E. histolytica*. It would be of value to carry out a serological survey in

conjunction with a parasitological one since antibodies, as indicated by such techniques as indirect haemagglutination and counterimmunoelectrophoresis, are usually elaborated only in individuals who have, or have had, invasive amoebiasis. Amongst other intestinal parasites, the prevalence of hookworm (*Necator americanus*) and ascariasis is notably high and the burdens reported to be generally heavy (Price *et al.*, 1963). Despite the high prevalence and burden of hookworm there is little anaemia, with few individuals showing a haemoglobin level <10 g/l although an earlier study (Van den Berghe, 1941), showed that haemoglobin levels were about 25 per cent less than in normal Europeans. Presumably their diet contains sufficient iron to compensate for the depletion. Van den Berghe (1938) observed a lower rate of intestinal parasitism in the pygmies than in Bantus dwelling at the forest edge while Price *et al.* (1963) found the converse to occur. Perhaps there was a shift in behaviour or sanitation during the 25 years between these two studies. The reasons for the high rate of intestinal parasitism are unknown and studies into their sanitary-behavioural practices need to be carried out.

TABLE 1. Results of health surveys of populations (per cent infected) of the tropical African forest zone.

Habitat:	Deep forest (hunter-gatherers)		Forest and forest settled agriculturalists		Village in forest
Population group:	Ituri forest pygmies (Zaire)	Cameroon: Babinga and other groups	Bantus of Congo[1]	Benaka langa[9] (Uganda)	Akufo, Nigeria[15]
Entamoeba histolytica	36[1]		12		
Giardia lamblia	7[1]		10		
Malaria	Children 12 *falciparum*[1] 17 *malariae*[1] 40[2] Adults 22[2]		Infant spleen rate 28 51[2]	24	51 parasite rate all *falciparum* with 14 mixed *malariae* 70 spleen rate 2–5 yr group 11 spleen rate 20 yr group
Trypanosomiasis	none reported	3.4 and 0.2[14]			none
Hookworm	86[1], 40[3]		80[3]		71
Ascaris	58[1], 22[3]		73[3]		70
Trichuris	70[1], 27[3]		11[3]		45
Onchocerca	Adults 43– 82[1]	4[10]	5[16] endemic	100	
Loa loa	2[1]	24[10]			4.1
Dipetalonema perstans	60–100	78[10]		35	1
Schistosomiasis	11[3]		28[3]		
Leprosy	7[1], 6–9[12]	rare[14]	endemic		rare
Syphilis, yaws and treponemal antibody	high prevalence syphilis, >50		yaws 20[11] syphilis 60[8]		
Rickettsiosis	typhus present in rain forest[13]				
Yellow fever			rare 0.1[11]		75 immune by 5 yrs
Sickle-cell trait	26[6]		26[6]	24	25

Sources
1. Price *et al.*, 1963
2. Duren, 1937
3. Van den Berghe, 1938
4. Degotte, 1940
5. Mann *et al.*, 1962
6. Van den Berghe and Janssen, 1950
7. Beghin, 1960
8. Ledent, 1944

9. Raper and Ladkin, 1950
10. Languillon, 1957a, b
11. Liégois *et al.*, 1948
12. Van Breuseghem, 1938
13. Barlovatz, 1940
14. Lalouel, 1950
15. Gilles, 1964
16. Geukens, 1950

There is a low prevalence (11 per cent) of Mansonian schistosomiasis; less than one third that of the Bantu peri-forest villagers (Van den Berghe, 1938). Planorbid snail vectors are rarely found in forest streams and it may be that the relatively few infections are acquired outside the forest. However, few investigations on the possible transmission of schistosomiasis in the African forests have been carried out (see also page 392).

Although the Babinga pygmies have no domestic cats the serological (dye test) positivity rates range from 20 to 50 per cent (Berengo et al., 1974). These authors suggest that the infection was acquired from their habit of eating raw or undercooked meat of wild animals which had become infected from ingesting the oocysts discharged by sylvatic Felidae. Such assumption seems doubtful as other authors (Jaeger, personal communication) observed that pygmies do not usually eat raw or undercooked meat.

The contribution of malaria as a cause of infant mortality is not known. The malariometric surveys of Duren (1937) and Price et al. (1963) showed little change in prevalence pattern to have occurred over 26 years. As with other African negroes, the common occurrence of sickle-cell trait in the pygmy would give a clinically protective effect against malaria. The percentage of sickle-cell trait (26 per cent, table 1) as reported in 1950 among Ituri pygmies by Van den Berghe and Janssen seems too high and may indicate cross-breeding with Bantu populations. The survey of Price et al. (1963) revealed that in children, the *Plasmodium malariae* rate was as high as that of *P. falciparum*. There is persuasive evidence that *P. malariae* is the etiological agent for nephrotic syndrome in African children but there are no reports of this clinical entity afflicting pygmy children.

There are considerable gaps in knowledge of the vectors and their bionomics. *Anopheles funestus* is said to be the main vector in the Ivory Coast. It is behaviourally exophilic spending little time indoors (Coz, 1966). In the Cameroon, *A. moucheti*, which breeds at the edge of slow moving streams, and *A. nili* have been identified as vectors (Languillon, 1957a; Mouchet and Gariou, 1966). Mattingly (1949) has carried out an excellent study on the seasonal distribution and biting habits of *A. hargreavesi* and *A. gambiae*, malaria vectors of the southern Nigerian swamp forests. *A. hargreavesi* is most numerous during the dry season while *A. gambiae* practically disappears during this period, being abundant and dispersed only during the heavy rains of July and August. Both species bite at ground level, a behavioural characteristic insuring man-vector contact (although clear-cut evidence was not brought for *A. hargreavesi* as a vector in Nigeria). *A. gambiae* is the most important African vector where man has made clearings, a process that provides a suitable habitat for its breeding requirements (Livadas et al., 1958). But under the forest cover, vectors are generally scarce and often absent.

Onchocerciasis is hyperendemic along some rivers. The clinical manifestations, i.e. loss of vision, thickening of the skin, associated with the infection in endemic zones of the savanna, has not been commented upon for forest people. It is probable that the forest form of *Onchocerca volvulus* does not generally induce severe ocular pathology. Forest peoples of Africa, including the pygmies, have very high infection rates of *Dipetalonema perstans*, *Acanthocheilonema streptocerca* and *Loa loa*, virtually every individual harbours these parasites. Infection with these filarial worms has been considered to be benign but this has been questioned by the recent WHO Expert Committee on Filariasis (1974).

The single report on leprosy by Van Breuseghem (1938) indicates that the pygmies are much infected (6–9 per cent prevalence) if not too severely affected. The majority of the infections were of the macular type, very few lepromatous cases being noted. Leprosy should be reinvestigated to determine its present status in pygmies.

Syphilis, but not gonorrhea, occurs with high frequency (Ledent, 1944; Mann et al., 1962). However unlike their Bantu neighbours the Nkundis, fetal wastage due to syphilis was not causing the rapid depopulation of the latter group. The ratio of Nkundi adults to children was 36:1 while the ratio in the pygmies was 1:1.5 despite a similar prevalence rate of syphilis (Ledent, 1944). He believed that the socio-sexual habits of the pygmies made them much less promiscuous than their Nkundi neighbours and the non-infected individuals were able to maintain population numbers. But according to recent investigations (Ciréra et al., 1977) no clear-cut clinical symptoms of syphilis have been found among the Bi-Aka pygmies living within the forest of lower Lobaye (Bokoka), Central African Empire; a high percentage of positive serological tests (80 per cent) among these pygmies is in fact due to endemic yaws (*Treponema pertenue*) and not to syphilis (*Treponema pallidum*); the latter may exist but more research is needed to show its real frequency among these groups.

Actually, very little is known about the demography of the pygmy peoples.

Virtually nothing is known regarding the viral infections, particularly arboviruses. Presumably the bionomic isolation of reservoirs and vectors prevents them from contracting yellow fever, as is the case for other dwellers of the Zaire high forest (Liégois et al., 1948).

Mann et al. (1962) studied the physiological health of the Ituri forest pygmy. As is the case of most other forest people who have been subject to similar study, it was found that there was little or no cardio-vascular disease; blood pressure, at all ages, being lower than for urbanized populations. The average serum cholesterol level of 106 mg/ml, was strikingly lower than in Europeans (which should not be considered as the standard reference). Their general assessment was one of poor health which they relate to the high frequency of hepatosplenomegaly. Nutritional status was relatively satisfactory although a few cases of overt malnutrition were noted amongst the children.

Since 1974 a more precise study on the biology of pygmies is performed by the Centre eurafricain de biologie humaine (Paris; Jaeger et al., 1977) in the forest clearings, south of Bokoka (forest of the lower Lobaye, Central African Empire). 1 200 Bi-Aka pygmies belonging to *ca.* 50 camps are examined and looked after. Anthropophysiological measurements and the analysis on the spot (glycemia and blood sedimentation speed) or in the laboratory (at Boukoko or in France) of blood, saliva, stool, urine and skin samples, of 460 individuals give a first approximation of the genetic stock and of the usual biological parameters of this nomadic people, when they come to stay annually at the forest fringe between the periods of hunting and of caterpillar or honey harvesting. The first peculiarity of such a study is to be based on the exact knowledge of the parenthood of the examined individuals (thanks to 70 blood-markers). A genealogical tree of these pygmies and of their close or remote relatives (4 500 individuals) was designed over 9 generations, 5 of living individuals and 4 of ancestors. Such a family census allows the demographic study of the population; the description of the genotype thanks to the probabilities of gene origin and to genetic markers; the qualitative and quantitative analysis of genetic flow from a generation to the other and from a family camp to another, according to a methodology set up for the study of another population of the Central African Empire, the Sara Kaba of Miamane (Jaeger, 1974). Table 2 summarizes the first results of this study. The second peculiarity of this research is to try to perceive the significance of the way of insertion of this pygmy population within an ecosystem which is still largely undisturbed and within modern society (see second part of chapter 19).

Settled villagers of the near forest region

While the pygmies and other hunter-gatherer settlers may be of special epidemiological interest, the majority of the population within the forest zone are inhabitants of villages, towns and cities. This is particularly true of the western zone where an ever-growing population had led to the penetration into and extensive settlement of, forest areas. Many if not most of the villagers are farmers with small holdings bordering the forest at some distance from their homes. As man becomes further removed from the forest the factors affecting his health and general well-being also alter. Not only will the villagers be affected by the usual problems of incipient urbanization e.g. a pure water supply, waste disposal, adequate housing, etc., but also the bordering forest remains a potential reservoir of pathogens. There have been several comprehensive investigations of communities that can serve as models for the others. One outstandingly thorough investigation is that made by Gilles (1964) of the people of Akufo, a village of about 1 500 inhabitants situated in the forest belt of western Nigeria and fairly typical of other villages throughout the forest region.

Children under the age of 14 years made up 43 per cent of Akufo village population. The broad base of the age pyramid indicated high wastage of child life. This was confirmed by a four year survey which revealed a mortality rate of 43 per cent/a over the first four years of life. Despite a list of conditions reading much like the index of Manson's 'Tropical Diseases', Gilles states 'The reason for this high death rate are multiple and not at all easy to dissociate'. Nutrition was considered marginally adequate. Cassava and yams were the main staples making for a high carbohydrate diet. Proteins were *ca.* 7 per cent of the total caloric intake; the subclinical protein malnutrition is reflected in the low average serum albumin of all age groups and the subnormal heights and weight of children. With the exception of some B vitamin deficiency no other vitamin deficient condition was noted. Vitamin A, in particular, was well provided by the high consumption of palm oil. The influence of low-grade protein calorie malnutrition on the clinical course of infectious diseases is not clear. However, current nutritional thought holds that provided energy requirements are met, a diet containing about 6 per cent of energy from protein would usually be adequate. Gilles observes that childhood mortality rates are as high as in western African areas where malnutrition is rare as in areas where nutrition is poor.

The summary of infectious disease findings is shown in table 1. The most striking feature is the extremely high rates of malaria and accompanying hepatosplenomegaly. The overall parasite rate was 50 per cent and in the 2–5 year group it was between 70–80 per cent. The difference in malaria rates in forest-dwelling pygmies (12 per cent) and settled Akufo villagers (50 per cent) is the most striking difference in disease pattern between these two populations. The actual toll taken by malaria is not known. Fatal parasite densities are not attained in sickle-cell trait (AS) carriers although the parasite rate in this group is the same as the AA segment of the population. Thus some of the children have a natural resistance against a lethal course of the infection; however the AS children are only 25 per cent of that age group. Entomological studies were not carried out in the Akufo study but the most likely explanation for the higher malaria rate in the villages than in the forest-dwelling population is that opening of the forest for agriculture produced suitable breeding sites for the principal vector, *A. gambiae*.

Hookworm rates and burdens were high in Akufo villagers of all age groups. The combination of malaria and hookworm would be expected to cause anaemia, and average haemoglobin levels were 2–3 g/l lower than for normal Europeans. However the Akufo villagers' mean Hb levels were rarely below 11 g/l which suggest a diet that adequately compensates for iron loss.

Diarrhoea commonly occurs in Akufo children and, although no figures are given, Gilles comments that hospital records indicate gastro-enteritis to be one of the commonest cause of death among young children in the region. The etiology is not clear. Many cases are due to *Shigella*, *Salmonella*, *Pseudomonas* and pathogenic *Escherichia coli*; but *E. histolytica* and other intestinal parasites can give rise to gastro-enteritis and it should be kept in mind that diarrhoea is a common presenting symptom in children with *Plasmodium falciparum* malaria. Nevertheless the high frequency of diarrhoea points to a major problem of people gathered into settled communities—obtaining a safe water supply. This is not a problem of the nomadic hunter-gatherers, but their high rate of intestinal parasites makes one consider possible contamination of their water supply. Dracontiasis found to be common in the Akufo villagers (23 per cent prevalence) was not reported in the Ituri pygmies. This would be another example of a water-borne

TABLE 2a. Immunogenetic study of 460 Bi-Aka pygmies of Bokoka, Central African Empire (3°44 N, 17°51 E) (Jaeger et al., 1977).

ERYTHROCYTE SYSTEMS[1]
ABO: A^1, 0.15; A^2, 0.02; B, 0.15; O, 0.68
Rhesus: R^0, 0.91; R^1, 0.02; R^2, 0.009; r, 0.06
MNSssu: MS, 0.03; Ms, 0.38; MS^u, 0.13; NS, 0.11; Ns, 0.28; NS^u, 0.07
Henshaw: He^+, 0.007
Antigen Vw: Vw, 1.0
P: P_1, 0.85; P_2, 0.15
Kell Cellano: K, 0.995; k, 0.005; kp^a, 0.0; kp^b, 1.0
Sutter: Js^a, 0.21; Js^b, 0.79
Duffy: Fy^a, 0.0; Fy^b, 0.004; Fy, 0.996
Kidd: JK^a, 0.83; JK^b, 0.17
Lutheran: Lu^a, 0.004; Lu^b, 0.996

ENZYME SYSTEMS[2]
G6PD: GdA^+, 0.136; GdA^-, 0.064
GdB, 0.80; Gds, 0.003
6PGD: 6PGDA, 0.98; 6PGDB, 0.02
PGM 1 et 2: PGM_1^1, 0.84; PGM_1^2, 0.16; PGM_2^1, 0.95; PGM_2^2, 0.004;
PGM_2^6 pygmée, 0.04
Acid phosphatase: P^a, 0.04; p^b, 0.87; p^r, 0.09
Adenylate kinase: AK^1, 1.0
Adenosine deaminase: ADA^1, 1.0
Superoxyde dismutase: SOD^1, 1.0
Malate dehydrogenase: MDH^1, 1.0
Pseudocholinesterase, C_5 est: E_1^u, 1.0; E_2^+, 0.09
Alkaline phosphatase: B, 98.1; C, 1.9

IMMUNOGLOBULINS[3]
Gm: *Gm1.17.5.6.10.11.14.27*, 0.002; *Gm1.17.5.6.11.24*, 0.19
Gm1.17.10.11.13.15, 0.02; *Gm1.17.5.10.11.13.14.27*, 0.79
Inv: $Inv^{1.2}$, 0.42; Inv^3, 0.58

HAPTOGLOBINS[4]
Hp0, 31.24%; Hp21M, 1.73%
$Hp1F$, 0.17; $Hp1^s$, 0.18; Hp^2, 0.65

TRANSFERRINS[4]
Tf^c, 0.84; Tf^b, 0.16

GROUP COMPONENT[4]
Gc^1, 0.88; Gc^2, 0.12; Gc^{Ab}, 0.03

HEMOGLOBINS[5]
Hb^A, 0.93; HbS, 0.07;
Hb A2, 0.97; A2 Babinga, 0.02; A2 Flatbush, 0.01

LEUCOCYTE AND PLATELET GROUPS[6]
Antigen frequencies % — HLA A1: 0.3; A2: 12.1; A3: 14.7; A9: 15.0; A10: 10.5;
A28: 7.1; A29: 8.7; AW19: 8.4; AW30: 10.3; AW32: 0.5; — HLA B5: 8.2; B7: 6.8;
B8: 0.3; B12: 7.6; B13: 1.1; B14: 1.3; BW15: 0.5; BW39: 7.9; BW17: 12.1; BW18: 2.4;
BW21: 6.3; BW22: 1.8; BW27: 3.9; BW35: 3.2; BW40: 5.8; BW40–13 pygmy: 2.1

Preferential gamete associations — HLA A3–B5: 7.1%; HLA A9–BW39: 6.0%;
HLA A9–B7: 4.7%; HLA A29–BW17: 4.5%; HLA A3–BW27: 3.4, i.e. 27.7% of
haplotypes determined by the family study.

1. Y. Marty; A. Muller.
2. H. Vergnes.
3. M. Blanc.
4. J. Constans.
5. M. L. Coquelet.
6. E. Ohayon; L. Halle.

TABLE 2b. Results of several biological tests and measurements of 350 pygmies over 18 years, examined in their forest camps, south of Bokoka (Jaeger et al., 1977).

ANTHROPOPHYSIOLOGY[1]		Men	Women
Height (cm)	m	154.2	144.2
	s	6.3	6.4
Size, seated (cm)	m	79.8	74.9
	s	2.8	3.9
Weight (kg)	m	48.1	42.3
	s	5.5	6.3
Blood systolic pressure (mm Hg)	m	144.3	135.3
	s	21.1	22.8
Blood diastolic pressure (mm Hg)	m	87.6	83.7
	s	11.8	14.2
Pulse (seated)	m	85.4	86.8
	s	13.1	13.2
Pulse (standing)	m	90.8	94.3
	s	13.3	13.3

HEMATOLOGY. BIOCHEMISTRY[1]		Men	Women
Blood sedimentation speed (mm)	m	31.0	38.1
	s	15.1	16.0
Hematocrit (%)	m	41.0	37.5
	s	4.8	5.2
Haemoglobin (g/1)	m	14.4	12.5
	s	1.4	2.7
Urea (g/1)	m	0.10	0.19
	s	0.05	0.06
Cholesterol (g/1)	m	1.52	1.53
	s	0.34	0.38
β-lipoprotein (units)	m	29.2	29.4
	s	7.6	7.9
Uric acid (mg/1)	m	50.8	45.3
	s	10.1	10.9
Glucose* (g/1)	m	1.05	1.18
	s	0.26	0.44
Bilirubin (mg/1)	m	7.5	5.1
	s	4.8	2.3

positive or abnormal qualitative results

		M%	W%
Reiter, Kolmer,[2] Kline +++		80.5	80.5
ASLO >200 u.		25.8	25.4
Hepatic test +++		94.5	96.9
Oxaloacetate transaminase		32.0	16.2

PROTEINS[3]		M%	W%
Total proteins (g/1)	m	91.5	85.2
	s	23.2	8.3
Albumin, A	m	39.4	49.7
	s	5.8	3.6
Globulins, G: Alpha 1 (%)	m	4.4	4.4
	s	0.7	0.8
Alpha 2 (%)	m	7.3	8.1
	s	1.4	1.5
Beta (%)	m	9.7	10.4
	s	1.3	1.6
Gamma (%)	m	39.3	36.3
	s	6.1	4.7
Ratio A/G	m	0.66	0.69
	s	0.15	0.10

Titration[5]	m		s	
IgG (g/1)	29.22		6.82	
IgA (g/1)	3.29		1.24	
IgM (g/1)	4.66		3.75	
IgE highly increased among 95% of people				
Transferrin	2.74		0.59	
Coeruloplasmin	0.474		0.177	
Alpha 2-macroglobulin	2.11		0.54	
Beta-1A-Beta 1C	1.12		0.30	

HEMATOMETRY[4] of 180 individuals			
White blood cells (x10³)	m	7.6	
	s	2.8	
Red blood cells (x10⁶)	m	4.63	
	s	0.56	
Platelets (x10³)	m	220.2	
	s	91.5	
Formula: neutrophils	m	30.6	
	s	13.5	
eosinophils	m	17.4	
	s	8.4	
lymphocytes	m	45.6	
	s	12.6	
monocytes	m	5.4	
	s	3.0	

PARASITE ANTIBODIES[5]
anti-Aspergillus fumigatus: 0.0; anti-nematods (Ascaris suum, Parascaris equorum): 0.20; anti-cestods (hydatic antigens, Taenia solium): 0.05.

TOXOPLASMOSIS[6]
92.0% of positive sera, of which 73% are natural antibodies.

AUSTRALIA ANTIGEN[7]
21 carriers of AgHBs (5.6%) of which 19 subtypes:
15 a_2yw; 3 a_1yw; 1 a_3yw.

ARBOVIRUSES[8] (antibody frequency)
A group: Sindbis, 4.7%;
Chikungunya, 9.4%;
Semliki Forest, 13.3%;
Bunyamwera group:
Bunyamwera, 10.7%;
B group: yellow fever, 10.9%;
West Nile, 3.4%;
Uganda S., 7.0%;
Zika, 0.8%.

1. M. J. Palisson; G. Pinerd; J. Saurois.
2. P. Cirèra.
3. M. L. Coquelet.
4. P. Colombies.
5. J. Petithory.
6. P. Bourée.
7. A. M. Courroucé.
8. P. Sureau.
* not necessarily fasting.
m average.
s standard deviation.

infection dependent on a contaminated community water supply, for contamination occurs through infected small crustaceans, *Cyclops*, which contain the larvae or embryos of the worm made free from wounds of people entering ponds.

Unfortunately, Gilles did not study other physiological aspects, such as blood pressure, but all other studies have shown that hypertension is prevalent in the settled communities.

The following statement by Gilles could well apply to any analysis of health in the tropical forest region: 'The causes of high mortality at Akufo, as in other parts of West Africa, are difficult to dissociate. The bewildering variety of infections from protozoa, helminths, bacteria and viruses, represents in our opinion the main obstacle to survival for the pre-school Nigerian child'.

Another study of the parasitic disease pattern of pygmies and settled population was made by Pampiglione and Ricciardi (1974) in the Central African Empire between 1968 and 1970. They have examined the Bantu

TABLE 3. Comparison of parasitic disease pattern (percentage of infected individuals) between two populations (*ca.* 300 individuals) living at the forest fringe (Bagandou-Bale Loko, Central African Empire), 1968–70 (Pampiglione and Ricciardi, 1974).

	Babinga pygmies %	Bantu village-dwellers %
BLOOD		
Presence of gametocysts	41.8	60.7
of which *P. falciparum*	35.9	58.3
P. malariae	8.2	5.3
P. ovale	1.1	0.3
Microfilariae		
of which *Dipetalonema perstans*	22.0	15.9
Loa loa	2.0	10.6
SKIN		
Dipetalonema streptocerca	27.8	29.5
Onchocerca volvulus	0.8	5.3
URINE		
Schistosoma haematobium	0.0	41.0
FAECES		
1 or several protozoans	85.9	77.3
of which *Entamoeba histolytica*	35.8	26.7
Entamoeba coli	73.6	59.5
Endolimax nana	29.1	19.0
Iodamoeba butschlii	13.0	6.5
Giardia intestinalis	11.4	16.5
Trichomonas intestinalis	18.7	22.2
Chilomastix mesnilii	3.7	1.7
Dientamoeba fragilis	0.3	
1 or several helminths	93.3	97.0
of which *Trichuris trichiura*	77.9	69.7
Ancylostoma sp.	73.6	80.2
Strongyloïdes fülleborni	23.7	11.7
Strongyloïdes stercoralis	5.0	25.7
Ascaris lumbricoïdes	16.7	53.0
Schistosoma mansoni	0.3	8.7

village-dwellers and their Babinga 'workers', and they have shown important epidemiological differences between these two populations (table 3).

From the epidemiological description given certain patterns of habitat relationships to health in African forest zone populations can be discerned. Semi-nomadic hunter-gatherers are relatively little affected by malaria and other vector-borne disease. Their nutritional status is also generally satisfactory. It should also be stressed that the African *forest-dwelling* pygmies (Bi-Aka, Central African Empire; Jaeger, personal communication) show obvious signs of freedom and absence of constraint which appear to many observers (see chapter 19) as an expression of environmental fitness and happiness. Settlement appears to bring a deterioration of health. Man-made changes in the forest create situations favouring transmission of malaria, intestinal parasites, schistosomiasis and other water and vector-borne infections.

South-East Asia

Within the forests of this region dwell diverse peoples; hill tribes, negrito and proto-Malay aborigines, as well as ethnic nationals settled in villages and plantations. Thai hill tribes, Malay aborigines and Borneo Muruts, have been used as models of the health problems of the southeastern Asian peoples. The region is undergoing rapid change with replacement of the forest by agriculture to provide food for increasing populations. Resettlement schemes at the forest edge have become a regionally popular policy to deal with surplus urban populations. The impact of war and political upheaval in many countries of the region has had an enormous, if still poorly recognized, effect upon the health of the forest-dwelling peoples.

Malaysian aborigines (Orang Asli)

More is known about the health of the aboriginal groups of peninsular Malaysia than many of the otherworld's forest-dwelling peoples. Even so there are some important deficiencies in knowledge, particularly in respect of serological studies for viral infections and demographic characteristics. The results of health surveys on the Orang Asli within and on the fringe of the Malaysian tropical rain forest are summarized in table 4.

It has been estimated that the population of all the Orang Asli in 1965 was about 46 000 (Dunn, 1972). Of these, there are approximately 2 000–5 000 negritos (Semang) who still pursue a nomadic hunter-gatherer existence deep in the forests of northern Malaysia and southern Thailand. The other Orang Asli groups also have a close association with the forests but practise some form of settled or semi-settled agriculture. This is usually of a swidden nature for the cultivation of cassava, bananas, millet and beans. This diet is supplemented by gathered roots, seeds and fruits. Hunting (mice to elephants) provides the animal proteins. This diet provides a generally satisfactory state of nutrition although Polunin (1953b) considers them to be somewhat underweight due to underfeeding. This was confirmed by the recent study on the nutritional status of deep forest-dwelling aboriginal children carried out by Robson *et al.* (1973). Using weight age/height age ratio as a standard correlate of nutritional status they found little indication of malnutrition, only 16 per cent of the children being below norm (see also chapter 16). There is no evidence of the vitamin A deficiency, prevalent in Malays, nor of vitamin B deficiency. Goitre, attributed to the low iodine content of the river water, is endemic; 40 per cent of the Orang Asli studied by Polunin (1953b) were found to have an enlarged thyroid. However the degree of thyroid enlargement is less than in Malays

probably owing to the fact that shifting agriculture in forest clearings does not deplete the soil of iodine. Despite the relative availability of protein-rich foods, cultural practices are reported to produce nutritional inadequacies for children and women of child bearing age (Bolton, 1972a). Children, breast-fed up to the age of 4 years, are not given some kinds of animal meat. Pregnant women also have a culturally restricted diet (the Negritos do not have similar food taboos for pregnant women). The difference in feeding habits between young women and men is reflected in the serum albumin level, the average being 4.18 per cent and 4.67 per cent respectively.

TABLE 4. Results of health surveys of populations (per cent infected) for peninsular Malaysia (Orang Asli).

Habitat:	Deep forest-dwellers			Settled and forest fringe-dwellers	
Population group:	Negrito	Tenure	Group not stated	Tenure	Group not stated
Entamoeba histolytica	1.2[1]	0.8[1]	Prevalent, in association with *Edwardsiella tarda*	5.7	6[6]
Giardia	9.3[1], 47[14] 50 spleen rate	3[1]	25–65[11]	12.4	
Malaria	2.5 *falciparum* 12 spleen rate[14]	15[3]	About 20, not varying with age, most *falciparum*[5]		endemic[15]
Hookworm	93[1], 65[14]	77.6[1]	3 high burden[11] 91[5]	78.9	5 high burden
Ascaris	11.6[1], 0[14]	2.7[1]	2[11]	59.3	
Trichuris	55[1], 14[14]	23[1]		91.3	
Brugia malayi	33[10]	16[3]	44 and 2 elephantiasis[5] 8.5[12]		15[9]
Wuchereria bancrofti		0[3]	16.7[14] 2.8[12]		10[9]
Tuberculosis			25 tuberculin +[5] 7[11]		2[15]
Venereal diseases	37[1]		none seen[5]		
Scrub typhus			< 20 yrs 23 > 20 yrs 56		
Haemoglobinopathies	no sickle-cell trait		HbE trait 22 G6PD deficiency 17[4]		

Sources
1. Dunn, 1972
2. Polunin, 1953a
3. Wharton *et al.*, 1963
4. Lie-Injo and Chin, 1964
5. Polunin, 1953b
6. Bolton, 1972a
7. Nevin, 1938
8. Cadigan *et al.*, 1972
9. Ramachandran *et al.*,1964
10. Polunin, 1951
11. Bolton, 1968
12. Onyah, 1967
13. Gilman *et al.*, 1971
14. Kinzie *et al.*, 1966
15. Bolton, 1972b

Those who carried out health studies on the Orang Asli during the 1950s showed a concern for their declining population. Resettlement of the deep forest-dwellers into camps resulted in a high mortality rate of 90 per 1 000 with a compensatory birth rate of only 23 per 1 000 (Bolton, 1972b). Reproductive histories were unobtainable but Polunin found that 28–40 per cent of the children die before their mothers. Infant mortality rather than sterility (as is the case for the Muruts of Borneo) is the major cause of population decline. The major cause of infant mortality appears to be malaria. Table 4 shows that the overall malaria rate is over 20 per cent and presumably the infant-toddler group must have the highest parasitaemia rates and

densities. The lack of sickle-cell trait, a protective selective adaptation of negroes of African origin, makes the aborigines particularly susceptible to *Plasmodium falciparum* malaria. There is a high incidence of HbE and G6PD (glucose-6-phosphate dehydrogenase) deficiency (Lie-Injo and Chin, 1964). These characters may be protective although this is still an unresolved question.

The main malaria vector is *Anopheles maculatus*. This mosquito breeds in small forest streams and seepages exposed to sunlight. Deforestation along the streams creates ideal conditions for its proliferation. The effect of more extensive deforestation on the transmission of malaria by *A. maculatus* will be discussed later. In contrast to Thailand

and Viet-Nam, *A. balabacensis* does not appear to be a significant vector of human malaria but its bionomic isolation makes it a vector of the many primate malarias present in the monkeys and gibbons of the Malaysian forests. Primate malaria is of a zoonotic potential, some species such as *Plasmodium knowlesi*, *P. cynomolgi* and *P. inui* being capable of infecting man. However, there is no evidence so far that forest-dwelling aborigines regularly become infected with primate malarias.

The prevalence of intestinal helminths, particularly hookworm, is surprisingly high in the Orang Asli of the forest. One would suspect that the shifting habit of these peoples would not result in the heavy seeding of their immediate habitat; although it is to be noted that the prevalence rate and burden of *Ascaris* and *Trichuris* are significantly lower in deep forest Negritos than semi-settled proto-Malay aborigines (Kinzie *et al.*, 1966; Bolton, 1968). However even the highly nomadic negritos have a hookworm prevalence of 93 per cent (Dunn, 1972). Perhaps further study of aboriginal behaviour in respect of their defecatory habits might help explain the high hookworm rate. Hookworm could also account for the low grade anaemia, Hb 9–10 g/l is common in the Orang Asli (Polunin, 1953b). Malaria does not seem to be a major contributory factor, the low Hb levels were also noted in the Senoi, a group that dwells at higher altitudes where malaria is hypoendemic. It is also possible that an iron-deficient diet plays a role in the general anaemic state. The Orang Asli are not prone to the cardio-vascular disorders. Their blood pressure is remarkably low at all ages, an average of 106 (systolic)/64 (diastolic) for the 30–59-year-old group and 104/67 for those over 60 years of age (Polunin, 1953b). Life in the forest does not produce the hypertensive anxieties of the civilized man, but rapid cultural change may have a serious impact. Kinzie *et al.* (1966) found a 16 per cent hypertension rate amongst a group of Negritos who had been settled in a rubber plantation. Amongst their proto-Malay neighbours hypertension was only 1.6 per cent of the group studied.

While the Orang Asli are not subject to such communicable diseases as smallpox or influenza (except for chance contacts with people from outside the forest when these epidemics may virtually annihilate them) there are some infectious diseases which are rife amongst them: tuberculosis, yaws, bronchial disorders and pneumonia are common and are leading causes of adult morbidity and mortality.

Filariasis is prevalent amongst the Orang Asli, particularly that caused by subperiodic *Brugia malayi* in groups living in or near the swamp forest bordering the large rivers where the *Mansonia* spp. vectors breed. Malayan filariasis due to subperiodic *Brugia malayi* is a zoonotic infection, the main reservoir host being the leaf-eating monkey (*Presbytis*) and the macaque (*Macaca fascicularis*). The presence of rural *Wuchereria bancrofti* infections in the Orang Asli of the forest is of interest for, as far as is known, it is a parasite of man only. It is difficult to understand how small bands of widely dispersed nomadic people can maintain the infection without the presence of a reservoir host. The sylvatic strain of *W. bancrofti* in Malayan aborigines

has not been adequately studied and merits further investigation into its epidemiology and biological characteristics. The forest strain is thought to be transmitted by *Anopheles whartoni*.

Scrub typhus, conventionally thought to be an infection of the scrub grassland area near the forest fringe, can be contracted in the forest and Cadigan *et al.* (1972) showed by the indirect fluorescent antibody technique that 56 per cent of the deep forest Orang Asli under the age of 20 years had antibodies to *Rickettsia tsutsugamushi*. The course of the disease is evidently much milder than in other situations and areas such as Japan and perhaps this sylvatic strain, with a probable forest rodent reservoir, has undergone humanization. Publications by Walker *et al.* (1973) and by Muul *et al.* (1974) also deal with this subject.

Hill tribes (northern Thailand)

Hill peoples are located in a belt extending from northern Burma, across Thailand and into Indo-China. Little information is available regarding the current health status of the Burmese hill tribes, while the Montagnards of Indo-China are now in such a perilous state as victims of recent events that an account of their health status would be of historical interest only. Relatively few health surveys have been carried out on Thai hill communities but the excellent demographic investigations by Kunstadter (1970) on the Lua and Karen living in the evergreen forests of northern Thailand give some insights into morbidity and mortality amongst these peoples. His findings for the Luas and Karens of Mae Sariang are presented here as a model for population and health characteristics of hill people. The hill people of northwestern Thailand are farmers of upland (swidden, slash and burn) evergreen forests. They live in small settlements separated by unoccupied forest. They occasionally come into the larger towns. The demographic findings are summarized below.

	Lua	Karen
Median age	17.1	16.1
Mean number of live births to women of completed fertility (A)	4.8	6.8
Mean number of surviving children born to women of completed fertility (B)	2.8	4.3
Survival ratio (B/A)	0.59	0.63

The above figures show that both populations are young and increasing rapidly despite low childhood survival (about 60 per cent of the children born to women of completed fertility survived to the time when the census was taken). Kunstadter has estimated that they were doubling every generation (see also chapter 15). The effect of this rapidly expanding population is everywhere evident—70 per cent of the evergreen forest has already been denuded. Why are these hill groups experiencing a population explosion while other isolated agriculturalists have, for the most part, stable or declining populations? It would seem that although infant mortality is high, in the absence of any other form of population control it is not of a magnitude to maintain stability. The health survey conducted by Kunstadter *et al.* (1968) indicated that modern medical care is not readily accepted or available and therefore mortality rates are essentially unaffected by medical treatment. The group studied was isolated from measles transmission (although outbreaks do occur from time to time) as well as the arboviruses causing dengue and chikungunya. The principal peridomestic vector of these infections, *Aedes aegypti*, was absent from their village. Japanese encephalitis, an infection occurring in lowland villages and towns surrounded by wet rice paddies, was also absent. Probably of prime importance is the very low malaria endemicity; no parasites were found by examination of thick-thin blood films. The spleen rate was only 6 per cent, giving further evidence of the existing hypoendemic situation. A malaria vector, *Anopheles maculatus*, was present but in very low density. It may be, that despite the presence of intestinal parasites, respiratory diseases (tuberculosis is common amongst the adults), gastro-enteritis as well as other prevalent infections, primitive agriculturalists' population numbers increase only when malaria is absent or hypoendemic.

Muruts of Sabah

A group who face imminent population extinction is the Muruts of Sabah due to female sterility superimposed upon a high infant mortality rate. The Muruts are a mongoloid tribe of hunter-shifting agriculturalists living in the forest interior. They live in semi-permanent houses collected in villages rarely exceeding 100 people. They grow cassava as their staple crop along with some hill rice and vegetables. This is supplemented with protein from animal meat of the forest and fish from the stream. Evidently this diet is either not completely adequate or customary habits withhold essential foods from children. Clarke (1951) observed malnutrition, rickets and xerophthalmia to be common amongst the children. On the other hand, Polunin (1958) did not find signs of nutritional deficiency, except enlarged thyroid in 20 per cent of the population.

Infant mortality, chiefly due to malaria transmitted by *Anopheles balabacensis*, is high. Respiratory diseases are the main cause of adult mortality (3.6/1 000). Periodic epidemics of influenza, measles and smallpox cause additional wastage (Clarke, 1951). Even though the mortality from infectious diseases is high the chief cause of population decline is considered to be female sterility (Polunin, 1958). Only 36 per cent of the population were under 20 years of age. Women completing their reproductive life had an average of less than 2 live births and 40 per cent of the married women were childless. Further investigation (Polunin, 1958) revealed the sterility was bacterial, probably anaerobic cocci. The customs relating to pregnancy and delivery are believed to cause pelvic sepsis, puerperal infection with consequent tubal blockage. During pregnancy it is the practice for untrained older women to make frequent manual examination of genitalia and delivery is carried out under highly unhygienic conditions. The Muruts

must have existed as a stable population for a thousand years or more. The reasons for the relatively sudden decline are not known; did customs relating to birth and pregnancy change? Did they move into a highly malarious area or alter their environment in a way that would favour transmission? Whatever the original causes, they represent a dwindling community facing extinction due to a combination of their customs and environment. The Muruts are now settled in forest fringe areas but no studies have been carried out to determine whether their health status has changed.

Conclusion

There are a wide variety of human associations with the tropical forests of South-East Asia. Deep-forest hunter-gatherers and incipient agriculturalists are reported to have a generally good health status. Malaria rates are relatively low, nutrition adequate and the diseases of affluence (e.g. cardio-vascular disease, diabetes) virtually absent. The infections contracted may be of an ecologically-dependent zoonotic nature, e.g. subperiodic Malayan filariasis and scrub typhus, or contaminant diseases such as water and food-borne gastro-enteritis and various intestinal parasites.

Settlement and opening of the forest often produces conditions suitable for the breeding of anopheline vectors of malaria. Resettlement schemes in particular seem to have been adversely affected. There is also a suggestion that general health deteriorates with settlement at the ecotone; there is a greater variety and intensity of infections and physiological diseases such as cardio-vascular abnormalities become evident. There is also evidence that settled farming in cleared forest areas may eventually deplete the soil of trace elements which in turn may cause the pathological conditions associated with such deficiencies.

Comprehensive health surveys for town-dwellers, in the forest zone, are notably lacking.

Melanesia

While the southeastern Asian and Melanesian forest ecosystems have an overall general common character, Wallace's biogeographical line divides the area into two distinct faunistic assemblages and has important epidemiological implications. The primate reservoirs of many arboviruses and *Brugia malayi* are absent from the Melanesian region as are the molluscan vectors and ungulate reservoirs of *Schistosoma japonicum*. Melanesia is not free from anophelines; *Anopheles farauti* and *A. punctulatus* are accompanied by many hyperendemic malaria foci. These vectors also transmit *Wuchereria bancrofti* and there is a high prevalence of filariasis in most areas at altitudes below 3 000 m (Desowitz *et al.*, 1966).

In the foot-hills of the central ranges of New Guinea are hunter-gatherer-incipient agricultural societies living in small scattered groups, completely within the forest. The bulk of the population dwell on the slopes of the upland valleys of the central cordillera; a dense population of rotational gardeners living in a grassland bordered by extensive primary rain forest. In addition, relatively small populations reside in areas where the forest has been replaced by monocultures of coco-nut, cocoa and coffee. However, these are increasing rapidly.

Very little information is available on the hunter-gatherers or agriculturalists. A demographic study of the Breri revealed that women marry at about the age of 16, have their first child at an average of 18.7 years and a child every three years thereafter with a mean of 8.4 children. After contact with a missionary group and the introduction of simple medical services such as malaria therapy the once stable population has been expanding at the rate of 2.3 per cent/a. Other forest peoples, particularly those living in the mid-montane and lower regions, without access to these services have decreased in population due to high infant mortality and female infertility (of presumably unknown etiology). Formerly warfare, which sometimes killed 5 per cent of a group's population in one raid, also acted as a regulatory factor. Demographic statistics for other New Guinea groups show great variability but the overall picture is one of high childhood mortality; only between 475 to 740 per 1 000 surviving to the age of 5 years. This high infant wastage is balanced by a fertility rate, at least in the Toricelli Range in the Sepik region, of 166 per 1 000 women aged 15–49 years and a crude birth rate of 36 per 1 000. The chief identifiable cause of infant mortality is malaria but other infections such as gastro-enteritis and pneumonia that leave no physical or serological residue would be difficult to assess by the point prevalence studies from which the data are obtained. Moreover, malaria cannot be disassociated from the interacting effects of malnutrition, intestinal parasitism and other intercurrent infections. Malaria in Melanesia is mainly a man-made disease. The vector *Anopheles farauti* breeds in the small areas of exposed water created by human activity. It would be of interest to obtain comparative malaria rates for nomadic hunter-gatherers living entirely within the intact forest and settled agriculturalists. Altitude is a crucial factor in the epidemiology of malaria and filariasis. In montane forest at heights above *ca*. 2 100 m these infections do not generally occur despite the presence of *A. farauti*. Presumably the ambient temperature is too low to allow completion of the intermediate developmental cycle.

A high prevalence of intestinal parasitism was found in virtually all populations surveyed. Here too altitude and associated ecological factors may influence the pattern. Lowland swidden agriculturalists have low rates and burdens of *Ascaris* and *Trichuris* but high infestations of hookworm while highlanders have high burdens of *Ascaris* and intestinal protozoa. Whether this is due to ecological factors or differences in behavioural customs related to defecation and fecal disposal is not known. Amoebiasis may be an important cause of morbidity in some communities. A high prevalence rate (>25 per cent) was noted as well as many cases of typical bloody diarrhoea in a stool survey carried out in a Cape Gloucester, New Britain, village bordered by the sea and forest (Desowitz, unpublished). New Guinea has foci of the highest balantidiasis rates in the world. This is obviously due to the close association between pig and man.

Little or no information is available on the presence of arbovirus infections and other viruses in New Guinea forest-dwellers, but Murray Valley encephalitis virus and dengue types 1 and 2 have been reported. Group C viruses are rare (Bell, 1973).

New Guinea rural populations too are free of the diseases of affluence e.g. hypertension, arteriosclerosis, coronary artery disease, stroke and gout. Maddocks and Vines (1966) speculate that the absence of hypertension may be the product of a barter with infectious disease, particularly malaria. They noted that New Guinea males with splenomegaly of malarial origin had lower blood pressure than males with normal size spleens and felt that this was consistent with the hypothesis that chronic infections lower blood pressure. A possibly related phenomenon has been reported by Desowitz and Langer (1968) who found that serum cholesterol levels were markedly below normal in non-human primates and rodents during malaria infection.

Hornabrook (personal communication) presents an account of the nutritional status of New Guinea forest-dwellers. He first comments on the lack of precise indicators of nutritional status. The bulk of the Melanesian's diet consists of protein-poor root foods e.g. sweet potato, taro, yam and cassava. The closer the association with the forest a community has the better able it is to supplement its diet with the protein-rich foods indigenous to the forest—meat of the wild birds and mammals, nuts and, for the Melanesians, certain insects. Highland-dwellers obtain only 6–7 per cent of the energy value of the diet from protein sources and fats provide 10 per cent. The Melanesian has adapted to this energy-poor diet. Norgan *et al*. (1974) found that the food eaten by New Guinea forest-dwellers had a low energy: protein ratio and a relatively low total energy content. Yet their way of life required moderate physical effort for several hours daily; their physique and muscular development were apparently admirable and there were few indications of nutritional inadequacy (see chapter 16).

Amerinds and other forest-dwellers of Central and South America

In the vast forest that extends from southern Mexico to eastern Peru, there is probably a greater variety of human associations with forests than in any other similar area of the world. The Amerinds are of many groups whose relation to the forest ranges from extremely isolated hunter-gatherers to settled agriculturalists living in villages of about 300 persons at the forest fringe. Tribes of more than a few hundred are unstable and tend to split, often violently. The Cayapo (Black, 1975) seem to have arisen by such a fission. If the population falls below 100, other pressures come into play, particularly the problem of finding marital partners not forbidden by incest taboos. The population of the Amazon periphery is very sparse, and 200 km or more of forest may separate neighbouring groups. Settlements of mestizos are distributed throughout much of the area, usually along the banks of the larger rivers. The mestizos are the harvesters of the forest, engaged in chicle and nut collection, wood cutting, etc., and are thus exposed to infections cycled within the forest ecosystem. In Surinam and the Guyanas the descendants of escaped slaves, the Bush Negroes, live deep within the forest. Large tracts of forest are being felled for cultivation. As in other parts of the world, the altered environment may cause new vectors and changes in epidemiological patterns.

Table 5 summarizes the findings of health surveys of forest peoples of Central and South America. A number of infections are intrinsic to the neotropical forest ecosystem. Notable amongst these are *Leishmania* of the *mexicana* and *brasiliensis* complexes, malaria transmitted by forest-breeding anophelines and arboviruses, e.g. yellow fever and Mayaro virus. The prevalence of herpes simplex, chicken-pox, Epstein-Barr (infectious mononucleosis), hepatitis B and cytomegalovirus diseases is higher in these communities than commonly found in more advanced cultures, and no group was found free of them; they seem to be well adapted to persistence in such groups for they caused little apparent morbidity and did not threaten the continuance of their hosts; the infectious agents are known to remain in infected persons for long periods of time and to be reactivable.

Introduced diseases, e.g. measles and tuberculosis, have caused great havoc. This group of diseases (measles, mumps, rubella, influenza, parainfluenza, poliomyelitis) causes acute infections which spread rapidly to the whole population and die out; children born after the epidemic remain free of disease. Because they do not persist in small populations, it is unlikely that they were a problem to ancient man; conversely, they are the diseases to which modern man may have had the least time to adapt (Black, 1975). There is no genetic deficiency amongst these peoples to respond immunologically (Neel *et al.*, 1970) but infection of immunologically virgin individuals causes severe clinical manifestations. An epidemic of measles was described by Neel *et al.* (1970) in the Yanomamo, a group of isolated agriculturalist-hunter-gatherers of the Brazilian-Venezuelan border. In some villages 17 per cent died. Tuberculosis seems to have been introduced recently; none of the Auca of the Ecuadorian Amazon give positive tuberculin tests, but the incidence of positive tests is high in the Suia, Txukharamai (Notels *et al.*, 1967) and Xikrin (Black *et al.*, 1970, 1974). Notels *et al.* (1967) described the devastation amongst Cayapo of Brazil after introduction of tuberculosis. Such events have been, and are being repeated and there is concern that they may destroy the Indian populations.

Investigators, such as Cabannes *et al.* (1964) who have observed the Amerinds in their undisturbed forest setting, have remarked how remarkably adapted they are to their forest setting. Hypertension and severe cardiac abnormalities are absent. The adults at least, are reported to be well nourished, the diet of protein foods of fish and game supplemented by fruit, and cassava and legumes from their slash-and-burn plots is nutritionally adequate. No vitamin deficiency states or other dietary disease as goitre have been reported.

Leishmaniasis, either in its more benign cutaneous or highly invasive muco-cutaneous forms, has not been reported in Amerind hunter-gatherers. Possibly the bionomic isolation of vectors and animals reservoirs in the deep forest is not conducive for transmission. Alternatively, one can speculate that selection has led to an immunologically highly competent group. The relationship of the immune response to the type of lesion produced by leishmania has been elegantly elucidated during this past decade. There appears to be some differences in susceptibility, in that leishmaniasis is prevalent and often severe in the mestizos and other non-Amerinds who enter the forests, often for extended periods, to collect chicle, rubber and other natural products. Indeed leishmaniasis constitutes a serious impediment to the exploitation of the neotropical forests.

The treponemal diseases, yaws and syphilis, are uncommon in forest Indians (as among the Brazilian Tiriyo or the Xavante) but in areas such as Surinam and the Guyanas they are highly prevalent in the Bush Negroes who also are deep forest-dwellers occupying the same general territory as the Amerinds (Sausse, 1951; Schaad, 1960). However, among the three Cayapo groups examined by Black (1975), a very high prevalence of positive tests was found; clinical examination revealed no evidence of venereal or congenital syphilis, pinta or yaws; the serological tests indicate that some treponema is very commonly encountered by these people and the absence of clinical signs suggests that it does minimal harm. Either the agent is of low virulence or the Cayapo possess unusual resistance; a stable relationship exists which permits continuance of both parasite and host.

There appears to be considerable clinical immunity to yellow fever despite a high percentage of Amerinds and other forest-dwellers having been exposed to the infection, as indicated by high serological positivity rates (Sneath, 1939; Janssen, 1961; Black *et al.*, 1970; Madalengoita *et al.*, 1973). The prevalence of antibody to this virus varies from one group to another, presumably depending on the availability of some other host or vector in the immediate area; however, where antibody was found in the human population, the positive proportion increased steadily with age. Apparently the risk of infection was proportional to duration of exposure, and the infections had not occurred in the form of major epidemics. It may be also that the forest strain is of relatively low virulence or that primary infection with the many other Group B arboviruses afford a protection against yellow fever.

Nevertheless there is considerable morbidity amongst the Amerinds and populations tend to be small or dying out, and as Notels *et al.* (1967) have commented in reference to the Amerinds of Brazil, 'little is known about the population structure and selective pressures to which groups of hunter-gatherers and incipient agriculturalists are submitted'. Certain patterns in morbidity, mortality and reproductive rates can, however, be discerned from the results of the health surveys. Malaria is undoubtedly a significant cause of mortality and morbidity in the Amerinds in whom glucose-6-phosphate dehydrogenase and the abnormal haemoglobin, particularly Hb–S, genes are absent (Tondo and Salzano, 1960; Arends, 1961; Arends and Gallango, 1965). As will be seen from table 5, spleen and parasite rates are high in most Amerinds studied. Sausse (1951) notes that the Amerinds of French Guyana suffer from malaria at all ages while their Bush Negro adult

TABLE 5. Results of health surveys of populations (per cent infected) of South American forest-dwellers.

Locality:	Brazil-Venezuela	Peru		Surinam, Guyanas	
Population group:	Amerinds	Amerind 8 (Machinguenes)	Indian and Mestizo forest-villagers	Indians	Bush Negroes
Entamoeba histolytica	23[1]	32			rare[11]
Giardia	12[1]				
Toxoplasma	47[5], 52[7] (serology)				
Malaria	14[6] spleen rate <19 hunter-gatherer children[6] 80 Alto Xingu[17]	cause of high infant mortality		spleen rate all ages 16[15], 89[9], 95[12]	children 89 spleen rate high 77[11, 14] Adults 32, spleen rate low 18[11, 14]
Leishmania	no+serology[17]		Muco-cutaneous[9] +cutaneous seen	common[15]	
Hookworm	95[1]	57	40[9]		
Strongyloides	26[1]	15	10[9]		
Ascaris	70[1]	62	60[9]		100 in children[11]
Trichuris	91[1]	53	40[9]		100 in children[11]
Onchocerca	47[2], 63[3] 4 ocular lesions[2, 3]				
Dipetalonema perstans				rare[12] 69[18]	
Mansonella ozzardi				rare[12] 45[18]	
Bacillary dysentery		cause of high infant mortality		Typhoid serology 22+	common in children[11] rare+Widal[12] little pulmonary disease[14, 15]
Tuberculosis	tuberculin+14 (Cayapo)[4], 80 hunter-gatherers[6]			tuberculin+ 40[12], 20[15]	
Arboviruses (serological+rates) and other virus	Group B: 18 yellow fever (Tiriyo)[5] yellow fever and Ilheus 50[6] Group A: Mayaro 80[5], 50, 90[6]	Ind[10] 70 85 45	Mes[10] 50 80 35 Indians: little or measles, 50 CMV, varicella, herpes[5, 6]	Yellow fever 28[13] 48 males, 8 females[16] 27[19]	Yellow fever 50 26 males, 1.9 females[13] >50[19]
Syphilis/yaws			No syphilis, yaws present in some tribes[6]	no VD, no yaws[14] 4 Wasserman[12]+	17 yaws[14], prevalent VD, 33 Wasserman[12]+

Sources
1. Knight and Prata, 1972
2. Baruzzi, 1970
3. Moraes et al., 1973
4. Notels et al., 1967
5. Black et al., 1970
6. Black et al., 1974
7. Baruzzi, 1970
8. Wieseke, 1968
9. Carrizales and Destombes, 1970
10. Madalengoita et al., 1973
11. Janssen, 1961
12. Schaad, 1960
13. Snijders et al., 1947
14. Sausse, 1951
15. Cabannes et al., 1964
16. Sneath, 1939
17. Baruzzi et al., 1971
18. Orihel, 1967
19. Wolff et al., 1958

Several workers have looked at the epidemiology of infectious disease in the people living in the periphery of the Amazon basin: serological and clinical evidence of disease in the Xavante of Brazil was the subject of a major study by Neel et al. (1970); Baruzzi and his colleagues (1970, 1971) have conducted various studies of the people of the Xingu Park; Schaad (1960) has studied two Surinam groups; Black et al. (1970, 1974) completed a study of seven Carib and Cayapo groups in Brazil.

neighbours (who have a high rate of sickle-cell trait) are relatively unaffected. There are endemic foci of malaria in settlements and towns near the forest throughout Central and South America. Malaria occurs where the principal vector, *Anopheles darlingi*, is abundant. This anopheline has the ability to breed under a range of conditions: reservoirs containing floating plants, rain collections, and forest streams. It is highly anthropophilic and endophilic, characters that further ensure its vectorial efficiency. Within the deep forest, malaria is transmitted chiefly by *A. cruzi*, which breeds in the tanks of bromeliads growing in the canopy.

Other diseases contribute to the health of forest-dwellers. Virtually all who have surveyed these people have commented on the high frequency of pulmonary disease, ranging from smoker's cough to tuberculosis. Dysentery causes high infant mortality (Wieseke, 1968). Adults also are subject to episodic bouts of diarrhoea, mainly of bacterial etiology. Schaad (1960) found only a low proportion of the sera from Amerinds and Bush Negroes of Surinam to give a positive Widal test for typhoid. Relatively little is known regarding drinking-water sources and the water purity. With settlements, the water supply will be one of the most crucial aspects affecting health. Amoebiasis is said to be rare, but one survey of Brazilian Amerinds revealed an *Entamoeba histolytica* prevalence of 23 per cent, and an indirect fluorescent antibody positivity rate also of 23 per cent. Since serological positivity is indicative of past or present invasive amoebiasis it would seem that clinically significant infections commonly occur in some groups. There was a high prevalence of intestinal helminthiasis in all communities studied. Where studied, the combination of malaria and hookworm, was shown to produce anaemia in children and women of child bearing age; 60 per cent of this group are reported to have haemoglobin levels below normal (Schaad, 1960; Cabannes *et al.*, 1964).

The neotropical forest-dwellers are generally unhealthy. All except the most isolated Amerind tribes are nearing extinction as a result of introduced diseases, such as tuberculosis, measles, cerebro-spinal meningitis and even malaria. Black (1975) suggested that, in addition to the highly endemic diseases which can persist in an individual for a prolonged period (treponemal infections, arbovirus diseases, leishmaniasis) there are other diseases which are infectious only in the acute phase and die out quickly after introduction. The latter diseases could not perpetuate themselves before the advent of advanced cultures and did not exert selective pressures on the people until relatively recently. The geographical features that separated the South American Indians from important sources of evolving disease agents and the filtering effect of the Bering and Panamanian land bridges may have protected them from a number of diseases; their small population concentrations would have protected them from many diseases which reached the area.

Other groups such as the Bush Negroes in the forest and Mestizo settlements adjacent to forest areas also present a rather dismal picture. However, few studies have been carried out on the demographic aspects except the obser-

vation of high infant mortality and the often relative paucity of children in the population composition. Janssen (1961) has provided one of the few examples of adequate demographic data. His study of the Bush Negroes of Surinam revealed an infant mortality of 113 deaths per 1 000 live births while there were only 39 births per 1 000 inhabitants/a. Moreover, the surviving children's growth was retarded, possibly due to a diet deficient in protein (estimated at 2 g/d). Here again is an example of an essentially benign environment in respect of the intrinsic inhabitants and the unhealthy conditions produced when they depart, even slightly, from their isolation as when settlers make inroads into the forests.

Epidemiology

Malaria

Malaria is a major cause of continuous morbidity and mortality of children in tropical forest zones throughout the world. The *Plasmodium* parasites of South-East Asia and South America have become resistant to the drugs that were the basis of therapy and prophylaxis. In many areas the anopheline vectors have become resistant to insecticides. Finally, authorities faced with these frustrations and other demands on their limited financial and professional resources have become resistant to further long-term expenditure required by an anti-malaria programme. Also, human activities have created many new endemic foci, for malaria is largely a man-made disease.

The transmission of malaria depends upon the presence of an efficient vector in sufficient density. Vector efficiency is related to being a suitable host, in which the parasite regularly completes its development, and behavioural characteristics such as the predilection to feed upon man (anthropophilism) and to either enter houses (endophily) or to be associated with human outdoor activity. Within the intact forest there are relatively few species of malaria vectors. Not only is there a paucity of water but also most anopheline vectors prefer breeding sites exposed to sunlight. In South America a group of mosquitoes has adapted to breeding in the water found in the tanks of bromeliads growing in the canopy. One member of this group, *Anopheles cruzi*, is an important vector of human and primate malaria (*Plasmodium simium* and *P. brasilianum*). This species also appears to have separated into behaviourally variant populations affecting its vectorial capacity; in some areas of Brazil it bites man at ground level while in other areas it remains in the canopy where it transmits simian malaria only (Deane *et al.*, 1966; Deane, 1971). Probably the only example of an artificially created forest leading to a malaria outbreak involved bromeliad anophelines. In Trinidad the establishment of the cocoa required planting *Erythrina glauca* and *E. micropteryx* to provide the shade. Epiphytic bromeliads colonized the tree canopy and *Anopheles bellator* and *A. homunculus* proliferated. *A. homunculus* remained confined to the forest and transmits malaria there but *A. bellator* entered clearings and moved to villages bordering the forest bringing malaria in its wake (Downs and Pittendrigh, 1946). In Africa, *A. moucheti* and *A. nili* are reputed to be vectors of malaria in the undisturbed forest (D'Haenens *et al.*, 1961; Mouchet and Gariou, 1966). The forest transmission of malaria in South-East Asia is less clear. The *umbrosus* and *leucosphyrus* groups are confined to the forest while the *hyrcanus* group will enter villages. Most of these anophelines, as well as canopy inhabiting species such as *A. hackeri*, are vectors of the malarias of monkeys and bite man only with reluctance, although *A. letifer* will bite monkeys in the canopy and comes to ground level to feed on man (Wharton and Eyles, 1961; Wharton *et al.*, 1963; Moorhouse and Wharton, 1965; Warren *et al.*, 1970). In Malaysia, *A. balabacensis* is zoophilic and an important vector of primate malaria only while in Thailand this species is anthropophilic and the major malaria vector in that country.

Small man-made alterations of the forest ecosystem may result in a great increase of malaria. Settlement in small clearings produces favourable habitats for highly efficient anopheline vectors by opening the forest to sunlight and by human activity making water for breeding. In Africa it is *A. gambiae*, in South America *A. darlingi* and in Melanesia *A. farauti* and *A. punctulatus* that breed, proliferate and transmit malaria under these conditions. Major alterations in the forest also may result in a physiography favourable for large mosquito populations. A classical example of this is that the destruction of the Malaysian forest for rubber plantations exposed the streams to sunlight and created the right conditions for the proliferation of *A. maculatus* and epidemic malaria amongst the newly settled (and non-immune) workers. Similarly Desowitz *et al.* (1974) noted that large areas of hill-side evergreen forest in southern Viet-Nam had been cleared by chemicals and the physical attrition of war. They hypothesized that conditions for *Anopheles maculatus* transmission of malaria may have been produced. They found that anopheles were not present in the intact dense mangrove forests of South-East Asia and the inhabitants of these forests were free of malaria. A similar observation had been made for Malaysian mangrove forests by Strickland (1936). The southern Viet-Nam mangroves were intensively sprayed with herbicide and the Rung Sat became a treeless wasteland. Under these conditions, and with the introduction of rice cultivation, *A. sinensis* and *A. lesteri* began to breed and malaria became endemic. In Thailand a large tract of mangrove forest was cleared; *A. subpictus* has proliferated and awaits the new inhabitants. In India the clearing of the forest for rice cultivation has resulted in malaria epidemics by paddy-breeding *A. fluviatilis* and *A. culicifacies* (White and Adhikari, 1940; Srivastava, 1955). In Africa the two major vectors are *A. gambiae*[1] and *A. funestus*. These species usually inhabit the humid savanna but will penetrate into the forest zone when it is suitably altered by human settlement especially in the clearings about the settlements, and particularly along the rutted roads with their pools of water. Road surfacing or grading would destroy these breeding sites. *A. nili* is a vector of local importance in forest villages situated near rapidly flowing water, while *A. moucheti* is a

vector limited to the central African forest zone and to the fringing forest derived from it. It is precisely to cleared areas that settlers are being sent, often with disastrous results. Until agricultural and administrative planners become cognizant of the health problems and make provision for safeguarding health it is difficult to see how the newly opened areas will ever be effectively settled.

Man is susceptible to infection with *Plasmodium simium* and *P. brasilianum* in South America, *P. cynomolgi*, *P. shortii*, *P. inui* and *P. knowlesi* in the Far East, and *P. schwetzi* in Africa. The occurrence of Asian and South American primate malarias in men who entered the forest gave rise to concern that the primate malaria reservoir would complicate control and eradication. However a number of facts tend to mitigate against this possibility:

— human cases of primate malaria are very rare; Warren *et al.* (1970) inoculated the blood of several hundred individuals living in a village situated in a Malaysian rain forest into rhesus monkeys; no infection in the sentinel monkeys occurred;
— most of the vectors of primate malaria remain in the canopy and are largely zoophilic;
— simian malaria in man gives rise to low grade infections with scanty or absent gametocytaemia.

Nevertheless, Warren *et al.* (1970) warn 'that man can only become infected with the simian malaria when he becomes involved in the normal mosquito-monkey cycles in the forest'. The form of involvement which would allow this contact is unknown. Destruction of the forest would reduce the primate population and the normally arboreal, zoophilic primate vectors might alter their behaviour and come to ground level seeking food, biting man and monkey alternately. The situation may be somewhat different in the areas of the African forest where man lives near areas inhabited by higher apes. *P. schwetzi* is congruent to *P. ovale* and may actually be the same species. Juminer (1970) believes *P. schwetzi* to be a zoonosis and Languillon *et al.* (1955) have found a somewhat higher prevalence rate of *P. ovale* in Cameroon forest zone populations than in populations living outside the forest.

It can be concluded that primate malarias hold little danger as zoonoses. However altering the forest may change this and continued surveillance is necessary. Reviews by Garnham (1969) and Coatney (1971) are comprehensive sources of information.

African trypanosomiasis

Trypanosomiasis does occur in the African rain forests. Our knowledge of the bionomics of the forest tsetse and the trypanosomes they transmit largely derives from the studies

1. The problems associated with *A. gambiae* transmission of malaria are complicated in that six sibling species are known to exist, each having somewhat different behavioural characteristics. The bionomics of each species are not fully understood but it appears that species A which is anthropophilic and endophilic is the most important vector in the forests, at least in western Africa. In the mangroves there are two vectors: *A. melas* in the west and *A. merus* in the east.

carried out in Cameroon, the People's Republic of the Congo, Zaire and southern Nigeria by entomologists of the West African (now Nigerian) Institute for Trypanosomiasis Research. The landscape epidemiology of Gambian trypanosomiasis (*Trypanosoma gambiense*) is also tree savanna through which run streams and rivers bordered by fringing forest, woodland or thicket. The vector, *Glossina palpalis*, is confined to these riparian thickets. *G. palpalis* is a relatively inefficient intermediate host and even after experimental infective blood meals only a relatively small percentage of flies harbour the metacyclic trypanosomes in the salivary glands. Thus even in epidemic circumstances only a few of the flies are infectious, but they remain infectious during their whole life of several months. However the concentration of the fly along the watercourses where there is so much human activity ensures a high degree of vector-host contact. *G. palpalis* is also abundant in the western African tropical forest, where the humid conditions permit wide dispersion. In the forest it was supposed not to concentrate at the streams or water-holes as is the case in the savannas; 'whereas man is dependent on the stream the fly is not' (Page and McDonald, 1959). This does not seem to be true and the fly concentrates along the rivers in the forests, but its dispersion is not known; all the historical trypanosomiasis foci are located in the forests (Cameroon, Congo, Gabon, Central African Empire). *G. palpalis*, which is implicated in these foci, is largely replaced by *G. fuscipes*, a tsetse of similar behaviour, east of 12°. *G. fuscipes* is present in the forest of Cameroon where it is responsible for endemic foci of human trypanosomiasis in settlements within the forest. Another example of a forest focus of Gambian trypanosomiasis exists in the M'Bomo region of the People's Republic of the Congo. In the epicentre of this focus, the village of Nzondo-Lebango, 33 per cent of the population were found to be serologically positive by the indirect immunofluorescent test (Frezil and Colum, in press). The situation is less clear in the forests of eastern and central Africa where the mainly savanna Rhodesian trypanosomiasis occurs. *Trypanosoma rhodesiense*, unlike *T. gambiense*, is a zoonotic infection of ungulate reservoir hosts, including bushbuck (*Tragelaphus scriptus*) which often frequents the forest but whether it, and other animals, maintain a sylvatic cycle in this eco-system is not known.

There is another condition in which the forest provides the habitat for the transmission of Gambian trypanosomiasis. Intensive settlement and cultivation of areas north of the Niger and Benue rivers resulted in almost complete deforestation (as in most of western Africa and was mentioned by Challier, Muraz, Richet, etc.). However it was the practice of many people to preserve a small forest grove. People retreated to these *tsafi* groves, often for days at a time, for various rites. *G. palpalis* also retreated into them and there was intense man-fly contact. Some travellers introduced the infection to the tsetse of the groves. Explosive outbreaks ensued and an excellent description of one such episode that almost destroyed the Rukuba tribe has been given by Duggan (1962). The same phenomenon was described by Vanel (1940) during an epidemic in southern Cameroon forest which killed 60 per cent or more of the villagers. Movements of population may therefore contribute to disease outbreaks.

It is the *fusca* group of tsetse that is associated with the rain forest ecosystem. Most of these species are not attracted to man but feed on ungulates and wild pigs (Nash, 1952; Jordan *et al.*, 1961). From these the tsetse acquire infections of trypanosomiasis of veterinary importance, *Trypanosoma vivax*, *T. congolense* and *T. simiae*. The infection rate in certain species may be very high. Jordan (1961) reported, from dissected flies from southern Nigeria and Cameroon forests, that 36 per cent of *G. caliginea*,[1] 15 per cent of *G. medicorum*, 15 per cent of *G. fusca* and 24 per cent of *G. nigrofusca* to be infected; about 75 per cent of the infections were *Trypanosoma vivax*. In the Irangi forest of Zaire, *Glossina vanhoofi*, which feeds mainly on wild pigs, were found to be infected with *Trypanosoma suis* (Van den Berghe and Zaghi, 1963). Thus when man destroys the forest to introduce animals, tsetse is already present to frustrate his efforts to produce animal protein. *Glossina medicorum*, *G. fusca* and *G. nigrofusca* have the particular ability to make forays from the forest to grazing areas. Jordan (1962) has written a comprehensive account of the forest *fusca* group of tsetse.

American trypanosomiasis (Chagas disease)

Chagas disease has been considered as typically occurring in settlements outside the forest (Garnham, 1971). Then Herrer (1960) reported cases from the high forest area of the Peruvian Amazon Basin suggesting a possible forest cycle. The vector was considered to be *Panstrongylus herreri* a triatomid bug that is not house-haunting and has little contact with man. At that time the reservoir hosts were unknown. This gap in the knowledge of the sylvatic cycle was filled by Deane (1958) and Coura *et al.* (1966) who found opposums and forest rats to be infected, and by Albuquerque *et al.* (1969) and Funayama and Barretto (1970) who discovered natural infections in *Alouatta caraya* and *Callithrix* monkeys. Some 40 species of vertebrates (bats, rodent marsupials, edentates) have been implicated as reservoirs to *Trypanosoma cruzi*; *Rattus rattus* may also have become infected. In Brazil the forest vectors were reported as being *Triatoma tibimaculata* and *Panstrongylus megistus*.

The extent of the sylvatic cycle and all the reservoir hosts are unknown. The threat of the sylvatic cycle to forest settlements remains conjectural but as Coura *et al.* (1966) note there is a real danger 'if forests near the city are destroyed or a large number of animal deaths occur with adaptation of the bugs to human dwellings and human blood such as occurs in endemic areas of Chagas disease'. Considering the large number of reservoirs of *Trypanosoma cruzi*, there is a good indication that the disease is well

---

1. *G. caliginea* belongs in fact to the *palpalis* group; it is found in the perhumid forest, bordering the mangroves in Cameroon; Roubaud and Rageau (1950) have found a high rate of infection but the role of the fly is still unknown.

established in the forests and the relationship to forest edge is intensified by a widely practised slash-and-burn agriculture, which increases the ecotone.

Leishmaniasis of the New World

The landscape epidemiology of Old World leishmaniasis is semi-desert country while leishmaniasis of the New World is essentially an infection of the tropical forest ecosystem. Leishmaniasis of the New World consists of a bewildering array of species and strains of parasites and as well as producing a variety of clinical manifestations. In two comprehensive reviews Lainson and Shaw (1971, 1973) have made a gallant attempt to bring order to the classification of the parasite. They divide New World leishmaniasis into complexes of two species, *Leishmania mexicana* and *L. brasiliensis*, but admit that as knowledge accrues on the biological and immunological characters, revision will be required. Leishmaniasis occurs from Mexico's Yucatan peninsula to Argentina. It is impossible to do more than present a synopsis of the salient features, particularly as they epidemiologically relate to the forest ecosystem.

For both *L. mexicana* and *L. brasiliensis*, as they occur within the forest, a number of common characters can be described: they are transmitted by sandflies of the *Lutzomyia* and *Psychodopygus* genera; they are zoonoses, the main reservoir hosts being forest rodents, although marsupials and primates have also been reported as reservoirs; the disease depends on the member of the complex involved; single cutaneous lesions, e.g. chiclero's ulcer in Mexican forest workers caused by *L. mexicana* group; metastatic lesions involving nasopharyngeal tissue, e.g. espundia of Brazil caused by the *L. brasiliensis* group. Lainson and Shaw (1973) give a synoptic description of the varieties of *Leishmania*.

It is possible that the parasite's long sojourn in the rodent host may have rendered it relatively non-invasive (cutaneous leishmaniasis) for man; only with human intrusion in the forest is the cycle altered with consequent man to man transmission through anthropophilic vectors and does the parasite produce the muco-cutaneous form of the disease. Garnham (1971) has envisaged such a process in Brazil: 'the major factor in creation of epidemics is the extensive deforestation of the country such as seen in Brazil. The animal reservoir disappears with the alteration in the landscape, and other species of *Phlebotomus* transmit the infection between man to man. The disease in the terrible form of muco-cutaneous leishmaniasis (espundia) becomes rife. . . . The mild chiclero's ulcer of zoonotic origin is replaced by the sometimes fatal purely human disease'. Unfortunately there is no experimental proof to confirm or disprove this concept. The situation appears to be much more complex than transformation in virulence. Lainson and Shaw have evidence that a number of strains exist in the rodents of the Brazilian forest, all producing discrete cutaneous lesions in the animal host. When man is infected some strains give rise to cutaneous lesions while other strains invade the nasopharyngeal tissues. There is also the possibility that primary exposure to the less patho-

genic strains or species gives rise to a protective immunity against the more virulent leishmaniasis. That cross protection may occur has been shown by Lainson and Shaw's finding that a previous infection by *L. brasiliensis panamensis* protects humans against *L. mexicana mexicana*. Amerind hunter-gatherers living deep in the forest often have antibodies to *Leishmania* but rarely show clinical evidence of the disease. It is possible that cross-protection afforded from exposure to relatively non-invasive animal *Leishmania* is also operative in this instance. If this is true then destruction of the forest with the disappearance of the animal reservoir would eliminate this process of immunization.

A large number of phlebotomine species have evolved in the New World tropics. Many species have been found infected with trypanosomes but how many of these were with leishmania promastigotes of species capable of infecting man is unknown. Knowledge of the vectors responsible for both transmission of the sylvatic and human cycles is incomplete. What is known is largely based upon epidemiological and behavioural inferences. There are considerable differences in predilection for man between species but it appears that zoophilic species are mainly arboreal while anthropophilic species are ground level feeders. In a Panamanian rain forest Chaniotis *et al.* (1971) found that of the species collected 38 per cent were ground biters, 12 per cent arboreal and 49 per cent were found resting on tree trunks. The mature forest contained more species than secondary forest but in both seasonal changes that caused drying or waterlogging of the ground reduced the density of sand-flies. Another study of Chaniotis and Correa (1974) revealed that the anthropophilic ground biting sand-flies, *Lutzomyia olmeca, gomezi, panamensis, sanguinaria, trapidoi*, and *ylephiletrix* were rarely found in forest clearings. It is conceivable that unless sand-flies with a greater tolerance to lower humidity such as *Lutzomyia longipalpis* become established in the open areas, deforestation would reduce or even eliminate leishmaniasis. It is interesting that as long ago as 1942 Beltran and Bustamante reported that leishmaniasis did not occur in the land cleared of the forests.

New World leishmaniasis is an infection of the forest but human exploitation of the forest seems to affect transmission in different ways from area to area. More information to explain these is required in order to devise logical schemes of protection.

Schistosomiasis

As noted earlier the prevalence of schistosomiasis in Africans living within the tropical rain forest is relatively low owing to the paucity of snail vectors in forest streams. However *Schistosoma intercalatum* seems to be confined to the forests of central Africa where it is transmitted by *Bulinus africanus* and *B. forskalli* (Deschiens *et al.*, 1972; Becquet et Decrocq, 1973). The epidemiology of *Schistosoma intercalatum* schistosomiasis is relatively unknown.

Where permanent bodies of water are present within the forest schistosomiasis may be endemic such as the focus in Danamé in Ivory Coast. Man-made water impoundment also greatly increases the risk of permitting transmission. This risk is often not appreciated by authorities responsible for such projects. In the Cameroon the creation of ponds for fish farming allowed the breeding of *Biomphalaria camarunensis* and induced new foci of *mansoni* schistosomiasis.

In Africa, at least, schistosomiasis does not appear to have a zoonotic cycle. Nevertheless the finding by De Paoli (1965) of a chimpanzee to be naturally infected with *Schistosoma haematobium* indicates that further study is desired.

Filariasis

Onchocerciasis

The clinical and social effects of onchocerciasis are sufficiently notorious as not to require restatement. The intensification of endemicity as a consequence of water impoundment schemes in Africa is also well known. The most severe effects occur in the Sudan zone. However, there is also a high endemicity in the forest belt; the biological character and epidemiological features of the forest cycle apparently being different from that in the savanna.

Human onchocerciasis caused by *Onchocerca volvulus* occurs in Africa and in the Americas. The distribution of the infection has been illustrated on maps in the WHO Expert Committee on Onchocerciasis report (1965). In Africa the disease is present in a wide belt south of the Sahara to Angola in the west and Tanzania in the east. In the Americas, important foci exist in Mexico, Guatemala, Colombia and Venezuela, and also among small human groups of the Amazonian forest—a discovery which questions the concept of this disease not being indigenous to America.

The difference between the African forest and savanna forms of onchocerciasis was largely discovered by Duke *et al.* (1966) in Cameroon. It has been observed that while the infection rates and densities are similar in both environments, onchocerciasis in forest communities rarely leads to severe ocular lesions whereas in the savanna ocular effects are severe and common, despite the forest inhabitants probably being exposed to more infective bites than savanna-dwellers (Duke, 1968). This led to the hypothesis that there are two strains of *O. volvulus* and Duke and Anderson (1972) provided some experimental confirmation; microfilariae from savanna-infected patients were more invasive and produced severer lesions in subconjunctivally inoculated rabbits than did microfilariae from forest-infected patients. Not only does the strain of parasite differ, but the western African vector, *Simulium damnosum*, also exhibits differences in reciprocal transmissability as well as morphology and cytotypes. For example, when *S. damnosum* from the savanna were fed on a patient from the forest, infection in the flies did not develop and they eliminated nearly all the parasites at the microfilarial stage (Duke, 1966; Duke *et al.*, 1966). Other differences between forest and savanna strains of *S. damnosum*, such as parous rates, biting time, and

possibly zoophily have been noted (Disney, 1972). Recent work by Dunbar and Vajime (1972) indicates that *S. damnosum* consists of a complex of species identifiable by their cytotypes. Further studies will have an important bearing on host-vector-parasite relationships in onchocerciasis. Although the blindness rates never attain the dramatic proportions encountered in the Sudan savanna, forest onchocerciasis can hardly be considered inocuous. In heavily infected villages eye lesions do occur and the skin lesions and other associated complications such as hanging groin are common, particularly in the older age groups (Duke, 1972). The distribution of African forest onchocerciasis is focal and undoubtedly related to the presence of breeding sites as well as the nature of man-fly contact. For example, the forest region around Feol, Zaire, had a prevalence of 5 per cent while the rate was 100 per cent in the surrounding savanna communities (Geukens, 1950). A comprehensive review of the epidemiology of onchocerciasis in the Congo basin has been written by Fain and Hallot (1965).

New World onchocerciasis has a somewhat different epidemiological pattern. The landscape epidemiology is generally the mountain slopes between 500 and 1 200 m altitude, a height where the streams provide the breeding conditions for the main vector, *S. ochraceum*. It is here that conditions are suitable for growing coffee and vast areas of the forest have been replaced by *Coffea arabica* plantations. The work force are the main victims of the infection which causes considerable loss in work efficiency. Little is known regarding the presence of onchocerciasis in forest-dwellers of the endemic regions but a high rate (47–62 per cent) has been noted in the Yanomamo and Waica Indians of the Brazil-Venezuela border (Moraes *et al.*, 1973; Moraes and Chaves, 1974). In Venezuela the principal vector is *S. metallicum*, a species that breeds in small streams and has the ability to travel some distance from the river-banks to adjacent villages. *S. ochraceum* has a tendency to bite on the upper part of the body with the result that the onchocercal nodules are usually located in this region. Ocular tropism of the American parasite is not well understood but it appears that it is akin to the African forest strain in that it does not produce eye lesions to the extent of African savanna strain (Waddy, 1960; WHO Expert Committee Report, 1965).

Whether or not onchocerciasis is a zoonosis is debatable. There are only two instances of natural infection in primates: one case of a monkey (*Ateles geoffroyi*) from Mexico and a gorilla from Zaire. Because of the rarity of the infections Nelson (1965) does not consider it to be a zoonosis. Nevertheless the very fact that some primates are naturally infected and that some forest peoples such as the Venezuelan Indians and Ituri forest pygmies do have a high prevalence rate leaves the question open; a zoonotic reservoir may exist in some areas.

Loa loa

The bionomic isolation of its several vectors and hosts has led to an incipient speciation into two distinct types of *Loa loa*, both occurring within tropical forests. Loasis is restricted to Africa where its vectors, various species of *Chrysops*, are found. *Loa loa* is a parasite of man and primates. In the forest, mandrills (*Mandrillus leucophaeus*), *Cercopithecus nictitans martini* and mona monkeys (*C. mona mona*) are infected with a variety whose microfilariae exhibit nocturnal periodicity. Approximately 26 m below the canopy habitat, man on the forest floor is also infected with *Loa loa* but the parasite of humans have diurnally periodic microfilariae smaller in size than the primate *L. loa* microfilariae. The degree of adaptation to their respective hosts is such that the parasite of man is poorly infective for monkey and *vice versa* (Duke and Wijers, 1958). Further evidence of the stability of these characters is that when monkeys are experimentally infected with *Loa* from man the small size and diurnal periodicity of the microfilariae are retained. The cycle of loasis from man to man is maintained by *Chrysops silacea* and *C. dimidiata*; flies that have diurnal activity and feed at ground level (Duke, 1958). The primate cycle is maintained by *C. centurionis* and *C. langi*, canopy-dwellers with biting peaks at dusk and early evening. Haddow *et al.* (1950) however have demonstrated that *C. centurionis* is not entirely restricted to the canopy and where there are seasonal climatic variations such as in the Bwamba forest of Uganda, it will descend to the ground level during the dry season, especially in the ecotones. This is yet another example where the behaviour of the arthropod vector is dependent on the microclimate and comes to inhabit another niche offering suitable conditions when its normal habitat is naturally or artificially altered.

Loasis is an infection only of inhabitants of the African forests. It is not a zoonosis although one can speculate that under altered forest conditions where the vector bites both man and monkey the primate strain could adapt to the human host, and *vice versa*. It is not an incapacitating infection although the allergic swelling produces some transient morbidity. However, according to Brumpt (personal communication), the eosinophilic fibroblastic endocarditis of Loeffler has often in Africa a loasis etiology; nephrosis and encephalitis do exist and they are not all due to diethylcarbamazine administration.

Brugia malayi and *Wuchereria bancrofti*

Brugia malayi is another example of diverging evolution with one coastal rice-plain strain cycled by anophelines and *Mansonia* mosquitoes and without a reservoir host, and a swamp forest strain as a zoonosis with a number of reservoir hosts and transmitted by *Mansonia*. The forest form of *Brugia malayi* occurs in Indonesia, Malaysia, in southern Viet-Nam (in Montagnards), Sabah and in the Philippines. Its habitat is forest with a river-bordered swamp forest. *Mansonia longipalpis*, *M. dives* and *M. bonneae* breed at the swamp forest verge while *M. uniformis* and *M. crassipes* select more exposed swamp areas. An efficient and important vector is *M. dives* which attacks man on the forest floor and the primate reservoirs, mainly *Presbytis* and other monkeys, in the canopy (Wilson *et al.*, 1958; Wharton, 1962). In peninsular Malaysia, with the exception of gibbons, most of the monkeys and leaf-monkeys are known to be reservoirs

of filariasis; the distributional study by Lim Boo Liat (1976) indicates that they are patchily distributed throughout the country and this reflects the epidemiology of filariasis.

It is not only the aboriginal hunter-gatherer and incipient agriculturalist who contract Malayan filariasis. The vector readily enters riverside habitations and the cycle of monkey to man and man is established. Cats become infected and constitute another source of infection. Clinical manifestations, fever, lymphadenopathies and ultimately in some instances, elephantiasis of the upper and lower limbs (but unlike Bancroftian filariasis, genital involvement is infrequent), are common and can result in significant loss in the work force. Non-immune transmigrant workers are particularly susceptible as has been described for rubber plantations in southern Kalimantan by Desowitz (1973).

The forest form of Malayan filariasis is a true zoonosis with feral and domestic reservoir host. Control is extremely difficult. The inaccessibility and vastness of the swamp makes vector control almost impossible. Mass drug administration with diethylcarbamazine gives relief where feasible but the presence of the reservoir requires a continuous programme. As Cheong (1974) has observed, the only prospect of eventual eradication or control is additional settlement of the area, with the filling or draining of the swamps to make the land usable for agriculture.

Other helminthiases

A number of other potentially infectious helminths exist within the tropical forest ecosystem. *Paragonimus westermani* occurs in Malaysia and western Africa. No human cases have been reported from Malaysia, the infection being confined to wild felines and crab-eating macaques (*Macaca irus*). A potential zoonotic hazard exists but undercooked crabs from forest streams are not eaten (Lee and Miyazaki, 1965). Some western African groups regularly collect crabs from streams, the Boksi of southern Cameroon have a paragonomiasis rate of 4 per cent (Zahra, 1952). *Paragonimus westermani* infection and the number of individuals with typical pulmonary manifestations due to food shortages during the Nigerian civil war, increased.

The extent of trichinosis as a sylvatically cycled infection of wild animals is not accurately known. Nelson *et al.* (1963) reported *Trichinella spiralis* in animals of the eastern African savanna such as hyaena and warthog but not in forest animals such as the giant forest hog or *Hylochaerus* (this is very rare and the infestation could be more frequent in the river-hog; infestation is often due to diseased warthog of the savanna areas and open woodlands). On the other hand, trichinosis is endemic in northern Thailand where the sylvatic reservoirs are the wild bush pig and the Himalayan bear. A number of outbreaks in northern Thai communities have occurred following eating undercooked meat of these (Doege *et al.*, 1969).

Paramanthan and Dissanaike (1961) believe *Echinococcus granulosus* to be present in Sri Lanka. The adult worms are reported in jackals and the hydatid cyst in deer. No human cases of echinococcosis have been observed but here too there is a zoonotic potential if the food habits of the people change or if dogs and domestic animals capable of initiating transmission to man are introduced.

One of the few well-documented examples of a zoonotic helminthic infection cycled within forest is strongyloidiasis due to *Strongyloides fülleborni*. *S. fülleborni* is normally a parasitic nematode of African monkeys and apes but it is also well adapted to man. It seems to affect a vast area in Africa comprising the forest belt of Cameroon and the Central African Empire between 2° and 4° N as well as forest areas in the mountains of Ethiopia. Infection rates as high as 24 per cent in the Babinga pygmies, 57 per cent in the Mingongol of the Cameroon and 78 per cent of the Dekkia of Ethiopia have been reported (Pampiglione and Ricciardi, 1971). Experimental infection of a human volunteer with *S. fülleborni* indicates that it can provoke a clinical syndrome of eosinophilia, rash, lymphangitis, bronchial symptoms and epigastric distress with diarrhoea (Pampiglione and Ricciardi, 1972).

Angiostrongylosis is a parasitic disease of mammals produced by various species of nematodes of the genus *Angiostrongylus*. *A. malaysiensis* is primarily a parasite of rats and from them it is spread indirectly to man. In man the parasite travels to the brain, where it develops to early maturity, soon dies, and produces an inflammatory reaction. Increased interest in this species has followed its implication as the probable agent of eosinophilic meningoencephalitis in Malaysia (Lim Boo Liat, 1976). A survey of *A. malaysiensis* infection in rodents and molluscs has been undertaken over the past several years in ten different habitats of peninsular Malaysia, including secondary forest and primary dipterocarp rain forest (at ground and arboreal levels).

Scrub typhus

The usual niche of scrub typhus has been considered to be *Imperata* grassland where trombeculid mites cycle the infection through rodents. The narrow zones at the periphery of the forest in small clearings are particularly rich areas for transmission (Audy, 1948, 1958). There is now convincing evidence that scrub typhus is also cycled, in some regions, within the tropical forests. *Rickettsia tsutsugamushi* has been isolated from the spleens of rats (*Rattus edwardsi*, *R. mulleri*, *R. rajah*) trapped deep in the primary Malaysian forest. In Malaysian and Sabah forests a high rate of isolates and serological positivity was obtained from terrestrial and tree-climbing mammals, particularly rodents, while arboreal species were not found to be involved in transmission (Walker *et al.*, 1973; Muul *et al.*, 1974). Muul *et al.* (1974) state: "Thus it seems possible that the forest habitats may serve as the enzootic source of the infection, at least for the natural mammalian hosts, and as forests are altered by man, the rickettsia may 'spill over' into such disturbed habitats as small animal populations establish themselves there". *Trombicula deliensis* was found but at a lower population density than in secondary growth areas (Traub *et al.*, 1950). On the other hand Mohr (1947) observed that rats taken deep in the New Guinea forest carried more *T. deliensis* than animals from an open area.

Scrub typhus may be an important infection in Ma-

laysian forest-dwellers; Cadigan *et al.* (1972) reported serological positivity rates to be higher in inhabitants of the deep forest than in populations living at the forest fringe or in settlements beyond the forest.

Yellow fever

Yellow fever, a group B arbovirus, occurs in tropical Africa and South and Central America, primarily as a sylvatically-cycled infection of primates. Transmission from host to host is accomplished via species of *Aedes* mosquitoes in Africa and *Haemagogus* and *Aedes* in South America.

The epidemiology of yellow fever in Africa was elegantly elucidated by investigators working in eastern Africa during the 1940s and 1950s and an account of their findings will be given as a general model of this zoonosis as it exists in the forest south of the Sahara. The purely sylvatic cycle is maintained by primates that are largely arboreal and remain within the forest. The black mangaby (*Cercocebus albicata johnstoni*) and the lowland colobus (*Colobus polykomos uellensis*) are in this category and 70 per cent of these monkeys may be serologically positive for yellow fever antibodies (Haddow *et al.*, 1947). They are territorial, usually sleeping in the same trees each night, even $5\frac{1}{2}$ years (Buxton, 1951). The mosquito responsible for arboreal transmission is *Aedes africanus* in most areas and the acrodendrophilic *Aedes deboeri* in the Langata forest of Kenya (Garnham, 1949). *Aedes africanus* breeds in the water in buttress roots and in tree-holes up to 15 m high. The essential link in the progression to man is the arboreal redtail monkey (*Cercopithecus nictitans mpangae*) which acquires the virus from the mangaby-colobus reservoir via *Ae. africanus*[1] (Buxton, 1952). The redtail is not restricted to the forest but emerges to raid the plantations. At the ecotone the vector link to man, *Ae. simpsoni*, becomes infected. It breeds in the axils of banana leaves and similar places. In these plantations *Ae. simpsoni* transmits the infection to man. Gillett (1951) notes that where yellow fever does not occur despite the presence of *Ae. simpsoni*, the strain of mosquito rarely feeds on man. He speculates that with cutting of the forests and the establishment of plantations and subsequent contact with man, *Ae. simpsoni* becomes domesticated and readily feeding on man. There are similar strain-behavioural differences for *Ae. aegypti*, the urban vector *par excellence*, of yellow fever and other Group B arboviruses such as dengue. *Ae. aegypti* is present in forests where it breeds in tree-holes. Here, *Ae. aegypti* feeds mainly on forest animals and bites man only with great reluctance[2] (Haddow, 1945; Garnham *et al.*, 1946). One may speculate that *Ae. aegypti* was originally a zoophilic sylvatic mosquito that has become humanized in the towns after man destroyed its forest habitat. *Ae. aegypti* provides the final step in the epidemiological

1. *Ae. africanus* positive for yellow fever have been found; in addition *Ae. africanus* readily bites man; thus it does not seem to be only associated with primates.
2. Recent work, and especially by the WHO Arbovirus Vector Research Unit, has shown that *Ae. aegypti* in forest relics bites man at a rather high rate.

cycle. The plantation workers having become infected from *Ae. simpsoni* visit the towns where they are bitten by *Ae. aegypti* and the urban cycle becomes established.

As noted earlier, yellow fever antibody is rarely found in African high forest-dwellers (Liégois *et al.*, 1948). This is presumably due to the bionomic isolation of vector and reservoirs, as the disease spreads mostly in the ecotones.

It seems that in western Africa (Mouchet, personal communication), where *Ae. simpsoni* does not bite humans, the yellow fever transmission does not follow the complicated cycle described by Haddow. *Ae. aegypti* does not play any role in the forest cases of the disease, as it is very rare there. But in the ecotones, fringing forests and plantations, the infected vectors descend from the canopy to the ground and infect humans. In western Africa, this seems to be more frequent than the infection of humans by monkeys. Such a transmission pattern explains the low titer of antibodies in the forest and its increase in the fringing forests and in the forest-savanna mosaics, where transmission occurs between the canopy-reservoir and men.

The virus reservoir is in fact the dynamic mosquito-monkey link; not a static concept as prevails in tick-borne diseases.

The situation appears to be different in South and Central America where serological studies have revealed a high percentage of yellow fever antibody positives in Indians and other forest inhabitants (Sneath, 1939; Snijders *et al.*, 1947; Laemmert *et al.*, 1949; Kumm, 1950; Black *et al.*, 1970, 1974; Madalengoita *et al.*, 1973). Since the virtual eradication of *Ae. aegypti* from South America, yellow fever has become a disease maintained solely within the forest cycle. There have been recent small outbreaks in Panamá, Colombia, Peru and Ecuador; in monkeys as well as humans. There are other important differences between the epidemiology of yellow fever in Africa and South America. In South America it is a wandering epizootic amongst primates (Kumm, 1950). In contrast to all the species of African monkeys which seem to be immune to yellow fever, there is considerable difference in susceptibility amongst neotropical primates. The spider (*Ateles*) and howler (*Alouatta*) monkeys, for example, are highly susceptible and large numbers die when the infection is introduced into their territory (Boshell and Bevier, 1958). The *Cebus* monkeys, and probably *Lagothrix* and *Saimiri* species as well, are much more tolerant. There is a high rate of serological positives amongst these (Burgher *et al.*, 1944) and they are the reservoirs primarily responsible for maintaining and rebuilding the sylvatic cycle when the numbers of susceptible species are reduced during an epizootic. Other animals may also be reservoirs; the common opossum (*Didelphis marsupialis*) is often found to have a positive serology for yellow fever antibody.

A number of mosquitoes are vectors of sylvatic yellow fever in the New World, the most important being *Haemagogus spegazzinii*. *H. spegazzinii* breeds in tree-holes and the adults are normally restricted to the forest canopy. However, where the forest has been felled it has been found breeding in tree stumps, fallen trees and cut bamboo. Under these conditions it will bite at ground level and is thus capable of transmitting the virus to humans (Komp, 1952). In many areas of South America, such as Minas Gerais, the forest is separated by pasture. The *Cebus* reservoir is highly territorial and does not cross these open spaces. Disseminative transmission is accomplished by the ability of *H. spegazzinii* and another forest vector *Aedes leucocelaenus* to traverse the open areas (Causey and Kumm, 1948; Causey *et al.*, 1948; Laemmert *et al.*, 1949). *Haemagogus spegazzinii* has been known to cross open spaces 11.5 km across (Causey *et al.*, 1950).

A number of physical variables influence the behaviour of the South and Central American mosquito vectors. The area north of Amazonia to Panamá is subject to seasonal variation in rainfall and the forests are an intermediate between deciduous and rain forest. Seasonal variations in humidity greatly influence the abundance and behaviour of the vector, as well as the abundance and behaviour of the human host in the forest. The majority of vectors normally stay in the canopy doing so because they prefer a sunlit environment. This restriction ceases in deforested areas where they will descend to the exposed clearings. Humidity also has an influence, *Haemagogus spegazzinii*, *H. equinus*, *H. lucifer* and *Aedes leucocelaenus* thrive under conditions of high humidity and their numbers are much reduced during the dry season. *Sabethes clarkii* however which is arboreal during the wet season, tolerates the lower humidity of the dry season when it descends to ground level where it comes in contact which the human hosts. *Haemagogus capricornii*, like *Sabethes clarkii*, is tolerant of lower humidity and exhibits similar behaviour but will also enter forest clearings. The ability to survive the dry season makes *Haemagogus capricornii* an important vector since the virus also persists and this mosquito becomes the reservoir host: it has been said that, because the viraemia in the primate host is so transient compared to the mosquito which harbours the virus its entire life, yellow fever is a zoonosis of the mosquito rather than primates (Burgher *et al.*, 1944; Downs and Pittendrigh, 1946; Kumm, 1950; Galindo *et al.*, 1956; Boshell and Bevier, 1958).

A number of workers such as Jorge (1938) have observed that forest yellow fever is less virulent for man than the urban form transmitted by *Aedes aegypti*. Certainly the high rate of serological positivity in Africans and Amerinds living in the forest would indicate that the infection is clinically negligible in these people. The reason for the relatively mild disease as it occurs in the forest in distinction to the virulent urban form is not clear. The virulence may have been enhanced by humanization as the virus comes from the monkey to the exclusively human cycle. Alternatively, or in addition, many other Group B arboviruses (as the Zika virus in Africa), found in the tropical forest ecosystem, are not highly pathogenic for man. Those viruses however share some common antigens with the yellow fever virus, as evidenced by serological techniques, and primary infection with these viruses may afford some immunological protection to subsequent infections with yellow fever virus.

Other arboviruses

There is a rich assortment of arboviruses in tropical forests. Some have been isolated only from arthropod vectors while others have been found only in wild mammals. These viruses produce disease in experimentally infected laboratory animals but their virulence for man remains largely unknown. Other arboviruses have the proven ability to cause overt disease in man. Kyasanur Forest Disease, caused by a tick-borne virus of the Russian spring summer virus complex is a haemorrhagic syndrom occurring in the deciduous and evergreen forests of the Western Ghats of India. Primates, rodents and birds act as reservoirs. The presence of antibodies in cattle suggests the possibility of domestication of the cycle with settlement of the forest for mixed agriculture (Boshell and Rao, 1963; Work, 1963). Another example is Semliki Forest virus of the forest and cultivated patches of western Uganda. It was originally isolated from the *Aedes abnormalis* group of mosquitoes and found to be neurotropic for experimental animals. It is related to the dengue and Chikungunya viruses but less pathogenic than either (Smithburn and Haddow, 1944). The Lassa virus is not a specific forest virus but is found after human modifications of forests, as in Sierra Leone; it is not an arbovirus, but an arenavirus with a reservoir of the rodent *Mastomys natalensis*.

Rudnick *et al.* (reported by Knudsen, 1974) have finally proved the primate sylvatic reservoir of the dengue virus. The presence of serological positive individuals to all four dengue serotypes in urban, rural and deep forest-dwellers was the initial clue to the presence of the virus in the forest ecosystem. Sentinel monkeys placed in the Malaysian forest canopy became infected with dengue types 1 and 2. The missing link of the forest cycle yet to be elucidated is the vector(s). Until this component is known it is difficult to predict the probability of man's contracting the infection. In keeping with many other arthropod transmitted infections it can be conjectured that in an altered environment the virus would somehow eventually find its way to man.

Conclusions: epidemiology and patterns of use

An appropriate starting point is to examine the condition of populations living in complete intimacy with their forest habitat—the remnant hunter-gatherers. These peoples are almost absolutely dependent on their habitat in which they subsist without causing permanent alteration. This steady state can prevail only if the population remains small. The forest hunter-gatherers are nomadic but wander within the territorial limits required by their needs and that imposed by neighbouring groups. The mechanisms by which these groups have maintained stability in numbers are not well understood. Childhood mortality certainly plays a major role in regulating population size but lack of precise information on pre-reproductive age mortality makes the effect of this factor difficult to assess. Childhood mortality due to infectious disease such as measles does not occur unless introduced from the outside and even then it is an episodic event since the critical mass is too small for the pathogens continued maintenance. Gastro-enteritis is a major killer of infants but presumably the water sources of nomadic peoples would be uncontaminated and food would be gathered and consumed quickly enough to prevent spoilage. Furthermore, the varieties and virulence of diseases in the intact forests are less than in the altered forests and at the forest border interface even for malaria. Social and customary practices could affect the group's size but little is known of this. It has been shown, for example, that the Muruts' practice of frequent examination of the pregnant woman's genitalia before and during birth leads to puerperal sepsis followed by tubal blockage sterility. Sterility and fetal wastage due not only to such practices but also to venereal diseases and other infections have not been carefully assessed as causes. Some infections that have been disregarded are now being recognized as of possible importance. For example, adult African males with Bancroftian filariasis hydrocele (which does not exist in forests) have fewer children than those with normal testicles. Vayda (1969) proposes that where shifting agriculture is practised customs have evolved to prevent overcropping. Since swidden agriculturalists can only maintain fertility of their plots by allowing long restorative fallow periods it follows that population numbers within their territorial limits must somehow be restricted. He suggests that witchcraft accusations have been the one of the means of expelling the surplus members of the group. As forest peoples evolve to become settled agriculturalists they too will feel the population pressures now being experienced by many other rural groups. Irreversible, exploitive destruction of their habitat can best be prevented if a logical means of population control is introduced. Such a scheme can best be devised if all the cultural and biological factors formerly operative in regulating their numbers have been identified.

The undisturbed forest ecosystem is probably more protective of its human component than are most other ecological settings. It is important to understand the means by which this protection is afforded in order to predict the hazards in altering the forest habitat.

1. Transmission of disease is limited because the group's size does not permit maintenance of some pathogens and, in the case of vector-borne zoonoses, the vector prefers not to feed on man and/or is confined to an area of the forest, such as the canopy, which prevents man-vector contact.

2. Bionomic isolation of the parasite in reservoir hosts has resulted in incipient speciation so that many taxa adapted to the animal reservoir either do not infect man or produce only transient subclinical infections. There is epidemiological and some experimental evidence that this has occurred in a wide variety of pathogens such as yellow fever, the rickettsia causing scrub typhus, some primate malarias such as *Plasmodium rhodaini*, and possibly some neotropical leishmaniasis, *Onchocerca volvulus* and *Loa loa*. It has also been shown that many arthropod vectors or species have

become divided into complexes whose members possess biological and behavioural variance affecting their transmission capability.

3. There is a growing body of epidemiological evidence, for which experimental proof is still largely lacking, that an antigenic experience with an animal adapted pathogen affords cross-protection against related pathogen species that are potentially pathogenic in man. This may be true for sylvatic B group viruses and yellow fever. Nelson (1965) has suggested that it may be operative for a number of helminth infections such as filariasis and schistosomiasis. There may be still another form of protective immunity, such as maternal immunity. Thus non-immune migrants and settlers are highly susceptible.

4. Forest-dwelling man has adapted behaviourally and physiologically to the demands and stresses of milieu. Some investigators believe this to be mainly a behavioural adaptation, i.e. dwellers of the humid forest adapt to the climate and their energy intake by expending little energy in their daily activities. Others have shown that there are physiological adaptations. For aborigines and pygmies their small size is an adaptive advantage giving large surface to mass ratio (Wyndham *et al.*, 1964). The absence of diseases of affluence, e.g. hypertension, in forest peoples is another characteristic, although chronic infectious disease may be the barter for low blood pressure. The non-immune arrival faces dangers to health; so do the physiologically and behaviourally non-adapted. Neel (1971) concludes ' . . . nutritional, infectious and parasitic diseases exact a lower toll in these groups than in the tropical agriculturists of the same areas'.

The manipulation of tropical forests proceeds at an ever-quickening pace; there is need for new agricultural land and the forest's natural resources are in greater demand. Settlement is proceeding apace but there is very little information on the precise patterns of development nor of the numbers of peoples involved. The impression is that the greatest proportion of disturbed forest is due to small settlements not to large development projects. These are mainly intended to be permanent. There is abundant evidence that these relatively modest disturbances of the forest often have a profound epidemiological consequence.

Many settlers are non-immunes from outside the forest. They are highly susceptible to the potential pathogens within the forest ecosystem. Examples of this are yellow fever and other arboviruses, filariases, scrub typhus and, possibly, leishmaniasis and Chagas disease. The problem are usually aggravated by impure and inadequate water supplies as well as lack of facilities for human waste disposal. Bacillary dysenteries and intestinal parasitoses are the consequence. Permanent settlement also results in a soil heavily seeded with nematode ova and larvae and which in turn produces high burdens of intestinal parasites.

While the emphasis has been put upon forest-related diseases some attention should be paid to veterinary infections: animal husbandry, on a modest scale, is a desirable adjunct to agricultural settlement. Domestic animals provide high quality protein lacking in some cereal and root crops as well as source of fertilizer. Virtually nothing is known regarding the health hazards to domestic animals when mixed farming is attempted in newly opened forest lands. In the African forest zone trypanosomes of veterinary importance are zoonotic and may prevent the rearing of pigs or domestic ungulates except for the tolerant dwarf breeds of cattle (Baoulé, Lagunes, N'Dama). Further research on veterinary epidemiological-ecological consequences of altering the forest is essential if the newly opened lands are to be fully productive.

Alteration of the forest may change the character of the pathogen and vector. Parasites which normally are cycled in animals often are slightly pathogenic for man. Destruction of their forest habitat either eliminates or greatly reduces much of the reservoir fauna and the parasites then enter a purely human cycle. In many instances man to man transmission ultimately enhances the virulence of the parasite. This phenomenon has been postulated for *Leishmania* and the yellow fever virus. There has been no convincing experimental proof despite its importance.

There are many instances of behavioural change in the arthropod vectors after disturbing or destroying their forest environment. The loss of both preferred hosts and habitat often brings a potential vector to feed on man. Examples can be drawn from the mosquito vectors of malaria and yellow fever and the sand-fly vectors of neotropical leishmaniasis normally participating in the sylvatic cycle. In their undisturbed habitat these arthropods are mainly confined to the canopy where they feed upon primate and rodent reservoirs but when the forest has been cleared they come to ground level and bite man. Little is known of the mechanism responsible. Is it a selection from a heterogeneous population or do these insects have the inherent physiological latitude to adapt to the new conditions?

In some instances the alteration of the forest ecosystem results in the replacement of non-vector by vector species, while in other instances it allows the proliferation of vector species which were at low densities in the intact environment. Species replacement generally only occurs where there has been extensive alteration of the environment. Man-made lakes may flood forest lands bringing malaria, schistosomiasis and arbovirus infections. Development schemes attract immigrants adding a further element to the complexities. Waddy (1975) has written an eloquent account of the medical problems associated with water impoundment projects. He strongly recommends that information be obtained on the planning of new lakes and that a central institution be created for the study of the biological and medical problems associated with such schemes, special attention being paid to epidemiology and environmental sanitation. There is actually a group in WHO assigned to oversee health problems associated with such developments.

Too little is known of the psychological and physiological stresses of people who come to live in an altered or new ecosystem. Many observers have commented on the absence of urban diseases such as hypertension in nomadic and incipient agriculturalists of the forest. There is at least one example, a Malaysian negrito group, of abnormally high blood pressure subsequent to a change from nomadism to wage-earning in a permanent settlement. The peoples who come to live in an altered environment must adopt a new way of life. For indigenous forest peoples the cost of the adaptive process is acculturation. The deterioration of health of forest Amerinds has been sohwn by Neel (1971) to be an almost inevitable accompaniment of acculturation. The abandonment of intercourse taboos for women with children under 12–18 months of age by the accultured society may result in a growth population in which the food resources of the community cannot accommodate.

Psychological-behavioural and physiological consequences of ecological change have not been adequately identified in assessing health. Nor have they been properly appreciated as interacting components of epidemiological processes. The perception of disease in indigenous peoples needs to be studied so that effective, acceptable control schemes can be devised.

Alteration of the forest almost invariably requires forest-dwellers to seek new sources of food. All too often there is a profound adverse nutritional effect due to dietary changes. For the hunter-gatherer the animal supply becomes exhausted. The diminished animal protein intake, in the absence of a proper balance of vegetable protein, may result in protein malnutrition. Similarly for the immigrant who raises mainly protein-poor cereal and root crops. Another hazard is that constant use may deplete the soil of its nutrients. The use of insecticides to control crop pests, initially at least, control arthropod vectors of medical importance. The danger is that ultimately these vectors will become insecticide resistant. Thus, it is important that the agricultural application of insecticides be done with the advice of biomedical experts and in concert with their public health plans (see also chapter 14).

During the war in Indo-China the forest was deliberately attacked with herbicides and much of the affected forest shows little sign of regeneration. Conditions in Viet-Nam are such that little is known of the epidemiological consequence of this destruction. In the new barren area, formerly mangrove forests, breeding sites for anophelines were created and malaria has become endemic. There is also evidence that the rat population has increased markedly. New breeding sites for other malaria vectors such as *Anopheles maculatus* have been created in the montane and valley forests cleared by herbicide and mechanical means. The numbers of vectors breeding in the numerous bomb craters can be conjectured. There has also been an enormous dislocation of peoples and undoubtedly there will be massive efforts to resettle these and others in forest areas with all the medical risks attendant upon such policy. Hopefully an assessment of the war-induced ecological-epidemiological changes can now be made and some of the more disastrous effects averted or rectified.

Physical changes caused by mechanization of forest industries may also produce health hazards. The water collections in the ruts made by vehicles make ideal breeding sites for many anopheline vectors of malaria.

Research needs and priorities

1. There is little doubt that the tropical forest ecosystem will undergo change at an ever-quickening pace. Developments will continue to be mainly small scale rather than enormous projects such as the water impoundment schemes that affect other ecosystems. Thus it is difficult to create project-oriented biomedical teams such as that envisaged by Waddy (1975) to assess and predict the health hazards created by ecological changes in the tropical forests. Nevertheless it is important that multidisciplinary advisory groups be established to give logical direction to health policies.

a. Expert groups consisting of epidemiologists, nutritionists, medical zoologists, demographers and medical anthropologists should be established on a regional basis to assess the medical problems related to the forest ecosystem and particularly to identify the problems inherent in ecological change. Priority should be given to the study and surveillance of resettlement schemes.

b. Where feasible this group should be drawn from national personnel resources. However, many countries facing such health problems do not have this expertise and an international panel of experts (possibly through appropriate UN agencies) should be established to collate information, make recommendations and provide periodic independent assessment following implementation. This obviously requires the establishment of an international office of medical ecology that cuts across organizational boundaries. This office, and its field teams, would have to work in close cooperation with, and possibly include, agricultural and forest scientists, sanitary engineers, and economic and health planners.

c. Although several sources of information do exist and need to be made available, it is necessary to pursue the investigations on ecological-epidemiological relationships in the tropical forest ecosystems at the regional or national levels. There is need for regionally specific information for many areas of the tropics.

2. The health status of forest peoples requires further study. Plans should be formulated and implemented, for their protection against disease, nutritional inadequacies and the stresses of acculturation as they come into more and more contact with other peoples (see chapter 19).

It is probable that for some time to come many of the studies of isolated forest peoples will be carried out by anthropologists. A simple medical research kit and questionnaire should be assembled so that a fundamental appreciation of health status can be obtained. Central specialized laboratories and specialists fully aware of the epidemiology and the medical geography of the regions should analyse this material.

3. Nutritional problems associated with the forest zone require further investigations (see chapter 16). Studies of this kind are linked to changes in agricultural practices. Recommendations need to be made on the most effective practicable farming methods to provide nutritionally adequate diets.

4. Reliable vital statistics for forest-dwellers are generally lacking. Strengthening and improving national health

services' statistical sections are recommended. Practical methods of data collections should be devised (see chapter 15).

5. Very few reports on epidemiological studies provide even a brief ecological description of the setting in which they were carried out. Epidemiologists and other investigators should be made mindful of the importance of giving an ecological account. Ecologists on the other hand should devise a relatively simple categorical system that can be understood and used by professionals who have had little or no training in ecology.

For example, in peninsular Malaysia (Lim Boo Liat, 1976), to study the arboreal mammals, an aerial transect *ca.* 305 m in length, built in the canopy of an undisturbed forest at heights between *ca.* 9 to 28 m from the ground was completed in 1969. This has been useful for observations and collections that are otherwise difficult to make from the ground because of the shielding effect of the sub-canopy formed by smaller trees. The transect walkway is constructed from aluminium ladders joined together and suspended from ropes like a suspension bridge. Associated with the walkway are observation platforms, which also support meteorological equipment. Temperature and humidity are measured at the various elevations as well as on the ground below to compare the micro-climate at these levels. In addition to collections and observations of vertebrate host species, the canopy transect serves for mosquito collections and sentinel bait studies. With the aid of this walkway it is possible for the first time to do cross-sectional studies of the hosts, vectors, and parasites that reside at different levels in the forest. Data are thus being collected for a variety of species of mammals to determine the habits of their activity and to determine the correlation of this aspect of their ecological niches with their involvement in zoonotic disease cycles. Studies on patterns of geographical and altitudinal distribution, temporal and spatial use of the environment, reproductive cycle, population dynamics, food habits, competition, parasitology, taxonomy and systemics are still continuing.

It should be stressed that wild animals, besides their important role as vectors and hosts of parasites, could be used as potential laboratory experimental animals, in order to collect very useful ecological data. In peninsular Malaysia it was shown for example (Lim Boo Liat, 1976) that several animals could be easily kept in captivity and adapt themselves to laboratory food; some of the primitive primates, *Tupaia glis, Nycticebus coucang,* field rats, *Rattus tiomanicus, R. annandalei,* and forest rats, *R. mulleri, R. sabanus,* do breed in captivity. With prolonged laboratory breeding of these species, there is also a possibility of establishing a new gene pool for various experimental laboratory and medical bioassay studies.

Such recommendations are not at all utopian, as is shown by the research done on the Bi-Aka pygmies, undertaken since 1974 by the Centre eurafricain de biologie humaine in Paris, at the request of the Government of the Central African Empire. Health problems, which had an obvious priority, were tackled under epidemiological, pharmacogenetic and ethno-social aspects. The discussion with the diseased persons, their medical examination, the analysis of biological samples and of water and foods, the study of arthropod vectors, serve to establish the frequency of diseases and the nature of the pathogenicity risk to which the population or its functional units (family, camp) are submitted. The results of the medical care provided in the forest and of the numerous biological analyses, allow for the determination of the most effective treatments, the least toxic according to the genome and the least expensive. The problems of demography and dynamics of the group were studied thanks to the indications given by the genealogical tree of the concerned population (9 generations, 4 500 individuals) as stored on magnetic tapes. The knowledge of the systems of choice of marital partners and of the wedding space is essential for the study of the genetic make-up of such a population, and of its variations as resulting from migrations, random drift, selection or mutation pressures.

Bibliography

ALBUQUERQUE, R. D. R.; BARRETTO, M. P. Estudos sobre reservatorios e vectores silvestro do *Trypanosoma cruzi.* XXVII. Infeccao natural do simio *Callithrix penicillata jordani* (Thomas, 1964) pelo *T. cruzi. Revta. Med. Trop.* (Sao Paulo), 11, 1969, p. 394–402.

ARENDS, T. El problema de las hemoglobinopatías en Venezuela. *Rev. Venezolana Sanidad y Asistencia Social,* 26, 1961, p. 61–68.

——; GALLANGO, M. L. Haemoglobin types and blood serum factors in British Guiana Indians. *British Journal of Haematology,* II, 1965, p. 350–359.

*AUDY, J. R. Some ecological effects of deforestation and settlement. *Malayan Nature Journal,* 8, 1948, p. 178–189.

*——. The localization of disease with special reference to the zoonoses. *Transactions of the Royal Society of Tropical Medicine and Hygiene,* 52, 1958, p. 308–328.

BARLOVATZ, A. Typhus exanthématique de forêt au Congo. *Annales Société belge de Médecine tropicale,* 20, 1940, p. 23–40.

BARUZZI, R. G. Contribution to the study of toxoplasmosis epidemiology. Serologic survey among Indians of the Upper Xingu River, central Brazil. *Revta. Med. Trop.* (Sao Paulo), 12, 1970, p. 43–164.

——; CAMARGO, M. E.; KAMEYAMA, I.; HOSHINO, S.; REBONATO, C.; d'ANDRETTA, C. Jr. Splenomegalia in Brazilian Indians from the "Alto Xingu" (Central Brazil). 1. Occurrence and results of serological tests for some parasitic diseases. *Annales Société belge de Médecine tropicale,* 51, 1971, p. 205–213.

BECQUET, R.; DECROCQ, J. Découverte d'un foyer de bilharziose intestinale à *Schistosoma intercalatum* en République Centrafricaine. *Bulletin Société Pathologie exotique* (Paris), 66, 1973, p. 720–727.

*BEGHIN, I. Enquête sur la nutrition et l'état de santé des enfants

*Major reference.

Warega (Congo belge). *Annales Société belge de Médecine tropicale*, 40, 1960, p. 253–288.

BELL, C. *The diseases and health services of Papua New Guinea*. Port Moresby, Department of Public Health, 1973, 647 p.

BELTRAN, C.; BUSTAMANTE, M. E. Datos epidemiológicos acerca de la "ulcera de los chicleros" (leishmaniasis americana) en Mexico. *Rev. Inst. Salubridad y Enfermedades Trop.*, 3, 1942, p. 1–28.

BERENGO, A.; PAMPIGLIONE, S.; DE LALLA, F. Serological studies on toxoplasmosis in some groups of Babinga pygmies in Central Africa. *Rivista di Parasitologia*, 35, 1974, p. 81–86.

*BLACK, F. L. Infectious diseases in primitive societies. *Science*, 187, 1975, p. 515–518.

——; WOODALL, J. P.; EVANS, A. S.; LIEBHABER, H.; HENLE, G. Prevalence of antibody against viruses in the Tiriyo, an isolated Amazon tribe. *American Journal of Epidemiology*, 91, 1970, p. 430–438.

*——; HIERHOLZER, W. J.; PINHEIRO, F.; EVANS, A. S.; WOODALL, J. P.; OPTON, E. M.; EMMONS, J. E.; WEST, B. S.; EDSALL, G.; DOWNS, W. G.; WALLACE, G. D. Evidence for persistence of infectious agents in isolated human populations. *American Journal of Epidemiology*, 100, 1974, p. 230–250.

BOLTON, J. M. Medical services to the aborigines in West Malaysia. *British Medical Journal*, 2, 1968, p. 818–823.

——. Food taboos among the Orang Asli in West Malaysia: a potential nutritional hazard. *American Journal of Clinical Nutrition*, 25, 1972a, p. 789–799.

——. The control of malaria among the Orang Asli in West Malaysia. *Medical Journal of Malaysia*, 27, 1972b, p. 10–19.

BOSHELL, J. M.; BEVIER, G. A. Yellow fever in the lower Motagua Valley, Guatemala. *American Journal of Tropical Medicine and Hygiene*, 7, 1958, p. 25–35.

——; RAO, T. R. Kyasanur forest disease. In: *Proceedings of the 7th International Congress on Tropical Medicine and Malaria* (Rio de Janeiro), 3, 1963, p. 304–305.

*BRENGUES, J. *La filariose de Bancroft en Afrique de l'Ouest*. Paris, ORSTOM, Mémoire n° 79, 1975, 299 p.

BRUMPT, L. C.; HO THI SANG; JAEGER, G.; RICOUR, A. Quelques réflexions à propos du parasitisme sanguin et intestinal dans deux villages d'Afrique centrale. *Bulletin Société Pathologie exotique* (Paris), vol. 65, n° 2, 1972, p. 263–270.

*BRUNHES, J. *La filariose de Bancroft dans la sous-région malgache (Comores, Madagascar, Réunion)*. Paris, ORSTOM, Mémoire n° 81, 1975, 212 p.

*BURGHER, J. C.; BOSHELL-MANRIQUE, J.; ROCA-GARCIA, M.; OSORNO-MESA, E. Epidemiology of jungle yellow fever in eastern Colombia. *American Journal of Hygiene*, 39, 1944, p. 16–51.

BUXTON, A. P. Further observations on the night-resting habits of monkeys in a small area in the edge of the Semliki Forest, Uganda. *Journal of Animal Ecology*, 20, 1951, p. 31–32.

——. Observations on the diurnal behaviour of the redtail monkey (*Cercopithecus ascarnius schmidti* Matschie) in a small forest in Uganda. *Journal of Animal Ecology*, 21, 1952, p. 25–58.

CABANNES, R.; LARRONY, G.; RUFFIÉ, J. Étude clinique et hématologique des Indiens du Haut-Oyapock et du Haut-Maroni (Guyane française), Oyampi, Emirillon et Oyana. *Bulletin Société Pathologie exotique* (Paris), 57, 1964, p. 307–325.

CADIGAN, F. C.; ANDRÉ, R. G.; BOLTON, M.; GAN, E.; WALKER, J. S. The effect of habitat on the prevalence of human scrub typhus in Malaysia. *Transactions of the Royal Society of Tropical Medicine and Hygiene*, 66, 1972, p. 582–593.

CARRIZALES, D.; DESTOMBES, P. Quelques particularités de la pathologie infectieuse et parasitaire du Pérou. *Bulletin Société Pathologie exotique* (Paris), 63, 1970, p. 597–606.

CAUSEY, O. R.; KUMM, H. W. Dispersion of forest mosquitoes in Brazil. Preliminary studies. *American Journal of Tropical Medicine and Hygiene*, 28, 1948, p. 469–480.

——; HAEMMENT, H. W. Jr.; HAYES, G. C. The home range of Brazilian Cebus monkeys in a region of small residual forests. *American Journal of Hygiene*, 47, 1948, p. 304–314.

——; KUMM, H. W.; HAEMMENT, H. W. Jr. Dispersion of forest mosquitoes in Brazil: further studies. *American Journal of Tropical Medicine and Hygiene*, 30, 1950, p. 301–312.

CAVALLI-SFORZA, L. L. Pygmies, an example of hunter-gatherers, and genetic consequence for man of domestication of plants and animals. In: *4th Int. Congress of Human Genetics* (September 1971), p. 79–95. Amsterdam, Excerpta Medica, 1972, 500 p.

CHANIOTIS, B. N.; NEELY, J. M.; CORREA, M. A.; TESH, R. B.; JOHNSON, K. M. Natural population dynamics of Phlebotomine sandflies in Panamá. *Journal of Medical Entomology*, 8, 1971, p. 339–352.

——; CORREA, M. A. Comparative flying and biting activity of Panamanian Phlebotomine sandflies in mature forest and adjacent open space. *Journal of Medical Entomology*, 11, 1974, p. 115–116.

CHEONG, W. H. The changing pattern of vector-borne disease transmission due to ecological changes resulting from development and human activity. Working paper. In: *IBP Synthesis Meeting* (Kuala Lumpur, 12–18 August 1974), multigr.

CHIPPAUX, A.; CHIPPAUX-HYPPOLITE, C. Immunologie des arboviroses chez les Pygmées Babinga de Centrafrique. *Bulletin Société Pathologie exotique* (Paris), vol. 58, n° 5, 1965, p. 820–833.

CIRÉRA, P.; PALISSON, M. J.; PINERD, G.; JAEGER, G. La sérologie tréponémique dans une population pygmée Bi-Aka centrafricaine. *Bulletin Société Pathologie exotique* (Paris), vol. 70, n° 1, 1977, p. 32–36.

*CLARKE, M. C. Some impressions of the Muruts of North Borneo. *Transactions of the Royal Society of Tropical Medicine and Hygiene*, 44, 1951, p. 453–464.

COATNEY, G. R. The simian malarias: zoonoses, anthropozoonoses, or both. *American Journal of Tropical Medicine and Hygiene*, 20, 1971, p. 795–803.

CORSON, J. F. Vertebrates and arthropods examined in investigations of the epidemiology of yellow fever. *Tropical Diseases Bulletin*, 42, 1945, p. 597–609.

COURA, J. R.; FERREIRA, L. F.; DA SILVA, J. R. Triatomineos no Estado da Guanabara e suas relacoes com o domicilio humano. *Revta. Med. Trop.* (Sao Paulo), 8, 1966, p. 162–166.

COZ, J. Études entomologiques sur la transmission du paludisme humain dans une zone de forêt humide dense, la région de Sassandra, République de Côte d'Ivoire. *Cahiers ORSTOM* (Paris), série Entomologie médicale, 4, 1966, p. 13–42.

DEANE, L. M. Novo hospedeiro de trepanosomos dos tipos *cruzi* e *rangeli* encontrado no Estado do Para: o marsupial *Metachirops opossum opossum*. *Revista Brasileira Malariologia* (Rio de Janeiro), 10, 1958, p. 531–541.

*——. On the transmission of simian malaria in Brazil. *Revta. Med. Trop.* (Sao Paulo), 13, 1971, p. 311–319.

——; DEANE, M. P.; NETO, J. F. Studies on transmission of simian malaria and on a natural infection of man with *Plasmodium simium* in Brazil. *Bulletin of the World Health Organization*, 35, 1966, p. 805–808.

DEGOTTE, J. Epidemiological leprosy survey in the Nepoko, Kibali-Ituri district, Belgian Congo. *International Journal of Leprosy*, 8, 1940, p. 421–444.

DE PAOLI, A. *Schistosoma haematobium* in the chimpanzee—a natural infection. *American Journal of Tropical Medicine and Hygiene*, 14, 1965, p. 561–565.

DESCHIENS, R.; VAUTHIER, G.; NORDAN, C. Observations écologiques et biologiques sur *Bulinus forskalli*, vecteur de la bilharziose à *Schistosoma intercalatum. Bulletin Société Pathologie exotique* (Paris), 65, 1972, p. 138–145.

DESOWITZ, R. S.; SAAVE, J. J.; SAWADA, T. Studies on the immuno-epidemiology of parasitic infections in New Guinea. I. Population studies on the relationship of a skin test to microfilaraemia. *Annals of Tropical Medicine and Parasitology*, 20, 1966, p. 257–264.

——; LANGER, B. W. Jr. Hypocholesterolemia in rodent malaria. *Journal of Parasitology*, 54, 1968, p. 1006–1008.

*——; BERMAN, S. J.; GUBLER, D. J.; HARINASUTA, C.; GUPTAVANIJ, P.; VASUVAT, C. The effects of herbicides in South Vietnam. Epidemiological-ecological effects: studies on intact and deforested ecosystem. In: *Report of the National Academy of Sciences Committee on the effects of herbicides in South Vietnam*, 1974, 398 p.

D'HAENENS, G.; LIPO, M.; MEYERS, H. Notice de la carte des zones malariologiques naturelles de la République du Congo et du Ruanda-Urundi. *Rivista di Parasitologia*, 22, 1961, p. 175–184.

DISNEY, R. H. L. Observations on chicken-biting blackflies in Cameroon with a discussion of parous rates of *Simulium damnosum. Annals of Tropical Medicine and Parasitology*, 66, 1972, p. 149–158.

DOEGE, T. C.; THIENPRASIT, P.; HEADINGTON, J. T.; PONGPROT, B.; TARAWNICH, S. Trichinosis and raw bear meat in Thailand. *Lancet*, March 1, 1969, p. 459–461.

*DOWNS, W. G.; PITTENDRIGH, C. S. Bromeliad malaria in Trinidad, British West Indies. *American Journal of Tropical Medicine and Hygiene*, 26, 1946, p. 47–66.

*DUGGAN, A. J. The occurrence of human trypanosomiasis among the Rukuba tribe of northern Nigeria. *Journal of Tropical Medicine and Hygiene*, 65, 1962, p. 151–163.

DUKE, B. O. L. Studies on the biting habits of *Chrysops*. V. The biting cycles and infection rates of *C. silacea, C. dimidiata, C. langi* and *C. centurionis* at canopy level in the rain forest at Bombe, British Cameroons. *Annals of Tropical Medicine and Parasitology*, 52, 1958, p. 24–35.

——. *Onchocerca-Simulium* complexes. III. The survival of *Simulium damnosum* after high intakes of microfilariae of incompatible strains of *Onchocerca volvulus*, and the survival of the parasites in the fly. *Annals of Tropical Medicine and Parasitology*, 60, 1966, p. 495–500.

——. Studies on factors influencing the transmission of onchocerciasis. VI. The infective biting potential of *Simulium damnosum* in different bioclimatic zones and its influence on the transmission potential. *Annals of Tropical Medicine and Parasitology*, 62, 1968, p. 164–170.

*——. Studies on factors influencing the transmission of onchocerciasis. VII. A comparison of the *Onchocerca volvulus* transmission potentials of *Simulium damnosum* populations in four Cameroon rain forest villages and the pattern of onchocerciasis associated therewith. *Annals of Tropical Medicine and Parasitology*, 66, 1972, p. 219–234.

——; WIJERS, D. J. B. Studies on loasis in monkeys. I. The relationship between human and simian *Loa* in the rain forest zone of the British Cameroons. *Annals of Tropical Medicine and Parasitology*, 52, 1958, p. 58–175.

*——; LEWIS, D. J.; MOORE, P. J. *Onchocerca-Simulium* complexes. I. Transmission of forest and Sudan-savanna strains of *Onchocerca volvulus* from Cameroon, by *Simulium*

damnosum from various West African bioclimatic zones. *Annals of Tropical Medicine and Parasitology*, 60, 1966, p. 318–336.

DUKE, B. O. L.; ANDERSON, J. A comparison of the lesions produced in the cornea of the rabbit eye by microfilariae of the forest and Sudan-savanna strains of *Onchocerca volvulus* from Cameroon. I. The clinical picture. *Zeitschr. Tropenmed. Parasit.*, 23, 1972, p. 354–368.

DUNBAR, R. W.; VAJIME, C. G. *Le complexe Simulium (Edwardsellum) damnosum: rapport sur les études effectuées jusqu'en avril 1972.* Genève, OMS, WHO/Oncho/72–100, 1972.

*DUNN, F. L. In: BAKER, P. T.; WEINER, J. S. (eds.). *The biology of human adaptability*, p. 539–563. Oxford, Clarendon Press, 1966, 541 p.

*——. Epidemiological factors: health and disease in hunter-gatherers. In: Lee, R. B.; De Vore, I. (eds.). *Man the hunter*, p. 221–228. Chicago, Aldine, 1968.

——. Intestinal parasitism in Malayan Aborigines (Orang Asli). *Bulletin of the World Health Organization*, 46, 1972, p. 99–113.

DUREN, A. *Un essai d'étude d'ensemble du paludisme au Congo belge.* Extrait de *Mémoire* publié par l'Institut Royal Colonial Belge (Section des Sciences naturelles et médicales), 1937.

FAIN, A.; HALLOT, R. *Répartition d'Onchocerca volvulus Leukart et de ses vecteurs dans le bassin du Congo et les régions limitrophes.* Bruxelles, Académie Royale des Sciences d'Outre-Mer, classe des Sciences naturelles et médicales, XVII-1, 1965.

*FENNER, F. In: Boyden, S. V. (ed.). *The impact of civilization on the biology of man*, p. 44–68. Toronto, University of Toronto Press, 1970.

FREZIL, J. H.; COLUM, J. Apport de l'immunofluorescence indirecte dans le dépistage et le contrôle de la trypanosomiase à *T. gambiense*. In: *10ᵉ Conf. Techn. OCEAC* (in press).

FUNAYAMA, G. K.; BARRETTO, M. P. Estudos sobre reservatorios e vectores silvestres do *Trypanosoma cruzi*. XLII. Infeccao natural do simio, *Alouatta caraya* (Humboldt, 1812) pelo *T. cruzi. Revta. Med. Trop.* (Sao Paulo), 12, 1970, p. 257–265.

GALINDO, P.; TRAPIDO, H.; CARPENTER, S. J.; BLANTON, F. S. The abundance cycles of arboreal mosquitoes during six years at a sylvan yellow fever locality in Panama. *Annals of the Entomological Society of America*, 49, 1956, p. 543–547.

GARNHAM, P. C. C. Acrodendrophilic mosquitoes of the Langata Forest, Kenya. *Bulletin of Entomological Research*, 39, 1949, p. 489–490.

*——. Malaria as a medical and veterinary zoonosis. *Bulletin Société Pathologie exotique* (Paris), 62, 1969, p. 325–332.

——. *Progress in Parasitology.* London, The Athlone Press, 1971.

——; HARPER, J. O.; HIGHTON, R. B. The mosquitoes of the Kaimosi Forest, Kenya Colony, with special reference to yellow fever. *Bulletin of Entomological Research*, 36, 1946, p. 473–496.

GEUKENS. Contribution à l'étude des filarioses dans le territoire de Feshi. *Annales Société belge de Médecine tropicale*, 30, 1950, p. 1483–1493.

*GILLES, H. M. *Akufo: an environmental study of a Nigerian village community.* Ibadan University Press, 1964, 80 p.

GILLETT, J. O. The habits of the mosquito *Aedes (Stegomyia) simpsoni* Theobald in relation to the epidemiology of yellow fever in Uganda. *Annals of Tropical Medicine and Parasitology*, 45, 1951, p. 110–122.

GILMAN, R. H.; MADASAMY, M.; GAN, E.; MARIAPPAN, M.; DAVIS, C. E.; KYSER, K. A. *Edwardsiella tarda* in jungle diarrhoea and a possible association with *Entamoeba histolytica*. *South-East Asian Journal of Tropical Medicine and Public Health*, 2, 1971, p. 186–189.

*HADDOW, A. J. On the mosquitoes of Bwamba County, Uganda: I. Description of Bwamba, with special reference to mosquito ecology. *Proceedings of the Zoological Society London*, 115, 1945, p. 1–13.

——; SMITHBURN, K. C.; MAHAFFY, A. F.; BUGHER, J. C. Monkeys in relation to yellow fever in Bwamba County, Uganda. *Transactions of the Royal Society of Tropical Medicine and Hygiene*, 40, 1947, p. 677–712.

——; GILLETT, J. D.; MAHAFFY, A. F.; HIGHTON, R. B. Observations on the biting habits of some Tabanidae in Uganda, with special reference to arboreal and nocturnal activity. *Bulletin of Entomological Research*, 50, 1950, p. 209–221.

HERRER, A. Distribución geografica de la enfermedad de Chagas y de sus vectores en Perú. *Bol. Oficina Sanitaria Panamericana*, 6, 1960, p. 572–581.

JACQUARD, A. *The genetic structure of populations biomathematics*. Berlin and New York, Springer Verlag, 1975, 569 p.

JAEGER, G. Étude hémotypologique d'une communauté Sara centrafricaine. *Cahiers d'Anthropologie et d'Écologie humaine* (Paris), vol. 2, n° 2, 1974, p. 19–124.

——; PINERD, G.; PALISSON, M. J. Les Pygmées Bi-Aka de Bokoka, étude médicobiologique et immunogénétique. 1977, in press.

JANSSEN, J. F. The health of Maroon children of Surinam. *Journal of Tropical Pediatrics*, 7, 1961, p. 91–99.

*JORDAN, A. M. An assessment of the economic importance of the tsetse species of southern Nigeria and the southern Cameroons based on their trypanosome infection rates and ecology. *Bulletin of Entomological Research*, 52, 1961, p. 431–441.

——. The ecology of the *fusca* group of tsetse flies (*Glossina*) in southern Nigeria. *Bulletin of Entomological Research*, 53, 1962, p. 356–393.

——; LEE-JONES, F.; WEITZ, B. The natural hosts of tsetse flies in the forest belt of Nigeria and the southern Cameroons. *Annals of Tropical Medicine and Parasitology*, 55, 1961, p. 167–179.

JORGE, R. La fièvre jaune sylvatique au Brésil. *Bulletin de l'Office international d'Hygiène publique*, 30, 1938, p. 54–68.

JUMINER, B. Le paludisme est-il une anthropozoonose? *Ann. Inst. Pasteur Tunis*, 47, 1970, p. 229–241.

KINZIE, J. D.; KINZIE, K.; TYAS, J. A comparative health survey among two groups of Malayan Aborigines. *Medical Journal of Malaya*, 21, 1966, p. 135–139.

KNIGHT, R.; PRATA, A. Intestinal parasitism in Amerindians at Coari, Brazil. *Transactions of the Royal Society of Tropical Medicine and Hygiene*, 66, 1972, p. 809–810.

KNUDSEN, B. A. The silent jungle transmission cycle of dengue virus and its tenable relationship to endemic dengue in Malaysia. Working paper. In: *IBP Synthesis Meeting* (Kuala Lumpur, 12–18 August 1974), multigr.

KOMP, W. H. W. The facultative breeding of *Haemagogus spegazzinii falco* Kumm et al., the vector of jungle yellow fever in Columbia. *American Journal of Tropical Medicine and Hygiene*, 1, 1952, p. 330–332.

KUMM, H. W. Seasonal variations in rainfall: prevalence of *Haemagogus* and incidence of jungle yellow fever in Brazil and Columbia. *Transactions of the Royal Society of Tropical Medicine and Hygiene*, 43, 1950, p. 673–682.

*KUNSTADTER, P. Cultural patterns, social structure and repro-

ductive differentials in northwestern Thailand. In: *Symposium on culture, family planning and human fertility*. San Diego, California, 1970.

KUNSTADTER, P.; CHIEWSLIP, D.; YUILL, T. M. *Social integration and ecological isolation: a report of a medical survey of a Lua village in Mae Hong Son Province, northwestern Thailand*. Report to the National Research Council of Thailand, 1968.

LAEMMERT, H. W.; HUGHES, T. P.; CAUSEY, O. R. The invasion of small forests by yellow fever virus as indicated by immunity in *Cebus* monkeys. *American Journal of Tropical Medicine and Hygiene*, 29, 1949, p. 555–565.

*LAINSON, R.; SHAW, J. J. Epidemiological considerations of the leishmanias with particular reference to the New World. In: Falles, A. M. (ed.). *Ecology and physiology of parasites*, p. 21–56. London, Adam Hilger, 1971.

——; ——. Leishmania and leishmaniasis of the New World, with particular reference to Brazil. *Bulletin of the Pan-American Health Organization*, 7, 1973, p. 1–19.

LALOUEL, J. Sur l'état sanitaire des Babinga. *Bulletin Société Pathologie exotique* (Paris), 43, 1950, p. 714–718.

*LANGUILLON, J. Carte épidémiologique du paludisme au Cameroun. *Bulletin Société Pathologie exotique* (Paris), 50, 1957a, p. 585–597.

——. Carte des filaires du Cameroun. *Bulletin Société Pathologie exotique* (Paris), 50, 1957b, p. 417–427.

——; MOUCHET, J.; RIVOLA, E. Contribution à l'étude du *Plasmodium ovale* (Stephens, 1922) dans les territoires français d'Afrique. Sa relative fréquence au Cameroun. *Bulletin Société Pathologie exotique*, 48, 1955, p. 819–823.

LEDENT, H. La dépopulation chez les Nkundo. *Rec. Travaux Science médicale au Congo belge*, 2, 1944, p. 130–140.

LEE, H. F.; MIYAZAKI, I. *Paragonimus westermani* infection in wild mammals and crustacean hosts in Malaysia. *American Journal of Tropical Medicine and Hygiene*, 14, 1965, p. 581–585.

LIE-INJO, Luan Eng; CHIN, J. Abnormal haemoglobin and glucose–6–phosphate dehydrogenase deficiency in Malayan aborigines. *Nature*, 104, 1964, p. 291–292.

LIÉGOIS, P.; ROUSSEAU, E.; CURTOIS, Ch. Complément d'enquête sur la distribution de l'immunité antiamarile naturelle chez les indigènes du Congo belge. *Annales Société belge de Médecine tropicale*, 28, 1948, p. 247–267.

LIM, B. L. *A review of published work on medical ecology in peninsular Malaysia*. Kuala Lumpur, Institute for Medical Research, 1976, 37 p. multigr.

——; MUUL, I. *Small mammal hosts of scrub typhus vectors*. Geneva, WHO Report VBC/SG/73-2, 1973.

*——, ——; CHAI KOH SHIN. Zoonotic studies of small animals in the canopy transect at Bukit Lanjan Forest Reserve, Selangor, Malaysia. In: *IBP Synthesis Meeting* (Kuala Lumpur, 12–18 August 1974), 36 p. multigr.

LIVADAS, G.; MOUCHET, J.; GARIOU, J.; CHASTANG, R. Peut-on envisager l'éradication du paludisme dans la région forestière du Sud Cameroun? *Rivista di Malariologia*, vol. 37, n° 4–6, 1958, p. 229–256.

MADALENGOITA, J.; FLORES, W.; CASALS, J. Arbovirus antibody survey of sera from residents of eastern Peru. *Bulletin of the Pan-American Health Organization*, 7, 1973, p. 25–34.

MADDOCKS, I.; VINES, A. P. The occurrence of chronic infection on blood pressure in New Guinea males. *Lancet*, July 30, 1966, p. 262–264.

*MANN, G. V.; ROEL, O. A.; PRICE, D. L.; MERRILL, J. M. Cardio-vascular disease in African Pygmies. A survey of the health status, serum lipids and diet of Pygmies in the Congo. *Journal of Chronic Diseases*, 15, 1962, p. 341–371.

MANTNER, H. W. Some aspects of the geographical distribution of parasites. *Journal of Parasitology*, 53, 1967, p. 3–9.

MATTINGLY, P. F. Studies on West African forest mosquitoes. Part I. The seasonal distribution, biting cycle and vertical distribution of four of the principal species. *Bulletin of Entomological Research*, 40, 1949, p. 149–168.

MOHR, C. O. Notes on chiggers, rats and habitats in New Guinea and Luzon. *Ecology*, 28, 1947, p. 194–199.

MOORHOUSE, D. E.; WHARTON, R. H. Studies on Malayan vectors of malaria; methods of trapping and observations of biting cycles. *Journal of Medical Entomology*, 1, 1965, p. 359–370.

MORAES, M. A. P.; FRAIHA, H.; CHAVES, G. Onchocerciasis in Brazil. *Bulletin of the Pan-American Health Organization*, 7, 1973, p. 50–56.

——; CHAVES, G. M. Onchocerciasis in Brazil: new findings among the Yanomamo Indians. *Bulletin of the Pan-American Health Organization*, 8, 1974, p. 95–99.

MOUCHET, J.; GARIOU, J. *Anopheles moucheti* au Cameroun. *Cah. ORSTOM* (Paris), série Entomologie médicale, 4, 1966, p. 71–81.

*MURDOCK, G. P. The current status of the world's hunting and gathering peoples. In: Lee, R. B.; De Vore, I. (eds.). *Man the hunter*. Chicago, Aldine-Atherton, 1968.

MUUL, I.; LIM, B. L.; GAN, E. Scrub typhus antibody in mammals in three habitats in Sabah. *South-East Asian Journal of Tropical Medicine and Public Health*, 5, 1974, p. 80–84.

——; ——; WALKER, J. S. Mammals and scrub typhus ecology in Peninsular Malaya. *Transactions of the Royal Society of Tropical Medicine and Hygiene* (in press).

NASH, T. A. M. Some observations on resting tsetse-fly populations, and evidence that *Glossina medicorum* is a carrier of trypanosomes. *Bulletin of Entomological Research*, 43, 1952, p. 33–42.

*NEEL, J. V. Genetic aspects of the ecology of diseases in the American Indian. In: Salzano, F. A. (ed.). *The on-going evolution of Latin American populations*. p. 561–590. Springfield, C. C. Thomas, 1971, 680 p.

*——; CENTERWALL, W. R.; CHAGNON, W. A.; CASEY, H. L. Notes on the effect of measles and measles vaccine in virgin-soil populations of South American Indians. *American Journal of Epidemiology*, 91, 1970, p. 418–429.

NELSON, G. S. Filarial infections as zoonoses. *Journal of Helminthology*, 39, 1965, p. 229–250.

——; GUGGIOBERG, C. W. A.; KUKUNDI, J. Animal hosts of *Trichinella spiralis* in East Africa. *Annals of Tropical Medicine and Parasitology*, 57, 1963, p. 332–346.

NEVIN, H. M. *Annual Report of the Institute for Medical Research*. F.M.S. for 1937, 1938, p. 145–147.

*NORGAN, N. G.; FERRO-LUZZI, A.; DURNIN, J. V. G. A. The energy and nutrient intake and the energy expenditure of 204 New Guinean adults. *Philosophical Transactions of the Royal Society of London*, 3, 268, 1974, p. 309–348.

NOTELS, N.; AYRES, M.; SALZANO, F. M. Tuberculin reactions, X-ray and bacteriological studies in the Cayapo Indians of Brazil. *Tubercle*, 48, 1967, p. 195–200.

ONYAH, Itam. Filariasis among Malayan Aborigines examined at the Gombak Hospital during the period 1961–1966. *Medical Journal of Malaya*, 21, 1967, p. 384–385.

ORIHEL, T. C. Infections with *Dipetalonema perstans* and *Mansonella ozzardi* in the aboriginal Indians of Guyana. *American Journal of Tropical Medicine and Hygiene*, 16, 1967, p. 628–635.

PAGE, W. A.; MCDONALD, A. W. An assessment of the degree of man-fly contact exhibited by *Glossina palpalis* at water-

holes in northern and southern Nigeria. *Annals of Tropical Medicine and Parasitology*, 53, 1959, p. 162–165.

PAMPIGLIONE, S.; RICCIARDI, M. L. The presence of *Strongyloides fülleborni* Von Linstow, 1905, in man in Central and East Africa. *Parasitologia*, 43, 1971, p. 257–269.

——. Experimental infestation with human strain *Strongyloides fülleborni* in man. *Lancet*, March 25, 1972, p. 663–665.

——; ——. Parasitological survey on pygmies in Central Africa in Babinga group (CAR). *Rivista di Parasitologia*, vol. 35, n° 3, 1974, p. 161–188.

PARAMANTHAN, D. C.; DISSANAIKE, A. S. Sylvatic hydatid infection in Ceylon. *Transactions of the Royal Society of Tropical Medicine and Hygiene*, 55, 1961, p. 483.

*PAVLOVSKY, Y. N. *Human diseases with natural foci*. Moscow, Foreign Languages Publishing House, 1963.

POLUNIN, I. Observations on the distribution of filariasis in the interior of the Malay Peninsula. *Medical Journal of Malaya*, 5, 1951, p. 320–327.

——. Racial-relationships. ABO blood-groups and the sickle-cell trait in Malayan negritos. *Journal of the Royal Anthropology Institute*, 1953a.

*——. The medical natural history of Malayan Aborigines. *Medical Journal of Malaya*, 8, 1953b, p. 56–174.

——. The Muruts of North Borneo and their declining population. *Transactions of the Royal Society of Tropical Medicine and Hygiene*, 53, 1958, p. 312–321.

*PRICE, D. L.; MANN, G. V.; ROELS, A. O.; MERRILL, J. M. Parasitism in Congo Pygmies. *American Journal of Tropical Medicine and Hygiene*, 12, 1963, p. 383–387.

RAMACHANDRAN, C. P.; HOO, C. C.; OMAR, A. H. Filariasis among Aborigines and Malayans living close to Kuala Lumpur. *Medical Journal of Malaya*, 18, 1964, p. 193–200.

RAPER, A. B.; LADKIN, R. G. Endemic dwarfism in Uganda. *East African Medical Journal*, 27, 1950, p. 339–359.

ROBSON, P.; BOLTON, J. M.; DUGDALE, A. E. The nutrition of Malaysian aboriginal children. *American Journal of Clinical Nutrition*, 26, 1973, p. 95–100.

*SAUSSE, A. Pathologie comparée des populations primitives noires et indiennes de la Guyane française. *Bulletin Société Pathologie exotique* (Paris), 44, 1951, p. 455–460.

SCHAAD, J. D. G. Epidemiological observations in Bush Negroes and Amerindians in Surinam. *Tropical Geographical Medicine*, 12, 1960, p. 38–46.

SMITHBURN, K. C.; HADDOW, A. J. Semliki Forest virus. I. Isolation and pathogenic properties. *Journal of Immunology*, 49, 1944, p. 141–157.

SNEATH, P. A. T. Yellow fever in British Guiana. *Transactions of the Royal Society of Tropical Medicine and Hygiene*, 33, 1939, p. 241–242.

SNIJDERS, E. P.; POLAK, M. F.; HOEKSTRA, J. Jungle yellow fever in Surinam. *Transactions of the Royal Society of Tropical Medicine and Hygiene*, 40, 1947, p. 861–868.

SRIVASTAVA, H. M. L. Anopheline fauna of Uttar Pradesh-India. Part I. Vectors of malaria. *Indian Journal of Entomology*, 17, 1955, p. 363–372.

STRICKLAND, C. Papers on malaria in Malaya. *Meded. Dienst Volksgezondheid in Nederl. Indie*, 25, 1936, p. 331–340.

TONDO, C.; SALZANO, F. M. Haemoglobin types of the Caingang Indians of Brazil. *Science*, 132, 1960, p. 1893–1894.

TRAUB, R.; FRICK, L. P.; DIERCKS, F. H. Observations on the occurrence of *Rickettsia tsutsugamushi* in rats and mites in the Malayan jungle. *American Journal of Hygiene*, 51, 1950, p. 269–273.

VAN BREUSEGHEM, R. La lèpre chez les Pygmées. *Annales Société belge de Médecine tropicale*, 18, 1938, p. 135–137.

Van den Berghe, L. Les parasites intestinaux des Pygmées Efe de l'Ituri (Congo belge). *Annales Société belge de Médecine tropicale*, 18, 1938, p. 293–296.

——. Contribution à la connaissance de l'hématologie normale des indigènes du Congo belge. Premier mémoire : le sang. *Annales Société belge de Médecine tropicale*, 21, 1941, p. 375–395.

——; Janssen, P. Maladie à *sickle cells* en Afrique noire. *Annales Société belge de Médecine tropicale*, 30, 1950, p. 1553–1566.

——; Zaghi, A. J. Wild pigs as hosts of *Glossina vanhoofi* Heurard and *Trypanosoma suis* Ochman in the central African forest. *Nature*, 197, 1963, p. 1126–1127.

——; Chardonne, M.; Peel, E. The filarial parasites of the eastern goulla in the Congo. *Journal of Helminthology*, 38, 1964, p. 349–368.

*Vayda, A. An ecological approach in cultural anthropology. *Bucknell Review*, vol. 17, no. 1, 1969, p. 112–119.

Waddy, B. B. *Report on a tour of onchocerciasis foci in Mexico, Guatemala, Venezuela and Colombia*. Geneva, WHO/ONCHO/67. 60., 1960.

*——. Research into the health problems of man-made lakes, with special reference to Africa. *Transactions of the Royal Society of Tropical Medicine and Hygiene*, 69, 1975, p. 39–50.

Walker, J. S.; Gan, E.; Chye, C. T.; Muul, I. Involvement of small mammals in the transmission of scrub typhus in Malaysia: isolation and serological evidence. *Transactions of the Royal Society of Tropical Medicine and Hygiene*, 67, 1973, p. 838–845.

Warren, M.; Cheong, W. H.; Fredericks, H. K.; Coatney, G. R. Cycles of jungle malaria in West Malaysia. *American Journal of Tropical Medicine and Hygiene*, 19, 1970, p. 383–393.

Wharton, R. H. The biology of *Mansonia* mosquitoes in relation to the transmission of filariasis in Malaya. *Bulletin of the Institute of Medical Research of Malaya*, 11, 1962, 114 p.

——; Eyles, D. E. *Anopheles hackeri*, a vector of *Plasmodium knowlesi*. *Science*, 134, 1961, p. 279–280.

*——; Laing, A. B. G.; Cheong, W. H. Studies on the distribution and transmission of malaria and filariasis among Aborigines in Malaya. *Annals of Tropical Medicine and Parasitology*, 57, 1963, p. 235–254.

White, R.; Adhikari, A. K. On malaria transmission in the eastern Satpura Ranges. *Journal of the Malaria Institute of India*, 3, 1940, p. 383–411.

*WHO. *Research on human population genetics*. Geneva, WHO Report Series no. 387, 1968.

Wieseke, N. M. Encuesta medica de dos aldeas machiguences. *Bol. Oficina Sanitaria Panamericana*, 6, 1968, p. 485–504.

Wilson, T.; Edeson, J. F. B.; Wharton, R. H.; Reid, J. A.; Turner, L. H.; Laing, A. B. G. The occurrence of two forms of *Wuchereria malayi*. *Transactions of the Royal Society of Tropical Medicine and Hygiene*, 52, 1958, p. 480–481.

Wolff, J. W.; Collier, W. A.; De Roever-Bonnet, H.; Hoekotra, J. Yellow fever immunity in rural population groups of Surinam. *Tropical and Geographical Medicine*, 10, 1958, p. 323–331.

Work, T. H Discussion of the paper of Boshell, J. and Rao, T. R., "Kyasanur forest disease". In: *Proceedings of the 7th Int. Cong. on Tropical Medicine and Malaria* (Rio de Janeiro), 3, 1963, p. 305–308.

Wyndham, C. H.; McPherson, R. K.; Munro, A. Reactions to heat of Aborigines and Caucasians. *Journal of Applied Physiology*, 19, 1964, p. 1033–1038.

Zahra, A. Paragonimiasis in the southern Cameroons: a preliminary report. *West African Medical Journal*, 1, 1952, p. 75–82.

Human adaptability and physical fitness

Introduction

The dominant pattern of use by man of the tropical forest ecosystems, until recent times, has been that of swidden agriculture. Demographic characteristics of forest people (see chapter 15) are related to the consequences of this, small groups requiring a large area. The density generally varies from 4 to 15/km² (Boyden, 1972), but in the Bandama valley of the Ivory Coast it is up to 60/km². Population increase is causing many to leave the forest and move to the cultivated lands and towns (see chapter 19, first part). In Malaysia, Pelzer (1945) considered that the forest could support up to 50/km². There are also hunter-gatherers, such as the pygmies of Zaire, as well as many who combine swidden agriculture with hunting and fishing, exemplified by the Aborigines of Malaysia. The small size of the community and its isolation means a small breeding unit resulting in genetic drift and considerable genetic diversity. Population increase leads to the splitting of settlements, and the founder principle can cause further genetic diversity (see chapter 17).

    The physiological characteristics of the peoples associated with forests will be described in a regional framework. Comparison will be made with hunter-gatherers from other ecosystems.

South America

The enormous area still covered by forest is inhabited by many different populations, but physiological investigations have been few.

Body size and build

Neel, Salzano *et al.* (in Salzano, 1971) have carried out extensive genetic surveys and have also made observations on health and body size. The anthropometric variation in the Aymara Indians of the Amazonian forest has been examined by Rothhammer and Spielman (1972). Baker (1966) measured 38 adult male Shibipo Indians of the Peruvian forest who averaged 159 cm tall and 59 kg in weight. Measurements reported by Newman (1962) for other tropical forest groups show that male stature averaged 154–159 cm and weight 55–58 kg. The people of the South American forests are, in general small but, as Baker points out, cannot be regarded as pygmies.

Physique and growth

Doornbos and Jonxis (1968) studied the Bush negroes (descendents from African slaves) who have lived for some 200 years in virtually isolated groups in the rain forests of Surinam. They practise swidden agriculture, fishing and hunting. The birth weight of 308 male infants averaged 3 040 g and of 338 females 2 940 g. The mean birth length was 48.7 cm for the males and 48.2 cm for the females. During the first six months, growth in height and weight was similar to that of European children, then the Bush negro children began to lag behind the Europeans. Weight differences were *ca.* 5.6 kg and the Bush negro children were about 9 cm shorter than their European counterparts at 8 years old. Doornbos and Jonxis suggest that the lag in growth is due to frequent severe illness, including malaria, which results in a 50 per cent mortality before the tenth birthday.

Energy expenditure and working capacity

The only measurements of working capacity appear to be those by Gardner (1973) who measured maximum oxygen consumption of the Warao Indians living in the forests of Venezuela and obtained a mean of 51.2 ml/kg body weight for males.

Energy expenditure of forest-dwellers is probably moderate in absolute terms, but relatively high in relation to body size and climate. Baker noted that the Shibipo women 'performed practically no hard work and spent much of their time in the shade of the open house'. Baker concludes that it is doubtful whether the women are ever exposed to more than very sporadic heat stress. The men carried out heavy work in the early morning and late evening, so avoiding the hottest time of the day.

Heat tolerance

Baker (1966) and Hanna and Baker (1974) assessed the capacity of 20 Shibipo Indian men to work in the heat by measuring their physiological responses to two levels of physical activity; these were compared with 20 Mestizo from a nearby town. Tests were carried with ambient temperatures ranging from 26.4–32.7° C dry-bulb and 22.1–26.2 wet-bulb. The subjects walked for either one or two hours on successive days at a speed of 5 km/h for the first 6 days and 8.3 km/h for the next 6 days. The results are shown in the following table:

Walking speed (km/h)	Group	Rectal temperature (° C)	Pulse rate (beats/min.)	Sweat (g/h)
5	Indians	37.78	93.6	670
	Mestizo	37.83	102.7	618
8.3	Indians	38.24	133.3	970
	Mestizo	38.24	129.9	913

Most of the Mestizo subjects worked as day labourers and probably in general worked harder than the Indians. However, the measurements show that there was little physiological difference between the two groups. Pulse rates might be considered to be relatively low for fast walking with a body temperature of over 38° C. Sweat rates might also be considered high, but as Baker points out 'we cannot seriously speculate on how Shibipo performance in the heat compares to other racial groups'.

Sweat rates may be affected by the number of functional sweat glands. Knip (1975) has been able to examine a small number of Bush negroes in Surinam. Their number of functioning sweat glands were intermediate between the Dutch and Hindu inhabitants of Surinam, and not significantly different from either.

Blood pressure and cardio-vascular disease

Some aspects of the health and disease of the Cayapo Indians of the Amazonian forests are described by Notels *et al.* (1967); few appear to be over 40 years old so the absence of atherosclerosis and coronary heart disease may not be of great significance.

Conclusion

'It seems we must await studies with better environmental controls and with better control populations before we can determine the physiological adaptation of the South American tropical forest dweller to his hot, wet environment.' Baker's statement (1966) still holds good.

Africa

The forest area of Africa is still very extensive, but population pressure is leading to the clearance of forest and bringing people into close contact with the forest-dwellers. Some findings will be described for non-forest inhabitants, since their culture and their activity probably resembles those of the forest people whose physiology has received scant study. Many of the relevant African studies were described at a conference held in Malawi in 1971 and published for the International Biological Programme (Vorster, 1972).

Body size and build

Hiernaux, Rudan and Brambati (1975) described the weight and height of 25 rain forest populations of central Africa and compared them with 44 open country populations. They conclude that 'the shorter stature of the rain forest populations seems to be largely genetic in origin: it probably results from selective pressure exerted by the thermal stress in this hot and wet biome where sweating is of low thermolytic efficiency. The amount of reduction of adult stature depends for a large part on the number of generations spent in the forest'.

The stature and weight of the male members of the 25 rain forest populations are given in table 1.

Stature has a highly significant negative correlation with rainfall ($r = -0.36$). Weight is not related to rainfall, unless stature is constant when the logarithm of weight

TABLE 1. Mean stature and body weight in adult males of rain forest populations (Hiernaux, Rudan and Brambati, 1975).

Population	Latitude and longitude	Stature (cm)	Weight (kg)
Anang (Ibibio)	5 N 7 E	161.0	52.6
Yambasa	5 N 10 E	169.5	62.5
Mangisa	5 N 11 E	165.5	56.0
Tanga	4 N 9 E	168.1	59.3
Ewondo	4 N 11 E	169.4	59.5
Jem	4 N 13 E	162.9	56.6
Zimu	4 N 13 E	163.4	57.7
Binga (Cameroon)	4 N 14 E	153.4	49.5
Mbimu	4 N 15 E	164.2	56.2
Binga (Lobaye)	4 N 15 E	152.3	46.4
Fang	3 N 12 E	166.4	61.7
Forest Bira	2 N 28 E	158.0	52.5
Mbuti	2 N 28 E	144.0	39.8
Binga (Mekambo)	1 N 13 E	157.9	50.1
Humu	1 N 29 E	157.8	49.6
Myuba	1 N 29 E	158.1	50.9
Nyanga	0 S 27 E	160.2	51.6
Oto Ekonda	1 S 19 E	166.2	54.4
Twa Ekonda	1 S 19 E	157.5	46.8
Tembo	1 S 28 E	159.0	49.9
Lega	2 S 27 E	162.2	57.2
Yans	3 S 17 E	162.6	49.7
Fubiru	3 S 29 E	159.1	47.8
Mbala	5 S 18 E	160.5	47.7
Mbun	5 S 19 E	164.5	49.9

had a highly significant partial correlation with rainfall ($r = +0.40$). Hiernaux *et al.* (1975) therefore argue that the shorter stature of the rain forest populations cannot be attributed to poor nutrition. This view is supported by the fact that holding log weight constant increases the correlation between stature and rainfall to $r = -0.48$. The Mbuti pygmies, who with a mean height of 144 cm are the shortest people in the world, have a body weight which is considerably above the average for their stature. According to Turnbull (1961), they have no signs of protein deficiency and apparently never experience famine.

Hiernaux's figures can be compared with those relating to southern Africa, as summarized by Tobias (1972). In 90 samples of Bantu-speaking negroes the median height was 167.1 cm; substantially taller than those living in the forests. Bushmen who are hunter-gatherers of arid grazing lands had a mean height of 159.4 cm for a sample of 292 males (Tobias, 1962); females were approximately 8.0 cm shorter. 23 Bushmen who had the same average height as Tobias' sample averaged 47.7 kg (Wyndham, 1972). These stature figures are similar to the forest-dwellers (apart from the true pygmies) but the body weight of the Bushmen appears to be lighter. Tobias has shown that there is a significant secular trend in height of the Bushmen and also an increased sexual dimorphism, and argues that these findings indicate that there has been an improvement in nutrition of the Bushmen (see chapter 16).

Austin (1974) has studied the Twa who are pygmoid and the Ntomba or Hutu, a Bantu population who live in the same area of rain forest close to Lake Tunba in Zaire. Their measurements are:

	Ntomba		Twa	
	Male	Female	Male	Female
Height (cm)	168.50	155.70	159.50	153.10
Weight (kg)	58.20	48.00	47.50	44.10
Surface area (m²)	1.66	1.44	1.46	1.38
Skinfolds—sum of 6 (mm)	30.40	40.20	27.80	34.50

The differences between the men, except for skinfold thickness, are significant, but the women in the two groups are not significantly different.

Physique and growth

A comprehensive, world-wide survey of growth based on the results of the Human Adaptability Section of the International Biological Programme has been prepared by Eveleth and Tanner (1976). These results include all the reported figures for African populations. Hiernaux (1972) has compared the growth curves of the Tutsi and the Hutu living in the rain forest of Rwanda and the adjacent people living outside the forest area. Birth sizes were not significantly different but growth rates were slower amongst the Hutu from an early age, certainly by 7 years:

Age (years):	7	9	11	13	15	Adult
Group:			Height (cm)			
Tutsi (Rwanda)	116.9	126.3	137.7	145.9	152.7	176.5
Hutu (Rwanda)	115.9	121.4	131.0	139.5	146.3	167.1
Zaire	121.9	131.2	141.8	151.8	165.5	—

Hiernaux *et al.* (1975) have drawn attention to the stature of adult forest-dwellers being heavier than of populations living in the open. Body fat of nomadic pastoralists is low, e.g. in the Turkana and the Dorobo, body fat was low at all ages, being *ca.* 8 per cent of body mass at 10 years, 15 per cent at 40 years and 11 per cent at 55 years. Body fat was also low in females; girls aged 10 had the same as boys; at the age of 20 this had increased to 18 per cent and remained the same in women aged 55 and over (Prampero and Ceretelli, 1969).

The Hadza hunter-gatherers of Tanzania have been studied by Barnicot *et al.* (1972). Both men and women were lean and had little body fat; the average height and weight was 161.3 cm and 55.8 kg for the men, 142 cm and 46.2 kg for the women.

Energy expenditure and working capacity

For those practising swidden agriculture, it is the low crop yields which make the community move, so there must be considerable year to year variation in nutrition. There may also be considerable seasonal variation. Fox (1953) studied the energy expenditure and food intake of a village in

Gambia for 15 months. These farmers did not practise shifting cultivation but their annual pattern may apply to swidden cultivators. After harvest, body weight rose since there were plentiful food and little heavy work. As food stocks ran down body weight stabilized and then fell as the land had to be prepared for the new crop. In the period leading up to the harvest, food supplies were becoming exhausted and work level remained high. Fox calculated that the villagers remained in balance over the year although

expenditure, were close to exhaustion. He attributes the undoubted fatigue to the inadequate food intake with an energy deficit and loss of weight. A temporary imbalance, even if severe, would not affect maximum working capacity.

Similar seasonal imbalances must be very common in forest-dwellers.

Ghesquiere (1972) obtained the following measured maximum oxygen consumptions for the Hutu and the Twa in Zaire:

	Number observed	Height (cm)	Weight (kg)	Vital capacity (l)	Oxygen consumption (ml/min./kg)
Hutu	27	169.0 ± 0.50	56.0 ± 0.70	3.82 ± 0.02	42.7 ± 0.54
Twa	23	160.0 ± 0.98	51.2 ± 0.80	3.17 ± 0.08	47.5 ± 0.80

they were not in energy balance at any particular time during the year. With little or no reserve there could be years when food intake was inadequate and some might die from starvation. In the absence of detailed nutrition studies of swidden cultivators, it may provisionally be assumed that the diet would, in favourable circumstances, be adequate in energy, proteins—including first-class animal protein—and vitamins, but might vary considerably in mineral content. Occasional natural hazards would lead to years when the crop yield was too low and there would be a danger of starvation.

Fox (1953) made many measurements of oxygen consumption during the performance of various tasks; examples are given in table 2.

TABLE 2. Oxygen consumption of Gambia villagers performing various tasks (Fox, 1953).

Task	Energy cost above basal (kcal/h)	O_2 consumption (l/min.)
Men		
Clearing	360	1.45
Ridging	509	1.93
Planting	154	0.76
Weeding	248	1.05
Hoeing	295	1.15
Women		
Pounding rice	247	1.00

These results show that many of the tasks probably exceed the 50 per cent physical work capacity of the villagers. They can only work in excess of an oxygen consumption of 1 l/min. for limited periods and then only in the relatively cool conditions early or late in the day. As Pirnay et al. (1969) have shown, hot and humid conditions limit physical working capacity, a fact which has been confirmed many times in climatic chamber experiments. The limitation is mainly due to the rise of body temperature.

Fox was not able to measure maximum oxygen consumption, but from the figures given for body weight it is probable that many of the villagers, at the peak of energy

All these differences are significant, and indicate that Twa as a group are fitter than the Hutu.

Other measurements of work capacity include the group of seven villagers who cultivated areas cleared from the forest near Lagos (Ojikutu, Fox, Davies and Davies, 1972). They had a mean maximum oxygen intake of 48.5 ml/min./kg body weight or 61.2 ml O_2/min./kg lean body mass.

The habitual activities of the Twa and the Ntomba in Zaire were studied by Austin (1974). The women are responsible for the crops; they plant and harvest as well as hoe. They also cut and take fire-wood to the villages. They prepare the plants they harvest for cooking and eating. The men clear and burn forest areas for new fields, hunt and fish, and build and maintain houses. The day begins between 05 00 and 06 00 h. After a small breakfast, men and women walk for 10–30 min. arriving at the fields ca. 07 00 h. The women leave the fields after 11 00 h, depending on the need to prepare food; all have left by 14 00 h. After cooking and eating, the women rest for ca. 2 h and then prepare the evening meal which is eaten ca. 19 00 h. The men work for 3–4 h in the morning, clearing away brush and burning undergrowth for new fields, on the average 3 days a week. They hunt, but the frequency varies greatly between individuals; some never, others up to 6 days a week. Hunting is not strenuous as it involves stalking and the use of bow and arrow. There was little difference in the activity patterns of the two groups. The Ntomba regard the Twa as serfs and expect work from them.

The results obtained by Ghesquiere (1972) and by Ojikutu et al. (1972) may be compared with the measurements made by Wyndham et al. (1966) on many different Bantu populations and a small number of Bushmen. The Bantu and the Bushmen are not forest-dwellers but they and their ancestors have lived in the same geographical and climatic environment, and make useful comparisons. The various Bantu groups had maximum oxygen consumptions ranging from 41.0–46.4 ml/min./kg body weight and the three Bushmen who were successfully studied averaged 47.1 ml/min./kg body weight. There is therefore no great difference between any of these groups although the forest-dwellers such as the Twa, the Nigerian villagers, and the three Bushmen appear to be the fittest.

Prampero and Ceretelli (1969) measured maximum oxygen consumption in the Turkana and the Dorobo and in the Masai, who herd cattle and have a high intake of meat and milk. Values were similar in the three groups and did not show much decline with age, e.g., boys of 10 years had a maximum oxygen consumption of 50 ml/min./kg and men of 55 years and over averaged 45 ml/min./kg.

Heat tolerance

There have been many studies of the physiological responses of Africans to high environmental temperatures, notably by Ladell at Oshodi in Nigeria and by Wyndham *et al.* at Johannesburg. However, none of the subjects have been forest-dwellers. Ojikutu *et al.* (1972) examined the response to heat exposure using the portable air-conditioned bed, designed by Fox, which was adopted for use in the International Biological Programme (IBP). This test involves raising the subject's body temperature to 38° C and holding it constant for 30–40 minutes whilst whole body sweat is measured. No physical exercise is involved and the test procedure is short. (One of the difficulties of the exercise tests is that many unsophisticated subjects fail to complete the procedure.)

The Nigerian subjects included male and female students, industrial workers and those living the traditional life in a forest village near Lagos. Sweat rates in the villagers were lower than in unacclimatized Europeans; those engaged in heavy industry had more than double the rates of the villagers. Comparing the measurements of oxygen consumption during work with these results suggests that moving from rural to urban sedentary or light industrial occupations may be accompanied by some physiological deterioration, increase in body weight in the form of fat and decrease in maximum aerobic power.

Wyndham (1972) has shown that Bushmen have higher sweat rates than unacclimatized Bantu when exposed to similar standardized climatic conditions, but in general it appears that Africans, unless they are specifically acclimatized to high environmental temperatures, have low sweat rates. The Bushmen's rates could not be directly compared with those of the Nigerian villagers as different techniques were used, although indirect calculations show they were fairly similar.

Knip (1975) quotes results obtained by Ojikutu who carried out sweat gland counts on 108 Nigerian soldiers. It is not clear if any of these were forest-dwellers. They had an average functional sweat gland density of 176/cm²; substantially higher than Knip recorded in Surinam in Bush negroes. Heat tolerance may be influenced by the surface area-body weight ratio. Hiernaux *et al.* (1975) have pointed out that the pygmies in the rain forests of Zaire are, in this sense, well adapted to their hot and humid climate, and that in forest-dwellers as a whole there were significant correlations between climate and height. Austin and Ghesquiere (1976), on the other hand, question the advantages of the pygmies compared with larger people, since those with a smaller body mass have a lower capacity for storing heat.

Austin (1974) measured body temperature and heart rate during the performance of various tasks. There was little difference between the members of the two groups. In the fields, the women had rectal temperatures ranging from 37.4–38.5° C with heart rates of 104–132, while the men had rather similar body temperatures of 37.5–38.7° C with heart rates of 96–160. These figures indicate that both men and women were exposed to moderately severe heat stress. A heat tolerance test was carried out with 10 men from each group who walked at 5 km/h for 100 min. on a level road without shade on 9 successive days, starting at 11 00 h. During the first 7 days the dry bulb temperature was 31–33° C and relative humidity 70–80 per cent; globe temperatures varied from 41 to 45°. The weather was much cooler during the last 2 days. Austin concluded that there were no significant differences between the two groups in heart rate (although this declined slightly during the 7 days), rectal temperature, sweat rate and sweat gland counts.

Blood pressure and cardio-vascular disease

Huizinga (1972) has prepared a table listing populations in which blood pressure does not increase with age in either one or both sexes. Amongst those included are the Ituri pygmies from the rain forests of Zaire, the Bushmen and the Hadza. Abraham *et al.* (1960) measured blood pressure in a rural community in a partially forested area of southwestern Nigeria and did not find any increase with increasing age.

Conclusion

In Africa, as elsewhere, there is a paucity of studies of populations actually living in the forests, but there is a considerable body of research on other rural and urban populations and in many cases on people who are probably genetically fairly similar to the forest-dwellers; where measurements have been made on the latter then useful comparisons can be made with rural and urban Africans.

It may be concluded that those who live in the forest are short people (and include pygmies, the shortest people in the world) who do not suffer severely, if at all, from under-nutrition but do have a high incidence of infectious disease, principally malaria, and hence have a high infant and child mortality (see chapter 17). They have a moderate physical working capacity, and probably have to do physically hard and exhausting work. They appear to be free of cardio-vascular disease and possibly other chronic degenerative diseases.

South-East Asia and Oceania

Body size and build

Although census returns (1952) indicated *ca.* 30 000 Aborigines in Malaysia, administrators in the area thought there were *ca.* 100 000. The greatest number cultivated cassava as a main crop. Sites were rapidly exhausted and the individual community soon had to move to a new site. In addition to planting and harvesting, they trapped animals and also used a blow-pipe with poisoned arrows to kill a great variety of animals, ranging from rats to elephants. Polunin (1953) comments that the Malaysian Aborigine should have had an excellent diet although he was unable to make any assessments of food intake.

Polunin (1953) determined the following heights and weights of various groups of Malaysian Aborigines:

	Group		
	Aborigines	Senoi	Negritos
Mean height (cm)			
Men	162.9	154.7–158.0	154.4
Women	150.4	144.3–147.8	144.0
Weight range (kg)			
Men		41.5– 45.0	
Women		34.0– 39.5	

An extensive and intensive study has been carried out under the auspices of the Human Adaptability Section of the IBP by a joint Australian-British team of the people on the island of Karkar (off the northern coast of New Guinea) and the highlanders living on the mainland near Goroka (Walshe, 1974). Karkar is forested and the Kaul people live in villages in clearings and cultivate land cleared from the forest. In the highlands, the Lufa also live in villages on the edge of the forest. Detailed anthropometric studies were carried out by Harvey (1974), who found some differences between the highlanders and the islanders. Adults in both groups were well developed, with no signs of malnutrition (see chapter 16). There was little difference in the average height and weight:

		Men		Women	
Locality	People	Height (cm)	Weight (kg)	Height (cm)	Weight (kg)
Karkar island	Kaul	165.0	58.3	156.0	52.4
Goroka highlands	Lufa	162.0	59.7	152.0	52.4

Physique and growth

Malcolm (1970) studied growth rates of Bundi children in the rugged forested mountain country of the New Guinea highlands. In both height and weight for age, the Bundi children were smaller and lighter than British children, and maturity was reached much later. He concludes that the Bundi children have the slowest growth rate in the world. Skinfold thickness was also low in children, falling on the 10th percentile for British children. The slow growth rate, together with late menarche and marked delay in skeletal age were attributed to the low protein intake.

Some aspects of bodily physique were examined by Clouse and Damon (1971) who carried out a radiological survey on the people living on the island of Malaita, one of the largest of the Solomon Islands group. In the wooded hinterland the Baegu practise shifting agriculture, living in small groups of 10–20. The main findings were: all adults smoked but emphysema was very rare; scoliosis was also rare; the females were *ca.* 3 months retarded in bone age and the males 6 months compared with white Americans; cortical thickness of hand wrist bones was less than in American children.

Energy expenditure and working capacity

In New Guinea, Cotes *et al.* (1974) measured lung function, maximum oxygen consumption and heart rate in Kauls living on Karkar island and Lufa highlanders. The cardiac frequency during submaximal exercise and the estimated maximal oxygen uptake of the Kauls was similar to that of factory workers in the United Kingdom; *ca.* 40 ml/min./kg body weight. The higlanders had significantly lower heart rates at the same level of submaximal exercise and higher values for estimated maximal oxygen uptake than the Kauls; their values were similar to those of the United Kingdom amateur athletes, —>50 ml/min./kg body weight.

Norgan, Ferro-Luzzi and Durnin (1974) measured the daily energy expenditure of the same subjects studied by Cotes *et al.* On Karkar island the men expended an average of 2 350 kcal daily and the women 1 800 kcal. The highland Lufa men expended 2 570 kcal and the women 2 250 kcal daily (see chapter 16). Although the details of the agricultural work differed between the two localities, there was much similarity in the pattern of work: 70 per cent of the time was spent lying, sitting or standing and these three activities accounted for 60 per cent of the energy expended; *ca.* 10 per cent of the total time was accounted for by walking but this involved *ca.* 25 per cent of the total energy expenditure; the men's oxygen consumption during lying was 1.3 kcal/min./65 kg subject at Karkar and 1.42 kcal/min./65 kg subject at Lufa. These are rather high figures when compared with those recorded in other groups, including United Kingdom subjects. Great care was taken by Norgan *et al.* (1974) to ensure that their subjects were fully relaxed —all were familiar with the apparatus and had used it during measurements of oxygen consumption during work.

In Malaysia, Polunin (1953) noted that although the men were small and not very muscular they could work very hard and efficiently at task such as tree-felling. This suggests that in relation to body size their working capacity was high and would have been reflected in a substantial maximum oxygen consumption.

Heat tolerance

Fox *et al.* (1974) studied the New Guinea highlanders and the population on Karkar island, using the same technique as in Nigeria. The climate on Karkar island is hotter and more humid than in the highlands and the sweat rates both of men and women were higher than in the highlanders, but substantially lower than that of Europeans living on Karkar. The sweat rates of the highlanders were very low.

Budd *et al.* (1974) carefully studied the climatic conditions which the same subjects examined by Fox *et al.* (1974) experienced during the day. They measured the dry and wet bulb, and the globe thermometer temperature together with air speed, in the immediate locality of the subjects. From these measurements, carried out over a period of a year on Karkar island and in the highlands, the degree of heat exposure was calculated; information was also available on their heat production. They concluded that, although the climatic conditions were fairly severe, the New Guineans had learnt how to avoid the worst conditions by organizing their work so that they did not have a high level of energy expenditure, and utilizing whatever shade was available. Behavioural adaptation to hot conditions appears therefore to be more important than a high sweat rate.

MacDonald (1974) found low sodium and high potassium in the sweat obtained from the Karkar islanders and the Lufa people from the highlands, reflecting the absence of sodium from the diet.

Blood pressure and cardio-vascular disease

The incidence of disease in the Aborigines of Malaysia was studied by Polunin (1953) who found that there was a high mortality of infants and children, much of it due to malaria (see also chapter 17). This was reflected in the age distribution of 513 Aborigines:

Age group	Men	Women
0– 9	81	66
10–19	64	43
20–29	18	36
30–39	25	47
40–49	32	10
50–59	29	19
60–69	22	17
70+	2	2
Total	273	240

28 per cent of the population were below 10 years old and 50 per cent below 20. Apart from malaria, there were some cases of filaria, but intestinal parasites were relatively uncommon. There were 61 cases of yaws (included old healed cases) out of 171 examined but venereal disease were rare. About 10 per cent of all adults had a fungal skin disease (*Tinea imbricata*).

A respiratory condition resembling chronic bronchitis was very common, but only one individual was found with a raised blood pressure; there were no cases of coronary heart disease or heart failure due to any other cause. There was no rise of blood pressure with age, as is generally accepted as a normal accompaniment of ageing. The following table shows the consistency of blood pressure and resting heart beat:

Age (years)	18–29	30–59	> 60
Blood pressure (mm Hg)	99/64	108/64	105/68
Heart rate (beats/min.)	82	64	76

The Malaysian Aborigine is not the only group without an age effect on blood pressure. In New Guinea, Hornabrook, Crane and Stanhope (1974) found no cases of hypertension or atherosclerosis in the inhabitants of Karkar island or the highlanders near to Goroka, nor did they find any increase of blood pressure with increasing age. Macfarlane *et al.* (1968) examined the blood pressure in a number of different Melanesian groups and related the incidence of rising blood pressure with increasing age to the degree of contact with Europeans. Sinnett and Whyte (1973) have carried out a detailed survey, including whole population of a highland village in New Guinea, finding little evidence of cardiovascular disease.

Clouse and Damon (1971) in their radiological survey of the Baegu people on the island of Malaita, found very few cases of raised blood pressure. Calcification of the aortic arch was found in 8 per cent of men and 13 per cent of women over the age of 40. In spite of this, coronary heart disease was very rare.

Conclusions

There are still many unanswered questions about the physiological characteristics, patterns of life and state of health of forest-dwellers. The people are, in general, small and hence with rather low maximal work capacity, suffer from many parasitic diseases (e.g. malaria, hookworm), and have high infant and child mortality (see chapter 17). Although they work in a hot and humid environment they have low sweat rates. There is only scanty information about their activity, but both Polunin and Ghesquiere refer to their capacity for very hard work, both mentioning felling of trees.

Changes when forest people leave the forests have been studied in northern Thailand by Kunstadter (1972). Here, as in much of South-East Asia, there are hill farms and paddyfields in the lowlands where the markets and the religious and administrative centres have developed. The hill people depend largely or exclusively on swidden agriculture and have never formed supra-village administrative organizations. According to Kunstadter 'there has been sustained and rapid growth of population both in the lowland and hill areas'. Since swidden agriculture imposes a strict limit to the numbers the forest can sustain, there has been a steady migration to the lowlands. Kunstadter has studied these migrants 'whose entire way of life is changed when they move to the lowland environment'. They usually move into the lowest economic layer; the have no land and no money to buy it, so they become 'labourers, servants or gatherers and sellers of forest produce'. Their diet changes from unpolished to polished rice and they cannot supplement it by hunting and gathering. Recent migrants tend to have higher still-births and miscarriages, and higher child mortality than the people who have lived in the valley for more than a generation. The children of recent migrants are smaller with lower haemoglobin levels. Nevertheless, within a few years of migration, improvement was observed in all these criteria.

Fox *et al.* (1974) suggested that migration from village

to town might be associated with physical deterioration. In New Guinea, when the highlanders move into towns such as Port Moresby, amongst the many changes to which they are exposed are alterations in diet—from one rich in potassium and poor in sodium to a salt-rich diet, which results in profound disturbance of water and electrolyte balance with a number of cases of renal failure. When the Australian Aborigine comes to live in the shanty towns, his diet becomes dominated by white flour and sugar, and some 25 per cent develop diabetes (Macfarlane *et al.*, 1968). In addition to these dramatic dietary diseases, the new environment has many deleterious social and economic factors, including poor housing, sanitation and water supply and exposure to many infectious illnesses. These may be partly responsible for the rise in blood pressure with age and high incidence of cardiovascular disease, including coronary thrombosis, as well as diabetes and gout. These were all clearly demonstrated in a study by Harvey and Prior (1972) of Polynesians transferred to New Zealand.

It appears that the physical working capacity of forest-dwellers is closely related to body weight (as is the case in other population groups) and, in general, is on the high side of average figures when expressed in terms of body weight or lean body mass. Weiner (1972) has drawn attention to the need to consider physical work capacity in determining population density and the area required to produce an adequate amount of energy. He also pointed out that the physical working capacity is likely to be limited by disease, specifically malaria or bilharzia. Allan (1967) estimated that a family unit of five would require 12 000 kcal daily, or an average of 2 400 kcal/day/head. A reasonable assumption is that a daily energy expenditure of 5 000 kcal would be necessary to produce the 12 000 kcal required (extraction ratio, 40 per cent) and that the most active worker in the family would need to contribute about half this work. This would mean an 8–hour day, average 5 kcal/min. or an oxygen consumption of 1 l/min.

Sustained work at a rate in excess of 50 per cent maximum working capacity leads to exhaustion. The maximum oxygen consumption as measured in forest-dwellers or those living in close proximity to the forests is relatively high when expressed as ml/kg body weight/min., i.e. of the order of 46–48 ml compared with levels of 43–45 for many groups in western Europe and North America. But body weight is so low the absolute values are of the order of 2.2–2.6 l O_2/min. for maximum oxygen consumption. So 1 litre O_2 consumption/min. which is of the order required is rather close to 50 per cent of maximum or exhaustion level. Since the level of health is, in general, considered to be low for forest-dwellers, with a high incidence of malaria and a varying but usually high level of intestinal parasites, it is likely that a significant proportion cannot maintain the necessary level of energy expenditure.

In the tropical forest at ground level there is little direct solar radiation except in clearings. Hunter-gatherers live in humid atmospheres with moderately high air temperatures and little air movement. There is little seasonal variation; in some areas there are wet seasons and relatively dry seasons. The diurnal variation in air temperature may be only

ca. 5° C, with a mean temperature close to 30–32° C. In this hot oppressive atmosphere clothes are an impediment to heat loss, although they may be a protection against thorns and biting insects, and humid forest-dwellers therefore usually have a minimum of clothing. The sweat rates of forest-dwellers is probably low, possibly due to genetically few sweat glands. Kuno (1956) found that sweat gland counts were lower in people living in the Philippines than in Japan, but this has not been confirmed, and Weiner has indicated that detailed sampling of the body surface is necessary if accurate counts of active sweat glands are to be achieved.

The problems facing the forest-dweller when he moves beyond the forest have been described by Kunstadter (1972). The opposite has been studied by Crocker (1971) for an Indian tribe of Canela, living in northwestern Brazilian savanna, who were resettled in forest country. They adapted badly and most eventually returned to the savanna. Unfamiliarity of the agricultural techniques required, although these were not so very different from those practised in the savanna, was the main reason for this failure.

Research needs and priorities

Forest-dwellers in different regions of the world are small people, with a very high infant and child mortality, mainly due to malaria, acute respiratory disease, diarrhoea and trauma (see chapter 17). They seem to be free of cardiovascular disease, and blood pressure does not increase with age. Although they may have a low incidence of chronic degenerative diseases, they suffer from a high incidence of infectious disease and disease due to parasites. They do not appear to have serious nutritional problems, but the physical work they have to do to raise sufficient crops may be near to their limit.

There is a need for many more studies of forest-dwellers, particularly the physiological characteristics of physical work capacity and heat tolerance, habitual activities and detailed measurements of the climatic conditions. The type of study carried out in Karkar island and at Goroka in New Guinea by the joint Australian-British team working under the auspices of the IBP may serve as an example for what is needed elsewhere.

The forests of South America are still immense and they probably contain more forest-dwellers than in the rest of the world. They have scarcely been examined by physiologists, although there have been substantial genetic and anthropometric studies by Neel, Salzano etc. (Salzano, 1971). There is a strong case for detailed physiological measurements to be made in this region. An attempt has been made above to summarize the physiological and related characteristics of forest-dwellers; it is important to know if the South American Indians fit this pattern or in what respects they diverge.

Another problem which deserves more intensive study is the effect on the forest-dweller of leaving the forest. There are increasing numbers of those migrants and little is known of their adjustement or physiological adaptations to the new conditions. Apart from the need to improve their conditions, long-term studies should be made to ascertain the effects of a new environment on cardio-vascular responses. Their diet is likely to change; the details need to be known and related to changes which may occur in teeth or the incidence of diseases such as diabetes.

This is urgent for the life-pattern of the forest people is threatened. The information gained could throw light on their freedom from cardio-vascular disease. One factor influencing the incidence of coronary heart disease is the lack

of physical exertion. Swidden agriculturalists and hunter-gatherers have periods of high energy expenditure but modern urban man seldom works hard physically. Boyden (1972) suggests that man is, evolutionarily, a hunter-gatherer and rising blood pressure and cardio-vascular failure are the consequences of modern urban living being so different from this. Although evidence in favour of this suggestion may be inadequate, further study of tropical forest-dwellers could be rewarding.

Bibliography

ABRAHAM, D. G.; ALELE, C. A.; BARNARD, B. G. The systemic blood pressure in a rural West African community. *West African Medical Journal*, 9, 1960, p. 45–58.

ALLAN, W. *The African husbandman*. 2nd edition. Edinburgh, Oliver and Boyd, 1967, 505 p.

AUSTIN, D. *Heat stress and heat tolerance in two African populations*. Dept. of Anthropology, Pennsylvania State University, Ph. D. thesis, 1974, 130 p.

——; GHESQUIERE, J. L. A. To be published in *Human Biology*.

BAKER, P. Ecological and physiological adaptation in indigenous South Americans. In: Baker, P.; Weiner, J. S. (eds.). *The biology of human adaptability*, p. 275–304. Oxford, Clarendon Press, 1966, 541 p.

BARNICOT, N. A.; BENNETT, F. J.; WOODBURN, J. C.; PILKINGTON, T. R. E.; ANTONIS, A. Blood pressure and serum cholesterol in the Hadza of Tanzania. *Human Biology*, 44, 1972, p. 87–116.

BOYDEN, S. Biological determinants of optimum health. In: Vorster, D. J. M. (ed.). *Human biology of environmental change*, p. 3–11. London, Taylor and Francis, 1972, 205 p.

BUDD, G. M.; FOX, R. H.; HENDRIE, A. G.; HICK, K. E. A field survey of thermal stress in New Guinea villagers. *Philosophical Transactions of the Royal Society of London*, B, 268, 1974, p. 393–400.

CLOUSE, M. E.; DAMON, A. Radiologic survey in the Solomon Islands, 1968: lungs, heart, spleen, bone age and dental developments. *Human Biology*, 43, 1971, p. 22–35.

COTES, J. E.; ANDERSON, H. R.; PATRICK, J. M. Lung function and the response to exercise in New Guineans; role of genetic and environmental factors. *Philosophical Transactions of the Royal Society of London*, B, 268, 1974, p. 349–361.

CROCKER, W. H. The non-adaptation of a savanna Indian tribe (Canela, Brazil) to forced forest relocation: an analysis of factors. In: *Seminario de Estudos Brasileiros*, p. 213–281. Brasil, Sao Paulo, Universidade de Sao Paulo, 1971, 370 p.

DOORNBOS, L.; JONXIS, J. H. P. Growth and Bush negro children on the Tapomahony River in Dutch Guyana. *Human Biology*, 40, 1968, p. 396–415.

EVELETH, P. B.; TANNER, J. M. *World wide variation in human growth*. Cambridge University Press, 1976, 498 p.

FOX, R. H. *A study of the energy expenditure of Africans engaged in various rural activities, with special reference to some environmental and physiological factors which may influence efficiency of their work*. University of London, Ph. D. thesis, 1953, 88 p.

——; BUDD, G. M.; WOODWARD, P. M.; HACKETT, A. J.; HENDRIE, A. G. A study of temperature regulation in New Guinea people. *Philosophical Transactions of the Royal Society of London*, B, 268, 1974, p. 375–391.

GARDNER, G. W. Physical fitness of primitive peoples—the

Warao Indians of Venezuela. In: Seliger, V. (ed.). *Physical fitness*, p. 158–163. Prague, Charles University Press, 1973, 230 p.

GHESQUIERE, J. L. A. Physical development and working capacity of Congolese. In: Vorster, D. J. M. (ed.). *Human biology of environmental change*, p. 117–120. London, Taylor and Francis, 1972, 205 p.

HANNA, J. M.; BAKER, P. T. Comparative heat tolerance of Shipibo Indians and Peruvian Mestizos. *Human Biology*, 46, 1974, p. 69–80.

HARVEY, R. G. An anthropometric survey of growth and physique of the populations of Karkar island and Lufa sub-district, New Guinea. *Philosophical Transactions of the Royal Society of London*, B, 268, 1974, p. 279–292.

HARVEY, H. P. B.; PRIOR, I. A. M. Cardio-vascular epidemiology in New Zealand and S. Pacific. In: Vorster, D. J. M. (ed.). *Human biology of environmental change*, p. 80–86. London, Taylor and Francis, 1972, 205 p.

HIERNAUX, J. La croissance des écoliers rwandais. *Mémoires Académie Royale Sciences d'Outre-mer, Classe Sci. Nat. Méd.*, 16, 1965, p. 1–204.

——. A comparison of growth and physique in rural, urban and industrial groups of similar ethnic origin: a few case studies from the Congo and Chad. In: Vorster, D. J. M. (ed.). *Human biology of environmental change*, p. 93–95. London, Taylor and Francis, 1972, 205 p.

——. *The people of Africa*. London, Weidenfeld and Nicolson, 1974, 217 p.

——; RUDAN, P.; BRAMBATI, A. Climate and the weight/height relationship in sub-Saharan Africa. *Annals of Human Biology*, 2, 1975, p. 3–12.

HORNABROOK, R. W. The demography of the population of Karkar island. *Philosophical Transactions of the Royal Society of London*, B, 268, 1974, p. 229–239.

——; CRANE, G. G.; STANHOPE, J. M. Karkar and Lufa: an epidemiological and health background to the HA studies of IBP. *Philosophical Transactions of the Royal Society of London*, B, 268, 1974, p. 240–250.

HUIZINGA, J. Casual BP in populations. In: Vorster, D. J. M. (ed.). *Human biology of environmental change*, p. 164–169. London, Taylor and Francis, 1972, 205 p.

KNIP, A. S. Acclimatization and maximum number of functioning sweat glands in Hindu and Dutch females and males. *Annals of Human Biology*, 2, 1975, p. 261–278.

KUNO, Y. *Human perspiration*. Springfield, Illinois, Charles C. Thomas, 1956, 416 p.

KUNSTADTER, P. Ecological change and human biology in northern Thailand. In: Vorster, D. J. M. (ed.). *Human biology of environmental change*, p. 21–30. London, Taylor and Francis, 1972, 205 p.

MacDonald, I. C. *The composition of human sweat with special reference to variation due to ethnic origin, acclimatization status and other factors.* University of London, Ph. D. thesis, 1974, 188 p.

Macfarlane, W. V.; Howard, B.; Scroggins, B.; Skinner, S. L. Water, electrolytes, hormones and blood pressure of Melanesians in relation to European contact. In: *Proceedings of the 24th International Congress of Physiological Sciences*, 7, 1968, p. 274.

Malcolm, L. A. Growth and development of the Bundi child of the New Guinea highlands. *Human Biology*, 42, 1970, p. 293–328.

Malhotra, M. Peoples of India including primitive tribes—a survey of physiological adaptation, physical fitness and nutrition. In: Baker, P.; Weiner, J. S. (eds.). *The biology of human adaptability*, p. 329–356. Oxford, Clarendon Press, 1966, 541 p.

Newman, M. T. Adaptations in the physique of American aborigines to nutritional factors. *Human Biology*, 32, 1962, p. 288–313.

Norgan, N. G.; Ferro-Luzzi, A.; Durnin, J. V. G. A. The energy and nutrient intake and the energy expenditure of 204 New Guinean adults. *Philosophical Transactions of the Royal Society of London*, B, 268, 1974, p. 309–348.

Notels, N.; Ayres, M.; Salzano, F. M. Tuberculin reactions, X-ray and bacteriological studies in the Cayapo Indians of Brazil. *Tubercle*, 48, 1967, p. 195–200.

Ojikutu, R. O.; Fox, R. H.; Davies, C. T. M.; Davies, T. W. Heat and exercise tolerance of rural and urban groups in Nigeria. In: Vorster, D. J. M. (ed.). *Human biology of environmental change*, p. 132–144. London, Taylor and Francis, 1972, 205 p.

Pelzer, K. J. Pioneer settlement in the Asiatic Tropics. *American Geographical Society Special Publication*, 29V, 1945, 93 p.

Pirnay, F.; Petit, J. M.; Deroanne, R. Consommation maximum d'oxygène et température corporelle. *Journal de Physiologie* (Paris), Suppl. 2, 1969, p. 376.

Polunin, I. The medical natural history of Malayan Aborigines. *Medical Journal of Malaya*, 8, 1953, p. 56–174.

Prampero, P. E. di; Ceretelli, P. Maximum muscular power (aerobic and anaerobic) in African natives. *Ergonomics*, 12, 1969, p. 51–59.

Rothhammer, F.; Spielman, R. S. Anthropometric variation in the Aymara. *American Journal of Human Genetics*, 24, 1972, p. 271–380.

Salzano, F. A. (ed.). *The on-going evolution of Latin American populations.* Springfield, Illinois, Charles C. Thomas, 1971, 680 p.

Sinnett, P.; Whyte, H. M. Epidemiological studies in a total highland population, Tukisarta New Guinea: cardio-vascular disease and relevant clinical, electrocardiographic, radiological and biochemical findings. *Journal of Chronic Disease*, 26, 1973, p. 265–290.

Tobias, P. V. On the increasing stature of the Bushmen. *Anthropos*, 57, 1962, p. 801–810.

——. The peoples of Africa south of the Sahara. In: Baker, P.; Weiner, J. S. (eds.). *The biology of human adaptability*, p. 111–200. Oxford, Clarendon Press, 1966, 541 p.

——. Growth and physique. In: Vorster, D. J. M. (ed.). *Human biology of environmental change*, p. 96–104. London, Taylor and Francis, 1972, 205 p.

Turnbull, C. M. *The forest people: a study of the pygmies of the Congo.* New York, Simon and Schuster, 1961.

Vorster, D. J. M. (ed.). *Human biology of environmental change.* London, Taylor and Francis, 1972, 205 p.

Walshe, R. J. Geographical, historical and social background of the people studied in IBP (in New Guinea). *Philosophical Transactions of the Royal Society of London*, B, 268, 1974, p. 223–228.

Weiner, J. S. Tropical ecology and population structure. In: Harrison, G. A.; Boyce, A. J. (eds.). *Biological and social structure of human populations*, p. 393–410. Oxford, Clarendon Press, 1972, 447 p.

——; Willson, J. O. C.; El-Neil, H.; Wheeler, E. F. The effect of work level and dietary intake on sweat nitrogen losses in a hot climate. *British Journal of Nutrition*, 27, 1972, p. 543–552.

Wyndham, C. H. Heat tolerance and work capacity of South African groups. In: Vorster, D. J. M. (ed.). *Human biology of environmental change*, p. 145–153. London, Taylor and Francis, 1972, 205 p.

——; Strydom, B. N.; Morrison, J. F.; Williams, C. G.; Bredell, G. A.; Heyns, H. The capacity for endurance effort of Bantu males from different tribes. *The South African Journal of Science*, 62, 1966, p. 259–263.

19 Populations, civilizations and human societies

First part: Populations densities

Introduction

In considering relationships between human societies and tropical forest ecosystems, a basic datum of the greatest importance is that of population numbers and their distribution in space. The number of people living within an area is a fairly good measure of the effect a society has on the environment. This relationship is even more accurate in the humid tropical zone where there are systems with moderate levels of productivity. Manual labour is largely predominant and there is only limited use of animal labour of a fairly low yield. This means that the amount of natural product taken by an individual varies between once, twice or three times for a given quantity of work. This is characteristic of rural economies that have been functioning over a long period as an almost closed system. A large part of the production goes to feed the producers or, more broadly, the producing society (or co-existing societies) within an exchange framework which is at most regional. It is therefore possible thanks to population numbers, to find a graduation, from natural forest to completely transformed ecosystems. There are no exceptions to the rule that low population densities take low harvests from the ecosystem and higher densities are linked to swidden cultivating societies that manipulate the ecosystem and redirect its energy flow to their own ends. Even higher density values result in a total reorganization of natural systems.

The relationship between an area and the population living on it (and living from it in the case of food-gatherers and cultivators) constitutes the population *density*. Population densities, expressed in numbers per km², have many advantages in making quantitative comparisons between areas or ecosystems. Many studies, usually undertaken by geographers, have led to the formulation of a number of definitions, precautions and rules.

General density takes into account the whole of a population, including town-dwellers. Rural density excludes the inhabitants of towns, using a definition of the threshold beyond which agglomerations become urban areas (such definitions vary according to the geographical areas and authors). Agricultural density measures the ratio of the rural population to the effectively cultivated area within a given year. The frequent presence of uncultivated land that nevertheless comes within the usable space of a human group (as fallow, and land allocated for activities which are more or less related to agricultural practices) makes the latter concept rather difficult to use.

Population density is subject to systematic mapping. At small scales it is likely that no national, international or world atlas exists that does not have a map of population density. A number of countries of the humid tropical zone have benefited from a much more refined level of presentation at scales ranging from 1/500 000 and 1/2 000 000. The maps of the provinces of Zaire (Gourou, 1960; de Smet, 1962, 1966, 1971), the sheets included in the regional atlas of Cameroon (Barral and Franqueville, 1969; Champaud, 1973; Franqueville, 1973), and the map of the Red River delta (Viet-Nam) by Gourou (1936) are all notable examples. The function of maps of population densities is to show inequalities in the distribution. Other maps achieve the same result by a different process. Inhabitants are shown in the form of a dot or other conventional sign, placed at the point at which they live; the sign represents a certain number of inhabitants such as 100 or 1 000. These maps of population distribution are an equally good expression of variations in human numbers in a form that is immediately perceptible. They are complementary to the preceeding type. One method permits inhabitants to be counted by adding up all the points present within any area; the other facilitates comparisons.

Population density is both one of the most simple and one of the most synthetic ways in which the nature of relationships between a given area and all the people making use of it can be understood. It also represents a constraining factor that governs the functioning and evolution of the systems of production and, in a more general sense, of the social system belonging to the human groups concerned.

This does not mean that study of human societies can be undertaken in the same way as for animal and plant communities. Many features are very characteristic of man's behaviour towards space and the natural environment. Man has been shown to be capable of reaching all areas, even when cut off by various obstacles. The ability of the human species to multiply whenever conditions permit is not controlled in an absolute way by the presence of other species. Lastly, no psycho-somatic type of regulation, as has been observed in various animal populations (Stott, in Vayda, 1969), controls effective increase within a given area above a certain level of accumulation.

An ecological analysis based on population density will have to take into account the specific character of human groups as well as the ecological environment into which man has intruded and which reacts to his presence. This operates at two levels. The human species shows an unquestionable autonomy towards the physical environment as well as towards several biological constraints that assign other species to a fixed place and limit their numbers. The distinctive cultural heritage of each human group is associated with its particular type of behaviour within the ecosystems. Hence the concept of civilization as 'a combination of techniques for production and management' (Gourou, 1973). Behind the social practices and productive processes of present day men is an old heritage that is firmly entrenched; thus civilization, understood in this way, is linked to the distinctive behaviour of each human group with regard to the surrounding natural system and the opportunities which it affords. Society applies rather to an assemblage located in time and space. It puts the emphasis on a functioning system which links ecological, social and economic factors. Many societies may exist or succeed each other within the framework of the same civilization. The distinction that is made between the two concepts of society and civilization that is used here, is far from being unanimously accepted, and these terms are currently used in a similar sense. The important fact is that there is not just the presence of the human species in front of natural ecosystems nor just a collection of individuals but of types of organized systems.

Groups of swidden cultivators, and especially groups of food-gatherers, undertake their activities largely within the framework of the natural ecosystem. However, their practices are never identical, nor mechanically determined. There are large differences in the types of crops, tools, the way in which fields are prepared, the presence or absence of successional crops and in the type of fallowing, between one society and another. Settlements may be mobile or permanent, of a village type or dispersed, and neither the agricultural practices nor the distinctiveness of the local environment can account for all the variations. Thus, a kind of independence from nature exists and this is never completely constraining. This margin of independence increases as groups having a higher technological level and operating a system of production that is completely unique in each case are considered. Each group operates within a compromise between two series of constraints: those of the physical and biological environment and those imposed by past history, habits acquired and transmitted, the mechanisms by which society reproduces and finally by the relationships with other societies.

Space, along the continuous chain from natural to purely cultural elements, imposes constraints in terms of distance and time. Mastering of space at different scales governs both the optimum use of the environment and the abilities of a group to defend itself or to dominate over others. Factors imprinted in space, such as land tenure or the type and form of settlement, are at a meeting point between ecology and social factors. Population densities themselves only have meaning with reference to space. More than that, they serve better than anything else to indicate the process that societies go through in moulding and differentiating space ('generate' it, to use Marxist terminology). In another way, densities reflect on societies some constraints inherent to space. Their inequalities in part show the spatial inertia of a population. This is able to accumulate more easily in some areas due to

the effects of demographic growth than it does spread out into adjacent areas. The basic reason for this is the distance factor.

Insular distribution of high densities

Inequalities in density distributions

There are vast areas in the humid tropical zone that are practically uninhabited and are adjacent to other rural areas with population densities of several hundreds or even a thousand people/km². The situation is one of islands with a more or less dense population, in the middle of areas of low density. The fall in density at the margins of one of these population concentrations is often very, or extremely rapid, and throws them into relief on a map. Some of these islands are isolated, many others form beads on a chain. On a smaller scale the high density areas also have a tendency to form larger groupings. In tropical America, the nuclei of population in Atlantic Brazil and those better demarcated of the Andean basins and some coastal sectors along the Pacific form two areas which are relatively well populated and which contrast to the empty interior of the continent, especially the larger part of Amazonia. In Asia, the geographical region of hot, wet climates is divided between the populated area of the Indian subcontinent and an extended area with a lower mean population density represented by the Indo-China and Indonesia. This second region has some remarkably well populated areas, for example a true island in the case of Java or figurative islands as in the Red River delta and other coastal plains of Viet-Nam. The intervening continental areas, which are more or less mountainous, and some islands or parts of islands have low densities. Lastly, in Africa, Trewartha and Zelinsky (1954) have underlined the way in which central Africa, from Chad to Rhodesia, appears to be an empty part of the continent lying between two higher populated regions of western and eastern Africa. At a more detailed level, a study of present day population densities in Zaire has shown that almost two thirds of the inhabitants live in three relatively well populated regions that together cover around 30 per cent of the country (Gourou, 1955). The higher populated area lies along the 5th parallel south and can be broken up into three principal areas of concentration. In the central area of Kwilu the density is 16/km². Five nuclei of population densities markedly higher than this mean, together have more than 40 per cent of the population on less than 20 per cent of the area (Nicolai, 1963).

This phenomenon of geographical isolation of population densities is partially masked in the usual numbering made according to administrative divisions as shown in official statistics. Boundaries created by frontiers and internal subdivisions rarely coincide with natural division of space based on population density and the inequalities tend to be lost at the level of the composite units. This 'island' distribution is characteristic of all developing countries and Europe formerly also had a comparable situation (Duby, 1962). It is evident that a phenomenon, which extends largely beyond the limits of the humid tropics, cannot be completely explained by the peculiar characteristics of this region. A more general explanation must take into account the specific role of demographic, technical and sociological factors that apply to all humanity.

It is not so easy to disassociate factors relating to man from those linked to the environment when considering the tropical forest ecosystem. The most fundamental fact is a climate with a minimum of 1 400–1 500 mm/a rainfall and at least 8–9 rainy months (Lauer, 1952). Temperature only becomes a discriminating factor on the eastern continental margins (Atlantic Brazil, Natal, southern China and Indo-China). According to Köppen, a mean temperature of 18° C for the coolest month is generally considered as marking the limits between tropical and subtropical climates in humid coastal regions. The terms equatorial and subequatorial are used to designate variations in the length of short relatively rainless equinotial periods. These two climatic types correspond to dense evergreen and dense semi-deciduous forest. But the association between a humid tropical climate and dense humid forest does not always occur. Disassociation may result from natural phenomena; for example, edaphic conditions (saturated soils) or, on a larger scale, a period of drought that may be recent enough for the results still to remain in evidence in the vegetation. Man has been capable of bringing about the disappearance or the retreat of the forest at the ecological limits of the dense humid forest or even slowing down forest expansion when this is initiated by an increase in rainfall. Even in the heart of the equatorial forest, high population densities can cause the disappearance of the original vegetation.

Favourable and unfavourable environments

The analysis of the relationships between natural ecosystems and human population density is classically undertaken in terms of favourable or unfavourable environments. This is a too generalized and dangerous way of approaching a geographical reality. It is better to ask: are there habitats that are intrinsically better adapted to supplying produce to be consumed or sold than others and that, as a result, can feed a greater number of people? The use of the term intrinsically implies at an equivalent technological level with the same amount of work. An affirmative reply to this question gives rise to another: do inequalities in population densities, as shown on the map, reflect land-capability differences between natural habitats?

Within the tropical zone, a major subdivision separates humid forest ecosystems from savanna ecosystems. In theory, the dense humid forest offers certain advantages. The long duration of vegetative growth enables the cultivation of plants such as the plantain banana or cassava the yield of which (in energy per unit area) and the ratio of produced quantities to time spent are particularly high when compared with that of cereals cultivated in savannas with their long dry season. It is also possible in a forest environment to obtain in a year two or three crops with a short growth cycle, especially maize. It remains to be seen to what extent these theoretical considerations are reflected in population

densities. Some calculations have been made for each continent and for the whole world, using the climatic classification developed by Köppen as a framework (Staszewski, 1961). The results for Africa (inhabitants/km²) are as follows: 9.6 for tropical humid regions, 7.6 for savanna and 5.5 for steppe climates. The equivalent results for the entire tropical world are: 18.4, 14.4 and 7.9 respectively. These figures destroy the image of an equatorial environment which is hostile to man that has been created in the minds of people because of the vast largely uninhabited forest areas in Amazonia, western and central Africa, or on islands such as Borneo. These exist but the forest is not necessarily the primary cause of their low population densities. The large empty spaces in the equatorial region of the population density map are largely compensated for by other areas that are intensively occupied. Many of these regions are found adjacent to the dense humid forest, or are enclaves within it, or occur in places which forest formerly occupied. For example, in western Africa a whole chain of concentrated population nuclei are located along the northern boundaries of the forest, from the Senegalese Casamanca, in the west, to the Bamiléké plateau (Cameroon) in the east, passing through the Kissi country and that of the Futa-Jallon (Guinea-Conakry), and the Yoruba and the Ibo-Ibibio (eastern Nigeria). In eastern Africa, concentrated areas of dense populations generally occupy the sites of former discontinuous forests. This is the situation in the highlands of Rwanda and Burundi as well as in Kikuyu country (Kenya) where the forest only remains in a residual form at high altitudes. Java and many of the Pacific and Caribbean islands provide other examples. Lastly, the rice-cultivated deltas of South-East Asia are an example of concentrations of people surrounded by forest, as well as, at much lower densities, the modest concentrations of population that occur around Belém (Amazonia) or in the Equatorial Province of Zaire.

In the Red River delta and in the coastal plains of Viet-Nam, the present rural population density is over 500/km² and it is higher in Java. The most populated regions of the wettest parts of tropical Africa (the Kikuyu country, central Rwanda, southeastern and southwestern Nigeria and the Bamiléké plateau) have in some places densities of around 300/km². Almost comparable values occur in parts of upland Madagascar. The difference between these figures and the mean density for the tropical humid forest zone as defined by Köppen is enormous. It is far greater than the differences between the principal subdivisions of the tropical climatic zone. The very low population densities of *ca.* 2/km² observed for large parts of the Congo basin or for the Amazonian forest cannot be explained in terms of the ecological conditions; even the poorest soils (when covered by forest) and agricultural practices the most wasteful of space cannot account for so low a population.

The tropical forest ecosystem represents a relatively homogeneous environment in terms of its ability to feed. At the slash-and-burn level of technology, the richest soils enable the period of cultivation to be extended without a des-cline in yield. This obviates the necessity for frequent move-of villages or of fields being too far from the settlement. It guarantees success with perennial crops. Nevertheless, many cases occur in which a move from an area of mediocre ecological potential to one of a higher potential is accompanied by local or regional higher densities, as shown on the map. For example, cocoa plantations were developed with government support in the north of the Congo, within the limits of the best soil formations and the population was partly redistributed in the villages having benefited from these conditions (Guillot, 1973). In cases where there is agreement between population densities and ecosystem potential, adjustment may have occurred by a series of trials and errors. For example, plots of equal size within an uninhabited forest were distributed by the Government to groups of settlers at the foot of the Bamiléké plateau (western Cameroon) without taking into account their unequal soil quality; thus, after several years some cultivators were rewarded for their efforts whilst others had nothing (Barbier, 1971).

There is only a general and systematic agreement between ecosystem quality and population density when there is a high population density. As soon as all available space, including unused or little used land, becomes rare, systematic efforts are made to obtain the most from all available space. When everything is exploited to the maximum limits of a given technology, the better endowed areas carry the highest demographic load. Inequalities in population distribution become more obvious once the forest has been eliminated or degraded and the true qualities of soils are no longer hidden.

In cases where agricultural systems are relatively advanced, natural fallowing is accompanied by an appropriate crop sequence, a differential use, through a variety of field types, of the local conditions and lastly by a painstaking work of the soil. Population density equivalents do not go above 30/km² on the poor sands of the southwestern region of Brazzaville but they reach around 60/km² in Kasai (southwestern Zaire) (Béguin, 1960) and *ca*. 150–200 in eastern Nigeria (Morgan, 1955). More or less the same level of equilibrium has been reached on the 'terre de barre' plateau area developed behind the coast of the Togo in western Africa. Regions subjected to permanent cropping *a fortiori* have the same or even higher density levels. On the Bamiléké plateau, sandy soils derived from granites feed a population which never rises above 100–150/km² and near by on soils of volcanic origin, densities are twice these values. In the highlands of Madagascar the dense Betsiléo population relies on an intensive technique based on irrigated rice and terracing slopes. Significant variations in density are due to the amount of water to supply the terraces and this depends on the geology. In Kikuyu country (Kenya), and the uplands of Rwanda, the equatorial climate becomes drier to the east and with decreasing altitude. The population density follows this variation. However, the significance of these observations must not be overstated. The most marked variations that can be directly attributed to ecological factors within the tropical forest ecosystem are of the order of 1 to 5 or 1 to 10 at the most. This is small in a gamut of densities which range from nearly zero to over 1 000/km².

Attractive and unattractive environments

An environment may be attractive for a number of reasons. For example, if it allows a group of cultivators to obtain equal production for less work. Land that is easy to work is thus preferred to heavy ground: in Rwanda, formerly only the hills were cultivated thereby excluding the markedly richer soils of low-lying areas. Generally, colonization of plains and rice-cultivating delta areas occurs from upstream downstreamwards and from the margins towards the centre. In both South-East Asia and the uplands of Madagascar marginal areas, where hydrological control is still relatively easy are preferred although these are rarely the most productive areas. At the junction of plains and hills an agricultural society may be able to draw on a greater variety of habitats and resources and make a better allocation of its labour force. This favourable effect is also present in regions with prominent mountains and in forest-savanna mosaics. That the resources to be used will be exploited by only one population or perhaps by two populations living in symbiosis through a series of more or less institutionalized exchanges is unimportant, as can be seen in the New Hebrides, especially in the islands of Aoba and Maewo where *Colocasia* and yam cultivators used to coexist until recently (Bonnemaison, 1974).

A final reason that may make an area attractive is a connection between the environment and the customs of the population. All populations have a tendency to reproduce their production system, not only in the physical sense but also in the sociological sense and this requires a search for suitable places. In both the Middle-West and the West of Madagascar the settlements of the populations on the high plateaux are directly related to suitable conditions for an advanced type of irrigated rice cultivation (Le Bourdiec, 1974). Parsons (1949) in his study of Antioquia (Colombia) has clearly shown how the pioneering movement associated with agriculture occurred along the ranges of the Cordillera and jumped from one range to another at heights suitable for the particular type of land use.

In cases where the environment has been more or less completely transformed by man, it is often difficult to visualize the conditions that the first inhabitants found. Present day judgments of attractiveness are very subjective. Thus, in the deltas of South-East Asia, the appearance of the area has become so favourable that what it looked like originally and the total effort required to transform it is forgotten.

Ecosystems and demographic accumulation

In all parts of the tropical humid world well endowed and/or attractive regions exist that only have insignificant populations. Natural factors may wield their influence in indirect ways and have effects that are only seen with the passage of time. It is therefore necessary to consider all the factors capable of accelerating, slowing down or hindering population growth within a given area and, in some cases, those which may lead to depopulation.

An initial type of control is liable to be exerted on the

natural population growth through fertility or mortality. In this respect the role of pathogenic complexes (Sorre, 1943) is important. These are diseases of many different kinds that are linked to a particular environment due to the requirements of the reservoir animals, the insect vectors or of the infective organisms themselves (see chapter 17). Some pathogenic complexes have been found to be associated with particular types of forest ecosystems, others are wide-spread but the importance and virulence of a disease may vary considerably from place to place. Contrary to an opinion that was wide-spread for a long time, it does not seem as if the tropical forest ecosystem is responsible for higher mortality levels than those in savanna ecosystems with their characteristic dry season. Inequalities in population distribution are difficult to explain on a large geographical scale in terms of differences in the distribution and character of pathogenic complexes. Tropical America has, for a long time, escaped serious infections such as yellow fever and some particularly active pathogenic vectors (e.g. *Anopheles gambiae*) that are wide-spread in Africa, but the Amazonian forest is more thinly populated than the Congo forest. In monsoon Asia, where population densities have means of *ca*. 100/km² as compared to densities of *ca*. 12/km² for the other two tropical continents, there are no natural advantages in terms of illnesses linked to heat and humidity. The differences have to be explained by something else.

Contrasts in population densities that are linked to the natural environment and to associated epidemiological features can be seen more clearly on a regional scale. One of the classic examples is that of northern Viet-Nam where the densely populated plains that suffer from a less virulent malaria can be contrasted with mountainous areas endowed with inverse characteristics. Whole regions of Africa have been depopulated or prevented from being populated by sleeping-sickness; this was present at a catastrophic level in the Congo basin from the latter part of the 19th century up to the end of the 1920s, when massive and systematic chemical treatments were applied. In the New Hebrides, the windward slopes are more malarial and less populated than the leeward slopes. Nevertheless, here again the situation is complex because of the inextricable interactions of natural factors and human geography. Epidemiological foci had a tendency to be located along rivers but have also occurred along caravan routes (for example in the Niari valley in the south of the Congo, where the exchanges between the coast and the interior used to take place along the paths made some distance from the river, in savannas completely stripped of trees).

Another more general explanation is that of auto-protection that can accompany population density once it exceeds certain levels: the infective power of vectors, as has been shown in the case of onchocerciasis, is diluted in the mass of the population, which as a result only feels diminished effects; the transformation of the environment that occurs above certain population densities can alter the ecology for transmitting insects; in the open dry forests of eastern Africa, population density above 40–50/km² breaks the continuity of forest cover by cultivation clearings and the movement of tsetse flies is no longer possible and human mortality

falls (Gillmann, 1936). Starting from a critical threshold, population densities can either evolve towards very low levels, or almost nothing can prevent an increase in population numbers. In the deltaic plains of Indo-China, high densities are correlated with an overall management of paddy-fields for this favours a mosquito that carries a relatively mild form of malaria. The shady watercourses of the mountains are the breeding places of another mosquito which carries a much more severe form of the illness. Here once more, pathogenic factors that are linked to a particular environment act as a secondary factor, the primary factor being a certain population level and its agricultural techniques. In other parts of South-East Asia, especially Java, where there are different mosquitos, paddy-fields are associated with malaria.

Ecosystems and the dynamics of populations

If two areas are initially populated with the same number of inhabitants and one of the two is subjected to more severe constraints affecting the health of the inhabitants, its population density will increase more slowly or not at all and in time, the density of this area will be lower than that of the other. But the control exerted by the natural environment on the growth of human numbers may be through another process, on the movement of people in space; the balance between immigrants and emigrants replaces that of births and deaths.

Some environments have been shown to be less easy to penetrate than others. For example, the central African forest has been capable of being crossed from one side to the other during the course of a very small number of generations of people without any modern technological aids. The Oubangui-Congo hydrographic axis was an aid to penetration because of its orientation. However, in other cases it is the interfluves that have been followed over long distances. The practice of large-scale fishing in the first case and hunting in the second, also on a large scale, could assist this movement, but only true agriculturalists have the ability to populate an area. The unique constraint to which they are subject is that of remaining close enough to harvest. Food supplies cannot be transported very far and the humid forest does not add any particular difficulties to these general limitations. In present day southern Cameroon and northern Gabon the historical migrations of the Fang show how much can be performed under such conditions by adding small moves over short distances (Alexandre and Binet, 1958).

The survival of immense areas of forest that are virtually devoid of agricultural peoples remains to be explained. An immediate answer lies in the scale of the areas. The equatorial forest in Africa or Amazonia or South-East Asia is a remarkably extended but uniform environment; the phenomena resulting from contact between the environments and different human populations, that are so often essential for developing a social and spatial framework and on which largely depend the population densities, only occur in the peripheral areas.

Another explanation, that may be more fundamental,

appears whenever there is a transition from one ecosystem to another. The environmental change tends to slow down or stop movement or to displace this laterally. Whether regarded objectively or subjectively, the true value of environments that are in contact count less than the direction of movement. This concept can be particularly well applied in the case of dense humid forest, when it is reached by populations coming from outside, i.e. from the savanna. It also explains the human concentrations in various parts of the savanna-forest contact areas in tropical Africa and which penetrate the margins of the large forest massifs. In this continent the dense forest has been reached by groups deprived of the tools that would have enabled them to make clearings without excessive work. These groups also initially lacked a supply of plants suitable for this new environment. The resources of the forest were unknown and the time required to learn imposed a delay in penetration of the forest. The Polynesians were in the same situation when they reached New Zealand. The absence of Bantu people in the Cape region when the first Europeans reached it may, perhaps, be explained in the same way. A migrating group will only settle in a new environment if it changes its habits and develops a new way of life.

The effect of the halt caused by natural obstacles encountered by migratory populations must not be exaggerated. Dense forests, swamps, mountains and plateaux isolated by high abrupt cliffs easily become refuges for societies whose existence is threatened. On the other hand, under present day conditions, the economic advantages of some environments may have completely cancelled out or even reversed the repulsive character that they had before. For example, in western Africa the populations living to the north of the large forest areas are often finding that the forest has become the best place for well paid work due to the economic success of cash crops (cocoa and coffee, but also oil-palm, bananas, pineapples and rubber, etc.). Permanent settlement in the forest by many people originating in the savanna has been a very substantial reinforcement to the population of regions where these crops have had the greatest spread (in the form of family orchards in large parts of the forest of the Ivory Coast and of Ghana or in company plantations as in western Cameroon).

There are thus cases in which the arrival of man has been prevented or filtered to such an extent that the population remains too spread out and does not offer conditions viable for society. In other cases, obstacles have been overcome through sufficient people and in a relatively small time formed a demographic nucleus. But from this moment, limits operate in a reverse sense, from the interior towards the exterior, and hinder or slow down emigration of excessive population. This trapping effect is particularly striking in the case of small, or even moderate sized, islands such as Java. Indonesia, the Philippines and many of the Pacific archipelagos had juxtapositions of very densely populated and relatively empty islands. Borneo and Mindanao (before the rural colonization during the last half century) are striking examples. The same kind of insular situation produces comparable effects in the case of isolated plateaux or intra-montane basins. A very good example is that of the

Andapa basin in the north of Madagascar which really only began to be populated through immigration from 1920 because of the high cost of vanilla; it took little more than one generation for the population to reach *ca.* 60 000 inhabitants on 300 km²; today almost all the area is occupied and the soil and economic conditions have declined, whilst all around the basin the planters of coffee and vanilla are constrained by relief to cultivate higher and higher slopes where the ecological conditions are less and less satisfactory (Portais, 1975).

Ecosystems and the social framework

Is the development of a social framework—a necessary precondition for all demographic progress towards high population densities—not hindered by some types of environment? Maquet (1966) has remarked that 'the dense and hostile forest' isolates villages and that this isolation 'made the development of political units that include more than several villages with similar culture more difficult'. The same reasoning can be applied to some types of very dissected relief and to archipelagos. However, in many cases, steep slopes, rugged mountains and sea passages have not hindered the formation of a dense network of social relationships. The Gurungs of the Nepalese Himalayas are a good example of a strongly structured society that occupies the broken slopes of a high mountain chain at fairly high densities (Pignède, 1966). In the New Hebrides, social networks have a tendency to form links between the facing slopes of different islands rather than those on opposite sides of the same island. Conditions appear to be very different in Amazonia from those in Africa: the pseudo-archaic phenomena of social regression as analysed by Lévi-Strauss (1958), refer to Indians from the savanna who are in a marginal situation. A dense forest with a tendency towards drought is the natural environment in which the political structures of Mayan society in Yucatan and the Khmers in Cambodia were developed.

Thresholds and cumulative processes

Many examples have shown the existence of density thresholds that are crucial for the evolution of the population. Below the critical density level the population increases slowly or may stagnate or regress. Above this level a cumulative process occurs, each increase creating conditions that are more favourable for further increase. The threshold effect first shows itself when agriculturalists have caused the disappearance of the forest and eliminated factors that are unfavourable to immigration or demographic growth or when management permits a change from extensive agriculture with low population densities to an intensive agriculture with high densities. In terms of social construction and the management of space, effective socio-political control is not possible except beyond a certain demographic density level and with a certain degree of settlement; thus conditions are suitable for a progressive accumulation of people within the geographical area having a relatively stable and solid social organization: violence is regulated inside the society and protection is given to the inhabitants from external undertakings, peace favours agricultural production and trade and large collective works become possible.

Change in the natural environment can alter the threshold level. Socio-political factors can cause changes elsewhere; for example accumulation in concentrations often causes decreases in population between them as people are attracted to the security and better livelihood of the politically strong centres. Lightly populated buffer zones separated the communities and were available for agriculture and other activities; thus the Middle-West of Madagascar, almost empty, was formerly devoted to grazing by zebu cattle belonging to the Merina kings. These various phenomena contribute to explain the insular distribution of human density in the tropical zone.

Conclusions

It seems to be impossible to establish a constant and simple relationship between tropical forest ecosystems and population density and two examples help in understanding this. One of the most densely populated places in tropical Africa is in eastern Nigeria and forms a zone nearly 200 km long and 40–50 km wide extending from near Onitsha in the north west to the Cross River estuary in the south east (Floyd, 1969). It has a mean population density of at least 350/km² and a maximum of *ca.* 700/km². Soils are not very fertile but very attractive because their sandy nature makes them easy to clear and to work for cultivators having only rudimentary tools. They are very suitable for growing oil-palms which play a very important role in food, economy and the social customs; yams—the preferred crop—do not require the large mounds which are needed in the clayey soils of lowlands that border the plateaux with high population density. One does not deal with an intrinsic advantage, but with a superiority with respect to the particular trends of the regional agricultural system; placed in similar conditions, cultivators who traditionally would cultivate rice or *Colocasia* would become concentrated on low-lying clayey lands, as in South-East Asia. Neither does this refer to a type of relationship with the natural environment that remains the same for ever. The plains of the upper basin of the Cross River, east of the Nsukka plateau (Nigeria), were abandoned for a long time but are now subject to a massive colonization by agriculturalists who work hard to take advantage of the availability of land and good soil quality to produce yams for urban centres (Floyd, 1969). Nevertheless, the 'historical concentration of human communities on sands' (Floyd, 1969) is not only due to agricultural reasons; the plateaux were free of trypanosomiasis following the transformation of the forest cover into savanna and cropland; and most importantly, the population found these areas safe from slave hunters who operated up the rivers; the dissected margins of the populated area afforded an additional protection. In modern economic and political conditions, it has been very easy on this flat open surface to convert the tangled network of paths linking human settlements into cycling tracks and roads. This has facilitated integration into national life and into a trading economy, and has thus contributed to the new demographic increases.

In the island of Aoba in the New Hebrides (Bonnemaison, 1972), the distribution of people has always matched to ecological factors to some extent. Three population zones extend from the shore to *ca.* 400–500 m altitude in the well populated sectors; in the others, the population is concentrated near the coast. Cultivation has never taken place above 800–900 m. The direct effects of low insolation and air saturated with humidity a few hundred metres in altitude can be seen in the concentration of the population and crops at the lower levels. The contrast between the windward eastern slopes and the leeward western slopes must also be underlined: higher densities occur on the latter as these have less malaria because there is less rainfall. Nevertheless, settlement was easier on the west coast where communication with the other islands was easier resulting in a very active system of exchange which certainly accounted for the demographic advantages of this slope of the island. Other contrasts in density are inexplicable in terms of environmental factors, e.g. those in seven small regions dissected at right angles to the coast. These formerly represented distinct social units. The competition between five different christian sects, having distinctive attitudes towards the traditional social and economic life, has maintained and even reinforced the divisions dating from the last century. In fact, the population density of Aoba has fluctuated irrespective of environmental factors. Three successive periods occurred. Initially there was a period of social division and insecurity which maintained density at a low level. A second period was the progressive organization and extension to the whole island of a competitive system called *hungwé* that was based on a complex scale of ranks to be reached through the accumulation of wealth in the form of pigs with artificially deformed teeth and tubers of *Colocasia* and yam of an exceptional size when the inhabited zone reached densities of 70–80/km² plus the pigs that were an essential part of the system and which also had to be fed. The last period followed the arrival of the Europeans: the infective organisms imported by them brought about a marked demographic crisis the effect of which was felt unequally in different places but the development of coco-nut plantations had more longlasting effects. The ecological requirements of this tree, which cannot produce fruit above 300–350 m, led to modifications of population distribution (a descent of many people towards the productive and accessible sea-shore zone and the saturation of land in the most densely populated area) and because of the requirements of trade and world market, i.e. of conditions outside the archipelago. A threshold appears here at around 30–35/km²: below this level the old social system and the old agricultural system have held good, while integrating the commercial crops; above it a true disintegration has resulted, with a variety of processes leading to the monopolization of land and the concentration of wealth to the benefit of a well-placed minority.

To conclude, the insular distribution of high population densities has nothing to do in its principle with the peculiar characteristics of the tropical forest ecosystems; the distribution of population nuclei within the tropical humid zone shows a preference for areas of forest of limited extent and for the margins of the large forest massifs. This can be explained by the extent and homogeneity of the forest environment; not by its intrinsic characters. The effect of scale plays a decisive role and implicates all kinds of material and social relationships among tropical people. Within the humid tropical forest, secondary biogeographical differentiations make some environments more attractive than others and have contributed towards the siting of some focal points for future demographic accumulation; these differentiations confer a higher agricultural potential to some areas and allow much higher levels of population densities with the same technical facilities. But the controls which are exerted on the accumulation process itself are largely unrelated with the ecosystem for they depend on the social structure and on the techniques available. Any substantial population increase within a given area or place, is subordinated to progress in both respects. This progress is not automatic but is dependent on historical changes and on the occurrence of physical or cultural communications between peoples. Discontinuities and threshold effects play a much more important role in the relationship between population density and technological and sociological effectiveness than in the relationship that links the level of population density to the natural characteristics of the ecosystem.

Problems of saturation

Saturation occurs when the production per head, of a population increasing within a limited area, is no longer limited by the labour force available but by space requirements.

Stages of agricultural intensification

From the technological viewpoint, systems of production can conveniently be classified into a coherent series in which each type has an efficiency that is higher than that of the preceeding one and which supports higher population density without space shortage. One of the most recent and most elaborate of such classifications is that developed by Boserup (1965) which identifies: forest-fallow cultivation, bush-fallow cultivation, short-fallow cultivation, annual cropping and multi-cropping. From a stage to the following one the diminution in length of the fallow period (and finally its complete suppression, even seasonally) corresponds, for the author, to the successive stages towards an intensification that is described as 'the gradual change towards patterns of land-use which make it possible to crop a given area of land more frequently than before'. The great advantage of Boserup's approach is that it shows the point at which technology, tools, land tenure and even social patterns are linked together in the evolution of agricultural systems. She also has clearly shown that increase in intensity is paid for by a reduction in work productivity as long as mechanization and chemical fertilizers, have not been introduced. Thus, people who are forced to obtain their subsistence from smaller areas must work harder. Hence the conclusion that intensification, with such an inconvenient burden, only occurs under demographic pressure. The stage of short-

fallow cultivation is closely linked to extensive use of the plough. In the tropical forest and in the surrounding savannas with a dense and perennial herbaceous layer this does not occur and agriculture is based on a variant of bush-fallow cultivation characterized by a lengthening of the period of cultivation up to eight years. Another interesting point of Boserup's system concerns the ways of moving from one stage to another. The substitution of continuous cultivation for discontinuous land use never happens immediately throughout all the land controlled by a group. It is the best located lands or those capable of yielding most versus the increase in labour corresponding to the intensification process, which are first transformed. Thus, a phase that can be considered as transitory begins during the course of which several chronologically unphased technological patterns will occur alongside each other. This type of mixed agriculture with forms of land use unequally intensive within the same living area occurs over a large part of the tropical world. The various types of cultivated land disposed concentrically in western Africa are one of the best examples and show striking similarities with organizational systems that formerly existed in western Europe (Sautter, 1962). In the humid tropics, for agricultures based on a long-term fallow, there is a diversity beyond the distinction between forest-fallow and bush-fallow cultivation (see chapter 20). It is possible to recognize three main evolutionary trends within the great variety of these agricultural systems, oriented towards a better use of the land and higher population densities. The first concerns settlements and has three stages: mobile settlements without any periodic return to, or near, the same sites; cyclical moves in settlement within a more or less defined area; and that in which the fields move and can be described as shifting whilst settlements move only in a very narrow range and in a way that is not directly related to agricultural requirements. The second has a bearing on the organization of agricultural space and occurs when there is some agricultural settlement. Sometimes the field rotation within the forest has a disorganized character and sometimes it obeys precise regulations. In some cases these rules express themselves in a division of the agricultural space into sections that are brought into cultivation in a particular order under a true rotational system. The terms shifting cultivation and rotational bush-fallow have been proposed to designate these two methods in western Africa. The third evolutionary trend relies on technological procedures. An orderly-patterned crop rotation permits the lengthening of the period of use without increasing the proportion of time under fallow. Forest fallow may be made more vigorous by protecting against grass invasion or by sowing species which restore soil fertility, as has been mentioned in southeastern Nigeria (Morgan, 1955) and the island of Timor (Ormeling, 1956). But more than anything else, it is work on the soil itself that increases the efficiency of crops. Africa provides two extreme examples of this. In Zambia and some neighbouring States, in the *chitimene* system the cultivators used to fell or lop the trees of the dry forest, collect combustible material into a heap, burn it and plant their crops in the bed of ashes thus obtained. At the other extreme, the specialized yam cultivators, of the region around Benin, obtain very high yields on plots in which the soil is very carefully hoed and formed into mounds. As a result, whereas in Zambia the Lala people of the Serenje plateau have saturated all available land with a density of 2.5/km² (Peters, 1950), the inhabitants of Umor in eastern Nigeria achieved, before 1940, a density of 92/km² (Forde, 1937). This gap cannot be explained only in terms of differences in soil quality, but by the techniques used.

Two situations at the junction between extensive systems and those in which the fallow period has been more or less eliminated need consideration. The first is that of permanently developed areas that are only used periodically and is found in Melanesia everywhere from the high valleys in the interior of New Guinea as far as New Caledonia (Barrau, 1956) and in some paddy-fields of the humid south of India. Here is an association of techniques for working the ground that demand much labour of the kind usually undertaken by truly intensive agriculturalists with a periodic cultivation characteristic of the most advanced type of long-fallow agriculture. The general trend towards intensification and high densities can thus never be reduced to a single-track process. The other situation concerns practices observed in various parts of Melanesia of the extraordinary care given to the cultivation of some varieties of yam and *Colocasia* with the object of obtaining gigantic tubers for ceremonial use. Deep holes are prepared and filled with compost to facilitate their growth. Comparable situations have been reported among the Abelam and Wosera from eastern New Guinea (Brookfield, 1971) and the New Hebrides (Bonnemaison, 1974). These are features of false intensification: much labour is spent but the cultivators are not seeking to increase the yield of the cultivated area and in some cases, there is no shortage of space. This behaviour is related to the considerable time and effort spent in raising pigs that in Melanesia are also for ceremonial use. The interpretation is obvious: extensive or semi-extensive cultivators use essentially for social aims the time which is left as a result of this type of less strenuous agriculture. An agricultural technical expertise that could lead to intensive cropping exists but in the present state of the system it is used for other ends.

Specific patterns of intensification

The patterns of agricultural intensification that occur within tropical forest ecosystems present a real originality as well as a certain degree of variety. Systems of agriculture associated with livestock-rearing are much less well represented in the aseasonal tropics than in regions with a marked dry season. In the forest zone and its margins, cattle are only occasionally capable of supporting high densities. Either the animals themselves, especially cattle, are absent, as, for example, over large parts of humid Africa due to trypanosomiasis. Or the animals are present, as in parts of Asia and Africa where rice is cultivated, but too few in number to play a definitive role apart from that of an additional labour force. Large herds have an important role in tropical America in the humid savannas around the edges of the Amazonian forests (especially the *llanos* of Venezuela) as well as in the savannas occurring inland from the Atlantic

forest. But these have never supported appreciable human population densities. In fact the reverse occurs as cattle are used to fill empty areas. When density increases, these extensive pastoral systems undergo a critical phase without giving rise to a mixed agricultural system. The studies of Pébayle (1974) in the subtropical south of Brazil show clearly that in this situation the gaucho continues to behave as stockman and not as a farmer. The slopes of Kilimanjaro that are cultivated by the Chagga show high densities that are associated with integrated cattle-rearing. Other examples include some peasant societies of eastern Africa that result from the fusion of two populations with different origins and agricultural and pastoral traditions. This is the situation in Rwanda and Burundi, where the climate is transitional between moist and dry, and also in the Sukuma country on the southern shores of Lake Victoria where agricultural land and pastures have replaced a dry forest (Malcolm, 1953). Nevertheless, even in these societies, the role of livestock is questioned today because of human population growth.

The only methods of land use that have been shown to be capable of feeding a high population density in the humid tropics are those that rely on rice and tree cultivation, as Gourou has shown in his *Pays tropicaux* (1947). Geertz (in Vayda, 1969) has opposed in the East Indian Archipelago swidden farming to *sawah* agriculture based on rice cultivation on irrigated terraces, as practised in Java. The first reproduces the natural forest ecosystem to man's benefit. In contrast, *sawah* cultivation marks a fundamental break with natural processes. A type of agriculture that is almost as intensive as controlled irrigation of rice-fields but whose landscape and characters are fairly close to the natural system, is that which is associated with the cultivation of food plants beneath a fairly dense canopy of trees and shrubs having several different layers, that contribute directly or indirectly to production. This was the appearance of the traditional Polynesian fields, other than the marshy areas devoted to *Colocasia* cultivation. For example, in Tahiti, mixed cropping of bananas and tubers is beneath useful trees (Ravault, 1970). The same situation is found in the Antilles as, for example, the gardens of the scattered settlements on the hills of Martinique. Bread-fruit, jack-fruit, avocado and various citrus trees make up these orchards, which have a stratification like forest. Identical features characterize compound farming in the most populated regions of eastern Nigeria where an ingenious technique of mixed and sequential cropping allows many different but superimposed levels of crops: sprawling plants, root crops, bananas, various fruit trees and oil-palm (Floyd, 1969), which plays a dominant role. A comparable landscape is present in the People's Republic of Benin (Dahomey) immediately north of Porto-Novo. The most densely populated parts of the Bamiléké plateau in Cameroon, just on the margins of the dense humid forest, where densities reach more than 300/km², show a combination of a dense array of man-made woodland and fields in which the enclosures are reinforced by lines of trees and of scattered stands of avocados, *Dacryodes edulis* and other trees species. The appearance is that of fairly dense forest, but it is an artificial woodland which develops together with permanent cropping above a certain density level. There is no relationship with fields of the swidden type, as described by Geertz (1969), and which are directly derived from natural forest. In this system, trees certainly play an essential role in fertility. The Kikuyu plateau, near Nairobi, is similar but is due to permanent banana plantations which are also present in a large number of other high population density areas in eastern Africa. In the south of the Ethiopian plateaux, the *Ensete* banana, is cultivated in more or less the same manner. In both cases banana plantations last for many years and form an extremely dense stand which maintains itself in the same way as natural forest (thanks to certain additions of manure and compost). In this form of intensive agriculture, the lower stratum supplies food and the trees and shrubs supply marketable products. The clove plantations of Zanzibar and Pemba, some kinds of coffee cultivation associated with food plants in both eastern and western Africa and in the wettest parts of Ethiopia and Madagascar, illustrate this agricultural stratification.

Levels of agricultural intensification

Of all the different classifications of agriculture ranked in order of intensity of labour and yield, that of Brookfield concerning Melanesia (1971) is one of the most detailed. It uses 48 features, each of which may be absent (0 score), present (1) or of high significance (2). The summation of the scores obtained enables each cultivated ecosystem to be ranked into one of the following categories: low intensity systems, subdivided into those dominated by swidden and partially intensive system; high intensity systems, subdivided according to the degree of control by terraces, irrigation or drainage into dominantly intensive systems with well-developed crop segregation and intensive systems of wider technological span. Geographers working in Nigeria have developed a comparable series: shifting cultivation, bush-fallowing, rudimentary sedentary cultivation, intensive sedentary cultivation or compound farming, and intensive sedentary cultivation or terrace farming. In the uplands of Madagascar, the predominant irrigated rice cultivation is practised at three different intensity levels (Le Bourdiec, 1974). The lower level comprises an association of rice-fields in valley bottoms and rearing of zebu cattle, which prepare the fields by their trampling, on the large pastures on the slopes. The only features of intensification are the managed rice-fields (but generally unirrigated) and the systematic planting out of rice seedlings. The intermediate level involves control of water to compensate for rainfall variations as well as an increasing care in cultivation (the ground is prepared by the Malagasy spade, the *angady*) and the use of fertilizers. The uppermost level is characterized by much more complex hydrological management, additional manuring and an attempt to make maximum use of all available ground (especially by growing rice on areas first used as nurseries). The degree of intensification reached in rice cultivation is in direct relation to the population density —*ca.* 10/km² on the northern and northwestern margins of the Tananarive area (Ankazobe and Anjazorobe) to more than 100/km² on the plains around the Malagasy capital.

It is inversely related to the size of family holdings: the area of rice-fields varies from 0.4 to over 2 hectares.

A graduation from extremely extensive to ultra intensive agricultural systems exists everywhere. In the southern parts of the uplands of Madagascar occupied by the Betsiléo, the cultivated landscapes reflect the following stages: (1) only scattered rice-fields, where development is recent, even in the valleys; (2) when the peasants have been in the area for at least four generations the valley floors are entirely managed but slope management in terraces (for rice or dry farming) remains discontinuous; (3) development of slopes is almost complete and only the tops of the hills and the highest slopes are used for pastures; (4) all the land is used and rice-fields cover all areas which can be supplied with water. Phase 3 requires between 5 and 6 generations of occupants and stage 4 requires even more. The methods of intensification found in the societies of upland Madagascar generally apply to paddy-fields and dry farming (maize, beans, *Xanthosoma*, cassava, etc.). The latter crops do not play a large role until population increase makes the area under rice for each family fall to 0.6–0.8 ha. The intensity of dry farming increases with the population but this is generally at a one-stage lower than the stage of rice cultivation. From a phase in which cropping alternates with fallow the stage of permanent cropping with manuring is gradually reached. It is this complex process with its many stages that has enabled the high densities found today in the central part of Imerina (in the southern and eastern surroundings of Tananarive) and in a certain number of nuclei scattered from north to south of the Betsiléo country to develop during the last three centuries. Progressive westwards colonization of the areas around populated regions is more and more recent as one goes far from the areas of high human concentration, and shows now itself as a series of rings that successively reproduce in inverse order the stages of evolution noted above.

The same processes of intensification in the centre and expansion at the periphery into unoccupied land have been observed in eastern Nigeria. For example, migrant tenant farmers sometimes go considerable distances to rent available land. In some cases organized communities choose a new site after having come to an agreement with groups that still have available land resources. But where population density exceeds 200/km², the change essentially consists of extending to all the land compound farming—the most intensive type of agriculture practised by the Ibo-Ibibio. The transition from intensification limited to compound farms occupying a central position to intensification over the whole area is very gradual. Farmland, a mixture of cropland, shrub or grass fallow, and secondary forest initially occupied 3 kilometres or more between settlements; this is progressively reduced and finally disappears. The countryside becomes one huge palm-grove in which the settlement and crops of one village can no longer be distinguished from those of another.

Although it is located at the margins of the tropical world, the example of the Gurungs in the Nepalese Himalaya gives a good illustration of the double process of intensification and expansion (Pignède, 1966). In one century the situation has changed completely. The slopes at 1 400–2 400 m have been progressively occupied by terraced crops. Livestock-rearing has ceased to be the principal source of wealth. It has become closely linked to cropland and the manuring of fields located downhill from the villages permits many harvests during the year. Agricultural expansion is only possible either upwards at an altitude at which only potatoes will grow, or much lower down in the valley floors which are progressively used for irrigated rice cultivation. These communities now have almost no free land and land concentration benefits those families that are the most favoured economically.

The Mekong delta was populated by rice cultivators that arrived from the north at a much later date than on the plains of central Viet-Nam and considerably later than in the Red River delta. This type of cultivation was extensive at the time of French colonization (Gourou, 1940). In a period of little more than one generation, the situation changed markedly: the most recent data show a density of 300/km² as compared with 540/km² for the Red River delta. The rice-fields giving a double harvest are expanding and around Ho Chi Minh (Saigon) one can see the change to intensive use with selected seed, chemical fertilizers, etc. A similar evolution can be noted in the central plain of Thailand.

Roles of technological innovation and demographic expansion

Technological progress and demographic expansion appear to be closely linked. Which is the driving force behind the processes of change? Boserup (1965) considers that it is demography. When the demographic load exceeds the production per unit area, it is rare that the human society concerned does not discover the technical knowledge, either from its own cultural resources or from contact with neighbouring groups that enables it to obtain higher yields from the soil and to progress to the next stage. Agriculturalists had to show some inventiveness in order to survive and it is difficult to otherwise explain the general progression towards intensification which everywhere was achieved by an increase in labour, until the introduction of modern equipment. This hypothesis appears to be reasonable in that each time agriculturalists have moved from densely populated areas to areas with available land, a regression towards extensive agriculture was observed.

Nevertheless, the response to the challenge from population growth that is made in terms of extra work and technological innovation is not automatic. Some groups of agriculturalists suffer from a break-down in spite of demographic pressure and this does support the opinion of Boserup. The force of inertia and weight of custom may maintain a population in a technological stage which has become incapable of satisfying its long-term needs. There are two other explanations of break-downs. Social resistance to change is often important and it takes a variety of forms. The ruin of the Maya empires may be explained by the incapacity of changing the agricultural system based on periodic clearing and long-fallow periods (see the second part of this chapter). Further, a socio-political structure

based on city-states makes it very difficult if not impossible for farmers to work fields a long way from their permanent settlement with the care required in intensive agriculture. The monopolizing of land holdings has had exactly the same effect in other societies. The same is true in the accumulation of taxes due from peasants to the authority in very hierarchical societies. How can cultivators be expected to expend the extra labour required by intensive agriculture when the surplus benefits landlords or higher levels of the power structure? For example, in tropical Africa and especially at the margins of the dense humid forest, it is the most egalitarian societies that have systematically been the most successful in intensification. It is only when inequality is so great so as to place some social categories in servitude that it ceases to remain an obstacle. For the same reasons colonization has generally resulted in a resistance to technological progress proposed by outsiders for their own benefit. Conversely, once certain conditions have been met, agrarian reform often clears the way for technological progress. One of the most convincing examples is Japan: the technological inability to improve rice yields that seemed to be evident before the second world war was followed by extraordinary improvements. The difficulties with the green revolution that have occurred because of a lack of land redistribution in India support this hypothesis.

Degradation of the natural ecosystem and overpopulation

Concentrations of population poses two series of problems. The first concerns the degradation of the natural ecosystem. Only a gathering economy leaves the ecosystem intact whilst profound changes are made to the vegetation even in the first stages of agriculture. Even long-term fallowing has the effect of making dense forest into secondary forest. The requirements of agriculture and forest exploitation are sometimes in acute competition. A moderate population density that is insufficient to cause problems for cultivators, or even to restrict their use of space, is sufficiently large to cause irreparable degradation of the growing stock of commercial tree species. One can see, for example, how the Government of the Ivory Coast has recently become aware of the economic risks that accompany the uncontrolled agricultural colonization of the southern part of the country. A detailed forest inventory has been decreed and safety measures declared: the forbidding of the creation of new orchards of coffee and cocoa (a difficult rule to respect), protected forests, etc. In general, forest officers have always reacted in the same way: preservation of areas and control of felling. Degradation and destruction have unquestionably been created by cultivators, but the pure stands of limba (*Terminalia superba*) present in Congo and in Zaire, those of okoumé (*Aucoumea klaineana*) in Gabon as well as the forest of *Lophira alata* in the coastal regions of Cameroon (Letouzey, 1968) are all the end result of an old period of slash-and-burn agriculture. Today there is only one solution to the conflicts between the general settlement of agriculturalists in the forset, under the double pressure exerted by the administration and economics, and forest management:

this is dividing the area permanently into restricted and intensified agricultural land and forestry evolving towards sylviculture.

Secondary forest offers big advantages to agriculture. A much higher proportion of trees that are easy to fell and burn. Some groups of African forest cultivators hold a right to old fallows and this bears witness to the large amount of work invested in vegetation management through a first hard clearing in order to reach this favourable condition (Jean, 1975). This is less significant when the vegetation is completely changed. The Betsimisaraka, of the Vavatenina region to the north of Tamatave where population density exceeds 50/km² in places, are traditionally dry rice farmers. Repeated clearing of the higher hills have resulted in reducing the forest to a scrub, *savoka*, or herbaceous vegetation. Today all the area has been irreversibly used. The farmers have reacted by progressively cultivating in a more intensive way the valley floors and lower slopes converted into terraces for irrigated rice cultivation (Dandoy, 1973). On the margins of the uplands, the Zafimaniry of Ambositra are maize and bean cultivators in a more difficult position in spite of their lower population densities—*ca.* 20/km² (Coulaud, 1973). Forest only occupies 220 km² out of 700 km² and it is much more fragile because of the dry season and winter frosts; the *tavy* (slash and burn) have transformed it into a low shrubby vegetation or a pseudo-steppe of *Aristida*. The damp cold that dominates over a large part of the year does not favour the use of artificial fertilizers; the narrow valleys are often rocky and unsuitable for productive permanent rice-fields. The Zafimaniry go to seek for outside resources by emigrating. Basic food requirements are only guaranteed for six months in an average year and less than three months in some years. Part of the money earned from outside serves to manage rice-fields but the population does not cease to grow. This crisis of the Zafimaniry country has been aggravated and accelerated by measures taken to safeguard the forest. For a long time these cultivators have been confined within authorized areas; they extend when they feel capable of doing so without risk. The rest of the safeguarded forest serves a small timber industry. But creation of savanna proceeds at a much faster pace within the portions devoted to agriculture.

At the northern limit of the last remnants of semi-deciduous forest in Cameroon, the savannas of the Adamawa plateau are occupied by Fulani livestock-rearers. They raise large herds of cattle and show seasonal transhumance movements. The pseudo-steppe which changes with altitude into a true grassland has replaced open or dense dry woodlands. This transformation was performed by a much more denser agricultural population that was largely destroyed during the 19th century and which only remains in residual groups. Livestock-rearing does not prevent the recolonization by trees. Woodlands have redeveloped and extended over the last ten years or so and an environment favourable for tsetse flies has re-established itself. Bovine trypanosomiasis has destroyed or caused the removal of herds from parts of Adamawa. The Fulani that want to remain must revert to agriculture of which they do a little. This example shows how environment and man interact; changes affecting the ecosystem may recur in cycles or in spirals and

the destruction of forest is not necessarily to the detriment of the inhabitants (Boutrais, 1974).

Areas of high population densities are accompanied by other effects on the natural ecosystems. Soil degradation and erosion occur each time that the cultivators are content to shorten the fallow period rather than fertilize, and/or lengthen the period of cropping. In the Anfouin area of lower Togo, population densities of *ca.* 200–300/km² or more have not been accompanied by any true intensification (Lecoq-Litoux, 1974). The cultivators just try to obtain the maximum possible yield by continuous cropping in space and time thanks to the very high natural fertility of the 'terre de barre', but this potential is being exhausted. Cassava, which requires little, has, for a long time and very extensively, replaced maize (now used in intercropping) and particularly yams; yields fell from *ca.* 10–12 t/ha to less than 6–7 t/ha and sometimes only 2 t/ha. An important migratory movement has occurred since the second world war towards land still available in the northern margins and especially towards coastal towns. A comparable evolution has occurred in the rural environment of the Yoruba towns of southwestern Nigeria. For example, Abeokuta, founded in 1830, has been the starting point for a true agricultural colonization by people who keep very strong links with the town in which many of them continue to live in for part of the time (Magobunje, 1959). The areas closest to the town that were the first to be farmed now have very impoverished soils which can only grow cassava. Yams, the much more favoured crop, and cash crops are, on the contrary, located at the periphery of the rural area dependent on Abeokuta, on soils that are less densely populated and less sought after. Even the most spectacular may create by downhill transport, soils which are more suitable for agriculture than the natural leached soils (Bourgeat, 1970). High density concentrations do not only correspond to degradation of the natural environment, they also can or may express themselves as an imbalance between the needs and resources of the concerned population. This imbalance can be easily corrected at the intermediate stages of agricultural intensification. But a fixed limit exists: that of the maximum human load that can be carried per unit area given existing techniques developed to their utmost limits. This limit may range from *ca.* 100/km² to more than 1 000/km². It is technically and financially possible to go beyond it with modern industrial techniques, but this is accompanied by a radical change in scale (see chapters 20 and 21). Relative overpopulation occurs when technical blocks are produced in the course of the general trend of evolution and true overpopulation at the end of this evolution. The term overpopulation only means that the lack of land explains or tries to explain insufficient resources, in relation to other factors that can also play an important role. The true test of overpopulation lies in the awareness of the scarcity of land, by the people concerned as a fact and not as a result of monopolization by particular social groups.

Most important is the bringing together of a high density area and of high population growth. Then pressure takes place towards technological change or the increasing degradation of the environment. In tropical Africa at least, a fairly good correlation exists between the highest population density nuclei and the highest rate of natural growth. Conversely, groups suffering from demographic difficulties are mainly located in places with the lowest densities: e.g. Mongo in the central Congolese basin, Zandé. Venereal diseases, especially gonorrhoea, are largely responsible in these concentrations of infertility (Retel-Laurentin, 1974). It is evidently not the natural ecosystem but the social conditions that are responsible for the transmission of these diseases. The generally fertile populations of high density areas are often at a disadvantage in terms of food supplies but this is compensated for by the elimination of some pathogenic complexes that occur in association with a densely occupied land area as well as the protection afforded by a strong social structure.

In those places where demographic accumulation occurs, increase in population and technical progress in agriculture are closely linked and settlement of the excess population takes place on the periphery. This peripheral occupation of available land tends to enlarge the areas of high population density fairly rapidly.

It is essential to consider the scale at which the shortage of land is felt by overpopulated agriculturalists. Concentrations of agricultural populations in very large village units or even in towns (as in the case of the ancient Maya and the Yoruba) may be sufficient to create pressure on land even without high regional densities. Some population redistributions enforced by administrative authorities resulted in a similar effect due to a failure to recognize the true nature of the relationship between a population, its land and its ecosystem. Another important concept is the chance of diffusing excess population produced in one area to other areas where demographic growth is less fast. The barriers may be natural or social. This concept of permeability and impermeability can be applied at various levels: from that of the family circle to that of the entire populated sector, through that of the village territory. Two African examples illustrate this concept. The Tiv of Benoue (Nigeria) studied by Bohannan (1954) have a process of social competition for land and arbitration that enables the strongest demographic group to drive back the weaker ones. Movement within the area is wide-spread and has repercussions extending to the limits of the Tiv country which thus expands progressively. The peasants of central Rwanda were for a long time maintained in a state of extreme social division by a foreign ruling class and they are incapable of implementing spatial adjustments to the needs of the people, even between restricted and related groups living among the hills (Meschi, 1974). Groups still having a relative abundance of systematically fallowed land live side by side in the same social restricted area as others that have very much below the minimum required for continuous cropping.

Many of the areas of high population density that occurred in the tropical humid world have disappeared and have left only archaeological evidence or memories within an oral tradition. This is the case with the Mambila, the Buté and other settled people of the Banyo area (Cameroon) that were destroyed due to contact with the Fulani that came from the north during the last century (Hurault, 1970).

Conversely, during recent decades, new areas of high density in economically attractive areas are being created.

The study of relationships between population density and resources provided by the ecosystem is delicate in the sense that some groups live within a rather narrow self-subsistence whilst others devote an important part of their time and their land to produce a surplus that can be exchanged for monetary items or desired goods having social values. By relentness work, the Lamet of the mountains of Indo-China can double their harvest of unirrigated rice in order to obtain bronze drums of distant origin (Izikowitz, 1951). The amount of energy used by a group of slash-and-burn cultivators in the interior of New Guinea to raise a large number of pigs under a rotational system is another example (Rappaport, 1967); the author tries to show that this has some rational explanation in terms of food but this is less obvious than the social justifications. The considerable development today of cash crops makes comparison more difficult both from group to group and from one high density area to another; some populations are in difficulties in terms of available land for cultivation because a large part of their land is devoted to commercial crops; others are maintained by the activities and income of their expatriated members, especially those working in urban areas.

Current trends

New world political and economic conditions have brought about important qualitative and quantitative changes in the relationships linking the tropical forest ecosystem to the people living within it. Before examining these it is advisable to underline the main points discussed so far. There is a great inequality in population densities within the tropical forest area. The variations are too large to be explained in terms of differences in the physical and biological conditions. They mainly reflect inequalities in technological ability for the production of food. There is a whole series of systems from food-gathering to permanent intensive continuous cropping. At the top of the scale, the more or less complete cultivation of the land is based on manuring that is not dependent on the existing vegetation and which requires complex management. Thus, a much greater amount of labour is required in order to obtain a given production per person. Intensification therefore involves an increasing amount of extraction from the ecosystem and a decreasing productivity per time and energy input, tools and equipment being at the same technological level. A second important fact concerns the juxtaposition of unequally populated and exploited areas, which does not express a fixed or static condition, but must be judged according to the impact of technological change and demographic growth closely linked together. The broad trend of evolution is towards technological progress and increasing population density, but there are variations due to local conditions. In many parts of the tropical forest world, this general trend can be reversed. The overall change can also be disturbed by transfers of people or techniques between different areas, where they will be considered as innovations that indirectly favour population growth. The demographic and technological progress of human societies results in the transformation of natural ecosystems.

The interplay of social factors determines significant thresholds in the progress of population densities. Does progress occur continuously in reaction to the number of mouths to feed or can it be broken down into a series of stable or semi-stable states separated by periods of change? A major break occurs between the food-gathering systems that are relatively stable both in terms of time and homogeneity, and agriculture. Within agriculture are a large number of weakly-differentiated stages and there may be a very gradual evolution. Once the process of intensification has been initiated and so long as no setbacks occur, nothing will hinder its progress to an advanced stage. In a large number of rural populations the cultivated space is occupied by both temporary fields and permanent fertilized fields. They are frequently arranged in an infield-outfield system concentrically around the settlement. In such mixed systems intensification increases in proportion to the increase in population. Difficulties do not generally arise until the intensive cultivation is applied to all available land. Obstruction may be due to saturation of the available labour force, competition for available space between man and animals for manure. European agriculturalists overcame this threshold by integrating fodder crops into the crop rotation. Pigs transform plant products more efficiently than cattle in the humid tropics. This helped overcome the food crisis due to a very rapid population growth in the Bamiléké country of Cameroon (Hurault, 1970). Other solutions are based on controlled irrigated rice cultivation and on some types of multi-layered horticulture. Whenever technological evolution meets obstacles, either at the early stages of agriculture, at the end of the intensification process or during the course of this process, a critical situation develops if the population continues to grow. A combination of technological stagnation and population growth is wide-spread today and could have been present in the past before the massive penetration of European influences. This threatens the balance of subsistence and the life-supporting environment. Even in the absence of any crisis, whenever technical progress and demographic growth are linked, the increasing pressure on the land as population density increases initiates a radical transformation of the natural environment. The forest gives way to regrowths and is finally eliminated. Creation of savanna is not a compulsory stage between a forest canopy with cultivated gaps and an integral agricultural occupation of the land. In many cases a bare landscape takes the place of dense forest.

The destruction of forest removes some direct or indirect obstacles to a rapid demographic growth; soil selection is developed and the various types of fields increase in number in response to the edaphic diversity. However, the identity of these societies, as rooted in their organization and culture is not basically affected. This is the case for the Zandé people (Zaire) and those of the Akan group (Ghana, Ivory Coast), these examples showing that a group whose geographical area is very unevenly populated within the forest or is divided between forest and savanna does not see its unity or continuity questioned except perhaps in the very long term. Differential adaptation to contrasted environments never affects anything but the periphery of the system: the technological organization of production, the social relationships that are most closely linked to it and the way in which the whole system is located in space. The central core that is really characteristic of the group and which consists of its true social structure, beliefs and values remains intact.

Demographic growth

Two fundamental facts relating to population density and its role in the interaction between the natural environment and agricultural societies are characteristic of the present times compared to the immediate past. The first is what is called the demographic explosion. The situation improved where populations were in stagnation or declining; those that were increasing at a moderate pace are now doing so more rapidly, so that to have 50 per cent of the population below 15 years of age and an annual growth increase of 3 per cent or more is no longer exceptional. The date of this rise in the demographic curve varies considerably. For example, in Africa some groups of peoples living within or in the neighbourhood of the rain forest were still experiencing a state of regression in 1960: these include the Nzakara of the upper Oubangui basin (Retel-Laurentin, 1974) and the Mongo of the Congo basin (Huysecom-Wolter, 1963); whilst the Ogooué populations of Gabon have only recently experienced a small surplus of births over deaths. The change came sooner in Madagascar for it was towards the end of the 1940s that the general stagnation in population growth gave way to the now very rapid increase (Chevalier, 1952). But at a world scale this great demographic growth generally took place much earlier.

In terms of its geographical distribution, it is the nuclei of high population densities that have been the first to produce substantial excesses and these remain the sites of the most vigorous demographic growth. It is above all within these nuclei that production of excesses of population has had maximum effects: the passage within one generation from 5 to 10/km² does not really change either the natural ecosystem or the place of man in this ecosystem, but the passage from 500 to 1 000/km² has upset all the established relationships between man and nature and those between men themselves. Societies that initially lacked land have no time to adjust their technology or their production relationships to rapidly increasing pressure. As a result it seems that setbacks become more numerous and tend to become the rule at the upper level of rural population densities. The overflow of excess populations, either around the margins of the populated area or beyond the natural obstacles which surround this area, has become more general and more massive. Examples of this are numerous and often spectacular: they include the population of Mindanao from the central and northern islands of the Philippines, the peasant migrations from Java to Sumatra and now to Borneo and the agricultural conquest of the Amazonian foot-hills of the Andes by the surplus population of the upland regions and Andean basins (Raison, 1968, 1973). In the present day demographic context the migration to virgin lands more and more often appears to be the only solution, but it is becoming more and more difficult to put into operation.

Development of cash crops

The other marked feature of the present times is the considerable development of so-called cash crops destined for export. Almost no peasant society exists that does not sell a fraction of its production. As a result there has been a considerable increase in land needs and this exerts an important pressure on the forests of the equatorial zone and on its margins. This is especially because the products required are ones which only the tropical forest ecosystem can supply: cocoa, coffee, bananas and other fruit, palm-oil and natural rubber. These products are more varied and less easy to replace (by synthetic products or those of temperate agriculture) than the products of savanna ecosystems (mainly cotton and ground-nuts). Generally they refer to woody perennial crops which can be grown within the natural ecosystem without radically changing it. In the most densely populated areas, whenever cash crops develop at the expense of food crops, a part of the money obtained is used to purchase food from outside sources. The area devoted to cash crops has a maximum limit. Nevertheless, it is remarkable to see the enormous extent of family cocoa farms in Yoruba country (southwestern Nigeria; Gourou, 1960) as well as the area devoted to *Coffea arabica* on the Bamiléké plateau of western Cameroon, where density is in the order of 100–200/km². In the case of Cameroon, the result is obtained through a close association of cash and food crops in mixed fields. In contrast, in Rwanda, the regression of coffee plantations can be seen, except in the sectors organized as paysannats (agricultural settlement schemes), because these plantations have become incompatible with densities that locally are as high as 500–600/km². The important expansion of cash crops orchards occurs in the areas of intact or little altered forest with a low or at least a moderate population. Sometimes, as in the case of the Yaoundé-Ebolowa region of Cameroon, there has been the diffusion of cocoa plantations in a region that was already occupied but given over to very extensive and patchy agriculture. In other places, the plant and its exploitation have become both the living basis and economic support of a pioneering population. The prototype of these advances of cash crops into forest can be found in the south of Brazil, especially in the coffee growing region between Sao Paulo and the Parana river (Monbeig, 1952). The most well known movement in Africa is that which has swept from east to west starting from the Akwapim range of forests of Ghana (Hill, 1961). The forests of the Ivory Coast have for over thirty years in turn been the site of comparable phenomena: e.g. the increase in numbers of orchards from the Dimbokro region towards the cocoa belt (Benveniste, 1974).

This spread into the forest immobilizes large areas of land. The areas used are very often away from the nuclei with the highest population densities. It is not generally into these areas that the excess population coming from those nuclei would tend to overspill. When this occurs it is because of a purely economic motivation that supersedes the simple hunger for land. The relatively low capability of receiving populations that is offered by the practice of tree cultivation in an extensive or semi-gathering form maintains intact the essential characteristics of the natural environment. The forest environment, the way in which the ecosystem functions and, up to a certain point, the floristic composition of larger tree species (kept in part for shade) are all preserved. All occurs as if, in relation to rather low population densities, dictated by this type of exploitation, large parts of the forest ecosystem were spared and provisionally withdrawn from other, much more destructive, forms of exploitation.

Intensification and occupation of virgin lands

To summarize, new circumstances born from an accelerating rate of demographic increase and growing pressure from dominant economic systems have had two main results: population density increases to such an extent that it far exceeds the capacities for production and the rate of progress of existing systems, thus resulting in an enormous potential number of migrants; an accelerated rate of occupation of the tropical forest space. Some solutions to these problems rely on the intensification of agriculture, others consist of providing a framework for the organization and planning of the occupation of vacant land by the people from overpopulated nuclei. This second approach was disappointing: the costs of supporting a migratory population are high and migrants do not always comply with the constraints judged to be necessary in terms of land distribution and agronomic practices; many failures occurred and the effective organizational framework concerns at best only a fraction of those looking for land (Raison, 1973). The colonization of virgin lands has never, or almost never, succeeded by itself in stopping demographic growth in the source areas. As for intensification, effective methods have been perfected, generally through the combination of a number of factors: dissemination of high yielding varieties of plants, livestock-rearing in close association with cultivation, distribution of chemical fertilizers, anti-erosion and water management practices, improvements in equipment and more rational use of space and time; all this occurring within the framework of the existing systems of exploitation (see chapter 20). In the tropical forest ecosystems, however, apart from the success in rice cultivation, solutions are still awaited. The systematic mixing of crops, the local ecosystem heterogeneity and the difficulties of clonal selection, all complicate the task of the innovators. If the rural populations are receptive, their poverty in high density areas and the low money value of an almost entirely subsistence production, mean that it is extremely difficult to purchase chemical fertilizers. The forest agro-ecosystems specializing in production for export pose different problems. Here the finance exists but the tree cultivators refuse to be deprived of the advantages and facilities of extensive cultivation, as has been shown in a large statistical study undertaken twenty years ago on the cocoa cultivators of the Yoruba area (southwestern Nigeria;

Gourou, 1960). It is possible to move to intensification, without working more for a given income, but this requires techniques that are not easy to acquire as well as a certain degree of concentration in terms of financial and technical means. One can therefore move to co-operative associations or enterprises in which the peasant-planters are organized around modern technical units, according to the model of cultivation in association. The prototype of this solution can be found in the West Indies at the end of the 19th century (Lasserre, 1961), after the revolution brought about by factories. The village planters, under the auspices of state capital in the Ivory Coast, represent a modern version of the same approach applied to oil-palm and its products. In 1972 there were more than 3 000 people grouped around a limited number of agro-industrial complexes arranged at intervals behind the coast within the framework of an operation mounted by the *Société pour le développement et l'exploitation du palmier à huile*. A system still functions in eastern Java in which the agricultural industry rents land to the peasants for sugar plantations over a discontinuous period. The landlord gets wages on his land during the period in which the area is occupied by sugar-cane. The reverse situation in which the peasants receive the land, but must deliver their harvest, still supplied 15 per cent of the crushed cane of Guadeloupe in 1960. All these examples deal with an enlargement of the large-scale capitalist exploitation by using the labour force available in a well populated area. Neither the economic role nor the place such a system holds in the tropical forest ecosystem compare to those of present day family tree crop cultivation.

Population movements

The movement of population towards virgin lands stems from an increasing demographic load and from needs which can no longer be met locally. Other movements are not motivated by the search for land to cultivate for food. The reasons for such movements are at a level largely above that of local or regional production systems. Their result is nevertheless that of moving pressure on land from one area to another and of thus modifying more or less deeply the map of population density.

A first type of movement is represented by migration from the savanna towards the forest. The most remarkable example in Africa is that originating in the north of the Ivory Coast and Ghana, and especially in Upper Volta and moving towards the forest regions of the first two countries that produce coffee and cocoa. In Ghana, where this phenomenon is fairly old, it led in the 1950s to an inversion in population density between the savanna and the forest with regard to the previous situation. In the Ivory Coast towards the end of the 1960s more than 100 000 people who came from Upper Volta lived in the forest environment; there were also more than 200 000 temporary or seasonal foreign workers mainly from Upper Volta or Mali and accompanied by a hundred thousand women and children. By a process of accumulation a number of nuclei of density 40–50/km² were created within the forest of the Ivory Coast which are spreading rapidly. In the Amazonian area of

Mato Grosso State the advance of peasant colonization is increasing and concentrations of population are forming in an economic context that is very different from that of western Africa.

The second form of migration is that from rural areas towards the towns, especially the large urban concentrations, many of which, especially in Africa, are concentrated in the forest zone. Africa south of the Sahara (excluding South Africa) had 5 towns with more than 100 000 inhabitants in 1940, 17 in 1955 and 56 or 57 in 1970. At the same time the total population of these towns increased from 700 000 to 3 million inhabitants and then to 14.5–15 million (Vennetier, 1972). This neo-urban population feeds on food imported from the temperate zone and continues to depend largely on the agricultural resources of the tropical zone itself. It is only rarely that the regions from which the migrants originate play a principal role in supplying them. Thus the need for land is displaced: in 1975 imported Chinese rice helped to feed the inhabitants of Tananarive and the countries of South-East Asia, especially Burma and Thailand, have, for a long period, played the role of granaries for countries deficient in food. On the other hand, as towns, and especially capital cities, develop so, regions orient their agriculture towards them. It could be the regions surrounding cities which thus benefit from the proximity factor, or more distant regions that have particular advantages (good soils, easy communications or available land). For example, Abidjan is supplied by yams cultivated in northwestern Ivory Coast where a speculative food agriculture has arisen; another region near Divo, in the forest sector, specializes in banana and plantain. Competition has arisen in many countries of the tropical forest zone between food crops for commercial markets and export crops and this has given rise to geographical divisions which are sometimes very clear-cut: the Bassa country of Cameroon, although located within the forest, has always refused to cultivate cocoa but has supplied the joint urban markets of Douala and Yaoundé (Champaud, 1970). With this type of demand coming from towns that are well supplied by a network of short or sometimes long communications, the agricultural response can be located with greater flexibility than in the case of village auto-consumption. Agriculture thus becomes relatively independent of the pre-existing rural population densities. The effects on the ecosystem are the opposite of those which are determined by demographic growth *in situ*: the pressure is distributed in a flexible manner over the area instead of reinforcing local or regional pressures.

A third type of movement concerns the evolution of human settlements in forest regions or Guinea savannas with a moderate population density. In a number of countries, especially in central Africa, a certain number of regrouping operations were undertaken by colonial administrations. These comprised the transfer of the population that was formerly scattered along the main tracks or roads into specially located villages of a certain size. Although originally undertaken under constraint, the populations finally accepted these operations. In Gabon the majority of villages located within the forest emerged to locate themselves along the lines of communication and today the population pattern is almost totally linear. On the other hand, in the 1950s, leaders at both regional and national level claimed for the creation of regrouped villages having a size of one thousand inhabitants or more, i.e. much larger than was usual until that time (Sautter, 1966). One of the reasons was to make it possible to build up infrastructures, especially schools and sanitary facilities, that require a minimum size concentration. In the Bamiléké plateau in Cameroon an attempt was made in the 1960s to create a certain number of peasant agglomerations reaching or exceeding 10 000 inhabitants in the place of dispersed settlements. In Tanzania the decision was taken at the end of 1973 to extend these regroupings and since 1974 regularly located places of permanent settlement demarcate the roads through the dry miombo forest that covers all the central part of the country. The creation of artificial villages means that, in the case of agriculturalists practising long-fallow cultivation, the fields will soon be located far from the settlement. There must thus be an intensification of agriculture in such a way as to indefinitely prolong soil use within the accessible area, or duplication of settlements by the creation of temporarily-occupied hamlets near to cultivated fields. The latter solution is that which has developed spontaneously in the Ivory Coast: the family planters scatter into isolated dwellings in the middle of their farms but all have some kind of residence in villages where considerable modernization and infrastructure are added by both the residents and the public authorities.

Thus progress, through activities brought about by concerted action or under the directive of authority, is towards an urbanization of the rural environment.

New types of exploitation

The large modern forms of agriculture in a forest environment doubly deserve the name plantations: they are not familiar to the traditional rural world (Gourou, 1966) and the cultivated plants are most often woody. The plantations are especially well represented in the old world, while livestock-rearing enterprises are mainly located in South America. The latter are nowadays the most dynamic form of economic use of space around the Amazonian basin (see chapter 20). A third type of large-scale use in the tropical humid environment is that for timber and a fourth type is for paper pulp production (see chapters 20 and 21). Of these four types of exploitation the first, that of plantation, occurs irrespective of whether it is in the heart of the forest ecosystem *sensu stricto* or on its margins within the humid savanna; the latter is a particularly suitable habitat for sugar-cane but tree and shrub crops need the true forest environment. Large-scale cattle-rearing finds ecological conditions at their best in savannas surrounding the forests, especially in South America where these habitats are free from tsetse fly.

These four types of use all have peculiar characteristics, compared to the usual rural activities. Both the surface area occupied per unit of production and the quantities supplied to markets are high. Considerable capital and competence with regard to technology and organizational capacity are required. These activities are foreign to the rural scene. More and more often the initial typical capitalist solution (in the sense that it is independent of the state) is taken up by governments looking for economic efficiency and thus draws on public investment. A second characteristic is that of economizing in the labour force, thanks to mechanization (see chapter 21). Thus these activities neither depend on nor create human densities and they can be devoted exclusively to supplying markets or factories.

Conclusion

In conclusion, it is necessary to stress the contrasts and contradictions that must be faced to achieve a better management of the tropical forest ecosystem. In the case of shifting cultivators these contradictory requirements concern work productivity (which needs extensive cultivation and mobility) and the access to concentrated infrastructures. The system of family plantations producing commercial crops is itself torn between the easy solution of a wide-spread extension at the risk of rapidly exhausting the soil reserves and that of intensification which is difficult but which enables the grouping of infrastructures and achieving of external economies. Many intact forest areas, such as those of zones 3 and 4 in Gabon, still constitute areas reserved for timber exploitation. But elsewhere the overpopulated nuclei are coming closer and are beginning to overspill their surplus. In other cases the pioneering peasant planters will dispute over their rights to the area with various kinds of capitalist exploitations. There are neither miracle nor definitive solutions to these problems and each country must find the appropriate ways and arrangements for its own geographic conditions and political options (see chapter 21).

Research needs and priorities

The first priority is to extend as rapidly as possible mapping the land according to its ecological potential for the population at a scale of 1/500 000 to 1/1 000 000, in all areas where the relationship between population and resources has reached a critical level. This should take into account the requirements of the population without referring to direct exploitation of the forest growing stock by modern technological methods for the world market. The criteria should be graded according to the way the cultivators judge the capabilities of the environment and conceive its use. The distinguished ecological units must be conceived so that each unit that is ecologically better or worse than its neighbouring one, nevertheless includes the diversity of natural aspects required for a good functioning of the agricultural system. Land cropping potential, which gives information on the yields to be expected is not required for it gives no information on the whole potential of the living spaces. It is appropriate to start by undertaking a general review on all work undertaken up until present in a scattered way and which meets these requirements; then to make an appraisal of this work to propose a method which is the most economical and easiest to generalize, and an essay on the ground at the level of at least a large region. The first project has never been undertaken but land capability studies undertaken since the 1950s on different regions of Melanesia do exist (Brookfield, 1971). The problems and techniques of land classification by English-speaking specialists, mainly but not exclusively in Africa have been reviewed by Thomas (1969). The work of Russian geographers on the classification of the natural environment into units easily related to human occupation and needs requires mention. An excellent study on *Les conditions géographiques de la mise en valeur agricole* (Bied-Charreton *et al.*, 1975) brings together the physical situation and the needs of utilizers in Madagascar; it includes a set of 12 maps at 1/500 000 representing the value of physical units which are classified into 24 categories.

The second priority is to map population densities. These maps are different from dot maps of population distribution but they complement each other. Three problems are: deciding the area of the census units; choosing a network for dividing the map into relatively homogenous units for calculating the population density; and choosing different intervals to represent density classes. The second is the most difficult and a variety of solutions have been used. Several publications have already dealt with many of the problems posed by density mapping (Sautter, 1966). These should be surveyed. It is important that, for maps designed to be used with maps of ecological potential for agriculture, the calculation should be made and the results presented within the same divisions used for physical land-use units. This is what the Madagascar geographers did in taking away from the surface area considered for the calculation the proportion of physical units whose natural character excluded any possibility of improvement. It is appropriate, as they did, to associate a density map with that of ecological potential in all cases where there is difficulty in surveying resources before national or regional development activities.

The third type of map should show maximum densities compatible with maintenance of a balance between the population and the local ecosystem. This concept of maximal density corresponds to the full use of the environment with the techniques and according to the concepts of the occupants of the area in question. It is the equivalent of the *human carrying capacity of land* as used by Allan (1967), who gave a didactic method for calculating it. The criteria showing overpopulation may also be identified empirically so as to indicate at which level the equilibrium is broken. This presupposes that a density map has already been prepared. The interest of this type of map lies in the comparisons possible for each spatial unit between effective density and maximal admissible density. As a result it would be very easy to prepare a map showing the positive or negative margin that separates the actual situation from the limiting situation for each unit. This margin may be expressed as a percentage, as an absolute density or, in an even more practical way, in number of inhabitants (so many inhabitants the local ecosystem cannot support over the long term without risk of transformation of the land-use system or, in contrast, so many additional inhabitants that could be received in complete safety without any change to existing practices). A last type of mapping is to make a similar document for the next 15 or 30 years (half or a complete generation ahead). For each basic unit the demographic trends could be

extrapolated from the last census and from a census about ten years older.

These maps would give the best available information on variations in space of the potential production, of the population density and of the relationship between these two. It would enable decision-makers to locate the critical sectors, realize the urgency of the problem in the different places, know the margin of time and action that are available in the least threatened places, appreciate the outlets for population excesses that may be offered by certain sectors (and for how long), effect a balance between agricultural intensification and the development of land that is still free, and judge what the eventual possibilities are for the adding or extending cash crops.

Another type of cartography at a much smaller scale would deal not only with the areas that are already fully occupied but with the whole of the occupied or unoccupied area whatever kind of activity is being undertaken. The key of the map would distinguish the following broad categories: 1) intact forest; 2) forest used for shifting cultivation at least one generation ago but now abandoned; 3) areas largely devoted to capitalistic types of production but without completely transforming the ecosystem (forest exploitation, extensive cattle-rearing); 4) areas transformed completely by agro- or sylvo-industrial enterprises; 5) areas regularly subjected to shifting cultivation but not to any type of permanent agriculture; 6) areas devoted partly to shifting cultivation and partly to permanent cropping; and 7) areas in which the ecosystem is totally artificial. Some of these categories should be subdivided. Thus 5 can be broken down into 5a and 5b, according to whether the settlement is permanent or not, and also 5c when there is a systematic organization in rotation; in the same way 7 can be divided into 7a (permanent cropping without soil fertilization), 7b (permanent cropping beneath a tree canopy) and 7c (many harvests during the year). A special design can be used to indicate the existence of management practices that modify the environment. An additional category could be created to indicate commercial orchards managed by families which might represent large surfaces. Lastly, superimposed hatching on categories 5 and 7 could indicate areas that are exploited at less than their capacity, areas in which the agro-ecosystem is just in balance and areas in which agricultural pressure exceeds the capacities of the forest environment. Such mapping would not duplicate the *World Atlas of Agriculture* (1969) which has different objectives. The principle is sufficiently broad and simple that it could be undertaken at the level of a continent, or even of the entire tropical zone, taking into account the peculiarities of each of the major regions (for example the important place that various forms of cattle-rearing have in South America).

The problem, with regard to practices that would contribute towards supporting more people on a given forest area or that would improve the standard of living of cultivators for the same amount of work, should not only be approached from the angle of intensification or of that of management techniques (see chapter 21); better ways of using the time and areas available to shifting cultivators practising long fallows must be looked at. Their complete survey has not been made up until present, although this work would cost little and could be completed in a short time. Thus it merits a high degree of priority. This information is found in the agricultural systems, described in a large number of studies, reports and experiments. Lengthening the cropping period and maintaining soil fertility by well organized successional crops (see chapter 20) has received much attention. Less attention has been given to the alternation between cultivation and fallow. It is only known that, after a limited rest period, an extra year does not add much in terms of reconstituted organic reserves. When shifting cultivators have a choice they prefer to clear fairly recent fallow rather than old forest: they look for less work and with softer trees they can work much faster. This advantage holds good even if a slightly larger area has to be cleared in order to compensate for the loss in yield. Comparative yields per unit for different lengths of fallowing have been the subject of many studies. A comparative balance sheet should be added giving the amount of work spent and the yield obtained in each case. Thus it would be possible to develop a rational model of forest agriculture-fallow and to adapt it to various situations, especially according to population density. It must be noted that many agriculturalists atlernate two phases of fallow: a short one and a long one, in the period between which cultivation resumes but with different crops. A rational element exists in this behaviour and should be taken into account as possibilities for progress exist. On the contrary, a certain amount of attention has been given to the arrangement of the land among cultivators and of the distribution of crops during a given year in relation to fallow. The advantages of such a defined pattern are obvious to agronomists, especially with regard to phytosanitary treatments and to the control of crop practices, but it runs against the principle of agriculture based on long fallowing which is to adapt closely to the forest ecosystem. The agriculturalists adapt the shape and area of their fields to the soil variations whereas a geometrically arranged subdivision, imposed from outside, creates inequalities between the plots: in the dry forest area of eastern Rwanda where the paysannats were installed, cases occurred in which unlucky peasants had mediocre sorghum crops but they were, in fact, victims of small-scale topographic variations (Silvestre, 1974).

A last priority is practices intended to facilitate and speed up the regrowth of forest after cultivation. These do not appear to have been either systematically enumerated or classified, to make them more wide-spread or to put them into use elsewhere. Reafforestation techniques or the simple prevention of invasion of the plot by herbaceous plants formerly existed in the lower Congo region. Processes consisting of enrichment of a young fallow by introducing fast-growing woody species with a view to creating an artificially shrubby fallow have been mentioned elsewhere. The transition from cultivation to forest regrowth occurs spontaneously with plantain bananas or cassava and this is even more rapid as weeding stops once these plants have reached a certain size. In the former Belgian Congo the agronomists took care to end the crop successions imposed on cultivators with the same plants having a long cycle of development. In the eastern region of Madagascar, action is being taken to promote forest regeneration after cultivation by planting *Grevillea* at the same time as the first year of rice so that when the field is abandoned the plot is covered with a dense shrubby canopy. This technique of *improved tavy* is close to the spontaneous practices of some groups living in eastern Nigeria. It is an example towards economy of land by cultivators using long fallows, i.e. a rise in the population density limit compatible with the system. This work of evaluation should give particular attention to all analogous attempts as well as to the response of the agriculturalists concerned.

Finally, it is appropriate to stress the danger of underestimating the real needs of shifting cultivators using long fallows. These have been calculated too tightly and often have only taken into account the portions of the forest being used or showing signs of recent cultivation. Since cycles of long-term use often alternate with shorter cycles, entire sections of the forest may appear to be unused whilst they are necessary to the long-term good functioning of the system. On the other hand, such tight calculations neglect the supplementary needs linked to the excess of population which may happen and do not consider the peculiar characters of the societies in question. An important part of their subsistence comes from non-agricultural resources and is the result of hunting, fishing or gathering. These resources presuppose free access to an area of forest that is normally much larger than that which meets their strict agricultural needs. One may create reserves if the area allocated the groups is very limited. Knowledge of these societies of the tropical forest ecosystem is thus indispensable for taking decisions on the distribution of land; it will be dealt with in the second part of this chapter.

Bibliography

ALEXANDRE, P.; BINET, J. *Le groupe dit Pahouin (Fang-Boulou-Beti)*. Paris, Presses Universitaires de France, 1958, 152 p.

ALLAN, W. *The African husbandman.* 2nd ed. Edinburgh, Oliver and Boyd and Barnes and Noble (1965), 1967, 505 p.

BARBIER, J. C. *Les villages pionniers de l'opération Yabassi-Bafang. Aspects sociologiques de l'émigration bamiléké en zone de forêt dans le département du Nkam (Cameroun).* Yaoundé, ORSTOM, 1971, 303 p. multigr.

BARRAL, H.; FRANQUEVILLE, A. *Atlas régional Sud-Est. République Fédérale du Cameroun.* Yaoundé, ORSTOM, 1969, maps 1/500 000, 53 p.

BARRAU, J. *L'agriculture vivrière autochtone de la Nouvelle-Calédonie.* Précédée de l'*Organisation sociale et coutumière de la population autochtone*, par Guiart, J. Nouméa, Commission du Pacifique Sud, 1956, 123 p.

BÉGUIN, H. *La mise en valeur agricole du sud-est du Kasaï.* Publications INEAC, sér. sci., n° 88, 1960, 289 p.

BENVENISTE, C. *La boucle du cacao, Côte-d'Ivoire. Étude régionale des circuits de transport*. Paris, ORSTOM, 1974, 221 p.

BIED-CHARRETON, M.; DANDOY, G.; RAISON, J. P. *Espaces naturels et développement rural. Un travail collectif de cartographie sur Madagascar: principes, méthodes, applications*. Paris, ORSTOM, 1975, 37 p. multigr., maps.

BOHANNAN, P. *Tiv farm and settlement*. London, His Majesty's Stationery Office, 1954, 87 p.

BONNEMAISON, J. Système de grades et différences régionales en Aoba (Nouvelles Hébrides). *Cahiers ORSTOM* (Paris), *sér. Sci. humaines*, vol. 9, n° 1, 1972, p. 87–108.

——. Espaces et paysages agraires dans le nord des Nouvelles Hébrides. L'exemple des îles d'Aoba et de Maewo. *Journal de la Société des Océanistes* (Paris, Musée de l'Homme), 1974, p. 163–281.

BOSERUP. E. *The conditions of agricultural growth; the economics of agrarian change under population pressure*. London, Allen and Unwin, 1965. Chicago, Aldine, 1966, 124 p. Traduit en français sous le titre: *Évolution agraire et pression démographique*. Paris, Flammarion, 1970, 222 p.

BOURGEAT, F. *Contribution à l'étude des sols sur socle ancien à Madagascar. Types de différenciation et interprétation chronologique au cours du Quaternaire*. Tananarive, ORSTOM, 1970, 2 vol., 310, 126 p. multigr.

BOUTRAIS, J. Les conditions naturelles de l'élevage sur le plateau de l'Adamaoua. *Cahiers ORSTOM* (Paris), *sér. Sci. humaines*, vol. 11, n° 2, 1974, p. 145–198, 2 maps.

BROOKFIELD, H. C.; HART, D. *Melanesia. A geographical interpretation of an island world*. London, Methuen, 1971, 464 p.

CHAMPAUD, J. Mom (Cameroun) ou le refus de l'agriculture de plantation. *Études Rurales* (Paris), n°s 37, 38, 39 (Terroirs africains et malgaches), 1970, p. 299–311.

——. *Atlas régional Ouest. II. République Unie du Cameroun*. Yaoundé, ORSTOM, 1973, maps 1/500 000 and 1/1 000 000, 113 p.

CHEVALIER, L. *Madagascar, populations et ressources*. Paris, Presses Universitaires de France, 1952, 212 p. (Travaux et Documents de l'Institut national d'études démographiques, cahier n° 15).

COULAUD, D. *Les Zafimaniry. Un groupe ethnique de Madagascar à la poursuite de la forêt*. Tananarive, Imprimerie Fanontam-Boky Malagasy, 1973, 385 p.

DANDOY, G. *Terroirs et économies villageoises de la région de Vavatenina (Côte orientale malgache)*. Paris, ORSTOM, 1973, 94 p., maps (Atlas des structures agraires à Madagascar. I.)

DUBY, G. *L'économie rurale et la vie des campagnes dans l'Occident médiéval (France, Angleterre, Empire, IX-XVᵉ siècles). Essai de synthèse et perspectives de recherches*. Paris, Aubier, 1962, 2 vol., 822 p.

FLOYD, B. *Eastern Nigeria. A geographical review*. London, Macmillan, 1969, 359 p.

FORDE, C. Daryll. Land and labour in a Cross River village, southern Nigeria. *Geographical Journal*, vol. 90, no. 1, 1937, p. 24–51.

FRANQUEVILLE, A. *Atlas régional Sud-Ouest. I. République Unie du Cameroun*. Yaoundé, ORSTOM, 1973, maps 1/500 000 and 1/1 000 000, 93 p.

GEERTZ, C. Two types of ecosystems. In: Vayda, A. P. (ed.). *Environment and cultural behavior; ecological studies in cultural anthropology*, p. 3–28. New-York, Natural History Press, 1969, 485 p.

GILLMANN, C. A population map of Tanganyika Territory. *Geographical Review*, vol. 26, no. 3, 1936, p. 353–375.

GOUROU, P. *Les paysans du delta tonkinois. Étude de géographie humaine*. Paris, Les Éditions d'Art et d'Histoire, 1936, 666 p. (Publications de l'École française d'Extrême-Orient). Paris, Mouton, 1965, 668 p.

——. *L'utilisation du sol en Indochine française*. Paris, Hartmann, 1940, 466 p.

——. *Les pays tropicaux. Principes d'une géographie humaine et économique*. Paris, Presses Universitaires de France, 1947, 199 p.; 4ᵉ ed., 1966, 271 p.; 1969, 272 p.

——. *La densité de la population rurale au Congo belge*. Bruxelles, Académie Royale des Sciences coloniales, Classe des Sciences naturelles et médicales, 1955, 168 p.

——. Cartes de la densité et de la localisation de la population dans la province de l'Équateur. In: *Atlas général du Congo*. Bruxelles, Académie Royale des Sciences d'Outre-Mer, 1960, maps, 22 p.

——. Les cacaoyers en pays Yoruba: un exemple d'expansion économique spontanée. *Annales Économies, Sociétés, Civilisations* (Paris), n° 1, 1960, p. 60–82.

——. *Pour une géographie humaine*. Paris, Flammarion, 1973, 388 p.

GUILLOT, B. *Projet de développement de la culture du cacaoyer dans la région de la Sangha. Études géographiques et sociologiques. Tome I. Géographie*. Brazzaville, ORSTOM, 1973, 91 p. multigr., maps.

HILL, P. The migrant cocoa farmers of southern Ghana. *Africa*, vol. 31, no. 3, 1961, p. 210–230.

HUNTER, G.; BUNTING, A. H.; BOTTRALL, A. F. *Change in agriculture*. Proceedings of the second international seminar on Change in agriculture (Reading, September 1974). London, Croom Helm (on behalf of the Overseas Development Institute), 1976.

HURAULT, J. L'organisation du terroir dans les groupements Bamiléké. *Études Rurales* (Paris), n°s 37, 38, 39 (Terroirs africains et malgaches), 1970, p. 232–256.

HUYSECOM-WOLTER, C. La démographie en Équateur (Congo). *Revue Belge de Géographie*, vol. 87, n° 2, 1963, p. 177–209.

International Association of Agricultural Economists (under the aegis of). *World Atlas of Agriculture*. Edited by the Committee for the World Atlas of Agriculture. 3 vol. I. Europe. USSR, Asia Minor. II. South and East Asia, Oceania. III. Americas. Novara (Italy), Istituto Geografico De Agostini, 1969, 527, 671 and 497 p.

IZIKOWITZ, K. G. *Lamet, hill peasants in French Indochina*. Göteborg, Etnografiska Museet, Etnologiska Studien, 17, 1951, 376 p.

JEAN, S. *Les jachères en Afrique tropicale. Interprétation technique et foncière*. Paris, Institut d'ethnologie — Musée de l'Homme, Mémoires de l'Institut d'ethnologie, 14, 1975, 168 p.

KNOWLES, R. L. Farming with forestry: multiple land use. *Farm Forestry*, 14, 1972, p. 61–70.

LASSERRE, G. *La Guadeloupe. Étude géographique*. Bordeaux, Union française d'Impression, 1961, 2 vol., 1 135 p.

LAUER, W. Humide und aride Jahreszeiten in Afrika und Südamerika und ihr Beziehung zu den Vegetationsgürteln. In: *Studien zur Klima und Vegetationskunde der Tropen*, p. 15–98. Bonner Geographische Abhandlungen, Heft 9, 1952.

LE BOURDIEC, F. *Hommes et paysages du riz à Madagascar. Étude de géographie humaine*. Doctorate thesis, 1974, 3 vol., 1 059 p. multigr.

LECOCQ-LITOUX, M. C. *Contribution à la connaissance régionale du sud-est du Togo. Surpeuplement et migrations, le village de Fiata*. Lomé (Togo), ORSTOM, 1974, 131 p. multigr., maps.

LETOUZEY, R. *Étude phytogéographique du Cameroun*. Paris, Lechevalier (Encyclopédie Biologique 69), 1968, 508 p.

LEVI-STRAUSS, C. *Anthropologie structurale*. Paris, Plon, 1958, 447 p.

MAGOBUNJE, A. L. The evolution of rural settlement in the Egba Division, Nigeria. *Journal of Tropical Geography*, 13, 1959, p. 65–77.

MALCOLM, D. W. *Sukumaland: an African people and their country. A study of land use in Tanganyika*. London, published for the International African Institute by the Oxford University Press, 1953, 224 p.

MAQUET, J. *Les civilisations noires. Histoire/techniques/arts/sociétés*. Nouvelle édition. Verviers (Belgique), Marabout Université, 1966, 319 p.

MESCHI, L. Évolution des structures foncières au Rwanda: le cas d'un lignage hutu. *Cahiers d'Études Africaines* (Paris), vol. 14, nᵒ 53, 1974, p. 39–51.

MESSERSCHMIDT, D. A. Ecological change and adaptation among the Gurungs of the Nepal Himalaya. *Human Ecology* (New York), vol. 4, no. 2, 1976, p. 167–185.

MONBEIG, P. *Pionniers et planteurs de São Paulo*. Paris, Armand Colin, 1952, 376 p.

MORGAN, W. B. Farming practice, settlement pattern and population density in southeastern Nigeria. *The Geographical Journal*, vol. 121, no. 3, 1955, p. 320–333.

NICOLAÏ, H. *Le Kwilu. Étude géographique d'une région congolaise*. Bruxelles, Édition CEMUBAC (Centre scientifique et médical de l'Université libre de Bruxelles en Afrique centrale), 69, 1963, 472 p.

OKIGBO, B. N. Fitting research to farming systems. In: Hunter, G.; Bunting, A. H.; Bottrall, A. F. *Change in agriculture*. London, Croom Helm (on behalf of the Overseas Development Institute), 1976.

ORMELING, F. S. *The Timor problem. A geographical interpretation of an underdeveloped island*. Djakarta, Gröningen, J. B. Wolkers, 1956, 284 p.

PARSONS, J. J. *Antioqueño colonization in Western Colombia*. Berkeley, University of California Press, 1949, 212 p.

PÉBAYLE, R. *Éleveurs et agriculteurs du Rio Grande do Sul (Brésil)*. Université de Lille III, thesis, 1974, 744 p.

PETERS, D. U. *Land usage in Serenje District*. Oxford University Press, 1950, 99 p.

PIGNÈDE, B. *Les Gurungs. Une population himalayenne du Népal*. Paris, La Haye, Mouton, 1966, 414 p.

PORTAIS, M. Les cultures commerciales dans un milieu géographique original: la cuvette d'Andapa (Madagascar). In: *Types de cultures commerciales paysannes en Asie du Sud-Est et dans le monde insulindien*, p. 327–355. Bordeaux, Centre d'études de géographie tropicale, CEGET (CNRS), Travaux et Documents de Géographie tropicale, nᵒ 20, 1975.

RAISON, J. P. La colonisation des terres neuves intertropicales. *Études Rurales* (Paris), nᵒ 31, 1968, p. 5–112.

——. La colonisation des terres neuves intertropicales d'après les travaux français. *Cahiers ORSTOM* (Paris), sér. Sci. humaines, vol. 10, nᵒ 4, 1973, p. 371–403.

——. Conditions et conséquences de l'intensification de l'agriculture sur les hautes terres malgaches. *Terre malgache* (Tananarive), nᵒ 15, 1973, p. 59–68.

RAPPAPORT, R. A. *Pigs for the ancestors: ritual in the ecology of a New Guinea people*. New Haven and London, Yale University Press, 1967. 5th edition, 1973, 311 p.

RAVAULT, F. La vie rurale dans un district de la côte ouest de Tahiti: l'exemple de Papeari. In: *Tahiti et Moorea. Études sur la société, l'économie et l'utilisation de l'espace*, p. 106–140. Paris, ORSTOM, 1970, 183 p., maps.

RETEL-LAURENTIN, A. *Infécondité en Afrique noire. Maladies et conséquences sociales*. Paris, Masson, 1974, 188 p.

ROCHE, L. *The practice of agri-silviculture in the tropics with specific reference to Nigeria*. Shifting cultivation and soil conservation in Africa. Rome, FAO, Soils Bulletin no. 24, 1973, 248 p.

RUELLAN, F. *La production du riz au Japon. Étude des conditions naturelles et historiques de la culture et des problèmes qui s'y rapportent*. Paris, Larose, 1938, 103 p.

RUTHENBERG, H. *Farming systems in the tropics*. Oxford, Clarendon Press, 1971, 313 p.

SAUTTER, G. A propos de quelques terroirs d'Afrique occidentale. Essai comparatif. *Études Rurales* (Paris), nᵒ 4, 1962, p. 24–86.

——. De l'Atlantique au fleuve Congo. Une géographie du sous-peuplement. République du Congo, République Gabonaise. Paris, La Haye, Mouton, 1966, 2 vol., 1 102 p.

SILVESTRE, V. Différenciation socio-économique dans une société à vocation égalitaire: Masaka dans le paysannat de l'Icyanya. *Cahiers d'Études Africaines* (Paris), vol. 14, nᵒ 53, 1974, p. 104–169.

SMET, R. E. de. *Carte de la densité et de la localisation de la population de la Province Orientale (Congo)*. Bruxelles, CEMUBAC, 1962, maps, 49 p.

——. Cartes de la densité et de la localisation de la population de l'ancienne province de Léopoldville (République démocratique du Congo). Bruxelles, CEMUBAC, 1966, maps, 46 p.

——. Cartes de la densité et de la localisation de la population de la Province du Katanga (République du Zaïre). Bruxelles, CEMUBAC, 1971, maps, 38 p.

SORRE, M. *Les fondements biologiques de la géographie humaine. Essai d'une écologie de l'homme*. Paris, Armand Colin, 1943, 440 p.

STASZEWSKI, J. Bevölkerungsverteilung nach den klimagebieten von W. Köppen. *Petermanns Geogr. Mitteilungen*, no. 2, 1961, p. 133–138.

STOTT, D. H. Cultural and natural checks on population growth. In: *Environment and cultural behavior; ecological studies in cultural anthropology*, p. 90–120. New York, Natural History Press, 1969, 485 p.

THOMAS, M. F. Geomorphology and land classification in tropical Africa. In: Thomas, M. F.; Whittington, G. W. (eds.). *Environment and land use in Africa*, p. 103–145. London, Methuen, 1969, 554 p.

TREWARTHA, G. T.; ZELINSKY, W. Population patterns in tropical Africa. *Annals of the Association of American Geographers*, vol. 44, no. 2, 1954, p. 135–162.

VENNETIER, P. La poussée urbaine en Afrique noire et à Madagascar. In: *La croissance urbaine en Afrique noire et à Madagascar*, vol. 1, p. 231–243. Paris, Éditions du Centre national de la recherche scientifique, 1972, 2 vol., 1 109 p. (Colloques internationaux du CNRS, Talence, 29 septembre–2 octobre 1970).

WHYTE, R. O. *Land and land appraisal*. The Hague, Junk, 1976, 370 p.

Second part: Civilizations and societies

Original nature of human societies

Introduction

Knowledge both of the civilizations and human societies of the tropical forest ecosystems and of their techniques for utilization of the natural resources of such ecosystems has for a long time suffered from a particular ethnocentric point of view often expressed by workers who are cultural and ecological strangers to such environments. It has only been during the last quarter of the century that a more accurate assessment has begun to appear of, for example, the hunter-gatherer societies that have persisted in tropical forests (as in other ecosystems). The Proceedings of a meeting held in Chicago in 1965 under the title *Man the hunter* (Lee and De Vore, 1968) represents this new trend. Previously, the ideal-ogical prejudgments of missionaries, as in the case of the otherwise interesting work of Schebesta (1928–1957) on African pygmies and Malaysian negritos, or colonial tech-nocrats have often concluded that the hunger-gatherers of the tropical forests are relicts of prehistoric savagery, suf-fering from a 'millenial stagnation' in a forest which 'masters them', and are incapable of managing their environment or drawing benefit from it (the expressions are quoted from Burkill, 1952). Such attitude is often shared by local non-forest societies that make use of this environment. This is a good explanation of the contrast shown in the ways in which the pygmies on one hand and the cultivators clearing the margins of the forest, on the other, perceive the forest as well as of the ambiguities present in the relationships between these two groups using part of the same ecosystem. This has been noted by Turnbull (1961, 1965). The cultivators, who are afraid of the forest, only have access to its resources through the pygmies whom they try to bind into their service and integrate to their culture. The pygmies accept this role to a greater or lesser extent, the final result being a kind of symbiosis based largely on mutual dupery, social and economic. The pygmies often regain their liberty by retreat-ing into the forest which remains the best guarantee of their relative independence since the cultivators regard it as formidable and consider it an obstacle to their activities.

The tropical forest ecosystem, with its diversity and complexity as well as with its impressive primary pro-ductivity, can at the same time fascinate and surprise, or even disturb those whose culture is rooted in different biomes, such as the Europeans whose civilization is based on cereal crops and animal husbandry born, 10 000 years ago, in the *saltus* of the Middle-East fertile crescent; such a civilization extended its landscapes in Europe while clearing the forests. It is also true for the agricultural civilizations which evolved in the African savannas.

This can in part explain judgments such as those of Haudricourt (1968) who stated: 'the forests of the equatorial region, the evergreen forest, in spite of its floristic richness,

does not appear to have constituted a very favourable environment for man . . . This environment which is favour-able for apes, especially for the present day anthropomorphic apes, does not appear to have had the same character for man himself!' This is not completely false but is not an attitude that will encourage an understanding of the inter-relationships between such societies and the ecosystems. Similar statements abound which refer to the types of cultivation practised in tropical forests (shifting cultivation, slash-and-burn agriculture or swidden cultivation; Barrau, 1972). These are commonly considered as a wastage of resources and a practice to be condemned because of its inefficiency and destructiveness.

It was only Conklin's (1957) *Hanunóo agriculture: a report on an integral system of shifting cultivation in the Philippines*, that led anthropological research to a better approach to the study of the agricultural civilizations of the tropical forest environment. Izikowitz (1951) showed the way with his study of Indo-Chinese swidden cultivators, *Lamet: hill peasants in French Indochina*. Condominas (1957), Carneiro (1960), Geertz (1963), Spencer (1966), Harris (1971) and others followed. This new positive attitude was devel-oped and reinforced in interdisciplinary studies of which that of the French Centre national de la Recherche scienti-fique on *Culture sur brûlis et évolution du milieu forestier en Amazonie du Nord-Ouest* is a good example (Slash-and-burn agriculture and evolution of the forest environment in north-western Amazonia; Centlivres *et al.*, 1975).

Yet the old prejudgments have not disappeared. The persistence of negative administrative attitudes, which re-press traditional ways of using tropical forests, is evidence of this. The better understanding of the ecology of tropical forests is not necessarily integrated with the study of the human elements which are part of it. The adoption of an ecological perspective in the study of human societies of the tropical forests is very recent and rather exceptional. The ma-jority of those undertaking such human studies have neither ecological knowledge nor the approach of the naturalist such as Conklin or Harris and Rappaport (1967, 1971).

A number of excellent contributions to our knowledge of tropical forest societies are somewhat lacking in ecological basis. Work that is as remarkable as that of Turnbull (1961, 1965) on the Mbuti pygmies of Zaire gives little precise information on their natural environment, or on the nature and hierarchy of the ecological constraints that these hunter-gatherers are subject to, or on their place and role in their ecosystem. These pygmies are a 'forest people'; an over-simplification since the ecological framework described by Turnbull does not go beyond this statement. It is sufficient to superimpose a map of the Mbuti territory on a vegetation map and a relief map in order to realize that such a territory includes different biomes. A rapid examination of available meteorological data shows a range of seasonal variations and botanical observations made in this region also show a diversity of resources which are not refered to by Turnbull. This shows clearly how necessary interdisciplinary approaches are in the study of such a vast and complex field in order to permit an acceptable interpretation of the inter-relationships between human societies and tropical forest ecosystems.

A further point is the relative lack of knowledge of human prehistory in tropical forest ecosystems. Very old plant domestications could have taken place and, in at least some parts of the tropical forest world, cultural and agri-cultural developments could have occurred in circumstances which the present conditions do not necessarily enlighten. Signs of former human presence have come to light in various tropical forests, often as an outcome of road-building and often accompanied by indications of agricultural activi-ties (for example, in Africa, see Letouzey, 1968). These witnesses of neolithic or post-neolithic activity are proof of a human occupation of the forest over a long period.

Lastly, the civilizations and societies that are peculiar to tropical forests appear to have induced two opposing reactions among western observers:
— those who only see archaic and primitive societies with no future, that can be suppressed or forgotten because they only serve as curbs to the profitable exploitation of the forest of which they belong;
— and others who adopt the romantic idealism of Thoreau and marvel at this way of life in the forests in which they rediscover the primitive golden age miraculously pre-served.

Neither of these attitudes permits an accurate appreciation of the present problem and both result in grave conse-quences. The first approach readily leads to ecocide and to ethnocide and the second to an unrealistic and utopic conser-vationism.

Transformation of tropical forest ecosystems

The historical character of the tropical forest must be re-membered; almost all remaining forests have traces of human activity which has considerably reduced their extent. Maps of Spencer (1954) for South-East Asia and the Indian Archipelago show this clearly. The transformation of tropical forest ecosystems will continue since they represent resources which cannot be left unused in the present demographic and economic state of the world. On the other hand it would also be an illusion to believe that the peoples of the forest can or wish to remain at their present level of development. They are not isolated and their civilizations and societies are the result of transformations which are still continuing. It seems likely, for example, that the pygmy communities of Zaire described by Turnbull (1961, 1965) have already been per-ceptibly changed by exterior contacts and pressures even during the course of the studies made by Turnbull, and that their home range has already been reduced by the activities of both African cultivators and Belgian colonizers.

This point should be stressed in order to again dispel the fixed illusion concerning virgin tropical forest where a few people live a pure and healthy life, as that of man's early days. A society of food-gatherers in tropical forest is not necessarily the relict of a pre-agricultural palaeotechnical period but can result from a society of agriculturalists who were driven from their own domesticated ecosystem and pushed into the forest, to which they re-adapt by adopting, for example, the gathering of spontaneous resources as a basis for subsistence. Some hunter-gatherer societies of neotropical forests could fall within this category (Lathrap, 1968; Godelier, 1974).

Tropical forests seem to be exploited by huge under-takings seeking rapid short-term profit with no thought for the human societies present, or the untouchability of such forests is proclaimed together with the inviolability of the populations which live in them. Some alternative is necessary to these over-simplified points of view.

The world-wide technological, economic and social change affects the tropical forest and its indigenous human populations. They and their forests are no longer untouched and can no longer remain so. These inevitable changes should occur through sensible ecological management and allocation of the resources, a management which takes into account both the rights and hopes of the human societies which dwell in them and the requirements of economic and social development. Once more it seems that a good know-ledge of the natural systems and of the human societies is necessary as a basis for management of these systems.

Contrasts between forest and non-forest civilizations

The humid tropical forests are generalized ecosystems in the sense used by Odum (1959), a term adopted by Geertz (1963) and Harris (1969). There are a large number of species each represented by a few individuals. Primary productivity is high, ecological niches are numerous and many alternative routes are available for the movement of matter and energy (see chapters 1–14).

These characteristics help understanding the reasons for the ethnobiological convergence which has occurred. The most important factor, in terms of cultural heritage and its role in these human societies, is the perception of the natural environment and in particular the relationships engendered between human beings and the other components of the biocenosis. One can, for example, compare a generalized ecosystem of the tropical forest type with a more special-ized ecosystem such as a savanna or steppe at the level of spontaneously-available resources for food-gatherers. In the first case, taking into account the diversity of the biocenosis, the perception and understanding of the environment are done individually and show themselves in a discriminating and detailed knowledge of the many components of the biocenosis. Thus Conklin (1957) has shown that the hunter-gatherers Hanunóo of Mindanao in the Philippines dis-tinguished 1 600 categories of plants in their forest area (in which *ca.* 1 200 plant species are known). The appropriation of such resources requires a knowledge of all of them. In contrast, in a more specialized ecosystem with a smaller number of species represented by a large number of indi-viduals, an inverse situation occurs, that is perception, understanding, appropriation, performed on a collective and heavy scale. The ancient mesolithic gatherers of the fertile crescent reaped wild cereals from meadows which were somewhat similar to our cereal fields and the flocks of gregarious herbivores present were similar to our herds. In contrast, the African pygmies or Asiatic negrito hunter-gatherers seek out plants individually and have to do so before their rapid decay.

Similar reasoning can be applied to production of domesticated plants in tropical forests. The gardens of swidden cultivators often reproduce the structure of the surrounding ecosystem: a high diversity of species and varieties and a complex structure with many layers. This small-scale replica of the forest is accompanied by an approach which gives individual attention to each plant, planted, tended and harvested on an individual basis. Haudricourt (1962, 1964) has quite rightly drawn attention to the contrast between this behaviour and that of the agriculturalists and herdsmen of more specialized ecosystems and has underlined the difference between them. He uses the societies and civilizations of Asia and tropical Oceania as a basis for his reasoning and he has also suggested that in these societies the relationship of man to the other components of the natural or domesticated ecosystem are echoed in the relationships between men and in the ideologies that they have developed.[1] It seems that these contrasts are founded in the nature of the ecosystems in which they evolved and developed the first stages of appropriation or production. Tropical forest ecosystems appear to have contributed to the development of the horticultural aspects which are expressed through limited assistance and by the integration of specific methods of appropriation and harvesting.

The restraining nature of tropical forest ecosystems may be stressed by pointing out that there is only one choice: either integrating into the ecosystem or destroying it. Thus, in Africa one can see cultivators originating from savanna regions several generations ago who have almost completely abandoned their traditional crops in order to clear gardens on the edge of the forest from which the pygmies also provided them with the products of hunting and gathering. But one can also see savanna cultivators who have not ceased to enlarge their agricultural clearings on the margins of the forest to the extent to which they are gradually replacing it by their own specialized environment, for they 'carry their ecology with them' (Rousseau, 1972). Societies born and raised in generalized tropical forest ecosystems usually remain deeply influenced by the type of relationship between man and natural resources which is peculiar to such an environment. 'Plant civilizations' (Gourou, 1948) in Asia and Oceania bear witness to this. The inverse is also true: civilizations and societies which are strange to the tropical forest integrate into it only when their technology does not allow them to escape from the constraints of the forest ecosystem. Otherwise they destroy the forest and substitute a more open environment which is less diversified and is similar to the ecosystem which was their cultural norm.

Some significant contrast exists especially when there is a symbiosis between a forest and a non-forest society which has moved into the margins of the forest. This is the case with Mbuti pygmies of Zaire and the Bantu cultivators with whom they maintain social and economic relationships on a slightly ambiguous basis, as was noted by Turnbull (1961, 1965).

Cultural factors are very important in the relationships between man and forest ecosystems. Cultivating cereals in a forest environment leads to conflicts, for which the solution is usually the destruction of the forest. Java and the other continental or island territories of South-East Asia give good examples of this (Geertz, 1963), since the development of cereal crops (millet, followed by rice) has eliminated forest and the swidden cultivation practised within it. It could also be said that the present day development of livestock-rearing and ranching in the Amazonian forests also illustrates this point. The history of agricultural development and livestock-rearing in Europe was similar.

Unity and diversity of traditional uses

It seems difficult to stick to the classic conflict between Nature and Culture, when considering human societies and civilizations of tropical forest ecosystems. Man can never be dissassociated from the influence of the environment: the hunter-gatherer of the tropical forest feels that he is part of the forest which is the provider of resources, his nurse and the overriding force. Much the same holds true for the shifting cultivator and even for any cultivator of former forest land: tropical forest has profoundly influenced both behaviours and ideas, even when it is considered as the land of the unexpected, the disturbing or of the wicked. Lombard (1974) has shown this clearly in his study of the attitudes towards the forest as shown in Java which is now almost totally deforested due to agricultural activity but where, nevertheless, the forest is still present throughout Javanese thought. It remains as a mythical environment, the abode of wicked genies and the site of the trials of strength of legendary heroes. In terms of cultural adaptation the tropical forest ecosystem can never be treated as a neutral or inert background against which physical activities, social organizations and idealogical conceptions of mankind can interplay with absolute freedom or real chance.

Despite the diversity of techniques and the resources within the various types of tropical forest there is an overriding unity. This is obvious in the case of hunter-gatherers. Comparison of the technical behaviours of the African pygmies and of the Malaysian negritos towards the natural environment and its resources shows this clearly as does the non-violence of these people living from harvesting the resources from a nature which they feel part of and which they only intervene into with moderation and care. The Yanomamö of Brazil and Venezuela (Chagnon, 1968; Lizot 1975) are considered as violent societies but they are cultivators who are also hunters and could have originated in another environment.

1. Haudricourt and Hedin (1943) drew the attention to the possible anteriority of a horticulture with perennial plants which were reproduced by vegetative means, contrasting with an agriculture of broadcast cereals. Such a horticulture was supposed to have originated in the humid tropics, especially in South-East Asia. This idea was developed later by Sauer (1952).

A similar convergence can be found in the case of swidden cultivators despite the variations of their agriculture as tabulated by Conklin (1957) and Spencer (1966) among others. Whatever these variations, three characteristics are always present:
— bringing into cultivation of a man-made clearing in the forest for a relatively short period and in which, the created domesticated ecosystem is often a small-scale reproduction of the surrounding ecosystem;
— using fire in preparing the clearing;
— after cultivation, the clearing is fallowed for a sufficiently long period for the forest cover to be re-established with a structure similar to that of untouched plant formations. There are two further points. Tools are certainly characterized by the adze or hatchet and the digging stick and its variations—a tool which is also used by gatherers. There is thus an overlap between gathering or collecting and swidden. The harvesting of spontaneous resources often persists in swidden societies and often contributes to subsistence in a far from negligible manner. Such examples are present in Melanesia and in New Guinea and it is common to find swidden cultivators relying on the resources provided by more than a hundred wild plant species (Blackwood, 1940; Barrau, 1962). There is a convergence in the types of plants suitable for this type of cultivation; they are predominantly plants with starchy organs such as yams or aroids, taro (*Colocasia*) in the Malayo-Oceanic region and *Xanthosoma* in the American tropics, cassava, banana, etc.

A striking feature is the convergence of adaptive techniques at the level of food production and by gathering. *Metroxylon* palms are used in the Indian Archipelago and New Guinea for the starch (sago) the pith of their trunks contains, by a tool which is very similar to that used by the Indians of the Amazon and Orinoco basins to harvest the starch present in *Mauritia* palms. This convergence of techniques reflects the common features of this ecosystem. The same could be said of convergences between different types of swidden in the various parts of the tropics.

The integration of techniques of appropriation or production into that of the natural environment is perhaps of greater significance. Geertz (1963) stresses this common feature when dealing with swidden cultivation in Indonesia and cites an expression of Kampto Utomo (1957) that it is 'a natural forest transformed into a harvestable forest'. This could equally appropriately be applied to various forms of gathering undertaken in tropical forest. There is a greater technological diversity in hunting (different techniques of trapping, variations in the use of poisons for hunting and fishing, differences in the use of hunting nets, the bow and its variations as well as the blowpipe, etc.). This diversity appears more in the detail than in the principles. An overriding care to safeguarding the natural regeneration of the all-providing ecosystem is apparent in all societies of gatherers living on the resources provided by the forest environment. This takes the form of various systems of prohibition with a magical and religious justification that ensure moderation in the use of resources and safeguard their regeneration (such

as selective hunting techniques, periodic prohibitions on harvesting which permit regeneration of vegetation thus protected; see Burkill, 1952). Hunter-gatherer societies have some kind of biocenotic behaviour and care for their subsistence while respecting the natural environment to which they feel they belong.

This behaviour also occurs in swidden culture despite the influence of domestication and the exercise of some control over the forest environment. But this agricultural system must remain integrated into the ecosystem and permit its regeneration if it is going to be able to perpetuate itself. Otherwise the forest would degenerate and could disappear and thus lead to the transformation of a society based on swidden. Such change is often due to factors outside the society and its ecosystem (introduction of new agricultural techniques, new crops, pressure from neighbouring populations, etc.).

This unity and convergence must not allow underestimation of the human and ecological diversity in the tropical forest world; the forests of South-East Asia can provide greater food sources than those of Africa because of their floristic richness. Therefore the major plant domestications occuring in the humid tropics of Asia and the resulting socio-cultural consequences have a magnitude and a style that are different from those in Africa. Sauer (1952) has stressed the possibility of an earlier date for domestication in tropical Asia, the region from which came cultivars of worldwide economic importance (the large yam, taro, banana, sugar-cane, rice, etc.) and which gave birth to the plant civilizations of Asia and Oceania.

Another danger is regarding the tropical forest as a homogeneous ecological formation and thus ignoring variations in man's adaptation to such natural habitats. Some idea of these variations will be gained by refering to the work of Robbins (1961) on the forests of New Guinea or that of Letouzey (1968) on the African forests.

It is essential not to neglect the way in which its inhabitants perceive the forest, draw in it ethno-ecological distinctions, to which correspond distinct human activities and behaviours. The value of such an approach is clearly shown in the work of Martin (1974) on the ethnophytogeography of Kampuchea. This work shows the significance, utility and refinements of local naturalist folkscience in the recognition of different categories of plant formations.

Although evidence of technological and cultural convergence among civilizations of tropical forest ecosystems has been stressed, it is also important to study these societies and civilizations in a micro-geographical or micro-ethnoecological way (Brookfield, 1973) so as to highlight the uniqueness of each example and the diversity of human adaptations to tropical forest habitats. It is also essential to consider the variations due to external factors; they may come from influences and contributions of other environments, other cultures or be derived from non-forest heritages. For example, the Asiatic banana introduced into Africa before colonial times has become a principal means by which the African forest is penetrated by agriculture and is thus a major factor in the ecological and economic transformation of this natural habitat.

Conservation, destruction or transformation?

The present day exploitation of tropical forests by large companies has aroused justifiable protests from scientists. Nevertheless, these have often been an ultra-conservative reaction which would like to see a return to maintenance of an integral natural and human *status quo*. There are no true virgin tropical forests just as there are no longer any human societies living within them that are still in a primitive golden age. Tropical forest ecosystems cannot escape from world-wide technological, economic and social change. Tropical forest societies have been able to maintain themselves over a long period through a subsistence economy based on gathering or swidden but all such societies have some evidence of transformation due either to internal evolution or to influences or contributions from outside. To believe that hunter-gatherer or swidden societies are capable of continuing to maintain and reproduce themselves indefinitely in the tropical forest would be to ignore the fact that these societies may be tempted to acquire new techniques capable of affecting the structure and even the ideology of such societies. They are coming into greater and greater contact with other cultures and their environment is already the result of transformations due to human activities.

The attitude of some conservationists either ignores or underestimates such a reality; it praises soft technology for the developing world as if one would limit its productive development. Thus one could end up with a situation in which the development that these young countries are aiming for is limited while at the same time their major resources remain free to exploitation by a few major foreign interests.

The tropical forest world is no longer sheltered from change and its transformation is inevitable. Aspirations for development are especially evident in the millenairist beliefs of the cargo cult type present in some parts of the tropical forest world (Worsley, 1957; Lanternari, 1960) and which can often be interpreted as a desire to inherit the riches and technology of colonialists. It is therefore advisable to look for ways by which development can be undertaken within the framework of wise ecological management and by respecting the aspirations of the societies and civilizations present.

There is no longer any question about advising strongly against activities leading to ecocide or ethnocide. But, in the majority of cases, science and technology do not have sufficient knowledge of the functioning of these ecosystems as to be able always to wisely guide their development. The main outcome is usually the appropriation of spontaneous resources by outsiders, and the destruction of forest eco-systems in order to try to replace them by artificial eco-systems. In this respect the knowledge of tropical forest people may be able to make significant contribution to the development of techniques better adapted to economic development of the forest. Perhaps a better knowledge of this ethnoscience should come first together with a better knowledge of the forest societies and civilizations before technical formulae perfected in completely different ecological and socio-cultural contexts are applied to such ecosystems. Reichel-Dolmatoff (1973) sums this up, while refering to Amerindian societies: 'The prime factor, the true infrastructure for description and analysis is . . . the meaning that the environment has for the indigenous person, his understanding of the habitat.'

Culture, economy and society

Ways of using natural resources

In this review of some of the technological and cultural characteristics of tropical forest societies, the two main ways of using natural resources were underlined: food-gathering and swidden cultivation. These often co-exist within one society which may, for example, have a subsistence garden in a cleared area but will largely rely on gathering in order to ensure its subsistence. This is frequently the case in the New Guinea forests. The economic symbiosis between hunter-gatherers and forest cultivators, as for example between the pygmies and the African swidden cultivators using the forest margins, must also be mentioned.

Food-gathering in tropical forest reflects a thorough appreciation of its ecological characteristics: the various resources are dispersed and decay is very rapid. Although usable resources are relatively abundant and their production is sustained, their harvesting requires mobility and they must be used rapidly, if not immediately. The forest hunter-gatherers, such as for example the African pygmies, do, nevertheless, make use of short-term conservation processes such as the smoking of meat and of plant products such as the seeds of *Irvingia* (Bahuchet, 1972).

A characteristic feature of hunter-gatherer societies appears to be the frequent practising of ritualized control of ecosystem resources through periodic prohibitions with a magic and religious justification that affect the harvesting of particular plants or the hunting of particular animals. This corresponds to the only form of authority which is recognized in such societies, an authority represented by a senior or a group of seniors taking decisions on place and time of prohibitions. Burkill (1952) has described this system very well in the case of the hunter-gatherers of the Andaman Islands where the most favoured resources (the tubers of *Dioscorea glabra*, the hearts of the *Caryota* palm and seeds of *Entada*) are protected from over-harvesting. However, a marked preference for some resource does not necessarily mean its disappearance and may in fact encourage its reproduction: the negritos of Malaysia search to locate wild durians, *Durio zibethinus*, and the change in habitat brought about during their temporary camps beneath such fruiting trees provides particularly suitable conditions for seed germination and regeneration. The same occurs with *Metroxylon* palms of the Malayo-Oceanic region where stands have sometimes been developed due to the sago-gatherers. These people have also often eliminated spiny forms of the palm by empirical selection for they made the starch-extraction more difficult. This is, however, without bringing about a real domestication. The specialized ecosystem of these palm-groves represents such a large collection of resources that the sago-gatherers are sedentary, living in villages and trading in this spontaneously available resource. The semi-nomadic people of the tropical forests have another interesting characteristic: their non-violence in a constant search for compromise as a solution to all conflicts or through recourse to other methods of 'de-fusing' such conflicts. Turnbull (1965) has given a good example of this in his description of the role of the jester in solving conflicts between African pygmies and Dentan (1968) has given another example of non-violence in the case of the Semai negritos of peninsular Malaysia, an approach which prevented their involvement in a colonial war which was in progress within their habitat.

Another striking feature among the African pygmies and Asiatic negritos is their collective method of taking decisions governing hunting and gathering. These are based on a consensus reached after debate between all the adults of a band.

The time spent in food-gathering in general does not exceed four hours daily by each member of a band. Sahlins (1968) saw these hunter-gatherer societies as the 'first societies of abundance'. There is generally an egalitarian society which safeguards the maintenance of the ecosystem into which it is integrated, using its resources moderately and without excessive labour in order to supply the needs of both the active and inactive members of the band which constantly tries to maintain its cohesion by a process of peaceful elimination of conflicts.

The principal features of swidden cultivation in tropical forest are:
— the use of a cleared area of forest for a crop (or a limited number of crops);
— the use of fire to create this temporary clearing;
— the observation of a sufficiently long fallow in order that tree cover can re-establish.

The practice of swidden modifies the floristic composition of the forest (in many cases the regrowth of certain fast-growing tree species considered as beneficial is promoted). Nevertheless, if the rest period is sufficiently long (more than 10 years in the aseasonal tropics), the structure of the forest which regenerates is not very different from that of the original forest. Such an agricultural system essentially permits the maintenance and regeneration of the ecosystem into which it is included. Swidden has many variations throughout the tropics and Spencer has made a complete description for South-East Asia (see chapter 20). These variations are primarily ecological and based on the ease of clearing and cultivation; they occur due to biocenotic constraints (structure and composition of the forest canopy) and biotopic constraints (climate and its seasonal variations, relief, altitude, soils, water, etc.). Thus the techniques of clearing evergreen equatorial forest will be different from those for semi-deciduous forest.

Many different plants are cultivated in such clearings. There is a striking contrast in the Malayo-Oceanic region between swidden cultivators who grow plants with starchy roots or fruit almost exclusively and those who have adopted cereals such as rice or millet as their basic food. The latter have a tendency towards specialization and uniformity within their domesticated ecosystem and therefore have a more marked impact on the forest ecosystem than the former who are better integrated into it. The introduction of exotic food crops into such agricultural systems is a source of potential ecological and technological change, which can affect the economic and social pattern. This also occurs whenever a new forest resource is exploited, or over-exploited, and with the provision of new techniques. Tropical swidden has a characteristic series of cultivated plants: perennial plants with starchy roots or fruit reproduced by vegetative means (yams, *Colocasia* and other aroids and bananas for Asia and Oceania; African yams and *Xanthosoma* and cassava in tropical America). During the first years of the colonial period there was an active exchange of plant material between tropical regions. As a result of this, African yams are cultivated in the neotropics where the Asiatic and Oceanic *Colocasia* has often replaced the native American *Xanthosoma*; and the American cassava and the Asiatic banana are found throughout tropical forest regions. Some of these diffusions took place before colonial times: the Asiatic large yam (*Dioscorea alata*), taro (*Colocasia esculenta*), bananas of the *Eumusa* section and sugar-cane (*Saccharum officinarum*) all reached Africa across the Kingdom of Sheba and the Indian Ocean. These are well suited to tropical forest swidden, and *Xanthosoma* and cassava have been factors in the transformation of ecosystems in Africa, Asia and Oceania while the introduction of the banana from Asia into Africa has facilitated the agricultural utilization of the forests of this region.

All agricultural specializations within the swidden system, i.e. simplifications of the structure and composition of the vegetation in the cultivated clearing, lead to the disruption of the system. It can maintain its relative stability in so far as the swidden garden remains a generalized and diversified domestic ecosystem.

Social organization

The two alternatives of food-gathering and swidden are the only ones that are both characteristic of tropical forest ecosystems and integrated into such ecosystems. The significant factors for the social organization of human societies practising these ways of exploitation are:
— the generalized and diversified character of the tropical forest biocenosis, especially the high level of primary production, the wide-spread distribution and large number of species that are utilizable;
— the strong degree of integration between man and the ecosystem;
— the consequent need for fluidity and mobility among human groups so as to permit adjustment to the distribution of resources in time and space (for hunter-gatherers) or to the regeneration of natural conditions safeguarding continued agricultural activities (in the case of swidden);
— the perception and management of resources on an individual basis which implies a detailed knowledge both of the biocenosis and the biotopes, and are based rather on the assistance of natural productivity than on its control and modification.

The two systems of food-gathering and swidden agriculture are, as far as the human populations forming the societies of the tropical forests are concerned, both based on indirect negative action as described by Haudricourt (1962, 1964). This is in marked contrast with the direct positive action shown by the cereal cultivators or pastoralists of savanna or steppe regions. This direct positive action, with the constant tendency to increase specialization within the ecosystem, is common to agriculturalists, sylviculturalists and livestock-rearers of the neo-technical era. The human societies which are characteristic of tropical forest ecosystems have only intervened with moderation and they remain integrated with the natural environment which they assist without excessive disturbance. Nevertheless the balance is very delicate and it is only necessary to depart from the normal constraints of the ecosystem to initiate an accelerating process of change which usually degrades the habitat. The spread of savanna due to an intensification of the swidden cycle shows this very clearly.

The tropical forest thus appears to be an environment with particular constraints that only offers two alternatives to man:
— either he accepts it and works within a biocenotic and homeostatic framework in bringing about transformations, as is the case when food-gathering or swidden is integrated into the tropical forest ecosystem; or
— the ecosystem is disposed of through complete transformation.

The tools for production are relatively simple: the digging stick, hatchet or adze, trapping nets, spear and bow and arrows, etc., and natural substances such as hunting or fishing poisons, and fire. Essentially the hunter can draw on a 'primitive arsenal' of working tools in a similar way to that in which the gatherer can draw from a 'primitive provision store'. Man relies on nature to provide his tools; moreover, he is a part of this forest nature:
— either he regards it, as do the majority of hunter-gatherers, as a kind mother which should be respected; or
— he regards it as a formidable power, as do the majority of swidden cultivators, a force with which man must come to a compromise since he lacks means to master it or suppress it. At this level of ecosystem integration and with only simple tools, it is not surprising that population densities are very low: the more this density increases, the more the regeneration of these ecosystems is constrained, their transformation is accelerated and their disappearance becomes inevitable. In addition, considerable experience is necessary in the harvesting of the spontaneous resources of the tropical forest ecosystem in order to support human life as well as for the cultivation of domestic resources. Gathering or swidden practices that are well integrated into the tropical forest ecosystem require a particularly well developed knowledge of natural history. Nevertheless man cannot act or subsist alone. The sanction of the Mbuti pygmies (Turnbull, 1961, 1965) against one of their member who is guilty of a grave offense consists of expelling him from the band and is the equivalent of a death sentence;

mutual assistance is necessary for netting game or in clearing an area. This reliance on mutual aid and collaboration appears to be one of the most important characteristics of human societies in forest ecosystems. Even when there is an apparent relationship in which one society is dominated by another, as for example in the case of the African swidden cultivators and 'their' pygmy hunter-gatherers (Turnbull, 1961, 1965; Godelier, 1974), more often than not this takes the form of a mutual dupery which appears as a kind of symbiotic relationship.

The small size of bands is a characteristic feature of hunter-gatherer societies (7–30 nuclear families in the case of the Mbuti pygmies (Turnbull, 1968) as is the fluidity and mobility of such bands (Godelier, 1974) so as to allow for adjustments to variations in spontaneous resources and to different requirements for their harvesting. Lee and De Vore (1968) have summarized the characteristics of the social organizations of various hunter-gatherer societies. Their fluidity and mobility enable such bands to respect the 'constraints of individual co-operation based on sex and age in the production process (game netting, etc.)'. Godelier also remarks that in such cases 'kinship relationships may function within the group as part of the social relationships which govern the production processes' and he adds that 'this means the function of deciding on access to and control of the means of production and of social product for the groups and individuals who make up a given type of society, and also the function of organizing the work and the distribution of products'. The socio-economic aspects of hunter-gatherer and swidden societies are those of non-capitalist societies in which 'the economy does not play the same role and, as a result, does not take the same form of development' as in capitalist societies. Therefore, the classic relationship between economy and society can never be the same in such cases and may even lead to 'hindering the understanding of the internal logic of such societies' (Godelier, 1974).

Social stratification is absent among hunter-gatherers where care has been taken to avoid the development of any individual authority. Turnbull (1961, 1965) has given a good description of this situation and attitudes among the Mbuti pygmies of Zaire. The same remarks apply to the hunter-gatherers of South-East Asia: Forde (1964) has noted that among the Malaysian negritos there is no individual authority or leadership system other than that of recognizing the experience and knowledge of the elders. Among the hunter-gatherers of the Andaman Islands, this simple acknowledgement of the ethnobiological knowledge of the elder members of a group gives them the right to prescribe the periodic prohibitions which protect some spontaneous resources from possible over-appropriation. In all these hunter-gatherer societies, decisions are taken collectively through a discussion in which all members participate. This also demonstrates the 'non-violent' attitude and a regard for social stability which is characteristic of these food-gatherers.

The situation is markedly different among swidden cultivators although some of their societies (especially in Melanesia) have developed some concern for equality. Once more it is difficult to generalize. The type of leadership system seems similar to that described by Hogbin (1943–44) for the Melanesians of northern Malaita in the Solomon Islands. The leader or centre-man is considered by the others 'like the banyan fig which although it is the largest tree, is still a tree like the others but due to its size and volume it can support more epiphytes and lianas, feed more birds and give better protection from the sun and rain'. This botanical imagery as used by the Melanesians appears to give a good definition of the *primus inter pares* of this type of leadership. Moreover, Sahlins (1963) has given an interesting description of this type of leadership and its development and functioning in Oceania. Harris (1972) summarizes the social organization of tropical swidden cultivators: 'The characteristic pattern of social organization among swidden cultivators is that of simple segmentary tribes living as decentralized autonomous communities in small dispersed settlements. The transition from this pattern to one of dependent peasantry under centralized control is a critically difficult one which populations wholly dependent on swidden cultivation appear unable, or at least most unlikely, to make . . . Other social factors reinforce the characteristic swidden pattern of decentralized autonomous communities and present the reorganization of simple kinship groups into conical clans or pyramidal chiefdoms that might permit the development of social stratification and centralized control.' Exceptions, or apparent exceptions, to this will be considered later.

Comparison of hunter-gatherers and swidden cultivators

The fundamental difference between the *appropriation of spontaneous resources* as practised by hunter-gatherers and the *production of domesticated resources* by swidden cultivation lies in a major change in the relationship between man and the other components of the ecosystem to which he belongs. The importance of this change in human history is obvious and Gordon Childe (1942) has used the phrase 'neolithic revolution' in the case of plant and animal domestications in the Near and Middle East. Nevertheless, these two means have similarities. Not all forest-based societies of the tropics are the result of a long period of association with the forest environment. Some human groups originated in other environments in spite of the fact that they now practise a food-gathering economy or a limited form of swidden cultivation or even a combination of the two. Where such people are relative newcomers, they may regard the forest as an area they are passing through. These groups often behave in a way that reflects the attitudes, and social and economic systems developed in surroundings that are ecologically different. This is the case with Amerindian societies in the neotropical forests as well as with some swidden cultivating groups in the mountains of South-East Asia. Their socio-economic heritage prevails against that of their present natural environment.

Thus it seems preferable to devote more attention to societies and civilizations that have had a long period of life in the forest as shown by their degree of integration into a relatively stable forest ecosystem. Societies that truly belong to tropical forest ecosystems do not cause problems concerning the maintenance and regeneration of the surrounding ecosystem.

The relationship between man and the ecosystem can justifiably be expressed by ecological concepts such as the home range, or territory and spacing. This is especially true of home range—the area within which an animal moves while engaged in its daily activities (Smith, 1966) and is without rigidity in terms of its uses, size or function. Perhaps this term is the best description of the characteristics of an area used for hunting and gathering or for swidden cultivation rather than conceptions based on sedentary agricultural situations.

Food-gathering and swidden cultivation are both accompanied by constraints that require fluidity and some divisibility within the human groups concerned. As soon as societies beome rigid there is a risk of breaking the relative stability which they express and which permits restrained assistance given to nature through indirect negative action (Haudricourt, 1968). The term territoriality can perhaps help to explain certain behavioural characteristics of hunter-gatherers and swidden cultivators and can justify the sociobiological viewpoint expressed recently by Wilson (1975) at least for the food-gathering societies (see also Peterson, 1975). This territoriality of the food-gatherers is in no way incompatible with their non-violent attitudes. In many cases, as for example among the African pygmies, an advertising behaviour rather than actual violence is developed to defend the territory (Odum, 1971). Social solutions to problems of adjustment to the environment (through the divisibility and fluidity of bands and of groups) do not enable these societies to be considered strictly in terms of animal behaviour. The home range covered by swidden cultivators in New Guinea during the course of their slash-and-burn cycle is not considered as falling within definitive limits and there is also an implication in terms of spacing between neighbouring human groups so that there is a kind of buffer zone within which they feel the effects of some kind of territorial competition. Although freedom of movement may be limited by other human groups, this is not accompanied by an acceptance of a relatively settled existence and the home range is considered as potentially capable of extension. The Tsembaga rituals in New Guinea (Rappaport, 1971) appear to express this kind of perception of the home range. It can also explain remarks about swidden cultivation in tropical forests under-exploiting an area (Carneiro, 1960; Harris, 1972). This also shows the difficulties in trying to define the carrying capacity in such situations (Brush, 1975). This outline of territorial behaviour implies that the food-gatherer or swidden cultivator plays his biocenotic role while maintaining himself beneath the highest possible threshold for use of resources. An example given by Carneiro (1960) illustrates this point. A Kinkuru village in the Upper Xingu region of Brazil has:

— a population of 145 that exist through swidden cultivation in an area that could support around 2 000 swidden cultivators;
— this population could settle itself by using only 7 per cent of the cultivable land that is relatively close to the village;

— it would take 400 years to cultivate and exhaust the soil of the total available cultivable land.

In spite of Carneiro's argument being based on incomplete data (see Harris, 1972), it is interesting to consider an example of tropical forest swidden that can use at least 25 year fallows without using all its area. This is not always the case and in the example of the Tsembaga (Rappaport, 1971) in New Guinea, there is a group of swidden cultivators that are always ready to conquer the land belonging to other groups and are always open to similar attacks from neighbours. In this case they have almost reached a threshold beyond which the balance between swidden cultivation and the ecosystem may be broken because of the pressure exerted on primary production by the Tsembaga population as well as that of their herds of pigs.

To summarize, it seems that the food-gathering and swidden are integrated into the tropical forest and man takes advantage of natural energy sources but does not control them. In the case of swidden he attempts to direct natural production but in neither case does he interfere in the functioning of the natural ecosystem so its regeneration is affected. If this should happen, it results in the long-term degradation of the natural environment so that the human activities which initiated it will no longer be possible or the ecosystem will be replaced by an artificial one controlled by man using different means.

Ecological constraints

For such an ecosystem to regenerate there must be a maintenance of its biocenotic diversity. The forms of swidden that are integrated into the tropical forest are always those with a large diversity of species and varieties and a structure similar to that of the surrounding ecosystem. Geertz (1963) has given a good description of this in the Indian Archipelago and Harris (1971) for the polycultural conuco in the forests of the upper Orinoco in Venezuela. As soon as there is any specialization in crops or a tendency towards monoculture, aggression against the forest ecosystem is increased and a plagiosere commences which implies the progressive disappearance of the forest environment.

The most important biotope constraint seems to be imposed by the soil. A very long fallow regenerates soil fertility. Undoubtedly it is the protection of the soil from insolation and from precipitation through the complex structure and species diversity of the domesticated swidden ecosystem, that is of major importance. This is especially important in preventing lateritization. Harris (1971) has given a good description of this from the Orinoco forest: 'The positioning of plants in the conuco followed no regular plan but was guided by the necessity of avoiding tree stumps, felled trunks and other forest debris that remained after clearing and burning. The effect of this apparently haphazard pattern of cultivation was to leave little bare soil exposed to the direct effect of insolation and raindrop impact . . . The interplanting of species with different growth habits and root systems—trees, shrubs, climbing and sprawling plants, root and fruit crops—also ensures effective vertical and lateral exploitation of available light, warmth, moisture and nu-

trients. In other words, by substituting a diverse assemblage of cultivated plants for the wild species of the forest this type of polycultural conuco simulates much more closely than the monocultural plots do the structure and dynamics of the natural forest ecosystem.'

Moreover, this biocenotic diversity in terms of species and varieties considerably limits the impact of predators and parasites. Barrau (unpublished) has shown this many times in Melanesia where all attempts at monoculture have resulted in pest outbreaks.

The climate of the forest causes rapid decomposition of collected or harvested products and this considerably limits storage. Continuous production of both cultivated and wild plants renders this unnecessary. The dispersion of available wild resources requires human groups to maintain mobility and fluidity so that they can adjust constantly to the distribution of such resources.

Other factors that may affect the movement of food-gatherers include the availability of running-water. For example, in Africa there are some areas that are not frequented by pygmies because they lack running-water or because the water that is present is considered to be of bad quality. In other regions the distribution of human populations appears more linked to the presence of watercourses which form a means of communication and a source of fish food. Sauer (1952) and Burkill (1960) have drawn attention to the role of watercourses and of the ecotones between the river-banks and the forests in the initial process of human settlement and plant domestication in South-East Asia.

The nature of the tropical forest therefore conveys to the food-gatherer or swidden cultivator a requirement for mobility over a relatively large home range in order to:

— be able to draw on resources scattered in a very diversified biocenosis, in the case of food-gatherers;
— safeguard the regeneration of the plant canopy between brief periods of swidden, in the case of swidden cultivators.

There are exceptions to this, especially the sago collectors of the *Metroxylon* swamp forests; the abundance and concentration of the resources permit a harvest of food that is easily above subsistence requirements and allow for trading of the surplus. This has enabled their settlement in large villages.

Social outcome of such constraints

Among food-gatherers, as for example the African pygmies, the area used for collecting defines a band and membership of such a band gives access to resources. But this requires mobility and fluidity. The pygmy bands that belong to the same group are subject to constant fission and fusion in such a way that none of them ever has the same composition. What is the magnitude of this mobility and fluidity? Turnbull (1961, 1965) described a group that was confined to a decreasing area of forest and the densities (bands of 230 people within 1 200 km²) appear high for a hunter-gatherer society.

In a large part of the African forest, in the frontier zone between the Central African Empire, the Congo and Cameroon, at least five pygmy groups exist. According to

Demesse (personal communication) these pygmies move over hundreds of kilometres and if there is a constant transfer between bands within the same pygmy group, there may be transfers between different groups. There remains this high degree of mobility and extreme fluidity resulting from the direct relationship with the natural environment and accessibility to its resources. Co-operation is necessary in harvesting; hunting with a net requires at least 7 and up to 30 nets, each operated by a married man, and this seems to be the characteristic means of production of these pygmy societies and implies the association of a number of nuclear families. According to Godelier (1974) this shows simultaneously the constraint of co-operation and that of flow and of the open nature of the bands. The last constraint is that members of a band have no right to a defined area and its resources. The constraints of flow, fluidity and openness allow for the adjustment of the numbers of the band to the variations in local resources; that of the membership of the band expresses the right to means of production.

This is a good example of the minimal size of a system of harvesting spontaneously occurring resources within a tropical forest ecosystem.

According to Harris (1972) swidden, integrated into an ecosystem, consists of relatively close hamlets or dispersed villages with populations ranging from 50 to 250 people. The population of swidden villages in tropical forests of South America ranges between 50 and 150 persons and the hamlets of the Isneg swidden cultivators of the Philippines average 85 people. The Akawaio swidden cultivators of Guyana live partly in a central village which is also the ceremonial centre and includes 20–60 people; in part they live in country houses distributed around this centre, each occupied by an extended family of about a dozen individuals. Membership of a swidden group can be interpreted in terms of the right to use a portion of the collective cultivated area. This implies mutual assistance and Condominas (1974) has given a good example of such a system in his study of the Mnong Gar in Viet-Nam. Matras-Troubetzkoy (1974) has given a detailed description of collective land management by the Brou swidden cultivators of Kampuchea.

Densities vary according to the region, the type of forest and its productivity. In the case of swidden societies they are most frequently lower than 40/km² and it is not exceptional to find densities of 1–5/km². These world-wide averages are confirmed by Harris (1972). The average density of hunter-gatherers in the generalized tropical forest ecosystems is *ca.* 0.005–0.12/km². All higher figures, as for example the Mbuti studied by Turnbull (1961), are undoubtedly an indication of a territory that has been considerably reduced due to external pressure. Population densities of settled collectors, such as those making use of the sago palm-groves, will inevitably be much higher.

The forest, an obstacle to social stratification?

Harris (1972) refers to a 'failure (of swidden) to support complex societies and concentrated settlements'. This remark also applies to food-gathering societies of generalized tropical forest ecosystems. It does not apply where gathering is practised in ecosystems with abundant and concentrated resources, as the sedentary sago-gatherers collecting from *Metroxylon* palm-groves in New Guinea. Harris also remarks: 'In view of all the ecological and social factors conspiring to limit the evolutionary potential of swidden it would seem right to conclude that, when practised as the only form of subsistence, it is incapable of sustaining civilization or indeed any level of socio-economic complexity above that of autonomous tribal communities in small dispersed settlements.' There are some notable exceptions to this, especially that quoted by Harris of the theocracy of the Maya who had successfully, if only temporarily, imposed and developed a social system with centralized control that ruled a swidden peasantry. It seems as if the latter had an agricultural activity based on a diversity of resources: cereal (maize), tubers (sweet potato and cassava?), and fruit (especially the *peji-baye* palm, *Guilielma gassipaes* and *ramon, Brosimum ali-castrum*). In refering to this Maya theocracy, Godelier (1974) writes: 'The example of the Maya enables to underline the large diversity in the productive capacity of a slash-and-burn agricultural system. It has been calculated, twenty years ago, that this system enabled a family of five to obtain its subsistence thanks to 65 days work during the year. A potential surplus of available labour therefore existed and the problem of mobilizing this was a social problem which depended on the social relationships of production and on the existence of dominant social classes that were not directly productive . . . In order to understand the dynamic relationships that exist between the societies and the various forest ecosystems that they exploit, it is necessary to analyse their various possibilities of intervention on these ecosystems, the social conditions under which such interventions take place as well as defining the social destination of usable products obtained from nature. Consequently there must be both an effort to establish what economists call the specific production functions of each system of productive forces and social relationships as well as that of analysing the conditions in which such systems could be maintained or fail to maintain themselves, taking into account both their internal structure and the internal variations in operating conditions.'

This appears relevant to all studies undertaken on societies of the tropical forest. In the example of the Kachin swidden cultivators of Burma that have been studied by Leach (1954), Godelier (1974) has drawn attention to the fact that 'one finds societies in the heart of which alternate a type of rural organization without internal hierarchies and another with hierarchical chiefs at many levels and in permanent competition. These encourage production of greater and greater surpluses with a result that there are increasing pressures on both natural and human resources. The Kachin that have been studied by Leach would appear to have a permanent fluctuation in social organization between a democratic (*gumlao*) form and an aristocratic form (*gumsa*), the development of which comes to a climax in crises in which the power of the nobles collapses and is replaced by a rural democratic society'. It could be questioned whether, in the case of the Mayan theocracy, a similar kind of process could explain its collapse: greater and greater pressure on natural and human resources could have led to a break in the system and thus its failure to reproduce. In fact, the Maya example is undoubtedly a more complex one since recent studies (Denevan, 1970; Siemens and Puleston, 1970) have brought to light traces of ridged fields in the region occupied by the Mayan theocracy and this implies the practice of an intensive agriculture which is completely different from swidden.

It cannot be stated that the tropical forest or the practice of swidden performed in it are obstacles to social stratification. In contrast, the most significant feature is the extent to which man is freed from natural constraints; if a human society is confined to a generalized forest ecosystem into which it is ecologically integrated through food-gathering or swidden practices that permit the essential regeneration of the ecosystem, the constraints imposed through a need for fluidity and mobility do, as a general rule, lead to a 'failure to support complex societies and concentrated settlements' (Harris, 1972). If, on the other hand, man causes the transformation of the generalized forest ecosystem into a domesticated ecosystem or if a society has access to a diversity of ecosystems that it uses in different fashions, there is then a strong possibility that social stratification will become more marked (see Sahlins, 1958, for the island ecosystems of Oceania). Mobility means also immigrants bringing technological, social and cultural habits. The immigrant either transforms the forest ecosystem or adapts to his new environment through adopting practices such as food-gathering or swidden that are relatively well integrated into the ecosystem. Here again the tropical forest environment only gives man two alternatives: to dispose of it or to fit in it.

The examples of the Maya and the Kachin have defined the limits of swidden to support a social system that requires an increasing surplus production. Nevertheless, to state this in purely quantitative terms is dangerous since everything will depend on the nature of such surplus and especially whether it leads to trade which would open the ecosystem to the exterior. In conditions that are certainly somewhat different from those of tropical forest, studies by MacNeish (1964) have shown the significance of spiritual power or temporal power in its relationships with the evolution of use of environmental resources and their allocation in Mexico: in the region studied, agglomerations grouped around ceremonial centres were followed by commercial cities secularized and the transformation of the earlier system continued. Possibly the Mayan theocracy found itself in some kind of sclerosis because it did not have adequate trade with the outside world or goods suitable for such a trade. This brings attention back to the example of the sago starch collectors of the specialized *Metroxylon* palm-grove ecosystem where the gathering of an abundant spontaneous resource as well as the nature of such a resource has enabled settlement and production of a surplus which has been exchanged for a long time with other regions of New Guinea. This shows the importance of the qualitative factor in considering the evaluation of the natural resources of a given environment. If the social, economic and cultural consequences of plant domestication in the fertile crescent of the Near and Middle East were very different from those that occurred in the tropical regions of Asia, this might be due

to the cereals being brought into cultivation in a specialized domesticated ecosystem that resembled the surrounding ecological conditions whereas in Asia the cultivation of tuberous and fruit-bearing plants had a similar structure to the surrounding forest. In the former the yield was a dry commodity capable of being subdivided, stored and traded; in the latter the yield was a vegetable which although well suited to a subsistence economy, could never be used for trade but could, on the other hand, be used in acquiring prestige (as is shown by the significance of accumulations of yams in some societies of Oceania).

Whatever the value of these hypotheses, the fact remains that, because of the way that human societies are integrated into and confined within the tropical forest ecosystem, it does not appear that they have ever reached the level of complexity of other societies. In cases in which complex societies have developed these have not been able to exceed a certain threshold imposed by the material foundation of the social structure, which then collapsed; otherwise, human mastery over the forest ecosystem was exercised to such an extent that the ecosystem is transformed and artificial or domesticated ecosystems are substituted and one is no longer dealing with a forest ecosystem *sensu stricto*.

Socio-cultural consequences of ecosystem transformation

Societies of food-gatherers and swidden cultivators that are integrated into the tropical forest ecosystem represent a baseline from which to measure the extent and nature of changes arising from other interventions. These changes lead from the natural ecosystem in which groups of men live by hunting and gathering or swidden cultivation to synthetic ecosystems (plantations or ranches). As soon as there is non-specialized collecting, human impact will be felt through assistance in propagating some harvested species. Thus, in the Melanesian forest, wild yam stands become gradually extended to form accidental gardens because of digging by collectors (Barrau, 1967) and the example of the durian in peninsular Malaysia has already been cited. In fact, as food-gatherers in tropical forests always practise some kind of discrimination between numerous resources of the biocenosis, they intervene in resource management and manipulate natural production to their own advantage, whether this by periodic prohibitions or by harvesting of particular components of the biocenosis. But this assistance does not modify the principal characteristics of the forest ecosystem (its high diversity, relative stability, sustained primary productivity, complex structure, etc.). This also holds true for swidden with long fallow. The swidden cultivations are synthetic ecosystems but are only temporary and allow the natural regeneration of the forest. In these examples, there is only a manipulation of the functioning of the natural ecosystem.

The critical threshold for transformation comes when there is a plagiosere which corresponds to a process of ecosystem specialization or when there is the sudden replacement of a generalized forest ecosystem by a specialized ecosystem (see chapter 20).

Historical aspects of transformations

It would be possible to prepare maps showing the inexorable reduction in extent of the tropical forest occurring everywhere that would be similar to those prepared by Spencer (1966) for South-East Asia and the Indian Archipelago (see chapter 1). Semi-deciduous or deciduous tropical forest or tree or grass savannas usually represent a 'plagio-climax community whose existence has been permitted by a degree of burning and human interference' (Eyre, 1963). The intrusion of techniques arising from industrialization and the increase in commercial demands during colonial times has considerably accelerated this process, while the justifications of technocracy affirm that it is necessary to replace such forest ecosystems by artificial ecosystems that are as profitable as they are productive. Also there is the recent destruction of forest for military reasons, especially in Indo-China.

Among the principal causes of the retreat of the forest was that of the demand for food, especially cereal cultivation. The socio-economic aspects must be considered, especially where complex, stratified societies required the production of a surplus in order to maintain and reproduce themselves. Commerce began to have its effect, even on food-gathering, long before colonial times, e.g. cardamom harvesting in Kampuchea (Martin, 1974) and its commercialization through a centralized power. Forest gathering for commerce only has a limited effect on the forest ecosystem as can be seen in the case of chicle in neotropical forests. Forest exploitation markedly increased since colonial times as has the development of plantations (see chapter 20).

Thus, the tranformation of tropical forest ecosystems only happens when the harvesting of spontaneous resources or production of domesticated resources is not limited to simple manipulation of the natural system. As soon as there is human pressure on such resources or demands for production of a surplus to serve requirements other than food or because of a combination of these reasons then such transformations occur.

Endogenous and exogenous transformations

Human societies that are peculiar to tropical forest ecosystems change more due to external pressure than because of their own dynamics. They are conservative societies as to their way of insertion in the natural environment (Harris, 1972) and their density remains stable as long as outside factors do not intrude. But they are not rooted in a permanent homeostatic situation. They develop and have developed techniques and, even in the case of food-gatherers, can have an impact on the ecosystem. The development of accidental or assisted orchards results in an increase in the harvest of this resource and this can increase its extent.

The rhythm of such change is slow and should not be compared to that arising from external change due to the acquisition of new tools for harvesting or production, the introduction of exotic food plants and the requirements for commercial surplus, the demands of work for others, or the imposition of new beliefs, etc.

For example the Central African pygmies were entrusted with fire-arms by the groups cultivating the margins of the forest so that they could hunt for them as part of their symbiotic relationship. The result was an overcropping that has considerably reduced the game resources of the pygmies and resulted in either a retreat to the heart of the forest in order to free themselves from these symbiotic constraints, or in the progressive destruction of their culture. Another example is that of Melanesian swidden cultivators converted to christianity by missionaries who resettled them in coastal villages near to the church; they were destroyed by malaria and their customary food sources were compromised because of the distance from the area that was traditionally cleared for cultivation. The introduction of metal tools provides other good examples of transformations brought about by the introduction of external technology that enables the environment to be used much easier than before. The same thing happened in the case of the banana which facilitated agricultural penetration of the African forest. Cassava and *Xanthosoma* from neotropical regions have played a similar role throughout the Old World tropics.

The greatest changes that affect the forest ecosystem and its human populations arise from external forces that intervene directly in the ecosystem in order to gain a profit. They can alienate the land for development or require the forest populations to produce a surplus or new crops. The recent eviction of the Dayak populations from the forests of Kalimantan now under exploitation is a typical case of what is also taking place within the neotropical forests where large ranches and plantations are being created. The recent military aspects of such transformations had effects that were as ecologically and economically disastrous as they are for health; Viet-Nam provides the most tragic example. The tropical forest and its human populations have rarely been treated with consideration by those desiring resources.

It would be injust to not recognize the agricultural and sylvicultural development which took place in, or at the expense of, the tropical forest. It must be regretted that a simple technical formula is generally applied, which consists of clearing the forest in order to replace it with a domesticated ecosystem, in which homogeneity makes exploitation easier, without looking for alternative solutions based on rational management of the natural ecosystem.

Immigration

The human populations present in tropical forests are generally inadequate to participate in the development of their natural environment and their low population density is another obstacle to such a development. The mobility, dispersion and fluidity within such people has often led to negative reactions from central authorities. There is a tendency to try to settle and regroup them.

Therefore new activities often draw on non-forest manpower. Thus, throughout the large forest of central Africa the role played by pygmies in the labour force is infinitely small. In the coastal plantations of New Guinea work was largely by mountain people whereas in New Britain this was performed by New Guineans from the Sepik valley. Amerindians never participate in the development of the

Amazonian forest, and the development of the forest areas of Indonesia has been largely based on attempts at colonization by displaced island peoples.

The social problems caused by such transplanted manpower, leaving aside autecological, physiological and health aspects (see chapters 16, 17 and 18), are relevant to the politics and sociology of the salaried or indented work. It is interesting to note the resulting tensions which arose from the polyethnic composition of the displaced and re-grouped people. The incidence of such tensions between immigrant labour and indigenous populations should also be noted. An example is the violent conflict between labour coming from the Sepik and the local Tolai population that took place in Rabaul in New Britain.

The inevitable effects of alienation of land for the establishment of new activities can no longer be hidden. This is bound to be seen by the forest population as a violation of their home range and as a threat to their livelihood and social structure. These alienations cannot but provoke changes in the human societies: when confronted by external aggression the pygmies either flee to the heart of the forest or gradually become included within the populations of the forest margins. This pattern recurs in many other cases of food-gatherers and swidden cultivators. The reduction of the home range of these cultivators often leads to increasing internal tensions and increase in their aggression against neighbouring groups. Many go to seek better luck in towns but continue to be supplied with food from the forest and may gradually begin trading with goods received from the forest, thus increasing pressure on its resources and sometimes accelerating the process of transformation of production techniques.

Research needs and priorities

The critical threshold beyond which the maintenance and regeneration of such natural systems and their human societies are no longer possible needs to be determined. So do the ways of transforming such societies, as a result of the constraints exerted on them, when the neo-technological exploitation of their natural environment is taking place.

It is therefore important to study societies of hunter-gatherers or swidden cultivators that have persisted up to the present day and the processes of their transformation. Societies of immigrants that were or are being recreated in the forest (for example the Boni of Guyana) may supply useful information on the processes of adaptation to a forest environment. In fact all such studies should take into account the societies in question, external interventions in the forest ecosystem for short-term gains and the long-term policy of development, management and protection of the resources.

It is essential to study the knowledge of the natural environment of these forest societies. This is also the best way of understanding the environmental perception by such populations. The major problem is that of integrating natural and social sciences. Useful information may be obtained from work completed by the American schools of

anthropology with their ethnoscientific and ecological approach (Berlin, Conklin, Rappaport, Vayda, etc.). Methodological adjustments would be necessary when considering the social relationships of production since the relationship between man and his environment cannot be understood without taking into account the relationships between men themselves. Inevitably this leads to consideration of the hierarchy of constraints that are exerted in terms of the maintenance, reproduction and transformation of such societies (Godelier, 1974).

It is indispensable to approach such studies from a micro-geographical or micro-ecological basis in order to avoid generalizations that do not take into account variations in the natural environment (in its resources and in traditional means of harvesting or production) which are important in understanding technological, economic and social phenomena. This leads to a refinement in ethnographic and ethnobiological observation that is the only way of taking into account the peculiarities of each case as well as in recommending an integrated approach by specialists of both the natural and human sciences.

It is first necessary to collect precise data on the infrastructure of the human societies. Too many anthropological studies have been based on insufficient knowledge of the natural environment and the ways in which man is inserted in it. The ethnoscientific approach, making use of the local knowledge, is highly recommended. It is essential to link the study of social relationships to that of development of productive forces: the evolution of the technology of a given society may clarify the process of transformation of the relationships between its component peoples and even supply an element of understanding of their beliefs.

The introduction of a new resource in the economy of a society, as for example, that of an exotic cultivated plant (e.g. the banana in Africa) may be a powerful factor in the human penetration and impact on the ecosystem. The adoption of fire-arms by tropical forest hunters will have similar effects which may be felt at the social level.

Conclusion

The possible, and often desirable, transformation of the economies of tropical forest societies requires the inclusion of new resources (improved food plants for example that are compatible with their habitual diet; tools or techniques should similarly be adapted to their customs and their natural environment). Such transformations would not suddenly or fundamentally disturb the personality of these societies nor that of their environment, and would allow progress to which they may legitimately aspire and which their countries consider desirable (Greenland, 1975).

A major role in the inevitable development of their habitat may be taken by the inhabitants of the tropical forest with beneficial results. The pygmies could become the shrewd managers for utilization of the forests in which they live and for which they undoubtedly have the best knowledge. Attempts have not been made to perfect techniques adapted to the various types of tropical forest and of human societies within it. It is not utopian to think that subsistence swiddening cultivators could gradually become planters. It is no more utopic than visualizing the transformation of hunter-gatherer societies into societies of cultivators or even to societies utilizing the commercial resources of their forests. To say the contrary would be to deny what has taken place elsewhere in the world, among other human groups.

These possibilities imply study and investigations since

the present characteristic of tropical forest development is to generalize the application of technical solutions which do not take into account either the natural or the human elements of the local conditions.

In this perspective, the inventory and evaluation of the resources of such environments need to rely on the knowledge of the forest peoples and this underlines the interest and importance of ethnoscientific studies.

Bibliography

BAHUCHET, S. Étude écologique d'un campement de pygmées Babinga. *Journal d'Agriculture tropicale et de Botanique appliquée* (Paris), vol. 19, n° 12, 1972, p. 509–559.
——. Rapport d'une mission effectuée en saison sèche en Lobaye (République Centrafricaine). *Journal d'Agriculture tropicale et de Botanique appliquée* (Paris), vol. 22, n° 4–5–6, 1975, p. 177–197.
——. Ethnozoologie des pygmées Babinga de la Lobaye (République Centrafricaine). In: *L'homme et l'animal* (comptes-rendus du 1er Colloque d'Ethnozoologie), p. 52–61. Paris, IES, 1975, 644 p.
BARRAU, J. *Les plantes alimentaires de l'Océanie, origines, distribution et usages*. Marseille, Musée Colonial de la Faculté des Sciences (fascicule unique des *Annales du Musée Colonial de Marseille*, 7e série, vol. 3–9, 1951–1961), 1962, 275 p.
——. De l'homme cueilleur à l'homme cultivateur. *Cahiers d'Histoire mondiale* (Neuchâtel), vol. 10, n° 2, 1967, p. 275–292.
——. Culture itinérante, culture sur brûlis, culture nomade, écobuage ou essartage? Un problème de terminologie agraire. *Études Rurales* (Paris), 45, 1972, p. 99–104.
——. *Environnements naturels, sociétés humaines et développement en Papua-Nouvelle Guinée. Guide bibliographique*. Paris, Maison des Sciences de l'Homme, Unité de documentation et de liaison sur l'écodéveloppement, 1975, 71 p.
BLACKWOOD, B. Use of plants among the Kukukuku of south-eastern central New Guinea. In: *Proceedings 6th Pacific Science Congress*, 1940, p. 111–119.
BROOKFIELD, H. C. (ed.). *The Pacific in transition: geographical perspectives on adaptation and change*. London, Arnold, 1973, 332 p.
——; BROWN, P. *Struggle for land: agriculture and group territories among the Chimbu of the New Guinea highlands*. London, Oxford University Press, 1963, 193 p.
BRUSH, S. B. The concept of carrying capacity for systems of shifting cultivation. *American Anthropologist*, 77, 1975, p. 799–811.
BURKILL, I. H. Habits of Man and the origin of the cultivated plants of the Old World. In: *Proceedings of the Linnean Society of London*, 164, 1952, p. 12–42.
——. The organography and the evolution of the Dioscoreaceae, the family of yams. *Journal of the Linnean Society* (London), *Botany*, 56, 1960, p. 319–412.
CARNEIRO, R. L. Slash-and-burn agriculture: a closer look at its implications for settlement patterns. In: Wallace A. F. C. (ed.). *Men and cultures*. Philadelphia, University of Pennsylvania Press, 1960, 810 p.
——. Slash-and-burn cultivation among the Kinkuru and its implications for cultural development in the Amazon basin. In: Wilbert, J. (ed.). *The evolution of horticultural systems in native South America*, p. 47–68. Caracas, Anthropologica, supplement 2, 1961.

CENTLIVRES, P.; GASCHE, J.; LOURTEIG, A. (eds.). *Culture sur brûlis et évolution du milieu forestier en Amazonie du Nord-Ouest* (Actes du Colloque de l'Institut d'ethnologie de Neuchâtel, 6–8 nov. 1975). Bulletin de la Société suisse d'Ethnologie (Basel), numéro spécial, 1975, 171 p.
CHAGNON, N. A. *Yanomamö: the fierce people*. New York, Holt, Rinehart and Winston, 1968, 142 p.
CHILDE, V. Gordon. *What happened in history*. Harmondsworth, Penguin Books, 1942; new edition 1964, 304 p.
CONDOMINAS, G. *Nous avons mangé la forêt*. Paris, Mercure de France, 1957, 491 p.
——. L'entraide agricole chez les Mnong-Gar (Proto-Indochinois du Vietnam central). *Études Rurales* (Paris), numéro spécial (*Agriculture et sociétés en Asie du Sud-Est*), 53–56, 1974, p. 407–420.
CONKLIN, H. C. An ethnoecological approach to shifting cultivation. *Transactions of the New York Academy of Science* (New York), vol. 2, no. 17, 1954, p. 133–142.
——. *The relation of Hanunóo culture to the plant world*. New Haven, Yale University, Ph. D. thesis, 1954, 411 p. Ann Arbor, Michigan, University Microfilms, 1967, n° 67–4119.
——. Shifting cultivation and the succession to grassland climax. In: *Proceedings 9th Pacific Science Congress* (Bangkok), 7, 1957, p. 60–62.
——. *Hanunóo agriculture: a report on an integral system of shifting cultivation in the Philippines*. Rome, FAO Forestry Development Papers no. 12, 1957, 209 p. Reedition, Northford, Connecticut, Elliot's Books, 1975.
——. The study of shifting cultivation. *Current Anthropology* (Chicago), 2, 1961, p. 27–61.
——. Ethnobotanical problems in the comparative study of folk taxonomy. In: *Proceedings 9th Pacific Science Congress* (Bangkok), 4, 1962, p. 299–301.
DENEVAN, W. M. Aboriginal drained field cultivation in the Americas. *Science*, vol. 169, no. 3946, 1970, p. 647–654.
DENTAN, R. K. *The Semai: a non-violent people of Malaya*. New York, Holt, Rinehart and Winston (Case studies in cultural anthropology), 1968, 110 p.
DOURNES, J. Bois-bambou (Köyau-ale): aspect végétal de l'univers Joraï. *Journal d'Agriculture tropicale et de Botanique appliquée* (Paris), vol. 15, n° 4–11, 1968, p. 89–156 et 369–498.
——. Chi-Ché: la botanique des Srê. *Journal d'Agriculture tropicale et de Botanique appliquée* (Paris), vol. 20, n° 1–12, 1973, p. 1–189.
——. Le milieu Joraï: éléments d'ethno-écologie d'une ethnie indochinoise. *Études Rurales* (Paris), numéro spécial (*Agriculture et sociétés en Asie du Sud-Est*), 53–56, 1974, p. 487–503.
DURHAM, W. H. The adaptive significance of cultural behavior. *Human Ecology* (New York), vol. 4, no. 2, 1976, p. 89–121.

EYRE, S. R. *Vegetation and soils: a world picture*. London, Arnold, 1963, 324 p.

FORDE, C. Daryll. *Habitat, economy and society: a geographical introduction to ethnology*. London, Methuen, 1964 (original edition, 1934), 500 p.

FOX, R. B. The Pinatubo negritos: their useful plants and material culture. *Philippine Journal of Science* (Manila), vol. 81, no. 3–4, 1971, p. 173–414.

GEERTZ, C. *Agricultural involution: the process of ecological change in Indonesia*. Berkeley and Los Angeles, University of California Press, 1963, XX, 176 p.

GODELIER, M. Anthropologie et biologie: vers une coopération nouvelle. *Revue Internationale des Sciences Sociales* (Paris), vol. 26, n° 4, 1974, p. 666–690.

GOUROU, P. La civilisation du végétal. *Indonésie*, vol. 1, n° 5, 1948, p. 385–394.

GREENLAND, D. J. Bringing the green revolution to the shifting cultivator. *Science* (Washington), vol. 190, no. 4217, 1975, p. 841–844.

HARLAN, J. R. *Crops and man*. Madison, Wisconsin, American Society of Agronomy, 1975, 295 p. Comprehensive bibliography.

HARRIS, D. R. Agricultural systems, ecosystems and the origin of agriculture. In: Ucko, P. J.; Dimbleby, G. W. (eds.). *The domestication and exploitation of plants and animals*, p. 3–16. London, Duckworth, 1969, 581 p.

——. The ecology of swidden cultivation in the upper Orinoco rain forest, Venezuela. *Geographical Review*, vol. 61, no. 4, 1971, p. 475–495.

——. Swidden systems and settlement. In: Ucko, P. J.; Tringham, R.; Dimbleby, G. W. (eds.). *Man, settlement and urbanism*, p. 245–262. London, Duckworth, 1972, 979 p.

HAUDRICOURT, A. G. Domestication des animaux, culture des plantes et traitement d'autrui. *L'Homme*, vol. 2, n° 1, 1962, p. 40–50.

——. Nature et culture dans la civilisation de l'igname: origines des clones et des clans. *L'Homme*, vol. 4, n° 1, 1964, p. 93–104.

——. La technologie culturelle: essai de méthodologie. In: Poirier, J. (ed.). *Ethnologie générale* (Encyclopédie de la Pléiade, 24), p. 731–822. Paris, Gallimard, 1968, 1 907 p.

——; HEDIN, L. *L'homme et les plantes cultivées*. Paris, Gallimard, 1943, 233 p.

HOGBIN, I. H. Native councils and courts in the Solomon Islands. *Oceania* (Sydney), 14, 1943–44, p. 258–283.

IZIKOWITZ, K. G. *Lamet, hill peasants in French Indochina*. Göteborg, Etnografiska Museet, Etnologiska Studien, 17, 1951, 376 p.

KAMPTO UTOMO. *Masjarakat transmigram spontan didaerah W. Sekampung (Lampung)*. Djarkarta, P. T. Penertiban Universitas, 1957.

LANTERNARI, V. *Movimenti religiosi di liberta e di salvessa dei popoli oppressi*. Milano, Feltrinelli, 1960, 39 p.

LATHRAP, D. W. The hunting economies of the tropical forest zone of South America: an attempt at historical perspective. In: Lee, R. B.; De Vore, I. (eds.). *Man the hunter*. Chicago, Aldine, 1968, 415 p.

LEACH, E. R. *Political systems of highland Burma: a study of Kachin social structure*. London, Bell, 1954; 2nd edition, 1964, 324 p.

LEE, R. B.; DE VORE, I. (eds.). *Man the hunter*. Chicago, Aldine, 1968, 415 p.

LETOUZEY, R. *Étude phytogéographique du Cameroun*. Paris, Lechevallier (Encyclopédie Biologique 69), 1968, 508 p.

LIZOT, J. Économie ou société? Les Yanomami. In: Cresswell, R. (ed.). *Éléments d'ethnologie*, vol. 1, p. 128–165. Paris, Armand Colin, 1975, 320 p.

——. *Le cercle des feux. Faits et dits des Indiens Yanomami*. Paris, Le Seuil, 1976, 256 p.

LOMBARD, D. La vision de la forêt à Java, Indonésie. *Études Rurales* (Paris), numéro spécial (*Agriculture et sociétés en Asie du Sud-Est*), 53–56, 1974, p. 473–486.

MACNEISH, R. S. Ancient mesoamerican civilization. *Science* (Washington), 143, 1964, p. 531–537.

MARTIN, M. A. *Introduction à l'ethnobotanique du Cambodge*. Paris, CNRS, 1971, 280 p.

——. De la cueillette à la culture. In: Thomas, J. M. C.; Bernot, L. (eds.). *Langues et techniques, nature et société*, vol. 2., p. 333–336. Paris, Klincksieck, 1972, 415 p.

——. Les Pear, agriculteurs-cueilleurs du Massif des Cardamomes, Cambodge. *Études Rurales* (Paris), numéro spécial (*Agriculture et sociétés en Asie du Sud-Est*), 53–56, 1974, p. 439–448.

——. Essai d'ethnophytogéographie khmère. *Journal d'Agriculture tropicale et de Botanique appliquée* (Paris), vol. 21, n° 7–8–9, 1974, p. 219–238.

MATRAS-TROUBETZKOY, J. L'essartage chez les Brou du Cambodge. Organisation collective et autonomie familiale. *Etudes Rurales* (Paris), numéro spécial (*Agriculture et sociétés en Asie du Sud-Est*), 53–56, 1974, p. 421–438.

ODUM, E. P. *Fundamentals of ecology*. Philadelphia and London, Saunders, 1959. 3rd ed., 1971, 574 p.

PETERSON, N. Hunter-gatherer territoriality: the perspective from Australia. *American Anthropologist*, vol. 77, no. 1, 1975, p. 53–68.

RAPPAPORT, R. A. *Pigs for the ancestors: ritual in the ecology of a New Guinea people*. New Haven and London, Yale University Press, 1967. 5th edition, 1973, 311 p.

——. The flow of energy in an agricultural society. *Scientific American* (New York), vol. 224, no. 3, 1971, p. 116–133.

REICHEL-DOLMATOFF, G. *Desana: le symbolisme universel des Indiens Tukano du Vaupes*. Paris, Gallimard, 1973, 341 p.

——. Cosmology as ecological analysis: a view from the rain forest. *The Ecologist*, vol. 7, no. 1, 1977, p. 4–11.

ROBBINS, R. G. The vegetation of New Guinea. *Australian Territories* (Canberra), 1, 1961, p. 21–32.

ROUSSEAU, J. Des colons qui apportent avec eux leur écologie. In: Thomas, J. M. C.; Bernot, L. (eds.). *Langues et techniques, nature et société*, vol. 2, p. 337–345. Paris, Klincksieck, 1972, 415 p.

SAHLINS, M. D. *Social stratification in Polynesia*. Seattle, University of Washington Press, 1958.

——. Poor man, rich man, big man, chief: political types in Melanesia and Polynesia. *Comparative studies in society and history*, vol. 5, no. 3, 1963, p. 285–303.

——. Notes on the original affluent society. In: Lee, R. B.; De Vore, I. (eds.). *Man the hunter*, p. 85–88. Chicago, Aldine, 1968, 415 p.

SAUER, C. O. *Agricultural origins and dispersals*. New York, American Geographical Society (Bowman Memorial Lecture II), 1952, 110 p.

SCHEBESTA, P. *Orang-utan: bei den urwaldmenschen Malayas und Samatras*. Leipzig, F. A. Brokhaus, 1928.

——. *Among the forest dwarfs of Malaya* (transl. Chambers, A.). London, Hutchinson, 1929, 288 p.

——. Erste mitteilungen über die ergebnisse meiner forschungsreise bei den pygmaen in Belgisch Kongo. *Anthropos* (Wien-Mödling), 26, 1931, p. 1–27.

SCHEBESTA, P. Die Bambute-pygmaen vom Huri. *Mémoires de l'Institut Royal Colonial Belge, section des Sciences morales et politiques* (Bruxelles), 1, 2, 4, 1938–1950.

——. Die Negrito Asiens. 1. Geschichte, Geographie, Umwelt, Demographie und Anthropologie der Negrito. Wien-Mödling, St Gabriel Verlag, *Studia Instituti Anthropos*, vol. 6, 1952, 496 p.

——. Die Negrito Asiens. 2. Ethnographie der Negrito, Religion and Mythologie. Wien-Mödling, St Gabriel Verlag, *Studia Instituti Anthropos*, vol. 13, 1957, 336 p.

SIEMENS, A.; PULESTON, D. Prehistoric ridged fields and related features in Campeche, Mexico. In: *Proc. Intern. Congress of Americanists* (Lima), 1970.

SIMMONDS, N. W. (ed.). *Evolution of crop plants*. London, New York, Longman, 1976, 339 p.

SMITH, R. L. *Ecology and field biology*. New York and London, Harper and Row, 1966.

SPENCER, J. E. Asia East by South: a cultural geography. New York, Wiley, 1954. 2nd ed. (Spencer, J. E.; Thomas, W. L.), 1971, 669 p.

SPENCER, J. E. *Shifting cultivation in southeastern Asia*. Berkeley and Los Angeles, University of California publications in geography no. 19, University of California Press, 1966, 247 p.

SUÁREZ, M. M. *Los Warao, Indígenas del delta del Orinoco*. Caracas, Instituto Venezolano de Investigaciones científicas (IVIC), Departamento de Antropología, 1968, 311 p.

TURNBULL, C. M. *The forest people: a study of the pygmies of the Congo*. New York, Simon and Schuster, 1961, 288 p.

——. *Wayward servants: the two worlds of the African pygmies*. London, Eyre, Spottiswoode, 1965, 390 p.

——. The importance of flux in two hunting societies. In: Lee, R. B.; De Vore, I. (eds.). *Man the hunter*, p. 132–137. Chicago, Aldine, 1968, 415 p.

WILSON, E. O. *Sociobiology: the new synthesis*. Cambridge, Mass., The Belknap Press of Harvard University Press, 1975, 697 p.

WORSLEY, P. *The trumpet shall sound: a study of cargo-cults [in Melanesia*. London, MacGibbon and Kee, 1957, 290 p.

The types of utilization

Introduction

*Biological limitations to the transformation
of tropical forest ecosystems*
    Critical characteristics
        Species diversity
        Genetic variability
        Reproduction
        Spatial distribution
        Atmospheric environment
        Water balance and biogeochemical cycles
    Prediction and evaluation of the impact of forestry operations on production and the environment
    Conclusion

Types of utilization
    Utilization without major modification
        Sylviculture
            Inducement of regeneration
            Felling and logging damage
            Diagnostic sampling and stocking levels
            Enrichment planting
            Selective felling
            Polycyclic systems
            Vegetative reproduction
            Conclusions
        Forest exploitation
        Wildlife management
            Existing uses of wild animals
            Forest management and wildlife
            Wildlife management
            Conclusions
        Cattle grazing
            Without forest modification
            With forest modification
    Simplified ecosystems
        Forest plantations
            Limiting factors
                Wind
                Fire
                Erosion
                Termites and other insect pests
                Grazing and browsing
            Plantation extension and tree improvement
            Conclusion

        Improved forest grassland
            Total clearance (Queensland)
            Pasture development from tropical rain forest
            Pasture development and afforestation on poor soils
    Shifting cultivation and other agri-sylvicultural systems
        Shifting cultivation
            Terminology
            Populations
            Stable systems
            Degenerative and accelerated systems
            Effects
            Food and energy
            Conclusions
        Other agri-sylvicultural systems
        Cattle pastures under plantations
        Conclusion
    Weed control
        Definition
        Weed problems
        Forest nurseries
        Plantations
        Tree weeds
        Parasites
        Fire-break maintenance

*The impact of forest operations
on the environment*
    Biocides
    Logging and transportation
    Man-made lakes
    Pulp effluents
    Industry, mining and human settlements

Research needs and priorities
    Biological limitations to the transformation of tropical forest ecosystems
    Types of utilization
        Sylviculture
        Plantations
        Wildlife management
        Agri-sylvicultural systems
    Weed control
    The impact of forest operations on the environment

Bibliography

Introduction

The likely choice of exploitation of tropical forest ecosystems will depend on the objectives; firstly and chiefly, it will be to increase the quantity and quality of resources demanded by the populations, trade and industry by modification or complete transformation of these ecosystems.

The possible types of exploitation have been examined in a great many international conferences and colloquia, but only those held after the second world war will be mentioned here. There was the United Nations Scientific Conference on the Conservation and Utilization of Resources (Lake Success, 1949), the Symposium organized by the Wenner Gren Foundation, on *Man's role in changing the face of the earth*, the United Nations Conference on the Application of Science and Technology to Development (Geneva, 1963), the Unesco Conference on Utilization and Conservation of the Biosphere (Paris, 1968), and finally, three conferences on food organized by FAO (Washington, 1963; The Hague, 1970), and by the United Nations (Rome, 1974). Mention should also be made of the Ninth Technical Meeting on the ecology of man in the tropical environment, organized in Nairobi in 1963 by IUCN.

Nevertheless, the literature on land use in the humid tropics remains exceedingly vague as regards long-term possibilities. Not enough distinction is made between the problems of regions which have a marked dry season and those which do not. In the former, simplification of the forest ecosystems by transforming them into tree plantations, permanent pastures, or intensively managed crops, has yielded positive results, not least due to the opportunities for controlling pests and diseases afforded by the existence of a dry season. In the non-seasonal tropics of evergreen forests, results have often been disappointing except on alluvial or volcanic soils, irrigated terraces or lands sufficiently close to villages for weed control to be practical. What is and is not possible in tropical land use is of the utmost importance and requires much attention, for great hopes continue to be placed on the possibilities of development of the humid tropics. For instance, the document on the state of world agriculture, presented to the United Nations Food Conference (1974), envisaged a potential increase in the area under cultivation in the developing countries from 737 million ha in 1970 to 900 million ha in 1985 and, provided trypanosomiasis can be solved, an increase of 120 million head of cattle, mainly in the humid tropics.

This underlines the importance of research by certain international institutes under the aegis of the Advisory Group for agricultural research and its technical Committee, such as the International Institute of Tropical Agriculture in Ibadan (IITA), the International Livestock Centre for Africa (ILCA) at Addis Ababa, The Maize and Wheat Improvement Centre in Mexico (CIMMYT), and the International Rice Research Institute in Los Baños (IRRI). However, the research projects of these institutes are chiefly concerned with the techniques and the socio-economic conditions of food production. Thus, recent work at IITA on agricultural systems and crop rotations is highly promising for the Sudan-Guinea region, but results cannot yet be applied to the strict Guinea region. A working party of IRRI is defining the areas in South-East Asia which are suitable for rice production from the agricultural and climatic point of view, where the Institute plans to establish research stations to study different methods of cultivation and the suitability of different varieties.

Infrastructures on the ground (roads, dams, etc.) as well as the measures undertaken to improve the harvesting of resources have a more or less long-term influence on the environment. These influences are often underestimated and it is essential to take them into account in any development either by trying to quantify their costs and benefits, or by planning measures to control their effects.

The purpose of this chapter is to examine different types of utilization of the forest ecosystems of the humid tropics, but excluding complete transformation to agro-ecosystems. Improved grassland will be discussed chiefly to demonstrate that it is possible to simplify the forest ecosystem to a certain degree, frequently in connection with tree crops, but without the intention of suggesting a research programme. The farming or cropping of large wild mammals will not be discussed but will be mentioned in connection with wildlife management. Harvesting and product transport and the socio-economic aspects of development schemes (though their direct or indirect influence on utilization is considerable) will not be considered except for the necessity of research.

Biological limitations to the transformation of tropical forest ecosystems

Critical characteristics

A comprehensive evaluation of the biological limits to the transformation of tropical forest ecosystems would require a study of all characteristics of such systems in order to ascertain the extent to which they would impede or facilitate the success of any manipulation. These attributes are not all entirely independent of each other; only those which are the most important limitations or constraints will be discussed in this chapter. It is not intended to deal with very general characteristics, such as entropy, or information which constitute a synthesis of other characteristics and whose relationships with the possibilities of manipulation would be rather difficult to establish in practice. Attributes connected with the energy balance, such as gross and net production, will also be excluded as these do not seem to constitute theoretical limitations in the tropics. Golley (personal communication) has proposed investigating the attributes of the different successional stages since these form the basis of stability; these are: the biomass, the number of functional options and the time taken to respond to environmental disturbances. 'Firstly, the mass of biological material which comprises the system has at least two aspects.

A large biomass can be thought to damp oscillations because it has inertia, as well as an associated zone of influence around it (mineral cycling). In addition, man can have a strong impact on the local climate. Secondly, the number of functional options in a community refer to the variety of tasks and opportunities in a system. Options can influence stability by providing alternative components and/or pathways in the event of a disturbance to the system. Finally, stability involves the timing of the various interactions between the members of a community. The timing of response may be exceedingly important in stability since feedback between units can amplify or damp an oscillation. For stability a community requires a repertoire of responses which fit the timing of environmental disturbance and the time lags of the unit interactions. For example, a rapid response of a unit might reduce time lag in response of other units and result in overall system adjustment' (Golley, personal communication).

In chapter 9, the characteristics of the different stages of succession and their land-use importance were mentioned. In this chapter species diversity, genetic variability, reproduction, spatial distribution, the atmospheric environment and the nutrient cycle will be considered.

Species diversity

The species diversity of tropical moist forests is well known (see chapters 4, 5 and 7). Species richness (the number of species) should be distinguished from species diversity (the relative importance of the different species); the term stability has different meanings for different authors, such as constancy, inertia, elasticity, directional stability, etc. Golley, Goodman, Orians, Whittaker and others have contributed to the definition of these concepts. A more detailed examination indicates that the species are highly susceptible to all kinds of disturbances because, having evolved under relatively constant environmental conditions they tend to have a low rate of reproduction and of propagule dispersion, and little or no seed dormancy. Moreover they tend to be highly specialized in their habitat requirements.

May (1975) introduced the concepts selection r and selection K (where r is the capacity of reproduction of each individual and K the maximum population capable of existing in a given ecosystem) which derive from the conventional parameters of the logistic equation $\frac{dN}{dt} = rN \frac{(K-N)}{(N)}$, where N is the number of individuals. This equation represents the familiar sigmoid curve of the growth of a population in a restricted environment (see also chapter 7). He concludes:

'(1) Complex natural ecosystems (such as the tropical rain forest), with their many species and rich interaction structure, are in general dynamically fragile. Although well adapted to persist in the relatively predictable environment in which they have evolved, they are likely to be much less resistant to disturbances wrought by man than are relatively simple and robust temperate ecosystems.

'(2) In the tropics, reproductive strategies for both plants and animals are adapted to a relatively constant, predictable environment. They are less well adapted to recover from large scale perturbations than are temperate zone organisms.

'(3) Much of the tropical moist forest grows on soils unsuited to sustained agriculture: commercial forestry is a more rational use of such areas. But, by virtue of (1) and (2), for sustained yield tropical forests are likely to require more sophisticated management techniques than temperate forests. Restraint, research, and avoidance of the uncritical use of inappropriate harvesting methods are needed if the tropical moist forests are not to suffer the fate of analogous K-selected organisms such as the bison or the blue whale.'

Hartshorn and Orians (1975) examine all the ways in which interactions between individuals can influence the number of species living together, and the way in which such interactions theoretically and practically affect stability in tropical forest ecosystems. They state that the concept of stability applies to the capacity of certain components of a system to maintain it close to a point of equilibrium even after a disturbance. These components are: constancy, persistence, inertia, elasticity, amplitude, cyclic stability and directional stability. The characteristics of tropical forests in relation to these seven components are analysed for trees in the different phases of the succession using the results of Budowski, Holdridge and Janzen. They stress the importance of rain and its seasonal distribution on the survival or decline of tropical forests, the lack of information concerning the amplitude (the range of states during which the system retains its stability) and the role of cyclical stability. This forms the basis of the regeneration mosaic theory as opposed to the traditional interpretation of stability in a directional sense during secondary successions.

Changes in the species composition or physical structure of a forest modifies favourably or adversely the environment of animals which depend on plants for their food and of their predators (Synnott, 1975). Changes in animal populations may have important effects on pollination and dispersal which may affect the species composition in all the later stages. The effects of structural modification and the removal of old trees on tree-inhabiting animals (such as, in particular, squirrels and bats) deserve special mention.

Genetic variability

Bouvarel (1975, see chapter 21) writes: 'As regards the competition which exists between species in closed forests, it is probable that its mechanism (within and between strata) is extremely complex and that the hazard of vicinity is of greater importance for the survival of an individual than its genetic constitution. What then, is the best system of adaptation under such conditions? In the first place, a system which ensures the survival of the species by a large quantity of seed, even when the seed bearer is far away from other trees of the same species; furthermore, a system which produces for each individual of the species a mean level of adaptation with only small fluctuations around that mean, that is to say, without strong individual variability. It is true that in a species with strong individual variability the best adapted genotypes may have an advantage over other species, but only if these genotypes find themselves, accidentally, in an ecological situation that is individually favourable for them. The risk that the opposite may be happening is great: the best genotypes may be disadvantaged and the poorer genotypes may be favoured, but without being able to dominate because of their initially poor genotypes. Both these conditions, high fertility of isolated trees and mean levels of adaptation are fulfilled in a reproductive system based on autogamy.' Bouvarel suggests that the species of dense closed forests are chiefly autogamous.

Reproduction

The tropical moist forest is easy to destroy and difficult to regenerate. This is due to limited seed dispersal and highly specialized requirements for germination and seedling establishment (apart from the effect of the removal of biological material and humus on the availability of nutrients). Light-demanding secondary species have seeds with long viability and dormancy, whilst the seeds of primary species are relatively short-lived with a limited dormancy and their seedlings require fairly dense cover. The difficulties experienced by methods of regeneration which are mostly derived from those used in temperate forests point to the need for more research on the mechanisms of reproduction. Chapters 8 and 9 contain information on seed viability and dormancy, dispersal mechanisms and on the little understood adaptation mechanisms of certain species to the opening of the canopy.

Spatial distribution

In addition to a vertical stratification of the vegetation there is a horizontal heterogeneity which is manifested by the existence of gaps of different age, discontinuity of emergent trees and in the herbaceous layer, and by a tendency towards aggregation among the smaller trees. This organization tends to ensure the best possible utilization of the space and to reduce competition. In addition to this spatial utilization there is also a temporal one since there is no seasonal interruption of the vegetation growth in dense moist evergreen forests (see also chapter 8). This spatial heterogeneity which has to be taken into account when sampling is equally important in connection with management and felling; it could make it impossible to apply any schematic schemes.

Atmospheric environment

A great deal is known about radiation, temperature, wind speed and other climatic factors and their vertical variation within the forest and the differences between these microclimatic factors and the external climate. Much less is known about the effects on these parameters of manipulations of the forest and it is this knowledge which ought to guide measures designed to achieve the regeneration or the growth of desirable species.

Compared with undisturbed secondary forests diurnal and seasonal variations in temperature, humidity and associated phenomena such as vapour tension, saturation deficit and evapotranspiration are considerably greater on large cleared areas, degraded sites and also (though less marked) in small gaps. Wind speed in the interior of a forest is very low near the ground—less than 1 per cent of that in the open. It increases from the ground upwards but even in the tree crowns it is no more than *ca.* 10 per cent of the speed above the canopy. This isolation of the forest causes considerable resistance to gas exchange on the surface of the leaves of the understorey and a stabilization of temperature and humidity regimes. It is clear that wind speed as well as the extent and intensity of turbulence increase in all strata as soon as the canopy is broken and this results in an increase of gas exchange and loss of moisture from plants and the soil.

The proportion of radiation penetrating to the forest floor varies according to the structure of the canopy and the pattern of gaps and sunflecks. The light reaching the forest floor represents *ca.* 1–4 per cent of the light falling on the canopy. Sunflecks pose an even greater problem for the characterization of the light quality than the light quantity. While the quality of bright sunflecks may be close to that of full sunlight, the quality of the diffuse radiation in the shade in forests is different, with relatively more in the green wave-bands and less in the blue and red (photosynthetically active) wave-bands. There is also an increase in the far red/red ratio which may be important for controlling seed germination and for the physiology of shade leaves through the phytochrome reaction. It is obvious that any operation which lets in more radiation from the sun and the sky will also bring the spectral composition of radiation in the forest closer to that outside. This may be partially responsible for changes in growth rates of understorey plants and in the composition of the regeneration after disturbance.

Carbon dioxide concentration may be greater inside the forest than outside because of an excess of respiration over photosynthesis during darkness. It was shown that the concentration may vary up to twice normal atmospheric values. The highest values were measured early in the morning and the lowest late in the afternoon. The destruction of a forest, by fire or by decay, returns to the atmosphere in the form of CO_2 the greater part of the carbon accumulated in it.

Water balance and biogeochemical cycles (see chapters 12 and 13)

In tropical rain forests more than half of all mineral nutrients (and sometimes as much as 80 per cent) is in the biomass and the superficial layers of the soil, and the rapid recycling allows such forests to thrive on poor soils and on ferrallitic soils (ferralsols), due to deep roots. Some run-off occurs between the litter layer and the soil proper. This means that there is neither complete protection against erosion nor against the loss of fertility by ferralitization, but this proceeds very slowly.

The forest canopy reduces run-off by the reduction of throughfall. Synnott (1975) mentions that interception may reach up to 50 per cent but that it is very much less where the canopy is open. Within a catchment area, forest cover evens out seasonal fluctuations of outflow by facilitating infiltration at the expense of run-off. Erosion is much more severe under degraded than under intact stands and it tends to increase with forest exploitation; but the degree of interference required to produce erosion varies with the slope, and the types of soil and rock. Agricultural use frequently causes erosion to accelerate. In a study in the Ivory Coast near Adiopodoumé it was ascertained that during 1956 the soil loss from secondary forest was 2.4 t/ha, that on land in the same locality which had been cleared and planted with cassava it had risen to 92.8 t/ha. In 1957, the respective values were 0.03 and 28.7 t/ha. Brünig (1975) gives the following annual rates of soil erosion:

Land form and vegetative cover type	Annual rate of erosion	
	mm	t/ha
Almost flat		
cotton	4	80
annual field frops	1.6	32
dense pasture	0.1 – 0.5	2 – 10
open pasture	1 –10	20 – 200
Undulating, moderate slope		
natural forest	0.01– 0.5	0.2– 10
teak plantation		
— wide spacing, mixed understorey	0.1 – 0.5	2 – 10
— dense, no understorey	1 – 8	20 – 160
Moderate to steep slope		
natural forest	0.5 – 2	10 – 40
shifting cultivation during cropping years	30 –60	600 –1 200

Hasan (1975, see chapter 21) provides information on the special role of bamboos: 'Due to tufts of rhizome and fibrous roots they conserve soil better (White and Chandler, 1945), protect river-bank embankments from flood (Urdo, 1960), are good for reclamation of ravines (Kaul, 1963) and are superior to other broad-leaved and coniferous species in returning nutrients but inferior in returning organic matter to soil (Seth, Kaul and Ramswarup, 1963). Within bamboo species Qureshi, Yadar and Prakash (1969) found that *Bambusa tulda* returns more calcium and *Nechouzeana dulloa* and *Oxytenanthera nigrociliata* more magnesium to soil under them.'

These aspects of mineral removal in connection with logging and erosion underline the need for caution in forest exploitation especially where whole trees, except for roots, are to be removed. Some maintain nevertheless that this problem may not be of immediate importance. In Queensland a normal logging operation of 70 m³/ha results in the loss per ha of 1 kg of phosphorus, 10 kg of potassium, 90 kg of calcium and 20 kg of magnesium; considering the total amounts of minerals contained in the majority of forest sites in the region this would be a relatively small loss. There is also the fact that the nutrients becoming available as the result of logging may be rapidly adsorbed by soil colloids and taken up again by secondary vegetation, or they may be leached by rain.

Finally, it is of interest to link the problems of the fertility of forest soils with those of primary production and stability. Recent studies suggest that on drier and less fertile soils mean increment (and even biomass) approaches and even exceeds that on soils of average humidity because the leaf surface is greater by reasons of a multistoreyed structure; but after logging operations or clear fellings, a restoration of the forest should be possible on fairly moist sites whilst on the drier sites succession would be deflected towards a regressive stage leading to further soil degradation.

Prediction and evaluation of the impact of forestry operations on production and the environment

Systems analyses and models have been used since 1968 to try and evaluate the consequences of forest operations on biological productivity. At the Pacific Science Congress Association in Canberra, Bethel (1971) stated that the most important basic data to be ascertained when a natural forest is turned into a managed forest were those relating to soil properties, biology of reproduction, tree form, structure and properties of timber, and the response of the forest to various sylvicultural treatments; and that such data are essential in order to evaluate actual and potential biological and economic productivity. He stressed that quantitative studies, such as those encouraged by the various activities of the IBP, can assist in forest management decisions and pointed out that the validity of models describing the behaviour of the natural forest systems can be transformed into models for the behaviour of manipulated ecosystems. He is convinced that in the extremely complex conversion of various broad-leaved tropical forests into managed forests, the use of models simulating the management plans could make a major contribution to decision-making. Studies of this type have made it possible to define more closely the influence of manipulations on certain ecosystem characteristics, in particular on nutrition, species richness and diversity, water cycling, etc., though mainly in the context of temperate forests. These have also been the object of models dealing with succession and erosion.

Goodall (1975), after enumerating the conditions necessary for satisfactory predictions and stressing the uncertainty of predictions that remain stochastic, analyses types of predictions based on observations, experiment and analysis. The last-named type implies systems analysis and the elaboration of models in which the dynamics and interactions inherent in the system are translated into mathematical and logical terms in such a fashion that the effects of any manipulation suggested can be followed. The author then discusses the problem of defining the state of a given system with the aid of the values of certain variables, while introducing exogenous variables in all cases (and they are the most frequent) which do not refer to closed systems. He concludes that the construction of models reflecting as far as possible the physical and biological mechanisms of ecosystems, is the best method available for predicting the effect of human interventions in tropical moist forests; such models would have, where necessary, to be subdivided into sub-models, but not without making sure that connections between the different sub-models were coherent.

Systems analysis and the construction of models appear to be suitable methods for predicting and evaluating the impact on the environment of possible interventions in tropical moist forests. Jeffers (1974) who is enthusiastic about the use of models states: 'Systems analysis, in this context of social responsability, may therefore be expected to achieve synthesis between data collection and modelling, in which five successive phases can be recognized: 1. setting of objectives and preliminary synthesis; 2. experimentation; 3. management; 4. evaluation; and 5. final synthesis.'

In fact this is what the Unesco programme on Man and the Biosphere (MAB) sets out to do; its essential elements are: analyses of ecosystems, studies of interactions between man and his environment, integration of such information at various spatial levels, and the use of the modelling technique in order to arrive at quantitative predictions.

Conclusion

The tropical forest ecosystem is highly complex. Though it is very stable under constant environmental conditions it is not well equipped to resist unexpected disturbances and particularly human interventions. The removal of individuals or species, excessive exploitation of part of the biomass, the exposure of the soil to light, etc., are events that often have irreversible consequences. Hence any intervention must be very carefully thought out as regards its intensity, nature and spatial extent, so as to avoid severe traumatic effects so that the damage caused becomes irreparable.

The role of interspecific relations is very important. The consequences of an imbalance between species could be analogous to such an imbalance recently found in marine ecosystems (Aubert, 1971), where biological life is regulated by chemicals secreted into the environment by the organisms inhabiting it. The concept appears to be particularly interesting for the understanding of interspecific relations and that of the triggering off or blocking of certain metabolic pathways. Webb *et al.* (1967) investigated the failure of the rain forest species, *Grevillea robusta*, in monoculture in eastern Australia and reported the presence of a factor associated with the actively growing roots of that species which inhibited the growth of its own seedlings. Moreover, some factor in the leaf drip had the same, though less significant result. Although no toxin was isolated or identified, the results indicated that a microbial factor in the rhizosphere, as well as a factor associated with the living roots and leaves of the parent, were causing this phenomenon. Since other rain forest species show similar behaviour both in monoculture and natural forests (the so-called non-gregarious species) a question arises as to the generality of these findings. Are some of the species in rain forest prevented from becoming more numerous because they actively exclude their own seedlings? Similarly, are some species prevented from becoming more numerous because of chemical exclusion by other species? The liberation of supposedly phytostatic and/or phytotoxic substances from plants has

been the subject of a number of comprehensive reviews and discussions, e.g. Bonner (1950), Börner (1960), Woods (1960), Aamisepp and Osvald (1961), Garb (1961), Muller (1966,1969), Rovira (1969) and Whittaker and Feeny (1971). Despite the fact that there is mounting evidence that allelopathic influences on vegetation patterns are a real possibility, there seem to be no published accounts (except that of Webb *et al.*, 1967) of investigations into these aspects (Connell, 1970). This is one aspect of tropical ecology which has been almost completely ignored. At present there is no conception of the practical effects of the interrelationships between soil microflora and fauna, plant exudates or decomposition products, and the success or otherwise of the plant species. The case of *Terminalia ivorensis* is also worth mentioning. This dense forest species does not seem to survive in monospecific stands. Research conducted by CTFT and ORSTOM shows that man-made stands disappear after 12–20 years; nitrogen mineralization in the litter is completely inhibited and a considerable amount of calcium accumulates in the rhytidoms. Bernhard-Reversat (ORSTOM) has shown for a restricted sample that a leaf extract of *T. ivorensis* inhibits nitrogen mineralization in the soil. It seems that this species poisons itself when it is the only species to occupy an area. In the unmanaged forests, it exists as isolated trees and there is no such mortality. This example provides a case of a biological limitation to the transformation of the ecosystem, as well as a possible biological explanation for the distribution of a forest species.

Types of utilization

It is possible to utilize tropical forest ecosystems as they are, or after only slight modification by sylviculture, logging, cattle grazing, or wildlife management; or special agricultural techniques such as shifting cultivation may be used, or they may be completely transformed into simplified ecosystems, especially forest plantations and sown pasture.

Utilization without major modification

Sylviculture

In tropical forests, selective felling of commercial species produces only small yields (e.g. 5–25 m³/ha in tropical Africa) whereas clear felling might produce as much as 450 m³/ha (especially in the dipterocarp forests of the Philippines). This is why foresters have long been trying to apply certain treatments (based on experience in temperate zones) to improve the composition and growth of marketable species. These include the methods for improving *Aucoumea klaineana* and *Terminalia superba* and certain forest stands in the Ivory Coast, the shelterwood system of Nigeria, the Malaysian uniform system, the selection system in Ghana, etc. There are also other methods for enriching the forest with valuable species, which involve the use of planting stock previously raised in nurseries, the killing of the existing forest trees by various processes (girdling, poisoning, etc.); the *taungya* method, etc. The present-day trend is to enrich the forest by planting at the time of exploitation. The FAO Committee on forest development in the tropics proposed the adoption (with certain modifications required to make it applicable to management in all humid tropical forests) of a system of management worked out by Catinot (1969) and Dawkins (1958b) which may be summarized as follows:

After having established the objectives of management, the working plan should decide on a programme of sylvicultural measures and exploitation fellings. Since sylvicultural techniques are mostly still in an experimental stage it will be necessary to establish various trials relating to:
— improvement in natural stands (thinning and clearing);
— increment and rotation;
— natural regeneration, for which an experimental layout is suggested consisting of exploitation of 100 m wide east to west strips alternating with similar strips left for cutting at the next rotation;
— artificial regeneration by the method of large strips.
It is important to base the exploitation on an accurate inventory (standard error less than 20 per cent for areas <30–35 000 ha) enumerating trees >20 cm DBH in three size classes (20–45 cm, 45–60 cm and >60 cm). Studies on forest exploitation (variations of cost price according to species, the nature of the terrain, and the size of the felling area), transport (effects on cost price of transport routes and their loading capacity) and on forest economics (calculating the maximum acceptable cost price on the felling site on the basis of prices paid FOB or at the factory) are necessary, in order to determine what timber can be produced at what price from a given area.

Forestry operations, unless they are on a very small scale, tend to have a long-term impact on the stands and their environment. These will affect species and genetic composition, general structure and growth; increased exposure to radiation, wind and rain will affect the micro-climate, and the removal of tree substance and increased tree mortality will interfere with the nutrient cycles, and the water balance will be affected by changes in interception, evaporation, etc. Synnott (1975) presents detailed information on the consequences of forestry operations and confirms the biological limitations which have already been stated.

It is not intended to give a historical account or describe in detail the various methods nor the socio-economic problems which are relevant to management (see chapter 21). The intention is to review the lessons which have been learnt and to discuss the problems that remain.

Inducement of regeneration

Measures taken to induce the regeneration of particular species only appear to aid the survival and growth of individuals already present by reducing the growth of other trees. This is the experience of the forest services of eastern and western Africa. Long-term observations of the sample plots at Sungei Keoh in peninsular Malaysia demonstrate how slow and dispersed is the rate of increase in desirable species even where there is an abundance of seedlings from the seedbearers that have been allowed to remain (Wyatt Smith, 1949, 1954, 1955, 1958; Kochummen, 1966). Considerably more knowledge about the behaviour of different species is required.

Felling and logging damage

Felling and logging damage is inevitable; at least half of the total height of the dominant trees lies within the crown and the ratio of crown diameter to stem diameter is high (15–25 according to Palmer, 20–40 according to Cati-not—compared with *ca.* 9–12 in conifers and *Eucalyptus* stands) (Palmer, 1975). Generally speaking, the severity of damage is correlated more with the number of trees felled than with basal area or volume removed. It is also possible to slightly reduce the damage by raising the girth limit of trees to be exploited, thereby reducing their number without substantially reducing the volume yield. The damage caused to the remaining stand has received less attention than that to the soil surface by the logging equipment.

Diagnostic sampling and stocking levels

Immediately after cutting a decision has to be made as to whether natural regeneration is adequate or whether a certain amount of artificial regeneration is necessary. This is the purpose of surveys to estimate the number of young stems of desirable species and to appraise their chance of survival until the next rotation. Many unknown factors are involved in such a decision; Palmer (1975) therefore recommends that any decision concerning the leading desirable species should be based only on the situation at the time of sampling and that essentially the relative values of one or more leading desirables should determine what should be taken into account per quadrat. According to Palmer (1975) values at the end of the rotation cannot be predicted with any degree of certainty, and therefore any species for which there is an actual or potential market should be treated as a desirable. For short rotations he would be prepared to accept sample surveys covering several species, on the pattern of those used in Queensland (Nicholson, 1972, 1974). However, instructions in such cases tend to become extremely complex, and the final picture and its evaluation will depend largely on the complete processing of the field data.

It is obvious that ideally a correct decision would depend on a knowledge of growth and mortality rates, and of the rate of replacement of the leading desirables, but such information is scanty or non-existent. Temporary empirical rules were therefore formulated (Dawkins, 1958b; Boerboom, 1966; Nicholson, 1972) for adequate density levels; and good reasons will have to be given for any changes in these. If, after a felling operation, the density of the leading desirable is less than 20 per cent, it will generally be necessary to use artificial regeneration on all or part of the area.

Enrichment planting

Natural regeneration of desirable species is slow and uncertain especially after logging activities which have caused soil compaction. Enrichment must start immediately after felling together with other sylvicultural treatments to open the canopy, for delay will cause the canopy to close and strong root competition to reassert itself. Various methods of enrichment have been described; they can be grouped into two major categories:

— enrichment planting in gaps where there is no natural young growth of desirable species; the future crop will then consist of the young desirables not yet harvested and their eventual natural regrowth supplemented by the planted trees;
— enrichment planting on cleared strips or in groups. The intensity of planting varies from a minor enrichment designed to supplement existing natural regeneration to a density sufficient for a complete final felling, thus merging into a conversion.

At its last session (Rome, 1974) the FAO Committee on forest development in the tropics surveyed the methods of enrichment and has studied a document prepared by Lamb on the basis of questionnaire enquiry and a literature review. This indicated that failures may be attributed to two causes: insufficient or belated opening of the canopy—the need for overhead illumination being one of the requirements most frequently indicated; and a choice of unsuitables species—the trees planted must be capable of rapid initial growth in order to escape the competition from herbs and natural regrowth—a mean height increment of 1.5 m/a has been suggested as the minimum.

Selective felling

If the objective is continuous timber production (i.e. sustained yield in the sense of production of logs of the same kind and sizes) selective felling is possible only in forests very rich in desirables. Furthermore, the dynamics of stand structure may change so that by the time of the next felling the structure as regards commercial species and sizes is equivalent to that at the time of the previous felling, or corresponds to the structure desired in accordance with the objectives of management.

In the early days of management it was noted that most forests had an all-species frequency distribution (stand table) shaped in the classical form of a reversed J; though the de Liocourt diminution coefficient tends to be rather variable (see chapter 5). Difficulties arose however when these were broken down into the distributions of their component species. Not only were the commercial species represented by a rather small number of stems, but their distributions quite often had irregular shapes. Most tropical forest services possess large collections of such distribution graphs derived from inventories of varying quality. Non-conformity with a de Liocourt graph does not in itself rule out the possibility of selective felling management, if the desirable species, possibly after some ecological grouping, can be encouraged by sylvicultural treatment. The following conditions must apply:

— a sufficiently large proportion of desirable species must pass from their size class at the time of the first logging operation into commercial sizes at the time of the next;
— these trees must not have suffered such severe logging

damage that they have been rendered unfit for the market by rot and insect infestation as a consequence of their injuries.

According to Palmer (1975), not many forests are rich enough to allow true selective working—the removal of each tree (of desirable species) as soon as it reaches commercial size. The present commercial sizes are rather large, > *ca.* 60 cm DBH, in these forests. Not only will each tree cause considerable damage when it falls, but the heavy logging equipment needed will cause further damage. To sum up, true selective felling is impracticable regardless of the structure, composition and dynamism of the original stands until minimum commercial sizes are appreciably reduced. Dawkins (1958b) using a slightly different argument, arrived at the same conclusion.

Polycyclic systems

Polycyclic logging is practicable where, notwithstanding the logging damage sustained during a logging operation, the forest is capable of producing a sustained volume and size yield from advance regeneration which is greater than seedling size. It is often thought that a positive, reverse J size distribution is essential (Dawkins, 1958b), but this is not necessarily so. A paucity in the lower and middle sizes is not a disadvantage if the desirable species can grow rapidly from poles to large sizes. In general, the commercially desirable species tend to be late secondary forest species in western Africa—such as the mahoganies, *Terminalia* spp., or *Triplochiton scleroxylon*, and early seral components in peninsular Malaysia (Cousens, 1965). These species tend to grow comparatively fast and have relatively light wood. Information is rather sparse, but there are indications that they can compete fairly successfully with undesirable species in regeneration after logging. On the basis of observations in Nigeria, Mervart (1972) suggests that once past the small-pole stage, mortality of fast-growing trees is rather less than that of slower-growing trees, so that the desirables increase their proportional representation by volume. Whether or not such improvement occurs will depend on the species present, their growth rates, and the felling damage. General indications from Malaysian and Philippine data tend to confirm this.

The Philippines are committed to polycyclic felling partly through a misunderstanding of the Malayan uniform system. The theoretical basis given in the *Handbook on selective logging* (Bureau of Forestry, 1965) is rather weak and has been criticised by Fox (1967). Suggestions for improvements were made by Nicholson (1970), who believes that the dipterocarp forests in all the climatic regions of the Philippines can probably sustain a 30–45-year felling cycle provided that the prescribed management is carried out which is not what happens at the moment. However, there are indications that the forest is rich enough in desirable species to tolerate considerable maltreatment, especially as fast-growing dipterocarp trees seem to be able to regenerate healthy crowns provided that at least one half of the original crown was left undamaged at the time of logging. There are however a few stands not rich in medium-sized advance growth, for example in areas liable to typhoon damage (parts of Samar and Negros), where large felling cycles may be necessary and relying on seedling advance regeneration like the Malayan uniform system.

The Queensland rain forest is so atypical that experience there cannot be transferred to other areas. A high proportion of the original stands consists of cabinet woods of high value and high quality. Strict logging rules are imposed and intensive tree marking is followed by intensive sylviculture. A very sharp price-size and price-quality gradient, together with a proper stumpage appraisal system, has allowed expensive forest management to take place. However, the post-logging sylvicultural operations are labour intensive, and information from the early 1970s suggests that the rapid rise in labour costs is resulting in major changes.

The arguments against the use of polycyclic logging are based on:

— the general financial desirability of removing the whole commercially saleable stand in one operation;
— the more or less uncontrollable damage to the larger residual stems;
— the frequent lack of sufficient stems in the middle size classes of desirable species;
— the obliteration of advance regeneration by tractor tracks, cableways and log landings which tend to cover larger and larger proportions of the forest.

Polycyclic logging implies the regular return to an area for a harvest at intervals less than the time taken by a desirable to grow from seedling to commercial size. For the same regular yield it requires a much larger annual coupe than monocyclic logging. Salvage logging, and two-stage conversion to uniform working, as discussed in Dawkins (1958b) are not types of true polycyclic working.

Baidoe (1972) shows that the so-called 15-year felling cycle (reduced from 25 years) in Ghana is nothing like a felling cycle. It was suggested to the government that a long cycle allows deterioration of the trees due for harvesting last. This might be true if the forest is in a strongly seral transition, but various calculations on the longevity of western African desirables (e.g. Redhead, 1960) provide no support for a change from 25 to 15 years. Indeed it is quite obvious that this is a commercial move to allow the removal of the biggest and best trees in larger quantities. The consequences are regrettably obvious.

Outside the Philippines and Queensland, only a few calculations have been published which attempt to justify short felling cycles. The study by Redhead (1960) on logging damage in Nigeria reported only that damage was not so great as to rule out polycyclic felling; changes in logging machinery and utilization intensity have made the calculations out of date. In peninsular Malaysia, Skapski (1971) suggested 5-year felling cycles as he thought a 20 per cent light intensity would stimulate lowland dipterocarp seedling growth, but this contradicts the fine body of literature of this forest type; furthermore he makes no allowance for felling damage. Fresh calculations are now being made in peninsular Malaysia to see if polycyclic fellings could be introduced (K. J. Sargent, personal communication; F. S. Walker, personal communication). It is difficult to

see how conclusions from previous work in Malaysia and in Sabah can be avoided, since the forests are so similar.

Vegetative reproduction

Vegetative reproduction, that is to say coppice systems have not been studied intensively in the humid tropics. In a report to FAO, Ramakrishna described a decline after two 40–60-year rotations: a vigorous mixed forest was transformed into a sea of weeds highly susceptible to fire (and hence threatened with further decline). Palmer (1975) estimates that most species of closed tropical forests do not easily regenerate stumps once these have reached a diameter of 5 cm. However, various forms of coppice with standards do exist in South-East Asia and with the help of both basic and applied research it ought to be possible to work out some rules for their treatment. Studies on the functioning of meristems are indispensable for the research into suitable methods of regeneration from cuttings.

Conclusions

The difficulties encountered and the failures experienced in the regeneration of natural forests have caused foresters to neglect research and to direct their efforts towards the complete replacement of natural forests by forest plantations. They have also led some to ask whether the problem should not be approached afresh in a different manner—the original trend having been (especially in Africa) quite simply to transfer methods developed on the basis of experience in temperate regions. This has led to a development of basic research on the structure and functioning of tropical forest ecosystems. Experience acquired on forest inventories and sample surveys, methods of regeneration and enrichment, plantation establishment and logging operations have certainly contributed to a better understanding of the functioning and structure of tropical forest ecosystems, and hence to an improvement of treatment possibilities—while waiting for the rules of a system of tropical sylviculture to emerge from basic studies now in progress and further field experience. Already it seems probable that the cost per m³ of such sylviculture will be rather high and that it will have to be justified either by the quality of the products or by the role of the tropical forest biome in the biosphere. Duvigneaud's statement (1971) is worth recalling: 'It is important to acknowledge the enormous amount of work accomplished in the last century by foresters within the field of forest productivity. Their aims may differ from ours, but they are sure to converge at the end of the *long road which still remains to be travelled.*' It is to be hoped that foresters and ecologists will quickly pool their knowledge and experience so as to arrive at that new science of sylviculture which is so much needed and so fervently desired.

Forest exploitation

The techniques of felling and skidding have made considerable progress owing to developments in the organization of work and in mechanization, but have caused some environmental problems. FAO (1974) has discussed these in two recent publications: *Exploitation and transport of logs in dense tropical forests* and *Logging and log transport in man-made forests in developing countries*. It also entrusted to the Centre technique forestier tropical (CTFT) a study on the cost of tropical forest roads. Ergonomic studies will be discussed in chapter 21. At present there is a tendency towards whole-tree utilization with complex equipment to increase labour productivity; under certain conditions this could result in harmful effects on the nutrient cycles. A symposium held in Sweden during 1975 under the sponsorship of FAO and EEC was devoted to this.

Wildlife management

Within the tropics are the world's richest fauna. This is particularly so for large mammals in Africa. This section will deal with ways of utilizing animals, the influence of forest management on such utilization, and trends towards the improved exploitation of wildlife.

Existing uses of wild animals

In all tropical forests, hunting is practised by local populations using traditional methods, professional commercial hunters and frequently by sportsmen. Fish, birds, small and large mammals are the usual quarry. The production of skins and other animal products such as ivory represent an important aspect of wildlife exploitation. Wild animals are a traditional source of protein for forest-dwellers; in recent years this form of land use has been intensified by the adoption of plans for the commercial exploitation of wild animals by their systematic regular cropping. In some forest regions it has been possible to improve the habitat so as to increase the productivity of animal populations.

In Ghana, wild animals account for 65 per cent of the proteins consumed in rural areas. In Nigeria the giant or pouched rat (*Cricetomys gambianus*) has been domesticated and is raised successfully; economic and ecological data have been collected to determine the feasibility of breeding them on an industrial scale; similar studies have been carried out on the giant edible snail (*Achatina* spp.).

Singapore, which is one of the two principal centres for the export of animal products of South-East Asia, exported the skins of crocodiles, snakes and lizards, aquarium fishes and wild birds, worth US dollars 10 million in 1966. Breeding deer is quite common in Viet-Nam for the production of venison and antlers for the Chinese market. Research has been carried out on the utilization of crocodiles and tortoises especially in South-East Asia; these are bred in the open or in batteries, or by a combination of both methods.

Little is known about the utilization of wild animals in Latin America. In Peru, fish and meat from wild animals from the region of Ucayali in the Amazon account for ≥85 per cent of the animal proteins consumed by low-income families. A start has been made with the utilization

of crocodiles and tortoises in the Amazon basin. Many rodents are used as human food (Pierret and Dourojeanni, 1966).

Especially in Latin America and Asia, the trade in primates has developed to a point which threatens the survival of certain species. The breeding of primates in captivity for laboratory experiments is a recent development.

Forest management and wildlife

Forest management affects the habitat of animals. The nature of management, for production or for protection, whether indigenous or introduced species are involved; thinning, selection or group selection fellings, etc., have a profound influence on the populations of animals. Forest management can change the damage of harmful animals by changing the habitat. Plantation of exotics tends to reduce the diversity of the vegetation and each kind of plantation in each region has its animal production potential. Grazing of animals modifies the habitat and the behaviour and productivity of wild animals.

Wildlife management

Major changes in the habitat can thus bring about major changes in animal populations. Examples of this have recently been given by Riney (1967). Wildlife biologists stress that the greatest wildlife potential occurs where there is the greatest diversity in habitats. Shifting cultivation increases the potential of the wildlife habitats, though the possibilities this offers have been ignored. Admittedly the peoples practising shifting cultivation tend to regard the animals as a supplementary source of food, and their, usually uncontrolled, hunting prevents the populations reaching the carrying capacity of the modified environment; with rational management the number of permissible kills could be considerably higher than now.

Efforts are being made to control the numbers killed, i.e. to exploit existing resources to obtain optimum sustained yields and avoid irreparable destruction. It is possible to achieve this by successive approximations but such a method is dangerous for it can lead to the extinction of a population (signs of over-exploitation usually appearing when the damage is already done). It is preferable to use direct computation, such as have been elaborated or are still under study, but such methods are by no means universally accepted and require improvement. Their application presupposes an improvement in the techniques of population inventories, carrying capacity determinations, and more information on population dynamics to allow greater selectivity in the choice of animals to be killed.

The exploitation, conversion and selling of products from wild animals need to be thoroughly improved. The transport and treatment of animals should also be made as efficient as possible to reduce losses and to increase revenue. The conservation of endangered animal species is essential.

Conclusions

Forest management will have to accord increasing consideration to aspects other than timber exploitation; wildlife management has not only increased the value of forested land, but is rapidly becoming a form of land use in its own right. Multiple use, originally no more than a rather vague and general aspiration, has become an important concept in which the various types of utilization combine to improve overall productivity. The FAO Committee on forest development in the tropics has nevertheless expressed its disquiet about the dearth of ecological data indispensable for a rational wildlife management. It has also expressed concern about the inconsiderate use of pesticides, the introduction of exotic animals and the danger of extermination of rare animals. The Committee recognized the importance of wild animals as a source of protein for many tropical areas. The Committee concluded that wildlife management should form an integral part in plans of forest management.

Cattle grazing

Without forest modification

Little use of tropical closed rain forest is made for grazing. In cases of extreme drought cattle may be given access to rain forests which usually contain a significant component of edible leguminous browse, but very little gramineous or herbaceous forage at ground level. In northeastern Thailand such a system is being practised. In Bali, Indonesia, cattle are sometimes tethered in rain forest. The situation in closed monsoon forests is similar. These contain some deciduous species and browse is obtained from their foliage and litter. In the wetter areas grazing is usually less nutritious due to lower soil fertility, leaching and a more rapid maturation of the gramineous species present. Inferior grasses such as *Imperata cylindrica* commonly occur as secondary growth following burning. Where the climate has a well-defined dry season annual burning of the ground cover is a normal practice.

With forest modification

Several methods of forest modification for grazing have been used. Before the advent of mechanical clearing equipment, trees were generally killed by ringbarking. Species differ in their reaction to ringbarking; some die quickly, others produce a succession of suckers from the cut surface, requiring follow-up treatment for their destruction. Killing the trees removed their competition for light, water and nutrients and thereby improved the production of grass and herbage in the ground flora. Carrying capacity of ringbarked land was at least twice that of the original forest. Intelligent ringbarking allowed selected trees to be left for shade, wind-breaks, stream-bank protection and animal browse, but there has often been a tendency to destroy too many trees, with the consequent danger of soil erosion.

Millions of hectares of grazing land in the sub-tropics have been created by ringbarking; these constitute the most important beef areas. As pressure for land increased, areas of greater fertility in more advantageous climates tended to be cultivated for cash crops. Thus much of the present ring-barked land is only devoted to breeding and the production of unfinished cattle—the animals being subsequently finished for market on sown pastures, fodder crops, or in feedlots. Ringbarking is now little practised because it is time-consuming and labour for such manual work may be in short supply; arboricides tend also to be used more frequently.

Simplified ecosystems

Forest plantations

Because of the failures and difficulties experienced with natural regeneration and the success achieved with certain plantations of *Pinus caribaea*, *Tectona grandis* and *Gmelina arborea* (especially in regions with a more pronounced dry season), many foresters have proposed that forests should be replaced by, usually monospecific, plantations. This development since 1945 and especially from 1960 to 1970 was helped by expeditions in search of suitable seed, exchange of information on the behaviour of species and provenances, and by research on species which had been grown in plantations for a sufficient length of time to furnish valuable information on their mechanical and pulping qualities. This tendency in favour of plantations was confirmed by the last International Forestry Congress (held at Buenos Aires in 1972) which, mindful of the difficulties of managing and utilizing humid tropical forests and the reluctance of international bodies to finance research into natural regeneration, recommended the reduction of pressure on humid tropical forests by concentrating research on and establishing plantations outside the humid tropical forest zone.

The establishment of plantations can be justified by their supply of wood and by protecting neighbouring forests provided they are limited in size and are properly managed. Considerable misgivings can be aroused by establishing large-scale plantations especially monocultures of ill-adapted species without the necessary control of origin of seed or planting stock. This is aggravated where the soil is poor, there is pollution from factories, and in climates without a pronounced dry season during which to control pests. Because of the longevity of forest species it is very difficult to breed resistant races. Weed infestation commences from the start of the plantation and the nutrient cycles are not always satisfactorily restored.

Limiting factors

Wind

Hurricanes, cyclones and typhoons are frequent in some areas. Other areas experience occasional violent storms. Perhaps the most notable example is the storm forest area in Kelantan, peninsular Malaysia, which is believed to have originated from a freak storm at about the time of the

Krakatau eruption in 1883 (Browne, 1949; Wyatt-Smith, 1954). Whitmore (1974) suggests that global atmospheric cooling is forcing the main zones of activity of atmospheric circulation towards the equator. Evidence from the Solomon Islands suggests that equatorial regions are becoming more liable to cyclone damage. Reliable records of cyclone paths are rare and insufficient to assist prediction (Gane, 1970). There is some evidence that local topography affects the storm path (Wadsworth and Englerth, 1959; Gane, 1970; Hindley, personal communication). Climber tangles over low uneven canopies usually indicate repeated strong disturbances.

Different types of forests, plantations and species react differently to hurricanes (Wadsworth and Englerth, 1959). Plantations generally receive some damage whilst damage to forests is more sporadic; large tracts of natural forest are sometimes destroyed by storms. Outside the hurricane belts, damage is more noticeable in certain forest types. Single or isolated dominant trees seem to be particularly damaged. The effects are well documented on aerial photographs of

Sarawak peat swamp forest (Anderson, 1964; Brünig, 1973). A somewhat similar effect seems to occur in plantations: heavily thinned areas suffer more damage than unthinned plantations, because thinning increases aerodynamic roughness at a time when the trees are slender and adapted to dense stand conditions. Some species are particularly resistant to storm damage; in Mauritius indigenous species are damaged less than exotic plantations (King, 1945). The crowns of the indigenous trees are low and umbrella-shaped, the leaves small and coriaceous, the timbers slow-growing and hard. Walker (personal communication) believes that *Agathis macrophylla* in the Santa Cruz islands in the Pacific is able to resist cyclones by shedding branches as wind pressure rises. Johnson (1971) observed not an single windthrown *A. obtusa* during a forest inventory on Erromango in the New Hebrides despite their 'large crowns spreading above the general top canopy of the forest', but Whitmore (1963) believes that such large trees are 'most likely to be damaged by wind'. The following table shows some estimates of wind damage frequency:

Locality	Period	Number of recorded storms	Remarks	References
Belize	1787–1961	14	1787–1925, once in 23 years. Since 1925, once in 6 years.	Lindo (1968)
British Solomon Islands	1950–1972	16	1966–72, once in six months.	Whitmore (1974)
Fiji	1900–1970	24	Some areas seem more likely to be hit than others.	Gane (1970)
Mauritius	1644–1960	12	Severe cyclones recently one in 11 years.	Brouard (1960)
New Hebrides			Damaging cyclones almost annually, total destruction of any one area once in 40–50 years.	Bennett (1974)
Puerto Rico	1509–1959	45	Severe hurricanes once every 10 years.	Wadsworth and Englerth (1959)
Queensland	72 years	10	Very severe cyclones.	Webb (1958)
Western Samoa	1961–1968	3	Also frequent high velocity wind storms.	Wood (1970)

Much more quantitative evaluation is required in order to assess the resistance of species to damage on particular sites and to recommend management. Three scales are tentatively proposed for the evaluation of damage to a particular tree. They are based on Nicholson's (1958) scales of logging damage. Associated observations should include notes on the soil, effective rooting depth and liability to weakening under the intense rainfall, aspect, tree and crop dimensions, topography, shelter, and the local characteristics of the storm.

Tentative damage evaluation scales are:
Stem
no damage 0
snapped 1
leaning 2
uprooted 3

Crown
no damage 0
defoliation 1
twigs and small branches broken 2
1 major branch broken 3
2 major branches broken 4
¼ crown lost 5
½ crown lost 6
¾ crown lost 7
whole crown lost 8
Bark
no damage 0
<30 cm of wood exposed 1
30–150 cm of wood exposed 2
150–300 cm of wood exposed 3
300–600 cm of wood exposed 4
>600 cm of wood exposed 5

In a hurricane zone, management must be strongly conditioned by the possibility of wind damage. Bennett (1974) divides cyclones into those that damage and those that destroy. He suggests that in the New Hebrides, severe destruction of any one area occurs seldom more than once in every 40–50 years, while damaging cyclones strike the islands almost annually. Tentatively, he suggests that plantations may be brought to financial maturity by:
— using wind-firm fast-growing species;
— close planting in small blocks surrounded by natural forests;
— line planting with retention of as much vegetation as possible between the lines;
— dispersing the plantation programme throughout the islands instead of concentrating on one or two islands.
Even after these precautions he suggests that overall timber production will have to be an estimated 20 per cent above the theoretical to allow for wind damage.

The limiting effect is summarized by Baur (1962): plantation forestry passes from the realms of sylviculture and economics to statistical probability on the likelihood of cyclone damage. The small and relatively poor countries liable to be affected by hurricanes and cyclones cannot afford to have their precious investments lost through inadequate techniques.

Fire

At forest margins with agricultural or savanna land, and increasingly towards seasonal forest, fire is a persistent threat. The prolonged effect of fire is being studied on the Olokemeji plots in Nigeria and in other areas in western Africa (Charter and Keay, 1960; Ramsay and Rose Innes, 1963). Mitigation of fire in plantations is largely a matter of how much money is spent on fire protection relative to the value of the growing product. Fire by itself has not been a limiting factor in the establishment or management of plantations in the rain forest zone, but it is serious when it destroys the ground cover protection against erosion.

Erosion

Steep slopes should generally be avoided for tree crop and timber plantations because of the problems and high cost harvesting. Even on relatively gentle slopes losses of soil may be considerable (see chapter 12). Little information on the rates of soil formation is available. Baillie (1971) suggested that the net accumulation rate of soil on a steep upper slope under mixed dipterocarp forest in the Ulu Balleh, Sarawak, could be *ca.* 0.2 mm/a, but that the rate of flow across this charcoal site might be as high as 1 mm/a unconsolidated depth. If one assumes that 1 mm depth of top soil is equivalent to 10 t/ha (Nye and Greenland, 1960) a mean loss of 2–10 t/ha/a might be balanced by the rate of pedogenesis.

It is clear that some of the soil losses are unlikely to be so matched and erosion should be controlled. The problem is clearly described in Bell's paper (1973) on erosion under teak plantations in Trinidad. Close-planted teak shades out the shrubs and undergrowth which spring up after taungya farming. At the beginning of the dry season, the leaves are shed and fires sweep through about 50 per cent of the plantation areas each year—sometimes twice in one season. The teak tolerates the fires; it coppices up to 3 years old, and though it becomes more sensitive later, this lasts only a few years until the bark thickens. However, shrubs and ground cover are eliminated by the combination of shade and fire. Teak does not produce new leaves until after the start of the rains, but there is only a slight indication that soil losses from early rains are proportionately greater than those at the end of the wet season. Bell also mentions that trials have been started in Trinidad to control erosion under teak. It is widely believed wide spacing leads to poor tree form, but evidence from planting in groups, lines and squares suggest that line planting might be best. These would result in the use of 40 per cent fewer trees and less weeding. It is possible that bamboo regrowth was the failure of line planting of teak in Burma, but this does not occur in Trinidad. Nye and Greenland (1960) is western Africa, and Freeman (1955) in Sarawak, point out that the soils most favoured by shifting cultivators are often on steep or relatively steep slopes. If systems of agri-sylviculture are to be introduced as alternatives, it is important to ensure an erosion-reducing understorey and the species chosen should not cast shade to eliminate undergrowth. Some reduction in yields and possibly of form or quality may have to be accepted if hill-sides are to remain stable.

Research should concentrate on the choice of species and planting intervals. Later, the degree and type of thinning and the mechanical aids used in harvesting will require careful attention.

Termites and other insect pests

For many years large areas, particularly in the drier zones, were almost uncultivable because of termites and it was considered fortunate if >30 per cent of planted eucalypts established themselves. The introduction of synthetic organic insecticides in the 1940s altered this. Trials in Brazil (Fonseca, 1949–50), Cameroon (Monnier, 1958), Nigeria (Kemp, 1956; Lowe, 1961) and Tanzania (Parry, 1959) showed that aldrin and dieldrin can give sufficient protection for the critical first growing season. It is now normal to add a small quantity of 2 per cent dieldrin dust to the potting mixture or transplant bed.

Termites are a problem to the establishment of plantations on some sites which recently carried rain forest. Exotic pines (*Pinus caribaea, P. oocarpa, P. patula*) on podzols in Sarawak were persistently attacked in spite of dieldrin sprays. Plantations on red-yellow podzols suffered less severely although no insecticide had been used in the potting mixtures. It is very likely that there were nutritional deficiencies in the pines. Termite damage also occurs in the more seasonal climate of Sabah on former forest sites, but this is not a major problem. Trouble may occur throughout the rotation, though it usually affects the less dominant trees and may be a secondary factor associated with stress symptoms (Freezaillah, 1966; Browne, 1968; Lamb, 1973).

Leaf-cutting ants (Attini) are a particular problem of the New World. Control requires the location and destruction of colonies before planting. On large schemes this is achieved by mechanical destruction and by high-power blowing of Mirex powder into the nests. On smaller areas, baits of Mirex plus other compounds have met with variable success (Cherrett, 1969). Jacobs (1972) noted the tendency of the afforestation companies in Brazil to neglect their control after the first few years. It is expensive in manpower and chemicals, but is indispensable for success. Better managed companies find that the expense is justified.

A number of valuable timber species are almost impossible to cultivate in plantations because of shoot borer or gall-maker attack. The tree family Meliaceae is especially liable to the depredations of the shoot borers of the genus *Hypsipyla* (see chapter 14).

Grazing and browsing

Plantations on sites formerly occupied by rain forest contain succulent growing points rich in minerals and, in many areas, grass. If herders arrive fires are bound to occur. Short-term prevention is by policing and legal means. In the long term, particularly if plantations lie accross migration routes, sylvicultural methods may be adopted in which a lower yield of wood is balanced by improved grazing. The substantial progress made in the United States and New Zealand may not be directly applicable to tropical sites, but the experience of Fiji is promising. It may well be profitable for a forest service to improve the grazing quality of its plantations so that graziers will not be inclined to burn the area.

In general wild mammals are not a limiting factor in large close-spaced plantations although in many areas potential timber production is not achieved because of persistent damage by monkeys, opossums, squirrels, rats, etc.

Expanding human populations force wild mammals into the remaining areas of forest for longer periods and at greater density than normal. Logged and enrichment areas, with succulent low growth, are attractive feeding grounds for many larger herbivorous mammals in Africa. In the Budongo forest in Uganda, elephants are an especial problem (Philip, 1965; Johnstone, 1968; Laws, Parker and Johnstone, 1975); their food species preferences are marked but they browse on a vast range of desirable and weed species and do indiscriminate damage to many more plants than they feed on. Control is difficult and expensive if not impossible; areas liable to large herbivore populations and especially palatable species should not be used for enrichment schemes (Foury, 1956; Dawkins in Lamb, 1969).

Plantation extension and tree improvement

Forest plantations and their industrial importance formed the subject of an international colloquium held at Canberra in 1967 under the sponsorship of FAO. Forest plantations covered 80 million ha in 1965 and it was estimated that their area would double in twenty years (FAO, 1967). FAO has published a series of monographs on *Tree planting practices in tropical Africa* (1956), *tropical Asia* (1957) and *Latin America* (1960), as well as on the choice of species (1952) and *Tree seed notes* (1955).

A report on tropical plantations was presented to the International Forestry Congress in Buenos Aires (1972) by Lamb, while Catinot and Wood dealt with plantations in the Guinea regions at the FAO Conference on the establishment of co-operative agricultural research programmes for African countries with comparable ecological conditions (Catinot, 1972). Groulez (1976) discussed conversion plantations in tropical humid forests. Conversion plantations have been chiefly established in Africa (for example, 26 000 ha with *Aucoumea klaineana* in Gabon, 6 000 ha with *Terminalia superba* in the People's Republic of the Congo, 16 000 ha of various species in the Ivory Coast), but also in tropical America (50 000 ha, chiefly with *Gmelina arborea*, in Brazil), in Asia and Australasia. Their advantage consists of concentrating optimum productivity on a selected site by fast-growing light-demanding species in relatively short rotations using simplified management. Difficulties arise principally from the preexisting forest in the form of falling or dying trees with disease hazards, the maintenance of soil fertility, and of costs, profits and financing.

Research is required on: the destruction of the natural forest (total utilization and studies of the forest structures to be destroyed); the ecology, biology and physiology of the species to be introduced (including their vegetative propagation); genetic improvement of planting stock or seed (provenance trials, progeny testing, clone isolations, hybridization, exploitation of heterosis, etc.); sylvicultural practices (planting, spacing, fertilizing, etc.); pests and diseases; and economic aspects.

There have been international consultations on forest tree improvement in Stockholm in 1962 and in Washington in 1969. A third (organized jointly by FAO and IUFRO took place in Canberra in 1977) will pay special attention to the improvement of fast-growing tropical species. An FAO panel of experts on forest gene resources was established by FAO in 1968 and has held three meetings. As the activities have, because of limited funds, concentrated on exploration and collection, and because this progress has shown the need for a more comprehensive programme, FAO in 1972 submitted to UNDP preliminary proposals for a global programme in forest genetic resources. The new International Board for Plant Genetic Resources (IBPGR) of the Consultative Group on International Agricultural Research (CGIAR), to which UNDP and UNEP are contributors, now deals with this. It aims to:
— identify species and field operations on which action is required during the next two decades;
— assign relative priority to the various combinations of species and field operations where possible;
— incorporate the high priority species and field operations in the action programme for 1975–1979.
An up-to-date list of species on which action is necessary is given by IBPGR, together with forest genetic resources priorities (by region, species and operations). The importance rating for a species is expected to remain constant over one to several decades in contrast with priority ratings for

individual operations which may change within a few years. The following tropical species were given first priority:

— Africa: *Aucoumea klaineana, Chlorophora excelsa, C. regia, Entandrophragma angolense, E. cylindricum, E. utile, Gymnostermum zaizou, Khaya anthotheca, K. grandiflora, K. ivorensis, Terminalia ivorensis, T. superba, Triplochiton scleroxylon, Turraeanthus africana.*
— Australia: numerous eucalypts.
— South-East Asia: *Araucaria cunninghamii, A. hunsteinii,* bamboo spp., *Bombax ceiba, Dipterocarpus* spp., *Eucalyptus deglupta, E. urophylla, Gmelina arborea, Pinus kesiya, Tectona grandis.*
— America: *Cedrela odorata, Pinus caribaea* vars. *hondurensis, caribaea* and *bahamensis, P. oocarpa, Swietenia macrophylla.*

The existence of ecotypes, of local races could play an important role in the development of plantations; they are in urgent need of research and conservation (see chapter 21).

Field operations refer mainly to botanical and genealogical exploration, collection for evaluation, provenance trials, conservation *in situ*, collection for conservation and selection *ex situ*, storage as seed, research on floral biology, seeds, utilization of bulk supplies, individual selection and breeding.

Collection of wide ranging samples should be followed by immediate provenance testing. In some cases, the seed collections may need storing for a few years until the necessary meticulous supervision of the trials can be assured. International assistance is desirable through the provision of standardized procedures for the design, assessments and analysis of results. Internationally accepted programmes for computerized storage and retrieval of the information will become imperative.

Co-operative programmes are carried out on *Cedrela odorata, Pinus caribaea* and *Pinus oocarpa,* by the Commonwealth Forestry Institute, Oxford, *Pinus kesiya* and *Eucalyptus* spp. by the Forest Research Institute, Canberra, *Pinus kesiya, P. merkusii, P. merkusiana* and *Tectona grandis,* by the Pine and Teak Centres in Thailand, established in co-operation with the government of Denmark, and *Terminalia superba* and *Aucoumea klaineana,* by the Forest Research Station in the Ivory Coast, in co-operation with the Centre technique forestier tropical, Paris. A manual on tropical species and provenances is in preparation by FAO.

Conclusion

It has been necessary to stress the difficulties in establishing plantations and the prospects they offer provided that the species and the provenances are well chosen. Like all enterprises, plantations need considerable investments of manpower and capital. Plantations should not be established on poor soils. It is necessary to quantify the gains resulting from breeding and fertilization, thinning and pruning, as well as from irrigation, associated crops, etc.

Before embarking on large-scale reforestation, the example of certain countries (for instance Nigeria) in devel-

oping plantations within the framework of agriculture and thereby make use of existing equipment and labour, might be followed.

Improved forest grassland

Total clearing was formerly achieved by the use of the axe and fire. The practice is still used, but since 1945 most has been accomplished by bulldozers. Some top soil removal may occur and scrub rakes, which allow most of the top soil to escape between the lines, are often preferred to the solid dozer blade.

Total clearance (Queensland)

In Queensland, Australia, *ca.* 7 million ha of *Acacia harpophylla* scrub and monsoon forest have been cleared for pasture for cattle-raising. This area lies mainly between the 600–750 mm isohyets. The soils are rather fertile brown self-mulching clays of good water-holding capacity and are not normally fertilized for pasture establishment. This scrub was cleared using two bulldozers equipped crawler tractors *ca.* 40–50 m apart pulling a heavy chain or steel cable between them. The scrub is totally flattened but strips of undisturbed scrub are left to provide shade, shelter and fence posts. The pulled vegetation is burnt when dry enough without windrowing or stocking as it easily ignites and usually burns very well to a white ash. Pulling is timed to allow burning just before the wet season. About a week after burning, *Chloris gayana, Panicum maximum* var. *trichoglume, Cenchrus ciliaris,* or a mixture of these, is sown in the ashes using aircraft operating some 100 m above ground level. The seeds germinate well in the weed free ash if adequate rain falls and a luxuriant pasture is normally established within two months. There is a good deal of accumulated nitrogen in this newly exposed soil from the *A. harpophylla* and pasture production is excellent for about ten years. *Chloris gayana* disappears when soil nitrogen level falls. The land is then ploughed for sorghum, sunflowers, ground-nuts and other crops in a rotation. Carrying capacities are one beast/10–15 ha on natural scrub, one/4 ha on improved pasture and one/2 ha under intensive management with the inclusion of a legume.

Establishment problems arise if rain does not fall soon after burning. The pulling and burning injure the lignotubers of *A. harpophylla* which quickly produce suckers. With good pasture establishment this timber regrowth can be suppressed, but if the germination is delayed the regrowth becomes a weed. This is sprayed with herbicides. Regrowth after a fire can be a problem up to twelve years after the initial felling and burning.

Pasture development from tropical rain forest

The clearing of rain forest for pastures is usually accomplished with bulldozers or tree-crushers, first pushing over the timber and later moving it into windrows for burning. Thorough cultivation after burning is now recommended for pasture establishment as seed and fertilizer are costly and subsequent weed control and cattle management are difficult if much timber remains (Teitzel, 1969). Regrowth of some of the original species is also suppressed by thorough land preparation.

Problems include excessive regrowth due to undertaking too large an area at one time, excessive weed growth, inadequate maintenance fertilizer, and poor grazing management. The decline in soil fertility after forest removal is extremely rapid due to insolation, oxidation and leaching; significant differences in yield usually occurring in 5–8 years. Soil fertility can be maintained under an adequately fertilized and managed grass and legume pasture. Phosphorus is the major nutrient required. Bruce (1965) has shown that land previously under rain forest at Innisfail, North Queensland, where the average rainfall is 3 750 mm/a, sown to a *Panicum maximum* and *Centrosema pubescens* pasture, had a total nitrogen in the top 7.5 cm of soil of 0.432 per cent, compared with the soil under adjacent undisturbed rain forest which contained 0.423 per cent. The live weight gain obtained on this pasture mixture was 610 kg/ha/a and 440 kg/ha/a on a pure *Panicum maximum* pasture (Grof, 1965). In the Amazon basin at Pucallpa, Peru, Santhirasegaram (personal communication) obtained a cattle live weight increase of 50 kg/ha/a when grazing on natural *Hyparrhenia rufa*. With adequately fertilized grass and legume pastures on similar soil the live weight gain was 650 kg/ha/a.

Pasture development and afforestation on poor soils

In coastal Queensland, there are *ca.* 2 million ha of poorly drained gley and humid peat soils supporting *Melaleuca* and *Banksia* scrub and forest. Much is seasonally waterlogged, the remainder being at altitudes up to 60 m. The soils are poor in nutrients and to establish grass and legume pastures inputs per hectare of *ca.* 620 kg each of lime and single superphosphate, 130 kg of muriate of potash, 8 kg each of copper and zinc sulphate and 280 g of sodium molybdate are required (Barr, 1971). The subsequent fertilizer requirement is 250 kg/ha/a of single superphosphate and 130 kg/ha/a of muriate of potash. Rainfall is 1 250–1 500 mm/a. Quite productive mixed pastures of tropical grasses and legumes have been established and live weight gains of 250–350 kg/ha/a have been obtained (Evans, 1971). In the same area the Queensland Department of Forestry has for some forty years been growing the exotic *Pinus elliottii* with the aid of some phosphorus fertilizer.

The best economic land use for this area has been examined by McCarthy *et al.* (1970): 'Given the underlying assumptions and the criteria used, it was found that a 40 year pulp-timber rotation was most profitable, closely followed by a beef breeding and yearling marketing alternative where replacements for the breeding herd are bought on the store market. Beef fattening of stores was superior to a 20 year pulp rotation, while beef breeding and yearly marketing with the maintenance of a breeding herd was the least profitable. Product price levels would have to increase for all alternatives by 5–10 per cent all other assumptions remaining unchanged, before the ranking in profitability would change in favour of beef alternatives. However, if all product prices fell, beef would become relatively more unprofitable than forestry.' With the present depressed world beef market (1973–74), forestry would no doubt be more profitable.

Shifting cultivation and other agri-sylvicultural systems

Agri-sylvicultural techniques are those combining agricultural and forest production in either time or space. They depend on forest fallow to restore the soil fertility in the case of shifting cultivation and on the forest environment in the case of tree crops which are also frequently combined with livestock husbandry.

Shifting cultivation

Since 1945 shifting cultivation has been the subject of many studies and discussions. The agricultural and forestry committees of FAO have discussed it in joint sessions. The FAO Committee on forest development in the tropics dealt with qualitative and quantitative aspects of shifting cultivation during its first two sessions. The limited success of intensive permanent agriculture in the humid tropics (except on alluvial and volcanic soils) causes shifting cultivation and its associated destruction and regression of forests to continue to be a focus of interest. It achieves high productivity per man-day with small capital investment. A regional seminar (FAO/SIDA/ARCN, 1974) reviewed the situation in Africa and FAO (1974) dealt with the management properties of ferralsols and fallows in humid tropical forests.

In most instances clearing with simple tools is incomplete; soil disturbance is limited (Nye and Greenland, 1960). Consequently, if cultivation is not prolonged, rapid regeneration of secondary forest vegetation occurs when the land is abandoned (see chapter 9). This will not happen if serious soil erosion occurs as on steep slopes, or with thorough mechanized clearing, or with prolonged cultivation so that the viable forest seeds are reduced and replaced by grasses, but if prolonged cultivation is associated with measures to maintain soil fertility, no permanent loss occurs. Thus the effects of cultivation on tropical forest ecosystems differ considerably depending on circumstances.

Terminology

The essential characteristics of shifting cultivation are that an area of forest is cleared, usually rather incompletely, the debris is burnt, and the land cultivated for a few years—usually less than five—and then allowed to revert to forest or other secondary vegetation before being cleared and used again. The system varies in detail from place to place. Consequently many names are used for it. Nye and Greenland (1960) and many others use the term shifting cultivation as a general term embracing many variations of natural fallow cultivation systems. Conklin (1957) discusses the revived term swidden farming, and proposes its use for the more general descriptions of shifting cultivation, leaving this last expression for the more specific types of agricultural practices. Ten variables are proposed to distinguish different types of swidden farming according to agronomic and cultural practices. Spencer (1966) lists numerous terms related to shifting cultivation, mostly from South-East Asia and proposes 19 basic qualitative elements including such factors as ecological adaptation, labour efficiency, erosion control, etc. Watters (1971) includes in the definition of shifting cultivation the use of primitive tools and the subsistence economy usually associated with its practice. In the FAO/SIDA (1974) Symposium, continuous and non-continuous cultivation systems were differentiated, the first referring to where some form of continuous management is practised. The non-continuous were subdivided into natural

fallow and, if the homes of the cultivators are also moved, shifting cultivation. The recommended terminology given in this report is shown in table 1. Greenland (1974), reviewing the evolution of shifting cultivation, recognized a four-phase development, based on intensity of land use and mobility of homes: I. simple shifting cultivation; II. recurrent shifting cultivation; III. recurrent cultivation with continuously cultivated plots; and IV. continuous cultivation (table 2). Phases II and III can be subdivided according to the intensity of land use, given by Allan's land-use factor which is expressed as:

$$L = \frac{C+F}{C}$$

TABLE 1. Recommended terminology (*FAO/SIDA Symposium on shifting cultivation and soil conservation in Africa*, 1974).

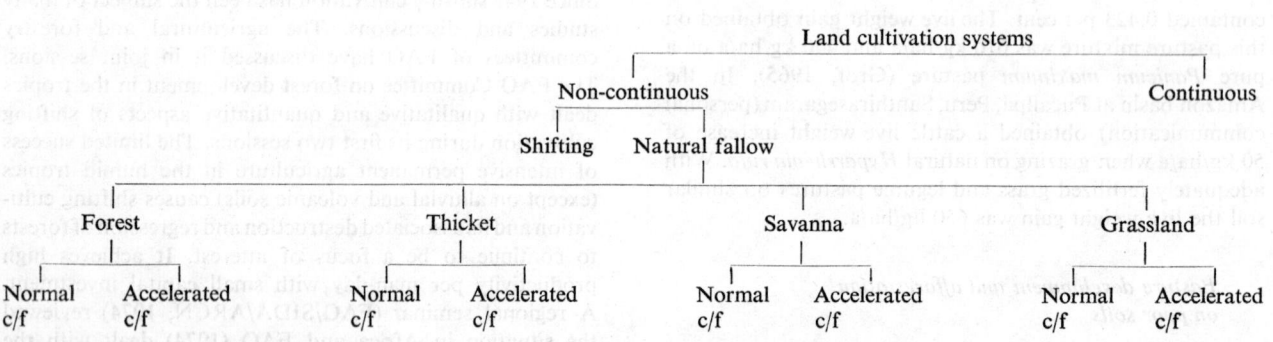

1. An initial division into *continuous* and *non-continuous* cultivation system.

2. *Continuous systems* include all where some form of management is continuously applied; e.g. planted fallows, planted tree crops, managed grassland and *taungya* cultivation.

3. *Shifting systems* are those in which the homes of the cultivators are moved on each occasion that the cultivated area of land is moved. *Natural fallow systems* are those in which the homes of the cultivators are not normally moved when the area of cultivated land is moved.

4. Natural fallow cultivation systems are very widely practised, and further subdivisions are necessary:

   a. The type of fallow vegetation:
     forest comprises woody vegetation with trunks and a closed canopy in which the trees are ecologically dominant;
     thicket comprises dense woody vegetation without trunks;
     savanna comprises a mixture of fire-tolerant trees and grasses in which the grasses are ecologically dominant;
     grassland comprises grasses without woody vegetation.

   b. The length of the fallow: *normal* fallow is adopted when land availability is unrestricted; if the length of the fallow period is less than would be voluntarily chosen because of population pressure or other factors, the term *accelerated* should be used.

   c. The lengths in years of the cropping and fallow periods; the end of the cropping period being taken as the time when the cropped land is no longer maintained. The lengths of crop and fallow periods should be shown as c/f where c is the length of the cropping period in years and f the length of the fallow period in years. Natural fallow systems shall thus be described in the following terms:

normal	forest	
or	or	
accelerated	thicket	
	or	fallow system, c/f
	savanna	
	or	
	grassland	

   e.g. an accelerated forest fallow cultivation system, $\frac{2}{5}$.

TABLE 2. The phases of land cultivation (Greenland, 1974).

Phase I	*Phase II*	*Phase III*	*Phase IV*
simple shifting cultivation	recurrent cultivation	recurrent cultivation with continuously cultivated plots	continuous cultivation
dwellings and cultivated area shift together	cultivated area shifts more frequently than dwellings	always complex; several field types	may involve alternate husbandry with planted and cultivated pastures or fallow crops
	may be complex; several field types		

TABLE 3. The categories of recurrent cultivation (Greenland, 1974, using Allan's land-use factor, 1965).

		land-use factor, L.
1. Intensive recurrent cultivation	only on highly fertile soils	1 to 2
2. Intermittent intensive recurrent cultivation	usually on highly fertile soils	2 to 4
3. a. Short-term recurrent cultivation	⎫	4 to 5
b. Medium-term recurrent cultivation	⎬ may be degraded by population pressure to	5 to 7
c. Long-term recurrent cultivation	⎭ shorter-term or more intensive system	7 to 10

where L: land-use factor, C: length of cropping period, F: length of fallow period (Allan, 1965). These categories (table 3) are useful in relation to ecological definition of land types. Watters (1971) differentiates two main categories: traditional shifting cultivation and that imposed by necessity which approximately correspond to normal and accelerated systems in the FAO classification.

The working group on terminology (FAO/SIDA, 1974) proposed four vegetation terms to describe the natural fallow vegetation: forest, thicket, savanna and grassland. The descriptions might be expanded using the unified system of classification and mapping of vegetation adopted by Unesco (1973) which deals with natural and semi-natural vegetation types including immature secondary growth and has proved useful in tropical regions (Kuechler and Montoya, 1971). Clarification should prevent duplication of research in similar agroecological zones whilst other zones are ignored. Ruthenberg (1971, 1974) classifies shifting cultivation on vegetation types, with further divisions related to migration, rotation practice, clearance methods, cropping sequence and tools used.

Terms such as migratory or pioneer agriculture are used mainly in tropical America to designate clearing forest, growing crops and leaving the area.

Populations

Some 200 million inhabitants of tropical countries were estimated to live by shifting cultivation (FAO, 1957). These have probably increased somewhat as little success has been achieved with continuous cultivation systems. The total population of tropical countries increased from 585 million in 1960 to 811 million in 1972. The total numbers of people engaged in or living by shifting cultivation must now be in the order of 240 million or more.

Their distribution in ecological zones is important. Cole (1965) gives the following table for tropical America:

Vegetation	Area $km^2 \times 1\,000$	Population $\times 10^6$	Persons per km^2	Per cent of area	Per cent of population
mountain	1 690	38.8	23	8.2	18.6
dry	5 226	41.4	8	25.3	19.7
forest	9 843	100.2	10	48.0	48.0
savanna	2 573	5.2	2	12.5	2.5
grassland	1 243	23.1	19	6.0	11.2

Comparable data for Africa and Asia have not been collected. Braun (1974) collected some data on Africa, but most refer to drier grassland areas, and uncertainties regarding whether natural fallow systems were included or whether the data refer only to phase I methods left considerable doubt of their value.

It is difficult to obtain an adequate estimate of the total area of the tropical forest currently subject to shifting cultivation. The areas where it is exploited by phase I type shifting cultivation are probably limited to small parts of the Ituri forests in Zaire, and some parts of the Amazonian forests. Natural fallow rotation systems of phases II and III are very much commoner. Such are practised throughout western Africa and support a population of *ca.* 80 million. In Zaire and neighbouring areas natural fallow systems are also wide-spread, although populations are lower. In the upland areas of Asia, phase III agriculture is wide-spread, with continuous cultivation in the wetter valley bottoms, and natural fallow rotations on higher ground (Spencer, 1966). In southern and central America some form of shifting cultivation is the predominant system being practised in most of the Amazon basin as well as in the more populated Andean Highlands, Atlantic coast of Central America, Antilles and in much of Mexico (Watters, 1971; Sanchez, 1972).

Population densities are much higher in Asia than in most of Africa or America. Although shifting cultivation played important roles in food production in the past in India and Pakistan, it is mostly in South-East Asia that it is now practised but with some pockets in the lower density areas of eastern and central India, the Himalayas, Burma and Sri Lanka. Much less information exists on the areas or people involved in shifting cultivation in Asia than for the rest of the tropics. Spencer (1966) estimated that in southeastern Asia *ca.* 50 million people live by shifting cultivation (including those practising phase III mixed systems) while 675 million live by permanent agriculture. These permanent agriculturists provide the bulk of food for the urban 200 million. The areas under each system are 100–112 million ha and 227–235 million ha respectively. Only about one fifth of the total area under shifting cultivation is under crops at a given time. The rate of land clearing is *ca.* 17 million ha/a; most is of lower quality than that used for permanent agriculture so their productivities are not comparable. Prothero (1972) collected papers on population and shifting cultivation in Africa.

Stable systems

De Schlippe (1956) gives probably the most detailed account of shifting cultivation as it is practised as a stable, phase III system, in derived savanna in the southern Sudan. Continuous cultivation is practised on old hut sites, and areas near the house—the compound farm; more distant areas are cultivated on a natural fallow rotation, with different intensities of use depending on soil type—riverine land being especially valued and more intensively used than other land. Cropping periods are two to three years, the land being abandoned when yields fall and weed competition becomes intense. Reopening of fallow land is determined by indicator species in the succession. Such systems have been described for all parts of the world (Cook, 1921; Conklin, 1957; Allan, 1965; Spencer, 1966; and others).

The adjustment of these systems to local ecological conditions reflects the equilibrium attained between man and the environment and that there has been sufficient time to achieve a balance. The system is commonest in areas where populations are not too high, and the rainfall not so great that leaching is serious i.e. it is confined to the drier forest and derived savanna areas with rainfalls of *ca.* 1 200–1 800 mm/a. The major soil types are ultisols, alfisols and entisols, not the more impoverished oxysols.

In some wetter areas the poorer nutrient levels of the soils are only compatible with infrequent use (land use factors >10). Consequently areas such as the Amazon basin (Sioli, 1973) and the Congo basin (Miracle, 1973) generally support very low population densities (<4 persons/km²). A good description of shifting cultivation in a similar area in Asia is given by Freeman (1955).

There are exceptions. In southeastern Nigeria, on poor sedimentary soils, with rainfalls >2 000 mm/a, populations of *ca.* 300 persons/km² are supported by shifting cultivation. The essential difference is the association of a permanent cover of oil-palms (Morgan, 1955). On volcanic soils (entisols) high population densities can be sustained as, in Indonesia, where rice cultivation is combined with a natural fallow system with a land-use factor close to 1. In intermediate areas, such as the moist semi-deciduous forests of Ghana and Nigeria with rainfalls of 1 400–1 800 mm/a, the cycling of nutrients through the vegetation is sufficient to maintain near neutral surface soils, although subsoils are highly acid. In these areas land-use factors of the order of 5 to 10 are common, and appear to maintain fertility (Nye and Greenland, 1960). Descriptions of practices in other parts of Africa are given by Allan (1965).

Degenerative and accelerated systems

Damage to tropical forest ecosystems arises in three situations: when the land type is unsuited to arable cultivation; when the agriculturists are unaware of the ease with which the ecosystem can be damaged, or are careless of such damage, and fail to adjust their practices accordingly; and when the agriculturists, although aware of the damage they may cause, are forced to exploit the ecosystem in order to survive.

Damage by cultivation of unsuitable soils is frequent on steep slopes, where clearing results in immediate severe erosion and exposure of a much less fertile subsoil. In many ultisols and oxysols, nutrients and organic matter are heavily concentrated in a thin A horizon, maintained by litter additions. Removal of the vegetation removes the source of soil enrichment. Removal of top soils leads to the disappearance of the nutrients and the seeds from which the forest may regenerate. Slope erosion also deposits sediments in other areas, affecting their fertility, and induces flooding because the slope soil is much less able to absorb rainfall and the valleys may be silted up. This is particularly important in catchment areas and where these are still forested cultivation should be forbidden; legislation passed and enforced to this effect. Failure to do so will result in no usable soils.

Where a group is in equilibrium with its environment, destructive shifting cultivation does not occur. Lack of adjustment is most common when people are displaced and enter a new and unknown situation. Many of the examples of shifting cultivation described by Watters (1971) in Latin America belong to this category. Where people are forced from urban communities by their inability to earn a living, they have to take up farming. Clearing forest is their only method and usually the poorest soils are available. A rapid decline in productivity almost invariably occurs (fig. 1) as the nutrients added to the soil when the forest is burnt are leached and weed competition becomes more intense. Whereas the traditional farmer has learnt to abandon the land early, allowing the forest to regenerate through his final plantings of cassava and plantains, the displaced cultivator will endeavour to obtain all he can from the land. Consequently fertility is further reduced, regeneration delayed and the liability to erosion increased. The land is abandoned only when real deterioration has occurred.

The reasons for the decline in fertility have been debated frequently. It is not due to removals of nutrients in crops, which are a small proportion of that added to the soil from the forest (Nye and Greenland, 1960). It may be due to leaching of nutrients out of the root range; this is almost certainly true in the higher rainfall zones. Loss of nutrients is only part of the reason for the abandonment; competition from weeds, particularly grasses, being at least equally important. This is well illustrated by data Kang (1975) has collected at the International Institute of Tropical Agriculture, Ibadan, Nigeria (fig. 2). Here on an oxic paleustalf, in the drier forest zone but near the savanna boundary, the decline in yield after clearing is clearly a combined effect of loss of nutrients and increasingly important weed competition.

The situation in which injudicious land use occurs, arises from increasing population densities. In traditional systems there is an adjustment of land use intensity to the inherent soil properties, longer fallow periods being required on poorer soils (Allan, 1965). As long as land is plentiful, the appropriate lengths of cropping and fallow can be respected. In phase II and phase III shifting cultivation, this involves use of land further and further from the fixed homes as population increases, and eventually a move of some homes to avoid excessive travel. Eventually land, even at considerable distances, is not available. The usual response is to shorten the fallow, reduce the land-use factor and initiate an accelerated system. These are becoming increasingly common in most areas of shifting cultivation, with the exception of the Amazon and Congo basins where population densities are still very low.

Increasing population pressure requires increased cultivation periods and decreased fallows, and threatens the balanced alternation of crop and fallow periods. The ability of the forest to regenerate is reduced to the point where grasses may invade and the succession diverted from forest to savanna. The introduction of relatively permanent tree crops may aggravate the situation by restricting the mobility and causing the over-exploitation of any natural fallow rotation.

Effects

Cultivation of forest affects the vegetation and soils. The data on viability of seeds of forest species after clearing and cultivation has received little attention, although Gómez-Pompa *et al.* (1972) have examined some South American rain forest soils and discussed the causes of deflected successions, and Brinkmann and Vieira (1971) have examined the effects of burning after clearing on seed vaibility. Kellman (1969) discusses species number and composition of the regenerating forest for Mindanao (see chapter 9). Keay (1960) found that most of the regeneration in Nigerian forests came from seeds lying in the soil before clearing.

The penetration of savanna into forest has been studied rather more. The practice of burning grass at the end of the dry season has received much attention, as it leads to suppression of many of the forest species, and dominance of pyrophilous savanna species. A very extensive bibliography on this topic has been prepared by Bartlett (1955, 1957, 1961). The effects of cultivation and fire on forest succession have been discussed by Budowski (1956, 1961), Bazilevič and Rodin (1966) and Daubenmire (1972). The general impact of shifting cultivation was also discussed by Donis (1975) and Denevan (1975) in two papers submitted to FAO. According to Donis, 'it favours light-demanding species, among them several species of economic importance; it makes the structure of stands less complex, but many forests are changed adversely and transformed into degraded secondary scrub without any production potential, and turns vast areas into savannas where annual burning causes intense erosion'. Denevan, after elaborating on the chemical and physical changes in soils under short and long-term fallows, discussed their environmental impact:

'Effects on wildlife are fairly obvious. If vegetation changes, then changes in kind and number of wildlife can be expected. Where long-fallow systems prevail, full recovery on disturbed sites is potentially very rapid. However, even if only sparse human populations exist in an area, they can modify wildlife by hunting even more than can temporary and scattered changes in vegetation. On the other hand, where protective forest cover survives around swiddens, certain species are encouraged by the availability of swidden crops or youthful successions of forest fallow. Many swidden farmers tolerate crop losses due to deer, monkeys, birds, rodents, and other animal pests because the hunting thereof is facilitated. With short-fallow cropping systems and where cropping is replaced by permanent pasture or scrub savanna, the effect on wildlife is devastating. Those species favouring open habitats may be intentionally destroyed to minimize competition with livestock for forage. Where forest clearing for swiddens is massive, hydrology is significantly modified. Without a forest cover, run-off is more rapid resulting in shorter but higher and more destructive flood levels. Soils dry out more rapidly affecting vegetation and crops, and lowered water tables cause lower stream levels during the dry season. Clearing of vegetation on slopes is particularly dangerous, as erosion may be greatly accelerated, and losses and damage can be catastrophic under heavy rains. Many of the deaths in the 1974 Honduran hurricane resulted from large mudflows from slopes denuded by shifting cultivators. The effect of deforestation from shifting cultivation on climate is poorly known. Microclimates are obviously changed, with greater surface heating. Haze from swidden fires at times covers vast areas and the smoke particles can reduce solar radiation reaching the ground. Extensive clearing must effect evapotranspiration locally, if not regionally, and while the ocean is the main source of moisture

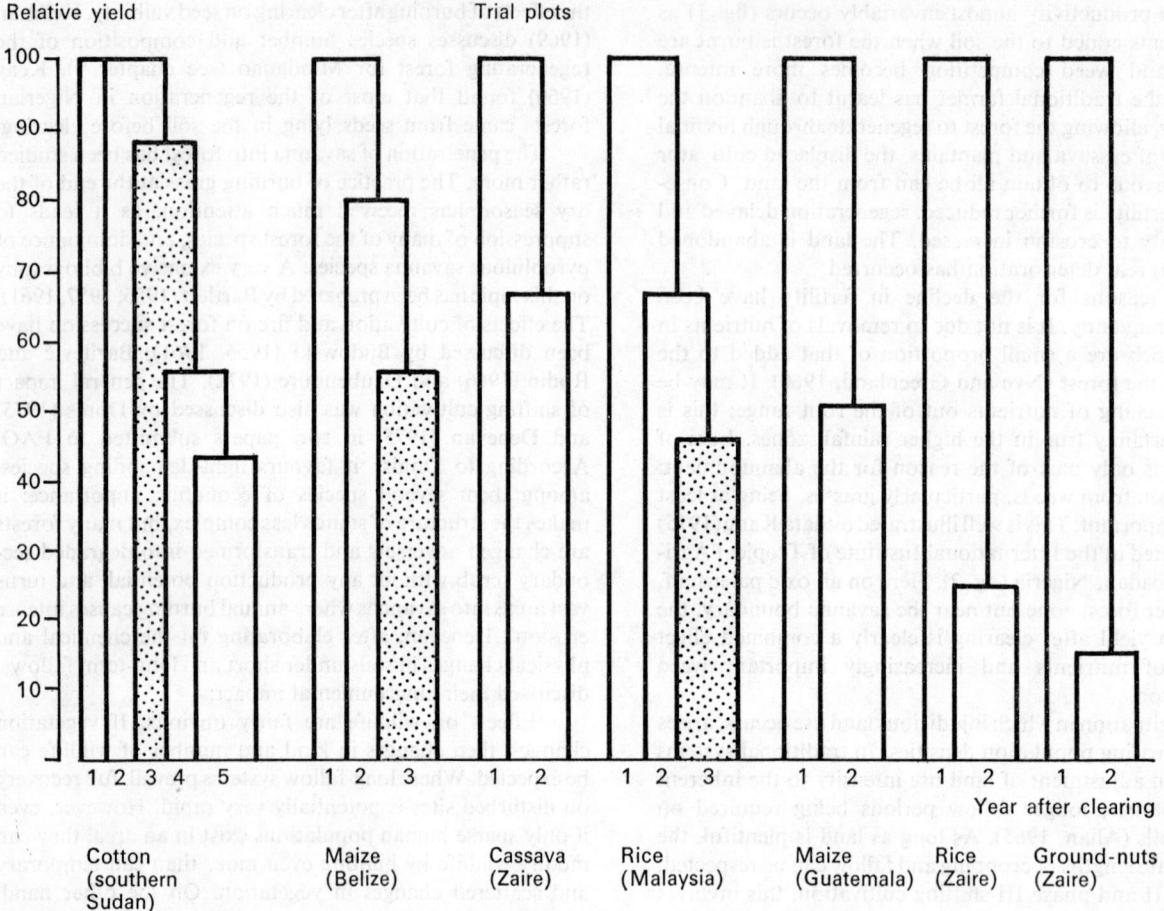

FIG. 1. Changes in yields following the clearing of natural fallow
(from Ruthenberg (1971) using data of Nye and Greenland, 1960).

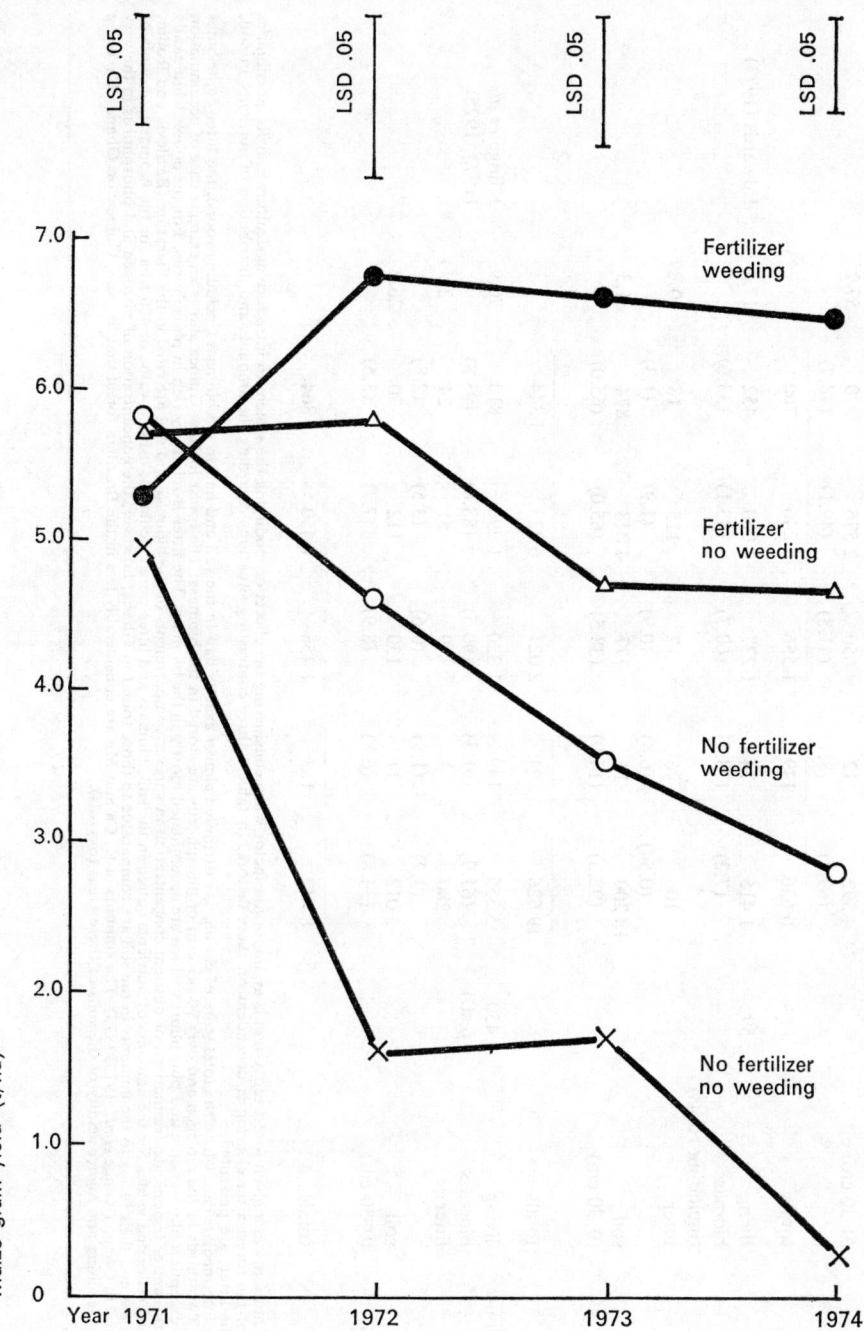

FIG. 2. Effect of cultural factors on successive maize yield on the same plots on Egbeda soil series at IITA (Kang, 1975, unpublished).

TABLE 4. Amounts and distribution of nutrients in soil, biomass, and the relative importance of litter, for three tropical forests.

Location	Forest and soil types	Ecosystem compartment	Biomass (t/ha)	Total nutrients stored (kg/ha) and (per cent)					Per cent total	References
				N	P	K	Ca	Mg		
Kade, Ghana	40 year mature secondary forest	living biomass (including roots)	361	2 009 (30.3)	136 (91.3)	896 (57.6)	2 625 (50.0)	384 (50.5)	42.2	Nye and Greenland (1960)
		litter		35 (0.5)	1 (0.7)	10 (0.7)	45 (0.9)	6 (0.8)	0.68	
	mid-slope brown loam over phyllite	soil (0–30 cm)		4 592 (69.2)	12 (8.1)	650 (41.8)	2 576 (49.1)	370 (48.7)	57.2	
	(ultisol?)	totals		6 636	149	1 556	5 246	760		
Marafunga, New Guinea	lower montane forest, 2 500 m	living biomass (including roots)	592	1 415 (7.2)	74 (78.7)	1 227 (60.7)	2 060 (33.1)	452 (33.6)	17.8	Edwards (1973)
		litter		106 (0.54)	6 (6.4)	17 (0.9)	115 (1.9)	18 (1.3)	0.89	
	humid brown clay developed from volcanic ash	soil (0–30 cm)		18 200 (92.3)	14 (14.9)	777 (38.5)	4 048 (65.0)	874 (65.0)	81.3	
	(entisol)	totals		19 626	94	2 021	6 223	1 344		
Manaus, Brazil	tropical ombrophilous alluvial forest	living biomass	450 (est.)	5 355 (61.9)	149 (91.4)	1 320 (90.5)	1 290 (88.8)	810 (93.8)	70.9	Klinge et al. (1973, 1975)
		litter		280 (3.2)	3 (1.8)	9 (0.6)	51 (3.5)	24 (2.8)	2.9	
	light yellow latosol (oxysol?)	soil (0–30 cm)		3 012 (34.8)	11 (6.8)	130 (8.9)	112 (7.7)	30 (3.5)	26.2	
		totals		8 647	163	1 458	1 450	864		

The distribution and the total amounts of nutrients in a given forest ecosystem is an important factor when considering the effects of modifying the system to introduce agriculture and its possibilities of success or failure. Also, the recovery of the forest after clearing is dependent on these factors. In table 4 three contrasting situations regarding the amounts and distribution of nutrients in soil, biomass and the relative importance of the litter, are presented.

The 40-year fallow in Ghana represents an intermediate case where the total store of cations is distributed almost exactly half in the soil and half in the biomass (including roots); the litter contributes less than one per cent of the nutrients. Two thirds of the nitrogen and over 90 per cent of phosphorus are stored in the biomass. The forest in New Guinea shows an extreme case of accumulation of nitrogen and cations in the soil with the exception of potassium and phosphorus which are accumulated mostly in the biomass. The litter is relatively rich in phosphorus. For the poorer white sandy soils which are found to support forest in the wetter regions, determinations of nutrient concentration in various compartments of the biomass and soil are cited in the literature (Bazilevič and Rodin, 1966; Stark, 1971). For Amazonia (Manaus) estimated values for the amounts of nutrients stored in the soil, biomass and litter were calculated, using a value of 450 t/ha for the biomass; amounts of both nitrogen and phosphorus in this biomass and their ratios to the amounts in the soil are comparable to those found in Ghana (Klinge et al., 1973, 1975). The amounts of K, Ca and Mg are considerably less in the Brazilian forest that in New Guinea or Ghana, Ca being much the lowest, in accord with the greater leaching and higher acidity of the soils of these areas (oxysols).

for rainfall, evapotranspiration is also a contributor. One preliminary study done of the water balance of Amazonia suggests that if the rain forest was completely destroyed rainfall in the interior would be reduced by up to about 25 per cent (Wise, no date).'

The effects of shifting cultivation on soil properties were reviewed by Nye and Greenland (1960). More recent work has been discussed by Vine (1968), Sanchez (1972), Fassbender (1974) and Greenland (1974). Measurements of changes in soil nutrient levels associated with shifting cultivation, subsequent to those discussed by Nye and Greenland (1960) are reported by Cunningham (1963), Nye and Greenland (1964), Sanchez (1972) and Brinkmann and Nascimento (1973). The general effects are now well established. The standing forest vegetation contains large quantities of plant nutrients (table 4; see also chapters 10 and 13). When the forest is felled and burnt, much carbon, nitrogen and sulphur are volatilized as oxides, while the remaining nutrients are added to the soil principally in the form of carbonates, with lesser amounts of silicates and phosphates. The carbonates cause the soil alkalinity to increase by as much as 2 pH units and there are substantial increases in the amounts of available nutrients (table 5). Although sulphur and nitrogen are largely lost during burning, the reserves of these elements in the soil are little if at all affected (Nye and Greenland, 1960; Sanchez, 1972; Brinkmann and Nascimento, 1973). The large release of nutrients immediately following burning can lead to a rapid loss of some of these by leaching. This is particularly important in the higher rainfall areas (Brinkmann and Nascimento, 1973) but in the semi-deciduous forest leaching is usually less intense and nutrient deficiencies may appear only after several years of cultivation (Greenland, 1974).

The study of the cycling of nutrients between forest and soil has received considerably increased attention in recent years (see chapter 13).

Changes in the physical properties of the soil which follow forest removal (Van der Weert, 1974; Wilkinson and Aina, 1976) consist of substantial losses of macropores if clearing was by heavy machinery resulting in decreases in infiltration rates and permeability. The change is considerably less if hand clearing is used. As a result of reduced infiltration rates, the run-off is considerably greater. If the soil is exposed, the detachment of soil particles by raindrop impact and the transport of detached material by run-off produces large erosion losses (Barnett *et al.*, 1972; Lal, 1974). However if the soil is protected by mulch or living vegetation detachment does not occur and erosion is much reduced (Lal, 1974).

Mulching is a common practice with many shifting cultivators (Allan, 1965) and the system of mixed cropping which is almost universally used helps to maintain cover. Nevertheless clearing and planting must lead to some soil exposure and the high rainfall will produce soil erosion loss. If steep slopes are cultivated, or poor practices used, the extent of erosion can be very severe.

Some substantial studies have recently been completed on the soil fauna and flora. Fittkau and Klinge (1973) review recent publications and provide data for the Amazonian rain

TABLE 5. Changes in the soil after clearing semi-deciduous forest at Kade, Ghana.

a. Following local practice.

		Before clearing	After clearing and burning	After 1 year's cropping	After 2 year's cropping
0–5 cm	pH	5.2	8.1	7.5	7.0
exchangeable	K	0.5	2.5	0.6	0.3
cations	Ca	7.2	21.2	20.5	14.5
meq/100 g	Mg	2.6	3.9	2.4	2.0
organic C (%)		2.86	2.34	2.67	2.19
organic N (%)		0.26	0.25	0.24	0.19
5–15 cm	pH	4.9	6.2	5.4	6.0
exchangeable	K	0.3	0.9	0.4	0.2
cations	Ca	2.0	5.0	3.2	5.2
meq/100 g	Mg	1.2	1.6	1.2	1.1
organic C (%)		1.06	1.18	1.01	1.06
organic N (%)		0.11	0.13	0.10	0.11
15–30 cm	pH	4.9	6.2	5.0	5.2
exchangeable	K	0.4	0.5	0.3	0.2
cations	Ca	1.7	3.8	2.5	3.2
meq/100 g	Mg	1.2	1.4	1.1	1.1
organic C (%)		0.84	0.85	0.63	0.71
organic N (%)		0.092	0.092	0.072	0.070

b. With tillage and crop succession of maize-cassava-maize.

		Before clearing	After clearing and burning	After 1 year's cropping	After 2 year's cropping
0–5 cm	pH	5.2	8.0	6.8	6.6
exchangeable	K	0.4	1.5	0.4	0.2
cations	Ca	4.0	12.7	7.2	6.8
meq/100 g	Mg	1.9	2.9	1.7	1.3
organic C (%)		2.26	2.06	1.64	1.60
organic N (%)		0.24	0.19	0.15	0.15
5–15 cm	pH	(4.7)*	7.4	5.5	5.5
exchangeable	K	(0.2)	0.8	0.4	0.2
cations	Ca	(1.2)	5.0	3.0	3.0
meq/100 g	Mg	(0.9)	1.5	1.2	1.0
organic C (%)		(0.78)	1.17	1.07	0.97
organic N (%)		(0.08)	0.13	0.10	0.094
15–30 cm	pH	(4.7)*	5.5	4.8	4.9
exchangeable	K	(0.2)	0.5	0.2	0.2
cations	Ca	(1.2)	2.5	1.2	1.5
meq/100 g	Mg	(0.9)	1.1	0.7	0.8
organic C (%)		(0.78)	0.98	0.61	0.59
organic N (%)		(0.08)	0.093	0.061	0.059

* 5–30 cm samples analysed together.

Data from Nye and Greenland (1960). See also Cunningham (1963).

The data refer to a ferrallitic soil at the University of Ghana research station, Kade (rainfall, 1 500 mm/a). Cultivation (a) according to local practice, with no treatments and seed planted with a dibble stick; maize was sown and interplanted with cassava, cocoyams and plantains; (b) with land-tillage followed by cropping to maize, cassava and maize in succession.

forest, and the group working at the IITA in Ibadan, Nigeria, supported by the Centre for Overseas Pest Research (COPR) have obtained much data on the soil populations and the changes which follow clearing and burning (IITA Annual Report, 1974). This group has also obtained information on the effects of insecticide (DDT) applications. In general the effects of clearing and burning are far more drastic than those of insecticides. The relation between the soil populations and soil productivity remains undefined, and it is difficult to know if the observed changes are important for forestry or agricultural production.

Food and energy

The productivity of shifting cultivation varies widely. Gathering from wild species in natural fallows, fishing and hunting often provide an important part of the diet of traditional shifting cultivators. In many cases most of the energy in their diet comes from a major staple crop particularly maize, taro, sweet potatoes and cassava. Of these cassava is one of the most efficient in the production of carbohydrates and has an additional advantage in that it can be stored in the ground up to two years before harvesting. Following the introduction of cassava in Africa and its wide-spread use by many shifting cultivators, some authors consider it as a cause not only of malnutrition due to low protein content (see chapter 16) but of some neurological disorders due to cyanogenic substances in the bitter varieties (Lowenstein, 1973). Cassava may produce three times the energy of maize due to 12 months growing period (versus 4 to 6 months for cereals), greater tolerance of poor soils, vegetative reproduction and a virtual elimination of storage losses (Miracle, 1973). Yields of 10–18 t/ha have been reported on poor soils in the Amazon basin (Meggers, 1971).

The energy flows in shifting cultivation and modern agriculture reveal some of the basic differences between the two systems. Shifting cultivation unlike modern agriculture makes little or no use of fossil fuel. In a very detailed account of the energy flows in a group of shifting cultivators in New Guinea, Rappaport (1971) compared the input in human labour with the yield produced by the plants. He found a 16 to 1 ratio of energy output to input. The per caput ingestion by the Tsembaga people was 2 200–2 600 kcal/day. The total calorie output of food was 22×10^6 kcal/ha of which *ca.* one third was used to feed ceremonial pigs. Stout (1974) points out that a large part of the success of American agriculture has come from applying fossil fuels but this system now requires about one calorie from fossil fuel for each edible calorie. Pimentel *et al.* (1973) give a more conservative ratio of one calorie from fossil fuel to 3 calories of food. Revelle (1974) has calculated that one hectare of maize in Iowa (6.4 t/ha/a) would provide 2 500 kcal/day for 24 people and for the population of the world (4×10^9 persons), only 158 million hectares would be required at this level of productivity to meet their energy needs. The world average agricultural efficiency is only one tenth of that of maize in Iowa. Comparison illustrates that for the same energy output in food per hectare, the input of energy is only 1/16th for the Tsembaga while for a farm in

Iowa it is $\frac{1}{8}$ or more. In the shifting cultivation system, relying only on manual labour, the natural tendency is to make it labour-efficient rather than economically efficient. In this respect, Conklin (1957), Nye and Greenland (1960), Spencer (1966) and Watters (1971) have found that shifting cultivation is labour-efficient however it is practised, but adapts poorly to the market-oriented economy imposed by large concentrations of people.

When shifting cultivation is forced to become more intensive it may lose its stability. More research is needed on the energetics of shifting cultivation and continuous management systems. Even if it were technically feasible to produce food for the estimated 4×10^9 people in the world with an efficiency similar to the best mechanized farming, the known reserves of petroleum would be finished in only 29 years—assuming that petroleum is the only source of energy and that it is only used to produce food (Pimentel *et al.*, 1973). Obviously, the kind of agriculture and human diet must be set at more realistic values if the goal is a stable system of production. Many of the international programmes to increase food production are aimed at high yields per unit area regardless of the energy input necessary to attain such production. *A new look at energy resources* (McCloud, 1974) critically reviews the situation and essentially proposes three solutions: a differential price of fuel for agriculture, a better use of the energy input on farms by recycling wastes as well as growing fuel crops, and improving the physiological conversion of solar energy into agricultural products. Pimentel *et al.* (1973) consider such alternatives as rotations and green manures to reduce the high demand for fossil fuel energy in the form of fertilizers and pesticides. Energy expenditure might be reduced rather than develop highly mechanized agriculture for different physical and economic environments. The consequences of this development of modern agronomy must be taken into account when planning for the development.

Conclusions

It can be concluded with Denevan (1975) that all forms of shifting cultivation have a destructive and modifying impact on tropical forests. However, the nature of this impact varies considerably, from dispersed and temporary to wide-spread and permanent. Systems of shifting cultivation can be classified as long fallow, short fallow, unstable fallow and migratory.

As is generally acknowledged (Conklin, 1957), long-fallow or forest-fallow shifting cultivation is ecologically sound, leaving small scars not too unlike those from natural disturbances (floods, land-slides, falling trees) which are rapidly healed (see chapter 8). The period of cultivation is short (1–2 years), the period of forest fallow comparatively long (>*ca.* 20 years), and the fields are usually widely dispersed. Small swiddens within the forest are protected from winds and erosion, and conserve moisture well. Regeneration following abandonment is rapid as result of immediately adjacent seed sources, and biotic diversity is assured (Gómez-Pompa *et al.*, 1972). These systems are characteristic of low density populations in Amazonia and to a lesser extent in very humid areas of central Africa and South-East Asia. They renew fertility but may not allow the development of mature forests which can take a century or longer. Early successional stages may contain relatively high percentages of economically useful and fast-growing species such as pine, teak, ceiba, mahogany and tropical cedar, as well as fruit- and other planted or feral domesticated trees, associated with abandoned swiddens. Long-fallow systems can support only sparse populations and small settlements since only a small portion of a total land area is utilized at one time.

Where fallows range from *ca.* 4 to 20 years, scrub and low forest regeneration may be adequate to sufficiently restore fertility and moisture, improve soil structure, and reduce weeds and pests to permit a renewal of cultivation. Under such a fallow the ecosystem will only partially recover in terms of tree size, species diversity and wildlife. Forest regrowth can be aided by interspersing swiddens with forest reserves which serve as seed and moisture sources and wind-breaks, as with the corridor system of the Congo Basin (Kellogg, 1963) or with the present Brazilian policy of requiring that half of all new land holdings in Amazonia remain in forest. The rate and composition of the regrowth can be influenced by planting trees, including timbers and plantation crops during cropping, as in the taungya system (Watters, 1971), nitrogen-fixing legumes or fast-growing trees such as *Casuarina*, as in the New Guinea highlands. The Tsembaga protect the tree seedlings which grow in their fields (Rappaport, 1971). Another option is the orchard-garden thicket, described by Gordon (1969) for Indians of Yucatan, Panamá and Colombia, where trees may be planted, but mainly the natural secondary species are hand-selected for usefulness by destroying the undesirables. Thus, shifting cultivation can rotate with arboriculture, or a managed forest that is both productive and protective, rather than with natural forest fallow. Stable short-fallow systems are very common in many parts of Asia and Africa among farmers who know the needs and limitations of their habitats and have adjusted their farming accordingly (Watters, 1971). Populations up to about 50/km² are possible but the systems break down when the population increases with either environmental deterioration and/or a shift to more intensive cultivation (see chapter 19).

Where the fallow is less than *ca.* 4 years it is dominated by grasses and shrubs and the recovery of soils and woody vegetation is insufficient. Two land-use developments can be differentiated, the determining factors being complex and little understood but population density is apparently critical. First, the natural fertility may be increased by mulches, composts, manures, ash and sometimes manufactured fertilizers, and the soil structure can be improved, and weeds eliminated by hoeing, mounding and ridging (Denevan, 1975). Environmental degradation may be severe, ultimately resulting in land abandonment, but can be compensated for by agrotechniques such as terracing. Cropping becomes permanent or near permanent. Population densities, labour inputs and production costs are high; consequently standards of living are low. There are no forests, but such agricultural intensification and population absorption in one area can potentially save other forest areas

from being destroyed by expanding shifting cultivation. Some Old World derived savannas are cultivated using shrub and grass-fallow techniques. Secondly, short-fallow cycles are accompanied by increasing soil deterioration, weeds and pests, and declining yields. The soil is likely to be so degraded that forest recovery is slow or it is replaced with shrub savanna maintained by a minimum of human disturbance (Budowski, 1956).

Migratory (pioneer) shifting cultivation without forest fallow may follow a series of short-fallow cycles and abandonment then or after only one or two years of cropping. This is common in Latin America, particularly near settlements and roads. Recent migrants have not learnt appropriate fertility maintaining techniques. They tend to crop too long and not fallow long enough. When production declines, rather than fertilize and control weeds, which demand high labour or capital inputs, the land is abandoned and new forest is sought. The abandoned fields may revert to a forest fallow or be used for plantation crops; usually they become cattle range. Land deterioration is permanent. The cover is grass and shrubs with little or no rotation with crops or forest. Forest recovery is intentionally prevented by burning and chopping and unintentionally by fires, invasion of agressive grasses or bracken, and by deterioration of soil structure and nutrients level. This savannization is often extensive and seeds for forest regeneration becomes unavailable.

The conversion to permanent grassland or shrub savanna provides the greatest threat to the survival of tropical moist forests. The process dominates the recent history of land use in tropical Latin America (Watters, 1971; Denevan, 1975), as well as in much of Asia and Africa. Clearing the forest reflects population pressure elsewhere, not locally. The failure to establish either a stable short fallow or permanent cultivation and the availability of new forest reflects this lack of local population pressure. Despite land availability, long forest fallows seldom occur. Abandoned swiddens become available to cattlemen at low cost who maintain the vegetation as grassland. Thus shifting cultivation is replaced by an even more extensive form of land use—cattle grazing. With poor soil, the pasture is poor and cattle quality and quantity are low. The beef and milk produced are too expensive for most people, but there is a market and the demand for even low quality grazing land is great. Thus, while shifting cultivation is the instrument of forest destruction in frontier regions, it is cattle ranching that impedes the forest recovery.

There are therefore several ways to carry out shifting cultivation according to the objectives, and much biological and socio-economic research remains to be carried out. The taungya system has worked well in many parts of the world (King, 1968). In such systems the cultivators grow crops for a few years and plant commercially useful timber species; after their harvest, the trash is burnt *in situ* and the cycle repeated. Taungya systems may be seen either as farm-forestry or as part of industrial timber development. As farm-forestry it has been described by Lowe (1975) in Nigeria where it has been usual for the forestry services to leave food production entirely to peasant farmers, who have cleared the forest in return for permission to use the land for arable farming and the right to all food crops grown. Some help might have been given with felling the largest trees or they might be left, though there might be an attempt to kill them by fire. The forestry staff surveyed, selected and allocated the land amongst the farmers. To some extent they regulated the farming activities by coordinating the brushing, felling, stacking and burning of the debris and the removal of logs. Tardy farmers were chivied, operations such as felling or burning being delayed until all farmers had completed the preliminary work. There was some regulation of crops in order to protect the tree crop. For example plantains were either prohibited or restricted to farm edges and it was usual to prohibit the planting of cassava at the beginning of the first season and to regulate its spacing. The unbranched varieties of cassava, which interfere less with the tree crop, might be favoured. The forestry services have accomodated the farmers by planting trees at a wide spacing (table 6) and by permitting farmers to prune side-branches from the young trees; in Benin State a gratuity was paid and was then discontinued. Farmers have also assisted in the planting operations without payment, but this may have been a mistake, as it can result in a poor quality and affect the percentage of trees established.

summary of farming cycles in three localities is given in table 7. Commonly raised crops are egusi melon (*Citrullus vulgaris*) intergrown with yams (*Dioscorea* spp.) or maize which may be followed by a second crop of maize. Alternatively they may all be intercropped. Intermixed with these are peppers (*Capsicum* spp.), tomatoes, okra (*Hibiscus esculentus*), spinach (*Amaranthus* spp.) and sometimes beans and other vegetables. It is usual for the farming cycle to terminate with cassava. The exact cycle depends on local conditions and customs, and also on the energy of the individual farmer. Yams, though much the most paying crop, require the greatest effort and the proportion covered by yams depends on the desire or need for cash. The Forest Department fits well into this pattern of labour and reward, organizing the farming and obtaining the cultivation of its trees which involves the farmer in little extra work.

Taungya can only be practised in limited areas usually on forest land near to existing communities. The opening of large areas of forest for farming could be unwise, because of eventual socio-economic problems. It is desirable to establish stable rural communities and each taungya series should be properly planned. It is unlikely that there will be an indefinite reservoir of farmers prepared to carry out taungya operations. In any event taungya schemes have a

TABLE 6. Spacings used for tree crops in farm forestry (taungya) in Nigeria (Lowe, 1975).

Tree crop	Planting stock	Planting spacing (m)	Trees/ha	State
Gmelina arborea	stumps	3 ×3	1 076	Western
		2.5×2.5	1 682	South-Eastern
Nauclea diderrichii and Meliaceae (mahoganies) $\frac{5}{1}$ mixtures	stumps	3.5×3.5	747	Mid-Western
Tectona grandis	stumps and/or polipots 7.5×12 cm	2.5×2.5	1 682	Western
Terminalia ivorensis	stumps	5 ×5	420	Western
		5.5×5.5	332	Mid-Western

Note: Polipots of 7.5×12 cm (flat) are also suitable for *Nauclea* and *Terminalia* giving superior survival and growth compared with stump plants.

Peasant farmers are prepared to clear and cultivate forest for two reasons: first, because of the fertility of newly cleared land, and secondly because much of their own land is covered by perennial tree crops so that, in some areas, there is a shortage of land for food production. Land hunger *per se* is not essential for successful taungya schemes, and where it exists (as in the East-Central State) the only land available for forest reservation is waste or at risk from erosion. The methods of farming used within taungya are essentially traditional, with an intimate mixture of crops in order to utilize the land as fully as possible. Taungya farmers are probably in general better than the average, and are frequently more orientated towards producing crops for sale. The need to clear high forest usually ensures that they are amongst the stronger farmers. Where areas larger than *ca.* 0.5 ha are allotted, the quality of the farming tends to deteriorate and the farmer has to hire assistance. A

finite life, as the farms must be reasonably close to the settlement; normally within 5 km, perhaps up to 10 km, if access were improved. By the time the accessible land is all converted to plantations, the forest estate should have developed to a point where sufficient employment can be supplied by forestry activities and wood-using industries. This might be about 10 years for *Gmelina* and about 30 years for *Tectona* or *Terminalia* or *Nauclea*/Meliaceae mixtures. The viability of taungya projects may depend on marketing provision for arable crops and the thinnings. When new schemes are launched all these factors should be carefully thought and written into a working plan.

Donis (1975) stressed that facilities for transport, exploitation and export should determine the location of timber production and regions suitable for such would be those already exploiting timber or those which could be opened up by the constructions of new outlets by road, railway or river. The regions destined to produce wood on an industrial scale will thus be the coastal zones, and areas served by large rivers, main roads or railways, for which transport distances of 100–200 km appear to be economically acceptable. Such projects should only be sited in regions where utilizable soils are ≥25–30 cm deep.

TABLE 7. Taungya farming cycle in Nigeria (Lowe, 1975).

State	Western	Mid-Western	South-Eastern
Annual rainfall (mm)	1 250–1 600	1 500–2 000	1 700–2 600
Geology	basement complex	sedimentary sand	basement complex
Farming agency	peasant farmer	peasant farmer	Forestry Department
Year 0			
October	—	—	demarcation
November	demarcate farms	} demarcate farms	brushing
December			felling
Year 1	} brushing		
January		} brushing	} burning and packing
February	felling		
March	burning and packing	felling	sow maize (1×1 m)
		burning and packing	
April	sow maize, egusi	sow maize, melon,	
	melon, okra	okra, *Amaranthus*	
	dig yam mounds	and peppers	
May, trees	plant yams when	dig mounds and	
planted	maize germinated	plant yams	
June, trees	harvest maize	—	harvest maize
planted			
July	harvest melon	} harvest maize	} plant cassava
		and melon	
August	plant cassava		
September	resow maize		resow maize
October			
November	harvest second		harvest second
	maize crop	} harvest yams	maize crop
December	harvest yams		
Year 2			
January	—	plant cassava	
April-August	harvest cassava		
July-December			harvest cassava
Year 3			
January-March		harvest cassava	

The objectives of farm forestry and industrial wood production could be combined; a development scheme for this was established in the Philippines. Harcharik (in Donis, 1975) discusses the role *Acacia falcata* could play in providing employment for rural people and in establishing tree farms within 100 km of the mill on private holdings in the Bislig region of Mindanao. Rather optimistic growth figures have been reported for an 8-year rotation with coppice regeneration: 290 m³/ha on poor sites, 570 m³/ha on typical sites, and 696 m³/ha on best sites.

Other agri-sylvicultural systems

Many other agricultural systems besides shifting cultivation have been and are used in the tropical forest regions (Duckham and Masefield, 1971; Ruthenberg, 1971). The most successful are those which involve perennial species, and so maintain a permanent vegetative cover. Plantation developments and the incorporation of perennial crops into systems including shifting cultivation have been successful (Coulter, 1972).

The increasing urban to rural population ratio, without the development of agricultural practices places great demands on shifting cultivators for food production. A smaller proportion of people are left with inadequate techniques to feed themselves and the town-dwellers (see chapter 19). The tendency to grow cash crops instead of more food aggravates the situation. The low yields usually force the farmer to decrease fallow periods and increase cropping periods causing yields to drop even more. There have been several recent attempts to determine more appropriate methods using continuous management systems.

For lowland hydromorphic soils, continuous rice cultivation is possible as is well established in the Far East; in many areas of Africa, as well as Asia, this form of continuous cultivation is combined with shifting cultivation of upper slope areas. The greatest reserve of land exists in the areas of ultisols and alfisols—the red earths and latosols of older classification systems—that are farmed with land-use factors of ⩾5. In western Africa and many other areas alternatives are not between its preservation as forest or use for agriculture, but between its exploitation by destructive agricultural techniques and its development by appropriately conservative farming methods. In the Congo and Amazon basins the low population densities afford an opportunity for a more carefully considered land-use policy, involving the preservation of some forest, agriculture and sylviculture.

The primary requirements of an appropriate agricultural technology is that soil fertility is preserved or enhanced. Under shifting cultivation soil conditions tend to be preserved where at least low productivity can be maintained. The difficulties with attempts at mechanization are that very efficient clearing is required to enable the ploughs, discs, harrows and other machinery to be used. This exposes the soil to erosion, compacts the surface soil and destroys the larger soil pores through which most water infiltrates the soil; in addition, the soil fauna is largely lost. Thus soils that have been stable become very prone to erosion, with the result that expensive measures of erosion control, such as contour banks and graded water channels, become necessary. With management of a very high standard such systems can prevent serious erosion and maintain fertility, especially if combined with appropriate fallow crops (Vine, 1953), but there have been many examples of failure.

The many failures to develop agriculture in the tropical forest areas of South America are discussed by Sioli (1973). A similar record of failure could be provided for Africa. In spite of the many examples of mechanized development, with adequate inputs of fertilizers and pesticides, giving high yields over several years on research stations, such techniques have not been commercially successful. This provides proof for their unsuitability. What is needed is the devising of methods specially adapted to the humid tropics (Greenland, 1975). Such methods must ensure that soil erosion is avoided and minimize the need for costly inputs in the form of mechanization, pesticides and (as far as possible) of fertilizers, and should control the tendency of soils to become highly acid.

Work of the Farming Systems Programme at the IITA in Nigeria has demonstrated how this may be achieved. Minimum or zero tillage (Rockwood and Lal, 1974) in which all crop residues are used as mulch are extremely effective in reducing run-off and lead to far less erosion than when land is ploughed and harrowed (see tables 8 and 9). Weed control can be achieved with herbicides or by carefully timed manual or mechanical surface cultivation carried out in a manner designed to produce a minimum of soil disturbance. A vegetative cover is provided by the mulch and by appropriate mixed and relay cropping methods comparable to those developed at the IRRI by Bradfield (1969). By using mixed crops, and improved materials that are resistant to pests and diseases, the need for chemical pesticides can be reduced. Application of fertilizer is unavoidable to replace the nutrients removed with the crops, and to improve the fertility status of the soil but the need for nitrogen can be reduced by including legumes possessing active nitrogen-fixing rhizobia in their roots: cow-pea (*Vigna*

TABLE 8. No tillage effects on soil and water loss under maize (Ibadan, first season, 1973; rainfall, 780 mm; plots 25 by 4 m). Unpublished data of Lal (IITA).

Slope (per cent)	Soil loss (t/ha)		Run-off (mm)	
	No tillage	Plowed	No tillage	Plowed
1	0.03	1.2	11.4	55.0
10	0.08	4.4	20.3	52.4
15	0.14	23.6	21.0	89.9

TABLE 9. Yields of cow-pea and intercropped cow-pea and maize under different tillage methods (Ibadan, second season, 1974). All plots received 30 kg/ha of nitrogen, 30 kg/ha of phosphorus and 30 kg/ha of potassium as fertilizers (on Egbeda series, paleustalf). Unpublished data of Nangju (IITA).

Tillage method	Grain yield (kg/ha)			
	Sole	Intercropping		
	Cow-pea	Cow-pea	Maize	Total
Plowed and ridged	1 185	665	1 705	2 370
Plowed, flat bed	1 274	725	1 675	2 300
Strip tillage	1 538	1 022	2 337	3 359
Zero tillage	1 649	941	2 809	3 750

unguiculata), lima bean (*Phaseolus lunatus*), winged bean (*Psophocarpus tetragonolobus*). As lime is not readily available, the most economic method of acidity control is likely to be by means of a deep-rooted fallow crop. Alternatively such a species could be grown elsewhere and the organic matter or ash transferred to the cultivated area. Hopefully agricultural systems based on these methods will combine the major virtues of traditional systems of shifting cultivation, stability, with high productivity. If they are combined with a proper land-use policy, and the preservation of forest on major and minor watersheds, development of agriculture should be compatible with proper conservation.

Merz (1975) provides information on the use of moist tropical forest in Ghana for cocoa. Cocoa was introduced there about 1883 after which it spread gradually until, by the 1950s, it had reached the northern limit of the moist forest. Originally the cocoa farms were spread in many forest areas as small trees which left much of the canopy intact; foodstuffs were imported very cheaply. With the increase of population the forest between the cocoa farms was gradually used for food production. When the timber industry was developed the farmers followed the extraction and took over the forest, sometimes for cocoa but mainly for food crops. Except for isolated pockets, the original tropical moist forest of Ghana has been replaced by transitional forest.

Watson (1973) presented the results of soil and plant nutrient studies in Malaysian rubber plantations. Post 1945 rehabilitation of the rubber industry necessitated replanting many impoverished and neglected areas. After forest clearing, mineralization of organic matter and cation leaching lead to a rapid impoverishment of the soil. In the young rubber plantations the mobilization of nitrogen and phosphorus by leguminous cover plants can have a beneficial effect on tree growth but progressive immobilization within the rubber trees can grossly deplete the nutrients of infertile latosols. If the first generation of trees is removed, nutrient deficiencies occur in the second generation. Soil and tree nutrient status are essential for diagnosing fertilizer requirements.

Cattle pastures under plantations

The practices of growing legumes as ground cover and a crop under plantation trees has been widely used. The legumes provide soil nitrogen and protect the soil against erosion, but compete with the tree crop. These legumes were found to be very palatable to grazing animals and in Brazil and Queensland they were adopted as a component of pasture for livestock grazing. It was then thought that a system of multiple land use might be initiated in which cattle could graze on such swards under the plantation crops; this would improve soil fertility nutrients from excreta and income from animal products. This has been subjected to a good deal of research, especially on the raising of cattle on pastures sown under coco-nuts. Such a multiple land use has social implications. Most coco-nuts are grown close to the sea and the people's carbohydrate diet is supplemented with marine fish. The people have to be motivated to cattle production and trained in cattle husbandry and pasture management. Cattle may involve additional inputs such as fencing, dipping facilities and yards.

Ohler (1969) and Whiteman (1974) have published reviews on pasture development under coco-nuts. Shading by the coco-nut trees generally suppresses pasture production but trees over twenty years old allow sufficient light to reach ground level to support a pasture. Competition for nutrients depends upon the fertility of the soil and the pasture species. Rajaratnam and Santhirasegaram (1963) found in Sri Lanka that fertilized *Panicum maximum* swards reduced coco-nut yields by 500 nuts/ha/a whereas *Brachiaria brizantha* and *B. milliformis* gave yield increases of 235 and 545 nuts/ha/a respectively. Generally, in mixed cropping with sufficient added fertilizer, there should be no yield reduction of either unit but the economics of the level of fertilizer must be assessed (Rodrigo, 1943). Competition for moisture may be apparent only in dry years (Krishna Marar, 1953, 1961). With young palms there is more competition than with older ones undersown with *Centrosema pubescens* (Fremond and Brumin, 1966). Improvement in soil structure due to the ramification of the roots of the pasture sward can lead to better water infiltration (de Silva, 1961).

Grazing management should aim at retaining the legume and keeping the pasture short enough so that the fallen coco-nuts can easily be found. In Tanzania *Alisycarpus vaginalis* is appreciated as a pasture legume because of its prostrate habit. Some recorded values for cattle carrying capacity under coco-nuts are:

— Rajaratnam and Santhirasegaram (1963), in Sri Lanka, grazed 0.5–2.0 head of Sinhala cattle/ha on pasture supplemented with 1.4 kg of coco-nut press cake/head/day;

— Fernandez (1968), in the Ivory Coast, grazed 0.75 head/ha/day on sandy soils growing *Centrosema pubescens*;

— Eden (1958), in the Solomon Islands, grazed 2–3 heads/ha. No live weight gains of the cattle were recorded.

With nearly one third of Sri Lanka's cattle in the coco-nut triangle, the prospects for a rapid development of the livestock industry based on improved pastures under coco-nut appear to be particularly bright. A rotational cross breeding involving the local Sinhala cattle and certain European breeds has been recommended and is being implemented.

Although oil-palms and rubber plantations use leguminous cover crops little grazing has been attempted mainly because of cattle management problems in young plantations and because a close tree canopy ultimately suppresses the undersown legume. Rombaut (1972, 1974) records trials with cattle-raising in association with oil-palm production in the Ivory Coast. He showed it to be quite feasible if:

— trypanotolerant cattle of the Baoulé type are used;

— the pastures are rotationally grazed with the cattle input limited to 125 kg cattle/ha or, in the case of Baoulé $\frac{1}{2}$ head/ha.

Maximum live weight gain without degradation of the pasture was 500 g/day. The cost of upkeep of the herd was compensated by the saving in weeding. The pasture was mainly the legume *Pueraria phaseoloides* and it was shaded out by the palms in their fifth year.

Conclusion

The goal of tropical land management should be the development of sustained-yield agroecosystems (Janzen, 1973). The most stable and environmentally least destructive are those which are integrated with perennials and livestock; that is, permanent field systems in which annual crops are rotated with and within a multistoried complex of trees and shrubs which include fruit, nuts, fibers, timbers, medicinals, beverage plants, shade and ornamentals, plus chickens and pigs and a few cattle. The natural physiognomy is reproduced as closely as possible, protecting the soil and permitting the maintenance of the nutrient cycle without manufactured fertilizers; a diversity is maintained which minimizes disease and pests. Such systems, which exist more in theory than in practice, except as relatively small house gardens, are only moderately productive and require a fair amount of labour but they are not dependent on fossil fuels. They can provide adequate subsistence and some commercial production. They are viable and maintain people on the land and hence can minimize short-fallow and migratory shifting cultivation which are so destructive of tropical forests and soils.

Weed control

Definition

The essential feature of weeds is that they interfere with man's utilization of land for a specific purpose, a characteristic reflected in the commonly accepted definition of a weed as 'a plant growing where it is not desired' (Shaw, 1956). This definition includes not only the wide range of common invasive, herbaceous weeds but, also the numerous forest trees which cannot at present be profitably exploited. With changing economic conditions and technological advances it is probable that many trees will lose their status as weeds and some be regarded as valuable (see chapter 21).

Weed problems

In natural forest the trees of low commercial value, climbers, parasites, etc., are in competition with the more valuable species, and the aim of management is to decrease these competitive effects by selective felling or poisoning. In plantations weeds are mainly important in the early stages of establishment, as few competing plants can survive after closure of the canopy. The problems and methods of weed control in the early stages of plantations are similar to those of transplanted agricultural crops. Similarly, the problems of weeds in forest nursery seed-beds have much in common with those in sown agricultural crops. Land with a long history of agricultural use tends to develop a weed flora related to the cropping system, but in the early stages of agricultural development the weed floras of crop land and forest plantations may be very similar and information obtained in one field may be relevant to the other.

In order to grow crops in cleared forest, regeneration must be kept in check for several years. Removal of the forest cover not only encourages the regrowth of many of the original tree species but also allows invasion by weeds from neighbouring cultivated ground. The present review is not intended to cover the control of weeds in tropical crops, which has been comprehensively summarized by Kasasian (1971). The problems of converting forest for agricultural development have much in common with those of land clearing for the establishment of forest plantations or improved pastures, and this aspect of crop production will be considered briefly.

The cost of controlling weeds is high compared with pest and disease control, but estimates of the magnitude of weed problems vary widely depending on how much unwanted vegetation is considered as weed growth. Weeding costs are generally high in forest nurseries and after planting. FAO (1974), for example, quotes figures showing that, where all operations establishing a plantation in an arid area were by hand, more than half the total labour cost in the first two years was for weeding. The costs of killing weed trees in natural forest are considerable but, if the clearing of forest prior to plantation establishment is also regarded as controlling weed trees, this must be accounted the most expensive weed control operation in the forest ecosystem.

Forest nurseries

Under certain conditions of forest nurseries, as in eastern Africa, seedlings only remain in seed-beds for 2–3 months and the soil of transplant beds is relatively seed free so that weed control is a minor problem (Dyson, 1964). Nevertheless, in a number of other areas there is interest in herbicides for nursery use, particularly for conifers and *Eucalyptus* spp.

The herbicides tested on conifers have mostly been residual chemicals of the triazine group and considerable variations in susceptibility are reported. In the West Indies, Kasasian (1965) found prometryne (4,6-bisisopropylamino-2-methylthio-1,3,5-triazine) to be safe and effective for *Cupressus lusitanica*, *Pinus caribaea* and other pines. In Argentina, Ortega (1963) records *Cedrus deodara* resistant to pre-emergence treatment with simazine or atrazine (2-chloro-4-ethylamino-6-isopropylamino-1,3,5-triazine), though several *Pinus* spp. were susceptible. Work on pines and eucalypts in Zambia is reviewed by Thomson (1968). Experiments in the United States (Ahrens, 1967) suggest that the tolerance of conifers to simazine can be greatly increased by dipping the roots in activated charcoal at transplanting and Peñaloza (1968) has used this technique effectively in Chile with *Pinus radiata*. Propazine (2-chloro-4,6-bisisopropylamino-1,3,5-triazine) appears to be one of the most selective triazines tested on conifers and is used successfully in New Zealand on a number of pines, including *P. radiata*.

The chemicals applied to *Eucalyptus* seed-beds before planting are mostly of the soil sterilant type, which kill fungi, nematodes, etc., in addition to weeds, but are non-persistent. With methyl bromide, the most effective material, good weed control has been reported with *E. saligna* in Brazil, though simazine reduced germination (Veiga, 1968). In South Africa, methyl bromide gave a 15-fold increase in germination of *E. grandis* (Knuffel, 1967) and, in Rhodesia, Barnes (1969) recorded increased germination and growth combined with effective weed control on *E. cloeziana*.

In addition to the above species, *Terminalia ivorensis* in the Ivory Coast has been successfully weeded by simazine which also increased germination and growth (Fabre and Brunck, 1971).

Herbicides can be used for the selective control of weeds in conifer seed-beds and transplant lines. Chemical control has shown advantages over hand weeding for a number of species but weeds are a minor problem under certain nursery conditions. Less is known about tropical hardwoods. *Eucalyptus* spp. present special problems, as the seedlings are liable to injury from residual herbicides.

Plantations

A few tropical forest trees, such as *Gmelina arborea*, make such rapid growth under favourable conditions that they are little affected by weed competition. With the majority of species, however, control of weed growth by soil preparation at planting, followed by hoeing, slashing or chemical treatment for several years is necessary. Weed competition is equally important whether trees are underplanted or planted on open sites, and there is often a succession of weed communities from predominantly annual to perennial and eventually woody species.

Weeds vary according to the site. In some southern African forest sites, *Pinus radiata* has shown no response to weeding (Donald, 1971). In Puerto Rico, Geary and Zambrana (1972) found little weeding necessary in certain areas planted with *P. caribaea*, though in others, up to five hand clearings plus two herbicide treatments in the two years after planting failed to prevent a growth check in the trees caused by weed competition. They suggested that broad-leaved species are more suitable than pines for planting in weedier sites.

Reports on plantation establishment mostly refer to the beneficial effects of controlling weeds which may be particularly important when planting grassland areas. With *Eucalyptus* spp. for example, complete ploughing has been found necessary in South Africa (Poynton, 1965) and strip ploughing in Ghana (Ankah and Amayaw, 1971). Strip cultivation has also allowed direct seedling of *Vernicia montana* (*Aleurites montana*) in Taiwan (Kan *et al.*, 1971) and cultivation before planting to reduce competition from *Imperata cylindrica* was needed for the successful introduction of *Araucaria cunninghamii* into a dry forest environment in Australia (Richards, 1967).

After planting, interrow weeding is essential for plantations in the savanna region of Nigeria (Allan, 1973b) and the need for regular weeding is similarly stressed for *Cedrela mexicana* and *Swietenia macrophylla* in Mexico (Rodriguez, 1963), *Shorea robusta* in India (Rajkhowa, 1966), *Pinus kesiya* in Zambia (Endean and Jones, 1972) and for teak in Tanzania (Bryant, 1968). Takle and Mujumdar (1958) consider the first weeding in teak to be critical for success or failure in establishment and recommend digging up weeds rather than slashing or pulling. Regular weeding for up to five years has been found equally important for the establishment of natural regeneration of *Shorea robusta* in India (Qureshi *et al.*, 1968), of *Newtonia buchananii* in Malawi (Foot, 1967a) and of *Araucaria angustifolia* in Argentina (Laharrague, 1967). The high costs of plantation weeding are illustrated by the estimate of Wang and Kuo (1971) that the 10–12 hand weedings needed in the first five years of plantations in Taiwan represent more than half the total cost of reforestation and by Baur's (1967) figures showing that plantation weeding costs in New South Wales were seven times those for nursery weed control.

Manual weed control is still widely practised, by hoeing or slashing. In suitable areas the ground between the tree lines can be cultivated (Allan, 1973b) or the weed growth cut by tractor-mounted machines.

Manual weed control is of special significance with the taungya system (Kadambi, 1958). The need to weed food crops ensures that the young trees make a good start. After the conclusion of taungya cropping the regrowth of weeds and crop residues (e.g. cassava) is generally rapid but the fast-growing tropical hardwoods normally planted (*Gmelina arborea, Shorea robusta, Tectona grandis, Terminalia ivorensis*) are sufficiently well grown by this stage for little additional weeding to be needed. The system has the great advantage that the critical early weeding is done at no cost to the forestry authority and is particularly well suited to areas where cropping practices are relatively non-intensive.

In other systems chemical weed control methods are being adopted increasingly. Sites can be kept free of weeds during the early stages of growth by the same soil treatments as those used in nursery transplant lines. In Taiwan, for example, Wang and Kuo (1971) quote successful results from application of a mixture of atrazine and 2,4-D

(2,4-dichlorophenoxyacetic acid) to the soil before planting China fir.

When the effects of residual herbicides have worn off and new weed growth appears, the herbicide most commonly applied is the non-selective paraquat, which must be carefully sprayed so as to avoid wetting the tree foliage. This was the most effective of a range of chemicals tested in Argentina by Revilla (1972). Because the trees can be injured if contacted by the spray solution, various methods of protection have been devised, such as the covering of the seedlings with plastic or paper bags (Queensland Department of Forestry, 1967) or by the use of specially shielded sprayers, which allow spray coverage of weeds growing close to the seedlings while preventing contact with the foliage (Bowers and Hawthorn, 1971).

Paraquat is effective in killing the majority of annual weeds but, as a result of being poorly translocated, has only temporary effects on perennials and other weeds with a capacity for regeneration from near or below-ground level. Where perennial grass weeds are a problem other chemicals must be employed. The grass-killer dalapon (2,2-dichloropropionic acid) is recommended to control *Panicum maximum* in Brazil as a means of reducing fire (da Silva, 1969). Good control of the rhizomatous grasses *Imperata cylindrica* and *Digitaria abyssinica* (*D. scalarum*), resulting in increased tree growth, has also been obtained with it for eucalypts and conifers in Uganda (Ball, 1970a), and, in Trinidad, applications in 2-year-old *Pinus caribaea* are safe as long as contact with the trees is avoided (University of West Indies, 1969). A related chemical tetrapion (2,2,3,3-tetrafluoropropionic acid) has proved successful against the most troublesome *Miscanthus sinensis* in Taiwan (Nakamura *et al.*, 1973).

Bamboos are particularly troublesome in certain areas and Kadambi (1958) refers, for the encouragement of teak regeneration, to a constant war against these woody grasses which need frequent cutting. Satisfactory chemical control of bamboo is difficult but promising results on one species in the West Indies have been obtained by sprays of dalapon, trichloroacetic acid or sodium chlorate (Kasasian, 1964). Encouraging results on several Malaysian species with various formulations of dalapon applied by injection or as a paste are reported by Lowe (1971). Dalapon paste has been used similarly in Malaysia to obtain control of a stemless palm hindering regeneration of *Shorea curtisii* (Burgess and Lowe, 1971).

For the control of perennial dicotyledons and woody weeds either 2,4-D or a 2,4-D+2,4,5-T (2,4,5-trichlorophenoxyacetic acid) mixture can be applied as sprays, directed so as to avoid the tree foliage. Successful trials with the mixture in 2-year-old *Pinus caribaea* and other species are reported from Trinidad (University of West Indies, 1968, 1969). Teak appears to be very liable to damage from these chemicals and applications should be confined to the dormant period (Murray, 1967). A practical guide to herbicides use for forest weeds in New South Wales is given by Forrest and Richardson (1965).

Several very aggressive introduced weeds have spread widely. The prickly shrub *Lantana camara* is troublesome in many African and Asian-Pacific areas and, in Bangladesh, has been observed to prevent natural regeneration after opening up mixed forest unless checked by frequent cutting (Ghani, 1958). Studies on biological control have been conducted in various places following the impressive success in Hawaii. The difficulties are illustrated by experiences in Australia, where various insects have been introduced in the past 50 years with limited success (Willson, 1968), and in South Africa, where the introduced moth, *Catabena esula*, has been badly affected by disease (Oosthuizen, 1964). In India much *Lantana camara* is being killed by the lace bug, *Teleonemia scrupulosa* which escaped from quarantine (Joshi, 1969). *Eupatorium odoratum* (Siam weed) is of similar growth habit and a serious problem of young oil-palm plantations, especially in Nigeria, but is regarded as less troublesome in forest plantings. In Malaysia it can replace the more undesirable *Imperata cylindrica*, if protected from fire (Landon, 1958), and in Nigeria plantations of *Gmelina arborea* is soon outgrown by the trees (Ivens, 1974). The climber, *Mikania micrantha*, whose twinning habit makes control by cutting or directed herbicide spraying difficult, is a particular problem in Assam (Choudhury, 1972) and a weed for which biological control would be very valuable.

Planting cover crops excludes weeds. Various legumes are employed extensively for this purpose in such tree crops as oil-palm, rubber, etc., but only a limited amount of information is available on forest plantations. The fast-growing trees, *Leucaena leucocephala* (*Leucaena glauca*) and *Anogeissus acuminata* have been sown with teak in Java, and Letourneux (FAO, 1957) states that such covers are able to suppress *Imperata* and do not interfere with tree growth as long as they are pruned a year after planting. The extra beneficial effect of *L. leucocephala* of enriching the soil nitrogen content is stressed by Van Alphen de Veer (1958). Herbaceous legumes have been tested in Australia and found to improve the growth of indigenous conifers, such as *Araucaria cunninghamii*, but to depress the yield of exotic pines (Richards and Bevege, 1967).

Thus control of weed growth may be necessary in preparing land for tree planting. It is essential for several years after planting on the majority of sites in order to release the young trees from competition, to lessen fire hazards and to reduce pests and diseases. Manual control is practised extensively. The superiority of chemical control is becoming evident and a variety of herbicides and equipment is available for many weed problems. Biological control has shown promise against a limited number of weeds and there may be scope for wider use of leguminous covers.

Tree weeds

The poisoning of non-commercial trees is an accepted operation in most tropical forest areas and its role in opening the canopy and encouraging better growth of desirable species is discussed in *Tropical silviculture* (FAO, 1958). A detailed account of the improved conditions for seedling regeneration in Uganda is given by Philip (1968). Under certain conditions, however, its value has been questioned, as in studies in Sabah (Meijer, 1970). For many years the poison was sodium arsenite, applied as an aqueous solution or paste to a circular cut or girdle which allowed access to the conducting tissues. Detailed recommendations for its application are given in several general sylvicultural works, such as Wyatt-Smith (1963) dealing with Malaysian, and Catinot (1965) with western African conditions. The majority of trees are killed rapidly. Trials on some of the more resistant western African species are reported by Hombert (1954), and a critical assessment of poison girdling in lowland Malaysian forest is made by Wong (1966). Sodium arsenite has the disadvantage of high toxicity towards man and animals and, with the introduction of the growth-regulating herbicides 2,4-D and 2,4,5-T, attempts were made to develop safer alternatives.

Pioneer work with these was started in Uganda by Dawkins (1953), who showed that the application of a mixture of 2,4-D and 2,4,5-T esters in used motor sump oil in a ring around the trunk, killed most trees without the need for girdling or frilling. Diesel oil was later found to be a better diluent and, with some of the more resistant species, the action was improved by application to frilled trunks (Dawkins, 1954). Apart from its reduced toxicity hazards, treatment with the 2,4-D and 2,4,5-T mixture was considered superior to sodium arsenite in that less labour was required and the treated trees died more slowly, and at different rates, resulting in a more gradual opening up of the canopy. Details of the methods used on a large scale in Uganda are given by Hughes and Lang Brown (1962).

Tests were later conducted in various parts of the world to compare the relative merits of sodium arsenite and 2,4,5-T (alone or mixed with 2,4-D) as arboricides. Trials in Malaysia are reported by Beveridge (1957) and Wong (1966), in Africa by Letourneux (1956) and Cebron (1957) and in South America by Boerboom (1964) and King (1965). Control of strangling figs (*Ficus* spp.) with 2,4,5-T in Sabah is reported by Liew and Charington (1972). Despite the marked superiority of 2,4-D and 2,4,5-T mixtures in Uganda, sodium arsenite is still preferred because of its lower cost in a number of countries where the use of arsenical pesticides has not been prohibited and where the more resistant trees requiring frilling predominate, as in Ghana (Amediwole, 1967). In Guyana the faster action of sodium arsenite has been considered an advantage (King, 1965). Arsenic was still regarded as the best chemical for routine tree poisoning in Malaysia in 1967, but the introduction of restrictions on its use has stimulated increased interest in less toxic alternatives.

More recent developments include the introduction of new application techniques and chemicals, in particular picloram (4-amino-3,5,6-trichloropicolinic acid, marketed as Tordon) (Huraux, 1970). This chemical, like 2,4-D and 2,4,5-T, has a hormone type of action. It is superior in being better translocated and in being effective against a wider range of woody species. It is most frequently used as a water soluble amine applied by injection, either by special tree injectors or by making axe cuts through the bark at intervals round the trunk and applying small volumes of concentrate. The injection technique has advantages in that only small volumes of herbicide solution have to be carried, is quicker than the spraying or brushing required with 2,4,5-T in oil and the expense of an oil diluent is avoided. Water soluble amine salts of 2,4-D and 2,4,5-T can be applied in the same way (Peevy, 1968) and a number of Tordon formulations consist of mixtures of one of these chemicals with picloram.

Organic arsenical compounds of low mammalian toxicity, such as cacodylic acid (dimethylarsinic acid) and MSMA (monosodium methylarsonate) have also been developed which are suitable for applying by injection and are extensively used in the United States. Details of the recommended techniques and the newer chemicals are summarized by Newton (1974). A number of suitable injecting tools employed in France and the United States are described by Leroy-Deval (1970) and tests with injectors under eastern African conditions are reported by Ivens (1970). A variety of Australian injectors are reviewed by Diatloff (1970). Other tools for achieving the same purpose include a hammer-shaped punch (MacConnell *et al.*, 1968) and herbicidal cartridges (Sterzig, 1970). Outside Australia little work has been done with the picloram injection technique in tropical forests. However, in Hawaii, Carpenter (1966b) has compared the effects of 2,4,5-T and picloram injected into *Metrosideros collina* and, in the Philippines, Zabala (1969) has recorded the susceptibility of a wide range of tree species to picloram absorbed via the root system after application to the ground as pellets.

The concept of utilizing weed trees for purposes other than timber, rather than killing them, appears to have received little attention. In most tropical forests, non-timber trees are valued for fire-wood or charcoal. In Uganda, Earl (1968) has developed a technique of combining fuel production with replanting, which could well have wider application. Under this system, weed trees are converted into charcoal and fast-growing, desirable species planted on the charcoal burning sites. The cost of refining is found to be considerably less with this method than where arboricides are used to kill the unwanted trees.

In plantations chemicals are used mainly for thinning. On occasion, however, it may be desirable to kill infected trees in order to prevent disease spread (Knutz and Nair, 1967). The methods used are generally similar to those employed in natural forest. Chemical thinning of plantations has been extensively studied in Australia, where the injection of picloram formulations has been found particularly effective on eucalypt trees (Kimber, 1967) and stumps (Jack, 1968). Comparative results of sodium arsenite, 2,4,5-T and picloram on standing trees, stumps and coppice growth are reported by Bachelard *et al.* (1965) and the increased growth of planted trees resulting from coppice control has been measured by the Queensland Department of Forestry (1967). In South Africa, chemicals have been tested on *Eucalyptus* plantations being converted to pine and the susceptibilities of 45 *Eucalyptus* spp. to a picloram and 2,4-D mixture have been recorded by Morze (1971). Unwanted coppice of *Gmelina arborea* has been killed in Malawi by stump or basal bark treatment with 2,4,5-T in diesel oil (Foot, 1967b).

One problem associated with the use of chemicals is the risk of desirable trees being injured through root uptake; because of the closer spacings this is more important in plantations. This has occurred in the United States but Kimber (1967) working on *Eucalyptus marginata* found that there was little movement of chemical between treated and untreated trees even though fused root systems were evident. Clearly, further information is needed on the dangers of herbicide transfer from one plant to another via the roots but few cases of injury have been reported.

The clearing of forest for tree planting is normally done by felling or mechanized clearing followed by burning. Where there is no objection to the presence of dead, standing trees, however, it may only be necessary to clear the undergrowth, the larger trees being killed by arboricides. In Ghana, for example, basal bark application of 2,4,5-T ester in diesel oil has been shown to be a considerably cheaper method of clearing than hand felling and burning (Liefsting, 1965). The use of herbicides based on 2,4,5-T and picloram to clear understorey vegetation in Hawaii is reported by Carpenter (1966a) while in Puerto Rico, high doses of picloram applied from the air have been found effective for the elimination of understorey vegetation in mixed forest in preparation for planting *Pinus caribaea*.

The relative merits of the various systems of site clearance in areas of varying rainfall are considered by Stuart Smith (1967) who concludes that, although arboricides are only employed on a limited scale at present, their use is likely to increase as chemical techniques improve and the cost of labour rises. Such improvements in technique include the newer methods of arboricide injection which have yet to be adequately tested under tropical conditions.

Stuart Smith also points out that weeds are of greater importance in the establishment of plantations in savanna and grassland than in forest areas. There is consequently a greater need for cultivation during site clearance, particularly for the control of highly competitive, rhizomatous grasses. An account of trials with various types of land clearing and ploughing equipment in the savanna of northern Nigeria is given by Allan (1973a).

The problems of killing unwanted trees are similar for weeds in forests and for thinnings in plantations. Tree poisoning techniques involve a choice between the older, cheap but highly toxic sodium arsenite and the newer, more expensive chemicals such as 2,4,5-T, picloram, etc., which are very much less toxic. Opinions on the comparative effectiveness of the two types of chemical vary, as do assessments of the hazards involved. Tree poisoning with sodium arsenite, however, is declining as increasing restrictions are placed upon its sale and as cheaper and more effective ways of employing the newer chemicals are developed. The same herbicides may be employed where land is cleared for replanting and, although hand or mechanical methods are still the most widely practised, chemical techniques are finding increasing acceptance as labour costs rise.

Parasites

Trees are frequently affected by parasitic flowering plants, chiefly members of the Loranthaceae, which grow on the trunk or branches and cause varying degrees of damage. In a survey in the Dominican Republic, Etheridge (1971) recorded up to 60 per cent of pines infected by several parasitic species and up to 50 per cent loss of merchantable *P. occidentalis* timber through the effects of dwarf mistletoe in Hispaniola. The high incidence of *Loranthus* in *Gmelina* plantations in Bangladesh is referred to by Ghani (1958). Herbicides show considerable promise for dealing with these parasites.

In Australia, for example, Brown and Greenham (1965) reported trials in which *Amyema* sp. growing on *Eucalyptus polyanthemos* was successfully controlled by injecting the stem of the host with 2,4-D or 2,4-DB (2,4-dichlorophenoxybutyric acid). Hartigan (1971) gives more general recommendations for controlling various mistletoes, based on application of 2,4-D as a spray. In the West Indies, injection of various chemicals was less successful in killing *Phthirusa adunca* on teak without injuring the trees (University of West Indies, 1964) but paraquat, applied with a long lance directly on to the parasites, was effective and non-injurious (University of West Indies, 1966). A similar technique using diquat (9,10-dihydro-8a,10a-diazoniaphenanthrene) to kill *Loranthus* growing on teak has been successfully applied in India (George, 1966).

Fire-break maintenance

Combustible vegetation on fire-breaks can be defined as weed growth, the desired vegetation being a fire tolerant type. After clearing fire-break lines, the aim is to prevent the growth of plants which would subsequently become inflammable, especially grasses, and this is normally done by periodic cultivation or cutting. Alternatively, the use of herbicides is available.

Non-selective persistent chemical total weed-killers can be applied to the soil to prevent the growth of all plants for long periods. Numerous effective products are marketed which have been developed for weed control on railways, industrial sites, etc., under varying environmental conditions. One of the most generally suitable is bromacil (5-bromo-6-methyl-3-(1-methyl-n-propyl)uracil), a herbicide active against a wide range of species for about a year. When mixed with such chemicals as paraquat, 2,4-D, dalapon or amitrole (3-amino-1,2,4-triazole), it can be used to kill back existing herbaceous vegetation in addition to maintaining bare ground. Where complete vegetation control is not necessary it may be sufficient to apply a grass-killer, such as dalapon. This chemical is recommended in Brazil, where da Silva (1969) has observed that it kills *Panicum maximum* and changes the predominantly grassy, inflammable vegetation to a community dominated by less inflammable dicotyledons.

An alternative approach is spraying with paraquat before the foliage dies back naturally and burning the fire-break while the surrounding unsprayed grass is still too green to burn (Connell and Cousins, 1969). Large areas of fire-break can be sprayed from the air in a very short time. Early burning is a recommended technique in Australia (Pearce, 1969; Freak, 1971), and has been successfully tested in Uganda (Ball, 1970b), where it was found to be cheaper

than hoeing and almost as cheap, but more effective, than slashing. On wide fire-breaks, chemical costs may be reduced by spraying and early burning strips along both edges and using these as subsidiary fire-breaks for carefully burning off the main fire-break area between the strips when the grass dries off naturally.

The impact of forest operations on the environment

Biocides

The majority of chemicals applied for weed control are of relatively low mammalian toxicity. The most widely used method of expressing toxicity is based on the orally administered dose causing death of 50 per cent of batches of laboratory animals (acute oral lethal dose or LD 50) and the doses applying to the herbicides mentioned are as follows:

Chemical	LD 50 dose (mg/kg body weight)
dalapon (2,2-dichloropropionic acid)	7 600–9 300
picloram (4-amino-3,5,6-trichloropicolinic acid)	8 200
bromacil (5-bromo-6-methyl-3-(1-methyl-n-propyl)uracil)	5 200
simazine (2-chloro-4,6-bisethylamino-1,3,5-triazine)	5 000
TCA (Na salt) (sodium trichloroacetate)	3 200–5 000
atrazine (2-chloro-4-ethylamino-6-isopropyl-amino-1,3,5-triazine)	3 100
MSMA (monosodium methylarsonate)	700
2,4-D (2,4-dichlorophenoxyacetic acid)	375– 666
2,4,5-T (2,4,5-trichlorophenoxyacetic acid)	500
diquat (9,10-dihydro-8a,10a-diazoniaphenanthrene)	400
paraquat (1,1'-dimethyl-4,4'-bipyridilium)	150
sodium arsenite	10

As these figures have been obtained from experiments on rats and mice they only provide an approximate guide to human toxicity, which may arise from other than oral ingestion. TCA, for example, has undesirable dermatitic effects which necessitate avoidance of contact. The human toxicity of paraquat is greater than suggested by the above list and small doses of concentrate can prove fatal. Further consideration must also be given to 2,4,5-T as a dioxin impurity (2,3,6,7-tetrachlorodibenzo-para-dioxin) is very highly toxic and may have teratogenic effects. The use of 2,4,5-T for defoliation in Viet-Nam gave rise to much publicity and details of its effects in this situation have been published by the National Academy of Sciences (1974). Since the discovery of the high toxicity of dioxin, the hazards of using 2,4,5-T have been greatly reduced by the restriction of the dioxin content to <0.1 ppm. In a symposium on the forestry uses of 2,4,5-T in Europe, it was concluded that the safety regulations existing in 1971 were adequate (Maier-Bode, 1971). In the United States the dangers of 2,4,5-T in the forest environment have been evaluated by Montgomery and Norris (1970), who concluded that the hazards are low when the chemical is used in accordance with tested procedures. Norris (1971) further reviewed the dangers of chemicals applied for woody plant control and stated that none is likely to cause acute or chronic hazards to non-target organisms (including mammals, birds and micro-organisms) if used properly.

With the normal precautions there appears to be little danger to man or animals in the use of weed-killers, though special care is required with such chemicals as sodium arsenite and paraquat. Precautions are needed to avoid undesirable effects on plants. Crops may be damaged by spray drift. Cotton, for example, is particularly susceptible to small quantities of 2,4-D, 2,4,5-T and picloram. Care is therefore needed in applying such chemicals in the vicinity of susceptible crops, especially when aerial applications are employed.

Persistent herbicide residues may also cause problems to tree seedlings, pasture legumes, etc., and the introduction of the relatively persistent and mobile picloram has aroused concern about possible contamination of water. Bachelard and Johnson (1969) tested the influence of 2,4,5-T and picloram residues on the growth of *Pinus radiata* seedlings in Australia. The effects of 2,4,5-T applied to the soil were no longer apparent after two months while, with picloram, survival of the pine seedlings was reduced until the sixth month. Water pollution aspects have been investigated by workers in the United States including Davis *et al.* (1968), who applied a high dose of picloram to a watershed area and found detectable residues in stream-water for up to 16 months. Patric (1971), however, in a review of American forests, finds no suggestion that proper herbicide use has limited the usefulness of stream-water and provides suggestions for minimizing the water pollution. Experiments were also conducted on regeneration with the aid of herbicides, especially sodium arsenite, in the dipterocarp forest of Sabah. Studies in progress deal with the influence of such treatments on soils, micro-climate, floristic composition, development of lianas, succession from secondary forest to climax forest, and on the fauna. Early results indicate: that the destruction of the understorey by sodium arsenite leads to soil degradation and high mortality among dipterocarp seedlings; that the eradication of undesirable species is not necessarily achieved because of their seed dormancy and the increased availability of light for their germination; and that succession towards a climax is not accelerated.

The effects of certain arboricides, such as picloram, 2,4-D and 2,4,5-T could be observed over 21 336 km² in Viet-Nam. Mangroves and plantations suffered most. In the region of Gia-Dinh, *Rhizophora* which occupied 51 per cent of the mangrove area, now occupies only 15 per cent. Moreover, the mangrove areas where malaria had been eradicated, are now infested with *Plasmodium vivax* and *P. falciparum*, and rat populations have greatly increased; regeneration is very slow and it is estimated that it will take more than 100 years before these communities are rehabilitated (National Academy of Sciences, 1974). The effects were different in natural forest. The opening of the upper canopy strata caused a high susceptibility to fire, which impeded regeneration, destroyed seedlings and favoured the expansion of bamboos. Teak plantations have suffered particularly.

The hazards of direct human toxicity from herbicides are small, apart from sodium arsenite (the toxic nature of which is well understood) and paraquat (oral ingestion of which in concentrated form has caused fatalities). Dangerous levels of a dioxin impurity suspected of teratogenicity have been found in some batches of 2,4,5-T, but permitted levels of the impurity are now extremely low and the safety regulations governing its use in Europe and the United States are considered adequate. The dangers to non-target organisms are low in temperate countries, except crops may be damaged by drift. Few assessments have been made in the tropics but, as long as similar doses and methods are employed, the hazards are unlikely to be higher. Picloram is more persistent than the other herbicides considered and residues have been detected in streams for a considerable period after application to a watershed at a high dosage.

Finally there is the possible effect of pesticides such as might be used for the control of animal trypanosomiasis in Africa; a long-term programme was recommended to FAO by the World Food Conference in Rome in 1974. The disease is more common in the savannas that have replaced forests and in the forests degraded by shifting cultivation, but this does not mean that it never occurs in the humid forests (FAO, 1974; see also chapter 17). Opening of forests for exploitation, plantation establishment, or grazing may increase contacts between forest populations of tsetse flies (*Glossina* spp.) and man and his livestock and may lead to the colonization of such areas with fly species from the savannas. One cannot foresee to what extent insecticides will be used inside forests. It is probable that the undesirable effects will be the same as those observed in the savanna, i.e. mainly the destruction of other insects, small mammals, birds and reptiles. Hence it ought to be possible to evaluate the effects in the forest. Research on the effects of pesticides ought to be directed on the number and abundance of parasitic species, possible changes in their and non-target species metabolism, reproduction, and behaviour, etc., also on the persistence, break-down and accumulation of the pesticides in various parts of the environment. Techniques for such studies exist and it is necessary to obtain the necessary information before pesticides are employed.

Logging and transportation

In certain areas, especially on slopes, severe erosion has been found to follow logging operations and road construction. This has engaged the attention of the FAO Committee on forest development in the tropics; in its third session (1974) the Committee emphasized that mechanization and the use of heavy equipment without due precautions and the construction of forest roads in steep terrain might cause damage to the environment—especially to soil stability. Apart from the removal of the top layers of soil (up to 60 per cent was removed in parts of the Tawau Hills Forest Reserve, Sabah) heavy felling and logging damage the remaining trees. It is a matter for debate how much of its crown a tree may lose and how much of its bark may be torn off before it is rendered useless for another crop; long-term research is needed to evaluate this. Measures to protect the remaining trees would seem a sensible precaution, but are not practical in Sabah because of a shortage of staff. As natural regeneration of desirable species is abundant there, it is suggested that the more logical course at present would be to lower the felling size, but it should be high enough to spare sufficient seed-bearers to allow fertile seeds of desirable species to reach all parts of the forest. Such a course would considerably increase current yields especially while there is a demand for relatively small logs, and there would be much less need for poisoning to open the canopy, but it would not prevent damage to the soil. Both in Sabah and in Sarawak damage to and loss of soil have been studied. The need for the kind of measures proposed and sometimes applied in North America is even more urgent in tropical rain forests. Implementation is largely a matter for political decision, the determination to insist on the necessary measures when drafting concession agreements.

In the Philippines construction of forest road-beds after the wet season caused soil losses. Spoil placed on the lower side of a cut, or left as fillings, is subject to erosion during the next wet season. Erosion is particularly severe on steep roads or where they have not been surfaced. On temporary roads drainage channels should be constructed at intervals after extraction has been completed to avoid serious erosion. In the humid regions of Sri Lanka exploitation is by selective felling in 20-year rotations followed by tractor skidding. Skidding by elephants in broken country might be more economic and less harmful to the environment than by tractors. Because of the undesirability of skidding methods which involve dragging heavy logs on the ground, other methods, especially helicopter or cable skidding, have been considered. Helicopter methods are too expensive except for logs of great value, but in steep country prone to erosion cable skidding is a practical alternative. It is however important to take into account the cost of equipment and the technical skills needed by the operators in this form of hauling.

Control measures to reduce soil damage include:
— prohibition of all logging operations on steep slopes, the critical gradient being related to the type of soil and the abundance of regeneration;
— laying of proper culverts at the base of every embankment and earth fill;
— prohibition of using hollow logs as culverts on major roads;
— maximum length limitations of skid hauls;
— minimum road density with prescribed standards of construction and maintenance;
— training to improve the standard of tractor use, and especially greater use of winching;
— all skidding tractors to be fitted with integral arches;
— encouraging the use of skidders with rubber types;
— destruction of earth fills after logging to restore natural free drainage.

Technical solutions exist and it ought to be possible to adapt them for use in humid tropical forests. North American technology is not free from environmental drawbacks especially as regards erosion and requires large capital investments. European equipment, which tends to be lighter, is generally unsuitable because of the size and weight of the logs in tropical forest. Hence a great deal of adaptation of existing technologies to the special problems of tropical forests remains to be carried out. The problems vary between regions. In Asia, e.g. in Sri Lanka, the Philippines and Thailand, where very large concessions are customary and where the soils are much more erosion-prone than in Africa or Latin America, the organization of forest operations should be within watersheds: sensitive areas should be left as reserves and care should be taken to adapt logging operations to those areas that can tolerate forest exploitation.

In addition to the direct effect of road building, the overall impact on the environment of the extension of transportation networks should be emphasized. The improvement of transportation networks, particularly roads, are indispensable for the successful development of agriculture, industry, tourism and forestry. Nevertheless most remaining forest areas have few roads or other transport systems. Recently States and international bodies concerned with assistance have given considerable attention to the construction of roads in these areas; the best known examples include the Carretera Marginal, now under construction along the eastern front of the Andes from Bolivia to Venezuela, and the Trans-Amazonian Highway being built across the Amazon basin. In 1960 the completion of Brazil's 2 000 km Belém-Brasilia Highway demonstrated the country's ability to use roads for the purpose of helping to colonize the country's interior. Currently Brazil is involved in an even more massive road-building effort planned to intersect the entire Amazon Basin, which will create vast changes in what is still the largest area of tropical forest in the world. If the equally vast biological resources of this region are not to be destroyed, an unprecedented effort to study the region's ecology and optimum resource use patterns must be mounted before rapid environmental change eliminates the options.

Research on the impact of large new transportation systems is limited. One case study was done by McNeal (in Farvar and Milton, 1972); it describes two rail-road-linked projects in Amazonian Brazil. Both resulted in forest destruction, soil degradation, deteriorating farm yields and eventually led to the establishment of scrub. In areas of more favourable soil, the replacement of forest by stable agriculture may be possible. However, in most cases new tropical transportation networks have led to spontaneous, unplanned human settlement, forest destruction, accelerated run-off and erosion, declining soil fertility and structure, and loss of forest products and wildlife.

It is important that the ecological impacts of transportation projects be carefully studied before a new project is started (and preferably at the pre-feasibility stage). So far, few precedents exist for the inclusion of such applied ecological research in road development.

Man-made lakes

Though the influence on the environment of artificial reservoirs or irrigation schemes in tropical countries has often been studied, especially in the context of epidemiology (see chapter 17), their influence on forest ecosystems has received much less attention. This may become important in connection with energy requirements and the construction of more dams (cf. the Third International Congress on Artificial Lakes, United States, 1971).

In 1969 there appeared a joint study of FAO and UNDP on the problems of planning and developing artificial lakes. It envisages especially studies on disease vectors, wildlife, human settlements, etc. The management of forests (to the extent as it affects erosion, sedimentation and the underground water resources) is sometimes discussed, but the influence of the impoundment on the forest is rarely considered. A recent study by Leentvaar (1974) on an artificial lake in Surinam contains interesting observations: the stagnant water soon led to a vigorous development of floating water-hyacinth which covered 41 200 ha two years after the dam was completed; this plant was eradicated by spraying 2,4-D; because of the absence of pollution, continuous discharge of water and its replenishment from oligotrophic rivers, the development of the lake is towards oligotrophy; the action of wind in the lake area has increased.

It is vital that in future projects for river basin and reservoir development, research and surveys on the role of forest resources be included at all levels of project planning and execution. The adequate protection of natural watersheds in which it is planned to construct reservoirs is particularly important in order to prevent erosion, lake sedimentation and floods which could shorten their operating life-spans (Milton, 1975). There is also a great scarcity of adequate biological and ecological studies or surveys to appraise the scientific, educational and recreational values of reservoir inundation sites and watersheds from the point of view of their potential use as natural reserves or national parks. Such studies are essential if the full costs and benefits of any project are to be properly evaluated. They are no less important for suggesting rational and ecologically sound alternatives for river basin developments.

Pulp effluents

Since plantations can supply large quantities of wood-pulp, and the prospects of pulping mixed hardwoods are good (see chapter 21), it is necessary to discuss the dangers to the environment represented by pulp and paper-mills. Under the sponsorship of the FAO Consultative Committee on Pulp and Paper, experts have undertaken research on effluent control, economic aspects of pollution control and on the prospects for the protection of the environment. The documents point out that the cost of pollution control depends on the degree of purification required, the techniques, the size of the installations and the capacity of the surrounding environment to absorb the damage. In its last session (May 1974) this Committee found that the

estimates for increasing production capacity to meet rising demands were inadequate because certain production costs had risen, and especially because of the high costs of purification to meet government standards. The regulations on the effect of pulp and paper-mills on the environment may be based on different principles, ranging from the absorption capacity of a given environment for pollution to uniform rules or to taxes based on the amount of pollutants emitted.

Industry, mining and human settlements

The impact of industry and human settlements on forest lands is both direct, from land clearance activities to provide space for new villages, towns and cities, and indirect from the accelerated demands for food, timber, water, power and other resources. A related demand is for fuel wood and charcoal for home heating and cooking—still important in many tropical and subtropical countries.

The impact of mining is of local significance in areas rich in minerals and fossil fuels such as the oil-bearing zones of eastern Colombia and Ecuador, the copper ore deposits of the Andes and central Africa, the silver mines of central Mexico, the iron ore areas of Brazil and the bauxite deposits of the Guyanas. In some cases, as with silver mining, the effect on forests has covered far greater areas than those immediately affected by mining; frequently forests have been devastated over wide areas to provide mine timbers and the charcoal used in ore reduction processes.

Although some research exists on the effects of industry, mining and human settlements on tropical and subtropical forests, an integrated world-wide programme of research is needed to examine the full range of ecological impacts and economic demands on this resource. Research should concentrate on project demands from rapid urbanization and the potential impact on local supplies. The role of high-production forest plantation to satisfy growing timber and fuel demands and in relieving pressure on marginal natural forest is also worthy of investigation.

Similarly, international research is needed to predict which forest areas will be directly threatened by expansions of human settlements, industry and mining. Studies to determine which of these are critical for recreation, science, gene preservation and watershed protection or are unusually fragile should then be mounted. This is vital to determine which forest sites should be protected in regions of fast growth of human populations, and it is crucial to provide planners with alternative growth choices at a stage when they can be seriously considered.

Research needs and priorities

Biological limitations to the transformation of tropical forest ecosystems

Priorities for research are:
— atmospheric factors and their influence on the manipulation of the plant cover;

— genetic variability of species and its quantification;
— competition and interrelations between species and the role of chemical telemediators;
— systems analysis and the use of models to ascertain the effects of various treatments. It is important to stress that, because of the great complexity of tropical forest ecosystems, such undertakings are far from simple and should not overshadow the uncertainties. To take into consideration most of the known factors and their interacting effects all possible solutions will have to be examined in order to find the best; any such attempt quickly leads to the unsurmountable obstacle which Dumas (personal communication) has called 'le mur du combinatoire'. To remain within reasonable limits of computer time, it is necessary to severely limit the number of hypotheses and the uncertainty persists.

Types of utilization

The problems are:
— Impact of various sylvicultural systems on soil structure, nutrient retention, run-off, micro-climate and subsequent land-use opportunities.
— Relationship between intensive resource development and the incidence of diseases, insects, and other pests.
— Long-term impact of different agricultural and forestry uses, including clear and partial cutting, shifting cultivation and replacement by plantations of exotic trees.
— Tree improvement including selection and breeding; the introduction and evaluation of improved trees; reproduction and seed physiology.
— Development and adaptation of efficient technologies for forest operations, geared to the special needs of tropical forests and labour.
— Reclamation and regeneration of soils damaged by logging and other activities.
— Relations between biological productivity, yield diversity, and types of management and investment.
— Development of prediction models for the main forest uses and land mosaics.

Sylviculture

The management protocole mentioned earlier aims at combining logging and enrichment operations and envisages research within the framework of field operations.

Studies should be started on ways to influence recovery. If this could be speeded up, by introducing key organisms, by adding nutrients, or other actions, it should be possible to reduce the harmful impacts of habitat modifications and to rehabilitate areas which have been subjected to excessive stress. Such studies should be integrated with studies on the inducement of natural regeneration, logging damage, diagnostic sampling and stocking levels, monocyclic and polycyclic felling systems, vegetative propagation, etc.

Studies need to be carried out on germination and its light requirements; the early growth of the more important desirable and undesirable species; on the optimum stocking level for maximum growth of various species at different

stages; on understanding the factors which influence production; and on establishing arboreta and sample plots to determine which species are likely to be of great potential.

The importance of research on vegetative propagation which might make it possible to develop coppice or coppice with standards systems needs stressing. This has been discussed in connection with the taungya system, but it is important to stress the importance of studies on meristems.

There seems to be an excessive duplication of effort in sylvicultural research. Treatments must be in accordance with ecological principles so the local peculiarities of each site have to be considered and the apparent repetition is justified. However there is need for regional co-operation and the exchange of information between ecologically similar areas.

Plantations

— Destruction of forest
Research aiming at more effective forest utilization will have a bearing on plantations—an increase in commercial species, or uses of wood for such purposes as carbonization, energy, ground wood or pulp.
Identification of forest types in terms of difficulty of eradication. Distribution of growing stock over diameter classes; potential uses of aerial photographs; more detailed studies on clearing costs as a function of structure; correlations of structures with costs. Trials concerning clearing the forest some time before planting; trends.
— Research on the ecology, biology and physiology of the principal species, including:
 • search for the best species;
 • basic research on their behaviour in relation to increased illumination, soil types, forest regrowth and competition;
 • natural rhythms of growth and sexual reproduction, and potential for vegetative propagation; grafting and propagation from cuttings.
— Genetic improvement:
To produce fast-growing and uniform plants it is necessary to make sure its source is reliable, local and concentrated in one place. This involves:
 • search, collecting and conservation of provenances;
 • provenance trials, selection for fast growth, quality and resistance to pests;
 • individual selection and progeny tests;
 • selection, conservation and isolation of clones;
 • crossing, hybrid propagation, exploitation of heterosis;
 • seed orchards.
— Sylvicultural research:
 • best type of planting stock for survival, speed of establishment, resistance to competition;
 • prolonged observation of various areas converted to plantations; development of new techniques; search for particularly suitable sites;
 • planting density in relation to risks from falling dead trees and thinning schedules;
 • economics of fertilizer use;

- regrowth after various methods of forest clearing and also in relation to the type and structure of the original forest; relations with tending problems;
- possibilities of establishing stands from cuttings.
— Pests and diseases:
- relative severity of certain attacks on plantations established by forest conversion as compared with other plantations;
- effect of the establishment of mixed stands with regard to the severity of borer attacks.
— Economics:
- cost of each stage of conversion and possibilities of cost reductions;
- cost elements in different plantation projects;
- profitability regarding all factors.

Wildlife management

There is a major need for investigations into animal population dynamics in tropical forest ecosystems, and particularly of the larger mammals which are valuable as food and possible tourist attractions. A successful culling programme depends on a sound knowledge of habitat carrying capacities and of the densities and population structures of the species concerned and how these are modified by:
— management for timber production;
— management for wild animal production;
— changes in the forest cover;
— plantations;
— shifting cultivation;
— introductions of exotic trees and animals;
— use of pesticides.

Agri-sylvicultural systems

Faced with the financial impossibility of an exhaustive research programme, the FAO Committee on forest development in the tropics has encouraged case studies within field projects (FAO/UNDP), especially in Guinea, Haiti, Indonesia, Paraguay, Peru, the Philippines, Thailand and Togo. There was the added advantage that these could approach the problem within the realistic framework of field operations. Shifting cultivation cannot be studied without taking into account the economic and social conditions prevailing, and clearly defined objectives and means.

Research in progress aims to rationalize the system of shifting cultivation, keep the land partly under forest by cultivating tree crops, form artificial grasslands for livestock husbandry, and to establish permanent agriculture. These include agroclimatic studies (especially for the development of rice cultivation in South-East Asia) designed to assist in the choice of suitable varieties and methods of culture. The possibilities of continuous crop cultivation in the lowland areas have been summarized by Greenland (1975): 'The potential farming systems for small farmers which are developing from this work involve zero tillage, and plant residue mulches; mixed crops of high yielding varieties that are disease and pest resistant; fertilizer to replace the phos-

phorus and possibly other nutrients removed in produce sold off the farm; legumes with highly active nitrogen fixing rhizobia to supply nitrogen to the soil and other crops, and control of acidity by means of ash or mulches of deep rooted species or by lime and trace elements where lime is readily available. The ingredients of this farming system have not yet been put together and tested as a whole, but fortunately this is not an essential preliminary to their use. As improved seeds of better yielding, pest and disease resistant materials are made available, as fertilizer is introduced, as legumes inoculated with highly active nitrogen fixing bacteria become more widely distributed and more widely used, yields will rise, and the shortening of fallow periods and the extension of cropping periods in shifting cultivation will have less and less detrimental effect. The evolution to continuous cultivation will follow naturally.'

These investigations include regional studies of integrated systems of land use, the effect on the soil of certain plants used in fallows, the effect on the environment of different types of shifting cultivation, and the demographic, social and economic factors of land-use systems.

The taungya system deserves a special mention because it may help to solve local food problems and to create the basis for a forest industry.

Much is still to be learnt on the restoration of soil fertility especially under short rotations, the loss of nutrients during cropping and the behaviour and response of ferrallitic soils to various intensities of use and methods of management. If a sustained agri-sylviculture programme is to be adopted for ferrallitic soils in western Africa the following questions must be posed and answered:
— the influence of various species of forest and arable crops on the physico-chemical properties of the soils;
— the fertility and productivity trends of these soils under combinations of tree and arable crops and cropping systems (including spacings and duration of cropping);
— the amount and rate of decomposition of organic matter (and the nutrients contained in it) added during the rotation and the extent to which the original soil structure and its attributes could be re-established at the end of a 5–10-year rotation of a wood-pulp species;
— the effect of wood and arable crop harvesting on nutrients;
— the effect of coppice as opposed to high forest regeneration on wood and arable crop production and nutrient drain;
— the effects of fertilizer dressing on the productivity of both arable and forest crops, litter turnover, and the continued use of the site for the production of these or other combinations of crops.

If wood is to be produced on an industrial scale, certain studies will have to be developed further. There will have to be more research into the autecology and requirements of local gregarious species. Selection trials and trials of vegetative propagation from cuttings will have to be carried out (bearing in mind the dangers inherent in a large degree of genetic homogeneity) similar to those already being conducted by the Centre technique forestier tropical in the People's Republic of the Congo. Mechanization of certain operations, especially in the Guinea savannas, and special treatments for soils with hard-pans will have to be studied.

Competition problems are of two kinds. First, that of the initial elements of the natural regrowth and the weeds which spread to form communities because of the greater uniformity. This might be relieved by mechanization or benefit from recent advances in chemical herbicides and arboricides. The main groups of troublesome competitive plants in Africa are the Mimosaceae (especially *Mimosa asperata*), the Scitamineae (species of *Aframomum, Palisota, Clinogyne*, etc.) and certain caespitous and rhizomatous savanna grasses.

Secondly, there is competition within the artificial stands themselves due to their density and the intensity of thinning which concern the understorey and stand productivity. In addition fire hazards, product uniformity and total yield require solutions.

In addition, it is of the greatest importance that during the preparation preceding the establishment of management plans, the initial efforts should be concerned with reviewing and updating aerial photographic surveys of regions considered suitable from the transport angle in order to ascertain the state of the plant cover; these should include the distribution of the major soil groups and their degree of utilization or degradation, and finally the necessary updating of information on the demography of the area.

Weed control

Research over the past twenty years has led to greatly increased knowledge of weeds and weed control although effective application of this has varied. Sufficient tropical experience has now been gained to provide good indications of the type of research needed. In many situations relatively minor research efforts may be sufficient to adapt existing techniques to local requirements.

Research on weeds is most advanced in Australia and the United States where modern control techniques have been widely adopted. A considerable body of information, especially relating to weed tree removal in natural forest, is also available from eastern and western Africa, the Indian region, and parts of South-East Asia (especially Malaysia). In these areas there is need to work out solutions to outstanding problems, but the continuation and expansion of existing programmes should produce the required information in a relatively short time. In many other parts of the tropics, however, including the Zaire basin region, much of central and southern America and a large part of South-East Asia, less progress has been made and there is an urgent need for the initiation of comprehensive research.

Some of the more important research requirements are listed and have been alloted degrees of priority in accordance with the following system:

— priority A, research of a high benefit/cost ratio, relating to major problems and expected to provide information of practical value in a relatively short time;

— priority B, research relating to more local problems or of a longer-term nature;

— priority C, research with smaller (though still substantial) chances of producing practical benefits.

More efficient and economical weed control in forest nurseries might be achieved by increased use of herbicides. For conifer seed-beds and transplant lines it should be relatively easy to adapt the techniques developed in temperate areas (priority A). More extensive development will be required for tropical hardwoods. Similar herbicides to those used in conifers are likely to prove successful on certain species, particularly triazines for large-seeded trees (priority B). For eucalypts a different type of selectivity is needed and preliminary screening of a wide range of chemicals is required (priority A). The possible use of activated charcoal for protecting some of the more sensitive species against soil acting herbicides may repay investigation (priority C).

In the establishment of forest plantations tending operations are necessary for varying periods to release young trees from the competitive effects of surrounding weed growth. Control by chemical methods is likely to be both quicker and cheaper than manual methods. Research is needed to develop selective treatments which are effective and safe when applied over both trees and weeds or, where sufficient selectivity is lacking, treatments directed so as to avoid the trees (priority A). In the reforestation of savanna areas, perennial grass weeds are best dealt with in the preparation of the land. Control can be achieved by cultivating under suitable conditions but research is needed to investigate herbicides which could reduce the amount of cultivation and afford more effective control at lower costs (priority A). The possibility of employing quick-growing and easily controlled leguminous trees to prevent weed growth and increase soil nitrogen would repay more investigation (priority C).

Little attention has been given to direct seeding; the elimination of the nursery stage could result in considerable savings but controlling weeds have hitherto made such an approach impracticable. Suitable herbicide treatments would overcome one of the principal obstacles and research is needed especially on the faster tree species (priority B).

Because of divergent opinions on the safety of sodium arsenite, careful reassessment of the toxicity hazards of arsenic treatment is needed in those countries where it is still in use (priority A). Less toxic chemicals, such as 2,4-D, 2,4,5-T, picloram, etc., are available as replacements, but research is needed to determine the relative susceptibilities of a wide range of tropical species to them so that they can be employed to the best advantage. Application by injectors should be widely tested for this will probably be the most economical technique (priority A). More investigation is required into the possibility of trees being injured through root associations (priority B).

Parasitic species of *Loranthus, Amyema* and *Phthirusa* cause appreciable loss of timber. Control with herbicides, found promising in India, Australia and the Carribean, should be tested further (priority B).

There is a lack of detailed ecological information about competition between weeds and trees. Certain perennial grasses are particularly strong competitors and thus among the most troublesome weeds. Little data are available for assessing the relative undesirability of other weed associations or for recognizing where plantations might be established without the need for expensive weed control. The potential benefits of being able to adjust weed control more closely to particular weed situations are likely to be considerable (priority B).

The increasing use of herbicides in forestry needs to be accompanied by more intensive study of their side effects. The evidence available suggests that the dangers to non-target organisms are slight. Continued research is needed to guard against water pollution and undesirable residues in the soil. Little is known about their effects on soil micro-organisms and such information should be obtained (priority B).

The impact of forest operations on the environment

There is need to develop and adapt forest technologies to the special needs of tropical forests and labour. The FAO Committee on forest development in the tropics has recommended at its third session (May 1974) that FAO should continue to provide practical guides for forest operations, including one for tropical swamp forests. It also recommended further studies on long-distance transport systems. This Committee recognized that forest roads play a key role not only in timber transport but also in forest

management and others forms of land use. Concern was expressed at the disturbance of soil resulting from road construction and it was recommended that soil protective measures should be treated in depth in the proposed FAO manual on forest roads. The Committee emphasized that cable transport should be considered as an alternative to road transport in steep terrain susceptible to erosion.

The approaches to environmental problems in the pulp and paper industries vary widely; many countries are still evolving legislative methodology and control techniques, while others are only considering them. Pollution control technology includes improving the washing efficiency, closing screening systems, reducing bleach plant water use, stripping condensates, installing equipment to reduce atmospheric emissions, external treatment plants, and the use of technology not yet commercially proven such as new pulping and bleaching processes, new equipment, and radically different treatment technologies. Several sulphate pulp-mills operate with almost closed water systems for washing and screening and bleaching technology is approaching the stage where the effluent may be returned to the recovery cycle. While much pollution control research and development is being undertaken by the industry, there is need for data exchange between countries; this would almost certainly decrease duplication of effort and some inefficiency of application.

Bibliography

Biological limitations to the transformation of tropical forest ecosystems

AAMISEPP, A.; OSVALD, H. Influence of higher plants upon each other-allelopathy. *Nova Acta Regiae Societatis Scientiarum Upsaliensis*, ser. IV, vol. 18, no. 2, 1961, p. 1–19.

AUBERT, M. Télémédiateurs chimiques et équilibre biologique océanique. Théorie générale. *Rev. Intern. Océanogr. Méd.*, 21, 1971, p. 5–16.

BARRETT, G. W.; VAN DYNE, G. M.; ODUM, E. P. Stress ecology. *Bio-Science*, vol. 26, no. 3, 1976, p. 192–194.

BETHEL, J. S. Problems in relating economic to biological production. In: *Records of Proceedings 12th Pacific Science Congress* (Canberra), vol. 1 (Abstracts of papers), 1971, p. 110.

BONNER, J. The role of toxic substances in the interactions of higher plants. *Bot. Rev.*, 16, 1950, p. 51–65.

BÖRNER, H. Liberation of organic substances from higher plants and their role in the soil sickness problem. *Bot. Rev.*, 26, 1960, p. 393–424.

CATINOT, R. Le présent et l'avenir des forêts tropicales humides. *Bois et Forêts des Tropiques*, n° 154, 1974, p. 3–26.

CONNELL, J. H. On the role of natural enemies in preventing competitive exclusion in some marine animals and in rain forest trees. In: *Proc. Adv. Study Inst. Dynamics Numbers Popul.* (Oosterbeek, 1970), p. 298–312.

DOBBEN, W. H. van; LOWE-MCCONNELL, R. H. (eds.). *Unifying concepts in ecology*. Report of the plenary sessions of the first international congress of ecology (The Hague, September 8–14, 1974). The Hague, W. Junk B. V. Publishers; Wageningen, Centre for agricultural publishing and documentation, 1975, 302 p.

Chapters on 'Flow of energy and matter between trophic levels'; 'Comparative productivity in ecosystems'; 'Diversity, stability and maturity in natural ecosystems'; 'Diversity, stability and maturity in ecosystems influenced by human activities'; 'Strategies for management of natural and man-made ecosystems'.

DUBOS, R. Humanizing the earth. *Science*, vol. 179, no. 4075, 1975, p. 769–772.

FAO. *Rapport de la Conférence FAO sur l'établissement de programmes coopératifs de recherche agronomique entre pays ayant des conditions écologiques semblables en Afrique. Zone guinéenne* (Ibadan, 1971). Rome, 1972, 313 p.

——. Management properties of ferralsols (by A. Van Wambecke). *Soils Bulletin*, 23, 1974, 129 p.

——. Committee on forest development in the tropics. Report of the 3rd session. Rome, May 1974, 65 p. + annexes.

FRANKLIN, J. F.; WARING, R. H. Predicting short and long term changes in the function and structure of temperate forest ecosystems. In: *Proceedings of the 1st International Congress of Ecology* (The Hague, September 1974), p. 228–232. Wageningen, Centre for agricultural publishing and documentation, 1974, 414 p.

GARB, S. Differential growth inhibitors produced by plants. *Bot. Rev.*, 27, 1961, p. 422–443.

GOLLEY, F. B. Structural and functional properties as they influence ecosystem stability. In: *Proceedings of the 1st International Congress of Ecology* (The Hague, September 1974), p. 97–102. Wageningen, Centre for agricultural publishing and documentation, 1974, 414 p.

GOODALL, D. W. *Predicting the results of human intervention in the moist tropics*. FAO, 1975, 6 p. multigr.

HARTSHORN, G. S.; ORIANS, G. H. *Diversity, stability and maturity in tropical forest ecosystems*. FAO, 1975, 26 p. multigr.

JACOBS, J. Diversity, stability and maturity in ecosystems influenced by human activities. In: *Proceedings of the 1st International Congress of Ecology* (The Hague, September 1974), p. 94–95. Wageningen, Centre for agricultural publishing and documentation, 1974, 414 p.

JEFFERS, J. N. R. Future prospects of systems analysis in ecology. In: *Proceedings of the 1st International Congress of Ecology* (The Hague, September 1974), p. 255–259. Wageningen, Centre for agricultural publishing and documentation, 1974, 414 p.

LABORIT, H. *Biologie et structure*. Paris, Gallimard, 1968, 187 p.

MAY, R. M. *Diversity, stability and maturity in natural ecosystems, with particular reference to the tropical moist forests*. FAO, 1975, 9 p. multigr.

MULLER, C. H. The role of chemical inhibition (allelopathy) in vegetational composition. *Bull. Torrey Bot. Club*, 93, 1966, p. 332–351.

——. Allelopathy as a factor in ecological process. *Vegetation*, 18, 1969, p. 348–357.

ODUM, E. P. The strategy of ecosystem development. *Science*, 164, 1969, p. 262–270.

PRESTON, F. W. Diversity and stability in the biological world. In: *Diversity and stability in ecological systems*, p. 1–12. Upton, N.Y., Brookhaven National Laboratory, Biol. Dept., Brookhaven Symposia in Biology (May 26–28, 1969), no. 22.

ROVIRA, A. D. Plant root exudates. *Bot. Rev.*, 35, 1969, p. 35–57.

SINGH, K. D. Spatial variation patterns in the tropical rain forest. *Unasylva* (FAO), vol. 26, no. 106, p. 18–23.

SYNNOTT, T. J. *The impact, short and long-term, of silvicultural, logging and other operations on tropical moist forest*. FAO, 1975, 18 p. multigr.

Unesco. *Use and conservation of the biosphere*. Paris, Unesco, 1970, 272 p.

——. *Expert panel on the rôle of systems analysis and modelling approaches in the Programme on Man and the Biosphere (MAB)*. MAB report series no. 2. Paris, Unesco, 1972, 50 p.

United Nations. *Proceedings of the United Nations Scientific Conference on the Conservation and Utilization of Resources* (17 August–6 September 1949, Lake Success, New York). New York, United Nations, Dept. of Economic Affairs, 8 vol., 1950.

WEBB, L. J.; TRACEY, J. G.; HAYDOCK, K. P. A factor toxic to seedlings of the same species associated with living roots of the non-gregarious subtropical rain forest tree *Grevillea robusta*. *J. Appl. Ecol.*, 4, 1967, p. 13–25.

WHITTAKER, R. H.; FEENY, P. P. Allelochemics: chemical interactions between species. *Science*, 171, 1971, p. 757–770.

WOODS, F. W. Biological antagonisms due to phytotoxic root exudates. *Bot. Rev.*, 26, 1960, p. 546–549.

Sylviculture. Enrichment planting. Forest exploitation

ANON. *Checklist of literature on ecological aspects of silviculture*. Bogor, Indonesia, BIOTROP, 1974, 76 p.

AUBRÉVILLE, A. La forêt coloniale: les forêts de l'Afrique occidentale française. *Ann. Acad. Sci. Colon.* (Paris), 9, 1938, p. 1–245.

BAIDOE, J. F. The management of the natural forests of Ghana. In: *7th World Forestry Congress* (Buenos Aires, Argentina), 1972, 7 CFM/C:1/4 E, 10 p.

BAUR, G. N. *The ecological basis of rain forest management*. Forestry Commission of New South Wales, Australia, 1961–62, 499 p. Rome, FAO, André Meyer Fellowship Programme Report, 1962, 499 p.

BAUR, G. N. Rain forest treatment. *Unasylva* (FAO), no. 72, 1964, p. 18–26.

BOERBOOM, J. H. A. Some remarks on the natural regeneration of tropical rain forest, with special reference to a method newly applied in Surinam. In: *Proceedings 6th World Forestry Congress* (Madrid, Spain), 3, 1966, p. 3187–3193.

BRITWUM, S. P. K. *Natural and artificial regeneration practices in the high forest of Ghana*. Rome, FAO, 1975, 8 p. multigr.

Bureau of Forestry, Philippines. *Handbook on selective logging*. Manila, 1965, 265 p.

BURGESS, P. F. The effect of logging on hill dipterocarp forests. *Malayan Nature Journal* (Kuala Lumpur), vol. 24, no. 3–4, 1971, p. 231–237.

——. Studies on the regeneration of the hill forests of the Malay peninsula—the phenology of dipterocarps. *Malayan Forester* (Kuala Lumpur), vol. 35, no. 2, 1972, p. 103–123.

CATINOT, R. Sylviculture tropicale en forêt dense africaine. *Bois et Forêts des Tropiques*, n°s 100, 101, 102, 103, 104, 1965, p. 5–18, 3–16, 3–16, 3–16, 17–30.

——. (in collaboration with LEPITRE, C.; CAILLIEZ, F.). Note condensée sur un protocole d'aménagement expérimental en forêt dense tropicale africaine. In: *Rapport 2e session Comité de la mise en valeur des forêts dans les tropiques*, p. 30–41. Rome, FAO, 1969.

——. Les éclaircies dans les peuplements artificiels de forêt dense africaine. Principes de base et application aux peuplements artificiels d'okoumé. *Bois et Forêts des Tropiques*, n° 126, 1969, p. 15–38.

——. Le présent et l'avenir des forêts tropicales humides. *Bois et Forêts des Tropiques*, n° 154, 1974, p. 3–26.

COUSENS, J. E. Some reflections on the nature of the Malayan lowland rain forest. *Malayan Forester*, vol. 28, no. 2, 1965, p. 122–128.

DAUBENMIRE, R. Phenology and other characteristics of tropical semi-deciduous forest in northwestern Costa Rica. *Journal of Ecology* (Oxford), vol. 60, no. 1, 1972, p. 147–170.

DAWKINS, H. C. *Felling damage*. Technical Note, Forest Department (Entebbe, Uganda), no. 6/58, 1958a, 2 p.

——. *The management of natural tropical high forest with special reference to Uganda*. Oxford, Commonwealth Forestry Institute, no. 34, 1958b, 155 p.

DAWKINS, H. C. Crown diameters: their relation to bole diameters in tropical forest trees. *Commonwealth Forestry Review* (London), vol. 42, no. 4, 1963, p. 318–333.

DONIS, C. La forêt dense congolaise et sa sylviculture. *Bulletin agricole du Congo belge* (Bruxelles), n° 2, 1956, p. 47.

DUVIGNEAUD, P. (ed.). *Productivity of forest ecosystems*. Paris, Unesco, 1971, 707 p.

FAO. *Tropical silviculture*, vol. I, II, III. Rome, 1957, 1958, 190, 415, 101 p.

——. *Exploitation and transport of logs in dense tropical forests*. Forest Development, no. 18. Rome, 1974, 100 p.

——. *Logging and log transport in man-made forests in developing countries*. Rome, FAO/SWE/TF 116, 1974, 134 p.

——. *Committee on forest development in the tropics. Reports of the 1st, 2nd, 3rd and 4th sessions*. Rome, October 1967, October 1969, May 1974, November 1976.

Fox, J. E. D. Selective logging in the Philippine dipterocarp forest. *Malayan Forester*, vol. 30, no. 3, 1967, p. 182–190.

——. Damage, defect and wastage. *Malayan Forester*, vol. 31, no. 3, 1968, p. 157–164.

Fox, J. E. D. Sylvicultural and economic aspects of re-logging. In: *Laporan 1969 Penyelidek Hutan, Negeri Sabah* (Sandakan, Sabah), 1972, p. 100–107.

Gilmour, D. A. The effects of logging on streamflow and sedimentation in a North Queensland rain forest catchment. *Commonwealth Forestry Review* (London), vol. 50, no. 1, 1971, p. 38–48.

IUFRO. *Preliminary report of ad hoc Committee on tropical forestry research.* June 18, 1975, 9 p.

Kio, P. R. O. What future for natural regeneration of tropical high forest? An appraisal with examples from Nigeria and Uganda. In: *Proceedings 6th annual Conference of the Forestry Association of Nigeria*, 1975.

Kochummen, K. M. Natural plant succession after farming in Sungei Keoh. *Malayan Forester*, vol. 29, no. 3, 1966, p. 170–181.

Lamb, A. F. A. Artificial regeneration within the humid lowland tropical forest. *Commonwealth Forestry Review*, vol. 48, no. 1, 1969, p. 41–53.

Liew, T. C. An analysis on staff, cost and labour in protective tree marking and climber cutting prior to logging. In: *Laporan 1971 Penyelidek Hutan, Negeri Sabah* (Sandakan, Sabah), 1973, p. 68–77.

Lowe, R. G. Unpublished cyclostyled research reports and lecture notes on the Tropical Shelterwood System. Ibadan, Nigeria, Federal Department of Forest Research, 1964–66.

——. *Some effects of stand density on the growth of individual trees of several plantation species in Nigeria.* University of Ibadan, Nigeria, Faculty of Agriculture, Forestry and Veterinary Science, Ph. D. thesis, 1971, 249 p.

——. *Nigerian experience with natural regeneration in tropical moist forest.* Rome, FAO, 1975, 14 p. multigr.

Lundgren, B. *Ecological comparison between softwood monoculture and natural forests in East Africa.* Stockholm, Royal College of Forestry, 1974, 33 p. multigr.

Mervart, J. Growth and mortality rates in the natural high forest of western Nigeria. *Nigeria Forestry Information Bulletin* (Ibadan), new series, 22, 1972, 28 p.

Moore, D. *Enrichment of the species composition in relation to management of the tropical moist forest.* Rome, FAO, 1975, 8 p. multigr.

Nicholson, D. I. Analysis of logging damage in tropical rain forest, North Borneo. *Malayan Forester*, vol. 21, no. 4, 1958, p. 235–245.

——. *Forest management; demonstration and training in forest, forest range and watershed management.* Unpublished technical report, UNDP/FAO project FO:SF/PHI 16. Rome, FAO, no. 3, 1970, 51 p.

——. Compartment sampling in North Queensland rain forests as a basis for silvicultural treatment. *Commonwealth Forestry Review* (London), vol. 51, no. 4, 1972, p. 314–326.

——. Specifying rain forest treatment with a wedge prism. *Commonwealth Forestry Review* (London), vol. 53, no. 3, 1974, p. 189–190.

Palmer, J. R. Towards more reasonable objectives in tropical high forest management for timber production. *Commonwealth Forestry Review* (London), vol. 54, no. 161–162, 1975, p. 273–289.

——; Dawkins, H. C. *Silvicultural research programme 1971–75.* Kuching, Sarawak, Silvicultural research section, Forest Research Branch, Forest Department, 1971.

Redhead, J. F. An analysis of logging damage in lowland rain forest, western Nigeria. *Nigeria Forestry Information Bulletin* (Ibadan), new series, 10, 1960, p. 5–16.

Rollet, B. *L'architecture des forêts denses humides sempervi-*

rentes de plaine. Nogent-sur-Marne, France, Centre technique forestier tropical, 1974, 298 p.

Sargent, K. J. *An analysis of problems affecting the development of forestry and forest industries in West Malaysia.* Technical Report, UNDP/FAO project FO:SF/MAL 16. Rome, FAO, no. 1, 1970, 229 p.

Skapski, K. *Intensive forest management.* Study paper, Pahang Tenggara Regional Master Planning Study, peninsular Malaysia, no. 11, 1971.

Synnott, T. J.; Kemp, R. H. *The relative merits of natural regeneration, enrichment planting and conversion planting in tropical moist forests, including agri-silvicultural techniques.* Rome, FAO (Committee on forest development in the tropics, 4th session, November 1976), FO:FDT/76/7(a), 1976, 12 p.

Tomlinson, P. B.; Gill, A. M. Growth habits of tropical trees: some guiding principles. In: Meggers, B. J.; Ayensu, E. S.; Duckworth, W. D. (eds.). *Tropical forest ecosystems in Africa and South America: a comparative review,* p. 129–143. Washington, D.C., Smithsonian Institution, 1973, 350 p.

Tran Van Nao. Forest resources of humid tropical Asia. In: *Natural resources of humid tropical Asia,* p. 197–215. Paris, Unesco, 1974, 456 p.

Troup, R. S. Silvicultural systems. In: *Oxford manuals of forestry.* Oxford, Clarendon Press, 1928, 199 p.

Wadsworth, F. H. Tropical forest regeneration practices. In: *Duke University tropical forestry Symposium.* Durham, North Carolina, Duke University, School of Forestry, April 1965, 29 p.

Wyatt-Smith, J. Survival of seedlings of meranti sarang punai (*Shorea parvifolia Dyer*) and kempas (*Koompassia malaccensis* Benth.) in belukar. *Malayan Forester,* vol. 12, no. 3, 1949a, p. 144–148.

——. Natural plant succession. *Malayan Forester,* vol. 12, no. 3, 1949b, p. 148–152.

——. Survival of isolated seedbearers. *Malayan Forester,* vol. 17, no. 1, 1954, p. 30–32.

——. Changes in composition in early natural plant succession. *Malayan Forester,* vol. 18, no. 1, 1955, p. 44–49.

——. Seedling/sapling survival of *Shorea leprosula, Shorea parvifolia* and *Koompassia malaccensis. Malayan Forester,* vol. 21, no. 3, 1958, p. 185–193.

——. *Manual of Malayan silviculture for inland forests.* Kuala Lumpur, Malayan Forest Record, no. 23, 1963, 400 p.

Wildlife management

Asibey, E. O. A. Wildlife as a source of protein in Africa south of the Sahara. In: *Report of the fourth session of the working party on wildlife management of the African Forestry Commission* (Nairobi, 1–3 February 1972). Rome, FAO, 1972.

Caughley, C. Sustained-yield harvesting. In: *Report of the fourth session of the working party on wildlife management of the African Forestry Commission* (Nairobi, 1–3 February 1972). Rome, FAO, 1972.

Charter, J. R. The economic value of wildlife in Nigeria. In: *Forestry Association of Nigeria, First Annual Conference* (Ibadan), 1970, p. 1–12.

Choudhury, S. R. Forestry and wildlife conservation in the tropics. *Indian Forester,* vol. 101, no. 1, 1975, p. 45–46.

Darling, F. F. *Wildlife in an African territory.* London, Oxford University Press, 1960, 166 p.

Dasmann, R. F. Biomass, yield and economic value of wild and domestic ungulates. In: *International Union of Game Biologists,* p. 227–235, 1965.

DEN HARTOG, A. P.; de VOS, A. The use of rodents as food in tropical Africa. *FAO Nutrition Newsletter*, vol. 11, no. 2, 1973, p. 1–14.

DE VOS, A.; JONES, T. *Proceedings Symposium on land use and wildlife management* (Nairobi, 1967). Special issue of the *East African Agriculture and Forestry J.*, 33, 1968, 297 p.

——; KAITTANY, K. M. *Selected bibliography on the economic uses of wildlife and wildlife products in Africa*. 1972, 10 p.

FAO. Le rôle de la faune sauvage et des parcs nationaux dans la foresterie tropicale. In: *Comité de la mise en valeur des forêts dans les tropiques. Rapport de la 2e session*, p. 96–105. Rome FAO, octobre 1969.

PIERRET, P. *Estudio de la importancia de la producción de la fauna en carne y pieles para las poblaciones rurales del Río Ucayali, Peru*. La Molina, Universidad Agraria, Instituto de Investigaciones forestales, 1967.

——; DOUROJEANNI, M. La caza y la alimentación humana en las riberas del Río Pachitea, Peru. *Turrialba*, vol. 16, no. 3, 1966, p. 271–277.

RINEY, T. The international importance of African wildlife. *Unasylva* (FAO), vol. 15, no. 2, 1961, p. 75–80.

——. The economic use of wildlife in terms of its productivity and its development as an agricultural activity. In: *First FAO Regional Meeting on Animal Production and Health* (Addis Ababa), 1964, 3 p.

——. *Conservation and management of African wildlife*. UNDP Technical Assistance Report. Rome, FAO, 1967, 35 p.

——; HILL, P. *Conservation and management of African wildlife. English-speaking country reports (Botswana, Ethiopia, Kenya, Malawi, Nigeria, Sierra Leone, Somali Republic)*. UNDP Technical Assistance Report. Rome, FAO, 1967, 145 p.

——; ——. *Conservation et aménagement de la faune et de son habitat en Afrique. Rapports sur les pays francophones (Burundi, Cameroun, Congo, Dahomey, Haute-Volta, Mali, République Centrafricaine, Sénégal, Tchad)*. UNDP Technical Assistance Report. Rome, FAO, 1967, 135 p.

SPILLET, J. J. Economic aspects of wildlife conservation; values of consumptive and non-consumptive uses of wildlife. In: *IUCN 11th Technical Meeting* (New Delhi), p. 121–129, 1970.

TALBOT, L. M. Wild animals as source of food. In: *Proceedings 6th International Congress Nutrition* (Edinburgh), p. 243–251, 1964.

—— et al. *The meat production potential of wild animals in Africa*. Edinburgh, Commonwealth Bureau of Animal Breeding and Genetics, 16, 1965, v+42 p.

Cattle grazing

BARR, N. C. Nutrition of grazing animals. *Tropical Grasslands*, vol. 5, no. 1, 1971, p. 50–53.

BRUCE, R. C. Effect of *Centrosema pubescens* Benth. on soil fertility in the humid tropics. *Queensland Journal of Agricultural and Animal Science*, 22, 1965, p. 221–226.

EDEN, D. R. A. Pacific copra production near possible serious decline. *South Pacific Commission Quarterly Bulletin*, 8, 1958, p. 1–32.

EVANS, T. R. Species for coastal pastures—their strengths and weaknesses. *Tropical Grasslands*, vol. 5, no. 1, 1971, p. 45–50.

FERNANDEZ, D. E. F. Effect of pasture on the yield of coconut. Annual Report of the Coconut Research Institute of Ceylon 1967. *Ceylon Coconut Quarterly*, vol. 19, no. 1–2, 1968, p. 54–56.

FREMOND, Y.; BRUMIN, C. Cocotier et couverture du sol. *Oléagineux*, vol. 21, n° 6, 1966, p. 361–364.

GROF, B. Establishment of legumes in the humid tropics of northeastern Australia. In: *Proceedings of the 9th International Grassland Congress*, vol. II, 1965, p. 1137–1142.

KRISHNA MARAR, M. M. Intercultivation in coconut gardens—its importance. *Indian Coconut Journal*, vol. 4, no. 4, 1953, p. 131–137.

——. Trial of intercultivation practices in coconut gardens. *Indian Coconut Journal*, vol. 14, no. 3, 1961, p. 87–99.

McCARTHY, W. D.; NUTHALL, P. L.; HIGHAM, C.; FERGUSON, D. Economic evaluation of land use alternatives for the Southern Wallum region, Queensland. *Tropical Grasslands*, vol. 4, no. 3, 1970, p. 195–212.

OHLER, J. G. Cattle under coconuts. *Tropical Abstracts*, vol. 24, no. 10, 1969, p. 639–645.

RAJARATNAM, D. T.; SANTHIRASEGARAM, K. Cultivation X pasture experiment, Ratmalagara Estate. *Ceylon Coconut Quarterly*, vol. 14, no. 1–2, 1963, p. 37–38.

——; ——. Intensity of grazing trial. *Ceylon Coconut Quarterly*, vol. 14, no. 1–2, 1963, p. 38–39.

RODRIGO, E. Fodder grass experiment (Lunuwila). In: *Annual Report Coconut Research Scheme*, 1943, p. 11.

ROMBAUT, D. *Élevage bovin sous palmiers*. Rome, FAO, 1972.

——. Étude sur l'élevage bovin dans les palmeraies de Côte-d'Ivoire. *Oléagineux*, vol. 29, n° 3, 1974, p. 121–125.

SILVA, M. A. T. de. Cover crops under coconuts. *Ceylon Coconut Planters' Review*, vol. 11, no. 1–2, 1961, p. 17–22.

TEITZEL, J. K. Pastures for the wet tropical coast. *Queensland Agricultural Journal*, vol. 95, 1969, p. 304–314, 380–388, 464–471, 532–537.

WHITEMAN, P. C. Pasture development under plantation and annual crops. In: Whiteman, P. C.; Humphreys, L. R.; Monteith, N. H. (eds.). *A course manual in tropical pasture science*, p. 52–57. Brisbane, Watson Ferguson, 1974.

Forest plantations

ANDERSON, J. A. R. Observations on climatic damages in *Shorea albida* forests in Sarawak attributed to lightning. *Commonwealth Forestry Review* (London), vol. 43, no. 2, 1964, p. 145–158.

AUBRÉVILLE, A. Érosion sous forêts et érosion en pays déforesté dans la zone tropicale humide. *Bois et Forêts des Tropiques* (Nogent-sur-Marne), n° 68, 1959, p. 3–14.

BAILLIE, I. C. *An occurrence of charcoal in soil under primary forest*. Kuching, Sarawak, Forest Department, unpublished report, 1971, 10 p.

BELL, T. I. W. Erosion in the Trinidad teak plantations. *Commonwealth Forestry Review* (London), vol. 52, no. 3, 1973, p. 223–233.

BENEDICT, W. V. *Protecting plantations of long-fibre tree species from loss by insects and diseases*. Technical Report, UNDP/SP project MAL/12. Rome, FAO, no. 4, 1971, 24 p.

BENNETT, R. M. *A forest plantation scheme for the New Hebrides*. Port Vila, Department of Agriculture, 1974, 16 p.

BOYCE, J. S. *Forest plantation protection against diseases and insect pests*. Rome, FAO, 1954, 41 p.

BROUARD, N. R. A brief account of the 1960 cyclones and their effects upon exotic plantations in Mauritius. *Empire Forestry Review* (London), vol. 39, no. 4, 1960, p. 411–416.

——. Damage by tropical cyclones to forest plantations, with particular reference to Mauritius. In: *9th Commonwealth Forestry Conference* (New Delhi, 1968). Port Louis (Mauritius), Government Printer, 1968, 8 p.

BROWNE, F. G. Storm forest in Kelantan. *Malayan Forester* (Kuala Lumpur), vol. 12, no. 1, 1949, p. 28–33.

BROWNE, F. G. *Pests and diseases of forest plantations: an anno-tated list of the principal species occurring in the British Commonwealth*. Oxford, Clarendon Press, 1968, 11+1330 p.

BRÜNIG, E. F. Some further evidence on the amount of damage attributed to lightning and wind-throw in *Shorea albida* forest in Sarawak. *Commonwealth Forestry Review* (London), vol. 52, no. 3, 1973, p. 260–265.

BURLEY, J.; KEMP, R. H. Centralised planning and international cooperation in the introduction and improvement of trop-ical tree species. In: *Second General Congress of the Society for the Advancement of breeding researches in Asia and Oceania* (New Delhi), 1972, 11 p.

——; NICKLES, D. G. (eds.). *Proceedings of a IUFRO meeting on tropical provenance and progeny research and international cooperation* (Nairobi, Kenya, 1973). Oxford, Common-wealth Forestry Institute, 1973, 597 p.

CATINOT, R. Plantation intensive des essences forestières sous les Tropiques humides. Environnement et principaux problèmes régionaux de recherches. In: *Rapport de la Conférence FAO sur l'établissement de programmes coopératifs de recherche agronomique entre pays ayant des conditions écologiques sem-blables en Afrique. Zone guinéenne* (Ibadan, 1971), p. 271–276. Rome, FAO, 1972, 313 p.

CHARTER, J. R.; KEAY, R. W. J. Assessment of the Olokemeji forest control experiment 28 years after institution. *Nigeria Forestry Information Bulletin* (Ibadan), new series, no. 3, 1960.

CHERRETT, J. M. Baits for the control of leaf-cutter ants. I. For-mulation. *Tropical Agriculture* (Trinidad), vol. 46, no. 2, 1969, p. 81–90. II. Toxicity evaluations of Mirex 450, Aldrin and Dieldrin to *Acromyrex octospinosus*. *Tropical Agriculture* (Trinidad), vol. 46, no. 3, 1969, p. 211–219. III. Waterproofing for general broadcasting. *Tropical Agriculture* (Trinidad), vol. 46, no. 3, 1969, p. 221–231.

DUGAIN, F.; FAUCK, R. Erosion and run-off measurements in middle Guinea. Relations with certain cultivations. In: *3rd Inter-African Soils Conference* (Dalaba), 1959.

FAO. *Catalogues de graines forestières*. Rome, FAO, 1956, 1961, 1975, 178, 523, 283 p.

——. *Tree seed notes. 2. Humid tropics*. Rome, FAO, 1955, p. 187–354.

——. *Tree planting practices in tropical Africa*. Rome, FAO, 1956, 302 p.

——. *Tree planting practices in tropical Asia*. Rome, FAO, 1957, 172 p.

——. *Prácticas de plantación forestal en América latina*. Roma, FAO, 1960, 499 p.

——. World Symposium on man made forest. *Unasylva*, vol. 21, no. 3–4, 1967, p. 1–116.

——. Second world consultation on forest tree breeding (Was-hington, 1969). *Unasylva*, vol. 24 (2–3), no. 97–98, 1970, p. 1–132.

——. *A manual on establishment techniques in man-made forests*. FO:MISC/73/3. Rome, FAO, February 1973, 108 p.

——. *Report of the 3rd session of the FAO Panel of experts on forest gene resources*. Rome, FAO, May 1974.

——. *Forest genetic resources. Information no. 4*. Rome, FAO, Forestry occasional paper 1975/1, 68 p.

FONSECA, J. P. da. Chemical control of subterranean termites in the Guarani Forestry Nursery. *Arquivos do Instituto de Biologia* (São Paulo, Brazil), 19, 1949–50, p. 57–58.

FOURY, P. Comparaison des méthodes d'enrichissement utilisées en forêt dense équatoriale. *Bois et Forêts des Tropiques* (Nogent-sur-Marne), n° 47, 1956, p. 15–25.

FREEZAILLAH BIN C. Y. *Some notes on Pinus caribaea Mor.*

grown in Malaya. Research pamphlet, Forest Research Institute (Kepong), no. 54, 1966, 24 p.

GANE, M. Hurricane risk assessment in Fiji. *Commonwealth Forestry Review* (London), vol. 49, no. 3, 1970, p. 253–256.

GROULEZ, J. *Conversion planting in tropical moist forests*. Rome, FAO (Committee on forest development in the tropics, 4th session, November 1976), FO:FDT/76/7 (b), 1976.

HOPKINS, B. Observations on savanna burning in the Olokemeji Forest Reserve, Nigeria. *J. Appl. Ecol.* (Oxford), vol. 2, no. 2, 1965, p. 367–381.

IUFRO. *Preliminary report of ad hoc Committee on tropical forestry research*. June 1975, 9 p.

JACOBS, M. R. *Research needs in silviculture and forest man-agement*. Technical report, UNDP/SF project BRA/45. Rome, FAO, no. 1, 1972, 85 p.

JOHNSON, M. S. *New Hebrides Condominium; Erromango forest inventory*. London, Land Resources Division (Overseas Development Administration), Land Resource Study no. 10, 1971, 70 p.

JOHNSON, N. E. *Biological opportunities and risks associated with fast growing plantations in the tropics*. Rome, FAO, 1975, 16 p. multigr.

JOHNSTONE, R. C. B. Elephant protection problems in Budongo tropical high forest. In: *9th Commonwealth Forestry Confer-ence* (New Delhi), 1968.

KEMP, R. H. *The control of root feeding nursery pests, with special reference to termites (especially those attacking Eucalyptus spp. in North Nigeria)*. Imperial Forestry Institute (Oxford), unpublished thesis, 1956, 39 p.

KING, H. C. Notes on three cyclones in Mauritius in 1945: their effect on exotic plantations, indigenous forest and on some timber buildings. *Empire Forestry Review* (London), vol. 24, no. 2, 1945, p. 192–195.

KOCHUMMEN, K. M. Natural plant succession after farming in Sungei Keoh. *Malayan Forester*, vol. 29, no. 3, 1966, p. 170–181.

LAMB, A. F. A. *Impressions of Nigerian forestry after an absence of twenty-three years*. Oxford, Commonwealth Forestry Institute, 1967, 42 p. multigr.

——. *Fast growing timber trees of the lowland tropics. No. 2. Cedrela odorata*. Oxford, Commonwealth Forestry Institute, 1968, 46 p.

——. Artificial regeneration within the humid lowland tropical forest. *Commonwealth Forestry Review* (London), vol. 48, no. 1, 1969, p. 41–53.

——. Tropical pulp and timber plantations, a brief account of forest plantations in the tropics. In: *Proceedings 7th World Forestry Congress* (Buenos Aires), 1972, p. 14.

——. *Fast growing timber trees of the lowland tropics. Pinus caribaea*, vol. 1. Oxford, Commonwealth Forestry Institute, 1973, 254 p.

LAWS; PARKER; JOHNSTONE, R. C. B. *Elephants and their habitats: the ecology of elephants in North Bunyoro, Uganda*. London, Oxford University Press, 1975, 376 p.

LEGGATE, J. *The resistance of certain tree species, at present on trial in the Solomons, to cyclone damage*. Honiara, British Solomon Islands Protectorate, Forestry Department tech-nical note no. 5/67, 1967, 3 p.

LINDO, L. S. The effect of hurricanes on the forests of British Honduras. In: *9th Commonwealth Forestry Conference* (New Delhi), 1968, 13 p.

LOWE, R. G. Control of termite attack on *Eucalyptus citriodora* Hook. *Empire Forestry Review* (London), vol. 40, no. 1, 1961, p. 73–78.

MONNIER, M. F. Eucalyptus et termites. In: *CCTA Conference* (Pointe Noire), 1958.

PARRY, M. S. Control of termites in *Eucalyptus* plantations. *Empire Forestry Review* (London), vol. 38, no. 3, 1959, p. 287–292.

PHILIP, M. S. *Working plan for Budongo Central Forest Reserve (including Budongo, Siba and Kitigo forests), third revision, for the period 1 July 1964 to 30 June 1974.* Entebbe, Uganda, Forest Department, 1965, 130 p.

REDHEAD, J. F. Taungya planting. *Nigeria Forestry Information Bulletin* (Ibadan), new series, no. 5, 1960, p. 13–16.

SELF, M. B. *The resistance of certain tree species, at present on trial in the Solomons, to cyclone damage.* Honiara, British Solomon Islands Protectorate, Forestry Department technical note no. 4/68, 1968, 2 p.

WADSWORTH, F. H.; ENGLERTH, G. H. Effects of the 1956 hurricane on forests in Puerto Rico. *Caribbean Forester* (Puerto Rico), vol. 20, no. 1–2, 1959, p. 38–51.

WARING, H. D. *Nutritional problems in Malaysian pine plantations.* Technical Report, UNDP/SF project MAL/12. Rome, FAO, no. 1, 1971, 59 p.

WEBB, L. J. Cyclones as an ecological factor in tropical lowland rain forest, North Queensland. *Australian Journal of Botany* (Melbourne), vol. 6, no. 3, 1958, p. 220–228.

WHITE, M. G. *The problem of the Phytolyma gall bug in the establishment of Chlorophora.* Oxford, Commonwealth Forestry Institute, paper no. 37, 1966, 52 p.

WHITMORE, J. L. Myths about the establishment of *Cedrela*. In: *Proceedings of the first symposium on integrated control of Hypsipyla* (IICA-CATIE, Turrialba, Costa Rica), 1973.

WHITMORE, T. C. *A botanist's notes on the Vanikoro kauri forests.* Honiara (British Solomon Islands Protectorate), Forestry Department, technical note no. 9/63, 1963, 3 p.

——. *Change with time and the role of cyclones in tropical rain forest on Kolombangara, Solomon Islands.* Oxford, Commonwealth Forestry Institute, paper no. 46, 1974, 92 p.

WOOD, P. J. Problèmes de développement forestier dans la zone guinéenne. In: *Rapport de la Conférence FAO sur l'établissement de programmes coopératifs de recherche agronomique entre pays ayant des conditions écologiques semblables en Afrique. Zone guinéenne* (Ibadan, 1971). p. 277–283. Rome, FAO, 1972, 313 p.

——. The evaluation of fast growing species in the tropics. In: *10th Commonwealth Forestry Conference*, 1974, p. 24.

WOOD, T. W. W. Wind damage in the forest of Western Samoa. *Malayan Forester*, vol. 33, no. 1, 1970, p. 92–99.

WYATT-SMITH, J. Storm forest in Kelantan. *Malayan Forester*, vol. 17, no. 1, 1954, p. 5–11.

Shifting cultivation and other agri-sylvicultural systems

AHN, P. M. Some observations on basic and applied research in shifting cultivation. In: *Report on the FAO/SIDA/ARCN Regional seminar on shifting cultivation and soil conservation in Africa* (Ibadan, Nigeria), p. 54–61. Rome, FAO, 1974.

ALEXANDER, E. B. A comparison of forest and savanna soils in northeastern Nicaragua. *Turrialba*, 23, 1973, p. 181–191.

ALLAN, W. *The African husbandman.* Edinburgh, Oliver and Boyd, 1965, 505 p.

ANAKWENZE, F. N.; ETTAH, A. F. The role of forestry in food production in Nigeria. In: *Proceedings 5th annual Conference of the Forestry Association of Nigeria* (Jos), 1974.

BARNETT, A. P.; CARREKAR, J. R.; ABRIENA, F.; JACKSON, W. A.; DOOLEY, A. E.; HOLLADAY, J. H. Soil and nutrient losses

in run-off with selected cropping treatments on tropical soils. *Soil Sci. Soc. Amer. Proc.*, 64, 1972, p. 391–394.

BARTLETT, H. R. *Fire in relation to primitive agriculture and grazing in the tropics. Annotated bibliography.* Ann Arbor, University of Michigan, vol. I, 1955, 568 p.

——. *Fire in relation to primitive agriculture and grazing in the tropics. Annotated bibliography.* Ann Arbor, University of Michigan, vol. II, 1957, 873 p.

——. *Fire in relation to primitive agriculture and grazing in the tropics. Annotated bibliography.* Ann Arbor, University of Michigan, vol. III, 1961, 216 p.

BATCHELDER, R. B.; HIRT, H. F. *Fire in tropical forests and grasslands.* United States Army Natick Lab. Techn. Rep., 1966, 380 p.

BAZILEVIČ, N. I.; RODIN, L. E. The biological cycle of nitrogen and ash elements in plant communities of the tropical and subtropical zones. *Forestry Abstr.*, vol. 27, no. 3, 1966, p. 357–368.

BRADFIELD, R. *Intensive multiple cropping.* Los Baños, International Rice Research Institute, 1969, multigr.

BRAUN, H. Shifting cultivation in Africa. In: *Report on the FAO/SIDA/ARCN Regional seminar on shifting cultivation and soil conservation in Africa* (Ibadan, Nigeria), p. 28–29. Rome, FAO, 1974.

BRINKMANN, W. L. F.; VIEIRA, A. N. The effect of burning on germination of seeds at different soil depths, of different tropical tree species. *Turrialba*, 21, 1971, p. 77–82.

——; NASCIMENTO, J. C. de. The effect of slash-and-burn agriculture on plant nutrients in the Tertiary region of Central Amazonia. *Turrialba*, vol. 23, no. 2, 1973, p. 248–290.

BRÜNIG, E. F. Taungya versus shifting cultivation. In: *Proceedings of International Seminar on Employment and transfer of technology in forestry*, p. 197–223. Berlin, 1974.

BUDOWSKI, G. Tropical savannas, a sequence of forest felling and repeated burnings. *Turrialba*, 6, 1956, p. 23–33.

——. Forest successions in tropical lowlands. *Turrialba*, 13, 1961, p. 42–44.

CARNEIRO, R. L. Slash-and-burn cultivation among the Kinkuru and its implications for cultural development in the Amazon basin. In: Wilbert, J. (ed.). *The evolution of horticultural systems in native South America*, p. 47–68. Caracas, Anthropologica, supplement 2, 1961.

CLARKE, W. C. Maintenance of agriculture and human habitats within the tropical forest ecosystem. *Human Ecology*, vol. 4, no. 3, 1976, p. 247–259.

COLE, J. P. *Latin America; economic and social geography.* London, Butterworths, 1965, 468 p.

CONKLIN, H. C. *Hanunóo agriculture: a report on an integral system of shifting cultivation in the Philippines.* Rome, FAO Forestry Development Papers no. 12, 1957, 209 p. Re-edition, Northford, Connecticut, Elliot's Books, 1975.

COOK, O. F. Milpa agriculture, a primitive tropical system. In: *Annual report* (Smithsonian Institute, Washington, D.C.), 1921, p. 307–326.

COULTER, J. K. Soil management systems. In: Drosdoff, M. (ed.). *Soils of the humid tropics*, p. 189–197. Washington, D.C., National Academy of Sciences, 1972, 219 p.

CUNNINGHAM, R. K. The effect of clearing a tropical forest soil. *J. Soil Sci.*, 14, 1963, p. 334–345.

DARBY, H. C. The clearing of the woodland in Europe. In: Thomas, W. L. (ed.) *Man's role in changing the face of the earth*, p. 183–216. Chicago, University of Chicago Press, 1956, 1 193 p.

DAUBENMIRE, R. Some ecological consequences of converting forest to savanna in northwestern Costa Rica. *Trop. Ecol.*, 13, 1972, p. 31–51.

DENEVAN, W. M. *The causes and consequences of tropical shifting cultivation*. Rome, FAO, 1975, 10 p. multigr.

DOBBY, E. H. C. *South-East Asia*. London, University of London Press, 1950, 415 p.

DOMMERGUES, Y. Les cycles biogéochimiques des éléments minéraux dans les formations tropicales. *Bois et Forêts des Tropiques*, 87, 1963, p. 9–25.

DONIS, C. *Agriculture itinérante et techniques sylvo-agricoles*. Rome, FAO, 1975, 7 p. multigr.

DOUGLAS, J. S.; HART, J. R. A. de. *Forest farming*. London, Watkins, 1976, 197 p.

DUCKHAM, A. N.; MASEFIELD, G. B., assisted by WILLEY, R. W. and DOWN, K. *Farming systems of the world*. London, Chatto and Windus, 1971, 542 p.

DUMOND, D. E. Swidden agriculture and the rise of Maya civilization. *Southwestern J. Anthropol.*, 17, 1961, p. 301–316.

ENABOR, E. E.; ADEYOJU, S. K. *An appraisal of departmental taungya as practised in the South-Eastern State of Nigeria*. Department of Forestry, Misc. Report no. 3, 1975, 80 p.

EYRE, S. R. *Vegetation and soils; a world picture*, 2nd edition. London, Arnold, 1968.

FAO. Shifting cultivation. *Tropical Agriculture* (Trinidad), 34, 1957, p. 159–164.

——. Management properties of ferralsols (by A. Van Wambecke). *Soils Bulletin* no. 23. Rome, FAO, 1974, 129 p.

——. Shifting cultivation and soil conservation in Africa. *Soils Bulletin* no. 24. Rome, FAO, 1974, 248 p.

FAO/SIDA. *Shifting cultivation and soil conservation in Africa. Summaries and Recommendations* (Seminar held at University of Ibadan, Nigeria). Rome, FAO, 1974.

FASSBENDER, H. W. Aspectos eco-pedológicos de la transformación de un ecosistema forestal a un ecosistema agrícola. *Sociedad Venezolana de la Ciencia del Suelo*, Publicación no. 10, 1974, p. 25–43.

FITTKAU, E. J.; KLINGE, H. On biomass and trophic structure of the Central Amazonian rain forest ecosystem. *Biotropica*, vol. 5, no. 1, 1973, p. 2–14.

FREEMAN, J. D. *Iban agriculture*. A report on the shifting cultivation of hill rice by the Iban of Sarawak. London, HM SO, 1955, 148 p.

GÓMEZ-POMPA, A.; VÁZQUEZ-YANES, C.; GUEVARA, S. The tropical rain forest: a non-renewable resource. *Science*, 177, 1972, p. 762–765.

GOUROU, P. The quality of land use of tropical cultivators. In: Thomas, W. L. (ed.). *Man's role in changing the face of the earth*. Chicago, University of Chicago Press, 1956, 1193 p.

GREENLAND, D. J. The maintenance of shifting cultivation versus the development of continuous management systems. In: *IITA Conference* (Ibadan, November 1970), 11 p.

——. Evolution and development of different types of shifting cultivation. In: *FAO/SIDA/ARCN Regional seminar on shifting cultivation and soil conservation in Africa* (Ibadan, Nigeria). Rome, FAO, 1974.

——. Intensification of agricultural systems with special reference to the role of potassium fertilizers. In: *10th Colloquium, International Potash Institute* (Budapest), 1974.

——. Bringing the green revolution to the shifting cultivator. *Science*, vol. 190, no. 4217, 1975, p. 841–844.

——; KOWAL, J. M. L. Nutrient content of a moist tropical forest of Ghana. *Plant and Soil*, 12, 1960, p. 154–174.

GRINNELL, H. R. *Agri-silviculture: a suggested research programm for West and Central Africa*. Report to International Development Research Centre (IDRC, Ottawa), 1975, 44 p. multigr.

HANEY, E. B. *The nature of shifting cultivation in Latin America*. Wisconsin, Land Tenure Center, University of Wisconsin LTC-45:29.

HARRIS, D. Venezuela's empty rain forests. *Geographical Magazine*, 41, 1968, p. 216–220.

HUGHILL, J. A. C. Brazil: development prospects for Amazonia. *Span*, vol. 16, no. 3, 1973, p. 123–124.

IUCN. *The ecology of man in the tropical environment*. Morges, Switzerland, Publication no. 4, 1964, 355 p.

JANZEN, D. H. Tropical agroecosystems. *Science*, 182, 1973, p. 1212–1219.

JONES, W. O. Manioc; an example of innovation in African economies. *Economic development and cultural change*, vol. 5, no. 2, 1957, p. 97–117.

JORDAN, C. F.; KLINE, J. R. Mineral cycling: some basic concepts and their application in a tropical rain forest. *Annual Rev. Ecol. Syst.*, 3, 1972, p. 33–49.

JURION, F.; HENRY, J. *De l'agriculture itinérante à l'agriculture intensifiée*. Bruxelles, INEAC, 1967, 498 p.

——; ——. *Can primitive farming be modernised?* ONRD/INEAC, hors série, 1969, 457 p.

KELLMAN, M. C. Some environmental components of shifting cultivation in upland Mindanao. *J. Trop. Geogr.*, 28, 1969, p. 40–56.

KELLOGG, C. E. Shifting cultivation. *Soil Sci.*, 95, 1963, p. 221–230.

——; ORVEDAL, A. C. Potentially arable soils of the world and critical measures for their use. *Adv. Agron.*, 21, 1969, p. 109–170.

KING, K. F. S. *Agri-silviculture (the taungya system)*. Bulletin no. 1, Department of Forestry, University of Ibadan, 1968, 109 p.

KLINGE, H.; RODRIGUES, W. A. Biomass estimation in a Central Amazonian rain forest. *Acta Científica Venezolana*, 24, 1973, p. 225–237.

——; ——; BRÜNIG, E. F.; FITTKAU, E. J. Biomass and structure in a Central Amazonian rain forest. In: Golley, F. B.; Medina, E. (eds.). *Tropical ecological systems: trends in terrestrial and aquatic research*, p. 115–122. Berlin, New York, Springer Verlag, Ecological Studies no. 11, 1975, 398 p.

KNOWLES, R. L. Farming with forestry: multiple land use. *Farm Forestry*, 14, 1972, p. 61–70.

KUECHLER, A. W.; MONTOYA-MAQUIN, J. M. The Unesco classification of vegetation: some tests in the tropics. *Turrialba*, 21, 1971, p. 98–109.

LAL, R. No-tillage, soil properties and maize yields. *Plant and Soil*, 40, 1974, p. 129–143, p. 321–331, p. 589–606.

——. Soil erosion and shifting cultivation. In: *FAO/SIDA/ARCN Regional seminar on shifting cultivation and soil conservation in Africa* (Ibadan, Nigeria). Rome, FAO, 1974.

LAUDELOUT, H. *Dynamics of tropical soils in relation to their fallowing techniques*. Rome, FAO, 1961, 111 p.

LOWE, R. G. *Farm forestry in Nigeria*. Rome, FAO, 1975, 12 p. multigr.

LOWENSTEIN, F. W. Some consideration of biological adaptations of aboriginal man to the tropical rain forests. In: Meggers, B. J. et al. (eds.). *Tropical forest ecosystems in Africa and South America: a comparative review*, p. 293–310. Washington, D.C., Smithsonian Institution, 1973, 350 p.

LUGO, A. Tropical ecosystem structure and function. In: Farnworth, E. G.; Golley, F. B. (eds.). *Fragile ecosystems*, p. 67–111. Berlin, New York, Springer Verlag, 1974, 258 p.

McCLOUD, D. E. (ed.). *A new look at energy resources*. American

Society of Agronomy (Madison), Special Publication no. 22, 1974, 50 p.

McGinnis, J. T.; Golley, F. B.; Clements, R. G.; Child, G. I.; Duever, M. J. Elemental and hydrological budgets of the Panamanian tropical moist forest. *Bio-Science*, 19, 1969, p. 697–700.

Meggers, B. J. Environmental limitations on the development of culture. *American Anthropologist*, 56, 1954, p. 801–824.

——. *Amazonia; man and culture in a counterfeit paradise*. Chicago, New York, Aldine Atherton, 1971, 182 p.

Merz, K. *Environmental deterioration in the area of the deciduous moist forest of Ghana caused by utilization*. Rome, FAO, 1975, 4 p. multigr.

Miracle, M. P. The Congo basin as habitat for man. In: Meggers, B. J. *et al.* (eds.). *Tropical forest ecosystems in Africa and South America: a comparative review*, p. 335–344. Washington, D.C., Smithsonian Institution, 1973, 350 p.

Morgan, W. B. Farming practice, settlement pattern, and population density in southeastern Nigeria. *Geog. J.*, 121, 1955, p. 320–333.

Nwoboshi, L. C. *The soil productivity aspects of agri-silviculture in the West African tropical moist forest zone*. Rome, FAO, 1975, 19 p. multigr.

Nye, P. H. Organic matter and nutrient cycles under moist tropical forest. *Plant and Soil*, 13, 1961, p. 333–346.

——; Greenland, D. J. *The soil under shifting cultivation*. Technical Comm. no. 51. Harpenden, Commonwealth Bureau of Soils, 1960, 156 p.

——. Changes in the soil after clearing tropical forest. *Plant and Soil*, 21, 1964, p. 101–112.

Okafor, J. C. Interim report on breeding of some Nigerian food trees. In: *Proceedings 2nd annual Conference of the Forestry Association of Nigeria* (Zaria), 1971.

——; Oholo, H. C. Potentialities of some indigenous forest trees of Nigeria. In: *Proceedings 5th annual Conference of the Forestry Association of Nigeria* (Jos), 1974, 13 p.

Petriceks, J. *Shifting cultivation in Venezuela*. SUNY College of Forestry (Syracuse, New York), Ph. D. thesis, 1968.

Phillips, J. F. V. *Agriculture and ecology in Africa*. Faber, 1959.

Pimentel, D.; Hurd, L. E.; Bellotti, A. C.; Forster, J. M.; Oka, I. N.; Sholes, O. D.; Whitman, R. J. Food production and the energy crisis. *Science*, 182, 1973, p. 443–449.

Popenoe, H. The influence of the shifting cultivation cycle on soil properties in Central America. In: *Proc. 9th Pacific Science Congress* (Bangkok), 1, 1957, p. 72–77.

——. Some soil cation relationships in an area of shifting cultivation in the humid tropics. In: *7th International Congress Soil Science* (Madison, Wisconsin), 1960, p. 303–311.

Prothero, R. M. *People and land in Africa south of the Sahara*. Oxford Univ. Press, 1972, 341 p.

Rappaport, R. A. The flow of energy in an agricultural society. *Scientific American*, 225, 1971, p. 116–132.

Revelle, R. Food and population. *Scientific American*, vol. 231, no. 3, 1974, p. 161–170.

Roche, L. *The practice of agri-silviculture in the tropics with specific reference to Nigeria*. Shifting cultivation and soil conservation in Africa. Rome, FAO, Soils Bulletin no. 24, 1973, 248 p.

Rockwood, W. G.; Lal, R. Mulch tillage; a technique for soil and water conservation in the tropics. *Span*, 17, 1974, p. 77–79.

Russell, W. M. S. The slash-and-burn technique. In: Gould, R. (ed.). *Man and man's ways*, p. 86–101. New York, National History Magazine, Harper and Row, 1973.

Ruthenberg, H. *Farming systems in the tropics*. Oxford, Clarendon Press, 1971, 313 p.

Ruthenberg, H. Agricultural aspects of shifting cultivation. *FAO Soils Bulletin*, 24, 1974, p. 99–111.

Salas, G. de las. Eigenschaften und Dinamik eines Waldstrandortes im Grenzbereich des immergrunen tropischen Regenwaldes in mittleren Magdalenatal (Kolumbien). *Gottinger Bodenkundliche Berichte*, 27, 1973, p. 1–206.

Sanchez, P. A. (ed.). *A review of soils research in tropical Latin America*. Raleigh, North Carolina State University, 1972, 263 p.

——. Soil management under shifting cultivation. In: Sanchez, P. A. (ed.). *A review of soils research in tropical Latin America*, p. 62–92. Raleigh, North Carolina State University, 1972, 263 p.

——; Buol, S. W. Properties of some soils of the Upper Amazon Basin of Peru. *Soil Sci. Soc. Am. Proc.*, 38, 1974, p. 117–121.

Schlippe, P. de. *Shifting cultivation in Africa: the Zande system of agriculture*. London, Routledge and Kegan Paul, 1956, 304 p.

Schultz, T. W. *Transforming traditional agriculture*. New Haven, Conn., Yale University Press, 1964, 212 p.

Shantz, M. L.; Marbut, C. F. The vegetation and soils of Africa. *Amer. Geog. Soc.* (New York), 1923, 263 p.

Sioli, H. Recent human activities in the Brazilian Amazon region and their ecological effects. In: Meggers, B. J. *et al.* (eds.). *Tropical forest ecosystems in Africa and South America: a comparative review*, p. 321–334. Washington, D.C., Smithsonian Institution, 1973, 350 p.

——; Klinge, H. Solos, tipos de vegetaçao e aguas na Amazonia. *Bol. Museu Paraense Emilio Goeldi*, 1, 1962, p. 27–41.

Spencer, J. E. *Shifting cultivation in southeastern Asia*. University of California (Berkeley) publications in geography, 19. University of California Press, 1966, 247 p.

Stark, N. Mycorrhizae and nutrient cycling in the tropics. In: *Proc. first North American conference on mycorrhizae*, p. 228–229. Washington, Misc. Publ. 1189 USDA-Forest Service, 1969.

——. Nutrient cycling. I. Nutrient distribution in some Amazonian soils. *Tropical Ecol.*, vol. 12, no. 1, 1971, p. 24–50.

——. Nutrient cycling. II. Nutrient distribution in Amazonian vegetation. *Tropical Ecol.*, vol. 12, no. 2, 1971, p. 177–201.

Sternberg, H. O. Man and environmental change in South America. In: Fittkau, E. J.; Illies, J.; Klinge, H.; Schwabe, G. H.; Sioli, H. (eds.). *Biogeography and ecology in South America*, vol. 1, p. 413–445. The Hague, Junk, Monographiae Biologicae no. 18, 19, 2 vol., 1968–69, p. 446, 449–946.

——. Development and conservation. *Erdkunde*, vol. 27, no. 4, 1973, p. 253–265.

Stout, P. R. Agriculture's energy requirements. In: McCloud, D. E. (ed.). *A new look at energy resources*, p. 13–22. American Society of Agronomy (Madison, Wisconsin), Special Publication no. 22, 1974, 50 p.

The Institute of Ecology. *Man in the living environment*. Madison, University of Wisconsin Press, 1972, 267 p.

Unesco. *International classification and mapping of vegetation*. Ecology and conservation 6. Paris, Unesco, 1973, 93 p.

Van der Weert. Influence of mechanical clearing on soil conditions and the resulting effects on root growth. *Tropical Agric.* (Trinidad), vol. 51, no. 2, 1974, p. 325–333.

Vine, H. Experiments on the maintenance of soil fertility at Ibadan, Nigeria, 1922–1951. *Emp. J. Expt. Agric.*, 21, 1953, p. 65–68.

Vine, H. Developments in the study of soils and shifting agriculture in tropical Africa. In: Moss, R. P. (ed.). *The soil resources of tropical Africa*, p. 89–119. Cambridge, The University Press, 1968.

WATSON, G. A. Soil and plant nutrient studies in rubber cultivation. In: *FAO/IUFRO Symposium on forest fertilization* [FOR:FAO/IUFRO/F/73/23]. Paris, Dec. 1973, 8 p.

WATTERS, R. F. *Shifting cultivation in Latin America.* FAO Forestry development paper no. 17. Rome, FAO, 1971, 305 p.

——; BASCONES, L. The influence of shifting cultivation on soil properties at Altamira-Calderas, Venezuelan Andes. In: Watters, R. F. *Shifting cultivation in Latin America*, p. 291–299. FAO Forestry development paper no. 17. Rome, FAO, 1971, 305 p.

WEBSTER, C. C.; WILSON, P. N. *Agriculture in the tropics.* London, Longmans, Green and Co., 1966.

WHITMORE, T. C. Wild fruit trees and some trees of pharmacological potential in the rain forest of Ulu Kelantan. *Malayan Nat. J.*, 24, 1971, p. 222–224.

WILKINSON, G. E.; AINA, P. O. Infiltration of water into two Nigerian soils under secondary forest and subsequent arable Cropping. *Geoderma* (Amsterdam), 15, 1976, p. 51–59.

YEN, D. E. Arboriculture in the subsistence of Santa Cruz, Solomon Islands. *Economic Botany*, 28, 1974, p. 247–281.

Weed control

AHRENS, J. F. Improving herbicide selectivity in transplanted crops with root dips of activated charcoal. In: *Proc. 21st Northeastern Weed Control Conf.*, 1967, p. 64–70.

*ALDHOUS, J. R. Review of practice and research in weed control in forestry in Great Britain. In: *Proc. 14th Cong. Int. Union For. Res. Org.* (Munich), Pt. IV, Sec. 23, 1967, p. 40–64.

ALLAN, T. G. *Land clearing and preparation trials using Caterpillar, Fleco and Rome equipment.* Res. pap. (Savanna ser.) Fed. Dept. For. Res. Nigeria, no. 17, 1973a, 22 p.

——. *Mechanized cultivation trials for forestry plantations in the savanna region of Nigeria.* Res. pap. (Savanna ser.) Fed. Dept. For. Res. Nigeria, no. 18, 1973b, 22 p.

*ALPHEN de VEER, E. J. van. Teak cultivation in Java. In: *Tropical silviculture*, vol. II (FAO Forestry and Forest Products Studies no. 13), 1958, p. 216–232.

AMEDIWOLE, E. K. Experiments on tree poisoning in Bobiri Research Centre using sodium arsenite and other arboricides. *Newsletter For. Prod. Res. Inst. Ghana*, no. 2, 1967, p. 8–10.

ANKAH, E. C.; AMAYAW, M. A. The effect of cultivation on the survival and growth of *Eucalyptus tereticornis* Sm. *Tech. Newsletter For. Prod. Res. Inst. Ghana*, no. 5, 1971, p. 6–14.

*ANNECKE, D. P.; KAMY, M.; BURGER, W. A. Improved biological control of the prickly pear *Opuntia megacantha* Salm-Dyck in South Africa through the use of an insecticide. *Phytophylactica*, 1, 1969, p. 9–13.

APLIN, T. E. H. Poison plants of western Australia. The toxic species of the genera *Gastrolobium* and *Oxylobium. J. Agric. W. Aust.*, 12, 1971, p. 12–18.

AUDUS, L. J. (ed.). *Herbicides. Physiology, biochemistry, ecology.* London, New York, Academic Press, 1976, vol. 1, 636 p.; vol. 2, 475 p.

AULD, B. A. Groundsel bush—a dangerous weedy weed of the far north coast. *Agric. Gaz. N.S.W.*, 81, 1970, p. 32–34.

——. Chemical control of *Eupatorium adenophorum*, Crofton weed. *Trop. Grassl.*, 6, 1972, p. 55–60.

*BACHELARD, E. P.; SARFATY, A.; ATTIWILL, P. M. Chemical control of eucalypt vegetation. *Aust. For.*, 29, 1965, p. 181–191.

BACK, P. V. Dawson gum control. *Qd. Agric. J.*, 98, 1972, p. 579–586.

BAILEY, D. R. Control of *Acacia flavescens* (A. Cunn. ex Benth.) in relation to the establishment of improved pastures in the wet tropical areas of Queensland. *J. Aust. Inst. Agric. Sci.*, 36, 1970, p. 302–303.

*——. Control of *Acacia flavescens* with herbicides. *Aust. J. Exp. Agric. Anim. Husb.*, 12, 1972, p. 441–446.

*BALL, J. B. Developments in herbicide research in Uganda forestry. In: *Proc. 4th E. Afr. Herbicide Conf.* (Arusha), p. 213–220, 1970a.

——. Early burning with paraquat. In: *Proc. 4th E. Afr. Herbicide Conf.* (Arusha), p. 228–232, 1970b.

BARNES, R. D. The use of methyl bromide soil sterilization and a coarse sand mulch in raising pines and eucalypts in the nursery. *Rhodesia Sci. News* (Salisbury), 3, 1969, p. 99–101.

*BARRONS, K. C. Some ecological benefits of woody plant control with herbicides. *Science*, 165, 1969, p. 465–468.

*BAUR, G. N. Economics of weeds in forestry. *Proc. Weed Soc. N.S.W.*, 1, pap. 6, 1967, 3 p.

BEMBRIDGE, T. J. Eradication of thorn trees. *Rhodesia Agric. J.*, 63, 1966, p. 86–88.

BEVERIDGE, A. E. Arboricide trials in lowland dipterocarp rain forest of Malaya. *Malayan Forester*, 20, 1957, p. 211–225.

BOERBOOM, J. H. A. *The natural regeneration of the mesophytic forest of Surinam after exploitation.* Part I. Wageningen, Landbouwhogeschool, 1964, 56 p.

BOVEY, R. W.; MILLER, F. R.; DIAZ-COLON, J. Growth of crops in soil after herbicidal treatments for brush control in the tropics. *Agron. J.*, 60, 1968, p. 678–679.

*——; HAAS, R. H.; MEYER, R. E. Daily and seasonal response of huisache and Macartney rose to herbicides. *Weed Sci.*, 20, 1972, p. 577–580.

BOWERS, A.; HAWTHORN, J. M. Development and use of an inverted cone tree releasing sprayer. In: *Proc. 24th N.Z. Weed and Pest Control Conf.*, p. 43–47, 1971.

BRYANT, C. L. The effect of weed control on the growth of young teak in Tanzania. *Silv. Res. Note Silv. Sect. For. Div. Lushoto*, 8, 1968, 2 p.

BURGESS, P. F.; LOWE, J. S. The control of bertam (*Eugeissonia tristis* Griff.) a stemless palm in hill forests of the Malay peninsula. *Malayan Forester*, 34, 1971, p. 36–46.

BURROWS, W. H. Studies in the dynamics and control of woody weeds in semi-arid Queensland. 1. *Eremophila gilesii. Qd. J. Agric. Anim. Sci.*, 30, 1973, p. 57–64.

CARPENTER, S. B. *Herbicides for site preparation; broadcast spray by mist blower tested against understorey in Hawaii rain forest.* U.S. For. Serv. Res. Note, Pac. S.W. For. Range Exp. Sta., PSW 115, 1966a, 8 p.

——. *Controlling cull ohia (Metrosideros collina) trees by injecting herbicides.* Res. Note, Pac. S.W. For. Range Exp. Sta., PSW 125, 1966b, 5 p.

*CATINOT, R. Sylviculture tropicale en forêt dense africaine. *Bois et Forêts des Tropiques*, n° 100, 101, 102, 103, 104, 1965, p. 5–18, 3–16, 3–16, 3–16, 17–30.

CEBRON, P. Les essais d'empoisonnement avec phyto-hormones en forêt de teck. Application des résultats aux travaux d'enrichissement en forêt dense. *Bois et Forêts des Tropiques*, 52, 1957, p. 9–15.

CHOUDHURY, A. Controversial *Mikania* (climber), a threat to the forests and agriculture. *Indian Forester*, 98, 1972, p. 178–186.

*CONNELL, C. A.; COUSINS, D. A. Practical developments in the use of chemicals for forest fire control. *Forestry*, 42, 1969, p. 119–132.

* Major reference.

Da Silva, S. A. F. Contribution to the study of capim-coloniao (*Panicum maximum* var. *maximum*). 2. Its spread and control. *Vellozia*, 7, 1969, p. 3–25.

*Daubenmire, R. The ecology of fire in grasslands. *Advanced Ecol. Res.*, 5, 1968, p. 209–266.

Davis, E. A.; Ingebo, P. A.; Pase, C. P. *Effect of a watershed treatment with picloram on water quality*. U.S. For. Serv. Res. Note RM 100, 1968, 4 p.

Dawkins, H. C. Trials of non-toxic arboricides in tropical forest. *Empire For. Rev.*, 32, 1953, p. 253–256.

——. Contact arboricides for rapid tree-weeding in tropical forest. In: *Proc. 4th World For. Cong.*, Sect. 5, 1954.

Delwaulle, J. C. The increasing unproductiveness of Africa south of the Sahara. *Bois et Forêts des Tropiques*, no. 149, 1973, p. 3–20.

Diatloff, G. Tree injectors and their use. *Qd. Agric. J.*, 96, 1970, p. 106–109.

*Donald, D. G. M. Cleaning operations in South African forestry. *For. in S. Africa*, 12, 1971, p. 55–65.

Donaldson, C. H. Control of blackthorn in the Molopo area with special reference to fire. *Proc. Grassl. Soc. S. Afr.*, 1, 1966, p. 57–62.

*Dougall, H. W.; Bogdan, A. V. Browse plants of Kenya with special reference to those occurring in South Baringo. *E. Afr. Agric. J.*, 23, 1958, p. 236–245.

Dutta, T. R.; Pandey, R. K. Brush-killers for *Sehima-Dichanthium* pastures in India. *Down to Earth*, 27, 1971, p. 16–18.

Dyson, W. G. Possibilities for the use of soil-applied herbicides in forest management in East Africa. In: *3rd Afr. Conf. on soil-acting herbicides*, 1964, 9 p.

*Earl, D. E. Latest techniques in the treatment of natural high forest in South Mengo district. In: *9th Commonwealth For. Conf.* (New Delhi), 1968, 26 p.

Egberink, J. *Study of Lantana spp. and their control; chemical control of Lantana camara, 1964–1967. Final Report*. Agric. Res. (Pretoria), part 1, 1970, p. 184–185.

Endean, F.; Jones, B. E. Clean cultivation and the establishment of *Pinus kesiya* in Zambia. *E. Afr. Agric. For. J.*, 38, 1972, p. 120–129.

*Ennis, W. B. Economic aspects of crop losses caused by weeds. In: *FAO Symp. on crop losses* (Rome), 1967, p. 127–145.

Etheridge, D. E. Dominican Republic. New pests and diseases of forest trees: mistletoes. *FAO Pl. Prot. Bull.*, 19, 1971, p. 21–22.

*Everist, S. L. *Use of fodder trees and shrubs*. Indooroopilly (Queensland), Dept. Primary Ind., 1969.

Fabre, J. P.; Brunck, F. Pre-emergence herbicide test on an Ivory Coast species used in afforestation: framiré—*Terminalia ivorensis. Bois et Forêts des Tropiques*, 136, 1971, p. 35–41.

FAO. *Tree planting practices in tropical Africa* (prepared by Parry, M. S.). FAO Forestry Div. paper no. 8. Rome, FAO, 1956.

——. *Tree planting practices in tropical Asia* (prepared by Letourneux, C.). FAO Forestry Div. paper no. 11. Rome, FAO, 1957.

*——. *Tropical silviculture*, vol. I (prepared by Haig, I. T.; Huberman, M. A.; U Aung Din). FAO Forestry and Forest Products Studies no. 13. Rome, FAO, 1958.

——. Land and Water Development Division. *Lectures presented at the FAO/SIDA Regional seminar on shifting cultivation and soil conservation in Africa* (Ibadan, Nigeria, 2–21 July 1973). Rome, FAO, 1974.

Feldman, J. Considerations concerning some problems of intrusive woody plants in the Argentine Republic. *Malezas y su Control*, 1, 1972, p. 4–27.

Fitzgerald, C. H.; McComb, W. H. Damage to pine released from hardwood competition by 2,4-D. *J. Forestry*, 68, 1970, p. 164–165.

Floyd, A. G. Effect of fire upon weed seeds in the wet sclerophyllous forests of northern New South Wales. *Aust. J. Bot.*, 14, 1966, p. 243–256.

Foot, D. L. *Mkwerenyani (Newtonia buchananii) regeneration trials R 102, R 104/7*. Silv. Res. Rec. Silv. Res. Sta. (Dedza, Malawi), no. 2, 1967a, 2 p.

——. *Arboricide treatment for unwanted coppice of Gmelina arborea: M 360*. Silv. Res. Rec. Silv. Res. Sta. (Dedza, Malawi), no. 4, 1967b, 2 p.

Forrest, W. G.; Richardson, R. R. *Chemical control of forest weeds*. Res. Note For. Comm. N.S.W., no. 16, 1965, 22 p.

Freak, R. H. T. Herbicides for firebreaks. *J. Agr. S. Australia*, 74, 1971, p. 181–185.

*Fryer, J. D.; Makepeace, R. J. *Weed control handbook*. Vol. 2, 7th ed. Oxford, Blackwell, 1972.

Geary, T. F.; Zambrana, J. A. Must Honduras pine be weeded frequently in Puerto Rico? For. Serv. Res. Pap. Inst. Trop. For. (Puerto Rico), 1972, 16 p.

George, K. Selective control of *Loranthus* on teak. *Curr. Sci.*, 35, 1966, p. 444.

*Ghani, Q. Heterogeneous types of tropical forests—Chittagong forests. In: *Tropical silviculture*, vol. II, p. 24–35. Rome, FAO, 1958, 415 p.

Harley, K. L. S. Biological control of *Lantana*. *P.A.N.S.*, 17, 1971, p. 433–437.

Hombert, J. Empoisonnement des arbres à l'aide de l'arsénite de soude. *Bull Inf. INEAC*, 3, 1954, p. 245–260.

Hughes, J. F.; Lang Brown, J. R. The planning and organization of current silvicultural treatments in the central forest reserves of S. Mengo District, Buganda Province. In: *8th Brit. Commonwealth For. Conf.*, 1962.

Huraux, M. J. Brush control with picloram in forestry and along rights-of-way. In: *Summs. Paps. 7th Int. Cong. Pl. Prot.* (Paris), 1970, p. 327–329.

Ivens, G. W. *Results of bush control experiments. 5. Application of chemicals by means of tree-injectors*. Report to Range Management Div., Min. Agric. and Animal Husbandry, Nairobi, April 1970 (unpublished).

——. *Results of bush control experiments. 8. Effects of grass competition on regrowth of Acacia species*. Report to Range Management Div., Min. Agric. and Animal Husbandry, Nairobi, July 1970 (unpublished).

*——. Seasonal differences in kill of two Kenya bush species after foliar herbicide treatment. *Weed Res.*, 11, 1971, p. 150–158.

Jack, J. B. Herbicides and woody growth control in Victorian state forests. In: *Proc. 1st Victorian Weeds Conf.*, 1968, p. 4–12.

*Johnson, R. W. *Ecology and control of brigalow (Acacia harpophylla) in Queensland*. Brisbane, Queensland, Dept. Primary Ind., 1964, 92 p.

Joshi, D. P. Eradication of *Lantana camara* by Lantana bug. *Indian Forester*, 95, 1969, p. 152–154.

*Kadambi, K. Methods of increasing growth and obtaining regeneration of tropical forests. In: *Tropical silviculture*, vol. II, p. 67–78. Rome, FAO, 1958, 415 p.

Kan, W. H.; Chang, K. C.; Chen, C. H. *Site preparation for direct sowing of Aleurites montana*. Bull. Taiwan For. Res. Inst. no. 207, 1971, 9 p.

KASASIAN, L. Bamboo (*Bambusa vulgaris*)—a progress report on its control by herbicides. *P.A.N.S.* (C), 10, 1964, p. 14–15.

*——. Chemical weed control in seedling *Pinus caribaea* var. *hondurensis*. *Commonw. For. Rev.*, 44, 1965, p. 139–142.

*——. *Weed control in the tropics*. London, Leonard Hill, 1971, 307 p.

KIMBER, P. C. Thinning jarrah with hormone herbicides. *Aust. For.*, 31, 1967, p. 128–136.

KING, K. F. S. The use of arboricides in the management of tropical high forest. *Turrialba*, 15, 1965, p. 35–39.

KNUFFEL, W. E. *Eucalyptus grandis* seed germination in soil sterilized with methyl bromide gas. *S. Afr. For. J.*, 62, 1967, p. 33–35.

KNUTZ, J. E.; NAIR, V. M. G. Control of forest tree diseases with herbicides. In: *Proc. N. Cent. Weed Control Conf.*, 1967, p. 31.

LAHARRAGUE, P. Natural regeneration in plantations of *Araucaria angustifolia*. *Rev. For. Argent.*, 11, 1967, p. 71–78.

*LANDON, F. H. Malayan tropical rain forest. In: *Tropical silviculture*, vol. II, p. 1–11. Rome, FAO, 1958, 415 p.

*LEROY-DEVAL, J. Putting new chemical weapons to use in tropical forestry. *Bois et Forêts des Tropiques*, no. 132, 1970, p. 22–29.

LETOURNEUX, C. Les dégagements par annélation et empoisonnement. *Bois et Forêts des Tropiques*, no. 46, 1956, p. 3–10.

LIEFSTING, C. Chemical clearing, a possibility. *Ghana Farmer*, 9, 1965, p. 8–14.

LIEW, T. C.; CHARINGTON, M. S. Chemical control of giant stranglers. *Malayan Forester*, 35, 1972, p. 13–16.

LITTLE, E. C. S. The control of bush by application of concentrated herbicide to stumps. In: *1st FAO Int. Conf. Weed Control* (Davis, California), 1970, 12 p.

LOWE, J. S. The control of bamboo. In: *Proc. 3rd Asian-Pacific Weed Sci. Soc. Conf.*, 1, 1971, p. 86–92.

MACCONNELL, W. P.; WHITNEY, L. F.; COSTA, A. J. A new tool for killing unwanted trees. *J. For.*, 66, 1968, p. 486–487.

*MAIER-BODE, H. The question of 2,4,5-T. *Anz. Schädlingsk.*, 45, 1971, p. 2–6.

MCKELL, C. M.; BLAISDELL, J. P.; GODIN, J. R. (eds.). *Wildland shrubs—their biology and utilization*. International Symposium (Utah State Univ., Logan, Utah, July 1971). USDA For. Serv. General Tech. Rep. INT–1, 1972, 494 p.

MEADLY, G. R. W. *Calotropis* or rubber tree (*Calotropis procera* (L.) Dryand). *J. Agric. W. Aust.*, 12, 1971, p. 69–71.

MEIJER, W. Regeneration of tropical lowland forest in Sabah, Malaysia, forty years after logging. *Malayan Forester*, 33, 1970, p. 204–229.

MONTGOMERY, M. L.; NORRIS, L. A. *A preliminary evaluation of the hazards of 2,4,5-T in the forest environment*. U.S. For. Serv. Res. Note Pacific N. West For. Range Exp. Sta., PNW–116, 1970, 9 p.

MOORE, R. M.; WALKER, J. *Eucalyptus populnea* shrub woodlands. Control of regenerating trees and shrubs. *Aust. J. Exp. Agric. Anim. Husb.*, 12, 1972, p. 437–440.

MORTON, H. L.; COMBS, J. A. Influence of surfactants on phytotoxicity of a picloram—2,4,5-T spray on three woody plants. *Abstr. Meet. Weed Sci. Soc. Amer.*, 1969, p. 65.

MORZE, J. Chemical control of *Eucalyptus* species. *For. S. Africa*, 12, 1971, p. 49–53.

MURRAY, C. H. Arboricides and clonal teak. *Commonw. For. Rev.*, 46, 1967, p. 133–137.

NAKAMURA, K.; YOSHIMURA, S.; KANG-HE, C.; YI, C.; LIN, C. Chemical control of weeds in the planted forest of Taiwan. *Weed Res.* (Japan), 15, 1973, p. 42–48.

*NORRIS, L. A. Chemical brush control: assessing the hazard. *J. For.*, 69, 1971, p. 715–720.

OAKES, A. J. Herbicide control of *Acacia*. *Turrialba*, 20, 1970a, p. 213–216.

——. Herbicidal control of *Croton*. *Turrialba*, 20, 1970b, p. 299–301.

ORTEGA, E. Weed control in conifer nurseries. In: *3ª Reunión nacional sobre malezas y su control* (Buenos Aires), 1963, 2 p.

*PEARCE, G. A. Chemical firebreaks. *J. Dept. Agric. W. Aust.*, 10, 1969, p. 166–172.

*PEEVY, F. A. Controlling upland southern hardwoods by injecting undiluted 2,4-D amine. *J. Forestry*, 66, 1968, p. 483–487.

PEÑALOZA WAGENKNECHT, R. *Triazines and charcoal for weed control in forest nurseries*. Publ. Cient. Univ. Austral Chile, no. 10, 1968, 12 p.

*PHILIP, M. S. The dynamics of seedling populations in a moist semi-deciduous tropical forest in Uganda. I. Interim report on research plot 441 Uganda Forest Department—survival of seedlings following destruction of the canopy with arboricide. In: *9th Commonw. For. Conf.* (New Delhi), 1968, 31 p.

*PHILLIPS, J. F. V. Fire as master and servant: its influence in the bioclimatic regions of trans-Sahara Africa. In: *Proc. Tall Timbers Fire Ecol. Conf.*, no. 4, 1965, p. 7–109.

*QURESHI, T. M.; SRIVASTAVA, P. B. L.; BORA, N. K. S. Sal (*Shorea robusta*) natural regeneration *de novo*. Effect of soil working and weeding on the growth and establishment. *Indian Forester*, 94, 1968, p. 591–598.

RAJKHOWA, S. The effect of rain weeding on the growth of sal seedlings. *Indian Forester*, 92, 1966, p. 75–78.

REVILLA, V. Chemical control of weeds in young conifer plantations. *Bol. Asoc. Plantadores Forestales de Misiones*, 7, 1972, p. 25–26.

RICHARDS, B. N. Introduction of the rain forest species *Araucaria cunninghamii* Ait. to a dry sclerophyll forest environment. *Plant and Soil*, 27, 1967, p. 201–216.

*RICHARDS, B. N.; BEVEGE, D. I. The productivity and nitrogen economy of artificial ecosystems comprising various combinations of perenial legumes and coniferous tree species. *Aust. J. Bot.*, 15, 1967, p. 467–480.

RODRIGUEZ, M. A. Some experiments relating to the growth of *Cedrela mexicana* and *Swietenia macrophylla*. *Bol. Inst. For. Lat.-Amer.* (Mérida), 13, 1963, p. 38–50.

*ROSE INNES, R. Fire in West African vegetation. In: *Proc. Ann. Tall Timbers Fire Ecol. Conf.*, 11, 1972, p. 147–173.

*STRANG, R. M. Bush encroachment and veld management in south-central Africa: the need for a reappraisal. *Biol. Conservation*, 5, 1973, p. 96–104.

*STUART SMITH, A. M. Practice and research in establishment techniques. In: *FAO World Symp. on man-made forests and their industrial importance* (Canberra), 1, 1967, p. 265–287.

——; BALL, J. B. Recent work on arboricides in Uganda forestry. In: *Proc. 4th E. Afr. Herbicide Conf.*, 1970, p. 221–227.

TAKLE, G. G.; MUJUMDAR, R. B. Increasing growth and natural regeneration of teak. In: *Tropical silviculture*, vol. II, p. 237–256. Rome, FAO, 1958, 415 p.

TOTHILL, J. C. Grazing, burning and fertilizing effects on the regrowth of some woody species in cleared open forest in south-east Queensland. *Trop. Grasslands*, 5, 1971, p. 31–34.

VAGELER, C. P. Practical application of 2,4,5-T based bush-killers in the Pantanal, Mato-Grosso. *Bol. Inst. Ecol. Exp. Agric.* (Rio de Janeiro), 23, 1962, p. 354–375.

*Veiga, R. A de. Effects of soil sterilizing chemicals on the germination and seedling development of *Eucalyptus saligna* and on weed control. *Rev. Agric. Piracicaba*, 43, 1968, p. 141–148.

Vogl, R. J. The role of fire in the evolution of Hawaiian flora and vegetation. In: *Proc. 9th Ann. Tall Timbers Fire Ecol. Conf.* (Tall Timbers Res. Sta., Tallahassee, Florida), 1969.

Wang, T. K.; Kuo, P. C. The effects of herbicides on the control of arrow bamboo sites. Tech. Bull. Dept. For., Nat. Taiwan Univ., no. 104, 1971, 22 p.

Ward, H. K.; Cleghorn, W. B. The effects of ringbarking trees in *Brachystegia* woodland on the yield of veld grasses. *Rhod. Agric. J.*, 61, 1964, p. 98–107.

*——; ——. The effects of grazing practices on tree regrowth after clearing indigenous woodland. *Rhod. J. Agric. Res.*, 8, 1970, p. 57–65.

*Wilde, S. A.; Shaw, B. H.; Fedkenheuer, A. W. Weeds as a factor depressing forest growth. *Weed Res.*, 8, 1968, p. 196–204.

Willson, B. W. Insects on trial in fight against *Lantana*. *Qd. Agric. J.*, 94, 1968, p. 748–751.

Wong, Y. K. Poison-girdling under the Malayan uniform system. *Malayan Forester*, 29, 1966, p. 69–77.

*Wyatt-Smith, J. *Manual of Malayan silviculture for inland forests*. Malayan For. Rec., no. 23, 1963, 400 p.

Young, N. D. Tordon for *Eucalyptus* control—a tool for land development in Australia. *Down to Earth*, 24, 1968, p. 2–6.

Zabala, N. Q. Researches on chemical control of forest weeds conducted in the U.P. College of Forestry. In: *Proc. 2nd Asian-Pacific Weed Control Interchange*, 1969, p. 378–400.

Impact of forest operations on the environment

Biocides

Bachelard, E. P.; Johnson, M. E. A study of the persistance of herbicides in soil. *Aust. For.*, 33, 1969, p. 19–24.

Davis, E. A.; Ingebo, D. A.; Pase, C. P. *Effect of watershed treatment with picloram on water quality*. U.S. Forest Service, Note RM 100, 1968, 4 p.

Dawkins, H. C. Trials of non toxic arboricides in tropical forest. *Empire For. Rev.*, 32, 1953, p. 253–256.

FAO. *Programme de lutte contre la trypanosom ase animale africaine*. Rome, FAO, 1974, 20 p.

Montgomery, M. L.; Norris, L. A. *A preliminary evaluation of the hazards of 2,4,5-T in the forest environment*. U.S. Forest Service Res. Note, Pac. S.W. For. Range Exp. Sta., PSW 116, 1970, 9 p.

Morton, H. L.; Combs, J. A. Influence of surfactants on phytotoxicity of a picloram—2,4,5-T spray on three woody plants. *Abstr. Meet. Weed Sci. Soc. Amer.*, 65, 1969.

National Academy of Sciences. *The effects of herbicides in South Vietnam. Part A. Summary and conclusion*. Washington, National Academy of Sciences Committee on the effects of herbicides in Vietnam, 1974, 398 p.

Orians, G. H.; Pfeiffer, E. W. Ecological effects of the war in Vietnam. *Science*, 168, May 1970, p. 544–554.

Riordan, K. *African forests in relation to the tsetse fly problem, utilization by man and the effects of pesticides*. Rome, FAO, 1975, 13 p. multigr.

Logging and transportation

FAO. *Exploitation and transport of logs in dense tropical forests*. Forest Development, no. 18. Rome, 1974, 100 p.

——. *Logging and log transport in man made forests in developing countries*. Rome, FAO/SWE/TF 116, 1974, 134 p.

FAO/UNDP. *Demonstration and training in forest range and watershed management. The Philippines*. Logging and transport Technical Report no. 5, FO:SF/PHI 16. Rome, 1971, 123 p.

Fontaine, R. G. *The impact upon the environment of forestry practices in tropical moist forests*. Rome, FAO (Committee on forest development in the tropics, 4th session, November 1976), FO:FDT/76/8(b), 1976, 20 p.

Goodland, R. J. A.; Irwin, H. S. *Amazon jungle: green hell to red desert? An ecological discussion of the environmental impact of the highway construction programme in the Amazon Basin*. The Cary Arboretum of the New York Botanical Garden, 1975, 205 p. multigr.

A selva amazônica: do inferno verde ao deserto vermelho? Tradução de Regina Regis Junqueira; revisão técnica, prefacio e notas de Mário Guimarães Ferri. São Paulo, Ed. Itatiaia, Ed. da Universidade de São Paulo, 1975, 156 p.

Poore, D. *The values of the tropical moist forest ecosystems and the environmental consequences of their removal*. Rome, FAO (Committee on forest development in the tropics, 4th session, November 1976), FO:FDT/76/8(a), 1976, 39 p.

Man-made lakes

FAO/UNDP. *Man-made lakes. Planning and development*. Rome, 1969, 71 p.

Freeman, P. H. *The environmental impact of tropical dams: guidelines for impact assessment based upon a case study of Volta lake*. Washington, D.C., Smithsonian Institution, 1975.

Leentvaar, P. Inundation of a tropical forest in Surinam. In: *Proceedings of the 1st International Congress of Ecology* (The Hague, September 1974), p. 348–354. Wageningen, Centre for agricultural publishing and documentation, 1974, 414 p.

Milton, J. P. The ecological effects of major engineering projects. In: *The use of ecological guidelines for development in the American humid tropics* (Proceedings of international meeting, 20–22 February 1974, Caracas), p. 207–222. Morges, IUCN Publications, new series no. 31, 1975, 249 p.

Pulp effluents

Easton, J. C.; McFarlane, M. M. *Economics of pulp and paper pollution abatement*. Rome, FAO, FO:Misc./79/9, April 1973, 34 p.

Ekono. *Study of pulp and paper industry's effluent treatment*. Helsinki, May 1973, 34 p.

FAO. *Guide for planning pulp and paper enterprises*. Rome, FAO, Forestry and Forest products Studies no. 18, 1973, 362 p.

OECD. *Pollution by the pulp and paper industry: present situation and trends*. Paris, 1973, 129 p.

Sikes, J. E. G. A clean piece of paper. *Unasylva* (FAO), vol. 27, no. 109, 1975, p. 11–16.

——. *A perspective on the environmental protection situation in the pulp and paper industry*. UNEP Industry Sector Seminars, Pulp and paper meeting (Paris, March 1975), 14 p.

Conservation and development

Introduction

Until recently, human activities in tropical and subtropical forests were usually confined to shifting cultivation, small-scale logging, hunting and fishing, the harvesting of wild plants and the limited extraction of minerals. In most areas, human populations were relatively small and there were no large permanent settlements; because population densities were low and populations mobile, forest residents could move easily according to the opportunities and requirements of the forest environment. At present, many of these areas are under increasing human pressure for intensive resource utilization.

As a result of development activities the original moist tropical forest area may have decreased roughly from 1.2 to 0.8×10^9 ha and there is evidence that the pace of clearing is accelerating (see chapters 1 and 2). Much of the remaining areas are of marginal or submarginal capability for intensive development. This is particularly true of humid tropical forests growing on old, highly leached soils or on slopes in excess of 20 per cent (Dasmann, Milton and Freeman, 1973). Attempts to clear and utilize these marginal or less than marginal lands have caused problems, but little work has been done to fully explore the importance of conserving and protecting the natural forest as an integral part of development planning. Considerable success has been achieved in setting aside parks and related reserves, but only rarely has this been achieved in the context of overall rational land-use planning on a local, national or regional basis.

It is necessary to clarify in what sense the words conservation and development will be used in this chapter. Conservation is frequently (and rightly) taken to mean a rational utilization of resources, i.e. a sustained utilization resting on ecological principles, and therefore including agricultural, forestry or pastoral development. But in this chapter the words conservation and protection, will be reserved for the specific meaning of maintaining tropical forest ecosystems in their natural state, i.e. without transformation or with only very slight changes, and the word development, or economic development, will be restricted to denote activities or operations designed to organize the production of goods and services from forests, to transform the forests into simplified ecosystems or to dedicate them to multiple use on the basis of combined ecological, economic and social considerations.

In view of the dangers with which forests, and the humid tropical forests in particular, are threatened it seems necessary to restate the principles of land management and land-use planning in the tropics, taking into consideration the various possibilities and constraints (chapter 20), the density and nature of local populations and the perspectives of the market. This has been the subject of many international conferences and the theme of many recommendations as to the data that ought to be collected and the methods that should be followed. Some recent work will be summarized. First, immediate measures will have to be taken to preserve certain tracts or representative areas of forest to safeguard a genetic pool which is threatened with extinction, as well as providing for essential observations and studies. Such a

policy, which aims principally at the establishment of a network of reserves, the issuing of laws and regulations for the protection of the environment, the promotion of education in these fields and informing the general public will be presented first. The degree to which such reserves have to be managed is a delicate question on which opinions are divided. Some would like them to be natural systems entirely under natural control, i.e. without any intervention from outside, while others would admit a certain amount of interference by man. The problem of buffer zones also requires examination.

It will no doubt be necessary to clear some forests for the cultivation of crops, or the raising of livestock, or in order to make way for certain infrastructures such as roads, air-fields, etc. Other forests will however remain in their present state and these will have to be the object of exploitation for forest products.

In order that the organization of the production of goods and services from forest areas will be in the right direction, it will be necessary to study several different problems. These include the situation of the market in forest products and market trends, then problems of employment and conditions of work in the forest and certain institutional aspects, especially the problem of forest concessions which is the subject of study in many countries.

In view of the richness of tropical forests on the one hand and the prevailing raw materials and energy crisis on the other, it seems clear that special attention will have to be given to the products of the forest, and wood in particular, as source of energy and raw material. It is important to explore the prospects of supply and demand for this material, existing and potential markets, to review research in this field and to examine the possibilities offered by new techniques.

Other factors to be taken into consideration are the employment of workers, their living and working conditions and the stresses imposed by forest work. Forest work has been recognized by the International Labour Organization (ILO) as one of the most strenuous occupations; it needs to be seen whether there are special limitations for this work in the tropics and to what extent mechanization can reduce these stresses without creating others (and also without creating adverse consequences for the environment in the long term). Furthermore it will be necessary to take into account the participation in the development of forest areas of local people, as well as immigrant or recently settled populations, and the possibility of conflicts between these groups (see chapter 19). These human aspects are closely related to certain institutional problems. After discussing problems of administration and education, the chapter will deal with concessions—the long-term agreements entered into by governments with large commercial firms for the exploitation of their forest resources.

The problem of the development and management of moist tropical forests will thus be considered on the basis of ecological, economic and social principles and within the general framework of the land-use planning of the whole country concerned and especially of a better distribution of land between agriculture, animal husbandry and forestry. The management of forests at the level of forest enterprises will have to conform with the national forest policy of the country and with a rational land-use policy at the provincial and local level.

Planning and land use

Since 1945, governments have paid considerable attention to spatial organization at national and regional levels and planning techniques have been designed. In the agricultural sector, the concept of rational land use, i.e. a sensible distribution of land between agriculture, forestry and animal husbandry, has been evolving rapidly in the direction of integrated rural management towards a condition in which the goods and services of the different subsectors combine to achieve the best possible result. At a more restricted local level, this management of space finds application in integrated watershed management, in specific natural resources management (forest, grazing land and farm management) and in community development, according to the priority given to physical, economic or social criteria. Unfortunately, there are few theories that really relate to the tropics and very often it has just been a case of transferring concepts and methods originally elaborated in industrialized countries.

In 1951, FAO organized a regional conference on land use in Asia and the Far East (FAO, 1952). The 9th Pacific Science Congress devoted a symposium to climate, vegetation and rational land utilization in the humid tropics (1958). Duke University, USA, organized a symposium on tropical forestry in April 1965. Two publications by FAO on the management of natural resources (FAO, 1972) and *Approaches to land classification* (Soils Bulletin no. 22, 1974), and *Land evaluation for rural purposes* (publication no. 17 of the International Institute for land reclamation and improvement) represent important contributions to this field.

Theoretically the methodology does not present any major difficulties; all that is necessary to construct a model that would optimize the yield of a system is to identify the various sites or habitats, assess their biological potential, establish the cost of producing certain goods and to localize the centres of consumption, so that transport costs can be computed. This has been tried in some countries, notably the United States but it is a costly exercise. Moreover it takes for granted the agricultural structures and sizes of exploitations, the availability of capital and labour as well as the technologies, whereas in the developing countries all these will often have to be carefully considered. Some authors have stressed the harmful consequences such models have on certain traditional systems which may embody valuable and irreplaceable experience acquired by observation and adjustment over many generations. Often such theoretical exercises take no account of difficult areas (mountain country, arid zones, etc.), though in practice it has often been proposed to set aside areas in need of protection, using as criteria the steepness of slopes, erodibility or poor productivity; such areas would then act as buffer zones between more productive, but also more modified and hence less stable ecosystems.

Whether the problem is the intensification of production in certain areas, or the development of areas that are still untouched or only slightly used by sparse local populations, the final decision on land use will always be a political one, even if this decision has to be taken with due regard to the existing ecological, economic and social possibilities and limitations. Two recent studies on rational land use, one large-scale and qualitative, the other more quantitative at small-scale, will be discussed in this chapter.

Ecological guide-lines for development

An effort to explore the partnership of conservation and development (Dasmann, Milton and Freeman, 1973) suggests that development planners must include ecological criteria, principles and guide-lines in all phases of regional economic development if the long-term well-being of regional populations is to be achieved. Where an undeveloped area of natural land exists, these authors mention six options open to development planners:

1. it can be left in a completely natural state and reserved for scientific study, educational use, watershed protection and for its contribution to landscape stability;
2. it can be developed as a national park or equivalent reserve, with the natural scene remaining largely undisturbed to serve as a setting for outdoor recreation and the attraction of tourism;
3. it can be used for limited harvest of its wild plant or animal life, but maintained in a wild state serving to maintain landscape stability, support certain kinds of scientific or educational uses, provide for some recreation and tourism, and yield certain commodities from its wild populations;
4. it can be used for more intensive harvest of its wild products, as in forest production, pasture production for domestic livestock, or intensive wildlife production; in this case its value as a wild area for scientific study diminishes, but it gains usefulness for other kinds of scientific and educational uses; its value for tourism and outdoor recreation diminishes; its role in landscape and watershed stability is changed, but may be maintained at a high level;
5. the wild vegetation and animal life having been partly removed, it can be intensively utilized for the cultivation of planted trees, pastures or food crops; or
6. the wild vegetation and animal life having been almost completely removed, it can be used for intensive urban, industrial or transportation purposes.

So long as the first three choices are taken, the option remains open to change from one of these uses to the other or to use the land for any of the latter three purposes. If the fourth choice is decided upon, the options for restoring the land to any of the first three categories are reduced but not eliminated. Selection of the latter two development possibilities largely prohibits any shift to other alternatives within a reasonable period of time. A rational and sensible choice from among the options available must be based on ecological and economic considerations, as well as on other grounds. In most developing nations numerous conservation and development options remain open. It is always easier to take into account conservation needs before intensive development begins. It is far less expensive to protect than to restore the environment, and some types of environmental damage are irreparable.

In tropical and subtropical forest areas the adequacy of ecological research and planning will largely determine the rational basis for selecting one or more of these six options. Unless such ecological studies and planning are well integrated with overall economic planning, the protection and conservation values of many natural forest areas will be threatened. However, there are still problems of integrating ecological and economic planning. Further research and demonstration projects are needed to overcome differences in time scales and in a quantitative versus qualitative approach to planning. For example, the economist is commonly concerned only with those resources which can be reduced to quantifiable monetary units of measurement. The ecologist, by contrast, is concerned with the nature and interrelationship of all resource factors affected by development, whether quantified by price, non-quantified (as is common with relatively free resources such as air and water), or quantified by non-monetary means (such as the measurement of energy flow, biogeochemical cycling, or the incidence of human disease). Similarly, the economist is likely to measure the resource potentials in terms of immediate, short-term, lowest cost resource supply responses to known, short-run demands. The ecologist, by contrast, is concerned with balancing short-term and long-term concerns and is unwilling to meet immediate resource demands if the result could lead to irreversible degradation of resources. Ultimately, the ecologist is concerned with stable, long-term, sustainable resource production and with an optimum spectrum of varying human uses based on rational management.

Some of the basic ecological aspects applicable to tropical and subtropical forest management include: first, the careful study of biogeochemical cycles, energy flow patterns, food webs, hydrological cycles, and biotic interrelationships within the ecosystem; secondly, the examination and definition of regulatory functions within the ecosystem such as succession, predator-prey relationships and the role of biotic factors in maintaining nutrient and energy balance; thirdly, the study and testing on demonstration areas of various patterns of probable and proposed resource use of the ecosystem, with careful attention to environmental destabilization and irreversible effects; and finally, the development of ecologically-sound approaches to agricultural, aquacultural, livestock and timber utilization based upon the results of the research. See Bene *et al.* (1977).

An excellent compilation of ecological principles for land-use planning appropriate to tropical and subtropical forest areas has been provided by Miller (1972): 'First, it was suggested that areas which are on or approaching ecological thresholds of irreversibility, i.e., which show accelerated erosion, extending areas of landslides and mass earth movements, uncontrolled stream flow, volcanic and seismic activity, rapid lateritization, etc., should be indicated on the planning maps as *critical zones*. These sites should be kept free for development until the problems can be studied,

the risk evaluated, and solutions given. Second, superlative examples of forest, fauna, scenery, archaeological or other natural or cultural values, should be classified as *unique zones*, until a detailed evaluation can determine the appropriate objectives to follow and the final management system to install. Third, permanent vegetative cover must be maintained on slopes, river catchments, swamps, lowlands, stream banks and highly erosive soils. In these areas, large *multipurpose zones* must be established to maintain permanent vegetative cover and to produce on a flexible basis a wide variety of such goods and services as timber, water, minerals, wildlife, hunting and fishing, tourism, forest industry, and other compatible uses. Fourth, areas which are distant from markets, which lack in critical or unique features, yet which appear to possess materials of high future value, should be designated as *holding zones* or government reserves. In the future, when more intensive resource evaluation is warranted by increasing demands and pressures, then the area can be allocated to permanent uses according to determined objectives. Finally, only those areas of high potential yields in agriculture, livestock and fast-growing fibre crops should be designated as *agriculture development zones*. There, risks of losses from floods, erosion, soil depletion, plant succession, animal damage and other biotic factors can be reduced and controlled.'

Whatever management goal is chosen, the primary concern must be to maintain the site's biological stability and resource productivity. If this is disregarded, the ecosystem may be exposed to serious and irreversible deterioration. This decline will close all productive options either for maintaining the area under current management or for shifting use to other worthwhile purposes.

Ecological stratification for land-use planning

In addition to such general guide-lines, it may be desirable to apply more analytical and more detailed methods which yield an ecological stratification and enable decisions to be made on the utilization of individual *terroirs*, i.e. particular sites or habitats, according to their potentials and limitations. But first one will have to agree that such *terroirs* exist. Whitmore (1975) states: 'most, if not all, species have highly precise habitat requirements and occupy distinct niches, so that variation in the forest mirrors the occurrence of these niches and can be interpreted once they have been elucidated. Forest cannot be considered as one extensive fluctuating association. Environment-correlated variation occurs as a result of the success in competition of those species within dispersal range which are best adapted to a particular habitat, and there may be several species equally well suited.'

Variation and ecological relationships

Greig-Smith (1974) in a review of the application of numerical methods to the classification of tropical forest states that 'the few investigations in which numerical methods have been used have already demonstrated not only that there is ordered variation in rain forest composition, but that this variation can be correlated with environmental differences'.

Environmental variation pattern at large scales of space has been related to broad *life zones* by Holdridge (1967) using three environmental parameters: mean annual temperature, mean annual precipitation and potential evapotranspiration. But in the tropics the relationship between actual climate and soil is often vague, and the annual temperature and rainfall alone are inadequate for an ecologically meaningful characterization of the life zone. At least seasonality of rainfall, radiation and water availability should even at this high level of classification be directly considered.

Ashton and Brünig (1975) point out that, while the variation in humid tropical forest is continuous along hypothetical environmental gradients, in the field variation is often discontinuous. More or less abrupt changes occur at well defined boundaries of site conditions such as changes in soil, exposure and aspect, or in current or past biotic or climatic impacts. The responses of the vegetation to recurrent fluctuations of environmental factors, such as rainfall and storm frequency and intensity, will generally depend on:

— the frequency and duration of the fluctuating states in relation to the life-span of the organism or community;
— the time pattern of occurrence of the various states, particularly its reliability and regularity;
— the effect of the various states on the fitness of the species to survive and compete;
— the effect of adaptation of a species to one state on its performance in another state.

Stratification, assessment and prediction

The choice of site in primary forest by the shifting cultivator is largely guided by the luxuriance and stature of the vegetation. Dense shady and tall forest is generally preferred. In some cases, the occurrence of certain trees, palms or herbs, is taken as an indication of the suitability of the soil for the cultivation of certain crops. Direct assessment of the soil is the exception. For example, Watters (in Brünig *et al.*, 1976) reports that farmers in Yukatan prefer rocky lands on low limestone hills for maize cultivation, presumably because the soils have initially a high organic matter content and better water-holding capacity than many other soils. Yet, in spite of regular dressings of chemical fertilizers, the yields have steadily dropped throughout an eight year observation period.

This attitude is rational and understandable because the first crop will thrive on the nutrient store of the primary forest and there is the option to shift later. The requirements for site determination and classification are more exacting if sustained and intensively producing land-use is the objective.

Classification and mapping of soils, vegetation cover and habitat must be coordinated and land-use oriented. The units of classification must relate to the natural potential as well as to the possibilities and limitations for improvements (fertilizing, biological and technical amelioration). Discrepancies occur and are difficult to avoid. The information value of a cover type map is chiefly as a timber stock inventory; it provides little information on habitat potential. Classification from air photographs can be refined in relation to moisture conditions and vigour of the standing crop if infra-red, 'false colour' (infra-red colour) films are used. Erodability and other ecological features of the forest in large tracts could be efficiently assessed by a combination of satellite imagery, and conventional air photography at small and large-scale (see chapter 1).

Various authors have then studied the contribution of the local climatic conditions, climatic factors variation (from season to season and from year to year), physiography and soil conditions, including water-holding capacity, to the complexity and variation of forest cover. The close relationship between land form or landscape unit (relief), parent material, soil type at fairly low level of classification (soil series) and certain soil features (depth of soil) is demonstrated by Burnham (in Brünig *et al.*, 1976) for Malaysia and between the resulting environmental pattern and the natural vegetation types by Ashton (1964), Whitmore and Burnham (in Whitmore, 1975), and Brünig *et al.* (1976).

The development of classificatory processes is progressing in two directions. Research concerns more objective methods in deciding the quantitative and qualitative criteria which would effectively allow agglomerative or divisive classifying. The other is the ecological significance of the classification scheme and its categories. The complexity and diversity of the humid tropical forest make both an extremely difficult task. The procedure is not unlike the schemes developed in Canada by Hills (1961) and in Australia by Christian and Stewart (in Brünig *et al.*, 1976) and now applied in many tropical countries. Classification of sample plots is not a classification of the vegetation as a whole, and should not be used in lieu of a standard forest type classification, but as a guide to the best means by which such classification can be constructed.

For local classification good use can be made of the following relationships between forest type and environmental features:

— within one geological formation exists a strong correlation of forest type with physiography;
— soil nutrient status, especially availability of phosphorus, is the major determinant of forest composition and to some degree of structure;
— soil physical properties and drainage are the principal determinants of forest physiognomy (architecture, aerodynamic and optical properties of the canopy, stature).

The usefulness of any ecological vegetation/site classification depends on the degree to which it provides information for a defined and specific objective. Thus the value of a classificatory criterion (e.g. phosphorus availability of the soil, or a structural or floristic element of the vegetation) depends on its significance in relation to the purpose of the classification which is usually the improvement of the prediction of the consequences when a particular option of land management is chosen.

Contribution of stratification

A major limitation is the small proportion of the area of the tropical moist forest which has yet been covered by land-use oriented surveys, and the often excessively low intensity of surveying. For example, little more than 10 per cent of Amazonian rain forest soils have been surveyed, including extremely extensive reconnaissance.

Another serious limitation is inadequate knowledge of the ecology of the various primary and modified tropical moist forest ecosystems. There are few data on the structure and functioning of natural forests, hardly any of modified forest, agricultural or combined agro-sylvicultural systems and their long-term changes under different management practices and environmental conditions on different sites.

Available knowledge does however suffice for the identification of soils with similar general characteristics, and this leads to a tendency to uniform land-use planning and practices. Chemical analysis allows effective planning of fertilizing. Combining data on nutrient status of soil and trees, physical soil characteristics and growth of species could produce the information needed for the delineating and mapping of soils in relation to productivity. Information on productivity rating of different plant genotypes in relation to different soils is, however, still inadequate and requires further research before satisfactory land capability maps can be produced. Practical experience in many countries has shown that careful and cautious allocation of adequately defined soil types to specific uses can at least reduce risks, but in practice allocation continues to be opportunistic rather than rational, by block of land rather than by soil type.

Fast-growing timber plantations almost invariably disappoint expectations and do not meet targets on soils which are too poor for subsistence agriculture. Short-rotation plantations of fast-growing light demanders such as species of _Gmelina_, _Tectona_, _Eucalyptus_ and _Anthocephalus_ spp. on light, sandy soil, may fail to maintain the fertility of the site unless a suitable understorey is encouraged. This also applies to other species such as _Aucoumea klaineana_ and _Nauclea diderrichii_ for tropical Africa and _Cordia alliodora_ and _Dididopomax morototonii_ for tropical America. Another risk in planting these fast-growing, light demanding species is that they are often biochemically less diverse and low in toxins. Consequently, they are more susceptible to insect and fungus damage, and this adds to the adverse effects of unsuitable site allocation.

Insufficient is known of the properties and possible uses of the native species and experience is lacking on the consequences of converting natural forest into mixed forest, plantation or into agriculture. Many of the species are of very doubtful potential as an industrial resource and their ecological properties are only vaguely known. With this lack of basic information it is nearly impossible to derive any kind of soils or habitat classification which could be meaningful in relation to economic, socio-economic and social objectives.

Plans must clearly define the type of human manipulation envisaged in order that the prediction may be expressed unambiguously, and understood correctly by those to whom it is addressed. If the prediction refers to clearing of forest and meant clearing by hand with the stumps left in place while the manager uses a bulldozer, the outcome may be disastrous. If the prediction is based on the weather of a single average season, and the manager assumes it applies to a series of years of which some may be wet and some dry, serious misjudgments may result.

Examples of adverse ecological changes as a result of the adoption of new techniques or crops are not restricted to more recent times. When the nomadic Temiar of Malaya adopted padi (hill rice) they often grew two crops instead of the one usual with traditionally grown millet. This and more intensive clearing eventually lead to the formation of degraded grasslands of _Imperata cylindrica_ as a fire disclimax (Wycherley, in Brünig _et al._, 1976). The introduction of soil working implements such as the hoe replacing the dibbling stick lead to more soil disturbance and hence to more erosion especially on slopes. Decision on the suitability of a soil and site for certain crop types must consider possible effects on the environment. For example, large areas of the most productive alluvial soils in Malaysia have been lost or degraded and their management made much more costly and difficult as a result of clearing and cultivation in the catchment areas causing erosion and excessive overland flow. Flooding of irrigated alluvial plains, silting of irrigation and drainage channels, burying of fertile fields with inferior silt load, are already common and rapidly spreading throughout the tropical world.

Another limitation is the lack of integration of ecological and economic aspects. As an example, the classification by land form-natural vegetative cover type of the Amazonian region (IBDF-PRODEPEF, in Brünig _et al._, 1976) defines the usual general preferential land uses for each unit, but gives no indication of risks and relative cost-benefits of the various options. Nor is the effect of forest stand and site variation within the units on the feasibility of alternative options with respect to risk and performance taken into account. This is a severe limitation of the usefulness of the classification for land-use planning.

Conclusions

Land management for production of food and raw materials can be sustained in humid lowland tropical areas only if four requirements are met:

— a vegetative cover must be maintained to provide effective protection against soil deterioration;
— the balanced nutrient cycle must be maintained;
— the crop's nutrient demands must be adjusted to the nutrient inputs from rain, dust, fixation and weathering;
— biotic diversity of physiognomy, trophic levels, life-forms, species composition and age distribution must be maintained above the level at which the activities of pests and other ecological risks become a serious factor ecologically and economically.

Tropical moist lowland forest vegetation is relatively well known in very broad and general terms. For a few selected sites, detailed studies of the vegetation have produced some useful information on forest structure, biomass, nutrient storage and cycles. Very little, however, is known of the interrelationships between structure and functioning of these ecosystems and their environment, the variation of this relationship, and their reaction to haphazard or planned interference by man.

Existing knowledge permits the stratification of the forests in some fashion into discrete but heterogeneous and unequal units at some usually very high level of the ecosystem hierarchy. For detailed project planning, information is necessary at smaller scale on the kind and amount of structural and functional variation within and between narrowly defined types of the natural and modified humid tropical forest. Knowledge in this field is generally rare, usually local. Consequently, stratifying, predicting and planning are still largely intuitive and subjective. Uncertainty of success due to the yet largely unpredictable reactions of the sensitive tropical moist forest ecosystem to interference makes any long-term land-use development project extremely risky. To reduce this risk to acceptable levels, very much more knowledge on the ecology of the forest is needed.

In conclusion, it seems that we have now at our disposal not only general guide-lines which enable decision makers to formulate a strategy for conservation and development but also more elaborate methods to decide on the most rational utilization of sites or habitats, provided that the necessary data are available. Much remains to be done to elaborate a simpler methodology, to reduce the amount of data required and to undertake specific research. One might first investigate practical methods that would perhaps allow a grouping of similar 'terroirs' and the choice of indicators fully representative of ecological, economical and social conditions.

Conservation

Conservation of habitats

Values and objectives

The values of protected natural forests have received much attention during the 20th century. A primary response of nations to the rapid loss of natural forest lands under various impacts has been to set aside at least some land in protected areas. The values and objectives of protected natural ecosystems are the following:

- the preservation of large, relatively self-contained natural areas in each of the world's major ecosystem types to safeguard the evolutionary processes;
- the protection of representative areas to safeguard the great diversity (biological and geological) of the biosphere's natural ecosystems, and to assure the preservation of the genetic pool they contain;
- the assurance that the normal regulatory functions of the biosphere can continue without irreversible disruption;
- the provision of representative and unique natural systems where basic and applied environmental research and education can be carried out, where baseline and monitoring studies can be initiated, and from which ecologically-sound planning and management of land and water resources can be derived and applied;
- the protection of watersheds, particularly from erosion and down-stream sedimentation: this will also help to maintain high quality of water and to prevent serious flooding;

- the protection and conservation of fish and wildlife;
- the protection and conservation of plant species and their values as recreation resources, timber products, sources of genetic material for plant breeding, medicinal plants, climatic enhancement and ecosystem regulation;
- the provision of a wide spectrum of undisturbed areas for aesthetic and recreation purposes, and the development of modest natural-area based local tourism economies.

These form another subject for critical research. For example, the role of forests as a provider of animal protein was mentioned as one possible objective for sustained-yield production (see chapter 20) but only limited studies have been carried out on its economic and nutritional significance. However, unless studied, such economic significance could easily be missed by economic development, and even by reserve planners.

The great resource use of tropical and subtropical systems by indigenous peoples needs to be heavily underscored. Many of these peoples are not part of national and international economic networks, and the value of the forest for its products is often left out of these broader socio-economic analyses. For example, the forest provides wild animals, unpolluted rivers for fishing, fertile soils for shifting cultivation, a wide variety of plant foods, medicines, construction materials, and many other resources not apart of larger market economies. In India and much of tropical and subtropical Asia, the forest can provide significant additions to the economies of village and rural peoples that have been quantified. For example, the Erenga Kharias of the Simplipal Hills in India used to collect 15 000 kg/a of wild honey from the forest, which they sold for Rs 5/kg in the forest and up to Rs 20/kg in the city; due to neglect by modern foresters this resource has now declined to under 2 000 kg/a. Another estimate from the Morena forest of west central India calculates that each hectare of *Acacia-Anogeissus* wooded grassland produces Rs 350 of fire-wood at every 20 year rotation, carries 1/2 unit of livestock and provided mule fodder, wild mammals and gallinaceous birds. Musk, feathers, ivory, hides, honey, green manures, rubber, forage, beverages, pest poisons, dyes, tannin, resins, spices, oils and medicines contribute in India about Rs 1 000 million/a.

This high productivity in the resource utilization of tropical forests by local peoples is due to the highly-evolved patterns of multiple and sustained yield management achieved over centuries of agricultural experimentation. Once destroyed, these self-sufficient peoples often undergo cultural disintegration from the loss of their forest economic base. This tragic process now threatens large parts of the Amazon Basin, and the forested areas of Central America, Africa and Asia. Instead of destroying these rich resources and values for the single-purpose, short-term production goals of timber companies, the resources use patterns of indigenous peoples should be studied, refined and propagated. Unfortunately, research on the ecological and economic values of such approaches has lacked support.

The values of forests in stabilizing micro-climate, pest populations and other ecosystem regulatory roles is a major field for research (see chapter 2). The crucial role of the tropical rain forest in maintaining the stability of biogeo-chemical cycling and soil nutrient levels is well known, but comprehensive studies have not yet been initiated on the exact functioning of this process in a representative range of ecosystems (see chapters 2 and 13). Similarly, little is known of the regulatory value of these forests in maintaining water flows and in preventing soil erosion, down-stream flooding and silting (see chapter 12). Such values should be quantified economically wherever applicable.

The values of tropical and subtropical forests for science and education have long been discussed, but only recently have major research programmes, such as the IBP and the MAB begun to fully explore the possibilities of these environments as living laboratories and education centers. Finally, the economic, social and psychological values for recreation and tourism are a rich field for investigation. In many parts of Latin America, Asia, Africa and the Pacific Region, natural forests and their associated wildlife, geo-logical and archaeological features, are playing primary roles in building up local economies from national and inter-national tourism, and providing major areas for recreation. Lastly, the social importance of protected areas as an antidote to the intense urban stresses and industrial pollution of modern cities is a major value of parks and reserves. Although it is difficult to quantify, it is clear that natural areas contrib-ute to human health and psychological well-being. Research into the social and economic significance of these trends is of critical importance to help elucidate the values and objectives of protection.

In summary, we know that protected natural forests assist gene flow preservation, preserve representative eco-systems for research, maintain hydrologic flows and soil stability, provide opportunities for recreation and tourism, and safeguard scenic resources. Nevertheless, it still requires considerable basic and applied research before the true sig-nificance of each of these objectives can be measured.

Existing status

During the last decades the current status of parks and related reserves in the world has been reviewed in great detail in one study of paramount importance and two equally important conferences. The First World Conference on National Parks, held in Seattle, Washington (1962), represented a major turning point in the evolution of global concern for the protection, establishment and sound management of national parks (Adams, 1962). The Second World Confer-ence on National Parks took place at Yellowstone and Grand Teton National Parks (United States) in 1972 (Elliot, 1974). More than 80 nations participated in a critical review of the more than 1 200 national parks or equivalent reserves that had been established by that date. The Conference recognized that some of the most serious problems in the development of parks in tropical zones were due to the lack of funds, materials, personnel and public support. Only 47 national parks or equivalent reserves in the wet tropical zone were recognized at that time by the United Nations list indicating 'serious underrepresentation in comparison with savanna and grasslands'. The third important effort concerning world

parks is the latest edition of *United Nations list of national parks and equivalent reserves* (IUCN International Com-mission on National Parks; 1971). An excellent popular summary is Curry-Lindahl and Harroy's *National parks of the world* (1972). The list, which reports on 1 204 areas covers 140 nations. It includes careful definitions for national parks and equivalent reserves as well as detailed criteria to be used for their selection. An excellent summary is given of the protection status, area, staff, budget, date of establish-ment and legal history, facilities for tourism, research history, and ecological and biological conditions of each park or reserve. In addition, the general picture of institutions and factors relating to conservation unique to each country is described in an introduction to the section for each nation.

The first national parks and equivalent reserves to be established in the tropics and subtropics were South Africa's Kruger Game Reserve, now National Park (created in 1898) and Natal National Park; Zaire's Virunga National Park (Albert, 1925); India's Kaziranga National Park (1908); Cambodia's Angkor National Park (1925); Burma's Pidaung Game Sanctuary (1927); Singapore's Kranji Reserve (1883); Royal National Park (1886); Ku Ring Gai Chase National Park (1894) and Australia's Lamington National Park (1915); Argentina's Iguazu National Park (1909); Guyana's Kaieteur National Park (1929) and Ecuador's Galapagos Islands National Park (1935).

From these important but modest beginnings, the estab-lishment of national parks and equivalent reserves has grown steadily throughout this region. In his survey of parks in wet tropical areas, Richards (1974) notes that by 1972, 47 nations had established national parks or equivalent re-serves in the wet tropics: 10 were in the Americas; 27 in tropical Africa, and 10 in tropical Asia. He then notes that a major goal of such parks should be to preserve represen-tative ecosystems, but that sufficient information was not available to assess how complete the coverage of protected areas was. Nevertheless he felt certain that the goal of protecting representative samples from each major tropical ecosystem type was far from being achieved. The United Nations list indicates that considerable levels of protection have already been achieved for grasslands and savannas (par-ticularly in Africa), but that protected lowland humid trop-ical rain forest areas are seriously under-represented. This discrepancy is largely due to the high priority given in the past to protecting the spectacular large mammals of African and Asian grasslands.

The emphasis on protecting species, rather than entire ecosystems, has caused other problems. Single species' habi-tats often include only a partial aspect of a regional eco-system. As a result, when parks are designed to meet the requirements for preserving the habitat of only a few species, or scenic features such as a waterfall, important other op-portunities for full ecosystem protection are frequently lost. The Kaieteur and Iguazu National Parks of tropical America, both built around waterfalls, are examples of such reserves originally intended to protect only one aspect of the land-scape. Many other reserves, too small or specialized to be listed by the United Nations, and often linked with univer-sities and research institutions, play important roles in forest ecosystem protection. The Organization of Tropical Studies' Finca La Selva, in Costa Rica, is an outstanding private reserve for research and education; the University of Malaya has protected an important forest site near Kuala Lumpur for research; and Panama's Barro Colorado Island now managed by the Smithsonian Institution presents fine opportunities for research.

National forest reserves are another type of at least partial protection. They are usually managed for a combination of timber, watershed protection, recreation, research and related aims. Because for many such areas the primary purpose is economic production, their very real contribution to nature protection has often been overlooked. However, such reserves often include parts specifically protected or allocated to research and education.

The range of natural forests under some form of protection extends from intensively-studied areas, usually situated close to major population centers and endowed with relatively good access, to others that are remote and little known. A good example of the former is the Luquillo Forest Park in Puerto Rico which is attracting over a million visitors a year and has become an important focal point for intensive studies of tropical forest ecosystems. On the other hand, many outstanding areas have been set aside that are little studied and rarely visited, such as the remote and beautiful Manu National Park which protects a vast area of lowland to montane tropical forest on the eastern slopes of the Andes in Peru.

Perusal of the 1973 edition of the United Nations list indicates that 62 countries whose territories are all or largely within the tropics and subtropics report they have established 406 national parks or equivalent reserves. If the parks and reserves of the tropical and subtropical parts of Australia (Northern Territory and Queensland), Japan (Kyushu), and the United States (Florida, South Georgia, Virgin Islands, Hawaii and Puerto Rico) are added, this figure becomes 460. The largest number of established areas are reported from: Australia (41), Brazil (20), India (16), Indonesia (48), Kenya (14), Malagasy Republic (31), Philippines (23), Rhodesia (20), South Africa and South-West Africa (33) and Zambia (17). However, the size and nature of the forest ecosystems protected is not known. Future surveys and research should obtain these data. Since the period of compilation of information for the 1973 edition, many new parks and reserves have been established (and, in all likelihood, others have ceased to exist).

For example, a new national park in Costa Rica, Parque Nacional de Corcovado in the Osa Peninsula, was created on 31 October 1975 by a presidential decree; the park includes the entire watershed of the Corcovado Basin, *ca.* 290 km², *ca.* 0.6 per cent of the area of Costa Rica. This is magnificent lowland tropical wet forest and *ca.* 95 per cent of it has never been cut. Venezuela has some magnificent areas represented in its system of 18 national parks and seven natural monuments. This system began in 1937 with the establishment of Henry Pittier (El Rancho Grande) National Park and it has grown quite consistently in number and area to where it now includes *ca.* 26 600 km² or 2.9 per cent of the country's area. A recent bulletin from the Division of National Parks in the

Directorate of Renewable Natural Resources gives the location, size and a brief description of each area as of August 1974. Those having substantial areas of tropical rain forest are: Henry Pittier, Sierra Nevada, Guatopo, El Avila, Canaima, Macarao, Yacambú and María Lionza. There is currently considerable sentiment for including Canaima within a much larger Gran Sabana Park which would include additional rain forest. Eichler (1973) has made a study of the park system—what it is and what it could be. Many of his proposed areas have since become reality: Morrocoy, Mochima, El Guácharo, Laguna de Tacarigua, Yacambú, Canaima, Médanos de Coro and Laguna de la Restinga have all become national parks and Cerro Santa Ana has become a natural monument. The Division of National Parks has proposals for additions or enlargements involving existing parks Canaima and Yacambú and now areas such as: Cerro La Neblina, Orinoquía, Tamá and Perijá. The most imminent of these seems to be an enlarged Canaima, with perhaps a name change to Gran Sabana. A comprehensive master plan for Gran Sabana has been prepared under the auspices of the Tourism Corporation and a study by IUCN aims at implementing as soon as possible the recommendations of this master plan. In his publication on reorganizing the structure and functions of the wildlife work, Mondolfi (1974) discusses the refuge system, and identifies the need for areas to protect the Andean bear, the jaguar and other animal wild species. Introductions of exotic wild species do not seem to be major problems in Venezuelan rain forest, nor has environmental contamination yet reached a point where the rain forest fauna is being significantly affected. The major problems are habitat destruction and pressure of hunting and trapping. Another problem already mentioned (in chapters 6 and 7) is the lack of knowledge of populations, life-histories and ecology of the animals. The Division of Wildlife is carrying out some excellent work at the biological station in Henry Pittier National Park. A research effort by the Smithsonian Institution, cooperating with the Consejo de Bienestar Rural and initiated in 1975, is focusing not only on the Llanos, but on Guatopo National Park as well. Also significant for the welfare of wildlife was the establishment of the Foundation for the Defense of Nature (FUDENA) in 1975 and the corresponding World Wildlife Fund National Appeal for Venezuela under the auspices of the United States National Academy of Sciences. Steyermark (in Hamilton, 1976) has identified 64 areas of tropical rain forest in Venezuela which he considered to be of such unusual biological value that they merit protection from disturbance. Some may become biological reserves or biosphere reserves.

In the Philippines, the Parks and Wildlife Office was merged into the forest department in 1972 as one of seven divisions of what is now the Bureau of Forest Development; the 30 km² Aurora Memorial National Park in Luzon covers the crest and flanks of an east-west running range of hills, clothed in forests from 100 to 600 m; a proposal has been made to reduce the 730 km² Mount Apo National Park in Mindanao to 130 km² and release the rest for logging and settlement; an extensive area of lowland dipterocarp forest is contiguous to the 30 km² St Paul Underground River

National Park in Palawan. In Papua New Guinea there are only three national parks; more will be needed if the various formations are to be adequately conserved. Five new parks covering the most important areas have been proposed. In Fiji, montane forests are perhaps adequately conserved in protected forests reserves and two national parks. Lowland and beach rain forest are probably inadequately represented, but are included in three proposed conservation areas.

It is difficult to determine from available information the exact number of national parks and equivalent reserves in tropical and subtropical forest areas. It would be most valuable to include a break-down of parks and reserves by major ecosystem and biome types in future editions of the United Nations list.

As mentioned earlier, the United Nations list excludes many protected areas because they do not comply with the criteria for selection. As well as forest reserves managed for economic production and natural areas of small size or specialized usage, there are other sites which are also not included. For example, many combined natural park and cultural reserves, such as the 22 000 km² Xingu National Park and Indian Reserve in the Brazilian Mato Grosso. However, good summaries of many such non-qualifying areas are included.

Another important initiative is the *Unesco Convention concerning the protection of the world's cultural and natural heritage*. This was designed to recognize and provide international assistance to national parks, cultural or historic sites and other areas of outstanding world interest. Examples of tropical and subtropical sites that have been suggested for inclusion in this system include: Tanzania's Serengeti Plains, Venezuela's Angel Falls, the historic ruins of Inca, Mayan and Aztec cities, Zaire's Virunga National Park, the proposed Mt. Everest National Park in Nepal and the Kaziranga Sanctuary in India (Elliot, 1974).

In November 1971, the International Co-ordinating Council for the Man and Biosphere Programme (MAB) initiated Project 8: *Conservation of natural areas and of the genetic material they contain*. Two reports (Unesco, 1973, 1974) explore the reasons for, and the means to achieve, protection for selected natural areas representative of the world's major ecosystems. The 1974 report describes the objectives of a proposed network of biosphere reserves, defines the criteria for reserve selection, and sets forth guidelines to aid in their establishment and management (see also di Castri and Loope, 1976). The objectives of this international network of biosphere reserves have been defined as:

— 'to conserve for present and future human use the diversity and integrity of biotic communities of plants and animals within natural ecosystems, and to safeguard the genetic diversity of species on which their continuing evolution depends;

— 'to provide areas for ecological and environmental research including, particularly, baseline studies, both within and adjacent to these reserves;

— 'to provide facilities for education and training.'

These reserves might often be established in concert with existing or planned national parks or equivalent reserves, but would be selected according to criteria under three categories: natural areas representative of biomes; unique ecosystems of special interest; man-modified landscapes. Each reserve would include both *core areas* and *buffer zones* with intensive human uses confined to the buffer zones. The guiding principle for multiple uses would be their compatibility with the primary goal of conservation. The Unesco MAB report (1974) suggests that the IUCN system for classification of natural regions might be a helpful basis for selection of representative biomes and ecosystem types (see Dasmann, 1972, 1973, 1974, for an explanation of this system and its relation to classifying protected areas; Udvardy, 1975). It is also emphasized that many countries will require programmes of international assistance in conducting surveys, inventories and evaluations needed to locate and establish suitable sites for biosphere reserves. Tropical forests, grasslands, coastal ecosystems and islands were deemed zones of critical need for more effective conservation.

Because of the major impact of substantial development schemes now under way in tropical and subtropical forests (Unesco, 1973, 1974), it is clear that particular emphasis has to be placed at this moment on the establishment of national parks, reserves and biosphere reserves in the major representative and unique ecosystems. Ecological base-line and monitoring studies in such natural parks and reserves could have a beneficial effect in modifying the excessively rapid and often destructive treatments frequently involved in present development practices. It is particularly important that such research be carried out both in protected reserves and in man-modified areas where development patterns are either stable and ecologically sound, or where examples of degraded ecosystems can be studied. As yet, such research is in its infancy in most tropical and subtropical forest areas.

Research and training needs

The goals of development and the goals of conservation and protection of forest areas must be made to coincide. Effective, long-range rational planning based on a balanced integration of sound social, ecological and economic principles is necessary if the interests of local and national populations, as well as those of the international community, are to be best served.

Sound planning requires sufficient information on the functioning and structure of natural areas to determine rational management that is ecologically sound. Similarly, inputs from planning and management are required to develop the most effective basic and applied research priorities. The ultimate effectiveness of planning and research programmes for forest conservation is dependent upon well-conceived programmes of public education. The conservation and protection of tropical and subtropical forest areas must be in agreement with the perceived values and problems of these areas.

The required basic and applied research programmes include: the data gathered for the United Nations list of parks and reserves; the regular evaluation of parks at the world conferences on national parks; the Unesco/MAB Programme's exploration of the ecological effects of human activities on tropical and subtropical ecosystems, and the development of the criteria and guide-lines for establishing biosphere reserves. Also, completion of IUCN's preliminary world provinces classification scheme is of crucial importance (Dasmann, 1972). In addition, there is a need for similar expansion of research to survey, inventory and classify the vegetation and animal communities (including rare and endangered species) in tropical and subtropical areas. Such inventories and classifications are essential to answer more fruitfully the following questions. What gaps exist in the region's system of parks and equivalent reserves? What important habitats may have been left out of existing protected areas that are vital to the stable functioning of a particular park or reserve? Similarly, what factors (such as air and water pollution) originating outside the boundaries of the protected area might affect the area's ecosystem?

It is also necessary to compile a register and to prepare maps on a national basis, of the existing parks, wildlife reserves, scientific reserves, regional protected areas and other equivalent reserves.

Another major field for research within national parks and other protected areas relates to their adequate design, planning and management. What are the minimum sizes of reserves capable of protecting a given ecosystem as a self-maintaining area? What opportunities exist to restore degraded ecosystems as natural areas and under what forms of management could this be achieved? What forms of economic utilization are compatible with natural area preservation goals?

Because of the importance of recreation and tourism for the economics of many developing countries and the vital role of forest parks and reserves in recreation and tourism, additional research is needed in the following areas: evaluation of the potential for reserves to support recreation and tourist industries; research on the past, present and potential future values and objectives of protected forests; the development of ecological principles and concepts for assessing the impact of recreation and tourism in such areas and formulating adequate management guide-lines; and research on the capability of legislative statutes, recreation infrastructure and enforcement systems to implement the protection and proper use of parks and reserves.

The need for education and training can be viewed in two ways: first, the preparation of adequate programmes to educate and train professional personnel in park and reserve survey, inventory, evaluation, establishment and management; and second the fullest possible utilization of protected forest areas to assist in programmes of environmental education. Duarte de Barrios and Strang (1970) and Miller (1974) have reviewed the training needs for park and reserve personnel in detail. In the past it was considered adequate to create parks and reserves to protect the physical resources of the areas involved and to provide a modest level of recreational facilities. Current problems and new trends, however, have led to accelerating demands for many other forms of management, including: applied ecological research, environmental education, sound planning for the pressures of mass recreation and commercial tourism, genetic preservation, safeguarding water quality and protection against

erosion, prevention of pollution, and the maintenance of regional ecological stability. Those new demands on management will require new approaches to training personnel for parks and reserves in many specialized fields; ideally, the personnel of parks and reserves should be composed of multi-disciplinary planning and management teams. Some of these fields include: management and protection; landscape architecture, architecture and engineering; facilities maintenance; administration and accounting; ecology, sociology and psychology; economics; resource law and economics; land tenure and acquisition; and a range of natural science specialities such as botany, zoology and anthropology, as well as archaeology and history (Miller, 1974).

There has been some progress in setting up regional and national programmes. The UNDP/FAO supported College of African Wildlife Management at Mweka, Tanzania, had by 1974 trained well over 400 individuals from over 13 nations to the medium level in wildlife and national park management. In Garoua (Cameroon), the Ecole de Faune was established in 1970 with the assistance of FAO and financed by UNDP; by 1975, over 100 candidates from 16 African countries have been trained in wildlife and national park management. At Dehra Dun and Coimbatore in India, forestry training now also includes training in wildlife, park and reserve management. Similarly, regional training programmes have been established at the Argentine Park Service School in Bariloche, the Ranger School in Conocoto, Ecuador, and training centers in Bolivia, Peru (La Molina) and Colombia (Piedras Blancas). FAO is currently engaged in developing programmes to support national park and wildlife management training in Asia and Latin America, and more recently in the Near East. Some national universities offer training at the professional level, UNDP/FAO assisted with the establishment of the Department of Forest Resources at the University of Ibadan, Nigeria, and the incorporation of wildlife management courses into the curriculum. In addition, many individuals from the developing countries of the tropics and subtropics have studied overseas at universities and other institutions under UNDP/FAO fellowships.

FAO is the United Nations Agency charged with the responsibility for executing field projects related to national park, wildlife and forestry projects. At the country level FAO is engaged in wildlife and national park training, survey, inventory, planning and management activities in such countries as Nigeria, Cameroon, Central African Empire, Gabon, Zaire, Nepal, Indonesia, Dominica Island, Ecuador, Brazil and Venezuela.

At the international level, the annual *International seminar on administration of national parks and equivalent reserves* in the United States has trained hundreds of park managers from all over the world since 1965; also, park professionals have been invited to attend programmes at the three National Park Service Training Centers. The Smithsonian Institution-Peace Corps Environment Programme has assisted in a broad range of educational and training assistance since 1972; it has also worked with the United States National Park Service in joint programmes. The United States Forest Service has offered courses in tropical forestry, including conservation, at its Institute for Tropical Forestry in Puerto Rico since 1953. The Latin American Committee on National Parks (CLAPN) has offered a wide variety of staff training and courses since 1965, and for over twenty years the Inter-American Institute of Agricultural Sciences (Costa Rica) has offered conservation training in tropical forestry, including a master's programme in wildland management to students throughout Latin America. In addition, FAO in cooperation with nine forestry and agronomy faculties in Latin America have developed a major series of training workshops for professors in wildland management. Unesco has developed a number of regional conferences, seminars and workshops (often in collaboration with IUCN) dealing at least in part, with park and reserve training; such sessions have been held in Africa, Latin America, South and South-East Asia. IUCN and the World Wildlife Fund (WWF) have also sponsored numerous efforts contributing to conservation training needs in tropical and subtropical countries. The two World Conferences on National Parks, the IUCN General Assemblies and Technical Sessions, and the IUCN regional and national meetings such as those held in Thailand, Argentina, Africa, the Pacific Islands and Madagascar have all provided significant assistance.

Despite these beginnings, conservation and management training programmes in most tropical and subtropical forest regions still lag behind those of most temperate regions. Little has been done to assess the overall future training and personnel requirements for the tropical forest region. Future research should include a study of such requirements and the factors upon which they will depend: growth of demand for recreation, tourism, biological research, public education, watershed protection and genetic protection. In concert with such studies, a major survey within the tropical and subtropical forest region is needed on future park and reserve planning and management problems such'as: conflicts with other needed land uses, environmental pollution, scarcities of funds and trained personnel, poor programmes of public information and lack of integration with key regional and national development programmes and policies.

Miller (1974) has established that, on the assumption that one manager is needed for each major park, 'approximately 1 000 managers will be required for the already existing parks of the approximately 100 Third World countries'. He adds that '600 designers, interpreters and planners, 800 support specialists, 10 000 guards and guides and 5 000 foremen, construction and maintenance crewmen will be required during the next decade' in these same nations. If one adds the many new parks that have recently been or will be established and the numerous other categories of protected natural areas (wildlife reserves, scientific areas, national forests, etc.) needing similar planning and management—a doubling or tripling of these estimates may be necessary.

Conservation of genetic resources

The conservation of forest genetic resources has been studied in detail during a meeting convened by FAO and UNEP, in

Rome, in 1975. The main conclusions of this meeting, as regards the humid tropics, will be summarized in this section.

Biological background

One of the most recent and comprehensive general statements on plant genetic resources, their exploration, conservation and utilization, is that contained in Frankel and Bennett (1970). The value of this document is enhanced by the fact that it also embraces forest genetic resources.

The principal components upon which a strategy of genetic conservation depends have been identified by Frankel (in Roche, 1975). They are the nature of the material and the objective and scope of conservation. 'The nature of the material is defined by the length of the life-cycle, the mode of reproduction, the size of individuals, and the ecological status—whether wild, weed or domesticated. The objective—research, introduction, breeding, etc.—may determine the degree of integrity which it is essential or desirable to maintain. The scope is the time scale over which preservation is projected, and the area, or space, to which it relates—a locality, a region, the world. The strategy will determine methodology, including the size of a population or sample which it is appropriate to preserve; in particular, whether to seek the preservation of a population as such, or of its genetic potential.'

There are a number of ways in which genetic resources may be conserved, and the methodology chosen will be determined by the factors referred to above. The ideal method is long-term conservation *in situ*. 'There can be little doubt that valuable gene pools of the wild plants we use in forests, pastures and elsewhere, and those which are related to our domesticated plants, not only should be preserved in perpetuity, but as far as possible with the genetic integrity of their natural state. A community in balance with a stable environment—the stability being subject to the general vagaries of natural environments—is the ideal model of long-term conservation' (Frankel and Bennett, 1970). The ideal model, however, is frequently not attainable, and there are numerous examples of forest tree species which are cultivated in many parts of the world as important commercial species while undergoing massive genetic impoverishment in their natural habitat. For a significant number of forest tree species of commercial importance, both hardwoods and softwoods, the centres of genetic diversity lie outside the areas where they are planted. For such species, *ex situ* conservation is frequently essential.

It is not inappropriate to discuss the concept of the niche under the broad heading of genetics, for the characteristics of the niche determine the genetic architecture of the species and its populations. The niche is defined as the environmental conditions that permit a population to survive permanently and with which this population interacts (Stern and Roche, 1974). The width of a niche may often be determined by investigations of one or more environmental components. An example is provided by the length of the growing season, which is taken to begin when the temperature sum reaches a certain level in the spring and to finish when a critical day-length is reached in the autumn. Adaptations to length of growing season may also depend on early and late frosts. Thus there are at least four niche factors which must be considered: date and size of the temperature sum needed to initiate the vegetative and reproductive cycle; frequency and distribution in time of spring frosts and date of the critical day-length in the autumn. Frost does not occur at low altitudes in the tropics, and is, therefore, not a niche factor, but there are numerous others which have allowed specialization for a multitude of narrow niches and a greater degree of speciation than in the north temperate zone.

Almost all species of forest trees which have been investigated have shown intraspecific variation. Nearly always the variation is striking and easily demonstrated. Its causes are less easily demonstrated, though often understood. Most experimental work has been carried out on species of the north temperate zone. Intraspecific variation poses a major problem for genetic resources conservation both *in situ* and *ex situ*.

The genetic structure of a population is determined by its environment and one or more individuals may be an adequate sample of that population, depending on its mating system which itself is under genetic control. In general terms three types of mating systems may be distinguished in forest trees: random mating where every individual mates with the same probability with every other individual of the opposite sex; genotypic assortative mating where the probability of mating is determined by the degree of relationship (negative genotypic assortative mating occurs in obligate cross-pollinators, positive genotypic assortative mating in self-pollinators); phenotypic assortative mating, positive and negative, where phenotypic characters are responsible for the deviation from random mating.

Among the accomodating species which have proved to be the most suitable for transfer are many local races or varieties. Not only do these differ in their tolerance of various climatic factors, or in resistance to particular pests and diseases, but they also differ in form. It is therefore not enough to select a species from a homoclime of the area in which it is to be tried but the varieties of the species which appear to be most particularly adapted to their proposed new homes and to possess the most desirable sylvicultural features, should also be selected (see also chapter 20).

Conservation in situ

Exclusion of felling and conservation *in situ* of representative samples of ecosystems in their natural state can best be done by establishing strict natural reserves within the larger units of forest reserves or national parks.

The range quoted for the minimum area likely to be needed for long-term conservation of samples of forest ecosystems is from 1 to 10 km². It is necessary to consider to what extent areas of this order would be adequate to conserve a viable local gene pool of the constituent species.

The minimum number of stems of an endangered population needed to form a viable gene pool will vary with the species. For example it is likely that this number will be relatively large for coniferous species of the north temperate zone, which are wind-pollinated and strongly outbreeding. A figure of 10 000 individuals has been suggested by Toda (in Roche, 1975). In other species a breeding population of much fewer individuals may be adequate. Dyson (in Roche, 1975) suggests 200 individuals, a figure based on experience in animal breeding. However, in view of lack of experimental data from forest trees, it would be prudent to double Dyson's figures in practice. It is likely, in any case, that the number required for tropical hardwood species, which are predominantly pollinated by insects, birds or bats and many of which are capable of self-fertilization, will be considerably less than for the northern anemophilous conifers.

If the environment varies rapidly within a relatively short distance, e.g. from valley bottom to ridge top, it may be possible to ensure the conservation of a range of ecological and genetic variation by establishing one large strict natural reserve comprising the whole watershed. Another reason for using a large strict natural reserve is to conserve seral stages in the succession, as well as the climax vegetation. Strict natural reserves should not be confined to undisturbed primary forest. It is of equal importance to establish them in disturbed secondary forest containing valuable gene resources, which makes up the bulk of the forest estate in many countries.

Strict natural reserves are normally established in forest reserves in Africa, though a number have been established in national parks. Strict natural reserves should be surrounded by a buffer zone of indigenous forest subjected to sustained yield management but not to clear-felling and replacement by plantations. In Nigeria, Kenya and Uganda, strict natural reserves are established well inside forest reserves and often in remote areas far from roads. They are thus surrounded by large tracts of reserved forest. This is the ideal situation. If forest gene resources scheduled for conservation are close to the boundary of the forest reserve then every effort must be made to ensure that a buffer zone of at least 300 m is established around the strict natural reserve.

More than one buffer zone may be required, as proposed for MAB Biosphere Reserves (Unesco, 1974). Gene pools must be managed and utilized as well as conserved and this implies periodic seed collection. An inviolate core area, from which human interference other than scientific observation is excluded, could be surrounded by an inner buffer zone of gene pool reserve, with outer buffer zones for tourism and commercial forest management.

It cannot be too strongly emphasized that if strict natural reserves are to play an important part in forest genetic resources conservation every attempt must be made to provide rigorous criteria both for their establishment and subsequent management. Areas of forest land set aside merely with the stated purpose of forest genetic resources conservation, and without clearly stated objectives incorporated in management plans are unlikely to remain inviolate. Furthermore, such static forms of conservation of forest genetic resources will not result in the accumulation of information about these resources as would result from dynamic conservation measures forming part of a forest management plan. In the long run, and if appropriately managed and studied, strict natural reserves should not only conserve forest genetic resources but should generate a continuous flow of information about these resources which will allow their eventual domestication, and in the case of tropical hardwoods the development of plans for the management of natural ecosystems of the species.

The maintenance of specific genetic resources within a strict natural reserve will require intervention in the ecosystem if the species concerned are seral forms which decrease or disappear as the ecosystem approaches a climax condition. For reasons such as these, management plans must be prepared.

Conservation *ex situ*

In some areas local pressures for total clearance of natural forest in favour of agriculture or other forms of land development are so great that destruction of the *in situ* gene resources is inevitable. In such cases steps must be taken to collect seed of endangered species of potential economic importance before it is too late, and to conserve them in seed banks or by planting artificial conservation stands *ex situ*, in localities where protection and management can be assured.

The problems of *ex situ* conservation have been fully discussed by the FAO panel of experts on forest gene resources and by the Unesco/MAB expert panel on conservation of natural areas and of the genetic material they contain. In the global programme for improved use of forest genetic resources (see chapter 20), various action proposals are included. They refer to early collection of *substantial* quantities of seed of an endangered provenance either for temporary storage or for immediate establishment of artificial stands on new sites. Similar procedures and quantities are required if selection and breeding rather than conservation, are the primary purposes of *ex situ* stands.

The establishment of artificial stands outside the natural range, but with a good prospect of long-term conservation, is a highly promising method of conserving gene pools. It requires careful siting and meticulous standards of site preparation, planting and tending. The cost per unit area will be appreciably higher than that of normal plantation establishment especially as the stands should be at least 10 ha.

A combination of *in situ* and *ex situ* conservation may be the solution for some species, certain provenances being suitable for permanent conservation in their natural ecosystems while others must be transferred to a new home if they are to survive.

An area of 10–30 ha per provenance or population on each site is appropriate. As an insurance against catastrophe, each population should be planted on at least two sites. Isolation from hybridizing provenances should be provided where possible, but practical consideration may render this difficult. In such cases, vegetative propagation or control pollination provide means of conserving a high degree of genetic integrity in the next generation.

It has been suggested by the FAO panel of experts on forest gene resources in its third session (FAO, 1975) to provide, within a *Global programme for improved use of forest genetic resources*, financial assistance for the establishment of provenance conservation and selection stands *ex situ* of two species of interest for humid tropics: *Pinus caribaea* var. *hondurensis* and *Pinus oocarpa*. The proposed host countries and regions are Congo, eastern Africa, Nigeria, Fiji, India, Thailand, Venezuela; and Congo, eastern Africa, Nigeria, Zambia, India, Thailand, Brazil, Mexico, respectively.

Forest development and management

Introduction

This section will deal chiefly with the history of development in tropical forests and will stress some recent ideas first on the development of the forest sector and its role, and later on forest management. Most developed countries have a framework of declared forest policies, institutionally secured or not; sometimes there is a programme of projects to be undertaken. In this respect foresters have been pioneers in the field of planning. Forest economics has really only been developed since the end of the 19th century, especially at the level of the enterprise or management unit; since then a large literature on forest management has been produced.

FAO, soon after its formation, issued a declaration of principles of forest policy, which was approved by the Conference of 1951, and a study entitled *Forest policy, administration and legislation*. A series of regional studies on the trends and prospects of production and consumption of wood were undertaken; they were designed to aid the formulation of national forest policies and to facilitate coordination within the regions recognized by the United Nations. A synthesis of these regional studies, brought up to date at regular intervals, was presented to the 6th World Forestry Congress (Madrid, 1966). The study on *Forest policy, administration and legislation* was revised to be reissued in 1976. Much has been written on the development of tropical forests and its economic implications. An article by Westoby (1963) played an important part in this for it clearly demonstrated the role of forest industries in the fight against underdevelopment especially in countries where forests abound or forest plantations can be created. It also showed the driving role of the forestry sector in a country's economy because of its links with the economic network.

Forest management and forest planning raise many problems in tropical countries and the last World Forestry Congress (Buenos Aires, 1972) gave them special attention. Many forest economists have studied this question and questioned certain concepts. King (1972) wrote that 'the rapid increase in the world's population, the new concern for economic development and growth, the rapid advances in technology and science, the recent predilection for the environment and the emancipation of large sections of the world from political domination by alien powers are but a few of the factors which have helped to make anachronistic many of the sacred tenets of the past, and to demand new approaches to the solution of the world's problems'. Leslie (1971), on the other hand pointed out that the situation now facing the utilization of forests in the tropics differed fundamentally from the situation which existed a hundred years ago in the now developed countries where forestry was born. In tropical countries well established forest departments now exist; forest products can be easily replaced by other materials; governments are inclined to interfere; foreign capital is small and sensitive to the political situation; local populations possess forest rights which are often more respected than during colonial days; there is heavy population pressure on the land; and the broad-leaved forests are heterogeneous and of low economic productivity. Whereas in the developed countries there was a hundred years ago, no restriction whatever on exploitation in the new territories; markets and outlets were assured to a large extent by the economic insignificance of substitute materials and the inadequacies of the transport system; capital was available for overseas, and there was not much population pressure on the land.

During its last session the FAO Conference (Rome, November 1975), taking note of the declaration and the programme for action concerning the inauguration of new international economic order, underlined the priority to be accorded to tropical forests. This means not only that the activities of FAO have to be given this direction, but the general outlines of a world forestry strategy for these forests have to be established.

In this survey of forest development and management it is not intended to deal with forest policy in the wider sense, for this would require the inclusion of institutional, administrative and legislative problems which are to be reviewed in a forthcoming FAO publication. The existence or desirability of such a policy within the framework of the institutions of each country is here taken for granted, and the discussion will concentrate on putting it into practice, in terms of time and space, especially as regards the product *timber*.

Some human and institutional problems

Employment and working conditions

The employment situation has changed considerably during the last 15 years. Whereas in the industrialized countries intensive mechanization has brought about an increase in labour productivity, labour productivity in developing countries has stagnated at a much lower level whilst the total number of workers in the wood producing sector has increased. An enquiry conducted by FAO shows that whereas in the northern countries the employment index (for logging and sylvicultural work) fell from 100 in 1960 to 60 in 1968, the index rose from 100 to 151 in Madagascar (sylvicultural work only) and western Malaysia (logging only).

In tropical countries employment in forestry is largely seasonal for planting work and a minority of workers are employed permanently, chiefly by large, well-equipped forestry companies. The turnover of personnel is considerable, doubtless because of the conditions of employment.

Mechanization is on the increase, particularly for the harvesting of tropical humid forests, where the building of roads, tractor skidding and lorry transport have made it possible to utilize large tracts which previously had been inaccessible. Power saws are being used more and more and heavy equipment plays an important role in forest clearing before planting. However there are cases where mechanization cannot be justified either socially or economically; a different approach, based on intermediate technology, is recommended.

The limitations to physical work brought about by climate have been the subject of a recent FAO (1974a) study. Heat stresses cause an additional strain on cardio-vascular activity because they induce a demand for a flow of blood towards the periphery of the body for the purpose of regulating body temperature. When the demands of thermoregulation (i.e. for heat dissipation) and those for oxygenated blood to be pumped to the muscles in action exhaust the capacity of the heart, the upper limits of tolerance are reached and work has to be reduced. But it is not just extra blood being diverted to the peripheral blood vessels; when the body temperature rises above 39° C there are several effects on the circulation, on the efficiency of sweat glands, and on the water balance of the body all of which reduce the capacity to perform muscular work. Therefore, the capacity to do hard physical work decreases quite significantly as the air temperature increases; an air temperature of *ca.* 26° C is the maximum tolerable when work has to be performed at levels that would be expected from forestry workers. This estimate implies a low humidity; if humidity is high working capacity is reduced by *ca.* 50 per cent. Working capacity in a hot environment depends also on the appropriate replacement of salt and water, as well as on individual characteristics, on acclimatization, etc.

This FAO study shows clearly that heat is a major problem in the organization and supervision of forest work in the tropics and for the well-being of workers; it stresses the need for further research into climatic conditions, and the performance and behaviour of workers, especially as regards their ability to replace the losses of water and salt resulting from transpiration. Workers in developing countries frequently differ in nutritional status, body weight, size and general health from workers in industrialized countries whose problems have been more thoroughly investigated, and this, too, makes further research a necessity (see chapter 16). Nor has the interaction of all causal factors been sufficiently investigated. There is no quantitative evaluation which is indispensable for the prediction of production costs and for the planning of development schemes. Moreover, since the tentative estimates used in this study to predict the reduction of output resulting from heat rely entirely on a model adapted from data found in the literature, it is quite possible that field observations would not agree with the predictions. Observations will have to be analysed for the modification of prediction methods which should be kept simple if they are to receive general application. It will also be necessary to study, for instance, the effect of air currents, the role of clothing and the relation between an increased heat radiation and reduction of output.

Whatever improvements are made, forestry work will always be hard and the risk of casualties will remain high. Though mechanization may reduce the physical effort involved, it brings other adverse effects on the cardio-vascular system (noise and vibration of power saws) and an increase in mental strain.

Apart from the pay of certain drivers of forest machines, the wages of forest workers are usually low; they equal the minimum wages of agricultural workers. Social benefits are very limited. Little has been carried out to adapt the length of the working day, the duration and distribution of rest periods, and the quantity and quality of food to tropical conditions. In the few cases which have been well documented, accident frequency appeared to be very high in comparison with industrialized countries. The constraints resulting from parasitic diseases should also be recalled (see chapter 17).

The present situation and trends of working conditions in developing countries were described in a report by the International Labour Organization (1973) to the second tripartite technical meeting for the timber industry, in Geneva. It can be summarized as follows:

In the developing countries, in spite of the low earning of timber workers, production costs still tend to be very high because of low productivity in all operations due to lack of training and efficient hand-tools; and because of extremely high transport costs due to lack of transport facilities. Frequently, these countries are also short of services such as public roads, railways or efficient water-ways. In some countries the climatic conditions make year-round forestry operations impossible. Yet forest resources are a real asset that could help to create jobs with little capital input. The timber industry offers unequalled opportunities, since it lends itself well to mass employment. The wood-consuming industries that are established after forest resources are tapped are estimated to generate three or four times as much employment as the timber industry. The most important investment is likely to be in the construction of forest roads. The next important step is the training of forest workers and the creation of the necessary conditions to attract permanent labour such as by planning for year-round employment, better wages, satisfactory housing and reasonable working conditions. Since the supply of labour generally is plentiful, it should be possible to select suitable workers.

Investments in exploiting forest resources and initiating an efficient timber industry are sound since forest land and standing timber are a solid security. It is essential to protect manpower and forests against unfair exploitation by appropriate and efficiently enforced laws concerning sustained yield, management of forest land, minimum wages, safety and health regulations, etc.

Certain forest operations can be carried out manually and many are mechanized only with difficulty. Below the level of mechanized work, there is ample room for large increases in productivity, safety and the saving of energy by improvements in working methods, tools and maintenance. Vocational training appears to be an essential step prior to mechanization. The latter may follow when rising wages make the use of machines economic, or where the terrain or the exceptional size of logs make manual labour too strenuous and transport is impossible. In general, the choice of technology, whether labour—or capital—intensive, has to be considered very carefully. If one of two technologies uses more labour with the same quantities of all other inputs, for a given output, that technique is clearly inferior. The real economic issue arises when, for the same level of output, the capital-output ratio can be reduced by increasing the ratio of labour to capital. Two principles have to be considered: the extent to which it is possible to choose labour-intensive products is limited by the composition of demand and the degree of substitution between products; and the freedom to choose alternative technologies is not present in all industries. In some, they can result in such an enormous difference in the quality of the product as to be of very limited use. But in some industries, there is a real choice of alternative technologies. The choice in the central production process may be restricted, but in subsidiary, peripheral operations, in particular those having little effect on the quality of the final product, labour intensive techniques can be used (e.g. the construction phase of a project, handling of raw materials, hand-tools versus power saws in logging, manual versus mechanical transport in sawmills). Before introducing operational methods based on a high degree of mechanization, it is imperative to investigate conditions thoroughly so as to be certain that the equipment which is to be introduced is adaptable to local conditions. The first step must be to teach the forest workers general working techniques (for manual work) through ordinary vocational training. A relatively high degree of mechanization must be arrived at gradually. However, it must be mentioned that in most cases the chain saw is now considered a hand-tool; it spread very rapidly as a result of the reduction in the number of loggers and a desire not to waste timber and time.

An intermediate technology is the use of elephants for timber extraction especially in south Asian forests. This has several advantages over manual labour and over large machines: trained operators are locally available; maintenance problems are few; damage to the environment is minimal; more rugged terrain can be harvested without erosion dangers. Animal-powered technology requires further study and probably more wide-spread introduction than oil-consuming machines.

Ergonomic investigations should be extended and should study the basic factors related to work. It is useful to make a distinction between ergonomic research (basic studies) and applied ergonomics (lay-out of work and environment); but both areas should be considered so that research will always be oriented towards the solution of practical problems. Basic research deals with the exploration of all relevant working, environmental and social factors with regard to human abilities, capacity and limitations. The most important factors relate to climate, and to the nutrition, clothing, physical capacity and endurance-limits of the workers. Such knowledge is needed as a basic information for the planning of work and training. Lay-out of work is essential at every working place and in every operational system and should deal with working procedure, organization and equipment with due consideration of the actual workers and this includes accident prevention.

Institutional problems

The major institutional questions which will have to be examined before a land-use policy and a forest policy can be formulated are: the ownership of forest lands, the status of the responsible administrations and the development of concessions.

In most cases forests are in public ownership (state, community, tribe, etc.) but there are privately owned forests and those where ownership is uncertain. In tropical America and in certain parts of Asia private ownership of forest land predominates (e.g. 93 per cent in Paraguay versus 13 per cent in Madagascar). State ownership is not always clearly established by real possession by the organs of public administration. Since decolonization there has even been a tendency to grant the claims by local communities to create communal forests or local authority forests. Some forests are protected temporarily or in a vague manner; their forest capability is recognized but they are also open to other uses; others are reserved for production, protection or as national parks whilst others are destined to be cleared. Finally there are forests planted by private companies or individuals; these are usually recognized as such, and their extent may be considerable, especially in tropical America (in Brazil, for example, they cover 4 000 km²).

The formulation of forest policy and the management of publicly owned forests is usually the task of a Forest Service which is normally attached to a Ministry of Agriculture, Natural Resources, or Rural Affairs. Sometimes the Forest Service has been combined with a Department of Tourism and raised to the rank of a separate ministry (e.g. in the Congo, the Ivory Coast, or Gabon). Wildlife and hunting may fall in its competence, but sometimes are dealt with by a State Secretariat for national parks, wildlife and nature conservation. The responsibilities of forest services have recently been increased. In countries with large forest resources, forest services need not only to improve their public functions and their assistance to private forestry but equally their capacity to manage and develop state forests where governmental activities in the past were rather limited. Many forest services have felt a need for decentralization which would not merely be of territorial organization but would ensure speedier and more efficient decision making at the field level. Some countries have created autonomous agencies or offices with responsibility for all forestry matters or for afforestation only. A tendency towards this form can be found in all regions (for example Indonesia, Malaysia, Nigeria and Gabon) but only in Latin America has it become a characteristic institution: a *Corporación Hondureña de Desarrollo Forestal*, a semi-autonomous state agency, was established in Honduras in 1974; an *Instituto Nacional Forestal* in Guatemala (1945); the Chilean *Corporación Nacional Forestal* is closely associated with regional planning; forestry plans have been successfully incorporated into the framework of the national plan of Colombia, etc.

Governments which need to intensify the exploitation of their resources can opt for giving timber harvesting concessions or forest utilization contracts. While the sale of wood at roadsides, and to a lesser extent the sale of standing trees, requires a considerable activity in logging, road construction and forest management on the part of the forest authority, utilization contracts involve the authority in much less technical work. Inadequate infrastructure, the lack of a sufficient number of technicians and professional foresters, the lack of a well-staffed national forest service and the considerable problems of skidding, road construction and marketing explain why the most common solution for forest resource allocation has been the granting of utilization contracts.

Basically, three types of contract can be distinguished: forest exploration, timber harvesting and forest management contracts. Such contracts do not free the government from all responsibility. It is necessary to determine appropriate provisions governing the rights and obligations of the contracting company which will implement the national wood disposal policy and safeguard future forest production. The government must inspect and supervise all operations of the company to ensure that the nation retains its full ownership of the land and receives its due share of the forest wealth. A national forest service able to undertake this task is therefore absolutely necessary.

A policy to enter into forest contracts has to be viewed in the context of the whole economic and social development of the country. Forests are not only a source of raw material for industry; they play an important role in the lives of people living in them (see chapter 19) and in the balance of the various ecosystems (chapter 2), and these facts must be appropriately covered in the provisions of a utilization contract. The responsible authorities must recognize the possibility of disagreement between local peoples and the contractor and ensure that these do not jeopardize the aims of the contract. In the last two decades, long-term utilization contracts have gained considerable importance. It is generally felt with good reason that they are an important way of using forest wealth and expanding wood industries and may attract new forest industries. With the benefits to national development that can flow from the establishment or expansion of an industry for which a forest contract is to be granted, there may be the temptation to settle for fairly easy terms in order to get an operation started quickly. Yet, because of the long-term and far-reaching effects of a utilization contract, it must be carefully and skilfully negotiated. It should be recognized that forests are becoming more valuable and, therefore, owners may be able to obtain more favourable contracts than past experience might suggest.

Negotiating and drawing up forest utilization contracts carries many responsibilities: for example ensuring that they are compatible with medium- and long-term management plans for the forest sector as a whole and with the long-term development interests of the nation, and that adequate institutional and staffing arrangements are made for the control of the execution of the contract.

This situation has caused FAO to be deeply concerned with forest utilization contracts. Its efforts have led to the publication of a handbook (FAO, 1972) which is intended to give forest administrations the information necessary for solving the major problems of forest utilization contracts. The first part of the handbook examines the nature of forest contracts and the various categories of public forest land on which they may be granted are described. The alternatives which are open to a government for utilization of the national forest resources and their relative merits are discussed. It explains why contracts form an important element of any wood disposal policy, why certain types of contracts (short-term timber harvesting contracts) should be considered as a temporary expedient, and why long-term contracts may be an efficient method of forest development and thus become an accepted form of forest organization, provided that the interests of the nation are safeguarded. It may be in a government's interest to use several methods of wood disposal and to adapt the conditions of contracts to the particular aims of forest management, the type of industry concerned, the legal status of the land and the type of forest. It also examines the major problems of legislation. The second part of the handbook is particularly concerned with contracts granted for 20 to 30 years including a government's aims, and the major matters to be included, the various fees which the grantees may have to pay and methods of assessing them. Measures of supervision and control, penalties and sanctions to ensure that the grantee will implement his obligations and responsibilities and comply with the clauses of the contract are mentioned. Finally, incentives that may be offered to invest international capital in forest industries are discussed. The third part lists the main conditions, requirements and provisions which should be included in long-term agreements or in legislation on forest utilization contracts and the appendices give three typical examples of long-term agreements.

At present about 90 per cent of all forest exploitation in tropical regions is carried out by forest concessions, but experience has not been entirely satisfactory because the conditions necessary for its successful functioning are far from universally present. Clear statements of national policies and general concepts of forest management are often lacking. There are still shortcomings in various aspects, such as land capability classifications, forest inventories, management plans, industrial planning and an adequate knowledge of exploitation and production costs indispensable for arriving at proper estimates of forest fees. Moreover experience in negotiating the clauses of concession agreements is still very insufficient.

Thus legislation on utilization contracts and individual agreements tend to be vague and imperfect because it is not known exactly what the forest can produce and at what cost, how it ought to be managed in terms of technology and economics, what should be the responsibilities of contractors and how the raw material could be processed in the country of origin.

Economics of forest products

Only wood and its derivatives will be considered; forest products provided by gathering will not be covered. In view of world market trends for wood, forest industries could play an important role in the development of those tropical countries that possess or can create forest resources by afforestation. A report by FAO (1967) has analysed the

driving role the forest sector can play in the economic growth of developing countries. This role is explained by the flexibility in the size of enterprises, the down-stream and up-stream economic linkages of the forest sector, the possibility of gradually employing new technologies and of profiting from a wide variety of qualified personnel. In addition, forest industries are supported by renewable national resources and are closely connected with the agricultural sector which plays an important role in developing countries.

In a subsistence economy, wood is often consumed locally and statistics of production and consumption are not always reliable. In the world-wide study presented by FAO to the World Forestry Congress of Madrid (1966) world consumption of wood was estimated as 2 131 million m³ of which fuel-wood alone accounted for 1 088 million m³. However the proportion of total consumption represented by fuel-wood varies considerably between regions. In Africa and Latin America, nine tenths of all wood consumed consist of fuel-wood; in Asia (excluding Japan) it is two thirds; in Europe and the USSR it is over a quarter, and in North America only one tenth. The use of wood as fuel remains the principal use in the developing countries. It was assumed that with the progress of urbanization and the spread of other sources of energy that are cleaner and easier to handle, the consumption of fuel-wood would decline; but the expansion of rural populations and the rise in the price of oil, will no doubt lead to an increase in world demand for fuel-wood for some time to come. The problem will be to ascertain how part of the wood that is now used as fuel can be made available for the manufacture of pulp or particle board, or timber.

Because they contain a high proportion of the world's wood reserves, and because the conditions are suitable for plantations, the tropical countries can hope to satisfy a large part of the future increase of the demand; this belief is plausible as (with the exception of the northern and eastern parts of the USSR and North America) the intensification of forest resources exploitation and their mobilization will be more expensive. The extent to which the tropical producer countries will be able to secure these markets will depend on raising production while keeping costs down. They will have to increase as far as possible the number of species they market, rationalize exploitation and transport, and export a greater proportion of their production in the form of finished products rather than logs. It also presupposes that the consumer countries will co-operate to the dissemination of new timber species, and by lowering customs barriers and other restrictions in order to facilitate the import of wood in the form of manufactured products. The United Nations conference held in New York in April 1974 stressed the need for co-operation between the two groups of nations. FAO, UNDP, UNIDO and UNCTAD considered the creation of a Bureau of Tropical Woods to contribute to a better knowledge of the resource and lesser-used species, and to sales promotion. The suggestion was taken up at the regional level and the formation of regional Bureaux, possibly with support from UNDP, is under study in Africa, tropical America and South-East Asia.

Trends and prospects of tropical broad-leaved woods

In view of the increasing importance of pulp and industrial wood for fibre and particle boards, saw and veneer logs represent a decreasing percentage of output. Total output increased by 2 per cent a year from 1954 to 1971, and the percentage of broad-leaved woods increased. Tropical broad-leaved (hardwood) logs represent a constantly rising proportion of the hardwood total, and the proportion of these which are either exported as such, or processed locally before being exported, is also rising.

Tropical broad-leaved wood

An estimate of wood from tropical moist forests is not readily available. However, an analysis of broad-leaved species removals of countries that are predominantly tropical closely approximates to the harvest of wood from the moist tropical forests, at least for industrial woods. Notable additions do, nevertheless, result from fuel-wood from the tropical savannas and logs from the dry deciduous forests.

In table 1, removals of tropical broad-leaved species during 1954–1973 are shown in relation to all broad-leaved species, conifers and total removals. It is clear that broad-leaved removals are becoming relatively more important and tropical woods are growing in importance. However, it must be recognized that a large portion of removals are for fuel, which is only crudely estimated, and that in this use category, particularly in tropical areas, broad-leaved species are much more important.

TABLE 1. World removals of wood (million m³) (Pringle, 1976)

	1954	1960
Total	1 745 (100%)	2 057 (100%)
Conifers	858 (49%)	947 (46%)
Broad-leaved	887 (51%)	1 110 (54%)
including tropical		
broad-leaved	472 (27%)	642 (31%)

	1970	1973
Total	2 389 (100%)	2 501 (100%)
Conifers	1 082 (45%)	1 122 (45%)
Broad-leaved	1 308 (55%)	1 379 (55%)
including tropical		
broad-leaved	829 (35%)	904 (36%)

Broad-leaved fuel-wood removals, for 1973, in predominantly tropical countries were estimated at 757 million m³, accounting for 30 per cent of all wood removals and its estimated share of all removals has been growing rapidly (by contrast coniferous fuel-wood removals were declining from 192 million m³ in 1954 to 174 million m³ in 1975). This may be largely the result of improvement in statistical estimating, but the figures do show how important this category is. The pressures in energy sources throughout the world as well as the fact that in many local areas fuel-wood is in critically short supply, stress the need to assess much more carefully the role of fuel-wood on both local and national bases.

Much more attention has been directed, at the world level, to industrial wood use and the role that wood from tropical regions plays in this. Table 2 parallels table 1 in showing, for industrial wood removals, a comparison of material by species group.

TABLE 2. World removals of industrial wood (million m³) (Pringle, 1976).

	1954	1960
Total	876 (100%)	1 036 (100%)
Coniferous	666 (76%)	769 (74%)
Broad-leaved	210 (24%)	267 (26%)
including tropical		
broad-leaved	57 (7%)	72 (7%)

	1970	1973
Total	1 276 (100%)	1 345 (100%)
Coniferous	909 (71%)	948 (70%)
Broad-leaved	367 (29%)	397 (30%)
including tropical		
broad-leaved	121 (9%)	147 (11%)

The much more important role of coniferous species is evident but it is clear that the broad-leaved species are growing in importance and that the tropical broad-leaved component is growing most rapidly. This is despite tropical broad-leaved species not yet playing a very important role in the most rapidly growing sub-category of industrial wood—raw material for pulping, particle board and fibre board. However, this is somewhat offset by the declining use of poles, pitprops, etc., in the temperate regions.

In the category of logs for sawn wood, veneer and plywood, shown in table 3, the increasing role of tropical broad-leaved species is more striking than the increase for industrial wood as a whole.

TABLE 3. World removals of saw logs and veneer logs (million m³) (Pringle, 1976).

	1954	1960
Total	557 (100%)	647 (100%)
Coniferous	415 (75%)	487 (75%)
Broad-leaved	142 (25%)	160 (25%)
including tropical		
broad-leaved	41 (7%)	49 (8%)

	1970	1973
Total	760 (100%)	832 (100%)
Coniferous	549 (72%)	594 (71%)
Broad-leaved	211 (28%)	238 (29%)
including tropical		
broad-leaved	87 (11%)	109 (13%)

TABLE 4. Wood removals in 1973 in relation to forest growing stock (Pringle, 1976).

Species group	Region	Growing stock (with bark) million m³	Removals			
			fuel-wood and industrial		industrial only	
			million m³	per cent of growing stock*	million m³	per cent of growing stock
Coniferous	World	107 000	1 122	1.1	948	1.0
	Japan	1 100	25	2.5	25	2.5
	Europe	9 500	212	2.5	200	2.3
	North America	27 000	380	1.6	377	1.6
	USSR	61 000	319	0.5	264	0.5
Non-coniferous (broad-leaved)	World	180 000	1 379	0.7	397	0.2
	Predominantly temperate countries	35 000	475	1.3	250	0.7
	Predominantly tropical countries	145 000	904	0.5	147	0.1
	Asia, Far-East	25 000	406	1.5	87	0.3
	Africa	42 000	269	0.4	31	0.1
	Latin America	78 000	229	0.3	29	0.04

* Coniferous growing stock reduced by 10 per cent to allow for bark; no allowance made for non-coniferous as most estimates exclude bark.

The reasons for these developments are well known. Among them are: a growing scarcity in temperate regions of large, evenly-shaped logs suitable for producing wide, clear and long pieces of sawn wood and particularly for slicing into decorative veneer or peeling into clear veneer sheets of a utility quality; and a rapidly growing deficient supply of any type in some heavily industrialized countries.

It is interesting to compare the 1973 wood removals by wood species groups with the corresponding estimated growing stock—the volume of standing timber: see table 4.

World broad-leaved wood

Table 5 provides an approximate model of the 1971 world supply and utilization of broad-leaved logs for sawn wood, veneer and plywood and their trade movements between major regions. A number of points are evident:
— about $\frac{4}{5}$ of the world's broad-leaved wood standing volumes are in the tropics;
— the temperate hardwood forests are much more intensively utilized and provide about $\frac{2}{3}$ of the world's industrial hardwood, but tropical hardwood logs now acount for more than $\frac{2}{5}$ of removals of logs of broad-leaved species;
— there was a net movement of 35 million m³ of hardwood logs from the tropical to temperate regions which was augmented by only 7 million m³ (in log equivalent) of exported processed sawn wood, veneer and plywood;
— imports of tropical logs were predominantly made by temperate countries of Asia while North America and Europe were the main recipients of processed tropical woods;
— as a result of these movements $\frac{3}{4}$ of all hardwood logs

were processed in temperate regions and nearly $\frac{4}{5}$ of sawn wood, veneer and plywood of hardwood species were consumed in temperate regions.

Movement and consumption of tropical broad-leaved logs

Tables 6 and 7 trace broadly the movement and consumption of tropical hardwood (broad-leaved) logs from producing countries to consuming regions with attention being given to countries which are indicated as consumer-processor-exporter countries. This last category includes countries which import logs for processing into sawn wood and veneer, substantial quantities of which are exported to other consumers. Producers of the tropics are composed primarily of countries of South-East Asia, west central Africa and the Amazon region. The consumer-processor-exporter countries in the tropics are predominantly Singapore and Hong Kong, while in the temperate region the most important are Japan, China and Korea. Predominant consuming areas are temperate Europe, North America, Australia and New Zealand. For completeness, this block also includes the Near East countries and the USSR as well as wood-deficient portions of Latin America, Africa and the Asian Pacific region.

Table 7 is shown in log equivalent so that the production column of this table coincides with the log consumption column of table 6.

Table 8 shows trade information on a similar basis but expanded to show detail for individual products and regions, as well as for the development from 1965 to 1971. It also postulates trade models for 1980 and 1990. The development of this model is based on a number of detailed

TABLE 5. Approximate model of world supply and use of broad-leaved logs, sawn wood, veneer and plywood, expressed in million m³ of logs or log equivalent in 1971 (Pringle, 1976).

Region	Growing stock (1)	Removals Industrial wood (2)	Removals total (3)	Logs only (4)	Net trade in logs (5)	Log supply to industry (6)	Production Sawn-wood (7)	Production Ply-wood (8)	Production Other veneer (9)	Net trade in processed wood Sawn-wood (10)	Net trade Ply-wood (11)	Net trade Other veneer (12)	Net trade Total (13)	Consumption Sawn-wood (14)	Consumption Ply-wood (15)	Consumption Other veneer (16)	Consumption Total (17)	Net trade total (18)
World	198 700	368	0.19%	212	—	212	169	36	7	—	—	—	—	169	36	7	212	—
Predominantly temperate countries	37 700	249	0.7 %	122	+35	157	120	32	5	+6	+1	—	+7	126	33	5	164	+42
Europe	4 500	77	1.7 %	36	+7	43	33	7	3	+2	+2	—	+4	35	9	3	47	+11
USSR	12 000	34	0.3 %	23	—	23	23	—	1	+1	—	—	+1	24	—	1	24	+1
North America	9 500	75	0.8 %	36	—	36	30	5	1	+1	+4	—	+5	31	9	1	41	+5
China	7 500	16	0.2 %	8	+2	10	8	2	—	—	-2	—	-2	8	—	—	8	—
Other temperate Asia	1 200	22	1.8 %	8	+26	34	17	16	1	+1	-3	—	-2	18	13	1	32	+24
Other temperate	3 000	25	0.8 %	11	—	11	9	2	—	+1	—	—	+1	10	2	—	12	+1
Predominantly tropical countries	161 000	119	0.07%	90	-35	55	49	4	2	-6	-1	—	-7	43	3	2	48	-42
Latin America	112 000	26	0.02%	21	—	21	20	1	—	-1	—	—	-1	19	1	—	20	-1
Africa	23 000	28	0.12%	16	-7	9	7	1	—	-1	—	—	-1	6	1	—	8	-8
Tropical Asia-Pacific	26 000	65	0.25%	53	-28	25	22	2	1	-4	-1	—	-5	18	1	1	20	-33

Net trade is shown by + for imports and — for exports.
Column (3) is calculated by expressing column (2) as a percentage of column (1).
Column (6) is derived by adding columns (4) and (5) and is then broken down to show the requirements for raw material in columns (7), (8) and (9) by the use of appropriate conversion factors; subsequently adjusted to total.
Column (13) is the sum of columns (10), (11) and (12).
Columns (14), (15) and (16) are derived by adding net trade to production for each product, e.g. column (14) = column (7) plus column (10).
Column (17) is the total of columns (14), (15) and (16) which is in turn equal to the sum of columns (4) and (18).
Column (18) is the sum of columns (5) and (13).

TABLE 6. Production, trade and consumption of tropical hardwood logs, in 1971 (million m³) (Pringle, 1976).

Region	Production	Exports	Imports	Consumption
Primary tropical producers	90	38	—	52
Consumer-processor-exporters	—	—	30	30
Consumers	—	—	8	8
Total	90	38	38	90

TABLE 7. Production, trade and consumption of processed wood from tropical hardwood logs, in 1971 (million m³, log equivalent) (Pringle, 1976).

Region	Production	Exports	Imports	Consumption
Primary tropical producers	52	7	—	45
Consumer-processor-exporters	30	7	2	25
Consumers	8	—	12	20
Total	90	14	14	90

studies listed in the bibliography but does not follow anyone of these in its pattern. The studies look at future demand and supply and the implicit trade flows. However, most of them concentrate on more than one or two regions and consequently do not provide a complete world model. The patterns shown for 1980 and 1990 must, therefore, be a compromise between the conflicting outlooks (sometimes already bypassed by developments more rapid than foreseen) of these various studies. Some of these studies did not go as far as 1990 or did not specifically identify tropical timber. Hence, a compromise, and frequent extrapolations, had to be made with subjective judgment. It should be noted that exports and imports for 1980 and 1990 have been made to balance roughly. This was done despite the fact that in the estimates based on demand and on supply considerations, discrepancies of as much as 10–15 per cent occurred in the world balances. However, it should also be noted that, even with the use of historical data, considerable imbalances may be found because of both time lags and the inaccuracies of the basic statistics.

A number of broad assumptions were made in deriving the outlook. It was recognized that utilization patterns must substantially change to a broader base of assortments of species and probably smaller sizes to permit the export and import of twice as much tropical log volume equivalent in the next two decades. It is assumed that a much larger portion of the material will be exported in the processed form and that not only consumers, but also the present consumer-processor-exporter group, will import less of

their requirements in the log form. Indeed, much of the group is foreseen as practically shifting into the consumer group.

The rapid current increases in price of most forest products, especially those for sawn wood, veneer and plywood, must have some effect on the relative position of timber and particularly of tropical timber in the future. Prices may well drop but not to earlier levels. Although these price developments are assumed to have some dampening impact on future demand, they will make harvesting possible in more remote areas and more particularly of species which have been less valuable. As increases in prices of temperate hardwoods, and especially softwoods, will continue, the possibility of these becoming substitutes for tropical woods is unlikely.

It is assumed that Japan, which since 1960 has reduced the proportion of broad-leaved species in her log imports from nearly 4/5 to roughly 1/2, will during the next few decades expand its imports of coniferous logs at approximately the same rate as that of broad-leaved logs, drawing on the resources of Siberia in this development, as the log exports of North America are not likely to increase substantially. It is assumed that additional regions of China will import tropical logs to a substantial degree and that a considerable expansion of the exporting base of this country will occur.

For 1971, domestic consumption of the primary tropical producing areas is estimated at 48 million m³ log equivalents. This figure is, however, subject to considerable error because of incomplete reporting. Indications of apparent consumption are derived by subtracting log and processed exports from estimated removals. During the first part of the 1960s this estimate of domestic consumption showed little change and declined slightly. However, from 1967 to 1971 it increased in all regions with a total gain of roughly 5 million m³, *ca.* 12 per cent in the four-year period. It is assumed that domestic consumption would increase to 60 million m³ by 1980 and 75 million m³ by 1990. This, together with suggested log exports, would increase the required output of tropical logs from 90 million m³ in 1970 to 124 and 162 million m³ in 1980 and 1990 respectively. A fundamental question underlying the trade estimates is whether the tropical forests can supply these increases. The key is the possible use of a wider range of species.

It is foreseen that exports from Latin America will expand very appreciably, but for processed wood rather than logs, because of both the complex forest structure and the proximity to sawn wood and plywood importing countries of North and South America. However, its total export volume by 1990 is still foreseen to be relatively small compared to that of other tropical forest areas. It is felt that the exports of western and central Africa must grow fairly modestly because of the growth of domestic demand and previous exploitation of much of its forests, especially the readily accessible portions. The most critical question raised by the model is whether South-East Asia, which already supplies 2/3 of the log material for the tropical hardwood products of importing areas, can continue to expand its output by some 25 million m³/a in each of the coming decades.

TABLE 8. Trade and trade prospects in tropical broad-leaved woods (in million m³) (after Pringle, 1976).

	Logs				Sawn wood				Veneer and plywood				Total log equivalent			
	1965	1971	1980	1990	1965	1971	1980	1990	1965	1971	1980	1990	1965	1971	1980	1990
IMPORTS																
Consumer areas	6.2	7.6	8.8	8.1	2.2	3.1	4.2	5.0	1.7	3.0	5.1	7.8	14.7	20.2	27.9	33.9
North America	0.2	0.1	0.1	0.1	0.4	0.5	0.8	1.0	1.5	2.3	3.5	5.0	4.8	6.3	9.0	12.1
Europe	5.4	6.8	7.5	6.0	1.1	1.6	2.0	2.4	0.2	0.5	1.0	1.5	8.1	10.7	13.5	13.8
Other temperate[1]	0.3	0.4	0.7	1.4	0.5	0.6	0.8	0.8	—	0.2	0.4	1.0	1.3	2.0	3.1	5.0
Wood importing areas of																
Latin America[2]	0.2	0.2	0.4	0.5	—	0.2	0.4	0.6	—	0.05	0.1	0.2	0.2	0.6	1.4	2.1
Africa[3]	0.1	0.1	0.1	0.1	0.2	0.2	0.2	0.2	—	0.05	0.1	0.1	0.5	0.6	0.9	0.9
Consumer-processor exporter areas	12.1	30.3	41.3	52.4	0.2	0.6	1.5	4.0	—	0.3	1.2	2.7	12.5	31.9	47.1	65.8
Japan	9.3	21.3	26.0	26.0	—	0.3	1.0	2.0	—	0.2	0.8	2.0	9.3	22.2	30.0	34.0
Other East Asia[4]	1.2	5.6	10.0	18.0	—	—	—	—	—	—	—	—	1.2	5.6	10.0	18.0
Near East[5]	0.1	0.2	0.3	0.4	—	—	—	—	—	—	0.1	0.2	0.1	0.2	0.5	0.8
Tropical Asia[6]	1.5	3.2	5.0	8.0	0.2	0.3	0.5	2.0	—	0.1	0.3	0.5	1.9	3.9	6.6	13.0
Total imports	18.3	37.9	50.1	60.5	2.4	3.7	5.7	9.0	1.7	3.3	6.3	10.5	27.2	52.1	75.2	99.7
EXPORTS																
Consumer-processor exporter areas	0.1	—	—	—	0.7	0.8	1.2	1.5	0.9	2.4	3.7	4.0	3.7	6.8	9.8	11.0
Japan	—	—	—	—	0.3	0.1	—	—	0.4	0.3	—	—	1.6	1.0	—	—
Other East Asia[4]	—	—	—	—	—	—	—	—	0.4	1.7	3.2	3.0	1.0	4.0	6.4	6.0
Near East[5]	—	—	—	—	—	—	—	—	—	—	—	—	0.2	0.1	—	—
Tropical Asia[6]	0.1	—	—	—	0.4	0.7	1.2	1.5	0.1	0.4	0.5	1.0	0.9	1.7	3.4	5.0
Primary tropical producers	19.1	37.7	50.3	60.5	2.1	2.9	4.5	7.5	0.8	1.1	2.6	5.0	24.9	45.1	64.5	87.9
Latin America	0.5	0.3	0.3	0.5	0.3	0.5	1.0	2.5	—	0.2	0.5	1.5	1.2	1.4	3.3	8.5
Africa	5.2	7.0	5.0	4.0	0.7	0.6	1.0	2.0	0.1	0.3	0.8	1.5	6.8	8.7	8.6	11.0
Asia-Pacific	13.3	30.4	45.0	56.0	1.1	1.8	2.5	3.0	0.7	0.6	1.3	2.0	16.9	35.0	52.6	69.0
Total exports	19.2	37.7	50.3	60.5	2.9	3.7	5.7	9.0	1.7	3.5	6.3	9.0	28.6	51.9	74.3	99.5

1. South Africa, Australia, New Zealand, North Africa, Near East, USSR.
2. Mainly Argentina, Uruguay and Cuba.
3. Mainly eastern Africa.
4. China and Republic of Korea.
5. Israel.
6. Singapore and Hong Kong.

The European case

So far as tropical hardwoods are concerned, the tentative estimate is that supplies to Europe will be 25–35 million m³ wood raw material equivalent (WRME) by the year 2 000, compared with around 10.5 million m³ WRME in 1969–71 and a temporary peak of about 16 million m³ WRME in 1973, from which the volume of imports has subsequently fallen back considerably. The rates of growth inferred by these forecasts are an average of 3–4 per cent/a between 1969–71 and 2 000, compared with 5.5 per cent/a between 1959–61 and 1969–71. During that decade, tropical hardwoods increased their share of Europe's apparent consumption of total industrial wood from 2.6 to 3.1 per cent. Based on consumption projections, the tentative estimates are that by 2 000 the share would, under the lower assumption, remain at the 1971–73 level of 3.4 per cent and under the more favourable assumption, continue to increase at the rate of the 1960s to attain about 4.5 per cent of the European market for industrial wood products by the end of the century.

The ability of the European market to absorb volumes even somewhat higher than the estimates presented above would probably not be a limiting factor under conditions of ready availability and competitive prices such as ruled during the 1960s. The more cautious estimates take into consideration certain constraints on the availability of supplies from the tropics. These include:

— the era of timber mining and creaming of the best commercial species and qualities, which was practised in at least some of the major tropical producing countries, is passing; more importance is being attached to the principles of sustained yield, involving a controlled rate of exploitation;
— unless the progressive degradation of the tropical forest from excessive shifting cultivation can be checked, the area of forest available for commercial wood exploitation could be reduced;
— as exploitation moves away from the most accessible forests, costs will rise;
— growing demand in the tropical countries and in other export markets could hold back the availability for European market.

Uses of tropical broad-leaved woods

Primary uses

Tropical hardwoods have widely differing properties which allow them to be used in a great many different ways, a number of which are quite specific. The relation between uses and wood properties is very close and has supported various marketing campaigns for tropical woods.

The greater part of all tropical woods is used in the round, chiefly for fuel. The estimate of 745 million m³ of tropical hardwoods used for fuel is almost certainly an underestimate. Whereas consumption of fuel-wood is decreasing in the industrialized countries, it is rising in the tropical countries, though at a slower rate than the rate of

population increase; this fact will have to be taken into account in future planning.

Of the *ca.* 125 million m³ (in the early 1970s) which appear in felling statistics as timber and industrial wood, *ca.* 2/3 (86 million m³) are used in their countries of origin either as wood in the round or after conversion (as sawn timber, sleepers, plywood and veneers), and the remaining third (*ca.* 39 million m³) is utilized in the industrialized countries usually after conversion. The situation is complex and difficult to analyse because the tropical countries export to the industrialized ones part of their sawn timber and sleepers (11 per cent) and of their plywood and veneers (45 per cent) but they also trade in roundwoods among themselves and with other developing countries and a proportion of that wood is then re-exported in the form of sawn timber and plywood.

The industrialized countries use the largest quantity of tropical wood (in terms of roundwood equivalent) in the form of plywood and veneers, with sawn wood use not much less. Relatively insignificant amounts of wood in the round or roughly squared are used in these countries. The tropical countries, on the other hand use relatively large quantities of wood in the round and relatively small quantities of plywood and veneers, a pattern which may be correlated with their stage of economic development. The contrast between the consumption patterns according to type of product is shown in table 9.

TABLE 9. Main product groups' shares (based on roundwood equivalent) in the utilization of industrial tropical hardwoods (Pringle, 1976).

	Tropical countries	Rest of world
	(percentages)	
Wood in the round or roughly squared	41	3
Sawn wood and sleepers	53	40
Plywood and veneers	6	57
Total	100	100

End uses

The end uses of tropical woods depend on their properties which determine their commercial classification. Three grades are sometimes distinguished: decorative, utility and construction woods, but for many species multiple uses are possible. Woods with decorative properties command a higher price than construction woods, which in turn are sold at a premium compared with utility woods. This classification applies chiefly to sawn timber, but also to plywood. In the case of the latter, the type of adhesive used is the decisive factor and in Japan, where all plywood is made from tropical woods, one distinguishes between exterior grades resistant to fresh and sea-water, interior grades resistant to water, and interior grades resistant to humidity. The development in the use of decorative species is associated with a higher standard of living and a desire for a high

TABLE 10. Summary of end-uses for tropical hardwoods in the mid-1960s (percentages).

	Furniture	Construction[1]	Transport	Ship- and boat-building (including repairs)	Packaging[2]	Hydraulic works	Other[3]	Total
Germany (Federal Republic of)	50–55	36–40	2	2	(........	1–2	5–7)	100
United Kingdom	40	40	3–5	5–7.5	(......................		7.5–10)	100
France	40	36–41.5	2–3	5–6	1.5–2	(......	7–10)	100
Netherlands	40	27	3	2–3	4–5	17	5–7	100
Belgium	45	35	2–2.5	4–5	2	(.......	8–11)	100
Switzerland	40	47–53	1–2	1	1–2	(.......	6)	100
Sweden	40	35–40	3–5	10	(.....................		8–10)	100
Norway	35–40	32–33	2	20	(.....................		5–7)	100
Weighted average (8 countries)	43–45	37–40	2–3	4–5	(.....................		9–12)	100

1. Includes door-making and parquet;
2. May include shuttering in some countries;
3. Where percentage is not shown for packaging or hydraulic works, it is included under other.

Source: Study on the consumption of tropical hardwoods in Europe, suppl. 9 to vol. 19 of the *FAO/ECE Timber Bulletin for Europe*, February 1967, p.29.

quality environment; these woods are however in constant competition with synthetic materials.

The methods employed to determine end uses depend not only on the objectives pursued (commercial introduction, marketing, products-research and development, planning of production, industrial development, utilization of lesser-known species) but also on the funds available.

Table 10, taken from a study of the FAO/ECE Timber Committee for Europe (1967) summarizes the percentages of end uses for tropical woods in the 1960s in eight countries.

Development of lesser-known species

It is estimated that there are 650 little known species (*ca*. 100 in the tropical moist forests of Africa, 250 in those of South-East Asia and 300 in those of South America) the wood of which could in all probability be used for commercial purposes. Since utilization potential depends closely on wood properties, and since the lesser-known species are at present arousing interest in the whole world, it would be desirable to establish an international standardized system for evaluating the utilization potential of these species.

The promotion of the various species is based on practices which have long been established in relation to utilization of and commerce in wood. Recently a start has been made in grouping together certain species with similar or identical characteristics for the purposes of trade. Measures envisaged to establish classification system and trading customs, or to improve existing ones are of two kinds: finding similar uses for useable woods so as to group together as many as possible of the lesser-known species which lend themselves to such uses, and preliminary screening to select species which appear to possess a higher potential which are subjected to special studies designed to optimize their utilization potential.

It will be possible to use the wood of many little-known species in a disintegrated form, such as chips and fibres. Hence it will be necessary to direct considerable efforts of commercial promotion towards a search of outlets for such products. Promotion of an integrated industrial system of utilization for lesser-known species will thus have to aim at a diversification of products and their acceptance by the trade, especially in local markets.

New uses for tropical broad-leaved woods

New production techniques are either in a stage of application or study and new uses for tropical woods are being introduced as a result of assays made and the development of new processes for the treatment of wood.

Wood-based panels

A world-wide consultation on wood-based panels was held during February 1975 in New Delhi. It studied the present situation and trends existing in the manufacture of veneer, plywood, fibre and particle board. Even though the consultation found that the consumption of panels ceased to increase or declined in many important countries in 1974 and 1975, the spectacular increase in the consumption of particle boards which rose from 6 million m³ in 1963 to 27 million in 1973, and the equally massive increase in the consumption of fibre boards which rose from 11 million m³ in 1963 to 17 million in 1973 should be noted. The increase in the consumption of panels is likely to continue provided that the economy of the world recovers; but the supply of wood might be insufficient in certain regions. The value of fibre and particle boards may be enhanced by certain techniques such as veneering and laminating.

An example was cited of exceptionally heavy wood (0.9 density) being successfully used to produce construction-grade plywood with superior strength properties. There was great need and scope for adapting existing techniques and developing new ones to fit operating conditions in developing countries. This task was recognized as being of prime importance, since a large share of raw material reserves are located in tropical countries and will increasingly be converted locally into finished products. On the other hand, there was full agreement that while there was a need for small-size wood-based panel operations, particularly for plants based on domestic markets only, economic criteria normally set the limit for the optimum minimum size of a plant. To improve this by the inclusion of export markets was often difficult because of high transportation costs. Manufacturers of equipment should therefore take steps to develop techniques and machinery suited for small-scale operations. It was agreed that the potential for smaller sized mills, particularly veneer and plywood for local consumption, and the possible integration of small-scale particle board and fibre board mills is not fully recognized. The consultation recommended that FAO investigates the question of small mills in developing countries taking into account markets, raw material supply and labour intensive technology. Veneer and block-board have been identified as products in their own right and small veneer plants present a logical first step, at lower investment than plywood, towards establishment of a wood-based panel industry.

Pulp and paper

The utilization of various tropical hardwoods for the manufacture of pulp and paper has become important chiefly because of the amounts of mixed hardwoods which became available with forest clearings. Already some tropical mills use certain tropical hardwood species for the integrated production of pulp and paper; but though a limited number of species are being utilized, there does not yet seem to exist any large-scale production based on the utilization of all available species. This was discussed by the FAO Committee on forest development in the tropics (FAO, 1976).

Research by the Centre technique forestier tropical (CTFT) conducted for some twenty years in the laboratory and recently also on an industrial scale, has yielded positive results. This is also true of trials carried out in Sweden on an industrial scale by the Swedish Stora Kopparberg which indicate that pulp made from mixed tropical hardwoods from Gabon is at least as good as pulp made from birch or other European hardwoods; the problem is an economic one, since it is difficult to estimate the harvesting costs in tropical humid forests. A statistical analysis of two pulps made from mixed Gabonese hardwoods derived from 11 exploitation units distributed over an area of 1 000 km², showed that day-to-day variability of their properties did not exceed 5 per cent of their mean values at a probability level of 0.95. A similar study has been carried out with mixed hardwoods from the Ivory Coast. The elimination of undesirable species (those with latex, high silica contents, etc.) and organization of harvesting are necessary to minimize the variability in pulp quality resulting from the diversity of the species used and the patchiness of their distribution. Comparative trials on undesirable species,

conducted by the CTFT, not only in the French-speaking countries of Africa but also in Surinam, New Guinea, Kalimantan, Malaysia and in French Guyana, have shown that these represented no more than 2–6 per cent of the total volume.

There appears to exist a potential demand for a short-fibred pulp which can be produced from a mixture of tropical hardwoods. It is of great importance to avoid the problems which would arise if, once such pulps are put on the market, they vary too greatly in quality. One solution would be to keep fairly large buffer stocks at the mill; this would make it possible to mix the various species in roughly constant proportions. Or exploitation could be so organized by different units working in different localities, so that overall deliveries would always remain fairly close to the average composition of the forest.

Projects of installing pulp-mills in Cameroon have been reported and there is already a mill working in Colombia. Gabon which has in 1976 the second world plywood factory (75 000 m³/a), has built at Kango, 100 km from Libreville, a cellulose-mill which is to produce 250 000–300 000 t/a of dry bleached pulp.

Forest development and management

In the preceding sections, the human, institutional and economic framework of forest development and management has been examined, as well as the new techniques which may aid the economic development of forests. The planning of forest development will now be treated at the national level and the management of forests at the level of the enterprise or the management unit.

Planning

It is of course undesirable that more resources for forestry mean fewer resources for other activities and that, because the forest sector is not the only venue towards the achievement of government objectives, forestry objectives cannot be defined in isolation even though, in many tropical countries, the forestry solution is the only one possible. With the economic aspects of growth, the availability of capital, land and labour, the industrialization in progress, the desirability of earning or saving foreign exchange, the situation and prospects of employment and unemployment, and the quality of the environment all aimed at, it will be necessary to:
— define the aims of the forestry sector;
— analyse the different policies and strategies available for achieving these aims;
— decide on the policy to adopt, and formulate a programme to put it into practice.
This requires a clear definition of the forestry sector (which may or may not include the forest industries); the construction of a framework for analysis, a system of accounting and the collection of the necessary information. Next there will be a need for analysing and predicting the needs of forest products, including exports, as well as the resources, and for adjusting supply and demand. Finally the feasibility of

possible choices will have to be examined and their contributions in terms of economic growth, employment, and savings or earnings of foreign exchange evaluated.

Environmental impact statements

It is clear that the limitations of development inherent in the socio-cultural situation and the attitudes of populations (who do not view all problems in the same light), and the protection of the environment have to be taken into account. This concern for the environment, and in some cases for the quality of life, is no more than a recognition of the indirect benefits of the forest which should be quantified and evaluated in order to determine the costs and benefits of any project.

In some countries each project now has to be accompanied by a statement on its environmental consequences. The idea was started in the United States in the form of *environmental impact statements*. In France, administrative measures have been taken in 1976 to define the content of such statements and to establish an *Atelier central d'environnement* composed of architects, town-planners, engineers and ecologists of high standing to examine the impact statements submitted. Each statement has to comprise four parts: justification of the project and alternatives, an analysis of the environment, consequences of the proposed management for the environment, and a financial estimate of ecological damage. Tropical environments which are so far only slightly industrialized have a much greater capacity to absorb harmful influences and pollution than industrialized environments, which are often already saturated.

Methodology of planning and demand forecasting

An *introduction to planning forestry development* (FAO, 1974) summarizes the principal aspects of development planning, and especially the procedures and organization; it deals with the formulation of objectives for forest development, with the analysis of the forest sector, and particularly with the forecasting of needs and resources, the evaluation of projects and with methods of estimating investments.

FAO and many individual countries have studied the methodology of demand forecasting and the methods that have been elaborated will be employed in studies on the future consumption and production of wood. This methodology is based on the following principles.

The usual approach to demand forecasting is to try and identify the factors which have caused demand to change, to quantify these relationships, and then extend them into the future with such modifications as are seen to be necessary. Export requirements have to be treated separately from domestic requirements, and, where much of the latter is subsistence demand (e.g. pole sand fuel-wood), it may be desirable to treat that part separately from commercial domestic requirements. The methods used to forecast requirements for forest products are not different from those used to analyse and forecast the requirements for other commodities. Usually the methods are simple,

because there are insufficient data available to warrant the use of sophisticated and elaborate methods. The key to producing reasonably accurate forecasts lies as much in the analyst having a sufficient knowledge of the product, and how it is used, to be able to interpret the evidence of what has happened in the past, and the likelihood of particular relationships being extended unchanged into the future, as in the particular mathematical or graphical methods used. It is important to remember that forecasts are no more than one element in the analysis, and quantitative projection techniques are only one tool for making forecasts. It is also important to remember that forecasts reduce the limits of the uncertainty that always attaches to the future, but do not eliminate it; forecasts will never be infallible. Because of this uncertainty it is desirable to present more than one forecast.

The most commonly used forecasting methods are indirect, and relate change in wood use to change in other factors which can be measured and predicted. A widely used method is to relate change in wood use to change in time; i.e. to study how much consumption grew over the last, say, ten years and to use this as a basis for predicting how much it will change over the next ten, or more, years. But it is not the passage of time that causes consumption to change; there are other factors which themselves tend to change through time. The principal demand shifting factors are changes in population, income, prices and the technology and tastes of the users. It is therefore preferable to attempt to predict future requirements by relating them to these factors. As the evolution of population and national income are nearly always estimated at the macro-planning level, this approach has the added advantage that the forecasts of demand for the products of the sector can be directly linked to what is being assumed, or planned, for the economy as a whole. In practice mathematical or graphical analyses and projections of forest products consumption tend to be confined to the relationship between change in consumption and the changes in population income, or a particular industry which uses a wood product. The effects of price changes have been largely ignored; most such forecasts thus have been based on the implicit or explicit assumption that forest product prices will not change relative to other prices. However, where it is clear that relative prices of forest products will change, these changes must be taken into account. The effects of changes in technology, habits and tastes on demand for wood are difficult to quantify, but are often very important determinants. Most forest products are inputs to other industries rather than final goods. Their demand is therefore very affected by usages and technologies in these other industries, as well as by changes in demand for the products of the latter. The use of sawnwood in housing, for example, has been enormously affected in industrially advanced countries by the shift from single houses to apartment blocks; by the shifts to building with concrete cast *in situ* and then to the use of dry pre-fabricated building unit systems; and by the application of engineering principles to the design of such structures as roof trusses. Where changes in technology and user practice become such important determinants of use, they can no longer be subsumed in the income-consumption, or price consumption, relationships; they must be taken into account separately, usually on the basis of subjective judgment.

Forestry planning is faced with certain problems such as the forecasting of demand and supply, the marketing of the products, the fixing of prices for standing timber and the analysis of investments; to these will have to be added the lack of data on productivity of labour, the difficulties in assessing the accessibility of forests, the complexity of the problems inherent in the manipulation of forest ecosystems, the degree of success of conversion plantations and forest enrichment schemes, the economies of scale of different operations, etc. All these are found in a more precise form at the level of management which should be concerned with the application of the national forest development plan at the level of the enterprise or management unit.

In view of the uncertainties of sylviculture and of natural and artificial regeneration techniques, the absence of an infrastructure for logging and transport, the difficulties inherent in estimating the costs of harvesting, and finally the multiplicity and variability of end uses, is the formulation of a development plan which can be translated into a system of sustained yield at the level of the different management units possible? Would it be better to organize production in the whole of the territory in a way which would ensure for the forest industries a continuous, or, where necessary, increasing flow of raw material, while safeguarding the production potential of the forest, rather than trying to organize a sustained-yield production at the level of management units, the life expectancy of which is not even fixed with regard to the medium-term and long-term forecasts of economic and social growth? On the other hand, the responsibility of the forest planner is not to propose a plan which relates merely to certain functions and productions of the forest; his duty is to present various possibilities while combining various assumptions and possible schemes of utilization, and assessing their influence on certain indicators of economic development.

The role of forests for the rural communities

Apart from the importance of forest development at the national level, the importance of forests for the daily life of rural communities needs consideration. Its chief purpose is the supply of fuel-wood (the main source of energy for rural populations), the supply of which may be regulated or free, accessory products gathered in the forest, and fish and game. The list of tubers, fruits and leaves used as food by local populations is very long, but much further research remains to be done (see chapters 16 and 19). The importance of wildlife was stressed in chapter 20; to this should be added fish production in freshwater mangrove swamp forests and the breeding of *Tilapia* fed on vegetable matter.

Farm forests, as being developed for instance in India, are a very important contribution to providing the whole agricultural needs of the community by integrating a variety of crop plants, domesticated animals including poultry, fish, and timber for fuel.

This vital contribution of forest products to the satis-

faction of the needs of people living in tropical forest ecosystems, that is to say the food-gatherers but also the shifting cultivators, has been discussed in chapter 19.

The Malaysian case

Among the few methodological studies dealing with the analysis of the forest sector, a model of forest development in Malaysia which has been elaborated within the framework of an FAO/UNDP project will be described. It starts from a highly complex situation since it was necessary to take into account both the pace of forest clearings for agriculture and the periodicity and/or intensity of fellings in the remaining forests.

The major objective was to prepare, separately for peninsular Malaysia and Sarawak, strategies for the long-term development of their forest sectors, with particular regard to the development of the primary wood-based industry. In order to facilitate the formulation of appropriate strategies, the project undertook intensive investigations into the quantity, quality and distribution of the forest resource, the cost and productivity of forest harvesting, the structure and operations of the forest industry, the marketing of forest products in both domestic and international markets, land-use policy, and into the institutional and infrastructural framework for the development of the sector.

The model itself was developed during 1971–73, but work on selected components (transportation model), revision of the input data, and refinement of the strategy selected for final adoption continued into 1975. Perhaps a succinct description of the model would define it as an *articulation of computational routines* which, for any given set of the values of selected policy variables (e.g. the rate of conversion of forest land to agriculture and the application of given forest management regimes), calculates for specified future periods such things like the timber output, its distribution to industries and markets, the current and investment costs of each operation, the sales revenues, etc. The model is computerized.

The analysis of alternative forest sector strategies with the aid of the model is carried out in the context of the wider perspective of overall economic and social development, with particular reference to the interrelationships between the forest sector and agriculture and transportation. First, the model provides estimates of the impact of alternative forest sector development strategies on a number of economic and social policy and planning criteria, e.g. employment, contribution to GNP, foreign exchange earnings, government revenue, etc. Secondly, the process of strategy evaluation and selection is based on a continuous dialogue and interaction between the policy-maker and planner-analyst. The model is used as an efficient aid in such dialogue. Several rounds of consultations are held, as alternative strategies are simulated on the computer and the outcomes are discussed and evaluated. This process of interaction between the policy-maker and the analyst (and the machine) continues until there emerges a consensus on the one strategy alternative which comes closest to the preferences of the policy-maker.

The objective is the formulation of a strategy for the rational exploitation of the forest resource. The method of analysis does not determine the *best* strategy. It provides certain indicators and criteria which can be used by the policy-maker in deciding which of many alternative strategies comes closest to his preferences. To fulfill this task the analysis estimates the effect of alternative strategies on a number of policy criteria such as employment, foreign exchange earnings, capital formation, government revenue, costs distinguished by cost category and operation, sale revenues, etc. The method of analysis is based on many runs of a sequence of computational operations which produce estimates of such effects. Each run simulates one alternative strategy. Fig. 1 provides a simplified flow chart of this sequence of operations. Any given strategy is defined as one combination of the rate at which forest land will be converted to agriculture and the intensity and frequency with which the permanent forest will be harvested. Once the values of these variables have been fixed, there follows a sequence of calculations to estimate the timber output, the costs of the different operations, the sales revenues and the effects on the other policy criteria, e.g. employment, balance of payments, government revenue, etc.

More concretely, values for the following policy variables are determined for each strategy alternative and used as inputs into the calculations:
— the rate of conversion of forest land to agriculture determines also the total area to be left as permanent forest;
— the harvesting intensities to be applied to the land earmarked for conversion and to the area to be harvested in the permanent forest; this variable is specified in terms of the species and size groups of the trees to be felled; therefore application of any given harvesting intensity implies the cutting of all trees belonging to the specified species and size groups;
— the cutting cycle to be applied to the permanent forest; this divides the total permanent forest area into the area to be logged each year.

The forest resource data available from the forest inventory and other sources are organized and stored in the computer. The basis for organizing the data is the use of a grid system which divides the entire forest area into equal squares (grid areas) of approximately 20 700 acres (8 280 ha) each. For each grid area several data are identified and stored in the computer, e.g. timber content distinguishing the species and size of trees, geographical location, distance from the nearest road or other transport facility (existing or planned), etc.

Application of any given combination of values of the three policy variables (land conversion rate, harvesting intensity, and cutting cycle) produces a number of computational results.
— The forest area to be converted to agriculture by 5-year period and location. The area to be converted in each 5-year period is given exogenously (value of the land conversion rate variable). The *location* of the areas to be converted in each 5-year period is derived from the calculations on the basis of a criterion of minimum distance from the roads, i.e. the grid areas nearest to

the roads are selected for conversion in the first 5-year period, the next nearest for the second period and so on. Only areas suitable for agriculture are selected.
— The total permanent forest and the part of it to be harvested in each 5-year period, together with the location of the areas to be logged. The *total permanent forest* is derived as the difference between the total forest area available for exploitaion and the area earmarked for conversion to agriculture. The permanent forest *acreage to be harvested* per year (or 5-year period) is obtained as the ratio of the total permanent area divided by the cutting cycle. The *location* of the permanent forest area to be logged in each 5-year period is derived on the basis of the minimum distance criterion.
— The total timber output to be obtained from the harvesting of both the areas earmarked for conversion and the permanent forest earmarked for logging. The *timber output* is derived from the application of the harvesting intensities (one for the land to be converted and one for the permanent forest) to the timber content data of the grid areas selected for logging and/or conversion per 5-year period. The timber production is in terms of size and species of timber to be produced in each 5-year period.

The results derived so far are subsequently used as inputs into a second phase of calculations which estimates for each alternative the costs and benefits to the forestry sector and to agriculture as well as the values of selected policy criteria (employment, foreign exchange, etc.).

The estimates of costs, revenues and profits of any given strategy represent only one part of the results which the policy-maker takes into account in the process of strategy evaluation and selection. The model was designed with the express purpose of estimating the values of a number of equally important indicators which are used by the policy-maker in assessing the implications of any strategy for overall economic and social development. These indicators are:
— employment,
— value added, total and per person employed,
— contribution to GNP,
— capital generation and requirements,
— foreign exchange earnings and expenditure,
— government revenue and expenditure,
— rate of return on investment,
— land made available to agriculture,
— quantities and prices of forest products in the domestic market,
— most favourable product mix in the forest industries.
Some of these are straightforward in their meaning. The nature of some others needs further clarification. The analytical framework used has the potential of identifying the effects of alternative strategies on these indicators with varying degrees of reliability. For example, it can identify readily the *land made available to agriculture* under each alternative strategy but it can provide only very vague answers to the question of what will be the employment created in other sectors or the rates of return on investment.

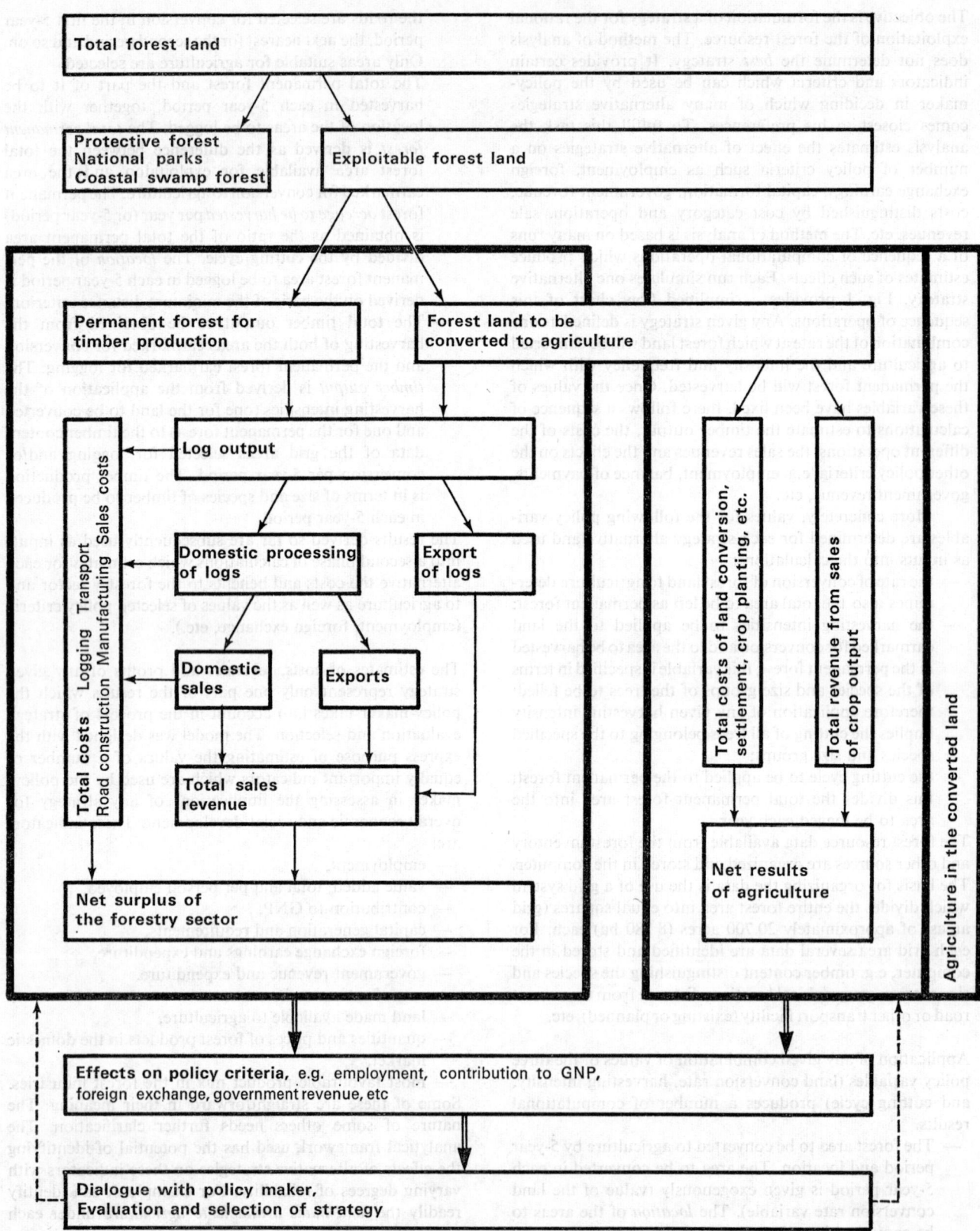

Fig. 1. Simplified flow chart of estimation procedure of one alternative strategy.

The rules which the policy-maker uses in utilizing the information contained in the above indicators for strategy evaluation and selection are not specified. Knowledge of such rules would certainly help the analyst produce values of the indicators consistent with such rules, e.g. if some of the indicators were to be used as inputs into an economy-wide analytical framework the values of the indicators should conform to the input specifications of such analysis. Current work is directed towards the development of such an economy-wide analytical framework. Therefore eventual revision of the forest sector analysis would benefit from coordination.

Many efforts have been made for the further elaboration of the results of the analysis and for making the strategy more concrete, bringing into the picture additional points of view and making suggestions for implementation in the context of the institutional and political framework of Malaysia. The following points could be gainfully explored in further model development work: time horizon of the analysis; selection of areas for logging; disaggregation of the results of analysis and the economics of the analysis. Time horizon is important to estimate the comparative rates of return on investment, or to examine the comparative merits or demerits of land use in agriculture or forestry.

Furthermore some decisions may require knowledge of the evolution of certain variables, notably the level of timber output at the national and regional levels over a period well beyond the 20-years time horizon. This is the case of decisions concerning the level and location of such long life investments as ports and industrial processing facilities. For the selection of areas for logging the analysis presumes that all grid areas which meet the minimum distance criterion in any given period will be selected for harvesting in that period, irrespective of their timber content, terrain morphology or distances from sales outlet. However the profitability of logging any given area is significantly affected by these factors, particularly timber content, and some improvement must be made to inject some more economic calculus into the area selection procedure. An economic criterion would need to be based on the timber removed from any area under the harvesting intensity of a given strategy rather than on the timber existing in the area. The results of the analysis are presented as agregates for the entire peninsular Malaysia and for all operations. However, the political and institutional framework suggests that disaggregation of the results in terms of regions or states and in terms of the actual structure of enterprises of the forest and forest industry sector, would enhance the value of the model as an aid to policy-making. As regional development is an important issue in economic and social planning and as the central planning authorities would certainly need to know the regional development implications of the strategies, there is a need for the model to explore the likely evolution of timber flows by region over the longer term.

The disaggregation of the forestry sector activities to correspond with the existing or foreseeable structure of the enterprise would be needed for the investigation of the levels of royalty rates which are imposed at the level of the logging enterprises rather than at the final point of sale and similar aspects. Finally the evaluation of the alternative strategies depends to a large measure on the values assumed by a set of indicators or policy criteria. One of the key indicators is the net surplus of the forestry sector over its own investment requirements. Such surplus is implicitly taken to represent a net benefit to society since all costs incurred for its production have been deducted. The approach employed for estimating the costs would correspond to a *financial* rather than *economic* analysis in a project evaluation context. Costs to society are assumed to be represented by the forecast market prices of the factors or production employed for the generation of this surplus. Economic analysis would require that such costs be estimated in terms of what it costs society to release these factors from other uses, i.e. their opportunity cost. Thus, in a situation of sizeable unemployment the opportunity cost of unskilled labour may be below its market price while management may have an opportunity cost higher than the market price. It is not suggested here that the analysis should have attempted to estimate shadow prices, since this may well be an impossible task. One needs, however, to interpret this estimate of surplus bearing in mind that it might have been different if socio-economic costs rather than private-market costs had been taken into account.

However, whatever improvements may be made, a certain imbalance between the various elements of the model is bound to persist. Certain functions or activities are studied in great detail because they are well known. Others are much less elaborately dealt with either because their representation in a model is more difficult or because there are insufficient data available. One of the excellent qualities of the model is its flexibility which makes it possible to take into account constructive criticisms. The concern for trying to represent reality should sustain or improve the model's predictive capacity. In any case the synthesis embodied in this model is an achievement of major importance. It succeeded in integrating a number of partial studies into a general understanding of the forest sector and a clarification of its links with the agricultural sector.

The Venezuelan case

Very different from the case of Malaysia is the Venezuelan case which shows the various kinds of problems and difficulties in rain forest exploitation and development (Hamilton, 1976).

It was estimated that *ca.* 78 per cent of Venezuela's forests were rain forest. Data for 1959 given in the *Forest Atlas of Venezuela* (1961) show 370 730 km^2 of rain forest. Of this, 314 400 km^2 were in the State of Bolivar and Amazonas Territory, that is, south of the Orinoco river where almost all the forest is in public ownership. Most of it is still relatively intact, although there has been some clearing. There are some large clearings along the boundary with Colombia south of the river. Some of this has become *de facto* private land by virtue of the clearing for agriculture and mining. Perhaps 3 per cent of the forest is privately owned, all in the State of Bolivar.

South of the Orinoco river large areas of primary

rain forest still exist, particularly in Amazonas Territory. There has been logging, and a significant amount of clearing, especially for grazing, in the State of Bolivar.

There are approximately 22 sawmills clustered primarily near the river; it is estimated that production is *ca*. 66 000 m³/a. Forest inventory for an area of northeastern Bolivar including the Imataca Forest Reserve (32 032 km²) has been carried out under FAO Project Venezuela 005. The current Project Venezuela 019 is building on this and developing new plans for this area. The recommendation of this FAO Project is to rely initially on timber salvaged from the enlarged Guri Reservoir to supply some of Venezuela's wood needs over the next 6–8 years. In the meantime, concessions in Imataca should be frozen in order to investigate sylvicultural management practices that would allow timber production without degradation of the forest. The knowledge from these should permit a rational utilization of this reserve from about 1983 onwards. At a later stage, the extensive pine plantations north of the Orinoco, and others to be established, might supply much of Venezuela's timber needs.

Mining activities south of the Orinoco are extremely important to the economy. Their impact on the forest is minimal in terms of the area physically occupied, even considering the fires that often accompany mining. But the roads to the mines open up the area to shifting slash-and-burn agriculture (the Venezuelan term for this small-scale shifting agriculture, often carried out on public lands, is *conuco*, and it is practised by *conuqueros*). Diamond miners have penetrated even to remote areas of Bolivar and Amazonas.

The large Guri Project on the Caroni river flooded a substantial area of forest which was not salvaged although there was an attempt to remove the wildlife. The enlarged Guri Reservoir will flood perhaps 4 000 km², containing 2 million m³ of wood. It is proposed to salvage the timber as a basis for the proposed wood industry, and thus take some of the pressure off the Imataca Forest Reserve until further management details can be worked out on regeneration possibilities.

But the pressure from shifting cultivators is the greatest.

In spite of all these activities there is still a lot of rain forest in this region although the rates of alteration or conversion to other uses are unknown, but surveys are proposed. Amazonas Territory is relatively intact, and the agency in charge of planning and development for the area, CODESUR (Commission for the Development of the South) now seems to be proceeding cautiously in its proposals for opening up the area. The State of Bolivar is losing so much forest by virtue of the greater accessibility that in June 1975, the Government prohibited deforestation and forest exploitation throughout the State although it exempted clearing whose objective was farming and livestock-raising. The prohibition therefore appears to be aimed primarily at timber harvesting and speculative land clearing.

North of the Orinoco statistics from the *Forest Atlas* suggest that *ca*. 44 per cent (approximately 73 930 km²) of the forest was rain forest, mostly in public ownership.

Logging has been going for a long time and a significant sawmill industry is established. Much has been converted into agriculture and grazing land.

Veillon (in Hamilton, 1976) aerially mapped the extent of forest in the west on both sides of the Andes. He had previously mapped the forest of the Andean piedmont, western Llanos, and, at ten years intervals, the States of Barinas and Portuguesa. These data show that there has been a 33 per cent reduction in forest area during 1950–1975, and that now only about 30 per cent of the total area bears forest. In the sub-regions, the greatest change has taken place in the State of Portuguesa whose forest area was reduced by half in the 25 years. Veillon suggests that only 21 per cent was forested in 1825 compared with 45 per cent in 1950 (the height of land clearing). This indicates that a large amount of the western Llanos forest was secondary in 1950; this includes most of the Forest Reserve of Turén. By the year 2000, he estimates that the forest area will be reduced to 16 per cent (14 320 km²). These maps show the disappearance of forest for only one major area north of the Orinoco. The same conversion forces are at work throughout the area, and one can presume that changes are taking place at somewhat similar rates in general, north of the Orinoco.

No accurate estimate of the rate of disturbance of rain forest by forest exploitation is available. Cutting has been fairly heavy in some semi-deciduous forests and in the tall cloud forest of the Andes where the number of commercial species is relatively high. Here the forest is substantially altered, but it is still forest, and with proper management could be even more commercially useful than the primary forest and it still provides essential watershed protection and wildlife habitat. In 1972 the forest was supplying 174 sawmills and 3 737 permits were granted for exploitation involving 421 187 m³ of wood. Over the years the number of permits has not fluctuated significantly, and has diminished slightly since 1964.

The national imperative for economic development includes increasing agricultural production and having more land under agricultural enterprise, in the hands of shifting cultivators. As well as continuing agrarian reform, there is much government activity in land development programmes. However, there is much illegal occupation of public lands. The three main pressures are: shifting cultivation; the extension of large land holdings, particularly for grazing; the government programmes of land clearing and settlements carried out by the National Agrarian Institute. The Agrarian Reform Law gives the National Agrarian Institute the authority and the responsibility, to move shifting cultivators from forest reserves, protected zones, national parks and other sensitive areas, and to relocate them in more appropriate locations preferably in the same region.

From 1960 to 1972, 176 465 permits were issued to large property owners for the removal of high and medium forest vegetation covering 12 070 km² (Ministry of Agriculture, Annuario estadístico agropecuario 1972). The number of permits and the area involved have been fairly constant each year; *ca*. 1 400 permits and 900 km² recently.

The forest reserve system of Venezuela consists of ten units, embracing originally 117 052 km². In actuality, three of these reserves have had their areas substantially reduced by conversion to agricultural use as the following data show:

Reserve	Original area (km²)	Area estimate (based on mapping by Veillon, 1975)
San Camilo	4 500	2 800
Caparo	1 744	1 690
Ticoporo	2 127	1 560
Turén	1 164	220
Guarapiche	3 700	same
Imataca	32 023	same
La Paragua	7 820	(needs surveying)
El Caura	51 340	(needs surveying)
Sipapa	12 155	same
Río Tocuyo	480	(needs surveying)
Total	117 053	

The Río Tocuyo reserve has little or no tropical rain forest whilst Guarapiche is mainly mangrove forest. The remaining eight reserves are essentially rain or somewhat seasonal and gallery forest.

They are not preserves but areas of public lands reserved and dedicated to timber production. They were established because of the valuable timber which remained. This is certainly the case north of the Orinoco river where the Turén, San Camilo, Ticoporo and Caparo are located. These are the lands which are most susceptible to penetration by agricultural activity: Turén has been the locale for one of the official IAN land settlement projects and San Camilo has been reduced in area by at least one third.

The forest reserve system is administered by the Directorate of Renewable Natural Resources within the Ministry of Agriculture. Professional foresters are in charge of management and cutting control, and forest technicians carry out the field operations. Orientation is clearly and exclusively towards wood production. Research is carried out on sylvicultural techniques and the Faculty of Forest Sciences in Mérida seems to emphasize underplanting and after-harvesting enrichment planting to try to increase the percentage of desirable species. On the Ticoporo Forest Reserve teak and saqui-saqui are being planted in cleared lanes after logging; 30-year contracts for three managements units have been set up: one unit is harvested by a private company; one is under joint management of the Corporation of the Andes, the Faculty of Forest Sciences at Mérida and a private company; the third is operated by a co-operative of shifting cultivators through the National Agrarian Institute. In the Caparo Reserve there has been a programme of research by the University of the Andes and the Corporation of the Andes which forms the basis of a management plan. Imataca Reserve comes within the working plan developed by FAO under projects 005 and 019. Some previous cutting removed only a few species in a light cut of probably no more than 10 per cent of the standing volume. The resulting forest seemed well stocked with quality stems and healthy trees. As small experimental plantations of fast-growing conifers and hardwoods established within the Imataca Reserve were not very successful it was considered undesirable to replace the forest with exotic plantations.

At present there are only about twelve species classified as having major commercial value. Total timber volumes of 80–200 m³/ha are reported. Thus cutting intensity has been fairly light until recently and not drastically disturbing. The increase of the number of species used could make the forest more economically productive, but FAO Project 019 proposed further investigation of sylvicultural management before taking any decision that may affect large areas.

Thus, within the forest reserve system there are some active programmes of controlled harvesting which need to be extended. But the idea of creating zones of protection and new reserves of deforested lands which would be reforested and become part of the production system is given more attention. This is the case with decreed protected zones, e.g. Turimiquire in eastern Venezuela, an area in the Andean piedmont and a buffer area eastward from El Avila Park in the Coastal Range. There are proposals for additional protected zones, and detailed ones for the Guasare system in the Perijá, and for Ríos Boconó, Negro, Anus and Tucupido in the Andes. These are excellent plans especially as they propose a monitoring system and management personnel.

Management planning

Introduction

Management planning is the implementation at enterprise or management unit level, of the national plan or programme, with due regard to regional or local problems. It is concerned with the organization of production of goods and services by the unit in order to achieve the targets fixed by the state planning authority or owner. The compliance with the principle of sustained yield is an added responsibility.

It might be surprising that management and sylviculture are not considered together, since management is not possible without sylviculture and a sylviculture which is not related to costs and benefits does not make sense. For a country which has a long forestry tradition such treatment would be difficult to accept, for it is not possible to deal with any treatment of the forest without reference to the socio-economic framework in which it takes place. But, in the case of the humid tropics, this separate treatment has been possible and indeed useful, for though there are situations (especially in South-East Asia) in which these two approaches can be combined, the principles of tropical sylviculture are far from being established, and they should therefore be discussed without reference to socio-economic factors. This view is not accepted by everybody. Whereas for Vannière (1975) 'management is a field that has hardly been touched', and Leslie (1971) is concerned with the lack of basic information, Palmer (1975, unpublished) thinks that 'the gaps in research for management guidance are local; the tactics for closing the gaps are in the literature' and 'there is generally enough information available from comparable countries to allow a state forest service to make a preliminary set of decisions'. Socio-economic considerations will have the last word in the choice of systems to adopt once the biological potentials and limitations of treatments are known.

However, a land-use plan, a clear and precise forest policy, and a more short-term programme for forest development, are often unavailable. Aided only by vague directives as to land use and forest policy, the manager has to decide on the best land use and define the forestry objectives of the tract he has to manage, bearing in mind market prospects as well as existing communications and their development potential, the density of populations, their needs and capacities. Where a forestry programme does exist, it must be remembered that planning is a continuous process with feedback procedures and information gained and problems encountered at the local level affect it. Moreover, modifications in the sylvicultural systems may become necessary by changes in demand. While an exploratory forest inventory is an aid towards deciding on the objectives of management, the manner of its execution depends on these objectives. The management of treated forests will be discussed in this section. Large-scale plantations are a different problem which is discussed in a document published by the FAO/UNDP (1975) project for the development of forest industries in Malaysia. This may serve as an example and a general guide; however, economic comparisons will be made where necessary. Leslie (1976) believes that plantations present the best solution and that in the long term, the management of tropical forests is generally doomed. It would be a very complicated and laborious exercise to present the state of knowledge on tropical forest management; the intention here is to simply give the views that exist where the advocates of natural regeneration confront those of plantations, and the upholders of sustained-yield management those of continuous supply.

Management objectives and regeneration methods

Although prediction of market demands and opportunities for the final crops must be uncertain at the time of the regeneration, an objective or objectives must be set and priorities be clearly established between them. This uncertainty places a premium on flexibility to accomodate changing demands, and this may influence the choice of the regeneration method. The more exactly the method is designed to meet a special market requirement the fewer will be the options for changes in management objectives later in the rotation. A major defect of most natural regeneration systems is the inability to predict production levels either of species or classes of timber, or of total merchantable wood. However, the natural variability of the forest is likely to assist flexibility to accomodate changing markets (Synnott and Kemp, 1976).

Some authors feel that the manager must produce a choice of alternatives as his plan. The important decision is determining the main uses for a tract. In the case of public lands the alternatives may be developed around optimum combinations of other uses which will cause a minimum reduction in the main uses. Many situations exist where the alternatives will provide a combination of minor uses at the expense of the main use without regard for the highest economic return.

Applications of modern computer techniques such as simulation can forecast the outputs of various combinations of inputs and restraints. Depending upon the constraints placed on the planning, an almost infinite number of options may be produced. This procedure rests on the establishment of a set of priorities and values for all uses. Monetary values cannot always be assigned and many that are assigned have a low degree of certainty; the judgment of the manager will be needed to weigh the values of intangibles against those to which monetary values can be assigned; hence, before reaching a decision there must be an assessment of the consequences of selecting that option. The assessment must look at environmental, social, political and economic consequences, and should be soundly based on available data and not 'a flight of fantasy into the future' (Arnold, 1971).

Another empirical approach is suggested by Palmer (1975, unpublished) who thinks that a simple flow-list could indicate the basic guide-lines for management, assuming that national aspirations framed by a development plan and based on demand studies have led to a forest policy, and that supply studies, forest inventories, market and land capability surveys have shown what could be obtained from the forest as it stands, and indicated what inputs of resources might be needed to supply other parts of the demands. His flow-list is:

1. Is there a projected demand for the same forest products as those obtained at the first cut, in terms of species, sizes and grades? If *yes*, go to 4; if *no*, go to 2.
2. No projected demand. Can the forest be modified by cutting to higher or lower limits, sylvicultural treatment to boost the growth of the leading desirable species, enrichment, so as to meet the demand for different forest products in whole or in part? If *yes*, go to 6; if *no*, go to 3.
3. The forest has no long-term use. Alternative methods of land use will supply the demand for large dimension timber, good quality water, gums, resins, fire-wood, house poles, flood control, fruit, wild animal meat. Management should be directed towards obtaining the maximum utilization of the forest crop before the change is made to other land use.
4. There is a projected demand for forest products. Is it possible to increase forest productivity towards these ends? If *yes*, go to 7; if *no*, go to 5.
5. After first trials of sylvicultural treatment and alterations to logging rules, it is not possible to improve productivity. If supply will fall short of demand, the difference can be made up by imports, or by compensatory plantations (though these will not supply large dimension timber profitably), or by overcutting in the hope that long-term demand for these products will fall. Management of the forest must concentrate on the prevention of diminution of the forest estate, and on further sylvicultural experiments. However it must be

realized that treatments must result in a massive and persistent positive effect to justify the compounded cost over long felling cycles.

6. There is a projected demand for forest products different from those obtained at the first cut of the forest, for example the use of secondary growth species in pulp and board products. Management methods approach those used in plantations and, because of the probably shorter felling cycle, allow both an increased first cut and greater expenditure on second felling cycle management. A continued research programme is needed to improve tactical management.

7. If there is capacity to utilize the increased productivity, the second felling cycle may be shortened (since crop trees will probably come to maturity earlier), but attempts to reduce the first cycle should be acceded to only if the improved productivity is certain to be achieved (to avoid a time of deficit between the ending of the first cycle and the commercial maturation of the second crop).

The impact of the various objectives of management on the choice of regeneration methods will be reviewed.

Wood production

For wood production, the aim can be either maximum volume or specialized timber. A major objective may be the maximum production of useful timber from a limited area of forest. In these circumstances, the operation of natural regeneration is not sufficiently productive and it is necessary to practise the interference through enrichment or conversion to plantation. Volume production in naturally regenerating tropical moist forest of mixed species is of the order of 0.5–2 m³/ha/a whereas plantations of fast-growing species can readily achieve ten times this rate or even higher rates by selection of species, provenances and individual trees and the production of improved material through breeding.

Almost all the valuable tropical hardwood timbers have been harvested from naturally regenerated forest and this is likely to be so for the rest of the present century. In many areas it has not been possible to increase or even maintain the stocking of commercial species and if demand continues they are likely to increase in value as the supplies diminish. Countries possessing tropical moist forests of valuable hardwoods may have therefore a great advantage in the future world market, provided that they can achieve their regeneration. Although a few major species, such as teak, are readily raised and managed in plantation, many are not, either because of insect damage or as slow initial growth rate, very short seed life, susceptibility to exposure, or other disturbances, etc. Nevertheless the stocking of valuable young trees before and after harvesting in the forest may be readily assessed and, if it is adequate to provide a final crop, there is a clear case for retaining natural regeneration even if the probable growth rate and ways of influencing the yield may be uncertain.

Soil and water conservation

The method of regeneration must take account of the need to protect the stability of the system in areas where disturbance may cause accelerated erosion, unfavourable changes in stream flow or loss of soil fertility. Clearance to create plantations causes maximum exposure to sun and rain, at least temporarily, and may entail burning and soil compaction. Although the plantations may later attain much of the stability of the original rain forest in regard to nutrient cycles, the type of forest canopy may be very different, particularly if the crop is deciduous and composed of a single species. Wherever soil and water resources may be vulnerable to changes in forest structure, the retention of an effective evergreen cover by the use of natural regeneration or enrichment methods may be preferred. If retention of an effective cover is the primary objective it can be most simply achieved in most areas by natural regeneration.

Social objectives

The provision of employment, food, recreation facilities and the protection of the human environment, particularly in regard to health hazards also influence the choice of regeneration methods.

Wherever the provision of employment is important the techniques employed in regeneration can be labour intensive. The more extensive systems, such as natural regeneration or enrichment, involving less interference with the forest, may be more readily adapted to this than the more intensive systems, in which the use of mechanical equipment for clearing, planting and weeding may have operational advantages. Nevertheless where an adequate labour force is available for such intensive operations to be carried out on schedule, the use of manual labour may reduce the dangers of adverse environmental impacts, such as unnecessary soil disturbance or soil compaction which might be associated with the use of mechanical equipment. If the objective is to provide continued work for a settled community then the more intensive systems such as conversion planting and agri-sylviculture, may be preferred since they provide a greater concentration of jobs and greater possibility for employment on work such as thinning and tending, at intermediate stages before harvesting.

The extent to which the regeneration of forests by agri-sylviculture may also contribute to food production is illustrated by the Nigerian experience where 100 km² are cleared annually for combined arable cropping and timber crop establishment, producing food crops worth over 5 million $ (Lowe, 1975). In Africa and tropical America, the soils are vulnerable to leaching and erosion and incapable of sustained arable farming without fallows and the combination of arable farming and timber crops offers considerable scope for sustained rural communities and appears more profitable than either separately.

At present, recreational and environmental benefits are minor objectives in most areas of tropical moist forests and unlikely to influence greatly the choice of regeneration systems. The ease of access, aesthetic appearance and the influence of the forest on wildlife vary at different stages in the life of the crop, and their value varies according to individual human judgments. In some circumstances the way in which the forest developed may affect populations of parasites or vectors of human diseases and could be an important consideration.

Management constraints

The choice of regeneration method has an important influence not only on the type of forest produced and its productivity but also on the way in which the nation's resources of land, forest vegetation, staff and finances are used to achieve the objective. When resources are severely restricted there may be strong social and political pressure for them to be used in ways that are profitable within a short time; that is short in comparison with a timber rotation period. This pressure has an important and sometimes decisive effect on the choice of regeneration method.

Land availability and capability

The greater the area of land available for forest production the less the pressure for intensive use. Competition may come from shifting cultivation, when farmers wish to move from soils already degraded by agricultural use, to the soils kept fertile by forest. Although natural regeneration might in the long term achieve maximum benefit from limited resources, the choice of conversion planting or agri-sylviculture may be the only way to resist the pressure for dereservation of the land.

The most fertile soils, capable of sustained arable cropping, have become the base for settled communities and in general it is the poorest soils which have been allowed to remain as forest (Fraser, 1975). The low human population leads to a low pressure on the land for uses other than forestry and limits labour for intensive operations. In these circumstances natural regeneration systems, which are least likely to upset the balance of nutrient cycles and which demand only low levels of financial and labour resources, can be an advantage.

Forest capability

Though the total vegetation productivity is fairly high, total production of wood is of the same order of magnitude as in temperate forests, that is *ca.* 3 m³/ha/a, and the growing stock associated with this is *ca.* 275–425 m³ of industrial wood and 250–300 m³ of stem-wood. Forests which have been subjected to treatment and which correspond to an early stage of succession achieve a growing stock of 600–750 m³/ha (*Aucoumea*, Dipterocarpaceae). Economic productivity is very much below biological productivity. It is estimated that in Africa the *ca.* 35 species which are at present merchantable and utilizable represent only 20–25 per cent of the total volume, *ca.* 15 non-commercial but technically promising species 20–40 per cent, and those which are neither 40–60 per cent of this total volume. Fig. 2, showing basal areas for different ages of artificial and natural stands in tropical and temperate regions, allows a

comparison between these climatic zones. The introduction of a greater number of valuable species, perhaps with faster growth, is clearly advantageous, provided that the anticipated increase in value of the crop will bear the cost of establishment. The use of selection and breeding methods may offer further possibilities for increasing the productivity of the forest in later rotations. By contrast the possibilities for increasing the productivity by natural regeneration alone are very limited. Climbers suppress or distort the regeneration of valuable species. Although the cost of weeding and climber control may be more readily borne by the more intensive plantation crops, the emergence of dense growth of weed species in dense uniform conversion planting has been a severe problem in some areas.

Financial resources

In many developing countries the lack of capital to invest in long-term projects such as timber production, where financial returns are relatively low and long-deferred, more severely limits the choice of regeneration methods than land availability. Natural regeneration is relatively cheap and an investment of a given size can be spread over a much wider area, thus retaining a larger forest estate with an assurance of continued production of wood even if growth rates and future market values may be uncertain. If unwanted trees can be cleared cheaply for charcoal production the cost of enrichment and selection weeding may be so low that a more predictable crop of higher value may be assured at a rate of return that compares favourably with more intensive methods such as conversion planting. However if the forest is to be used for wood using industries with greater predictability in the quality and yield, a greater production per unit area may be required. In such a case the conversion to high yielding plantations may be possible, perhaps with the aid of the private capital of the companies concerned in the initial investment.

The initial cost, through its effect on the discounted value of the final crop, is a major factor influencing the choice of regeneration method which is influenced by the present low value of most tropical timber. Nevertheless there is a high price differential between the more valuable hardwoods in the forests and the fast-grown plantation crops which may replace them. In some cases, as in Queensland, Australia, and in Trinidad, for example, this difference in value may compensate for the longer rotations and lower yields per unit area. A slightly higher market price for present commercial species, and acceptance by the international market of some presently non-commercial species, could improve the financial attraction of natural regeneration and enrichment methods in comparison with intensive plantations of fast-growing species of lower unit value.

Agri-sylviculture, as Lowe (1975) points out, assists not only in the amortization of the costs of establishment of the tree crop, but also in achieving greater financial profitability than either arable farming or timber crops separately. However Lowe also indicates the problems that arise in operating such a system, and the dangers that threaten if there is inadequate control of the system in practice.

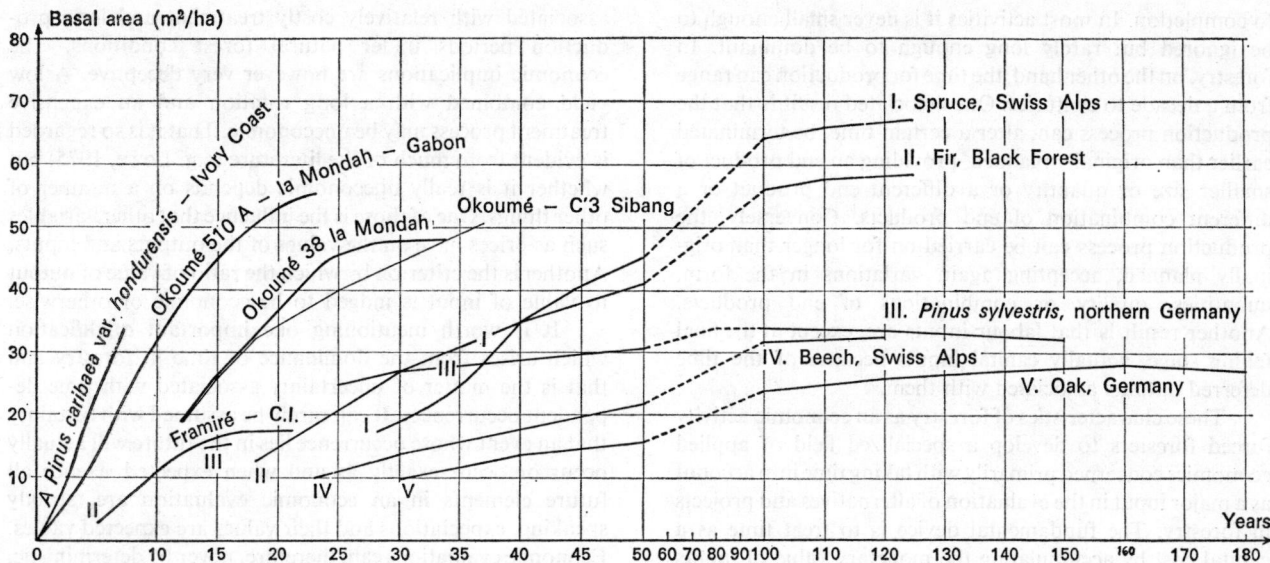

FIG. 2. Basal areas (m²/ha) at different ages of various artificial and natural stands in tropical and temperate regions (after Catinot, 1965).

Human and technical resources

Any method of regeneration presents technical and operational problems related to the complexity of the ecosystems, with many species and site differences over small areas, the difficulties of access and movement in the forest and the frequently large extent of the area to be treated. In many developing countries there is a shortage of trained staff and sometimes of labour available for work in the relatively remote and often uncomfortable conditions of the forest. The more extensive systems of regeneration, if they also demand close attention, present the greatest problem in control and supervision. The administrative difficulties are greatest if frequent visits need to be made to each area of forest for a sequence of relatively minor operations, as in the original tropical shelterwood system developed in Nigeria.

The more intensive systems avoid some of these problems but create others if they involve repeated tending operations and lead on to later thinning, pruning and protection from pests, fire or other dangers. The more intensive the system and the greater the initial investment, the greater the potential losses from failures in control and supervision. Nevertheless there is in general a better understanding of the techniques involved in the management of artificially regenerated forests, particularly close-planted stands, than those involved in the management of naturally regenerated tropical moist forests. Assuming that continued research may improve our knowledge and understanding of the dynamics of the natural system, and that education and training programmes will produce more skilled staff, it may be preferable to concentrate the present resources on the intensive regeneration of a relatively limited area provided that the danger of dereservation of other areas through apparent lack of management is not increased.

The possibilities and limitations of forest manipulation together with the impact of human and natural disturbances have been discussed in chapter 20. The deterministic or probabilistic character of regeneration pattern, succession and end product has also been discussed (chapters 8 and 9). Many problems remain: how to induce natural regeneration of desirable species, how to clear and tend the stand to improve it, and the limits of these interventions. Hence the need for research in management planning.

Economic implications

The economic implications of any activity arise from evaluations of its products (outputs) relative to what was used (inputs) to produce the outputs. Forest management is deliberate intervention designed to change the natural production process into some specific combination of outputs considered to be more useful or more satisfying and to maintain that output for some specified time. So long as no permanent drastic change in the forest's composition and structure is induced by the manipulation the system is regarded as natural management.

There are several complications about the forestry production process which make its analysis as an economic activity somewhat different in practice, but not in principle, from most other activities. One is that the forest itself is simultaneously product, producer and an input. To some degree that is true of production in general, but for most processes it is possible to distinguish the three aspects well enough for them to be treated as separate entities. In forestry the distinctions are much more blurred and hence analysis and management less clear-cut. A second complication is that time in forestry is a major and perhaps the major input. Again the difference between forestry and other production processes is one of degree. Time is needed for the process of combining and transforming inputs to run from initiation to completion. In most activities it is never small enough to be ignored but rarely long enough to be dominant. In forestry, on the other hand, the time for production can range from a decade to centuries. One associated result is that the production process can, after a certain time, be terminated earlier than originally planned, providing an end product of smaller size or quantity or a different end product or a different combination of end products. Conversely, the production process can be carried on for longer than originally planned, accepting again variations in the form, quantities, quality or combinations of end products. Another result is that labour inputs are, except at the final felling stage, actually capital inputs because of the time deferred outputs associated with them.

These characteristics of forestry as an economic activity forced foresters to develop a specialized field of applied economics concerned primarily with taking time into account as a major input in the evaluation of alternatives and projects in forestry. The fundamental device is to treat time as a capital cost by accumulating the monetary value of inputs and outputs exponentially at some specified or implied rate of compound interest. But the solution to the problem of time as an input created another problem—that of the so-called burden of compound interest—which has troubled forestry ever since.

The device of compound interest for measuring time as an input automatically gets over the problem of comparing inputs and outputs that occur at different times. They are, in effect, reduced to the same point in time. But there is still another problem of incommensurability to overcome before an evaluation can be made. This one arises from the differences in the physical nature of inputs and outputs. In their basic form physical quantities of different items cannot be added, subtracted, multiplied or compared.

The transformation to a common measurement is usually effected by expressing the quantities as values in terms of money. It is a simplification which only works readily for those items for which market prices exist and it works well only for those items whose market prices are fairly close approximations to their social values. All of this adds up to the fact that to make an economic evaluation in forestry requires a great deal of quantitative information regarding:
— the quantities and prices of the inputs (costs) and their timing;
— the quantities and prices of the outputs (returns) and their timing;
— the way in which outputs change or are affected as given inputs are varied;
— the time between an input and its output or its effect on output;
— the interest rate.
When these requirements are considered in relation to the management of the moist tropical forests two things are obvious. The first is a paucity of quantified information in many aspects; this approaches complete ignorance in the case of the relationships between intermediate treatments and final outputs. The second is that what is known or can be guessed at, points to a relatively small wood volume output

associated with relatively costly treatment and long production periods under natural forest conditions. The economic implications are however very deceptive. A low yield combined with a long rotation and an expensive treatment process may be uneconomic. That it is so regarded is evident from much of the literature (e.g. Lowe, 1975) but whether it is really uneconomic depends on a number of other things. One of those is the influence that other variables such as prices have on the values of the outputs and inputs. Another is the criterion by which the ratio of value of output to value of input is judged to be economic or otherwise.

It is worth mentioning one important qualification which arises from the dominance of time in forestry and that is the matter of uncertainty associated with time dependent occurrences. It can rarely be assumed with certainty that an event whose occurrence lies in the future will actually occur or occur exactly as and when expected. Hence all future elements in an economic evaluation are, strictly speaking, expectations and their values are expected values. Economic evaluations can, therefore, never be deterministic, except in retrospect. That applies with double force in forestry where the stochastic nature of measurements of biological relationships is compounded by the uncertainty associated with the occurrence and level of future events. It is impossible to exaggerate the importance of uncertainty in forest management. If we consider that the utilization of moist tropical forests for their wood and other products (in such a way that the initial ecological structure is perpetuated without drastic or permanent change) is either ecologically impossible, as some ecologists now suspect, or so difficult as to be not worth to attempt, as many tropical foresters now feel, their reservation on a large scale without commercial utilization or conversion to agriculture, simply to preserve them as a major biome is hardly a serious option and they will continue to disappear as a result of the economic situation in the countries of their main occurrence.

One of the potentially most damaging consequences would be global or regional climatic imbalance associated with the disappearance of the major blocks of moist tropical forest. Although the possibility was fairly strongly mooted a few years ago, global climatic deterioration now seems to be discounted as a factor which would justify the type of world-wide action needed to halt the destruction of tropical forests. That something would be lost to mankind as a whole by the reduction or extinction of the many complex biological systems that make up the moist tropical forests is obvious enough. Exactly what would be lost is rather difficult to specify and what it would mean is almost unknowable. There would be a psychic loss to some people resulting from the disappearance of some speciality timbers. But what it would all add up to is anybody's guess. The case for retaining the moist tropical forests, if they could not be managed as such, depends therefore on doubts about the future that are too speculative to match the urgency of demands on them. The uncertainty may point to a need for caution, but it is hardly strong enough to justify a halt.

The lack of information which inhibits preservation of moist tropical forests, if they are a non-renewable resource, handicaps even more so action to justify their management as natural forests if they are renewable. It is far from certain how to re-establish naturally most types of moist tropical forests after or in conjunction with harvesting (Catinot, 1974; Nwoboshi, 1975). Even if that were not so, then the lack of knowledge of stand dynamics in managed or unmanaged forests virtually reduces sylvicultural treatments aimed at maintaining productivity, let alone increasing it, to an act of sheer faith. Yet it is only through increased productivity that moist tropical forests can compete with the alternative land uses.

However some forest ecologists, mainly with experience in the Far East, remain convinced that through a better ecological stratification it will be possible to identify areas which could be profitably managed, even if the calculation of allowable cut and yield prediction, as a result of lack of production table, remains a critical issue.

Furthermore the road to increased productivity is not confined to solutions of biological management problems. Increased utilization of secondary species would give an automatic and substantial increase in output. Adjustments that raised the standing timber prices for the presently preferred and easily marketed species relative to the less preferred and difficult to market species, could give a substantial lift to the value of output. Such economic measures possibly offer more scope for achieving an increase in value productivity than the biological approaches presently being pursued, and they would do it with much less uncertainty and at considerably lower investment cost. Equally significant is that they do not involve waiting on solutions to the problems of the biology of stand development. But they do involve an understanding of markets and marketing, particularly in foreign countries.

There is some scope for improving the ratio of output to input by reductions in the cost of management. It is not a promising field unless the present crop can be utilized much more intensively. Even then, it is hard to see the improvement in economic performance resulting from cost reduction measures being great enough to have a really significant effect. The major factor affecting the value-cost relationship is the interest rate and any case for the application of a special low rate of interest in forestry investments is likely to be general for all forestry rather than specific to the natural management systems. A lower rate would help to improve the relative economic performance of natural management, but it would need to be very low and combined with a substantial reduction in rotation lengths for it to make a significant impact on the gap between natural management and its competitors. The shortening of rotations could be envisaged when an end use has been found for small dimension logs. When this is reached it will be opportune to reconsider the objectives of management. If the chief objectives can then be declared as the production of logs from trees of 40–50 cm DBH minimum size instead of 60–80 cm DBH, there are potentialities for the use of lighter logging equipments and less logging damage.

Some overall improvement in the economic performance of natural management of moist tropical forests is possible; but it remains to be seen if improvement in sylviculture or marketing will have a big enough impact on the performance of natural management relative to other forms of land use. There is, however, one very important qualification. It holds only so long as the performances are judged in terms of commercial output valued from the point of view of the organization responsible for the forests. But that way of evaluating alternatives can be misleading if not completely erroneous. It tends, for instance, to leave too many things out, and to take too narrow a view of effects and their significance. Once however an attempt is made to extend the economic implications of natural management to a social level, the inadequacies of data on wood production are compounded enormously by the inadequacies of data on the environmental and social side. An evaluation that was somewhat speculative but at least fairly clear-cut could be replaced by one much more speculative and anything but clear-cut. At the extreme, a social evaluation of natural management of moist tropical forests could show it to be or not to be a viable alternative to agriculture or plantations according to the answer that was wanted. In such circumstances, financial analysis may be a wrong guide to decision but social analysis could be no guide. It is hard to say which is worse.

In part the difficulties can be traced to the data; they are not good enough to provide definitive benefit-cost relationships. But a large part of the trouble also lies with the decision analysis techniques. They are either too narrow in the range of interest, or too arbitrary in their valuation procedures, or too narrow in the range of objectives they include in their criteria to provide definitive benefit-cost relationships. But the major problem lies in uncertainty about the future. No analysis based on what is known now of the future can ever be definitive. Foresters almost everywhere are forced to look too far into the future for them to be so sure about their assumptions that they can be dogmatic about their decisions. Some of the strongest arguments for the natural management of tropical forests lie therefore in the conceptual weaknesses of the case against it. But they are not yet the sort of arguments that would carry much weight in practice. The best reasons for retaining part of tropical moist forest under natural management lie in the insurance it provides against a short-sighted and irrevocable narrowing of land-use options.

Long-lease contracts

The forest service responsible for the administration of public forests must take measures for the preparation of management plans of different units, such as forest inventories, communication network, the organization of the area into compartments, etc. The majority of forest services are not fully equipped to carry out these tasks so the beneficiaries of long-lease contracts should be obliged to carry out their coupes according to a general plan approved by the forest services and under their supervision and control. Hence the problem of concessions is closely linked with the formulation of a general forest policy and general principles of management by a central authority. This chiefly implies that contractors should accept harvesting programme covering 20–30 years so as to provide a secure basis for the development of forest industries in the region. The programme should provide for revisions every 5 or 10 years and it should take into consideration the objectives, potentials and limitations previously discussed. Independent of any theoretical considerations such agreed rules of exploitation must, in the basis of actual knowledge, foresee the areas to be cleared for agriculture or those to be protected; they should state the allowable annual cut, the method of felling and skidding, and the number, sizes and species of trees to be felled; they should also specify the necessary communication infrastructure, and contain instructions about forest improvement work to be carried out.

Management and working plans

Where no national plan exists each area should be managed in such a way that it can supply a unit or be handed over for exploitation to a contractor, provision being made for the availability of the necessary equipment, the construction of the necessary infrastructure, etc. Where necessary, distinct management objectives may have to be determined for the different compartments, and the plan should incorporate studies and practical trials which are needed to put it into effect and to provide for its future orientation.

African high forest

The sylvicultural aspects of this protocole have been discussed in chapter 20; they include the determination of growth rates of the leading species in natural stands either by reference to permanent sample plots laid out according to statistical principles to study the increment of all trees of commercial interest contained therein (Dawkins' permanent sample plots), or by the study of randomly distributed sample plots in the forest, or by the analysis of growth rings which in several species exhibit an annual pattern. There is also a need for accurate inventories; utilization studies to determine the variation of the cost per m^3 according to species, terrain, size of logging area, etc.; studies on transport costs taking into account the major routes and the loads they will have to carry; and for studies in economics, calculating the maximum admissible cost for timber to be harvested by working backwards from the FOB price or the price at the factory gate as a means of determining the assortments of wood that can be produced from a given managed tract of forest taking into account transport distances.

By synthesizing the results of these studies and trials, it should be possible to arrive at figures for the number and intensity of felling operations, their periodicity and size (usually determined on an area—and only rarely on a volume basis), the nature and costs of infrastructural works and forest operations. Such a synthesis will also enable predictive balance sheets to be drawn up on the basis of high or low assumptions as regards expenses and receipts. All management plans are subject to revision at fixed intervals to incorporate the results of research (especially as regards the rotation) and to make adaptations demanded by changes in circumstances.

Peninsular Malaysia

A development strategy for the forest sector of peninsular Malaysia has been compiled up to 1990. It emphasizes the urgent need to establish a firm management control over the entire forest resource of mixed dipterocarp hill forest so as to optimize the commercial potential and to create a basis for its treatment as a perpetually renewable resource. A decision has therefore been taken by the Forest Department to control all harvesting operations through the consolidation of licence areas. The previous pattern of numerous small licences of short tenure is being replaced by long-term licences which will be subject to comprehensive planning and strict supervision.

In order to enable the Forest Department to coordinate log flow from the various classes of forest into a properly planned and phased basis for industrial development and to implement these plans through strictly supervised control so as to provide a basis for resource renewal, it is necessary to obtain additional data. The basic data collection and management programme which were initiated in 1974 are summarized in order to indicate the type of action which is being taken to obtain data on stock composition, log yields, growth rates, development of regeneration and weed growth, in response to a variety of different intensity periods of harvesting under a wide range of stand and topographic situations.

The programme envisages the establishment of at least 200 permanent observation samples, 35 series of experimental management plots and systematic stock surveys covering approximately 1 million ha over 5 years in 7 forest management units. All samples and plots will be subject to detailed observation and analysis for a minimum of 25 years, while large-scale application of controlled harvesting will also provide data on the extent of felling damage and on the methods to minimize it. Priority is being given to ranking species and species groups according to their potential commercial productivity; identifying safe minimum diameter limits for species or species groups in specific types of stock composition so as to facilitate adequate regeneration with minimum weed invasion; establishing the optimum harvesting period between successive cuts; determining the average yield for defining forest regions; and investigating the possibility of large-scale plantations of commercially valuable fast-growing indigenous hardwoods. Interim results will provide a reasonable realistic basis for the formulations of interim management procedures so as to provide the planning framework for regulated log flows. These results will be applied within designated forest management units with the object of developing practical management procedures based upon adequate experimental data and tested by large-scale application.

The systematic stock surveys are of two types. The commercial cruise undertaken by licensees on large long-term concessions covers primary forest and is limited to a minimum diameter of 15 cm whereas the regeneration cruise within harvested forests, as undertaken by the Forest Department, makes an assessment of small-sized stock and includes an evaluation of weed occurrence.

Permanent observation samples include:
— management sample plots in adequately regenerating, treated forests;
— random inventory plots in all types of harvested forests which, in most cases, have not received any form of treatment since harvesting.

Treatments in the permanent management sample plots consisted mainly of the poison-girdling of certain trees and the cutting of climbers.

Periodic remeasurements concentrate on diameter growth of individual trees, on mortality rates, and on growth in stand basal areas. Additional information is being collected to strengthen final analysis and comparisons; these include tree dominance classifications and crown diameters, light assessment, bole and total heights of selected trees on management plots, with regeneration, weed and stump surveys on inventory plots. Particulars of the plot locations including altitude, slope, soil type and information on date of logging, log outturn, date of treatment, type and costs of treatment, are also recorded for use in analysis. Volume data will be derived from basal area, clear bole height, and from factor data as a final step in analysis. Estimates of mortality rates and assessments of development of small size regeneration and weed growth, will follow from periodic counts.

Experimental plots are being laid out in specially selected forest areas. These plots are reserved for at least 25 years: they include replicated cutting and growth experiments in blocks of primary forest. A more simple trial series of different cutting, replanting and treatment methods, is being established in harvested forests. These field experiments were initiated by the Forest Department in 1974.

Each forest management unit will be designed as a natural resource-flow catchment area and, wherever possible, will contain a minimum area of 1 000 km² of forest, the greater part of which will be managed as permanent forest. Under existing circumstances it is inevitable that each forest management unit will contain forest land which is scheduled for conversion to agriculture and substantial areas of already harvested forest in addition to primary forest. The first forest management unit has been selected and field surveys within it are well advanced; these include detailed assessments of the forest resource, existing pattern of harvesting and log flows, plans for agricultural development and their phasing, soil and terrain classification and road locations. Stock surveys and the collection of basic planning data are also in progress in three additional forest management units, the boundaries of which have yet to be finalized.

Maps at 1/25 000 are being prepared to show forest category and stock type details, felling schedules within licence areas, clearing schedules in agricultural development areas, alignments of existing and planned roads and the locations of forest industries. Statistics of numbers of trees and volumes per hectare are being computed as sample averages for stock types and/or categories by blocks of forest considered logical units for detailed planning. Areas of stock types and categories per block will be multiplied with sample averages of stock, to give estimates of total resources per block. As a final step, several alternative block yield totals, by year, will be computed and the most desirable harvesting schedule selected.

Initially, the harvesting regime for the permanent forests will be based upon rather high minimum diameter limits for the more valuable commercial species. This conservative method of harvesting control is a temporary expedient pending confirmation, by research, of the most appropriate management system.

In order to maintain adequate log flows to meet the requirements of industry during the period of transition towards a proven practicable management system, emphasis is being placed upon maximizing the utilization of the full commercial potential of the agricultural conversion areas. Until recently only a small proportion of this potential has been utilized by industry; much of value has been burnt. It is, therefore, possible during the next ten years, to reduce the pressure on the permanent forest and yet maintain an adequate flow of logs to industry. This phasing of log flows from the different categories of forest will be a major aspect of planning within each forest management unit.

A major feature of the forest management units will be the alignment and construction of roads to high standards to provide an efficient permanent network capable of meeting the requirements of high-production low-cost logging and log transportation within the permanent forest. Major access roads within the agricultural conversion areas will be planned by the land development authority concerned, and the phasing of forest harvesting in these areas will require full integration with agricultural planning.

The forest management units represent a new approach to organized forest sector development. Together with the extensive work now being undertaken by the Forest Department on the measurement of the productive potential of the mixed dipterocarp hill forests, and the large-scale applied research on the development of appropriate regimes of forest management, they will ensure that the forest sector will continue to develop progressively as a major contributor to the national economy.

Surinam

There are over 14.8 million ha of land carrying some form of forest in Surinam, representing 90 per cent of the land area. The first attempt to bring the forests under management was made in 1904 with the setting up of a Forestry Service. However, the Service was abandoned in 1925 and it was not until 1947 that the present Forest Service was reinstalled with the directives 'to manage the country's forest to yield in perpetually the maximum benefits for the community'. Forest policy must ensure, *inter alia*, a regular wood supply to local small and middle-sized enterprises, while maintaining the possibility of large integrated industries created in joint venture with the government.

Concessions are awarded by the Governor who hears the advice of the Forest Concession Advisory Committee which includes the Forest Service. The latter has no power to decide who should have the concessions, or how large they should be, and it cannot limit the total area under concessions at any one time. Only in managed concessions can it regulate the amount of timber cut. Managed concessions have been introduced in areas which have been inventoried or opened up by road building; but transportation difficulties for the Forest Service are making it impossible to keep up with the inspection load.

The main instruments for controlling the annual cut is the management plan and an outline plan for one unit has been prepared. The following procedure has been followed:

— the external boundaries of the unit, the limits of the proposed storage basin, the limits of proposed agricultural development, and a zone of protection forest, have been marked on 1/40 000 maps;
— the remaining forest was considered to be potentially productive, and was divided into blocks and compartments; simultaneously a road network was designed and drawn on the map; using the estimated standing volumes/ha derived from the FAO inventory (40–50 m³/ha of merchantable timber), the optimum road density with current harvesting techniques was estimated at about 8 m/ha;
— the gross area of each compartment was measured on the map, and the area of savanna forest, open forest and marsh forest within each compartment measured; the length of creek shown on the map in each compartment was also measured; the area of high forest in each compartment was subdivided as a result of inventory on the basis of 80, 17.5, 2.5 per cent to dry land (*terra firme*), creek and liana forest respectively;
— the average skidding distance for each compartment was also measured and an estimate of the total volume of commercial species obtained by multiplying the areas of the forest types by the following average volume/ha obtained from the inventory (dry land 50 m³/ha, swamp and creek 20 m³/ha, liana 12 m³/ha);
— the data for each compartment in block I were recorded in full on forms which will be available in the project record. For other blocks in the unit general estimates were made from the gross area of the block.

One chapter of the plan deals with the regulation of the yield. Criteria must be established for calculating the allowable cut and then different ways examined of achieving it. No data are available about increment following treatments; and it is assumed that increment with no cutting is nil; increment with clear cutting and natural regeneration is 2 m³/ha/a; increment with clear cutting and planting of eucalypt is 15 m³/ha/a.

The volumes quoted represent only the commercial species. The allowable cut should take into account the available volumes in as many other species as possible. Contracts to supply volumes should be written to include all species which are known to be usable. If concessionaires are unable to use some species in their own factory they may be forced to cut them and endeavour to sell them to other mills. Only by doing this can the forest be made to contribute fully to development.

The system of control used should be by area rather than by volume as at present. By this is meant that the volume to be felled as calculated for the allowable cut is

converted to an area which will be expected to carry the required volume as derived from inventory data. The area would then be granted on the basis that it contained a specified volume. It would be up to the purchaser to satisfy himself that the volume was actually present before making a bid for the area, because once a price is agreed there would be no adjustment of the area. The Forest Service would therefore merely have to ensure that the operator did not stray beyond the boundaries of the area. Demarcation of compartments on the ground would further facilitate control, as all timber could be sold by compartments.

Conclusions

In the present state of knowledge of tropical forest ecosystems and on the basis of experience acquired on their treatment and transformation it can be confidently stated that it is possible to carry out the various stages of their development, provided that present research continues and new studies are undertaken, though there may be cases in which such development is not feasible immediately. It is now possible to envisage a land-use policy for the development of the forest sector and medium or long-term programmes specifying various operations in time and space; and guide-lines on general management which allow working plans for large forest units (*ca.* 1 000 km²) to be prepared.

This presupposes that one recognizes and can estimate and quantify, the role of forests in the ecological environmental equilibrium at a global, regional and local level, the significance of the forest sector as a driving force in economic and social development, and finally the social significance of forests both as regards the benefits they provide to the rural populations and those experienced by urban populations. Hence the inclusion in national balance sheets, and thence into development plans, of benefits in kind and in services not usually the object of trade activity which forests render to the nation as a whole and to the rural populations in particular.

The debate on forest development is sometimes distorted by different concepts and leads to unnecessary friction. The conflicts are between the advocates of natural regeneration and the advocates of plantations; and between insisting on the adherence to sustained yield and recommending a more flexible formula. The confusion between capital and revenue, as well as the use of compound interest (a factor which is often neglected) is a peculiar problem characteristic of the management of renewable natural resources.

But apart from certain difficult areas there is really no conflict between plantations and natural forests (with or without aided natural regeneration). Plantations are the obvious choice where there are available good soils nearby centres of consumption, qualified technical personnel, adequate financial resources, a communication infrastructure, and other institutions which make it possible to protect the plantations against harm. In contrast, the management of natural forests is the inescapable choice for not easily accessible forests on vulnerable soils and where financial and human resources are limited. For with this method investments are relatively small (including the research necessary) and the output per man-day in harvesting may be high. All kinds of intermediate forms may be adopted in accordance with local circumstances and restraints. The failures of intensive agriculture and the fact that experience with intensively managed plantations is relatively recent should lead to caution in the choice of options; they suggest the desirability of research into simple methods of ecological stratification of tropical environments and the elaboration of principles for stimulating natural regeneration of leading commercial species as well as for treatments for the improvement of stands.

The term sustained yield must be used correctly. Forest lands designated for agriculture are exploited without any thought of sustained yield; countries with vast forest resources can exhaust the forests of the first zone in expectation of forests of the second or third zone becoming accessible and providing time for the forests of the first zone to recover and countries with extensive afforestation programmes can anticipate future yields in order to regulate the distribution of age classes. However, when the forest capability of an area has been recognized, at least for a long period, utilization has to be guided firmly by the principle of sustained yield. Even if all working plans are based on the principle of sustained yield, this does not necessarily mean that local officers must never be allowed to anticipate future

yields or to stock according to market conditions. On the practical level, the only feasible method is the sustained yield from a forest assumed to have a normal structure; all other models are too complicated and costly in relation to the objectives aimed at.

In conclusion, the knowledge of structure and functioning of tropical forests ecosystems, the experience acquired in the regeneration and treatment of these ecosystems, the improvement of inventory techniques and data processing, of planning methods at national level and of management at enterprise level, allow an optimistic view of the prospects of developing and managing tropical humid forests within an interdisciplinary setting, i.e. with full consideration of ecological, economic and social factors. Forest management plans have as a compulsory consequence to postpone an immediate exploitation, but such a financial sacrifice preserves the future of the forest growing stock and its long-term development. But as long as there is insufficient information available to make such an approach possible, it is necessary to preserve areas that are unique and representative and to proceed systematically with the management of large enough tracts to support an industry, while making reasonable assumptions as regards desirable national objectives.

Bibliography

Country planning and land use

ASHTON, P. S. *Ecological studies in the mixed dipterocarp forests of Brunei State*. Oxford Forestry Memoirs, 25. Oxford, Clarendon Press, 1964a, 110 p.

——. A quantitative phytosociological technique applied to tropical mixed rain forest. *Malayan Forester*, vol. 27, no. 3, 1964b, p. 304–317.

——; BRÜNIG, E. F. *The variation of tropical moist forest in relation to environmental factors as key to ecologically oriented land-use planning*. Rome, FAO, 1975, 34 p. multigr.

AUBRÉVILLE, A. *Étude écologique des principales formations végétales du Brésil et contribution à la connaissance des forêts de l'Amazonie brésilienne*. Nogent-sur-Marne, Centre technique forestier tropical, 1961, 268 p.

AUSTIN, M. P.; GREIG-SMITH, P.; WHITMORE, T. C. The application of quantitative methods to vegetation survey. I. Association analysis and principle component ordination of rain forest. *J. Ecol.*, 55, 1967, p. 483–503.

——; ASHTON, P. S.; GREIG-SMITH, P. The application of quantitative methods to vegetation survey. III. Re-examination of rain forest data from Brunei. *J. Ecol.*, vol. 60, no. 2, 1972, p. 305–324.

BENE, J. G.; BEALL, H. W.; CÔTÉ, A. *Trees, food and people: land management in the tropics*. Ottawa, International Development Research Centre, IDRC-084e, 1977, 52 p.

BOON, D. A. Some aspects on plant ecology in the tropics in connection with the use of aerial photography. *Turrialba*, 2, 1965, p. 132–134.

BRÜNIG, E. F. Forestry on tropical podzols and related soils. *Trop. Ecol.*, vol. 10, no. 1, 1969a, p. 45–58.

——. On the seasonality of droughts in the lowlands of Sarawak (Borneo). *Erdkunde* (Bonn), vol. 23, no. 2, 1969b, p. 127–133.

BRÜNIG, E. F. Stand structure, physiognomy and environmental factors in some lowland forests in Sarawak. In: *Abstr. Proc. 11th Int. Bot. Congress* (24.8–2.9.1969, Seattle), 1970, 24 p. *Trop. Ecol.*, vol. 11, no. 1, 1970, p. 26–43.

——. On the ecological significance of drought in the equatorial wet evergreen (rain) forest of Sarawak (Borneo). In: Flenley, J. R. (ed.). *The water relations of Malesian forests. Transactions of the first Aberdeen-Hull symposium on Malesian ecology*, p. 66–97. Univ. Hull, Dept. Geography, Misc. ser. no. 11, 1971, 97 p.

——; BUCH, M. Von; HEUVELDOP, J.; PANZER, K. F. *Stratification of the tropical moist forests for land-use planning*. Rome, FAO (Committee on forest development in the tropics, 4th session, November 1976), FO:FDT/76/8(c), 1976, 37 p.

BUDOWSKI, G. The opening of new areas and landscape planning in tropical countries. In: *12th Congress of the International Federation of Landscape Architects* (Lisbon, September 8, 1970).

CATINOT, R. Le présent et l'avenir des forêts tropicales humides. *Bois et Forêts des Tropiques*, n° 154, 1974, p. 3–26.

CONWAY, G.; ROMM, J. *Ecology and resource development in South-East Asia*. New York, Ford Foundation, 1973.

CURRY-LINDAHL, K. *Conservation for survival: an ecological strategy*. New York, William Morrow, 1972, 335 p.

DARLING, F.; MILTON, J. P. (eds.). *Future environments of North America: transformation of a continent*. New York, Natural History Press, Doubleday, 1966, 767 p.

DASMANN, R. F.; MILTON, J. P.; FREEMAN, P. H. *Ecological principles for economic development*. London, New York, Wiley, 1973, 252 p.

DENEVAN, W. M. Development and the imminent demise of the Amazon rain forest. *Prof. Geog.*, 25, 1973, p. 130–135.

DUVIGNEAUD, P. *La synthèse écologique.* Paris, Douin, 1974, 296 p.

FAO. *Land utilization in the tropical areas.* Rome, 1952, 10 p.

——. *Conservation and management of African wildlife.* Rome, 1967, 35 p.

——. *Environmental aspects of natural resources management. Agriculture and soils.* Agriculture services Bulletin no. 14. Rome, 1972, 39 p.

——. *The environmental aspects of forest land use.* Report on the FAO/SIDA Seminar on forest social relations for English-speaking countries in Africa and the Caribbean (April 1974). Rome, 1975, 184 p.

FARNWORTH, E. G.; GOLLEY, F. B. (eds.). *Fragile ecosystems. Evaluation of research and applications in the Neotropics.* Berlin, New York, Springer Verlag, 1974, 258 p.

FARVAR, M. T.; MILTON, J. P. *The careless technology: ecology and international development.* New York, Natural History Press, Doubleday, 1972, 1 030 p.

GÓMEZ-POMPA, A.; VÁZQUEZ-YANES, C.; GUEVARA, S. The tropical rain forest: a non-renewable resource. *Science,* 177, 1972, p. 762–765.

GOUROU, P. *The tropical world.* London, Longmans, Green, 1966, 196 p.

GREIG-SMITH, P. Application of numerical methods to tropical forests. In: *Statistical ecology,* vol. 3 (*Populations, ecosystems and systems analysis*), p. 195–206, 1974.

HILLS, G. A. *The ecological base for land-use planning.* Ontario Dept. Lands Forests, Res. Rep. no. 46, 1961.

HOLDRIDGE, L. R. *Life zone ecology.* San José, Costa Rica, Trop. Science Center, 1967, 206 p.

——; GRENKE, W. C.; HATHAWAY, W. H.; LIANG, T.; TOSI, J. A. Jr. *Forest environments in tropical life zones. A pilot study.* London, Pergamon Press, 1971, 731 p.

INTERNATIONAL INSTITUTE FOR LAND RECLAMATION AND DEVELOPMENT. *Land evaluation for rural purposes* (summary of an expert consultation). Wageningen, October 1972.

JACKSON, D. S. Soil factors in land use. *N.Z. J. Forestry,* vol. 18, no. 1, 1973, p. 55–62.

LANLY, J. P. Use of recent techniques in the evaluation of forest resources. *Bois et Forêts des Tropiques,* n° 147, 1973, p. 35–45.

——. *The inventory of tropical moist forests for industrial investment decisions.* Rome, FAO (Committee on forest development in the tropics, 4th session, November 1976), FO:FDT/76/9(d), 1976, 11 p.

LEEUW, P. N. de; TULEY, P. The land resources of North-East Nigeria. Vol. I. The environment: vegetation. In: *Land Resource Study no. 9,* p. 121–155. Land Resources Division, Ministry of Overseas Development (Tolworth Tower, Surbiton, Surrey, United Kingdom), 1972.

MEGGERS, B. J.; AYENSU, E. S.; DUCKWORTH, W. D. (eds.). *Tropical forest ecosystems in Africa and South America: a comparative review.* Washington, D.C., Smithsonian Institution, 1973, 350 p.

MILLER, K. Conservation and development of tropical rain forest areas. In: *11th General Assembly and 12th Technical Meeting of International Union for Conservation of Nature* (Banff, Canada), 1972.

MILTON, J. P. The ecological effects of major engineering projects. In: *The use of ecological guidelines for development in the American humid tropics* (Proceedings of international meeting, 20–22 February 1974, Caracas), p. 207–222. Morges, Switzerland, IUCN Publications, new series no. 31, 1975, 249 p.

MOORE, D. *Assessment of land capability as a prerequisite to*

national forest management of the tropical moist forests. Rome, FAO, 1975, 5 p. multigr.

MUELLER-DOMBOIS, D. Planned utilization of the lowland tropical forests. *Nature and Resources* (Unesco, Paris), vol. 7, no. 4, 1971, p. 18–22.

PARSONS, J. J. The changing nature of New World tropical forests since European colonization. In: *The use of ecological guidelines for development in the American humid tropics* (Proceedings of international meeting, 20–22 February 1974, Caracas), p. 28–38. Morges, Switzerland, IUCN Publications, new series no. 31, 1975, 249 p.

PHILLIPS, J. *The development of agriculture and forestry in the tropics: patterns, problems and promise.* Faber and Faber, 1961.

POORE, D. *Ecological guidelines for development in tropical forest areas of South-East Asia.* Morges, Switzerland, IUCN occasional paper no. 10, 1975, 33 p. IUCN Publications, new series no. 32, 185 p.

——. *Ecological guidelines for development in tropical rain forests.* Morges, Switzerland, IUCN, 1976, 39 p.

Proceedings of the ninth Pacific Science Congress (1957). Vol. 20, Special Symposium on climate, vegetation and rational land utilization in the humid tropics. Bangkok, Department of Science, 1958, 168 p.

SHULTZ, J. P. *Ecological studies on rain forest in northern Surinam.* North-Holland, 1960.

SMITH, A. P. Stratification of temperate and tropical forests. *The American Naturalist,* vol. 107, no. 957, 1973, p. 671–683.

STERNBERG, H. O. La percepción cambiante de los recursos naturales y la región amazónica. *Revista Florestal Venezolana* (Facultad de Ciencias forestales, ULA, Mérida), 23, 1973.

THOMAS, W. L. (ed.). *Man's role in changing the face of the earth.* Chicago, Univ. of Chicago Press, 1956, 1 193 p.

THORNTHWAITE, C. W.; HARE, F. K. Climatic classification in forestry. *Unasylva* (FAO), 9, 1955, p. 51–59.

UNESCO. *Ecological effects of increasing human activities on tropical and sub-tropical forest ecosystems.* Paris, Unesco, MAB report series no. 3, 1972, 35 p.; no. 16, 1974, 96 p.

WADSWORTH, F. H. Natural forests in the development of the humid American tropics. In: *The use of ecological guidelines for development in the American humid tropics* (Proceedings of international meeting, 20–22 February 1974, Caracas), p. 129–139. Morges, Switzerland, IUCN Publications, new series no. 31, 1975, 249 p.

WEBB, W. L. Development of incentive control of shifting cultivation. In: *Trans. National Conf. on the Kaingin problem* (Manila, Philippines, March 12–13, 1964).

WHITMORE, T. C. *Modes of variation in the composition of tropical lowland evergreen forest in the South-East Archipelago and their correlation with the environment.* Rome, FAO, 1975, 28 p. multigr.

Conservation of natural forest ecosystems

ADAMS, A. B. (ed.). *First World Conference on National Parks.* U.S. Dept. of Interior, National Park Service, 1962, 471 p.

ALLEN, P. H. *The rain forests of Golfo Dulce.* Gainsville, Univ. of Florida Press, 1956.

ANON. *Comparative studies of faunistic communities of Ujung Kulon Nature Reserve, S.W. Java.* Bogor, Indonesia, BIOTROP, 1973, 30 p.

——. *Un parque nacional en la República Dominicana.* Washington, D.C., Benchmarks, 1973, 52 p.

AUBERT DE LA RUE, E.; BOURLIÈRE, F.; HARROY, J. P. *The tropics*. New York, Knopf., 1957.

BATES, M. *The forest and the sea*. New York, Random House, 1960.

BISWAS, A. S.; BISWAS, M. R. State of the environment and its implications to resource policy development. *Bio-Science*, vol. 26, no. 1, 1976, p. 19–25.

BOZA, M. A. *Plan de manejo y desarrollo para el parque nacional Volcán Poas, Costa Rica*. Turrialba, Costa Rica, Instituto Interamericano de Ciencias Agrícolas (IICA), M.Sc. thesis, 1968.

BROWN, L. *Africa: a natural history*. New York, Random House, 1965, 300 p.

BUDOWSKI, G. *The classification of natural habitats in need of preservation in Central America*. Turrialba, Costa Rica, Instituto Interamericano de Ciencias Agrícolas, 1964, 35 p. multigr.

CONSERVATION FOUNDATION. *Dominica: a chance for a choice*. Washington, D.C., 1970.

CURRY-LINDAHL, K.; HARROY, J. P. *National parks of the world*. New York, Golden Press, 1972, vol. 1, 217 p., vol. 2, 240 p.

DARLING, F. F.; EICHHORN, N. D. *Man and nature in the national parks*. Washington, D.C., The Conservation Foundation, 1969, 86 p.

DASMANN, R. F. Towards a system for classifying natural regions of the world and their representation by natural parks and reserves. *Biological Conservation*, 4, 1972, p. 247–255.

——. *Classification and use of protected natural and cultural areas*. Morges, Switzerland, IUCN occasional paper no. 4, 1973.

——. Development of a classification system for protected natural and cultural areas. In: Elliott, H. (ed.). *Second World Conference on National Parks*, p. 338–396. Morges, Switzerland, IUCN, 1974, 504 p.

DE VOS, A. *Africa, the devastated continent*. The Hague, Junk, 1975, 236 p.

DI CASTRI, F.; LOOPE, L. *Thoughts on the biosphere reserve concept and its implementation*. Paris, Unesco, Division of Ecological Sciences, 1976, 13 p. multigr.

DORST, J. *South America and Central America: a natural history*. New York, Random House, 1967, 300 p.

DUARTE DE BARRIOS, W.; STRANG, H. E. *Training of national park personnel*. Quito (Ecuador), paper presented at FAO Latin American Forestry Commission, 1970.

ECKHOLM, E. P. *Losing ground: environmental stress and world food prospects*. Norton, 1976, 223 p.

EICHLER, A. *Parques nacionales y reservas afines. Política y planificación*. Mérida, Universidad de los Andes, Instituto de Investigaciones económicas, 1973, 221 p.

ELLIOTT, Sir Hugh (ed.). *Second World Conference on National Parks*. Morges, Switzerland, IUCN, 1974, 504 p.

FAO. Segundo taller internacional sobre el manejo de areas silvestres. *Proposición para el manejo del Parque y de la Reserva Nacional Iguazú, Argentina*. Puerto Iguazu, Argentina, 1973, 83 p. multigr.

——. *Planning interpretive programme in national parks. Manual*. Rome, 1976, 22 p.

——. *National parks planning. Manual with annotated examples*. Rome, 1976, 42 p.

——. *Report of the fifth session of the working party on wildlife management and national parks of the African Forestry Commission* (Bangui, 17–19 March 1976). Rome, FAO, FO:AFC/WL/76/Rep., 1976, 9 p.+annexes.

FORSTER, R. R. *Planning for man and nature in national parks: reconciling perpetuation and use*. Morges, Switzerland, IUCN Publications, new series no. 26, 1973.

FOSBERG, R. (ed.). *Man's place in the island ecosystem*. Honolulu, Bishop Museum Press, 1963.

GALINDO, H.; GABALDÓN, M. *Venezuela 1974 Parques nacionales y Monumentos naturales*. Caracas, Ministerio de Agricultura y Cría, Dirección de Recursos naturales renovables, 1974.

GONDELLES, A. R. *Parque nacional Canaima*. Caracas, Corporación de Turismo de Venezuela y Ministerio de Agricultura y Cría, 1974.

HART, W. J. *A systems approach to park planning*. Morges, Switzerland, IUCN, 1966.

HOFMANN, R.; PONCE DEL PRADO, C. *El Gran Parque Nacional del Manu*. Lima, Ministerio de Agricultura, Oficina de Información Técnica, informe no. 17, 1971, multigr.

HOOPER, M. D. The size and surroundings of nature reserves. In: Duffey, E.; Watt, A. S. (eds.). *The scientific management of animal and plant communities for conservation*, p. 555–561. Oxford, Blackwell, 1971.

IUCN. *Conservation of nature and natural resources in modern African States*. Morges, Switzerland, IUCN Publications, new series no. 1, 1963, 367 p.

——. *The ecology of man in the tropical environment*. Morges, Switzerland, IUCN Publications, new series no. 4, 1964, 355 p.

——. *Proceedings of the Latin American Conference on the Conservation of renewable natural resources* (Bariloche, Argentina). Morges, Switzerland, IUCN Publications, new series no. 13, 1968.

——. *United Nations list of national parks and equivalent reserves*. Second edition. Morges, Switzerland, IUCN Publications, 1971, 601 p.

——. *Ecological guidelines for the use of natural resources in the Middle East and South West Asia*. Morges, Switzerland, IUCN, 1976.

KEAST, A. *Australia and the Pacific Islands: a natural history*. New York, Random House, 1966, 300 p.

KROPP, G. *Wildlife and national park legislation in Latin America*. Rome, FAO Legislative Studies no. 2, 1971.

LAMPREY, H. *The distribution of protected areas in relation to the needs of biotic community conservation in eastern Africa*. Morges, Switzerland, IUCN occasional paper no. 16, 1976, 85 p.

LAZARTE, P. B. *La región de Guayacán, Costa Rica, y sus posibilidades como reserva biológica*. Turrialba, Instituto Interamericano de Ciencias Agrícolas, M. Sc. thesis, 1967, 140 p. multigr.

LUNA, L. A. Problemática de las reservas forestales en Venezuela. *Revista Florestal Venezolana* (Facultad de Ciencias forestales, ULA, Mérida), 23, 1973, p. 21–32.

MEGGERS, B. J. *Amazonia: man and culture in a counterfeit paradise*. Chicago, New York, Aldine Atherton, 1971, 182 p.

MILLER, K. R. *Estrategia general para un programa de manejo de parques nacionales en el norte de Colombia*. Turrialba, IICA, 1968, 67 p.

——. *El programa de manejo y desarrollo de los parques nacionales de la CVM, Colombia*. Turrialba, IICA, 1968.

——. *Scheduling the development of national parks*. Santiago, Chile, FAO, 1971, multigr.

——. Development and training personnel. The foundation of national park programs in the future. In: Elliott, H. (ed.). *Second World Conference on National Parks*, p. 326–347. Morges, Switzerland, IUCN, 1974, 504 p.

——. Ecological guidelines for the management and development of national parks and reserves in the American humid tropics. In: *The use of ecological guidelines for development in the American humid tropics* (Proceedings of international meeting, 20–22 February 1974, Caracas), p. 91–108. Morges, Switzerland, IUCN Publications, new series no. 31, 1975, 249 p.

MILLER, K. R.; BORSTEL, K. R. Von. *Parque nacional histórico Santa Rosa, Guanacaste, Costa Rica.* Turrialba, Costa Rica, ICT/IICA/FAO, 1968, 76 p.

MONDOLFI, E. *Proyecto de reestructuración de los servicios y programas de fauna silvestre del Ministerio de Agricultura y Cría.* Caracas, Consejo de Bienestar Rural, 1974, 120 p.

MUELLER-DOMBOIS, D. *Natural area system development for the Pacific region, a concept and symposium.* U.S. IBP, technical report no. 26, 1973.

MYERS, N. An expanded approach to the problem of disappearing species. *Science,* 193, 1976, p. 198–202.

NICHOLSON, E. M. *Handbook of the conservation section of the International Biological Programme.* IBP Handbook no. 5. Oxford, Blackwell, 1968.

PFEFFER, P. *Asia: a natural history.* New York, Random House, 1968, 300 p.

PURI, G. S. *Indian forest ecology.* New Delhi, Oxford Book and Stationery, 1960. London, George Allen and Unwin, 1962.

RICHARDS, P. W. National parks in wet tropical areas. In: Elliott, H. (ed.). *Second World Conference on National Parks,* p. 219–227. Morges, Switzerland, IUCN, 1974, 504 p.

ROUTLEY, R. and V. *The fight for the forest.* Canberra, Australian National University, Research School of Social Sciences, 1974, 407 p.

RUSSELL, E. W. *Management policy in the national parks.* Arusha, Tanzania, Tanzania National Parks, 1970, 24 p.

SARTORIUS, P. *Sociological and environmental consequences of the reduction or elimination of primary tropical forests.* Rome, FAO, 1975, 15 p. multigr.

SLATYER, R. O. Ecological reserves; size, structure and management. In: Fenner, F. (ed.). *A national system of ecological reserves in Australia,* p. 22–38. Canberra, Reports of the Australian Academy of Science, no. 19, 1975, 114 p.

SOEGENG REKSODIHARDJO, W.; ANDERSON, J. A. R.; PHUNG TRUNG NGAN. *Preliminary report on investigation of the Kutei Nature Reserve, East Kalimantan, Indonesia.* Bogor, Indonesia, BIOTROP, 1974, 33 p.

STERNBERG, H. O. Man and environmental change in South America. In: Fittkau, E. J. *et al.* (eds.). *Biogeography and ecology in South America,* vol. 1, p. 413–445. The Hague, Junk, Monographiae Biologicae no. 18, 1968, 446 p.

STONE, B. C. National parks as a national resource. In: Stone, B. C. (ed.). *Natural resources in Malaysia and Singapore* (Proceedings of the Second Symposium on Scientific and Technological Research in Malaysia). Kuala Lumpur, 1967, 265 p.

TOSI, J. A. *Zonas de vida natural en el Perú. Memoria explicativa sobre el mapa ecológico del Perú.* Boletín Técnico no. 5, Zona Andina, Proyecto 39. Turrialba, Instituto Interamericano de Ciencias Agrícolas, 1960, 271 p.

TOWLE, E. L.; HANIF, M. *National parks in the Caribbean area.* St. Thomas, U.S. Virgin Islands, Occasional Paper no. 2, June 1973, 19 p.

TRZYNA, T. C.; COAN, E. V. (eds.). *World directory of environmental organizations.* 2nd edition. Sequoia Institute (Claremont, P.O.B. 30, California 91711), 1976, 288 p.

UDVARDY, M. D. F. *A classification of the biogeographical provinces of the world.* Morges, Switzerland, IUCN occasional paper 18, 1975, 48 p.

UNESCO. *Use and conservation of the biosphere.* Paris, 1970, 272 p.
——. *Expert panel on Project no. 8: Conservation of natural*

areas and of the genetic material they contain. Paris, Unesco, MAB report series no. 12, 1973, 64 p.

UNESCO. *Task force on: Criteria and guidelines for the choice and establishment of biosphere reserves.* Paris, Unesco, MAB report series no. 22, 1974, 61 p.

UNITED STATES NATIONAL PARK SERVICE. *Kilimanjaro: a survey for the proposed Mount Kilimanjaro National Park, Tanzania.* Washington, D.C., 1970.

——. *A master plan for the preservation and use of Tikal National Park* (USAID/Government of Guatemala). Washington, D.C., 1971.

Conservation of forest genetic resources

BOUVAREL, P. *Report on forest genetic resources.* Rome, FAO, International Board for Plant Genetic Resources (IBPGR), second meeting, 1975, 17 p. multigr.

FAO. *Forest genetic resources information no. 3.* Rome, Forestry Occasional Paper 1975/1, 68 p.

——. *Report of the third session of the FAO panel of experts on forest gene resources* (Rome, May 1974). Rome, 1975, 90 p.

——. *Report on the FAO/DANIDA training course on forest seed collection and handling* (Chiang Mai, Thailand, February-March 1975). Rome, 1975, vol. 1, 80 p.; vol. 2, 453 p.

——/UNEP. *The methodology of conservation of forest genetic resources* (Rome, 1974). Rome, 1975, FO:MISC/78/8, 127 p.

FRANKEL, O. H.; BENNETT, E. (eds.). *Genetic resources in plants: their exploration and conservation.* Oxford, Blackwell, 1970.

——; HAWKES, J. G. *Crop genetic resources for today and tomorrow.* Cambridge University Press, 1975, 491 p.

GULDAGER, P. *Ex situ* conservation stands in the tropics. In: *The methodology of conservation of forest genetic resources,* p. 85–92. Rome, FAO/UNEP, 1975, 127 p.

JONG, K. Malaysian tropical forest; an unexploited genetic reservoir of edible fruit tree species. In: *Proceedings Symp. Biol. Res. and Nat. Dev.,* 1973, p. 113–121.

MIKSCHE, J. P. (ed.). *Modern methods in forest genetics.* Berlin, New York, Springer Verlag, 1976, 288 p.

ROCHE, L. Biological background. In: *The methodology of conservation of forest genetic resources,* p. 5–18. Rome, FAO/UNEP, 1975, 127 p.

——. Tropical hardwoods. In: *The methodology of conservation of forest genetic resources,* p. 65–78. Rome, FAO/UNEP, 1975, 127 p.

——; OLA-ADAMS, B. A. *Gene resources conservation: IUFRO Working Party S2.02.2. Progress Report 1972–1975.* Bangor, North Wales, United Kingdom, University College of North Wales, Department of Forestry and Wood Science, 13 p. multigr.

STERN, K.; ROCHE, L. *Genetics of forest ecosystems.* Berlin, New York, Springer Verlag, Ecological Studies no. 6, 1974, 330 p.

UNESCO. *Conservation of natural areas and of the genetic material they contain.* Paris, MAB report series no. 12, 1973, 64 p.

Human and institutional problems

ADEYOJU, S. K. *Land tenure problems and tropical forestry development.* Rome, FAO (Committee on forest development in the tropics, 4th session, November 1976), FO:FDT/76/5(b), 1976, 36 p.

BELDING, H. S. Symposium 5: Work in hot environment. In: *XVI International Congress on Occupational Health* (Tokyo), 1969.

FAO. *Contrats d'exploitation forestière sur domaine public. Manuel de référence.* Rome, 1972, 173 p.

——. *Heat stress in forest work* (by AXELSON, O.). Rome, TF-INT 74 (SWE), 1974a, 31 p.

——. *Logging and log transport in man-made forests in developing countries.* Rome, FAO/SWE/TF 116, 1974, 134 p.

——. *Employment in forestry. Report on the FAO/ILO/SIDA Consultation on employment in forestry* (Chiang Mai, Thailand, February-March 1974). Rome, FAO/SWE/TF 126, 1974, 27 p.+annex 486 p.

——. *The state of food and agriculture. La situation mondiale de l'alimentation et de l'agriculture* (1974). Rome, 1975, 196 p.

——/ILO/ECE. *Symposium on ergonomics applied to forestry* (Joint Committee on forest working techniques and training of forest workers), vol. 1. Geneva, 1971, 131 p.

HANSSON, J. E.; LINDHOLM, A.; BIRATH, H. *Men and tools in Indian logging operations. A pilot study in ergonomics.* Stockholm, Royal College of Forestry, Research notes no. 29, 1966, 27 p.

ILO. Second tripartite technical meeting for the timber industry. Geneva, 1973. Report I. *General report: effect given to the conclusions of the first meeting*, 65 p. Report II. *Conditions of work and life in the timber industry*, 115 p. Report III. *The training of managers and workers in the wood working industries*, 36 p.

KING, K. F. S. A plan of action for the next 6 years. A summary of the revised FAO study on forest policy, law and administration. In: *7th World Forestry Congress* (Buenos Aires, 4–18 October 1972), 76 p.

LEITHEAD, C. S.; LIND, A. R. *Heat stress and heat disorders.* London, Cassel, 1964.

MUELLER-DARSS, H. *Ergonomics of forest work in the tropical moist forests.* Rome, FAO, 1975, part I, 9 p., part II, 11 p. multigr.

SCHMITHÜSEN, F. *Forest utilization contracts on public land in the humid tropics: experiences, problems and trends.* Rome, FAO (Committee on forest development in the tropics, 4th session, November 1976), FO:FDT/76/5(c), 1976, 29 p.

SVANQVIST, N. *Employment opportunities in the tropical moist forests under alternative silvicultural systems, including agri-silvicultural techniques.* Rome, FAO (Committee on forest development in the tropics, 4th session, November 1976), FO:FDT/76/6(b), Add. 1, 1976, 107 p.

VELAY, L. *Administrative organization of forestry in the developing countries.* Rome, FAO (Committee on forest development in the tropics, 4th session, November 1976), FO:FDT/76/5(a), 1976, 28 p.

Economics of forest products

BRAZIER, J. D. *Defining end use property requirements as a contribution to the more efficient use of tropical forest resources.* Rome, FAO, 1975, 9 p. multigr.

DSE/FAO. *Properties, uses and marketing of tropical timber.* Proceedings of an international meeting organized by the German Foundation for international development (DSE) in collaboration with FAO (Berlin-Tegel), 1973, vol. 1, Final report, 37 p.; vol. 2, Meeting papers, 236 p.

EARL, D. E. *Forest energy and economic development.* Oxford, Clarendon Press, 1975, 128 p.

ECE. *Study of timber trends and prospects in the ECE region 1950–2000.* Timber Committee, 33rd session (Geneva, October 1974). Chapter 4 (The special role of tropical hardwoods in the region's trade and consumption of forest products), 53 p. Chapter 9 (International trade), 37 p.

ERFURTH, T. International trade and trade flows of tropical wood products. In: *Proceedings of an international meeting on properties, uses and marketing of tropical timber*, organized by the German Foundation for international development in collaboration with FAO (Berlin-Tegel), p. 99–108; vol. 2, 1973, 236 p.

——. *Product development and the choice and effective application of promotional measures to advance the wider use of products from the tropical moist forests.* Rome, FAO (Committee on forest development in the tropics, 4th session, November 1976), FO:FDT/76/10(b), 1976, 10 p.

FAO. *Le bois. Evolution et perspectives mondiales.* Rome, 1967, 133 p.

——. Seventh World Forestry Congress (Buenos Aires, 1972). Rome, *Unasylva* (FAO), no. spécial 104, 1972, 108 p.

——. *The marketing of tropical wood. A. Wood species from African tropical moist forests* (by ERFURTH, T.; RUSCHE, H.). *Commercialisation des bois tropicaux. A. Essences des forêts tropicales humides d'Afrique* (par ERFURTH, T.; RUSCHE, H.). Rome, 1976, 60 p. (72 p.). B. *Wood species from South American tropical moist forests* (in preparation). C. *Wood species from South-East Asian tropical moist forests* (in preparation).

——/ECE. *Study on the consumption of tropical hardwoods in Europe.* Suppl. 9 to vol. 19 of the FAO/ECE Timber Bulletin for Europe. Geneva, February 1967, 33 p.

——/——. *European timber trends and prospects, 1950–1980, an interim review.* Suppl. 7 to vol. 21 of the Timber Bulletin for Europe. Geneva, 1969, vol. 1, 182 p.; vol. 2, 112 p.+27 charts.

HASAN, S. M. *Bamboo: the most important constituent of tropical moist forests; a critical assessment of research done and suggestions for future improvements.* Rome, FAO, 1975, 11 p.

KYRKLUND, B.; ERFURTH, T. *The future of mixed tropical hardwoods: an important renewable natural resource.* Rome, FAO (Committee on forest development in the tropics, 4th session, November 1976), FO:FDT/76/9(a), 1976, 16 p.

LESLIE, A. J. Targets, policies and inputs in forestry. *Unasylva* (FAO), 95, 1969, p. 3–18.

LONGWOOD, F. R. *Puerto Rican woods, their machining, seasoning and related characteristics.* U.S. Department of Agriculture, Forest Service, Agriculture Handbook no. 205, 1961, 98 p. (Institute of Tropical Forestry, Río Piedras, Puerto Rico).

——. *Present and potential commercial timbers of the Caribbean, with special reference to the West Indies, the Guianas and British Honduras.* U.S. Department of Agriculture, Forest Service, Agriculture Handbook no. 207, 1962, 167 p.; reprinted 1971 (Institute of Tropical Forestry, Río Piedras, Puerto Rico).

PECK, T. J. Worldwide information on uses and consumption patterns: surveys of end-uses for tropical hardwood. Objectives and result. In: *Proceedings of an international meeting on properties, uses and marketing of tropical timber*, organized by the German Foundation for international development in collaboration with FAO (Berlin-Tegel), p. 3–14; vol. 2, 1973, 236 p.

PRINGLE, S. L. Hardwoods. World supply and demand. *Unasylva* (FAO), 93, 94, 1969, p. 24–33, p. 34–39.

——. Tropical hardwood products: world summary of trends and prospects in demand, supply and trade. In: *Proceedings of an international meeting on properties, uses and marketing of tropical timber*, organized by the German Foundation for international development in collaboration with FAO (Berlin-Tegel), p. 85–90; vol. 2, 1973, 236 p.

——. *The role of the tropical moist forests in world demand, supply and trade of forest products.* Rome, FAO (Committee on forest development in the tropics, 4th session, November 1976), FO:FDT/76/10(a), 1976, 22 p.

TAKEUCHI, K. *Tropical hardwood trade in the Asia-Pacific region.* World Bank Staff occasional paper no. 17. Baltimore, London, John Hopkins University Press, 1974.

VAKOMIES, P. Basic requirements for principal industrial uses. In: *Proceedings of an international meeting on properties, uses and marketing of tropical timber,* organized by the German Foundation for international development in collaboration with FAO (Berlin-Tegel), p. 19–22; vol. 2, 1973, 236 p.

New techniques

CENTRE TECHNIQUE FORESTIER TROPICAL (CTFT). *Recherches et essais effectués sur les bois tropicaux par divers organismes de recherche.* Nogent-sur-Marne, France, 1972, 291 p.

COLLARDET, J. *Improvements and adjustments in industrial processing to encounter problems specific to the use of wood species with varying properties from the humid tropics.* Rome, FAO (Committee on forest development in the tropics, 4th session, November 1976), FO:FDT/76/9(c), 1976, 22 p.

FAO. *The production, handling and transport of wood chips.* Rome, TF-INT 55 (NOR), 1973, 86 p.

——. *Committee on forest development in the tropics. Report of the 4th session* (November 1976). Rome, 1977.

——. *Pulping and paper making properties of fast growing plantation wood species.* Rome, 1975, 466 p.

——. *Proceedings of the World Consultation on wood-based panels* (New Delhi, February 1975). Brussels, Miller Freeman Publications, 1976, 442 p.

——/ECE. *Recherches sur les bois feuillus tropicaux.* Suppl. 5 au vol. 22 du Bulletin du Bois pour l'Europe. Genève, avril 1970, 39 p.

KING, K. F. S. It's time to make paper in the tropics. *Unasylva* (FAO), vol. 27, no. 109, 1975, p. 2–5.

MARRA, G. G. Le bois, élément de synthèse. *Unasylva* (FAO), vol. 27, no. 108, 1975, p. 2–9.

PETROFF, G. La production de cellulose dans les pays de l'OCAM. État actuel des projets. *Bois et Forêts des Tropiques* (Nogent-sur-Marne, France), n° 143, 1972, p. 35–44.

SIKES, J. E. G. A clean piece of paper. *Unasylva* (FAO), vol. 27, no. 109, 1975, p. 11–16.

SIMEON DE JESUS. How to make paper in the tropics. *Unasylva* (FAO), vol. 27, no. 109, 1975, p. 6–10.

Forest development and management planning

ANDEL, S. *The development of forest management regimes in peninsular Malaysia.* Rome, FAO, 1975, 11 p. multigr.

ANON. *Planned utilization of the lowland tropical forests* (symposium Cipayung, Bogor, Java). Bogor, Indonesia, BIOTROP, 1973, 263 p.

ARNOLD, R. K. Forest management by alternatives rather than systems. In: *Records of Proceedings 12th Pacific Science Congress* (Canberra), vol. 1 (Abstracts of papers), 1971, p. 109.

AUSTRALIAN UNESCO COMMITTEE FOR MAN AND THE BIOSPHERE (MAB). *Report of symposium on Ecological effects of increasing human activities on tropical and sub-tropical forest ecosystems* (University of Papua–New Guinea, 28 April–1 May 1975). Chapters on 'Use of the ecosystem for wood production', 'Use of the ecosystem for food production and

as a habitat', 'Monitoring and conserving the ecosystem and its role in development', 'Case study—The Gogol wood chip project'. Canberra, Australian Unesco Committee for Man and the Biosphere, Publication no. 3, Australian Government Publishing Service, 1976, 214 p.

CATINOT, R. Sylviculture tropicale en forêt dense africaine. *Bois et Forêts des Tropiques,* n° 100, 1965, p. 5–18; n° 101, 1965, p. 3–16; n° 102, 1965, p. 3–16; n° 103, 1965, p. 3–16; n° 104, 1965, p. 17–30.

CHAUVIN, H. *Factors conditioning the systems and costs for opening up and harvesting the tropical moist forests.* Rome, FAO (Committee on forest development in the tropics, 4th session, November 1976), FO:FDT/76/6(a), 1976, 17 p.

DONIS, C. La forêt dense congolaise et sa sylviculture. *Bulletin agricole du Congo belge* (Bruxelles), 2, 1956, p. 261–360.

FAO. *Tropical silviculture,* vol. I, II, III. Rome, 1957, 1958, 190, 415, 101 p.

——. *Committee on forest development in the tropics. Report of the first session* (October 1967). Rome, 1968, 170 p. *Report of the second session* (October 1969). Rome, 1970, 160 p. *Report of the third session* (May 1974). Rome, 1974, 65 p.+annexes.

——. *Simulation studies of forest sector development alternatives in West Malaysia.* Working paper no. 17 (FO:SF/MAL/72/009). Rome, 1972.

——. *An introduction to planning forestry development.* Rome, FAO/SWE/TF 118, 1974, 86 p.

——. *Forestry for community development. Proposal for FAO/SIDA programme.* Rome, FAO, 1975, 18 p.

——. *Forestry for community development: background paper.* FAO Regional Conference for Asia and Far East. Rome, FAO, 1976, 15 p.

——. *Report of the fourth session of the African Forestry Commission* (Bangui, 22–27 March 1976). Rome, FAO, FO:AFC/76/Rep., 1976, 14 p.+annexes.

——/UNDP. *Forestry development in Surinam. Forest management.* Project working document no. 5 (FO:SF/SUR/71/506). Rome, October 1973, 105 p.+maps.

——/——. *A development strategy for the forest sector of peninsular Malaysia.* Technical report no. 9 (FO:DP/MAL/72/009). Rome, 1974, 156 p.+topographical map of peninsular Malaysia.

FAO/UNDP. *An economic model for the planning of forest sector development strategy in peninsular Malaysia.* Rome, 1975, 78 p.

——/——. *Plantation management procedures for large scale plantations in peninsular Malaysia.* Working paper no. 36 (FO:DP/MAL/72/009). Kuala Lumpur, September 1975, 205 p.

FRASER, A. I. *Technical and economic implications of the management systems in moist tropical forests in Asia.* Rome, FAO, 1975, 13 p. multigr.

HAMILTON, L. S. *Tropical rain forest use and preservation: a study of problems and practices in Venezuela.* San Francisco, Sierra Club Special Publication, International Series no. 4, 1976, 72 p.+annexes.

LEE PENG CHOONG. Multi-use management of West Malaysia's forest resources. In: *Proceedings Symp. Biol. Res. and Nat. Dev.,* 1973, p. 93–101.

LESLIE, A. J. *Economic problems in tropical forestry.* Rome, FAO, 1971, 201 p.

——. *Economic implications of the management systems applied to the tropical moist forests.* Rome, FAO (Committee on forest development in the tropics, 4th session, November 1976), FO:FDT/76/7(c), Add. 1, 1976, 31 p.

NWOBOSHI, L. C. *Problems and prospects of natural regeneration in the future management of the tropical moist forest for timber production*. Rome, FAO, 1975.

ROCHE, L. *Priorities for forestry research and development in the tropics*. Report to International Development Research Centre (IDRC, Ottawa), first draft for a workshop (University of Reading, Plant Science Laboratories, 11 June 1976), 105 p. multigr. (Department of Forestry and Wood Science, University College of North Wales, Bangor, United Kingdom).

SARGENT, K. *Factors in the planning of wood based industrial development in the context of South-East Asia*. Rome, 1975, 17 p. multigr.

SYNNOTT, T. J.; KEMP, R. H. *The relative merits of natural regeneration, enrichment planting and conversion planting in tropical moist forests, including agri-silvicultural techniques*. Rome, FAO (Committee on forest development in the tropics, 4th session, November 1976), FO:FDT/76/7(a), 1976, 12 p.

VANNIÈRE, B. *Les possibilités d'aménagement de la forêt dense africaine*. Nogent-sur-Marne, CTFT, 1974, 75 p.

——. *Influence de l'environnement économique sur l'aménagement en Afrique tropicale*. Rome, FAO, 1975, 17 p. multigr.

WESTOBY, J. C. Les industries forestières dans la lutte contre le sous-développement économique. *Unasylva* (FAO), vol. 16, n° 4, 1963.

Part III

Some regional case studies

Structure and functioning of evergreen rain forest ecosystems of the Ivory Coast

by F. Bernhard-Reversat[1], C. Huttel[2] and G. Lemée[3]

1. Centre ORSTOM (Laboratoire de Botanique), B.P. 1386, Dakar, Sénégal.
2. Centre ORSTOM (Laboratoire de Botanique), B.P. V 51, Abidjan, Côte-d'Ivoire.
3. Laboratoire d'Écologie, Faculté des Sciences, Université de Paris XI, 91405 Orsay.

Introduction

Research on the evergreen forest ecosystems of the lower Ivory Coast has been carried out within the framework of the International Biological Programme (IBP) by the Office de la Recherche scientifique et technique outre-mer (ORSTOM) with financial support from the IBP. This research at the ORSTOM Centre of Adiopodoumé, near Abidjan, began in 1966 and involved many scientific disciplines. This team benefited from the experience of the Centre technique forestier tropical (CTFT) in the Ivory Coast.

Ecological systems as complex as those of the tropical rain forests pose complicated methodological problems and some, especially those on faunal aspects, are still being evolved. Observations have had to be limited to the essential aspects of ecosystems' functioning: the quantitative and structural characterization of the phytocenosis, the measurement of primary productivity and the decomposition of the organic matter produced, and the characterization of the cycles of water and the major mineral elements.

Geographical setting

Forest still covers approximately a third of the Ivory Coast; the remainder being cultivated areas and savannas. Two main types of forest may be distinguished each occupying about half the forest area: the evergreen forest in the south and south west (the ombrophilous sector), and the semi-deciduous forest (the mesophilous sector) to the north of this. Studies were made on two different geological formations in the evergreen forest.

Localities studied

The *Banco National Park* is a few miles west of Abidjan, in the sublittoral zone of Tertiary sands. This type of forest has been largely cleared for cultivation. One of the remaining parts, the Banco forest, was made into a reserve in 1924. It has an area of 30 km² and approximately occupies the drainage basin of the Banco river. The central part has been conserved. Although it is located in a well-populated area, the forest is little influenced by human activity but the larger animals have largely disappeared.

The *Yapo forest* is 45 km north of the Banco forest on

TABLE 1. Climatic characteristics of localities near the Banco and Yapo forests.

	January	February	March	April	May	June	July	August	September	October	November	December	Year
Rainfall (mm)													
Banco[1] (1935–1973)													
minimum	0	0	46	21	118	181	6	5	9	16	41	15	1 369
maximum	132	161	273	332	538	1 059	868	216	338	627	385	155	3 287
mean	*41*	*55*	*106*	*138*	*282*	*602*	*267*	*61*	*102*	*191*	*166*	*84*	*2 095*
Yapo[2] (1933–1973)													
minimum	0	0	25	8	84	167	2	2	16	30	74	4	1 213
maximum	203	162	321	294	394	524	417	215	402	412	418	169	2 456
mean	*30*	*54*	*138*	*153*	*236*	*315*	*163*	*64*	*122*	*218*	*174*	*72*	*1 739*
Shade temperature (°C)													
Adiopodoumé[3] (1950–1972)													
mean minimum	22.0	22.8	23.1	23.2	22.9	22.4	21.7	21.2	21.8	22.5	22.4	22.1	22.3
mean maximum	31.2	32.1	32.3	31.9	31.1	28.8	27.8	27.4	28.0	29.3	30.6	30.5	30.0
mean	*26.6*	*27.5*	*27.7*	*27.5*	*27.0*	*25.6*	*24.7*	*24.3*	*25.0*	*25.9*	*26.5*	*26.3*	*26.2*
Length of insolation (hours)													
Adiopodoumé (1956–1972)	161	176	196	182	172	84	87	74	84	157	182	166	1 722
Global solar radiation (cal/cm²)													
Adiopodoumé (1968–1972)	10 759	12 022	14 486	13 759	13 332	9 641	8 506	8 110	9 430	12 497	11 881	11 537	135 960
Potential evapotranspiration (mm, Turc formula)													
Adiopodoumé (1956–1972)	105	113	128	121	113	77	79	77	81	108	112	105	1 219

1. Banco School of Forestry.
2. Forest House at South-Yapo.
3. ORSTOM bioclimatology laboratory.

Birrimian schist. It is an example of a forest type that is widely distributed in the ombrophilous sector and is a relatively large area containing forest plantations. It is a *forêt classée*, so limited forestry is authorized. The locality studied shows traces of fairly recent exploitation, but, villages being distant, human intervention remains negligible.

Climate

The climate of the Ivory Coast was described by Eldin (1971, 1974). Local data were provided by ASECNA publications for Banco, by CTFT archives for Yapo, and by the ORSTOM bioclimatology laboratory at Adiopodoumé (Gosse and Eldin, 1973). Table 1 gives the monthly and annual averages of climatic factors at stations near the forests studied.

There is an equatorial pattern with two wet seasons: the heaviest, from March to July, has a maximum in June; the lighter, from September to November, has a maximum in October. Thus two drier seasons are defined, with the long dry season in January and the other in August (little dry season). Rain intensity differs at the two stations in June and July, being higher at Banco, which is nearer to the coast; the difference is found in the annual totals.

This pattern represents an average established over 40 years, but variations from one year to the next may be very large, the annual total varying from the normal to the double at Yapo and being even greater at Banco. The length and intensity of the four defined seasons are also very variable from year to year, since the lesser dry season may be completely non-existent (Yapo in 1973).

Atmospheric humidity is always very high, never below a relative humidity of 65 per cent on the sunniest days. Monthly averages vary between 80 and 86 per cent except in August when the minimum is 77 per cent. In the long dry season, short periods of dry north *harmattan* wind may lower the relative humidity to 45 per cent or less over several days, particularly in January.

The annual average *temperature* cycle shows one maximum, at the end of the long dry season, and one minimum, in the little dry season. The annual variation is less than 4° C, although the daily average variation is between 6° in the long wet season and 9° in the long dry season.

The length of insolation and the global radiation flow show two maxima; one in March at the end of the long dry season, the other, less pronounced, in October-November, in the short wet season.

The monthly values of potential evapotranspiration (PET), shown in table 1, have been calculated by the Turc formula:

$$PET = 0.4 \frac{t}{t+15} \times \left[\left(0.62 \frac{h}{H} + 0.18 \right) I_{ga} + 50 \right]$$

where t is the average temperature, H the astronomic length of the day (in hours), h the daily length of insolation (in hours) and I_{ga} the solar radiation energy which would reach the earth in the absence of atmosphere (in cal/cm^2/d). The measurement of PET by an evapotranspirometer under a

lawn of *Paspalum notatum* at the ORSTOM bioclimatology laboratory gave values close to those calculated. Its seasonal variation parallelled solar radiation.

The water shortage is measured by the difference between the PET and the rainfall over a given period. The normal long dry season occurs during the months of December, January, February and part of March, with an average cumulative water shortage of 170 mm. This is underestimated and Eldin (1971) proposes a correction of 20 per cent; this increases it to 200 mm. The little dry season only occurs during August, with a small average shortage. Important differences in the length and severity of the shortage occur between years, and this variability constitutes in itself a very important climatic characteristic.

Therefore the evergreen forest climate of the lower Ivory Coast has four seasons:

— a long dry season of 3–4 months (December to March), sunny, hot with a high PET;
— a long wet season, maximum in June, sunny, temperature and PET progressively decreasing;
— a short dry season in August-September, very cloudy, cool, with low PET;
— a little wet season in October-November, less cloudy and with a higher PET.

Geology and geomorphology

The sedimentary zone of Tertiary sands occurs along a strip some 30 km wide on the central and eastern coast of the Ivory Coast. These are deposits of detritic sands, clay patches being found in many places: this zone forms a plateau at nearly 100 m altitude, intersected by valleys 50 m deep at Banco, whose slopes may reach 50 per cent. Of the two Banco plots, one is located on the plateau and the other on the flat valley bottom; they are 400 m apart with a difference of 50 m in altitude. The Tertiary sands are very permeable and show a preponderance of water displacement by percolation. Conditions are optimal for good infiltration and the formation of a deep water-table.

The Yapo forest is located on a schistous metamorphic parent rock, mainly of arkosic or clay schist with veins of quartz. This formation covers the entire south east of the Ivory Coast north of the Tertiary sands. The topography consists of a succession of hills with rounded summits and valleys with gentle slopes. The plot studied is on a hill-top.

Soils

Under the French classification, the soils of these two forests come into the category of largely desaturated ferrallitic soils, in the impoverished group (Perraud, 1971). Desaturation involves the B layer, and the largely desaturated soils of the lower Ivory Coast are defined by the sum of the exchangeable bases inferior to 1 milliequivalent/100 g, the saturation rate of the absorbant complex inferior to 20 per cent and a pH less than 5.5. The impoverished group is characterized by the loss of clay in at least the upper 40 cm.

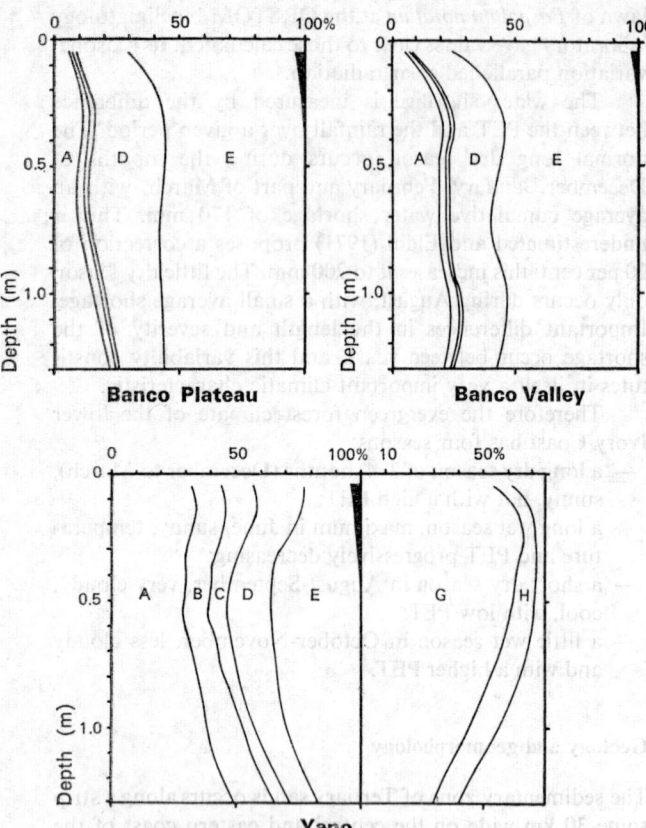

FIG. 1. Soil texture at the three stations (C. Huttel, 1975).

A: clay (<2 m)
B: fine silt (2 to 20 μm)
C: coarse silt (20 to 50 μm)
D: fine sand (50 to 200 μm)
E: coarse sand (0.2 to 2 mm)
G: all fine elements (up to 2 mm)
H: coarse elements (>2 mm)
In dark: organic matter

A to E: in percentages of earth sieved at 2 mm (fine fraction).
G and H: in percentages of total sample.

Fig. 1 shows the textures of the soils on the three plots studied. The Banco soils are very sandy. The valley soil, a colluvial deposit, is of sandy clay over the whole profile. The plateau soil shows a light clayey sand surface resting on a thick clay-sand layer. At Yapo the soils are characterized by a gravelly surface 100–150 cm deep. The content of clay and silt of the fine fraction is higher than at Banco, the soil being clayey-sand at the surface and clayey below 120 cm in depth.

Composition and structure

Mangenot (1955), applying the analytical method of the Zurich-Montpellier school to all the vascular vegetation except epiphytes, defines three associations in the *Uapace-*

talia class (ombrophilous forests at low altitudes on well-drained soils):

— *Turraeantho-Heisterietum*, a psammohygrophilous forest linked to heavy rainfall and a sandy soil;
— *Diospyro-Mapanietum*, a pelohygrophilous forest linked to a drier climate but on a soil rich in clay;
— *Eremosphatho-Mabetum*, a subhygrophilous forest present in the same climatic region as Diospyro-Mapanietum, but on soils containing less clay.

This classification was adopted and mapped for the entire Ivory Coast by Guillaumet and Adjanohoun (1971). The Banco forest belongs to the first of these associations and the Yapo to the second.

Floristic composition

The floristic composition of tree species was established from complete inventories of 50×50 cm quadrats of all individuals >40 cm GBH. Sampling 20–40 cm GBH trees and of the shrub and herb strata was carried out on some quadrats.

Table 2 shows the tree species commonly found in at least one of the forests studied. Only ten of over 100 determined species occur frequently in only one forest and may be considered as characteristic. Many species are not numerous enough to allow their preferences to be charac-

TABLE 2. Average density of trees >40 cm GBH (individuals/ha) of the most numerous species in the forests of Banco nad Yapo.

Species	Banco	Yapo
Dacryodes klaineana	26	85
Strombosia glaucescens	25	37
Allanblackia floribunda	14	25
Coula edulis	13	18
Diospyros sanza-minika	8	22
Blighia welwitschii	15	4
Trichoscypha arborea	6	8
Combretodendron africanum	7	2
Cola nitida	7	2
Carapa procera	3	6
Vitex micrantha	4	3
Turraeanthus africana	15	2
Chrysophyllum albidum	6	+
Berlina confusa	4	
Tabernaemontana crassa	4	
Cola lateritia var. *maclaudii*	4	+
Baphia bancoensis	3	+
Monodora myristica	3	
Scottelia coriacea		15
Scottelia chevalieri		14
Drypetes aylmeri		14
Scytopetalum thieghemii		13
Tarrietia utilis	+	8
Coelocaryon oxycarpum	1	7
Garcinia gnetoides		5
Anthostema aubryanum		4

terized. The rarity of species may be an absolute rarity but may be due to:

— species that rarely reach 40 cm GBH but which are numerous in the smaller sizes, e.g. *Discoglypremna caloneura, Cleistanthus polystachyus, Maesobotrya barteri, Memecylon guineense*, etc.;

— species characteristic of other associations: from secondary or semi-deciduous forests such as *Alstonia congensis, Ceiba pentandra, Funtumia elastica, Musanga cecropioides, Macaranga* spp., *Lophira alata*; species from swampy lowland forests: *Symphonia globulifera, Uapaca esculenta, U. heudelotii, Cynometra ananta*.

99 species of trees were identified at Banco and 124 at Yapo on an area of *ca.* 5 ha per forest. To these 5–10 indeterminate species must be added. These show clearly the relative floristic poverty of African forests. Higher values are obtained in tropical America, and especially in Indo-Malesia. This difference is illustrated by the species/area curves compared with those from Surinam and Malaysia (fig. 2).

The differences in floristic diversity at the three plots are in agreement with Mangenot's (1950) findings that the pelohygrophilous association (Yapo) is the richest and the psammohygrophilous (Banco), the poorest. This difference is significant: the number of species counted per 0.25 ha survey is 25 at Banco and 29 at Yapo.

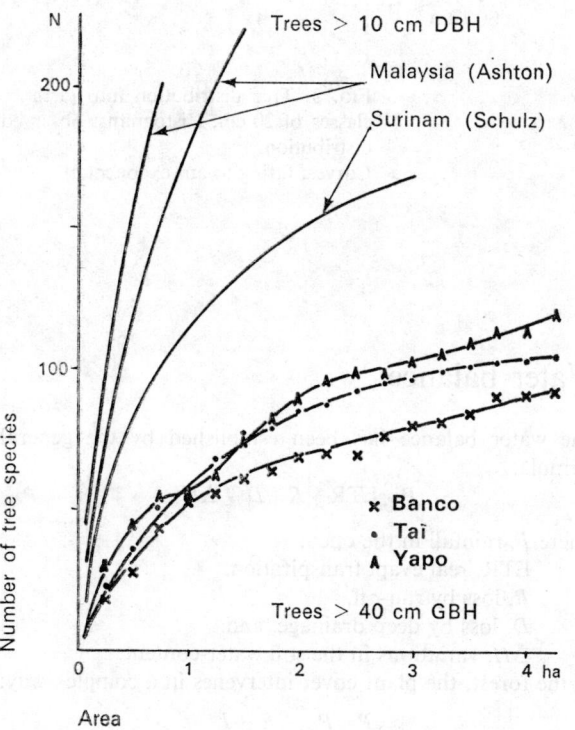

FIG. 2. Curve showing the relationship between area and numbers of tree species in three forests of the Ivory Coast, compared to three forests outside Africa.

Structure

Although densities (table 2) are very variable within the same forest, Yapo has a significantly higher density than Banco, although the density of trees over 120 cm GBH is the same in the two forests.

The basal areas (trunk areas at 1.3 m above ground level in m²/ha of ground) are shown in table 3. The values per plot vary from 21 to 44, with an average of *ca.* 31; there is no difference between the two forests.

The girth measurements were grouped into classes of 10 or 20 cm. There are many theoretical distributions attempting to fit such data (see chapter 5). Two were calculated: the hyperbole method of Pierlot (1968) and the exponential according to Pernes (personal communication).

TABLE 3. Density and basal area of the trees.

	Banco	Yapo
Density (trees/ha)		
GBH > 40 cm	265 ± 21	427 ± 44
GBH > 120 cm	66 ± 10	68 ± 4
Basal area (m²/ha)	30 ± 2.6	31 ± 2.3

In fig. 3 the observed distributions are shown by a histogram and the exponential theoretical curve. The fit is good at Banco, but at Yapo there are too many trees in the girth class 40–60 cm. The hyperbolic distributions cannot fit for either of the two forests.

Tree mortality at Banco seems to be constant and independent of size. This factor is found at Yapo only for individuals with a girth > 60 cm, the mortality rate at the passage from the first to the second class being higher than at the larger classes.

The distribution for the most abundant species shows differences. Exponential smoothing is suitable for small trees (*Diospyros sanza-minika* and *Strombosia glaucescens*) and for taller species, such as *Coula edulis* and *Allanblackia floribunda*, whereas *Turraeanthus africana, Blighia welwitschii, Dacryodes klaineana* have half-bell shaped curves that differ significantly from an exponential distribution. The exponential curves show a strong sciaphilous tendency of the species (as for small overshadowed trees) whilst the bell-shaped curves express a heliophilous tendency (larger emergent trees). The tallest trees are not numerous enough to allow a theoretical distribution to be calculated.

The most sciaphilous behaviour is found at Yapo and the most heliophilous at Banco. This tendency is evident in fig. 3: the very erect L-curves (Yapo) and the flattened L-curves (Banco) (Caussinus and Rollet, 1970).

The curve relating tree height to girth has a classic form with the small girths being represented by a steep slope. The points scatter and the curve flattens with increasing girth, but no plateau was reached. The logarithmic transformation gives a straight line of correlation (fig. 4). The correlation is highly significant ($r = 0.99$) and a test confirms that the relationship is linear.

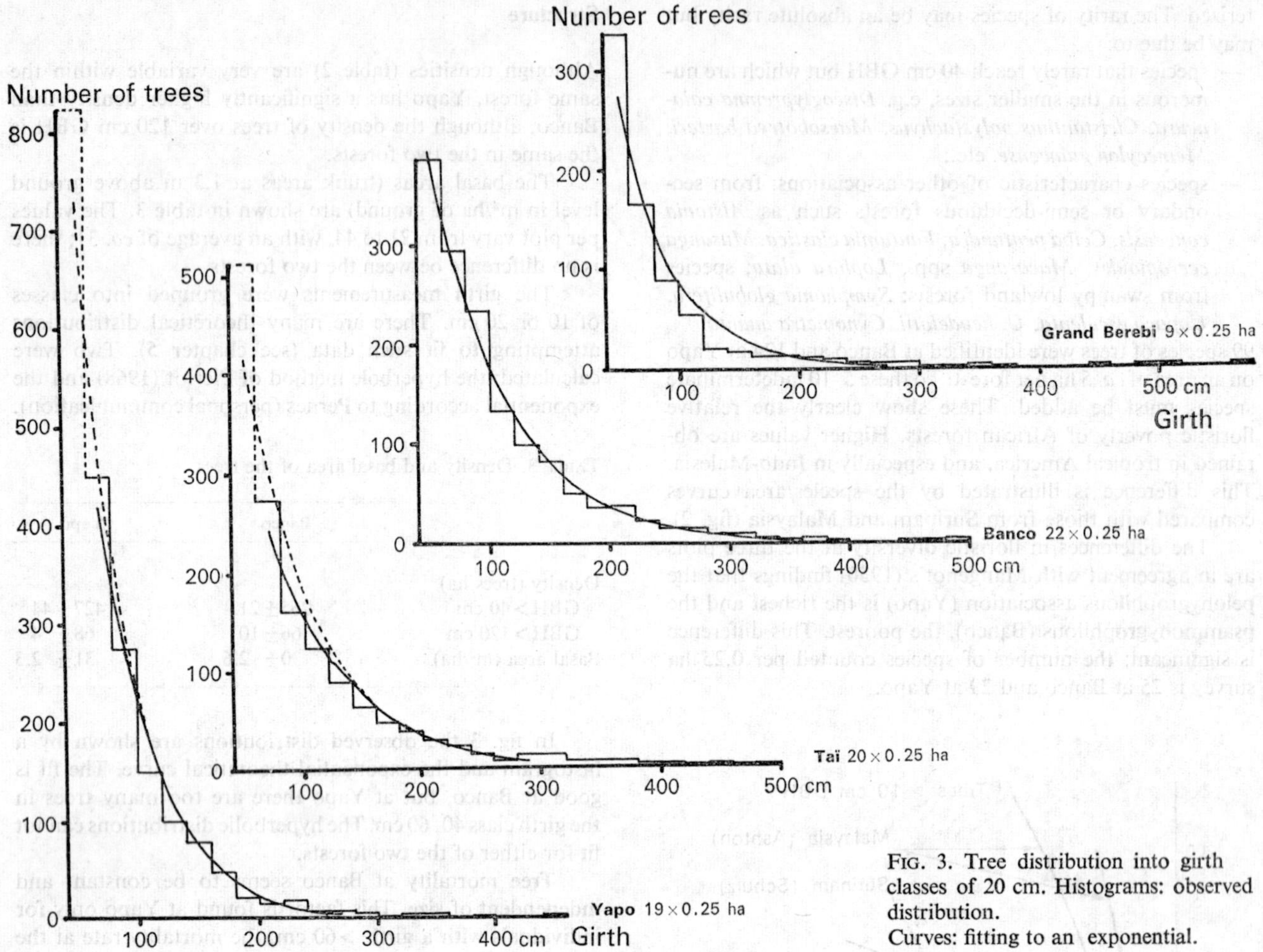

Number of trees

Number of trees

Number of trees

Grand Berebi 9 × 0.25 ha

Girth

Banco 22 × 0.25 ha

Taï 20 × 0.25 ha

Yapo 19 × 0.25 ha   Girth

FIG. 3. Tree distribution into girth classes of 20 cm. Histograms: observed distribution.
Curves: fitting to an exponential.

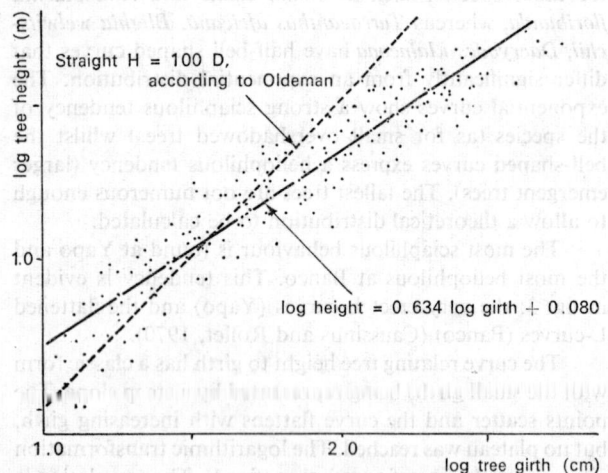

Straight H = 100 D, according to Oldeman

log height = 0.634 log girth + 0.080

log tree height (m)

log tree girth (cm)

FIG. 4. Relationship of tree height to girth in logarithmic coordinates.

Water balance

The water balance has been established by the general formula:

$$P = ETR + R + D \pm \Delta H$$

where P, rainfall in the open,

ETR, real evapotranspiration,

R, loss by run-off,

D, loss by deep drainage, and

ΔH, variations in the soil water-content.

In the forest, the plant cover intervenes in a complex way:

$$P = P_{sol} + E_t + I$$

where P_{sol}, throughfall,

E_t, stemflow, and

I, real interception.

The *rainfall* (P) was discussed in the climatic section above.

The *stemflow* (E_t) was measured for over a year in the Banco forest on a plot of 300 m², where all the 16 trees were fitted with a plastic drain spiral at the base of the trunk. This flow represented less than 1 per cent of the throughfall.

The *throughfall* (P_{sol}) was collected in random rectangular troughs each with a collection surface of 500 cm² and a profile that avoided splashing losses. These rain-meters were installed 50 cm above the ground and the watre collected in 25 l containers was measured weekly. The precision was equivalent to 0.1 mm of rain. The number of rain-meters was increased from 3 to 12, and later to 24. The precision obtained from 12 averages 10 per cent when rainfall is more than 5 mm.

Total interception ($E_t + I$) may be obtained directly or indirectly. A small clearing of wind-fallen trees on the plateau station allowed rainfall measurements to be taken in the open and the interception was estimated at 15 per cent. The small size of this clearing could modify locally the precipitation amount by aerodynamic effects. The indirect method by which throughfall is compared with the rainfall data at the nearest meteorological posts only allows a limited comparison to be made over fairly long periods of time due to the inequality of rainfall distribution. The 1969–1971 data for the School of Forestry at Banco, 2.2 km from the plots, give an interception of 10–12 per cent. The CTFT data at South-Yapo, 3.5 km from the plot, gives an interception of 22 per cent. This large difference corresponds to the denser intermediate and lower strata at Yapo.

Other work has produced very variable results: 38 per cent in a subtropical forest in Brazil (Freise, 1936), 28 per cent in a dense plantation of *Shorea robusta* in Bengal (Dabral and Rao, 1969), 17 per cent in Panamá (McGinnis *et al.*, 1969), 12–26 per cent in Puerto Rico (Kline and Jordan, 1968; Odum *et al.*, 1970), 15 per cent in Ghana (Nye, 1961), 5 per cent in Costa Rica (McColl, 1970), 3 per cent in Nigeria (Hopkins, 1965).

Run-off (R) was not measured for the plots on flat areas where the diffused run-off only plays a minor role and Roose (1967) showed that even on steep slopes the run-off does not reach 1 per cent of the rainfall.

Drainage (D) measurement is either impossible to implement in the forest, or too imprecise to establish a balance (Roose and Henry des Turreaux, 1970). It was thus impossible to calculate a weekly balance when the soil humid line descended below the deepest soil humidity measurement. On the other hand, it is possible to estimate for an annual balance the losses by drainage.

Variations of soil water content (ΔH) were measured weekly by the slowing up of rapid neutrons method. The equipment was provided and maintained by the Radio-Isotope laboratory of the Commissariat à l'Énergie atomique at the ORSTOM Centre of Adiopodoumé. A neutron probe HP 110 (source Ra-Bé of 5 mCi), linked to an IP 110 were used until December 1970; in January 1971 this was replaced by the neutron probe HP 310 (source Am-Bé of 50 mCi) linked to a counter-scale EC 310 or ECP 511, measurements being taken in hard aluminium tubes 2.50 m long planted vertically in the soil (5 tubes per station).

The measurements are taken at 10 cm intervals to 1.4 m depth, and at 20 cm intervals below this depth. By this method, the measurements are not pinpointed, the neutrons spreading over a 'sphere of action' having a radius of 20 to 30 cm in our case. The use of a neutronic reflector (Moutonnet *et al.*, 1967) allows the first measurement to be taken at 12 cm depth. For practical purposes, the measurements were grouped into four soil layers: 0–27, 27–67, 67–127 and 127–232 cm.

Scaling in the field was accomplished by relating the neutron counting with soil volume humidity, which was obtained by measuring the weighable humidity and the apparent dry density of the soil. The weighable humidity was obtained on specimens taken by drilling. A density meter with diaphragm was used to ascertain the apparent soil density at Banco, on samples taken from several sections in pits. At Yapo, due to the large proportion of coarse elements, other methods had to be used. The agronomy laboratory drill of the ORSTOM Centre at Adiopodoumé (Bonzon and Picard, 1969) allows soil samples of known volume to be taken on the spot. The apparent densities up to 1.30 m depth on 50 profiles were measured. Below this depth, which is the drill's limit, a gamma ray density meter was employed. This apparatus allows the measurement of the apparent soil density in the same tubes used for neutronic measurements.

Table 4 gives the maximum and minimum water-content

TABLE 4. Values of soil water reserves observed at different depths.

Depth (cm)		Banco, plateau 1969–71		Banco, valley 1969–71		Yapo 1972–73	
		mm	percentage	mm	percentage	mm	percentage
0– 27	Maximum	43.5	100	67.2	100	76.4	100
	Minimum	20.3	47	40	59	29.4	37
27– 67	Maximum	55.6	100	91.6	100	108.4	100
	Minimum	28.8	52	56	61	69.2	64
67–127	Maximum	111	100	140.4	100	234	100
	Minimum	70.8	64	86.4	62	167.4	72
127–230	Maximum	276	100	247.8	100	491.4	100
	Minimum	200	72	164.8	66	379	77
Total	Maximum	486	100	547	100	910	100
(0–230)	Minimum	320	66	347	63	645	71

in mm and its variation amplitude. At the Banco valley plot where the soil texture is homogeneous over the whole profile, the amplitude decreases regularly with depth and only depends on root activity. At the plateau plot, the decrease is masked by the increase of the soil clay content with depth. At Yapo the gravels disturb the pattern. The superficial layer may yield up to half of their water, but at greater depth this fraction is not more than a third.

The variations follow a cycle with two annual minima or one only when the little dry season is not very marked (1971 at Banco, 1975 at Yapo).

Very fragmentary values of the *evapotranspiration from the soil* (ETR—I) were obtained. They varied between 0.5 and 8.4 mm/d. The highest values were at the onset or at the end of the wet seasons and the lowest in the middle of dry season. The averages are 2.5 mm/d at Banco and 3.2 mm/d at Yapo. There is no difference between the two Banco stations.

Annual water balance

By complementing the measurements with meteorological observations, an annual balance may be established. During the periods when it is not possible to measure the real evapotranspiration at the plots, the potential evapotranspi-

ration calculated by Turc's (1961) formula was substituted. There is a close concordance between these two evapotranspiration values in the wet season (Bernhard-Reversat, Huttel and Lemée, 1972). Once this evapotranspiration, including the intercepted rain-water that evaporated directly from the tree tops, was calculated, a correction to separate the water that evaporated from the soil (ET) and the intercepted water (I) was necessary. The drainage is calculated by the difference between the two. The water retention in the vegetation by increase in the biomass was not calculated, but may be estimated at 0.3 mm/a. Table 5 gives the average measurements during three years at Banco, and during two years at Yapo. Depending on the plot, the drainage represents from a quarter to a third of the rainfall. These values should not be considered as representative averages: measurements occurred during the three years of low rainfall at Banco (38 year average 2 140 mm) and during a normal year a drainage of *ca*. 900 mm, i.e. 40 per cent of the rainfall, may be expected. A representative average is harder to estimate at Yapo. The two years of study were very wet (40 year average 1 740 mm), but were characterized by more sunny periods during the wet season than the average. The drainage may not exceed 550 mm, i.e. 30 per cent of the rainfall.

TABLE 5. Elements of the annual water balance.

	Banco, plateau 1969–71		Banco, valley 1969–71		Yapo 1972–73	
	mm	percentage	mm	percentage	mm	percentage
Rainfall in the open (P)	1 800	100	1 800	100	1 950	100
Throughfall (P_{sol})	1 615	90	1 555	86	1 510	77
Stemflow (E_t)	15	1	15	1	15	1
Real interception (I)=P—($P_{sol}+E_t$)	170	10	230	12	425	22
Water evaporated from the soil (ETR—I)	975	54	965	54	1 000	51
Real evapotranspiration (ETR)	1 145	64	1 196	66	1 425	73
Drainage ($D=P$—ETR)	655	36	605	34	525	27

Plant biomass, productivity and organic matter decomposition

The annual primary productivity of the forest is divided between additions to the woody biomass, an input of organic matter to the heterotrophs, and, in the case of utilization, output due to human activity. This latter was negligible in the plots studied so the complete organic matter balance of the ecosystem only concerns stocks in the different compartments and the flow between these.

Biomass

The method of CTFT (1968) allows the volume of trunk and branches (>7 cm diameter) of standing trees to be estimated. These tables having an admission type $V=aD^b$ were established from 2 614 felled trees of 120 species. Twelve different relations were calculated; they are applicable for one species, one family, or a group of species of the same

habit. Having established these in forests of the lower Ivory Coast, they were applied to Yapo and Banco where the trees had the same characteristics.

The average values obtained for trees with a GBH >40 cm on 20 plots of 0.25 ha were 560 m³/ha at Banco, and 500 m³/ha at Yapo. These are not significantly different. The wider dispersion of results at Banco (280–875 m³/ha) than at Yapo (340–660 m³/ha) simply indicates its greater structural heterogeneity.

Two wood samples per tree were taken with a Pressler drill. Disc samples from felled trees in timber yards showed the absence of systematic variations according to the height, and allowed measurement of densities of trees too hard for the drill. For certain types the data of CTFT (1968) were used. The densities ranged from 0.33 (*Hannoa klaineana*) to 1.07 (*Lophira alata*). Measurements were taken on the species which accounted for three quarters of the total volume. The average density of these species was applied to the others.

Observations made in other tropical forests have led to the adoption of the factor 1.3 for converting the volume of branches and boles to the total volume of wood (Dawkins, 1967). Biomass measurements of trunks less than 40 cm GBH were taken in a natural clearing at Banco. This also allowed the biomass of climbers to be estimated at 5 per cent of the tree biomass.

The estimation of leaf biomass was made from the annual production of litter, assuming that leaves have an average life-span of one year and applying the correction of 20 per cent proposed by Bray and Gorham (1967) for the dry matter loss before their fall. These calculated values were confirmed on a sample taken in a clearing at the Banco plateau plot as *ca.* 9 t/ha.

The herb layer and the epiphytes, which only represent a very small fraction of the biomass, were not measured.

All the foregoing refers to aerial biomass. The vertical distribution of roots and their biomass were described by Huttel (1974) (fig. 5). This was evaluated by taking soundings with a special drill (Bonzon and Picard, 1969) and also by uncovering roots on one plot at Banco.

Table 6 shows the results and those of Müller and Nielsen (1965), who, on a limited sample of felled trees in a degraded forest near Banco, obtained much lower results. The data obtained here rank among the highest established, those for forests in South-East Asia (Ogawa *et al.*, 1965; Kira and Ogawa, 1971).

TABLE 6. Plant biomass (t/ha) in evergreen forests of the Ivory Coast.

	Banco	Yapo	Anguédédou (Müller and Nielsen, 1965)
Woody biomass:			
Trees (girth > 40 cm)			
trunks	360	330	
branches	105	95	240
Shrubs (girth < 40 cm)	15	—	
Roots	49	—	
Leaves	9	8	2.5
Climbers	24	—	—
Total	562		290.5

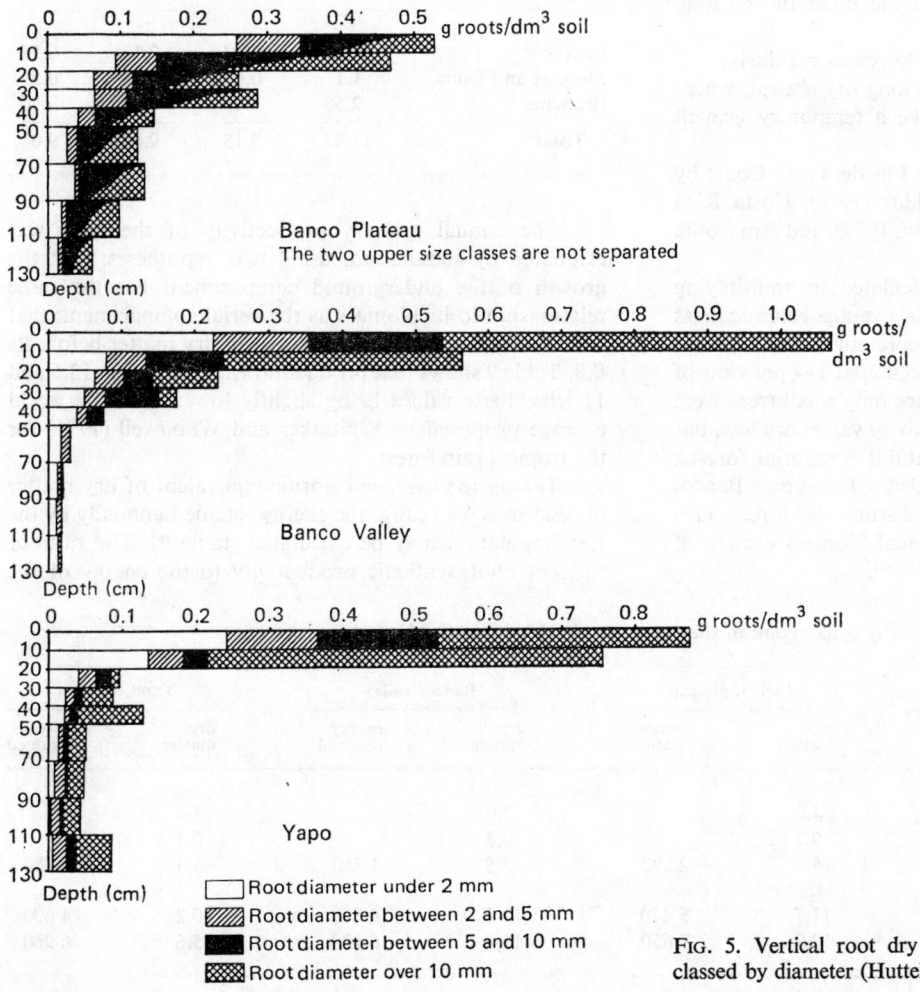

Root diameter under 2 mm
Root diameter between 2 and 5 mm
Root diameter between 5 and 10 mm
Root diameter over 10 mm

FIG. 5. Vertical root dry matter distribution (g/dm³ of soil) classed by diameter (Huttel, 1975).

Net primary productivity

The net primary productivity of a forest ecosystem is divided between an input to the woody biomass (increment or current growth) and to the production of deciduous organs, essentially leaves and reproductive organs.

The growth in girth was measured every three weeks with metal ribbon dendrometers on selected trees divided into two series. A first series of measurements was on trees of different heights of the most common species at the three plots (*Allanblackia floribunda, Dacryodes klaineana, Strombosia glaucescens* and *Turraeanthus africana*, this latter species only at Banco). A second series was taken in plots of 800 m² (Banco) or 400 m² (Yapo) where all trees having a GBH >30 cm were measured. More than 250 trees were observed over periods of 5–7 years.

A growth minimum for all trees occurred in January. Maximum growth occurred at different times according to the year, the species and the individual tree; on some trees two growth maxima per year were observed. Many authors, quoted by Hopkins (1970), have noticed this great intra-specific variability. It is less clear-cut when one observes a group of trees, or the growth rate variations of the basal area. The following conclusions may be drawn:

— the greatest growth occurs at the onset of the long wet season;
— the average growth rate then decreases regularly;
— the least growth occurs in the long dry season, where accidental heavy rains provoke a temporary growth renewal.

Similar rhythms have been observed in the Ivory Coast by Mariaux (1967), Nigeria by Hopkins (1970), Costa Rica (Lojan, 1967), Puerto Rico (Murphy, 1970), and Amazonia (Moraes, 1970).

Increases in volume were calculated by multiplying the CTFT (1968) cubage data by the average volumic mass (0.65). The data shown in table 7 represent a clear biomass increase. Losses by death only represented 1–4 per cent of the average annual production, since only small trees were studied on the plots. These productivity values are low, but very similar to those for other natural equatorial forests. Measurements made in the Anguédédou forest near Banco, but where the frequent clearings disturb the forest, give higher values: 7.5 t/ha, i.e. an annual biomass growth of 3.1 per cent (Müller and Nielsen, 1965).

TABLE 7. Annual growth of the above-ground ligneous biomass

	Measurement period	Average t/ha/a	Extreme values t/ha/a	Average percentage of biomass	Mortality losses t/ha/a
Banco, plateau	1969–74	4.60	2.89–5.43	0.96	0.05
Banco, valley	1969–74	3.05	1.63–4.87	0.65	0.12
Yapo, plateau	1968–74	4.65	2.86–6.12	1.06	0.11

Table 8 gives the litter data for two years (3 years for the leaves at Banco) (Bernhard, 1970). The values obtained correspond closely to most results found in similar forests. Leaf-fall has an annual rhythm, a maximum occurring in the long dry season and a minimum in the long wet season. The comparison of rhythms for certain species at different stations shows that these are specific and independent of site factors.

TABLE 8. Litter production, in t/ha/a (dry matter).

	Banco, plateau	Banco, valley	Yapo, plateau	Yapo, slope
Leaves	8.19	7.43	7.12	6.25
Flowers and fruits	1.1	0.66	1.05	0.53
Branches	2.58	1.09	1.45	2.26
Total	11.87	9.18	9.62	9.04

The annual primary productivity of the trees was estimated by summation, using two hypotheses: that the growth of the underground compartment has the same relationship to its biomass as the aerial compartment; that fresh leaf litter has lost 20 per cent of dry matter before its fall. Table 9 shows that production varied between 13.6 and 17 t/ha, these values being slightly lower than the world average proposed by Whittaker and Woodwell (1971) for the tropical rain forest.

Taking the average calorific equivalent of dry matter produced as 4.5 kcal/g, the energy retained annually by the tree population may be calculated (table 9). The ratio of this net photosynthetic productivity to the energy of the

TABLE 9. Evaluation of the net productivity (t/ha/a) and of the energy retained by trees (kcal/m²/a).

	Banco, plateau		Banco, valley		Yapo, plateau	
	dry matter	energy retained	dry matter	energy retained	dry matter	energy retained
(1) Growth of boles and branches	4.6		3.0		4.6	
(2) Growth of roots	0.7		0.5		0.7	
(1+2) Biomass increase	5.3	2 380	3.5	1 560	5.3	2 380
(3) Production of leaves and reproductive organs	11.7	5 270	10.1	4 540	10.2	4 600
(1+2+3) Total annual production	17.0	7 650	13.6	6 100	15.6	6 980

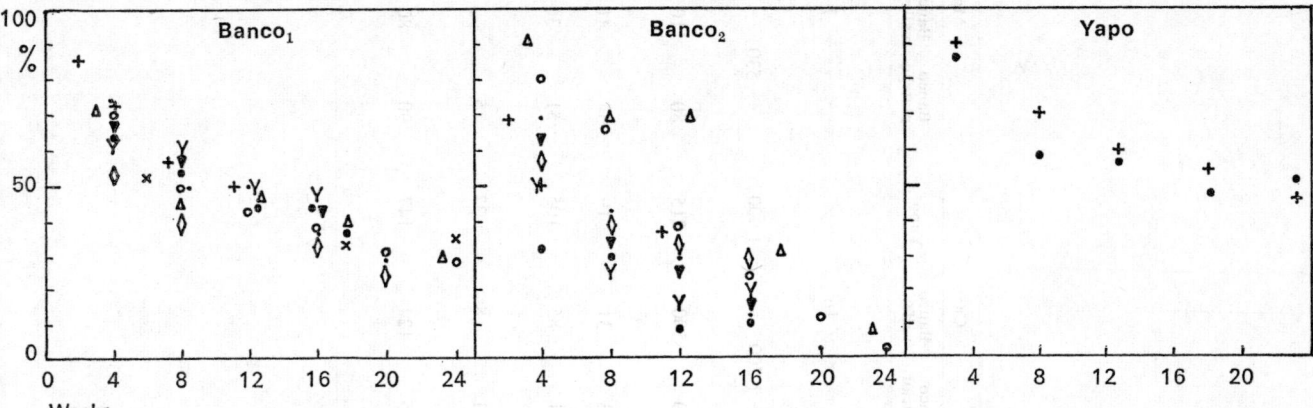

FIG. 6. Leaf litter decomposition *in situ*: changes in dry matter weight shown as percentages of the initial weight. The different signs represent trials made at different times of the year.
Banco$_1$: plateau; Banco$_2$: valley.

incident visible solar radiation estimated at 45 per cent of the total radiation, i.e. 61 180 cal/cm²/a, would be:

Banco, plateau plot — 1.25 per cent
Banco, valley plot — 1.00 per cent
Yapo, plateau plot — 1.15 per cent.

This photosynthetic efficiency is clearly higher than the value of 0.8 per cent calculated by Müller and Nielsen in the degraded forest of Anguédédou, near Banco.

Litter and its decomposition

The quantity of litter on the ground was measured monthly. It reaches a maximum of 3–4 t/ha in the long dry season, and a minimum of *ca.* 1 t/ha or less from July to October. Although this seasonal variation follows that of litter production, they are not entirely parallel, due to seasonal variations in the decomposition rate. This was measured by two methods (Bernhard, 1970; Bernhard-Reversat, 1972).

The measurement of the average quantity of litter on the ground at different times allows the calculation of the decomposition coefficient β established by Jenny *et al.* (1949) and applied to tropical forests by Nye (1961) and Olson (1963); see chapter 13. During one year it was found to equal 3.3 on the Banco plateau, 4.2 in the valley and 3.6 at Yapo.

The monthly values of the decomposition rate β were calculated to discover the variations during the year. The decomposition rate showed a clear correlation with rainfall outside the long dry season, and with fresh litter input at the onset of this season.

A direct study of decomposition was made by measuring the weight loss of samples of leaves placed in wide-meshed plastic nets. The results (fig. 6) show the differences between the two Banco plots. When litter was exchanged between these, it was seen that site influence was preponderant,

despite differences in chemical composition of the litters. It seems that the greater amount of exchangeable bases and higher pH in the bottom of the valley may cause this by an intermediary influence on the detritivores. Total litter decomposition takes *ca.* 9 months on the Banco plateau, 5 months in the valley, and 11 months at Yapo. This last result, obtained from one series of measurements only, is no doubt too high. Madge in Nigeria (1965, 1969) and Nye in Ghana (1961) found decomposition rates corresponding to these.

Soil organic matter and its decomposition

The data given in table 10 show a humiferous surface horizon on the Banco plateau and its lack in the valley; the Yapo plot is intermediate.

Carbon mineralization was measured by the CO_2 that escaped from soil kept during seven days at 30° C at a humidity near to its material retention capacity. The data given in table 10 show that CO_2 production is very low from Banco soils below 10 cm in depth. At Yapo it is high; this could help explain the low organic matter stocks in the latter forest soil. Measurements made on Banco soil samples exposed to different humidities showed that the latter become unfavourable to respiratory activity at very low levels (<7–12 per cent) exceptionally reached *in situ*.

TABLE 10. Carbon of soil organic matter and its potential capacity for mineralization.

	Depth (cm)	Banco plateau	Banco valley	Yapo
Carbon, t/ha	0–10	50	29	37
	10–50	120	70	33
C mineralized in 7 days of incubation (mg/g of total C)	0–10	4.6	3.5	6.0
	10–50	1.2	3.1	14.4

Activity of soil fauna

The role played by fauna in the cycling of organic matter is important, but little known in the tropical forest; the approaches which have been made are very limited; they

TABLE 11. Reserves and flows of the major elements.

	N			P			K			Ca			Mg		
	Banco plateau	Banco valley	Yapo	Banco plateau	Banco valley	Yapo	Banco plateau	Banco valley	Yapo	Banco plateau	Banco valley	Yapo	Banco plateau	Banco valley	Yapo
1. Input by rainfall (kg/ha/a)	21			0.5			5.5			16			7		
2. Storage in the total above-ground biomass (kg/ha)	1 400		1 000	100		70	600		350	1 200		1 900	530		180
3. Soil reserves (0–50 cm)* (kg/ha)	6 500	5 800	2 600	50	330	25	80	160	115	100	200	215	80	110	95
4. Rain-leaching contribution to the soil (kg/ha/a)	60	50	13	1.5	9.5	5.5	60	170	82	23	31	19	34	41	16
5. Litter contribution to the soil (kg/ha/a)	170	158	113	8	14	4	28	81	26	61	85	105	51	36	23
6. Retention in the woody biomass (kg/ha/a)	12	8	10	1	1	1	6	4	5	11	8	23	5	3	2
7. Absorption (4+5+6) (kg/ha/a)	242	226	136	11	25	11	94	255	114	95	124	147	90	80	41

* Total N; assimilable P; exchangeable K, Ca, Mg.

will however allow evaluation of the importance of certain elements of the fauna.

The Banco valley plot is characterized by a large number of earthworm casts which are very rare both at the plateau plot and at Yapo. These are approximately cylindrical, 5–12 cm high by 1–3 cm wide. All recognizable casts were collected on ten 1 m² plots. The results show large variations; the average was 2 600 g/m², with a significant standard deviation of 450 g/m².

Analyses of these casts showed that they contained a little more clay and silt and a little less coarse sand than the surface soil (0–10 cm). Their apparent density is very high: 2.20 ± 0.1, whereas that of the soil is 1.3 between 0 and 10 cm deep and 1.5 near 100 cm deep. This explains why even the casts which are no longer active are not penetrated by roots in spite of their richness in certain elements.

The overall biological activity, measured as the production of CO_2, is not significantly different in the casts and in the upper soil. On the contrary, nitrogen mineralization is more active in the casts: since this nitrogen is not absorbed by roots and since the great compactness of the rejects prevents its leaching by the rains, their mineral nitrogen content is very high.

The casts thus have a dual role: materials found at depth are raised to the surface, and an active mineralization takes place with accumulation of mineral nitrogen.

Only overall counting of the soil micro-arthropods in the upper humiferous soil layer was carried out. Samples were taken by the Berlese method on 10 cm² square plots 3 cm in depth. The numbers per m² were 54 000 for the Banco plateau soil, 26 000 at Yapo, and 17 500 for the Banco valley. These values are directly related to the organic matter content.

Many authors have proved the importance of termites in the pedogenesis of tropical soils and reviews have been devoted to them (Boyer, 1971; Bachelier, 1973), but little quantitative data about their action on the organic matter of forest soils are available.

Above-ground termite hills were counted on 20 plots of 50 m² at each site. 57 ± 8.4/m² were found on the Banco plateau, 75 ± 8.6/m² in the valley and 23 ± 4.6/m² at Yapo. A study made by Maldague (1970) in the Zairian rain forest gave a density of 87 termite hills per 1 000 m², with humivorous species predominating (*Cubitermes fungifaber* and *Thoracotermes* sp.).

Biogeochemical cycles

The circulation of nutrients in an ecosystem is characterized by inputs, transfers between the different storage sites, and outputs. Measurements were taken of these phases, table 11 giving the overall view.

Inputs

Input of elements is essentially by rainfall, since alterable minerals exist in negligible quantities in the upper soil layers. However, the nitrogen input by fixation of atmospheric nitrogen remains an unknown factor. Measurements of the element content of rain-water have been made by Roose (1974) at Adiopodoumé. Table 11 shows the abundance of calcium and nitrogen, the latter mainly in organic form. Large inputs of potassium have been found in the African tropical zone by Nye (1961), Genevois (1967) and Boyer (1973), but they are scanty at Adiopodoumé.

The vegetation

The quantity of nutrients stored in the woody aerial biomass was estimated by tree-felling in forests similar to those studied, with samples of sawdust taken from different heights on the trunk. From the data on the contribution to the biomass of the main species, it was calculated that the mineral mass of these species represented 60 per cent of the total. Average amounts obtained were extrapolated to other species. The amount of mineral elements in the roots was not measured; the proportion of 10 per cent of the total value was adopted, as proposed by other workers (Greenland and Kowal, 1960; Golley *et al.*, 1969). The total values are shown in table 11. Notable differences are observed between the two forests, the calcium stored being greater and the other elements less at Yapo than at Banco. The quantity of elements contained in the litter during one year was measured in order to estimate the mineral mass of the leaves. For nitrogen a correction factor of 1.3 was applied, this ratio having been observed on several occasions between the leaves on the tree and leaves recently fallen.

Transfers between vegetation and soil occur mainly through litter fall, root decomposition (not measured) and rain-leaching.

Samples of recently fallen fine litter containing leaves, reproductive organs, twigs and small branches, were analysed separately during one year (Bernhard, 1970). Table 11 shows the differences between the plots; at Banco the litter is richer in the valley, especially in potassium, than on the plateaut with the exception of magnesium; at Yapo, the values of nitrogen, phosphorus and magnesium are lower than a, Banco, but that of calcium is much higher. These differences follow those of the biomass accumulation and depend on the nature of the soil.

Only leaching by throughfall was measured, stemflow only contributing slightly to the total. Rain-water was collected every week in 12 to 20 rain-meters per plot on the three sites. The grouped samples were analysed every four weeks during two years. The amounts of nitrogen, potassium and magnesium were proportional to the amounts in the litter, but the relative accumulation of calcium in the leaves at Yapo does not correlate with its greater leaching and thus seems to concern fixed calcium. The quantity of phosphorus found in the throughfall is relatively large at Yapo, when compared with the amount in the leaves. As shown in fig. 7, a relationship exists between the amount of rainfall and the amount of mineral elements it contains. Under the forest canopy the relationship type depends on the element considered. In all cases the leached amount increases rapidly with the increase in rainfall, up to 100–200 mm of rainfall

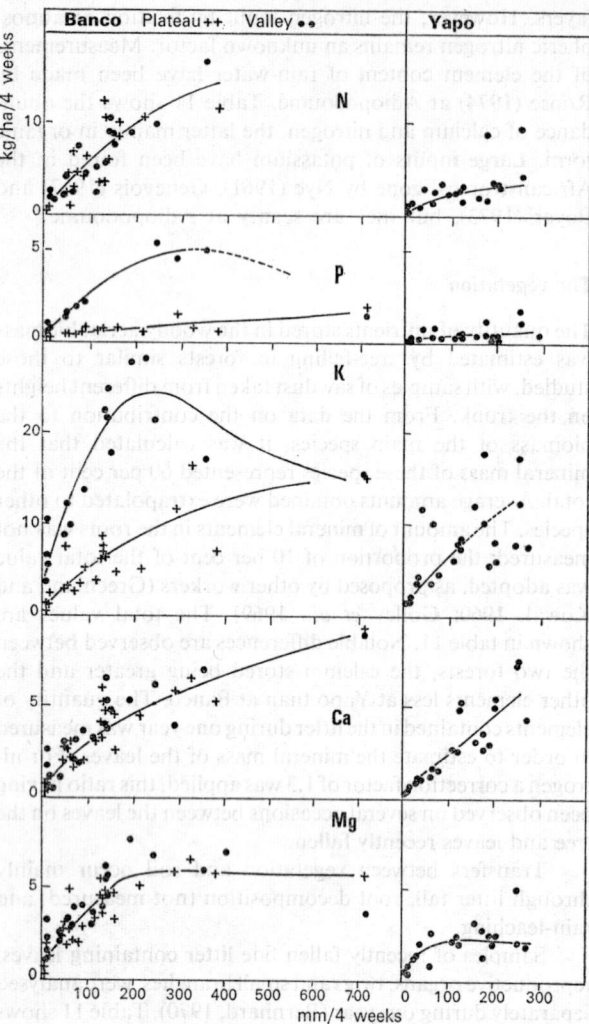

FIG. 7. Relationships between throughfall intensity and the quantity of mineral elements leached to the ground per hectare during 4-week periods.

The soil

The changes in chemical composition of the leaf litter were followed during its decomposition *in situ* (Bernhard-Reversat, 1972). In spite of the variation of results the following conclusions may be drawn. Potassium is rapidly leached, 70–80 per cent having disappeared after two weeks, 80–90 per cent after four weeks. Thus potassium contributed by litter fall in the dry season may be used by the vegetation before the onset of the wet season, and the possible losses by leaching are thus diminished. Magnesium and phosphorus disappear noticeably faster than organic matter, but slower

than calcium. The behaviour of nitrogen is complex. In one case an increase in the absolute value of the nitrogen quantity present was noted and it has been shown that a fairly important net nitrogen mineralization could already have taken place in the litter (Bernhard-Reversat, 1974). Elements mineralized in the litter may be used directly by the vegetation through the hair-fine root network that penetrates the litter.

The analysis of exchangeable bases was performed several times during different seasons to determine the exchangeable cations in the soil. The variability is high and the average results obtained represent rough approximations. On the other hand, the soil reserves capable of mobilization contain, besides the exchangeable cations, labile forms that have not been studied. Table 11 shows that the greatest exchangeable reserves are found at the bottom of the Banco valley.

The total stock of soil nitrogen is less at Yapo than at Banco. The quantity of mineral nitrogen is always small despite an active net mineralization, because it is rapidly used by the vegetation.

The assimilable phosphorus was extracted by the method described by Duchaufour. There was a greater disparity between the three plots than for the other elements.

Nitrogen mineralization was studied in the laboratory and in the field (Bernhard-Reversat, 1974). The laboratory measurements showed a high production of nitrate only in the upper layer (0–10 cm); in lower layers production is very small or non-existent; ammonization also decreases with depth, but more slowly. Nitrification shows a high sensitivity to different humidities; it is non-existent at low humidities, reaches an optimum which varies according to the site, then decreases at high humidities. Very rapid denitrification occurs in saturated soils. Measurements of mineralization in the field were made every four weeks over 18 months to 2 years. No characteristic seasonal variations were noticed. The annual production is given in table 12.

TABLE 12. Net mineralization of nitrogen *in situ* (kg/ha/a).

	Banco, plateau	Banco, valley	Yapo
Litter	8	12	3
Soil (0–10 cm)	167	156	125
Litter and soil	175	168	128
N mineralized in the soil (percentage of the total N)	9.8	12.5	8.0

The amount of mineralization in the Yapo forest is relatively small despite high potential mineralization in oven conditions. It is possible that nitrogen exists here in a form less easily decomposed than at Banco, perhaps because it is protected by clayey colloids. At Banco the difference between the two sites corresponds to a smaller mineralization capacity on the plateau, in accordance with a lower pH and a higher C/N ratio.

Lysimeters, composed of troughs 50 cm long and 10 cm wide, were placed at each plot. They were filled with coarse-grained siliceous sand and placed horizontally at 40 cm depth, in such a way that the profile and the vegetation above them were not disturbed. Because of an insufficient number of lysimeters, the collected solution was compared to that of two rain-meters placed in the immediate vicinity. Phosphorus and potassium were only collected in minimal proportions below 40 cm, these two elements being retained or reabsorbed in the upper soil layers. The element least retained was calcium. Nitrogen, whose behaviour was intermediary, was collected less at Banco than at Yapo.

In the absence of measurements of the quantities of elements absorbed annually by the vegetation, an evaluation of the flow between soil and vegetation may be made by the sum of elements retained in the woody biomass and restored with the litter, in the dead roots and in the water leaching the vegetation. Only its restoration by dead roots has not been evaluated. The input of elements by rain no doubt contributes to absorption, but the systems studied being in a stable state, it was considered that this input was balanced by the elements' output. The absorption thus evaluated (table 11) shows large differences in the turnover of the elements from one plot to another; in the Yapo forest the absorption rates of N and Mg were smallest, and that of Ca largest; the Banco valley showed a much higher absorption of P and K than at the other stations.

Outputs

In the absence of exploitation, the element losses are limited to their leaching to the water-table and to watercourses. Although it has not been possible to determine these outputs quantitatively, some qualitative data are available. The water specimens taken in a stream whose source is situated below the Banco valley plot showed a mineral composition that confirmed the results obtained through the lysimeters: traces of phosphorus, little potassium, higher concentrations of magnesium and total nitrogen (of which only half is in mineral form), and calcium having the highest amounts.

Overall aspects

The nitrogen cycle is characterized by low soil stock amounts relative to those of the total biomass: the ratio of these respective values being 4 at Banco and only 2 at Yapo; these values are lower than those in forests in other climatic zones (Ellenberg, 1971). The reason is the high mineralization rate of organic nitrogen, which undergoes nearly total nitrification in spite of the acidity of the soil (de Rham, 1971; Bernhard-Reversat, 1974). Moreover, the annual mineralization in the litter and the upper soil layer is equal to the quantity contributed in the same period by the total fine litter (table 12).

Phosphorus is present in the soil in extremely variable quantities. Its cycle is nearly closed: the contribution by rains is very small, its leaching by percolating water and the losses by drainage are almost non-existent. Experiments made with radio-active phosphorus (Luse, 1970) in a tropical forest in Puerto Rico show that the root systems in the upper soil layer are remarkably efficient in absorbing mineralized phosphorus as the litter decomposes. On the other hand, in soils containing little phosphorus, the competition between the vegetation and the soil micro-organisms, which shows as a lack of net mineralization, prevents any loss of mineral phosphorus from the ecosystem.

The potassium cycle is characterized by flows that are larger than the soil reserves of exchangeable elements. These reserves only form a small part of the potassium in biogeochemical cycles. Without being as closed as the phosphorus cycle, the potassium cycle shows few exchanges with the outside. Mathieu (1972) came to the same conclusions through the study of a watershed covered with a semi-deciduous forest in the Ivory Coast. Magnesium and particularly calcium have a more open cycle. The soil reserves of exchangeable calcium are slightly larger than those of other cations, but much lower than the quantities stored in the vegetation.

Conclusions

Studies on two types of evergreen rain forests in the lower Ivory Coast allow several essential characteristics of these ecosystems to be distinguished, but also show vast areas in which our knowledge is still very insufficient.

1. Complete floristic inventories on sample plots, necessary to characterize the phytocenosis, are possible in these forests as elsewhere, as Mangenot, Miège, Schnell, Guillaumet have shown in western Africa, and which led these authors to distinguish phytosociological units that indicate differing ecological conditions. However, a certain heterogeneity exists within these units, linked to topographical factors (landforms), floristic factors (a tendency of many lower strata species to agregate), structural factors (discontinuity of the crowns of emergent trees), successional factors (recent wind-fall clearings and regrowths of varying age). This heterogeneity raises the problem of plot selection that is as representative as possible of the ecosystem. In this work, it was after making an inventory of the flora and a structural approach to the whole phytocenosis that the plots were chosen, representing an average stage of an old stand.

2. To the spatial variability must be added a *time variability* on different scales:
— an annual climatic cycle to which the biological processes adjust: cambial activity, seed production, leaf-fall, the speed of litter break-down;
— a variability from year to year, that was noticed particularly for primary production and that might be partly linked to meteorological variations but which also depends on other factors, as shown by the lack of parallelism between the annual production of wood in the different plots;
— a variability of a secular nature linked to the biological cycle of the trees, whose initial phase originates in a clearing.

3. The *functioning* of forest ecosystems is dominated by two trophic levels, that of primary producers and that of the decomposers. The latter, which form extremely complex food webs where both micro-organisms and detritivorous fauna are active, have an importance in the ecosystem's functioning that is out of proportion to their biomass. The very active mineralization of carbon and nitrogen, the numerical importance of micro-arthropods in the humus horizon, the abundance of termite hills, and locally, of Oligochaeta, are evidence of this importance.

As a link between producers and decomposers, the pool of *dead organic matter* produced by the former may be divided into several sub-compartments. Only the leaf litter, flowers and fruits, fallen dead wood of small dimensions and organic matter incorporated to the soil were measured. Standing dead trunks and branches, together with those that have fallen, which in an unfelled forest form a large part of the dead biomass, were not measured.

The *annual water balance* and its seasonal climatic cycle, as well as the biogeochemical cycles are also basic characteristics of tropical forest ecosystems, which regulate the water cycle and conserve soil fertility.

4. These studies on the Ivory Coast forest have shown several *problems* peculiar to this type of ecosystem, which, in spite of their importance, have been little explored because of the technical difficulties involved.

— Firstly the representativity of the sample plots, with regard to the structural and floristic heterogeneity of the phytocenosis. This heterogeneity may be either analysed on small numerous plots, or on fewer larger areas. Studies on the number and size of the representative plots have still to be made.

— Variations over a very long period linked to the *internal dynamics* of the natural forest, beginning with the fall of dead trees, have hardly been studied. There is a wide-spread tendency to consider the forest as a stable formation, whilst it is rather a mosaic of young, mature and senescent stages. This approach relates to methods of demographic analysis of populations.

— *Knowledge of the components of the net photosynthetic balance and of the photosynthetic efficiency* of radiation using assimilation and respirometer devices *in situ*, or using the micrometeorological method of measurement of CO_2 flows, has hardly been used. The dimensions and lateral heterogeneity of the canopy make such attempts particularly difficult and necessitate considerable and expensive replication.

— The overall balance of mineral element inputs and outputs in forest ecosystems, which is indispensable in order to perceive the evolutionary trends of the biogeochemical cycles, will only be achieved by the choice of watersheds that are appropriate for such a study.

— The role of fauna in primary consumption, its consequences on the production of the vegetation and the regeneration of trees have hardly been studied.

5. The demographic growth and economic development of the Ivory Coast are the cause of many pressures on the forest. Studies in the natural forest will serve as a base-line for similar studies which should be performed in derived artificial formations.

— The great industrial *plantations* (rubber, oil-palm, cocoa) and, on a lesser scale, plantations of local or introduced forest species, form simplified ecosystems whose productivity and whose action on the soil organic matter and on the circulation of mineral elements should be compared with those of the climax forest.

— *Traditional cultivation* causes a more serious disturbance in the forest by clearing, burning and modifications of the micro-climate. In particular, it induces modifications of the water balance, the amount of organic matter and biogeochemical cycles, about which only very fragmentary data are available.

— *Secondary formations* occur after cultivation has been abandoned and consist of a series of stages whose floristic description has been partially made, but where the amounts of primary production remain to be established.

— Besides its productive function, the equatorial forest has a *protective function*: soil conservation, regulation of flow of the watercourses, maintenance of the local climatic conditions, protection of the flora and fauna associated with the forest biotopes. These point to the necessity of conserving the large forest areas that still exist in the Ivory Coast.

The results of these studies and the gaps of knowledge they stressed allowed to define the research in plant ecology to be undertaken within the interdisciplinary project of the Tai forest. This project, in southwestern Ivory Coast, aims to recommend methods of rational and planned use of the tropical humid forest ecosystem, which are compatible with the conservation of the productive potential of this ecosystem. The research to be undertaken within such a project will therefore concern the natural forest, the utilized forest (timber extraction and traditional cultivation) and the reconstitution of the forest environment during the fallow period. The research, foreseen for a period of 5 years, will be performed on a research site, which will be selected to be representative of the intact forest areas and according to protection and accessibility criteria. There will be a very close relationship between the research of the ecologists and those in the human sciences. The results obtained and their recommendations will be extended to the whole region. Seven programmes constitute the project; each programme includes several research operations on change in the physical environment, soil properties, vegetation and animal populations under the influence of man's activities, the trends in utilizing the ecosystem (organization of space and types of forest utilization), and the evolution of the forest landscape (which will be a synthesis of the results of the six other programmes). Several research operations commenced in 1976. The project is sponsored by the University Institute of Tropical Ecology (IUET) of Abidjan and is linked to the Man and Biosphere Programme Project no. 1 of Unesco. It deals with a series of integrated research topics, which are complementary between the natural and the social sciences, between research and management, which take place in an environment which is not only representative of the Ivory Coast but also of western and central Africa. Consequently, the results obtained will be useful to other countries of this region.

Selective bibliography

BACHELIER, G. Faune des sols et termites. In: *Les sols ferralli-tiques*, t. IV, p. 107–142. ORSTOM (Paris), *Init. Doc. Techn.*, n° 21, 1973.

BARTHOLOMEW, W. V.; MEYER, J.; LAUDELOUT, H. *Mineral nutrient immobilization under forest and grass fallow in the Yangambi region, with some preliminary results on the decomposition of plant material on the forest floor*. Bruxelles, Publ. INEAC (série scientifique), n° 57, 1953, 27 p.

BEAUFORT, W. H. J. de. *Distribution des arbres en forêt semper-virente de Côte-d'Ivoire*. ORSTOM (Adiopodoumé), 1972, multigr.

BERNHARD, F. Étude de la litière et de sa contribution au cycle des éléments minéraux en forêt ombrophile de Côte-d'Ivoire. *Oecol. Plant.*, 5, 1970, p. 247–266.

BERNHARD-REVERSAT, F. Décomposition de la litière de feuilles en forêt ombrophile de basse Côte-d'Ivoire. *Oecol. Plant.*, 7, 1972, p. 279–300.

——. L'azote du sol et sa participation au cycle biogéochimique en forêt ombrophile de Côte-d'Ivoire. *Rev. Écol. Biol. Sol*, 11, 1974, p. 263–282.

——; HUTTEL, C.; LEMÉE, G. Quelques aspects de la périodicité écologique et de l'activité végétale saisonnière en forêt ombrophile sempervirente de Côte-d'Ivoire. In: Golley, F.B.; Golley, P. M. (eds.). *Tropical ecology with an emphasis on organic production*, p. 217–234. Athens, Univ. Georgia, 1972, 418 p.

BONZON, B.; PICARD, D. Matériel et méthodes pour l'étude de la croissance et du développement en pleine terre des systèmes racinaires. *Cahiers ORSTOM, sér. Biol.*, 9, 1969, p. 3–18.

BOYER, J. Cycles de la matière organique et des éléments miné-raux dans une cacaoyère camerounaise. *Café, Cacao, Thé*, 17, 1973, p. 3–23.

BOYER, P. Les différents aspects de l'action des termites sur les sols tropicaux In: Pesson, P. (ed.). *La vie dans les sols*, p. 279–334. Paris, Gauthier-Villars, 1971, 472 p.

CAUSSINUS, H.; ROLLET, B. Sur l'analyse au moyen d'un modèle mathématique des structures par espèces des forêts denses humides sempervirentes de plaine. *C.R. Acad. sci. Paris*, sér. D, 1970, p. 1341–1344.

CENTRE TECHNIQUE FORESTIER TROPICAL. *Étude sur l'approvi-sionnement en bois de l'usine de pâte cellulosique de Yaou. 1re partie: étude des potentiels en bois disponible*. Nogent-sur-Marne, CTFT, 1968, 126 p. multigr.

DAWKINS, H. C. Wood production in tropical rain forest. *J. Ecol.*, 55, 1967, 20P–21P.

ELDIN, M. Le climat. In: *Le milieu naturel de la Côte-d'Ivoire*, p. 73–108. ORSTOM (Paris), Mémoire n° 50, 1971.

ELLENBERG, H. Nitrogen content, mineralization and cycling. In: *Productivity of forest ecosystems* (Proc. Symp. Brussels, 1969), p. 509–514. Paris, Unesco, 1971, 707 p.

GENEVOIS, L. L'alimentation minérale des végétaux par la pluie. Cas des régions tropicales. *J. Agric. Trop. Bot. Appl.*, 14, 1967, p. 582–597.

GOLLEY, F. B.; McGINNIS, J. T.; CLEMENTS, R. G.; CHILD, G. I.; DUEVER, M. J. The structure of tropical forests in Panamá and Colombia. *Bio-Science*, 19, 1969, p. 693–696.

GOSSE, G.; ELDIN, M. *Données agroclimatologiques recueillies à la station ORSTOM d'Adiopodoumé, 1948–1972*. ORSTOM (Adiopodoumé), 1973, 23 p. multigr.

GREENLAND, D. J.; KOWAL, J. L. M. Nutrient content of the moist tropical forest of Ghana. *Plant and Soil*, 12, 1960, p. 154–174.

GUILLAUMET, J. L.; ADJANOHOUN, E. La végétation de la Côte-d'Ivoire. In: *Le milieu naturel de la Côte-d'Ivoire*, p. 157–263. Paris, ORSTOM, Mémoire n° 50, 1971.

HOPKINS, B. Vegetation of the Olokemeji Forest Reserve, Nigeria. III. The microclimate, with special reference to their seasonal changes. *J. Ecol.*, 53, 1965, p. 125–138.

——. Vegetation of the Olokemeji Forest Reserve, Nigeria. VI. The plants on the forest site, with special reference to their seasonal growth. *J. Ecol.*, 58, 1970, p. 765–793.

HUTTEL, C. Estimation du bilan hydrique dans une forêt semper-virente de basse Côte-d'Ivoire. In: *Proc. Symp. Vienne*, p. 439–452. AIEA, 1972.

——. Root distribution and biomass in three Ivory Coast rain forest plots. In: Golley, F. B.; Medina, E. (eds.). *Tropical ecological systems: trends in terrestrial and aquatic research*, p. 123–130. Berlin, New York, Springer Verlag, Ecological Studies no. 11, 1975, 398 p.

JONES, E. W. Ecological studies on the rain forest of southern Nigeria. IV. The plateau forest of the Okomu Forest Re-serve. *J. Ecol.*, 43, 1955, p. 564–594.

JORDAN, C. F.; KLINE, J. R. Mineral cycling: some basic concepts and their application in a tropical rain forest. *Ann. Rev. Ecology Systematics*, 3, 1972, p. 33–50.

KIRA, T.; OGAWA, H.; YODA, K.; OGINO, K. Comparative ecological studies on three main types of forest vegetation in Thailand. IV. Dry matter production with special refer-ence to the Khao Chong rain forest. *Nature and Life in South-East Asia*, 6, 1967, p. 149–174.

——; ——. Assessment of primary production in tropical and equatorial forests. In: *Productivity of forest ecosystems* (Proc. Symp. Brussels, 1969), p. 309–321. Paris, Unesco, 1971, 707 p.

KLINE, J. R.; JORDAN, C. F. Tritium movement in soil of tropical rain forest (Puerto Rico). *Science*, 160, 1968, p. 550–551.

LOJAN, L. The tendencies of the radial growth of 23 tropical forestal species. *Turrialba*, vol. 18, no. 3, 1967, p. 275–281.

LONGMAN, K. A.; JENÍK, J. *Tropical forest and its environment*. London, Longman, Tropical Ecology Series, 1974, 196 p.

LUSE, R. A. The phosphorus cycle in a tropical rain forest. In: Odum, H. T.; Pigeon, R. F. (eds.). *A tropical rain forest. A study of irradiation and ecology at El Verde, Puerto Rico*, H 161–H 166. Division of Technical Information, U.S. Atomic Energy Commission (USAEC, Oakridge, Tenn.), 1970, 1 678 p.

MADGE, D. S. Leaf fall and litter disappearance in a tropical forest. *Pedobiologia*, 5, 1965, p. 272–288.

MALDAGUE, M. E. *Rôle des animaux édaphiques dans la fertilité des sols forestiers*. Bruxelles, Publ. INEAC (série scienti-fique), n° 112, 1970, 245 p.

MANGENOT, G. Études sur les forêts des plaines et plateaux de la Côte-d'Ivoire. *IFAN, Études éburnéennes*, 4, 1955, p. 5–61.

——; MIÈGE, J.; AUBERT, G. Les éléments floristiques de la basse Côte-d'Ivoire et leur répartition. *C.R. Soc. Biogéogr.*, 212–214, 1948, p. 30–34.

MARIAUX, A. Les cernes dans les bois tropicaux africains. Nature et périodicité. *Bois et Forêts des Tropiques*, 114, 1967, p. 23–37.

MATHIEU, P. *Apports chimiques par les précipitations atmosphé-riques en savane et sous forêt. Influence du milieu forestier intertropical sur la migration des ions et sur les transports solides (Bassin de l'Amitioro, Côte-d'Ivoire)*. Nice, thesis, 1972, 454 p.

McCOLL, J. G. Properties of some natural waters in a tropica wet forest of Costa-Rica. *Bio-Science*, 20, 1970, p. 1096–1100.

McGINNIS, J. T.; GOLLEY, F. B.; CLEMENTS, R. G.; CHILD, G. I.; DUEVER, M. J. Elemental and hydrologic budgets of the Panamian tropical moist forest. *Bio-Science*, 19, 1969, p. 697–700.

MORAES, V. H. F. Periodicity in stem growth of trees of the Amazonian forest. *Pesquis. Agropec. Brasil (Ser. Agr.)*, 5, 1970, p. 315–320.

MOUTONNET, P.; BUSCARLET, L. A.; MARCESSE, J. Emploi d'un humidimètre à neutrons de profondeur associé à un réflecteur pour la mesure de la teneur en eau des sols au voisinage de la surface. *Ann. I.T.B.T.P.*, 233, 1967, p. 1–5.

MÜLLER, D.; NIELSEN, J. Production brute, pertes par respiration et production nette dans la forêt ombrophile tropicale. *Det Forstlige Forssvaesen i Danmark*, 29, 1965, p. 69–160.

MURPHY, P. G. Tree growth at El Verde and the effects of ionizing radiations. In: Odum, H. T.; Pigeon, R. F. (eds.). *A tropical rain forest*, D 141–D 171. Div. Techn. Inf. (USAEC), 1970, 1 678 p.

NYE, P. H. Some soil forming processes in the humid tropics. IV. The action of soil fauna. *J. Soil Sci.*, 6, 1955, p. 73–83.

——. Organic matter and nutrients cycles under moist tropical forest *Plant and Soil*, 13, 1961, p. 333–346.

ODUM, H. T.; MORE, A. M.; BURNS, L. A. Hydrogen budget and compartments in the rain forest. In: Odum, H. T.; Pigeon, R. F. (eds.). *A tropical rain forest*, H 105–H 122. Div. Techn. Inf. (USAEC), 1970, 1 678 p.

OGAWA, H.; YODA, K.; OGINO, K.; KIRA, T. Comparative ecological studies on three main types of forest vegetation in Thailand. II. Plant biomass. *Nature and Life in South-East Asia*, 4, 1965, p. 51–81.

PERRAUD, A. Les sols. In: *Le milieu naturel de la Côte-d'Ivoire*, p. 265–291. ORSTOM (Paris), Mémoire n° 50, 1971.

PIERLOT, R. *Structure et composition des forêts denses d'Afrique centrale, spécialement celles du Kivu*. Bruxelles, Mémoires de l'Académie Royale des Sciences Outre-Mer, vol. 16, n° 4, 1966, 367 p.

RHAM, P. de. *L'azote dans quelques forêts, savanes et terrains de culture d'Afrique tropicale humide (Côte-d'Ivoire)*. University of Lausanne, thesis; Zurich, Buchdruckerei Berichthaus, 1971, 124 p.

RICHARDS, P. W. *The tropical rain forest: an ecological study*. Cambridge, Cambridge University Press, 1952, 450 p. 4th reprint with corrections, 1972.

ROLLET, B.; CAUSSINUS, H. Sur l'utilisation d'un modèle mathématique pour l'étude des structures des forêts denses humides sempervirentes de plaine. *C.R. Acad. sci. Paris*, sér. D., 268, 1969, p. 1853–1855.

ROOSE, E. J. Quelques exemples des effets de l'érosion hydrique sur les cultures. In: *C.R. Colloque Fertilité des sols* (Tananarive), II, 1967, p. 1385.

——. Influence du type de plante et du niveau de fertilisation sur la composition des eaux de drainage en climat tropical humide. In: *Comm. XIIIe Journ. Hydraulique*. Paris, 1974.

——; HENRY DES TURREAUX, P. Deux méthodes de mesure du drainage vertical dans un sol en place. *Agr. Trop.*, 25, 1970, p. 1079–1087.

SCHNELL, R. *Introduction à la phytogéographie des pays tropicaux. 2. Le milieu. Les groupements végétaux*. Paris, Gauthier-Villars, 1971, 452 p.

TURC, L. Évaluation des besoins en eau d'irrigation, évapotranspiration potentielle. *Annales Agr. INRA*, 12, 1961, p. 13–49.

WHITTAKER, R. H.; WOODWELL, G. M. Measurement of net primary production of forests. In: *Productivity of forest ecosystems* (Proc. Symp. Brussels, 1969), p. 159–175. Paris, Unesco, 1971, 707 p.

WIEGERT, R. G. Effect of ionizing radiations on leaf fall, decomposition, and litter micro-arthropods of a montane rain forest. In: Odum, H. T.; Pigeon, R. F. (eds.). *A tropical rain forest*, H 89-H 100. Div. Techn. Inf. (USAEC), 1970, 1 678 p.

The forest ecosystems of Gabon: an overview

by R. Catinot[1]

Biogeography

Gabon lies on the Atlantic coast of central Africa. It has an area of 267 000 km² and a coast-line of *ca.* 1 000 km. Altitudes vary from sea level to 300 m in the western sedimentary basin, from 300 to 600 m in the central part, and from 600 to 900 m in the mountainous region, where isolated peaks slightly exceed 1 000 m.

The climate is equatorial with a very even mean temperature, varying only between 23 and 26° C, a relative humidity fluctuating between 75 and 95 per cent and a rainfall between 1 500 and 3 500 mm/a. A more than 50-year-old network of meteorological stations operates in each prefecture, sous-prefecture and all major centres (mines, plantations, air fields, missions, etc.). The results from this very clearly demonstrate a coastal climate (in the western half of the country) characterized by a high precipitation (2 000–4 000 mm/a), a dry season of 2.5 to 3.5 months, a permanently cloudy sky and a mean temperature varying from 18 to 23° C which seems to have given rise to a very special forest flora (*Aucoumea, Sacoglottis,* etc.). There is a falling precipitation gradient from the west to the east, where precipitation averages 1 700–1 800 mm/a, and drier local climates (Mouila, Franceville: 1 200 mm/a) which correspond to savanna zones. The peculiarity of these climates is due to the Benguela marine current, etc. (Aubréville, 1948). The climatic equator crosses the country in a NW-SE direction and places four fifths of its area in the southern hemisphere.

There is a crystalline and metamorphic shelf from the east to the centre, composed of the original ancient shelf, the Francevillean, the schistous limestone and the schistous sandstone formations. In the west, the sedimentary coastal basin is composed of overlying formations from the Mesozoic to the Quaternary. Tectonic has given rise to two gulfs in the south and east which are filled with old sediments.

A soil map is being prepared.

The axis of the hydrographic system of the country is the river Ogooué the drainage basin of which covers more than half its area and terminates in a large delta opposite Port-Gentil. There are also some important coastal rivers, e.g. the Noya, the Como and the Nyanga. The system comprises a large number of lakes, the Ogooué lakes, in the sedimentary basin, and *ca.* 2 000 km² of coastal lagoons (the total area of all lagoons on the Atlantic coast of Africa being 2 500 km²).

The protection provided by the forests which cover

1. Centre technique forestier tropical (CTFT)
   45 bis, avenue de la Belle Gabrielle
   94130 Nogent-sur-Marne, France.

80 per cent of the country is considerable. They regulate the fluctuations of river flow and reduce the amount of solid matter carried as a result of erosion. The profiles of the rivers are only slightly inclined, except at the point of the rift between the crystalline formations and the sedimentary basin, where rapids and falls interrupt the monotony of their course and prevent navigation. Hydrological measurements are taken continuously; special attention being given to the Monts de Cristal and the rapids in the interior with a view to their management and control.

The population is small and there is a high dispersion of inhabitants, and a strong tendency towards concentration at urban centres (Libreville, Port-Gentil, Franceville, Mouila). This tendency must have been greater in the past, as shown by the presence of large stands of *Aucoumea klaineana* in areas that are now practically uninhabited.

There are many human groups, but three predominate: the Fangs who came from the north, the Bapounous south of the Ogooué, and the Bakotas in the south east. The peoples of Gabon are well adapted forest-dwellers, without fear of their environment. They are excellent hunters and fishermen.

Forest ecosystems

More than 80 per cent of the area (220 000 km²) is covered with closed forest. This is one of the highest proportions anywhere in the tropics. Savannas are chiefly found in the south (Mouila-Tchibanga), in the east (Batéké plateaux), and in the centre of the west (coastal plains between Port-Gentil and Libreville).

Types of forest ecosystems

De Saint-Aubin (1963) distinguishes the following ecosystems:
— coastal sedimentary basin;
  • *Aucoumea klaineana – Dacryodes buttnerii – Desbordesia glaucescens* forest;
  • *Aucoumea – Sacoglottis* forest;
  • a type intermediate between these.
— south:
  • *Aucoumea – Dacryodes* forests of Monts Tandou;
  • *Aucoumea – Dacryodes* forests with some *Terminalia superba* and Meliaceae, similar to the Mayombe of the Congo;
— centre and east (south of the Ogooué):
  • *Aucoumea – Dacryodes* forests with *Scyphocephalium ochocoa* and *Paraberlinia bifoliolata*;
— centre and east (north of the Ogooué):
  • *Aucoumea – Dacryodes* with *Monopetalanthus* on the Monts de Cristal;
  • between the Monts de Cristal and the M'Voung with *Gossweilerodendron balsamiferum*, *Scyphocephalium ochocoa* and *Paraberlinia bifoliolata*;
  • east of M'Voung without *Aucoumea klaineana* but with *Scyphocephalium ochocoa, Paraberlinia bifoliolata,* and, locally, *Gilbertiodendron dewewrii*;

— north: no *Aucoumea klaineana* or *Scyphocephalium ochocoa* but with *Terminalia superba* and *Triplochiton scleroxylon*.

Inventories

There are many inventories which constitute exceptionally rich information compared with other tropical countries.

There are numerous qualitative data on the flora of the forests, both old (Klaine, Pellegrin, Walker) and modern (CNRS, CTFT, and in particular, Muséum national d'histoire naturelle of Paris). *The forest Flora of Gabon* (Aubréville, ed.; Muséum national d'histoire naturelle, Paris) has reached its 23rd volume and is nearing completion.

Quantitative forest data are so numerous that the closed forests of Gabon are among the best known in the world. The Research Department of the Gabon Forest Service, with assistance from France, has made 17 inventories in the sedimentary basin, since 1950–1955. The CTFT, with the assistance of UNDP and FAO, surveyed over 7 million ha of closed forest between 1969 and 1974 and made a detailed inventory of 270 000 ha in the region of Kango. This makes it possible to assess the biomass of the forest of the sedimentary basin. These studies show that while the forest ecosystems are very varied, there is a predominance of Burseraceae in the west, and of Myristicaceae and Leguminosae in the whole of the country.

Functioning

Knowledge of plant physiology is superficial. Nevertheless, phenology of the principal species, the germination requirements of *Aucoumea klaineana* and *Musanga cecropioides*, and the growth rhythm of the former are well known.

The primary production of the natural forest is 200–300 m³/ha. That of *Aucoumea klaineana* plantations can reach 600 m³/ha of stem volume and 750 m³ of total volume (stems and branches) for a basal area *ca*. 50 m²/ha.

The fauna of Gabon comprises most of the animals of the high African forest: primates (gorillas, chimpanzees, mandrills, etc.), elephants (probably the largest herd in African forests which is maintaining its numbers), buffaloes, antelopes (bongo, situtunga, duikers, etc.). Because of the sparse human population the fauna remains stable. Elephants destroy a considerable proportion of *Aucoumea klaineana* plantations. The Centre of Makokou is studying ecological balances and primates.

Utilization and modification

Shifting cultivation

Traditional fellings are of small extent, except in the north and the south, because the population is so sparse. They are made in order to plant food crops.

In the sedimentary basin, where the *Aucoumea klaineana* is natural, a large proportion of cleared areas which had been planted with cassava and bananas, return to forest. The wind-dispersed *Aucoumea klaineana* forms very dense thickets and result eventually in a few trees per ha. These are not always very straight because of the excessive light demands, which cause the terminal shoot to change direction. Thus these small-scale clearings regenerate the forest and enrich it with an economically important species.

In other ecological zones cleared areas are invaded by other species which are not commercially important (*Alchornea, Macaranga, Harungana, Musanga*, etc.). They tend to be short-lived and are progressively replaced by other species of high natural forest except where due to repeated clearings, a herbaceous layer establishes itself.

The ecological consequences of these fellings are not very disquieting, because the human population density is so low; during a long period up to *ca.* 1930 these traditional fellings were very beneficial economically, since they led to an enrichment of the forest with *Aucoumea klaineana*. Since the people have become more sedentary, they tend to cultivate the same land several times and to destroy the young stands they had accidentally created.

Industrial felling and management

Industrial fellings produce a similar result to fellings made to clear land for crops. One can often recognize old skidding trails and landings by the existence of exceptionally large numbers of young *Aucoumea klaineana*.

Until recently exploitation was essentially based on minimum felling diameters which varied with species. It was hoped that a regeneration potential would be maintained and the existing young growth would survive. More complex management methods are now being studied experimentally, to provide a sustained yield within the framework of autonomous management units. Regeneration operations are planned to fit in with felling in both time and space; this is an extrapolation of the classical systems of management as practised in countries with an old forestry tradition.

Management types

In view of the strong regeneration potential of *Aucoumea klaineana* in its natural range it was tempting to base the management of the forests purely on this species and to rely largely on its natural regeneration. Unfortunately, experience has shown that such a method is beset with very high risks, is of purely local importance and much more costly than artificial regeneration by planting. Hence the establishment of pure plantations of *A. klaineana* on well-chosen sites.

The sylviculture of *A. klaineana* is very complex because of the exceptionally high light requirement, its tendency for branch growth to continue without natural pruning, and its susceptibility to certain pests (psyllids, scale insects, *Botryodiplodia, Pestallozia*). It has taken over 30 years to develop a perfected method. From the rain forest remaining after logging the operations are:
- suppression of all trees less than 35–40 cm DBH;
- progressive destruction of all other trees by girdling or poisoning;
- introduction of *A. klaineana* at 4×4 m (or 5×4 m)

spacings by direct planting or by planting in 15–20 cm high containers;
- manual tending 2–3 times a year for 5–6 years so that the natural regrowth is allowed to surround the young plants but not to overtop them;
- thinning in one operation at 10–12 years or in two, leaving 100–120 stems/ha;
- harvesting at 45–50 years when the trees will have a diameter of 70–75 cm.

To date 250 km² of such plantations have been established. However, progress is temporarily at a standstill because of a change of priorities by the Government.

Other species have also been tried. The most promising of these were *Nauclea diderrichii, Terminalia superba* (in the south) and *Tarrietia utilis*.

With a view to supplying a very large pulp-mill (250 000–300 000 t/a of dry pulp) more trials were carried out. These showed that the following species could be used successfully:
- *Aucoumea klaineana* (15–18 m³/ha/a) in coppice rotations of 10 years;
- *Gmelina arborea* (25–35 m³/ha/a);
- *Pinus caribaea* and *P. oocarpa* (15–18 m³/ha/a).

Research continues on all these projects.

Suggested forms of management

Management is based on the priority to be given to *Aucoumea klaineana*, a species of great commercial interest which is admirably adapted to its country of origin; on the employment of additional species of great local interest (douka, bilinga, ilomba); on the results of the forest inventories; on autonomous management units, in which logging and regeneration operations are to be planned in harmony so as to ensure sustained yield, and to be entrusted either to an exploiting contractor or to a forest service, or to large commercial forest industries (pulp-mills, board manufactures, etc.). These latter require large forest areas, long-term concessions and would have to be responsible for regeneration and maintenance of the resource at their own expense, in exchange for paying low fees, but accompanied by strict and precise schedules of obligations. Alternatively forest zones can be allocated to agriculture according to land-use planning recommendations.

Research needs and priorities

In many respects Gabon offers an exceptional array of information on rain forest ecosystems. Many valuable efforts have been made to describe the natural environment and the gaps which remain relate chiefly to the functioning of these ecosystems. Hence priority should be given to the following problems.
1. Plant physiology
   Studies of the rhythms of vegetation growth, basic biological (respiration, transpiration, assimilation) and nutritional phenomena (specific absorption of elements, energy budgets, etc.).

2. Animal physiology

Biological studies of the forest fauna and its environmental interactions, especially its integration into the ecosystem and its role in the nutrient and energy budget.

3. Regeneration

Studies on the ecology of most forest species, their genetic improvement and technique of regeneration.

4. Composition and structure

Preparation of much more detailed inventories, especially if certain new techniques can be applied. These are the subject of mathematical and field studies (theory of regionalized variables) for which new means would have to be found, but the application of which might be of far-reaching importance.

Conclusions

Gabon has rain forest ecosystems which have been little disturbed by man and are rich in animal life. They possess certain characteristics which are rather uncommon in Africa, and have been the object of numerous studies on the natural environment (vegetation, fauna, forest inventories) and the human element, as well as the potentialities for forest management (plantations of *Aucoumea klaineana* and research on other species). Moreover, the Centre of Makokou has started a research project on the balance within these ecosystems.

Bibliography

ANON. World wood review 1974. *World Wood* (Bruxelles, San Francisco), vol. 15, no. 6, 1974, 99 p.

AUBRÉVILLE, A. Les brousses secondaires en Afrique équatoriale. *Bois et Forêts des Tropiques* (Nogent-sur-Marne), n° 2, 1947, p. 24–49.

——. *Richesses et misères des forêts de l'Afrique noire: Mission forestière 1945–46*. Agronomie Tropicale (Nogent-sur-Marne), 1948, 251 p.

——. *Climats, forêts et désertification de l'Afrique tropicale*. Paris, Société d'éditions géographiques, maritimes et coloniales, 1949, 351 p.

——. *Contribution à la paléohistoire des forêts de l'Afrique tropicale*. Paris, Société d'éditions géographiques, maritimes et coloniales, 1949, 99 p.

——. A la recherche de la forêt en Côte-d'Ivoire. *Bois et Forêts des Tropiques* (Nogent-sur-Marne), n° 56, 1957, p. 17–32, n° 57, 1958, p. 12–28.

——. L'érosion sous forêt et érosion en pays déforesté dans la zone tropicale humide. *Bois et Forêts des Tropiques* (Nogent-sur-Marne), n° 68, 1959, p. 3–14.

——. Aperçus sur la forêt de la Guyane française. *Bois et Forêts des Tropiques* (Nogent-sur-Marne), n° 80, 1961, p. 3–12.

——. Principes d'une systématique des formations végétales tropicales. *Adansonia* (Paris), tome 5, fasc. 2, 1965, p. 153–196.

——. La destruction des forêts et des sols en pays tropical. *Adansonia* (Paris), tome 11, fasc. 1, 1971, p. 5–39.

BAILLY, C.; BENOIT DE COIGNAC, G.; HUEBER, R.; MALVOS, C.; RAMANAHADRAY. Essai d'aménagement des terres dans la zone forestière de l'Est de Madagascar. Expérience des villages de Morolafa et Andranamody. *Bois et Forêts des Tropiques* (Nogent-sur-Marne), n° 52, 1973, p. 3–18.

BELLOUARD, P. La situation forestière de l'Afrique occidentale française. *Bois et Forêts des Tropiques* (Nogent-sur-Marne), n° 39, 1955, p. 3–23.

CATINOT, R. Sylviculture tropicale en forêt dense africaine. *Bois et Forêts des Tropiques* (Nogent-sur-Marne), n° 100, 1965, p. 5–18; n° 101, 1965, p. 3–16; n° 102, 1965, p. 3–16; n° 103, 1965, p. 3–16; n° 104, 1965, p. 17–30.

——. Les éclaircies dans les peuplements artificiels de forêt dense africaine. Principes de base et applications aux peuplements artificiels d'okoumé. *Bois et Forêts des Tropiques* (Nogent-sur-Marne), n° 126, 1969, p. 15–38.

——. Le présent et l'avenir des forêts tropicales humides. *Bois et Forêts des Tropiques* (Nogent-sur-Marne), n° 154, 1974, p. 3–26.

CENTRE TECHNIQUE FORESTIER TROPICAL. Économie forestière des pays d'Afrique tropicale de l'Ouest. *Études Scientifiques* (Le Caire), numéro spécial, juin 1971, 78 p.

CHEVALIER, A. La décadence des sols et de la végétation en Afrique occidentale française et la protection de la nature. *Bois et Forêts des Tropiques* (Nogent-sur-Marne), n° 16, 1950, p. 335–353.

CLÉMENT, J.; GUELLEC, J. Utilisation des photographies aériennes au 1/5 000 en couleurs pour la détection de l'okoumé dans la forêt dense du Gabon. *Bois et Forêts des Tropiques* (Nogent-sur-Marne), n° 153, 1974, p. 3–22.

DEVRED, R. La végétation forestière du Congo et du Ruanda-Urundi. *Bulletin de la Société Forestière Belge* (Bruxelles), vol. 65, n° 6, 1958, p. 409–468.

DONIS, C. La forêt dense congolaise et sa sylviculture. *Bulletin Agricole du Congo Belge* (Bruxelles), n° 2, 1956, p. 261–320.

FAO. *Politique, législation et administration forestières*. Rome, FAO, 1950, 240 p. (Études des Forêts et des Produits forestiers n° 2).

——. *Forêt et pâturage*. Rome, FAO, 1952, 185 p. (Études des Forêts et des Produits forestiers n° 4).

——. *L'agriculture nomade. Volume 1: Congo Belge et Côte-d'Ivoire*. Rome, FAO, 1956, 166 p. (Études des Forêts et des Produits forestiers n° 1).

——. *Tropical silviculture*. Rome, FAO, 1958, 190 p.

——. *Consommation, production et commerce du bois en Afrique, évolution et perspectives*. Rome, FAO, 1967, 100 p., 83 tabl.

——. *Committee on forest development in the tropics*. Report of the first session (october 1967). Rome, FAO, 1968, 170 p.

——. *Annuaire des produits forestiers*. Rome, FAO, 1970, 216 p.

——. *Exploitation and transport of logs in dense tropical forests*. Rome, FAO, 1974, 100 p.

GAZEL, M. Le développement de l'exploitation forestière en Afrique de l'Ouest. *Présence Africaine* (Paris), n° 86, 1971, p. 38–67.

GROULEZ, J. Le reboisement des savanes pauvres de la ceinture brazzavilloise. *Bois et Forêts des Tropiques* (Nogent-sur-Marne), n° 50, 1956, p. 9–15.

GUIGONIS, G. République Centrafricaine. Association pour l'étude taxonomique de la Flore d'Afrique. *Acta Phytogeographica Suecica* (Uppsala), n° 54, 1968, p. 107–111.

LANLY, J. P. La forêt dense centrafricaine. *Bois et Forêts des Tropiques* (Nogent-sur-Marne), n° 108, 1966, p. 43–55.

——. Régression de la forêt dense en Côte-d'Ivoire. *Bois et*

Forêts des Tropiques (Nogent-sur-Marne), n° 127, 1969, p. 45–59.

LANLY, J. P. Use of recent techniques in the evaluation of forest resources. *Bois et Forêts des Tropiques* (Nogent-sur-Marne), n° 147, 1973, p. 35-45.

LEBRUN, J.; GILBERT, G. *Une classification écologique des forêts du Congo*. Bruxelles, Publ. INEAC (série scientifique), n° 63, 1954, 89 p.

LEMAIGNEN, G. Les industries du bois en Afrique noire franco-phone, un secteur en mutation. *Europe Outremer* (Paris), n° 525, 1973, p. 5.

LETOUZEY, R. *Étude phytogéographique du Cameroun*. Paris, Lechevalier, Encyclopédie Biologique 69, 1968, 508 p.

NANSON, A.; GENNART, M. Contribution à l'étude du climax et en particulier du pédoclimax en forêt équatoriale congolaise. *Bull. Inst. Agron. de Gembloux*, vol. 28, n° 3, 1960, p. 287–342.

NORMAND, D. *Forêts et bois tropicaux*. Paris, Presses Universitaires de France, 1971, 128 p.

PERRIER DE LA BATHIE. *La végétation malgache*. Annales du Musée colonial de Marseille; Paris, Challamel, 1921, 270 p.

PETROFF, G. La production de cellulose dans les pays de l'OCAM. *Bois et Forêts des Tropiques* (Nogent-sur-Marne), n° 143, 1972, p. 35-44.

PIERLOT, R. *Structure et composition des forêts denses d'Afrique centrale, spécialement celles du Kivu*. Bruxelles, Mémoires de l'Académie Royale des Sciences Outre-Mer, vol. 16, n° 4, 1966, 367 p.

RICHARDS, P. W. *The tropical rain forest: an ecological study*. Cambridge, Cambridge Univ. Press, 1952, 450 p.; 4th reprint with corrections, 1972.

ROLLET, B. *Nord Congo: Introduction à l'inventaire forestier*. Rome, FAO, n° 1782, 1964, 116 p.

SAINT-AUBIN, G. de. *La forêt du Gabon*. Nogent-sur-Marne, Centre technique forestier tropical, 1963, 208 p.

SARLIN, P. *Bois et forêts de la Nouvelle-Calédonie*. Nogent-sur-Marne, Centre technique forestier tropical, 1954, 303 p., 131 pl.

——. Répartition des espèces forestières de la Côte-d'Ivoire. *Bois et Forêts des Tropiques* (Nogent-sur-Marne), n° 126, 1969, p. 3–14.

SCHMID, M. *Aperçu sur les forêts du Sud-Ouest de la Mélanésie (Nouvelle-Calédonie, Iles Loyauté, Nouvelles-Hébrides)*. Paris, ORSTOM, 1975, 32 p. multigr.

SCHNELL, R. Aperçu préliminaire sur la phytogéographie de la Guyane. *Adansonia* (Paris), tome 5, fasc. 3, 1965, p. 308–355.

The management and regeneration of some Nigerian high forest ecosystems

by R. M. Lawton[1]

1. Ministry of Overseas Development,
   Land Resources Division,
   Tolworth Tower, Surbiton, Surrey, KT6 7DY,
   United Kingdom.

Introduction

It is estimated that Nigeria has over 95 000 km² of lowland rain forest of which nearly 20 000 km² is forest reserve, and over 25 000 km² of freshwater swamp communities of which only 256 km² is reserved (Progress Report 1966–72).

Most of the lowland forest areas have been cleared at some time for cultivation. Fragments of pottery and pieces of charcoal in soils under tropical high forests, suggest that they have been used for shifting cultivation. A number of historical factors have determined the pattern of human settlement: the effects of slave raiding probably caused people to disperse throughout the forest; and the early states with conflicts between them, led to the creation of inter-tribal buffer zones of uninhabited tracts of forest. Disease and epidemics have also caused settlements to be abandoned.

The forest canopy is broken by natural causes, such as the fall of dead or dying trees, and in some forests, elephants play a role in creating and maintaining gaps (Jones, 1956).

Most natural forests are therefore usually in some stage of dynamic succession following disturbance.

Composition and structure

Species lists of the main trees are of little value, but profile diagrams give some idea of structure as well as composition (Jones, 1948; Keay, 1949; Richards, 1952; Jones, 1956; Baur, 1962; Onyeagocha, 1962). Quantitative enumerations give the distribution of species by girth (or diameter) classes throughout the forest which is essential for the preparation of management plans.

Numerous enumeration and other quantitative data are available for Nigerian forests. Two examples will be used as illustrations: the Akure Forest Reserve with a mean annual rainfall of 1 523 mm (see table 1) and the somewhat moister Sapoba area (near Benin City) with a rainfall of *ca.* 2 200 mm/a; both areas have two months with <30 mm rainfall.

A 1 per cent enumeration of the Akure Forest Reserve (Jones, 1948) is given as an example in annex 1. The permanent 'Natural Forest Inviolate Plot', of *ca.* 32 ha, having a similar structure and composition to that recorded in the enumeration, was set aside (Jones, 1948). One of the soil pits dug in the plot contained fragments of pottery—evidence of previous human occupation.

Immediately above the main or upper tree canopy is a discontinuous emergent layer at a height of at least 46 m. Below the main canopy there are lower canopies of shade tolerant tree and shrub species. The canopies are continuous; some species are growing through to reach the upper canopy, others remain within the dense shade of the lower canopies. Where the canopies are broken, dense climber tangles may develop, or communities of fast-growing light demanding species may become established in the gaps.

TABLE 1. Climatic data for Akure Forest Reserve (from Bamgbala and Oguntala, 1973).

Month	Mean monthly rainfall (mm)	Number of rainy days
January	20	1.3
February	38	3.3
March	135	8.2
April	146	10.7
May	159	11.9
June	204	14.6
July	188	15.7
August	157	17.5
September	204	19.3
October	179	13.2
November	66	6.2
December	29	2.4
Total	1 525	124.3

The dry season lasts from about mid-November to March, with low rainfall and raised day temperatures. The wet season lasts from late March to early November with peaks in June and September. This is typical of southern Nigeria.

Species of the emergent and main canopies mainly belong to four families:
Meliaceae (*Entandrophragma* spp., *Guarea* spp., *Khaya* spp., *Lovoa trichilioides*)
Sterculiaceae (*Cola* spp., *Sterculia* spp., *Mansonia altissima*, *Nesogordonia papaverifera*, *Triplochiton scleroxylon*)
Leguminosae (*Afzelia* spp., *Brachystegia* spp., *Cylicodiscus gabunensis*, *Distemonanthus benthamianus*, *Gossweilerodendron balsamiferum*, *Piptadeniastrum africanum*)
Combretaceae (*Terminalia* spp.).
The lower canopies are rich in species from many families, in particular Annonaceae, which are well represented by *Anonidium mannii*, *Monodora myristica*, *Polyalthia suaveolens*, *Uvariodendron angustifolium* and *Xylopia* spp., in the Akure Forest Reserve enumeration. Species of Ebenaceae are also common in the lower canopies.

The rich high forests of the Sapoba area, described by Onyeagocha (1962), have an emergent layer of *Cylicodiscus gabunensis*, *Gossweilerodendron balsamiferum* and *Entandrophragma* spp. at 76 m, with a main canopy of *Antiaris africana*, *Ceiba pentandra*, *Cylicodiscus gabunensis*, *Distemonanthus benthamianus*, *Entandrophragma candollei*, *Entandrophragma cylindricum*, *Entandrophragma macrophyllum*, *Guarea cedrata*, *Guarea thomsonii*, *Gossweilerodendron*

balsamiferum, *Khaya ivorensis*, *Lovoa trichilioides* and *Strombosia pustulata*.

Large *Brachystegia* spp. are common by water. The light demanders, *Triplochiton scleroxylon*, *Terminalia* spp. and *Chlorophora excelsa* occur where the canopy has been broken or disturbed. *Lophira alata* and *Nauclea diderrichii* are common at the swampy edges of the forest. The important middle canopy in the Benin forests contain the shade tolerant *Anonidium mannii*, *Diospyros confertiflora* and *Strombosia grandifolia*, etc. Climbers are common and form dense impenetrable tangles where the canopy is open or where a dead tree lets in light. *Acacia pennata* and *A. ataxacantha* are two common species. The lower storey contains *Randia acuminata* (*Massularia acuminata*), *Diospyros* spp., *Enantia chlorantha* and *Trichilia prieuriana*. Where light reaches the forest floor, colonies of *Aframomum* spp. and other herbs may occur. In many places a secondary forest dominated by *Musanga cecropioides*, follows a period of cultivation.

Some ecological observations

Knowledge of the autecology of the main species and the dynamic relationships between the species is fragmentary. Richards (1952) noted that *Musanga cecropioides* will not colonize bare ground, but it will become established once a vegetative cover has been formed. It grows rapidly and dies within 15–20 years. It does not regenerate under its own canopy, but by the time it dies, other secondary species have become established and they form the forest canopy.

Jones (1956) has observed that *Khaya ivorensis*, *Lovoa trichilioides*, *Guarea cedrata* and *Lophira alata* are frequently described as shade tolerant and their seedlings will survive for several seasons in fairly dense shade, but much more light than is normally present under an unbroken forest canopy is necessary if the seedlings are to grow. Once past the sapling stage these species need abundant light. *Entandrophragma* spp. also requires large gaps and full light. Some of the middle or lower storey species are truly shade tolerant, in particular *Anonidium mannii*, *Enantia chlorantha* and *Diospyros* spp. Seedlings of *Ricinodendron* sp., *Ceiba* sp., *Canarium* sp. and *Triplochiton scleroxylon* only occur on open ground where there is no shade. Jones (1956) noted that where the light demander *Terminalia superba* was abundant as a mature tree, it did not regenerate under its own canopy, but shade tolerant seedlings of *Khaya ivorensis*, *Entandrophragma* spp., *Guarea* spp. and *Lovoa trichilioides* were present where the canopy had been opened.

Onyeagocha (1962) has made similar observations and he added that the emergent tree species will grow rapidly through the middle girth classes to reach the main canopy, provided they are given sufficient light. An *Entandrophragma cyclindricum* at Sapoba reached a girth of 122–152 cm in 16 years, and two *Khaya ivorensis* of *ca.* 36 years old attained girths of 305 cm and 267 cm.

From this sort of information it is clear that the pattern of natural regeneration is complex and there is still a lack of knowledge on the autecology of many of the species and on the synecology of the forest.

An investigation into the ecology of *Terminalia ivorensis* and *Triplochiton scleroxylon*, two important indigenous timber trees that are being grown in plantations, has shown that *T. scleroxylon* can be propagated from young vegetative cuttings. Different genotypes have been recognized and this study by Jones and Howland (1974) could provide a basis for similar ecological studies on some of the other important indigenous timber trees.

Changes in markets and exploitation

The export of mahogany logs to the United Kingdom started in the 1880s (Pollard, 1955) and the market expanded with minor set-backs until the early 1960s when exports declined in value from £ 11 443 025 in 1962 to £ 4 164 151 in 1972 (Progress Report 1966–72). At the same time there has been an increase in demand by the local market, and imports of raw pulp, paper, newsprint and packing materials have risen from £ 5 782 667 in 1962 to £ 22 380 667 in 1971 (Progress Report 1966–72). Obviously inflation is partly responsible for this fourfold increase in imports.

In the early days companies were granted concessions to selectively log areas of 500–800 km². The lease ran for 25 years and a minimum girth limit, usually of 3 m, was the only method of control. The licensee was required to plant a number of seedlings to replace the exploited timber, but this clause was rarely honoured, or enforced. Without subsequent tending, it is unlikely that the seedlings would have survived. There was no control, the method was very destructive and wasteful, and only the best trees were removed.

In 1944 it was decided to control felling by area as well as girth, and new licences for a period of 25 years were issued. A felling cycle of 100 years was adopted, and each annual coupe was usually one compartment of 259 ha. The felling cycle (or working circle) was divided into four blocks, each of 25 coupes and the licensee was allowed to exploit a block of 5 annual coupes, over a 5 year period. A new list of scheduled trees was introduced in an attempt to increase the utilization of the timber resources. Later the felling cycle was reduced to 50 years.

With the increase in local markets, almost all species are now utilized. The list of 24 economic indigenous tree species in 1953 was increased to 50 in 1969; it is given in annex 2.

It was assumed that the forest would be replaced by natural regeneration, but with the introduction of more intensive methods of exploitation, it became necessary to improve methods of regeneration. The changes in exploitation have influenced and caused changes to be made in methods of management and regeneration.

Early work on regeneration

Kennedy started to work on methods of regeneration at Sapoba in the Benin high forest in 1927, and MacGregor worked on drier types of forest. Kennedy (1935) experimented with various methods of regeneration, including one method in which he cleared areas of *ca.* 0.8 ha around the stumps of exploited trees, but regeneration usually failed to colonize the gaps, and it was necessary to establish groups by planting, or by broadcasting seed. He then selected trees of valuable timber species under the exploitable girth limit, noted when they were in flower, and then cut the climbers, undergrowth and some of the lower canopy species, down wind of the trees, so that the seeds would be blown into the opened area. This method was fairly successful, although sometimes it had to be supplemented with dibbled seed, or planted seedlings. The light demanding species like *Triplochiton scleroxylon* were particularly successful, and *Khaya* spp. and *Entandrophragma* spp. survived in mixtures with other species, where some of the seedlings escaped damage by the shoot borer. The next step was to try an adaptation of the European uniform system, which involves the establishment of regeneration before exploitation. Climber cutting was carried out in year 1, followed by girdling uneconomic lower canopy species in year 2 and 3. Exploitation followed in year 4 and 5. Where regeneration failed, seed was dibbled into gaps, or seedlings were planted. It is noted that the lower canopy trees were only girdled, not poisoned; this would not kill all species and may have led to dense coppice regrowth in some cases. Nevertheless the tropical shelterwood system (TSS) evolved from this adaptation of the uniform system.

Another method of regeneration is the agri-sylvicultural system known as taungya. This system has been practised successfully in Europe and the Far East. An area of exploited forest is allocated to farmers. They clear it to grow food crops and at the same time the Forestry Department plant tree seedlings. The farmers cultivate the area for one to two years and during that time they tend the tree crops. The farms are then abandoned and a new area is allocated to the farmers. The Forestry Department may need to weed, or cut back the climbers, for a few years after the farms have been abandoned. The taungya system succeeds where there is a shortage of land for cultivation, but it usually fails in remote sparsely populated forest areas where there is no pressure on the land. The present developments in the taungya system have been discussed by Olawoye (1975). See also chapter 20.

Line sowing is reported to have succeeded with *Khaya* spp. where germination was good, and the young seedlings were freed from weed competition (Lancaster, 1961a). Enrichment by line planting has given satisfactory results with light demanders e.g. *Triplochiton scleroxylon* and *Terminalia* spp. but failed with *Khaya ivorensis* (Igugu and Bamgbala, 1972).

These methods are intensive and often require considerable supervision so it is perhaps not practical to apply them to large areas. With the introduction of restrictions on the area felled, in 1944, it was estimated that there would be about 52 km² of forest for regeneration each year in the Benin high forests (Lancaster, 1961b). The taungya system could not cope with such a large area and it was therefore decided to introduce the TSS.

The tropical shelterwood system (TSS)

The presence of a number of experienced forest officers from Malaysia, who were serving in Nigeria during the war, helped to formulate the sylvicultural prescriptions of the TSS. The initial operations and subsequent alteration are discussed in detail by Lancaster (1961b) and Baur (1962); see also Lowe, 1975. The first instructions (1944) were:

Year 1		— demarcation of compartment
	any time	— climber cutting
	dry season	— removal of middle storey by poisoning.
Year 2	wet season	— 2nd climber cutting
	dry season	— 2nd canopy opening by poisoning.
Year 3	wet season	— 1st and 2nd cleaning.
Year 4	wet season	— 3rd and 4th cleaning.
Year 5	wet season	— 5th cleaning.
Year 6		— exploitation
	wet season	— 1st post-exploitation cleaning.
Year 9	wet season	— 2nd post-exploitation cleaning.
Year 14	wet season	— 3rd post-exploitation cleaning.
Year 19	wet season	— 4th post-exploitation cleaning.

The aim is to produce a large number of saplings of valuable species before exploitation, and to establish them after exploitation. It was considered that the first climber cutting would be necessary to allow freedom of movement, but it also let in light to the forest floor and enhanced seed germination. The removal of climbers is therefore important.

Poisoning the middle and lower storey trees is to let light in gradually to encourage the regeneration of timber trees; felling lets in light rapidly and leads to an invasion of weeds and the formation of dense climber tangles. Where three is a market for fuel or poles, some of the trees could be felled and removed instead of being poisoned.

The amount of weeding to be carried out leaves plenty of scope for individual interpretation; it ranges from cutting weeds just around the seedlings of desirable species to slashing all the weeds. In practice heavy cleaning leads to the accidental slashing of many desirable species, so light cleanings are less destructive. The post-exploitation cleanings were rarely carried out, apart from an early cleaning to repair damage following exploitation.

Two regeneration counts, including advanced growth, were taken over the whole area. Only one, of two or more, individual plants of economic species, within 2 m of each other, were counted. This method of sampling gives an under-estimate of actual stocking, but includes the important factor of the spatial distribution of regenerating plants. A figure of 100 stocked sample areas (2 m × 2 m) per hectare is considered satisfactory.

By 1953 it was decided to issue revised instructions. Many of the cleaning or weeding operations were dropped, and the most important addition was the removal of the shelterwood of uneconomic and malformed trees a few years after exploitation, by poisoning. After a few more years of experience, final instructions (1961) were:

Year 1	A milliacre assessment.
	Demarcation.
	Climber cutting and cutting back saplings of unec-

onomic species, if stocking of valuable species is low. Poisoning of middle and lower storey if there is sufficient advanced growth of valuable species.

Year 2	Poisoning the middle and lower storey if the assessment showed insufficient advanced growth of valuable species.
Year 6	Exploitation.
Year 8	Repair damage following exploitation, e.g. coppice damaged saplings of valuable species. Post-exploitation climber cutting and weeding. Removal of shelterwood of uneconomic and malformed tree species.
Year 15	Final enumeration (i.e. count of established regeneration).

The assessment is usually a 1 per cent enumeration by linearly arranged 2 × 2 m quadrats.

The TSS instructions cannot be followed rigidly. One of the chief problems is to control the invasion of climbers and, at the same time allow sufficient light to reach the forest floor to favour the germination and establishment of a diverse tree crop. Conditions vary from season to season and it is often impossible to predict the time of seed-fall or to forecast its species composition and the opening of the middle and lower storeys should be co-ordinated with seed-fall. The timing of all operations is important and requires skill and experience. The tropical forest ecosystem is complex, and only experienced field staff can hope to make a success of the TSS.

Reasons for the abandonment of TSS

TSS was abandoned in 1966 and replaced by monoculture plantations of fast-growing indigenous and exotic species. The reasons for this were partly political: there was pressure to release forest reserves for food cultivation; and plantations were visual evidence that the land was being used to grow timber. The practical problems of TSS have been discussed, but it was by no means a failure. In an evaluation of the system, Lowe (1966, 1975) recorded the maximum stocking of *ca.* 885 2×2 m quadrats per ha containing species between 1 m in height and 1.5 m in girth—twice the stocking in untreated forest.

Onyeagocha (1962) observed that there are plenty of seedlings in untreated forests but few reach 3 m in height or 30 cm in girth. This point is illustrated by the following data on the number of small trees of three height classes per hectare:

| | height (m) | | |
	0–1	1–3	>3 (or >30 cm girth)
untreated forest	877	187	50
TSS treated forest	315	150	260.

There is a fivefold increase in the saplings over 3 m in height (or 30 cm girth) in the TSS treated sample as compared with the untreated sample; 67 per cent of the saplings recorded belonged to the valuable Meliaceae family.

An assessment of potential timber stocks was made in the Akure Forest Reserve, fifteen years after exploitation and following TSS treatment (Bamgbala and Oguntala, 1973). The summarized results are:

girth (cm)	30–60	60–120	>120
trees per hectare	102	42	32.

The most important species in order of abundance are: *Nesogordonia papaverifera, Sterculia rhinopetala, Celtis* spp., *Pterygota bequaertii, Mansonia altissima, Triplochiton scleroxylon*.

This potential timber crop is different in composition from the original crop as recorded in 1 per cent enumeration (annex 1), where large *Khaya* spp., *Entandrophragma* spp. and *Brachystegia* spp. form an important part of the emergent layer. The new crop is mainly of light demanding species; a successively younger type of forest.

Present trends and prospects

Fast-growing timber plantations are needed to meet the increasing demands of local markets and every effort should be made to grow them successfully. The natural forest will have to meet these requirements for the next 2–3 decades. Estimates of its yield vary: exploitable timber is generally *ca.* 30 m³/ha (Bamgbala and Oguntala, 1973; Lowe, 1975) although 70 m³/ha is estimated for the Akure Forest Reserve (Bamgbala and Oguntala, 1973); Lowe (1975) considers plantations yield about three times the amount of natural forests. It is possible to establish plantations and manage the natural forest. The natural forest is a renewable resource and more intensive utilization is required to build up a thriving rural economy. Instead of poisoning unwanted trees, they could be used to make charcoal and other by-products (Earl, 1975).

Some of the fast-growing pioneer and secondary species, like *Musanga cecropioides*, may be suitable.

An FAO/UNDP High Forest Development Project, in collaboration with the Nigerian Federal Department of Forestry, is carrying out an inventory of the remaining high forest areas. The inventory has three parts.

1. Indicative planning inventory.
   This is a semi-random tract system sample. Although the size of sample is not known, all trees are recorded by species and measured by relascope down to 40 cm DBH (diameter at breast height). The report unit covers 388 km².
2. Pre-harvesting inventory.
   This is a more intensive inventory covering units of 13 km².
3. Post-harvesting inventory.
   Although this type of inventory has not yet been designed, it will sample the remaining cover and determine the floristic composition and species frequency, following exploitation.

The results of this project will form the basis for any future management and exploitation of the Nigerian high forest ecosystems. It should provide guidance for the re-introduction of some modified form of TSS.

The Federal Department of Forest Research is actively pursuing research concerned with the management of the natural high forest.

Bibliography

BAMGBALA, E. O.; OGUNTALA, A. B. *Merchantable yields and stand projection in Akure Forest Reserve, Western Nigeria.* Ibadan, Federal Department of Forest Research, Research paper (Forest Series), no. 21, 1973, p. 1–22.

BAUR, G. N. *The ecological basis of rain forest management.* Forestry Commission of New South Wales, Australia, 1961–62, 499 p. Rome, FAO, André Meyer Fellowship Programme Report, 1962, 499 p.

EARL, D. E. *Forest energy and economic development.* Oxford, Clarendon Press, 1975, 128 p.

IGUGU, C. O.; BAMGBALA, E. O. Enrichment planting in tropical high forest of Western and mid-Western States of Nigeria. In: Onochie, C. F. A.; Adeyoja, S. K. (eds.). *The development of forest resources in economic advancement of Nigeria*, 1972, p. 213–227.

JONES, A. P. D. *The Natural Forest Inviolate Plot.* Ibadan, Nigerian Forest Department, 1948, 33 p.

JONES, E. W. Ecological studies on the rain forest of southern Nigeria. IV. The plateau forest of the Okomu Forest Reserve. *J. Ecol.* (Oxford), 44, 1956, p. 83–117.

JONES, N.; HOWLAND, P. Research notes: Federal Department of Forest Research, Ibadan, West African Hardwood Improvement Project. *Commonwealth Forestry Review* (London), vol. 53 (3), no. 157, 1974, p. 190–194.

KEAY, R. W. J. *An outline of Nigerian vegetation.* Lagos, Government Printer, 1949, 52 p.

KENNEDY, J. D. The group method of natural regeneration in the rain forest at Sapoba, Southern Nigeria. *Empire Forestry Review* (London), 14, 1935, p. 19–24.

LANCASTER, P. C. Experiments with natural regeneration in the Omo Forest Reserve. *Nigerian Forestry Information Bulletin* (New Series), 1961a, no. 13, p. 5–16.

——. *History of TSS.* Ibadan, Federal Department of Forest Research, Technical Note no. 12, 1961b, 11 p.

LOWE, R. G. TSS investigations in moist semi-deciduous forest of southern Nigeria. In: King, K. F. S.; Iyamabo, D. E. (eds.). *The role and practice of forestry in the national economy* (Proceedings of the second Nigerian Forestry Conference, Enugu), 1966, p. 13–37.

——. *Nigerian experience with natural regeneration in tropical moist forest.* Rome, FAO, 1975, 14 p. multigr.

OLAWOYE, O. O. The agri-silvicultural system in Nigeria. *Commonwealth Forestry Review* (London), vol. 54 (3–4), no. 161–162, 1975, p. 229–236.

ONYEAGOCHA, S. C. *The development of the system of natural regeneration in Malaya compared with Nigeria.* Oxford, Commonwealth Forestry Institute, special study, 1962, unpublished.

OSENI, A. M.; ABAYOMI, J. O. Development trends of Nigerian silvicultural practice. In: Onochie, C. F. A.; Adeyoja, S. K. (eds.). *The development of forest resources in economic advancement of Nigeria*, 1972, p. 127–140.

POLLARD, J. F. A history of the efforts to export timber from Nigeria. *Empire Forestry Review* (London), vol. 34, no. 3, 1955, p. 285–293.

Progress Report 1966–72, prepared for the *10th Commonwealth Forestry Conference, 1974.*

RICHARDS, P. W. *The tropical rain forest: an ecological study.* Cambridge University Press, 1952, 450 p., 4th reprint with corrections, 1972.

Annex 1

1 per cent enumeration data for Akure Forest Reserve

A 1 per cent enumeration of the whole reserve was carried out in 1934 by Lancaster. 105 ha were enumerated. The total figures for this enumeration are given because they provide an interesting comparison and show that the Natural Forest Inviolate Plot is representative of the reserve as a whole, lacking only the strictly river-bank species.

EMERGENTS Girth classes (cm):	6–122	122–183	183–244	244–305	305–366	366 and over	Total
Afzelia bipindensis	15	7	4	5	3	4	38
Alstonia congensis	19	14	16	5	3	3	60
Brachystegia sp. (2)	21	19	12	13	6	10	81
Brachystegia nigerica (3)	47	22	19	22	14	38	162
Canarium schweinfurthii	1	—	—	—	—	—	1
Ceiba pentandra	33	11	6	6	11	37	104
Chlorophora excelsa	12	6	2	2	3	2	27
Combretodendron africanum	13	6	12	1	—	—	32
Cylicodiscus gabunensis	1	2	2	2	6	5	18
Daniellia ogea	2	2	—	—	2	—	6
Entandrophragma candollei	8	4	3	2	—	6	23
Entandrophragma cylindricum	5	—	2	8	4	11	30
Entandrophragma utile	21	4	7	6	2	8	48
Gossweilerodendron balsamiferum	1	—	2	3	—	1	7
Guarea cedrata	10	11	6	4	2	7	40
Guarea thomsonii	1	—	1	2	1	—	5
Holoptelea grandis	14	12	14	10	6	5	61
Khaya sp. (1)	25	9	4	12	4	9	63
Klainedoxa gabonensis	20	13	6	5	4	8	56
Piptadenia africana (4)	7	4	2	3	3	4	23
Pterygopodium oxyphyllum (5)	—	—	1	—	1	1	3
Terminalia ivorensis	—	1	3	8	3	3	18
Terminalia superba	41	30	29	37	30	36	203
Triplochiton scleroxylon	129	120	115	73	51	65	553
Totals	446	297	268	229	159	263	1 662

Density per ha, all sizes, 15.64
Density per ha over 183 cm GBH, 8.6.

UPPER STOREY Girth classes (cm):	6–122	122–183	183–244	244–305	305–366	366 and over	Total
Afzelia africana	2	—	—	2	1	—	5
Afzelia sp. (6)	28	32	18	7	6	3	94
Albizia ferruginea	18	5	2	2	—	—	27
Albizia gummifera	5	4	4	2	2	—	17
Albizia zygia	26	7	2	2	—	—	37
Amphimas pterocarpoides	23	12	5	3	—	1	44
Antiaris africana	3	3	3	1	1	1	12
Aningeria robusta	36	12	14	6	9	8	85
Berlinia sp.	17	9	5	3	—	1	35
Bombax brevicuspe	7	6	2	1	—	1	17
Bombax buonopozense	4	7	5	7	1	2	26
Bosquiea angolensis	35	6	1	1	1	—	44
Casearia dinklagei	1	—	1	—	1	—	3
Celtis soyauxii	429	204	70	49	13	6	771
Chrysophyllum africanum (7)	71	56	13	6	2	—	148
Cistanthera papaverifera (10)	184	168	112	53	18	14	549
Cleistopholis patens	11	11	4	—	—	—	26
Cola cordifolia	385	238	108	58	27	29	845
Cordia millenii	46	45	44	32	12	10	189

Discoglypremna caloneura	45	12	2	1	1	—	61
Distemonanthus benthamianus	1	4	2	1	1	2	11
Diospyros confertiflora	191	9	—	—	—	—	200
Duboscia viridiflora	28	3	—	—	—	—	31
Enantia sp.	99	1	—	—	—	—	100
Erythropsis barteri (11)	116	54	7	—	—	—	177
Hannoa klaineana	15	3	2	—	—	—	20
Hexalobus crispiflorus	225	152	67	26	9	2	481
Homalium alnifolium	42	19	4	—	—	—	65
Hylodendron gabunense	2	7	2	3	—	—	14
Irvingia gabonensis	79	33	11	4	3	—	130
Lannea acidissima	45	18	11	15	4	2	95
Lovoa klaineana (12)	—	—	—	1	—	—	1
Mansonia altissima	292	270	119	24	5	—	710
Mitragyna sp.	1	3	—	—	1	—	5
Ongokea gore	11	15	12	7	—	1	46
Parinari sp. (8)	8	10	4	3	3	5	33
Pausinystalia spp. (?)	17	1	—	—	—	—	18
Pentaclethra macrophylla	11	6	—	1	—	1	19
Phialodiscus unijugatus (9)	16	3	2	1	—	1	23
Phyllanthus discoideus	10	2	—	—	—	—	12
Pterocarpus osun	41	12	6	7	—	2	68
Pterygota macrocarpa	70	79	76	46	17	5	293
Pycnanthus angolensis	16	3	2	2	—	—	23
Ricinodendron africanum (13)	271	146	55	29	9	5	515
Sapium ellipticum	20	12	3	—	—	—	35
Sarcocephalus pobeguinii (14)	2	—	—	—	—	—	2
Scotellia coriacea	208	35	2	—	—	—	245
Spathodea campanulata	13	3	1	—	—	—	17
Spondianthus preussii	17	9	—	—	—	—	26
Staudtia stipitata	5	3	5	—	—	—	13
Sterculia oblonga	39	18	15	3	—	—	75
Sterculia rhinopetala	305	202	118	65	3	—	693
Stereospermum acuminatissimum	42	20	3	—	—	—	65
Strombosia pustulata	619	26	—	—	—	—	645
Treculia africana	19	8	7	9	1	—	44
Totals	4 272	2 026	951	483	151	102	7 985

Density per ha, 76.6.

LOWER STOREY Girth classes (cm): 6–122	122–183	183–244	244–305	305–366	366 and over	Total
Diospyros cauliflora	24	6	2	—	—	—	32
Diospyros monbuttensis	6	—	—	—	—	—	6
Diospyros piscatoria	108	—	—	—	—	—	108
Allanblackia sp.	4	1	1	—	—	—	6
Anonidium mannii	430	76	1	—	—	—	507
Anthocleista sp.	2	1	—	—	—	—	3
Baphia pubescens	3	—	—	—	—	—	3
Bridelia atroviridis	12	2	—	—	—	—	14
Carapa sp.	2	—	—	—	—	—	2
Celtis scotellioides	410	84	14	—	—	—	508
Cola acuminata (*sensu lato*)	20	—	2	—	—	—	22
Cola togoensis	95	2	—	—	—	—	97
Cola sp. (very doubtful)	397	41	1	—	—	—	439
Cola sp. (?)	130	—	—	—	—	—	130
Desplatzia sp.	72	1	—	—	—	1	73
Dracaena sp.	26	5	—	—	—	—	31
Drypetes rowlandii	11	1	—	—	—	—	12
Elaeophorbia drupifera	—	1	—	—	—	—	1
Erythrina sp.	11	2	—	—	—	—	13
Fagara spp.	56	4	—	—	—	—	60

Ficus spp.	8	5	1	—	—	—	14
Funtumia spp.	78	6	—	—	—	—	84
Gardenia imperialis	5	—	—	—	—	—	5
Isolona sp. (?)	16	2	—	—	—	—	18
Lecaniodiscus cupanioides	7	—	—	—	—	—	7
Macaranga barteri	5	—	—	—	—	—	5
Macaranga sp.	3	—	—	—	—	—	3
Macrolobium dewevrei (15)	3	—	2	—	3	—	8
Mannia africana	4	2	1	—	—	—	7
Monodora myristica (?)	15	2	—	—	—	—	17
Musanga cecropioides	24	10	1	—	—	—	35
Myrianthus arboreus	62	2	—	—	—	—	64
Newbouldia laevis	9	—	—	—	—	—	9
Octolobus angustatus	51	1	—	—	—	—	52
Pachylobus spp.	18	4	1	—	—	—	23
Picralima nitida	3	—	—	—	—	—	3
Picralima umbellata	160	—	—	—	—	—	160
Polyalthia suaveolens	41	—	—	—	—	—	41
Randia cladantha (16)	1	—	—	—	—	—	1
Randia genipaeflora (17)	158	3	—	—	—	—	161
Rauvolfia spp.	156	50	3	—	—	—	209
Strombosia grandifolia	118	1	—	—	—	—	119
Tetrapleura teraptera	3	7	—	—	—	—	10
Tetrastemma (or *Uvaria*)	14	6	4	2	—	—	26
Trema guineensis (?)	14	2	1	—	—	—	17
Trichilia heudelotii and *prieuriana*	96	5	8	7 (also 18 large trees, possibly *T. mildbraedii*)	—	—	116
Uapaca staudtii (?)	5	2	—	—	—	—	7
Uvariodendron angustifolium	80	—	—	—	—	—	80
Vitex sp.	79	4	—	—	—	—	83
Xylopia aethiopica	4	—	—	—	—	—	4
Xylopia quintasii	48	—	—	—	—	—	48
Totals	3 107	341	43	9	3	—	3 521
Density per ha over 61 cm GBH, 34.6							
Unidentified trees	188	22	7	3	6	6	233

1. Enumerated as *K. ivorensis*, but it is probably all *K. grandifoliola*.
2. Enumerated as *B. eurycoma*, but probably includes allied species.
3. Enumerated as *B. leonensis*, but corrected to the recently described *B. nigerica*.
4. Now *Piptadeniastrum africanum*.
5. Now *Oxystigma oxyphyllum*.
6. Enumerated as *A. bella*, but probably mostly *A. caudata*.
7. Including also no doubt *C. perpulchrum*.
8. Including probably *P. glabra* and *P. excelsa*.
9. Including probably *Eriocoelum* spp., *Pseudospondias* and other in Sapindaceae and Anacardiaceae.
10. Now *Nesogordonia papaverifera*.
11. Now *Hildegardia barteri*.
12. Now *Lovoa trichilioides*.
13. Now *R. heudelotii*.
14. Now *Nauclea pobeguinii*.
15. Now *Gilbertiodendron dewevrei*.
16. Now *Porterandia cladantha*.
17. Now *Aidia genipaeflora*.

Annex II

List of economic indigenous tree species (1969)

Aformosia elata	*Erythrophleum ivorense*
Afzelia africana	*Gossweilerodendron balsamiferum*
Afzelia bipindensis	*Guarea cedrata*
Afzelia pachyloba	*Guarea thomsonii*
Albizia spp.	*Guibourtia ehie*
Alstonia spp.	*Holoptelea grandis*
Antiaris spp.	*Khaya anthotheca*
Berlinia spp.	*Khaya grandifoliola*
Bombax spp.	*Khaya ivorensis*
Brachystegia spp.	*Lophira alata*
Canarium schweinfurthii	*Lovoa trichilioides*
Ceiba pentandra	*Mansonia altissima*
Celtis spp.	*Mitragyna ciliata*
Chlorophora excelsa	*Nauclea diderrichii*
Coelocaryon preussii	*Nesogordonia papaverifera*
Combretodendron macrocarpum	*Piptadeniastrum africanum*
Cordia millenii	*Pterygota bequaertii*
Cylicodiscus gabunensis	*Pycnanthus angolensis*
Daniellia ogea	*Ricinodendron heudelotii*
Diospyros spp.	*Sterculia oblonga*
Distemonanthus benthamianus	*Sterculia rhinopetala*
Entandrophragma angolense	*Terminalia ivorensis*
Entandrophragma candollei	*Terminalia superba*
Entandrophragma cylindricum	*Tieghemella (Mimusops) heckelii*
Entandrophragma utile	*Triplochiton scleroxylon*

Source: Oseni, A. M.; Abayomi, J. O. Development trends of Nigerian silvicultural practice. In: Onochie, C. F. A.; Adeyoja, [S. K. (eds.). *The development of forest resources in economic advancement of Nigeria*, 1972, p. 127–140.

The Miombo ecosystem

by F. Malaisse[1]

1. Université nationale du Zaire, Campus de Lubumbashi, Lubumbashi, Zaire.

Introduction

Miombo is a type of woodland dominated by *Brachystegia-Julbernardia-Isoberlinia*. The common Bemba term miombo is also used for various component species (*Brachystegia*, in Malawi, Rhodesia, Tanzania and Zambia; *Julbernardia* and *Isoberlinia* in Zaire).

Woodland was defined at the Symposium of phytogeography at Yangambi (CCTA, 1956) and a Symposium was later held in Ndola (CCTA, 1959). Forest is a formation where the highest layer is formed of trees, the crowns of which are not necessarily interlocking but their density is sufficient to cause a herbaceous layer that is distinct from that of savannas. This definition permits the distinction between woodlands and tree and shrub savannas. This difference may be observed when the tree cover exceeds 60 per cent. Aubréville (1957) described the Shabian (a previous province of Katanga) woodland as a 'mixed formation with a sparse graminaceous layer, trees being 15 to 20 m high, that looks like a real forest. Few intermediates between the forest and the herbaceous layer. The trees have interlocking or nearly interlocking crowns, but the foliage is light, the branches being most often extended in umbrella fashion, so that the whole is light and airy, it is an open forest ("forêt claire"); the expression seems correct'.

Miombo has been considered as open deciduous microphyllous forest (Peterken, 1967), rain green forest (Lieth, 1974), tropophilous forest (Lebrun and Gilbert, 1954) and heterothermic forest (Streel, 1963). The miombo type woodlands are similar to the dondo, mikondo, tenda and tumbi of various authors.

An IBP Miombo Project (Malaisse, 1973) has been developed at the National University of Zaire. This has given rise to numerous publications which, with original results, are summarized here.

Climate

The macroclimate of the Lubumbashi region is characterized by a wet season (November to March), a dry season (May to September) and two transition months (October and April). The annual precipitation is *ca.* 1 270 mm, but large between year differences occur (716 to 1 551 mm). Very rarely a rainfall may exceed 100 mm (160 mm in 2 hours 40 min in 1923). Half the daily precipitations are less than 5 mm, but these represent less than 10 per cent of the total

Table 1. Climatic data measured at the Kaspa University Campus Station (Lubumbashi) from 1964–1970.

	J	F	M	A	M	J	J	A	S	O	N	D
T̄M̄ (maximum: 32.3)	28.1	28.2	28.5	29.3	28.6	27.1	26.9	29.1	31.9	32.8	29.6	27.9
TM (maximum: 37.3)	36.4	31.0	31.7	33.0	34.2	30.5	35.0	35.0	37.8	36.6	35.5	32.5
T̄m̄ (minimum: 4.2)	16.3	16.6	15.6	12.3	7.5	5.3	4.2	7.0	10.0	13.2	15.7	16.4
Tm (minimum: −1.5)	12.4	13.5	10.3	6.0	1.5	−1.0	−1.5	0.9	4.2	6.0	11.6	11.8
T (mean annual: 20.3)	22.2	22.4	22.1	20.8	18.1	16.2	15.6	18.1	21.0	23.1	22.7	22.0
P (annual total: 1 275.4)	299.1	247.4	174.9	36.5	3.7	9.2	0	0	4.1	36.1	186.2	278.2
R.H. (mean annual: 77.5)	89	90	89	82	76	76	73	63	58	53	82	87
Piche (mean annual: 3.5)	1.8	1.8	1.9	2.5	3.2	3.6	4.4	5.6	6.2	5.8	2.8	1.9
A (mean annual: 17.3)	11.8	11.9	12.8	16.9	21.1	21.7	22.2	22.2	21.9	19.6	13.8	11.9
R (mean annual: 460.5)	423	414	442	472	478	436	469	493	533	516	456	394
I (mean annual: 7.7)	4.9	4.8	5.9	8.3	10.0	9.7	10.3	9.9	9.8	8.9	6.1	4.2

T̄M̄:mean daily maximum temperatures (°C)
TM:absolute maximum temperature (°C)
T̄m̄:mean daily minimum temperatures (°C)
Tm:absolute minimum temperature (°C)
T:mean monthly temperature (°C)

Abbreviations

P:mean monthly rainfall (mm)
R.H.:relative humidity at 08.00 h (%)
Piche:Piche evaporimeter (mm/d)
A:mean daily temperature amplitude (°C)
R:mean radiation (cal/cm²/d)
I:mean daily sunshine (h)

rainfall; daily rainfalls exceeding 50 mm represent 24 per cent of the total and occur on 4.3 days a year. On average there are 118 rain days.

The average temperature is *ca.* 20° C. It is lowest at the beginning of the dry season (mid-May to mid-July). The night minimum is most frequently *ca.* 06.00 h. Absolute minimums of 0° C are very rare but occur some years. October, or sometimes November, is the hottest month with a daily maximum of 31–33° C. The daily thermal range is small in the wet season and large in the dry season. The humidity of the air depends on the rainfall; it is minimal in October and maximal in February. The annual global radiation is 16.8×10^9 kcal/ha, of which 61 per cent is as direct radiation. Table 1 recapitulates the regional climate calculated from observations made at the meteorological station of the University campus of Lubumbashi.

The climate belongs to Koppen's Cw type with a dry season lasting an average of 186 days and an average rainfall of 1 270 mm.

Soils

The IBP sites are located on zonal soils of the Kaponda series (soil type A-2; Sys and Schmitz, 1959). These are deep, reddish-yellow lateritic soils, of fine clay texture, impermeable, but which drain well. The upper organic horizon is thin and does not exceed 2 cm; the A 1 horizon is also thin, less than 3 cm in general and exceptionally reaching 5 cm; the surface horizons overlie deeper redder horizons which are poorly differentiated. The average values of pH of 25 soil samples, over a year of observation and for each horizon level, are given in the following table (Luiswishi miombo site); the soils of the Kaponda series are characterized by a fairly pronounced acidity; on the other hand, it is likely that the higher values measured at the upper humiferous level are a direct consequence of bush fires. The C/N values were measured in January and September, the total carbon according to the modified method of Anne (1945) and the total nitrogen by the Kjeldahl method. The granulometric analyses made at various depths show a high predominance of fine elements (0–2 µ):

Grain size distribution (per cent)

Depth (cm)	Clay (0–2 µ)	Fine silt (2–20 µ)	Coarse silt (20–50 µ)	Fine sand (50–250 µ)	Coarse sand (350–2 380 µ)
5	36.5	13.9	24.3	14.3	10.9
10	36.4	14.1	26.0	13.1	10.4
20	36.7	15.1	24.9	12.4	9.7
50	40.3	16.7	21.6	11.3	9.4
100	41.8	14.6	26.1	10.5	6.9
200	35.1	12.3	34.4	11.4	6.5

Horizon	pH	C/N January 1969	C/N September 1969
L		20.6	17.5
F+H			
A 1 (−3 cm)	5.3	11.8	13.9
−10 to −20 cm	4.6	11.7	13.0

Structure and function

Distribution

According to the map of African vegetation published in 1959 by AETFAT (Association pour l'étude taxonomique de la flore d'Afrique tropicale) woodlands represent 12.1 per cent of Africa, i.e. *ca.* 3.765×10^6 km². According to Ernst (1971), this corresponds to the limits of resistance to heat and cold of the main trees ($+48°$ C for the leaves and buds, $-2°$ C for the stems of *Brachystegia spiciformis*). The area is subdivided into two blocks, a narrow strip situated north of the equator and a massive ensemble in the Zambezian domain. It is in the latter that the Zairian miombo is found occupying 11 per cent of the area of the Republic of Zaire.

Recent maps giving more detail of the distribution of woodlands are rare; examples include Angola (Grandvaux Barbosa, 1970), areas of Shaba (Sys and Schmitz, 1959; Streel, 1963; Malaisse, 1975) and the Selous Game Reserve in Tanzania (Rodgers, 1973).

The examination of new aerial photographs of Shaba allows a rapid detection of miombo, which is easily differentiated from the other main types of vegetation. However the dense thickets covering the termite hills occasionally modify its appearance.

The absence of a detailed map of Shaba vegetation is an important lack. Its construction from aerial photographs and, when they exist, controlled mosaics, does not present any important difficulties. Reconnaissances on the ground along carefully chosen itineraries are a necessary preliminary. It is desirable that such a map be drawn soon. It would not only give information on the distribution of the main types of vegetation but also allow a general evaluation of the vegetation biomass.

Physiognomy, architecture and structure

General aspect

The tree layer is characterized by its light covering and slight density. The trunks are short and twisted; there are no buttresses except in the case of *Marquesia macroura*, so that well-formed logs are rare. The upper branches are twisted and spread out; the canopy often an umbrella shaped, especially in *Brachystegia* spp. The leaves are small; most often they are pinnate or bi-pinnate with short leaflet stalks. Only a few species of undergrowth have large leaves (*Uapaca* spp. and *Protea* spp.). The lamina is slightly flexible, with variable hairiness, and usually with an entire margin. The species tend to be heliophiles. The miombo flowers all year round. Cauliflory is rare but occurs on some *Ficus* spp., among others. Fruiting is abundant, but often occurs every two years for species or individuals. Dispersal is especially by animals.

Epiphytism is fairly wide-spread in the unburnt areas where carpets of heliophile orchids are found: *Bulbophyllum* spp., *Polystachya* spp. and *Calyptrochilum christianum*.

Crustose and foliose lichens are more abundant than fruticose ones which only occur in the upper part of the crowns, but all lichens and bryophytes are scarce. Epiphyte biomass distribution is illustrated in table 2 for a 16.7 m tall *Marquesia macroura*. Lianas are found but are not well developed because of the fires, some strangling species of *Ficus* are present. There are also a number of species of hemiparasitic Loranthaceae as well as the true parasites *Pilostyles aethiopica* on tree branches and *Thonningia sanguinea* on tree roots.

TABLE 2. Distribution of the biomass (g of dry matter) of epiphytes on a 16.7 m tall *Marquesia macroura*. Values per ha are forty times greater, *ca.* 15 kg.

Height on tree (m)	Mosses	Liver-worts	Lichens foliose	Lichens crustose	Lichens fruti-cose	Vascular plants
0–1						
1–2	0.3					
2–3	1.3	0.6		0.3	6.5	2.6
3–4	2.1			8.5	15.2	
4–5	2.9	0.1		5.5	0.6	
5–6	6.1	0.4		4.4	5.7	
6–7	1.0			0.2	5.5	
7–8	1.2	0.1		0.5	2.1	
8–9	1.5	0.3		11.5	25.6	6.9
9–10	5.3	0.6		13.0	9.5	3.5	5.1
10–11	1.3	0.1		4.3	54.8	0.2
11–12	2.2			17.4	10.5	3.2
12–13		0.1		6.9	20.2	8.9
13–14				2.5	18.4	4.3
14–15	0.3			2.2	35.1	0.9
15–16				9.0	3.3	10.4
16–17	0.1			16.8	31.3	4.4
Total	25.6	2.3		103.0	244.3	42.7	7.7
Percentage	6.0	0.5		24.2	57.4	10.0	1.8

Clumps spared by fire develop rapidly into a dense structure, sometimes called *forêt claire muhuluteuse*, characterized by the exuberant understorey and lianas, the accumulation of litter and the poor development of the herb layer. The understorey is open, scattered with small trees and rare shrubs. The grass layer is more or less continuous and well developed. Sucker shoots are abundant and mix with the grasses which dominate in the gaps where the water-table is close to the soil surface after the last heavy rains. At the end of the wet season flower stalks exceed one meter in height.

Tall termitaria

Mounds built by *Macrotermes falciger* and reaching 8 m in height and 14–15 m basal diameter are characteristic; their density varies from 2.7 to 4.9/ha with an average cover of 6 per cent.

The termitophilous flora is formed of a mosaic of ecological groups amongst which xerophilous and eutrophic tendencies dominate. The xerophily is due to the high clay content of the termite mounds and their low rain-water penetration because of their conical form. The eutrophy is

because their soil has a higher pH than the surrounding ground. The existence of limestone concretions at their centre has been noticed by many authors. Their detailed study is necessary, dealing with regional and local variations, and the presence and absence of termite activity (fossil termitaria do exist there).

Floristics and adaptations

A phytogeographical analysis of 235 characteristic species gives the following distribution (Schmitz, 1971);

6.6 per cent	endemic species in the neighbourhood of Lubumbashi
13.6 per cent	Shabano-Zambian species
30.0 per cent	Zambezian species
9.9 per cent	Zambezian and Oriental species
3.3 per cent	Zambezian and Austral-African or Sahel-Sudan species
6.6 per cent	Zambezian, Oriental and Somalo-Ethiopian, Austral-African, Sahel-Sudan or species of Madagascar
9.5 per cent	omni-Sudano-Zambezian species
2.5 per cent	Zambezian and Guinea species
2.9 per cent	Sudano-Zambezian and Guinea species
7.0 per cent	panafrican to cosmopolitan species
7.0 per cent	African species with varied distribution.

A preliminary phytogeographic analysis on 336 species whose presence has been observed on termite mounds, showed the clear preponderance of the Sudano-Zambezian type (Malaisse and Anastassiou-Socquet, 1977): 18.5 per cent of the plants belong to the Shabano-Zambian distribution type, 7.4 per cent to the Zambezian, 12.5 per cent to the Zambezian and Oriental, 12.2 per cent to the omni-Sudano-Zambezian, 11.3 per cent to other distribution types within the Sudano-Zambezian. 16.1 per cent of the plants are bridge species of the Sudano-Zambezian Guinea-Zairian type, whilst other bridge types (2.7 per cent), pluriregional (8.3 per cent), palaeotropical (6.2 per cent), pantropical (2.1 per cent), Afro-American (0.9 per cent) and cosmopolitan (1.8 per cent) elements play a minor role.

These analyses show a high proportion of Zambezian which spread little outside the limits of the Zambezian zone and few wide-ranging species (Schmitz, 1963).

The life-form spectra show a preponderance of phanerophytes:

	Phanero-phytes	Chamae-phytes	Hemicrypto-phytes	Geo-phytes	Thero-phytes
Gross spectrum (percentage)	41.3	16.6	7.2	27.2	7.7
Balanced spectrum (percentage)	47.1	6.3	19.1	26.2	1.6

Tropical regions with a clear-cut dry season have a typically tropophilous flora and a large number of tropophytes are deciduous (Schnell, 1970). During the dry season the relative humidity decreases to 10 per cent (daily average minimum, 32 per cent), whilst the soil hydration of the

upper 50 cm falls to 11 per cent. The existence of certain xeromorphic tendencies is not surprising. The leaves show:
— thick cuticles;
— reduced lamina surface, with rapid deciduousness; the lamina is sometimes lacking;
— lamina rolling;
— leathery laminas;
— development of supporting tissue;
— protection of the stomata by a thick indumentum, and sunken stomata;
— succulence (aquiferous parenchyma and viscous substances);
— sclerenchymatous sheath around vascular bundles.
Stems are frequently thorny, more woody or succulent. The roots are laterally or vertically extensive; xylopods are frequent.

These various morphological and anatomical adaptations especially characterize the flora of termite mounds. Prickles and fleshiness are the most spectacular.

Variation

The regional differences in the composition of the woodlands of Shaba were studied in detail by Schmitz (1963, 1971) who, using the Zurich-Montpellier phytosociological method, noted many associations. For this author, these should be grouped into three alliances.
— the Berlinio-Marquesion (Lebrun and Gilbert, 1954);
— the Mesobrachystegion (Schmitz, 1950);
— the Xerobrachystegion (Schmitz, 1950).
The first alliance comprises the deciduous and semi-evergreen miombos located mainly on the edge of the Guinea region. These are mainly tall woodlands, largely dominated by *Marquesia macroura* which has a short leafless period. The second alliance comprises the miombo on relatively fertile clay soils. The third alliance corresponds to miombo on drier, often shallow soils. These three alliances form an order which includes all the Zambezian woodlands, the *Julbernardio-Brachystegietalia spiciformis* (Duvigneaud, 1949). Above this order is the *Erythrophleetea africani circumguineensia* class (Schmitz, 1963), i.e. woodlands largely formed of Caesalpiniaceae, spread over the Guinea-Zairian region.

In Zambia, Fanshawe's (1969) analysis distinguished four types of woodlands, including miombo. Other divisions have been proposed by Trapnell (1937, 1943) for Zambia and Lees (1962) for the Copperbelt. Fanshawe (1969) quotes Savory (1963) who had established the soil preferences and rooting characteristics of the main species:
'*Julbernardia paniculata* is very adaptable and occurs on all sites;
'*Brachystegia spiciformis* prefers really deep soils and cannot stand waterlogging;
'*Brachystegia longifolia* prefers deep sandy soils, cannot penetrate murram but can live in a permanent (moving) water-table;
'*Brachystegia utilis* prefers deep loams and cannot penetrate hard murram;

'*Brachystegia floribunda* prefers heavy-textured soils and can penetrate murram;
'*Brachystegia boehmii* prefers clay loams and can penetrate murram;
'*Isoberlinia angolensis* is very adaptable and can live within a permanent (moving) water table'.
In the sense of Schlenker, miombo forms a regional association, including a large number of local variations which are closely connected with edaphical conditions. Lawton (1972) distinguishes several ecological groups in an intensive study of Zambian miombo. He first recognizes two groups: the fire-hardy chipya ecological group; and the fire-sensitive dry evergreen forest ecological group or muhulu (mateshi in Zambia). In addition there are the *Brachystegia-Julbernardia* woodland canopy group, the *Uapaca* group and the ubiquitous group of species with ecological wide range. These groups are not discrete but, as his principal components analysis shows, form a vegetation continuum. The groups are related to soil differences, especially of depth and moisture availability in the dry season, and to the severity or fire which causes some of them to be dynamically related.

In general, the flora on termite mounds is different from that of the surrounding miombo. Moreover, the termite mound flora varies within the Zambezian region. Wild (1952) found 72 termitophilous species, including 10 Capparidaceae in Rhodesia. Only 17 of these species were found on the Shabian termite mounds; only one near the Rio Queve in Cela, Angola (Diniz and Aguiar, 1972); and none spread to the edge of the Chambesi plain in northern Zambia (Lawton, 1972). The number of termite mound species is large: 208 woody species in Zambia (Fanshawe, 1969), 212, including grasses, in Shaba (Schmitz, 1963).

Many authors have emphasized the differences of this flora from the surrounding vegetation. Thus Mullenders (1954) points out that, in the Kaniama region (lower Shaba) the vegetation of the (almost all uninhabited) termite mounds is extremely poor and varies according to the association in which they occur. Fanshawe (1969) distinguishes five habitats (miombo, Kalahari, mopane, munga and riparian) where termite mounds are found and each contains particular species. Another reason for this variation is the presence or absence of termite activity. Only active mounds would have a characteristic vegetation. There are several stages from an active termitarium to a non-active, eroded ones which are often called fossilized. A succession and a zonation occur from typical termite mound vegetation to the ordinary miombo or savanna. This change towards miombo is more rapid at the base of the termite mound and thus it is possible to observe differences on the same termitarium.

Fire

Natural fires have been seen in Shaba at the beginning of the wet season, but these are very rare. The rain that normally follows lightning generally extinguishes these natural fires. Fire is almost always started by man. Generally it takes place once a year during the dry season. Early burning (before the end of June) is less intense and less destructive to woody plants than late burning when the grass is drier and there is more tree litter on the ground and when regrowth may have commenced. Some areas may remain unburnt for one or several years; young trees and climbers grow in these.

Burning experiments carried out at Ndola (Trapnell, 1959; Lawton, 1972) and Lubumbashi (Delvaux, 1958; Symoens and Bingen-Gathy, 1959) show the following species groups:

— fire-tolerant species: *Parinari curatellifolia, Erythrophleum africanum, Pterocarpus angolensis, Anisophyllea boehmii, Diplorhynchus condylocarpon, Combretum* spp., *Ochna schweinfurthiana, Ochthocosmus lemaireanus, Strychnos innocua, S. cocculoides, S. spinosa, Maprounea africana, Hymenocardia acida, Syzygium guineense* subsp. *macrocarpum* and *Uapaca nitida*;

— semi-tolerant species: *Baphia bequaertii, Pseudolachnostylis maprouneifolia, Strychnos pungens, Isoberlinia angolensis, Uapaca kirkiana, Bridelia cathartica, Hexalobus monopetalus, Xylopia odoratissima, Uapaca pilosa*;

— fire-tender species: the woodland canopy dominant species of *Brachystegia* and *Julbernardia, Chrysophyllum bangweolense, Garcinia huillensis, Bridelia duvigneaudii.* Trapnell (1959) also listed intolerant species but these, mainly forest climbers and evergreen trees such as *Entandrophragma delevoyi, Artabotrys monteiroae, Uvaria angolensis*, etc., occur in Shaba in the dry evergreen forest.

Architecture

The density of the strata varies considerably and it is difficult to give average heights. On the main IBP research site, Malaisse distinguished three layers:

— a dominant tree layer, 14–18 (–21) m tall, with a density of *ca.* 65 trees/ha;
— a secondary tree layer, 8–12 (–14) m tall, with a density of *ca.* 80 trees/ha;
— a shrub layer, <8 m tall with 375–500 stems/ha.

The secondary tree layer may be locally absent. Late fires and cutting cause shrubs to branch at or near the base: thus one plant of *Diplorhynchus condylocarpon*, subsp. *mossambicensis* may have *ca.* 20 stems at 1.3 m and the understorey remains open in spite of the large number of stems.

This is why the basal area measured at 1.3 m above ground level is an excellent way to evaluate the density of miombo. It varies from 12 to 25 m²/ha. The dominant tree layer represents *ca.* 35 per cent of the total basal area. When the basal area is less than 10 m²/ha, the composition of the grass layer is different and the vegetation is wooded savanna. Frequency by diameter classes were (IBP site, on 0.5 ha plots):

Diameter classes (cm)	Trees/ha
0–<5	358
5–<10	264
10–<20	174
20–<30	54
30–<40	18
40–<50	14
50–<60	4
100–<110	2

Regular burning hastens the death of weak and small trees. The following results indicate the amount of dead standing trees:

Diameter classes (cm)	Dead trees/ha	Basal area (m²/ha)
0–<5	18	0.02
5–<10	16	0.05
10–<20	16	0.2
20–<30	4	0.2

Miombo has several local types and a relative floristic richness; at least 480 species with an average of 138/ha (table 3).

TABLE 3. Floristic diversity of a miombo woodland.

Family	Number of species	Stems ≥10 cm DBH (percentage)
Tree and shrub layers		
Apocynaceae	1	22
Caesalpiniaceae	4	15
Combretaceae	1	6
Dipterocarpaceae	3	13
Euphorbiaceae	5	12
Papilionaceae	4	11
Rosaceae	1	2
Rubiaceae	4	3
17 other families	20	15
	43	
Grass layer		
Acanthaceae	4	
Commelinaceae	5	
Compositae	11	
Cyperaceae	4	
Gramineae	14	
Iridaceae	4	
Lamiaceae	5	
Papilionaceae	14	
Rubiaceae	12	
26 other families	39	
	112	

At the research station, numerous branched *Diplorhynchus* spp. swell the numbers of Apocynaceae; elsewhere Caesalpiniaceae often represent 40 per cent or more of the tree population.

There remain several gaps in knowledge. The list of miombo species is richer than the reported statistics due to incomplete identifications. A key based on leaves and seedlings would be most valuable, even for the genera such as *Brachystegia*. Studies of variation in density and basal area should be undertaken to aid understanding miombo dynamics, change towards a dense dry forest or wooded savanna and regeneration.

Function

Phenology

Miombo shows well-defined seasonal variations correlated with rainfall and temperature. Phenologically there

TABLE 4. Miombo diaspores, weights and areas.

Species	Number of diaspores observed	Fresh weight of disseminating units (g)			Maximum surface of diaspore's projection (cm²)			Ratio $x:y$ (g/cm²)
		minimum	mean (x)	maximum	minimum	mean (y)	maximum	
Afrormosia angolensis	16	0.65	1.23	1.99	10.7	19.31	31.2	0,0637
Albizia antunesiana	30	0.36	1.23	2.05	17.4	48.43	80.7	0,0254
Brachystegia spiciformis	100	0.34	0.70	0.97	1.6	2.74	3.4	0,2544
Combretum zeyheri	25	1.04	1.71	2.85	13.1	25.14	36.0	0,0682
Dalbergia boehmii	75	0.05	0.10	0.23	3.9	5.87	13.7	0,0174
Dioscorea bulbifera	100	—	0.007	—	1.4	2.01	2.5	0,0035
Monotes africanus	25	0.31	0.58	0.93	4.0	7.81	12.5	0,0747
Ochna schweinfurthiana	150	0.07	0.299	0.57	0.14	0.49	0.79	0,6102
Parinari curatellifolia	77	6.12	13.94	24.34	3.2	6.45	10.0	2,1612
Pseudolachnostylis maprouneifolia	66	1.95	3.12	4.28	1.8	2.28	2.9	1,3672
Pterocarpus angolensis	16	2.60	3.97	6.34	56.0	82.93	118.0	0,0478
Strophanthus welwitschii	100	—	0.0158	—	3.46	10.75	15.76	0,0015
Strychnos innocua	10	42.45	72.41	119.75	16.5	24.49	36.6	2,9567
Uapaca kirkiana	25	3.86	8.10	15.84	2.9	4.92	7.7	1,6444
Vitex mombassae	80	1.51	3.34	5.75	0.84	2.16	3.28	1,5461

are five seasons: cold dry season (May–July), hot dry season (August–September), early wet season (October–November), main wet season (December–February) and late wet season (March–April) (Malaisse, 1974).

The Zairian miombo flowers all year round but the different layers have their own rhythms of flowering. The tree and shrub layers show a flowering maximum during the dry hot season although some shrubs flower as soon as the rains end. The grass layer has two maxima: the largest, especially of hemicryptophytes, in the main wet season and the second, especially of geophytes and geofrutex, in September.

Fruiting shows a seasonal rhythm with a maximum during the hot dry season. The quantity of fruit produced annually varies enormously: from 1968 to 1971, only 160 kg/ha (dry matter) compared with 2 t/ha in 1972; 1970–1971 was a wet year and 1971–1972 a very sunny dry year. Individual trees tend to fruit abundantly in alternate years; in the intermediate years they fruit partially or not at all. For the dominant trees, the maximum recorded values are in the range of 25–30 kg (dry matter) per tree of which, for certain Caesalpiniaceae, 84 per cent accounts for the valves of the pods and the rest for the seeds. Tables 4 and 5 summarize fruit and seed dimensions and production.

TABLE 5. Total production of fruit or seeds for miombo species.

Species	Number of fruit or seeds per tree	Height (m)	Crown projection (m²)	Girth at 1.3 m (cm)	Total production of fruit or seeds (g)		Individual seed dry weight (g)	Maximum dispersal (m)
					fresh weight	dry weight		
Albizia adianthifolia	10 620	11	86.9	101.4	444	397	0.037	103
Brachystegia spiciformis (valve of pod)	4 036	14	56.1	96.3		22 446	5.561	13
Brachystegia spiciformis (seed)	7 921	13	78.2	112.4	4 647	4 196	0.530	20
Chrysophyllum bangweolense	1 250	7.2	27.1	68.3	54 071	—	—	6
Combretum zeyheri	293	5.8	20.7	37.8	502	291	0.993	42
Hymenocardia acida	7 101	8.4	4.2	66.4	538	239	0.034	28
Ochna schweinfurthiana	17 057	6.1	12.5	41.8	5 100	3 151	0.185	6
Pseudolachnostylis maprouneifolia	2 877	6.8	47.8	91.2	8 976	3 087	1.073	9
Pterocarpus chrysothrix	2 530	13	173.0	164.5	8 475	5 027	1.987	20
Strophanthus welwitschii (seed)	2 338	4.4	2.3	6.8	37	33	0.014	>250
(follicle)	7	4.4	2.3	6.8		72	10.27	2
Strychnos innocua	177	5.8	39.2	48.0	12 817	9 733	54.989	6
	462	8.3	51.6	78.9	—	—		7
Swartzia madagascariensis	298	8.2	27.1	59.4	3 327	1 273	4.272	7
Uapaca kirkiana	3 998	8.7	49.6	74.7	5 168	2 067	0.517	8
Vitex doniana	1 326	7.2	28.3	70.1	10 389	3 173	2.393	5
Vitex mombassae	1 978	4.6	30.7	67.3	5 616	2 532	1.33	5

TABLE 6. Relative frequency (in percentage) of fruit dissemination types for some Zairian forests.

Vegetation type	Equatorial dense forest	Semi-deciduous forest	Transition forest	Mountain forest	Forest on waterlogged soil	Miombo
Area	Befale-Equator	Befale-Equator	Kivu	Kivu	Tshuapa-Equator	Southern Shaba
Source	Nanson and Gennart, 1960 (reported by Evrard, 1968)		Liben, 1962		Evrard, 1968	Malaisse (unpublished)
Autochory	37.9	33.5	14	8	12.0	30.4
Zoochory	54.7	54.5	57	49	59.4	38.5
Anemochory	7.0	11.5	24	37	22.2	31.1
Hydrochory	0.4	0.4	5	6	6.4	0

An analysis of the dissemination of 135 miombo specesi shows a slight emphasis on zoochory; this contrasts with other types of forests in Zaire (table 6).

Hydration has not been studied in the Zairian miombo but Ernst and Walker (1973) found a slight seasonal variation in Rhodesia. The young leaves of the main species in October give values between 230 and 280 per cent (of dry matter) whilst the older leaves remain within narrower limits of 100 to 130 per cent from December to leaf-fall in July. Osmotic pressure shows a marked seasonality. In young tree leaves it is less than 10 atmospheres, but increases rapidly to reach a first maximum (17–18 atmospheres) at the end of October; in the middle of the main wet season it decreases to 12–16 atmospheres and then rises progressively until the onset of the dry season; these high values remain until leaf-fall.

More information on the phenology of fruiting is required especially of the main woody species, particularly as so many fruits are edible. Hydration of the different organs of the characteristic species in each layer should be followed throughout the year.

Regeneration dynamics

Lawton (1972) provides information on regeneration. When an area is cleared for cultivation it is replaced by *Diplorhynchus condylocarpon*, *Hymenocardia acida*, *Pericopsis angolensis*, *Pterocarpus angolensis*, *Syzygium guineensis* subsp. *macrocarpum* and *Vitex doniana*. The two last species form patches of canopy at a height of 4 m which partially suppresses the tall grasses and bracken. *Uapaca* species (mainly *U. benguelensis*, *U. kirkiana*, *U. nitida* and *U. sansibarica*) established themselves under these conditions. When their canopy is 4–12 m tall, the herb stratum is reduced to sparse short grass mostly covered with leaf litter. Under these conditions fires creep along the ground and the *Brachystegia-Julbernardia* spp. as well as *Marquesia macroura* may regenerate. They overtop and shade out the *Uapaca* although some *Uapaca* trees survive, while the first group of colonizing species forms a coppice; only if the canopy is opened again will they regrow.

Vegetation biomass and primary production

The completion of observations on the biomass and productivity of woody strata should be one of the primary objectives of miombo research. An indication of the average growth of three trees in the neighbourhood of Lubumbashi is given by Schmitz (1971) (table 7).

TABLE 7. Mean growth of three miombo trees in the neighbourhood of Lubumbashi.

Age (years)	Girth (cm)		
	Julbernardia paniculata	*Brachystegia spiciformis* var. *latifoliolata*	*Parinari curatellifolia* subsp. *mobola*
5	10	18	12
10	19	32	25
15	28	44	37
20	34	55	48
25	42	65	60
30	49	74	71
35	56	83	82
40	63	91	94
45	69	99	106
50	76	106	118
60	89	119	142
70	103	130	166
80	117	140	192
90	130	148	218
100	144		
110	157		
120	171		
130	185		
140	198		
150	212		
160	226		

The analysis of grass and shrub productivity is complete. Annual production and biomass of the grass layer in the miombo have been studied by Freson (1973) who showed a highly positive correlation ($r=0.72$) between rainfall and organic matter production. The above-ground biomass increases rapidly from September to a maximum in January.

The dead biomass appears in December and increases rapidly from January to April (table 8). The losses due to herbivores also begin in December and are highest in March.

TABLE 8. Productivity (dry matter, g/m²) of the grass layer of miombo (modified from Freson, 1973).

Month	Biomass	Monthly change in biomass	Dead biomass including litter fall	Consumption loss	Net production
1969					
September	0	—	0	0	0
October	25.1	25.1	0	0	25.1
November	69.9	44.8	0	0	44.7
December	129.0	59.1	12.4	0.6	72.1
1970					
January	221.8	92.8	17.1	1.2	111.2
February	180.4	—41.5	57.2	3.6	19.3
March	151.6	—28.7	48.4	5.4	25.1
April	141.8	— 9.8	20.3	4.6	15.1
May	135.5	— 6.4	14.4	3.0	11.0
June	133.7	— 1.7	5.9	1.1	5.3
July	Burnt on 2 July				

The annual primary production of the above-ground part of the grass layer of the miombo is 3.3 t/ha/a, which corresponds to 15.4×10^6 kcal/ha/a. The losses by fire were estimated at 71 per cent of the total organic matter present when the fire started.

For the shrub layer, observations were made on *Baphia bequaertii*, a species with a relatively straight trunk and rarely branching below 1.5 m height (table 9).

TABLE 9. Dry weight biomass of leaves and wood of *Baphia bequaertii* by height classes.

Height (m)	DBH (cm)	Leaves (g)	Wood (g)
2–3	3.0	252	1 091
3–4	6.2	1 249	8 051
4–5	9.4	2 865	26 747
5–6	12.2	3 212	54 255
6–7	15.7	4 841	70 236
7–8	18.3	5 552	96 701

The dry weight (g) biomass of two *Baphia bequaertii* shrubs was as follows:

Height of shrub (m)	5.5	5.7
Root depth (cm)		
0–50	6 872	13 142
50–100	5 281	1 529
>100	55	35
Total	12 208	14 706

This is only 15 per cent of the above-ground biomass.

Four years observations on a 1/8 ha plot with an initial basal area for all woody plants of 13.335 m²/ha showed an average growth of 0.388 m²/ha/a. This growth was spread irregularly between the various size categories; girths between 30 and 40 cm showed the maximum growth. The annual dead biomass was 0.037 m²/ha and newly fallen trunks represented 0.018 m²/ha.

These results emphasize the necessity of continuing research on productivity to obtain more precision especially on above-ground biomass, monthly growth in girth of the main species and root productivity of the different layers.

Secondary production

Studies dealing with the secondary production have recently started and the only results available are for numbers and biomass of certain taxa (table 10). The two most important herbivore taxa are the Acridians and the caterpillars. The Acridians show the highest numbers during the main wet season.

Caterpillars are among the largest forest herbivores. Miombo caterpillars seem mostly to eat one species (*Ekebergia benguelensis* for *Rhenea mediata*), or one family (Caesalpiniaceae for *Elaphrodes lactea*, Passifloraceae for *Acraea natalica*, Myrtaceae for *Charaxes druceanus*, Annonaceae for *Xanthopan morgani*, Moraceae for *Pseudoclanis postica*, Lamiaceae for *Precis octavia*, etc.). Less common are caterpillars eating only one genus (*Ficus* for *Myrina silenicus*). On the other hand, several caterpillars seem to feed on plants belonging to two or three unrelated families; they might have a phytochemical relationship. Thus *Hippotion celerio* and *H. eson* feed on Vitaceae, Balsaminaceae and Araceae. Table 11 presents information on the diet of characteristic caterpillars.

Elaphrodes lactea (Lepidoptera, Notodontidae) feeds on Caesalpiniaceae and because of their predominance in the tree stratum is capable of spectacular destruction and total defoliation over large areas. The rapid multiplication of this insect was reported in the Lubumbashi region in 1934 and in 1969–1970 when observations were made. The main predators are birds, notably the black-headed oriolo, Hemiptera, spiders, chameleons and lizards. Parasites have been observed at various stages: Chalcidians on eggs, Ichneumonids and Tachinids on caterpillars. In 1970 82.3 per cent of the eggs hatched; half of the young caterpillars became adult and 74 per cent of these made cocoons, but only 51 per cent became chrysalids; only 8.3 per cent of the eggs gave rise to butterflies. With a sex-ratio (♂/♀) of 11.5, there is effective population control by parasites and predators in the second year of its multiplication.

Rearing experiments showed that, at their peak density, their monthly leaf consumption was 733 m²/ha (a dry weight of 98 kg/ha) and the dry weight of their faeces was 90 kg/ha. This represents an enormous transfer of green leaf material to the soil; its effect on tree growth is unknown but new leaves appear in May or June.

Other species have also high densities (table 12). All the *Pericopsis angolensis* trees were defoliated in March 1971 when the *Uapaca* spp. were attacked by a *Parasa* sp.

TABLE 10. Comparison of the monthly average number and live weight biomass (g/1 000 m^2) of the main animal groups in the herb layer for the dry and wet seasons in burnt and unburnt miombo during 1973–1974.

	Unburnt				Burnt 5 July 1973					
	Dry season		Wet season		Dry season				Wet season	
					before fire		after fire			
	Number	Biomass	Number	Biomass	Number	Biomass	Number	Biomass	Number	Biomass
Araneidea	236	28	308	108	147	35	39	10	257	84
Acridoidea	227	582	118	253	296	884	26	43	124	240
Tettigonioidea	33	32	29	41	41	37	0	0	29	32
Grylloidea	55	36	164	166	65	41	19	18	190	240
Blattodea	96	24	203	79	137	32	89	14	123	63
Mantodea	25	39	57	50	28	36	5	3	29	31
Heteroptera	131	51	208	135	201	67	10	6	86	56
Homoptera	82	12	58	17	92	20	16	1	71	22
Coleoptera	22	12	87	90	32	18	2	—	74	79
Caterpillars	5	6	50	75	20	11	1	1	34	60

TABLE 11. Diet of some miombo caterpillars.

Caterpillar family and species	Thunbergia lathyroides	Syzygium guineense	Stephania abyssinica	Pterocarpus angolensis	Ochna schweinfurthiana	Marquesia macroura	Markhamia obtusifolia	Julbernardia paniculata	Flacourtia indica	Ficus persicifolia	Fagaria chalybea	Erythrophleum africanum	Erythrina tomentosa	Ekebergia benguelensis	Dalbergia boehmii	Ceropegia meyeri-johannis var. verdickii	Canthium crassum	Bridelia ferruginea	Brachystegia spiciformis var. latifoliolata	Brachystegia microphylla	Brachystegia longifolia	Brachystegia boehmii	Baphia bequaertii	Annona senegalensis	Amorphophallus abyssinicus	Allophyllus africanus	Albizia antunesiana	Adenia gummifera
ACRAEIDAE																												
Acraea natalica																												×
ARCTIIDAE																												
Aganaïs baumanniana					×																							
ATTACIDAE																												
Bunaea caffraria														×														
Cirina forda												×																
Tagoropsis hanningtoni																											×	
DANAIDAE																												
Danaida chrysippus																×												
EUPTEROTIDAE																												
Camerunia flava																	×											
LASIOCAMPIDAE																												
Batalebeda strandi			×									×																
Epitrabela argyrostigma							×																					
Pachymeta robusta																				×								
LIMACODIDAE																												
Parasa viridissima							×														×							
NOCTUIDAE																												
Diva casta			×																									
Naurilia arcuata																												
Oraesia excavata			×																									
Rhanidophora cinctigutta	×																											
NOTODONTIDAE																												
Antheua tricolor																									×			
Cerurina marshalli																									×			
Desmocraera interpellatrix rileyi		×																										
Elaphrodes lactea		×					×												×	×		×						×
NYCTEMERIDAE																												
Hibrildes ansorgei																	×											
NYMPHALIDAE																												
Charaxes bohemanni																									×			
PAPILIONIDAE																												
Papilio demodocus											×																	
SPHINGIDAE																												
Coelonia fulvinotata						×																						
Hippotion osiris																											×	
Libyoclanis punctum					×																							
Platysphinx stigmatica piabilis				×								×													×			
Xanthopan morgani																									×			

Studies on population dynamics of the main species of Lepidoptera should be undertaken as soon as possible. Certain Hemiptera and plant-eating Coleoptera also show rapid multiplication and their study is desirable.

Table 12. Main Lepidoptera of the Kaspa miombo ($+++ > 2\ 500$ caterpillars/100 m²; $++ > 250$; $+ > 25$).

Season	1968–1969	1969–1970	1970–1971	1971–1972	1972–1973	1973–1974
Attacidae						
Cyrtogone sp.	+	—	—	+	—	—
Lasiocampidae						
Pachymeta robusta	—	—	—	—	—	+
Limacodidae						
Parasa cf. *urda*	—	+	—	++	—	—
Lymantriidae						
Dasychira goodi	—	—	—	+	+	—
Notodontidae						
Elaphrodes lactea	+++	—	+++	—	—	—
Thaumetopoeidae						
Paradrallia rhodesi	+	—	+	++	—	—
Tortricidae						
unidentified species	—	—	—	+	—	—

Rodents have not been studied in the Zairian miombo. For a Zambian miombo on hydromorphic soils Sidorowicz (1974) found miombo to be poorer in small mammals than *Loudetia* savannas, alluvial savannas with *Acacia*, or thickets on halomorphic soils. He noted in the miombo the presence, in decreasing importance, of *Praomys natalensis*, *Tatera leucogaster*, *Saccostomus campestris* and *Beamys major*, and considered that carnivorous mammals, birds and reptiles and parasites controlled the density of rodents; they were well adapted to fire. Dowsett (1966) in the Ngoma area of the Kafue National Park, in Zambia, has studied, during the wet season, the game population structure and biomass in the miombo; the study area was a self-contained ecological unit, comprising 31 km² of woodlands. The following table gives for each species the density and biomass per km² of suitable habitat:

	Density per km²	Average weight (kg)	Biomass (kg/km²)
Wildebeest			
(*Connochaetes taurinus*)	1.62	195	315.6
Waterbuck			
(*Kobus defassa*)	1.47	147	216.4
Hartebeest			
(*Alcelaphus lichtensteini*)	1.47	140	206.4
Reedbuck			
(*Redunca arundinum*)	2.31	52	120.3
Sable			
(*Hippotragus niger*)	0.66	180	119.6
Zebra			
(*Equus burchelli*)	0.42	250	104.6
Elephant			
(*Loxodonta africana*)	0.02	4 980	99.7
Kudu			
(*Tragelaphus strepsiceros*)	0.58	140	81.4
Roan			
(*Hippotragus equinus*)	0.23	140	32.3
Warthog			
(*Phacochoerus aethiopicus*)	0.39	68	26.5
Eland			
(*Taurotragus oryx*)	0.04	362	14.5
Lion			
(*Panthera leo*)	0.08	154	12.3
Common duiker			
(*Sylvicapra grimmia*)	0.01	14	11.0
Bushbuck			
(*Tragelaphus scriptus*)	0.31	32	9.8
Spotted hyaena			
(*Crocuta crocuta*)	0.15	59	8.8
Sharpe's grysbok			
(*Raphicerus sharpei*)	0.85	7	6.2
Impala			
(*Aepyceros melampus*)	0.08	59	4.7
Cheetah			
(*Acinonyx jubatus*)	0.08	27	2.2
Baboon			
(*Papio ursinus*)	0.23	14	3.1
Vervet monkey			
(*Cercopithecus aethiops*)	0.50	3	1.6
Jackal			
(*Canis adustus*)	0.23	5	1.2

Tertiary production

Studies of the carnivores are still scanty, but a preliminary study of spiders has been completed. The spiders living in the tree and shrub and grass strata can be divided into six behaviour patterns which correspond to their ecological niches:

— sedentary spiders that build large vertical webs on the periphery of the tree crowns;
— spiders that are commensal to the first;
— sedentary spiders that build horizontal webs, mainly at branch level;
— sedentary spiders that spin small vertical webs in trunk irregularities;
— hunting spiders that move quickly;
— mimic hunting spiders with long immobile waits.

Nephila pilipes pilipes is very abundant and has a relatively permanent web with long sticky threads. Its life-cycle has been studied: copulation occurs in April and eggs are laid inside cocoons stuck to the leaves and branchlets at the beginning of the dry season; after activity inside the cocoon, the young spiders leave them from mid-August to mid-October for a community life; in October-November each establishes an individual web and they have several instars before becoming adults. The diet of the young spiders is 59 per cent Hymenoptera, 22 per cent Coleoptera and 11 per cent Diptera. Adults eat Acridians, Heteroptera, adult Lepidoptera and bees. During May the dry weight of food captured daily by the population is 7.9 g/ha compared with 0.97 g/ha in mid-December. Amongst predators are rodents for eggs, the lizard *Mabuya* spp. and spiders for the young, and chameleons and *Ammophila* (Sphecidae) for the adults. 87 per cent of the eggs hatch and most losses are due to parasites, especially Chalcidians. Thus the numbers per hectare vary from 188 000 young spiders in October, 36 320 in November, 24 160 in December, 8 320 in March, 1 610 in May and 790 in June just before burning.

The variation in numbers and biomass of spiders in the herb stratum has been followed over a year, on burnt and unburnt sites (table 10). They are two maxima (April with 335 and

TABLE 13. Relative frequency (percentage) of spiders in the grass layer of the miombo.

(March 1973)

Lycosidae	18.5
Salticidae	10.5
Oxyopidae	10.1
Drassidae	9.4
Clubionidae	9.1
Thomisidae	9.1
Pisauridae	7.0
Ctenidae	5.6
Araneidae	5.6
Sparassidae	4.5
Theridiidae	3.1
Asamiidae	2.8
Other families which represent less than 2 per cent	4.7

November with 347 individuals/100 m²) and a minimum in September (unburnt 80/100 m² and burnt 34/100 m²). The maximum fresh weight biomass occurs in November (16.9 g/ha) and the minimum just after burning (0.4 g/ha). The relative importance of the different families at the end of the wet season has been estimated (table 13).

These results indicate future research on carnivorous groups or those with a mixed diet; ants and Heteroptera are most in need of study.

Water balance and biogeochemical cycles

Water balance

Rainfall interception and stemflow in the tree and shrub strata on a 625 m² plot (with 52 woody plants; equivalent to 832/ha and a basal area, at 1.3 m above-ground level, of 19.6 m²/ha) are shown in table 14. The proportion intercepted appears to be related to the duration and intensity of the rainfall rather than the leaf canopy. The dense herb layer will, no doubt, considerably reduce the amount of water reaching the ground surface.

Alexandre (1973) tried to establish a global water balance. Of the *ca.* 1 200 mm/a rainfall, scarcely 50 mm/a flow away. This mainly occurs during the second half of the wet

TABLE 14. Water balance in the Kaspa miombo during 1973 and 1974 on a 625 m² plot.

Period	Rainfall (mm)	Penetration						Interception	
		Throughfall		Stemflow		Total			
		mm	Percentage	mm	Percentage	mm	Percentage	mm	Percentage
1973									
January[1]	239.0	201.6	84.4	3.3	1.4	204.9	85.8	34.1	14.2
February[1]	228.5	198.3	86.8	5.0	2.2	203.3	89.0	25.2	11.0
March	95.8	82.0	85.6	1.5	1.5	83.5	87.1	12.3	12.9
April	118.3	88.0	74.4	2.2	1.9	90.2	76.3	28.1	23.7
May-August	0.0	—	—	—	—	—	—	—	—
September	21.5	18.9	88.0	0.3	1.4	19.2	89.4	2.3	10.6
October	10.1	7.2	72.4	0.0	0.2	7.2	72.6	2.9	27.4
November	119.6	95.3	79.7	2.0	1.7	97.3	81.4	22.3	18.6
December	272.4	198.8	73.0	1.8	0.7	200.6	73.7	71.8	26.3
Total	1 105.2	890.1	80.5	16.1	1.5	906.2	82.0	199.0	18.0
1974									
January	363.6	294.6	81.0	3.8	1.0	298.4	82.0	65.2	18.0
February	217.7	185.4	85.2	2.9	1.3	188.3	86.5	29.4	13.5
March	370.5	284.6	76.8	5.8	1.6	290.4	78.4	80.1	21.6
April	52.7	40.7	77.2	0.3	0.6	41.0	77.8	11.7	22.2
May	90.9	75.5	83.0	0.5	0.6	76.0	83.6	14.9	16.4
June	0.0	—	—	—	—	—	—	—	—
July	0.0	—	—	—	—	—	—	—	—
August*	(3.4)								
September	0.0	—	—	—	—	—	—	—	—
October	2.4	1.9	78.2	0.0	0.0	1.9	78.2	0.5	21.8
November	63.4	48.6	76.6	1.2	1.9	49.8	78.5	13.6	21.5
December	301.9	217.1	71.9	6.0	2.0	223.1	73.9	78.8	26.1
Total	1 463.1	1 148.4	78.5	20.5	1.4	1 168.9	79.9	294.2	20.1

* Throughfall not measured; rainfall exceptional (observed only twice in 53 years) and not taken into consideration.
1. Observations on 400 m².

season, when the water-table is near the ground surface in the dembo savannas that periodically flood. The infiltration water in the first half of the wet season restores the soil to field capacity; this represents an average of 220 mm for the soil above the capillary fringe. A similar volume feeds the water-table. With the growth of vegetation, the interception increases to 35 per cent (19 per cent by the tree and shrub layers). Infiltration of stemflow is *ca.* 18 mm, and direct evaporation 200 mm. During the dry season, the actual evapotranspiration (ETR) is equal to 360 mm, and 680 mm during the wet season except in October and November when plant growth is high. Over the whole year evapotranspiration is 1 050 mm, of which 200 mm evaporate directly and 850 mm transpire. Transpiration was 3 mm/a during the wet season and 1.5 mm during the dry season. Rivers flow at a rate equivalent to 160 mm/a which represents a run-off coefficient of only 13 per cent.

Litter fall and decomposition

92 per cent of the miombo species are deciduous and the annual leaf-fall of the tree and shrub layers is 2.5–3.4 t/ha of dry matter, with an average of 2.9 t/ha. This represents 68 per cent of the total litter (table 15).

TABLE 15. Variations in litter fall (dry matter in g/m^2) in the Kaspa miombo (1968–1973).

Year	Leaves	Flowers and fruits	Wood	Total
1968	261.0	9.9	96.7	367.6
1969	338.5	26.8	78.7	444.2
1970	295.0	19.5	84.9	399.4
1971	307.5	11.4	86.1	405.0
1972	255.5	201.0	90.6	546.8
1973	271.1	37.3	85.6	394.0
Average	288.1	51.0	87.1	426.2

In 1973, a study of the litter fall gave the following percentages:

leaves	68.1	scales and buds	0.45
fruits	6.7	wood (diameter <1.5 cm)	17.3
flowers	2.2	wood (diameter >1.5 cm)	2.7
lichens	0.03	bark	1.5
fungi	0.04	dead animals	0.4
mosses	0.01	excrements	0.7

The contribution made by large branches is being studied. Flowers and scales contribute little, but a large number of species have heavy fruits. The litter on the ground is highest during the early wet season (4.4 t/ha) and minimal during the cold dry season just after burning (1.6 t/ha) and before the leaves dried by the fire begin to fall; the average is 3.3 t/ha/a.

Litter decomposition is performed by three principal agents: micro-organisms, termites and fire. Soil micro-organisms show a relative homogeneous spatial distribution and a well-defined seasonal periodicity. However, the rate of decomposition of the leaves of different species varies greatly; those which are slowly decomposed protect the soil from the first showers at the beginning of the wet season and slow down erosion; they seem to occur more frequently in the miombo than in tree savannas.

Termites (table 16) are very active except during the hot dry season, when the fires are most destructive. The activity of the latter takes over that of insects.

Three diet patterns can be distinguished among termites: humivorous, lignivorous and mixed. The humivorous genera (*Cubitermes*, *Anoplotermes*, *Crenitermes* and *Basidentermes*) excrete organic compounds that are more resistant to degradation by physico-chemical agents and bacteria. The lignivorous termites (*Microcerotermes*, *Amitermes* and the Macrotermitinae in general) eat dead intact plant material; their superficial attacks are particularly spectacular and their densities may exceed 15 000/m².

The coefficient of decomposition k (see chapter 13) varies between 1.32 and 1.11 according to whether or not fire has occurred. The cumulative effect of termites and fire doubles the speed of litter decomposition.

Biogeochemical cycles

Biogeochemical cycles have not been studied but preliminary chemical analyses suggest:
— the leaves of the main species show little difference in macro-elements composition (in per cent of dry weight): N=1.5–2.9; K=0.6–1.2; Ca=0.16–0.90; P=0.07–0.15; Mn=0.05–0.32; Fe=0.013–0.027; Na, Cu and Zn = traces;
— one-year-old branchlets and roots have fairly similar amounts (N=1.3–1.6) which are higher than those of the trunks and large branches (N=0.4–0.6).
Ernst (1975) has studied the variation of the mineral content of tree leaves in a Rhodesian miombo located at Warren Hill (17° 50′ S, 30° 57′ E), during the main wet season. Mean results for the dominant trees (*Brachystegia* spp. and *Julbernardia* spp.) were 3.89 of ash (per cent of dry weight) and 524 (Ca), 150 (Mg), 78 (K), 19 (Na), 10 (Mn), 52 (P) and 28 (Cl) μg/g of dry weight Three dominant trees were surveyed from the time of flushing to that of shedding. With the exception of iron, manganese and sodium, they seemed to be able to similarly accumulate mineral elements, but young leaves were richer in nitrogen, phosphorus and potassium than mature ones, whilst the concentrations of aluminium, calcium, iron, manganese and sodium increased until abscission.

Freson (1974) has shown that the mineral output by rivers was *ca.* 124 kg/ha/a.

The study of biogeochemical cycles should be undertaken as soon as possible; a separate and detailed study should be made of the nitrogen cycle, because of its agricultural importance.

TABLE 16. Abundance and biomass of the main groups of soil fauna in the Zairian miombo (Goffinet, 1973a and b, 1975).

	Density (individuals/m²)		Biomass (dry weight, mg/m²)
Protozoa (testaceous Rhizopoda)	1.44×10^6		—
Acarida	81 514		30.92
Collembola	14 531		14.70
Protura	213		—
Diplura	206		—
Thysanura	1		—
Pauropoda	1 882		—
Symphylans	90		—
Homoptera (larvae)	1 985		—
Oligochaetae (worms)	10		225.69
Molluscs	rare		negligible
Isopoda	rare		—
Pseudo-scorpions	5.21		—
Araneidae	11.83		47.55
Chilopoda	4.71		10.04
Diplopoda	4.17		71.24
Isoptera { hypogeous termites	469.92	lignivorous	60
epigeous termites		humivorous	250
— lignivorous (*Macrotermes*)*	± 200		950
— humivorous (*Cubitermes*)	860		490
Embioptera	0.54		0.39
Dictyoptera	5.17		29.41
Dermaptera	0.46		—
Orthoptera	2.17		22.66
Formicidae	390.92		31.23
Homoptera (adults)	9.17		45.17
Thysanoptera	38.46		0.09
Psocoptera	10.25		0.07
Holometabolic larvae	69.17		263.97
Coleoptera (adults)	21.29		71.37
Carabidae			7.47
Tenebrionidae			1.02
Lagriidae			5.12
Elateridae			0.14
Curculionidae			0.09
Scolytidae			0.38
Staphylinidae			2.33
Scarabaeidae			49.23
Other families			5.59

* Approximative value on the basis of 2 million individuals per nest of *Macrotermes falciger*.

Man-made modifications

The regressive succession

The miombo is generally not considered as a climatic climax, but is a fire climax. The climatic climax is the very rare and disappearing dense dry forest called muhulu. Its detailed study is therefore urgent. Wood cutting and fire have changed it into miombo woodland; and this fire climax still covers 85 per cent of Upper Shaba but it is being replaced by savannas.

The dense dry forest is a closed vegetation type with several strata; most of the trees in the upper layers lose their leaves for a very short period; the understorey is evergreen or deciduous and the grass cover is discontinuous. The savanna consists of grasses reaching 1.1 m in height, with scattered shrubs.

The characteristics of the three stages of such a regressive series (dense dry forest—open woodland—savanna), often called the muhulu—miombo—savanna regression, are shown in table 17.

Savanna formation increases the mean annual temperature, its mean daily amplitude, and decreases the relative humidity. Such changes, on a large scale, may modify the regional climate and the water balance. Moreover, the savanna protects the soil less than the forest and woodland. For a territory mainly covered by woodland, the minerals exported are *ca.* 125 kg/ha/a; they could increase as a result of the progressive extension of anthropic savannas.

The comparative study of the three stages in the regressive series would provide very useful information for the management of dry tropical regions. The role of litter, the evolution of the soils, the water balance and the biogeochemical cycles, should be the primary themes. In spite of their small area, the remaining muhulu require immediate study and protection.

E. umbellata, E. camaldulensis, E. citriodora, etc.) have been used in experimental plantations and although the best species may not yet be clear, the beneficial results of *Eucalyptus* on a carbonized surface is confirmed. The plantations always yield >300 m³/ha at 25 years and sometimes

TABLE 17. Ecological characteristics of dry dense forest, woodland and savanna in Upper Shaba (modified from Freson *et al.*, 1974).

	Dry dense forest	Woodland	Savanna
CLIMATE			
Mean annual temperature (°C)*	19.2	20.6	22.1
Mean daily amplitude (°C)*	10.4	16.5	20.8
Solar radiation at 1.3 m (per cent of external)	2.3	26.8	100.0
Throughfall (per cent of rainfall)	57.7	78.8	100.0
Mean annual relative humidity (per cent)	81.7	71.8	64.0
SOIL			
Depth of A_1 (cm)	5–10	2–3	0–1
pH (A_1)	4.2	5.3	5.9
Mean annual humidity (per cent) at:			
10 cm	27.6	16.7	18.7
25 cm	24.4	17.9	17.2
50 cm	21.8	18.0	18.6
100 cm	21.3	19.2	19.4
VEGETATION			
Height (m)	18–22	14–17	1.1–5
Species diversity**	105	480	330
Number of woody plants/ha	8 500	500–900	30–70
Basal area at 1.3 m (m²/ha)	35–45	15–25	0.5
Total dry weight biomass (t/ha)	320	150	10

* soil, screened temperature.
** number of phanerogams in whole flora.

Forestry

The soils of woodlands are relatively poor and are unsuited for permanent agriculture, and require considerable vegetation cover to protect them from erosion and degradation; they should remain under forest land-use. Foresters have therefore tried to adapt techniques in order to manage such immense areas.

The natural forest may be enriched by the introduction of indigenous or introduced species (Schmitz, 1959). The experiments on indigenous species generally failed and only two exotic species (*Cupressus lusitanica* and *Callitris* sp.) gave acceptable results and these required protection against fire for several years.

Coppice growth may be stimulated by mechanically clearing at ground level and burning the material as small heaps in order not to kill the stumps. This encourages the production of dense regeneration except on the small burning sites. Early burning is carried out until the coppice regrowth is high enough to shade out the grasses. Later thinning improves the production of the best stems. After 20 years girth increase decreases. Further experiments on this method are necessary.

Several species of *Eucalyptus* (*E. saligna, E. maculata,*

exceed 500 m³/ha (Schmitz, 1969). The natural coppice, which needs the same care as the stands enriched with *Eucalyptus*, only produces a quarter of this. At twelve years of age, one hectare of seedlings can produce 6 tons of charcoal.

Pinus kesiya shows rapid growth, a relatively dense cover in young formations and reaches adult height in 4–5 years. It has a fairly straight trunk, but often shows a curve at the base. *P. insularis* has equal growth, straighter trunks and no basal curve, but its cover is less dense. Together with the very fire sensitive *Cupressus lusitanica*, these are the three conifers that are best adapted to local conditions.

The role of fire and protection against it are controversial topics. Experiments made by Delvaux (1958) and observations of Symoens and Bingen-Gathy (1959) suggest that the productivity of protected plots is higher than that of early burnt plots. But an accidental fire in protected areas causes considerable loss and the yield is then less than that of early burnt area. Moreover, early burning is better than protection for the rapid selection of the best elements of the stand. It is always possible to change from early burning to protection; the reverse is not possible. The high tolerance of certain pines and eucalypts to fire could be improved by early burning; the obsession of fire would thus be eliminated and an acceptable rate of wood production would be obtained.

Charcoal, which has been produced from miombo for a long time has given rise to industry and commerce. The needs of the mining towns of Shaba grow and thousands of uncontrolled people are deforesting the region. All charcoal for domestic use is provided by traditional exploitation using the forest stack method. The stack capacity is normally 20 m³ but may reach 60 m³. In spite of its low yield (a maximum of 17.5 per cent of the wood weight is carbonized), it meets the consumer's demand which is much more for easy use than quality.

Unfortunately, traditional scattered cutting and defective regrowth on huge areas is disastrous. After cutting, the forest is scattered with clearing where grass and a few worthless trees grow. Thus the derived savannas spread progressively and these at best, under present conditions, could evolve only into wooded savannas.

Pastoralism

The best fodder species are the grasses *Setaria thermitaria* and *Brachiaria brizantha*, and several leguminous species of *Indigofera*, *Crotalaria*, *Eriosema* and *Vigna*. These species are rarely abundant and the low palatability would dictate a carrying capacity of one head of cattle per fifteen hectares at least.

During the dry season the cellulose content of the grass increases whilst the protein content decreases. Hence, the need for early burning and supplementary feeding by silage. For a Zambian miombo, Rees (1974) suggests cutting down the trees and shrubs at ± 1.2 m height (the chitimene system). The stumps produce numerous shoots during the dry season and their young leaves, containing four times the protein

of the grass layer, are browsed by cattle. Grazing will change the flora and experiments are needed for the introduction and extension of cattle-raising.

Glossina morsitans, the vector of trypanosomiasis is not found in the grass savannas of the high Shabian plateaux which are the zones of cattle-breeding. The present prohibition of hunting, in as much as it is respected, should allow large mammals to re-establish and trypanosomiasis to spread, although most workers now conclude that the control of large wild mammals has no effect on tsetse in savanna areas.

Any livestock scheme necessitates dips twice a month to protect against ticks which carry anaplasmosis, piroplasmosis, etc. Hence cattle should not wander for more than 10 km from a central point.

Miombo contains several plants that are poisonous to cattle and poisoning is fairly frequent at the beginning of the wet season. A list of these plants is required, and botanical and toxicological research is urgent. *Buphane disticha* and *Gnidia kraussiana* are the two main species responsible for lethal poisoning. Poisoning was also observed after the cattle have eaten *Gloriosa superba*, *Peucedanum wildemanianum* and *Urginea altissima*, and various *Moraea* spp. The elimination of underground organs of toxic plants is the only way of avoiding fatal accidents; the uprooting of bulbs and rhizomes is a lengthy task, but possible over small areas (the case of a society that destroyed 20 000 bulbs in Shaba has been recorded). Such an operation remains problematical for large areas.

Traditional use

Miombo also provides food, medicinal, ichthyotoxic plants and fibres; meat and livestock products. The miombo of the

TABLE 18. Food value of some Zairian miombo products (after Heymans *et al.*, 1970; Thoen *et al.*, 1974).

	Water (%)	Dry matter (%)	Percentage of the dry weight					
			Ash	Lipids	Nitrogen	Protein	Fibre	Calcium
Cantharellaceae								
Cantharellus spp.	89.2	10.8	12.0	6.6	3.0	19.0	4.9	—
Amanitaceae								
Termictomyces spp.	87.8	12.1	8.8	4.1	6.3	39.1	7.0	—
Boletaceae								
Boletus granulatus	85.5	14.5	4.5	4.7	3.1	19.5	9.6	—
Tricholomataceae								
Schizophyllum commune	35.0	65.0	3.0	0.5	2.7	17.0	4.0	—
Russulaceae								
Lactarius spp.	89.3	10.7	7.0	6.7	3.6	22.7	7.2	—
Acrididae								
Homoxyrrhepes punctipennis	74.6	25.4	4.5	4.9	11.3	70.6	—	0.13
Notodontidae								
Elaphrodes lactea (dried caterpillars)	14.6	85.4	3.4	29.6	9.6	60.1	—	0.06
Termitidae								
Macrotermes sp. (soldiers)	66.0	34.0	6.8	5.3	9.8	61.3	—	0.17
Stenogyridae								
Achatina sp.	80.5	19.5	6.6	4.1	8.7	54.4	—	1.44

Lubumbashi region produces more than fifty edible plant species: berries of *Strychnos* and *Chrysophyllum bangweolense*; drupes of *Vitex*, *Parinari* and *Uapaca*; false fruit of *Ficus*, berry-like and fleshy fruit of *Syzygium* and *Aframomum*; flowers and seeds of *Sphenostylis*, aggregated fruit of *Annona*, tubers of *Dioscorea* spp. and certain orchids, bulbs of *Cyanastrum johnstonii*, young shoots of *Adenia gummifera* and *Pteridium aquilinum*. In addition there are more than twenty mushrooms. Besides wild meat, caterpillars, grasshoppers and termites are the insects that are most often eaten. Their food value (table 18) and their importance (Lambrecht and Bernier, 1953) have been defined.

Medicinal plants in Shaba have been listed and described by Schmitz (1967).

The miombo also produces ichthyotoxic plants (although some are not used as such in Shaba): *Pterocarpus angolensis*, *Strychnos innocua*, *Syzygium guineense*, *Ziziphus mucronata* subsp. *rhodesica* and *Balanites aegyptiaca*, the use of which has been noticed in Zambia and in Cameroon. *Diospyros mweroensis*, *Neorautanenia pseudopachyrhiza*, *Gnidia kraussiana* and *Euphorbia* cf. *ingens* are used in the Zairian miombo. *Diospyros mweroensis* (katula) is the wild plant that is most used, with the cultivated *Tephrosia vogelii* (buba), second.

The inventory and study of the food value of animal and plant products should be continued. A list of medicinal and ichthyotoxic plants should be made giving the vernacular and scientific names as well as determining the active chemicals. This would be a valuable guide for selecting species on which a thorough phytochemical study should be undertaken.

Research needs and priorities

Woodlands cover 12 per cent of Africa and represent one of the most important of its plant formations. Miombo is the major woodland type. This ecosystem has been the subject of multidisciplinary study in Zaire but several gaps remain and it is necessary to continue research.

The first area of study should be its structure and function, the degree of modification it can tolerate, and its ability to reach equilibrium. It is necessary to determine the productivity of the main plant species, including the roots; the composition and role of the various groups of animals, notably the ants, grasshoppers, birds and termites. The incidence of various types of fire should be further specified as should the biogeochemical cycles which will indicate ways of improving the ecosystem.

A second area of study is towards a more rational land use. It is therefore necessary to map the different types of vegetation in Upper Shaba and to determine more precisely the factors which control its various facies, and especially the relationships between soil and vegetation. Research is necessary on the two possible fundamental options: forestry and grazing. Forest development implies controlled felling and a better knowledge of natural regeneration and planting of indigenous and exotic species, including fruit trees like mango and guava. Grazing implies the increase of cattle health, improving the food during the dry season, reducing the frequency of dip treatments and delimiting areas cleared of toxic plants.

The last area of study should be the history of vegetation and human activities in Upper Shaba. The long controversy as to whether dry forest is the climatic climax of Upper Shaba or an edaphic climax with a limited distribution, might be solved if areas of woodlands on various soils and at different stages of evolution could be protected and observed over fifteen years. An ecological study of the few remaining dry forest patches should be undertaken immediately. It should include the micro-climate, water balance, leaf-fall, accumulation and decomposition of litter, and soil characteristics. The flora inventory of Schmitz (1963) should be extended by phenological data. These dry forests should be protected as natural reserves.

The survey and evaluation of natural foods should include phenological and ecological observations which would allow an estimation of the maximum human carrying capacity in undisturbed miombo. This would tie in with research by the anthropologists of the National Museums of Zaire. In the same way an inventory of the medicinal and ichthyotoxic plants should be completed, giving their vernacular and scientific names, and their active chemicals when these have been elucidated. This work would be a valuable guide for selecting more detailed phytochemical studies. These two inventories should be completed rapidly, as the traditional knowledge is quickly being lost because of changing customs and migrations.

Selective bibliography

ALEXANDRE, J. Le bilan hydrique du miombo. In: Malaisse, F. *Semaine Étude Problèmes Intertropicaux*, p. 144–150. Gembloux, 1973, 839 p.

AUBRÉVILLE, A. Muhulus, termitières fossiles géantes et forêt claire katanguiens. *Bois et Forêts des Tropiques* (Nogent-sur-Marne), 51, 1957, p. 33–39.

COLONVAL-ELENKOV, E.; MALAISSE, F. Contribution à l'étude de l'écosystème forêt claire (miombo). Note 20. Remarques sur l'écomorphologie de la flore termitophile du Haut-Shaba (Zaïre). *Bull. Soc. Roy. Bot. Belg.*, vol. 108, n° 2, 1975, p. 167–181.

ERNST, W. Zür Ökologie der Miombo-Wälder. *Flora*, 160, 1971, p. 317–331.

——; WALKER, B. Studies on the hydrature of trees in miombo woodland in south central Africa. *J. Ecol.*, 61, 1973, p. 667–673.

FANSHAWE, D. The vegetation of Zambia. *For. Res. Bull.*, 7, 1969, 67 p.

FRESON, R. Contribution à l'étude de l'écosystème forêt claire (miombo). Note 13. Aperçu de la biomasse et de la productivité de la strate herbacée au miombo de la Luiswishi. *Ann. Univ. Abidjan*, E, vol. 6, n° 2, 1973, p. 265–277.

FRESON, R.; GOFFINET, G.; MALAISSE, F. Ecological effects of the regressive succession muhulu-miombo-savannah in Upper Shaba (Zaire). In: *Proc. 1st Int. Congr. Ecol.* (The Hague, September 1974), p. 365–371. Wageningen, Centre for agricultural publishing and documentation, 1974, 414 p.

GOFFINET, G. *Synécologie comparée des milieux édaphiques de quatre écosystèmes caractéristiques du Haut-Shaba (Zaïre).* University of Liège, Ph. D. thesis, 1973a, 332 p.

——. Recherches préliminaires sur les fluctuations saisonnières des peuplements d'Acariens et de Collemboles au niveau de la litière du miombo. *Ann. Univ. Abidjan*, E, vol. 6, n° 2, 1973b, p. 257–263.

——. Écologie édaphique des milieux naturels du Haut-Shaba (Zaïre). 1. Caractéristiques écotopiques et synécologie comparée des zoocénoses intercaliques. *Rev. Écol. Biol. Sol* (Paris), vol. 12, n° 4, 1975, p. 691–722.

——; FRESON, R. Recherches synécologiques sur la pédofaune de l'écosystème forêt claire (miombo). *Bull. Soc. Écol.*, vol. 3, n° 2, 1972, p. 138–150.

HEYMANS, J. C.; EVRARD, A. Contribution à l'étude de la composition alimentaire des insectes comestibles de la Province du Katanga. *Problèmes Sociaux Congolais* (Lubumbashi), 90–91, 1970, p. 333–340.

LAWTON, R. M. *An ecological study of miombo and chipya woodland with particular reference to Zambia.* Oxford, Ph. D. thesis, 1972, multigr.

MALAISSE, F. Caractéristiques climatiques et écologiques du Shaba (Zaïre). In: *Semaine Études Problèmes Intertropicaux*, p. 144–150. Gembloux, 1973, 839 p.

——. Contribution à l'étude de l'écosystème forêt claire (miombo). Note 8. Le Projet Miombo. *Ann. Univ. Abidjan*, E, vol. 6, n° 2, 1973, p. 227–250.

——. Phenology of the Zambezian woodland area, with emphasis on the miombo ecosystem. In: Lieth (ed.). *Phenology and seasonality modelling*, p. 269–286. Berlin, New York, Springer Verlag, Ecological Studies no. 8, 1974, 444 p.

——. Conséquences écologiques de certaines modifications récentes de l'environnement rural au Shaba méridional (Zaïre). In: *Semaine Études Agriculture Environnement*, p. 343–352. Gembloux, 1974, 726 p.

MALAISSE, F. De l'origine de la flore termitophile du Haut-Shaba (Zaïre). *Boissiera*, 24b, 1976, p. 505–513.

——. Quelques méthodes d'étude de la structure en forêt. Exemple d'application au miombo zaïrois, écosystème tropical. In: *La pratique de l'écologie*, p. 104–118. Bruxelles, Administration générale de la coopération au développement, AGCD, 1976, 140 p.

——; MALAISSE-MOUSSET, M. Contribution à l'étude de l'écosystème forêt claire (miombo). Phénologie de la défoliation. *Bull. Soc. Roy. Bot. Belg.*, vol. 103, n° 1, 1970, p. 119–124.

——; VERSTRAETEN, C.; BULAIMU, T. Contribution à l'étude de l'écosystème forêt claire (miombo). Note 3. Dynamique des populations d'*Elaphrodes lactea* (Gaede) Lepidoptera, Notodontidae). *Rev. Zool. Afr.*, vol. 88, n° 2, 1974, p. 286–310.

——; ALEXANDRE, J.; FRESON, R.; GOFFINET, G.; MALAISSE-MOUSSET, M. The miombo ecosystem: a preliminary study. In: Golley, P. M.; Golley, F. B. (eds.). *Tropical ecology with an emphasis on organic production*, p. 363–405. Athens (Georgia, USA), Univ. of Georgia, 1972, 418 p.

——; FRESON, R.; GOFFINET, G.; MALAISSE-MOUSSET, M. Litterfall and litter breakdown in miombo. In: Golley, F. B.; Medina, E. (eds.). *Tropical ecological systems: trends in terrestrial and aquatic research*, p. 137–152. Berlin, New York, Springer Verlag, Ecological Studies no. 11, 1975, 398 p.

——; ANASTASSIOU-SOCQUET, F. Contribution à l'étude de l'écosystème forêt claire (miombo). Note 24. Phytogéographie des hautes termitières du Shaba méridional (Zaïre). *Bull. Soc. Roy. Bot. Belg.*, vol. 110, n° 1, 1977.

SCHMITZ, A. Aperçu sur les groupements végétaux du Katanga. *Bull. Soc. Roy. Bot. Belg.*, 96, 1963, p. 233–447.

——. L'utilisation des plantes du Haut-Katanga. *Africa-Tervuren*, vol. 13, n° 2, 1967, p. 41–54.

SCHNELL, R. *Introduction à la phytogéographie des pays tropicaux.* 2 vol. Paris, Gauthier-Villars, 1970, 951 p.

THOEN, D.; PARENT, G.; TSHITEYA, L. L'usage des champignons dans le Haut-Shaba. *Problèmes Sociaux Zaïrois*, 100–101, 1974, p. 69–85.

WHITE, F. The savanna woodlands of the Zambezian and Sudanian domains. An ecological and phytogeographical comparison. *Webbia*, vol. 19, no. 2, 1965, p. 651–681.

The forest ecosystems of the Brazilian Amazon: description, functioning and research needs

by João Murça Pires[1]

Introduction

Very little has been written in the form of an integrated treatment of the national or regional ecosystems of Brazil. The studies which show the greatest appreciation of the region as a whole are those published or in preparation by the Brazilian Institute of Geography and Statistics (IBGE) and by the RADAM Project of the Ministry of Mines and Energy. The IBGE, through its National Council of Geography, published the *Atlas Geral do Brasil* (1966) and five volumes, each dedicated to one of the great geographical regions (1959–65). The RADAM Project is based on a series of radar photographs. The series has been completed for the Amazonian area and RADAM is presently publishing monographs based on their interpretation and on field observations. Seven monographs have been published so far.

Broad-based surveys of forest resources have been carried out by the FAO Mission (1956–61) and are under the jurisdiction of the RADAM Project while studies on a smaller scale have been carried out by the Brazilian Forest Development Institute (IBDF), the Brazilian Agricultural Research Enterprize (EMBRAPA), the Institute for the Social and Economic Development of Pará (IDESP) and the private sector.

Besides Martius' *Flora brasiliensis*, now out-of-date, there exist scattered studies by a great number of botanists, zoologists and ethnologists.

Brazilian vegetation and climate

South America possesses a very diversified vegetation of which more than half is forest (Hueck, 1972). The Brazilian forests outside of the Amazon extend from the east coast of the northern frontier to Rio Grande do Sul.

The non-forest part of Brazil includes the xerophytic thorn-scrub *caatinga* of the Northeast; the *cerrados* (xeromorphic grasslands, savannas, scrubs and woodlands) of central Brazil; the subtropical grasslands of Rio Grande do Sul; the Pantanal complex (inundated *campos cerrados* and forests) of Mato Grosso; savanna areas within Amazonia (Territories of Roraima, Rondônia and Amapá and the State of Pará) and areas of edaphic non-forest vegetation known as caatingas, *campinas* or *chavascais* of various sizes scattered throughout Amazonia; these caatingas have nothing to do with the dry thorn-scrub caatingas of the Northeast. See fig. 1.

1. Museu Paraense Emilio Goeldi,
   Belém, Pará, Brazil.

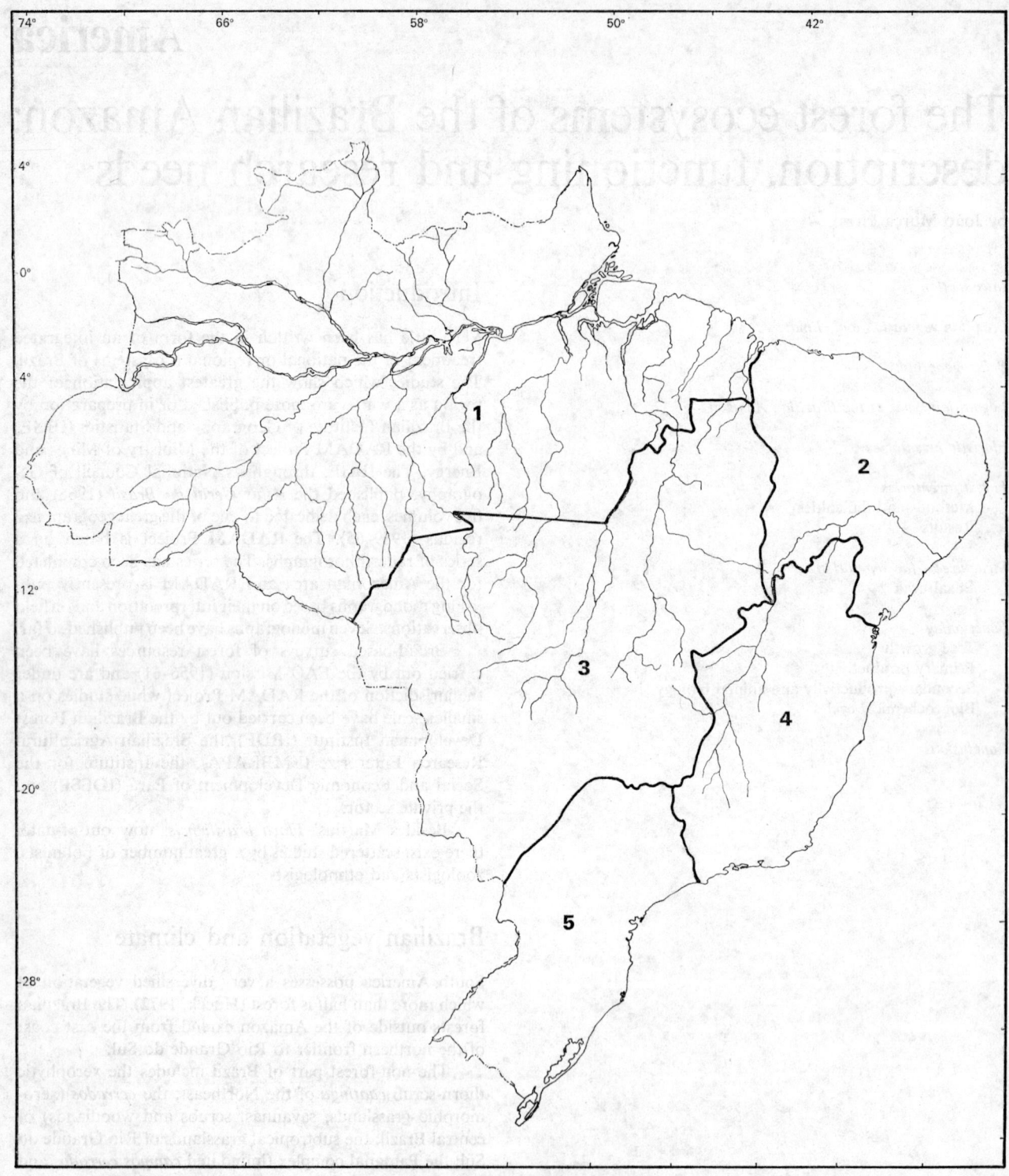

FIG. 1. Principal physiographic regions of Brazil, according to the Brazilian National Council of Geography:
1, North; 2, North-east; 3, Central-west; 4, East; 5, South.

Brazilian territory thus includes:

— a forested region, estimated as *ca.* 60 per cent of Brazilian territory, made up of:
- the forests of the Amazon region, *ca.* 40–50 per cent;
- the forests of the Atlantic eastern coast region, *ca.* 12–18 per cent (tropical forest to the north; temperate forest to the south: mixed forest without *Araucaria* and *Araucaria* forest);
- other types of forests, *ca.* 5–10 per cent;

— a non-forested region, *ca.* 40 per cent of the Brazilian territory, made up of:
- semi-arid xerophytic caatingas of the Northeast, *ca.* 10 per cent;
- savannas of Central Brazil, *ca.* 20 per cent;
- subtropical prairie-like *campos* of southern Brazil, *ca.* 5 per cent;
- Pantanal complex of Mato Grosso, *ca.* 5 per cent;
- other types of vegetation, *ca.* 5 per cent.

See fig. 2.

The Amazon region is characterized by tall forests and an aseasonal climate with high temperatures and rainfall.

The Atlantic coastal forests occupy a narrow strip along the Atlantic especially in its northern part between the States of Bahia and Pernambuco. These areas contain tall forests, with a physiognomy similar to that of the Amazon forests although the species composition differs and there are a number of common species. The climate varies with latitude. To the south there is a cool season which becomes more accentuated in the temperate forests of the south. The rainfall is in general less intense than in the Amazon, but in some places it may be of equal intensity as in the mountainous Serra do Mar near Santos where the annual precipitation can reach 4 500 mm. Between 20 and 30° S, the strip of coastal forest widens considerably to the west and in certain areas *Araucaria angustifolia* (Paraná pine) appears and can reach a considerable concentration within the mixed forest forming the so-called pine region.

The Amazon region, the cerrado, the caatinga and the Atlantic coastal forest, have an annual average of temperatures of 22° C–27° C. To the south of latitude 20° S the transition to the temperate climate with a cooler season from May to August becomes apparent. In the coastal forest this differentiation begins from the State of Bahia. In the south, the minimum temperature may be less than 0° C.

In the northern part of Brazil temperatures are uniform, thus simplifying the classification of climate which chiefly depends on rainfall. The mean temperature is 25° C. The classification used by Aubréville (1961) is quite practical. It is based on a pluviometric index of three numbers corresponding to the number of months with rainfall >100 mm, 30–100 mm, and <30 mm respectively.

For the Amazon, this pluviometric index varies from 12–0–0 to 8–4–0 and 7–4–1 (with the exception of the Territory of Roraima where the index is 5–6–1 or 5–7–0).

The lengths of the wet and dry season vary considerably. In the Amazon the wet season lasts from December or January until May (with the exception of the Territory of Roraima which, differs from the rest of Brazil, is humid from May to September, with two well-defined seasons, a wet and a dry season in reversed position in comparison to the rest of Amazonia). The climate of the Amazon can be subdivided into three according to dry season: the western part has no dry season; the central Amazon basin, the coast of Amapá and a band from central Brazil to the Amazon have a very moderate dry season; in northern Roraima there is a moderately accentuated dry season.

Palaeogeography

The Amazon region originated from two ancient shields, the Guiana and the Brazilian (Mendes, 1967), between which the ocean was replaced by sediments which in some places exceed 4 000 m depth, taking in the Acre basin, the Amazon basin, the Marajó basin and the Parnaíba (or Maranhão) basin.

According to certain authors (Ab'Saber, 1957; Vanzolini, 1970; Vuilleumier, 1971) the Amazon region has undergone great fluctuation in climate, the last of which was between 3 500 and 2 000 BP (Vanzolini, 1970). There might have been shrinkages and expansions of the forested area, with the formation of refuge areas which would later function as nuclei of colonization (Vanzolini, 1970; Meggers and Evans, 1973; Prance, 1973).

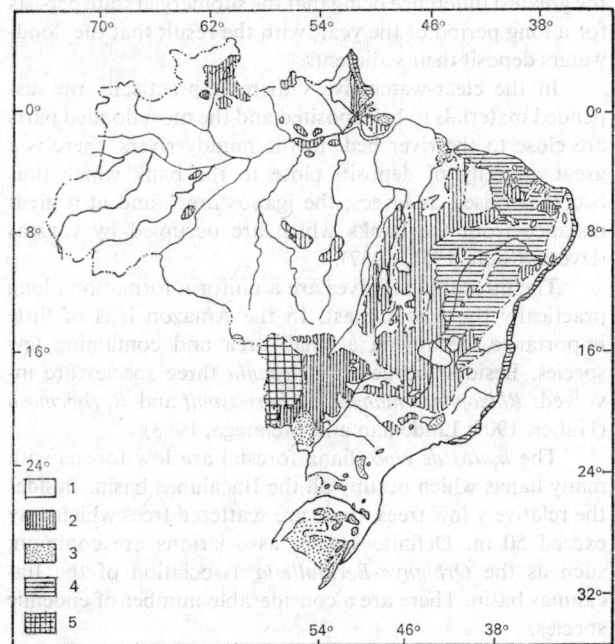

FIG. 2. Main vegetation types of Brazil: 1, forests; 2, savanna region mainly in central Brazil, savanna parkland, savanna woodland; 3, open savanna, grasslands; 4, caatinga region of northeastern Brazil, dry savanna; 5, Pantanal of Mato Grosso, wet grasslands (according to the Brazilian National Council of Geography).

Vegetation types of the Brazilian Amazon

The Amazon region covers about 6 million km² of which about 3.5 million are in Brazil. Due to the lack of data, it is very difficult to make an estimate of the area occupied by each of the types of vegetation described below. A much more critical evaluation may be made when the RADAM Project is complete. Also the types of vegetation are not sharply separated; intermediate and transitional stages are sometimes difficult to interpret as they do not always fit within any classification.

Tall upland forest
It covers 65–70 per cent of the Amazonian area.

Low upland forest
Less biomass; frequently with *Orbignya* (babaçu) or other species of palm; may or may not contain an excessive quantity of lianas; covers *ca.* 10–15 per cent of the Brazilian Amazon, principally south of the Amazon between the Tocantins and Madeira rivers and also in the valley of the Rio Branco in the Territory of Roraima.

Vegetation of mountains and slopes
Very diversified, from bush and scrub at greater altitude to open forest or dense forest at low altitude. It occupies perhaps <0.5 per cent of the total area.

Lowland and swamp forest
Moderate to small biomass, occasionally high forest. Soils with excessive moisture content and lack of aeration. Covering *ca.* 2 per cent of the Amazon area (Gourou, 1949) excluding the white-sand savanna.

White-sand savannas (campinas, caatingas or chavascais)
Very diversified vegetation, arboreal or shrubby, on poor deep sandy soils. During the wet season excess moisture induces waterlogging. Soils of the regosol or podzol type; many endemics; occupy an enormous area in small disjunct patches throughout the Amazon and as large patches in some regions such as in the Rio Branco basin or in the middle or upper Rio Negro.

Littoral vegetation
Medium tall forest to scrub, along the coasts; beach, dunes, mangrove, swamps and xeromorphic vegetation known as *restingas*. They occupy a very small area.

Upland savannas (*campos de terra firme*)
Grasses and sedges predominate in almost pure stands or between shrubs or small trees. Soil not subject to excess moisture. Light reaching the lowest plants and the soil. Most of these campos are pure grass fields (*campos limpos*), sometimes savannas with scattered small trees (*campos cobertos*).

The tall forest
It includes areas of flooded lowlands (*várzea alta*) as occur in the southwest of Marajó Island. The distribution of tall dense forest is not continuous because of vegetation of shorter or dwarf trees or shrubs, or campos. The species in the forest shade are generally medium-sized or small trees but there are exceptions (*Manilkara huberi* is a large tree). Shade tolerant species have a high number of young small individuals, their number diminishes gradually as the diameter class increases. The light demanding species have the greatest number of individuals in the intermediate size class. For example, the greatest number of *Bertholletia excelsa* (Brazil nut tree) in the forest of the Jari basin are in the class of 3.5 m GBH (or a DBH >1 m), while in the class of very small diameters the frequency is nil in the dense shaded forest. See fig. 3.

There is a great variation among the várzea bottomland forest caused chiefly by the degree of inundation. Wet season flooding can last for months; tidal flooding is twice daily. The várzeas of the lower Amazon basin (Monte Alegre, Santarém, Oriximiná, Parintins) are always associated with robust grasses (*canaranas*), while the várzeas of the Amazon estuary are forests, except for Marajó Island where the grasslands are lower and of the littoral savanna type. Also, in the lower Amazon basin palms are practically absent while várzeas of the estuary region are very rich in palms of the genera *Mauritia, Euterpe, Iriartea, Manicaria, Raphia, Bactris, Astrocaryum, Geonoma* and *Desmoncus* (Huber, 1903, 1909; Sioli, 1951; Rodrigues, 1961; Pires and Koury, 1959).

The *igapó* (swamp) forests are similar to the várzeas, the greatest difference being that the submerged state persists for a long period of the year, with the result that the floodwaters deposit their sediments.

In the clear-water rivers there is practically no suspended materials to be deposited and the most flooded parts are close to the river bed. In the muddy rivers, there is a great quantity of deposits close to the bank which thus becomes raised as levees; the igapós are found at a great distance from the banks which are occupied by várzeas (levees) (Sioli, 1951, 1957).

The littoral mangroves are a uniform formation along practically the whole coast. In the Amazon it is of little importance, occupying a small area and containing few species. Besides the genus *Avicennia* three species are involved: *Rhizophora mangle, R. harrissonii* and *R. racemosa* (Huber, 1909; Lindeman and Mennega, 1963).

The *matas de cipó* (liana forests) are low forests with many lianas which occupy all the Itacaiunas basin. Besides the relatively low trees, there are scattered trees which may exceed 50 m. Definite species associations are common, such as the *Orbignya-Bertholletia* association of the Itacaiunas basin. There are a considerable number of endemic species.

The diminution in the height of the liana forest does not seem to be due to a drier climate or soil impoverishment. The greater part of the area of the Transamazon highway, between Marabá and Altamira, is covered by liana forests on all types of soils, including the eutrophic *terra roxa* considered to be of high fertility. The causes which determine the appearance of liana forest have not yet been clarified.

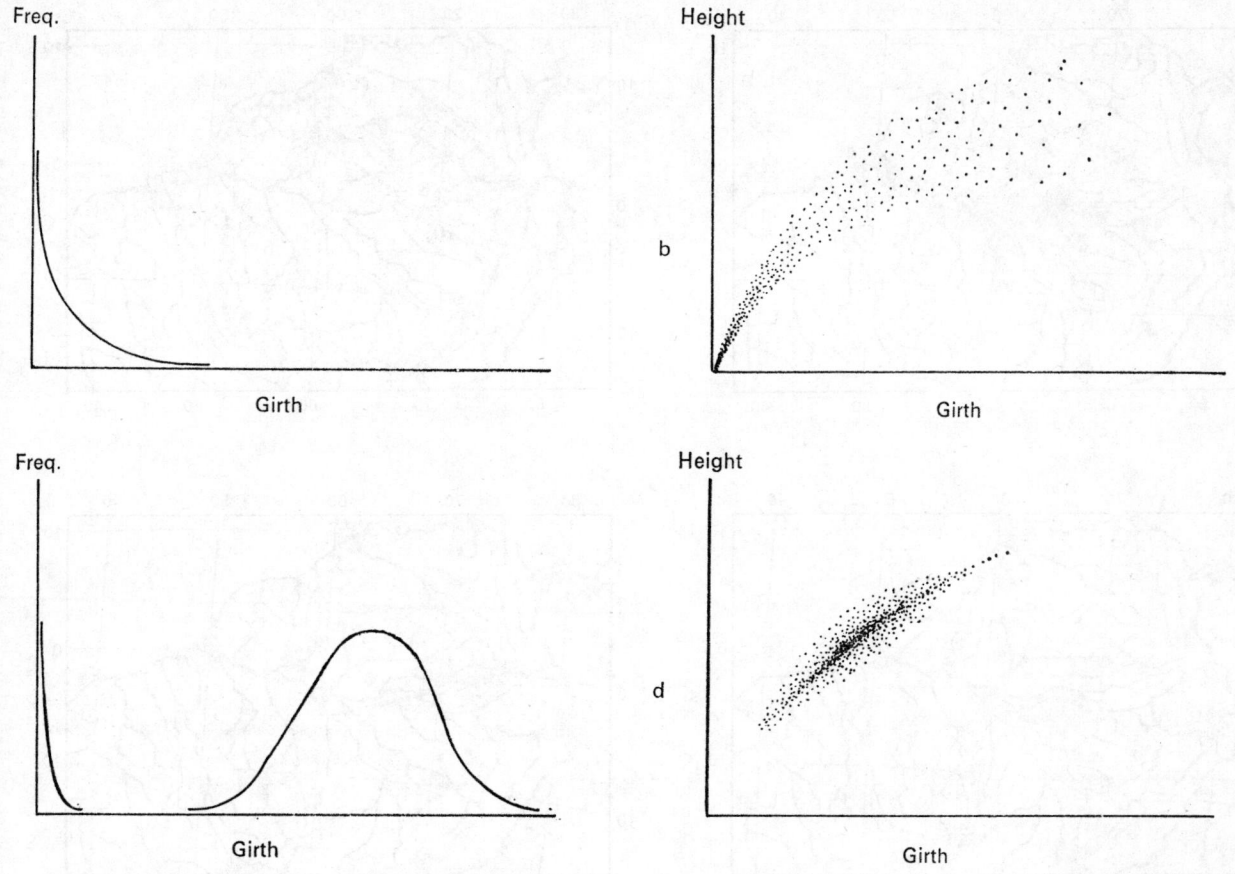

FIG. 3. Frequency of distribution of plant individuals in the forest.
a, b: each shade tolerant species or all species mixed together.
c, d: each light demanding species. (In b and d each individual plant is represented by a point.)

Floristic inventories

As examples of the floristic variation within short distances, the data from forests located on the upland (*terra firme*), várzea and igapó, all less than 2 km apart, near Belém, are given below. The upland soil is always dry and moderately sandy; the várzea and igapó soils are hydromorphic and clayey, with high moisture and periodic inundation. The várzea receives daily flooding by the muddy Guama river, the igapó is flooded for most of the year by clean rain-water.

Each sampled area was 5 ha; 345 species with 8 311 individuals (GBH ⩾30 cm) were recorded on the total 15 ha. 83 species were common to all three; the exclusive species to each area were 73 *terra firme*, 26 igapó and 48 várzea:

  terra firme: 52 families, 136 genera, 224 species,
               2 607 individuals
    igapó   : 47 families, 118 genera, 180 species,
               2 792 individuals
    várzea : 45 families, 124 genera, 196 species,
               2 912 individuals.

On the *terra firme* the 5 most common species accounted for 31 per cent of the trees; in the igapó for 34 per cent and in the várzea, 43 per cent.

The *terra firme* five characteristic species are: *Eschweilera odora*, 9.9 per cent of the trees sampled; *Eschweilera amara*, 9.8; *Tetragastris trifoliolata*, 6.0; *Vouacapoua americana*, 2.8 and *Goupia glabra*, 2.7. The várzea species are: *Euterpe oleracea*, 23.9 per cent; *Pentaclethra macroloba*, 8.8; *Carapa guianensis*, 4.8; *Theobroma subincana*, 3.3 and *Pterocarpus officinalis*, 2.4.

In the *terra firme* forest the Leguminosae was the most abundant family in terms of numbers of genera and species and the Lecythidaceae in numbers of individuals. In the várzea forest Leguminosae and Palmae were the most important. In the *terra firme* the 21 most common species contributed 50.4 per cent of the individuals; in the igapó, the 11 most common contributed 50.3; and in the várzea, the 8 most common contributed 49.8 per cent. These data indicate that the *terra firme* forest tree flora is most diverse and the várzea least so.

In a similar survey made in the Serra de Buritirama, at the Itacaiunas river, a tributary of the Tocantins river, the

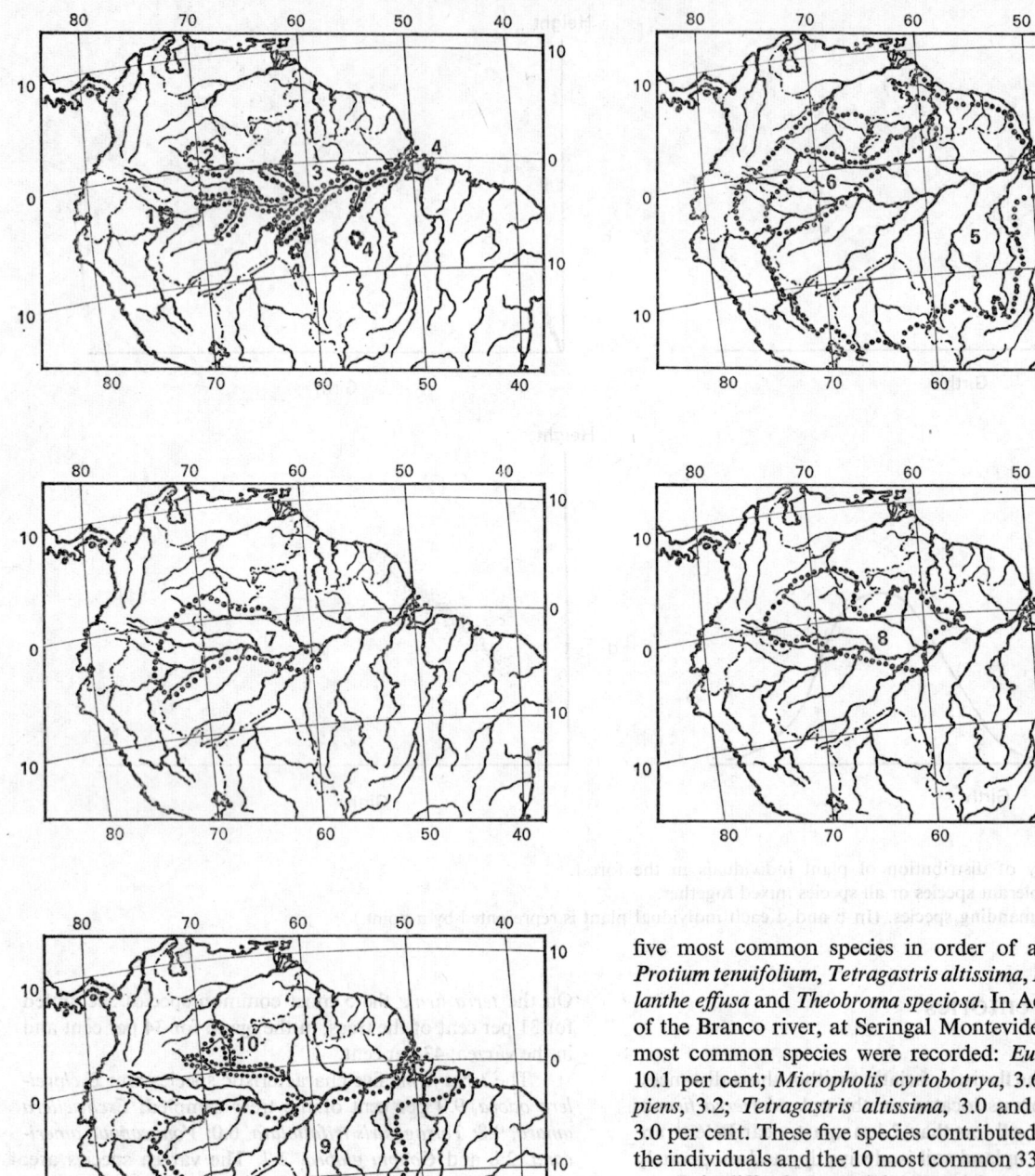

FIG. 4. Distribution of the species of *Hevea*: 1, *H. paludosa*;
2, *H. rigidifolia*; 3, *H. spruceana*; 4, *H. camporum*; 5, *H. guianensis*; 6, *H. pauciflora*; 7, *H. nitida*; 8, *H. benthamiana*; 9, *H. brasiliensis*; 10, *H. microphylla*.

five most common species in order of abundance were:
Protium tenuifolium, Tetragastris altissima, Neea spp., *Poeci-
lanthe effusa* and *Theobroma speciosa*. In Acre, in the region
of the Branco river, at Seringal Montevideu, the following
most common species were recorded: *Euterpe precatoria*,
10.1 per cent; *Micropholis cyrtobotrya*, 3.6; *Siparuna deci-
piens*, 3.2; *Tetragastris altissima*, 3.0 and *Quiina juruana*,
3.0 per cent. These five species contributed 22.9 per cent of
the individuals and the 10 most common 33.8 per cent.

It is difficult to explain why a species is abundant or
rare. One of the reasons for the rarity could be isolation in
face of the competition from adverse organisms. Certain
species when planted close together meet many problems
with disease organisms as does *Hevea*.

Each species has an index of abundance and an area
of dispersion which may be either continuous or disjunct.
Within its area a species possesses localities of greater or
lesser density. See fig. 4 for the distributions of species of
Hevea.

The described composition of the *terra firme*, várzea
and igapó forests demonstrates the polymorphism (or mixture of species) of the Amazon forest and of the predominance of certain taxa.

Forest inventories

In recent years, the Government has established various incentive plans for regional production and colonization by means of financing, land concession and the elimination of taxes. The projects have to obey certain requirements for the better zoneing, protection of the environment and the preservation of a certain percentage of the forests. In order to be approved, projects must present forest inventories. These inventories are generally carried out by planning and assessment offices, a private enterprise which has proliferated recently, and leave much to be desired.

Another purpose of the inventories is to provide additional information to the large-scale programmes which are being organized by the Government such as the RADAM Project; opening of the great road system (Transamazon highway, North-perimetral highway, Cuiabá-Santarém highway); implantation of a free port in Manaus; plan for settlement under the responsability of INCRA; regional plans organized by SUDAM, SUDENE, etc. Such inventories have been partly executed by government institutions and partly by the participation of the planning offices.

Methods and difficulties

An analysis of the results of forest inventories already available shows the existence of important methodological deficiencies. This must be improved.

Practically all of the studies carried out were based on common names given by local inhabitants. The names are compared with reference lists which give the Latin names. Such lists are general for Brazil or for the whole Amazon region; they are not specific for each locality. Studies on this basis contain errors of identification which may be as much as 90 per cent and are therefore unacceptable.

The introduction of refinements such as statistical interpretation and computation is desired, but no statistical method or computer can correct or improve data. It is better to use fewer samples and make sure the trees are accurately identified.

In the published inventories the commercial volumes are generally calculated taking into consideration the discarding of defective parts—stumps, hollow or twisted parts of the trunk, etc. The conception of what is a defect depends on personal criteria; both the total result and the volume of defects are necessary.

Generally a minimum girth is established. FAO utilized 25 cm diameter; the RADAM Project 100 cm girth (31.8 cm diameter); certain private planning offices 45 cm diameter.

In order to improve the inventories, and floristic and forest prospectings the following are recommended: incentivate the training of university and secondary level personnel in plant identification; training of field personnel to recognize plants and to collect samples for identification; re-evaluate the principles relative to planning and methodology.

Results

There is a considerable mass of inventories made in the Amazon since 1956, when broad inventories were carried out by the FAO mission in co-operation with the SPVEA (now SUDAM). The results of this project were published in twelve reports of FAO and translated into Portuguese (SUDAM, 1974).

The project was carried out on 225 000 km² principally in strips 150 km×1 500 km, parallel to and south of the Amazon river. Several thousand kilometers of 10 m wide transects and more than 120 000 trees were recorded. Unfortunately, the identification of the trees was based on common names supplied by local field men.

Another broad inventory is that of the RADAM Project which is broader and more expensive than the FAO Project, being based on radar photographs. All of Brazilian Amazonia has been photographed and these are being interpreted by specialists in vegetation, soil, geology, geomorphology and land use. The partial results are being edited as *Monografias*, seven of which have been published.

The volume of wood varies according to the density of the forest and the valuable species vary from place to place. Nevertheless in any stand a certain percentage of useful species is always found. Thus data on the gross volume of the whole forest would give a good idea of the true economic value of a certain forest.

Structure of the vegetation

Heinsdijk (1960, 1965) and Schulz (1960) did not detect strata in Amazonian forests.

There is a stratification of micro-organisms, herbaceous plants, shrubs, lianas, mosses and small pteridophytes which

TABLE 1. Basal areas (m²/ha) of some Amazonian vegetation types.

Vegetation types	Locality	Minimal trunk diameter (cm)					
		5	10	20	30	40	50
Campo (savanna)	West of Estreito (near the Tocantins river)	—	3.6	1.6	0.3	0.0	—
Liana forest	Between Estreito and Marabá	—	14.9	12.7	9.5	5.4	3.8
Tall forest	River Jari, southwest of Santo Antônio waterfall	38.0	34.6	30.4	25.9	20.6	13.4

live on the bark, and groups of organisms which locate themselves at various heights. But the graph of tree frequency distribution by height classes (fig. 5) is continuous.

Various authors stress the importance of stratification in the Amazonian forests (Davis and Richards, 1933–34; Aubréville, 1961; Rodrigues, 1961). However, the supposed discreteness of the strata results from a distorted interpretation of the profiles.

Basal area

The basal area gives a very good indication of the type of vegetation, its size, biomass and economic value as timber. The basal area is easy to calculate from trunk measurements, avoiding the calculation of volume which involves the subjective estimation of height. Results expressed in terms of

basal area are more comparable than those expressed in terms of volume. Basal areas for some Amazonian vegetation types are shown in table 1.

Functioning

Maximum flowering occurs at the beginning of the wet season and at the beginning of the dry season, although there are always some species in flower.

Fruits are eaten by animals and disseminated; some very large flowers (*Tabebuia* and *Caryocar* spp.) also serve as food for animals and probably become pollinated. *Bertholletia excelsa* (the Brazil nut tree) is pollinated by insects (*Bombus* for example) which have sufficient strength to raise the androphore which covers the stigma. The Brazil nut tree

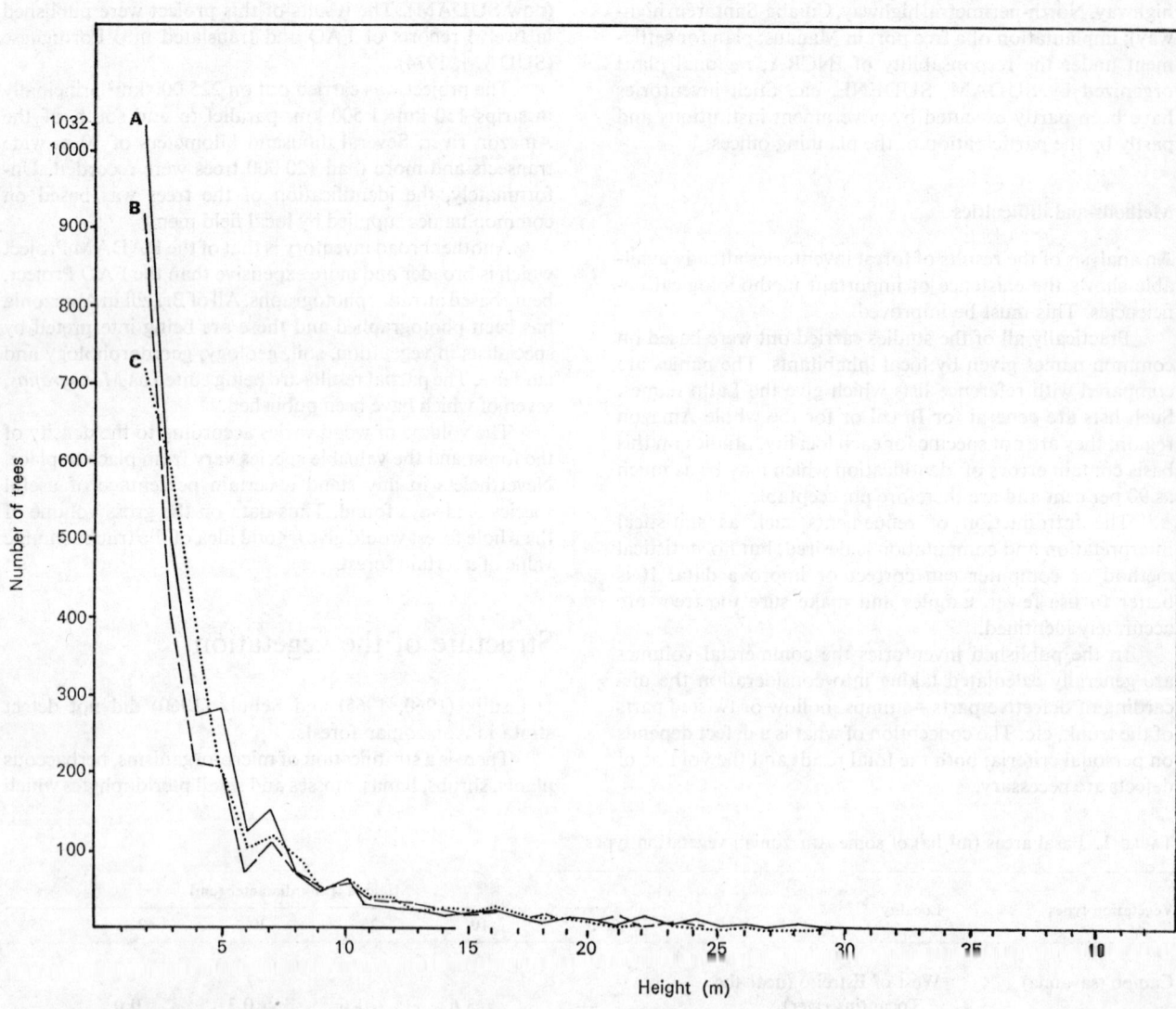

FIG. 5. Distribution of trees ≥2 m high
in height classes from three different types
of vegetation in the vicinity of Belém.
A: *terra firme* forest; B: várzea; C: igapó.

only flowers at a certain time of the year; other species, which flower at other times of the year, must provide food for the pollinating insects.

Tree growth

Natural clearings exist in the undisturbed forest and are caused chiefly by the fall of large trees by storms. A considerable number of trees die because of parasites, strangler trees, etc.

Some authors consider there is a suppression period in the relatively young phase. This suppression period does not seem to exist under Amazonian conditions and it is probable that data obtained elsewhere were incorrectly interpreted, having mixed observations from clearings with observations from mature forests when dealing with light demanding species.

Pires (unpublished) took six girth measurements during 15 years on trees (>30 cm GBH) on two hectares of forest in Belém. Some had no growth during the 15 year period. Large, small and zero increments occurred in all diameter classes and in all species. Each tree behave differently. These results might be explained by the effect of position: light is very important and varies according to the position a tree occupies in relation to its neighbours.

The average girth increase was <1 cm/a; a 3-year-old *Ochroma lagopus* on poor soils, in open plantation, may surpass 150 cm/a.

As tree growth fluctuates it is very difficult to estimate tree age. The number of growth rings in the trunk does not give a good indication. A tentative indication of age could be obtained from the maximum growth of each species, assuming that the maximum increment corresponds to the development of those individuals whose growth was not inhibited. Another possibility is analysing the individuals which die. If within a certain period all the individuals were to die and the population totally renewed, this would be the period corresponding to the limiting age of the species. Each species should have an annual index of mortality.

Heinsdijk (1965) analysing the diameter distribution of more than 120 000 trees, concluded that the ideal average tree of diameter class 25–35 cm would have a diameter increase in the order of 8 mm/a; between 25 and 155 cm, the increment would be 8–3.7 mm/a; and the average age of such ideal trees would range between 27 for trees 25–35 cm diameter and 418 years for those 145–155 cm diameter. Such theoretical predictions must be taken with some reservation.

According to some authors, longevity of trees in tropical rain forest generally does not surpass 400 years but there are exceptions: a *Bertholletia excelsa* in the Jari river region had a girth of 14 m and its age might exceed 1 400 years.

Primary productivity

Studies made in the tall forest near Manaus (Klinge and Rodrigues, 1968, 1971; Klinge, 1973; Fittkau and Klinge, 1973) give a biomass of *ca.* 1 000–1 100 t/ha (fresh weight); *ca.* 730 t above the soil surface, *ca.* 225 t of roots and 59 t of non-living organic matter.

Pires (unpublished) in a tall forest below the Santo Antônio waterfall on the Jari river has found 584 m³/ha of plants (excluding roots); 61 per cent trunks and 39 per cent branches and leaves. The volume of the roots and of non-living organic matter was not estimated.

There are some litter fall data from Manaus (Klinge and Rodrigues, 1968, 1971) and Belém (Klinge and Pires, in press). In tall *terra firme* forest in Manaus the oven dry litter fall was 7.9 (1963) and 6.7 (1964) t/ha/a. In Belém, samples taken between 1969 and 1971 in tall *terra firme* forest (Mocambo), igapó-forest (Catú) and várzea-forest (Aurá) gave litter falls of 9.8, 7.8 and 9.0 t/ha/a (oven dry material).

In an equilibrium, the production of organic matter could be estimated at *ca.* 10 t/ha/a. If the forest were to be completely cut down, it would involve the death of *ca.* 500–1 200 t/ha. The utilizable timber is much less. According to Heinsdijk (1960) the economic production of wood is *ca.* 4 m³/ha/a; in commercial plantings, for the enterprise to be economic, this production must attain 10–15 m³/ha/a.

Secondary productivity

The animal biomass of the Amazon forest is very small; Fittkau and Klinge (1973) estimated only *ca.* 200 kg/ha about half of which was soil fauna. However the fauna plays a very important role in pollination and dissemination. The destructive effect of animals on the flora is small. More intense effects occur in plantings, when the equilibrium is disturbed. An example is the great difficulty of introducing some species into cultivation. A clone of *Hevea brasiliensis* which is very resistant to the fungus *Dothidella ulei* later becomes susceptible to the fungus, which being free to undergo genetic variation takes advantage of the selected clone whose genotype is stabilized.

In *terra firme* forest near Belém, Elton (1973) found that the invertebrate fauna was not very numerous, was composed principally of many different species of insects less than 3 mm. Elton sought to explain the low density and great diversification by the equilibrium between prey and predators. According to Domiciano Dias, the swamp forest (várzea and igapó) is incomparably richer in small animals than the upland forest.

An intensive research programme starting in the 1950s between the Rockefeller Foundation and the Evandro Chagas Institute (Belém Virus Laboratory) is studying arboviruses; a survey of the virus population animals placed in the forest to become infested by virus, principally transmitted by blood-sucking animals. From 1966 to 1968, the IPEAN and the Smithsonian Institution organized a supplementary project at IPEAN, in Belém, known as the study of the *Area de Pesquisas Ecológicas do Guamá* (APEG). It included studies on botany, birds (behaviour, populations and biology), insects (social insects, leaf-cutter ants, army ants), bats, epidemiology (mosquitoes, arbovirus), climatology and soils. Enormous quantities of data were collected. For example, the birds and other animals captured and recaptured for studies of population and behaviour, also served for haematological tests related to virus. In this way the reserve areas of IPEAN became of great scientific value.

Some ecological studies of the poorly known fauna have been developed with the virus project. The best known group is the birds. Among the invertebrates, the insects are the best studied, especially those which transmit diseases, principally mosquitoes.

Sioli, after developing field research for about 20 years in Amazonia founded a branch of the Max Planck Institute in Manaus and developed an *Ecology of Amazon landscape* programme in co-operation with the National Institute of Amazon Research (INPA). The programme is so important that the journal *Amazoniana* was created to publish its results.

Another subject of great importance is the protection of the fauna. With the recent intensive development the most urgent measures to be taken include the localization of reserves of typical vegetation, control of settlement with a view to avoiding biological degradation, in order to guarantee not only present studies but also those which can be made in the future. Several parks and reserves exist in the Brazilian Amazonia (see table 2).

TABLE 2. List of national parks and national reserves in the Brazilian Amazonia.

Localization	Area (ha)	Year of creation
National parks		
Araguaia, Goiás	460 000	1959 and 1971
Neblina, Amazonas	—	1961
Tocantins, Mato Grosso	624 994	1961
Xingu, Mato Grosso	2 200 000	1961
National reserves		
Mundurucânia, Pará	2 375	1961
Tumucumaque, Pará	1 793 000	1961
Pedras Negras, Ter.		
Rondônia	1 761 000	1961
Rio Jauru, Ter. Rondônia	1 085 000	1961
Juruena, Mato Grosso	1 808 000	1961
Parima, Ter. Roraima	1 756 000	1961
Gorotire, Pará	1 843 000	1961
Rio Negro, Amazonas	3 790 000	1961
Caxiuaná, Pará	200 000	1961
Adolfo Ducke, Amazonas	10 000	1963
Walter Egler, Amazonas	750	1968

Biogeochemical cycles

The tropical rain forest which covers *ca.* 70 per cent of the Amazonian area is a closed system. The Amazon forest generally has poor soils. The enormous biomass is the product of a biological equilibrium. Almost all of the available nutrients are contained in the vegetation. The dead parts undergo rapid decomposition, are mineralized and reabsorbed by the roots. There is a very efficient mechanism to avoid losses through leaching. The soil organic matter protects the soil against leaching. When a leaf falls, it is only a short time until it is fixed to the forest floor by fungal mycelia and rootlets and its elements return to circulation.

The small nutrient losses are compensated for by the addition of some nutrients in rain-water.

According to Fittkau and Klinge (1973), in a tall forest near Manaus, the biomass of living plants is 1 000 t/ha, the biomass of the herbivorous animals 30 kg/ha and the soil fauna 165 kg/ha.

Another important element in this process is represented by the litter fall, of which fine litter is the most important. For Manaus it has been estimated as 7–8 t/ha/a and for Belém as 8–10 t/ha/a.

Conclusions

The primeval ecosystem of Amazonia has suffered insignificant interference by man. The amount of this interference is now increasing very much, principally as a result of encouragement by the Government.

The population of Brazilian Amazonia is a little over four million, *ca.* 30 per cent of it living in the two main capitals, Belém and Manaus. In addition, part of the population is dispersed in a number of small villages so that the interior of the country is very sparsely populated, having *ca.* one inhabitant/km².

The first projects with all the intention of promoting colonization were not successful. There is one successful example—the settlement of Tomé Açu, State of Pará, where is now the principal center of black pepper (*Piper nigrum*) production organized by Japanese immigrants and where the farmers are making considerable economic progress. The Agricultural Colony of Tomé Açu, with *ca.* 30 000 ha, is the most important nucleus of Japanese colonization in Amazonia. This colony was pioneered in 1929 and is located in the State of Pará, between the valley of the Acará-Mirim and its affluent, the Mari igarapé. It is at 250 km from Belém. The municipality which was formed due to the great progress of its principal colony, was established in 1961, covers 5 828 km² and has a population of 9 487 representing a density of 1.63/km², greater than the Amazonian mean of 1.03/km². It is a completely planned, directed and based on small properties with an efficient co-operative system.

The experience of Tomé Açu, which is still growing constitutes the first example in Amazonia history of what rational intensive agriculture can do.

Most of the inhabitants of the interior of Amazonia are nearly self-sufficient, living from what can be found in the forest and are little different from the native Indian population. The very few articles they have to import from outside the forest, such as salt, sugar, clothes and some medicines, are usually exchanged for forest products such as gums, latex, seeds, fibers, etc. The inland merchant, using a special kind of boat, the *regatão*, to store and transport the goods which are exchanged, plays an important role in the economy of the region.

With the opening of new roads, the arrival of settlers and the increase of population, the inhabitants of the interior are not prepared to support competition and a large proportion of them are not adjusted to the new situation, principally the ones who migrate to the cities.

At present, the Government is dedicating special attention to the conquest of Brazilian Amazonia and is spending a considerable amount of the national economic resources on it. Several official organizations and programmes have been created known as SUDAM, INCRA, IBDF, EMBRAPA, INPA, etc. The programmes are oriented primarily to the following aims: survey of natural resources (in which the RADAM Project is an important element); communications including roads and telecommunication; colonization. Forest reserves are already under legislation (fig. 6) or suggested by SUDAM, as well as forests for management (fig. 7).

The SUDAM (*Superintendencia do Desenvolvimento da Amazônia*) replaces the SPEVEA (Superintendency for the Economic Valuation of Amazonia) and deals with the supervision of government and private development. It is supported by 3 per cent of all the Brazilian taxes.

The INCRA (*Instituto Nacional de Colonização e Reforma Agraria*) deals with colonization and especially with problems involving the occupation of the interior of the country where the land has no owners. The IBDF (*Instituto Brasileiro do Desenvolvimento Florestal*) deals with the protection of the flora and the fauna, and with forestry research (FAO Project BRA–45). The FUNAI (*Fundação Nacional do indio*) deals with protection of the Indian populations and problems envolved. The IDESP (*Instituto do Desenvolvimento Economico-Social do Pará*) deals with problems for the development of the State of Pará, including the private ones. IDESP is also carrying out research on forest inventories, mineral prospection and social studies.

The EMBRAPA (*Empresa Brasileira de Pesquisa Agropecuária*) is a national institution supported by the Federal Government, operating in different regions; EMBRAPA-Pará and EMBRAPA-Amazonas are state branches of it. This institution has changed its name three times. It was initiated in 1940 as IAN (*Instituto Agronômico do Norte*), then replaced by IPEAN (*Instituto de Pesquisa Agropecuária*

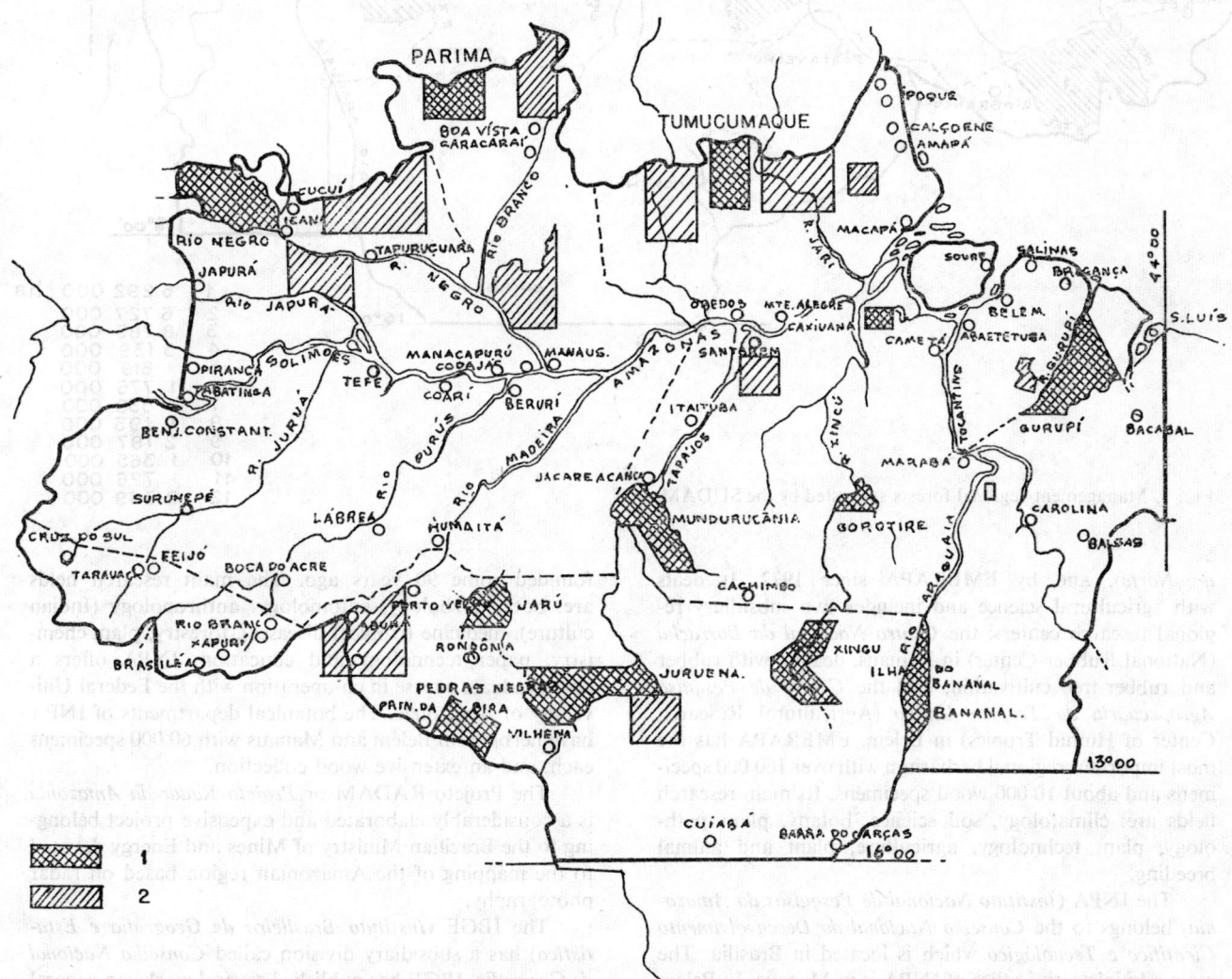

Fig. 6. Forest reserves. 1: already under legislation. 2: areas suggested by the SUDAM.

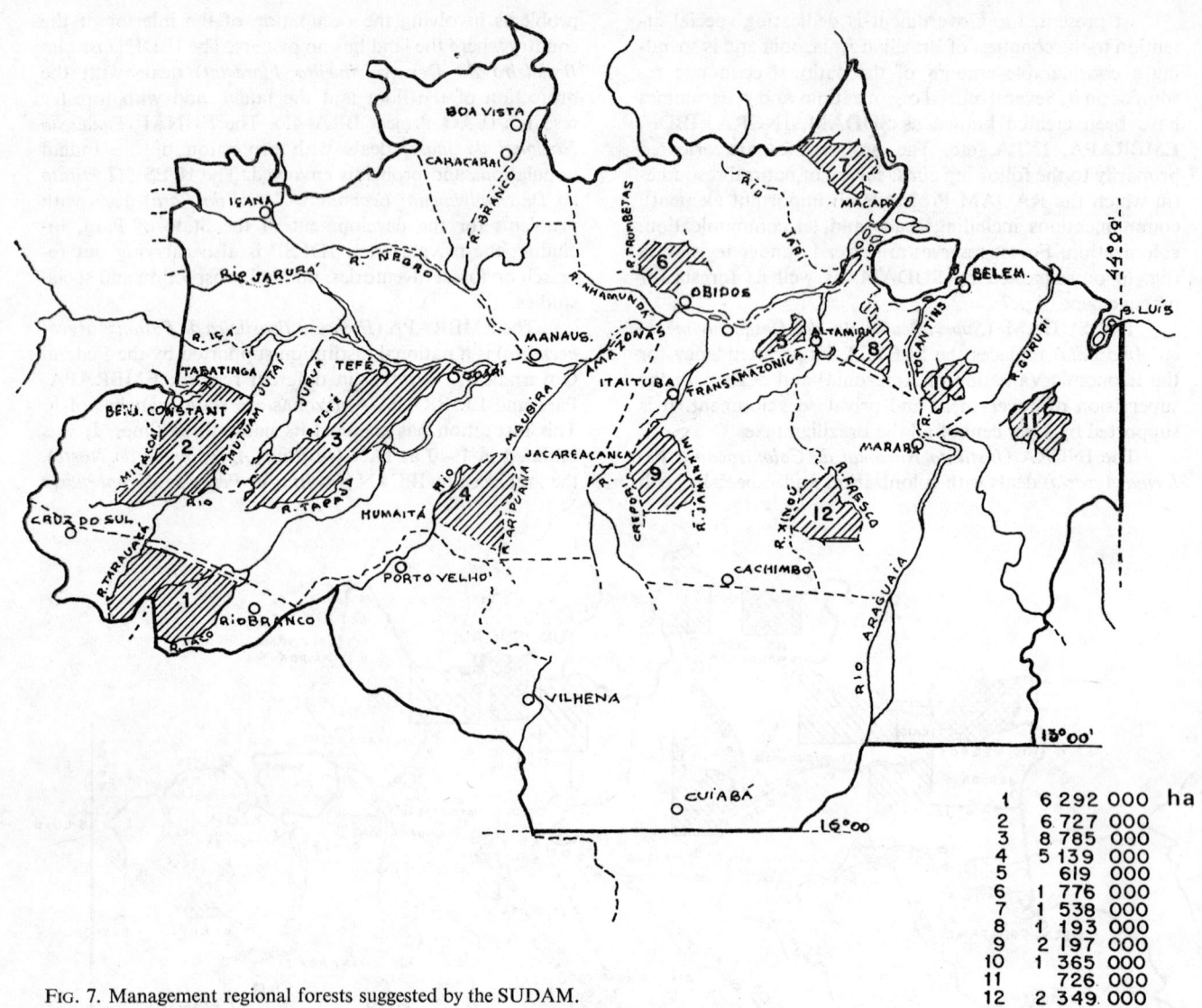

Fig. 7. Management regional forests suggested by the SUDAM.

1	6 292 000	ha
2	6 727 000	
3	8 785 000	
4	5 139 000	
5	619 000	
6	1 776 000	
7	1 538 000	
8	1 193 000	
9	2 197 000	
10	1 365 000	
11	726 000	
12	2 349 000	

do Norte), and by EMBRAPA since 1972. It deals with agricultural science and includes two subsidiary regional research centers: the *Centro Nacional da Borracha* (National Rubber Center) in Manaus, dealing with rubber and rubber-tree cultivation, and the *Centro de Pesquisa Agropecuária do Tropico Umido* (Agricultural Research Center of Humid Tropics) in Belém. EMBRAPA has the most important regional herbarium with over 160 000 specimens and about 10 000 wood specimens. Its main research fields are: climatology, soil science, botany, plant pathology, plant technology, agriculture, plant and animal breeding.

The INPA (*Instituto Nacional de Pesquisas da Amazonia*) belongs to the *Conselho Nacional do Desenvolvimento Científico e Tecnológico* which is located in Brasilia. The main administrative office of INPA is in Manaus. In Belém INPA is a continuation of the Museu Paraense Emilio Goeldi (Pará Museum Emilio Goeldi), a State institution

founded some 90 years ago. The main research fields are: botany, zoology, entomology, anthropology (Indian culture), medicine (tropical diseases), forestry, plant chemistry, paper technology and education. INPA offers a post-graduate course in co-operation with the Federal University of Amazonas. The botanical departments of INPA have herbaria in Belém and Manaus with 60 000 specimens each, and an extensive wood collection.

The Projeto RADAM or *Projeto Radar da Amazonia* is a considerably elaborated and expensive project belonging to the Brazilian Ministry of Mines and Energy devoted to the mapping of the Amazonian region based on radar photography.

The IBGE (*Instituto Brasileiro de Geografia e Estatística*) has a subsidiary division called *Conselho Nacional de Geografia*. IBGE has published several works on general physiognomy of Brazil including, *A Grande Região Norte* which is especially devoted to the Amazonian region.

Several periodicals are concerned with the publication of the research on Amazonia:
— *Boletim Técnico da EMBRAPA*, which is the continuation of the *Boletim Técnico do IPEAN* and is the continuation of the older *Boletim Técnico do Instituto Agronômico do Norte*.
— The EMBRAPA publishes also the review *Pesquisa Agropecuaria Brasileira*.
— *Acta Amazonica*, edited by the INPA.

— *Boletim do Museu Paraense Emilio Goeldi* has ceased since *Acta Amazonica* was created.
— *Boletim do INPA* ceased since the creation of *Acta Amazonica*.
— *Amazoniana* belongs to the Max Planck Institute of Hydrobiology.
— *Boletim da Universidade Federal do Pará*.
— *Boletim do IDESP*.
— *Monografias of the RADAM Project*.

Bibliography

ABREU, S. F. O Solo da Amazônia. *Rev. Bras. Geogr.* (Rio de Janeiro), vol. 4, no. 2, 1942, p. 299–312.

AB'SABER, A. N. Conhecimento sobre as flutuações climáticas do Quaternário do Brazil. *Bol. Soc. Bras. Geogr.* (Rio de Janeiro), vol. 6, no. 1, 1957, p. 41–48.

ACOSTA, A. C. A preliminary study of organic detritus in a tropical forest ecosystem. *Rev. Biol. Trop.* (Univ. Costa Rica), vol. 12, no. 2, 1964, p. 175–184.

ALENCAR, J. C.; BARROS, J. C. M.; VIEIRA, A. N. *Inventário florestal de área do Projeto piloto agropecuário da zona franca de Manaus*. Manaus, PROFLAMA, 1970, 55 p. multigr.

——; ——; ——. *Inventário florestal do Baixo e Médio Curuçá. Cia Industrial do Norte 'CIA-NORTE'*. Manaus, PROFLAMA, 1970, 63 p. multigr.

——; ——. *Inventário florestal do Alto Ituí e Baixo Itacoaí. 'PRAMA'—Produtos da Amazônia S.A*. Manaus, PROFLAMA, 1970, 81 p. multigr.

——; ——; ——. *Plano integrado de exploração—Industrialização e Reflorestamento. Cia. Siderúrgica da Amazônia. 'SIDERAMA'*. Manaus, PROFLAMA, 1970, 75 p. multigr.

ALMEIDA, L. C. Distribuição das quantidades de fungos leveduriformes em diferentes profundidades nos solos do Estado do Pará. In: *Atas do Instituto de Micologia* (Recife), 2, publ. no. 458, 1965, p. 397–406.

——. Distribuição das quantidades de fungos filamentosos e leveduriformes nas profundidades nos solos do Estado do Pará. In: *Atas Inst. Micologia* (Recife), no. 459, 1965, p. 407–416.

——. Distribuição das quantidades de bactérias, em diferentes profundidades, nos solos do Estado do Pará. In: *Atas Inst. Micologia* (Recife), no. 460, 1965, p. 417–427.

——. Distribuição das quantidades de fungos e bactérias nas diferentes profundidades nos solos do Estado do Pará. In: *Atas Inst. Micologia* (Recife), no. 461, 1965, p. 429–439.

ALTEMULLER, H. J.; KLINGE, H. Mikromorphologische Untersuchungen über die Entwicklung von Podsolen in Amazonas becken. In: Jongerius, A. (ed.). *Soil micromorphology*, p. 295–305. Amsterdam, Elsevier, 1964.

ALVIM, P. T. El suelo como factor ecológico en el desarrollo de la vegetación en el centro-oeste del Brazil. *Turrialba* (Costa Rica), vol. 2, no. 4, 1952, p. 153–160.

——. Los Trópicos bajos de América Latina: recursos y ambiente para el desarrollo agricola. In: *Simposio sobre el potencial de los Trópicos bajos de América Latina* (Cali, Colombia), 1973, 25 p.

——. Potencial agrícola da região Amazônica. *Rev. Amazonense de Desenvolvimento* (Manaus), 1, 1973, p. 9–18.

——; SANTANA, C. J. L. Diagnóstico das deficiências minerais em solos da região amazônica pelo método das microparcelas. In: *Atas Simpósio Bioto Amazônica*, vol. 1 (Geociências), 1967, p. 69–73.

ANON. Comissão brasileira junto à missão oficial norte-americana de estudos do vale amazônico. *Geologia, physiografia e solos (Valle do Amazonas). Relatório*. Rio de Janeiro, Min. Agricultura, Indústria e Comércio, pt. II, 1924, p. 341–416.

——. Conselho nacional de geografia. *Geografia do Brazil: Grande Região Norte*. Rio de Janeiro, IBGE, Biblioteca Geográfica Brasileira, vol. 1, publ. no. 15, 1959, 422 p.

——. Instituto de pesquísas e experimentação agropecuárias do norte. *Principais culturas da Amazônia*. Belém, IPEAN, 1965.

——. *Solos da estação experimental de Porto Velho, Território de Rondônia*. Belém, IPEAN, Solos da Amazônia, 1, 1967, 99 p.

——. Instituto brasileiro de geografia. *Paisagens do Brasil*. Rio de Janeiro, publ. no. 2, 1968, 286 p.

——. *Atlas climatológico do Brasil*. Rio de Janeiro, Ministério da Agricultura, Escritório de Meteorologia, 1969, 100 mapas.

——. Instituto de pesquísas e experimentação agropecuárias do norte. *Os solos da área Manaus-Itacoatiara*. Rio de Janeiro, Série Estudos e ensaios, 1, 1969, 116 p.

——. *Levantamento de reconhecimento dos solos da Colônia Agrícola Paes de Carvalho, Alenquer-Pará*. Belém, IPEAN, Solos da Amazônia, 2, 1970, 150 p.

——. I. *Levantamento de reconhecimento dos solos da zona Iguatemi, Mato Grosso*. II. *Interpretação para uso agrícola dos solos da zona Iguatemi, Mato Grosso*. IPEAN, Rio de Janeiro, 1970, 99 p.

——. Superintendência de desenvolvimento da Amazônia. Departamento de recursos naturais. *O extrativismo do pau-rosa: Aniba duckei* Kosterm., *A. rosaeodora* Ducke. Belém, SUDAM Documenta, 3(1/4), 1971–72, p. 5–58.

——. Instituto brasileiro de desenvolvimento florestal. *Códico florestal*. 2 ed. Rio de Janeiro, J. di Giorgio, 1972, 65 p.

——. *Levantamentos florestais realizados pela FAO na Amazônia (1956–1961)*. Belém, SUDAM, 2 vol., 1973.

——. Departamento de produção mineral. Projeto RADAM. *Folha SA. 23 São Luiz e parte da folha SA. 24 Fortaleza; geologia, geomorfologia, solos, vegetação e uso potencial da terra*. Rio de Janeiro, Levantamento de Recursos naturais, 3, 1973.

——. *Folha SB. 22 Araguaia e parte da folha SC. 22 Tocantins; geologia, geomorfologia, solos, vegetação e uso potencial da terra*. Rio de Janeiro, Levantamento de Recursos naturais, 4, 1974.

——. *Folha SA. 22 Belém; geologia, geomorfologia, solos, vegetação e uso potencial da terra*. Rio de Janeiro, Levantamento de Recursos naturais, 5, 1974.

——. *Folha NA/NB 22 Macapá; geologia, geomorfologia, solos, vegetação e uso potencial da terra*. Rio de Janeiro, Levantamento de Recursos naturais, 6, 1974.

ARAUJO, V. C. A reserva florestal Ducke (Manaus): características e principais elementos florísticos e faunísticos protegidos.

In: *Atas Simpósio Biota Amazônica*, vol. 7 (Conservação da natureza e recursos naturais), 1967, p. 57–68.

——. Fenologia de essências florestais amazônicas. I. *Bol. Inst. Nac. de Pesquísas da Amazônia, sér. Pesquísas florestais*, 4, 1970, p. 1–25.

——. The factor light as a basic element in tree growth in the Amazonian forest. In: *Proc. Symposium on Environment in Amazonia* (Manaus), part 1, 1970, p. 67–77.

ARENS, K. Considerações sobre as causas do xeromorfismo foliar. *Bol. Fac. Filos. Ci. Letr. USP Bot.* (São Paulo), 15, 1958, p. 25–56.

——. O cerrado como vegetação oligotrófica. *Bol. Fac. Filos. Ci. Letr. USP Bot.* (São Paulo), 15, 1958, p. 59–77.

ASHTON, P. S. Light intensity measurements in rain forest near Santarem, Brazil, *J. Ecol.*, vol. 46, no. 1, 1958, p. 65–70.

ASKEW, G. P.; MOFFATT, D. J.; MONTGOMERY, R. F.; SEARL, P. L. Soils and soil moisture as factors influencing the distribution of the vegetation of the Serra do Roncador, Mato Grosso. In: *Simposio sobre o cerrado*, 3, p. 150–160. São Paulo, Edgard Blucher, USP, 1971.

AUBLET, J. B. C. F. *Histoire des plantes de la Guyane française.* Londres, 4 vol., 1775.

AUBRÉVILLE, A. La forêt coloniale: les forêts de l'Afrique occidentale française. *Ann. Acad. Sci. Colon.* (Paris), 9, 1938, p. 1–245.

——. Le Vénézuéla forestier. *Bois et Forêts des Tropiques* (Nogent-sur-Marne), 45, 1956, p. 3–14.

——. As florestas do Brasil—Estudo fitogeográfico e florestal. *Anuário Bras. Ecón. Florestal*, 11, 1959, p. 201–232.

——. Étude comparée de la famille des Légumineuses dans la forêt équatoriale africaine et dans la flore de la forêt amazonienne. *Comptes Rendus Soc. Biogéographie* (Paris), 314/316, 1959, p. 43–57.

——. Études écologiques sur la forêt dense humide du Surinam. *Bois et Forêts des Tropiques* (Nogent-sur-Marne), 77, 1961, p. 61–64.

——. *Étude écologique des principales formations végétales du Brésil.* Nogent-sur-Marne, Centre technique forestier tropical, 1961, 268 p.

——. Savanisation tropicale et glaciations quaternaires. *Adansonia* (Paris), vol. 2, n° 1, 1962, p. 15–84.

——. Principes d'une systématique des formations végétales tropicales. *Adansonia*, vol. 5, n° 2, 1965, p. 153–196.

——. Conceptions modernes en bioclimatologie et classification des formations végétales. *Adansonia*, vol. 5, n° 3, 1965, p. 297–306.

AZEVEDO, A. *et al. Brazil—A Terra e o Homen.* 2 ed. São Paulo, Companhia Editora Nacional, 1972, 607 p.

BASTOS, A. M. A floresta do Amapari-Matapi-Cupixi (inventário florestal). *Bol. Serviço Florestal* (Rio de Janeiro), 2, 1960, p. 1–54.

BASTOS, T. X. O estado atual dos conhecimentos das condições climáticas da Amazônia brasileira. Zoneamento agrícola da Amazônia-Parte II. *Bol. Técn. Inst. Pesquísas e Experimentação Agropecuárias do Norte* (IPEAN, Belém), 54, 1972, p. 68–122.

——; PEREIRA, F. B.; DINIZ, T. D. A. S. Contribuição ao conhecimento da ecologia da floresta equatorial úmida. In: *Reunión técnica de Programación sobre investigaciones económicas para el Trópico americano* (Maracaibo, Venezuela), 1973. Fac. de Agron., Univ. Zulia, 1973. IICA, Informes de Conferencias, Cursos y Reuniones, Documento no. 8, 1973.

BATISTA, A. C.; LIMA, J. A. Aspectos microbiológicos dos solos do Território de Roraima. In: *An. 14 Congresso Soc. Bot. Brasil* (Manaus, 1963), 1964, p. 304–309.

BATISTA A. C. *et al.* Espécies fúngicas dos solos do Estado do Maranhão. *Atas Inst. Micologia* (Recife), no. 442, 1965, p. 309–317.

——; UPADHYAY, H. P. Soil fungi from Northeast Brasil. I. In: *Atas Inst. Micologia* (Recife), no. 443, 1965, p. 319–350.

BATISTA, D. Inventário científico da Amazônia. In: *Problemática da Amazônia*, p. 221–244. Rio de Janeiro, Ed. Casa do Estudante do Brasil, 1969.

BAZAN, R. *et al. Estudio comparativo sobre la productividad de ecosistemas tropicales bajo diferentes sistemas de manejo.* Turrialba, Costa Rica, IICA, 1973, 17 p.

BEARD, J. S. Climax vegetation in Tropical America. *Ecology*, 25, 1944, p. 127–158.

——. Los climax de vegetación en la América tropical. *Rev. Fac. Nac. Agric. Medellin*, vol. 6, no. 23, 1946, p. 225–293.

——. The savana vegetation of northern tropical America. *Ecological Monographs*, vol. 23, no. 2, 1953, p. 149–215.

——. The classification of tropical American vegetation types. *Ecology*, 36, 1955, p. 89–100.

——. Some vegetation types of tropical Australia in relation to those of Africa and America. *J. Ecol.*, vol. 55, no. 2, 1967, p. 271–290.

BÉGUÉ, L. Les formations végétales de la Colombie. *Bois et Forêts des Tropiques* (Nogent-sur-Marne), 102, 1965, p. 63–70.

BENA, P. *Essences forestières de Guyane.* Bureau agricole et forestier guyanais, Paris, Imprimerie Nationale, 1960, 488 p.

BENACCHIO, S. Investigaciones ecológicas en los trópicos húmedos en Venezuela. In: *Seminario sobre Ecología tropical* (Itabuna, Brasil, 1972). Turrialba, Costa Rica.

——. Investigaciones ecológicas de Venezuela. In: *Reunión técnica de Programación sobre investigaciones ecológicas para el Trópico americano* (Maracaibo, Venezuela), 1973. Fac. de Agron., Univ. Zulia, 1973. IICA, Informes de Conferencias, Cursos y Reuniones, Documento no. 8, 1973.

BENAVIDES, S. T. *Caracterización y classificación de suelos de la Amazonia colombiana.* Tesis de grado. Raleigh, North Carolina State Univ., thesis, 1973.

BENNEMA, J. *Report to the Government of Brazil on classification of Brazilian soils.* Rome, FAO-EPTA/2197, 1966.

BENOIST, R. La végétation de la Guyane française. *Bull. Soc. Bot. France*, 71, 1924, p. 1169–1177.

BERRY, E. W. Tertiary flora of Trinidad. *Stud. Geol.* (Baltimore), 6, 1925, p. 71–161.

BIARD, J.; WABENAAR, G. A. W. *Report to the Government of Brazil on crop production in selected areas of the Amazon Valley.* Rome, FAO Report no. 1254, 1960, 47 p.

BLACK, G. A. Os capins aquáticos da Amazônia. *Bol. Técn. Inst. Pesquisas e Experimentação Agropecuárias do Norte* (IPEAN, Belém), 19, 1950, p. 53–94.

——; DOBZHANSKY, T.; PAVAN, C. Some attempts to estimate species diversity and population density of trees in Amazonian forests. *The Botanical Gazette* (Chicago), vol. 3, no. 4, 1950, p. 413–425.

BLAIR, F. Problemas ecológicos da América Latina. *Lavoura* (Brasil), 74, 1971, p. 14–17.

BOUILLENNE, R. Notes sur les savanes équatoriales du bas Amazone. *Soc. Biogéographie* et Congrès de l'Association française pour l'avancement des sciences (Liège), 1974, p. 957–964.

——. Un voyage botanique dans le bas Amazone. In: Massart, J. (ed.). *Une mission biologique belge au Brésil*, p. 1–185. Bruxelles, Impr. médicale et scientifique, 2, 1930.

BRAUM, E. H.; RAMOS, J. R. A. Estudo agrogeológico dos campos

Puciari-Humaitá, Estado do Amazonas e Território Federal de Rondônia. *Rev. Bras. Geogr.* (Rio de Janeiro), vol. 21, no. 4, 1959, p. 443–497.

BRAUN, W. A. G. Contribuição ao estudo da erosão no Brasil e seu controle. *Rev. Bras. Geogr.*, vol. 23, no. 4, 1962, p. 591–642.

BRINKMANN, W. L. F. Light environment in tropical rain forest of Central Amazonia. *Acta Amazonica* (Manaus), vol. 1, no. 2, 1971, p. 37–49.

——. Air temperatures in Central Amazonia. II. The effect of near-surface temperatures on land-use in Tertiary region of Central Amazonia. *Acta Amazonica* (Manaus), vol. 1, no. 3, 1971, p. 27–32.

——; SANTOS, A. Natural waters in Amazonia. III. Ammonium molybdate reactive silica. *Amazoniana*, vol. 2, no. 4, 1970, p. 443–448.

——; ——. Natural waters in Amazonia. V. Soluble magnesium properties. *Turrialba*, vol. 21, no. 4, 1971, p. 459–464.

——; ——. Natural waters in Amazonia. VI· Soluble calcium properties. In: *Seminario sobre Ecologia tropical* (Itabuna, Brasil, 1972). Turrialba, Costa Rica.

——; WEIMAN, J. A.; RIBEIRO, G. M. N. Air temperatures in Central Amazonia. I. The daily record of air temperatures in a secondary forest near Manaus under cold front conditions. (July 4th to July 13th, 1969.) *Acta Amazonica* (Manaus), vol. 1, no. 2, 1971, p. 51–56.

——; NASCIMENTO, J. C. The effect of slash and burn agriculture on plant nutrients in the Tertiary region of Central Amazonia. *Turrialba*, vol. 23, no. 3, 1973, p. 284–290. *Acta Amazonica*, vol. 3, no. 1, 1973, p. 55–61.

BUDOWSKI, G. Forest succession in tropical lowlands. *Turrialba*, vol. 13, no. 1, 1963, p. 42–44.

——. Holdridge's world classification of life zones—a reappraisal. *Turrialba*, vol. 14, no. 1, 1964, p. 96–110.

——. Plant succession and the distribution of northern tropical American rain forest species. In: *Abstracts 10th Intern. Bot. Congr.* (Edinburgh), 1964, p. 274.

——. A review of studies on the distribution of tropical American tree species. *Bois et Forêts des Tropiques* (Nogent-sur-Marne), 95, 1964, p. 3–13.

——. Distribution of tropical American rain forest species in the light of successional processes. *Turrialba*, vol. 15, no. 1, 1965, p. 40–42.

——. The distinction between old secondary and climax species in tropical Central American lowland forest. *Tropical Ecology*, vol. 11, no. 1, 1970, p. 44–48.

CAIN, S. A.; CASTRO, G. M. O.; PIRES, J. M.; SILVA, N. T. Application of some phytosociological techniques to Brazilian rain forest. *Amer. J. Bot.*, 43, 1956, p. 911–941.

——; ——, G. M. O. *Manual of vegetation analysis.* New York, Harper, 1959, 325 p.

CALDERON, U. C.; FIGUEROA, Z. R. Estudios ecológicos en la Amazonia peruana. Proyecto: Desarrollo económico integral del area amazónica Madre de Dios Inambaru. In: *Seminario sobre Ecología tropical* (Itabuna, Brasil, 1972). Turrialba, Costa Rica.

CAMARGO, F. C. Report on the Amazon Region. In: *Proceedings of the Symposium on humid tropics vegetation* (Tjiawi, Indonesia), p. 11–24. Council for Sciences of Indonesia and Unesco Science Co-operation Office for South-East Asia, 1958, 312 p.

——. Agricultura na América do Sul. In: Fittkau, E. J. *et al.* (eds.). *Biogeography and ecology in South America*, p. 302–328. The Hague, Junk, 2 vol., 1968–69, 946 p.

CAMPOS, J. C. Ch. Considerações sobre o sistema de classificação ecológia proposto por Holdridge. *Rev. Ceres* (Brasil), vol. 20, no. 108, 1973, p. 87–96.

CAMPUZANO, G.; AREVALO, A. Situación de la investigación ecológica en el Ecuador. In: *Reunión técnica de Progamación sobre investigaciones ecológicas para el Trópico americano* (Maracaibo, Venezuela, 1973). Fac. de Agron., Univ. Zulia, 1973. IICA, Informes de Conferencias, Cursos y Reuniones, Documento no. 8, 1973.

CARNEIRO, L. R. S. *Os solos do Território Federal do Amapá.* Macapá (s. ed.), 1953, 107 p.

——. *Os solos do Território Federal do Amapá (Contribuição ao seu estudo).* Belém, Superintendência do Plano de Valorização Econômica da Amazônia, Setor de Coordenação e Divulgação, 1955, 110 p.

CARTER, G. S. Results of the Cambridge Expedition to British Guiana, 1933. The freshwaters of the rain forest areas of British Guiana. *J. Linn. Zool. Soc. London*, vol. 39, no. 264, 1934, p. 147–193.

CASTELLANOS, A. Introdução à Geobotânica. *Rev. Bras. Geogr.*, vol. 22, no. 4, 1960, p. 585–617.

COIMBRA FILHO, A. F. O gênero *Neomarica* no combate à erosão. *Rodriguésia* (Rio de Janeiro), 24, 1949, p. 189–196.

——; MAGNANINI, A. Considerações sobre a *Mimosa pudica* no combate à erosão superficial. *An. Bras. Econ. Florestal* (Rio de Janeiro), 6, 1953, p. 131–136.

——; MARTINS, H. F. Soluções ecológicas para problemas hidráulico-florestais. *Brasil Florestal* (Rio de Janeiro), vol. 4, no. 13, 1973, p. 4–19.

COTT, H. B. The natural history of the lower Amazon. *Ann. Rept. and Proc. Bristol Nat. Soc.*, vol. 7, no. 3, 1930–31, p. 181–188.

COURET, P. Observaciones sobre las mirmecofitas venezolanas, part. 1. *Mem. Soc. Ci. Nat. La Salle*, vol. 26, no. 73, 1966, p. 5–40.

COUTINHO, L. M.; LAMBERTI, A. Respiração edáfica e produtividade primaria numa comunidade amazônica de mata de terra firme. *Ciência e Cultura* (São Paulo), vol. 23, no. 3, 1971, p. 411–419.

DANSEREAU, P. A distribuição e a estrutura das florestas brasileiras. *Bol. Geogr.* (Rio de Janeiro), 61, 1948, p. 34–44.

DAVIS, T. A. W.; RICHARDS, P. W. The vegetation of Moraballi Creek, British Guiana. *J. Ecol.* (London), 21, 1933, p. 350–384; 22, 1934, p. 106–155.

DAY, T. H. *Guide to classification of the late Tertiary and Quaternary soils of the lower Amazon Valley.* Rome, FAO, J-6685, 1959, 56 p.

——. *Guia preliminar para a classificação dos solos do vale do baixo Amazonas.* Belém, SUDAM, 1959, 38 p.

——. *Relatório do levantamento expedido dos solos da área Caeté-Maracassumé.* Belém, FAO-SPVEA, 1959, 29 p. multigr.

——. *Report for the reconnaissance soil survey of the Caeté-Maracassumé area.* Belém, FAO-SPVEA, 1959, 22 p. multigr.

——. *Report to the Government of Brazil in soil investigation conducted in the lower Amazon Valley.* Rome, FAO Report no. 1935, 1961, 34 p.

DENEVAN, W. E. The campos cerrados vegetation of central Brazil. *Geogr. Rev.*, vol. 55, no. 1, 1965, p. 112–115.

DONSELAAR, J. Water and marsh plants in the artificial Brokopondo lake (Surinam, S. America) during the first three years of its existence. *Mededel. Bot. Mus. Herb. Rijksuniv. Utrecht*, 299, 1968, p. 183–196. *Acta Bot. Neerl.*, vol. 17, no. 3, 1968, p. 183–196.

——. Phytogeographic notes on the savanna flora of southern

Surinam (South America). *Mededel. Bot. Mus. Herb. Rijksuniv. Utrecht*, 306, 1968, p. 393–404.

DONSELAAR, J. Observations on savanna vegetation—types in the Guianas. *Mededel. Bot. Mus. Herb. Rijksuniv. Utrecht*, 326, 1969, p. 271–312.

DORNEY, R. S. Present status of existing and proposed national parks and national reserves in the Amazon basin. In: *Atas Simpósio Biota Amaz.*, vol. 7 (Conservação da natureza e recursos naturais), 1967, p. 105–114.

DUBOIS, J. Relatório do programa silvicultural no centro de Curuá-Una, Estado do Pará (SPVEA-FAO). In: *An. 14 Congr. Soc. Bot. Brasil* (Manaus, 1963), 1964, p. 379–397.

——. Programa de pesquisas silvícolas na Amazônia. *Rev. Escola Nac. de Florestas* (Curitiba), 1, 1966, p. 28–46.

——. A floresta amazônica e sua utilização face aos princípios modernos de conservação da natureza. In: *Atas Simpósio Biota Amaz.*, vol. 7 (Conservação da natureza e recursos naturais), 1967, p. 115–146.

——. *Considerações sobre o reflorestamento na Amazônia.* Belém, SUDAM, sér. Recursos naturais, 1969, 18 p.

——. *Desenvolvimento de uma economia florestal na Amazônia; análise focalizando especialmente os aspectos silviculturais.* Rio de Janeiro, Superintendência do Plano de Valorização Econômica da Amazônia, 1969, 36 p.

——; HALLEWAS, P. H.; KNOWLES, O. H. *A Amazônia brasileira como fonte de produtos madeireiros.* Belém, SUDAM, sér. Recursos naturais, 1966, 17 p.

DUCKE, A. A flora do Curicuriari, afluente do rio Negro, observado em viagem com a Comissão Demarcadora des Fronteiras do Setor Oeste. *An. Primeira Reunião Sul-Americana de Botânica* (Rio de Janeiro), vol. 3, no. 6, 1938, p. 389–398.

——. A Amazônia brasileira. *An. Bras. Economia Florestal* (Rio de Janeiro), 1, 1948, p. 28–37.

——; BLACK, G. A. Phytogeographical notes on the Brazilian Amazon. *An. Acad. Bras. Ciência* (Rio de Janeiro), 25, 1953, p. 1–46.

——; ——. Notas sobre a fitogeografia da Amazônia brasileira. *Bol. Técn. IPEAN* (Belém), 29, 1954, p. 1–62.

EGLER, E. G. A zona Bragantina no Estado do Pará. *Rev. Bras. Geogr.* (Rio de Janeiro) vol. 23, no. 3, 1961, p. 527–555.

EGLER, W. A. Contribuição ao conhecimento dos campos da Amazônia. I. Os Campos de Ariramba. *Boletim do Museu Paraense Emilio Goeldi, n. ser. Bot.* (Belém), 4, 1960, p. 1–40.

EITEN, G. Vegetation forms. *Bol. Inst. Bot.* (São Paulo), 4, 1968, p. 1–88.

——. The cerrado vegetation of Brazil. *Bot. Rev.*, vol. 38, no. 2, 1972, p. 201–341.

ELTON, C. S. The structure of invertebrate populations inside neotropical rain forest. *J. Anim. Ecol.* (Cambridge), 42, 1973, p. 55–104.

ESPINAL, L. S. T.; MONTENEGRO, E. M. *Formaciones vegetales de Colómbia y mapa ecológico.* Bogotá, Inst. Geogr. Ag. Codazzi, 1963, 201 p.

FALESI, I. C. Os solos da colônia agrícola de Tomé-Açu. *Bol. Técn. IPEAN* (Belém), 44, 1964, p. 7–93.

——. Levantamento de reconhecimento detalhado dos solos trecho 150–171 da estrada de ferro do Amapá. *Bol. Técn. IPEAN* (Belém), 45, 1964, p. 1–53.

——. O estado atual dos conhecimentos sobre os solos da Amazônia brasileira. In: *Atas Simpósio Biota Amaz.* (Belém, 1966), vol. 1 (Geociências), 1967, p. 151–168.

——. *Solos de Monte Alegre.* Belém, IPEAN, Solos da Amazônia, vol. 2, no. 1, 1970, 127 p.

——. O estado atual dos conhecimentos sobre os solos da Amazônia brasileira. Parte I. In: Zoneamento agrícola da

Amazônia (1ª aproximação). *Bol. Técn. IPEAN* (Belém), 54, 1972, p. 17–67.

FALESI, I. C. Os solos da rodovia Transamazônica. *Bol. Técn. IPEAN* (Belém), 55, 1972, 196 p.

—— et al. Levantamento de reconhecimento dos solos da Região Bragantina, Estado do Pará. *Pesquisa Agropecuária Brasileira* (Rio de Janeiro), 2, 1967, p. 1–63.

—— et al. Contribuição ao estudo dos solos de Altamira (Região fisiográfica do Xingu). *Circular do Inst. Pesq. Exp. Agropecuárias do Norte* (Belém), 10, 1967, p. 8–47.

FANSHAWE, D. B. *The vegetation of British Guiana, a preliminary review.* Imperial Forestry Institute (Oxford), paper no. 29, 1952, 96 p.

——. Forest types of British Guiana. *Carib. For.* (Rio Piedras), 15, 1954, p. 73–111.

FERRI, M. G. Transpiração de plantas permanentes dos 'cerrados'. *Bol. Fac. Filos. Ci. Letr. USP Bot.* (São Paulo), 4, 1944, p. 159–224.

——. Balanço de água de plantas da caatinga. In: *Congresso nacional da sociedade botânica do Brasil* (Recife, 1953), p. 314–332.

——. Ecological information on the 'Rio Negro' caatinga. In: *Comptes Rendus 9ᵉ Congrès International de Botanique* (Montréal), 1959.

——. Contribution to the knowledge of the ecology of the Rio Negro caatinga (Amazon). *Bull. Research Council of Israël* (Jerusalem), 80 (3/4), 1960, p. 195–207.

——. Aspects of the soil-water-plant relationships in connection with some Brazilian types of vegetation. In: *Tropical soils and vegetation* (Proc. of the Abidjan Symposium, 1959), p. 103–109. Paris, Unesco, 1961, 115 p.

FITTKAU, E. J. On the ecology of Amazonian rain forest streams. In: *Atas Simpósio Biota Amazônica*, vol. 3 (Limnologia), 1967, p. 97–108.

——. Limnological conditions in the head-water region of the Xingu river, Brasil. *Tropical Ecology*, vol. 11, no. 1, 1970, p. 21–25.

——. Esboço de uma divisão ecológico-paisagístico da região amazônica. In: Idrobo, J. M. (ed.). *II Simpósio y Foro de Biologia tropical amazônica, Florencia (Caquetá) y Letícia (Amazonas)*, 1969. *Amazônia Brasileira em Foco* (Rio de Janeiro), 9, 1974, p. 17–23.

——; ILLIES, J.; KLINGE, H.; SCHWABE, G. H.; SIOLI, H. (eds.). *Biogeography and ecology in South America.* The Hague, Junk, 2 vol., 1968–69, 946 p.

——; KLINGE, H. On biomass and trophic structure of the Central Amazonian rain forest ecosystem. *Biotropica*, vol. 5, no. 1, 1973, p. 2–14.

——; JUNK, W.; KLINGE, H.; SIOLI, H. Substrate and vegetation in the Amazon region. In: Dierschke, H. (ed.). *Vegetation und Substrat* (Berichte der internationalen Symposien der internationalen Vereinigung für Vegetationskunde herausgegeben von Reinhold Tüxen), p. 73–90. Vadüz, J. Cramer, 1975.

FLOR, H. M. Levantamento florestal de uma área de 500 hectares destinada a parcelas dos cortes rasos para futuro aproveitamento agrícola no municipio de Monção, Estado do Maranhão. In: *Atas Simpósio Biota Amazônica*, vol. 7 (Conservação da natureza e recursos naturais), 1967, p. 147–163.

FOSBERG, F. R. Ecological notes on the upper Amazon. *Ecology*, vol. 31, no. 4, 1950, p. 650–653.

FOUGEROUZE, J. Quelques problèmes de bioclimatologie en Guyane française. *Agron. Trop.*, vol. 21, nᵒ 3, 1966, p. 291–346.

FURON, R. Introduction à l'étude paléogéographique de l'Amérique du Sud. *C. R. Soc. Biogéogr.* (Paris), 272–274, 1954, p. 46–49.

GACHOT, R. *et al. Desenvolvimento florestal no vale do Amazonas.* Rio de Janeiro, Superintendência do Plano de Valorização Econômica da Amazônia, FAO, ETAP, report no. 171, 1966, 87 p.

GALVÃO, M. V. Clima da Amazônia. In: *Geografia do Brasil*, vol. 1, p. 61–111. Rio de Janeiro, 1959.

GLEASON, H. A. Botanical results of the Tyler-Duida expedition. *Bull. Torrey Bot. Club* (New York), 58, 1931, p. 277–506.

GLERUM, B. B. *Forest inventory (Caeté-Maracassumé).* Rome, FAO Report no. 1250, 1960, 67 p.

——. *Forest inventory (Ucuuba region, Tocantins river).* Rome, FAO Report no. 1492, 1962, 7 p.

——. *Forest inventory (Mohogani region, States of Goiás and Pará).* Rome, FAO Report no. 1562, 1962.

——. *Pesquisa combinada floresta-solo no Pará-Maranhão (área: margens da Rodovia Belém-Brasilia entre São Miguel do Guamã e Imperatriz).* Rio de Janeiro, SPVEA (Inventário florestal na Amazônia, 9), 1965, 113 p.

——; SMITH, G. *Forest inventory (Santarém).* Rome, FAO Report no. 1271, 1960.

——; ——. *Forest inventory (São Miguel—Imperatriz).* Rome, FAO Report no. 1483, 1962.

GOMES, P. F. Inconvenientes do uso do valor médio do diâmetro para determinação da área basal. *Escola Sup. Agr. 'Luiz de Queiroz'*, 22, 1965, p. 111–116.

GOODLAND, R. J. A. On the savanna vegetation of Calabozo, Venezuela and Rupununi, British Guiana. *Bol. Soc. Venez. Cienc. Nat.*, vol. 24, no. 110, 1966, p. 341–359.

——. *Glossario de ecologia brasileira.* Manaus, Instituto Nacional de Pesquisas da Amazônia, 1975, 96 p.

GOUROU, P. Observações geográficas na Amazônia (2 parte). *Rev. Bras. Geogr.*, vol. 12, no. 2, 1950, p. 171–350.

GRUBB, P. I. *et al.* A comparison of montane and lowland rain forest in Ecuador. I. The forest structure, physiognomy and floristics. *J. Ecol.*, vol. 51, no. 3, 1963, p. 567–601.

GUERRA, A. T. Aspectos gerais da vegetação do Amapá. *An. Bras. Economia Florestal* (Rio de Janeiro), vol. 6, no. 6, 1953, p. 227–232.

GUERRA, F. C. U. *A explotação da floresta amazônica e seu significado econômico.* Belém, SUDAM, 1971, 29 p.

——. *Colônias agróflorestais.* 2 ed. Belém, SUDAM, 1973, 17 p.

GUERRA, I. A. L. T. Tipos de clima da região norte. In: *Enciclopédia dos Municípios Brasileiros.* Rio de Janeiro, Inst. Bras. Geogr. Estatística, 14, 1957.

GUERRA, S. W. A method of making enumeration surveys in the Amazonian forests of Peru. *Rev. For. Peru*, vol. 1, no. 1, 1967, p. 33–41.

GUERREIRO, R. Seminario de FAO sobre territorios amazónicos de Brasil, Venezuela y Colombia (Manaus, Brasil, 1972). Bogotá, Instituto Colombiano Agropecuario, 1972.

HEGEN, E. E. Man and the tropical environment. Problems of resource use and conservation. In: *Atas Simpósio Biota Amazônica*, vol. 7 (Conservação da natureza e recursos naturais), 1967, p. 165–175.

HEINSDIJK, D. *Forest inventory (Tapajós—Xingú).* Rome, FAO Report 601, 1957, 135 p.

——. *Forest inventory (Xingú—Tocantins).* Rome, FAO Report 949, 1958, 93 p.

——. *Forest inventory (Madeira—Tapajós).* Rome, FAO Report 969, 1958, 83+24 p.

——. *Forest inventory (Guamá—Capim).* Rome, FAO Report 992, 1958, 72 p.

HEINSDIJK, D. *Dryland forest on the Tertiary and Quaternary south of the Amazon river.* Rome, FAO Report 1284, Part I: 1–2; Part II: 1–30; Part III: 1–21; Part IV: 116; Part V: 1–24; Part VI: 1–24, 1960.

——. A distribuição dos diâmetros nas florestas brasileiras. *Bol. Depto. Rec. Nat. Renov.* (Brasília), 11, 1965, p. 1–56.

——. As parcelas zero em inventários florestais. *Bol. Sec. Pesos Florestais* (Brasília), 8, 1965, p. 1–54.

——; MIRANDA BASTOS, A. de. Inventários florestais na Amazônia. *Bol. Setor Inventários Florestais* (Brasília), 6, 1963, p. 1–100.

HEYLIGERS, P. C. Vegetation and soil of a white-sand savanna in Suriname. *Mededel. Bot. Mus. Herb. Rijksuniv. Utrecht*, 191, 1963, p. 1–148.

HOLDRIDGE, L. R. Determination of world · plant formations from simple climatic data. *Science*, vol. 105, no. 2727, 1947, p. 367–368.

——. *Life zone ecology* (provisional edition). San José, Costa Rica, Tropical Science Center, 1967, 124 p. Revised edition, 206 p.

—— *et al. Forest environments in tropical life zones: a pilot study.* New York, Pergamon Press, 1971, 747 p.

HORN, E. F. The Amazon hylea, some notes on its development. *Carribean Forester*, 9, 1948, p. 316–366.

——. A exploração racional das florestas da Amazônia. The rational exploring of the Amazonian forests. Die rationalle der Amazonianwälder. *O Papel* (São Paulo), 12, 1961, p. 1–9.

HUBER, J. Contribuição à Geografia botânica do litoral da Guiana entre o Amazonas e o Oiapoque. *Bol. Mus. Par. Hist. Nat. Etnogr.* (Belém), vol. 1, no. 1/4, 1894–96, p. 381–402.

——. Sur la végétation du cap Magoary, île de Marajó. *Bull. Herb. Boissier* (Genève), vol. 2, n° 1, 1900, p. 86–107.

——. Sur les campos de l'Amazone inférieure et leur origine. In: *Congrès international de Botanique* (Paris, 1900).

——. Contribuição à geografia physica dos furos de Breves e da parte occidental de Marajó. *Boletim Museu Paraense Emilio Goeldi* (Belém), 3, 1902, p. 447–498.

——. La végétation de la vallée du rio Purus (Amazone). *Bull. Herb. Boissier* (Genève), 2 sér., 6, 1906, p. 249–276.

——. Matas e madeiras amazônicas. *Boletim Museu Paraense Emilio Goeldi* (Belém), 6, 1909, p. 91–225.

HUECK, K. Las regiones forestales de sur-America. *Inst. For. Latino-Amer. Bol.*, 2, 1957, p. 1–40.

——. *As Florestas da América do Sul: ecologia, composição e importância econômica* (tradução de H. Reichardt). Univ. de Brasilia (ed.), São Paulo, Poligono, 1972, 466 p.

HUTCHINSON, I. D. *Report on a preliminary study of the forest of Bolivia.* Rome, FAO Report AT 2323, 1967.

JANZEN, D. H. Tropical blackwater rivers, animals, and mast fruiting by the Dipterocarpaceae. *Biotropica*, vol. 6, no. 2, 1974, p. 69–103.

JOLY, A. B. *Conheça a vegetação brasileira.* Univ. São Paulo, Poligono, 1970, 181 p.

JUNK, W. Investigations on the ecology and production-biology of the 'floating meadows' (Paspalo-echinochloetum) on middle Amazon. I. The floating vegetation and its ecology. *Amazoniana*, vol. 2, no. 4, 1970, p. 449–495.

KERFOOT, O. The root systems of tropical forest trees. *Commonw. For. Rev.*, vol. 42, no. 3, 1963, p. 19–26.

KLINGE, H. Podzol soils in the Amazon basin. *J. Soil Sci.*, 16, 1965, p. 95–103.

——. Podzol soils: a source of blackwater rivers in Amazonia. In: *Atas Simpósio Biota Amazonica*, 3, 1967, p. 117–125.

KLINGE, H. *Report on tropical podzols*. Rome, unpubl. FAO Report, 1968, 88 p.

——. Climatic conditions in lowland tropical podzol areas. *Tropical Ecology*, 10, 1969, p. 222–239.

——. *Review of research on tropical podzols*. Report to FAO and Unesco, 1969, 249 p. multigr.

——. Biomasa y materia orgánica del suelo en el ecosistema de la pluviselva centro-amazónica. *Acta Cient. Venez.*, 24, 1973, p. 174–181.

——. Root mass estimation in lowland tropical rain forests of Central Amazonia, Brazil. I. Fine root masses of a pale yellow latosol and a giant humus podzol. *Tropical Ecology*, vol. 14, no. 1, 1973, p. 29–38.

II. 'Coarse root mass' of trees and palms in different height classes. *An. Acad. Brasil. Ciênc.*, vol. 45, no. 3/4, 1973, p. 595–609.

III. Nutrients in fine roots from giant humus podzols. *Tropical Ecology*, vol. 16, no. 1, 1975, p. 28–38.

——; RODRIGUES, W. A. Litter production in an area of Amazonian terra firme forest. Part. I Litter fall, organic carbon and total nitrogen contents of litter. *Amazoniana*, vol. 1, no. 4, 1968, p. 287–302. Part II. Mineral nutrient content of the litter. *Amazoniana*, vol. 1, no. 4, 1968, p. 303–310.

——; ——. Matéria orgânica e nutrientes na mata de terra firme perto de Manaus, *Acta Amazonica* (Manaus), vol. 1, no. 1, 1971, p. 69–72.

——; ——. Phytomass estimation in a Central Amazonian rain forest. In: Young, H. E. (ed.). *IUFRO biomass studies*, p. 339–350. Orono, Univ. of Maine, 1974.

——; PIRES, J. M. Fine litter production in three forest stands of eastern Amazonia. I. The fall of leaves, twigs and small flowers and fruits. *Amazoniana* (in press).

KNOWLES, O. H. *Relatório ao Governo do Brasil sobre produção e mercado de madeira na Amazônia*. Belém, SUDAM, Projeto do Fundo Especial, 52, 1966, 169 p.

KUECHLER, A. W.; MONTOYA-MAQUIN, J. M. The Unesco classification of vegetation; some tests in the tropics. *Turrialba* (Costa Rica), vol. 21, no. 1, 1971, p. 98–109.

KUHLMANN, E. Tipo de vegetação. Cap. IV. In: *Geografia do Brasil, Grande Região Norte*. Rio de Janeiro, Bibl. Geogr. Bras., vol. 1, publ. no. 15, 1959, p. 112–127.

LANGEHEIM, J. H. Leguminous resin-producing trees in Africa and South America. In: Meggers, B. J. *et al.* (eds.). *Tropical forest ecosystems in Africa and South America: a comparative review*, p. 89–104. Washington, D.C., Smithsonian Institution, 1973, 350 p.

LANJOUW, J. Studies on the vegetation of the Suriname savannas and swamps. *Ned. Kruid. Arch.* (Leiden), 46, 1936, p. 823–851.

——. The vegetation and origin of the Suriname savannas. In: *Congrès international de Botanique* (Paris, 1954), sect. 7, p. 45–48.

LA RUE, E. A. Sur l'origine naturelle probable de quelques savannes de la Guyane française et de l'Amazonie brési-lienne. *Comptes Rendus de la Société de Biogéographie* (Paris), 305/307, 1958, p. 50–53.

LECHTHALER, R. Inventário das árvores de um hectare de terra firme da zona Reserva Florestal Ducke, Município de Manaus, Amazônia. Rio de Janeiro, *INPA, Bot.*, 3, 1956, p. 1–10.

LEDOUX, P.; LOBATO, R. C. Investigações de bio-ecologia experimental sobre uma população de *Minquaria guianensis* Aublet (Fam. Olacaceae). In: *Simposio internacional sobre plantas de interés económico de la flora amazonica* (Belém, 1972). Turrialba, Costa Rica.

LEITÃO, C. M. Fauna amazônica. *Rev. Bras. Geogr.* (Rio de Janeiro), vol. 5, no. 3, 1943, p. 343–370.

LIMA, D. A. Viagem aos campos de Monte Alegre, Pará; contribuição para o conhecimento de sua flora. *Bol. Técn. IPEAN* (Belém), 36, 1959, p. 49–149.

——. Contribuição ao estudo do paralelismo da flora amazônica-nordestina. *Bol. Técn. Inst. Pesq. Agron. Pernambuco* (Recife), n. sér., 19, 1966, p. 1–30.

——. A agricultura nas várzeas do estuário do Amazonas. *Bol. Técn. IPEAN* (Belém), 33, 1956, p. 1–164.

——. Os efeitos das queimadas sobre a vegetação dos solos arenosos da região da Estrada de Ferro Bragançe. *Bol. Inpetoria Regional do Fomento Agrícola do Estado do Pará* (Belém), 8, 1958, p. 23–35.

LINDEMAN, J. C. The vegetation of the coastal region of Suriname. *Mededel. Bot. Mus. Herb. Rijksuniv. Utrecht*, 113, IV, 1953, 135 p.

——; MOOLENAAR, S. P. Preliminary survey of the vegetation types of northern Suriname. *Mededel. Bot. Mus. Herb. Rijksuniv. Utrecht*, 159, 1959, p. 1–45.

——; MENNEGA, A. M. W. Bomenbock voor Suriname. *Mededel. Bot. Mus. Herb. Rijksuniv. Utrecht*, 200, 1963, p. 62–63.

LÜTZELBURG, P. F. Die pflanzengeographischen Verhältnisse im Amazonas-gebiet. *Berricht der Deutschen Bot. Gesellschaft* (Berlin), 57, 1939, p. 247–262.

MAGNANINI, A. Chuvas e erosão dos solos. *An. Bras. Econ. Florestal* (Rio de Janeiro), 12, 1960, p. 404–417.

——. Notas sobre vegetação—climax e seus aspectos no Brasil. *Rev. Bras. Geogr.* (Rio de Janeiro), vol. 23, no. 1, 1961, p. 235–243.

——. Aspectos fitogeográficos do Brasil-Áreas e características no passado e no presente. *Rev. Bras. Geogr.* (Rio de Janeiro), vol. 23, no. 4, 1961, p. 681–690.

——; JORGE, M. T. Situação dos Parques Nacionais do Brasil. *Bol. Informativo Fundação Bras. Conserv. Natureza* (Rio de Janeiro), 4, 1969, p. 38–58.

MAGUIRE, B. On the flora of the Guyana highland. *Biotropica*, vol. 2, no. 2, 1970, p. 85–100.

MARBUT, C. F.; MANIFOLD, C. B. The soil of the Amazon basin in relation to their agricultural possibilities. *Geogr. Rev.*, 16, 1924, p. 414–442.

MARLIER, G. Ecological studies on some lakes of the Amazon Valley. *Amazoniana*, vol. 1, no. 2, 1967, p. 91–115.

——. Limnology of the Congo and Amazon rivers. In: Meggers, B. J. *et al.* (eds.). *Tropical forest ecosystems in Africa and South America: a comparative review*, p. 223–238. Washington, D.C., Smithsonian Institution 1973, 350 p.

MARTIUS, C. F. P. Tabulae physiognomicae Brasiliae regiones iconibus expressas. In: *Flora brasiliensis*. Vol. 1, pt. 1, 1840–1869, p. 22–29, tab. IX.

McGRATH, K.; GACHOT, R.; GALLANT, N. M. El valle del Amazonas. *Unasylva* (FAO), vol. 7, no. 3, 1953, p. 109–115.

MEGGERS, B. J. *Amazonia: man and culture in a counterfeit paradise*. Chicago, New York, Aldine Atherton, 1971, 182 p.

——; EVANS, C. A reconstituição pré-histórica. *Publ. av. Museu Paraense Emilio Goeldi* (Belém), 20, 1973, p. 51–69.

——; AYENSU, E. S.; DUCKWORTH, W. D. (eds.). *Tropical forest ecosystems in Africa and South America: a comparative review*. Washington, D.C., Smithsonian Institution, 1973, 350 p.

MENDES, J. C. Evolução geológica da Amazônia; breve histórico das pesquisas. In: *Simpósio Biota Amazônica* (Belém, 1966), vol. 1 (Geociências), 1967, p. 1–9.

MIRANDA, V. C. Os campos de Marajó. *Boletim Museu Paraense Emilio Goeldi* (Belém), vol. 5, no. 1, 1907, p. 96–151.

——. Os campos de Marajó e sua flora, considerados sob o ponto de vista pastoril. *Bol. Mus. Hist. Nat. Ethnogr.* (Belém), vol. 5, no. 1, 1908, p. 96–151.

MONTEITH, J. L. Methods and instruments for measuring the response of plants to weather. In: *Seminario sobre Ecología Tropical* (Itabuna, Brasil, 1972). Turrialba, Costa Rica.

MONTOYA, L. Resumen de la situación actual de los programas de investigación ecológica en el Trópico americano. In: *Reunión técnica de Programación sobre investigaciones ecológicas para el Trópico americano* (Maracaibo, Venezuela, 1973). Fac. Agron., Univ. de Zulia, 1973. IICA, Informes de Conferencias, Cursos y Reuniones, Documento no. 8, 1973.

MORAES, V. H. F. Periodicidade de crescimento do tronco em árvores da floresta amazônica. *Pesquisa Agropecuária Brasileira* (Rio de Janeiro), 5, 1970, p. 315–320.

——; PIRES, J. M. Estudo sobre regeneração natural na mata amazônica. In: *Áreas de Pesquisas Ecológicas do Guamá; um programa integrado de colaboração científico-educacional na Amazônia. Relatório anual* (Belém, IPEAN), 1966, multigr.

——; BASTOS, T. X. Viabilidade e limitações climáticas para as culturas permanentes, semipermanentes e anuais com possibilidade de expansão na Amazônia. Zoneamento Agrícola da Amazônia (Belém). *Bol. Técn. IPEAN*, 54, 1972, p. 123–153.

MOREIRA, A. Investigaciones ecológicas en los trópicos humedos de Bolívia. In: *Seminario sobre Ecología tropical* (Itabuna, Brasil, 1972). Turrialba, Costa Rica.

MYERS, J. G. Savannah and forest vegetation of the interior Guiana plateau. *J. Ecol.*, 24, 1936, p. 162–184.

NAVEZ, A. La forêt équatoriale brésilienne, I. *Bull. Soc. Bot. Belgique*, 57, 1924, p. 7–17.

NIEUWENHUIS, W. H. *Report to the Government of Brazil on the development of grazing and fodder resources in the Amazon Valley*. Rome, FAO Report no. 1238, 1960, 23 p.

OLIVEIRA, A. B. Considerações sobre a exploração de castanha no baixo e médio Tocantins. *Rev. Bras. Geogr.* (Rio de Janeiro), vol. 2, no. 1, 1940, p. 3–15.

OLIVEIRA, A. I.; LEONARDOS, D. H. *Geologia do Brasil*. Rio de Janeiro, Ministério da Agricultura, 1943.

Organização dos Estados Americanos (OEA). Departamento de desenvolvimento regional. *Marajó—Um estudo para e seu desenvolvimento*. Washington, D.C., 1974, 124 p.

PANDOLFO, C. *A atuação da SUDAM na presenvação do patrimônio florestal da Amazônia*. Belém, SUDAM, 1972, 14 p.

——. *Estudos básicos e estabelecimento de uma política de desenvolvimento dos recursos florestais e o uso racional das terras na Amazônia*. 2 ed. Belém, SUDAM, 1974, 54 p.

PEDROSO, L. M. Informaçoes sobre o atual comportamento de espécies exóticas na região do médio Amazonas. *Brasil Florestal* (Rio de Janeiro), vol. 4, no. 16, 1973, p. 64–68.

—— et al. *Informaçoes preliminares sobre a silvicultura de 38 espécies florestais da Extação Experimental de Curuá-Una*. Belém, Superintendência de Desenvolvimento da Amazônia, 1971.

PENTEADO, A. R. *Problemas de colonização e de uso da terra na região bragantina do Estado do Pará*. Belém, Univ. Fed. Pará, 2 vol.; 1967.

PEREIRA, A. P.; PEDROSO, L. M. *Experimentos de silvicultura tropical*. Belém, SUDAM, 1972, 79 p.

PIRES, J. M. Noções sobre ecologia e fitogeografia da Amazônia. *Norte Agronômico* (Belém), vol. 3, no. 3, 1957, p. 37–54.

PIRES, J. M. Sobre as necessidades de Reservas florestais. *Norte Agronômico* (Belém), vol. 5, no. 5, 1959, p. 120–124.

——. Exploração botânica no Território do Amapá (Rio Oiapoque). In: *Anais 13 Congr. Soc. Bot. Brasil* (Recife, 1962), 1964, p. 164–199.

——. The estuaries of the Amazon and Oiapoque rivers and their floras. In: *Proceedings Symposium on: scientific problems of the humid tropical zone deltas and their implications* (Dacca, 1964), p. 211–218. Paris, Unesco, 1966.

——; DOBZHANSKY, T.; BLACK, G. A. An estimate of the number of species of trees in an Amazonian forest community. *Bot. Gaz.* (Chicago), 114, 1953, p. 467–477.

——; KOURY, R. M. Estudo de um trecho de várzea próximo a Belém. *Bol. Técn. IPEAN* (Belém), 36, 1959, p. 1–44.

——; RODRIGUES, J. S. Sobre a flora das caatingas do Rio Negro. In: *Anais 13 Congr. Soc. Bot. Brasil* (Recife, 1962), 1964, p. 242–262.

——; KLINGE, H. Fine litter production in three forest stands of eastern Amazonia. II. Vegetation. *Amazoniana* (in press).

PITT, G. J. W. *Report to the Government of Brazil on the application of silvicultural methods to some of the forests of the Amazon*. Rome, FAO Report no. 1337, 1961, 139 p.

——. *Amazon forests. Possible methods of regeneration and improvement. Unasylva* (FAO), vol. 15, no. 2, 1961, p. 63–69.

PRANCE, G. T. Phytogeographic support for the theory of Pleistocene forest refuges in the Amazon Basin, based on evidence from distribution patterns in Caryocaraceae, Chrysobalanaceae, Dichapetalaceae and Lecythidaceae. *Acta Amazonica* (Manaus), vol. 3, no. 3, 1973, p. 5–28.

PRESTON, F. W. The commonness and rarity of species. *Ecology*, 29, 1948, p. 254–283.

RAMOS, A. A. *et al*. Inventário florestal do Distrito agropecuário da Zona franca de Manaus. *Revista Floresta* (Curitiba), vol. 4, no. 1, 1972, p. 40–53.

RATTER, J. A. Some notes on two types of cerradão occurring in northeastern Mato Grosso. In: *Simposio sobre o cerrado*, 3, p. 100–102. São Paulo, Edgard Blucher, USP, 1971.

——; RICHARDS, P. W.; ARGENT, G.; GIFFORD, D. R. Observations on the vegetation of northeastern Mato Grosso. I. The woody vegetation types of the Xavantina-Cachimbo Expedition Area. *Philosophical Transactions of the Royal Society of London, Biological Sciences*, vol. 266, no. 880, 1973, p. 449–492.

RAWITSCHER, F. *Elementos básicos de botânica*. 3 ed. São Paulo, Melhoramentos, 1953.

REIS, A. C. F. *O seringal e o seringueiro*. Rio de Janeiro, Serv. Informação Agrícola, Min. Agricultura, 1954, 149 p.

RICE, E. L. The future of the tropical rain forest. In: *Atas Simpósio Biota Amazonica*, vol. 7 (Conservação da natureza e recursos naturais), 1967, p. 49–56.

RIZZINI, C. T. Nota prévia sobre a divisão fitogeográfica do Brasil. *Rev. Bras. Geogr.* (Rio de Janeiro), vol. 25, no. 1, 1963, p. 1–64.

——. Delimitação, caracterização e relações da flora silvestre hileiana. In: *Atas Simpósio Biota Amazonica*, vol. 4 (Botânica), 1967, p. 13–36.

——; PINTO, M. M. Áreas climato-vegetacionais do Brasil segundo os métodos de Thornthwaite e de Mohr. *Rev. Bras. Geogr.* (Rio de Janeiro), vol. 26, no. 4, 1964, p. 523–547.

RODRIGUES, W. A. Estudo preliminar da mata de várzea alta de uma ilha do baixo rio Negro de solo argiloso e úmido. *Bol. Inst. Nac. Pesquisas da Amazônia, sér. Bot.* (Manaus), 10, 1961, p. 1–50.

——. Aspectos fitosociológicos das caatingas do rio Negro.

Boletim Museu Paraense Emilio Goeldi, n. ser., Bot. (Belém), 15, 1961, p. 1–41.

RODRIGUES, W. A. Aspects phytosociologiques des pseudo-caatingas et forêts de várzea du rio Negro. In: Aubréville, A. *Étude écologique des principales formations végétales du Brésil*, p. 209–265. Nogent-sur-Marne, Centre technique forestier tropical, 1961, 268 p.

——. Estudo de 2,6 hectares de mata de terra firme da Serra do Navio-Amapá. *Boletim Museu Paraense Emilio Goeldi, n. ser., Bot.* (Belém), 19, 1962, p. 1–22.

——. Vegetação aquática dos campos alagáveis de Quatipuru, Estado do Pará. In: *An. 15 Congr. Bras. Bot.* (Porto Alegre), 1964, p. 221–224.

——. Inventário florestal piloto ao longo da estrada Manaus-Itacoatiara, Estado do Amazonas. Dados preliminares. In: *Atas Simpósio Biota Amazonica*, vol. 7 (Conservação da natureza e recursos naturais), 1967, p. 257–267.

——. Plantas dos campos do Rio Branco (Território de Roraima). In: *Simpósio sobre o cerrado*, 3, p. 180–193. São Paulo, Edgard Blucher, USP, 1971.

——. A situação atual das pesquisas ecológicas na Amazônia brasileira. In: *Reunión técnica de Programación sobre investigaciones ecológicas para el Trópico americano* (Maracaibo, Venezuela). Fac. Agron. Univ. Zulia, 1973. IICA, Informes de Conferencias, Cursos y Reuniones, Documento no. 8, 1973.

ROMERO, V. Efeito de la vegetación sobre la fertilidad natural de los suelos del trópico húmedo. In: *Seminario sobre Ecología tropical* (Itabuna, Brasil, 1972). Turrialba, Costa Rica.

ROXO, M. G. D. Terras sulamericanas emersas nos tempos permo-carboníferos. *Rev. Bras. Geogr.* (Rio de Janeiro), vol. 5, no. 1, 1943, 43 p.

SAMPAIO, A. J. Os campos gerais de Cuminá e a phytogeografia do Brasil. *Bol. Mus. Nacional* (Rio de Janeiro), vol. 5, no. 2, 1929, p. 25–29.

——. A flora do rio Cuminá (Estado do Pará-Brasil). Resultados botânicos da Expedição Rondon à Serra Tumuc-Humac em 1928. *Arq. Mus. Nacional* (Rio de Janeiro), 35, 1933, p. 9–206.

——. *Fitogeografia do Brasil*. São Paulo, Cia. Ed. Nacional (Biblioteca Pedagógica Brasileira, sér. 5: Brasiliana, 35), 1945, 372 p.

SANTOS, H. M. Balança hídrico de Manaus, Amazonas. *Inst. Nac. Pesq. Amazônia sér. Avulsa*, 1, 1968, p. 1–16.

SANTOS, W. H.; FALESI, I. C. Contribuição ao estudo dos solos da ilha de Marajó. *Bol. Técn. IPEAN* (Belém), 45, 1964, p. 56–161.

SCHMIDT, J. C. J. O clima da Amazônia. *Rev. Bras. Geogr.* (Rio de Janeiro), vol. 4, no. 3, 1947, p. 465–500.

SCHMIDT, M. Anotações sobre as plantas de cultivo e os métodos da agricultura dos indígenas sul-americanos. *Bol. Geogr.* (Rio de Janeiro), vol. 20, no. 168, 1962, p. 258–267.

SCHMIDT, P. B.; VOLPATO, E. Aspectos silviculturais de algumas espécies nativas da Amazônia. *Acta Amazonica* (Manaus), vol. 2, no. 2, 1972, p. 99–122.

SCHNELL, R. Problèmes biogéographiques comparés de l'hylaea amazonienne et de la forêt dense tropicale d'Afrique. In: *Atas Simpósio Biota Amazonica*, vol. 4 (Botânica), 1967, p. 229–239.

SCHULZ, J. P. Ecological studies on rain forest in northern Suriname. *Mededel. Bot. Mus. Herb. Rijksuniv. Utrecht*, 163, 1960, p. 1–267.

——; RODRIGUES, L. Plantaciones forestales en Surinam. *Rev. Forestal Venez.*, 14, 1966, p. 5–36.

SIOLI, H. Alguns resultados e problemas da limnologia amazônica. *Bol. Técn. IPEAN* (Belém), 24, 1951, p. 3–44.

——. Sobre a sedimentação na várzea do Baixo Amazonas. *Bol. Técn. IPEAN* (Belém), 24, 1951, p. 45–65.

——. Valores de pH, de águas amazônicas. *Boletim Museu Paraense Emilio Goeldi* (Belém), *n. sér., Geologia*, 1, 1957, p. 1–37.

——. A limnologia e a sua importância em pesquisas da Amazônia. *Amazoniana*, vol. 1, no. 1, 1965, p. 11–35.

——. Studies in Amazonian waters. In: *Atas Simpósio Biota Amazonica*, vol. 3 (Limnologia), 1967, p. 9–50.

——. The Cururu Region in Brazilian Amazonia, a transition zone between hylaea and cerrado. *J. Bot. Society*, vol. 46, no. 14, 1967, p. 453–462.

——. Ecologia da paisagem e agricultura racional na Amazônia brasileira. *Amazônia Brasileira em Foco* (Rio de Janeiro), 8, 1973, p. 25–26.

——; KLINGE, H. Solos, tipos de vegetação e águas na Amazônia. *Boletim Museu Paraense Emilio Goeldi* (Belém), *n. ser., Avulsa*, 1, 1962, p. 27–41.

——; ——. Anthropogene Vegetation am brasilianischen Amazonasgebiet. In: Tüxen, R. (ed.). *Anthropogen Vegetation*, p. 357–363. Bericht über das Internationale Symp. in Stolzenau, Weser, 1961. The Hague, 1966.

SOARES, L. C. Delimitação da Amazônia para fins de planejamento econômico. *Rev. Bras. Geogr.* (Rio de Janeiro), vol. 10, no. 2, 1948, p. 163–210.

——. Limites meridionais e orientais da área da ocorrência da floresta amazônica em território brasileiro. *Rev. Bras. Geogr.* (Rio de Janeiro), vol. 15, no. 1, 1953, p. 2–95.

SOARES, R. O. Inventários florestais na Amazônia. *Brasil Florestal* (Rio de Janeiro), vol. 1, no. 1, 1970, p. 4–9.

SOMBROEK, W. G. *Amazon soils. A reconnaissance of the soils of the Brazilian Amazon region*. Wageningen, Centre for agricultural publishing and documentation, 1966, 292 p.

STARK, N. M. Mycorrhiza and nutrient cycling in the tropics. In: *Proc. First North American Conference on Mycorrhizae* (April 1969), p. 228–229. US. Dept. of Agric., Forest Service, Misc. Publ. 1189, 1969.

——. The nutrient content of plants and soils from Brazil and Surinam. *Biotropica*, 2, 1970, p. 51–60.

——. Nutrient cycling. I. Nutrient distribution in some Amazonian soils. *Trop. Ecol.*, 12, 1971, p. 24–50.

——. Nutrient cycling. II. Nutrient distribution in some Amazonian vegetation. *Trop. Ecol.*, 12, 1971, p. 177–201.

STERNBERG, O'REILLY, H. *A água e o homem na várzea do Careiro.* Univ. do Brasil (Rio de Janeiro), tese, 1956.

STEYERMARK, J. Flora del Auyan-tepui. *Acta Bot. Venez.* (Caracas), 5/8, 1967, p. 5–370.

TAKEUCHI, M. A estrutura da vegetação na Amazônia. I. A mata pluvial tropical. *Boletim Museu Paraense Emilio Goeldi* (Belém), *n. ser., Bot.*, 6, 1960, p. 1–17.

——. A estrutura da vegetação na Amazônia. II. As savanas do norte do rio Negro. *Boletim Museu Paraense Emilio Goeldi* (Belém), *n. ser., Bot.*, 7, 1960, p. 1–14.

——. A estrutura da vegetação na Amazônia. III. A mata de campina na região do rio Negro. *Boletim Museu Paraense Emilio Goeldi* (Belém), *n. ser., Bot.*, 8, 1960, p. 1–13.

——. The structure of the Amazonian vegetation. I. Savanna in Northern region. *J. Fac. of Science, Univ. of Tokyo, section III, Bot.*, vol. 7, no. 12, 1960, p. 523–533.

——. The structure of the Amazonian vegetation. II. Tropical rain forest. *J. Fac. Sci., Univ. Tokyo, section III, Bot.*, vol. 8, no. 1/3, 1961, p. 1–26.

——. The structure of the Amazonian vegetation. III. Campina

forest in the rio Negro region. *J. Fac. Sci., Univ. Tokyo, section III, Bot.*, vol. 8, no. 4, 1961, p. 27–35.

TAKEUCHI, M. The structure of the Amazonian vegetation. IV. High campina forest in upper rio Negro. *J. Fac. Sci., Univ. Tokyo, section III, Bot.*, vol. 8, no. 5, 1962, p. 279–288.

——. The structure of the Amazonian vegetation. V. Tropical rain forest near Uaupés. *J. Fac. Sci., Univ. Tokyo, section III, Bot.*, vol. 8, no. 6, 1962, p. 289–296.

——. The structure of the Amazonian vegetation. VI. Igapó. *J. Fac. Sci., Univ. Tokyo, section III, Bot.*, vol. 8, no. 7, 1962, p. 297–304.

TATE, G. H. H. Life zones at Mount Roraima. *Ecology*, vol. 13, no. 3, 1932, p. 235–257.

TEREZO, E. F. M. *et al. O extrativismo do pau-rosa (Aniba duckei* Kosterm; *A. rosaeodora* Ducke); *aspectos sócio-econômicos; a silvicultura da espécie*. Belém, SUDAM, 1971, 40 p.

TORTORELLI, L. A. The Amazonian forest in the Goiás-Pará region. *An. Bras. Econ. Florestal* (Rio de Janeiro), 17, 1965, p. 13–30.

TUNDISI, J. G. Produção primária em ecosistemas lacustres da região tropical. In: *Seminario sobre Ecología tropical* (Itabuna, Brasil, 1972). Turrialba, Costa Rica.

UHART, E. *A floresta amazônica, fonte de energia*. Belém, SUDAM, 1971, 91 p.

ULE, E. Die Pflanzenformationen des Amazonas-Gebietes. *Bot. Jb.*, 40, 1908, p. 114–172, 398–443.

VANZOLINI, P. E. *Zoologia sistemática, geografia e origem das espécies*. São Paulo, Inst. Geografia, USP, 1970, 56 p.

——. Paleoclimates, relief and species multiplication in equatorial forests. In: Meggers, B. J. *et al.* (eds.). *Tropical forest ecosystems in Africa and South America: a comparative review*, p. 255–257. Washington, D.C., Smithsonian Institution 1973, 350 p.

VELOSO, H. P. Os grandes clímaces do Brasil. II. Considerações gerais sobre a vegetação da região amazônica. *Memórias Inst. Oswaldo Cruz* (Rio de Janeiro), vol. 60, no. 3, 1962, p. 393–403. *Bol. Geogr.* (Rio de Janeiro), 192, 1966, p. 311–318.

——. *Atlas florestal do Brasil*. Rio de Janeiro, Ministério da Agricultura, Serv. Informação Agrícola, 1966, 82 p.

VIEIRA, L. S.; SANTOS, W. H. Contribuição ao estudo dos solos de Breves. *Bol. Técn. IPEAN* (Belém), 42, 1962, p. 33–45.

——; OLIVEIRA FILHO, J. P. S. As caatingas do rio Negro. *Bol. Técn. IPEAN* (Belém), 42, 1962, p. 7–32.

VILLEGAS, C. *Ecologia del trópico americano; uma bibliografia parcialmente anotada*. Turrialba, Costa Rica, IICA, Programa Cooperativo para el Desarrollo del Trópico Americano, 1974, 64 p. (IICA, Documentación e Información Agrícola, no. 33).

VOLTAPO, E.; SCHMIDT, P. B.; ARAÚJO, V. C. Situação dos plantios experimentais na Reserva Florestal Ducke. *Acta Amazonica* (Manaus), vol. 3, no. 1, 1973, p. 71–82.

VUILLEUMIER, B. S. Pleistocene changes in the fauna and flora of South America. *Science* (Washington), 173, 1971, p. 171–180.

WENT, F. W.; STARK, N. The biological and mechanical role of soil fungi. *Proc. Nat. Acad. Sci. USA* (Washington), 60, 1968, p. 497–504.

WILLIAMS, W. R.; LOOMIS, R. S.; ALVIM, P. T. Environments of evergreen rain forests on the lower rio Negro, Brazil. *Tropical Ecology*, vol. 13, no. 1, 1972, p. 65–78.

WIMJSTRA, T. A.; HAMMEN, T. van der. Palynological data on the history of tropical savannas in Northern South America. *Leid. Geol. Mededel.* (Leiden), 38, 1966, p. 71–90.

ZARUR, J. Um comentário sobre a classificação de Koeppen. *Rev. Bras. Geogr.* (Rio de Janeiro), vol. 5, no. 2, 1943, p. 250–266.

Tropical forest ecosystems of India: the teak forests
(as a case study of sylviculture and management)

by S. K. Seth[1] and O. N. Kaul[1]

Introduction

Geographical distribution

Description
    Typology
    Size class distributions

Autecology

Ecosystem functioning
    Flowering and seed production
    Seed origins and provenances
    Germinative capacity and establishment
    Seed dormancy and pre-treatment
    Biomass and productivity
    Water balance
    Nutrient cycling
    Protection

Management
    Natural regeneration
    Regeneration techniques and management practices

Research needs and priorities

Conclusions

Bibliography

1. Forest Research Institute and Colleges New Forest, Dehra Dun, India.

Introduction

As the most valuable tree of the countries of South-East Asia in which it grows, teak (*Tectona grandis* Linn.) sylviculture and management has dominated the forestry practice of Burma, the Indian peninsula, Indonesia and Thailand since the advent of scientific forestry. Though most attention has been paid to artificial regeneration, considerable work has been carried out on natural regeneration. Troup (1921) and later publications (Anonymous, 1958; Kadambi, 1972; Mathur, 1973) present most of the information on this species.

This case study is restricted to India.

Geographical distribution

Natural teak forests are practically confined to areas south of 24° N. The distribution is discontinuous and the most important teak forests are found in the States of Madhya Pradesh, Karnataka, Tamil Nadu, Maharashtra, Andhra Pradesh, Orissa, Gujarat and Rajasthan. Outside its natural range, teak has been planted in the moist deciduous forests of Assam, Bihar, Orissa, Andamans, Uttar Pradesh and West Bengal.

Description

Teak forests cover 8.9 million ha; over half of this being in Madhya Pradesh. Other important States are Maharashtra, Andhra Pradesh, Karnataka, Kerala and Gujarat (MOA, 1974).

Typology

The species occurs naturally in the southern tropical moist deciduous, and southern tropical dry deciduous forests (Champion and Seth, 1968). In the teak-bearing areas of southern tropical moist deciduous forests, *Tectona grandis* is the most characteristic species with excellent development wherever the soil permits. It is generally associated with *Terminalia* spp., *Pterocarpus* spp. and *Lagerstroemia* spp. *Adina* is often present though tending to indicate drier conditions, and *Dalbergia latifolia* is characteristic; *Xylia* is also very generally present, and *Schleichera* and *Ca-*

reya spp. are common, mainly in the second storey. The typical bamboo is *Bambusa arundinacea*. The following three sub-types are differentiated:

— *Very moist teak forest*

Rainfall >2 500 mm/a; deep alluvial often clayey soils; low (<10) percentage of teak; very dense evergreen undergrowth; little natural regeneration; no fires; negligible grazing.

The floristics correspond to the moister end of the moist deciduous, approaching the semi-evergreen. Site quality is high (I to II). Some of the forests are evidently of secondary origin on semi-evergreen sites, stabilized as a subclimax. Characteristic species are: *Grewia tiliaefolia*, *Lagerstroemia lanceolata* (top canopy), *Dillenia pentagyna*, *Kydia calycina* (second storey), *Bambusa arundinacea*, *Glycosmis pentaphylla*, *Clerodendrum viscosum* (shrubs).

— *Moist teak forest*

Rainfall 1 600–2 500 mm/a; deep loamy soils; fair to medium (10–25) percentage of teak; dense undergrowth; fair but patchy natural regeneration; fires rare; light grazing.

Deciduous associates predominate in this sub-type. Teak is usually of site quality II to III. Some of the forests may be secondary in origin. Characteristic species are: *Terminalia tomentosa*, *Dalbergia latifolia* (top canopy), *Xylia xylocarpa* (second storey), *Bambusa arundinacea*, *Dendrocalamus strictus*.

— *Slightly moist teak forest*

Rainfall 1 200–1 600 mm/a; moderately deep loamy soils; medium to high (20–60) percentage of teak; moderate undergrowth; fair natural regeneration; occasional fires; moderate grazing.

The proportion of dry deciduous associates is higher than in the last sub-type. Site quality is III or lower. Characteristic species are: *Terminalia tomentosa*, *Adina cordifolia* (top canopy), *Schleichera oleosa* (second storey), *Dendrocalamus strictus*.

The southern tropical dry deciduous forests can be divided into two classes according as teak is present or not. The most characteristic tree of the teak-bearing type is *Tectona grandis* and its most typical associates *Anogeissus latifolia* and *Terminalia* spp. In the non-teak-bearing forests, the dominant genera are again *Anogeissus* and *Terminalia* accompanied by *Diospyros*, *Boswellia* and *Sterculia*. Representatives of many families are included among the co-dominants, their chief characteristic being perhaps their wide distribution and general adaptability such that most are equally common in the cooler northern forest and in the moist deciduous types. Dipterocarpaceae are virtually absent (only *Shorea talura* and *S. tumbuggaia* very locally). The main bamboo species is *Dendrocalamus strictus*. The grasses are of medium height, *Heteropogon*, *Themeda* spp., *Saccharum spontaneum*, etc., the tall-tufted *Erianthus* spp., etc., not occurring. The following two sub-types are differentiated:

— *Very dry teak forest*

Rainfall <900 mm/a; dry infertile soils; teak in low to fair amount mixed with dry deciduous species; ground cover scanty; seedling regeneration practically absent; fires annual; grazing heavy.

Very open forest of very poor quality on stony, detrital, and shallow soils, usually derived from crystalline rocks or trap. Characteristic species with teak are: *Boswellia serrata*, *Anogeissus latifolia*, *Sterculia urens*, *Cochlospermum religiosum*, *Acacia catechu* and occasionally *Anogeissus pendula*, especially in transition zone to *Anogeissus pendula* edaphic community.

— *Dry teak forest*

Rainfall 900–1 300 mm/a; shallow or porous or clayey soils; teak present in high proportion, sometimes practically pure; undergrowth light and patchy; seedlings in groups and patches; fires frequent; grazing heavy.

Mixed dry deciduous forest with teak usually forming the major proportion of the crop on shallow porous or stiff clayey soils. Characteristic species are *Anogeissus latifolia*, *Diospyros tomentosa*, *Hardwickia binata* and other common dry deciduous trees.

Size class distributions

Some size class distributions are shown in tables 1 and 2 (FRI, 1964; Champion and Seth, 1968a). These figures are all for C-grade thinning[1] at usual rotation age. The height for site quality I is usually 1.5 times that for site quality III.

TABLE 1. Site quality and crop height of teak.

Site quality class	Crop age (years)	Crop height (m)
I	80	39.3
II	80	32.3
III	80	25.0
IV	80	18.0

TABLE 2. Distribution of diameter classes in even-aged crops of teak.

Diameter class (cm)	Percentage in each class
25–30	1.5
31–40	17.5
41–50	49.0
51–60	28.0
61–70	4.0

1. Heavy thinning (C-grade) consists in the removal of dead, dying, diseased and suppressed trees; of defective dominated stems and whips, of branchy advance growth, which is impracticable or not desirable to prune or lop; and of the further removal of the remaining dominated stems and such of the defective co-dominants as can be removed without making lasting gaps in the canopy.

The distribution of trees in diameter classes, even in even-aged crops, shows a wide range, from 25–70 cm of which 81, 32 and 4 per cent are above 40, 50 and 60 cm respectively. These figures represent crops in which many trees mainly in the lower diameter classes have been removed in thinning. The constitution of uneven-aged teak ecosystems has not been studied to any extent.

Autecology

Tectona grandis thrives naturally in the tropics with a warm and moist climate. It appears to grow best in localities where the annual rainfall is 1 250–3 750 mm/a, although it is found in areas with annual rainfall range of 500–5 000 mm. The temperature over its range varies from 2.2° C in winter to 48° C in summer, but it thrives best in the moist localities of the West Coast under an equable climate, with an absolute minimum of 13–17° C and an absolute maximum of 39–43° C.

Teak forests are largely on hilly and undulating terrain. The growth is better on the cooler northern and eastern aspects and well-drained alluvial deposits carry the finest stands. The occurrence of *Tectona grandis* appears to be correlated with the geology; it is prominent on traps, basalt and other lime and base rich rocks but insignificant on sandstones, quartzites and other Gondwana rocks which are usually non-calcareous. It also occupies the alluvial soils especially if they are derived from traps and lime-bearing rocks.

Seth and Yadav (1959) found soil moisture to be an important factor for its growth; it avoids waterlogged or very dry habitats. There is a correlation between its growth and pH, exchangeable calcium, magnesium and phosphorus; but not with nitrogen, organic matter and C/N ratio (Bhatia, 1954). Thus trap soils which have a high calcium and phosphate content support a better quality teak forests. Studies on the correlation of teak distribution with the subsoil acidity in Madhya Pradesh show that the species usually occurs on soils having pH range of 6.5–7.5; below 6.0 it is practically absent and above pH 8.5 its growth is retarded. Soils supporting good teak forests are characterized by an increase in the pH of the subsoil, while those supporting a poorer quality teak maintain uniformly acid conditions. In plantations, however, no distinct correlation between growth and soil pH has been found. Most good *Tectona grandis* soils in Madhya Pradesh usually contain more than 0.3 per cent exchangeable calcium which seems to be the critical minimum. In this area alluvial soils have a higher level of exchangeable calcium and are fairly moist. Soils under teak plantations in Nilambur Valley also have a higher percentage of calcium. Where laterite outcrops occur in the alluvial deposit, teak growth deteriorates. Natural regeneration of *Tectona grandis* has been found in Madhya Pradesh on soils with exchangeable calcium above 0.3–0.4 per cent and magnesium of *ca.* 0.2–0.25 per cent. Good regeneration was found on soils with a phosphorus range of 5–7 mg/100 g soil. Other factors which inhibit teak regeneration are waterlogging, dense shade, excessive grazing and frequent fires.

Griffith and Gupta (1947) found that four factors

(SiO_2/R_2O_3, silicaoxyde/iron and aluminium oxides ratio, dispersion coefficient, depth of permanent moisture availability, and aspect) were a reliable guide in deciding the suitability of a particular site for plantation and for forecasting the quality of teak expected. In general, a low SiO_2/R_2O_3 ratio, a low dispersion coefficient, a southerly or westerly aspect and a very low or extra high water-table are unfavourable. The high quality of teak in Nellikutta is believed to be due mainly to a high SiO_2/R_2O_3 ratio, alluvial site, good moisture availability, well-drained sandy loam soil and high content of bases like calcium and magnesium. Of the various mineral elements needed for its proper development, calcium is the most important. The positive relationship between soil calcium and teak growth in both natural stands and plantations appears to be confirmed by foliar analysis (Seth and Yadav, 1959). In mineral deficiency studies on the seedlings, sulphur deficiency produced chlorosis, curling of leaves, premature defoliation and restricted shoot growth. As sulphur and calcium are frequently associated in nature and teak is a calcicole, it is possible that the good growth of the species on calcareous soils may be partly due to sulphur (Kaul, Gupta and Negi, 1972).

Ecosystem functioning

Flowering and seed production

Teak flowers appear within about a month of the onset of the rains, according to season and locality (in July-August with the southwest monsoon and in December-January where only northeast monsoon causes rain). The fruits ripen from November to January and fall gradually, some remaining on the tree through part of the hot season. The data on phenological behaviour of the species are indicated in table 3 (Krishnaswamy and Mathauda, 1954).

TABLE 3. Teak phenology at Dehra Dun (latitude 30° 19′ N; altitude 680 m); based on observations from over a decade (days counted from 1 to 365 from January 1 to December 31).

	First new leaves	Last new leaves	First flowers open	Maximum flowers open
Mean date	144	189	223	245
Range of dates	115–153	159–209	185–247	217–285

	Last flowers open	First ripe fruit	Maximum ripe fruit	First fruit or seed-fall
Mean date	268	352	37	53
Range of dates	235–325	317– 14	6– 69	12– 94

	Maximum fruit or seed-fall	Last fruit or seed-fall	First leaf-fall	Last leaf-fall
Mean date	81	118	82	135
Range of dates	46–116	78–142	68–106	108–155

Teak produces abundant seeds almost every year and starts flowering and seeding at a young age: 20 years from seedlings and *ca.* 10 years from coppice. The seed crop may be partially destroyed by storms between flowering and fruiting and insects also often destroy much of the seed. It is estimated that a large tree produces *ca.* 31 000 seeds/a (Jain, 1962) and there are about 1.6 seed/g (Sen Gupta, 1937).

Seed storage does not present any serious problems in gunny bags, sealed tins or even by heaping it on the ground. Germination is not adversely affected by storage for at least 2 years; storage for one year is generally considered to facilitate germination.

Seed origins and provenances

Attempts have been made to distinguish the varieties available on the basis of morphological and other characters but this work cannot be considered as complete. Some are superior in stem-form, fluting, twist, shape of crown, branching habit, ease with which they can be regenerated from seed, growth rates, etc. It is believed that some of these characters (e.g., bole-form and branching habit) are hereditary.

Teak growing in Allapali Range in Maharashtra has a high reputation for wood quality (Kedharnath and Matthews, 1962). A variety called *Teli* in North Kanara (Karnataka) has smoother, more shiny and less heavy but darker leaves than the standard variety. The boles of *Teli* trees are more cylindrical, and the bark smoother and duller, the trees are believed to grow faster, leaf earlier and to have stronger timber. All these characters are believed to be heritable (Kadambi, 1972). It is believed to be resistant to leaf skeletonizer, *Pyrausta machaeralis* (Kedharnath and Matthews, 1962).

Two varieties of teak were reported in 1895 from South Kanara (Karnataka): *Kallu-theku* is confined to the interior, rocky or gravelly lateritic regions, its trees being stunted, crooked and with harder and heavier timber; *Theku* is found on deeper, richer gneissic soils (Kadambi, 1972).

In his preliminary studies on delimiting the provenance regions, Gopal (1972) suggested the classification of teak forests into 30 regions of provenance as a basis for creation of selected stands and seed orchards and provenance and progeny tests ultimately leading to production of certified seed.

Little experimental work has been carried out to investigate the inheritance of desirable characters. A set of experimental plots was established in 1931 in seven different localities with the object of studying the comparative behaviour of the important seed origins tried in each locality. This has given some useful information (Mathauda, 1954) though no conclusive results have been obtained.

Not all varieties have been recognized and their economic importance has not been assessed. Further, the various desirable qualities that some of these possess have seldom proved to be heritable. It is impossible to predict the behaviour of a particular seed origin in a new set of surroundings.

Germinative capacity and establishment

The germinative capacity has been found to vary within wide limits for seed from different areas but the percentage is generally higher for selected seed; the range of germination generally being 30–50 per cent, though it may be higher. The germinative capacity and plant per cent are 44 and 25 per cent respectively (Champion and Seth, 1968a) and *ca.* 660 plants are likely to be obtained from 1 kg of seeds (*ca.* 1 600).

Experience in Tamil Nadu has indicated that large seeds give a somewhat higher germination (Anonymous, 1956), but sorting out of seed is not economically justifiable. This factor influences vigour of the resulting plants. Possibly only large seed should be used for raising nursery stock. Seed from immature trees produces poor offspring; otherwise the age of mother trees has no effect. There is no difference between seed collected from plantations and natural forest.

Seed dormancy and pre-treatment

A considerable portion of the fresh seed remains dormant during the first year, whilst seed from dry localities has been found to show more persistant dormancy. Seeds are known to lie dormant in the natural forest for years (Troup, 1921).

The hard thick pericarp prevents easy germination especially in seeds from dry regions. A large number of presowing treatments aimed at softening or breaking of the pericarp through physical or physico-chemical action have been tried with varying success. A comparative study of the germination behaviour of five provenances of the species has indicated a differential response to presowing treatments. Whereas, Manantavady (Kerala) and Belgaun (Orissa) provenances responded best to nutrient treatment, Nandiyal (Andhra Pradesh) provenance required the removal of felty pericarp while the Top Slip (Tamil Nadu) provenance needed no presowing treatment (Gopal, Pattanath and Kumar, 1972).

Biomass and productivity

Data on biomass and productivity of teak forest ecosystems are wanting though plantation yield tables showing the growing stock and increment in terms of standard volume down to 5 cm diameter are available (Laurie and Sant Ram, 1940; FRI, 1964). The only study from which biomass and productivity of moist deciduous forests where teak has been introduced can be obtained is by Seth, Kaul and Gupta (1963) the results of which are presented in table 4.

A 33-year-old teak plantation has a biomass of >86 500 kg/ha (excluding underground parts) and 137 kg/tree; the productivity of non-photosynthetic biomass is >2 460 kg/ha/a and 4 kg/tree/a. Reference has been made to yield tables, which provide data for standing volume on 1 ha of fully stocked more or less even-aged crops on different site qualities. Table 5 shows timber (over bark diameter ⩾20 cm) volumes of plantation teak on site quality II at different ages (Champion and Seth, 1968a), while table 6 indicates the maximum standing volume/ha

TABLE 4. Biomass and productivity (oven dry weight) of a 33-year-old teak plantation at Dehra Dun.

Age (years)	33
Number of trees/ha	630
Biomass (kg/ha)	
Wood	67 500
Bark	13 709
Leaves	5 329
Total	86 538
Biomass (kg/tree)	
Wood	107
Bark	22
Leaves	8
Total	137
Productivity	
Non-photosynthetic biomass (kg/ha/a)	2 461
Non-photosynthetic biomass (kg/tree/a)	4

TABLE 5. Stand timber volume of teak even-aged crops (site quality II).

Age (years)	40	60	80
Timber volume (m³/ha)	115	186	241

TABLE 6. Maximum crop volume of teak even-aged crops.

Crop average after thinning	
Height (m)	36
Diameter (cm)	64
Age (years)	62
Volume before thinning	
Timber (m³/ha)	367
Timber and small wood (m³/ha)	381
Volume after thinning	
Timber (m³/ha)	305
Timber and small wood (m³/ha)	317

taken from sample plot records. Few large areas carry more than about half the full stocking possible, and most considerably less (Champion and Seth, 1968a).

Table 7 illustrates the rate of growth of even-aged crops (C-grade thinning) (Champion and Seth, 1968a). Maximum mean annual increment in even-aged crops is obtained at 50 years and *ca.* 70 years for site qualities I and II respectively.

TABLE 7. Annual increment in even-aged crops of teak.

Site quality	Maximum CAI* of timber (m³/ha)	Maximum MAI** of timber (m³/ha)	Age of maximum MAI of timber (years)
I	10.5	7.1	50
II	5.8	4	60–80

* Current annual increment.
** Mean annual increment.

Growing stock and increment in even-aged crops (site quality III, 80 years age) are shown in table 8 which also illustrates the great differences between mixed and more or less pure uneven-aged and pure even-aged crops (Mathauda, 1953, 1956, 1958; Champion and Seth, 1968a). The tropical evergreen forest carries the largest socking in numbers as well as basal area but it is not optimum for increment and diameter growth is slow in young and middle ages; only when the trees reach the emergent layer does the diameter increment accelerate. In the mixed moist deciduous forest the stocking is next highest in numbers but not in growing stock due to poor growth form of the trees as is evident by comparing it with uneven-aged or even-aged *Shorea robusta*. In comparable even-aged *Shorea robusta* and *Tectona grandis* crops (*ca.* 23–26 m dominant height at maturity) there is a progressive fall in growing stock and increment as compared to mixed uneven-aged stands, although the economic returns are perhaps higher due to the valuable nature of the timber.

TABLE 8. Growing stock and increment in different types of crops.

Type of forest	Number of trees/ha (>20 cm)	Basal area (m²/ha)	Growing stock stem timber (m³/ha)	Mean annual increment	
				Sem timber (m³/ha)	Stem timber and small wood (m³/ha)
Tropical wet evergreen (uneven-aged)	289	35.4	—	—	—
Tropical moist deciduous (uneven-aged)	167	14.7	77	2.3	3.4
Uneven-aged *Shorea robusta*	126	13.3	104	2.0	2.9
Even-aged *Shorea robusta* (125 years age, site quality III)	138	15.6	85	1.5	3.9
Even-aged *Tectona grandis* (80 years age, site quality III)	111	11.9	46	1.0	5.0

In view of the paucity of data on productivity, it is impossible to make any realistic estimates of biomass and productivity.

Water balance

Very little research work has been undertaken on water balance.

Interception studies carried out in 35-year-old plantations at Dehra Dun have indicated that 6, 73 and 21 per cent of rainfall were accounted for by stemflow, throughfall and interception respectively. For rainfalls of <50 mm the corresponding figures were 7, 64 and 29 per cent.

Rainfall amount and intensity have varying effects upon stemflow (Dabral, 1967): stemflow decreased with increased duration of rainfall. Sunshine influences the storage of rain on the foliage and evaporation and hence stemflow, especially during milder storms. Air temperature and humidity also influence stemflow.

Rainfall interception by leaf litter is governed by the amount of rain and its intensity as well as by the amount of litter. Litter induced 9 per cent interception; values were higher with lower amounts and intensities of rainfall and *vice-versa* (Dabral, Premnath and Ramswarup, 1963). In another study the litter interception was *ca.* 29 per cent (Pradhan, 1973).

Data on evapotranspiration are extremely meagre. Champion and Seth (1968a) computed the potential evapotranspiration for Dehra Dun using different formulae; see table 9. The calculated values differ widely during the hotter months. On the basis of one year's study at Dehra Dun to a soil depth of 122 cm under a 35-year-old teak plantation Dabral, Yadav and Sharma (1965) estimated evapotranspiration as *ca.* 840 mm.

TABLE 9. Monthly potential evapotranspiration at Dehra Dun (Champion and Seth, 1968a).

Month	Precipitation (mm)	Potential evapotranspiration (mm)		
		Thornthwaite	Rohwer	Leeper
January	59	19	25	29
February	63	26	24	24
March	32	65	62	40
April	17	120	124	57
May	37	177	165	79
June	217	179	120	110
July	668	166	39	131
August	731	147	36	130
September	270	121	45	112
October	32	85	59	67
November	9	43	32	37
December	26	24	22	21
Annual total	2 161	1 172	753	827

Data on soil moisture changes are not available. Seth and Khan (1960) have computed the general soil moisture regime at Dehra Dun according to Leeper's (1950) modification of Thornthwaite's climatic formula (table 10). In these calculations a maximum storage of 102 mm is assumed for a loamy type of soil. Studies (to 122 cm depth) were undertaken during 1961 and 1962 by Dabral, Yadav and Sharma (1965), indicated soil moisture accretion and depletion rates for every 30 cm depth.

TABLE 10. Soil moisture regime, Dehra Dun.

Month	Precipitation (mm)	Evapo-transpiration (mm)	Soil moisture (mm)	
June	217	144	+ 73	soil moisture
July	668	124	+102	recharge
August	731	117	+102	
September	270	112	+102	maximum storage
October	32	88	+ 46	soil moisture utilization
November	9	59	— 5	restricted
December	26	56	— 30	moisture
January	59	34	+ 25	soil moisture
February	63	41	+ 48	recharge
March	32	72	+ 7	soil moisture utilization
April	17	109	— 86	water deficiency
May	37	139	—102	

Champion and Seth (1968a) computed the monthly water budget for Dehra Dun according to Thornthwaite's scheme and showed a very good agreement with observed values so that the procedure can be used for diagnosing critical moisture conditions in relation to drought damage and mortality. Soil moisture storage for each soil layer and cumulative depths (to 120 cm), have also been indicated which are important from the point of view of seedlings which exploit shallower depths of soil.

Nutrient cycling

Very little work has been carried out on nutrient cycling. Seth, Kaul and Gupta (1963) reported an annual litter production of 5 930 kg/ha (table 11) in a 33-year-old teak plantation at Dehra Dun; leaf litter constituting nearly 90 per cent of the total. The high component of fruit

TABLE 11. Annual return of organic matter in a teak plantation.

Average number of trees/ha	630
Litter (oven dry weight, kg/ha)	
Leaf	5 329
Twigs	100
Fruits	500
Total	5 929
Leaf litter as per cent of total litter	89.8
Total litter/tree (kg)	9.4
Total leaf litter/tree (kg)	8.5

litter is also examplified. In another study carried out during 1961–62 (Dabral and Sagar, 1967) the average yearly litter in a 37-year-old plantation with 472 trees/ha was 4 541 kg/ha. The average leaf litter production in a *Tectona grandis* community of a tropical deciduous forest is 5.02±0.12 t/ha/a; the contribution of teak leaf litter being 95 per cent (Singh, 1968). The leaf litter production was 8.4 kg/tree/a. Leaf litter production figures from other parts of India are 6.1–17.3 t/ha/a (Singh, 1968).

Srivastava, Kaul and Mathur (1972) reported the monthly leaf litter production and its nutrient content in a teak plantation at Dehra Dun (table 12). Maximum leaf-fall occurs from February to July, prior to the onset of the monsoon.

TABLE 12. Monthly leaf litter-fall and return of nutrients in a teak plantation (kg/ha) (Srivastava, Kaul and Mathur, 1972).

Month	Leaf litter-fall	Nutrient content		
		N	P	K
January	198	2.4	0.4	0.7
February	621	6.6	0.9	1.9
March	1 318	12.9	2.0	3.7
April	2 488	17.2	3.0	10.0
May	673	5.2	0.9	2.2
June	221	1.7	0.3	1.2
July	718	8.4	1.2	4.2
August	221	2.4	0.3	1.4
September	396	3.8	0.5	2.0
October	428	4.5	0.6	2.1
November	257	3.1	0.4	1.5
December	150	1.9	0.2	0.8
Annual total	7 689	70.1	10.7	31.7

Rao, Dabral and Pande (1972) studied total litter production at four weekly intervals for four years at Dehra Dun and concluded that average yearly accumulation of litter was *ca.* 7 772 kg/ha, there being significant differences between years. Density and age of the crop have a significant effect on litter production (table 13).

There are no data on the contribution of understorey vegetation.

The main inorganic constituents of leaf litter have been reported by Seth, Kaul and Gupta (1963). The values on an oven dry weight basis for ash, nitrogen, phosphorus and potassium are 13.2, 1.0, 0.2 and 0.4 per cent respectively.

TABLE 13. Effect of stand density and age on litter production in teak plantations.

Age (years)	Stand density (trees/ha)	Average litter production (kg/ha)	Litter production for 100 trees	Source
35–37	472	4 541	962	Dabral and Sagar (1967)
39–43	666	7 772	1 167	Rao, Dabral and Pande (1972)

The annual return of these nutrients to the soil through leaf-fall is 705, 52, 11 and 20 kg/ha respectively. The main nutrients contained in the standing above-ground biomass of this plantation are given in tables 14 and 15.

TABLE 14. Mineral constituents of green leaf, wood and bark of a teak plantation at Dehra Dun (age, 33 years; percentage oven dry weight) (Seth, Kaul and Gupta, 1963).

Sample	Ash	N	P	K
Leaf	9.3	1.9	0.3	0.8
Wood	1.1	0.1	0.1	0.1
Bark	11.9	0.4	0.2	0.7

TABLE 15. Total quantity of standing nutrients in leaf, wood and bark of a teak plantation at Dehra Dun (age, 33 years; kg/ha) (Seth, Kaul and Gupta, 1963).

Sample	Total standing biomass	Ash	N	P	K
Leaf	5 329	497	102	14	41
Wood	67 500	770	88	88	88
Bark	13 709	1 628	55	21	101
Total	86 538	2 895	245	123	230

The nutrient concentrations of leaf litter as reported by Singh (1968, 1969) were: nitrogen, 0.8 per cent; phosporus, 0.2 per cent; and potassium, 0.2 per cent. On this basis, the annual release of nutrients in a *Tectona grandis* community of a dry deciduous forest is: nitrogen, 36 kg/ha; phosphorus, 8 kg/ha; and potassium, 20 kg/ha.

The monthly variation in the release of nutrients through leaf litter are presented in table 12 (Srivastava, Kaul and Mathur, 1972). As the maximum litter fall occurs from February to July, maximum return of nutrients is also confined to these months.

There is practically no data available for nutrient inputs through precipitation, dust, soil weathering and nitrogen fixation or for outputs through run-off, harvesting, etc. The rates of litter decomposition and nutrient release are poorly known.

Protection

Fire is a major cause of injury, especially in the drier forests. As teak regenerates profusely by sprouting and large trees are insulated by a fire-protective bark, the species is able to tolerate burning better than most of its associates. For the same reasons, it is possible to employ controlled burning in management. Losses from uncontrolled fire, however, can be excessive. Intense crown fires can destroy mature trees; hot ground fires can kill the phloem and cambium of younger thin-barked trees, causing wounds which succeeding fires enlarge until the butt log or the entire tree may be destroyed; intense ground fires in young plantations may kill the trees back to the ground. Fires also accelerate erosion and may cause epidemic defoliation (Kadambi, 1972). Fire protection measures are, therefore, followed in all plantations.

Heart-rot disease is caused by a non-decay fungus (*Phialophora* spp.) followed by decay fungi (*Fomes lividus* and *Polyporus zonalis*); up to 50 per cent of trees may show unsoundness. Other diseases are root-rot caused by *Polyporus zonalis* and wide-spread leaf rust caused by *Olivea tectonae* causing severe damage in young plantations (Bakshi, Reddy, Puri and Singh, 1972).

There are periodic outbreaks of two defoliators, *Pyrausta machaeralis* (teak skeletonizer) and *Hyblaea puera* where the leaves of the trees are seriously damaged. The loss has been estimated to be *ca.* 13 per cent of the annual increment. Complete defoliation of young plants may cause shoot die-back and a change in the feeding habits of the larvae from leaf to boring of tender growing leading shoots which results in forking of the main stem. Repeated annual defoliation is believed to stimulate epicormic branching. *Sahyadrassus malabaricus* (shoot borer) occurs in southern India and is injurious to young plantations (Beeson, 1941; Mathur, 1964; Chatterjee and Sen Sarma, 1968).

Loranthus is regarded as a serious problem in parts of Maharashtra State, but occurs all over the country. Regular eradication measures are undertaken in most plantations by lopping infected branches and cutting out of heavily infested trees during thinning. Preliminary investigations indicate that it could probably be controlled by injecting selective phytocides such as copper salts (Anonymous, 1956).

Teak is not readily browsed. Heavy grazing is, however, harmful in regeneration areas because the young shoots are easily broken or trampled by cattle.

Damage by elephants, gaurs, deer, porcupines, pigs, monkeys, hares, rats, etc., have been reported. Wild elephants may break down and uproot seedlings and saplings in plantations, strip off the bark during the dry season and uproot young teak trees to feed on tender branches. Gaurs and deer may browse tender seedlings, strip off bark, or break off the leading shoots.

Frost is the most probable factor limiting its distribution and it causes varying degree of damage. Certain races are regarded to be more frost hardy; of the various seed origins tried at Dehra Dun, Burma was found to be the best from this aspect.

Teak is sensitive to severe droughts and although it often recovers well, the damage may be fatal, especially in young plantations.

Teak is wind-firm owing to its well developed tap-root and subsidiary root system, although the 1938 cyclone damaged plantations at Puri (Orissa) very badly (Kadambi, 1972).

Management

On account of its high value and light demanding character there is a pronounced tendency to manage teak as a pure even-aged crop. There are, however, still vast areas where it occurs in uneven-aged mixtures. In mixed forests the predominant trend is to increase its proportion by selective cuttings or by raising pure plantations.

Natural regeneration

The comparatively drier types of deciduous forests fall under two broad categories on the basis of natural regeneration: the moister type, where the canopy is complete or nearly so, and teak forms but a small proportion of the crop; tall, well-formed scattered clumps of *Bambusa arundinacea* are generally present; natural regeneration of teak is poor or wanting; and the drier type, where teak is generally associated with *Anogeissus latifolia* and the two form a high proportion of the crop; *Dendrocalamus strictus* is the characteristic bamboo, and the ground is well illuminated under a fairly open canopy, but not so as to allow grass; the natural regeneration of teak is generally adequate, if not often abundant, e.g. various forests of central India (Madhya Pradesh) and the dry types of moist deciduous forests of Tamil Nadu, Karnataka, and Maharashtra.

Natural regeneration of teak has received little attention. Regeneration is profuse in some localities; it fails to do so in others. The problem is complicated by some easy regeneration outside its natural range, e.g., Gorakhpur Division (Uttar Pradesh) and Puri Division (Orissa). The ease with which the tree can be regenerated artificially, its hardiness and tenacity once it has established itself, and it being perhaps the first forest tree in India to be established on a large scale in plantations, have resulted in comparatively little attention being paid to its natural regeneration.

There is, generally, a sharp alternation of dry season and wet season. This is beneficial to its regeneration because the fruit which is shed at the beginning of the dry season weathers in that season, soaks in the premonsoon showers of April-May, dries out in the monsoon break that follows and finally soaks again in the monsoon. However, if the break between the premonsoon and monsoon is too long the seed, whose germination would start in the premonsoon rain, will die. The hard seed coat beneath the spongy layer of the seed protects it from premature germination during premonsoon rains. If the first, good premonsoon showers are followed by a long dry spell, hardly any germinating seedlings can survive.

The variety of rock formations and resultant soils indicate that it is the physical structure of the soil, more especially its porosity, that is responsible for success of natural reproduction. Granite, gneiss and trap produce moderately favourable soils; basalt and sandstone develop soils that are adequately porous, but on the latter humus disintegrates too rapidly. Teak regeneration does not thrive on laterite because humus is rapidly leached out.

Teak usually seeds abundantly, but the floor of a natural teak forest of the moist type often has a dense undergrowth which prevents the radicle from reaching the soil. Excessive opening of the forest canopy to secure regeneration may stimulate the growth of grass and other weeds which smother many of the seedlings.

Some warmth is required for germination; in cool shady places the seeds may lie dormant for years. Sowing seed in shaded and exposed plots resulted in 1 and 17 per cent germination respectively. When light was admitted to the shaded plot after two years, the dormant seeds germinated.

In natural forests, the annual ground fires scorch the seeds and facilitate germination by removing a portion of the seed coat. In a burnt plot in Betul Division (Madhya Pradesh) there was 26 per cent new seedlings against none in the unburnt plot. Fire is beneficial in heavy rainfall areas only. It is unfavourable in low rainfall areas as it destroys humus and makes the ground surface hard.

The establishment of a teak seedling depends especially on light. In moister localities the seedling must have light from its beginning, but in dry, hot localities, e.g., in parts of Madhya Pradesh, a sudden influx of light on young seedlings may cause their death for it requires protection against the sun. Drip from trees adversely affects the young plants. The very early development of a strong tap-root requires a soft friable soil or the seedling will remain weak and small and incapable of surviving the ensuing dry season.

Seedlings rot in excessively moist situations. In dry forests drought is one of the principal enemies to establishment. Soil working has a marked beneficial effect on the development of young plants, and in dry localities, this is an excellent method of conserving soil moisture and enabling the seedlings to survive. Excessive rains and grazing when teak seedlings are very young, also prevent the establishment of regeneration. However, limited grazing may help in keeping down heavy grass and thus favour reproduction.

In the moister type of teak forests indiscriminate fire protection has an a adverse effect on reproduction by encouraging weeds. The forests also progress to the evergreen type. In the dry type of teak forests, the seasonal fires are harmful and may kill the seedlings. Light burning, however, is considered advantageous.

Seedling shoots are killed by frost, but the root-stock is not killed and a stronger shoot is formed each season until it is able to rise above the level of frost.

The various kinds of bamboo often play a vital role in the natural reproduction of teak. Little regeneration survives under vigorous bamboo stands. Gregarious flowering affords an opportunity for establishment of teak from dormant seed. Cutting and burning the bamboo may provide similar conditions. Cutting or burning of young competing bamboo is necessary to ensure survival of a satisfactory stand of teak.

The status of natural regeneration in different moisture types of teak forests could be summarized as follows (Seth and Khan, 1958).

Very dry type. These forests receive the minimum possible rain for teak growth. The soil layer is thin (and usually murramy) with rock outcrops. The annual fires and grazing largely preclude natural regeneration. Ground cover is very scanty.

Dry type. Fires and grazing are serious but the greater rainfall is more favourable to natural regeneration. Stagnating seedlings usually occur in sheltered places. Effective protection results in a gradual progression towards a moister type. The amount of regeneration increases with the increase in moisture conditions and depth of soil. Ground cover is patchy.

Semi-moist type. These areas are much less burnt than the dry type, although grazing continues to be fairly heavy. The chief effect of protection from fire is the survival and establishment of some tree and shrub seedlings which grow to maturity. Whilst the canopy is open sufficient light reaches the ground to permit the development of an understorey of bamboo, *Lantana*, *Petalidium* spp., etc. Teak seems to regenerate with ease in the moister areas. Moderate growth of bamboos eliminates grass and other herbs of which teak is intolerant; teak seedlings survive under moderate bamboo shade for a considerable time and then grow rapidly when gaps appear in the canopy. Thus the second storey plays an important role in the natural regeneration of teak; it is inhibitory only when it is too dense. Generally speaking, in this type, the ground cover is fair and natural regeneration of teak is fairly adequate.

Moist type. This is well protected from fire and grazing is low. Towards the drier end the understorey is suitably dense but not too dense to favour the abundance of natural regeneration of teak provided the soil is not heavy. It, therefore, appears that as the upper canopy starts closing up and before bamboo or other shrubby growth becomes dense, the conditions are most favourable for natural regeneration of teak. In the more humid part of this type, where the overwood is comparatively denser, canopy manipulation does increase regeneration but the progress is slow and cannot be relied upon. As a rule, the ground cover is dense and adequate natural regeneration occurs only in patches.

Very moist type. In this type, fires have been excluded for a long period. This prolonged protection from fire has resulted in a dense understorey of semi-evergreen shrubs which inhibit the reproduction of teak and its valuable associates. In addition, the thick undergrowth prevents the teak seedlings from growing. Also the understorey consists of a dense mass of bamboos which seriously decreases the amount of light. Consequently, this type is particularly devoid of natural regeneration, and presents the greatest difficulty in achieving it. The percentage of teak trees is generally quite low.

Regeneration techniques and management practices

The successful artificial regeneration of teak in the moist types of forests results in hardly any reliance being placed on natural regeneration in some States (Kerala, Tamil Nadu, Maharashtra, etc.). Planting is the regular method of re-stocking the forests after exploitation. There is, however, little doubt that natural regeneration would be much more advantageous, were it practicable in a reasonable time and with reasonable costs. The method of artificial regeneration, involving clear-felling and planting, results in considerable loss of soil and soil fertility. This could be eliminated by natural regeneration, which would also result in greater mixture of species and the retention of the natural character of vegetation thereby reducing the risks to which mono-cultures are open.

Artificial regeneration cannot, however, be used over the greater and drier part of the teak zone for various reasons where there is no alternative to natural regeneration.

The methods of restocking may be divided into three categories:

— methods of natural regeneration such as coppice or clear-felling with conversion to uniform, or selection, including improvement fellings;[1]
— methods of aided natural regeneration, with provision to supplement stocking by planting;
— methods of artificial regeneration with no dependence on natural regeneration.

Very dry type. In this type there is the heaviest demand for fuel, small constructional timber and grazing. As natural regeneration is absent, these forests are managed under some coppice system. In Maharashtra and Madhya Pradesh, the system adopted is coppice-with-standards/reserves, the most common rotation being *ca.* 40–50 years. Areas with adequate advance growth are regenerated in one operation, well-grown patches of poles being retained; whereas areas unfit to be regenerated naturally are worked under improvement fellings. In this manner, the failure of natural regeneration is guarded against. In Maharashtra, the system employed is coppice-with-standards, with planting in patches. Rotation varies from 30 to 60 years. In steep eroded and degraded areas only light improvement fellings, amounting to protection, are carried out.

Dry type. This type covers extensive areas and natural regeneration is usually inadequate and occurs in patches only in sheltered places. In a few areas in Madhya Pradesh and Maharashtra the usual system is high forest with concentrated regeneration fellings, except in areas bordering settlements and low quality forests, where the coppice-with-reserves system is adopted. Such areas are usually of small extent. Here again the sylvicultural system takes advantage of the presence of advance growth. The major part of areas of this type are worked under coppice-with-standards/reserves systems. The relatively moister forests are worked

1. Clear-felling system where the crop is clear-felled by compartments or sub-compartments; regeneration is usually artificial.
   Uniform system where the felling is uniform over a compartment or sub-compartment; regeneration mainly natural, often largely assisted by artificial regeneration; even-aged crop.
   Selection system where trees are removed singly in selection fellings; regeneration is natural and the type of resultant crop uneven-aged.
   Improvement fellings system which consists of removal of inferior growing stock of all ages and tending the better elements of the crop; regeneration is natural and the resultant crop uneven-aged.
   Coppice system where the crop is removed by compartments by clear-felling, with no reservation of a shelterwood; regeneration is of coppice origin often assisted by artificial regeneration in strips, lines or large patches; the resultant crop is even-aged.
   Coppice-with-standards system in which the part of the crop is reserved to form an uneven-aged overwood; regeneration is mainly of coppice origin.
   Coppice-with-reserves system where part of the crop (sapling and poles) is reserved singly or in groups for part or whole of the second rotation; regeneration is mainly of coppice origin; underwood even-aged and overwood uneven-aged.

under improvement fellings systems. Artificial methods of restocking are also resorted to, but only to a limited extent.

Semi-moist type. Natural reproduction occurs rather frequently in this type, and the system largely adopted is the high forest with concentrated regeneration fellings. Planting is practised but only over limited areas very deficient in natural regeneration in order to restock the annual coupes completely. Some of these forests, having inadequate natural reproduction and occurring on erodible terrain, are worked under selection-cum-improvement fellings systems.

Moist type. Here natural regeneration is adequate on the well-drained soils and clear-felling, followed by intensive cultural operations results in a fully stocked second-growth forest. Artificial methods of restocking, however, have generally to be resorted to in order to restock the areas fully. Thus, clear-felling is adopted with conversion to uniform. Occasionally, preparatory cultural operations, consisting in the clear-felling of bamboos, are also carried out.

Very moist type. Due to the rank vegetation of semi-evergreen nature, natural regeneration is completely wanting, except locally in the zone merging with the moist teak type. Artificial regeneration is the only remedy, although it is costly and may be only partially successful. Selection-cum-improvement fellings, with planting restricted to a small suitable area, is the usual system of management.

Research needs and priorities

Available evidence suggests an increasing need for wood for any foreseeable future. This is specially true of a prized species like teak (covering nearly 12 per cent of the forest area of India) which is of importance in the national economy and by reason of its high price and multiple use, even a slight improvement in the quantity and quality of production will have a significant effect. It thus seems prudent to consider how best these demands can be met in the light of the pressures on forest resources and competition for land between forests and other uses, without deterioration of the environment and with rational management of the resource. As a result, great attention is to be paid to the production of maximum quantity of teak timber of all sizes, of the highest quality which the sites are capable of producing, from all existing and potential teak areas of the country, in the shortest possible time consistent with sound management and environmental stability.

The following broad fields of research are outlined for these forests. They only indicate some major priority trends and are by no means exhaustive.

1. In terms of long-term planning and management strategies to be adopted for the rational use of teak forests, detailed studies on their typology have to be undertaken along with soil-vegetation surveys to provide basic data. The classification into forest types and teak sub-types has to be based on climatic, floristic, edaphic and biotic characters and attempts have to be made to indicate some definite relations with appropriate sylvicultural and management practices. It is also essential to study the synecology and autecology of the species in natural as well as artificial

communities, in order to evolve suitable sylvicultural techniques and management practices for better management, regeneration, utilization and development of teak forests while at the same time maintaining and improving the site. Teak soils need to be studied in detail to correlate their properties with the suitability for growing teak; to evolve reliable indices, based on easily ascertainable coefficients relating to the soil, the soil moisture and the climate for predicting the quality expected to be attained and evolving suitable techniques for correcting soil deficiencies.

2. Sufficient emphasis has not been placed on the proper assessment of factors governing *natural regeneration*, even in areas where natural reproduction is being relied upon as the principal method of regenerating the forests. There are large areas which could possibly be regenerated by natural means, if suitable techniques were evolved. Detailed studies and experiments have to be carried out in semi-moist teak forests and in the suitable parts of dry and moist types, to determine the conditions for inducing natural regeneration and to work out techniques suitable for large-scale application. This would also result in great mixture of species and the retention of the natural character of vegetation thereby reducing the risks of monocultures.

Very little information is available on the physiology of teak in order to appraise the effects of various factors like light, soil air, soil moisture, nutrient availability, etc., on its regeneration and growth. Such physiological investigations need immediate attention.

Teak forms characteristic associations with some species of bamboo, and the latter is also a crop of rapidly growing economic importance. In view of the great ecological significance of these associations, especially in relation to natural regeneration, preservation of soil fertility, etc., detailed studies need to be carried out both in natural communities and in teak plantations, in order to make use of this knowledge in scientific management.

In some types of teak forests there is a dearth of undergrowth; in other places weeds like *Lantana, Eupatorium, Petalidium, Imperata* spp., *Sorghum halepense*, etc., invade to the almost total exclusion of other species, while elsewhere a mixed understorey develops naturally. The effects of different types of natural or introduced undergrowth on the soil and development of teak need to be investigated in detail with a view to determine the conditions which favour the development of particular types of undergrowth, and desirable mixtures of understorey species to be encouraged or introduced in teak areas.

3. Teak suffers from a number of diseases and pests which cause considerable damage, through outright killing and reducing quality and usefulness of the wood produced. Research to determine the cause, evaluate the status of pests and diseases, determine actual and potential threats, assess damage, study factors leading to epidemics and take timely action to organize control measures and maintain the health of forest, is necessary.

4. As clear-felling and planting is the sylvicultural system in vogue in extensive areas of teak forests, intensive research on the ecological requirements of the species, physiological aspects and forest genetics and tree improvement is needed for commercial planting to ensure faster rate of growth, better wood properties and resistance to diseases and pests. Genetical and tree improvement research would consist of classification of various provenances of teak according to important features, selection, propagation, testing and finally production of seed and plants of the improved varieties. As an interim measure seed production areas need to be established in the best stands of teak.

5. It is also necessary to maintain large areas of teak forests as biosphere reserves for scientific, educational and aesthetic purposes; to serve as control points against which the effects of other practices can be judged and to serve as gene reservoirs. There are at present only 37 (32 in natural forests and 5 in plantations) small preservation plots throughout the nearly 9 million ha of teak forests.

6. Wide-scale modification or destruction of tropical forests including teak forests has become a matter of scientific concern. The capacity of an ecosystem to withstand modification, adaptation and change needs to be known. It is also necessary to know as to how disturbed ecosystems can be repaired.

The impact of land-use alternatives on the production and fertility of teak forest ecosystems needs investigation. This would include studies on biological productivity and budgets of minerals and materials critical to productivity and stability in natural undisturbed, partially disturbed and man-made ecosystems; impacts of clear-cutting on soil structure, nutrient retention, regeneration capacity, run-off, micro-climate and subsequent land-use opportunities, and long-range effects of different forestry practices and agricultural uses including clear-felling, partial felling, and teak monocultures.

Studies on the effects of human settlements on teak forest ecosystems, would focus attention on the pressing problem created by existing or proposed settlement or development schemes. The epidemiological impacts of teak forest land usage on man, domesticated animals and agriculture pests should also be studied.

The effects of manipulation of teak forests on the socio-cultural and behavioural characteristics of human population living there need to be studied, as also the demographic changes that are taking place and the relations between these changes and the manipulations of forest areas. This would also involve studies on the immigration of population into altered tropical environments (monoculture, etc.) as well as emigration of population as a result of deterioration of the habitat.

7. Within the various research approaches, simulation, optimization and prediction models may be developed as an integral part of the work plan. These models could be used to describe the cycles and budgets of nutrients and to help devise procedures for minimizing the nutrient loss.

Conclusions

The systematic management of teak forests according to the present pattern of development will lead to profound floristic and ecological changes in these ecosystems. As teak is very easy to raise artificially, and mostly manual means are used thus facilitating its planting even on moderate gradients, it will be more easy to raise the species in the more equable part of its range. It is visualized that, if demand for teak timber continues as at present, most areas will be converted into teak plantations and only isolated pockets of natural vegetation will be left. These pockets will not be representative of the original forest because they will either occupy inhospitable terrain or suffer from some other limitation to prevent their being converted into plantations.

At the two extreme ends of its distribution the changes will tend towards elimination rather than concentration of teak. In the very dry teak forests, elements of thorn forests will gradually gain preponderance because of heavy grazing and frequent fires. They will also not be normally planted due to low survival. The shallow and infertile nature of the soil will also preclude much plantation activity in these areas. At the very moist end of its range, the tendency will be to gradually exploit the species on a selective basis and as most of these areas will be unfit for conversion into pure teak crops, teak may form a progressively decreasing element in the floral complex and may ultimately disappear altogether.

Thus there will be a contraction in the range of teak, but there will be an increase in its concentration in the core areas where conditions are conducive for it to be planted and grown relatively easily and at high economic benefit.

The possible ecological repercussions of this large-scale conversion of natural teak forest ecosystems into monocultures of high economic value are hard to predict in the absence of experience. The understorey, especially the undergrowth characteristic of the area, will invade the initial plantations. However, when plantations form continuous belts of teak monocultures, it is quite likely that the undergrowth may also undergo a significant change; possibly aggressive exotics like *Lantana, Eupatorium, Mikania* spp., etc., will become increasingly conspicuous elements and may profoundly change the floristic composition of the undergrowth and affect teak regeneration.

The consequent changes of these monocultures on the soil are hard to predict, but soil fertility may deteriorate leading even to lateritization. Soil morphology and physicochemical properties of the soil will definitely undergo a change and there could be changes in profile development also.

The long-term effects of teak introduced outside its natural range will be similar to those which might take place under plantations of exotic origin and may even lead to an accelerated change in the biogeocenotic features of the areas.

Bibliography

ANON. *National progress report on teak, India*. First Session, Teak Sub-Commission (FAO, Bangkok), 1956, 20 p.

——. *Proceedings of All India Teak Study Tour and Symposium*. Dehra Dun, Forestry Research Institute (FRI), 1958, 196 p.

——. Indian forest and forest products terminology. Part I. Forests. *Indian For. Rec. (new series), Silviculture*, vol. 10, no. 6, 1960, p. 125–215.

BAKSHI, B. K.; REDDY, M. A. R.; PURI, Y. N.; SUJAN SINGH. *Forest disease survey*. Dehra Dun, FRI, 1972, 117 p.

BEESON, C. F. C. *The ecology and control of the forest insects of India and the neighbouring countries*. Dehra Dun, FRI, 1941, 767 p.

BHATIA, K. K. *Factors in the distribution of teak (Tectona grandis Linn.) and a study of teak forests of Madhya Pradesh*. Saugar University, Madhya Pradesh, Ph. D. thesis, 1954, multigr.

CHAMPION, H. G.; SETH, S. K. *A revised survey of the forest types of India*. Delhi, Manager of Publications, 1968, 404 p.

——; ——. *General silviculture for India*. Delhi, Manager of Publications, 1968a, 511 p.

CHATTERJEE, P. N.; SEN SARMA, P. K. Important current problems of forest entomology in India. *Indian Forester*, vol. 94, no. 1, 1968, p. 112–117.

DABRAL, B. G. Preliminary observations on the stemflow behaviour in teak. *J. Soil and Water Conservation in India*, vol. 15, no. 1 and 2, 1967, p. 91–97.

——; PREMNATH; RAMSWARUP. Some preliminary investigations on the rainfall interception by leaf litter. *Indian Forester*, vol. 89, no. 2, 1963, p. 112–116.

——; YADAV, J. S. P.; SHARMA, D. R. Soil moisture studies in chir pine, teak and sal plantations at New Forest, Dehra Dun. *Indian Forester*, vol. 91, no. 10, 1965, p. 701–713.

——; VIDYA SAGAR. Litter studies in forest plantations at New Forest. In: *Proc. 11th Silvicultural Conference* (FRI, Dehra Dun), vol. 2, 1967, p. 365–371.

DABRAL, B. G.; SUBBA RAO, B. K. Interception studies in chir and teak plantations, New Forest. *Indian Forester*, vol. 94, no. 7, 1968, p. 541–551.

FOREST RESEARCH INSTITUTE (FRI, Dehra Dun). Yield and stand tables for plantation teak (*Tectona grandis* Linn.). *Indian For. Rec. (new series), Silviculture*, vol. 9, no. 4, 1964, p. 151–216.

GOPAL, M. Delimiting regions of provenance of teak for seed improvement and certification. In: *Proc. and Tech. Papers Symp. man-made forests in India* (Society of Indian Foresters, Dehra Dun), 1972, p. III B17–III B22.

——; PATTANATH, P. G.; KUMAR, A. A comparative study of germination behaviour of *Tectona grandis* of some Indian provenances. In: *Proc. and Tech. Papers Symp. man-made forests in India* (Society of Indian Foresters, Dehra Dun), 1972, p. III B22–III B27.

GRIFFITH, A. L.; GUPTAL, R. S. Soil in relation to teak with special reference to laterisation. *Indian For. Bull.*, no. 141, 1947, p. 1–57.

JAIN, N. K. Physical characters and output of sal seeds. *Tropical Ecology*, vol. 3, no. 1 and 2, 1962, p. 133–138.

KADAMBI, K. *Silviculture and management of teak*. School of Forestry, Stephen F. Austin State University (Nacogdoches, Texas), Bull. 24, 1972, 137 p.

KAUL, O. N.; GUPTA, A. C.; NEGI, J. D. S. Diagnosis of mineral deficiencies in teak (*Tectona grandis*) seedling. *Indian Forester*, vol. 98, no. 3, 1972, p. 173–177.

KEDHARNATH, S.; MATTHEWS, J. D. Improvement of teak by selection and breeding. *Indian Forester*, vol. 88, no. 4, 1962, p. 277–284.

KRISHNASWAMY, V. S.; MATHAUDA, G. S. Phenological behaviour of a few forest species at New Forest, Dehra Dun. *Indian For. Rec. (new series), Silviculture*, vol. 9, no. 2, 1954, p. 89–134.

LAURIE, M. V.; BAKSHI SANT RAM. Yield and stand tables for teak (*Tectona grandis* Linn.) plantations in India and Burma. *Indian For. Rec. (new series)*, *Silviculture*, vol. 4-A, 1940, no. 1, p. 1–115.

MATHAUDA, G. S. The tree species of the tropical evergreen Ghat forests of Kanara (Bombay) and their rate of growth. *Indian For. Bull.*, no. 169, 1953, p. 1–23.

——. The All-India teak seed origin sample plots. *Indian For. Bull.*, no. 177, 1954, p. 1–14.

——. The constitution and rate of growth of a tropical moist deciduous forest in South Chanda Division, Madhya Pradesh. *Indian For. Rec. (new series)*, *Silviculture*, vol. 9, no. 3, 1956, p. 135–150.

——. The unevenaged sal forests of Ramnagar Forest Division, U.P. *Indian Forester*, vol. 84, no. 5, 1958, p. 255–269.

MATHUR, K. B. L. *Teak bibliography*. Dehra Dun, FRI, 1973, 320 p.

MATHUR, R. N. Forest entomology. *Indian J. of Ent.*, 25, 1964, p. 437–455.

MINISTRY OF AGRICULTURE. *Progress report, 1966–72*. Tenth Comm. For. Conf. (London), 1974, 41 p.

PRADHAN, I. P. Preliminary study on rainfall interception through leaf litter. *Indian Forester*, vol. 99, no. 7, 1973, p. 440–445.

RAO, B. K., SUBBA; DABRAL, B. G.; PANDE, S. K. Litter production in forest plantations of chir (*Pinus roxburghii*), teak (*Tectona grandis*) and sal (*Shorea robusta*) at New Forest, Dehra Dun. In: Golley, P. M.; Golley, F. B. (eds.). *Tropical ecology with emphasis on organic production*, p. 235–243. Athens, University of Georgia, 1972, 418 p.

SEN GUPTA, J. N. Seed weights, plant per cent, etc., for forest plants in India. *Indian For. Rec. (new series)*, *Silviculture*, vol. 2, no. 5, 1937, p. 175–221.

SETH, S. K.; KHAN, M. A. W. Regeneration of teak forests. In: *Proc. All India Teak Study Tour and Symposium* (FRI, Dehra Dun), 1958, p. 107–120.

——; YADAV, J. S. P. Teak soils. *Indian Forester*, vol. 85, no. 1, 1959, p. 2–16.

——; KHAN, M. A. W. An analysis of soil moisture regime in sal (*Shorea robusta*) forests of Dehra Dun with reference to natural regeneration. *Indian Forester*, vol. 86, no. 6, 1960, p. 323–335.

——; KAUL, O. N.; GUPTA, A. C. Some observations on nutrient cycle and return of nutrients in plantations at New Forest. *Indian Forester*, vol. 89, no. 2, 1963, p. 90–98.

SINGH, K. P. Litter production and nutrient turnover in deciduous forests of Varanasi. In: *Proc. Symp. on Recent advances in tropical ecology*, 2, 1968, p. 655–665.

——. Nutrient concentration in leaf litter of ten important tree species of deciduous forests at Varanasi. *Tropical Ecology*, vol. 10, no. 1, 1969, p. 83–95.

SRIVASTAVA, P. B. L.; KAUL, O. N.; MATHUR, H. M. Seasonal variation of nutrients in foliage and their return through leaf litter in some plantation ecosystems. In: *Proc. and Tech. Papers Symp. man-made forests in India* (Society of Indian Foresters, Dehra Dun), 1972, p. III 66–III 71.

TROUP, R. S. *The silviculture of Indian trees*. 3 vol. London, Oxford University Press, 1921, 1 195 p.

The forest ecosystems of Malaysia, Singapore and Brunei: description, functioning and research needs

by T. C. Whitmore[1]

1. Commonwealth Forestry Institute,
   South Parks Road, Oxford OX1 3RB,
   United Kingdom.

Introduction

There are no up to date maps of the vegetation. The best older maps are those of Malaysia by Van Steenis (1958a) and of W. Malaysia by Wyatt-Smith (1964); both have explanatory notes. Some details of the distribution and extent of the various forest types were given in chapter 1. See also Ashton (1976).

A number of regional floras are in preparation. Two of the four volumes of the *Tree flora of Malaysia* (Whitmore, 1972a, 1973a) have been published as well as a book on its palms (Whitmore, 1973b). The Dipterocarpaceae of Sarawak, Brunei and Sabah have been described (Meijer and Wood, 1964; Ashton, 1965, 1968). Manuals of the main trees of Sabah are being prepared.

The species composition of the forests broadly reflects the regional geological history.

W. Malaysia has an extremely rich flora and fauna; the rain forests here are more complex and species-rich than anywhere else. The flora becomes progressively more depauperate eastwards and southwards (Whitmore, 1969). The most important difference is that Dipterocarpaceae are the dominant family in most lowland forest formations in the Sunda region whereas they are only of very local occurrence in the Sulu region. Another example is the climbing palms or rattans (subfamily Lepidocaryoideae) which have a strongly Sundaic centre of diversity and richness, with only a slight extension to the Sulu region (Whitmore, 1973b). Nearly all monographs of large plant genera show a distinction between Sundaic and Sulu groups, as do the mammals except for the bats (Medway, 1972b).

Major forest formations

Formations differ from each other in structure, physiognomy and floristic composition. Despite floristic differences, the same formation can generally be recognized from its structure and physiognomy in different parts of the tropics. Different formations occupy different habitats; they may have sharp boundaries or diffuse ecotones.

The formations can be grouped according to habitat and arranged roughly into a hierarchy as shown in table 1 (Whitmore, 1975). Availability of water plays the most important role in this hierarchy. An important classification for tropical and subtropical Asia was introduced by

TABLE 1. The main forest formations of the tropical Far East (Whitmore, 1975).

Formation		Climate	Soil water	Localities	Soils	Altitude (m)
Tropical rain forests	lowland evergreen rain forest	everwet	water-table not high	inland	zonal	1 200
	lower montane rain forest					(750) 1 200–1 500
	upper montane rain forest					(600) 1 500–3 000 (3 350)
	subalpine forest					3 000 (3 530) to tree line
	heath forest				podzolized sands	mostly lowland
	forest over limestone				limestone	mostly lowland
	forest over ultra-basic rocks				ultra-basic rocks	mostly lowland
	beach vegetation			coastal		
	mangrove forest		high water-	salt water		
	brackish water forest		table (at	brackish water		
	peat swamp forest		least	freshwater	oligotrophic peats	
	freshwater swamp forest		periodically)		eutrophic almost permanently wet	
	seasonal swamp forest				eutrophic periodically wet	
	semi-evergreen rain forest	seasonally dry	moderate annual shortage			
Monsoon forests	moist deciduous forest		marked annual shortage			
	other formations of increasingly					
	dry seasonal climates					

Champion (1936). Burtt Davy (1938) extrapolated it to Africa and his types were used by Symington (1943) in his masterly account of the forests of Malaya.

Tropical lowland evergreen rain forest

This is the most extensive lowland rain forest formation. It occurs in two great blocks which more or less coincide with the Sunda and Sulu shelves. The western block is the dipterocarp rain forest block. It covers virtually all Malaysia, Kalimantan (Borneo) and Sumatra and is patchy on south and west Java. The eastern block is centred on New Guinea.

Montane forest types

Considerable thought has been given to the factors which determine the altitudinal changes in forests. Three types are widely represented and sufficiently distinct to warrant formation rank: lower montane rain forest, upper montane rain forest and subalpine forest. The old simplistic view that decreasing temperature or increasing cloud were the determining factors are now realized to be inadequate; see Flenley (1974) and Whitmore (1975). Several factors can cause the diminished stature, biomass, leaf size, increased stunting and gnarling, peat accumulation and epiphytes. It is necessary to make sophisticated observations on climate, radiation balance, water balance and mineral status in order to decide which factors are at work.

In the series of montane rain forest formations there is some understanding of why structure and physiognomy alter with environment. Parkhurst and Loukes (1972) constructed a theoretical model of optimum leaf size and predicted its form in different environments: their prediction more or less fitted observation in tropical rain forest formations.

Heath forest

Heath forest is very distinct. It is found throughout the region on coastal sand terraces. Only in Borneo is it extensive inland, where it occurs on flat bedded Tertiary sandstones. The soil is siliceous sand, usually freely draining and nearly always podzolized. It has been extensively degraded by burning, grazing and cultivation to open savanna or low scrub.

Heath forest has been very thoroughly studied by Brünig (1968, 1969b, 1971, 1974) in Sarawak. He suggests that a major factor determining its occurrence is marked seasonal water stress, although he does not dismiss the earlier idea that oligotrophy was the cause. The practical relevance of Brünig's work is that their important timber tree is *Agathis* which is only locally present. If foresters can solve the problems of growing trees on such sites (Brünig, 1969a) they have great economic possibilities because there are no competing land uses.

Forest over limestone

Limestone is of limited extent, its flora is distinctive and the vegetation shows signs of seasonal water stress. The flora is rich in endemics and many species are of horticultural interest. A Unesco project studied the Southeast Asian limestone, and expeditions were run in both Malaysia and Sarawak, which served to train several young regional botanists.

Anderson (1965) described limestone habitats in Sarawak, whilst Chin (1973) has produced a detailed account of its flora in W. Malaysia.

Forest over ultra-basic rocks

Forest over ultra-basic rocks is of limited extent and of very diverse composition: in W. Malaysia and near Ranau, Sabah, it is only floristically distinct from evergreen rain forest, whereas on the slopes of Mt. Kinabalu it is a low scrub of distinct structure and physiognomy as well as flora.

Mangrove forest

Mangrove forest has been very fully described by Watson (1928), Van Steenis (1958b), McNae (1968) and Berry (1972). The physiological mechanisms which enable mangroves to tolerate salt water conditions have been investigated by Scholander *et al.* (1962), whilst Van Steenis (1958b) pointed out that mangroves are only facultative halophytes.

Mangrove has long been intensively managed for poles, fire-wood and charcoal, and cutch for tanning. Watson (1928) includes this aspect for Malaysia, from where there are numerous Forest Departmental reports. Its importance has declined with the advent of kerosene and other tanning agents. But in 1973 mangrove in east Sabah was clear-felled for wood chips for rayon manufacture. Careful control of felling intensity is needed to ensure regeneration.

Peat swamp forest

Amongst the lowland forest formations on wet soils, peat swamp forest is the most studied. The early work is reviewed and incorporated in the very full studies by Anderson (1958, 1961, 1963, 1964a, 1964b) in Sarawak, where peat swamp forest occupies 12.5 per cent of the land. He showed that the concentric series of forest types on domed surfaces several km across represents a time series, as peat swamp forest develops behind advancing mangrove forest. The communities become more stunted inland due to increasing oligotrophy and periodic water stress. From the Kapuas to the Tutong rivers in northwestern Borneo several of the peat swamp forests are dominated by the gregarious *Shorea albida* Anderson worked out its autecology. Deep peat is useless for agriculture and should remain part of the forest estate. All except the most stunted types contain high volume of valuable timber per hectare. The more stunted types in Sarawak have a high volume of small trees which has potential as a source of chips or fibre. Sylvicultural systems have to be perfected: it is not commercially possible to await felling of *Shorea albida* in Sarawak until regeneration is present so this valuable forest type thus does not regenerate.

Tropical semi-evergreen rain forest

Tropical semi-evergreen rain forest is of very limited extent, being marginal to tropical lowland evergreen rain forest where there is a seasonal water shortage and is therefore probably wide-spread. It occurs in intermontane valleys of the major mountain ranges in Sumatra, southeastern Borneo and New Guinea.

Monsoon forests

The monsoon forest formations are of only limited extent within the Malay archipelago, which is bisected by a north-south belt of seasonally dry climate running from west Luzon to the Lesser Sunda Islands. This belt separates the evergreen rain forest zone into western and eastern blocks and is occupied by semi-evergreen rain forest and the various monsoon forest formations which have never been precisely mapped. The vegetation depends on the water available during the dry season, so there are complex interactions between topography, soil and altitude. Furthermore, seasonally dry vegetation is inflammable; man and lightning have caused fires so most of the monsoon forests have been degraded to savanna. Teak is a monsoon forest species, favoured by fire which has been successfully planted in Perlis, northernmost W. Malaysia, at the southern limit of semi-evergreen rain forest, but performs very poorly.

The indigenous *Pinus* spp., *P. merkusii* are pioneer species of seasonally dry climates. Their natural range on landslips and stony ridges has been greatly extended by fire, which they can tolerate once past the small seedling stage. Without fire broad-leaved forest encroaches and huge relict trees can be found in rain forest (Van Steenis, 1972). The other subtropical and tropical *Pinus* spp. all have similar ecology. Much effort has been expended since *ca.* 1966 by FAO and the Peninsular Malaysian Forest Department to establish *Pinus* plantations. There have been great difficulties because *Pinus* is, in the tropics, basically a monsoon forest genus.

Floristic diversity

The Far East tropical rain forests have a richer flora and probably fauna too than any others. The most extensive and species-rich formation is the lowland evergreen tropical rain forest—the climatic climax. The flora becomes poorer with increasing altitude though many montane forest types have a wealth of plant species. In general, the more extreme the habitat, the poorer in species is the vegetation. This is exemplified in the peat swamp forests of Sarawak discussed above.

A very full analysis of this variation is given by Whitmore (1975). Variation is due to many factors which can be grouped roughly into a hierarchy from the gross geographical patterns downwards. Most species do not occupy the full geographical range of suitable habitat and the historical causes for this are now understood at least in outline.

The second important cause of variation is major disturbance, for example by wind (e.g. cyclones or typhoons), man or earthquake. Disturbed forests are dominated by light-demanding pioneer species which do not regenerate in their own shade and eventually change to forests dominated by shade bearers if disturbance is not repeated. The time scale is likely to be several centuries. Important studies of forests originating from disturbance are by Browne (1949) and Wyatt-Smith (1954b) on the Kelantan storm forest in northeastern Malaysia, now nearly a century old.

Next in the hierarchy, is variation correlated with topography. An analysis of forest data for Ulu Kelantan, Malaysia (26 628 trees ⩾4ft girth on a 676 ha sample from 10 000 km²) revealed that most species were present on all of four major rock and topographic types but that their abundance differed revealing site preferences (Whitmore, 1973c). However, several common species were about equally abundant on all sites. Soil and topography both reflect the parent material, and operate partly at the stage of seedling establishment and survival and partly because large trees probably show differential sensitivity to soil creep and slip which is always occurring on slopes in hot wet climates.

Shorea curtisii in Malaysia is an example; its marked general restriction to certain ridge crests and its gregarious occurrence has been correlated with habitat features. Burgess has made the most detailed study of the autecology of any tropical tree in his work on *S. curtisii* (Burgess, 1969a, 1969b, 1970, 1972, unpublished; Whitmore, 1975). He concluded that *S. curtisii* probably succeeds in competition because of its superior tolerance of periodic water stress. His investigations are a model and show how extremely profitable it is to study a single species in great detail. High priority should be accorded to similar very detailed studies of other valuable economic species.

An earlier similar study of *Dryobalanops aromatica*, another very important Malaysian dipterocarp, was undertaken by Lee (1967) who clearly defined the problem of its gregarious occurrence. It would be valuable to restudy *D. aromatica* in the light of the insight Burgess has given to the importance of mineral and water relations and using his methods.

The most subtle kinds of variation are associated with reproductive behaviour or with soil variation within homogeneous topography. At Pasoh forest, Malaysia, Wong and Whitmore (1970) detected groves of several dipterocarp species. At Jengka forest, Malaysia, Poore (1968) found clustering of several dipterocarps into family groups (a large tree with smaller ones of various girths around it), and that the degree of aggregation reflected the efficiency of the species dispersal mechanism. By contrast at the Andulau forest, Brunei, Austin, Ashton and Greig-Smith (1972) found associations between species which appeared correlated with soil factors, which they were unable to define.

The occurrence of family groups seems to be due to reproductive pressure; diaspores and seedlings of some species are present more than those of others at the times and places for establishment of a new tree. Lee concluded that the success of *Dryobalanops aromatica* was probably due to its more frequent fruiting and longer-lived seedlings than other dipterocarps.

Modern numerical techniques of phytosociological analysis have worked well at the grosser level where major discontinuities occur between and within forest formations. Such techniques are inappropriate for the investigation of lesser, more subtle variation and it has been shown, using data from Sepilok Forest, Sabah, that the more subtle the variation the more likely for particular numerical techniques to affect the results (Austin and Greig-Smith, 1968). Plot lay-out is also important and perhaps accounts for the different variation patterns detected at Malaysia-Pasoh and Brunei-Andulau.

Enough is now known about modes of variation to realize how carefully questions should be posed. Static studies of rain forest phytosociology are no longer likely to advance understanding, which will more likely come from investigations of dynamics and autecology.

Dynamics

The dynamics of tropical forests must be based on the concept of a dynamic continuously changing ecosystem. Whitmore (1975) provided a detailed description of the growing canopy based partly on detailed observations in Malaysia at Sungei Menyala and Bukit Lagong. Dynamics includes regeneration in gaps which in central W. Malaysian dipterocarp evergreen rain forest occupied 10 per cent of the area (Poore, 1968). Regeneration in these gaps will be discussed under the stages of seed production, germination, establishment and growth.

Flowering and fruiting

The evidence for seasonal variations within perhumid tropical climates is reviewed by Whitmore (1975) and is mainly based on the work of Nieuwolt (1965, 1966) in W. Malaysia and Singapore, and Brünig (1969a, 1971) and Baillie (1972) in Sarawak. Seasonal water shortage is ecologically the most important variation.

There are marked seasonal cycles in leaf change, flowering and fruiting and considerable evidence that these can be correlated with changes in water supply. This is counter to the classic concept that seasonality in climate is non existent or negligible in the perhumid lowland tropics. See also chapters 2, 8 and 20. Early studies of phenology in the Singapore Botanic Garden (e.g. Corner, 1940; Holttum, 1940, 1953; Koriba, 1958) showed no regular seasonality in trees. It now seems that this was because the specimens studied were isolated trees whose behaviour differs from that of trees in the forest. Whitmore (1975) states: 'An extremely important study at Ulu Gombak, Malaya, monitored flowering and fruiting for a ten year period 1960–1969 (McClure, 1966; Medway, 1972a). The sample as a whole showed a single distinct and marked annual peak of flowering between March and July then fruiting from about July to October. A minimum of 44 per cent of the 45 species flowered each year, and a minimum of 27 per cent fruited. Flowering follows the early year dry season. The highest incidence of flowering was in 1963 and 1968 when that was most pronounced, and included gregarious flowering of the Dipterocarpaceae. The species which flowered varied from year to year, the aggregate never falling below 20. This pattern of a peak of flowering followed by a fruiting peak was found to be generally true, of Malaya as a whole from analysis of nearly 4 000 herbarium collections in lowland and lower montane rain forest throughout the country made during 1966–1970 by the Forest Research Institute, Kepong. There was a distinct indication of a second peak at the end of each year, not shown in the Ulu Gombak study, but reflected in the availability then of durians and other fruits in the towns and villages from itinerant merchants.'

Besides seasonality amongst fruit trees an important practical aspect is the reproductive behaviour of the predominant timber tree family, Dipterocarpaceae. Whitmore (1975) analyses the data as follows: 'Flowering of Dipterocarpaceae is notorious for its infrequency and gregariousness. 1963 and 1968 saw a general flowering of the family throughout Malaya including Ulu Gombak. Poore (1968) has published figures for Temerloh in central Malaya showing that in 1963 evaporation exceeded rainfall in every month except October and November, and attributed to this drought the gregarious flowering of dipterocarps nearby. Burgess (1972) has recently analysed all available data for Malaya in Forest Department files, these cover the period 1925–1970. He concluded that dipterocarps in general fruit heavily every two to three years with occasional intervals of up to five years with some difference between species. Even in the best years only 40–50 per cent of mature trees in a given area are fertile, and in some cases groups of trees flowered together. Careful observations on *Shorea curtisii* and *S. platyclados* revealed that with one exception none of the trees which flowered in 1968 did so again in 1970. This suggests that a given tree may need a prolonged period of physiological preparation before it is ready to respond to an external stimulus to flower (cf. Wood, 1956; Fox, 1972). Wood noted that in the 1956 gregarious flowering in Sabah lowland species flowered about two weeks earlier than those in the hills. In addition many dipterocarp species flower sporadically during any month of the year, and this has tended to obscure the existence of a single regular maximum of flowering in May. *Dryobalanops aromatica* is unusual, it has a first peak in April and a second one in October, it also tends to flower more often than other species. From examination of the rainfall figures Burgess concluded that there is considerable circumstantial evidence that gregarious flowering is in some way connected with drought periods occurring 3–5 months before flowering, at a time when axillary buds are developing on a new flush of leafy shoots, and in order to turn them into flower buds. Similarly Whitehead (1959) found in Malaya that for the rambutan (*Nephelium lappaceum*) heavy rainfall prior to flower bud differentiation promoted subsequent vegetative growth instead of flowering. The nutritional status of the tree was also important. Burgess concluded that further progress will only come from study of individual trees and monitoring the weather they experience.

'Not all droughts are followed by flowering. Medway noted at Ulu Gombak that flowering was more copious in 1968 when a short dry season was followed by wetter weather, than in 1963 when the drought was prolonged. Baillie (1972) found a good correlation between drought in 1968 at Kuching in Sarawak and a subsequent heavy crop of illipe nuts from an experimental plantation. No such correlation was found in 1970, when there was a good crop but no drought. Flowering of the illipe nut trees, as indeed of other dipterocarps, may not be followed by fruiting, heavy rain or high winds may destroy the crop. (In Sabah in 1951 high winds and rain damaged flowers over a wide area and little fruit was set: Fox, 1972.) The size of the illipe crop harvested, which for most areas is the only measure available, further depends on what other source of income is available to the Iban gatherers at the time the fruit ripens. Fox (1972) noted that in Sabah high forest emergent dipterocarps seldom flower until they have grown to full height, whereas non emergent species flower at small size. Ng (1966) recorded flowering at young age and low height of planted dipterocarps in Malaya. Anderson (1961) made the very interesting discovery that the dipterocarps in the peat swamp forest of Sarawak flower out of phase with those of dry land, i.e. not at the dry season.

'In conclusion, there is now a substantial body of circumstantial evidence which indicates that flowering of many lowland rain forest tree species is promoted by water stress, but the relationship is not simple. The part played by endogenous controls is not clear, that such controls exist is definitely shown where single limbs are out of phase, as in the tree *Gluta renghas* observed at Ulu Gombak (Medway, 1972) and in species, such as some figs which flower at regular intervals during the year.'

Stimuli other than water stress which are well known to trigger blossoming were reviewed by Whitmore (1975), and include temperature shock (Corner, 1940; Kerling, 1941; Holttum, 1954a) and photoperiod (Garner and Allard, 1923; Dore, 1959).

An important distinction can be made between more or less regular periodic flowering, which is probably the case for most species of primary rain forest, and continuous flowering which occurs amongst many pioneer species which are discussed more fully later.

Gregarious flowering at much longer intervals than in the Dipterocarpaceae is known. It has been detected amongst the common Malayan hill forest bamboos *Dendrocalamus pendulus* and *Gigantochloa scortechinii* (Burgess), *Schizostachyum* sp. in entire river systems, and a few palms, for example *Plectocomia griffithii* (Whitmore, 1973b).

'It can be argued that it is advantageous for a species that individuals blossom together. The evolutionary strategy of long-lived trees is to maintain high variability (Stebbins, 1950) and one means to this end is by outbreeding, by cross pollination between individuals. Many species commence flowering in response to the external stimulus of water stress, which is why they come in to flower more or less simultaneously, though. . . related species often blossom a few days apart. Simultaneous flowering will tend to lead to simultaneous fruiting and simultaneous germination, this might be advantageous in some cases in ensuring that at least some seeds and seedlings avoid being eaten. Wood (1956) noted that in Sabah the fruit of late-flowering dipterocarp species matured faster.

'The simultaneous, periodic flowering and fruiting of forest trees at long intervals can have important practical implications for forestry. As dormancy is short or nil seed supply is sporadic. This is the main reason why promising dipterocarp timber species, such as the Light Red Meranti group of *Shorea* have never been planted even experimentally outside the region. Secondly, seedlings will appear on the forest floor in occasional immense populations.

'Contemporaneous flowering and fruiting of many canopy trees results in seasonal peaks in food for herbivorous animals. Recent work in Malaya and Sarawak, mainly on birds indicates that there are marked peaks in breeding activity at periods of abundant food (Ward, 1969; Chivers, 1971; Fogden, 1972; Murphy, 1973).'

Work at Pasoh forest, peninsular Malaysia, is investigating breeding systems, pollination and genetic diversity of selected tree species.

Dormancy and germination

Many species of primary rain forest trees (including all Dipterocarpaceae) have seeds which have no dormancy and lose their viability within a few weeks if stored. There has been very little work to attempt to extend viability. Tang (1971) extended the viability of *Shorea curtisii* and *S. platyclados* from 1 to 3–4 weeks by reducing moisture content. Tang and Tamari (1973) experimented with drying and cooling on various dipterocarp species and achieved modest success. Burgess (reviewed in Whitmore, 1975) found that *S. curtisii* seeds had to be kept wet. The viability of *Agathis* seed can be extended from one week to several months by storing fully imbibed at low temperature (Whitmore *et al.*, unpublished).

There is no doubt that this work is of fundamental importance for the choice of tree species for sylvicultural trial and plantation growth is limited by fruiting infrequency and absence of dormancy. The same problems occur with indigenous fruit trees (*Durio, Nephelium, Baccaurea*, etc.). Moreover, it is impossible to establish seed stores for genetic conservation unless dormancy can be induced.

Some species show dormancy; some have a staggered germination over several months; *Dyera costulata* and *Koompassia malaccensis* germinated between 24–130 and 13–56 days after sowing (Wyatt-Smith, 1963).

There are numerous claims that all light demanding pioneers possess seeds which have a dormant period. This is certainly true of *Anthocephalus chinensis* (Fox, 1972). These species are desirable on other grounds for aforestation, and dormancy is a valuable characteristic which has enabled seed to be stored and transported.

Establishment

The most detailed work on establishment is that by Burgess (unpublished) on *Shorea curtisii*; it is summarized by Whitmore (1975).

Many Dipterocarpaceae are able to establish themselves in dense shade, but the seedlings die unless the shade is reduced within a few years. Sylvicultural systems for dipterocarp forest (see e.g. Wyatt-Smith, 1963, on the Malayan Uniform System) utilize this behaviour which has been studied in most detail in Sabah by Fox (1972) following important earlier work (especially Nicholson, 1960).

Rates of photosynthesis and respiration in seedlings growing in dense shade would be of considerable interest for species of different shade tolerances. Sunderland (unpublished) worked at Pasoh Forest on the CO_2 exchange of seedlings of the light demander *Shorea leprosula* and the shade tolerant *S. maxwelliana*.

Light and shade trees

The concept of light and shade tree species has been reviewed by Whitmore (1975):

'As a first generalisation, it is found that established seedlings and saplings most often grow up to maturity in small gaps. The drastic change in microclimate which follows on the formation of a big gap is followed by the rapid death of young trees which became established in the cool shady humid microclimate of the closed canopy. Big gaps are found to be colonised by a group of species, rare or absent in the undergrowth of high forest which are equipped successfully to exploit the very different microclimate of open sites.'

For the Far East rain forest region Whitmore (1975) states:

'Rain forest trees can crudely be divided into two groups, those which regenerate in situ, in shade and those which regenerate in gaps, these groups are respectively shade tolerant and light demanding in their early life.

'Species which require gaps for regeneration cannot grow up even under their own shade and are commonly called pioneers.

'Foresters have long recognised that the degree to which species are shade bearers or light demanders is an important key to other aspects of their behaviour. It is found that pioneers (light demanders) characteristically have very rapid height growth in youth and, rapid girth growth, at least at first. Rapid height growth clearly has high selective value, as a pioneer tree which can overtop others in the same gap will suppress them by shading and by physical occupation of a volume of species, and so preempt that gap for itself for its whole remaining life. Wood of low density is a feature correlated with rapid growth, and it is usually pale in colour... Pioneer species characteristically flower and fruit copiously and continuously, without regard to season which is unusual in the flora as a whole. Individuals become fertile at an early age and dispersal is efficient; and in many cases by birds. For all these characters also there is strong selective pressure... *Macaranga* is the genus par excellence of pioneer trees; if is found throughout the tropical Far East, where most o. its 280 species occur. In Borneo for example there are *ca.* 50 species. In Malaya there are 27 species, of which 20 can be found along roadsides, 18 of them gregariously (Whitmore, 1967, 1973); like other pioneers *Macaranga* has been able enormously to expand its populations as man has provided an ever larger area of suitable habitats, especially over the last century or so. One Malayan species, *M. constricta*, which was first discovered in 1967 on the slopes of Gunung Benom, was at that place moving for the first time from natural landslips and forest gaps on to the sides of newly built logging roads (Whitmore, 1969; Whitmore and Airy Shaw, 1971). The enormous expansion in range of *Macaranga* species presents an interesting situation for biosystematic study. In fact, in Malaya, man's intervention appears to have had little effect on species evolution and no signs of introgressive hybridisation or polyploidy could be detected (Whitmore, 1969; Whitmore, Soh and Jones, 1970). Biosystematic studies of tropical groups especially of trees have scarcely yet begun, despite the high degree of activity there has been in temperate countries. Such work is commended to the attention of taxonomists resident in the region. It has important implications for the study of species evolution and hence floristic richness of rain forest... It is more usually the case that a few species of a big genus have pioneering properties, for example *Mallotus* (notably *M. paniculatus*), which is mainly a genus of C storey small trees of high forest (Whitmore, 1973d).

'There are several reports that the seeds of pioneers have a long dormancy and germinate on the stimulation of either bright light or high temperature or both when a gap forms (Symington, 1933, Malaya; Nicholson, 1970; Fox, 1972, Sabah).

'The observations that pioneers usually occur in pure stands, and that adjacent gaps created at different times, as in logging areas, commonly have different species are indicative that invasion on a first come first served basis after gap formation may in fact prevail.

'By contrast with light demanders, seeds of shade bearing species are able to germinate and establish in the gloom of the high forest, and some have substantial food reserves which they utilise to become established... Seedlings can commonly persist for several years, growing only very slowly. As a generalisation, shade bearers are commonly slow growing in girth and height and have dense, dark timber. These features all contrast with light demanding species though a very few hard-wooded light demanders are known...

'The population structure of light demanding and shade bearing species in a rain forest stand is markedly different. The former, have a preponderance of big stems, simply because they cannot regenerate in shade. The latter have most stems in the smallest classes. Foresters refer to these population structures as negative (i.e. not reproducing) and positive stands respectively.

'Extreme light demanding and shade bearing species show the two contrasting ecological strategies of forest trees with respect to perpetuation of the species. It must not be thought that all tree species fall neatly into one group or the other. Rather, there is a complete spectrum of responses to light and these are the two extremes. The conditions which stimulate seedlings of shade bearing species to commence upwards growth are not well under-

stood, and indeed are likely to vary from species to species. Some species probably need some increase in light, such as in a small gap, this has now been shown for several dipterocarps. Others may merely need a volume of space above ground such as will develop after a small tree dies, and may be stimulated to grow by the reduced root competition which is concomittant with such a death.

'In an interesting early experiment apparently never repeated Kramer (1926, 1933) made artificial gaps in high forest on Gunung Gede, Java. Small gaps of 0.1 ha were soon filled by surviving young individuals of primary forest tree species. In larger gaps of 0.2–0.3 ha these were suppressed by a lush growth of secondary forest pioneers.

'Two groups of pioneers can now be distinguished, those which are short-lived and long-lived respectively.

'Short-lived pioneers which are common in the tropical Far East are *Commersonia bartramia*, *Trichospermum*, *Trema*, *Mallotus* and most *Macaranga*. Kochummen (1966) found that a stand of *Macaranga gigantea* on an area farmed for one crop at Sungei Keoh, Malaya had grown to 60–65ft tall and 4ft girth in fifteen years and was beginning to die...

'Light demanding species which have been observed locally to colonise clearings, especially those made by man, but which are less widespread and common than those just considered, are *Symingtonia populnea* and *Weinmannia blumei* (for example at Cameron Highlands and Fraser's Hill, Malaya); *Sehima wallichii* locally in lowland and lower montane forest; and *Leptospermum flavescens* in upper montane forest. This group of species has some pioneering characters, but lacks the aggression of the true pioneer species. In the Sarawak peat swamp forest *Cratoxylon arborescens*, *C. glaucum*, *Dactyloclados stenostachys*, *Macaranga caladiifolia* and *Quercus sundaicus* exhibit pioneer features. *Dyera costulata*, a giant emergent tree of lowland forest is a very strong light demander. It occurs as scattered trees never in stands, despite its frequently produced wind borne seeds. In Malaya it has been shown that its seeds take from 24–130 days to germinate (Wyatt-Smith, 1963; Whitmore, 1973e). Why it exhibits only some pioneering characteristics remains to be ascertained.

'Once a tree has grown to occupy a space in the forest it will remain there until its death, and that may be several centuries hence. It follows that an understanding of the conditions which lead to the establishment of one species rather than another are essential to an unravelling of the nature of plant communities in the rain forest.'

The concept of pioneer light demanders and climax forest shade bearers is relevant to the nature of species associations. Where big gaps occur light demanders will predominate; in other areas shade bearers will be predominant. Forests of the northeastern Kelantan (W. Malaysia) result from major disturbance and differ in composition from other forests (Browne, 1949; Wyatt-Smith, 1954b). Relatively minor changes in species composition were detected in virgin lowland evergreen dipterocarp tropical rain forest at Sungei Menyala (Wyatt-Smith, 1966) during 1947–1959; these could be correlated with the disturbance factor, in particular a destructive episode in 1917.

Permanent plots on which these studies at Sungei Menyala, and a similar one in hill dipterocarp forest at Bukit Lagong, were carried out, still exist. The value of such long-continuing observations cannot be over emphasized and the longer the studies are continued the more valuable they become.

Productivity

Primary productivity

The major difficulties to obtaining accurate measures of growth and production, with estimates of their error, for tropical forests are:
— estimating production and biomass of roots;
— the necessity to define the stage of the regeneration cycle being studied as productivity differs between the gaps, growing and mature phases;
— seasonal fluctuations in litter fall and growth rates;
— the unreliability of estimating biomass by allometry based on diameter and height measurements of a few individuals of a few species;
— difficulties in measuring respiration of leaves and axial parts;
— the loss due to herbivores.
The major study of productivity in the region is that carried out under the auspices of the IBP at Pasoh Forest, W. Malaysia, during 1969–1974. The Pasoh Forest study area lies 2° 59′ N and 102° 18′ E at an altitude of 75–150 m. Rainfall is only *ca.* 1 900 mm/a with maxima in December/January and April. The predominant canopy species are dipterocarps.

Plant biomass was measured by clearing and weighing two 0.1 ha plots and by applying the formula of Ogawa, Yoda, Ogino and Kira (1965) to diameter and height measurements (see chapter 10). A total aerial biomass of 366 t/ha was estimated. Root biomass was calculated similarly as 30 t/ha or, by the often used method of 10 per cent of the aerial biomass, as 37 t/ha. The observed 20 t/ha for roots ≤1 cm diameter suggests these are underestimates.

Leaf litter fall was 6.5 t/ha/a, small branch fall 1.5 t/ha/a and large branch fall was estimated as 1.9 and 3.7 t/ha/a. Estimates of tree fall varied from 4.4. to 6.9 t/ha/a; these are not reliable.

Leaf respiration was in the order of 375 mg CO_2/kg (fresh weight)/h. Total respiratory losses were variously estimated as 45–74 t/ha/a.

Grazing insect faeces in litter traps amounted to 245 kg/ha/a suggesting an ingestion of plant material in the order of 0.3 t/ha/a.

CO_2 flux measurements suggested a gross production of 97–110 t/ha/a and a net production of 37–50 t/ha/a.

Despite the enormous amount of work carried out at Pasoh, the values for roots and large tree fall are little more than guesses (see chapter 10).

Secondary productivity

The amount of secondary productivity is unknown. The preliminary data from Pasoh suggest that insect larvae account for a small but significant leaf loss which reduces photosynthesis. There are other incomplete studies concerned which the birds and Hexapoda. See chapter 11.

Decomposition

Almost all the decomposition data also come from Pasoh. The litter on the ground averages 3.2 t/ha and ranges from 1.5 to 5.0 t/ha. The input of (non-trunk and branch) litter is 8–10 t/ha/a so that it disappears in *ca.* 4 months on average. Soil respiration studies indicate that *ca.* 80 per cent of the litter break-down occurs in the Ao horizon and *ca.* 20 per cent is mineralized. The rate of soil respiration is *ca.* 600 mg CO_2/m²/h, equivalent to 32 t/ha/a of organic matter. The difference of 21.5 t/ha/a is mainly due to living and dead roots and to mineralized trunks and branches. These decomposition rates are high.

Little work has been carried out on decomposer organisms apart from the Isoptera. 36 species of termites were recorded on 2 ha, but only *Homallotermes foraminifer* and *Macrotermes carbonarius* predominated. Termite biomass was *ca.* 6 kg/ha with an organic matter respiratory turnover of *ca.* 60 kg/ha/a.

Nutrient cycling data from Pasoh and elsewhere were discussed in chapter 13.

Animal ecology

Most of the work on animal ecology in rain forests has been carried out in Malaysia and Kalimantan (Borneo), much of it at the University of Malaysia by Lord Medway and his associates and in the Game Department by Mohammed Khan. Murphy (1973), in Singapore, has published extensive and elegant studies on the ecology of invertebrates, whilst Start (1974) has completed a detailed study of the three pollinating Microchiropteran bats of peninsular Malaysia.

There is marked seasonality in the breeding and moulting behaviour of some (e.g. birds) but not all animal groups. No evidence of seasonal trends in various reproductive features of six species of frog in lowland rain forest in Sarawak was found (Inger and Bacon, 1968), or in one associated with man in Kuching (Inger and Greenberg, 1963), or in ten species of lizards (Inger and Greenberg, 1967). See also chapters 6 and 7.

The major differences in structure, physiognomy and flora might be expected to be reflected in the fauna. This is still at a very early state of development (see Whitmore, 1975). Mangrove forest has an avifauna distinct from dry-land forest, and several examples of species replacements are known from Malaysia (Nisbet, 1968; Wells, 1974).

The primary rain forest faunas appear, on the whole, to have been unable to take advantage of the vast new areas of secondary forest. Stevens (1968) gives a useful computation for Malaysia which clearly demonstrates how few animals survive destruction of primary lowland rain forest there (Harrison, 1965). The only way in which the majority of native mammals and birds can be preserved is in the undisturbed lowland rain forest in which they evolved. The only way it can be retained is by establishing large permanent reserves.

Medway (1972b) distinguished between lowland and montane vertebrate faunas on Gunung Benom in Malaysia, the boundary being approximately that between the lower and upper montane forest formations.

Most birds and mammals and certainly most of the arboreal animals species are essentially lowland-dwellers. They will only survive if adequate tracts of the lowland climatic climax formation of evergreen rain forest are conserved. This is the formation which contains most commercial timber for montane forests are inaccessible and have few species of commercial value.

Logging operations largely destroy the structure of the canopy and therefore have a serious impact on arboreal animals (see chapter 20). The problem is not simple; canopy opening stimulates flowering and fruiting so that fig trees for example (which are a very important food source and which are never felled for timber) become more useful to animals. Lush herbaceous growth develops in gaps; this actually encourages ground-living browsing animals.

A project should be established to monitor changes in animal populations following logging with a view to designing logging methods compatible with continued existence of animals (see chapter 20).

Water balance

Regular annual water stress occurs within this perhumid rain forest region. Important papers include the Aberdeen-Hull Symposium on the *Water relations of Malesian forests* (Flenley, 1971), the work of Guha (1969) on rainfall interception and penetration under rubber plantations in Malaysia, and of Low and Goh (1972) on run-off under different vegetation covers in Malaysia. See also chapter 12. Since evaporation pans were established in Malaysia during the 1950s and 1960s, it has been discussed that much of the peninsula has 1–4 months per year when evaporation exceeds precipitation (Nieuwolt, 1965). The consequences of this for plant life have been discussed by Whitmore (1973b, 1975) and by Prance and Whitmore (1973). Burgess (1969 a) established the importance of water relations in determining the sites occupied by *Shorea curtisii*. A scheme of climatic types developed for Indonesia by Schmidt and Ferguson (1951) has been extended to the rest of the region by Whitmore (1975). Unlike the earlier ones of Köppen and Thornthwaite, it reflects vegetation types and is thus of high value to biologists.

Use of forest ecosystems

Until the early 1960s some 80 per cent of peninsular Malaysia and of Sarawak, Brunei and Sabah remained under forest, much of it primary lowland evergreen diptero-carp tropical rain forest. Later, Government agencies gave national parks more priority in land-use development plans (Soepadmo and Singh, 1973; Lee, 1974) and by 1975 the following main national parks existed:

Peninsular Malaysia
— *Taman Negara*, 4 340 km², containing most inland habitat types and ecosystems except high elevation granite.
— *Templer Park*, a small area of mainly open land and secondary forest near Kuala Lumpur; most valuable as a recreation area.

Sarawak
— *Bako National Park*, mainly primary and secondary heath forest; some riverine swamp forest, mangrove and mixed dipterocarp forest.
— *Gunung Mulu National Park*, a large area including most inland ecosystems except peat swamp forest.

Sabah
— *Kinabalu National Park*, including Gunung Kinabalu (4 100 m) plus surrounding lower montane rain forest and lowland rain forest.

In peninsular Malaysia there is also a network of Virgin Jungle Reserves although many have been logged and some have disappeared. Two, Sungei Menyala and Bukit Lagong, which have been under continual observation since 1947–1949 to monitor long-term changes in forest structure and composition are of prime importance. In Sarawak and Sabah there are also long-term observation plots in primary and sylviculturally treated lowland forest. These too are becoming increasingly valuable as they age and should be given high priority for conservation.

Plans for land use of lowland rain forest exist in varying detail for all territories (see chapters 20 and 21). It is vital that these plans should include areas adequate to conserve representative samples of all ecosystems. The following areas need consideration:

Peninsular Malaysia
— An area of high elevation granite, such as Gunung Benom in the Krau Game reserve provides. This area is also important as the focus of studies into primate ecology since 1968.
— The Gunung Besar massif on the Johore/Pahang border, and to include some stands of lowland red meranti forest despite its high commercial value.
— Gunung Belumut, Johore, plus a skirt of lowland forest.

Sarawak
A peat swamp forest including all six forest types.

Brunei
An extensive tract of lowland forest, such as Ulu Temburong, which is very rugged and in which the main timber stands are inaccessible.

Sabah
— A lowland area within the range of the orang utan and also to conserve the extremely tall, high biomass lowland rain forests of the east coast—probably the grandest representation of this formation in the world; the Ulu Telom valley would fulfill these functions.
— The Klias peninsula in the south west. This is an area of swamp and seasonal swamp forest, including mangrove, rich in species and also the spawning ground for the fish, etc., shoals of Brunei Bay, which are important as a food source.

Suggested research priorities

A revised version of the excellent Unesco vegetation map of Malesia and a map of the vegetation of continental South-East Asia would be useful as syntheses and pedagogic aids.

Plant monographic studies are a background for all applied work. Plant and animal collecting is urgently needed in Celebes, the Moluccas and the Lesser Sunda Islands before the forests there are seriously depleted by logging.

Sophisticated ecophysiological studies on heath forest to establish the limits within which tree crops can be successfully grown and to increase basic understanding of mineral and water relations in rain forests in general. As heath forest sites are likely to be permanently available for production forestry, solution of these problems would help to relieve pressure on the lowland evergreen rain forest.

Very high priority should be given to developing sylvicultural systems for peat swamp forest, and especially to perfecting one for the *Shorea albida* consociation of Sarawak and Brunei.

Experiments on tree growth to elucidate the basic factors controlling crown construction. Trees with monopodial crowns have potential as plantation trees. Basic growth rate data are lacking for most of the many species which have this construction. Some might prove valuable as plantation trees for cellulose or fibres.

Detailed autecological studies of valuable economic tree species in all countries to provide the basis for sylvicultural recommendations. Water and mineral relations are likely to be important. In particular, *Dryobalanops aromatica* in Malaysia could profitably be investigated.

Long-term observation plots coupled with meteoro-logical stations are required to study the effect of weather on flowering and fruiting. Water stress and flowering in rain forest and monsoon forest species.

Modes of inducing tree seed dormancy and simultaneous germination.

Basic physiological studies on respiration and photosynthetic rates, partly in growth chambers, are needed to increase understanding of seedling physiology.

Critical experiments to discover whether secondary forest pioneer species colonize from seeds dormant in the soil or from seeds invading the clearing after its creation.

The physiological differences between light demanding and shade bearing species.

Further study of periodicity in animal breeding.

As a very high priority project, an inventory of all long-term observation plots and their conservation.

Very high priority should be given to a project to monitor the changes which occur in animal populations following a logging, with a view to modifying, if necessary, the logging practices so that species are not totally driven out of an area when the timber is extracted.

Evaporation pans should be added to meteorological stations. Synecological studies, especially those on the effects of disturbance on the forest, should incorporate investigations of soil erosion. Very high priority should be given to a project, possibly at Gogol, to determine the amount of minerals in different parts of the ecosystem and their cycling before forest removal and after, and to monitor developments as vegetation regrows. Removal of timber for wood chips may result in serious depletion of minerals and failure of forest to re-establish itself.

Important fields for investigation are whether diaspores arrive after a gap is created or are there and are released from dormancy; nutrient cycling and partitioning through the forest; autecological studies of particular key species; changes in soil, which are especially probable in monsoon forests. Very high priority should be given to the study of secondary forest which is increasing in extent very rapidly and remains poorly investigated.

A series of reserves is needed to provide representation of adequate samples of all ecosystems, including those of commercially valuable lowland rain forest. These could be on a regional basis. A graded series of sanctuaries with different degrees of protection for special purposes is also needed.

High priority should be given to research into the problems which arise from the multiple use of rain forest.

Bibliography

ANDERSON, J. A. R. Observations on the ecology of the peat-swamp forests of Sarawak and Brunei. In: *Proceedings of the Symposium on humid tropics vegetation* (Tjiawi, Indonesia). Council for Sciences of Indonesia and Unesco Science Co-operation Office for South-East Asia, 1958, 312 p.

——. *The ecology and forest types of the peat-swamp forests of Sarawak and Brunei in relation to their silviculture.* Edinburgh University, Ph. D. thesis, 1961, 191 p. multigr. and annexes.

ANDERSON, J. A. R. Research on the effects of shifting cultivation in Sarawak. In: *Proceedings of the Symposium on the impact of man on humid tropics vegetation* (Goroka, New Guinea), p. 203–206. Djakarta, Unesco Science Co-operation Office for South-East Asia, 1962, 402 p.

——. The flora of the peat-swamp forests of Sarawak and Brunei including a catalogue of all recorded species of flowering plants, ferns and fern allies. *Gard. Bull. Sing.,* 20, 1963, p. 131–228.

ANDERSON, J. A. R. The structure and development of the peat-swamps of Sarawak and Brunei. *J. Trop. Geogr.*, 18, 1964a, p. 7–16.

——. Observations on climatic damage in peat-swamp forest in Sarawak. *Commonwealth Forestry Review*, 43, 1964b, p. 145–158.

——. Limestone habitat in Sarawak. In: *Proceedings of the Symposium on ecological research in humid tropics vegetation* (Kuching, Sarawak). Government of Sarawak and Unesco Science Co-operation Office for South-East Asia, 1965, 376 p.

ASHTON, P. S. *Manual of the dipterocarp trees of Brunei State.* London, 1965.

——. *A manual of the dipterocarp trees of Brunei State and Sarawak. Supplement.* Kuching, Sarawak, 1968.

ASHTON, P. and M. *The classification and mapping of Southeast Asian ecosystems. Transactions of the fourth Aberdeen-Hull Symposium on Malesian ecology.* Univ. of Hull, Dept. of Geography, miscellaneous series no. 17, 1976, 103 p.

AUSTIN, M. P.; GREIG-SMITH, P. The application of quantitative methods to vegetation survey. II. Some methodological problems of data from rain forest. *J. Ecol.*, 56, 1968, p. 827–844.

——; ASHTON, P. S.; GREIG-SMITH, P. The application of quantitative methods to vegetation survey. III. Reexamination of rain forest data from Brunei. *J. Ecol.*, 60, 1972, p. 305–324.

BAILLIE, I. C. *Further studies on the occurrence of drought in Sarawak.* Sarawak, Kuching, Forest Dept., Soil Survey Report F7, 1972, 12 p.

BERRY, A. J. The natural history of west Malaysian mangrove faunas. *Malay. Nat. J.*, 25, 1972, p. 135–162.

BRAY, J. R.; GORHAM, E. Litter production in forests of the world. In: Cragg, J. (ed.). *Advances in ecological research*, 2, p. 101–157. Academic Press, 1964.

BROWNE, F. G. Storm forest in Kelantan. *Malayan Forester*, 12, 1949, p. 28–33.

BRÜNIG, E. F. Some observations on the status of heath forests in Sarawak and Brunei. In: Misra, R.; Gopal, B. (eds.). *Proc. Symp. Recent Adv. Trop. Ecol.* (Varanasi), 2, 1968, p. 451–457.

——. On the seasonality of droughts in the lowlands of Sarawak (Borneo). *Erdkunde* (Bonn), vol. 23, no. 2, 1969a, p. 127–133.

——. Forestry on tropical podzols and related soils. *Trop. Ecol.*, 10, 1969b, p. 45–58.

——. On the ecological significance of drought in the equatorial wet evergreen (rain) forest of Sarawak (Borneo). In: Flenley, J. R. (ed.). *The water relations of Malesian forests. Transactions of the first Aberdeen-Hull Symposium on Malesian ecology*, p. 66–96. University of Hull, Dept. of Geography, miscellaneous series no. 11, 1971, 97 p.

——. *Ecological studies in the kerangas forests of Sarawak and Brunei.* Kuching, Borneo Literature Bureau, 1974, 237 p.

BURGESS, P. F. Preliminary observations on the autecology of *Shorea curtisii* Dyer ex King in the Malay peninsula. *Malayan Forester*, vol. 32, no. 4, 1969a, p. 438.

——. Ecological factors in hill and mountain forests of the States of Malaya. *Malay. Nat. J.*, 22, 1969b, p. 119–128.

——. An approach towards a silvicultural system for the hill forests of the Malay peninsula. *Malayan Forester*, vol. 33, no. 2, 1970, p. 126–134.

——. Studies on the regeneration of the hill forests of the Malay peninsula: the phenology of dipterocarps. *Malayan Forester*, 35, 1972, p. 103–123.

BURKILL, I. H. The composition of a piece of well-drained Singapore secondary jungle thirty years old. *Gard. Bull. Str. Settl.*, 2, 1919, p. 147–157.

BURTT DAVY, J. *The classification of tropical woody vegetation types.* Oxford, Imp. For. Inst. Paper no. 13, 1938, 85 p.

CHAMPION, H. G. A preliminary survey of the forest types of India and Burma. *Indian For. Rec. (New Series)*, vol. 1, no. 1, 1966, p. 1–286.

CHIN, C. S. *On limestone flora and vegetation of Malaya.* University of Malaya, M. Sc. thesis, 1973, multigr.

CHIVERS, D. J. Spatial relations within siamang group. In: *Proc. 3rd Cong. Primatology* (Zurich, 1970), 3, 1971, p. 14–21.

CORNER, E. J. H. *Wayside trees of Malaya.* 2nd ed. Singapore, 1952, vol. 1, 772 p.; vol. 2, 228 pl.+5 p., 359 fig.

DORE, J. Responses of rice to small differences in length of day. *Nature* (London), 183, 1959, p. 413–414.

ECAFE. *Report of the study mission on hardwood resources in the Philippines, Indonesia and Malaysia.* Asian Industrial Development Council, 1974.

FLENLEY, J. R. (ed.). *The water relations of Malesian forests. Transactions of the first Aberdeen-Hull Symposium on Malesian ecology* (1970). University of Hull, Dept. of Geography, miscellaneous series no. 11, 1971, 97 p.

——. *Altitudinal zonation in Malesia. Transactions of the third Aberdeen-Hull Symposium on Malesian ecology* (1973). University of Hull, Dept. of Geography, miscellaneous series no. 16, 1974, 119 p.

FOGDEN, M. P. L. The seasonality and population dynamics of equatorial forest birds in Sarawak. *Ibis*, 114, 1972, p. 307–343.

FOX, J. E. D. *The natural vegetation of Sabah and natural regeneration of the dipterocarp forests.* University College of North Wales, Bangor, Ph. D. thesis, 1972, 477 p. multigr.

FOXWORTHY, F. W. The size of trees in the Malay Peninsula. *J. Mal. Br. Roy. As. Soc.*, 4, 1926, p. 382–384.

GARNER, W. W.; ALLARD, H. A. Further studies in photoperiodism, the response of the plant to relative length of day and night. *J. Agr. Res.*, 23, 1923, p. 871–920.

GAUSSEN, H.; LEGRIS, P.; BLASCO, F. *Bioclimats du Sud-Est asiatique.* Inst. français de Pondichéry, Trav. Sect. scientifique et technique, 3, 1967, 114 p.

GUHA, M. M. A preliminary assessment of moisture and nutrients in soil as factors for production of vegetation in West Malaysia. *Malayan Forester*, 32, 1969, p. 423–433.

HALLÉ, F.; OLDEMAN, R. A. A. *Essai sur l'architecture et la dynamique de croissance des arbres tropicaux.* Paris, Masson, 1970, 178 p. (translation by Stone, B. C., Univ. of Malaya, Kuala Lumpur).

HARRISON, J. L. The effect of forest clearance on small mammals. In: *Bangkok Conference of IUCN*, 1965, multigr.

HO, T. H.; SOEPADMO, E.; WHITMORE, T. C. (eds.). National parks of Malaysia. *Malay. Nat. J.*, 24, 1971, p. 111–259.

HOLTTUM, R. E. Periodic leaf change and flowering of trees in Singapore. *Gard. Bull. Str. Settl.*, 11, 1940, p. 119–175.

——. Evolutionary trends in an equatorial climate. *Symp. Soc. Experimental Biology*, 7, 1953, p. 154–173.

——. *Plant life in Malaya.* London, 1954a.

——. *Adinandra* belukar. *J. Trop. Geog.*, 3, 1954b, p. 27–32.

——. Growth habits of Monocotyledons—variation on a theme. *Phytomorphology*, 5, 1955, p. 299–413.

INGER, R. F.; GREENBERG, B. The annual reproductive pattern of the frog *Rana erythraea* in Sarawak. *Physiological Zoology* (Chicago), vol. 36, no. 1, 1963, p. 21–23.

——; ——. Annual reproductive patterns of lizards from a Bornean rain forest. *Ecology* (Durham), vol. 47, no. 6, 1967, p. 1007–1021.

INGER, R. F.; BACON, J. P. Jr. Annual reproduction and clutch size in rain forest frogs from Sarawak. *Copeia* (Washington, D.C.), 3, 1968, p. 602–606.

KENWORTHY, J. B. Water and nutrient cycling in a tropical rain forest. In: Flenley, J. R. (ed.). *The water relations of Malesian forests. Transactions of the first Aberdeen-Hull Symposium on Malesian ecology*, p. 49–65. University of Hull, Dept. of Geography, miscellaneous series no. 11, 1971, 97 p.

KERLING, L. C. The gregarious flowering of *Zephyranthes rosea* Lindl. *Ann. Bot. Gard. Buitenzorg*, 51, 1941, p. 1–42.

KOCHUMMEN, K. M. Natural plant succession after farming in Sungei Keoh. *Malayan Forester*, vol. 29, no. 3, 1966, p. 170–281.

KORIBA, K. On the periodicity of tree growth in the tropics, with reference to the mode of branching, the leaf-fall and the formation of the resting bud. *Gard. Bull. Sing.*, vol. 17, no. 1, 1958, p. 11–81.

KRAMER, F. Onderzoek naar de naturlijke verjonging in uitkap in Preanger gebergtebosch. *Med. Proefst. Boschw. Bogor.*, 14, 1926.

——. De naturlijke verjonging in het Goenoeng Gedeh complex. *Tectona*, 26, 1933, p. 156–185.

LEE, P. C. *Ecological studies on Dryobalanops aromatica* Gaertn. f. University of Malaya, Ph. D. thesis, 1967, multigr.

——. Forestry and land use. In: *Tenth Commonwealth Forestry Conference*, 1974.

LOW, K. S.; GOH, K. C. The water balance of five catchments in Selangor, West Malaysia. *J. Trop. Geog.*, 35, 1972, p. 60–66.

McCLURE, H. E. Flowering, fruiting and animals in the canopy of a tropical rain forest. *Malayan Forester*, 29, 1966, p. 182–203.

MEDWAY, Lord. Phenology of a tropical rain forest in Malaya. *Biol. J. Linn. Soc.*, 4, 1972a, p. 117–146.

——. The Gunung Benom expedition 1967. 6. The distribution and altitudinal zonation of birds and mammals on Gunung Benom. *Bull. Br. Mus. Nat. Hist. (Zoology)*, 23, 1972b, p. 105–154.

MEIJER, W.; WOOD, G. H. S. Dipterocarps of Sabah. *Sabah For. Rec.*, 5, 1964.

MURPHY, D. H. In: Chuang, C. H. (ed.). *Animal life in Singapore*. Singapore, 1973.

NG, F. S. P. Age at first flowering in dipterocarps. *Malayan Forester*, 29, 1966, p. 290–295.

NICHOLSON, D. I. Light requirements of seedlings of five species of Dipterocarpaceae. *Malayan Forester*, 23, 1960, p. 344–356.

——. Forest management. In: *Demonstration and training in forest, forest range and watershed management. The Philippines*. Rome, FAO, Technical Report 3, 1970.

NIEUWOLT, S. Evaporation and water balances in Malaya. *J. Trop. Geog.*, 20, 1965, p. 34–53.

——. A comparison of rainfall in the exceptionally dry year 1963 and average conditions in Malaya. *Erdkunde* (Bonn), 20, 1966, p. 169–181.

NISBET, I. C. T. The utilisation of mangroves by Malayan birds. *Ibis*, 110, 1968, p. 348–352.

PARKHURST, D. F.; LOUKES, O. L. Optimal leaf size in relation to environment. *J. Ecol.*, 60, 1972, p. 505–537.

POORE, M. E. D. Studies in Malaysian rain forest. I. The forest on Triassic sediments in Jengka Forest Reserve. *J. Ecol.*, 56, 1968, p. 143–196.

PRANCE, G. T.; WHITMORE, T. C. Rosaceae. In: Whitmore, T. C. (ed.). *Tree Flora of Malaya*, vol. 2. Kuala Lumpur and London, Longman, 1973.

RICHARDS, P. W. *The tropical rain forest: an ecological study*. Cambridge University Press, 1952, 450 p.; 4th reprint with corrections, 1972.

ROBBINS, R. G. A prerequisite to understanding tropical rain forest. *Malayan Forester*, 32, 1969, p. 361–367.

——; WYATT-SMITH, J. Dryland forest formations and forest types in the Malayan peninsula. *Malayan Forester*, vol. 27, no. 3, 1964, p. 188–217.

SCHMIDT, F. H.; FERGUSON, J. H. A. Rainfall types based on wet and dry period ratios for Indonesia with western New Guinea. *Verhan. Djawatan Meteorologi dan Geofisik* (Djakarta), 42, 1951.

SCHOLANDER, P. F.; HAMMEL, H. T.; HEMMINGSEN, E.; GAREY, W. Salt balance in mangroves. *Plant Physiology*, 37, 1962, p. 722–729.

SOEPADMO, E.; SINGH, K. G. (eds.). *Proceedings of the Symposium on biological resources and national development*. Kuala Lumpur, Malayan Nature Society, 1973.

START, A. G. *Pollination by bats in Malaya*. University of Aberdeen, Ph. D. thesis, 1974, multigr.

STEENIS, C. G. G. J. Van. *Vegetation map of Malaysia*, 1958a, 1 sheet, 8 p. commentary.

——. Rhizophoraceae. In: *Flora Malesiana*, ser. I., 5, 1958b, p. 431–436.

——. *The mountain flora of Java*. Leiden, 1972.

STEVENS, W. E. *The conservation of wildlife in West Malaysia*. Seremban, 1968.

SYMINGTON, C. F. The study of secondary growth on rain forest sites in Malaya. *Malayan Forester*, 2, 1933, p. 107–117.

——. Forester's manual of dipterocarps. *Malay. For. Rec.*, 16, 1943.

TANG, H. T. Preliminary tests on the storage and collection of some *Shorea* spp. seeds. *Malayan Forester*, 34, 1971, p. 84–98.

——; TAMARI, C. Seed description and storage tests of some dipterocarps. *Malayan Forester*, 36, 1973, p. 38–53.

WARD, P. The annual cycle of the yellow-vented bulbul *Pycnonotus goiavier* in a humid equatorial environment. *J. Zool.* (London), vol. 157, part 2, 1969, p. 25–45.

WATSON, J. G. The mangrove swamps of the Malay peninsula. *Malay. For. Rec.*, 6, 1928, 275 p.

WEE, W. C. Weed succession observations on arable peat land. *Malayan Forester*, 33, 1970, p. 63–69.

WELLS, D. R. Resident birds. In: Medway, Lord; Wells, D. R. (eds.). *Birds of the Malay peninsula*. London and Kuala Lumpur, 1974.

WHITEHEAD, C. The rambutan, a description of the characteristics and potential of the more important varieties. *Malay. Agric. J.*, 42, 1959, p. 53–75.

WHITMORE, T. C. Studies in *Macaranga*, an easy genus of Malayan wayside trees. *Malay. Nat. J.*, 20, 1967, p. 89–99.

——. First thoughts on species evolution in Malayan *Macaranga* (studies in *Macaranga*. III.). *Biol. J. Linn. Soc.*, 1, 1969, p. 223–231.

——. Geography of the flowering plants. *Phil. Trans. Roy. Soc. London*, B, 255, 1969, p. 549–566.

——. Wild fruit trees and some trees of pharmacological potential in the rain forest of Ulu Kelantan. *Malay. Nat. J.*, 24, 1971, p. 222–224.

—— (ed.). *Tree Flora of Malaya*, vol. 1. Kuala Lumpur, London, Longman, 1972a, 471 p.

——. Leguminosae. In: Whitmore, T. C. (ed.). *Tree Flora of Malaya*, vol. 1. Kuala Lumpur, London, Longman, 1972b, 471 p.

—— (ed.). *Tree Flora of Malaya*, vol. 2. Kuala Lumpur, London, Longman, 1973a.

WHITMORE, T. C. *Palms of Malaya*. Kuala Lumpur, Oxford University Press, 1973b, 129 p.

——. Frequency and habitat of tree species in the rain forest of Ulu Kelantan. *Gard. Bull. Sing.*, 26, 1973c, p. 195–210.

——. Euphorbiaceae. In: Whitmore, T. C. (ed.). *Tree Flora of Malaya*, vol. 2. Kuala Lumpur, London, Longman, 1973d.

——. Apocynaceae. In: Whitmore, T. C. (ed.). *Tree Flora of Malaya*, vol. 2. Kuala Lumpur, London, Longman, 1973e.

——. *Tropical rain forests of the Far East*. Oxford, Clarendon Press, 1975, 278 p., 550 references.

——; SOH, K. G.; JONES, B. M. G. Studies in *Macaranga*. II. Chromosome counts. *Taxon*, 19, 1970, p. 255–256.

——; AIRY SHAW, H. K. Studies in *Macaranga*. IV. New and notable records for Malaya. *Kew Bulletin*, 25, 1971, p. 237–242.

WONG, Y. K. *Some indications of the total volume of wood per acre in lowland dipterocarp forest*. Malayan For. Dept., Research Pamphlet 53, 1967.

WONG, Y. K.; WHITMORE, T. C. On the influence of soil properties on species distribution in a Malayan lowland dipterocarp rain forest. *Malayan Forester*, 33, 1970, p. 42–54.

WOOD, G. H. S. The dipterocarp flowering season in North Borneo. *Malayan Forester*, 19, 1956, p. 193–201.

WYATT-SMITH, J. Suggested definitions of field characters (for use in the identification of tropical forest trees in Malaya). *Malayan Forester*, 17, 1954a, p. 170–183.

——. Storm forest in Kelantan. *Malayan Forester*, 17, 1954b, p. 5–11.

——. *Manual of Malayan silviculture of inland forests*. Kuala Lumpur, Malayan Forestry Rec., 23, 1963, 2 vol., 400 p.

——. A preliminary vegetation map of Malaya with descriptions of the vegetation types. *J. Trop. Geog.*, 18, 1964, p. 200–213.

——. *Ecological studies on Malayan forest. I*. Malayan For. Dept., Research Pamphlet 52, 1966.

The Melanesian forest ecosystems
(New Caledonia, New Hebrides, Fiji Islands and Solomon Islands)

by M. Schmid[1]

1. Centre de l'Office de la recherche
scientifique et technique
outre-mer (ORSTOM)
B.P. A.5, Nouméa, Nouvelle-Calédonie.

Introduction

The regions lying to the east and north of Australia between the Equator, the Tropic of Capricorn and the 130° and 185° meridians have long been occupied by Melanesian peoples who, despite considerable heterogeneity, are clearly distinguished by both physical features and culture from the other ethnic groups of the Pacific. Melanesia includes New Guinea and its adjoining islands, some of which are of considerable size (New Guinea is 772 000 km², the Bismarck archipelago, 53 000 km², total population: *ca*. 3 millions); the Solomon Islands (28 000 km², 200 000 inhabitants) to which Bougainville (9 000 km², 80 000 inhabitants) belongs geographically but not politically; the New Hebrides (12 000–13 000 km², 80 000 inhabitants); the Fiji Islands (18 000 km², 500 000 inhabitants) and New Caledonia and its dependencies (19 100 km², 150 000 inhabitants). See fig. 1.

New Caledonia and New Guinea have a long geological history; the Fiji Islands begin in the Tertiary; the Solomon Islands and the New Hebrides are of more recent and almost entirely volcanic origin. All these territories have a hot and humid climate, but precipitation varies considerably with the aspect. The decrease of temperature and increase in cloud cover with altitude cause considerable differences in structure and floristic composition between the lowland and mountain vegetation types. The highest peaks of New Guinea are over 5 000 m and are always snow covered. Apart from this area, climatic conditions are suitable to forest growth and it is probable that before the advent of man, forests covered the whole of Melanesia, with the exception of certain ultra-basic areas in New Caledonia and the Solomon Islands and the immediate surroundings of some active volcanoes.

New Guinea has an exceptional diversity of biotopes and an unusually rich flora of 20 000–25 000 species. This case study concerns the other Melanesian territories, distinguishing two main ensembles: the New Caledonian region which is the oldest and is remarkable for the distinctiveness of its flora—though there are undeniable affinities with Australia and New Zealand, and in certain respects, also with New Guinea; the region of the Fiji, the Solomon Islands and the New Hebrides which might be called an eastern, if somewhat impoverished, annex of the Malesian floristic region to which it is linked via New Guinea (although the flora of the Fijis contains some special elements).

The environment

It is now accepted that the archipelagos of Melanesia owe their existence to horizontal movements of the earth's crust which affected the bottom of the Pacific Ocean and the landmasses corresponding to the old continent of Gondwana, on the eastern edge of which New Caledonia, the Norfolk Islands and New Zealand would have been situated. These landmasses might have been under the oceanic plate in their progression eastwards at the point of the deep trench which separates New Caledonia from the New Hebrides. This might provide the biogeographical explanation for the marked isolation of the New Caledonian region. This isolation is shown in the differences between the flora of New Caledonia and of the New Hebrides, which lie less than 250 km away, and in the close affinities between the flora of the New Hebrides and of the Fiji Islands, which are separated by 800 km.

Relief and climate

In each archipelago there are some islands larger than 3 000 km² where the great variation in environmental conditions together with insularity favour speciation. Except in some smaller islands (Loyalty, Renell) the relief is young and broken but altitude never prevents forest growth. The highest peaks are in the Solomon Islands (on Bougainville, Mt Balbi, 2 765 m; on Guadalcanal, Mt Popomanaseu, 2 330 m). The highest point in the New Hebrides is 1 880 m on the Island of Santo, that of New Caledonia *ca*. 1 640 m (Mt Panié and Mt Humboldt) and that of the Fiji Islands only 1 323 m: Mt Victoria (Tomanivi) on Viti Levu. The extent of coastal plains and deltas is small and the plateaux, which correspond to raised coral reefs or basaltic formations are intersected by deep valleys. The narrowness of ridges and steepness of slopes, especially in the forested zones which include the most mountainous and humid regions, make forest exploitation more difficult and more threatening for the environment.

There are two majors seasons, one relatively cool, extending from May to September or October, when the wind blows regularly from south-east, and another hot and humid, when the region is covered with equatorial air masses, irregular winds and frequent cyclones. The less humid regions are those protected by mountains from the winter trade-winds: in all islands they occur in the west and northwest, but both in the east and in the west the volume of rainfall may vary considerably over short distances, according to the exposition of the main slopes, and, up to a certain level depending on the altitude and the size of an island. On the coast, the mean highest figures have been recorded in the Solomon Islands where rainfall can exceed 6 000 mm/a; in the northeast of New Caledonia it reaches 4 400 mm/a near the coast and 8 000 mm at Mt Panié (1 600 m) whereas in the northwest, less than 50 km away, it falls to 700 mm/a. Apart from certain sites in New Caledonia where the rainfall regime is particularly contrasted, the islands of Melanesia have a very humid climate (with

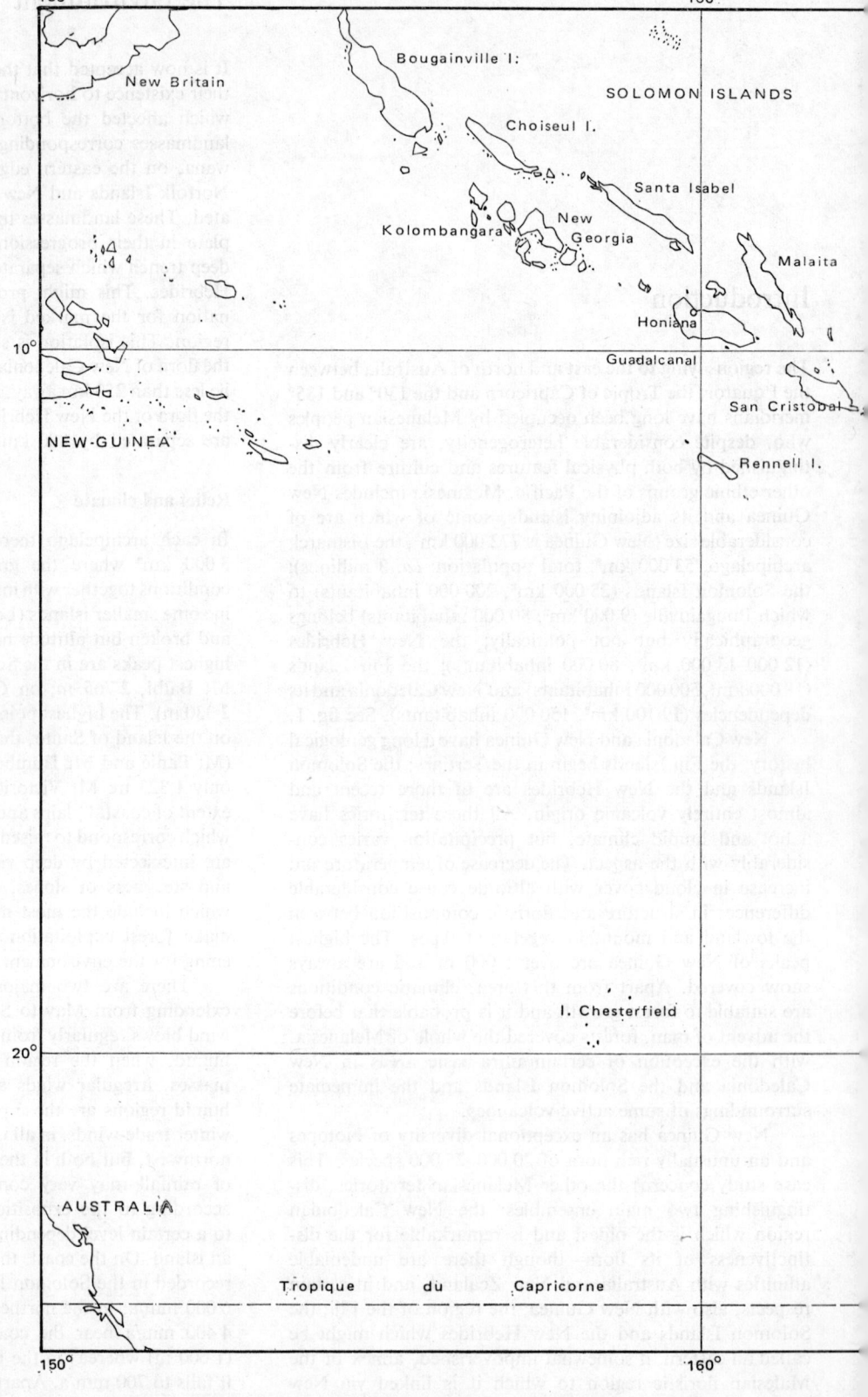

FIG. 1. Situation of the main archipelagos of Melanesia.

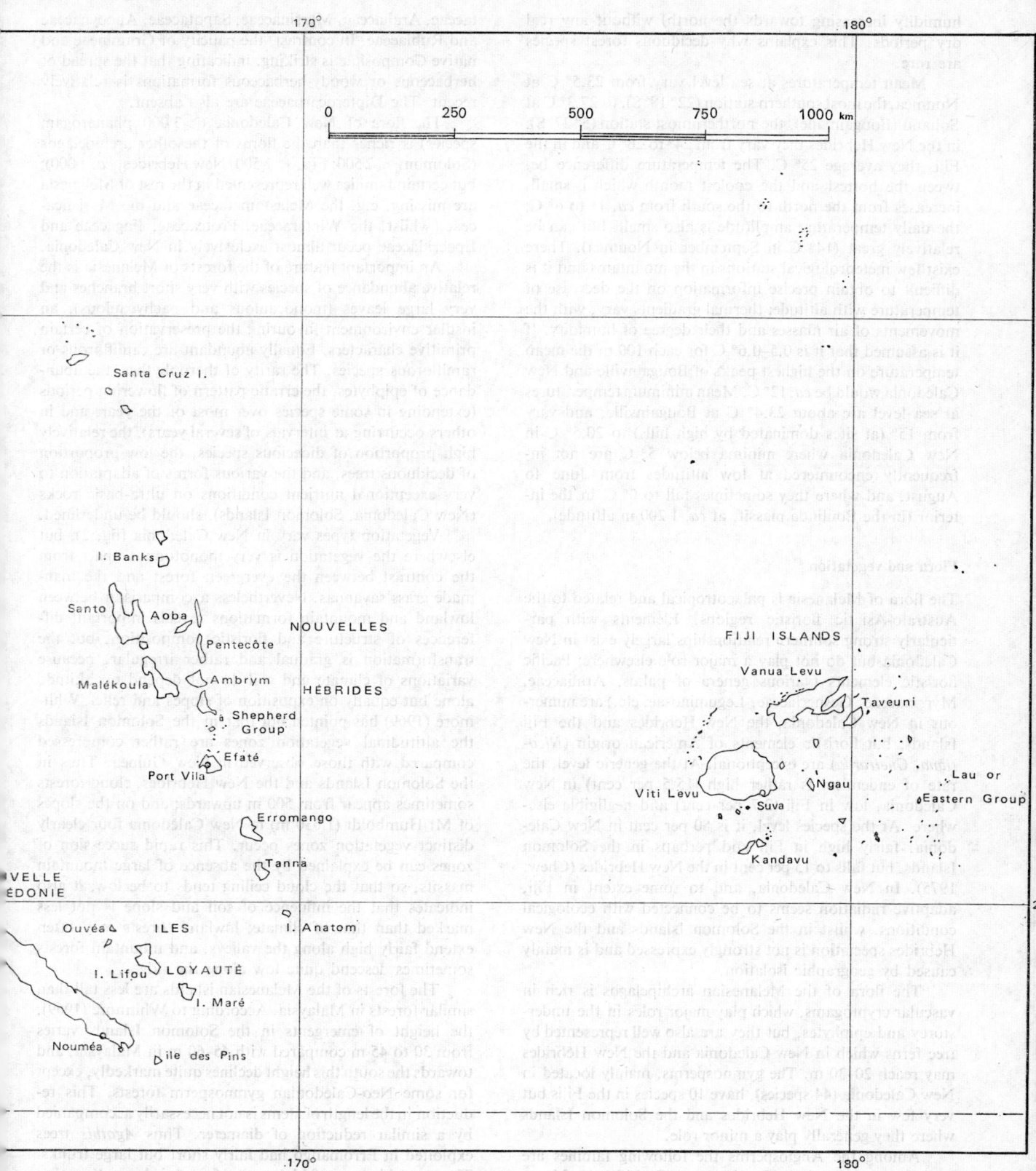

humidity increasing towards the north) without any real dry periods. This explains why deciduous forest species are rare.

Mean temperatures at sea level vary from 23.5° C at Nouméa, the most southern station (22° 18′ S), to 27.3° C at Sohano (Bougainville), the northernmost station (5° 27′ S); in the New Hebrides they vary from 24° to 26° C and in the Fijis they average 25° C. The temperature difference between the hottest and the coolest month which is small, increases from the north to the south from *ca.* 1° to 6° C; the daily temperature amplitude is also small, but can be relatively great (14° C in September in Nouméa). There exist few meteorological stations in the mountains and it is difficult to obtain precise information on the decrease of temperature with altitude; thermal gradients vary, with the movements of air masses and their degree of humidity. If it is assumed that it is 0.5–0.6° C for each 100 m the mean temperature on the highest peaks of Bougainville and New Caledonia would be *ca.* 12° C. Mean minimum temperatures at sea level are about 23.4° C at Bougainville, and vary from 15° (at sites dominated by high hills) to 20.5° C in New Caledonia where minima below 5° C are not infrequently encountered at low altitudes from June to August, and where they sometimes fall to 0° C in the interior (in the Boulinda massif, at *ca.* 1 200 m altitude).

Flora and vegetation

The flora of Melanesia is palaeotropical and related to the Australo-Asiatic floristic regions. Elements with particularly strong southern relationships largely exist in New Caledonia but do not play a major role elsewhere; Pacific floristic elements (various genera of palms, Araliaceae, Myrsinaceae, Gesneriaceae, Leguminosae, etc.) are numerous in New Caledonia, the New Hebrides and the Fiji Islands; but floristic elements of American origin (*Nicotiana*, *Guettarda*) are exceptional. At the generic level, the rate of endemism is rather high (15.5 per cent) in New Caledonia, low in Fiji (2.7 per cent) and negligible elsewhere. At the species level, it is 80 per cent in New Caledonia, fairly high in Fiji and perhaps in the Solomon Islands, but falls to 15 per cent in the New Hebrides (Chew, 1975). In New Caledonia, and to some extent in Fiji, adaptive radiation seems to be connected with ecological conditions, whilst in the Solomon Islands and the New Hebrides speciation is not strongly expressed and is mainly caused by geographic isolation.

The flora of the Melanesian archipelagos is rich in vascular cryptogams, which play major roles in the understorey and epiphytes, but they are also well represented by tree ferns which in New Caledonia and the New Hebrides may reach 20–30 m. The gymnosperms, mainly located in New Caledonia (44 species), have 10 species in the Fijis but very few in the New Hebrides and the Solomon Islands where they generally play a minor role.

Among the Angiosperms the following families are best represented: Palmae, Pandanaceae, Orchidaceae, Moraceae (*Ficus*), Lauraceae, Meliaceae, Sapindaceae, Elaeocarpaceae (*Elaeocarpus*), Euphorbiaceae, Cunoniaceae, Myr-

taceae, Araliaceae, Myrsinaceae, Sapotaceae, Apocynaceae and Rubiaceae. In contrast, the paucity of Gramineae and native Compositae is striking, indicating that the spread of herbaceous or woody-herbaceous formations is relatively recent. The Dipterocarpaceae are also absent.

The flora of New Caledonia (>3 000 phanerogam species) is richer than the floras of the other archipelagos (Solomon, <2 500; Fiji, <1 500; New Hebrides, *ca.* 1 000); but certain families well represented in the rest of Melanesia are missing, e.g. the Melastomataceae and the Myristicaceae, whilst, the Winteraceae, Proteaceae, Fagaceae and Epacridaceae occur almost exclusively in New Caledonia.

An important feature of the forests of Melanesia is the relative abundance of species with very short branches and very large leaves (monocaulous and pachycaulous), an insular environment favouring the preservation of certain primitive characters. Equally abundant are cauliflorous or ramiflorous species. The rarity of therophytes, the abundance of epiphytes, the erratic pattern of flowering periods (extending in some species over most of the year, and in others occurring at intervals of several years), the relatively high proportion of dioecious species, the low proportion of deciduous trees, and the various forms of adaptation to very exceptional nutrient conditions on ultra-basic rocks (New Caledonia, Solomon Islands), should be underlined.

Vegetation types vary in New Caledonia (fig. 2), but elsewhere the vegetation is very monotonous apart from the contrast between the evergreen forest and the man-made grass savannas. Nevertheless a comparison between lowland and mountain formations reveals important differences of structure and floristic composition, but the transformation is gradual and rather irregular, because variations of climate and soil do not depend on altitude alone but equally on exposition of slopes and relief. Whitmore (1969) has pointed out that in the Solomon Islands the altitudinal vegetation zones are rather compressed compared with those observed in New Guinea. Thus in the Solomon Islands and the New Hebrides, cloud forests sometimes appear from 500 m upwards; and on the slopes of Mt Humboldt (1 630 m) in New Caledonia four clearly distinct vegetation zones occur. This rapid succession of zones can be explained by the absence of large mountain massifs, so that the cloud ceiling tends to be low; it also indicates that the influence of soil and slope is not less marked than that of climate; lowland forests may often extend fairly high along the valleys, and mountain forests sometimes descend quite low along the ridges.

The forests of the Melanesian islands are less tall than similar forests in Malaysia. According to Whitmore (1969), the height of emergents in the Solomon Islands varies from 30 to 45 m compared with 45–60 m in Malaysia, and towards the south this height declines quite markedly, except for some Neo-Caledonian gymnosperm forests. This reduction in the length of stems is not necessarily accompanied by a similar reduction of diameter. Thus *Agathis* trees exploited in Erromango had fairly short but large trunks. The mean biomass of Melanesian forests is lower than that of the other high tropical forests, but their relative paucity of species favours exploitation: according to Whitmore, a survey of 3.6 ha in the Solomon Islands gave 104 tree species with stems >20 cm girth, compared with 383 species on 4 ha in Malaysia. In New Caledonia, where the flora is much more varied, the figures would approach the Malaysian ones, but the majority of woody species are only shrubs and many of them have a local distribution. Several Melanesian species of economic interest are gregarious (Araucariaceae, *Arillastrum*).

Outside the ultra-basic areas, the climax vegetation is forest and if all human activity ceased the recovery of the original vegetation would be possible almost everywhere, except perhaps in some relatively dry regions where certain rather agressive exotics (*Leucaena*, *Aleurites*) are well established. Whilst the large extension of savannas appears to be due essentially to fires lit by man, the vast areas of thickets and very low forests laden with lianas might, at least partly, be attributable to cyclones which are particularly frequent in the New Hebrides (one per five years on average) and the Solomon Islands but also affect Fiji and New Caledonia, though damage to forests there tends to be less severe. The destruction caused by repeated cyclones are particularly serious because they are mostly accompanied by downpours which weaken the stability of upper soil layers and increase the weight of the foliage. Gaps are frequent on deep soils and in areas which receive heavy rainfalls; on slopes land-slides occur. Certain species with deep tap-roots do, however, resist even very violent gales; this is the case of the Araucariaceae which are so numerous in New Caledonia. Volcanic eruptions also affect the evolution of the natural vegetation; falls of ash may sometimes occur at great distances from craters, but the effects are not really spectacular except in the neighbourhood of erupting volcanoes (Ambrym, Tanna, in the New Hebrides).

The herbaceous or woody-herbaceous formations, described as savannas or, in isolated cases (Erromango) as grasslands, occur in relatively dry areas, though in New Caledonia they extend to the humid zones. They should be considered as secondary formations, since they are mainly composed of pantropical Gramineae, and include sometimes recently introduced species, such as *Pennisetum polystachyon* in Viti Levu (Fiji) a species from tropical Africa. *Miscanthus floridulus*, the most important species of the savannas in the Fijis and also well represented in the New Hebrides and New Caledonia, is probably indigenous. *Imperata cylindrica*, *Heteropogon contortus*, *Themeda gigantea* and *Chrysopogon aciculatus* are particularly numerous in New Caledonia where they existed before the advent of Europeans. The secondary nature of these formations is confirmed by the absence in the tree layer of fire-tolerant species. *Melaleuca quinquenervia* (niaouli) which occurs only in New Caledonia is the only exception.

The herbaceous element is also present in the maquis of New Caledonia which cover a large part of the ultra-basic soils and are also found in highly siliceous areas. The structure and floristic composition of these varies according to the types of soil and site; there are lowland and mountain maquis. Their flora contains many distinctive species and some of them could be climax formations; it is, however, poor on siliceous terrain. Their physiognomy is due to a woody layer consisting of bushy small trees or shrubs often having an ombelliform or candelabra habit; the herbaceous layer is often well developed (on ferrallitic soils without hard-pan) or sometimes almost completely absent (on eutrophic brown soils); it mainly consists of Cyperaceae. Formations with such a physiognomy are occasionally but rarely found outside New Caledonia; they appear to be a terminal stage of a long process of degradation, and, except on ultra-basic rocks in the Solomon Islands, seem related to low open shrub formations with an understorey of ferns (Gleicheniaceae) occurring on heavily leached soils, which several authors described as landes (*Vaccinium* heath).

It is also possible to link the maquis with certain shrub formations growing on steep mountain slopes or narrow ridges, generally described as thickets, a term which also applies to very dense secondary formations a few meters high. Secondary thickets cover considerable areas in the New Hebrides and the Solomon Islands and are mainly composed of *Hibiscus tiliaceus*, various Euphorbiaceae (*Macaranga*), Musaceae (*Heliconia*), Zingiberaceae, Pandanaceae, Bambuseae (in the Solomon Islands) and ferns. Climbers are often abundant. *Leucaena leucocephala* forms almost monospecific thickets in comparatively dry areas.

It is possible to distinguish several forest types according to the density of the dominant stratum, the height of the canopy, the abundance of epiphytes (in cloud forest), the density or composition of the understorey, the forms of the trunks (proportion of species with large buttresses), or the floristic composition of the dominant storey. The vast majority of the forest are evergreen; but stands of deciduous species (*Pterocarpus*, *Gyrocarpus*) occur in restricted localities of the New Hebrides and the Solomon Islands.

There also exist swamp formations and mangroves. The former are of some importance in the Solomon Islands where they consist sometimes of *Terminalia brassii*, *Campnosperma brevipetiolatum*, *Eugenia tierneyana*, or of tall grass (*Phragmites*, *Saccharum*) or sedge (*Baumea*, *Rynchospora*) communities. Mangroves are wide-spread on the western coast of New Caledonia; they are also well developed on certain islands of the Solomon group (Santa Isabel, Malaita). In the Fiji Islands they are of more local importance (northwestern parts of Viti Levu and Vanua Levu) as well as in the New Hebrides (Mallicolo). Their composition is the usual one of Rhizophoraceae, *Lumnitzera*, *Avicennia* and *Sonneratia*. The palm *Nypa fruticans* occurs in the Solomon Islands.

Fauna

The unity of the great Melanesian archipelagos is more clearly apparent in their fauna, at least their vertebrate fauna, than in their flora; New Caledonia shows a higher degree of endemism, and the Solomon group a greater richness (attributable to the size of their territory and their proximity to New Guinea).

Most indigenous terrestrial bird species have a forest habitat. This is particularly so for the endemic species; but their ecological specialization is not very strong: many birds can be observed both in the lowlands and the mountains, in closed forest or open woodlands. It seems that the diversification of species in each genus, which is often not very marked, depends on geographic isolation.

There are 68 indigenous species of terrestrial birds in New Caledonia (4 genera and 20 species are endemic), 56 species in the New Hebrides (1 genus and 7 species endemic), 57 species in the Fijis (1 genus and 13 species endemic), and >150 species in the Solomon Islands (the bird fauna is incompletely known). A number of species are common to all four areas; some of them (including some predators) have a world-wide distribution, while others are represented by one or more distinct sub-species in each archipelago. There are close affinities between the ornithological faunas of the New Hebrides and the Solomon Islands, and also between the latter and that of New Guinea, which has a much richer fauna (520 species); 27 bird species in New Caledonia may be of Australian origin, and some of New Zealand kinship.

Because of the length of fructification periods, the forests provide fruit-eating and seed-eating species with abundant food during most part of the year (figs, wild olives, fruits of Palms or of *Couthovia* especially appreciated by pigeons, kauri seeds consumed by parrots, etc.).

Apart from a phalanger, in the Solomon Islands, and some rats (whose origin is doubtful) the mammals are exclusively represented by bats. The Megachiroptera (flying foxes) which are fruit and flower eaters represent a characteristic element of the fauna, and include a number of species found only in the South West Pacific. The Microchiroptera are numerous and have larger distribution areas extending as far as the Asian continent.

Apart from indigenous mammals there are animals descended from domestic animals, pigs (*Sus papuensis*) being the first introduced. They are now common in a feral state in the forests of nearly all the large islands. In New Caledonia the great Indo-Malaysian deer, *Rusa unicolor*, was introduced a hundred years ago.

There are many lizards, mainly belonging to the genus *Emoia*, geckos (some of these in New Caledonia possessing striking primitive characters), an *Iguana* occurring only in the Fiji Islands, and some snakes. In the main island of New Caledonia snakes are only represented by a *Typhlops*, whilst 3 species, including a fairly large one, live in the Loyalty Islands.

There are no indigenous amphibia in New Caledonia or the New Hebrides, but there are several in the Solomon and Fiji Islands.

The freshwater fishes have been little studied, apart from the *Galaxias* of New Caledonia, which represent a primitive group belonging to the austral regions.

The terrestrial molluscs have Asian affinities with the exception of the large Bulimulideae, snails of the forests which live in the four archipelagos and the northern part of New Zealand.

The arthropod faunas are not yet well known; the degrees of endemism and their affinities may be similar to those of the corresponding floras: New Caledonia being clearly distinct (several endemic families) and closer with Australia. According to Gross (1975) the arthropod fauna

of the New Hebrides is comparatively rich, having only weak affinities with New Caledonia, the Fiji and Solomon Islands.

The forest ecosystems

New Caledonia

Distribution

New Caledonia and its surrounding islands, the Île des Pins and the Belep islands, differ from the other Melanesian islands by the great extension of ultra-basic rocks overlaid by very poor soils having a peculiar flora. Forests occur on sedimentary or metamorphic rocks (micaschists), but there are also large stands in the peridotitic areas. Little forest now exists at low altitudes (such as the coastal forests of the region of Bourail and forests of the Île des Pins on calcareous sites); most of these forests are in the east where the country is more mountainous and, because of exposure to the prevailing winds, is more humid. In the Loyalty Islands, forest occupies the cliffs and the coral plateaux.

The surveys made so far do not give the precise determination of the areas covered by forests and by the other major vegetation types—the climax vegetation being very heterogeneous. The formations growing in the valleys and on the lower slopes are frequently quite tall, whilst the steep slopes and narrow ridges are covered with low formations which sometimes are reduced to thickets. Altitude does not appear to be the direct cause of this, since on the top of Mt Panié, the highest point of the area (1 640 m), there is a fine stand of kauris. The following estimates have only indicative value (see fig. 2).

Area of the main island	16 750 km²
Area of the surrounding islands and islets	17 135 km²
— Formations on massifs of ultra-basic volcanic rock (with some intrusions of more or less acid rocks)	*ca.* 6 000 km²
Forests (including low cloud forests and thicket forests on narrow ridges and steep slopes)	750 km²
Various types of maquis and woodlands	5 250 km²
— Formations on sedimentary or metamorphic terrain	*ca.* 11 135 km²
Climax forests (including low formations on steep slopes or ridges, and formations that are more or less degraded by exploitation) of which 100 km² are on calcareous soils (Île des Pins, Koumac region)	2 000 km²
Secondary forests	600 km²
Maquis on highly siliceous rocks	500 km²
Melaleuca quinquenervia (niaouli) savannas (very heterogeneous)	5 500 km²

Grass savannas and natural grass-lands	1 000 km²
Mangroves (on the western coast and the north east extremity of the main island)	250 km²
Various formations (*Lantana* and *Leucaena* thickets, plantations, artificial grasslands, urban zones, etc.)	1 285 km²

Total surface of the Loyalty Islands (entirely calcareous and flat, with the exception of two basaltic elevations of a few km² at Maré) *ca.* 1 970 km².

Climax forests	550 km²
Thickets on rocks, low secondary forests, man-made thickets	1 220 km²
Mangrove (Ouvea)	a few km²
Formations with a dominant herbaceous layer (especially on Maré, on substratum rich in Ca and Mg)	200 km²

In the inventories of the Forest Service, the land under forest is estimated as 10 per cent of the total area. This does not agree with ecological reality but can be considered reasonable if only dense forests are taken into account. The *Melaleuca* savannas and the grass savannas include 3 500 km² of natural pastures which are partly used for livestock husbandry; these areas are the most interesting for reforestation.

Flora

The distinctiveness and richness of the New Caledonian flora has long attracted the attention of botanists. The mean rate of specific endemism is *ca.* 80 per cent; it is certainly >90 per cent of the angiosperms in the climax formations of the interior of the main island and nearly 100 per cent in the formations growing on ultra-basic rocks. But the floras of the coastal or calcicolous forests, as well as that of the secondary forests, contain a large proportion of species which extend considerably beyond New Caledonia.

The dominant stratum of a particular forest is often composed of a small number of species, but its composition varies over the territory; the flora of the dominated layers is richer and more varied.

In the tree layers there is an abundance of gymnosperms, especially Araucariaceae, and of Guttiferae, Myrtaceae, Sapotaceae, Proteaceae, Lauraceae, Araliaceae, Meliaceae, Elaeocarpaceae, Moraceae, Cunoniaceae, Euphorbiaceae and Fagaceae (*Nothofagus*); among the Monocotyledons, Palmae are abundant, and among the vascular cryptogams, the Cyatheaceae. The Casuarinaceae occur in the maquis, but mainly in secondary forests on ultra-basic soils, and the Sapindaceae in secondary forests on sedimentary rocks. The Leguminosae, which play a comparatively minor role, are represented by *Albizia* spp., *Storckiella* (occurring only on New Caledonia and the Fiji Islands) and, on limestone, by *Afzelia bijuga*; the Rhizophoraceae are represented by the

Melanesian genus *Crossostylis*, the Icacinaceae by endemic genus *Anisomallon*. The epiphytes are mainly composed of ferns; the semi-epiphytes are mainly Pandanaceae (*Freycinetia*), Piperaceae (*Piper*), Araceae (*Epipremnum*) and ferns (*Teratophyllum*); and lianas of Verbenaceae (*Oxera*), Apocynaceae (*Alyxia, Parsonsia*), Linaceae (*Hugonia*), Leguminosae (*Caesalpinia, Mucuna*), Rhamnaceae (*Ventilago*), Violaceae (*Agatea*), Menispermaceae and Philesiaceae. In the shrub layers, the Rubiaceae, Myrsinaceae, Apocynaceae, Myrtaceae, Euphorbiaceae, Flacourtiaceae, Sapindaceae, Pandanaceae and the Palmae dominate. There are also some characteristic elements such as the endemic families Amborellaceae and Phellineaceae, and the Winteraceae. Stranglers, such as figs, exist at low altitudes, and *Metrosideros* in the interior forests. With the exception of *Afzelia bijuga* all species that are regularly felled are endemic. The most sought after is the *Montrouziera cauliflora* which is peculiar to the main island. The *Kermadecia* and *Sleumerodendron* of the Proteaceae also yield valuable timber.

On the ultra-basic sites and in the mountains competition favours species which are adapted to harsh conditions rather than highly dynamic species having stricter ecological requirements, and the rugged relief contains innumerable refugial sites. Hence many species which had never been recorded or were considered extinct have been found or rediscovered.

Types of forest

Because the climatic and edaphic conditions vary greatly over very short distances it is rare to find communities that are homogeneous in structure and composition over large areas—except in the Loyalty Islands: this heterogeneity is more marked on ultra-basic than on sedimentary sites. However, the very rugged relief presents a certain uniformity, as does the influence of aspect and altitude on climate. Hence, the study of the forest cover by sampling techniques remain a valid method—for detecting the distribution of species of general occurrence and for determining the volume of usable timber.

The abundance of species with a very special habit (Araucariaceae, pachycaulous plants, including many monocaulous shrubs in the understorey) gives certain communities a characteristic physiognomy.

Within species the dimensions of individuals vary greatly with soil conditions, and it is not unusual to find tall trees in valley bottoms of species which on hard-pan soils or on steep slopes are shrubs.

Structure and stratification are not always distinct.

Apart from certain communities associated with rather rich soils and relatively dry conditions, the forests of New Caledonia are closed and exotic species can not penetrate so long as they are intact.

There are open and closed forests, but no New Caledonian formation even on ultra-basic sites really corresponds to the descriptions generally given of open tropical forests.

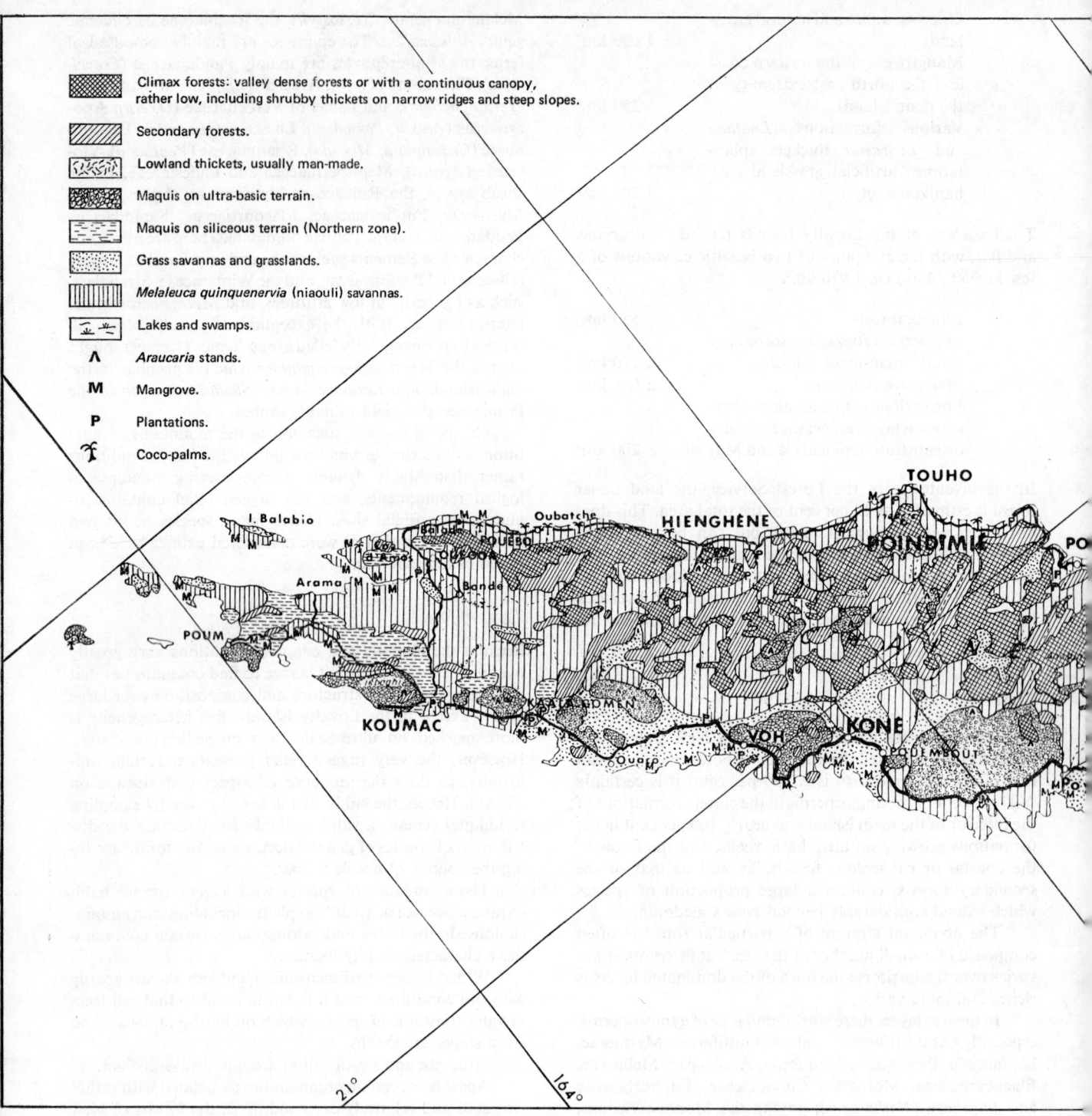

Climax forests: valley dense forests or with a continuous canopy,
rather low, including shrubby thickets on narrow ridges and steep slopes.

Secondary forests.

Lowland thickets, usually man-made.

Maquis on ultra-basic terrain.

Maquis on siliceous terrain (Northern zone).

Grass savannas and grasslands.

Melaleuca quinquenervia (niaouli) savannas.

Lakes and swamps.

Λ   *Araucaria* stands.

M   Mangrove.

P   Plantations.

    Coco-palms.

Fɪɢ. 2. Main vegetation types of New Caledonia.

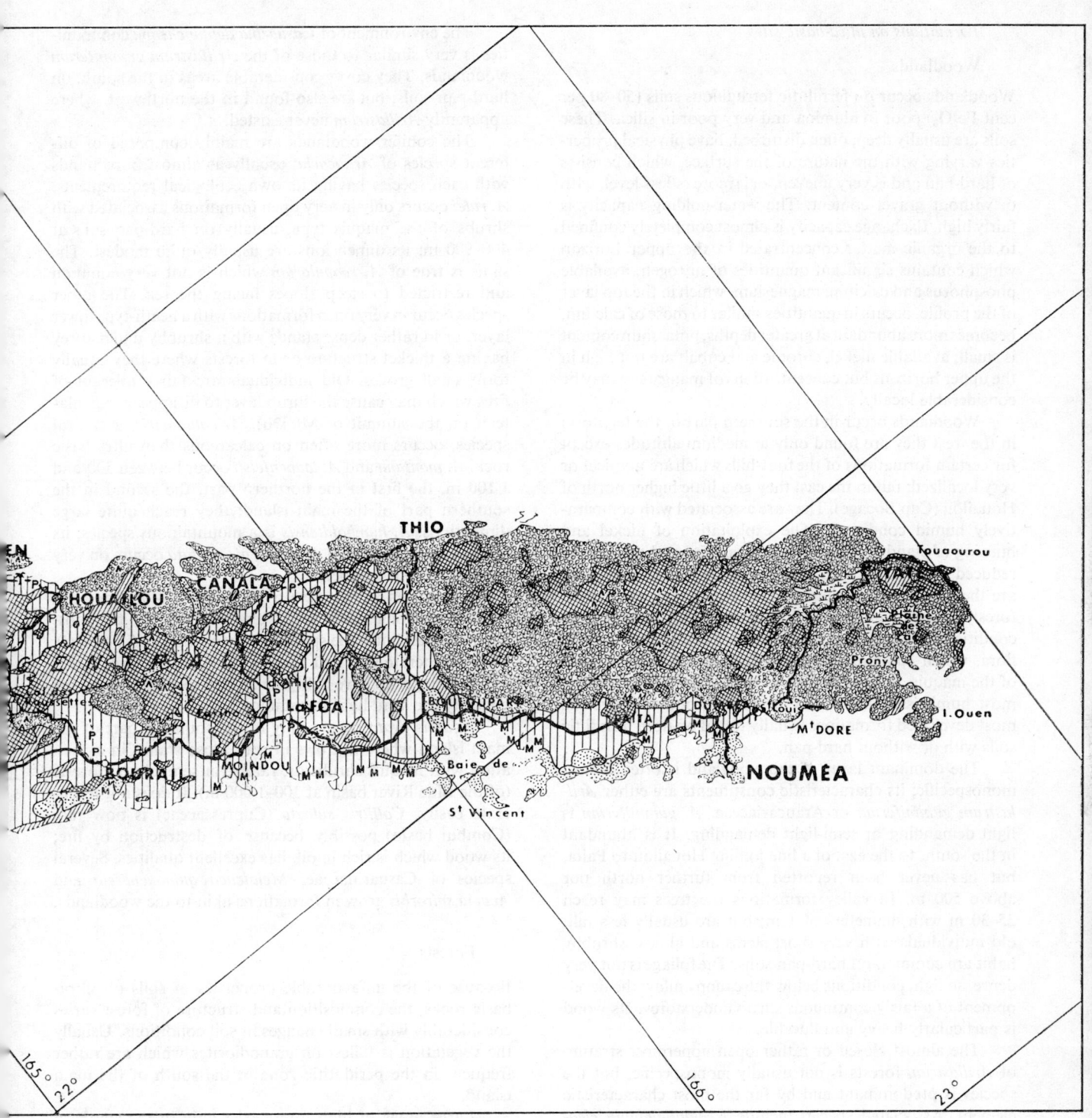

Formations on ultra-basic sites

Woodlands

Woodlands occur on ferrallitic ferruginous soils (50–80 per cent Fe_2O_3) poor in alumina and very poor in silica. These soils are usually deep, often disturbed, have physical properties varying with the nature of the surface, which consists of hard-pan and is very uneven, or is more or less level, with or without gravel content. The water-holding capacity is fairly high. Exchange capacity is almost completely confined to the organic matter concentrated in the upper horizon which contains significant quantities of nitrogen, available phosphorus and calcium; magnesium, which in the top layer of the profile, occurs in quantities similar to those of calcium, becomes more abundant at greater depths; potassium content is small; available nickel, chrome and cobalt are not high in the upper horizons but concentrations of manganese may be considerable locally.

Woodlands occur in the southern part of the territory; in the west they are found only at medium altitudes except for certain formations of the foot-hills which are atypical or very localized; but in the east they go a little higher north of Houailou (Cap Bocage). They are associated with comparatively humid conditions. The exploitation of nickel and burning of stands of *Arillastrum gummiferum*, have greatly reduced their area. Their origin remains to be elucidated: are they climax formations or formations replacing closed forest? The answer may well be different for different site conditions. Considering the distinctive composition of the flora, which contains few forest species but many species of the maquis, it is reasonable to think that, apart from the most humid sites and valley bottoms, woodlands are the most developed formations usually found on deep ferrallitic soils with or without hard-pan.

The dominant layer of the woodland is often almost monospecific; its characteristic constituents are either *Arillastrum gummiferum* or Araucariaceae. *A. gummiferum* is light demanding or semi-light demanding. It is abundant in the south, to the east of a line joining Houailou to Paita, but has never been reported from further north nor above 500 m. In valley formations the trees may reach 25–30 m with diameters of 1 m; but are usually less tall; old individuals with very short stems and almost shrubby habit are common on hard-pan soils. The foliage is not very dense, so light conditions below the canopy allow the development of a fairly continuous shrub understorey. Its wood is particularly heavy and durable.

The almost closed or rather open uppermost stratum of *Arillastrum* forests is not usually monospecific; but the species is predominant and by far the most characteristic element. Associated species include *Neoguillauminia cleopatra* and *Agathis lanceolata* in the semi-closed formations. The flora of the understorey is rich and varies with the soil conditions.

Arillastrum regenerates well on various types of ferrallitic soils but its growth is slow. It responds positively to fertilizers and, at least in its natural area, could be the most interesting species for reforestation in the zones of mineral exploitation. Unfortunately, it is not fire-tolerant.

The environment of *Casuarina deplancheana* communities is very similar to those of the *Arillastrum gummiferum* woodlands. They cover considerable areas in the south, on hard-pan soils; but are also found in the northwest, where apparently *Arillastrum* never existed.

The conifer woodlands are mainly composed of different species of *Araucaria*; usually as almost pure stands with each species having its own ecological requirements. *A. rulei* occurs only in very open formations associated with shrubs of the maquis type, usually on hard-pan soils at 400–900 m; its dimensions are usually quite modest. The same is true of *A. scopulorum* which is not very common and restricted to steep slopes facing the sea. The other species occur in very open formations with a heath-type lower layer, or in rather dense stands with a shrubby understorey having a thicket structure or in forests where they usually form small groves. Old individuals are fairly tolerant of fires which may cause the shrub layer to disappear (e.g. plateau on the summit of Mt Do). *A. columnaris*, a coastal species, occurs more often on calcareous than ultra-basic rock. *A. montana* and *A. laubenfelsii* occur between 500 and 1 100 m, the first in the northern part, the second in the southern part of the main island; they reach quite large dimensions. *A. humboldtensis* is a mountainous species; its height rarely exceeds *ca.* 10 m. *A. biramulata* occurs on very steep and eroded slopes.

Araucarias which have a very powerful tap-root can resist the strong winds and, generally occupy the most exposed sites, but some species grow in valleys (*A. bernieri*). They regenerate well and become fire-tolerant when rather old. Some species have light demanding seedlings whilst the majority are semi-light demanding.

Agathis ovata occurs only in the southern part of the main island where it grows in very loose stands in association with maquis shrubs or, rarely, in semi-closed forests (of the Blue River basin at 300–1 000 m). It does not regenerate easily. *Callitris sulcata* (Cupressaceae) is now rare (Combui basin) possibly because of destruction by fire; its wood which is rich in oil, has excellent qualities. Several species of Casuarinaceae, *Melaleuca quinquenervia* and *Acacia spirorbis* grow in formations akin to the woodlands.

Forests

Because of the unfavourable properties of soils on ultra-basic rocks, the composition and structure of forest varies considerably with small changes in soil conditions. Usually the vegetation is tallest on granodiorites which are rather frequent in the peridotitic zone in the south of the main island.

Large areas of high forest exist only in the southern part of the main island; in the centre (Mt Maoya) and in the west, except in the valleys, only low forests occur on strongly exposed slopes and near the peaks. This distribution could be attributable to mining, which has long been much more active in the western parts; however, high concentrations of minerals are rare under forest which occupies loose ferrallitic soils of medium depths on rocky slopes.

The most characteristic species is *Agathis lanceolata*

(kauri) which may reach 35–40 m in height with sometimes 20 m to the first branch and over 2 m diameter. Such trees, which are found in valley communities, are exceptional. Associated are columnar araucarias (*A. bernieri, A. subulata*) which usually occur in patches and sometimes reach 50 m in height but rarely more than 1 m in diameter. There are a considerable number of other associated species (Sapotaceae, Guttiferae, Myrtaceae, Leguminosae, Elaeocarpaceae, Cunoniaceae, Podocarpaceae); many, including several valuable ones, occur also on sedimentary sites where some reach their largest dimensions. The lower tree storey and the shrub understorey possess a very rich flora (Myrtaceae, Araliaceae, Lauraceae, Rubiaceae, Apocynaceae, Flacourtiaceae and Palmae), and *Cyathea novae-caledoniae*.

Agathis lanceolata is never found above 850 m; the large araucarias occur a little higher. At higher altitudes the leaves become more coriaceous and smaller. Myrtaceae (*Metrosideros*), Podocarpaceae and tree ferns (*Dicksonia*) become more abundant; the palms decrease.

The slopes and ridges between 1 100 and 1 300 m are covered with cloud or moss forest, composed of low branching trees covered with epiphytes. In New Caledonia this formation is associated with a loose soil rich in organic matter, resting on hardly weathered parent rock and is found only on peridotitic mountain ranges (Mt Mou, Humboldt, Boulinda). Scrub and a high altitude maquis with a very special flora (Mt Humboldt) occur above it.

Among the communities found on peridotitic sites, the *Cocconerion* forests (Euphorbiaceae) are particularly common in the northwest between 500 and 900 m. The *Nothofagus* forests (Fagaceae) contain several species of this genus, some of which grow also on sedimentary soils. These semi-closed formations are 6–20 m high and have an almost monospecific top stratum; they occur along valleys.

The forests of the ultra-basic zone are unstable. Their destruction causes the loss of valuable minerals from the soil either by erosion or leaching, and consequently a lasting depletion because of the slowness or impossibility of replenishing nutrients from the rock. The trees of the upper layer are very old; the mean annual diameter increment of the *Agathis* is a few mm; trees with diameters of 1 m are 250–300 years old, and the giant individuals perhaps more than a thousand years. The small isolated gaps are invaded by palms and a fairly complex flora of dicotyledons; conditions are quite favourable for the regeneration of Araucariaceae; yet their seedlings are rarely abundant. In larger gaps and in areas once covered with forests, *Pteridium esculentum* propagation is facilitated by fire, and occasionally bamboos (*Greslania*) occur.

Formations on metamorphic or non-calcareous sedimentary terrains

These formations differ from the forests on ultra-basic rocks by their greater structural homogeneity and by the lesser importance of the gymnosperms although the largest trees are *Agathis* species but not the same as those growing in the mining areas. In the upper stratum are Guttiferae (*Montrouziera, Calophyllum*) which also occur in mining

areas but which here are more abundant and reach up to 3 m in diameter for *Montrouziera caulifora*. Proteaceae (*Kermadecia, Sleumerodendron*), Myrtaceae (*Piliocalyx, Syzygium*), Cunoniaceae, Leguminosae (*Albizia*), Rhizophoraceae (*Crossostylis*), Meliaceae (*Dysoxylum*), Icacinaceae (*Anisomallon*) and Lauraceae (*Cryptocarya*) also occur. The Sapotaceae (*Ochrothallus sarlinii*) are less numerous than on peridotites. The flora of this stratum is relatively varied and the canopy more continuous and regular than in similar formations on peridotite terrain. In the dominated strata are many palms; tree ferns play a major role in the forest margins and in secondary successions. The flora of the understorey is rich in Myrtaceae, Rubiaceae, Winteraceae, Amborellaceae.

The most beautiful forests can be seen in the valleys or on their sides. According to studies made by the CTFT, the high forests though localized in the humid zones, are nevertheless mainly located on the sheltered southwest slopes. The influence of exposure declines above 600 m. On steep slopes, the high forest becomes less dense and the palms are more abundant; on eroded or very siliceous soils and on very exposed high ridges the forest is replaced by thicket. There is no altitudinal forest limit comparable to that on peridotite sites at *ca.* 1 200 m. Thus the upper part of Mt Panié (1 200–1 640 m), which looks like a succession of terraces, is occupied by a forest of *Agathis montana ca.* 15 m high with an open understorey. The flora of the mountains which is very rich (Myrtaceae, Cunoniaceae, Myrsinaceae, Symplocaceae, Paracryphiaceae, Chloranthaceae) appears closer to the flora on ultra-basic areas than the flora of the lowlands.

Among the formations particular to sedimentary terrains there are the forests of *Aleurites moluccana*, a species which was presumably introduced a long time ago, and is now well adapted on the (often stony and relatively dry) slopes of the western coast. There are also Myrtaceae, Lauraceae, Celastraceae (*Elaeodendron*), Sapindaceae, Anacardiaceae, and *Elaeocarpus* spp., several of which occur also on calcareous sites. Neither palms nor tree ferns are found in these forests.

The forests on sedimentary or metamorphic sites form fairly stable, well-balanced ecosystems and, in the absence of human intervention, tend to encroach on adjoining savannas, at least in fairly humid areas. The pioneer species are Sapindaceae, Myrtaceae (*Metrosideros demonstrans*), Leguminosae (*Albizia*) and ferns, the floristic composition of the colonizing front varying greatly with site conditions.

Formations on calcareous terrains

On the main island, forests on calcareous terrains occur only locally, chiefly in the region of Koumac. However, most of the coastal communities belong to this type (sectors of Bourail and Kuébini). They cover a large part of the Loyalty Islands and about half the Île des Pins, and some small islands to the west and south. Their flora is rather poor and contains few characteristic species; it nevertheless varies from one sector to another.

The forests of Lifou and Maré (Loyalty Islands) are

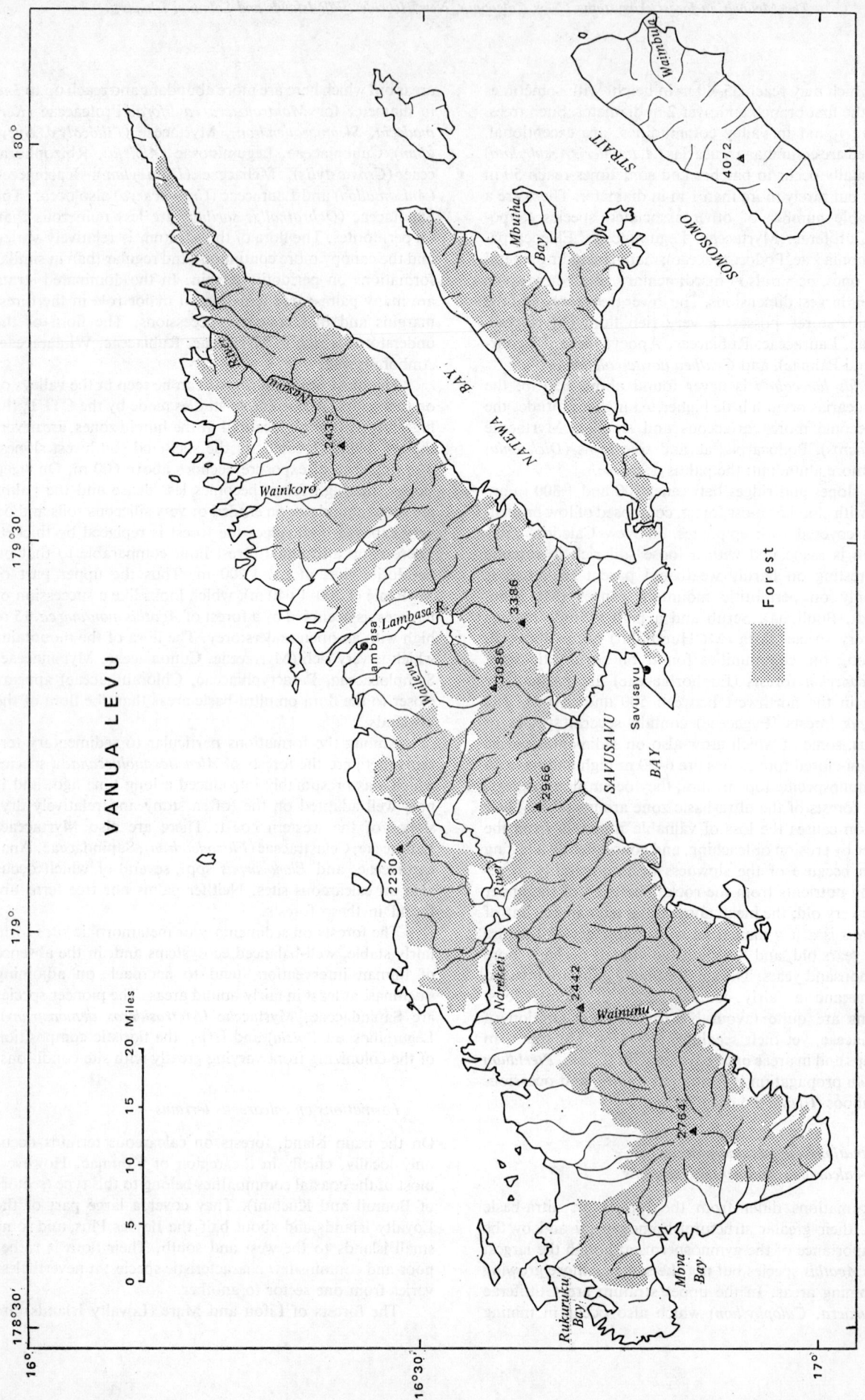

FIG. 3. Distribution of forests on Vanua Levu (Fiji Islands; after Parham, 1972).

the largest. Their upper layer, consisting entirely of evergreens at a height of 15–20 m, is overtopped in places by banian figs with very wide spreading branches. There are also patches of 25–30 m tall trees (Burseraceae, Celastraceae, Sapotaceae) and on the cliffs along the coast to the south and east, *Araucaria columnaris*. The most frequent species on the plateau of Lifou, *Schefflera golip* (Araliaceae), has a spherical habit and its trunk rarely exceeds 70–80 cm in diameter. There are also many Myrtaceae, Ebenaceae, Sapindaceae, one *Olea* sp. and, in secondary formations, two *Elaeocarpus* spp.; on the cliffs, *Hernandia cordigera*, *Bischofia javanica* and some Sapotaceae (*Mimusops*, *Manilkara*) occur. Leguminosae are rare (*Serianthes*). The understorey is often open, includes Euphorbiaceae (*Codiaeum*, *Cleidion*), Rubiaceae, Violaceae (*Hybanthus*); and the discontinuous herbaceous layer is formed by ferns (*Microsorium*). There are many lianas (Menispermaceae, Moraceae, *Ventilago*).

The forests of the Loyalty Islands are located on shallow, loose, permeable ferrallitic soils probably formed on volcanic materials laid as very thin layers on calcareous substrates from which they are enriched in P and Ca. They form fairly stable ecosystems. They seem taller on limestone rocks and all the large species possess roots which penetrate into fissures in the rocks. The extensive thickets of Euphorbiaceae and Leguminosae (*Acacia spirorbis*) are due to shifting cultivation. The existence in the subsoil of fairly compacted sediment layers richs in Ca and Mg originating from former lagune bottoms, impedes the establishment of forests (savannas of Maré); and on some of the cliffs with skeletal soils there are low shrub communities which are climax formations.

On Lifou and Maré, the groundwater-table is too low for the trees to reach it. In the Île des Pins, it is nearer the surface; this may explain the different composition of the forest there, which is richer floristically and larger structurally—though the stout trunks remain fairly short. The islands owes its name to the large stands of *Araucaria columnaris*, some of which were over 60 m heigh; but the most characteristic species ecologically is a legume, *Afzelia bijuga* which is also found further north in the forests of Uvea (where, however, it remains rather small) and in some sites on or near the coast of the main island (Kuébini, Golone).

On the main island, the calciphile forests are characterized by *Gyrocarpus americanus*, a deciduous species, *Spondias* (likewise deciduous), Euphorbiaceae (*Aleurites*) and Celastraceae (*Elaeodendron*). On raised beaches (Bourail region) are low forests of *Mimusops* and *Cycas*. On the islets off the western coast, with low precipitation, the tree layer, which is low and rather open, consists of *Elaeodendron*, *Acacia* and *Planchonella* spp. (Sapotaceae).

Coastal formations

Apart from the mangroves, usually low and without great economic interest, the coastal forests are limited to open stands including *Calophyllum inophyllum*, which reaches large dimensions, *Casuarina equisetifolia*, *Acacia simpli-*

cifolia, *Hernandia peltata*, etc. All these, as the mangrove species, have large distribution areas through the tropical Pacific.

Other archipelagos

Distribution

The vegetation cover of the Solomon and Fiji Islands (fig. 3 and 4) has been analysed in some detail and vegetation maps showing the distribution of the major formations have been published, but the distinctions were based on the interpretation of aerial photographs; from the botanical point of view they are too numerous and do not always show a clear distinction in flora or in environmental factors. In the New Hebrides only the vegetation of Vaté and that of the southern part of the archipelago have been studied in detail; in the north, exploratory studies have been carried out at Santo, Mallicolo and Pentecost.

There are no woody savannas comparable with the *Melaleuca* formations of New Caledonia, but there exist extensive herbaceous savannas, at least in the Fijis. The stands of *Acacia spirorbis* in the New Hebrides and the associations of *Casuarina papuana* and Cyperaceae on ultra-basic rock in the Solomon Islands have similarities with the woodlands and maquis of New Caledonia, but their flora is much less varied. In the Solomon Islands, mangroves and freshwater swamp formations are of some importance; elsewhere they are negligible in extension. Lowland formations have largely been cleared and replaced by anthropogenic vegetation; nevertheless the forest remains the dominant element of the landscape on most of the islands.

Where herbaceous savannas and grasslands exists, they are found in the northwestern sectors where conditions are drier, and they are associated with relatively compact soils of high base content. Quantin (1975) has observed that in the most humid northern part of the New Hebrides the finest forests occur towards the west, on medium altitude slopes protected from the prevailing winds. He believes that heavy continuous rains are not favourable to the growth of large trees, and that the forests of the eastern sectors which are more directly exposed suffer severe damage from cyclones.

Solomon (without Bougainville) and Santa Cruz Islands (Hansell and Wall, 1976)	28 000 km²	
— Forests	24 200 km²	
Mangrove		650 km²
Swamp forests		970 km²
Forests on ultra-basic rocks (open formations of *Casuarina*)		527 km²
Formations peculiar to calcareous terrain		1 150 km²
Various communities at low and medium altitudes, on basalts or andesites		20 000 km²

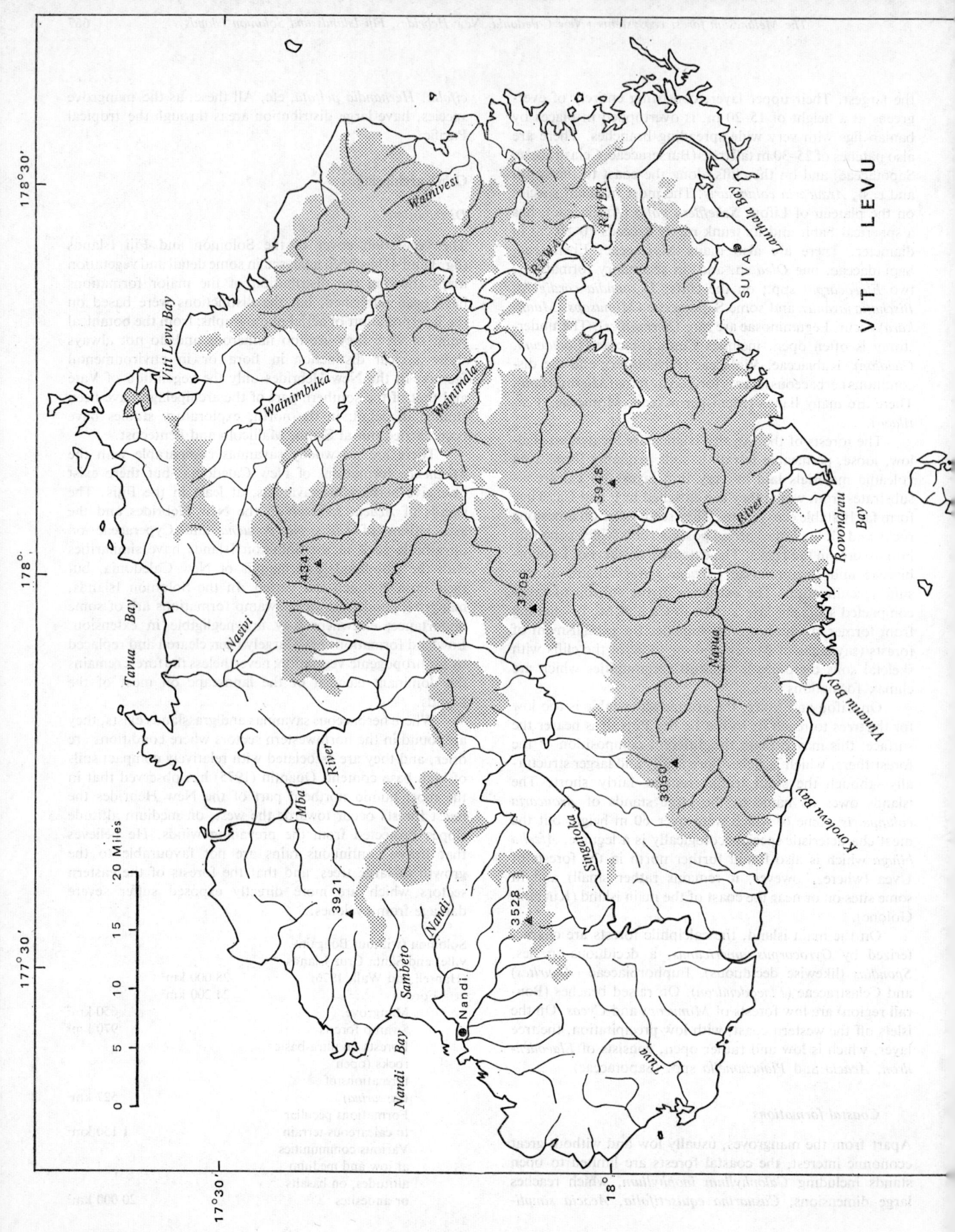

VITI LEVU

Waimbala

Rewa RIVER

RIVER

SUVA

Laucala Bay

Wainivesi

Wainimbuka

Wainimala

Viti Levu Bay

3948

3709

River

Rovondrau Bay

Nasivi

Tavua Bay

4341

Napua

Vunauu Bay

Korolevu Bay

Mba River

3060

Singatoka

Sambeto

Nandi

3921

3528

Tinu

Nandi Bay

Nandi

0 5 10 15 20 Miles

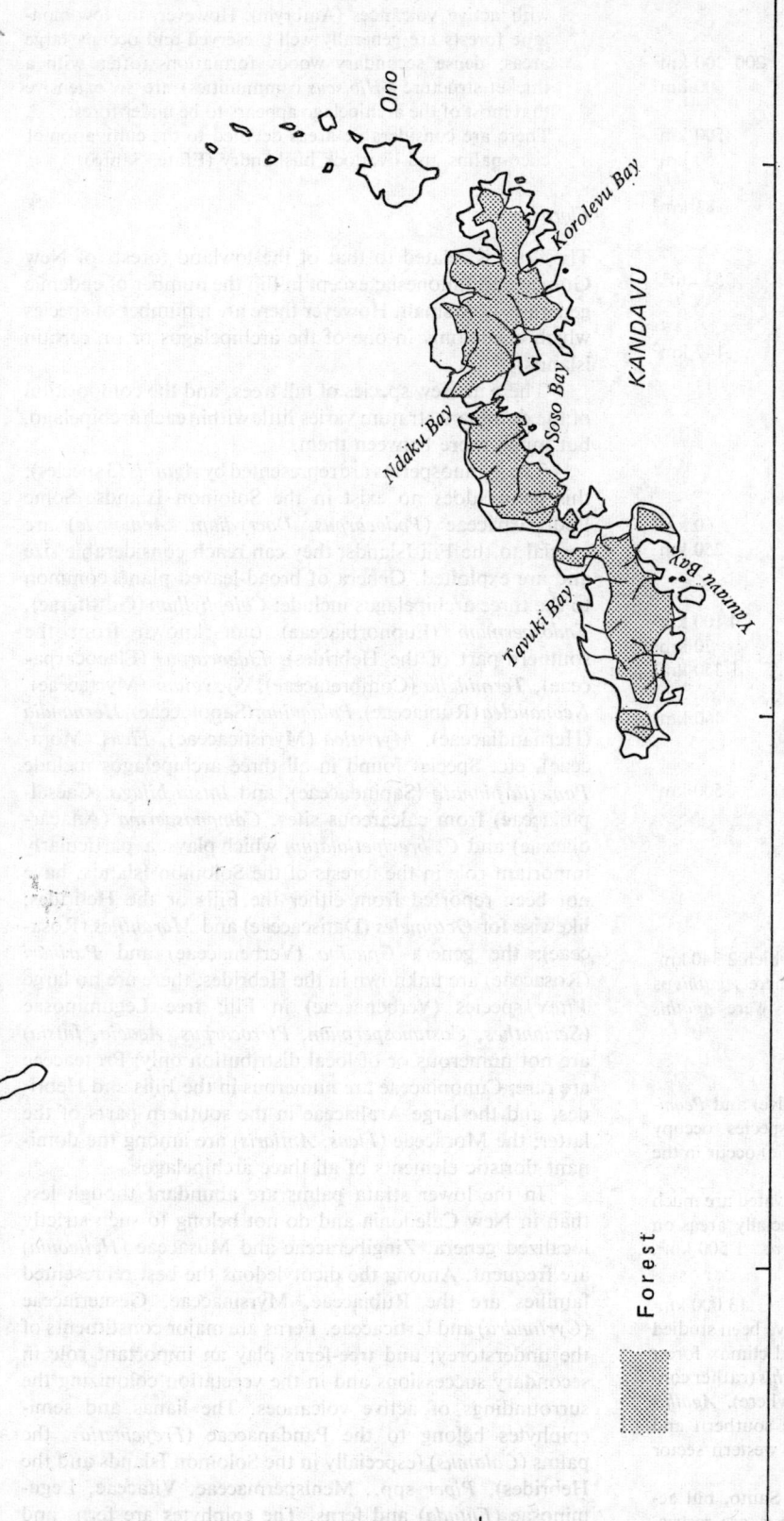

Ono

Korolevu Bay

KANDAVU

Soso Bay

Ndaku Bay

Tavuki Bay

Yanuvan Bay

Vatulele

Forest

19°

Fig. 4. Distribution of forests on Viti Levu and Kandavu (Fiji Islands; after Parham, 1972).

Forests containing *Agathis* (Santa Cruz Islands)		200–300 km²
Low mountain forests		600 km²
— Climax shrub formations	800 km²	
On calcareous sites		500 km²
On ultra-basic rocks		50 km²
On swampy ground (*Pandanus*)		83 km²
— Grass savannas	200 km²	
Reed swamps (*Phragmites*)		53 km²
Savannas on well-drained soils (*Themeda*)		147 km²
— Heathlands (fern communities)	50 km²	
— Crops and bush-fallow	2 500–3 000 km²	
Bougainville (Heyligers, 1967)	9 000 km²	
— Forests	5 850 km²	
Mangroves		60 km²
Swamp forests		250 km²
Various climax formations on well-drained soils		4 100 km²
Mountain forests		70 km²
Secondary forests		1 350 km²
— Climax shrub formations	520 km²	
Montane formations		460 km²
— Savannas	625 km²	
Grass formations on swampy ground		500 km²
— Crops and anthropogenic formations	2 000 km²	
Fiji Islands	18 000 km²	

(Parham, 1972; Berry and Howard, 1973), see fig. 3 and 4.

— Forests	8 390 km²

Surveys were carried out on 3 840 km², of which 2 340 km² are exploitable forests including 360 km² where *Agathis* is the predominant genus, and 1 500–2 000 km² were *Agathis* is more scattered.

— Thickets (climax or anthropogenic)	2 000–2 500 km²

— Grass savannas of *Miscanthus floridulus* (native) and *Pennisetum polystachyon* or other introduced species occupy *ca.* 6 000 km²; fern heathlands (Gleicheniaceae) occur in the eastern part of Vanua Levu and in Taveuni.

— Areas permanently or semi-permanently cultivated are much greater than in the other archipelagos, especially areas on or near the coast, and large valleys,   *ca.*   1 500 km².

New Hebrides	13 000 km²

Only the forests of Erromango (850 km²) have been studied in detail; *ca.* 180 km² are occupied by closed climax forest (Johnson, 1971); this includes stands of *Agathis* (rather continuous over 50 km², more scattered elsewhere). *Agathis* stands are also found at Aneityum (in the southern and southeastern sectors on 40 km²) and in the western sector of Santo.

There are forests in Erromango, Efate and Santo, but according to Quantin's (1975) maps high forests are restricted on most of the islands, especially those that are densely populated (Pentecost, Aoba, Tanna, Sheperd) and those with active volcanoes (Ambrym). However, the low montane forests are generally well preserved and occupy large areas; dense secondary woody formations (often with a thicket structure—*Hibiscus* communities) are so extensive that most of the archipelago appears to be under forest.

There are considerable areas devoted to the cultivation of coco-palms and livestock husbandry (Efate, Santo).

Flora

The flora is related to that of the lowland forests of New Guinea and Indonesia; except in Fiji the number of endemic genera is very small. However there are a number of species which occur only in one of the archipelagos or on certain islands.

There are few species of tall trees, and the composition of the dominant stratum varies little within each archipelago, but much more between them.

The gymnosperms are represented by *Agathis* (3 species); this genus does no exist in the Solomon Islands. Some Podocarpaceae (*Podocarpus*, *Dacrydium*, *Acmopyle*) are special to the Fiji Islands; they can reach considerable size and are exploited. Genera of broad-leaved plants common to the three archipelagos include: *Calophyllum* (Guttiferae), *Endospermum* (Euphorbiaceae) (not known from the southern part of the Hebrides), *Elaeocarpus* (Elaeocarpaceae), *Terminalia* (Combretaceae), *Syzygium* (Myrtaceae), *Neonauclea* (Rubiaceae), *Palaquium* (Sapotaceae), *Hernandia* (Hernandiaceae), *Myristica* (Myristicaceae), *Ficus* (Moraceae), etc. Species found in all three archipelagos include *Pometia pinnata* (Sapindaceae), and *Intsia bijuga* (Caesalpiniaceae) from calcareous sites. *Campnosperma* (Anacardiaceae) and *C. brevipetiolatum* which plays a particularly important role in the forests of the Solomon Islands, have not been reported from either the Fijis or the Hebrides; likewise for *Octomeles* (Datiscaceae) and *Maranthes* (Rosaceae); the genera *Gmelina* (Verbenaceae) and *Parinari* (Rosaceae) are unknown in the Hebrides; there are no large *Vitex* species (Verbenaceae) in Fiji; tree Leguminosae (*Serianthes*, *Castanospermum*, *Pterocarpus*, *Acacia*, *Intsia*) are not numerous or of local distribution only; Proteaceae are rare; Cunoniaceae are numerous in the Fijis and Hebrides, and the large Araliaceae in the southern parts of the latter; the Moraceae (*Ficus*, *Antiaris*) are among the dominant floristic elements of all three archipelagos.

In the lower strata palms are abundant though less than in New Caledonia and do not belong to such strictly localized genera. Zingiberaceae and Musaceae (*Heliconia*) are frequent. Among the dicotyledons the best represented families are the Rubiaceae, Myrsinaceae, Gesneriaceae (*Cyrtandra*) and Urticaceae. Ferns are major constituents of the understorey; and tree-ferns play an important role in secondary successions and in the vegetation colonizing the surroundings of active volcanoes. The lianas and semi-epiphytes belong to the Pandanaceae (*Freycinetia*), the palms (*Calamus*) (especially in the Solomon Islands and the Hebrides), *Piper* spp., Menispermaceae, Vitaceae, Leguminosae (*Entada*) and ferns. The epiphytes are ferns and orchids (*Dendrobium*) the latter being more abundant than in New Caledonia.

The montane flora is comparatively poor, at least in the Solomon Islands and the Hebrides; it includes many Myrtaceae (*Metrosideros*, *Syzygium*), also Melastomataceae (*Astronia*), Cunoniaceae (*Weinmannia*, *Geissois*), some Ternstroemiaceae (*Eurya*) and Ericaceae (*Vaccinium*) in the Hebrides and Solomon Islands, *Rhododendron* in the Solomon, and *Paphia* in the Fiji Islands. At higher altitudes trees are covered with epiphytes—large mosses (*Spiridens*), Hymenophyllaceae and orchids.

Types of forest

Climate is not as contrasted as in New Caledonia. With the exception of the northern part of Guadalcanal and the Florida Islands, the Solomon Islands, as well as the northern part of the New Hebrides and a great part of the Fiji Islands, may be considered as perhumid. It is an excess of rainfall which, by leaching the soils and making the trees prones to cyclone damage, might be unfavourable for the development of a very tall vegetation. Even the least humid sectors (the western part of Viti Levu, the northwest of Mallicolo and Efate) which are little forested, receive more rainfall than the western part of New Caledonia. The decrease of temperature with altitude, the rainfall variations and the increase of cloud density cause the distinct vegetation zones, the altitudinal limits of which vary with the size and geographical position of the islands, and with relief and aspect.

Soil conditions are likewise not varied; the parent rock is usually basic or slightly acid (basalts, andesites). In the Hebrides and the Solomon Islands calcareous bed-rock is the main reason of soil diversification. In the Fiji Islands, quartzite-containing rocks occur in the centre and the western part of Viti Levu and the northeastern part of Vanua Levu; the acid soils developing on these rocks suit gymnosperms. In the Solomon Islands and the Hebrides, there are also some ultra-basic areas, and the important role played by recent deposits of volcanic ash in Bougainville and the Hebrides should be stressed.

The forest landscape is therefore rather monotonous. Yet the gregarious tendency of certain species (*Campnosperma*, *Terminalia*, *Agathis*), the occasional occurrence of very tall emergents (*Terminalia*, *Ficus*, *Agathis*) and the patchy distribution of palms, introduce some variety in the physiognomy. The climax communities of low and medium altitudes contain many trees with large buttresses and very long slender trunks, except where the gymnosperms predominate; stratification is often marked, and the understorey quite distinct: the tropical environment is more obvious than in New Caledonia.

Coastal and swamp forests

Mangroves

Mangroves and adjoining inland communities on sediments permeated by sea-water which usually rest on coral terraces, are not wide-spread except in the Solomon Islands. The floristic composition is similar to those of New Caledonia, but the trees are much taller (up to 25 m).

The areas regularly flooded by the sea are occupied by Rhizophoraceae; where the tidal influence is less marked, *Sonneratia* and *Xylocarpus* occur, and on the landward border of the mangroves species which belong to the coastal forests: *Calophyllum*, *Heritiera*, *Intsia*. The palm *Nypa fructicans*, which reaches its eastern distribution limits in the Solomon Islands, forms small stands along some estuaries.

Coastal forests

In the Hebrides and the Solomon Islands coastal forests are of some importance and are associated with calcareous soils. Their composition differs between raised beaches and rocky terraces, but in both cases their flora contains species which are widely distributed in the Pacific: *Intsia bijuga*, *Calophyllum inophyllum*, *Terminalia catappa*, *Barringtonia asiatica*, *Casuarina equisetifolia*, *Ochrosia oppositifolia*, *Hernandia peltata*, *Acacia simplicifolia* (the two last named on sandy soils) and *Pandanus* spp. The tallest trees reach 25–30 m.

Freshwater swamp forests

Freshwater swamp forests are very extensive in the Solomon Islands where they occupy the coastal plains and the low-lying parts of valleys and grow in almost constantly flooded areas on soils with an horizon of more or less fibrous peat overlying alluvial substrates.

The upper stratum may be composed of a number of species or may be almost monospecific. The most characteristic species are *Eugenia tierneyana*, which is also very common in the Hebrides in riparian communities, *Inocarpus fagiferus*, *Barringtonia racemosa*, various *Calophyllum* and *Ficus* species, and in the understorey *Quassia indica* and *Horsfieldia spicata*. The trunks are covered with semi-epiphytes, Araceae (*Raphidophora*) and ferns (*Stenochlaena*). The herbaceous stratum consists of ferns. Where there is a tendency towards monospecificity, the dominant species may be *Terminalia brassii* (Combretaceae) or *Campnosperma brevipetiolatum* (Anacardiaceae), which is also a major constituent of forests on well-drained soils. Finally there exist on ultra-basic sites swamp forests of *Casuarina papuana*.

The stands of *Terminalia* reach a height of *ca.* 40 m. The tree layer is more or less open, the shrub layer is rather dense and epiphytes abound. Pandanaceae are numerous, but palms are rare. The only riparian species is a *Metroxylon* which is only found where the canopy is very open.

Closed forests without Agathis at low and medium altitudes

Closed forests without *Agathis* at low and medium altitudes are the most important forests, both as regards area and economic interest. They are more extensive and taller in the Solomon Islands than in the Hebrides (though these often grow on better soils) and the Fiji Islands. Their composition varies between the archipelagos.

The closed forests of the Solomon Islands are usually located on loose acid, very deep and strongly leached ferrallitic soils, which originate from basalts and andesites. On limestone, the soils do not always differ from those on eruptive rocks, because they have developed from volcanic material; but where the relief is rugged, the influence of the parent rock becomes obvious and the composition of the vegetation changes because of the disappearance of lime-intolerant species (*Campnosperma*). On basalts or andesites, P and Ca deficiencies are particularly marked; on calcareous sites, there are K deficiencies (in Bougainville, the nutrient reserves are greater, because of recent considerable deposits of ashes). Because of the poverty of the soil, regeneration of vegetation on large desnuded areas is difficult, whilst the climate is favourable for regeneration of forest in small gaps especially as many species of the upper stratum are semi-light demanding (*Campnosperma, Endospermum, Gmelina*). Site plays a major role in the relations between soil and vegetation (Lee, 1969).

The upper stratum reaches 30–40 m, but it is rarely continuous; the forest is strewn with gaps or intersected by corridors occupied by shrubby thickets due to natural disasters or human destruction. The only emergents are *Terminalia calamansanai* and some banian figs; these are frequent in some places (Whitmore, 1969).

Some 60 species of tall trees have been identified; those most abundant belong to the genera *Campnosperma, Calophyllum, Canarium, Dillenia, Endospermum, Elaeocarpus, Gmelina, Maranthes, Parinari, Pometia, Schizomeria, Terminalia* and *Vitex*. The number is much smaller than in Malaysia, but greater than in the other Melanesian archipelagos. Certain species predominate over large areas, particularly *Campnosperma brevipetiolatum* which is not known in Guadalcanal and San Cristobal and does not occur on coral terraces where *Pometia pinnata, Neonauclea* sp., *Eugenia* sp. and *Dysoxylum* sp. are abundant.

In the dominated strata pachycaulous species (*Boerlagiodendron, Tapeinosperma*) and many palms (*Areca, Licuala, Heterospathe*) are found, and in the herbaceous discontinuous layer, cryptogams (*Selaginella*, ferns) and Urticaceae (*Elatostema*). Where the density of the canopy decreases, rattans (*Calamus*) appear, also Scitaminales (Zingiberaceae, *Heliconia*) and bamboos. Lianas and epiphytes are more abundant than in the forests of Malaysia; these are Araceae, Pandanaceae (*Freycinetia*), Rubiaceae (*Hydnophytum*) and ferns.

Forests on alluvial deposits have a different composition from forests on slopes, with some peculiar species (Leguminosae, Sapotaceae and, in Bougainville, *Octomeles*); but the vegetation of the alluvial areas has generally been modified by man and low formations of *Hibiscus, Kleinhovia* and *Macaranga* occur in association with large Monocotyledons (Palmae, Zingiberaceae, Maranthaceae, *Heliconia*).

One forest type described by Heyligers (1967) in his study of Bougainville may be considered as intermediate between forest of medium altitudes and montane forest. It is a community in which Guttiferae (*Garcinia*) and Elaeocarpaceae (*Elaeocarpus, Sloanea*) predominate, with a rather

open upper stratum 25–30 m high and a comparatively dense intermediate stratum at *ca*. 15 m. It contains palms with stilt roots and among the lianas, a *Rubus*; in the herbaceous layer there are grasses, many ferns, *Begonia, Elatostema* and *Cyrtandra* spp. This type has been observed between 800 and 1 300 m. Certain elements characteristic of it (Palmae, *Rubus*, Gramineae) also occur in the New Hebrides at similar altitudes.

The closed forest of the New Hebrides is usually located on poorly developed soils which at medium altitudes resemble andosols, or on weakly desaturated ferrallitic soils covering calcareous plateaux, or on strongly desaturated ferrallitic soils over volcanic rocks. But there are also slope or valley forests on clayey or rendzina brown soils. Except for the desaturated ferrallitic soils, which carry the forests rich in gymnosperms, the forest soils of the New Hebrides are better supplied with nutrients than those of the Solomon Islands, but they are less deep.

The height of the canopy is 20–35 m; the only emergent trees are banian figs, on the plateaux and the valleys. The trunks, which frequently have strong buttresses, are relatively large and short; there are no trees with tall slender boles similar to *Terminalia* of the Solomon Islands.

On the calcareous plateaux, *Antiaris toxicaria* (Moraceae) is frequently the main component of the upper stratum; there are also some banian figs, Myrtaceae (*Syzygium, Piliocalyx*), and Leguminosae (*Castanospermum*), usually in small stands, *Hernandia* sp., *Myristica* sp., *Dendrocnide* sp. (Urticaceae), at Efate, *Evodia speciosa* (Rutaceae), and in the centre and the north, *Pangium* sp. (Flacourtiaceae). The subcoastal and valley formations, on more clayey soils richer in calcium, include *Gyrocarpus americanus, Aleurites moluccana, Bischoffia javanica, Spondias dulcis, Pometia pinnata, Dracontomelon vitiense, Kleinhovia hospita, Pterocarpus indicus* (often in small stands) and various Sapotaceae; several of these species are deciduous (*Gyrocarpus, Pterocarpus, Spondias*). Among the large trees common on the coral cliffs there are *Ficus* spp., *Cryptocarya* spp., *Neonauclea* spp. (at Erromango) and *Intsia bijuga* (Pentecost).

On volcanic rocks, the Guttiferae are well represented by *Calophyllum*; Myrtaceae, Sapotaceae (*Palaquium* at Erromango), *Kermadecia* (Proteaceae), Cunoniaceae, *Elaeocarpus* and *Hernandia* spp., *Schefflera* (Araliaceae) and *Myristica* sp. are also common.

The understorey is more open on volcanic than on calcareous rocks. There are Rubiaceae, Euphorbiaceae (especially on calcareous sites), Ebenaceae, Sapindaceae, Araliaceae and *Corynocarpus* sp. In the centre and the north palms (*Licuala, Veitchia, Calamus*) are abundant in places, but they are not very common at Erromango, and rare at Aneityum in the extreme south. In more open sites tree ferns are common. The herbaceous layer mainly includes Urticaceae and cryptogams (*Selaginella*, ferns), whereas the large monocotyledons (*Alpinia, Heliconia*) play an important role in secondary successions. Lianas and epiphytes become more abundant towards the north; they belong to the same taxa as those in the Solomon Islands, but species are often different.

The forests of the Fiji Islands, located in the humid sectors, are as in the Solomon Islands on deep leached soils, which are either ferrallitic on neutral or basic volcanic rocks, or fersiallitic (red and yellow podzols) on acid rocks; but as in the Hebrides and Bougainville, some forests also occur on weakly desaturated andosols, especially in the small islands in the east (Taveuni); less frequently, forests occur on clayey soils rich in calcium of the less humid sectors where herbaceous formations predominate.

The gymnosperms dominate in a large part of the formations at medium altitudes. Where *Agathis* is absent, the forests (which resemble physiognomically the broad-leaved formations of the New Hebrides) include mainly *Endospermum* (Euphorbiaceae), *Calophyllum* (Guttiferae), *Myristica*, *Gonystylus* and Myrtaceae; also found are *Palaquium* (Sapotaceae), *Canarium* sp., *Parinari* sp. and *Elaeocarpus* spp. In the valley formations or those on calcareous terrain (coastal sectors), *Pometia pinnata*, *Parinari insularum*, *Bischofia javanica* or *Intsia bijuga* occur; in relatively dry areas and on poor soils, *Casuarina nodiflora* is frequenty associated with Podocarpaceae.

Closed gymnosperm forests

Closed gymnosperm forests are wide-spread in the Fiji Islands and restricted to the islands of Santa Cruz, and to western parts of Santo, Erromango and Aneityum in the Hebrides. In very humid areas they grow at low and medium altitudes on deep soils of various types (fersiallitic soils in the Fiji Islands, leached ferrallitic soils in the south of New Hebrides, and in ferrallitic soils, only weakly leached, in Santa Cruz); they are never found in calcareous soils. Their predominant species is the kauri (*Agathis*). In each archipelago the genus is represented by a different species. Certain Podocarpaceae—*Dacrydium* in the Santa Cruz and Fiji Islands, two species of *Podocarpus*, s.l. in the New Hebrides and three in the Fijis—are found in association with the kauris, but they are more scattered and never reach comparable dimensions. The upper stratum of these forests contains the same broad-leaved species as the neighbouring forests not containing gymnosperms, e.g. *Calophyllum* spp. and Myrtaceae, and in Santa Cruz, *Campnosperma brevipetiolatum*, and in the Fiji Islands and the New Hebrides, *Palaquium* sp.

The kauris of the forests of New Caledonia occur mainly in the valleys, whilst the large ones of this part of Melanesia are mainly found on slopes and ridges. They grow scattered (western part of Santo) or in groups, and normally emerge above the broad-leaved canopy; their trunks are relatively large and short, and though their total heigth may reach 40 m, it does not generally exceed *ca*. 30. Diameters of 2 m are not exceptional in the New Hebrides (Erromango) and even trees with *ca*. 3 m DBH have been recorded.

Whitmore (1965), who studied the forests of the Santa Cruz Islands, states that in the absence of human intervention, the kauris regenerate naturally in the forest outside the zones affected by natural catastrophes. However, seedlings in the forest suffer from the competition by broad-leaved species and their initial growth is extremely slow: it can be greatly speeded up by thinning. At Erromango seedlings are abundant near logging areas being favoured by light conditions that are more intense than in the understorey but they may be suffocated by weeds.

Other forests

Woodlands of *Acacia spirorbis* occur in the driest parts of the large islands of the New Hebrides. On relatively rich soils (saturated fersiallitic soils, on calcareous bed-rocks) the woodland is rather dense and the *Acacia* is associated with Myrtaceae, Leguminosae, and Meliaceae; on more or less desaturated soils the tree layer is much more open, over a very discontinuous shrub layer and an open herbaceous layer of Gramineae and Cyperaceae (*Gahnia*).

The woodlands on ultra-basic rocks closely resemble those in New Caledonia under similar conditions; they are rather extensive in Choiseul, Santa Ysabel and San Jorge (Solomon Islands). They grow on ferrugineous ferrallitic soils which are not deep and become richer in organic matter with increasing altitude. The tree layer includes *Casuarina papuana*, which is usually the predominant species, *Dillenia crenata* and various Myrtaceae, and a gymnosperm, *Dacrydium beccari*, some palms (*Gulubia*) and *Pandanus* spp. The shrub layer is fairly dense (Myrtaceae); the herbaceous layer is composed of sedges and ferns. These woodlands are very susceptible to fire.

Montane forest occurs at various altitudes depending on the relief, exposure and geographic situation, but it generally starts at relatively low altitudes; for instance at 500 m at Aneityum in the extreme south of the New Hebrides. The distribution of montane communities depends more on variations in cloud density (perhaps also on wind speed) than on variations of temperature or the rainfall. On none of the mountain ranges does the forest reach the upper forest limit; however, some peaks are covered with very low shrubby or even herbaceous vegetation. The soils are very rich in organic matter and real rankers can be found.

Whitmore (1969) thinks that the montane flora of the Solomon Islands is poor. Nevertheless, it contains some peculiar species (*Rhododendron*) or some which are found at similar altitudes in the other archipelagos (*Gunnera* and *Vaccinium* in the Hebrides, *Machaerina*). The only species of the Vacciniaceae in Fiji, *Paphia vitiensis*, is found at high altitudes. But the Myrtaceae play the most important role in montane formations. The Podocarpaceae are fairly well represented, at least in the Solomon and Fiji Islands, also the Melastomataceae (*Astronia*), the Araliaceae, Cunoniaceae and Gesneriaceae. There are also some palms (*Clinostigma*, *Gulubia*) and Pandanaceae and, in the Solomon Islands, bamboos. Ferns are particularly abundant. Among the epiphytes are, in addition to cryptogams, many orchids and some *Astelia* spp. (Liliaceae). The transition from close lowland forest to montane forest is sometimes abrupt but mostly takes place through intermediate physiognomic and floristic types, with large trees occurring along deep valley or in sheltered sites. The genus *Metrosideros* (Myrtaceae) which is equally represented at rather low altitudes, plays an important role in the New Hebrides, where individuals with enormous trunks but a low branching habit occur.

The cloud forest consists of large shrubs or low-branching trees heavily laden with epiphytes (*Spiridens*, Hymenophyllaceae); the ground is covered with mosses and ferns.

Forest utilization

The Melanesians obtain from the forest several products for their crafts; certain timbers to build huts and boats, and edible plants. However, shifting cultivation is the main cause for the reduction of forest area on most islands. The establishment of plantations, the increase of sugar-cane in the Fijis, and the development of livestock husbandry all are permanent forms of land use which have greatly modified the landscape. Logging is being rapidly developed and greatly affects the forest. Fires and mining (at least in New Caledonia) cause serious damage. In the Fijis and New Caledonia efforts are being made to protect the forest, and reafforestation is carried out. In the New Hebrides and the Solomon Islands, where the degradation of the environment and particularly of the plant cover is less obvious, very few conservation measures are utilized.

Traditional uses

The human population density of the Melanesian islands remains low (less than 10/km² except in the Fiji Islands where it reaches *ca*. 30/km²) but it is very unevenly distributed. Some islands are heavily populated, others have been in the past, but their populations have declined since the arrival of the Europeans due to epidemics; except in New Caledonia, the total rural population exceeds the urban population.

Forest products

The forests of New Caledonia offer few resources to rural communities except perhaps for (overhunted) game and medicinal plants; in the other archipelagos many products of the forests are widely used, especially as building materials, and gathering sometimes has a considerable effect on the physiognomy of the understorey. Certain plants are edible (palms, *Barringtonia*) but the fruits gathered in the forest are mainly those of introduced species planted along paths. The overall damage is not severe.

Shifting cultivation

Before the Europeans came, shifting cultivation was not threatening the environment. Steep slopes were not cleared and forests on poor soils were spared. The periods of fallow were long enough to allow the nutrients of the soil to be replenished. The development of a cash crop agriculture, leading to permanent cropping on a large part of the coastal

plains, may locally induce a shortening of rotations and an extension in altitude of food crops (Bonnemaison, 1973). However, the areas required per inhabitant are modest (*ca*. 0.2 ha) because the traditionally planted tubers give high yields.

In the New Hebrides land shortage becomes a problem in the most densely populated islands; in addition there are misgivings about the extension of plantations and pastures in the forest areas on the plateaux of Efate and Santo. The Solomon and Fiji Islands, the soils of which are less fertile, are faced with the same problem, perhaps even more acutely.

Forest regenerates very slowly on abandoned plantations which are rapidly colonized by dense scrub which remains for several decades. Considering the enormous areas of secondary formations (low forests or thickets of Euphorbiaceae, *Hibiscus tiliaceus*, Sapindaceae and Meliaceae) which now occur in almost deserted areas, the same process probably, occurred in the past.

Wildlife

Most of the native animal species live in the forest. Some are threatened by the destruction or degradation of the habitat; several are endangered by hunting which is difficult to control. The introduction of cats and dogs, which very easily become feral, is a serious danger for reptiles and flightless birds (rails, cagou, megapodes). In the Fiji Islands at least four bird species have become extinct, including a screech owl and the Asian cock which had been introduced a long time ago; three other species have become very rare, including the red-throated parrot (*Vina amabilis*). In New Caledonia, where the hunters are particularly active, the flying foxes have become rare, but the main concern is the rapid decline of the large pigeons (*Ducula goliath*), peculiar to the main island, and the white-collar (*Columba vitiensis hypoenochroa*); in addition, the latter which nests in the savannas frequently falls victim to fires. The cagou (*Rhynochetos jubatus*) of New Caledonia is rare but its population dynamics are not well known since it lives in the most remote forests. Such a flightless species is very easy prey for dogs and, although its capture is forbidden, it is much sought after for keeping as pets or for zoos. The endemic parakeet of Ouvea is equally threatened by the demands of pets. The great *Rallus* of Lafresnaye, which was thought to be extinct since 1880, is now believed to have found a refuge in the forests of the western slopes of Mt Panié; the Asian cock still exists in the Île des Pins, and *Columba vitiensis* has found a refuge on the almost inaccessible cliffs of the Loyalty Islands. In the Hebrides and the Solomon Islands, the scattered distribution of the islands, the extension of almost impenetrable closed forests, and the primitivity of hunting weapons all contribute to making the animal populations relatively secure; but little is known about distribution and numbers of the most valuable species.

Timber

Existing reserves

Surveys and inventories are far from complete. In New Caledonia and the Fiji Islands a large part of the forests have been surveyed, but no serious study has been made on the regeneration of the most valuable species. In the New Hebrides detailed inventories have been carried out at Erromango where the finest forests have been exploited. For the other islands the available information, mainly provided by botanists, is rather vague. The estimates given for the Solomon Islands are very uncertain.

New Caledonia

Sarlin's (1954) *Bois et forêts de la Nouvelle-Calédonie* contains much information on the technological properties of woods and on the ecology and distribution of *ca*. 130 species. However, the information is incomplete. The inventory carried out by the Centre technique forestier tropical (CTFT) from 1973 to 1975 is virtually limited to the formations on sedimentary or metamorphic areas of the main island; some sampling has been carried out in forests on ultra-basic rocks of the great peridotitic massif in the south, but no satisfactory statistical analysis was made; the forests on calcareous terrain in the Loyalty Islands have not been surveyed. In the area studied, which includes the richest sectors, the closed forest is estimated at 734 km², of which 402 km² are fairly easily accessible. There is little forest below 200 m; *ca*. 50 per cent of exploitable formations are between 400 and 600 m; above 800 m, the forest is of no commercial interest and should be classed as protection forest, with the possible exception of some stands of *Montrouziera* occupying gentle slopes or flat uplands at *ca*. 900 m on the southwestern slopes of the Panié-Ignambi range.

Two species (*Montrouziera cauliflora*, houp, and *Calophyllum caledonicum*, mountain tamanou, both Guttiferae) scattered over the whole forest area of the main island, including sites on ultra-basic rocks, represent the largest volume of growing stock. Their diameter distribution (for trees >20 cm DBH) is well balanced and the stands seem healty. The distribution of tamanous is apparently little influenced by aspect; it is determined by altitude—it is common between 400 and 600 m and scarce above 800 m. The reverse is true of houp, which more often occurs on western slopes and remains abundant between 800 and 1 000 m. In the areas surveyed, the standing volume of trees with a DBH >40 cm is estimated at 2.4 million m³ for tamanou and 1.5 million for houp. The wood of tamanou which is pink with red venations resembles that of some African mahogany (*Entandrophragma*) but is heavier (density >0.8). It is a cabinet wood which takes a very fine polish but with the serious drawback of spiral grain, so that it should be used as veneer. The wood of houp is yellow when first cut but later of a greyish tint; it is of a medium density (0.75) and good dimensional stability; it is a durable wood suitable for exterior carpentry work.

The Myrtaceae are as important as the Guttiferae; they include the genera *Piliocalyx* and *Syzygium*, species of *Metrosideros* including *M. demonstrans* particularly frequent at higher altitudes, and *Arillastrum gummiferum* in the ultra-basic zone in the south. The gross volume of *Piliocalyx* and *Syzygium* (DBH >40 cm) is estimated at 3.3 million m³. Their wood is greyish in colour and very homogeneous, easy to saw and durable, and of medium density (0.75); the wood of *Arillastrum gummiferum* is red and very heavy (d=1), hard, strong and exceptionally durable; it is suitable for heavy construction work and other heavily exposed carpentry (bridges, boats, etc.).

Melaleuca quinquenervia (Myrtaceae) is not a forest tree and never reaches large dimensions; its abundance might make it suitable for pulp production, but the product is medium quality. The leaf oil, extracted by distillation, is used in the preparation of gomenol.

The best known wood is probably kauri, the five species of *Agathis* which occur only on the main island of New Caledonia and those of the New Hebrides and Santa Cruz Islands; *Agathis* of the Fiji Islands is known locally by the vernacular name dakua makadre. Until recently three species were exploited in New Caledonia, but *A. lanceolata* which occurs only on ultra-basic sites is now protected. The *Agathis* reserves of the sedimentary and metamorphic areas are estimated at 188 000–435 000 m³; the large margin of uncertainty is due to comparative rarity of the two used species, the large area over which they are spread and their gregarious habit. The reserves of *A. lanceolata*, a species now mainly found in areas difficult of access, may still be considerable. The kauris, the trunks of which reach large diameters, yield a reddish, rather light (0.6) wood of poor resistance against insects and the inclemencies of the weather, unless treated; it is very homogeneous and suitable for veneers. Foresters distinguish red and white kauri. The wood of the white kauri is lighter in colour but less easy to work. Red kauri, which yields the best construction timber, occurs on ridges and would belong to *A. corbassonii*; white kauri would be *A. moorei*, which, like *A. lanceolata*, grows on more humid sites. Certain users prefer the wood of *A. lanceolata* the qualities of which are associated with the slow growth rate of the species on ultra-basic sites.

The wood of 'beech', a name given to various Proteaceae (*Kermadecia*, *Sleumerodendron*) is also much appreciated. These species are abundant in the centre and the northeast of the main island; they are rare on ultra-basic sites. Their wood has a density of 0.5, it is light in colour, with a pinkish tint and a pattern of medullary rays which gives it a certain distinction. It is used mainly for interior joinery work.

The species regularly exploited on a small scale or occasionally include: *Intsia bijuga* (Caesalpiniaceae), a limestone species exploited in the Île des Pins, which yields a beautiful cabinet wood, reddish in colour and heavy (d=0.76); *Albizia granulosa* (Mimosaceae), occurring on all site types, yielding a brownish or light-brown, fairly dense wood (d=0.75); *Hernandia cordigera* (Hernandiaceae), occurring in most places and *Schefflera* spp. (Araliaceae), the wood of which, soft and light, is suitable for box making; *Cunonia*, *Geissois*, *Pancheria* and *Weinmannia* species (Cunoniaceae), the wood of which is heavy, hard and strong, suitable for exterior carpentry and cabinet making; araucarias which today are protected, yield woods of medium or light density (mountain *Araucaria*, with a density of 0.65, and *A. columnaris* with a density of 0.5), and long fibres, which would be suitable for pulp production; *Cocconerion* (Euphorbiaceae) and *Acacia spirorbis* yield woods of small diameter but great durability which are used as pit props; *Bischofia javanica*, and some Sapotaceae (*Manilkara* or buni, *Mimusops*), also have woods of high quality which are mainly used by the inhabitants of the Loyalty Islands.

Fiji Islands

The forest resources of the Fiji Islands have been surveyed by the Land Resources Division (Ministry of Overseas Development, 1973). The inventories covered 384 000 ha and dealt with the most important parts of the forest (estimated at 839 000 ha for the three main islands). 234 000 are considered suitable for exploitation and 88 000 should be designated as protection forests; the remaining 62 000 ha are not of commercial interest. Total standing volumes for the 40 most important species (DBH >40 cm) in these three forest types are estimated as follows (in thousand m³):

Islands	Exploitable forests	Protection forests	Forests without commercial interest
Viti Levu	7 018	1 122	478
Vanua Levu	5 442	1 398	303
Kandavu	262	174	3
Total	12 722	2 694	784

The lowland forests contain *ca.* 45 per cent of broad-leaved species, with soft, light wood (*Endospermum*, *Myristica*), 40 per cent of broad-leaved species with heavy or semi-heavy woods (*Calophyllum*, Myrtaceae, *Palaquium*, *Intsia*, *Gmelina*) and only 10 per cent gymnosperms. The medium altitudes or montane forests contain 45 per cent gymnosperms, 30 per cent broad-leaved species with heavy woods, and only 20 per cent with light woods. Some 90 species are being exploited but only a small number are of commercial importance, such as the conifers, Euphorbiaceae, Guttiferae, Myrtaceae and Myristicaceae.

Agathis vitiensis, Fijian kauri, is by far the most exploited species, although reserves are limited. Its wood which has excellent technological properties and lends itself to many uses (carpentry, boat making and the manufacture of various wooden objects) is comparable to that of Douglas fir; it is also similar to that of the other kauris, having the same good qualities and the same defects (low resistance to the weather and to termites). The other exploited conifers are all Podocarpaceae: *Podocarpus vitiensis* and *Dacrydium nidulum*. The wood of the first species is similar to that of kauri but it is lighter in weight and darker in colour; it is used mainly for interior carpentry. The wood of *Dacrydium* is harder, more durable and red

in colour; it is highly valued for its decorative properties (cabinet making, interior carpentry). These two species are more and more in demand, but the reserves are small.

Among the dicotyledons, *Endospermum macrophyllum* (Euphorbiaceae) yields a light-coloured, light-weight wood that is easy to work and is used for interior carpentry or light construction work; it is however highly susceptible to the attacks of insects and for this reason needs a previous preservative treatment. Kaudamu, *Myristica* spp. (Myristicaceae) have a light or semi-heavy wood of a brownish red colour which is used in interior carpentry and light constructions; damanu, *Calophyllum* spp. (Guttiferae) yield a semi-heavy wood of a brownish red colour, used in carpentry and cabinet making; yasi yasi, *Syzygium* or *Cleistocalyx* spp. (Myrtaceae) have a heavier wood with similar uses. All these are exploited regularly. In addition *Heritiera ornithocephala* (Sterculiaceae) which has a brownish-red wood, heavy and fairly durable, is fairly common; sacau, *Palaquium hornei* (Sapotaceae), yields a red, heavy, very durable wood; bauvudi, *Palaquium* spp., with a much lighter and much less resistant wood than sacau; mavota, *Gonystylus punctatus* (Thymeleaceae), with a semi-heavy, bright yellow wood susceptible to insect attack, is still quite common; and *Intsia bijuga* (Caesalpiniaceae), with a heavy, very durable heart-wood, mainly used in boat building, are commonly utilized although the last one is becoming scarce.

Solomon and Santa Cruz Islands

The forest resources of the Solomon and Santa Cruz Islands are greater than the other archipelagos; they are being rather heavily exploited and have not yet been properly surveyed.

Total standing volume of all species could be *ca.* 50 million m³ including some 15 million in Bougainville, for which more detailed information is available. Half the forest area is very difficult of access. Apart from the *Agathis* of Santa Cruz which represented a reserve of *ca.* 150 000 m³ (*ca.* 1960) the exploitable species are all dicotyledons: *Terminalia* (Combretaceae), *Campnosperma* (Anacardiaceae), *Calophyllum* (Guttiferae), *Endospermum* (Euphorbiaceae), *Dillenia* (Dilleniaceae), *Vitex* (Verbenaceae) and *Pometia* (Sapindaceae). The Myrtaceae are less important than in the more southern islands lying more to the south.

New Hebrides

Johnson (1971) has given detailed information on the volume of timber available in the forests of Erromango before the exploitation (1969–1974) of the sectors richest in kauris and tamanous.

Of *ca.* 180 km² of closed forest, 141 km² were considered suitable for exploitation; total volume for all species was estimated at 694 000 m³, including 118 000 m³ of kauri, 418 000 m³ of tamanous (*Calophyllum*), 75 000 m³ of *Hernandia* and 22 400 m³ of *Palaquium*. The reserves in Myrtaceae, which do not generally reach large diameters and have a different distribution pattern from that of the other species, may have been under-estimated. No estimate is given for the volume of gaiac (*Acacia spirorbis*) wood which occurs in open formations which in the west of the island colonize abandoned plantations and pastures. *Santalum* is still found in the south. The woods of *Agathis*, *Hernandia* and *Calophyllum* have properties which do not differ greatly from those of the same genera in New Caledonia and the Fiji Islands; however, according to Sarlin (1954), the tamanous of the New Hebrides have a wood of lower density (0.6) with a less twisted fibre. The wood of *Palaquium* is semi-heavy, polishes well and is suitable for interior carpentry; it is not resistant to termite attack The gaiacs of Erromango and the other islands of the archipelago reach much larger dimensions than in New Caledonia; their wood is lighter in weight, less hard, but very durable.

In Aneityum, in the extreme south of the archipelago, exploitation has been practised for 50 years on a number of species, but mainly on kauri. Reserves are now very reduced. *Intsia bijuga* is commercially exploited in Efate.

Extent of exploitation

The volume of timber extracted per annum is moderate compared with the enormous amounts exported from Africa or Malaysia. But in relation to the size of the various islands they are nevertheless important, especially in view of the difficulty of exploitation due to the nature of the topography which makes logging more costly and destructive. Melanesia profits to some extent from the relative poverty of its forest flora compared with that of other tropical regions, because the commercially interesting species are less widely scattered than elsewhere.

In New Caledonia, kauris and araucarias have long been exploited by the local people; the same is true of the hardwoods buni (*Manilkara*) and kohu (*Intsia*) of the lowland forests. However the quantities felled were small. In the other archipelagos, apart from kauri, the species most used in the past were different from those now heavily exploited. Since the beginning of the 19th century the Europeans were particularly interested in sandal wood which was relatively abundant, especially in the Loyalty Islands, in the south of the New Hebrides and in the Fiji Islands; this exploitation which went on for some hundred years led to an exhaustion of reserves. The exploitation of kauri became an important industry in the beginning of the 20th century; the wood was exported from New Caledonia, the islands of Santa Cruz, Aneityum and the Fiji Islands, to Australia. As timber surveys provided information on the distribution of the principal species and technological studies revealed their utilization potentials, the logging programme was gradually diversified.

In New Caledonia total annual timber production in 1974 was 17 714 m³. In 1962 it had been only 7 663 m³. Since then it has increased, but differing according to the species. The volume of *Araucaria* was 753 m³ in 1962, it then declined and reached zero in 1974 when felling was prohibited. The production of kauri timber increased from 901 m³ in 1962 to 2 607 in 1970 but declined to

1 824 m³ in 1974, because stands have become scarce and consumers showed less interest. The volumes of houp rose from 2 303 m³ in 1962 to 4 363 m³ in 1974, those of 'beech' from 1 172 to 3 155 m³, and those of tamanou from 607 to 5 223 m³. Ralia (*Schefflera*) appeared on the market in 1974 (304 m³); *Arillastrum* is no longer being exploited.

The total production of the Fiji Islands during the last 10 years was *ca.* 1 100 000 m³; a large part of this is used locally but certain timbers (*Endospermum*) are exported. The exploitation of kauri is the most important: *ca.* 30 per cent of the total; *Endospermum* accounts for 20 per cent, *Myristica ca.* 10 per cent, *Calophyllum* 9 per cent, *Syzygium* 7 per cent, Podocarpaceae 5–6 per cent, Sapotaceae 5 per cent, *Gonystylus* 3.8 per cent and *Heritiera* 2 per cent.

In the Solomon Islands production was *ca.* 250 000 m³ in 1968, nearly all for export. In the New Hebrides, towards 1970 when the forests of Erromango were exploited, production reached several ten thousands of m³, most of which (kauri and tamanou) were exported. But apart from that period, production carried out by small enterprises working for the home market, has been small.

The forests are exploited by private firms which own sawmills near the logging areas. Their activities are supervised by the forest administration. This supervision is very strict in New Caledonia where the forests belong to the state and where no tree may be felled without having first been marked by an officer of the Forest Service. In Fiji, too, exploitation is well supervised but only *ca.* 15 per cent of the forested area is state-owned. In the Hebrides and the Solomon Islands the concessionaires have more freedom. In New Caledonia *ca.* a dozen firms hold defined three year renewable concessions. The largest company produced *ca.* 7 000 m³ of timber, i.e. more than a third of total production in 1974. In the Fiji Islands, there are many more logging firms; they employ *ca.* 2 000 people and operate *ca.* 12 sawmills each with a capacity of over 3 000 m³/a. In the Solomon Islands, *ca.* one third of the known resources have been leased to four foreign companies.

Reforestation and protection

Management institutions

The administrations of the various territories have taken some measures to protect the forests and have undertaken the planting of trees on some deforested lands. Reforestation with exotic species aims at providing enough wood to satisfy local demand and supporting an export trade and to protect the soil. Regeneration and enrichment operations have also been undertaken in natural forests but, because of insufficient ecological knowledge, have not always proved successful.

The Forest Service of New Caledonia has only been in existence for *ca.* twenty years and is now very active. It is assisted by the Centre technique forestier tropical in ecological research and exotic trials. A commission to prevent mining damage has been recently established. The Forest Department of the Fiji Islands was established in 1938. In the Solomon Islands and the New Hebrides, the Forest Services do not have the necessary means to achieve their various tasks.

The Forest Service of New Caledonia is in a privileged position because most of the forests belong to the state; the private sector owns deforested land allocated for agricultural or pastoral uses. However, the reserves belonging to Melanesian communities are inalienable and include important forested areas in the main island and all the forests of the Loyalty Islands and Île des Pins; moreover most of the ultra-basic areas have been handed over to mining. In the Fiji Islands, the state owns less than 20 per cent of all forests, and land tenure problems with enrichment or reforestation are serious on some islands. In the Solomon Islands and the New Hebrides, the land belongs either to the native people or to settlers.

Protection

In New Caledonia it is essential that all vegetation above 800 m is protected; about 8 per cent of the areas surveyed should therefore be reserved. But *ca.* 40 per cent of the total forest is on steep slopes and very poor soils, so that exploitation would be costly and environmentally damaging. It is also important to protect many sites and a great variety of biotopes are of special interest because they harbour ecosystems of rare and unusual endemic species. Some classification has been carried out and regulations defining the status of various kinds of reserves and mapping their areas are to be published. The total area is *ca.* 50 000 ha. It was aimed to safeguard the most immediately threatened forests and those of most valuable biologically, particularly those on ultra-basic terrain and those situated in the region of Nouméa where the forest of the Thy park has been created for city-dwellers. The reserves also protect other wildlife.

In the Fiji Islands it is estimated that of the 384 000 ha of surveyed forest, 88 000 ha should be managed for protection. Of the 50 000 ha so far reserved, 20 000 ha are absolutely protected and 30 000 may be exploited. The protection measures appear adequate for the mountain regions; although they seem insufficient for the lowland and coastal forests, and the creation of three reserves in these areas has been recommended.

In the New Hebrides, various reserves have been projected but none have so far been established. The two most important ones would cover a sector of 2 000 ha in the southern part of Erromango (the Happy Land) which include kauri stands and 2 000–3 000 ha in the centre of Efate, where a fine tropical forest which harbours a rich bird fauna exists.

In the Solomon Islands the creation of a reserve in the Kalombangara island has been recommended; this would include the whole catchment area, which is covered with climax vegetation. It would be equally desirable to establish reserves at Ndeni (Santa Cruz) where the exploitation of the kauri stands has begun, and in Rennell, where bauxite deposits may be leased to a mining company in the near future.

Reforestation

Large-scale reforestation programmes have been carried out or are being pursued in New Caledonia and the Fiji Islands. In the New Hebrides trial plantings started some years ago have given remarkable results particularly at Efate on relatively rich brown-red soils (*Agathis* and various broad-leaved species), and at Aneityum on very poor ferrallitic soils (*Pinus*); but they are not extensive. In the Solomon Islands trials are in progress with broad-leaved species (*Campnosperma*, *Terminalia*). Except in the Solomon Islands, reforestation is carried out with exotics though some local kauris have been successfully planted.

In New Caledonia, the first plantations were made in 1962 in the south of the main island, on ferrallitic soils developed from ultra-basic rocks. These plantations, using native species (*Agathis*, *Araucaria*) not suitable to local conditions, proved disappointing; pines (especially *Pinus caribaea*) introduced later did better though growth was uneven.

In the Île des Pins, 410 ha of plantations of *P. caribaea*, also on ultra-basic terrain but on ferrallitic soils less deficient in P and Ca, are much better.

Other plantations (*ca.* 800 ha) were established on shale and sandstone terrains (fersiallitic leached soils). The growth of pines was good on relatively humid sites; that of various *Eucalyptus* species seemed to be better on very poor soils and in drier conditions, or in more seasonal climates (northern part of the main island). Kauri (*Agathis moorei*) is a promising species for fairly rich moist valley soils.

Trials carried out with local species have not been very conclusive; the ecophysiological characters of the species were not taken sufficiently into consideration in the choice of sites and planting techniques. However, kauri is an exception. The growth of *Arillastrum* sp. planted on ferrallitic soils in the Ploum pass under climatic and edaphic conditions suitable to the species, has also been most encouraging.

The Forest Service has large areas of government-owned land where it can establish plantations, but these areas are on steep slopes with very poor soils or in the ultra-basic regions most of which are exploited for mining so that the protection of young plantations presents difficulties. Hence land is used as well in the private sector or in the native population lands, where large areas, at present unused or devoted to a very extensive form of livestock husbandry, are suitable for economically viable reforestation projects. Seedlings have been supplied to owners and arrangements have been made with the representatives of several Melanesian communities. On the main island, fires have become less frequent in the planted sectors; the Melanesians are of the opinion that the trees they have planted grant their right of possession to the land.

In the Fiji Islands, reforestation operations have been pursued vigorously for *ca.* 20 years in closed forest with *Swietenia macrophylla*, where all valuable species have been extracted, or in savanna with *Pinus*, *P. elliottii* and chiefly

P. caribaea. More recently trials have been made with other species (including *Eucalyptus citriodora*) which may occupy an important place in the future.

The extent of state-owned land is less than 8 per cent of the archipelago; over 80 per cent of the land is subject to traditional tenure. Hence it has been necessary to purchase land or to enter into long-term agreements with the people in order to establish protected forest areas and to define areas for reforestation.

Swietenia macrophylla was introduced to the Fiji Islands in 1911; its use for reforestation was started *ca.* 20 years ago with planting stock raised from local seed harvested on trees aged *ca.* 30 years. No selection was practised. The plantations were established in a forest environment on 3–4 m wide lines. The original forest cover, which provided shelter in the first phase, was removed as soon as the plants were firmly established. Attacks by insects (*Xyloborus*) and fungi (*Fomes*) can cause severe losses. Reforestation projects of this type have been established on 15–20 per cent of exploited forest lands (*ca.* 6 000 ha). A rotation of 30–40 years is envisaged.

The introduction of pines is recent, but some plantations are over 10 years old. A major place in the reforestation programme has been reserved for *Pinus caribaea* which can adapt itself better than *P. elliottii* to different conditions. The results obtained are excellent. The plantations at Latauka and Nausori are situated in the western part of Viti Levu (a region of medium humidity) between 350 and 600 m, on ferrallitic soils developed from andesites and basalts which in natural conditions are covered by a rather poor *Pennisetum polystachyon* savanna. Outside the plantations, the humus horizon is thin and of poor structure and there is sheet erosion. Under the pines the flora becomes more varied and the herbaceous layer increases in density; guavas and other shrubby weed species appear; but there is a marked improvement in the soil properties due to an enrichment in organic matter and a reduction of erosion (Latham, 1975).

In the New Hebrides and the Solomon Islands, growth of various species trials are in progress. In the Solomon Islands are plantations of certain indigenous species of *Terminalia* and *Campnosperma*. In the New Hebrides the techniques used in the Fijis are adopted. It is as yet too early to evaluate the strip planting trials in Aneityum, but pines seem to grow very well on the leached ferrallitic soils in the cleared areas in the southwestern part of the island, whilst the rich soils of Tanna do not seem to suit them. Useful information can also be gained from trial plantings at Efate on the plateau that dominates Port Havannah; there, most of the introduced broad-leaved species (*Terminalia*, *Gmelina*, *Swietenia* spp., etc.) have grown very well on weakly desaturated soils, as well as *Agathis obtusa*, kauri of Erromango, which under natural conditions is found only on much poorer soils and in a more humid climate.

Research needs

The knowledge of the floras of each of the large archipelagos is satisfactory, especially New Caledonia and Fiji; there is also adequate information on the factors controlling the distribution of the most common species but hardly anything is known about the internal dynamics of the forest ecosystems and the effect of human actions on their development.

Natural forests

Research is needed on the forest environment and the biology of valuable timber species.

New Caledonia

There are numerous rainfall stations, but their distribution has often been governed by the requirements of hydrological studies and few records are available for the main forest areas. Little is known of the variations in temperature and humidity in the mountains, and nothing about wind speeds, insolation or cloud density. The distribution of the major soil types has been mapped, but forest soils have not been studied; research on organic matter has only just begun and (apart from some isolated observations) the soil fauna is unknown. Chemical properties and available soil moisture play an important role in the distribution of species but the only data on these, relate to the maquis or cultivated areas. The same remarks apply to run-off and erosion studies.

Because of the extreme diversity of site conditions the choice of study areas is of considerable importance; it is desirable to give priority to the large forest areas in the central part of the main island (col des Roussettes, col d'Amieu) where the reserves of timber could be exploited more easily, and to the forests on ultra-basic terrains in the west and south west (Nouméa region), because of their hydrological importance and the special problems associated with the adaptation of the flora to their climatic and soil conditions.

Fires cannot be controlled effectively whilst nothing is known about the conditions affecting them or their effects.

Relations of association and competition within forest ecosystems have not been seriously studied. No studies exist of re-establishment of forest in gaps, nor of the succession towards forest in savanna or maquis areas once human influence has ceased. Is the closed forest in mining areas a relict or a climax, or both? Are the majority of the existing maquis secondary formations? Are the stands of Casuarinaceae on iron hard-pan soils (*Gymnostoma deplancheana*) or on brown soils with magnesium and nickel contents (*G. chamaecyparis*) secondary communities and, if so, are they of a transitory nature? Does the appearance of palms in a maquis or a thicket indicate that the ecological conditions have become favourable to forest, where their increase seems to be associated with a degradation of the dominant layer? Are the patches of *Melaleuca* within forest the result of natural factors (local differences in soil conditions) or human intervention?

There is an almost complete lack of data on the functioning of the forests of Melanesia (water and energy balance, primary and secondary productivity, and biogeochemical cycles).

Recent inventories have enabled approximate estimates of the amount of timber available, but nothing is known about the time required for these reserves to be restored after exploitation. Any management trial to favour the regeneration of the most valuable species requires a knowledge of their biological characters. The species, the wood of which is mostly used, are not the only ones of interest; others, which yield high quality wood and grow to suitable sizes are little exploited for no other reason than that they are rare; in addition there are others, also infrequent, which contain pharmacological substances. Hence it would be useful to investigate whether regeneration of such species could be improved by some relatively simple treatment. The studies pursued for some twenty years by the Forest Service, and also by the CTFT and ORSTOM (Verlière, 1974) on *Arillastrum* and certain Araucariaceae (*Araucaria columnaris*, *Agathis* spp.) are important. The environments which suit these trees are known and good results have been achieved with plantations whilst enrichment planting of young plants in lines in the forest have not been very encouraging, perhaps because suitable thinning techniques were not available. Grafted seed orchards have been established to obtain seed of *Agathis lanceolata*, which produces few fertile cones in its natural habitat. Exploratory studies have also been made on the sylviculture of *Montrouziera cauliflora*—the species which is at present the most highly valued—but difficulties remain.

Other archipelagos

The problems involved in successful management of the forests in other archipelagos are similar to those discussed for New Caledonia but are less because the environment is more homogeneous.

The knowledge of climate and soils of the Fiji Islands is similar and in certain respects superior to that in New Caledonia. The climatological data available for the Solomon Islands and the New Hebrides are very incomplete; nevertheless enough is known to make it probable that in low and medium altitudes, except very locally, the establishment of closed high forest is not limited by humidity and temperature in the Solomon Islands and the northern part of the Hebrides. Cloud density may be excessive at certain levels. It is important to have precise information on the frequency and violence of cyclones in each sector. Quantin (1975) reported on the properties of the soils of the New Hebrides and good soil maps are available. Detailed soil maps for the Solomon Islands were published by the Land Resources Division. But the relations between soil and vegetation should be examined more closely.

Fires still cause serious damage in the west and north of the islands of Viti Levu and Vanua Levu. The forest borders are perhaps more stable than in New Caledonia, but the behaviour of the species which are most strongly exposed to fire, as well as the effect of the frequency of fires on the physiognomy and floristic composition of savannas, should be studied. In the Solomon Islands it is the vegetation of the ultra-basic areas which appears to be chiefly threatened, and it would be useful to compare the findings made in this respect in both territories. In the New Hebrides, it is important to ascertain the damage caused by fires in the south of the archipelago, especially at Aneityum and Erromango.

There is a considerable extent of secondary woody formations in areas now uninhabited, where the environmental conditions are clearly suitable for the development of closed high forest. Many of these occupy sites which were cultivated when the population of the islands was less concentrated near the coast. It would be interesting to find out the reason for this apparent delay in the restoration of the climax vegetation. The studies should aim at developing treatment techniques which would accelerate the re-establishment of species belonging to the original canopy.

The field for autecological studies is wider than in New Caledonia. Interesting work has been done on the biology of *Agathis* (Whitmore, 1965 ; Beveridge, 1975).

Plantations

It is desirable to make trials with indigenous species, and in the natural state, in the nursery, and after planting out before establishing large-scale plantations for protection on ultra-basic sectors of New Caledonia. Such difficulties, together with economic considerations, have led to reforestation projects using a small number of exotic species of pines, eucalypts and *Swietenia* in the Fijis, the sylviculture of which was already well known.

Plantations of exotics may modify the environment in a manner not easy to predict; they do not usually offer suitable shelter to species which are threatened by habitat destruction. Plantations are highly susceptible to severe attacks from parasites. It is therefore desirable to intensify research on the behaviour of indigenous species and to introduce an increasing number of exotic species from a greater variety of taxa. Such trials do not demand heavy technical means.

Once the plantation site has been fixed, detailed studies of soil conditions must precede planting.

Timber

Numerous timbers have been studied, but the properties of many are not well known, especially species from the New Hebrides and the Solomon Islands. Research should include wood preservation and possible diversification of uses.

Human activities and the future of forests

More data on the migrations of people and on the different agricultural systems are required. Such data, together with the inventory and mapping operations, would help to explain the distribution of the main ecosystem types, predict their long-term changes and determine treatments to counteract vegetation degradation.

In the New Hebrides the region of Dillon Bay, west of Erromango, has long been a centre of sheep-rearing. Since the settlers left, the extensive grasslands have been invaded by *Acacia spirorbis*, which in 20–30 years has formed woodlands on the calcareous plateaux but has remained rare on basaltic sites. It would be relatively easy to determine its mode of establishment and growth rate, which is much faster than in New Caledonia.

Savannas and grasslands chiefly occur in dry sectors but on the southern islands (Erromango, Tanna) they also occur in humid regions; their composition varies with the soil. The role played by man in their genesis and maintenance needs to be determined.

Outside the villages are many fruit trees (*Mangifera, Citrus, Inocarpus, Artocarpus, Barringtonia*); sometimes, as in Tanna, they form groves. Can these propagate without human intervention?

The vegetation of the mountains of Santo is rather degraded; is this due to agricultural operations? On the island of Aoba the development of coco-palm plantations in the lowlands causes an extension of food crops in the mountains: does this threaten the equilibrium of the whole insular environment? At Efate, the extension of livestock husbandry on forest land may trigger off accelerated erosion; a comparative study of two catchments, the Colle which is largely deforested, and the Rentapao which is almost entirely wooded, would test this.

A study of plant succession on areas abandoned by the white settlers (Epi) and areas where the native people have greatly diminished (Maewo, Aneityum) should be undertaken; such studies are particularly important on islands with dense populations (Sheperds), where it is desirable to determine the changes in soil conditions according to the data on the present and past state of the vegetation.

Conclusions

Though knowledge of the forest ecosystems of Melanesia contains important gaps, it is adequate for management. The rugged relief, the climate with its frequent contrasts, and the low fertility and instability of the soils make it imperative to create large protected zones; the establishment of reserves is essential for the survival of many endemic plant and animal species. The species which compose the higher strata are few; many are gregarious. Hence, considerable selectivity in logging is desirable in order to preserve the equilibrium. The heterogeneity of the environment makes the task of the forester more difficult than in other tropical regions.

Adequate management of the forest is not possible without considering the state of agriculture and the prospects of economic development; these differ greatly between the archipelagos.

In New Caledonia, the standard of living is high due to nickel mining which has brought about a decline in agro-pastoral activities. Though the present recession in mining has led to rehabilitation of rural sectors, these efforts are directed towards intensive agriculture; large areas of savanna and scrub on sites which are not suitable for intensive cultivation could be used for forest plantations. The management of the natural forests, which are less extensive than in the other archipelagos, should give priority to protection.

In the Fiji Islands forest production is more important than in New Caledonia. The great rate of population increase and the improvement in agro-pastoral techniques will lead to the encroachment of agriculture on the forested sectors, particularly on the savannas where they may take over pine and eucalypt plantations. Enrichment planting brings about a decline of natural forests, but is necessary for the maintenance of productivity.

In the New Hebrides, the forest areas are large but mainly secondary and poor in valuable timber species. Considering the quality of the soils, their extension seems excessive; however the situation differs greatly between islands. Within the general framework of land-use planning for the whole archipelago, stressing agricultural development, the stabilizing and regulating influence of the forest on the environment is more important than its economic productivity.

The Solomon Islands make large profits from timber exports; forest resources are considerable, but renewal under natural conditions is very slow. It is important to ensure that excessive exploitation does not lead to a serious decrease in productivity of the richest and most accessible forests, and to extend trials to re-introduce valuable species in areas now covered with thickets; moreover, crop cultivation should be concentrated in the most favourable sectors.

Selective bibliography

This list quotes the recent publications, some of which contain comprehensive lists of references (as those published by the Land Resources Division). The major references are indicated by an asterisk.

AUBRÉVILLE, A. Sapotacées. In: *Flore de la Nouvelle-Calédonie et dépendances*, n° 1. Paris, Muséum national d'histoire naturelle, 1967, 168 p.

BALGOOY, M. M. J. van. Plant geography of the Pacific. *Blumea*, suppl. 6, 1971, p. 1–222.

*BERRY, M. J.; HOWARD, W. J. *Fiji forest inventory*. Land Resources Study no. 12. Surbiton, Land Resources Division, Overseas Devel. Adm., 1973. Vol. 1. The environment and forest types, 98 p., 1/50 000 maps. Vol. 2. Catchment groups of Viti Levu and Kandavu, 62 p. Vol. 3. Catchment groups of Vanua Levu, 88 p.

BEVERIDGE, A. E. Kauri forest in the New Hebrides. *Phil. Trans. Roy. Soc. London*, B, 272, 1975, p. 369–383.

BONNEMAISON, J. *Espaces et paysages agraires dans le nord des Nouvelles-Hébrides*. Nouméa, ORSTOM, 1973, 108 p. multigr.

BRAITHWAITE, A. F. The phytogeographical relationships and origin of the New Hebrides fern flora. *Phil. Trans. Roy. Soc. London*, B, 272, 1975, p. 293–313.

BROWNLIE, G. Ptéridophytes. In: *Flore de la Nouvelle-Calédonie et dépendances*, n° 3. Paris, Muséum national d'histoire naturelle, 1969, 307 p.

CENTRE TECHNIQUE FORESTIER TROPICAL (CTFT). *Inventaire des ressources forestières de la Nouvelle-Calédonie*, Nogent-sur-Marne, 1975, 227 p. multigr.

CHEW, W. L. The phanerogamic flora of the New Hebrides and its relationships. *Phil. Trans. Roy. Soc. London*, B, 272, 1975, p. 315–328.

CORNER, E. J. H. *Ficus* in the Solomon islands. *Phil. Trans. Roy. Soc. London*, B, 253, 1967, p. 23–159.

——. Mountain flora of Popomanaseu, Guadalcanal. *Phil. Trans. Roy. Soc. London*, B, 255, 1969, p. 575–577.

——. *Ficus* subgen. *Pharmacosycea* with references to the species of New Caledonia. *Phil. Trans. Roy. Soc. London*, B, 259, 1970, p. 383–433.

——. *Ficus* in the New Hebrides. *Phil. Trans. Roy. Soc. London*, B, 272, 1975, p. 343–367.

DELACOUR, J. *Guide des oiseaux de la Nouvelle-Calédonie*. Neuchâtel, Delachaux et Niestlé, 1966, 172 p.

FIJI DEPARTMENT OF FORESTRY. *Some timbers of Fiji*. Suva, 1968, 60 p.

GORMAN, M. L. Habitats of the land-birds of Viti Levu, Fiji Islands. *The Ibis*, vol. 117, no. 2, 1975, p. 152–162.

GROSS, G. F. The land invertebrates of the New Hebrides and their relationships. *Phil. Trans. Roy. Soc. London*, B, 272, 1975, p. 391–421.

GUILLAUMIN, A. Contributions to the flora of the New Hebrides. Plants collected by S. F. Kajewski in 1928 and 1929. *J. Arnold Arbor.*, 13, 1932, p. 1–29, 81–126; 14, 1933, p. 53–61.

——. Compendium de la flore phanérogamique des Nouvelles-Hébrides. *Annales du Musée Colonial de Marseille*, 6, 1948, p. 5–56.

——. *Flore analytique et synoptique de la Nouvelle-Calédonie*. Paris, ORSTOM, 1948, 369 p.

HADLEY, C. J. The kauri forests of the Santa Cruz islands. *For. Soc. Journal* (Oxford), vol. 5, no. 7, 1959, p. 11–15.

*HANSELL, J. R. F.; WALL, J. R. D. *Land resources of the Solomon islands*. Land Resources study no. 18. Surbiton, Land Resources Division, Overseas Devel. Adm., 1976, 8 vol.

HEYLIGERS, P. C. Vegetation and ecology of Bougainville and Buka islands. In: *Land Research Series* (CSIRO, Melbourne), no. 20, 1967, p. 121–145, 1/600 000 map.

JAFFRÉ, T.; LATHAM, M. Contribution à l'étude des relations sol-végétation sur un massif de roches ultrabasiques de la côte ouest de la Nouvelle-Calédonie, le Boulinda. *Adansonia*, vol. 2, n° 14, 1974, p. 311–336.

JOHNSON, M. S. *New Hebrides Condominium, Erromango forest inventory*. Land Resources study no. 10. Surbiton, Land Resources Division, Overseas Devel. Adm., 1971, 91 p., 1/50 000 maps.

KOSTERMANS, A. J. G. H. Lauracées. In: *Flore de la Nouvelle-Calédonie et dépendances*, n° 5. Paris, Muséum national d'histoire naturelle, 1974, 123 p.

*LAND RESOURCES DIVISION, OVERSEAS DEVELOPMENT ADMINIS-

TRATION (Surbiton, England). *Land resources bibliography*: 4. *Fiji*, 1973, 138 p.; 7. *Solomon islands*, 1975, 125 p.

LATHAM, M. *Compte rendu de mission aux îles Salomon avec référence spéciale aux sols issus de roches ultrabasiques*. Nouméa, ORSTOM, 1973, 17 p. multigr.

——. *Étude du complexe sol-végétation dans les îles de l'est des Fidji (rapport préliminaire)*. Nouméa, ORSTOM, 1975, 22 p. multigr.

LAUBENFELS, D. J. de. Gymnospermes. In: *Flore de la Nouvelle-Calédonie et dépendances*, n° 4. Paris, Muséum national d'histoire naturelle, 1972, 168 p.

LEE, K. E. Some soils of the British Solomon Islands Protectorate. *Phil. Trans. Roy. Soc. London*, B, 255, 1969, p. 211–257.

——. Introductory remarks (Discussion on the results of the 1971 Royal Society expedition to the New Hebrides). *Phil. Trans. Roy. Soc. London*, B, 272, 1975, p. 269–276.

MAYR, E. *Birds of the southwest Pacific*. New York, Macmillan, 1945, 316 p.

McALPINE, J. R. Climate of Bougainville and Buka islands. In: *Land Research Series* (CSIRO, Melbourne), no. 20, 1967, p. 62–70.

MEDWAY, Lord; MARSHALL, A. G. Terrestrial vertebrates of the New Hebrides: origin and distribution. *Phil. Trans. Roy. Soc. London*, B, 272, 1975, p. 423–459.

MOORE, H. E. A preliminary analysis of the palm flora of the British Solomon islands. *Phil. Trans. Roy. Soc. London*, B, 255, 1969, p. 589–593.

O'REILLY, P. *Bibliographie de la Nouvelle-Calédonie*. Paris, Musée de l'Homme, Soc. des Océanistes, 1955, 361 p.

PARHAM, J. W. The grasses of Fiji. *Bull. Depart. Agric.* (Suva), no. 30, 1955, 166 p.

——. *Plants of the Fiji islands*. Suva, Government Press, 1972, 490 p.

QUANTIN, P. Soils of the New Hebrides islands. *Phil. Trans. Roy. Soc. London*, B, 272, 1975, p. 287–292.

——. *Cartes des sols, des formes de relief et de la végétation, avec notices, des îles Vaté, Sheperd, Aoba, Ambrym, Maewo, Pentecôte* (en cours de publication pour les autres îles). Paris, ORSTOM, 1971–1975.

SARLIN, P. *Bois et forêts de la Nouvelle-Calédonie*. Nogent-sur-Marne, CTFT, 1954, 303 p.

SAUNDERS, J. C. Forest resources of Bougainville and Buka islands. In: *Land Research Series* (CSIRO, Melbourne), no. 20, 1967, p. 146–156.

SCHMID, M. *Note sur la végétation des îles Loyauté*. Nouméa, ORSTOM, 1967, 70 p. multigr.

——. La flore et la végétation de la partie méridionale de l'archipel des Nouvelles-Hébrides. *Phil. Trans. Roy. Soc. London*, B, 272, 1975, p. 329–342.

SCOTT, R. M. Soils of Bougainville and Buka islands. In: *Land Research Series* (CSIRO, Melbourne), no. 20, 1967, p. 105–120, 1/600 000 map.

THORNE, R. F. Floristic relationships between New Caledonia and the Solomon islands. *Phil. Trans. Roy. Soc. London*, B, 255, 1969, p. 595–601.

TWYFORD, I. J.; WRIGHT, A. C. S. *The soil resources of the Fiji islands*. Suva, Government Press, 1965, 2 vol., 570 p., 23 maps.

UNESCO. *Population, resources and development in the eastern islands of Fiji: information for decision-making*. Paris, A report by the Unesco/United Nations Fund for Population Activities (UNFPA); Population and Environment Project in the eastern islands of Fiji; Man and Biosphere (MAB) Programme, Project 7: Ecology and rational use of island ecosystems; march 1977, 407 p. multigr.

VEILLON, J. M. *Architecture végétative de quelques arbres de la Nouvelle-Calédonie*. Montpellier, thesis, 1976, 300 p.

VERLIÈRE, G. *Étude de la croissance et de la nutrition minérale du chêne-gomme (Arillastrum gummiferum) sur quelques sols calédoniens*. Nouméa, ORSTOM, 1974, 18 p. multigr.

*VIROT, R. *La végétation canaque*. Paris, Mém. Muséum national d'histoire naturelle, B. 7, 1956, 400 p.

——. Protéacées. In: *Flore de la Nouvelle-Calédonie et dépendances*, n° 2. Paris, Muséum national d'histoire naturelle, 1968, 254 p.

WALL, J. R. D.; HANSELL, J. R. F. *Land resources of the Solomon Islands*. 8 vol. Land Resources Division, Ministry of Overseas Development (Tolworth Tower, Surbiton, Surrey, United Kingdom).

WHITMORE, T. C. A kauri forest in the Solomon islands. In: *Proc. Symp. Ecol. Res. Humid Trop. Veg.* (Kuching, Sarawak), p. 58–64. Paris, Unesco, 1965.

——. *Guide to the forests of the British Solomon islands*. London, Oxford Univ. Press, 1966, 208 p.

——. The vegetation of the Solomon islands. *Phil. Trans. Roy. Soc. London*, B, 255, 1969, p. 259–270.

[B.] SC.77/xii.14/A.